ANALYTIC COMBINATORICS

Analytic combinatorics aims to enable precise quantitative predictions of the properties of large combinatorial structures. The theory has emerged over recent decades as essential both for the analysis of algorithms and for the study of scientific models in many disciplines, including probability theory, statistical physics, computational biology and information theory. With a careful combination of symbolic enumeration methods and complex analysis, drawing heavily on generating functions, results of sweeping generality emerge that can be applied in particular to fundamental structures such as permutations, sequences, strings, walks, paths, trees, graphs and maps.

This account is the definitive treatment of the topic. In order to make it self-contained, the authors give full coverage of the underlying mathematics and give a thorough treatment of both classical and modern applications of the theory. The text is complemented with exercises, examples, appendices and notes throughout the book to aid understanding. The book can be used as a reference for researchers, as a textbook for an advanced undergraduate or a graduate course on the subject, or for self-study.

PHILIPPE FLAJOLET is Research Director of the Algorithms Project at INRIA Rocquencourt.

ROBERT SEDGEWICK is William O. Baker Professor of Computer Science at Princeton University.

ANALYTIC COMBINATORICS

PHILIPPE FLAJOLET

Algorithms Project
INRIA Rocquencourt
78153 Le Chesnay
France

&

ROBERT SEDGEWICK

Department of Computer Science
Princeton University
Princeton, NJ 08540
USA

CAMBRIDGE
UNIVERSITY PRESS

CAMBRIDGE UNIVERSITY PRESS

Cambridge, New York, Melbourne, Madrid, Cape Town, Singapore, São Paulo, Delhi

Cambridge University Press
The Edinburgh Building, Cambridge CB2 8RU, UK

Published in the United States of America by Cambridge University Press, New York

www.cambridge.org
Information on this title: www.cambridge.org/9780521898065

First published 2009

Printed in the United Kingdom at the University Press, Cambridge

A catalogue record for this publication is available from the British Library

ISBN 978-0-521-89806-5 hardback

Contents

Preface

ANALYTIC COMBINATORICS aims at predicting precisely the properties of large structured combinatorial configurations, through an approach based extensively on analytic methods. Generating functions are the central objects of study of the theory.

Analytic combinatorics starts from an exact enumerative description of combinatorial structures by means of generating functions: these make their first appearance as purely formal algebraic objects. Next, generating functions are interpreted as analytic objects, that is, as mappings of the complex plane into itself. Singularities determine a function's coefficients in asymptotic form and lead to precise estimates for counting sequences. This chain of reasoning applies to a large number of problems of discrete mathematics relative to words, compositions, partitions, trees, permutations, graphs, mappings, planar configurations, and so on. A suitable adaptation of the methods also opens the way to the quantitative analysis of characteristic parameters of large random structures, via a perturbational approach.

THE APPROACH to quantitative problems of discrete mathematics provided by analytic combinatorics can be viewed as an *operational calculus* for combinatorics organized around three components.

> *Symbolic methods* develops systematic relations between some of the major constructions of discrete mathematics and operations on generating functions that exactly encode counting sequences.
>
> *Complex asymptotics* elaborates a collection of methods by which one can extract asymptotic counting information from generating functions, once these are viewed as analytic transformations of the complex domain. Singularities then appear to be a key determinant of asymptotic behaviour.
>
> *Random structures* concerns itself with probabilistic properties of large random structures. Which properties hold with high probability? Which laws govern randomness in large objects? In the context of analytic combinatorics, these questions are treated by a deformation (adding auxiliary variables) and a perturbation (examining the effect of small variations of such auxiliary variables) of the standard enumerative theory.

The present book expounds this view by means of a very large number of examples concerning classical objects of discrete mathematics and combinatorics. The eventual goal is an effective way of quantifying metric properties of large random structures.

Given its capacity of quantifying properties of large discrete structures, *Analytic Combinatorics* is susceptible to many applications, not only within combinatorics itself, but, perhaps more importantly, within other areas of science where discrete probabilistic models recurrently surface, like statistical physics, computational biology, electrical engineering, and information theory. Last but not least, the analysis of algorithms and data structures in computer science has served and still serves as an important incentive for the development of the theory.

★ ★ ★ ★ ★ ★

Part A: Symbolic methods. This part specifically develops *Symbolic methods*, which constitute a unified algebraic theory dedicated to setting up functional relations between counting generating functions. As it turns out, a collection of general (and simple) theorems provide a systematic translation mechanism between combinatorial constructions and operations on generating functions. This translation process is a purely formal one. In fact, with regard to basic counting, two parallel frameworks coexist—one for unlabelled structures and ordinary generating functions, the other for labelled structures and exponential generating functions. Furthermore, within the theory, parameters of combinatorial configurations can be easily taken into account by adding supplementary variables. Three chapters then form Part A: Chapter I deals with unlabelled objects; Chapter II develops labelled objects in a parallel way; Chapter III treats multivariate aspects of the theory suitable for the analysis of parameters of combinatorial structures.

★ ★ ★ ★ ★ ★

Part B: Complex asymptotics. This part specifically expounds *Complex asymptotics*, which is a unified analytic theory dedicated to the process of extracting asymptotic information from counting generating functions. A collection of general (and simple) theorems now provide a systematic translation mechanism between generating functions and asymptotic forms of coefficients. Five chapters form this part. Chapter IV serves as an *introduction to complex-analytic methods* and proceeds with the treatment of *meromorphic functions*, that is, functions whose singularities are poles, *rational functions* being the simplest case. Chapter V develops *applications of rational and meromorphic asymptotics of generating functions*, with numerous applications related to words and languages, walks and graphs, as well as permutations. Chapter VI develops a general theory of *singularity analysis* that applies to a wide variety of singularity types, such as square-root or logarithmic, and has consequences regarding trees as well as other recursively-defined combinatorial classes. Chapter VII presents *applications of singularity analysis* to 2–regular graphs and polynomials, trees of various sorts, mappings, context-free languages, walks, and maps. It contains in particular a discussion of the analysis of coefficients of algebraic functions. Chapter VIII explores *saddle-point methods*, which are instrumental in analysing functions with a violent growth at a singularity, as well as many functions with a singularity only at infinity (i.e., entire functions).

★ ★ ★ ★ ★ ★

Part C: Random structures. This part is comprised of Chapter IX, which is dedicated to the analysis of multivariate generating functions viewed as deformation and perturbation of simple (univariate) functions. Many known laws of probability theory, either discrete or continuous, from Poisson to Gaussian and stable distributions, are found to arise in combinatorics, by a process combining symbolic methods, complex asymptotics, and perturbation methods. As a consequence, many important characteristics of classical combinatorial structures can be precisely quantified in distribution.

★ ★ ★ ★ ★

Part D: Appendices. Appendix A summarizes some key elementary concepts of combinatorics and asymptotics, with entries relative to asymptotic expansions, languages, and trees, among others. Appendix B recapitulates the necessary background in complex analysis. It may be viewed as a self-contained minicourse on the subject, with entries relative to analytic functions, the Gamma function, the implicit function theorem, and Mellin transforms. Appendix C recalls some of the basic notions of probability theory that are useful in analytic combinatorics.

★ ★ ★ ★ ★

THIS BOOK is meant to be reader-friendly. Each major method is abundantly illustrated by means of concrete *Examples*[1] treated in detail—there are scores of them, spanning from a fraction of a page to several pages—offering a complete treatment of a specific problem. These are borrowed not only from combinatorics itself but also from neighbouring areas of science. With a view to addressing not only mathematicians of varied profiles but also scientists of other disciplines, *Analytic Combinatorics* is self-contained, including ample appendices that recapitulate the necessary background in combinatorics, complex function theory, and probability. A rich set of short *Notes*—there are more than 450 of them—are inserted in the text[2] and can provide exercises meant for self-study or for student practice, as well as introductions to the vast body of literature that is available. We have also made every effort to focus on *core ideas* rather than technical details, supposing a certain amount of mathematical maturity but only basic prerequisites on the part of our gentle readers. The book is also meant to be strongly problem-oriented, and indeed it can be regarded as a manual, or even a huge algorithm, guiding the reader to the solution of a very large variety of problems regarding discrete mathematical models of varied origins. In this spirit, many of our developments connect nicely with computer algebra and symbolic manipulation systems.

COURSES can be (and indeed have been) based on the book in various ways. Chapters I–III on *Symbolic methods* serve as a systematic yet accessible introduction to the formal side of combinatorial enumeration. As such it organizes transparently some of the rich material found in treatises[3] such as those of Bergeron–Labelle–Leroux, Comtet, Goulden–Jackson, and Stanley. Chapters IV–VIII relative to *Complex asymptotics* provide a large set of concrete examples illustrating the power

---

[1]Examples are marked by "*Example* · · · ■".
[2]Notes are indicated by ▷ · · · ◁.
[3]References are to be found in the bibliography section at the end of the book.

of classical complex analysis and of asymptotic analysis outside of their traditional range of applications. This material can thus be used in courses of either pure or applied mathematics, providing a wealth of non-classical examples. In addition, the quiet but ubiquitous presence of symbolic manipulation systems provides a number of illustrations of the power of these systems while making it possible to test and concretely experiment with a great many combinatorial models. Symbolic systems allow for instance for fast random generation, close examination of non-asymptotic regimes, efficient experimentation with analytic expansions and singularities, and so on.

Our initial motivation when starting this project was to build a coherent set of methods useful in the analysis of algorithms, a domain of computer science now well-developed and presented in books by Knuth, Hofri, Mahmoud, and Szpankowski, in the survey by Vitter–Flajolet, as well as in our earlier *Introduction to the Analysis of Algorithms* published in 1996. This book, *Analytic Combinatorics*, can then be used as a systematic presentation of methods that have proved immensely useful in this area; see in particular the *Art of Computer Programming* by Knuth for background. Studies in statistical physics (van Rensburg, and others), statistics (e.g., David and Barton) and probability theory (e.g., Billingsley, Feller), mathematical logic (Burris' book), analytic number theory (e.g., Tenenbaum), computational biology (Waterman's textbook), as well as information theory (e.g., the books by Cover–Thomas, MacKay, and Szpankowski) point to many startling connections with yet other areas of science. The book may thus be useful as a supplementary reference on methods and applications in courses on statistics, probability theory, statistical physics, finite model theory, analytic number theory, information theory, computer algebra, complex analysis, or analysis of algorithms.

Acknowledgements. This book would be substantially different and much less informative without Neil Sloane's *Encyclopedia of Integer Sequences*, Steve Finch's *Mathematical Constants*, Eric Weisstein's *MathWorld*, and the *MacTutor History of Mathematics* site hosted at St Andrews. We have also greatly benefited of the existence of open on-line archives such as *Numdam, Gallica, GDZ* (digitalized mathematical documents), *ArXiv*, as well as the *Euler Archive*. All the corresponding sites are (or at least have been at some stage) freely available on the Internet. Bruno Salvy and Paul Zimmermann have developed algorithms and libraries for combinatorial structures and generating functions that are based on the MAPLE system for symbolic computations and that have proven to be extremely useful. We are deeply grateful to the authors of the free software Unix, Linux, Emacs, X11, TEX and LATEX as well as to the designers of the symbolic manipulation system MAPLE for creating an environment that has proved invaluable to us. We also thank students in courses at Barcelona, Berkeley (MSRI), Bordeaux, Caen, Graz, Paris (École Polytechnique, École Normale Supérieure, University), Princeton, Santiago de Chile, Udine, and Vienna whose reactions have greatly helped us prepare a better book. Thanks finally to numerous colleagues for their contributions to this book project. In particular, we wish to acknowledge the support, help, and interaction provided at a high level by members of the *Analysis of Algorithms (AofA)* community, with a special mention for Nicolas Broutin, Michael Drmota, Éric Fusy, Hsien-Kuei Hwang, Svante Janson, Don Knuth, Guy Louchard, Andrew Odlyzko, Daniel Panario, Carine Pivoteau, Helmut Prodinger, Bruno Salvy, Michèle Soria, Wojtek Szpankowski, Brigitte Vallée, Mark Daniel Ward, and Mark Wilson. In addition, Ed Bender, Stan Burris, Philippe Dumas, Svante Janson, Philippe Robert, Loïc Turban, and Brigitte Vallée have provided insightful suggestions and generous feedback that have led us to revise the presentation of several sections of this book and correct many errors. We were also extremely lucky to work with David Tranah, the mathematics editor of Cambridge University Press, who has been an exceptionally supportive (and patient) companion of this book project, throughout all these years. Finally, support of our home institutions (INRIA and Princeton University) as well as various grants (French government, European Union, and NSF) have contributed to making our collaboration possible.

Philippe Flajolet, INRIA-Rocquencourt, France
Robert Sedgewick, Princeton University, USA

An Invitation to Analytic Combinatorics

διὸ δὴ συμμειγνύμενα αὐτά τε πρὸς αὑτὰ καὶ πρὸς ἄλληλα τὴν
ποικιλίαν ἐστὶν ἄπειρα· ἧς δὴ δεῖ θεωροὺς γίγνεσθαι τοὺς
μέλλοντας περὶ φύσεως εἰκότι λόγῳ

— PLATO, The Timaeus[1]

ANALYTIC COMBINATORICS is primarily a book about *combinatorics*, that is, the study of finite structures built according to a finite set of rules. *Analytic* in the title means that we concern ourselves with methods from mathematical analysis, in particular complex and asymptotic analysis. The two fields, combinatorial enumeration and complex analysis, are organized into a coherent set of methods for the first time in this book. Our broad objective is to discover how the continuous may help us to understand the discrete and to *quantify* its properties.

COMBINATORICS is, as told by its name, the science of combinations. Given basic rules for assembling simple components, what are the properties of the resulting objects? Here, our goal is to develop methods dedicated to *quantitative* properties of combinatorial structures. In other words, we want to measure things. Say that we have n different items like cards or balls of different colours. In how many ways can we lay them on a table, all in one row? You certainly recognize this counting problem—finding the number of *permutations* of n elements. The answer is of course the factorial number

$$n! = 1 \cdot 2 \cdot \ldots \cdot n.$$

This is a good start, and, equipped with patience or a calculator, we soon determine that if $n = 31$, say, then the number of permutations is the rather large quantity

$$31! = 8222838654177922817725562880000000, .$$

an integer with 34 decimal digits. The factorials solve an enumeration problem, one that took mankind some time to sort out, because the sense of the "$\cdots$" in the formula for $n!$ is not that easily grasped. In his book *The Art of Computer Programming*

---

[1] *"So their combinations with themselves and with each other give rise to endless complexities, which anyone who is to give a likely account of reality must survey."* Plato speaks of Platonic solids viewed as idealized primary constituents of the physical universe.

1

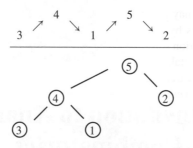

Figure 0.1. An example of the correspondence between an alternating permutation (top) and a decreasing binary tree (bottom): each binary node has two descendants, which bear smaller labels. Such *constructions*, which give access to *generating functions* and eventually provide solutions to counting problems, are the main subject of Part A.

(vol III, p. 23), Donald Knuth traces the discovery to the Hebrew *Book of Creation* (c. AD 400) and the Indian classic *Anuyogadvāra-sutra* (c. AD 500).

Here is another more subtle problem. Assume that you are interested in permutations such that the first element is smaller than the second, the second is larger than the third, itself smaller than the fourth, and so on. The permutations go up and down and they are diversely known as up-and-down or zigzag permutations, the more dignified name being *alternating* permutations. Say that $n = 2m + 1$ is odd. An example is for $n = 9$:

$$\nearrow \overset{8}{} \searrow \underset{6}{} \nearrow \overset{7}{} \searrow \underset{5}{} \nearrow \overset{9}{} \searrow \underset{1}{} \nearrow \overset{3}{} \searrow \underset{2}{}$$

The number of alternating permutations for $n = 1, 3, 5, \ldots, 15$ turns out to be

$$1, 2, 16, 272, 7936, 353792, 22368256, 1903757312.$$

What are these numbers and how do they relate to the total number of permutations of corresponding size? A glance at the corresponding figures, that is, $1!, 3!, 5!, \ldots, 15!$, or

$$1, 6, 120, 5040, 362880, 39916800, 6227020800, 1307674368000,$$

suggests that the factorials grow somewhat faster—just compare the lengths of the last two displayed lines. But how and by how much? This is the prototypical question we are addressing in this book.

Let us now examine the counting of alternating permutations. In 1881, the French mathematician Désiré André made a startling discovery. Look at the first terms of the Taylor expansion of the trigonometric function $\tan z$:

$$\tan z = 1\frac{z}{1!} + 2\frac{z^3}{3!} + 16\frac{z^5}{5!} + 272\frac{z^7}{7!} + 7936\frac{z^9}{9!} + 353792\frac{z^{11}}{11!} + \cdots.$$

The counting sequence for alternating permutations, $1, 2, 16, \ldots$, curiously surfaces. We say that the function on the left is a *generating function* for the numerical sequence (precisely, a generating function of the *exponential* type, due to the presence of factorials in the denominators).

André's derivation may nowadays be viewed very simply as reflecting the construction of permutations by means of certain labelled binary trees (Figure 0.1 and p. 143): given a permutation σ a tree can be obtained once σ has been decomposed as a triple $\langle \sigma_L, \max, \sigma_R \rangle$, by taking the maximum element as the root, and appending, as left and right subtrees, the trees recursively constructed from σ_L and σ_R. Part A of this book develops at length *symbolic methods* by which the construction of the class $\mathcal{T}$ of all such trees,

$$\mathcal{T} \;=\; \textcircled{1} \;\cup\; (\mathcal{T}, \max, \mathcal{T}),$$

translates into an equation relating generating functions,

$$T(z) \;=\; z \;+\; \int_0^z T(w)^2 \, dw.$$

In this equation, $T(z) := \sum_n T_n z^n / n!$ is the exponential generating function of the sequence (T_n), where T_n is the number of alternating permutations of (odd) length n. There is a compelling formal analogy between the combinatorial *specification* and its generating function: Unions ($\cup$) give rise to sums ($+$), max-placement gives an integral ($\int$), forming a pair of trees corresponds to taking a square ($[\cdot]^2$).

At this stage, we know that $T(z)$ must solve the differential equation

$$\frac{d}{dz} T(z) = 1 + T(z)^2, \qquad T(0) = 0,$$

which, by classical manipulations[2], yields the explicit form

$$T(z) = \tan z.$$

The generating function then provides a simple *algorithm* to compute the coefficients recurrently. Indeed, the formula,

$$\tan z = \frac{\sin z}{\cos z} = \frac{z - \frac{z^3}{3!} + \frac{z^5}{5!} - \cdots}{1 - \frac{z^2}{2!} + \frac{z^4}{4!} - \cdots},$$

implies, for n odd, the relation (extract the coefficient of z^n in $T(z) \cos z = \sin z$)

$$T_n - \binom{n}{2} T_{n-2} + \binom{n}{4} T_{n-4} - \cdots = (-1)^{(n-1)/2}, \quad \text{where} \quad \binom{a}{b} = \frac{a!}{b!(a-b)!}$$

is the conventional notation for binomial coefficients. Now, the exact enumeration problem may be regarded as solved since a very simple algorithm is available for determining the counting sequence, while the generating function admits an explicit expression in terms of well-known mathematical objects.

ANALYSIS, by which we mean mathematical analysis, is often described as the art and science of *approximation*. How fast do the factorial and the tangent number sequences grow? What about *comparing* their growths? These are typical problems of analysis.

---

[2]We have $T'/(1 + T^2) = 1$, hence $\arctan(T) = z$ and $T = \tan z$.

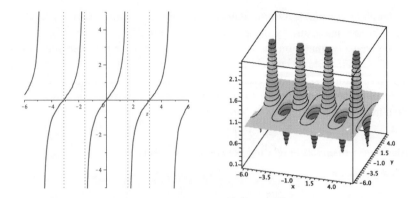

Figure 0.2. Two views of the function $z \mapsto \tan z$. Left: a plot for real values of $z \in$ $[-6, 6]$. Right: the modulus $|\tan z|$ when $z = x + iy$ (with $i = \sqrt{-1}$) is assigned complex values in the square $\pm 6 \pm 6i$. As developed at length in Part B, it is the nature of *singularities* in the *complex domain* that matters.

First, consider the number of permutations, $n!$. Quantifying its growth, as n gets large, takes us to the realm of *asymptotic analysis*. The way to express factorial numbers in terms of elementary functions is known as Stirling's formula[3]

$$n! \sim n^n e^{-n} \sqrt{2\pi n},$$

where the $\sim$ sign means "approximately equal" (in the precise sense that the ratio of both terms tends to 1 as n gets large). This beautiful formula, associated with the name of the Scottish mathematician James Stirling (1692–1770), curiously involves both the basis e of natural logarithms and the perimeter 2π of the circle. Certainly, you cannot get such a thing without analysis. As a first step, there is an estimate

$$\log n! = \sum_{j=1}^{n} \log j \sim \int_{1}^{n} \log x \, dx \sim n \log \left(\frac{n}{e}\right),$$

explaining at least the $n^n e^{-n}$ term, but already requiring a certain amount of elementary calculus. (Stirling's formula precisely came a few decades after the fundamental bases of calculus had been laid by Newton and Leibniz.) Note the utility of Stirling's formula: it tells us almost instantly that 100! has 158 digits, while 1000! borders the astronomical 10^{2568}.

We are now left with estimating the growth of the sequence of tangent numbers, T_n. The analysis leading to the derivation of the generating function $\tan(z)$ has been so far essentially algebraic or "formal". Well, we can plot the graph of the tangent function, for real values of its argument and see that the function becomes infinite at the points $\pm\frac{\pi}{2}, \pm 3\frac{\pi}{2}$, and so on (Figure 0.2). Such points where a function ceases to be

---

[3]In this book, we shall encounter five different proofs of Stirling's formula, each of interest for its own sake: (*i*) by singularity analysis of the Cayley tree function (p. 407); (*ii*) by singularity analysis of polylogarithms (p. 410); (*iii*) by the saddle-point method (p. 555); (*iv*) by Laplace's method (p. 760); (*v*) by the Mellin transform method applied to the logarithm of the Gamma function (p. 766).

"smooth" (differentiable) are called *singularities*. By methods amply developed in this book, it is the local nature of a generating function at its "dominant" singularities (i.e., the ones closest to the origin) that determines the asymptotic growth of the sequence of coefficients. From this perspective, the basic fact that tan z has dominant singularities at $\pm\frac{\pi}{2}$ enables us to reason as follows: first approximate the generating function tan z near its two dominant singularities, namely,

$$\tan(z) \underset{z \to \pm\pi/2}{\sim} \frac{8z}{\pi^2 - 4z^2};$$

then extract coefficients of this approximation; finally, get in this way a valid approximation of coefficients:

$$\frac{T_n}{n!} \underset{n \to \infty}{\sim} 2 \cdot \left(\frac{2}{\pi}\right)^{n+1} \qquad (n \text{ odd}).$$

With present day technology, we also have available *symbolic manipulation* systems (also called "computer algebra" systems) and it is not difficult to verify the accuracy of our estimates. Here is a small pyramid for $n = 3, 5, \ldots, 21$,

(T_n)	$(T_n^\star)$
2	*1*
16	1*5*
272	27*1*
7936	793*5*
353792	35379*1*
22368256	2236825*1*
1903757312	1903757*267*
209865342976	20986534*2434*
29088885112832	290888851*04489*
4951498053124096	495149805*2966307*

comparing the exact values of T_n against the approximations $T_n^\star$, where (n odd)

$$T_n^\star := \left\lfloor 2 \cdot n! \left(\frac{2}{\pi}\right)^{n+1} \right\rfloor,$$

and discrepant digits of the approximation are displayed in bold. For $n = 21$, the error is only of the order of one in a billion. Asymptotic analysis (p. 269) is in this case wonderfully accurate.

In the foregoing discussion, we have played down a fact—one that is important. When investigating generating functions from an analytic standpoint, one should generally assign *complex* values to arguments not just real ones. It is singularities in the complex plane that matter and complex analysis is needed in drawing conclusions regarding the asymptotic form of coefficients of a generating function. Thus, a large portion of this book relies on a *complex analysis* technology, which starts to be developed in Part B dedicated to *Complex asymptotics*. This approach to combinatorial enumeration parallels what happened in the nineteenth century, when Riemann first recognized the deep relation between complex analytic properties of the *zeta* function, $\zeta(s) := \sum 1/n^s$, and the distribution of primes, eventually leading to the long-sought proof of the Prime Number Theorem by Hadamard and de la Vallée-Poussin in 1896. Fortunately, relatively elementary complex analysis suffices for our purposes, and we

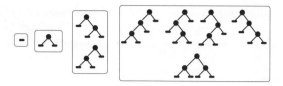

Figure 0.3. The collection of binary trees with $n = 0, 1, 2, 3$ binary nodes, with respective cardinalities $1, 1, 2, 5$.

---

can include in this book a complete treatment of the fragment of the theory needed to develop the fundamentals of analytic combinatorics.

Here is yet another example illustrating the close interplay between combinatorics and analysis. When discussing alternating permutations, we have enumerated binary trees bearing distinct integer labels that satisfy a constraint—to decrease along branches. What about the simpler problem of determining the number of possible *shapes* of binary trees? Let C_n be the number of binary trees that have n binary branching nodes, hence $n + 1$ "external nodes". It is not hard to come up with an exhaustive listing for small values of n (Figure 0.3), from which we determine that

$$C_0 = 1, \quad C_1 = 1, \quad C_2 = 2, \quad C_3 = 5, \quad C_4 = 14, \quad C_5 = 42.$$

These numbers are probably the most famous ones of combinatorics. They have come to be known as the *Catalan numbers* as a tribute to the Franco-Belgian mathematician Eugène Charles Catalan (1814–1894), but they already appear in the works of Euler and Segner in the second half of the eighteenth century (see p. 20). In his reference treatise *Enumerative Combinatorics*, Stanley, over 20 pages, lists a collection of some 66 different types of combinatorial structures that are enumerated by the Catalan numbers.

First, one can write a combinatorial equation, very much in the style of what has been done earlier, but without labels:

$$\mathcal{C} \quad = \quad \Box \quad \cup \quad (\mathcal{C}, \bullet, \mathcal{C}).$$

(Here, the $\Box$–symbol represents an external node.) With symbolic methods, it is easy to see that the *ordinary generating function* of the Catalan numbers, defined as

$$C(z) := \sum_{n \geq 0} C_n z^n,$$

satisfies an equation that is a direct reflection of the combinatorial definition, namely,

$$C(z) \quad = \quad 1 \quad + \quad z\, C(z)^2.$$

This is a quadratic equation whose solution is

$$C(z) = \frac{1 - \sqrt{1 - 4z}}{2z}.$$

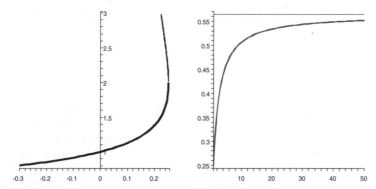

Figure 0.4. Left: the real values of the Catalan generating function, which has a square-root singularity at $z = \frac{1}{4}$. Right: the ratio $C_n/(4^n n^{-3/2})$ plotted together with its asymptote at $1/\sqrt{\pi} \doteq 0.56418$. The correspondence between *singularities* and *asymptotic* forms of *coefficients* is the central theme of Part B.

Then, by means of Newton's theorem relative to the expansion of $(1 + x)^\alpha$, one finds easily ($x = -4z$, $\alpha = \frac{1}{2}$) the *closed form* expression

$$C_n = \frac{1}{n+1}\binom{2n}{n}.$$

Stirling's asymptotic formula now comes to the rescue: it implies

$$C_n \sim C_n^\star \qquad \text{where} \quad C_n^\star := \frac{4^n}{\sqrt{\pi n^3}}.$$

This last approximation is quite usable[4]: it gives $C_1^\star \doteq 2.25$ (whereas $C_1 = 1$), which is off by a factor of 2, but the error drops to 10% already for $n = 10$, and it appears to be less than 1% for any $n \geq 100$.

A plot of the generating function $C(z)$ in Figure 0.4 illustrates the fact that $C(z)$ has a *singularity* at $z = \frac{1}{4}$ as it ceases to be differentiable (its derivative becomes infinite). That singularity is quite different from a pole and for natural reasons it is known as a square-root singularity. As we shall see repeatedly, under suitable conditions in the complex plane, a square root singularity for a function at a point ρ invariably entails an asymptotic form $\rho^{-n} n^{-3/2}$ for its coefficients. More generally, it suffices to estimate a generating function near a singularity in order to deduce an asymptotic approximation of its coefficients. This correspondence is a major theme of the book, one that motivates the five central chapters (Chapters IV to VIII).

A consequence of the complex analytic vision of combinatorics is the detection of *universality phenomena* in large random structures. (The term is originally borrowed from statistical physics and is nowadays finding increasing use in areas of mathematics such as probability theory.) By universality is meant here that many quantitative

---

[4]We use $\alpha \doteq$ d to represent a numerical approximation of the real α by the decimal d, with the last digit of d being at most ± 1 from its actual value.

properties of combinatorial structures only depend on a few global features of their definitions, not on details. For instance a growth in the counting sequence of the form

$$K \cdot A^n n^{-3/2},$$

arising from a square-root singularity, will be shown to be universal across *all* varieties of trees determined by a finite set of allowed node degrees—this includes unary–binary trees, ternary trees, 0–11–13 trees, as well as many variations such as non-plane trees and labelled trees. Even though generating functions may become arbitrarily complicated—as in an algebraic function of a very high degree or even the solution to an infinite functional equation—it is still possible to extract with relative ease *global asymptotic laws* governing *counting sequences*.

RANDOMNESS is another ingredient in our story. How useful is it to determine, exactly or approximately, counts that may be so large as to require hundreds if not thousands of digits in order to be written down? Take again the example of alternating permutations. When estimating their number, we have indeed quantified the proportion of these among all permutations. In other words, we have been predicting the *probability* that a random permutation of some size n is alternating. Results of this sort are of interest in all branches of science. For instance, biologists routinely deal with genomic sequences of length 10^5, and the interpretation of data requires developing enumerative or probabilistic models where the number of possibilities is of the order of 4^{10^5}. The language of probability theory then proves of great convenience when discussing characteristic parameters of discrete structures, since we can interpret exact or asymptotic enumeration results as saying something concrete about the likelihood of values that such parameters assume. Equally important of course are results from several areas of probability theory: as demonstrated in the last chapter of this book, such results merge extremely well with the analytic–combinatorial framework.

Say we are now interested in runs in permutations. These are the longest fragments of a permutation that already appear in (increasing) sorted order. Here is a permutation with 4 runs, separated by vertical bars:

$$2\,5\,8\,|\,3\,9\,|\,1\,4\,7\,|\,6.$$

Runs naturally present in a permutation are for instance exploited by a sorting algorithm called "natural list mergesort", which builds longer and longer runs, starting from the original ones and merging them until the permutation is eventually sorted. For our understanding of this algorithm, it is then of obvious interest to quantify how many runs a permutation is likely to have.

Let $P_{n,k}$ be the number of permutations of size n having k runs. Then, the problem is once more best approached by generating functions and one finds that the coefficient of $u^k z^n$ inside the *bivariate* generating function,

$$P(z, u) \equiv \frac{1 - u}{1 - u e^{z(1-u)}} = 1 + zu + \frac{z^2}{2!} u(u + 1) + \frac{z^3}{3!} u(u^2 + 4u + 1) + \cdots,$$

gives the desired numbers $P_{n,k}/n!$. (A simple way of establishing the last formula bases itself on the tree decomposition of permutations and on the symbolic method; the numbers $P_{n,k}$, whose importance seems to have been first recognized by Euler,

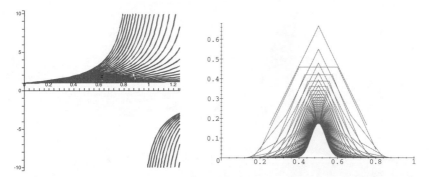

Figure 0.5. Left: A partial plot of the real values of the Eulerian generating function $z \mapsto P(z, u)$ for $z \in [0, \frac{5}{4}]$, illustrates the presence of a movable pole for A as u varies between 0 and $\frac{5}{4}$. Right: A suitable superposition of the histograms of the distribution of the number of runs, for $n = 2, \ldots, 60$, reveals the convergence to a Gaussian distribution (p. 695). Part C relates systematically the analysis of such a collection of singular behaviours to *limit distributions*.

---

are related to the *Eulerian numbers*, p. 210.) From here, we can easily determine effectively the mean, variance, and even the higher moments of the number of runs that a random permutation has: it suffices to expand blindly, or even better with the help of a computer, the bivariate generating function above as $u \to 1$:

$$\frac{1}{1-z} + \frac{1}{2} \frac{z(2-z)}{(1-z)^2} (u-1) + \frac{1}{2} \frac{z^2 (6 - 4z + z^2)}{(1-z)^3} (u-1)^2 + \cdots .$$

When $u = 1$, we just enumerate all permutations: this is the constant term $1/(1 - z)$ equal to the exponential generating function of all permutations. The coefficient of the term $u - 1$ gives the generating function of the *mean* number of runs, the next one provides the second moment, and so on. In this way, we discover the expectation and standard deviation of the number of runs in a permutation of size n:

$$\mu_n = \frac{n+1}{2}, \qquad \sigma_n = \sqrt{\frac{n+1}{12}}.$$

Then, by easy analytic–probabilistic inequalities (Chebyshev inequalities) that otherwise form the basis of what is known as the second moment method, we learn that the distribution of the number of runs is concentrated around its mean: in all likelihood, if one takes a random permutation, the number of its runs is going to be very close to its mean. The effects of such quantitative laws are quite tangible. It suffices to draw a *sample of one element* for $n = 30$ to get, for instance:

13, 22, 29|12, 15, 23|8, 28|18|6, 26|4, 10, 16|1, 5, 27|3, 14, 17, 20|2, 21, 30|25|11, 19|9|7, 24.

For $n = 30$, the mean is $15\frac{1}{2}$, and this sample comes rather close as it has 13 runs. We shall furthermore see in Chapter IX that even for moderately large permutations of size 10 000 and beyond, the probability for the number of observed runs to deviate

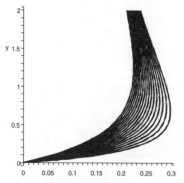

Figure 0.6. Left: The bivariate generating function $z \mapsto C(z, u)$ enumerating binary trees by size and number of leaves exhibits consistently a square-root singularity, for several values of u. Right: a binary tree of size 300 drawn uniformly at random has 69 leaves. As shown in Part C, *singularity perturbation* properties are at the origin of many randomness properties of combinatorial structures.

by more than 10% from the mean is less than 10^{-65}. As witnessed by this example, much regularity accompanies properties of large combinatorial structures.

More refined methods combine the observation of singularities with analytic results from probability theory (e.g., continuity theorems for characteristic functions). In the case of runs in permutations, the quantity $P(z, u)$ viewed as a function of z when u is fixed appears to have a pole: this fact is suggested by Figure 0.5 [left]. Then we are confronted with a fairly regular *deformation* of the generating function of all permutations. A parameterized version (with parameter u) of singularity analysis then gives access to a description of the asymptotic behaviour of the Eulerian numbers $P_{n,k}$. This enables us to describe very precisely what goes on: in a random permutation of large size n, once it has been centred by its mean and scaled by its standard deviation, *the distribution of the number of runs is asymptotically Gaussian*; see Figure 0.5 [right].

A somewhat similar type of situation prevails for binary trees. Say we are interested in leaves (also sometimes figuratively known as "cherries") in trees: these are binary nodes that are attached to two external nodes ($\square$). Let $C_{n,k}$ be the number of trees of size n having k leaves. The bivariate generating function $C(z, u) := \sum_{n,k} C_{n,k} z^n u^k$ encodes all the information relative to leaf statistics in random binary trees. A modification of previously seen symbolic arguments shows that $C(z, u)$ still satisfies a quadratic equation resulting in the explicit form,

$$C(z, u) = \frac{1 - \sqrt{1 - 4z + 4z^2(1 - u)}}{2z}.$$

This reduces to $C(z)$ for $u = 1$, as it should, and the bivariate generating function $C(z, u)$ is a deformation of $C(z)$ as u varies. In fact, the network of curves of Figure 0.6 for several fixed values of u illustrates the presence of a smoothly varying square-root singularity (the aspect of each curve is similar to that of Figure 0.4). It is possible to analyse the *perturbation* induced by varying values of u, to the effect that

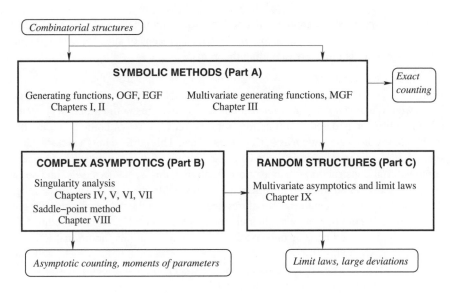

Figure 0.7. The logical structure of *Analytic Combinatorics*.

$C(z, u)$ is of the global analytic type

$$\sqrt{1 - \frac{z}{\rho(u)}},$$

for some analytic $\rho(u)$. The already evoked process of singularity analysis then shows that the probability generating function of the number of leaves in a tree of size n is of the rough form

$$\left(\frac{\rho(1)}{\rho(u)}\right)^n (1 + o(1)).$$

This is known as a "quasi-powers" approximation. It resembles very much the probability generating function of a sum of n independent random variables, a situation that gives rise to the classical Central Limit Theorem of probability theory. Accordingly, one gets that *the limit distribution of the number of leaves in a large random binary tree is Gaussian*. In abstract terms, the deformation induced by the secondary parameter (here, the number of leaves, previously, the number of runs) is susceptible to a *perturbation analysis*, to the effect that a singularity gets smoothly displaced without changing its nature (here, a square root singularity, earlier a pole) and a limit law systematically results. Again some of the conclusions can be verified even by very small samples: the single tree of size 300 drawn at random and displayed in Figure 0.6 (right) has 69 leaves, whereas the expected value of this number is $\doteq 75.375$ and the standard deviation is a little over 4. In a large number of cases of which this one is typical, we find *metric laws* of combinatorial structures that govern large structures with high probability and eventually make them highly predictable.

Such randomness properties form the subject of Part C of this book dedicated to *random structures*. As our earlier description implies, there is an extreme degree of generality in this analytic approach to combinatorial parameters, and after reading this book, the reader will be able to recognize by herself dozens of such cases at sight, and effortlessly establish the corresponding theorems.

A RATHER ABSTRACT VIEW of combinatorics emerges from the previous discussion; see Figure 0.7. A combinatorial class, as regards its enumerative properties, can be viewed as a *surface in four-dimensional real space*: this is the graph of its generating function, considered as a function from the set $\mathbb{C} \cong \mathbb{R}^2$ of complex numbers to itself, and is otherwise known as a Riemann surface. This surface has "cracks", that is, *singularities*, which determine the asymptotic behaviour of the counting sequence. A combinatorial construction (such as those freely forming sequences, sets, and so on) can then be examined through the effect it has on singularities. In this way, seemingly different types of combinatorial structures appear to be subject to *common laws* governing not only counting but also finer characteristics of combinatorial structures. For the already discussed case of universality in tree enumerations, additional universal laws valid across many tree varieties constrain for instance height (which, with high probability, is proportional to the square root of size) and the number of leaves (which is invariably normal in the asymptotic limit).

What happens regarding probabilistic properties of combinatorial parameters is this. A parameter of a combinatorial class is fully determined by a bivariate generating function, which is a deformation of the basic counting generating function of the class (in the sense that setting the secondary variable u to 1 erases the information relative to the parameter and leads back to the univariate counting generating function). Then, the *asymptotic distribution* of a parameter of interest is characterized by a collection of surfaces, each having its own singularities. The way the singularities' locations move or their nature changes under deformation encodes all the necessary information regarding the distribution of the parameter under consideration. Limit laws for combinatorial parameters can then be obtained and the corresponding phenomena can be organized into broad categories, called *schemas*. It would be inconceivable to attain such a far-reaching classification of metric properties of combinatorial structures by elementary real analysis alone.

Objects on which we are going to inflict the treatments just described include many of the most important ones of discrete mathematics, as well as the ones that surface recurrently in several branches of the applied sciences. We shall thus encounter words and sequences, trees and lattice paths, graphs of various sorts, mappings, allocations, permutations, integer partitions and compositions, polyominoes and planar maps, to name but a few. In most cases, their principal characteristics will be finely quantified by the methods of analytic combinatorics. This book indeed develops a coherent theory of random combinatorial structures based on a powerful analytic methodology. Literally dozens of quite diverse combinatorial types can then be treated by a logically transparent chain. You will not find ready-made answers to all questions in this book, but, hopefully, *methods* that can be successfully used to address a great many of them.

Bienvenue! Welcome!

Part A

SYMBOLIC METHODS

I

Combinatorial Structures and Ordinary Generating Functions

Laplace discovered the remarkable correspondence between
set theoretic operations and operations on formal power series
and put it to great use to solve a variety of combinatorial problems.

— GIAN–CARLO ROTA [518]

This chapter and the next are devoted to enumeration, where the problem is to determine the number of combinatorial configurations described by finite rules, and do so for all possible sizes. For instance, how many different words are there of length 17? Of length n, for general n? These questions are easy, but what if some constraints are imposed, e.g., no four identical elements in a row? The solutions are exactly encoded by *generating functions*, and, as we shall see, *generating functions are the central mathematical object* of combinatorial analysis. We examine here a framework that, contrary to traditional treatments based on recurrences, explains the surprising efficiency of generating functions in the solution of combinatorial enumeration problems.

This chapter serves to introduce the *symbolic* approach to combinatorial enumerations. The principle is that many general set-theoretic *constructions* admit a direct translation as operations over generating functions. This principle is made concrete by means of a dictionary that includes a collection of core constructions, namely the operations of union, cartesian product, sequence, set, multiset, and cycle. Supplementary operations such as pointing and substitution can also be similarly translated. In this way, a *language* describing elementary combinatorial classes is defined. The problem of enumerating a class of combinatorial structures then simply reduces to finding a proper *specification*, a sort of computer program for the class expressed in terms of the basic constructions. The translation into generating functions becomes, after this, a purely mechanical symbolic process.

We show here how to describe in such a context integer partitions and compositions, as well as many word and tree enumeration problems, by means of *ordinary generating functions.* A parallel approach, developed in Chapter II, applies to labelled objects—in contrast the plain structures considered in this chapter are called *unlabelled*. The methodology is susceptible to multivariate extensions with which many characteristic parameters of combinatorial objects can also be analysed in a unified manner: this is to be examined in Chapter III. The symbolic method also has the great merit of connecting nicely with complex asymptotic methods that exploit analyticity properties and singularities, to the effect that precise asymptotic estimates are usually available whenever the symbolic method applies—a systematic treatment of these aspects forms the basis of Part B of this book *Complex asymptotics* (Chapters IV–VIII).

I. 1. Symbolic enumeration methods

First and foremost, combinatorics deals with *discrete objects*, that is, objects that can be finitely described by construction rules. Examples are words, trees, graphs, permutations, allocations, functions from a finite set into itself, topological configurations, and so on. A major question is to *enumerate* such objects according to some characteristic parameter(s).

Definition I.1. *A combinatorial class, or simply a* class, *is a finite or denumerable set on which a* size *function is defined, satisfying the following conditions:*

(*i*) *the size of an element is a non-negative integer;*
(*ii*) *the number of elements of any given size is finite.*

If $\mathcal{A}$ is a class, the size of an element $\alpha \in \mathcal{A}$ is denoted by $|\alpha|$, or $|\alpha|_{\mathcal{A}}$ in the few cases where the underlying class needs to be made explicit. Given a class $\mathcal{A}$, we consistently denote by $\mathcal{A}_n$ the set of objects in $\mathcal{A}$ that have size n and use the same group of letters for the counts $A_n = \text{card}(\mathcal{A}_n)$ (alternatively, also $a_n = \text{card}(\mathcal{A}_n)$). An axiomatic presentation is then as follows: a combinatorial class is a pair $(\mathcal{A}, |\cdot|)$ where $\mathcal{A}$ is at most denumerable and the mapping $|\cdot| \in (\mathcal{A} \mapsto \mathbb{Z}_{\geq 0})$ is such that the inverse image of any integer is finite.

Definition I.2. *The* counting sequence *of a combinatorial class is the sequence of integers* $(A_n)_{n \geq 0}$ *where* $A_n = \text{card}(\mathcal{A}_n)$ *is the number of objects in class $\mathcal{A}$ that have size n.*

***Example* I.1.** *Binary words.* Consider first the set $\mathcal{W}$ of binary words, which are sequences of elements taken from the binary alphabet $\mathcal{A} = \{0, 1\}$,

$$\mathcal{W} := \{\varepsilon, 0, 1, 00, 01, 10, 11, 000, 001, 010, \ldots, 1001101, \ldots\},$$

with ε the empty word. Define size to be the number of letters that a word comprises. There are two possibilities for each letter and possibilities multiply, so that the counting sequence (W_n) satisfies

$$W_n = 2^n.$$

(This sequence has a well-known legend associated with the invention of the game of chess: the inventor was promised by his king one grain of rice for the first square of the chessboard, two for the second, four for the third, and so on. The king naturally could not deliver the promised $2^{64} - 1$ grains!) .... ∎

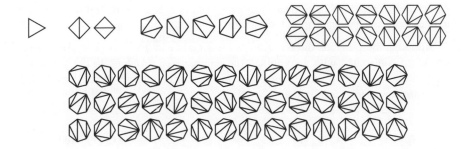

Figure I.1. The collection $\mathcal{T}$ of all triangulations of regular polygons (with size defined as the number of triangles) is a combinatorial class, whose counting sequence starts as $T_0 = 1$, $T_1 = 1$, $T_2 = 2$, $T_3 = 5$, $T_4 = 14$, $T_5 = 42$.

***Example* I.2.** *Permutations.* A permutation of size n is by definition a bijective mapping of the integer interval[1] $\mathcal{I}_n := [1 \mathinner{.\,.} n]$. It is thus representable by an array,

$$\begin{pmatrix} 1 & 2 & \cdots & n \\ \sigma_1 & \sigma_2 & & \sigma_n \end{pmatrix},$$

or equivalently by the sequence $\sigma_1 \sigma_2 \cdots \sigma_n$ of its distinct elements. The set $\mathcal{P}$ of permutations is

$$\mathcal{P} = \{\ldots, \; 12, 21, 123, 132, 213, 231, 312, 321, 1234, \ldots, 532614, \ldots\},$$

For a permutation written as a sequence of n distinct numbers, there are n places where one can accommodate n, then $n - 1$ remaining places for $n - 1$, and so on. Therefore, the number P_n of permutations of size n satisfies

$$P_n = n! = 1 \cdot 2 \cdot \ldots \cdot n \,.$$

As indicated in our *Invitation* chapter (p. 2), this formula has been known for at least fifteen centuries. .... ∎

***Example* I.3.** *Triangulations.* The class $\mathcal{T}$ of triangulations comprises triangulations of convex polygonal domains which are decompositions into non-overlapping triangles (taken up to smooth deformations of the plane). We define the size of a triangulation to be the number of triangles it is composed of. For instance, a convex quadrilateral $ABCD$ can be decomposed into two triangles in two ways (by means of either the diagonal AC or the diagonal BD); similarly, there are five different ways to dissect a convex pentagon into three triangles: see Figure I.1. Agreeing that $T_0 = 1$, we then find

$$T_0 = 1, \qquad T_1 = 1, \qquad T_2 = 2, \qquad T_3 = 5, \qquad T_4 = 14, \qquad T_5 = 42.$$

It is a non-trivial combinatorial result due to Euler and Segner [146, 196, 197] around 1750 that the number T_n of triangulations is

(1)
$$T_n = \frac{1}{n+1}\binom{2n}{n} = \frac{(2n)!}{(n+1)!\,n!},$$

a central quantity of combinatorial analysis known as a *Catalan number*: see our *Invitation*, p. 7, the historical synopsis on p. 20, the discussion on p. 35, and Subsection I. 5.3, p. 73.

---

[1]We borrow from computer science the convenient practice of denoting an integer interval by $1 \mathinner{.\,.} n$ or $[1 \mathinner{.\,.} n]$, whereas $[0, n]$ represents a real interval.

Following Euler [196], the counting of triangulations is best approached by generating functions: see again Figure I.2, p. 20 for historical context. .... ∎

Although the previous three examples are simple enough, it is generally a good idea, when confronted with a combinatorial enumeration problem, to determine the initial values of counting sequences, either by hand or better with the help of a computer, somehow. Here, we find:

(2)

n	0	1	2	3	4	5	6	7	8	9	10
W_n	1	2	4	8	16	32	64	128	256	512	1024
P_n	1	1	2	6	24	120	720	5040	40320	362880	3628800
T_n	1	1	2	5	14	42	132	429	1430	4862	16796

Such an experimental approach may greatly help identify sequences. For instance, had we not known the formula (1) for triangulations, observing unusual factorizations such as

$$T_{40} = 2^2 \cdot 5 \cdot 7^2 \cdot 11 \cdot 23 \cdot 43 \cdot 47 \cdot 53 \cdot 59 \cdot 61 \cdot 67 \cdot 71 \cdot 73 \cdot 79,$$

which contains all prime numbers from 43 to 79 and no prime larger than 80, would quickly put us on the track of the right formula. There even exists nowadays a huge *On-line Encyclopedia of Integer Sequences (EIS)* due to Sloane that is available in electronic form [543] (see also an earlier book by Sloane and Plouffe [544]) and contains more than 100 000 sequences. Indeed, the three sequences (W_n), (P_n), and (T_n) are respectively identified[2] as *EIS* **A000079**, *EIS* **A000142**, and *EIS* **A000108**.

▷ **I.1.** *Necklaces.* How many different types of necklace designs can you form with n beads, each having one of two colours, ∘ and •, where it is postulated that orientation matters? Here are the possibilities for $n = 1, 2, 3$,

This is equivalent to enumerating circular arrangements of two letters and an exhaustive listing program can be based on the smallest lexicographical representation of each word, as suggested by (20), p. 26. The counting sequence starts as 2, 3, 4, 6, 8, 14, 20, 36, 60, 108, 188, 352 and constitutes *EIS* **A000031**. [An explicit formula appears later in this chapter (p. 64).] What if two necklace designs that are mirror images of one another are identified?   ◁

▷ **I.2.** *Unimodal permutations.* Such a permutation has exactly one local maximum. In other words it is of the form $\sigma_1 \cdots \sigma_n$ with $\sigma_1 < \sigma_2 < \cdots < \sigma_k = n$ and $\sigma_k = n > \sigma_{k+1} > \cdots > \sigma_n$, for some $k \geq 1$. How many such permutations are there of size n? For $n = 5$, the number is 16: the permutations are 12345, 12354, 12453, 12543, 13452, 13542, 14532 and 15432 and their reversals. [Due to Jon Perry, see *EIS* **A000079**.]   ◁

It is also of interest to note that words and permutations may be enumerated using the most elementary counting principles, namely, for finite sets $\mathcal{B}$ and $\mathcal{C}$

(3)
$$\begin{cases} \operatorname{card}(\mathcal{B} \cup \mathcal{C}) &= \operatorname{card}(B) + \operatorname{card}(C) \quad \text{(provided } \mathcal{B} \cap \mathcal{C} = \emptyset\text{)} \\ \operatorname{card}(\mathcal{B} \times \mathcal{C}) &= \operatorname{card}(B) \cdot \operatorname{card}(C). \end{cases}$$

---

[2]Throughout this book, a reference such *EIS* **Axxx** points to Sloane's *Encyclopedia of Integer Sequences* [543]. The database contains more than 100 000 entries.

We shall see soon that these principles, which lie at the basis of our very concept of number, admit a powerful generalization (Equation (19), p. 23, below).

Next, for combinatorial enumeration purposes, it proves convenient to identify combinatorial classes that are merely variants of one another.

Definition I.3. *Two combinatorial classes A and B are said to be (combinatorially) isomorphic, which is written $A \cong B$, iff their counting sequences are identical. This condition is equivalent to the existence of a bijection from A to B that preserves size, and one also says that A and B are* bijectively equivalent.

We normally identify isomorphic classes and accordingly employ a plain equality sign ($A = B$). We then confine the notation $A \cong B$ to stress cases where combinatorial isomorphism results from some non-trivial transformation.

Definition I.4. *The* ordinary generating function *(OGF) of a sequence (A_n) is the formal power series*

$$(7) \qquad A(z) = \sum_{n=0}^{\infty} A_n z^n.$$

The ordinary generating function *(OGF) of a combinatorial class A is the generating function of the numbers $A_n = \text{card}(A_n)$. Equivalently, the OGF of class A admits the combinatorial form*

$$(8) \qquad A(z) = \sum_{\alpha \in A} z^{|\alpha|}.$$

It is also said that the variable z marks *size in the generating function.*

The combinatorial form of an OGF in (8) results straightforwardly from observing that the term z^n occurs as many times as there are objects in A having size n. We stress the fact that, at this stage and throughout Part A, generating functions are manipulated algebraically as formal sums; that is, they are considered as *formal power series* (see the framework of Appendix A.5: *Formal power series*, p. 730)

Naming convention. We adhere to a systematic *naming convention*: classes, their counting sequences, and their generating functions are systematically denoted by the same groups of letters: for instance, A for a class, $\{A_n\}$ (or $\{a_n\}$) for the counting sequence, and $A(z)$ (or $a(z)$) for its OGF.

Coefficient extraction. We let generally $[z^n] f(z)$ denote the operation of extracting the coefficient of z^n in the formal power series $f(z) = \sum f_n z^n$, so that

$$(9) \qquad [z^n] \left(\sum_{n \geq 0} f_n z^n \right) = f_n.$$

(The coefficient extractor $[z^n] f(z)$ reads as "coefficient of z^n in $f(z)$".)

1. On September 4, 1751, Euler writes to his friend Goldbach [196]:

Ich bin neulich auf eine Betrachtung gefallen, welche mir nicht wenig merkwürdig vorkam. Dieselbe betrifft, auf wie vielerley Arten ein gegebenes polygonum durch Diagonallinien in triangula zerchnitten werden könne.	I have recently encountered a question, which appears to me rather noteworthy. It concerns the number of ways in which a given [convex] polygon can be decomposed into triangles by diagonal lines.

Euler then describes the problem (for an n–gon, i.e., $(n-2)$ triangles) and concludes:

Setze ich nun die Anzahl dieser verschiedenen Arten $= x$ [...]. Hieraus habe ich nun den Schluss gemacht, dass generaliter sey

$$x = \frac{2.6.10.14....(4n-10)}{2.3.4.5....(n-1)}$$

Let me now denote by x this number of ways [...]. I have then reached the conclusion that in all generality

$$x = \frac{2.6.10.14....(4n-10)}{2.3.4.5....(n-1)}$$

[...] Ueber die Progression der Zahlen $1, 2, 5, 14, 42, 132$, *etc. habe ich auch diese Eigenschaft angemerket, dass* $1 + 2a + 5a^2 + 14a^3 + 42a^4 + 132a^5 + etc. = \frac{1-2a-\sqrt{1-4a}}{2aa}$.

[...] Regarding the progression of the numbers $1, 2, 5, 14, 42, 132$, and so on, I have also observed the following property: $1 + 2a + 5a^2 + 14a^3 + 42a^4 + 132a^5 + etc. = \frac{1-2a-\sqrt{1-4a}}{2aa}$.

Thus, as early as 1751, Euler knew the solution as well as the associated **generating function**. From his writing, it is however unclear whether he had found complete proofs.

2. In the course of the 1750s, Euler communicated the problem, together with initial elements of the counting sequence, to Segner, who writes in his publication [146] dated 1758: "The great Euler has benevolently communicated these numbers to me; the way in which he found them, and the law of their progression having remained hidden to me" [*"quos numeros mecum beneuolus communicauit summus Eulerus; modo, quo eos reperit, atque progressionis ordine, celatis"*]. Segner develops a recurrence approach to Catalan numbers. By a root decomposition analogous to ours, on p. 35, he proves (in our notation, for decompositions into n triangles)

$$\text{(4)} \qquad T_n = \sum_{k=0}^{n-1} T_k T_{n-1-k}, \qquad T_0 = 1,$$

a recurrence by which the Catalan numbers can be computed to any desired order. (Segner's work was to be reviewed in [197], anonymously, but most probably, by Euler.)

3. During the 1830s, Liouville circulated the problem and wrote to Lamé, who answered the next day(!) with a proof [399] based on recurrences similar to (4) of the explicit expression:

$$\text{(5)} \qquad T_n = \frac{1}{n+1}\binom{2n}{n}.$$

Interestingly enough, Lamé's three-page note [399] appeared in the 1838 issue of the *Journal de mathématiques pures et appliquées* ("Journal de Liouville"), immediately followed by a longer study by Catalan [106], who also observed that the T_n intervene in the number of ways of multiplying n numbers (this book, §I. 5.3, p. 73). Catalan would then return to these problems [107, 108], and the numbers $1, 1, 2, 5, 14, 42, \ldots$ eventually became known as the **Catalan numbers**. In [107], Catalan finally *proves* the validity of Euler's generating function:

$$\text{(6)} \qquad T(z) := \sum_n T_n z^n = \frac{1 - \sqrt{1-4z}}{2z}.$$

4. Nowadays, *symbolic methods* directly yield the generating function (6), from which both the recurrence (4) and the explicit form (5) follow easily; see pp. 6 and 35.

Figure I.2. The prehistory of Catalan numbers.

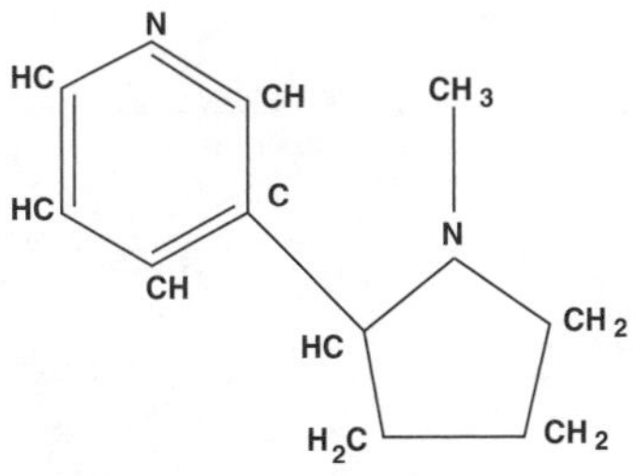

Figure I.3. A molecule, methylpyrrolidinyl-pyridine (nicotine), is a complex assembly whose description can be reduced to a single formula corresponding here to a total of 26 atoms.

The OGFs corresponding to our three examples $\mathcal{W}, \mathcal{P}, \mathcal{T}$ are then

$$
(10) \quad
\begin{cases}
W(z) & = & \displaystyle\sum_{n=0}^{\infty} 2^n z^n & = & \dfrac{1}{1-2z} \\[2ex]
P(z) & = & \displaystyle\sum_{n=0}^{\infty} n!\, z^n \\[2ex]
T(z) & = & \displaystyle\sum_{n=0}^{\infty} \dfrac{1}{n+1}\binom{2n}{n} z^n & = & \dfrac{1-\sqrt{1-4z}}{2z}.
\end{cases}
$$

The first expression relative to $W(z)$ is immediate as it is the sum of a geometric progression. The second generating function $P(z)$ is not clearly related to simple functions of analysis. (Note that the expression still makes sense within the strict framework of formal power series.) The third expression relative to $T(z)$ is equivalent to the explicit form of T_n via Newton's expansion of $(1+x)^{1/2}$ (pp. 7 and 35 as well as Figure I.2). The OGFs $W(z)$ and $T(z)$ can then be interpreted as standard analytic objects, upon assigning values in the complex domain $\mathbb{C}$ to the formal variable z. In effect, the series $W(z)$ and $T(z)$ converge in a neighbourhood of 0 and represent complex functions that are well defined near the origin, namely when $|z| < \frac{1}{2}$ for $W(z)$ and $|z| < \frac{1}{4}$ for $T(z)$. The OGF $P(z)$ is a purely formal power series (its radius of convergence is 0) that can nonetheless be subjected to the usual algebraic operations of power series. (Permutation enumeration is most conveniently approached by the exponential generating functions developed in Chapter II.)

Combinatorial form of generating functions (GFs). The combinatorial form (8) shows that generating functions are nothing but a reduced representation of the combinatorial class, where internal structures are destroyed and elements contributing to size (atoms) are replaced by the variable z. In a sense, this is analogous to what chemists do by writing linear reduced ("molecular") formulae for complex molecules (Figure I.3). Great use of this observation was made by Schützenberger as early as the 1950s and 1960s. It explains the many formal similarities that are observed between combinatorial structures and generating functions.

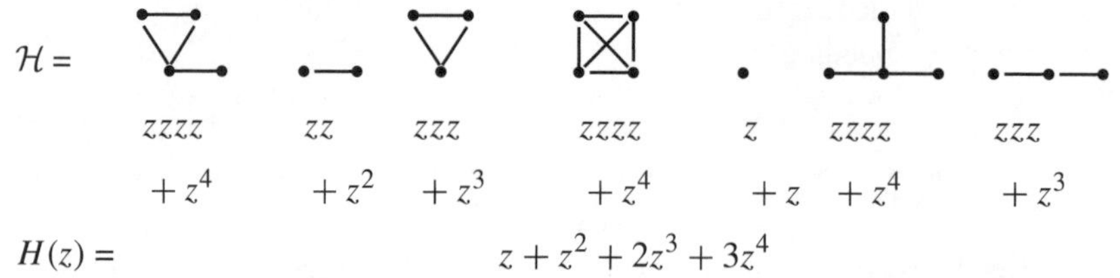

$$\mathcal{H} =$$

$$\begin{array}{ccccccc} zzzz & zz & zzz & zzzz & z & zzzz & zzz \\ +z^4 & +z^2 & +z^3 & +z^4 & +z & +z^4 & +z^3 \end{array}$$

$$H(z) = \qquad\qquad z + z^2 + 2z^3 + 3z^4$$

Figure I.4. A finite family of graphs and its eventual reduction to a generating function.

Figure I.4 provides a combinatorial illustration: start with a (finite) family of graphs $\mathcal{H}$, with size taken as the number of vertices. Each vertex in each graph is replaced by the variable z and the graph structure is "forgotten"; then the monomials corresponding to each graph are formed and the generating function is finally obtained by gathering all the monomials.

For instance, there are 3 graphs of size 4 in $\mathcal{H}$, in agreement with the fact that $[z^4]H(z) = 3$. If size had been instead defined by number of edges, another generating function would have resulted, namely, with y marking the new size: $1+y+y^2+2y^3+y^4+y^6$. If both number of vertices and number of edges are of interest, then a bivariate generating function is obtained: $H(z, y) = z+z^2y+z^3y^2+z^3y^3+z^4y^3+z^4y^4+z^4y^6$; such multivariate generating functions are developed systematically in Chapter III.

A path often taken in the literature is to decompose the structures to be enumerated into smaller structures either of the same type or of simpler types, and then extract from such a decomposition *recurrence relations* that are satisfied by the $\{A_n\}$. In this context, the recurrence relations are either solved directly—whenever they are simple enough—or by means of *ad hoc* generating functions, introduced as mere technical artifices.

By contrast, in the framework of this book, classes of combinatorial structures are built *directly* in terms of simpler classes by means of a collection of elementary combinatorial *constructions*. This closely resembles the description of formal languages by means of grammars, as well as the construction of structured data types in programming languages. The approach developed here has been termed *symbolic*, as it relies on a formal specification language for combinatorial structures. Specifically, it is based on so–called *admissible constructions* that permit direct translations into generating functions.

Definition I.5. *Let Φ be an m–ary construction that associates to any collection of classes $\mathcal{B}^{(1)}, \ldots \mathcal{B}^{(m)}$ a new class*

$$\mathcal{A} = \Phi[\mathcal{B}^{(1)}, \ldots, \mathcal{B}^{(m)}].$$

The construction Φ is admissible *iff the counting sequence (A_n) of $\mathcal{A}$ only depends on the counting sequences $(B_n^{(1)}), \ldots, (B_n^{(m)})$ of $\mathcal{B}^{(1)}, \ldots, \mathcal{B}^{(m)}$.*

For such an admissible construction, there then exists a well-defined operator Ψ acting on the corresponding ordinary generating functions:

$$A(z) = \Psi[B^{(1)}(z), \ldots, B^{(m)}],$$

and it is this basic fact about admissibility that will be used throughout the book.

As an introductory example, take the construction of cartesian product, which is the usual one enriched with a natural notion of size.

Definition I.6. *The* cartesian product construction *applied to two classes $\mathcal{B}$ and $\mathcal{C}$ forms ordered pairs,*

$$(11) \qquad \mathcal{A} = \mathcal{B} \times \mathcal{C} \quad \text{iff} \quad \mathcal{A} = \{\alpha = (\beta, \gamma) \mid \beta \in \mathcal{B}, \ \gamma \in \mathcal{C}\},$$

with the size of a pair $\alpha = (\beta, \gamma)$ being defined by

$$(12) \qquad |\alpha|_{\mathcal{A}} = |\beta|_{\mathcal{B}} + |\gamma|_{\mathcal{C}}.$$

By considering all possibilities, it is immediately seen that the counting sequences corresponding to $\mathcal{A}, \mathcal{B}, \mathcal{C}$ are related by the convolution relation

$$(13) \qquad A_n = \sum_{k=0}^{n} B_k C_{n-k},$$

which means admissibility. Furthermore, we recognize here the formula for a product of two power series:

$$(14) \qquad A(z) = B(z) \cdot C(z).$$

In summary: *the cartesian product is admissible and it translates as a product of OGFs.*

Similarly, let $\mathcal{A}, \mathcal{B}, \mathcal{C}$ be combinatorial classes satisfying

$$(15) \qquad \mathcal{A} = \mathcal{B} \cup \mathcal{C}, \qquad \text{with} \quad \mathcal{B} \cap \mathcal{C} = \emptyset,$$

with size defined in a consistent manner: for $\omega \in \mathcal{A}$,

$$(16) \qquad |\omega|_{\mathcal{A}} = \begin{cases} |\omega|_{\mathcal{B}} & \text{if } \omega \in \mathcal{B} \\ |\omega|_{\mathcal{C}} & \text{if } \omega \in \mathcal{C}. \end{cases}$$

One has

$$(17) \qquad A_n = B_n + C_n,$$

which, at generating function level, means

$$(18) \qquad A(z) = B(z) + C(z).$$

Thus, *the union of disjoint sets is admissible and it translates as a sum of generating functions.* (A more formal version of this statement is given in the next section.)

The correspondences provided by (11)–(14) and (15)–(18) are summarized by the strikingly simple dictionary

$$(19) \qquad \begin{cases} \mathcal{A} = \mathcal{B} \cup \mathcal{C} \implies A(z) = B(z) + C(z) & (\text{provided } \mathcal{B} \cap \mathcal{C} = \emptyset) \\ \mathcal{A} = \mathcal{B} \times \mathcal{C} \implies A(z) = B(z) \cdot C(z), \end{cases}$$

to be compared with the plain arithmetic case of (3), p. 18. The merit of such relations is that they can be stated as general purpose translation rules that only need to be established once and for all. As soon as the problem of counting elements of a union of disjoint sets or a cartesian product is recognized, it becomes possible to dispense altogether with the intermediate stages of writing explicitly coefficient relations or recurrences as in (13) or (17). This is the spirit of the *symbolic method* for combinatorial enumerations. Its interest lies in the fact that several powerful set-theoretic constructions are amenable to such a treatment, as we see in the next section.

▷ **I.3.** *Continuity, Lipschitz and Hölder conditions.* An admissible construction is said to be *continuous* if it is a continuous function on the space of formal power series equipped with its standard ultrametric distance (Appendix A.5: *Formal power series*, p. 730). Continuity captures the desirable property that constructions depend on their arguments in a finitary way. For all the constructions of this book, there furthermore exists a function $\vartheta(n)$, such that (A_n) only depends on the first $\vartheta(n)$ elements of the $(B_k^{(1)}), \ldots, (B_k^{(m)})$, with $\vartheta(n) \le Kn + L$ (Hölder condition) or $\vartheta(n) \le n + L$ (Lipschitz condition). For instance, the functional $f(z) \mapsto f(z^2)$ is Hölder; the functional $f(z) \mapsto \partial_z f(z)$ is Lipschitz. ◁

I. 2. Admissible constructions and specifications

The main goal of this section is to introduce formally the basic *constructions* that constitute the core of a specification language for combinatorial structures. This core is based on disjoint unions, also known as combinatorial sums, and on cartesian products that we have just discussed. We shall augment it by the constructions of sequence, cycle, multiset, and powerset. A class is *constructible* or *specifiable* if it can be defined from primal elements by means of these constructions. The generating function of any such class satisfies functional equations that can be transcribed systematically from a specification; see Theorems I.1 (p. 27) and I.2 (p. 33), as well as Figure I.18 (p. 93) at the end of this chapter for a summary.

I. 2. 1. Basic constructions. First, we assume we are given a class $\mathcal{E}$ called the *neutral class* that consists of a single object of size 0; any such object of size 0 is called a *neutral object* and is usually denoted by symbols such as ϵ or **1**. The reason for this terminology becomes clear if one considers the combinatorial isomorphism

$$\mathcal{A} \cong \mathcal{E} \times \mathcal{A} \cong \mathcal{A} \times \mathcal{E}.$$

We also assume as given an *atomic class* $\mathcal{Z}$ comprising a single element of size 1; any such element is called an atom; an atom may be used to describe a generic node in a tree or graph, in which case it may be represented by a circle (● or ○), but also a generic letter in a word, in which case it may be instantiated as a, b, c, ... Distinct copies of the neutral or atomic class may also be subscripted by indices in various ways. Thus, for instance, we may use the classes $\mathcal{Z}_a = \{a\}$, $\mathcal{Z}_b = \{b\}$ (with a, b of size 1) to build up binary words over the alphabet $\{a, b\}$, or $\mathcal{Z}_\bullet = \{\bullet\}$, $\mathcal{Z}_\circ = \{\circ\}$ (with ●, ○ taken to be of size 1) to build trees with nodes of two colours. Similarly, we may introduce $\mathcal{E}_\square, \mathcal{E}_1, \mathcal{E}_2$ to denote a class comprising the neutral objects $\square, \epsilon_1, \epsilon_2$ respectively.

Clearly, the generating functions of a neutral class $\mathcal{E}$ and an atomic class $\mathcal{Z}$ are

$$E(z) = 1, \qquad Z(z) = z,$$

corresponding to the unit 1, and the variable z, of generating functions.

Combinatorial sum (disjoint union). The intent of *combinatorial sum* also known as *disjoint union* is to capture the idea of a union of disjoint sets, but without any extraneous condition (disjointness) being imposed on the arguments of the construction. To do so, we formalize the (combinatorial) sum of two classes $\mathcal{B}$ and $\mathcal{C}$ as the union (in the standard set-theoretic sense) of two *disjoint* copies, say $\mathcal{B}^\square$ and $\mathcal{C}^\diamond$, of $\mathcal{B}$ and $\mathcal{C}$. A picturesque way to view the construction is as follows: first choose two distinct colours and repaint the elements of $\mathcal{B}$ with the first colour and the elements of $\mathcal{C}$ with the second colour. This is made precise by introducing two distinct "markers", say $\square$ and $\diamond$, each a neutral object (i.e., of size zero); the disjoint union $\mathcal{B}+\mathcal{C}$ of $\mathcal{B}, \mathcal{C}$ is then defined as a standard set-theoretic union:

$$\mathcal{B} + \mathcal{C} := (\{\square\} \times \mathcal{B}) \cup (\{\diamond\} \times \mathcal{C}).$$

The size of an object in a disjoint union $\mathcal{A} = \mathcal{B} + \mathcal{C}$ is by definition inherited from its size in its class of origin, as in Equation (16). One good reason behind the definition adopted here is that the combinatorial sum of two classes is *always* well defined, no matter whether or not the classes intersect. Furthermore, disjoint union is equivalent to a standard union whenever it is applied to disjoint sets.

Because of disjointness of the copies, one has the implication

$$\mathcal{A} = \mathcal{B} + \mathcal{C} \implies A_n = B_n + C_n \quad \text{and} \quad A(z) = B(z) + C(z),$$

so that disjoint union is admissible. Note that, in contrast, standard set-theoretic union is not an admissible construction since

$$\mathrm{card}(\mathcal{B}_n \cup \mathcal{C}_n) = \mathrm{card}(\mathcal{B}_n) + \mathrm{card}(\mathcal{C}_n) - \mathrm{card}(\mathcal{B}_n \cap \mathcal{C}_n),$$

and information on the internal structure of $\mathcal{B}$ and $\mathcal{C}$ (i.e., the nature of their intersection) is needed in order to be able to enumerate the elements of their union.

Cartesian product. This construction $\mathcal{A} = \mathcal{B} \times \mathcal{C}$ forms all possible ordered pairs in accordance with Definition I.6. The size of a pair is obtained additively from the size of components in accordance with (12).

Next, we introduce a few fundamental constructions that build upon set-theoretic union and product, and form sequences, sets, and cycles. These powerful constructions suffice to define a broad variety of combinatorial structures.

Sequence construction. If $\mathcal{B}$ is a class then the *sequence* class $\mathrm{SEQ}(\mathcal{B})$ is defined as the infinite sum

$$\mathrm{SEQ}(\mathcal{B}) = \{\epsilon\} + \mathcal{B} + (\mathcal{B} \times \mathcal{B}) + (\mathcal{B} \times \mathcal{B} \times \mathcal{B}) + \cdots$$

with ϵ being a neutral structure (of size 0). In other words, we have

$$\mathcal{A} = \big\{ (\beta_1, \ldots, \beta_\ell) \;\big|\; \ell \geq 0, \; \beta_j \in \mathcal{B} \big\},$$

which matches our intuition as to what sequences should be. (The neutral structure in this context corresponds to $\ell = 0$; it plays a rôle similar to that of the "empty" word in formal language theory.) It is then readily checked that the construction $\mathcal{A} = \mathrm{SEQ}(\mathcal{B})$ defines a proper class satisfying the finiteness condition for sizes if and only if $\mathcal{B}$ *contains no object of size* 0. From the definition of size for sums and products, it follows that the size of an object $\alpha \in \mathcal{A}$ is to be taken as the sum of the sizes of its components:

$$\alpha = (\beta_1, \ldots, \beta_\ell) \qquad \Longrightarrow \qquad |\alpha| = |\beta_1| + \cdots + |\beta_\ell|.$$

Cycle construction. Sequences taken up to a circular shift of their components define cycles, the notation being $\text{CYC}(\mathcal{B})$. In precise terms, one has[3]

$$\text{CYC}(\mathcal{B}) := (\text{SEQ}(\mathcal{B}) \setminus \{\epsilon\}) / \mathbf{S},$$

where $\mathbf{S}$ is the equivalence relation between sequences defined by

$$(\beta_1, \ldots, \beta_r) \, \mathbf{S} \, (\beta_1', \ldots, \beta_r')$$

iff there exists some *circular shift* τ of $[1 .. r]$ such that for all j, $\beta_j' = \beta_{\tau(j)}$; in other words, for some d, one has $\beta_j' = \beta_{1 + (j-1+d) \bmod r}$. Here is, for instance, a depiction of the cycles formed from the 8 and 16 sequences of lengths 3 and 4 over two types of objects (a, b): the number of cycles is 4 (for $n = 3$) and 6 (for $n = 4$). Sequences are grouped into equivalence classes according to the relation $\mathbf{S}$:

$$(20) \qquad 3\text{-cycles} : \begin{cases} aaa \\ aab\ aba\ baa \\ abb\ bba\ bab \\ bbb \end{cases}, \qquad 4\text{-cycles} : \begin{cases} aaaa \\ aaab\ aaba\ abaa\ baaa \\ aabb\ abba\ bbaa\ baab \\ abab\ baba \\ abbb\ bbba\ bbab\ babb \\ bbbb \end{cases}.$$

According to the definition, this construction corresponds to the formation of directed cycles (see also the necklaces of Note I.1, p. 18). We make only a limited use of it for unlabelled objects; however, its counterpart plays a rather important rôle in the context of labelled structures and exponential generating functions of Chapter II.

Multiset construction. Following common mathematical terminology, *multisets* are like finite sets (that is the order between elements does not count), but arbitrary repetitions of elements are allowed. The notation is $\mathcal{A} = \text{MSET}(\mathcal{B})$ when $\mathcal{A}$ is obtained by forming all *finite* multisets of elements from $\mathcal{B}$. The precise way of defining $\text{MSET}(\mathcal{B})$ is as a quotient:

$$\text{MSET}(\mathcal{B}) := \text{SEQ}(\mathcal{B}) / \mathbf{R} \quad \text{with} \quad \mathbf{R},$$

the equivalence relation of sequences being defined by $(\alpha_1, \ldots, \alpha_r) \, \mathbf{R} \, (\beta_1, \ldots, \beta_r)$ iff there exists some *arbitrary permutation* σ of $[1 .. r]$ such that for all j, $\beta_j = \alpha_{\sigma(j)}$.

Powerset construction. The *powerset* class (or set class) $\mathcal{A} = \text{PSET}(\mathcal{B})$ is defined as the class consisting of all *finite* subsets of class $\mathcal{B}$, or equivalently, as the class $\text{PSET}(\mathcal{B}) \subset \text{MSET}(\mathcal{B})$ formed of multisets that involve no repetitions.

We again need to make explicit the way the size function is defined when such constructions are performed: as for products and sequences, the size of a composite object—set, multiset, or cycle—is defined to be the sum of the sizes of its components.

▷ **I.4.** *The semi-ring of combinatorial classes.* Under the convention of identifying isomorphic classes, sum and product acquire pleasant algebraic properties: combinatorial sums and cartesian products become commutative and associative operations, e.g.,

$$(\mathcal{A} + \mathcal{B}) + \mathcal{C} = \mathcal{A} + (\mathcal{B} + \mathcal{C}), \qquad \mathcal{A} \times (\mathcal{B} \times \mathcal{C}) = (\mathcal{A} \times \mathcal{B}) \times \mathcal{C},$$

while distributivity holds, $(\mathcal{A} + \mathcal{B}) \times \mathcal{C} = (\mathcal{A} \times \mathcal{C}) + (\mathcal{B} \times \mathcal{C})$. ◁

---

[3]By convention, there are no "empty" cycles.

▷ **I.5.** *Natural numbers.* Let $\mathcal{Z} := \{\bullet\}$ with $\bullet$ an atom (of size 1). Then $\mathcal{I} = \text{SEQ}(\mathcal{Z}) \setminus \{\epsilon\}$ is a way of describing positive integers in unary notation: $\mathcal{I} = \{\bullet, \bullet\bullet, \bullet\bullet\bullet, \ldots\}$. The corresponding OGF is $I(z) = z/(1 - z) = z + z^2 + z^3 + \cdots$. ◁

▷ **I.6.** *Interval coverings.* Let $\mathcal{Z} := \{\bullet\}$ be as before. Then $\mathcal{A} = \mathcal{Z} + (\mathcal{Z} \times \mathcal{Z})$ is a set of two elements, $\bullet$ and $(\bullet, \bullet)$, which we choose to draw as $\{\bullet, \bullet\!-\!\bullet\}$. Then $\mathcal{C} = \text{SEQ}(\mathcal{A})$ contains

$$\bullet, \ \bullet\bullet, \ \bullet\!-\!\bullet, \ \bullet\bullet\!-\!\bullet, \ \bullet\!-\!\bullet\bullet, \ \bullet\!-\!\bullet\bullet\!-\!\bullet, \ \bullet\bullet\bullet\bullet, \ \ldots$$

With the notion of size adopted, the objects of size n in $\mathcal{C} = \text{SEQ}(\mathcal{Z} + (\mathcal{Z} \times \mathcal{Z}))$ are (isomorphic to) the *coverings* of $[0, n]$ by intervals (matches) of length either 1 or 2. The OGF

$$C(z) = 1 + z + 2z^2 + 3z^3 + 5z^4 + 8z^5 + 13z^6 + 21z^7 + 34z^8 + 55z^9 + \cdots,$$

is, as we shall see shortly (p. 42), the OGF of Fibonacci numbers. ◁

I. 2.2. The admissibility theorem for ordinary generating functions.

This section is a formal treatment of admissibility proofs for the constructions that we have introduced. The final implication is that any specification of a constructible class translates directly into generating function equations. The translation of the cycle construction involves the Euler totient function $\varphi(k)$ defined as the number of integers in $[1, k]$ that are relatively prime to k (Appendix A.1: *Arithmetical functions*, p. 721).

Theorem I.1 (Basic admissibility, unlabelled universe). *The constructions of union, cartesian product, sequence, powerset, multiset, and cycle are all admissible. The associated operators are as follows.*

Sum: $\qquad\qquad\quad \mathcal{A} = \mathcal{B} + \mathcal{C} \quad\Longrightarrow\quad A(z) = B(z) + C(z)$

Cartesian product: $\quad \mathcal{A} = \mathcal{B} \times \mathcal{C} \quad\Longrightarrow\quad A(z) = B(z) \cdot C(z)$

Sequence: $\qquad\qquad \mathcal{A} = \text{SEQ}(\mathcal{B}) \quad\Longrightarrow\quad A(z) = \dfrac{1}{1 - B(z)}$

Powerset: $\qquad\quad \mathcal{A} = \text{PSET}(\mathcal{B}) \quad\Longrightarrow\quad A(z) = \begin{cases} \displaystyle\prod_{n\geq 1}(1 + z^n)^{B_n} \\[2em] \displaystyle\exp\left(\sum_{k=1}^{\infty} \frac{(-1)^{k-1}}{k} B(z^k) \right) \end{cases}$

Multiset: $\qquad\quad \mathcal{A} = \text{MSET}(\mathcal{B}) \quad\Longrightarrow\quad A(z) = \begin{cases} \displaystyle\prod_{n\geq 1}(1 - z^n)^{-B_n} \\[2em] \displaystyle\exp\left(\sum_{k=1}^{\infty} \frac{1}{k} B(z^k) \right) \end{cases}$

Cycle: $\qquad\qquad\quad \mathcal{A} = \text{CYC}(\mathcal{B}) \quad\Longrightarrow\quad A(z) = \displaystyle\sum_{k=1}^{\infty} \frac{\varphi(k)}{k} \log \frac{1}{1 - B(z^k)}.$

For the sequence, powerset, multiset, and cycle translations, it is assumed that $\mathcal{B}_0 = \emptyset$.

The class $\mathcal{E} = \{\epsilon\}$ consisting of the neutral object only, and the class $\mathcal{Z}$ consisting of a single "atomic" object (node, letter) of size 1 have OGFs

$$E(z) = 1 \quad \text{and} \quad Z(z) = z.$$

Proof. The proof proceeds case by case, building upon what we have just seen regarding unions and products.

Combinatorial sum (disjoint union). Let $\mathcal{A} = \mathcal{B} + \mathcal{C}$. Since the union is *disjoint*, and the size of an $\mathcal{A}$–element coincides with its size in $\mathcal{B}$ or $\mathcal{C}$, one has $A_n = B_n + C_n$ and $A(z) = B(z) + C(z)$, as discussed earlier. The rule also follows directly from the combinatorial form of generating functions as expressed by (8), p. 19:

$$A(z) = \sum_{\alpha \in \mathcal{A}} z^{|\alpha|} = \sum_{\alpha \in \mathcal{B}} z^{|\alpha|} + \sum_{\alpha \in \mathcal{C}} z^{|\alpha|} = B(z) + C(z).$$

Cartesian product. The admissibility result for $\mathcal{A} = \mathcal{B} \times \mathcal{C}$ was considered as an example for Definition I.6, the convolution equation (13) leading to the relation $A(z) = B(z) \cdot C(z)$. We can also offer a direct derivation based on the combinatorial form of generating functions (8), p. 19,

$$A(z) = \sum_{\alpha \in \mathcal{A}} z^{|\alpha|} = \sum_{(\beta,\gamma) \in (\mathcal{B} \times \mathcal{C})} z^{|\beta|+|\gamma|} = \left(\sum_{\beta \in \mathcal{B}} z^{|\beta|} \right) \times \left(\sum_{\gamma \in \mathcal{C}} z^{|\gamma|} \right) = B(z) \cdot C(z),$$

as follows from distributing products over sums. This derivation readily extends to an arbitrary number of factors.

Sequence construction. Admissibility for $\mathcal{A} = \text{SEQ}(\mathcal{B})$ (with $\mathcal{B}_0 = \emptyset$) follows from the union and product relations. One has

$$\mathcal{A} = \{\epsilon\} + \mathcal{B} + (\mathcal{B} \times \mathcal{B}) + (\mathcal{B} \times \mathcal{B} \times \mathcal{B}) + \cdots ,$$

so that

$$A(z) = 1 + B(z) + B(z)^2 + B(z)^3 + \cdots = \frac{1}{1 - B(z)},$$

where the geometric sum converges in the sense of formal power series since $[z^0]B(z) = 0$, by assumption.

Powerset construction. Let $\mathcal{A} = \text{PSET}(\mathcal{B})$ and first take $\mathcal{B}$ to be finite. Then, the class $\mathcal{A}$ of all the finite subsets of $\mathcal{B}$ is isomorphic to a product,

$$(21) \qquad \qquad \text{PSET}(\mathcal{B}) \cong \prod_{\beta \in \mathcal{B}} (\{\epsilon\} + \{\beta\}),$$

with ϵ a neutral structure of size 0. Indeed, distributing the products in all possible ways forms all the possible combinations (sets with no repetition allowed) of elements of $\mathcal{B}$; the reasoning is the same as what leads to an identity such as

$$(1 + a)(1 + b)(1 + c) = 1 + [a + b + c] + [ab + bc + ac] + abc,$$

where all combinations of variables appear in monomials. Then, directly from the combinatorial form of generating functions and the sum and product rules, we find

$$(22) \qquad \qquad A(z) = \prod_{\beta \in \mathcal{B}} (1 + z^{|\beta|}) = \prod_n (1 + z^n)^{B_n}.$$

The *exp–log transformation* $A(z) = \exp(\log A(z))$ then yields

$$
\begin{aligned}
A(z) &= \exp\left(\sum_{n=1}^{\infty} B_n \log(1 + z^n)\right) \\
&= \exp\left(\sum_{n=1}^{\infty} B_n \cdot \sum_{k=1}^{\infty} (-1)^{k-1} \frac{z^{nk}}{k}\right) \\
&= \exp\left(\frac{B(z)}{1} - \frac{B(z^2)}{2} + \frac{B(z^3)}{3} - \cdots\right),
\end{aligned}
$$

(23)

where the second line results from expanding the logarithm,

$$
\log(1 + u) = \frac{u}{1} - \frac{u^2}{2} + \frac{u^3}{3} - \cdots,
$$

and the third line results from exchanging the order of summations.

The proof finally extends to the case of B being infinite by noting that each A_n depends only on those B_j for which $j \le n$, to which the relations given above for the finite case apply. Precisely, let $B^{(\le m)} = \sum_{k=1}^{m} B_j$ and $A^{(\le m)} = \text{PSET}(B^{(\le m)})$. Then, with $O(z^{m+1})$ denoting any series that has no term of degree $\le m$, one has

$$
A(z) = A^{(\le m)}(z) + O(z^{m+1}) \qquad \text{and} \qquad B(z) = B^{(\le m)}(z) + O(z^{m+1}).
$$

On the other hand, $A^{(\le m)}(z)$ and $B^{(\le m)}(z)$ are connected by the fundamental exponential relation (23), since $B^{(\le m)}$ is finite. Letting m tend to infinity, there follows in the limit

$$
A(z) = \exp\left(\frac{B(z)}{1} - \frac{B(z^2)}{2} + \frac{B(z^3)}{3} - \cdots\right).
$$

(See Appendix A.5: *Formal power series*, p. 730 for the notion of formal convergence.)

Multiset construction. First for finite B (with $B_0 = \emptyset$), the multiset class $A = \text{MSET}(B)$ is definable by

$$
\text{MSET}(B) \cong \prod_{\beta \in B} \text{SEQ}(\{\beta\}).
$$

(24)

In words, any multiset can be sorted, in which case it can be viewed as formed of a sequence of repeated elements β_1, followed by a sequence of repeated elements β_2, where $\beta_1, \beta_2, \ldots$ is a canonical listing of the elements of B. The relation translates into generating functions by the product and sequence rules,

$$
\begin{aligned}
A(z) &= \prod_{\beta \in B} (1 - z^{|\beta|})^{-1} = \prod_{n=1}^{\infty} (1 - z^n)^{-B_n} \\
&= \exp\left(\sum_{n=1}^{\infty} B_n \log(1 - z^n)^{-1}\right) \\
&= \exp\left(\frac{B(z)}{1} + \frac{B(z^2)}{2} + \frac{B(z^3)}{3} + \cdots\right),
\end{aligned}
$$

(25)

where the exponential form results from the exp–log transformation. The case of an infinite class $\mathcal{B}$ follows by a limit argument analogous the one used for powersets.

Cycle construction. The translation of the cycle relation $\mathcal{A} = \text{CYC}(\mathcal{B})$ turns out to be

$$A(z) = \sum_{k=1}^{\infty} \frac{\varphi(k)}{k} \log \frac{1}{1 - B(z^k)},$$

where $\varphi(k)$ is the Euler totient function. The first terms, with $L_k(z) := \log(1 - B(z^k))^{-1}$ are

$$A(z) = \frac{1}{1}L_1(z) + \frac{1}{2}L_2(z) + \frac{2}{3}L_3(z) + \frac{2}{4}L_4(z) + \frac{4}{5}L_5(z) + \frac{2}{6}L_6(z) + \cdots.$$

We reserve the proof to Appendix A.4: *Cycle construction*, p. 729, since it relies in part on multivariate generating functions to be officially introduced in Chapter III. ∎

The results for sets, multisets, and cycles are particular cases of the well-known *Pólya theory* that deals more generally with the enumeration of objects under group symmetry actions; for Pólya's original and its edited version, see [488, 491]. This theory is described in many textbooks, for instance, those of Comtet [129] and Harary and Palmer [129, 319]; Notes I.58–I.60, pp. 85–86, distil its most basic aspects. The approach adopted here amounts to considering simultaneously all possible values of the number of components by means of bivariate generating functions. Powerful generalizations within Joyal's elegant theory of species [359] are presented in the book by Bergeron, Labelle, and Leroux [50].

▷ **I.7.** *Vallée's identity.* Let $\mathcal{M} = \text{MSET}(\mathcal{C})$, $\mathcal{P} = \text{PSET}(\mathcal{C})$. One has combinatorially:

$$M(z) = P(z)M(z^2).$$

(Hint: a multiset contains elements of either odd or even multiplicity.) Accordingly, one can deduce the translation of powersets from the formula for multisets. Iterating the relation above yields $M(z) = P(z)P(z^2)P(z^4)P(z^8)\cdots$: this is closely related to the binary representation of numbers and to Euler's identity (p. 49). It is used for instance in Note I.66 p. 91.   ◁

Restricted constructions. In order to increase the descriptive power of the framework of constructions, we ought to be able to allow restrictions on the number of components in sequences, sets, multisets, and cycles. Let $\mathfrak{K}$ be a metasymbol representing any of SEQ, CYC, MSET, PSET and let Ω be a predicate over the integers; then $\mathfrak{K}_\Omega(\mathcal{A})$ will represent the class of objects constructed by $\mathfrak{K}$, with a number of components constrained to satisfy Ω. For instance, the notation

(26) $\text{SEQ}_{=k}$ (or simply SEQ_k), $\text{SEQ}_{>k}$, $\text{SEQ}_{1..k}$

refers to sequences whose number of components are exactly k, larger than k, or in the interval $1..k$ respectively. In particular,

$$\text{SEQ}_k(\mathcal{B}) := \overbrace{\mathcal{B} \times \cdots \times \mathcal{B}}^{k \text{ times}} \equiv \mathcal{B}^k, \qquad \text{SEQ}_{\geq k}(\mathcal{B}) = \sum_{j \geq k} \mathcal{B}^j \cong \mathcal{B}^k \times \text{SEQ}(\mathcal{B}),$$

$$\text{MSET}_k(\mathcal{B}) := \text{SEQ}_k(\mathcal{B})/\mathbf{R}.$$

Similarly, SEQ_{odd}, SEQ_{even} will denote sequences with an odd or even number of components, and so on.

Translations for such restricted constructions are available, as shown generally in Subsection I. 6.1, p. 83. Suffice it to note for the moment that the construction $\mathcal{A} = \mathrm{SEQ}_k(\mathcal{B})$ is really an abbreviation for a k-fold product, hence it admits the translation into OGFs

$$(27) \qquad \mathcal{A} = \mathrm{SEQ}_k(\mathcal{B}) \qquad \Longrightarrow \qquad A(z) = B(z)^k.$$

I. 2.3. Constructibility and combinatorial specifications. By composing basic constructions, we can build compact descriptions (specifications) of a broad variety of combinatorial classes. Since we restrict attention to *admissible* constructions, we can immediately derive OGFs for these classes. Put differently, the task of enumerating a combinatorial class is reduced to *programming* a specification for it in the language of admissible constructions. In this subsection, we first discuss the expressive power of the language of constructions, then summarize the symbolic method (for unlabelled classes and OGFs) by Theorem I.2.

First, in the framework just introduced, the class of all binary words is described by

$$\mathcal{W} = \mathrm{SEQ}(\mathcal{A}), \qquad \text{where} \quad \mathcal{A} = \{a, b\} \cong \mathcal{Z} + \mathcal{Z},$$

the ground alphabet, comprises two elements (letters) of size 1. The size of a binary word then coincides with its length (the number of letters it contains). In other terms, we start from basic atomic elements and build up words by forming freely all the objects determined by the sequence construction. Such a combinatorial description of a class that only involves a composition of basic constructions applied to initial classes $\mathcal{E}, \mathcal{Z}$ is said to be an *iterative* (or *non-recursive*) *specification*. Other examples already encountered include binary necklaces (Note I.1, p. 18) and the positive integers (Note I.5, p. 27) respectively defined by

$$\mathcal{N} = \mathrm{CYC}(\mathcal{Z} + \mathcal{Z}) \qquad \text{and} \qquad \mathcal{I} = \mathrm{SEQ}_{\geq 1}(\mathcal{Z}).$$

From this, one can construct ever more complicated objects. For instance,

$$\mathcal{P} = \mathrm{MSET}(\mathcal{I}) \equiv \mathrm{MSET}(\mathrm{SEQ}_{\geq 1}(\mathcal{Z}))$$

means the class of multisets of positive integers, which is isomorphic to the class of integer partitions (see Section I. 3 below for a detailed discussion). As such examples demonstrate, a specification that is iterative can be represented as a single term built on $\mathcal{E}, \mathcal{Z}$ and the constructions $+, \times, \mathrm{SEQ}, \mathrm{CYC}, \mathrm{MSET}, \mathrm{PSET}$. An iterative specification can be equivalently listed by naming some of the subterms (for instance, partitions in terms of natural integers $\mathcal{I}$, themselves defined as sequences of atoms $\mathcal{Z}$).

Semantics of recursion. We next turn our attention to recursive specifications, starting with trees (cf also Appendix A.9: *Tree concepts*, p. 737, for basic definitions). In graph theory, a tree is classically defined as an undirected graph that is connected and acyclic. Additionally, a tree is *rooted* if a particular vertex is specified (this vertex is then kown as the root). Computer scientists commonly make use of trees called *plane*[4] that are rooted but also embedded in the plane, so that the ordering of subtrees

---

[4]The alternative terminology "planar tree" is also often used, but it is frowned upon by some as incorrect (all trees are planar graphs). We have thus opted for the expression "plane tree", which parallels the phrase "plane curve".

attached to any node matters. Here, we will give the name of *general plane trees* to such rooted plane trees and call $\mathcal{G}$ their class, where size is the number of vertices; see, e.g., reference [538]. (The term "general" refers to the fact that all nodes degrees are allowed.) For instance, a general tree of size 16, drawn with the root on top, is:

$$\tau = \quad$$

As a consequence of the definition, if one interchanges, say, the second and third root subtrees, then a different tree results—the original tree and its variant are not equivalent under a smooth deformation of the plane. (General trees are thus comparable to graphical renderings of genealogies where children are ordered by age.). Although we have introduced plane trees as two-dimensional diagrams, it is obvious that any tree also admits a linear representation: a tree τ with root ζ and root subtrees $\tau_1, \ldots, \tau_r$ (in that order) can be seen as the object $\zeta \boxed{\tau_1, \ldots, \tau_r}$, where the box encloses similar representations of subtrees. Typographically, a box $\boxed{\cdot}$ may be reduced to a matching pair of parentheses, "$(\cdot)$", and one gets in this way a linear description that illustrates the correspondence between trees viewed as plane diagrams and functional terms of mathematical logic and computer science.

Trees are best described recursively. A plane tree is a root to which is attached a (possibly empty) sequence of trees. In other words, the class $\mathcal{G}$ of general trees is definable by the recursive equation

$$(28) \qquad\qquad \mathcal{G} = \mathcal{Z} \times \mathrm{SEQ}(\mathcal{G}),$$

where $\mathcal{Z}$ comprises a single atom written "•" that represents a generic node.

Although such recursive definitions are familiar to computer scientists, the specification (28) may look dangerously circular to some. One way of making good sense of it is via an adaptation of the numerical technique of iteration. Start with $\mathcal{G}^{[0]} = \emptyset$, the empty set, and define successively the classes

$$\mathcal{G}^{[j+1]} = \mathcal{Z} \times \mathrm{SEQ}(\mathcal{G}^{[j]}).$$

For instance, $\mathcal{G}^{[1]} = \mathcal{Z} \times \mathrm{SEQ}(\emptyset) = \{(\bullet, \epsilon)\} \cong \{\bullet\}$ describes the tree of size 1, and

$$\mathcal{G}^{[2]} \;=\; \Big\{ \bullet,\; \bullet\boxed{\bullet},\; \bullet\boxed{\bullet\ \bullet},\; \bullet\boxed{\bullet\ \bullet\ \bullet},\; \ldots \Big\}$$

$$\mathcal{G}^{[3]} \;=\; \Big\{ \bullet,\; \bullet\boxed{\bullet},\; \bullet\boxed{\bullet\ \bullet},\; \bullet\boxed{\bullet\ \bullet\ \bullet},\; \ldots,$$

$$\bullet\boxed{\bullet\boxed{\bullet}},\; \bullet\boxed{\bullet\boxed{\bullet\ \bullet}},\; \bullet\boxed{\boxed{\bullet\ \bullet}\bullet},\; \bullet\boxed{\bullet\boxed{\bullet\ \bullet}\boxed{\bullet\ \bullet}},\; \ldots \Big\}.$$

First, each $\mathcal{G}^{[j]}$ is well defined since it corresponds to a purely iterative specification. Next, we have the inclusion $\mathcal{G}^{[j]} \subset \mathcal{G}^{[j+1]}$ (a simple interpretation of $\mathcal{G}^{[j]}$ is the class of all trees of height $< j$). We can therefore regard the complete class $\mathcal{G}$ as defined by the limit of the $\mathcal{G}^{[j]}$; that is, $\mathcal{G} := \bigcup_j \mathcal{G}^{[j]}$.

▷ **I.8.** *Lim-sup of classes.* Let $\{\mathcal{A}^{[j]}\}$ be any increasing sequence of combinatorial classes, in the sense that $\mathcal{A}^{[j]} \subset \mathcal{A}^{[j+1]}$, and the notions of size are compatible. If $\mathcal{A}^{[\infty]} = \bigcup_j \mathcal{A}^{[j]}$ is a combinatorial class (there are finitely many elements of size n, for each n), then the corresponding OGFs satisfy $A^{[\infty]}(z) = \lim_{j \to \infty} A^{[j]}(z)$ in the formal topology (Appendix A.5: *Formal power series*, p. 730). ◁

Definition I.7. *A specification for an r–tuple $\vec{\mathcal{A}} = (\mathcal{A}^{(1)}, \ldots, \mathcal{A}^{(r)})$ of classes is a collection of r equations,*

$$(29) \qquad \begin{cases} \mathcal{A}^{(1)} & = & \Phi_1(\mathcal{A}^{(1)}, \ldots, \mathcal{A}^{(r)}) \\ \mathcal{A}^{(2)} & = & \Phi_2(\mathcal{A}^{(1)}, \ldots, \mathcal{A}^{(r)}) \\ & \ldots & \\ \mathcal{A}^{(r)} & = & \Phi_r(\mathcal{A}^{(1)}, \ldots, \mathcal{A}^{(r)}) \end{cases}$$

where each Φ_i denotes a term built from the $\mathcal{A}$ using the constructions of disjoint union, cartesian product, sequence, powerset, multiset, and cycle, as well as the initial classes $\mathcal{E}$ (neutral) and $\mathcal{Z}$ (atomic).

We also say that the system is a specification of $\mathcal{A}^{(1)}$. A specification for a combinatorial class is thus a sort of formal grammar defining that class. Formally, the system (29) is an *iterative* or *non-recursive* specification if it is strictly upper-triangular, that is, $\mathcal{A}^{(r)}$ is defined solely in terms of initial classes $\mathcal{Z}, \mathcal{E}$; the definition of $\mathcal{A}^{(r-1)}$ only involves $\mathcal{A}^{(r)}$, and so on; in that case, by back substitutions, it is apparent that *for an iterative specification, $\mathcal{A}^{(1)}$ can be equivalently described by a single term involving only the initial classes and the basic constructors.* Otherwise, the system is said to be *recursive*. In the latter case, the semantics of recursion is identical to the one introduced in the case of trees: start with the "empty" vector of classes, $\vec{\mathcal{A}}^{[0]} := (\emptyset, \ldots, \emptyset)$, iterate $\vec{\mathcal{A}}^{[j+1]} = \vec{\Phi}\big[\vec{\mathcal{A}}^{[j]}\big]$, and finally take the limit.

There is an alternative and convenient way to visualize these notions. Given a specification of the form (29), we can associate its *dependency (di)graph* Γ to it as follows. The set of vertices of Γ is the set of indices $\{1, \ldots, r\}$; for each equation $\mathcal{A}^{(i)} = \Xi_i(\mathcal{A}^{(1)}, \ldots, \mathcal{A}^{(r)})$ and for each j such that $\mathcal{A}^{(j)}$ appears explicitly on the right-hand side of the equation, place a directed edge $(i \to j)$ in Γ. It is then easily recognized that a class is iterative if the dependency graph of its specification is acyclic; it is recursive is the dependency graph has a directed cycle. (This notion will serve to define irreducible linear systems, p. 341, and irreducible polynomial systems, p. 482, which enjoy strong asymptotic properties.)

Definition I.8. *A class of combinatorial structures is said to be* constructible *or* specifiable *iff it admits a (possibly recursive) specification in terms of sum, product, sequence, set, multiset, and cycle constructions.*

At this stage, we have therefore available a specification language for combinatorial structures which is some fragment of set theory with recursion added. Each constructible class has by virtue of Theorem I.1 an ordinary generating function for which functional equations can be produced systematically. (In fact, it is even possible to use computer algebra systems in order to compute it *automatically*! See the article by Flajolet, Salvy, and Zimmermann [255] for the description of such a system.)

Theorem I.2 (Symbolic method, unlabelled universe). *The generating function of a constructible class is a component of a system of* functional equations *whose terms* *are built from*

$$1, \ z, \ + \ , \ \times \ , \ Q \ , \ \mathrm{Exp} \ , \ \overline{\mathrm{Exp}} \ , \mathrm{Log},$$

where

$$\begin{cases} Q[f] &= \dfrac{1}{1-f}, & \mathrm{Log}[f] &= \displaystyle\sum_{k=1}^{\infty} \dfrac{\varphi(k)}{k} \log \dfrac{1}{1-f(z^k)}, \\[4mm] \mathrm{Exp}[f] &= \exp\left(\displaystyle\sum_{k=1}^{\infty} \dfrac{f(z^k)}{k}\right), & \overline{\mathrm{Exp}}[f] &= \exp\left(\displaystyle\sum_{k=1}^{\infty} (-1)^{k-1}\dfrac{f(z^k)}{k}\right). \end{cases}$$

Pólya operators. The operator Q translating sequences (SEQ) is classically known as the *quasi-inverse*. The operator Exp (multisets, MSET) is called the *Pólya exponential*[5] and $\overline{\mathrm{Exp}}$ (powersets, PSET) is the *modified Pólya exponential*. The operator Log is the *Pólya logarithm*. They are named after Pólya who first developed the general enumerative theory of objects under permutation groups (pp. 85–86).

The statement of Theorem I.2 signifies that iterative classes have explicit generating functions involving compositions of the basic operators only, while recursive structures have OGFs that are accessible indirectly via systems of functional equations. As we shall see at various places in this chapter, the following classes are constructible: binary words, binary trees, general trees, integer partitions, integer compositions, non-plane trees, polynomials over finite fields, necklaces, and wheels. We conclude this section with a few simple illustrations of the symbolic method expressed by Theorem I.2.

Binary words. The OGF of binary words, as seen already, can be obtained directly from the iterative specification,

$$\mathcal{W} = \mathrm{SEQ}(\mathcal{Z} + \mathcal{Z}) \qquad \Longrightarrow \qquad W(z) = \frac{1}{1-2z},$$

whence the expected result, $W_n = 2^n$. (Note: in our framework, if a, b are letters, then $\mathcal{Z} + \mathcal{Z} \cong \{a, b\}$.)

General trees. The recursive specification of general trees leads to an implicit definition of their OGF,

$$\mathcal{G} = \mathcal{Z} \times \mathrm{SEQ}(\mathcal{G}) \qquad \Longrightarrow \qquad G(z) = \frac{z}{1-G(z)}.$$

From this point on, basic algebra[6] does the rest. First the original equation is equivalent (in the ring of formal power series) to $G - G^2 - z = 0$. Next, the quadratic equation

---

[5]It is a notable fact that, although the Pólya operators look algebraically "difficult" to compute with, their treatment by complex asymptotic methods, as regards coefficient asymptotics, is comparatively "easy". We shall see many examples in Chapters IV–VII (e.g., pp. 252, 475).

[6]Methodological note: for simplicity, our computation is developed using the usual language of mathematics. However, *analysis* is not needed in this derivation, and operations such as solving quadratic equations and expanding fractional powers can all be cast within the purely algebraic framework of *formal power series* (p. 730).

is solvable by radicals, and one finds

$$G(z) = \frac{1}{2}\left(1 - \sqrt{1 - 4z}\right)$$
$$= z + z^2 + 2z^3 + 5z^4 + 14z^5 + 42z^6 + 132z^7 + 429z^8 + \cdots$$
$$= \sum_{n \geq 1} \frac{1}{n}\binom{2n-2}{n-1} z^n.$$

(The conjugate root is to be discarded since it involves a term z^{-1} as well as negative coefficients.) The expansion then results from Newton's binomial expansion,

$$(1+x)^\alpha = 1 + \frac{\alpha}{1}x + \frac{\alpha(\alpha-1)}{2!}x^2 + \cdots,$$

applied with $\alpha = \frac{1}{2}$ and $x = -4z$.

The numbers

$$(30) \qquad C_n = \frac{1}{n+1}\binom{2n}{n} = \frac{(2n)!}{(n+1)!\,n!} \qquad \text{with OGF} \quad C(z) = \frac{1 - \sqrt{1 - 4z}}{2z}$$

are known as the Catalan numbers (*EIS* **A000108**) in the honour of Eugène Catalan, the mathematician who first studied their properties in geat depth (pp. 6 and 20). In summary, *general trees are enumerated by Catalan numbers:*

$$G_n = C_{n-1} \equiv \frac{1}{n}\binom{2n-2}{n-1}.$$

For this reason the term *Catalan tree* is often employed as synonymous to "general (rooted unlabelled plane) tree".

Triangulations. Fix $n + 2$ points arranged in anticlockwise order on a circle and conventionally numbered from 0 to $n + 1$ (for instance the $(n + 2)$th roots of unity). A triangulation is defined as a (maximal) decomposition of the convex $(n + 2)$-gon defined by the points into n triangles (Figure I.1, p. 17). Triangulations are taken here as abstract topological configurations defined up to continuous deformations of the plane. The size of the triangulation is the number of triangles; that is, n. Given a triangulation, we define its "root" as a triangle chosen in some conventional and unambiguous manner (e.g., at the start, the triangle that contains the two smallest labels). Then, a triangulation decomposes into its root triangle and two subtriangulations (that may well be "empty") appearing on the left and right sides of the root triangle; the decomposition is illustrated by the following diagram:

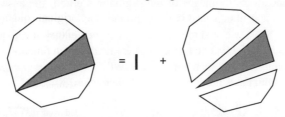

The class $\mathcal{T}$ of all triangulations can be specified recursively as

$$\mathcal{T} = \{\epsilon\} + (\mathcal{T} \times \nabla \times \mathcal{T}),$$

provided that we agree to consider a 2-gon (a segment) as giving rise to an "empty" triangulation of size 0. (The subtriangulations are topologically and combinatorially equivalent to standard ones, with vertices regularly spaced on a circle.) Consequently, the OGF $T(z)$ satisfies the equation

$$(31) \qquad T(z) = 1 + zT(z)^2, \qquad \text{so that} \quad T(z) = \frac{1}{2z}\left(1 - \sqrt{1 - 4z}\right).$$

As a result of (30) and (31), *triangulations are enumerated by Catalan numbers*:

$$T_n = C_n \equiv \frac{1}{n+1}\binom{2n}{n}.$$

This particular result goes back to Euler and Segner, a century before Catalan; see Figure I.1 on p. 17 for first values and p. 73 below for related bijections.

▷ **I.9.** *A bijection.* Since both general trees and triangulations are enumerated by Catalan numbers, there must exist a size-preserving bijection between the two classes. Find one such bijection. [Hint: the construction of triangulations is evocative of binary trees, while binary trees are themselves in bijective correspondence with general trees (p. 73).]   ◁

▷ **I.10.** *A variant specification of triangulations.* Consider the class $\mathcal{U}$ of "non-empty" triangulations of the n-gon, that is, we exclude the 2-gon and the corresponding "empty" triangulation of size 0. Then $\mathcal{U} = \mathcal{T} \setminus \{\epsilon\}$ admits the specification

$$\mathcal{U} = \nabla + (\nabla \times \mathcal{U}) + (\mathcal{U} \times \nabla) + (\mathcal{U} \times \nabla \times \mathcal{U})$$

which also leads to the Catalan numbers via $U = z(1 + U)^2$, so that $U(z) = (1 - 2z - \sqrt{1-4z})/(2z) \equiv T(z) - 1$.   ◁

I.2.4. Exploiting generating functions and counting sequences.

In this book we are going to see altogether more than a hundred applications of the symbolic method. Before engaging in technical developments, it is worth inserting a few comments on the way generating functions and counting sequences can be put to good use in order to solve combinatorial problems.

Explicit enumeration formulae. In a number of situations, generating functions are explicit and can be expanded in such a way that explicit formulae result for their coefficients. A prime example is the counting of general trees and of triangulations above, where the quadratic equation satisfied by an OGF is amenable to an explicit solution—the resulting OGF could then be expanded by means of Newton's binomial theorem. Similarly, we derive later in this chapter an explicit form for the number of integer compositions by means of the symbolic method (the answer turns out to be simply 2^{n-1}) and obtain in this way, through OGFs, many related enumeration results. In this book, we assume as known the elementary techniques from basic calculus by which the Taylor expansion of an explicitly given function can be obtained. (Elementary references on such aspects are Wilf's *Generatingfunctionology* [608], Graham, Knuth, and Patashnik's *Concrete Mathematics* [307], and our book [538].)

Implicit enumeration formulae. In a number of cases, the generating functions obtained by the symbolic method are still in a sense explicit, but their form is such that their coefficients are not clearly reducible to a closed form. It is then still possible to obtain initial values of the corresponding counting sequence by means of a symbolic manipulation system. Furthermore, from generating functions, it is possible systematically to derive recurrences that lead to a procedure for computing an arbitrary number of terms of the counting sequence in a reasonably efficient manner. A typical example of this situation is the OGF of integer partitions,

$$\prod_{m=1}^{\infty} \frac{1}{1 - z^m},$$

for which recurrences obtained from the OGF and associated to fast algorithms are given in Note I.13 (p. 42) and Note I.19 (p. 49). An even more spectacular example is the OGF of non-plane trees, which is proved below (p. 71) to satisfy the infinite *functional equation*

$$H(z) = z \exp\left(H(z) + \frac{1}{2} H(z^2) + \frac{1}{3} H(z^3) + \cdots \right),$$

and for which coefficients are computable in low complexity: see Note I.43, p. 72. (The references [255, 264, 456] develop a systematic approach to such problems.) The corresponding asymptotic analysis constitutes the main theme of Section VII. 5, p. 475.

Asymptotic formulae. Such forms are our eventual goal as they allow for an easy interpretation and comparison of counting sequences. From a quick glance at the table of initial values of W_n (words), P_n (permutations), T_n (triangulations), as given in (2), p. 18, it is apparent that W_n grows more slowly than T_n, which itself grows more slowly than P_n. The classification of growth rates of counting sequences belongs properly to the asymptotic theory of combinatorial structures which neatly relates to the symbolic method via complex analysis. A thorough treatment of this part of the theory is presented in Chapters IV–VIII. Given the methods expounded there, it becomes possible to estimate asymptotically the coefficients of virtually any generating function, however complicated, that is provided by the symbolic method; that is, implicit enumerations in the sense above are well covered by complex asymptotic methods.

Here, we content ourselves with a few remarks based on elementary real analysis. (The basic notations are described in Appendix A.2: *Asymptotic notation*, p. 722.) The sequence $W_n = 2^n$ grows exponentially and, in such an extreme simple case, the exact form coincides with the asymptotic form. The sequence $P_n = n!$ must grow faster. But how fast? The answer is provided by Stirling's formula, an important approximation originally due to James Stirling (*Invitation*, p. 4):

$$(32) \qquad n! = \left(\frac{n}{e} \right)^n \sqrt{2\pi n} \left(1 + O\left(\frac{1}{n} \right) \right) \qquad (n \to +\infty).$$

(Several proofs are given in this book, based on the method of Laplace, p. 760, Mellin transforms, p. 766, singularity analysis, p. 407, and the saddle-point method, p 555.) The ratios of the exact values to Stirling's approximations

n	1	2	5	10	100	1 000
$\dfrac{n!}{n^n e^{-n} \sqrt{2\pi n}}$	1.084437	1.042207	1.016783	1.008365	1.000833	1.000083

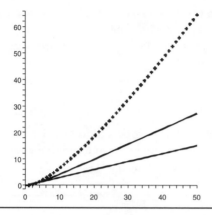

Figure I.5. The growth regimes of three sequences $f(n) = 2^n, T_n, n!$ (from bottom to top) rendered by a plot of $\log_{10} f(n)$ versus n.

show an *excellent quality* of the asymptotic estimate: the error is only 8% for $n = 1$, less than 1% for $n = 10$, and less than 1 per thousand for any n greater than 100.

Stirling's formula provides in turn the asymptotic form of the Catalan numbers, by means of a simple calculation:

$$C_n = \frac{1}{n+1} \frac{(2n)!}{(n!)^2} \sim \frac{1}{n} \frac{(2n)^{2n} e^{-2n} \sqrt{4\pi n}}{n^{2n} e^{-2n} 2\pi n},$$

which simplifies to

$$(33) \qquad\qquad C_n \sim \frac{4^n}{\sqrt{\pi n^3}}.$$

Thus, the growth of Catalan numbers is roughly comparable to an exponential, 4^n, modulated by a subexponential factor, here $1/\sqrt{\pi n^3}$. A surprising consequence of this asymptotic estimate in the area of boolean function complexity appears in Example I.17 below (p. 77).

Altogether, the asymptotic number of general trees and triangulations is well summarized by a simple formula. Approximations become more and more accurate as n becomes large. Figure I.5 illustrates the different growth regimes of our three reference sequences while Figure I.6 exemplifies the quality of the approximation with subtler phenomena also apparent on the figures and well explained by asymptotic theory. Such asymptotic formulae then make comparison between the growth rates of sequences easy.

The interplay between combinatorial structure and asymptotic structure is indeed the principal theme of this book. We shall see in Part B that the generating functions provided by the symbolic method typically admit similarly simple asymptotic coefficient estimates.

▷ **I.11.** *The complexity of coding.* A company specializing in computer-aided design has sold to you a scheme that (they claim) can encode any triangulation of size $n \geq 100$ using at most $1.5n$ bits of storage. After reading these pages, what do you do? [Hint: sue them!] See also Note I.24 (p. 53) for related coding arguments.   ◁

n	C_n	$C_n^\star$	$C_n^\star/C_n$
1	1	2.25	2.25675 83341 91025 14779 23178
10	16796	18707.89	1.11383 05127 52445 89437 89064
100	$0.89651 \cdot 10^{57}$	$0.90661 \cdot 10^{57}$	1.01126 32841 24540 52257 13957
1 000	$0.20461 \cdot 10^{598}$	$0.20484 \cdot 10^{598}$	1.00112 51328 15424 16470 12827
10 000	$0.22453 \cdot 10^{6015}$	$0.22456 \cdot 10^{6015}$	1.00011 25013 28127 92913 51406
100 000	$0.17805 \cdot 10^{60199}$	$0.17805 \cdot 10^{60199}$	1.00001 12500 13281 25292 96322
1 000 000	$0.55303 \cdot 10^{602051}$	$0.55303 \cdot 10^{602051}$	1.00000 11250 00132 81250 29296

Figure I.6. The Catalan numbers C_n, their Stirling approximation $C_n^\star = 4^n/\sqrt{\pi n^3}$, and the ratio $C_n^\star/C_n$.

▷ **I.12.** *Experimental asymptotics.* From the data of Figure I.6, guess the values[7] of $C_{10^7}^\star/C_{10^7}$ and of $C_{5\cdot10^6}^\star/C_{5\cdot10^6}$ to 25D. (See, Figure VI.3, p. 384, as well as, e.g., [385] for related asymptotic expansions and [80] for similar properties.) ◁

I. 3. Integer compositions and partitions

This section and the next few provide examples of counting via specifications in classical areas of combinatorial theory. They illustrate the benefits of the symbolic method: generating functions are obtained with hardly any computation, and at the same time, many counting refinements follow from a basic combinatorial construction. The most direct applications described here relate to the additive decomposition of integers into summands with the classical combinatorial–arithmetic structures of partitions and compositions. The specifications are iterative and simply combine two levels of constructions of type SEQ, MSET, CYC, PSET.

I. 3.1. Compositions and partitions. Our first examples have to do with decomposing integers into sums.

Definition I.9. *A composition of an integer n is a sequence $(x_1, x_2, \ldots, x_k)$ of integers (for some k) such that*
$$n = x_1 + x_2 + \cdots + x_k, \qquad x_j \geq 1.$$
A partition of an integer n is a sequence $(x_1, x_2, \ldots, x_k)$ of integers (for some k) such that
$$n = x_1 + x_2 + \cdots + x_k \qquad and \qquad x_1 \geq x_2 \geq \cdots \geq x_k \geq 1.$$
In both cases, the x_i are called the summands *or the* parts *and the quantity n is called the* size.

By representing summands in unary using small discs ("•"), we can render graphically a composition by drawing bars between some of the balls; if we arrange summands vertically, compositions appear as ragged landscapes. In contrast, partitions appear as staircases, also known as Ferrers diagrams [129, p. 100]; see Figure I.7. We

---

[7]In this book, we abbreviate a phrase such as "*25 decimal places*" by "25D".

Figure I.7. Graphical representations of compositions and partitions: (left) the composition $1 + 3 + 1 + 4 + 2 + 3 = 14$ with its "ragged landscape" and "balls-and-bars" models; (right) the partition $8 + 8 + 6 + 5 + 4 + 4 + 4 + 2 + 1 + 1 = 43$ with its staircase (Ferrers diagram) model.

let $\mathcal{C}$ and $\mathcal{P}$ denote the class of all compositions and all partitions, respectively. Since a set can always be presented in sorted order, the difference between compositions and partitions lies in the fact that the order of summands *does* or *does not* matter. This is reflected by the use of a sequence construction (for $\mathcal{C}$) against a multiset construction (for $\mathcal{P}$). From this perspective, it proves convenient to regard 0 as obtained by the empty sequence of summands ($k = 0$), and we shall do so from now on.

Integers, as a combinatorial class. Let $\mathcal{I} = \{1, 2, \ldots\}$ denote the combinatorial class of all integers at least 1 (the summands), and let the size of each integer be its value. Then, the OGF of $\mathcal{I}$ is

$$(34) \qquad\qquad I(z) = \sum_{n \geq 1} z^n = \frac{z}{1 - z},$$

since $I_n = 1$ for $n \geq 1$, corresponding to the fact that there is exactly one object in $\mathcal{I}$ for each size $n \geq 1$. If integers are represented in unary, say by small balls, one has

$$(35) \qquad \mathcal{I} = \{1, \ 2, \ 3, \ \ldots\} \cong \{\bullet, \ \bullet\bullet, \ \bullet \ \bullet \ \bullet, \ \ldots\} = \mathrm{SEQ}_{\geq 1}\{\bullet\},$$

which constitutes a direct way to visualize the equality $I(z) = z/(1 - z)$.

Compositions. First, the specification of compositions as sequences admits, by Theorem I.1, a direct translation into OGF:

$$(36) \qquad\qquad \mathcal{C} = \mathrm{SEQ}(\mathcal{I}) \qquad \Longrightarrow \qquad C(z) = \frac{1}{1 - I(z)}.$$

The collection of equations (34), (36) thus fully determines $C(z)$:

$$
\begin{aligned}
C(z) &= \frac{1}{1 - \frac{z}{1-z}} = \frac{1 - z}{1 - 2z} \\
&= 1 + z + 2z^2 + 4z^3 + 8z^4 + 16z^5 + 32z^6 + \cdots.
\end{aligned}
$$

From here, the counting problem for compositions is solved by a straightforward expansion of the OGF: one has

$$C(z) = \left(\sum_{n \geq 0} 2^n z^n \right) - \left(\sum_{n \geq 0} 2^n z^{n+1} \right),$$

0	1	1
10	1024	42
20	1048576	627
30	1073741824	5604
40	1099511627776	37338
50	1125899906842624	204226
60	1152921504606846976	966467
70	1180591620717411303424	4087968
80	1208925819614629174706176	15796476
90	1237940039285380274899124224	56634173
100	1267650600228229401496703205376	190569292
110	1298074214633706907132624082305024	607163746
120	1329227995784915872903807060280344576	1844349560
130	1361129467683753853853498429727072845824	5371315400
140	1393796574908163946345982392040522594123776	15065878135
150	1427247692705959881058285969449495136382746624	40853235313
160	1461501637330902918203684832716283019655932542976	107438159466
170	1496577676626844588240573268701473812127674924007424	274768617130
180	1532495540865888583583470271503091836187391221183602176	684957390936
190	1569275433846670190958947355801916604025588861116008628224	1667727404093
200	1606938044258990275541962092341162602522202993782792835301376	3972999029388
210	1645504557321206042154969182557350504982735865633579863348609024	9275102575355
220	1684996666696961498716668844293872691710232152640878578006897564576	21248279009367
230	1725436586697640946858688965569256363112777243042596638790631055949824	47826239745920
240	1766847064778384329583297500742918515827483896875618958121606201292619776	105882246722733
250	1809251394333065553493296640760748560207343510400633813116524750123642650624	230793554364681

Figure I.8. For $n = 0, 10, 20, \ldots, 250$ (left), the number of compositions C_n (middle) and the number of partitions P_n (right). The figure illustrates the difference in growth between $C_n = 2^{n-1}$ and $P_n = e^{O(\sqrt{n})}$.

---

implying $C_0 = 1$ and $C_n = 2^n - 2^{n-1}$ for $n \geq 1$; that is,

$$(37) \qquad C_n = 2^{n-1}, \quad n \geq 1.$$

This agrees with basic combinatorics since a composition of n can be viewed as the placement of separation bars at a subset of the $n - 1$ existing places in between n aligned balls (the "balls-and-bars" model of Figure I.7), of which there are clearly 2^{n-1} possibilities.

Partitions. For partitions specified as multisets, the general translation mechanism of Theorem I.1, p. 27, provides

$$(38) \quad \mathcal{P} = \text{MSET}(\mathcal{I}) \quad \Longrightarrow \quad P(z) = \exp\left(I(z) + \frac{1}{2}I(z^2) + \frac{1}{3}I(z^3) + \cdots\right),$$

together with the product form corresponding to (25), p. 29,

$$(39) \quad \begin{aligned} P(z) &= \prod_{m=1}^{\infty} \frac{1}{1 - z^m} \\ &= \left(1 + z + z^2 + \cdots\right)\left(1 + z^2 + z^4 + \cdots\right)\left(1 + z^3 + z^6 + \cdots\right)\cdots \\ &= 1 + z + 2z^2 + 3z^3 + 5z^4 + 7z^5 + 11z^6 + 15z^7 + 22z^8 + \cdots \end{aligned}$$

(the counting sequence is *EIS* **A000041**). Contrary to compositions that are counted by the explicit formula 2^{n-1}, no simple form exists for P_n. Asymptotic analysis of the OGF (38) based on the saddle-point method (Chapter VIII, p. 574) shows that $P_n = e^{O(\sqrt{n})}$. In fact an extremely famous theorem of Hardy and Ramanujan later improved by Rademacher (see Andrews' book [14] and Chapter VIII) provides a full expansion of which the asymptotically dominant term is

$$(40) \qquad P_n \sim \frac{1}{4n\sqrt{3}} \exp\left(\pi\sqrt{\frac{2n}{3}}\right).$$

There are consequently appreciably fewer partitions than compositions (Figure I.8).

▷ **I.13.** *A recurrence for the partition numbers.* Logarithmic differentiation gives

$$z \frac{P'(z)}{P(z)} = \sum_{n=1}^{\infty} \frac{n z^n}{1 - z^n} \qquad \text{implying} \quad n P_n = \sum_{j=1}^{n} \sigma(j) P_{n-j},$$

where $\sigma(n)$ is the sum of the divisors of n (e.g., $\sigma(6) = 1 + 2 + 3 + 6 = 12$). Consequently, $P_1, \ldots, P_N$ can be computed in $O(N^2)$ integer-arithmetic operations. (The technique is generally applicable to powersets and multisets; see Note I.43 (p. 72) for another application. Note I.19 (p. 49) further lowers the bound to $O(N\sqrt{N})$, in the case of partitions.)   ◁

By varying (36) and (38), we can use the symbolic method to derive a number of counting results in a straightforward manner. First, we state the following proposition.

Proposition I.1. *Let $\mathcal{T} \subseteq \mathcal{I}$ be a subset of the positive integers. The OGFs of the classes $\mathcal{C}^{\mathcal{T}} := \text{SEQ}(\text{SEQ}_{\mathcal{T}}(\mathcal{Z}))$ and $\mathcal{P}^{\mathcal{T}} := \text{MSET}(\text{SEQ}_{\mathcal{T}}(\mathcal{Z}))$ of compositions and partitions having summands restricted to $\mathcal{T} \subset \mathbb{Z}_{\geq 1}$ are given by*

$$C^{\mathcal{T}}(z) = \frac{1}{1 - \sum_{n \in T} z^n} = \frac{1}{1 - T(z)}, \qquad P^{\mathcal{T}}(z) = \prod_{n \in \mathcal{T}} \frac{1}{1 - z^n}.$$

Proof. A direct consequence of the specifications and Theorem I.1, p. 27.   ∎

This proposition permits us to enumerate compositions and partitions with restricted summands, as well as with a fixed number of parts.

Example I.4. *Compositions with restricted summands.* In order to enumerate the class $\mathcal{C}^{\{1,2\}}$ of compositions of n whose parts are only allowed to be taken from the set $\{1, 2\}$, simply write

$$\mathcal{C}^{\{1,2\}} = \text{SEQ}(\mathcal{I}^{\{1,2\}}) \qquad \text{with } \mathcal{I}^{\{1,2\}} = \{1, 2\}.$$

Thus, in terms of generating functions, one has

$$C^{\{1,2\}}(z) = \frac{1}{1 - I^{\{1,2\}}(z)} \qquad \text{with} \quad I^{\{1,2\}}(z) = z + z^2.$$

This formula implies

$$C^{\{1,2\}}(z) = \frac{1}{1 - z - z^2} = 1 + z + 2z^2 + 3z^3 + 5z^4 + 8z^5 + 13z^6 + \cdots,$$

and the number of compositions of n in this class is expressed by a Fibonacci number,

$$C_n^{\{1,2\}} = F_{n+1} \text{ where } F_n = \frac{1}{\sqrt{5}} \left[\left(\frac{1 + \sqrt{5}}{2} \right)^n - \left(\frac{1 - \sqrt{5}}{2} \right)^n \right],$$

of daisy–artichoke–rabbit fame In particular, the rate of growth is of the exponential type φ^n, where $\varphi := \dfrac{1 + \sqrt{5}}{2}$ is the golden ratio.

Similarly, compositions all of whose summands lie in the set $\{1, 2, \ldots, r\}$ have generating function

(41) $$C^{\{1,\ldots,r\}}(z) = \frac{1}{1 - z - z^2 - \cdots z^r} = \frac{1}{1 - z \frac{1 - z^r}{1 - z}} = \frac{1 - z}{1 - 2z + z^{r+1}},$$

and the corresponding counts are generalized Fibonacci numbers. A double combinatorial sum expresses these counts

$$
(42) \qquad C_n^{\{1,\dots,r\}} = [z^n] \sum_j \left(\frac{z(1 - z^r)}{(1 - z)} \right)^j = \sum_{j,k} (-1)^k \binom{j}{k} \binom{n - rk - 1}{j - 1}.
$$

This result is perhaps not too useful for grasping the rate of growth of the sequence when n gets large, so that asymptotic analysis is called for. Asymptotically, for any fixed $r \geq 2$, there is a unique root ρ_r of the denominator $1 - 2z + z^{r+1}$ in $(\frac{1}{2}, 1)$, this root dominates all the other roots and is simple. Methods amply developed in Chapter IV and Example V.4 (p. 308) imply that, for some constant $c_r > 0$,

$$
(43) \qquad C_n^{\{1,\dots,r\}} \sim c_r \rho_r^{-n} \quad \text{for fixed } r \text{ as } n \to \infty.
$$

The quantity ρ_r plays a rôle similar to that of the golden ratio when $r = 2$. .... ∎

▷ **I.14.** *Compositions into primes.* The additive decomposition of integers into primes is still surrounded with mystery. For instance, it is not known whether every even number is the sum of two primes (Goldbach's conjecture). However, the number of compositions of n into prime summands (*any* number of summands is permitted) is $B_n = [z^n] B(z)$ where

$$
\begin{aligned}
B(z) &= \left(1 - \sum_{p \text{ prime}} z^p \right)^{-1} = \left(1 - z^2 - z^3 - z^5 - z^7 - z^{11} - \cdots \right)^{-1} \\
&= 1 + z^2 + z^3 + z^4 + 3z^5 + 2z^6 + 6z^7 + 6z^8 + 10z^9 + 16z^{10} + \cdots
\end{aligned}
$$

(*EIS* **A023360**), and complex asymptotic methods make it *easy* to determine the asymptotic form $B_n \sim 0.30365 \cdot 1.47622^n$; see Example V.2, p. 297. ◁

Example I.5. *Partitions with restricted summands (denumerants).* Whenever summands are restricted to a finite set, the special partitions that result are called denumerants. A denumerant problem popularized by Pólya [493, §3] consists in finding the number of ways of giving change of 99 cents using coins that are pennies (1 cent), nickels (5 cents), dimes (10 cents) and quarters (25 cents). (The order in which the coins are taken does not matter and repetitions are allowed.) For the case of a finite $\mathcal{T}$, we predict from Proposition I.1 that $P^{\mathcal{T}}(z)$ is always a *rational* function with poles that are at roots of unity; also the $P_n^{\mathcal{T}}$ satisfy a linear recurrence related to the structure of $\mathcal{T}$. The solution to the original coin change problem is found to be

$$
[z^{99}] \frac{1}{(1 - z)(1 - z^5)(1 - z^{10})(1 - z^{25})} = 213.
$$

In the same vein, one proves that

$$
P_n^{\{1,2\}} = \left\lceil \frac{2n + 3}{4} \right\rfloor \qquad P_n^{\{1,2,3\}} = \left\lceil \frac{(n + 3)^2}{12} \right\rfloor ;
$$

here $\lceil x \rfloor \equiv \lfloor x + \frac{1}{2} \rfloor$ denotes the integer closest to the real number x. Such results are typically obtained by the two-step process: (*i*) decompose the rational generating function into simple fractions; (*ii*) compute the coefficients of each simple fraction and combine them to get the final result [129, p. 108].

The general argument also gives the generating function of partitions whose summands lie in the set $\{1, 2, \dots, r\}$ as

$$
(44) \qquad P^{\{1,\dots,r\}}(z) = \prod_{m=1}^r \frac{1}{1 - z^m}.
$$

In other words, we are enumerating partitions according to the value of the largest summand. One then finds by looking at the poles (Theorem IV.9, p. 256):

$$(45) \qquad P_n^{\{1,\dots,r\}} \sim c_r n^{r-1} \qquad \text{with} \quad c_r = \frac{1}{r!(r-1)!}.$$

A similar argument provides the asymptotic form of $P_n^{\mathcal{T}}$ when $\mathcal{T}$ is an arbitrary finite set:

$$P_n^{\mathcal{T}} \sim \frac{1}{\tau}\frac{n^{r-1}}{(r-1)!} \qquad \text{with } \tau := \prod_{n \in \mathcal{T}} n, \quad r := \text{card}(\mathcal{T}).$$

This last estimate, originally due to Schur, is proved in Proposition IV.2, p. 258. .... ∎

We next examine compositions and partitions with a fixed number of summands.

Example I.6. *Compositions with a fixed number of parts.*   Let $\mathcal{C}^{(k)}$ denote the class of compositions made of k summands, k a fixed integer ≥ 1. One has

$$\mathcal{C}^{(k)} = \text{SEQ}_k(\mathcal{I}) \equiv \mathcal{I} \times \mathcal{I} \times \cdots \times \mathcal{I},$$

where the number of terms in the cartesian product is k. From here, the corresponding generating function is found to be

$$C^{(k)}(z) = \left(I(z)\right)^k \qquad \text{with} \qquad I(z) = \frac{z}{1-z}.$$

The number of compositions of n having k parts is thus

$$C_n^{(k)} = [z^n]\frac{z^k}{(1-z)^k} = \binom{n-1}{k-1},$$

a result which constitutes a combinatorial refinement of $C_n = 2^{n-1}$. (Note that the formula $C_n^{(k)} = \binom{n-1}{k-1}$ also results easily from the balls-and-bars model of compositions (Figure I.7)). In such a case, the asymptotic estimate $C_n^{(k)} \sim n^{k-1}/(k-1)!$ results immediately from the polynomial form of the binomial coefficient $\binom{n-1}{k-1}$. .... ∎

Example I.7. *Partitions with a fixed number of parts.*   Let $\mathcal{P}^{(\leq k)}$ be the class of integer partitions with at most k summands. With our notation for restricted constructions (p. 30), this class is specified as

$$\mathcal{P}^{(\leq k)} = \text{MSET}_{\leq k}(\mathcal{I}).$$

It would be possible to appeal to the admissibility of such restricted compositions as developed in Subsection I.6.1 below, but the following direct argument suffices in the case at hand. Geometrically, partitions, are represented as collections of points: this is the staircase model of Figure I.7, p. 40. A symmetry around the main diagonal (also known in the specialized literature as conjugation) exchanges number of summands and value of largest summand; one then has (with earlier notations)

$$\mathcal{P}^{(\leq k)} \cong \mathcal{P}^{\{1,\dots k\}} \qquad \Longrightarrow \qquad P^{(\leq k)}(z) = P^{\{1,\dots k\}}(z),$$

so that, by (44),

$$(46) \qquad P^{(\leq k)}(z) \equiv P^{\{1,\dots,k\}} = \prod_{m=1}^{k} \frac{1}{1-z^m}.$$

As a consequence, the OGF of partitions with *exactly* k summands, $P^{(k)}(z) = P^{(\leq k)}(z) - P^{(\leq k-1)}(z)$, evaluates to

$$P^{(k)}(z) = \frac{z^k}{(1-z)(1-z^2)\cdots(1-z^k)}.$$

Given the equivalence between number of parts and largest part in partitions, the asymptotic estimate (45) applies verbatim here. .... ∎

▷ **I.15.** *Compositions with summands bounded in number and size.* The number of compositions of size n with k summands each at most r is expressible as

$$[z^n] \left(z\frac{1-z^r}{1-z}\right)^k,$$

which reduces to a simple binomial convolution (the calculation is similar to (42), p. 43). ◁

▷ **I.16.** *Partitions with summands bounded in number and size.* The number of partitions of size n with at most k summands each at most ℓ is

$$[z^n] \frac{(1-z)(1-z^2)\cdots(1-z^{k+\ell})}{\big((1-z)(1-z^2)\cdots(1-z^k)\big)\cdot\big((1-z)(1-z^2)\cdots(1-z^\ell)\big)}.$$

(Verifying this by recurrence is easy.) The GF reduces to the binomial coefficient $\binom{k+\ell}{k}$ as $z \to 1$; it is known as a Gaussian binomial coefficient, denoted $\binom{k+\ell}{k}_z$, or a "q–analogue" of the binomial coefficient [14, 129]. ◁

The last example of this section illustrates the close interplay between combinatorial decompositions and special function identities, which constitutes a recurrent theme of classical combinatorial analysis.

Example I.8. *The Durfee square of partitions and stack polyominoes.* The diagram of any partition contains a uniquely determined square (known as the Durfee square) that is maximal, as exemplified by the following diagram:

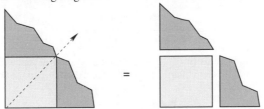

This decomposition is expressed in terms of partition GFs as

$$\mathcal{P} \cong \bigcup_{h \geq 0} \left(\mathcal{Z}^{h^2} \times \mathcal{P}^{(\leq h)} \times \mathcal{P}^{\{1,\dots,h\}}\right),$$

It gives automatically, via (44) and (46), a non-trivial identity, which is nothing but a formal rewriting of the geometric decomposition:

$$\prod_{n=1}^{\infty} \frac{1}{1-z^n} = \sum_{h \geq 0} \frac{z^{h^2}}{\big((1-z)\cdots(1-z^h)\big)^2}$$

(h is the size of the Durfee square, known to manic bibliometricians as the "H-index").

Stack polyominoes. Here is a similar case illustrating the direct correspondence between geometric diagrams and generating functions, as afforded by the symbolic method. A *stack polyomino* is the diagram of a composition such that for some j, ℓ, one has $1 \leq x_1 \leq x_2 \leq$

$\cdots \le x_j \ge x_{j+1} \ge \cdots \ge x_\ell \ge 1$ (see [552, §2.5] for further properties). The diagram representation of stack polyominoes

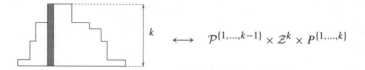

$$k \quad \longleftrightarrow \quad \mathcal{P}^{\{1,\dots,k-1\}} \times \mathcal{Z}^k \times P^{\{1,\dots,k\}}$$

translates immediately into the OGF

$$S(z) = \sum_{k \ge 1} \frac{z^k}{1 - z^k} \frac{1}{\left((1-z)(1-z^2)\cdots(1-z^{k-1})\right)^2},$$

once use is made of the partition GFs $P^{\{1,\dots,k\}}(z)$ of (44). This last relation provides a *bona fide* algorithm for computing the initial values of the number of stack polyominoes (*EIS* **A001523**):

$$S(z) = z + 2z^2 + 4z^3 + 8z^4 + 15z^5 + 27z^6 + 47z^7 + 79z^8 + \cdots .$$

The book of van Rensburg [592] describes many such constructions and their relation to models of statistical physics, especially polyominoes. For instance, related "q–Bessel" functions appear in the enumeration of parallelogram polyominoes (Example IX.14, p. 660). .... ■

▷ **I.17.** *Systems of linear diophantine inequalities.* Consider the class $\mathcal{F}$ of compositions of integers into four summands (x_1, x_2, x_3, x_4) such that

$$x_1 \ge 0, \quad x_2 \ge 2x_1, \quad x_3 \ge 2x_2, \quad x_4 \ge 2x_3,$$

where the x_j are in $\mathbb{Z}_{\ge 0}$. The OGF is

$$F(z) = \frac{1}{(1-z)(1-z^3)(1-z^7)(1-z^{15})}.$$

Generalize to $r \ge 4$ summands (in $\mathbb{Z}_{\ge 0}$) and a similar system of inequalities. (Related GFs appear on p. 200.) Work out elementarily the OGFs corresponding to the following systems of inequalities:

$$\{x_1 + x_2 \le x_3\}, \quad \{x_1 + x_2 \ge x_3\}, \quad \{x_1 + x_2 \le x_3 + x_4\}, \quad \{x_1 \le x_2, x_2 \ge x_3, x_3 \le x_4\}.$$

More generally, the OGF of compositions into a *fixed* number of summands (in $\mathbb{Z}_{\ge 0}$), constrained to satisfy a linear system of equations and inequalities with coefficients in $\mathbb{Z}$, is rational; its denominator is a product of factors of the form $(1 - z^j)$. (Caution: this generalization is non-trivial: see Stanley's treatment in [552, §4.6].) ◁

Figure I.9 summarizes what has been learned regarding compositions and partitions. The way several combinatorial problems are solved effortlessly by the symbolic method is worth noting.

I.3.2. Related constructions.

It is also natural to consider the two constructions of cycle and powerset when these are applied to the set of integers $\mathcal{I}$.

	Specification	OGF	coefficients	
Compositions:				
all	$\mathrm{SEQ}(\mathrm{SEQ}_{\geq 1}(\mathcal{Z}))$	$\dfrac{1-z}{1-2z}$	2^{n-1}	(p. 40)
parts $\leq r$	$\mathrm{SEQ}(\mathrm{SEQ}_{1..r}(\mathcal{Z}))$	$\dfrac{1-z}{1-2z+z^{r+2}}$	$\sim c_r \rho_r^{-n}$	(pp. 42, 308)
k parts	$\mathrm{SEQ}_k(\mathrm{SEQ}_{\geq 1}(\mathcal{Z}))$	$\dfrac{z^k}{(1-z)^k}$	$\sim \dfrac{n^{k-1}}{(k-1)!}$	(p. 44)
cyclic	$\mathrm{CYC}(\mathrm{SEQ}_{\geq 1}(\mathcal{Z}))$	Eq. (48)	$\sim \dfrac{2^n}{n}$	(p. 48)
Partitions:				
all	$\mathrm{MSET}(\mathrm{SEQ}_{\geq 1}(\mathcal{Z}))$	$\displaystyle\prod_{m=1}^{\infty}(1-z^m)^{-1}$	$\sim \dfrac{1}{4n\sqrt{3}}e^{\pi\sqrt{\frac{2n}{3}}}$	(pp. 41, 574)
parts $\leq r$	$\mathrm{MSET}(\mathrm{SEQ}_{1..r}(\mathcal{Z}))$	$\displaystyle\prod_{m=1}^{r}(1-z^m)^{-1}$	$\sim \dfrac{n^{r-1}}{r!(r-1)!}$	(pp. 43, 258)
$\leq k$ parts	$\cong \mathrm{MSET}(\mathrm{SEQ}_{1..k}(\mathcal{Z}))$	$\displaystyle\prod_{m=1}^{k}(1-z^m)^{-1}$	$\sim \dfrac{n^{k-1}}{k!(k-1)!}$	(pp. 44, 258)
distinct parts	$\mathrm{PSET}(\mathrm{SEQ}_{\geq 1}(\mathcal{Z}))$	$\displaystyle\prod_{m=1}^{\infty}(1+z^m)$	$\sim \dfrac{3^{3/4}}{12n^{3/4}}e^{\pi\sqrt{n/3}}$	(pp. 48, 579)

Figure I.9. Partitions and compositions: specifications, generating functions, and coefficients (in exact or asymptotic form).

Cyclic compositions (wheels). The class $\mathcal{D} = \mathrm{CYC}(\mathcal{I})$ comprises compositions defined up to circular shift of the summands; so, for instance $2 + 3 + 1 + 2 + 5$, $3 + 1 + 2 + 5 + 2$, etc, are identified. Alternatively, we may view elements of $\mathcal{D}$ as "wheels" composed of circular arrangements of rows of balls (taken up to rotation):

a "wheel" (cyclic composition)

By the translation of the cycle construction, the OGF is

$$(47) \quad D(z) = \sum_{k=1}^{\infty} \frac{\varphi(k)}{k} \log\left(1 - \frac{z^k}{1-z^k}\right)^{-1}$$

$$= z + 2z^2 + 3z^3 + 5z^4 + 7z^5 + 13z^6 + 19z^7 + 35z^8 + \cdots.$$

The coefficients are thus (*EIS* **A008965**)

$$(48) \qquad D_n = \frac{1}{n} \sum_{k \mid n} \varphi(k)(2^{n/k} - 1) \equiv -1 + \frac{1}{n} \sum_{k \mid n} \varphi(k) 2^{n/k} \sim \frac{2^n}{n},$$

where the condition "$k \mid n$" indicates a sum over the integers k dividing n. Notice that D_n is of the same asymptotic order as $\frac{1}{n} C_n$, which is suggested by circular symmetry of wheels, but there is a factor: $D_n \sim 2C_n/n$.

Partitions into distinct summands. The class $\mathcal{Q} = \mathrm{PSET}(\mathcal{I})$ is the subclass of $\mathcal{P} = \mathrm{MSET}(\mathcal{I})$ corresponding to partitions determined as in Definition I.9, but with the strict inequalities $x_k > \cdots > x_1$, so that the OGF is

$$(49) \quad Q(z) = \prod_{n \geq 1} (1 + z^n) = 1 + z + z^2 + 2z^3 + 2z^4 + 3z^5 + 4z^6 + 5z^7 + \cdots.$$

The coefficients (*EIS* **A000009**) are not expressible in closed form. However, the saddle-point method (Section VIII. 6, p. 574) yields the approximation:

$$(50) \qquad Q_n \sim \frac{3^{3/4}}{12n^{3/4}} \exp\left(\pi \sqrt{\frac{n}{3}}\right),$$

which has a shape similar to that of P_n in (40), p. 41.

▷ **I.18.** *Odd versus distinct summands.* The partitions of n into odd summands ($\mathcal{O}_n$) and the ones into distinct summands ($\mathcal{Q}_n$) are equinumerous. Indeed, one has

$$Q(z) = \prod_{m=1}^{\infty} (1 + z^m), \qquad O(z) = \prod_{j=0}^{\infty} (1 - z^{2j+1})^{-1}.$$

Equality results from substituting $(1 + a) = (1 - a^2)/(1 - a)$ with $a = z^m$,

$$Q(z) = \frac{1 - z^2}{1 - z} \frac{1 - z^4}{1 - z^2} \frac{1 - z^6}{1 - z^3} \frac{1 - z^8}{1 - z^4} \frac{1 - z^{10}}{1 - z^5} \cdots = \frac{1}{1 - z} \frac{1}{1 - z^3} \frac{1}{1 - z^5} \cdots,$$

and simplification of the numerators with half of the denominators (in boldface).   ◁

Partitions into powers. Let $\mathcal{I}^{\mathrm{pow}} = \{1, 2, 4, 8, \ldots\}$ be the set of powers of 2. The corresponding $\mathcal{P}$ and $\mathcal{Q}$ partitions have OGFs

$$
\begin{aligned}
P^{\mathrm{pow}}(z) &= \prod_{j=0}^{\infty} \frac{1}{1 - z^{2^j}} \\
&= 1 + z + 2z^2 + 2z^3 + 4z^4 + 4z^5 + 6z^6 + 6z^7 + 10z^8 + \cdots \\
Q^{\mathrm{pow}}(z) &= \prod_{j=0}^{\infty} (1 + z^{2^j}) \\
&= 1 + z + z^2 + z^3 + z^4 + z^5 + \cdots.
\end{aligned}
$$

The first sequence $1, 1, 2, 2, \ldots$ is the "binary partition sequence" (*EIS* **A018819**); the difficult asymptotic analysis was performed by de Bruijn [141] who obtained an estimate that involves subtle fluctuations and is of the global form $e^{O(\log^2 n)}$. The function

$Q^{\mathrm{pow}}(z)$ reduces to $(1-z)^{-1}$ since every number has a unique additive decomposition into powers of 2. Accordingly, the identity

$$\frac{1}{1-z} = \prod_{j=0}^{\infty}(1 + z^{2^j}),$$

first observed by Euler is sometimes nicknamed the "computer scientist's identity" as it reflects the property that every number admits a unique binary representation.

There exists a rich set of identities satisfied by partition generating functions— this fact is down to deep connections with elliptic functions, modular forms, and q–analogues of special functions on the one hand, basic combinatorics and number theory on the other hand. See [14, 129] for introductions to this fascinating subject.

▷ **I.19.** *Euler's pentagonal number theorem.* This famous identity expresses $1/P(z)$ as

$$\prod_{n\geq 1}(1-z^n) = \sum_{k\in\mathbb{Z}}(-1)^k z^{k(3k+1)/2}.$$

It is proved formally and combinatorially in Comtet's reference [129, p. 105] and it serves to illustrate "proofs from THE BOOK" in the splendid exposition of Aigner and Ziegler [7, §29]. Consequently, the numbers $\{P_j\}_{j=0}^{N}$ can be determined in $O(N\sqrt{N})$ integer operations. ◁

▷ **I.20.** *A digital surprise.* Define the constant

$$\varphi := \frac{9}{10}\frac{99}{100}\frac{999}{1000}\frac{9999}{10000}\cdots.$$

Is it a surprise that it evaluates numerically to

$$\varphi \doteq 0.8900100999989990000001000099999998999990000000000010\cdots,$$

that is, its decimal representation involves only the digits 0, 1, 8, 9? [This is suggested by a note of S. Ramanujan, "Some definite integrals", *Messenger of Math.* XLIV, 1915, pp. 10–18.] ◁

▷ **I.21.** *Lattice points.* The number of lattice points with integer coordinates that belong to the closed ball of radius n in d-dimensional Euclidean space is

$$[z^{n^2}]\frac{1}{1-z}(\Theta(z))^d \qquad \text{where} \qquad \Theta(z) = 1 + 2\sum_{n=1}^{\infty}z^{n^2}.$$

Estimates may be obtained via the saddle-point method (Note VIII.35, p. 589). ◁

I.4. Words and regular languages

Fix a finite *alphabet* $\mathcal{A}$ whose elements are called *letters*. Each letter is taken to have size 1; i.e., it is an atom. A *word*[8] is any finite sequence of letters, usually written without separators. So, for us, with the choice of the Latin alphabet ($\mathcal{A} = \{a,\ldots,z\}$), sequences such as `ygololihp`, `philology`, `zgrmblglps` are words. We denote the set of all words (often written as $\mathcal{A}^\star$ in formal linguistics) by $\mathcal{W}$. Following a well-established tradition in theoretical computer science and formal linguistics, any subset of $\mathcal{W}$ is called a *language* (or formal language, when the distinction with natural languages has to be made).

---

[8]An alternative to the term "word" sometimes preferred by computer scientists is *"string"*; biologists often refer to words as *"sequences"*.

	OGF	*coefficients*	
Words:	$\dfrac{1}{1-mz}$	m^n	(p. 50)
a–runs $< k$	$\dfrac{1-z^k}{1-mz+(m-1)z^{k+1}}$	$\sim c_k \rho_k^{-n}$	(pp. 51, 308)
exclude subseq. $\mathfrak{p}$	Eq. (55)	$\approx (m-1)^n n^{\lvert \mathfrak{p} \rvert -1}$	(p. 54)
exclude factor $\mathfrak{p}$	$\dfrac{c_{\mathfrak{p}}(z)}{z^{\lvert \mathfrak{p} \rvert} + (1-mz)c_{\mathfrak{p}}(z)}$	$\sim c_{\mathfrak{p}} \rho_{\mathfrak{p}}^{-n}$	(pp. 61, 271)
circular	Eq. (64)	$\sim m^n/n$	(p. 64)
regular language	[rational]	$\approx C \cdot A^n n^k$	(pp. 56, 302, 342)
context-free lang.	[algebraic]	$\approx C \cdot A^n n^{p/q}$	(pp. 80, 501)

Figure I.10. Words over an m–ary alphabet: generating functions and coefficients.

From the definition of the set of words $\mathcal{W}$, one has

$$(51) \qquad\qquad \mathcal{W} \cong \text{SEQ}(\mathcal{A}) \qquad \Longrightarrow \qquad W(z) = \frac{1}{1-mz},$$

where m is the cardinality of the alphabet, i.e., the number of letters. The generating function gives us the counting result

$$W_n = m^n.$$

This result is elementary, but, as is usual with symbolic methods, many enumerative consequences result from a given construction. It is precisely the purpose of this section to examine some of them.

We shall introduce separately two frameworks that each have great expressive power for describing languages. The first one is iterative (i.e., non-recursive) and it bases itself on "regular specifications" that only involve the constructions of sum, product, and sequence; the other one, which is recursive (but of a very simple form), is best conceived of in terms of finite automata and is equivalent to linear systems of equations. Both frameworks turn out to be logically equivalent in the sense that they determine the same family of languages, the *regular languages*, though the equivalence is non-trivial (Appendix A.7: *Regular languages*, p. 733), and each particular problem usually admits a preferred representation. The resulting OGFs are invariably rational functions, a fact to be systematically exploited from an asymptotic standpoint in Chapter V. Figure I.10 recapitulates some of the major word problems studied in this chapter, together with corresponding approximations[9].

---

[9]In this book, we reserve "$\sim$" for the technical sense of "asymptotically equivalent" defined in Appendix A.2: *Asymptotic notations*, p. 722; we reserve the symbol "$\approx$" to mean "approximately equal" in a vaguer sense, where formulae have been simplified by omitting constant factors or terms of secondary importance (in context).

I. 4.1. Regular specifications. Consider words (or strings) over the binary alphabet $\mathcal{A} = \{a, b\}$. There is an alternative way to construct binary strings. It is based on the observation that, with a minor adjustment at the beginning, a string decomposes into a succession of "blocks" each formed with a single b followed by an arbitrary (possibly empty) sequence of as. For instance $aaabaababaabbabbaaa$ decomposes as

$$[aaa]\, baa \mid ba \mid baa \mid b \mid ba \mid b \mid baaa.$$

Omitting redundant[10] symbols, we have the alternative decomposition:

$$(52) \qquad \mathcal{W} \cong \text{SEQ}(a) \times \text{SEQ}(b\,\text{SEQ}(a)) \quad \Longrightarrow \quad W(z) = \frac{1}{1-z}\frac{1}{1-z\frac{1}{1-z}}.$$

This last expression reduces to $(1 - 2z)^{-1}$ as it should.

Longest runs. The interest of the construction just seen is to take into account various meaningful properties, for example longest runs. Abbreviate by $a^{<k} := \text{SEQ}_{<k}(a)$ the collection of all words formed with the letter a only and whose length is between 0 and $k-1$; the corresponding OGF is $1 + z + \cdots + z^{k-1} = (1 - z^k)/(1 - z)$. The collection $\mathcal{W}^{\langle k \rangle}$ of words which do not have k consecutive as is described by an amended form of (52):

$$\mathcal{W}^{\langle k \rangle} = a^{<k}\, \text{SEQ}(ba^{<k}) \quad \Longrightarrow \quad W^{\langle k \rangle}(z) = \frac{1 - z^k}{1 - z} \cdot \frac{1}{1 - z\frac{1-z^k}{1-z}} = \frac{1 - z^k}{1 - 2z + z^{k+1}}.$$

The OGF is in principle amenable to expansion, but the resulting coefficients expressions are complicated and, in such a case, asymptotic estimates tend to be more usable. From the analysis developed in Example V.4 (p. 308), it can indeed be deduced that the longest run of a's in a random binary string of length n is on average asymptotic to $\log_2 n$.

$\triangleright$ **I.22.** *Runs in arbitrary alphabets.* For an alphabet of cardinality m, the quantity

$$\frac{1 - z^k}{1 - mz + (m-1)z^{k+1}}$$

is the OGF of words without k consecutive occurrences of a designated letter. $\triangleleft$

The case of longest runs exemplifies the utility of nested constructions involving sequences. We set:

Definition I.10. *An iterative specification that only involves atoms (e.g., letters of a finite alphabet $\mathcal{A}$) together with combinatorial sums, cartesian products, and sequence constructions is said to be a* regular specification.

A language $\mathcal{L}$ is said to be S–regular ("specification–regular") if there exists a class $\mathcal{M}$ described by a regular specification such that $\mathcal{L}$ and $\mathcal{M}$ are combinatorially isomorphic: $\mathcal{L} \cong \mathcal{M}$.

An equivalent way of expressing the definition is as follows: a language is S–regular if it can be described *unambiguously* by a regular expression (Appendix A.7:

---

[10]When dealing with words, especially, we freely omit redundant braces "$\{,\}$" and cartesian products "$\times$", for readability. For instance, $\text{SEQ}(a + b)$ and $a\,b$ are shorthand for $\text{SEQ}(\{a\} + \{b\})$ and $\{a\} \times \{b\}$.

Regular languages, p. 733). The definition of a regular specification and the basic admissibility theorem (p. 27) imply immediately:

Proposition I.2. *Any S–regular language has an OGF that is a* rational function. *This OGF is obtained from a regular specification of the language by translating each letter into the variable z, disjoint unions into sums, cartesian products into products, and sequences into quasi-inverses,* $(1 - \cdot)^{-1}$.

This result is technically shallow but its importance derives from the fact that regular languages have great expressive power devolving from their rich closure properties (Appendix A.7: *Regular languages*, p. 733) as well as their relation to finite automata discussed in the next subsection. Examples I.9 and I.10 below make use of Proposition I.2 and treat two problems closely related to longest runs.

Example **I.9.** *Combinations and spacings.* A regular specification describes the set $\mathcal{L}$ of words that contain exactly k occurrences of the letter b, from which the OGF automatically follows:

(53)
$$\mathcal{L} = \mathrm{SEQ}(a)\,(b\,\mathrm{SEQ}(a))^k \quad \Longrightarrow \quad L(z) = z^k/(1 - z)^{k+1}.$$

Hence the number of words in the language satisfies $L_n = \binom{n}{k}$. This is otherwise combinatorially evident, since each word of length n is characterized by the positions of its letters b; that is, the choice of k positions among n possible ones. Symbolic methods thus give us back the well-known count of combinations by binomial coefficients.

Let $\binom{n}{k}_{<d}$ be the number of combinations of k elements among $[1, n]$ with constrained spacings: no element can be at distance d or more from its successor. The refinement of (53)

$$\mathcal{L}^{[d]} = \mathrm{SEQ}(a)\,(b\,\mathrm{SEQ}_{<d}(a))^{k-1}\,(b\,\mathrm{SEQ}(a)) \quad \Longrightarrow \quad \sum_{n\geq 0} \binom{n}{k}_{<d} z^n = \frac{z^k(1 - z^d)^{k-1}}{(1 - z)^{k+1}},$$

leads to a binomial convolution expression,

$$\binom{n}{k}_{<d} = \sum_j (-1)^j \binom{k-1}{j}\binom{n - dj}{k}.$$

(This problem is analogous to compositions with bounded summands in (42), p. 43.) What we have just analysed is the *largest* spacing (constrained to be at most d) in subsets. A parallel analysis yields information regarding the *smallest* spacing. .... ∎

Example **I.10.** *Double run statistics.* By forming maximal groups of equal letters in words, one finds easily that, for a binary alphabet,

$$\mathcal{W} \cong \mathrm{SEQ}(b)\,\mathrm{SEQ}(a\,\mathrm{SEQ}(a)\,b\,\mathrm{SEQ}(b))\,\mathrm{SEQ}(a).$$

Let $\mathcal{W}^{\langle\alpha,\beta\rangle}$ be the class of all words that have at most α consecutive as and β consecutive bs. The specification of $\mathcal{W}$ induces a specification of $\mathcal{W}^{\langle\alpha,\beta\rangle}$, upon replacing $\mathrm{SEQ}(a)$, $\mathrm{SEQ}(b)$ by $\mathrm{SEQ}_{<\alpha}(a)$, $\mathrm{SEQ}_{<\beta}(b)$ internally, and by $\mathrm{SEQ}_{\leq\alpha}(a)$, $\mathrm{SEQ}_{\leq\beta}(b)$ externally. In particular, the OGF of binary words that never have more than r consecutive identical letters is found to be (set $\alpha = \beta = r$)

(54)
$$W^{\langle r,r\rangle} = \frac{1 - z^{r+1}}{1 - 2z + z^{r+1}} = \frac{1 + z + \cdots + z^r}{1 - z - \cdots - z^r},$$

after simplification. (This result can be extended to an arbitrary alphabet by means of "Smirnov words", Example III.24, p. 204.)

Révész in [508] tells the following amusing story attributed to T. Varga: "A class of high school children is divided into two sections. In one of the sections, each child is given a coin which he throws two hundred times, recording the resulting head and tail sequence on a piece of paper. In the other section, the children do not receive coins, but are told instead that they should try to write down a 'random' head and tail sequence of length two hundred. Collecting these slips of paper, [a statistician] then tries to subdivide them into their original groups. Most of the time, he succeeds quite well."

The statistician's secret is to determine the probability distribution of the maximum length of runs of consecutive letters in a random binary word of length n (here $n = 200$). The probability that this parameter equals k is

$$\frac{1}{2^n} \left(W_n^{\langle k,k \rangle} - W_n^{\langle k-1,k-1 \rangle} \right)$$

and is fully determined by (54). The probabilities are then easily computed using any symbolic package: for $n = 200$, the values found are

k	3	4	5	6	7	8	9	10	11	12
$\mathbb{P}(k)$	$6.54\,10^{-8}$	$7.07\,10^{-4}$	0.0339	0.1660	0.2574	0.2235	0.1459	0.0829	0.0440	0.0226

Thus, in a randomly produced sequence of length 200, there are usually runs of length 6 or more: the probability of the event turns out to be close to 97% (and there is still a probability of about 8% to have a run of length 11 or more). On the other hand most children (and adults) are usually afraid of writing down runs longer than 4 or 5 as this is felt as strongly "non-random". The statistician simply selects the slips that contain runs of length 6 or more as the true random ones. *Voilà!* .... ∎

▷ **I.23.** *Alice, Bob, and coding bounds.* Alice wants to communicate n bits of information to Bob over a channel (a wire, an optic fibre) that transmits $0,1$-bits but is such that any occurrence of 11 terminates the transmission. Thus, she can only send on the channel an encoded version of her message (where the code is of some length $\ell \geq n$) that does not contain the pattern 11.

Here is a first coding scheme: given the message $m = m_1 m_2 \cdots m_n$, where $m_j \in \{0, 1\}$, apply the substitution: $0 \mapsto 00$ and $1 \mapsto 10$; terminate the transmission by sending 11. This scheme has $\ell = 2n + O(1)$, and we say that its *rate* is 2. Can one design codes with better rates? with rates arbitrarily close to 1, asymptotically?

Let $\mathcal{C}$ be the class of allowed code words. For words of length n, a code of length $L \equiv L(n)$ is achievable only if there exists a one-to-one mapping from $\{0, 1\}^n$ into $\bigcup_{j=0}^{L} \mathcal{C}_j$, i.e., $2^n \leq \sum_{j=0}^{L} C_j$. Working out the OGF of $\mathcal{C}$, one finds that necessarily

$$L(n) \geq \lambda n + O(1), \qquad \lambda = \frac{1}{\log_2 \varphi} \doteq 1.440420, \qquad \varphi = \frac{1 + \sqrt{5}}{2}.$$

Thus no code can achieve a rate better than 1.44; i.e., a loss of at least 44% is unavoidable. (For this and the next note, see, e.g., MacKay [427, Ch. 17].)   ◁

▷ **I.24.** *Coding without long runs.* Because of hysteresis in magnetic heads, certain storage devices cannot store binary sequences that have more than four consecutive 0s or more than four consecutive 1s. We seek a coding scheme that transforms an arbitrary binary string into a string obeying this constraint.

From the OGF, one finds $[z^{11}] W^{\langle 4,4 \rangle}(z) = 1546 > 2^{10} = 1024$. Consequently, a substitution can be built that translates an original 10-bit word into an 11-bit block that does not have five consecutive equal letters. When 11-bit blocks are concatenated, this may however give rise to forbidden sequences of identical consecutive letters at the junction of two blocks. It then suffices to use "separators" and replace a substituted block of the form $\alpha \cdot X \cdot \beta$ by the longer block $\overline{\alpha} \alpha \cdot X \cdot \beta \overline{\beta}$, where $\overline{0} = 1$ and $\overline{1} = 0$. The resulting code has rate $\frac{13}{10}$.

Extensions of this method show that the rate 1.057 is achievable (theoretically). On the other hand, by the principles of the previous note, any acceptable code must use asymptotically at least $1.056n$ bits to encode strings of n bits. (Hint: let α be the root near $\frac{1}{2}$ of $1 - 2\alpha + \alpha^5 = 0$, which is a pole of $W^{(4,4)}$. One has $1/\log_2(1/\alpha) \doteq 1.05621$.) $\triangleleft$

Patterns. There are many situations in the sciences where it is of interest to determine whether the appearance of a certain *pattern* in long sequences of observations is significant. In a genomic sequence of length $100\,000$ (the alphabet is A, G, C, T), is it or is it not meaningful to detect three occurrences of the pattern TAGATAA, where the letters appear consecutively and in the prescribed order? In computer network security, certain attacks can be detected by some well-defined alarming sequences of events, although these events may be separated by perfectly legitimate actions. On another register, data mining aims at broadly categorizing electronic documents in an automatic way, and in this context the observation of well-chosen patterns can provide highly discriminating criteria. These various applications require determining which patterns are, with high probability, bound to occur (these are *not* significant) and which are very unlikely to arise, so that actually observing them carries useful information. Quantifying the corresponding probabilistic phenomena reduces to an enumerative problem—the case of double runs in Example I.10 (p. 52) is in this respect typical.

The notion of pattern can be formalized in several ways. In this book, we shall principally consider two of them.

(*a*) *Subsequence pattern*: such a pattern is defined by the fact that its letters must appear in the right order, but not necessarily contiguously [263]. Subsequence patterns are also known as "hidden patterns".

(*b*) *Factor pattern*: such a pattern is defined by the fact that its letter must appear in the right order *and* contiguously [312, 564]. Factor patterns are also called "block patterns" or simply "patterns" when the context is clear.

For a given notion of pattern, there are then two related categories of problems. First, one may aim at determining the probability that a random word contains (or dually, excludes) a pattern; this problem is equivalently formulated as an existence problem—enumerate all words in which the pattern exists (i.e., occurs) independently of the number of occurrences. Second, one may aim at determining the expectation (or even the distribution) of the number of occurrences of a pattern in a random text; this problem involves enumerating enriched words, each with one occurrence of the pattern distinguished.

Such questions are amenable to methods of analytic combinatorics and in particular to the theory of regular specifications and automata: see Example I.11 below for a first attempt at analysing hidden patterns (to be continued in Chapter V, p. 315) and Example I.12 for an analysis of factor patterns (to be further extended in Chapters III, p. 211, IV, p. 271, and IX, p. 659).

Example I.11. *Subsequence (hidden) patterns in a text.* A sequence of letters that occurs in the right order, but not necessarily contiguously in a text is said to be a "hidden pattern". For instance the pattern *"combinatorics"* is to be found hidden in Shakespeare's Hamlet (Act I, Scene 1)

Dared to the | comb |at; | in | which our v| a |lian| t | Hamlet–

F| or | so th| i |s side of our known world esteem'd him–

Did slay this Fortinbras; who by a seal'd | c |ompact,

Well ratified by law and heraldry,
Did forfeit, with hi⎡s⎤ life, all those his lands [...]

Take a fixed finite alphabet $\mathcal{A}$ comprising m letters ($m = 26$ for English). First, let us examine the language $\mathcal{L}$ of all words, also called "texts", that contain a given word $\mathfrak{p} = p_1 p_2 \cdots p_k$ of length k as a subsequence. These words can be described unambiguously as starting with a sequence of letters not containing p_1 followed by the letter p_1 followed by a sequence not containing p_2, and so on:

$$\mathcal{L} = \text{SEQ}(\mathcal{A} \setminus p_1) p_1 \, \text{SEQ}(\mathcal{A} \setminus p_2) p_2 \cdots \text{SEQ}(\mathcal{A} \setminus p_k) p_k \, \text{SEQ}(\mathcal{A}).$$

This is in a sense equivalent to parsing words unambiguously according to the left-most occurrence of $\mathfrak{p}$ as a subsequence. The OGF is accordingly

(55)
$$L(z) = \frac{z^k}{(1 - (m-1)z)^k} \frac{1}{1 - mz}.$$

An easy analysis of the dominant simple pole at $z = 1/m$ shows that

$$L(z) \underset{z \to 1/m}{\sim} \frac{1}{1 - mz}, \qquad \text{so that} \quad L_n \underset{n \to \infty}{\sim} m^n.$$

Thus, a proportion tending to 1 of all the words of length n *do* contain a fixed pattern $\mathfrak{p}$ as a subsequence. (Note I.25 below refines this estimate.)

Mean number of occurrences. A census (Note I.26, p. 56) shows that there are in fact $1.63 \cdot 10^{39}$ occurrences of "combinatorics" as a subsequence hidden somewhere in the text of Hamlet, whose length is 120 057 (this is the number of letters that constitute the text). Is this the sign of a secret encouragement passed to us by the author of Hamlet?

To answer this somewhat frivolous question, here is an analysis of the expected number of occurrences of a hidden pattern. It is based on enumerating enriched words, where an enriched word is a word together with a distinguished occurrence of the pattern as a subsequence. Consider the regular specification

$$\mathcal{O} = \text{SEQ}(\mathcal{A}) \, p_1 \, \text{SEQ}(\mathcal{A}) \, p_2 \, \text{SEQ}(\mathcal{A}) \cdots \text{SEQ}(\mathcal{A}) \, p_{k-1} \, \text{SEQ}(\mathcal{A}) \, p_k \, \text{SEQ}(\mathcal{A}).$$

An element of $\mathcal{O}$ is a $(2k + 1)$–tuple whose first component is an arbitrary word, whose second component is the letter p_1, and so on, with letters of the pattern and free blocks alternating. In other terms, any $\omega \in \mathcal{O}$ represents precisely one possible occurrence of the hidden pattern $\mathfrak{p}$ in a text built over the alphabet $\mathcal{A}$. The associated OGF is simply

$$O(z) = \frac{z^k}{(1 - mz)^{k+1}}.$$

The ratio between the number of occurrences and the number of words of length n then equals

(56)
$$\Omega_n = \frac{[z^n] O(z)}{m^n} = m^{-k} \binom{n}{k},$$

and this quantity represents the expectation of the number of occurrences of $\mathfrak{p}$ in a random word of length n, assuming all such words to be equally likely. For the parameters corresponding to the text of Hamlet ($n = 120\,057$) and the pattern *"combinatorics"* ($k = 13$), the quantity Ω_n evaluates to $6.96 \cdot 10^{37}$. The number of hidden occurrences observed is thus 23 times higher than what the uniform model predicts! However, similar methods make it possible to take into account non-uniform letter probabilities (Subsection III.6.1, p. 189): based on the frequencies of letters in the English text itself, the expected number of occurrences is found to be $1.71 \cdot 10^{39}$—this is now only within 5% of what is observed. Thus, Shakespeare did not

(probably) conceal any message relative to combinatorics—see Example V.7, p. 315, for more
on this topic. .... ∎

▷ **I.25.** *A refined analysis.* Further consideration of the subdominant pole at $z = 1/(m - 1)$
yields, by the methods of Theorem IV.9 (p. 256), the refined estimate:

$$1 - \frac{L_n}{m^n} = O\left(n^{k-1}\left(1 - \frac{1}{m}\right)^n\right).$$

Thus, the probability of *not* containing a given subsequence pattern is exponentially small. ◁

▷ **I.26.** *Dynamic programming.* The number of occurrences of a subsequence pattern in a text
can be determined efficiently by scanning the text from left to right and maintaining a running
count of the number of occurrences of the pattern as well as all its prefixes.   ◁

I. 4.2. Finite automata.
We begin with a simple device, the *finite automaton*,
that is widely used in the study of models of computation [189] and has wide descriptive power with regard to structural properties of words. (A systematic treatment of
automata and paths in graphs, combining both algebraic and asymptotic aspects, is
given in Part B, Section V. 5, p. 336.)

Definition I.11. *A finite automaton is a directed multigraph whose edges are labelled
by letters of the alphabet $\mathcal{A}$. It is customary to refer to vertices as* states *and to denote
by Q the set of states. One designates an initial state $q_0 \in Q$ and a set of final states
$\overline{Q} \subseteq Q$.*

The automaton is said to be deterministic *if for each pair (q, α) with $q \in Q$ and
$\alpha \in A$ there exists at most one edge (one also says a* transition*) starting from q, which
is labelled by the letter α.*

A finite automaton (Figure I.11) is able to process words, as we now explain.
A word $w = w_1 \dots w_n$ is *accepted* by the automaton if there exists a path in the
multigraph connecting the initial state q_0 to one of the final states of $\overline{Q}$ and whose
sequence of edge labels is precisely $w_1, \dots, w_n$. For a deterministic finite automaton,
it suffices to start from the initial state q_0, scan the letters of the word from left to right,
and follow at each stage the only transition permitted; the word is accepted if the state
reached in this way after scanning the last letter of w is a final state. Schematically:

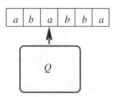

A finite automaton thus keeps only a finite memory of the past (hence its name) and
is in a sense a combinatorial counterpart of the notion of Markov chain in probability
theory. In this book, we shall only consider deterministic automata.

As an illustration, consider the class $\mathcal{L}$ of all words w that contain the pattern
abb as a factor (the letters of the pattern should appear contiguously). Such words are
recognized by a finite automaton with four states, q_0, q_1, q_2, q_3. The construction is
classical: state q_j is interpreted as meaning *"the first j characters of the pattern have*

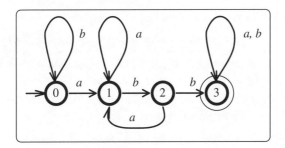

Figure I.11. Words that contain the pattern *abb* are recognized by a four-state automaton with initial state q_0 and final state q_3.

just been scanned", and the corresponding automaton appears in Figure I.11. The initial state is q_0, and there is a unique final state q_3.

Definition I.12. *A language is said to be A–regular (automaton regular) if it coincides with the set of words accepted by a deterministic finite automaton. A class $\mathcal{M}$ is A–regular if for some regular language $\mathcal{L}$, one has $\mathcal{M} \cong \mathcal{L}$.*

▷ **I.27.** *Congruence languages.* The language of binary representations of numbers that are congruent to 2 modulo 7 is *A*–regular. A similar property holds for any numeration base and any boolean combination of basic congruence conditions. ◁

▷ **I.28.** *Binary representation of primes.* The language of binary representations of prime numbers is neither *A*–regular nor *S*–regular. [Hint: use the Prime Number Theorem and asymptotic methods of Chapter IV.] ◁

The following equivalence theorem is briefly discussed in Appendix A.7: *Regular languages*, p. 733.

Equivalence theorem (Kleene–Rabin–Scott). *A language is S–regular (specification regular) if and only if it is A–regular (automaton regular).*

These two equivalent notions also coincide with the notion of regularity in formal language theory, where the latter is defined by means of (possibly ambiguous) regular expressions and (possibly non-deterministic) finite automata [6, 189]. As already pointed out, the equivalences are non-trivial: they are given by algorithms that transform one formalism into the other, but do not transparently preserve combinatorial structure (in some cases, an exponential blow-up in the size of descriptions is involved). For this reason, we have opted to develop independently the notions of *S*–regularity and *A*–regularity.

We next examine the way generating functions can be obtained from a deterministic automaton. The process was first discovered in the late 1950s by Chomsky and Schützenberger [119].

Proposition I.3. *Suppose that G is a deterministic finite automaton with state set $Q = \{q_0, \ldots, q_s\}$, initial state q_0, and set of final states $\overline{Q} = \{q_{i_1}, \ldots, q_{i_f}\}$. The generating function of the language $\mathcal{L}$ of all words accepted by the automaton is a rational function that is determined under matrix form as*

$$L(z) = \mathbf{u}(I - zT)^{-1}\mathbf{v}.$$

Here the transition matrix T is defined by

$$T_{j,k} = \text{card}\left\{\alpha \in \mathcal{A} \text{ such that an edge } (q_j, q_k) \text{ is labelled by } \alpha\right\};$$

the row vector $\mathbf{u}$ is the vector $(1, 0, 0, \ldots, 0)$ and the column vector $\mathbf{v} = (v_0, \ldots, v_s)^t$ is such that[11] $v_j = [\![q_j \in \overline{Q}]\!]$.

In particular, by Cramer's rule, the OGF of a regular language is the quotient of two (sparse) determinants whose structure directly reflects the automaton transitions.

Proof. The proof we present is based on a "first-letter decomposition", which is conceptually analogous to the Kolmogorov backward-equations of Markov chain theory [93, p. 153]. (Note I.29 provides an alternative approach.) For $j \in \{0, \ldots, s\}$, introduce the class (language) $\mathcal{L}_j$ of all words w such that the automaton, when started in state q_j, terminates in one of the final states of $\overline{Q}$, after having read w. The following relation holds for any j:

$$(57) \qquad \mathcal{L}_j \cong \Delta_j + \left(\sum_{\alpha \in \mathcal{A}} \{\alpha\} \mathcal{L}_{(q_j \circ \alpha)}\right);$$

there Δ_j is the class $\{\epsilon\}$ formed of the word of length 0 if q_j is final and the empty set ($\emptyset$) otherwise; the notation $(q_j \circ \alpha)$ designates the state reached in one step from state q_j upon reading letter α. The justification is simple: a language $\mathcal{L}_j$ contains the word of length 0 only if the corresponding state q_j is final; a word of length ≥ 1 that is accepted starting from state q_j has a first letter α followed by a word that must lead to an accepting state, when starting from state $q_j \circ \alpha$.

The translation of (57) is then immediate:

$$(58) \qquad L_j(z) = [\![q_j \in \overline{Q}]\!] + z \sum_{\alpha \in \mathcal{A}} L_{(q_j \circ \alpha)}(z).$$

The collection of all the equations as j varies forms a linear system: with $\mathbf{L}(z)$ the column vector $(L_0(z), \ldots, L_s(z))$, one has

$$\mathbf{L}(z) = \mathbf{v} + zT\,\mathbf{L}(z),$$

where $\mathbf{v}$ and T are as described in the statement. The result follows by matrix inversion upon observing that the OGF of the language $\mathcal{L}$ is $L_0(z)$. ∎

▷ **I.29.** *The forward equations.* Let $\mathcal{M}_k$ be the set of words, which lead to state q_k, when the automaton is started in state q_0. By a "last-letter decomposition", the $\mathcal{M}_k$ satisfy a system that is a transposed version of (58). ◁

The pattern abb. Consider the automaton recognizing the pattern abb as given in Figure I.11. The languages $\mathcal{L}_j$ (where $\mathcal{L}_j$ is the set of accepted words when starting from state q_j) are connected by the system of equations

$$\begin{array}{rcll}
\mathcal{L}_0 & = & a\mathcal{L}_1 & + b\mathcal{L}_0 \\
\mathcal{L}_1 & = & a\mathcal{L}_1 & + b\mathcal{L}_2 \\
\mathcal{L}_2 & = & a\mathcal{L}_1 & + b\mathcal{L}_3 \\
\mathcal{L}_3 & = & a\mathcal{L}_3 & + b\mathcal{L}_3 & + \epsilon,
\end{array}$$

---

[11] It proves convenient at this stage to introduce Iverson's bracket notation: for a predicate P, the quantity $[\![P]\!]$ has value 1 if P is true and 0 otherwise.

which directly reflects the graph structure of the automaton. This gives rise to a set of equations for the associated OGFs

$$
\begin{aligned}
L_0 &= zL_1 &+ zL_0 \\
L_1 &= zL_1 &+ zL_2 \\
L_2 &= zL_1 &+ zL_3 \\
L_3 &= zL_3 &+ zL_3 &+ 1.
\end{aligned}
$$

Solving the system, we find the OGF of all words containing the pattern *abb*: it is $L_0(z)$ since the initial state of the automaton is q_0, and

(59)
$$
L_0(z) = \frac{z^3}{(1-z)(1-2z)(1-z-z^2)}.
$$

The partial fraction decomposition

$$
L_0(z) = \frac{1}{1-2z} - \frac{2+z}{1-z-z^2} + \frac{1}{1-z},
$$

then yields

$$
L_{0,n} = 2^n - F_{n+3} + 1,
$$

with F_n a Fibonacci number (p. 42). In particular the number of words of length n that do *not* contain *abb* is $F_{n+3} - 1$, a quantity that grows at an exponential rate of φ^n, with $\varphi = (1+\sqrt{5})/2$ the golden ratio. Thus, all but an exponentially vanishing proportion of the strings of length n contain the given pattern *abb*, a fact that was otherwise to be expected on probabilistic grounds. (For instance, from Note I.32, p. 61, a random word contains a large number, about $\sim n/8$, of occurrences of the pattern *abb*.)

▷ **I.30. Regular specification for pattern abb.** The pattern *abb* is simple enough that one can come up with an equivalent regular expression describing $\mathcal{L}_0$, whose existence is otherwise granted by the Kleene–Rabin–Scott Theorem. An accepting path in the automaton of Figure I.11 loops around state 0 with a sequence of b, then reads an a, loops around state 1 with a sequence of a's and moves to state 2 upon reading a b; then there should be letters making the automaton passs through states 1-2-1-2-·····-1-2 and finally a b followed by an arbitrary sequence of as and bs at state 3. This corresponds to the specification (with $X^\star$ abbreviating $\mathrm{S{\small EQ}}(X)$)

$$
\mathcal{L}_0 = (b)^\star\, a(a)^\star b\, (a(a)^\star b)^\star\, b(a+b)^\star \quad\Longrightarrow\quad L_0(z) = \frac{z^3}{(1-z)^2(1-\frac{z^2}{1-z})(1-2z)},
$$

which gives back a form equivalent to (59).   ◁

Example I.12. *Words containing or excluding a pattern.* Fix an arbitrary pattern $\mathfrak{p} = \mathfrak{p}_1\mathfrak{p}_2\cdots\mathfrak{p}_k$ and let $\mathcal{L}$ be the language of words containing *at least* one occurrence of $\mathfrak{p}$ as a factor. Automata theory implies that the set of words containing a pattern as a factor is A–regular, hence admits a rational generating function. Indeed, the construction given for $\mathfrak{p} = abb$ generalizes in an easy manner: there exists a deterministic finite automaton with $k+1$ states that recognizes $\mathcal{L}$, the states memorizing the largest prefix of the pattern $\mathfrak{p}$ just seen. As a consequence: *the OGF of the language of words containing a given factor pattern of length k is a rational function of degree at most $k+1$.* (The corresponding automaton is in fact known as a Knuth–Morris–Pratt automaton [382].) The automaton construction however provides the OGF $L(z)$ in determinantal form, so that the relation between this rational form and the structure of the pattern is not transparent.

Autocorrelations. An explicit construction due to Guibas and Odlyzko [313] nicely circumvents this problem. It is based on an "equational" specification that yields an alternative linear system. The fundamental notion is that of an *autocorrelation vector*. For a given $\mathfrak{p}$, this vector of bits $c = (c_0, \ldots, c_{k-1})$ is most conveniently defined in terms of Iverson's bracket as

$$c_i = [\![p_{i+1} p_{i+2} \cdots p_k = p_1 p_2 \cdots p_{k-i}]\!].$$

In other words, the bit c_i is determined by shifting $\mathfrak{p}$ right by i positions and putting a 1 if the remaining letters match the original $\mathfrak{p}$. Graphically, $c_i = 1$ if the two framed factors of $\mathfrak{p}$ coincide in

$$\mathfrak{p} \equiv \quad p_1 \cdots p_i \boxed{p_{i+1} \cdots p_k}$$
$$\boxed{p_1 \cdots p_{k-i}} \, p_{k-i+1} \cdots p_k \quad \equiv \mathfrak{p}.$$

For instance, with $\mathfrak{p} = aabbaa$, one has

a a b b a a	
a a b b a a	1
a a b b a a	0
a a b b a a	0
a a b b a a	0
a a b b a a	1
a a b b a a	1.

The autocorrelation is then $c = (1, 0, 0, 0, 1, 1)$. The *autocorrelation polynomial* is defined as

$$c(z) := \sum_{j=0}^{k-1} c_j z^j.$$

For the example pattern, this gives $c(z) = 1 + z^4 + z^5$.

Let $\mathcal{S}$ be the language of words with *no* occurrence of $\mathfrak{p}$ and $\mathcal{T}$ the language of words that end with $\mathfrak{p}$ but have no other occurrence of $\mathfrak{p}$. First, by appending a letter to a word of $\mathcal{S}$, one finds a non-empty word either in $\mathcal{S}$ or $\mathcal{T}$, so that

(60) $$\mathcal{S} + \mathcal{T} = \{\epsilon\} + \mathcal{S} \times \mathcal{A}.$$

Next, appending a copy of the word $\mathfrak{p}$ to a word in $\mathcal{S}$ may only give words that contain $\mathfrak{p}$ at or "near" the end. In precise terms, the decomposition based on the left-most occurrence of $\mathfrak{p}$ in $\mathcal{S}\mathfrak{p}$ is

(61) $$\mathcal{S} \times \{\mathfrak{p}\} = \mathcal{T} \times \sum_{c_i \neq 0} \{p_{k-i+1} p_{k-i+2} \cdots p_k\},$$

corresponding to the configurations

The translation of the system (60), (61) into OGFs then gives a system of two equations in the two unknowns S, T,

$$S + T = 1 + mzS, \qquad S \cdot z^k = T c(z),$$

which is then readily solved.

Proposition I.4. *The OGF of words* not *containing the pattern* $\mathfrak{p}$ *as a factor is*

(62)
$$S(z) = \frac{c(z)}{z^k + (1 - mz)c(z)},$$

where m is the alphabet cardinality, $k = |\mathfrak{p}|$ the pattern length, and $c(z)$ the autocorrelation polynomial of $\mathfrak{p}$.

A bivariate generating function based on the autocorrelation polynomial is derived in Chapter III, p. 212, from which is deduced, in Proposition IX.10, p. 660, the existence of a limiting Gaussian law for the number of occurrences of any pattern. . . . . ∎

▷ **I.31.** *At least once.* The GFs of words containing at least once the pattern (anywhere) and containing it only once at the end are

$$L(z) = \frac{z^k}{(1 - mz)(z^k + (1 - mz)c(z))}, \quad T(z) = \frac{z^k}{z^k + (1 - mz)c(z)},$$

respectively. ◁

▷ **I.32.** *Expected number of occurrences of a pattern.* For the *mean* number of occurrences of a factor pattern, calculations similar to those employed for the number of occurrences of a subsequence (even simpler) can be based on regular specifications. All the occurrences (contexts) of $\mathfrak{p} = p_1 p_2 \cdots p_k$ as a factor are described by

$$\widehat{\mathcal{O}} = \mathrm{SEQ}(\mathcal{A})\,(p_1 p_2 \cdots p_k)\,\mathrm{SEQ}(\mathcal{A}), \quad \Longrightarrow \quad \widehat{O}(z) = \frac{z^k}{(1 - mz)^2}.$$

Consequently, the expected number of such contiguous occurrences satisfies

(63)
$$\widehat{\Omega}_n = m^{-k}(n - k + 1) \sim \frac{n}{m^k}.$$

Thus, the mean number of occurrences is proportional to n. ◁

▷ **I.33.** *Waiting times in strings.* Let $\mathcal{L} \subset \mathrm{SEQ}\{a, b\}$ be a language and $S = \{a, b\}^{\infty}$ be the set of infinite strings with the product probability induced by $\mathbb{P}(a) = \mathbb{P}(b) = \frac{1}{2}$. The probability that a random string $\omega \in S$ starts with a word of $\mathcal{L}$ is $\widehat{L}(1/2)$, where $\widehat{L}(z)$ is the OGF of the "prefix language" of $\mathcal{L}$, that is, the set of words $w \in \mathcal{L}$ that have no strict prefix belonging to $\mathcal{L}$. The GF $\widehat{L}(z)$ serves to express the expected time at which a word in $\mathcal{L}$ is first encountered: this is $\frac{1}{2}\widehat{L}'(\frac{1}{2})$. For a regular language, this quantity must be a rational number. ◁

▷ **I.34.** *A probabilistic paradox on strings.* In a random infinite sequence, a pattern $\mathfrak{p}$ of length k first occurs on average at time $2^k c(1/2)$, where $c(z)$ is the autocorrelation polynomial. For instance, the pattern $\mathfrak{p} = abb$ tends to occur "sooner" (at average position 8) than $\mathfrak{p}' = aaa$ (at average position 14). See [313] for a thorough discussion. Here are for instance the epochs at which $\mathfrak{p}$ and $\mathfrak{p}'$ are first found in a sample of 20 runs:

$$\mathfrak{p}: \quad 3, 4, 5, 5, 6, 6, 7, 8, 8, 8, 8, 9, 9, 10, 11, 14, 15, 15, 16, 21$$

$$\mathfrak{p}': \quad 3, 4, 8, 8, 9, 10, 11, 11, 11, 12, 17, 22, 23, 27, 27, 27, 44, 47, 52, 52.$$

On the other hand, patterns of the same length have the same expected number of occurrences, which is puzzling. *Is analytic combinatorics contradictory?* (Hint. The catch is that, due to overlaps of $\mathfrak{p}'$ with itself, occurrences of $\mathfrak{p}'$ tend to occur in clusters, but, then, clusters tend to be separated by wider gaps than for $\mathfrak{p}$; eventually, there is no contradiction.) ◁

▷ **I.35.** *Borges's Theorem.* Take any fixed finite set Π of patterns. A random text of length n contains all the patterns of the set Π (as factors) with probability tending to 1 exponentially fast as $n \to \infty$. Reason: the rational functions $S(z/2)$ with $S(z)$ as in (62) have no pole in $|z| \leq 1$; see also Chapters III (p. 213), IV(p. 271), V(p. 308). This property is sometimes called "*Borges's Theorem*" as a tribute to the famous Argentinian writer Jorge Luis Borges (1899–1986) who, in his essay "*The Library of Babel*", describes a library so huge as to contain:

> "Everything: the minutely detailed history of the future, the archangels' autobiographies, the faithful catalogues of the Library, thousands and thousands of false catalogues, the demonstration of the fallacy of those catalogues, the demonstration of the fallacy of the true catalogue, the Gnostic gospel of Basilides, the commentary on that gospel, the commentary on the commentary on that gospel, the true story of your death, the translation of every book in all languages, the interpolations of every book in all books."

Strong versions of Borges's Theorem, including the existence of limit Gaussian laws, hold for many random combinatorial structures, including trees, permutations, and planar maps (see Chapter IX, p. 659 and pp. 680–684).   ◁

▷ **I.36.** *Variable length codes.* A finite set $\mathcal{F} \subset \mathcal{W}$, where $\mathcal{W} = \text{Seq}(\mathcal{A})$ is called a *code* if any word of $\mathcal{W}$ decomposes in at most one manner into factors that belong to $\mathcal{F}$ (with repetitions allowed). For instance $\mathcal{F} = \{a, ab, bb\}$ is a code and $aaabbb = a|a|ab|bb$ has a unique decomposition; $\mathcal{F}' = \{a, aa, b\}$ is not a code since $aaa = a|aa = aa|a = a|a|a$. The OGF of the set $\mathcal{S}_{\mathcal{F}}$ of all words that admit a decomposition into factors all in $\mathcal{F}$ is a computable rational function, irrespective of whether $\mathcal{F}$ is a code. (Hint: use an "Aho–Corasick" automaton [5].) A finite set $\mathcal{F}$ is a code iff $S_{\mathcal{F}}(z) = (1 - F(z))^{-1}$. Consequently, the property of being a code can be decided in polynomial time using linear algebra. The book by Berstel and Perrin [55] develops systematically the theory of such variable-length codes.   ◁

In general, automata are useful in establishing *a priori* the rational character of generating functions. They are also surrounded by interesting analytic properties (e.g., Perron–Frobenius theory, Section V. 5, p. 336, that characterizes the dominant poles) and by asymptotic probability distributions of associated parameters that are normally Gaussian. They are most conveniently used for proving existence theorems, then supplemented when possible by regular specifications, which are likely to lead to more tractable expressions.

I. 4.3. Related constructions. Words can, at least in principle, encode any combinatorial structure. We detail here one situation that demonstrates the utility of such encodings: it is relative to set partitions and Stirling numbers. The point to be made is that some amount of "combinatorial preprocessing" is sometimes necessary in order to bring combinatorial structures into the orbit of symbolic methods.

Set partitions and Stirling partition numbers. A *set partition* is a partition of a finite domain into a certain number of non-empty sets, also called blocks. For instance, if the domain is $\mathcal{D} = \{\alpha, \beta, \gamma, \delta\}$, there are 15 ways to partition it (Figure I.12). Let $\mathcal{S}_n^{(r)}$ denote the collection of all partitions of the set $[1 \mathinner{\ldotp\ldotp} n]$ into r non-empty blocks and $S_n^{(r)} = \text{card}(\mathcal{S}_n^{(r)})$ the corresponding cardinality. The basic object under consideration here is a *set partition* (not to be confused with integer partitions considered earlier).

It is possible to find an encoding of partitions in $\mathcal{S}_n^{(r)}$ of an n-set into r blocks by words over a r letter alphabet, $\mathcal{B} = \{b_1, b_2, \ldots, b_r\}$ as follows. Consider a set partition ϖ that is formed of r blocks. Identify each block by its smallest element called the block *leader*; then sort the block leaders into increasing order. Define the index of a block as the rank of its leader among all the r leaders, with ranks conventionally starting at 1. Scan the elements 1 to n in order and produce sequentially n letters from the alphabet $\mathcal{B}$: for an element belonging to the block of index j, produce the letter b_j.

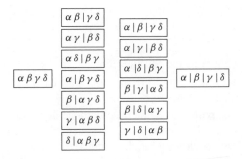

Figure I.12. The 15 ways of partitioning a four-element domain into blocks corresponds to $S_4^{(1)} = 1$, $S_4^{(2)} = 7$, $S_4^{(3)} = 6$, $S_4^{(4)} = 1$.

For instance for $n = 6$, $r = 3$, the set partition $\varpi = \{\{6, 4\}, \{5, 1, 2\}, \{3, 7, 8\}\}$, is reorganized by putting leaders in first position of the blocks and sorting them,

$$\varpi = \{\overbrace{\{1, 2, 5\}}^{b_1}, \overbrace{\{3, 7, 8\}}^{b_2}, \overbrace{\{4, 6\}}^{b_3}\},$$

so that the encoding is

$$\begin{pmatrix} 1 & 2 & 3 & 4 & 5 & 6 & 7 & 8 \\ b_1 & b_1 & b_2 & b_3 & b_1 & b_3 & b_2 & b_2 \end{pmatrix}.$$

In this way, a partition is encoded as a word of length n over $\mathcal{B}$ with the additional properties that: (i) all r letters occur; (ii) the first occurrence of b_1 precedes the first occurrence of b_2, which itself precedes the first occurrence of b_3, etc. Graphically, this correspondence can be rendered by an "irregular staircase" representation, such as

$$\begin{array}{ccccccccc} & & & \mathbf{4} & - & 6 & - & - \\ & & 3 & - & - & - & 7 & 8 \\ \mathbf{1} & 2 & - & - & 5 & - & - & - \end{array}$$

where the staircase has length n and height r, each column contains exactly one element, each row corresponds to a class in the partition.

From the foregoing discussion, $S_n^{(r)}$ is mapped into words of length n in the language

$$b_1 \, \text{SEQ}(b_1) \cdot b_2 \, \text{SEQ}(b_1 + b_2) \cdot b_3 \, \text{SEQ}(b_1 + b_2 + b_3) \, \cdots \, b_r \, \text{SEQ}(b_1 + b_2 + \cdots + b_r).$$

The language specification immediately gives the OGF

$$S^{(r)}(z) = \frac{z^r}{(1 - z)(1 - 2z)(1 - 3z) \cdots (1 - rz)}.$$

The partial fraction expansion of $S^{(r)}(z)$ is then readily computed,

$$S^{(r)}(z) = \frac{1}{r!} \sum_{j=0}^{r} \binom{r}{j} \frac{(-1)^{r-j}}{1 - jz}, \quad \text{so that} \quad S_n^{(r)} = \frac{1}{r!} \sum_{j=1}^{r} (-1)^{r-j} \binom{r}{j} j^n.$$

In particular, one has

$$S_n^{(1)} = 1, \quad S_n^{(2)} = \frac{1}{2!}(2^n - 2), \quad S_n^{(3)} = \frac{1}{3!}(3^n - 3 \cdot 2^n + 3).$$

These numbers are known as the Stirling numbers of the second kind, or better, as the Stirling partition numbers, and the $S_n^{(r)}$ are nowadays usually denoted by $\left\{ {n \atop r} \right\}$; see Appendix A.8: *Stirling numbers*, p. 735.

The counting of set partitions could eventually be done successfully thanks to an encoding into words, and the corresponding language forms a constructible class of combinatorial structures (indeed, a regular language). In the next chapter, we shall examine a flexible approach to the counting of set partitions that is based on labelled structures and exponential generating functions (Subsection II. 3.1, p. 106).

Circular words (necklaces). Let $\mathcal{A}$ be a binary alphabet, viewed as comprised of beads of two distinct colours. The class of *circular words* or *necklaces* (Note I.1, p. 18, and Equation (20), p. 26) is defined by a CYC composition:

$$(64) \qquad \mathcal{N} = \text{CYC}(\mathcal{A}) \quad \Longrightarrow \quad N(z) = \sum_{k=1}^{\infty} \frac{\varphi(k)}{k} \log \frac{1}{1 - 2z^k}.$$

The series starts as (*EIS* **A000031**)

$$N(z) = 2z + 3z^2 + 4z^3 + 6z^4 + 8z^5 + 14z^6 + 20z^7 + 36z^8 + 60z^9 + \cdots,$$

and the OGF can be expanded:

$$(65) \qquad\qquad N_n = \frac{1}{n} \sum_{k \mid n} \varphi(k) 2^{n/k}.$$

It turns out that $N_n = D_n + 1$ where D_n is the wheel count, p. 47. [The connection is easily explained combinatorially: start from a wheel and repaint in white all the nodes that are not on the basic circle; then fold them onto the circle.] The same argument proves that the number of necklaces over an m–ary alphabet is obtained by replacing 2 by m in (65).

▷ **I.37.** *Finite languages.* Viewed as a combinatorial object, a *finite language* λ is a set of distinct words, with size being the total number of letters of all words in λ. For a binary alphabet, the class of all finite languages is thus

$$\mathcal{FL} = \text{PSET}(\text{SEQ}_{\geq 1}(\mathcal{A})) \quad \Longrightarrow \quad FL(z) = \exp\left(\sum_{k \geq 1} \frac{(-1)^{k-1}}{k} \frac{2z^k}{1 - 2z^k} \right).$$

The series is (*EIS* **A102866**) $1 + 2z + 5z^2 + 16z^3 + 42z^4 + 116z^5 + 310z^6 + \cdots$. ◁

I. 5. Tree structures

This section is concerned with basic *tree enumerations*. Trees are, as we saw already, the prototypical recursive structure. The corresponding specifications normally lead to *nonlinear* equations (and systems of such) over generating functions, the Lagrange inversion theorem being exactly suited to solving the simplest category of problems. The functional equations furnished by the symbolic method can then conveniently be exploited by the asymptotic theory of Chapter VII (pp. 452–482). As we

	Specification	OGF	coefficient
Trees:			
plane general	$\mathcal{G} = \mathcal{Z} \times \text{SEQ}(\mathcal{G})$	$\frac{1}{2}(1 - \sqrt{1 - 4z})$	$\frac{1}{n}\binom{2n-2}{n-1} \sim \frac{4^{n-1}}{\sqrt{\pi n^3}}$
— binary	$\mathcal{B} = 1 + \mathcal{Z} \times \mathcal{B} \times \mathcal{B}$	$\frac{1}{2z}(1 - \sqrt{1 - 4z})$	$\frac{1}{n+1}\binom{2n}{n} \sim \frac{4^n}{\sqrt{\pi n^3}}$
— simple	$\mathcal{T} = \mathcal{Z} \times \text{SEQ}_\Omega(\mathcal{T})$	$T(z) = z\phi(T(z))$	$\sim c\rho^{-n}n^{-3/2}$
non-plane gen.	$\mathcal{H} = \mathcal{Z} \times \text{MSET}(\mathcal{H})$	$H(z) = z\,\text{Exp}(H(z))$	$\sim \lambda \cdot \beta^n/n^{3/2}$
— binary	$\mathcal{U} = \mathcal{Z} + \text{MSET}_2(\mathcal{U})$	Eq. (76), p. 72	$\sim \lambda_2 \cdot \beta_2^n/n^{3/2}$
— simple	$\mathcal{V} = \mathcal{Z}\,\text{MSET}_\Omega(\mathcal{V})$	Eq. (73), p. 71	$\sim \bar{c}\bar{\rho}^{-n}n^{-3/2}$

Figure I.13. Rooted trees of type either plane or non-plane and asymptotic forms. There, $\lambda \doteq 0.43992$, $\beta \doteq 2.95576$; $\lambda_2 \doteq 0.31877$, $\beta_2 \doteq 2.48325$. References for asymptotics are pp. 452–482 of Chapter VII.

shall see there, a certain type of analytic behaviour appears to be "*universal*" in trees, namely the occurrence of a $\sqrt{}$-singularity; accordingly, most tree families arising in the combinatorial world have counting sequences obeying a universal asymptotic form $C A^n n^{-3/2}$, which widely extends what we obtained elementarily for Catalan numbers on p. 38. A synopsis of what awaits us in this section is given in Figure I.13.

I. 5.1. Plane trees. Trees are commonly defined as undirected acyclic connected graphs. In addition, the trees considered in this book are, unless otherwise specified, *rooted* (Appendix A.9: *Tree concepts*, p. 737 and [377, §2.3]). In this subsection, we focus attention on *plane trees*, also sometimes called ordered trees, where subtrees dangling from a node are ordered between themselves. Alternatively, these trees may be viewed as abstract graph structures accompanied by an embedding into the plane. They are precisely described in terms of a sequence construction.

First, consider the class $\mathcal{G}$ of general plane trees where all node degrees are allowed (this repeats material on p. 35): we have

$$\text{(66)} \qquad \mathcal{G} = \mathcal{Z} \times \text{SEQ}(\mathcal{G}) \qquad \Longrightarrow \qquad G(z) = \frac{z}{1 - G(z)},$$

and, accordingly, $G(z) = \dfrac{1 - \sqrt{1 - 4z}}{2}$, so that the number of general trees of size n is a shifted Catalan number:

$$\text{(67)} \qquad G_n = C_{n-1} = \frac{1}{n}\binom{2n-2}{n-1}.$$

Many classes of trees defined by all sorts of constraints on properties of nodes appear to be of interest in combinatorics and in related areas such as formal logic and computer science. Let Ω be a subset of the integers that contains 0. Define the class $\mathcal{T}^\Omega$ of Ω–*restricted trees* as formed of trees such that the outdegrees of nodes are constrained to lie in Ω. In what follows, an essential rôle is played by a characteristic function that encapsulates Ω,

$$\phi(u) := \sum_{\omega \in \Omega} u^\omega.$$

Thus, $\Omega = \{0, 2\}$ determines binary trees, where each node has either 0 or 2 descendants, so that $\phi(u) = 1 + u^2$; the choices $\Omega = \{0, 1, 2\}$ and $\Omega = \{0, 3\}$ determine, respectively, unary–binary trees ($\phi(u) = 1+u+u^2$) and ternary trees ($\phi(u) = 1+u^3$); the case of general trees corresponds to $\Omega = \mathbb{Z}_{\geq 0}$ and $\phi(u) = (1 - u)^{-1}$.

Proposition I.5. *The ordinary generating function $T^\Omega(z)$ of the class T^Ω of Ω–restricted trees is determined implicitly by the equation*

$$T^\Omega(z) = z\,\phi(T^\Omega(z)),$$

where ϕ is the characteristic of Ω, namely $\phi(u) := \sum_{\omega \in \Omega} u^\omega$. The tree counts are given by

(68) $$T_n^\Omega \equiv [z^n] T^\Omega(z) = \frac{1}{n}[u^{n-1}]\phi(u)^n.$$

A class of trees whose generating function satisfies an equation of the form $y = z\phi(y(z))$ is also called a *simple variety of trees*. The study of such families (in the unlabelled and labelled cases alike) is one of the recurrent themes of this book.

Proof. Clearly, for Ω–restricted sequences, we have

$$\mathcal{A} = \text{SEQ}_\Omega(\mathcal{B}) \quad \Longrightarrow \quad A(z) = \phi(B(z)),$$

so

$$T^\Omega = \mathcal{Z} \times \text{SEQ}_\Omega(T^\Omega) \quad \Longrightarrow \quad T^\Omega(z) = z\phi(T^\Omega(z)).$$

This shows that $T \equiv T^\Omega$ is related to z by functional inversion:

$$z = \frac{T}{\phi(T)}.$$

The Lagrange Inversion Theorem precisely provides expressions for such a case (see Appendix A.6: *Lagrange Inversion*, p. 732 for an analytic proof and Note I.47, p. 75, for combinatorial aspects):

Lagrange Inversion Theorem. *The coefficients of an inverse function and of all its powers are determined by coefficients of powers of the direct function: if $z = T/\phi(T)$, then one has (with any $k \in \mathbb{Z}_{\geq 0}$):*

(69) $$[z^n]T(z) = \frac{1}{n}[w^{n-1}]\phi(w)^n, \qquad [z^n]T(z)^k = \frac{k}{n}[w^{n-k}]\phi(w)^n.$$

The theorem immediately implies (68).   ∎

The form relative to powers T^k in (69) is known as "Bürmann's form" of Lagrange inversion; it yields the counting of (ordered) k–forests, which are k–sequences of trees. Furthermore, the statement of Proposition I.5 extends trivially to the case where Ω is a multiset; that is, a set of integers with repetitions allowed. For instance, $\Omega = \{0, 1, 1, 3\}$ corresponds to unary–ternary trees with two types of unary nodes, say, having one of two colours; in this case, the characteristic is $\phi(u) = u^0 + 2u^1 + u^3$. The theorem gives back the enumeration of general trees, where $\phi(u) = (1 - u)^{-1}$, by

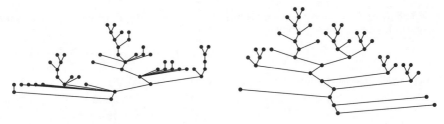

Figure I.14. A general tree of $\mathcal{G}_{51}$ (left) and a binary tree of $\mathcal{T}_{51}^{\{0,2\}} \cong \mathcal{B}_{25}$ (right) drawn uniformly at random among the C_{50} and C_{25} possible trees, respectively, with $C_n = \frac{1}{n+1}\binom{2n}{n}$, the nth Catalan number.

way of the binomial theorem applied to $(1-u)^{-n}$. In general, it implies that, whenever Ω comprises r elements, $\Omega = \{\omega_1, \ldots, \omega_r\}$, the tree counts are expressed as an $(r-1)$-fold summation of binomial coefficients (use the multinomial expansion). An important special case detailed in the next two examples below is when Ω has only two elements.

***Example* I.13.** *Binary trees and Catalan numbers.* A *binary tree* is a rooted plane tree, in which every node has either 0 or 2 successors (Figure I.14). In this case, it is customary to consider *size* to be the number of internal "branching" nodes, and we shall do so in most of the analyses to come. (By elementary combinatorics, if such a tree has ν internal nodes, it has $\nu + 1$ external nodes, hence it comprises $2\nu + 1$ nodes in total.) The specification and OGF of the class $\mathcal{B}$ of binary trees are then

$$\mathcal{B} = \mathbf{1} + (\mathcal{Z} \times \mathcal{B} \times \mathcal{B}) \quad \Longrightarrow \quad B(z) = 1 + zB(z)^2$$

(observe the structural analogy with triangulations in (31), p. 36), so that

$$B(z) = \frac{1 - \sqrt{1-4z}}{2z} \quad \text{and} \quad B_n = \frac{1}{n+1}\binom{2n}{n},$$

again a Catalan number (with a shift of index when compared to (67)). In summary:

The number B_n of plane binary trees having n internal nodes, i.e., $(n+1)$ external nodes and $(2n+1)$ nodes in total, is the Catalan number $B_n = C_n \equiv \frac{1}{n+1}\binom{2n}{n}$.

If one considers all nodes, internal and external alike, as contributing to size, the corresponding specification and OGF become

$$\widehat{\mathcal{B}} = \mathcal{Z} + (\mathcal{Z} \times \widehat{\mathcal{B}} \times \widehat{\mathcal{B}}) \quad \Longrightarrow \quad \widehat{B}(z) = z\left(1 + \widehat{B}(z)^2\right),$$

and the Lagrangean form is recovered (as well as $\widehat{B}_{2n+1} = B_n$), with $\phi(u) = (1+u^2)$.

Alternatively, consider the class $\overline{\mathcal{B}}$ of *pruned binary trees*, which are binary trees stripped of their external nodes (Appendix A.9: *Tree concepts*, p. 737), where only trees in $\mathcal{B} \setminus \mathcal{B}_0$ are taken. The corresponding class $\overline{\mathcal{B}}$ satisfies (upon distinguishing left- and right-branching unary nodes of the pruned tree)

$$\overline{\mathcal{B}} = \mathcal{Z} + (\mathcal{Z} \times \overline{\mathcal{B}}) + (\mathcal{Z} \times \overline{\mathcal{B}}) + (\mathcal{Z} \times \overline{\mathcal{B}} \times \overline{\mathcal{B}}) \quad \Longrightarrow \quad \overline{B}(z) = z\left(1 + \overline{B}(z)\right)^2$$

which is now Lagrangean with $\phi(u) = (1+u)^2$. These calculations, all with a strongly similar flavour, are explained by natural bijections in Subsection I.5.3, p. 73. . . . . ∎

▷ **I.38.** *Forests.* Consider ordered k–forests of trees defined by $\mathcal{F} = \text{SEQ}_k(\mathcal{T})$. The general form of Lagrange inversion implies

$$[z^n]F(z) \equiv [z^n]T(z)^k = \frac{k}{n}[u^{n-k}]\,\phi(u)^n.$$

In particular, one has for forests of general trees ($\phi(u) = (1-u)^{-1}$):

$$[z^n]\left(\frac{1-\sqrt{1-4z}}{2}\right)^k = \frac{k}{n}\binom{2n-k-1}{n-1};$$

the coefficients are also known as "ballot numbers".   ◁

Example I.14. *"Regular" (t–ary) trees.* A tree is said to be t–regular or t–ary if Ω consists only of the elements $\{0, t\}$ (the case $t = 2$ gives back binary trees). In other words, all internal nodes have degree t exactly. Let $\mathcal{A} := \mathcal{T}^{\{0,t\}}$. In this case, the characteristic is $\phi(u) = 1 + u^t$ and the binomial theorem combined with the Lagrange inversion formula gives

$$
\begin{aligned}
A_n &= \frac{1}{n}[u^{n-1}](1+u^t)^n \\
&= \frac{1}{n}\binom{n}{\frac{n-1}{t}} \quad \text{provided } n \equiv 1 \bmod t.
\end{aligned}
$$

As the formula shows, only trees of total size of the form $n = t\nu + 1$ exist (a well-known fact otherwise easily checked by induction), and

$$(70) \qquad A_{t\nu+1} = \frac{1}{t\nu+1}\binom{t\nu+1}{\nu} = \frac{1}{(t-1)\nu+1}\binom{t\nu}{\nu}.$$

As in the binary case, there is a variant of the determination of (70) that avoids congruence restrictions. Define the class $\overline{\mathcal{A}}$ of "pruned" trees as trees of $\mathcal{A} \setminus \mathcal{A}_0$ deprived of all their external nodes. The trees in $\overline{\mathcal{A}}$ now have nodes that are of degree at most t. In order to make $\overline{\mathcal{A}}$ bijectively equivalent to $\mathcal{A}$, it suffices to regard trees of $\overline{\mathcal{A}}$ as having $\binom{t}{j}$ possible types of nodes of degree j, for any $j \in [0, t]$: each node type in $\overline{\mathcal{A}}$ plainly encodes which of the original $t - j$ subtrees have been pruned. With Ω now being a multiset, we find $\overline{\phi}(u) = (1+u)^t$ and $\overline{A}(z) = z\overline{\phi}(\overline{A}(z))$, so that, by Lagrange inversion,

$$\overline{A}_\nu = \frac{1}{\nu}\binom{t\nu}{\nu-1} = \frac{1}{(t-1)\nu+1}\binom{t\nu}{\nu},$$

yet another form of (70), since $\overline{A}_\nu = A_{t\nu+1}$. .... ■

▷ **I.39.** *Unary–binary trees and Motzkin numbers.* Let $\mathcal{M}$ be the class of unary–binary trees:

$$\mathcal{M} = \mathcal{Z} \times \text{SEQ}_{\leq 2}(\mathcal{M}) \qquad \Longrightarrow \qquad M(z) = \frac{1-z-\sqrt{1-2z-3z^2}}{2z}.$$

One has $M(z) = z + z^2 + 2z^3 + 4z^4 + 9z^5 + 21z^6 + 51z^7 + \cdots$. The coefficients $M_n = [z^n]M(z)$, known as Motzkin numbers (*EIS* **A001006**), are given by

$$M_n = \frac{1}{n}\sum_k \binom{n}{k}\binom{n-k}{k-1},$$

as a consequence of the Lagrange Inversion Theorem.   ◁

▷ **I.40.** *Yet another variant of t–ary trees.* Let $\widetilde{\mathcal{A}}$ be the class of t–ary trees, but with size now defined as the number of external nodes (leaves). Then, one has

$$\widetilde{\mathcal{A}} = \mathcal{Z} + \text{SEQ}_t(\widetilde{\mathcal{A}}).$$

The binomial form of $\widetilde{A}_n$ follows from Lagrange inversion, since $\widetilde{A} = z/(1 - \widetilde{A}^{t-1})$. Can this last relation be interpreted combinatorially? ◁

Example I.15. *Hipparchus of Rhodes and Schröder.* In 1870, the German mathematician Ernst Schröder (1841–1902) published a paper entitled *Vier combinatorische Probleme.* The paper had to do with the number of terms that can be built out of n variables using non-associative operations. In particular, the second of his four problems asks for the number of ways a string of n identical letters, say x, can be "bracketed". The rule is best stated recursively: x itself is a bracketing and if $\sigma_1, \sigma_2, \ldots, \sigma_k$ with $k \geq 2$ are bracketed expressions, then the k–ary product $(\sigma_1\sigma_2\cdots\sigma_k)$ is a bracketing. For instance: $(((x\,x)x(x\,xx))((x\,x)(x\,x)x))$.

Let S denote the class of all bracketings, where *size* is taken to be the number of variable instances. Then, the recursive definition is readily translated into the formal specification (with $\mathcal{Z}$ representing x) and the OGF equation:

$$(71) \qquad \mathcal{S} = \mathcal{Z} + \text{SEQ}_{\geq 2}(\mathcal{S}) \qquad \Longrightarrow \qquad S(z) = z + \frac{S(z)^2}{1 - S(z)}.$$

Indeed, to each bracketing of size n is associated a tree whose external nodes contain the variable x (and determine size), with internal nodes corresponding to bracketings and having degree at least 2 (while not contributing to size).

The functional equation satisfied by the OGF is not *a priori* of the type corresponding to Proposition I.5, because *not all* nodes contribute to size in this particular application. Note I.41 provides a reduction to Lagrangean form; however, in a simple case like this, the quadratic equation induced by (71) is readily solved, giving

$$\begin{aligned} S(z) &= \frac{1}{4}\left(1 + z - \sqrt{1 - 6z + z^2}\right) \\ &= z + z^2 + 3z^3 + 11z^4 + 45z^5 + 197z^6 + 903z^7 + 4279z^8 + 20793z^9 \\ &\quad + \underline{103049}z^{10} + \underline{518859}z^{11} + \cdots, \end{aligned}$$

where the coefficients are *EIS* **A001003**. (These numbers also count series–parallel networks of a specified type (e.g., serial in Figure I.15, bottom), where placement in the plane matters.)

In an instructive paper, Stanley [553] discusses a page of Plutarch's *Moralia* where there appears the following statement:

> "Chrysippus says that the number of compound propositions that can be made from only ten simple propositions exceeds a million. (Hipparchus, to be sure, refuted this by showing that on the affirmative side there are $\underline{103\,049}$ compound statements, and on the negative side $\underline{310\,952}$.)"

It is notable that the tenth number of Hipparchus of Rhodes[12] (*c.* 190–120BC) is precisely $S_{10} = 103\,049$. This is, for instance, the number of logical formulae that can be formed from ten boolean variables $x_1, \ldots, x_{10}$ (used once each and in this order) using and–or connectives in alternation (no "negation"), upon starting from the top in some conventional fashion[13], e.g, with

---

[12]This was first observed by David Hough in 1994; see [553]. In [315], Habsieger *et al.* further note that $\frac{1}{2}(S_{10} + S_{11}) = 310\,954$, and suggest a related interpretation (based on negated variables) for the other count given by Hipparchus.

[13]Any functional term admits a unique tree representation. Here, as soon as the root type has been fixed (e.g., an $\wedge$ connective), the others are determined by level parity. The constraint of node degrees ≥ 2 in the tree means that no superfluous connectives are used. Finally, any monotone boolean expression can be represented by a series–parallel network: the x_j are viewed as switches with the *true* and *false* values being associated with closed and open circuits, respectively.

$$(x_1) \wedge (x_2 \vee (x_3 \wedge x_4 \wedge x_5) \vee x_6) \wedge ((x_7 \wedge x_8) \vee (x_9 \wedge x_{10}))$$

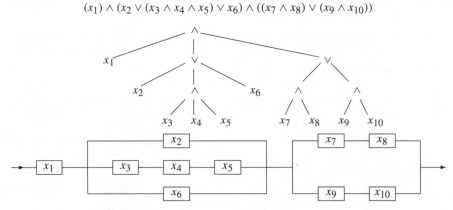

Figure I.15. An and–or positive proposition of the conjunctive type (top), its associated tree (middle), and an equivalent planar series–parallel network of the serial type (bottom).

an and-clause; see Figure I.15. Hipparchus was naturally not cognizant of generating functions, but with the technology of the time (and a rather remarkable mind!), he would still be able to discover a recurrence equivalent to (71),

$$(72) \qquad S_n = [\![n \geq 2]\!] \left(\sum_{n_1 + \cdots + n_k = n} S_{n_1} S_{n_2} \cdots S_{n_k} \right) + [\![n = 1]\!],$$

where the sum has only 42 essentially different terms for $n = 10$ (see [553] for a discussion), and finally determine S_{10}. .... ∎

▷ **I.41.** *The Lagrangean form of Schröder's GF.* The generating function $S(z)$ admits the form

$$S(z) = z\phi(S(z)) \quad \text{where} \quad \phi(y) = \frac{1 - y}{1 - 2y}$$

is the OGF of compositions. Consequently, one has

$$\begin{aligned} S_n &= \frac{1}{n} [u^{n-1}] \left(\frac{1 - u}{1 - 2u} \right)^n \\ &= \frac{(-1)^{n-1}}{n} \sum_k (-2)^k \binom{n}{k+1} \binom{n+k-1}{k} = \frac{1}{n} \sum_{k=0}^{n-2} \binom{2n-k-2}{n-1} \binom{n-2}{k}. \end{aligned}$$

Is there a direct combinatorial relation to compositions?   ◁

▷ **I.42.** *Faster determination of Schröder numbers.* By forming a differential equation satisfied by $S(z)$ and extracting coefficients, one obtains a recurrence

$$(n+2)S_{n+2} - 3(2n+1)S_{n+1} + (n-1)S_n = 0, \qquad n \geq 1,$$

that entails a fast determination, in linear time, of the S_n. (This technique, which originates with Euler [199], is applicable to any algebraic function; see Appendix B.4: *Holonomic functions*, p. 748.) In contrast, Hipparchus's recurrence (72) implies an algorithm of complexity $\exp(O(\sqrt{n}))$ in the number of arithmetic operations involved.   ◁

I. 5.2. Non-plane trees. An *unordered tree*, also called *non-plane* tree, is just a tree in the general graph-theoretic sense, so that there is no order between subtrees emanating from a common node. The unordered trees considered here are furthermore rooted, meaning that one of the nodes is distinguished as the root. Accordingly, in the language of constructions, a rooted *unordered* tree is a root node linked to a *multiset* of trees. Thus, the class $\mathcal{H}$ of all unordered trees, admits the recursive specification:

$$(73) \qquad \mathcal{H} = \mathcal{Z} \times \text{MSET}(\mathcal{H}) \implies \begin{cases} H(z) = z \displaystyle\prod_{m=1}^{\infty} (1 - z^m)^{-H_m} \\ = z \exp\left(H(z) + \tfrac{1}{2}H(z^2) + \cdots \right). \end{cases}$$

The first form of the OGF was given by Cayley in 1857 [67, p. 43]; it does not admit a closed form solution, although the equation permits one to determine all the H_n recursively (*EIS* **A000081**):

$$H(z) = z + z^2 + 2z^3 + 4z^4 + 9z^5 + 20z^6 + 48z^7 + 115z^8 + 286z^9 + \cdots .$$

The enumeration of the class of trees defined by an arbitrary set Ω of node degrees immediately results from the translation of sets of fixed cardinality.

Proposition I.6. *Let $\Omega \subset \mathbb{N}$ be a finite set of integers containing 0. The OGF $U(z)$ of non-plane trees with degrees constrained to lie in Ω satisfies a functional equation of the form*

$$(74) \qquad U(z) = z\Phi(U(z), U(z^2), U(z^3), \ldots),$$

for some computable polynomial Φ.

Proof. The class of trees satisfies the combinatorial equation,

$$\mathcal{U} = \mathcal{Z} \times \text{MSET}_\Omega(\mathcal{U}) \qquad \left(\text{MSET}_\Omega(\mathcal{U}) \equiv \sum_{\omega \in \Omega} \text{MSET}_\omega(\mathcal{U}) \right),$$

where the multiset construction reflects non-planarity, since subtrees stemming from a node can be freely rearranged between themselves and may appear repeated. Anticipating on what we shall see later, we note that Theorem I.3 (p. 84) provides the translation of $\text{MSET}_k(\mathcal{U})$:

$$\Phi(U(z), U(z^2), U(z^3), \ldots) = \sum_{\omega \in \Omega} [u^\omega] \exp\left(\frac{u}{1}U(z) + \frac{u^2}{2}U(z^2) + \cdots \right).$$

The statement then follows immediately. ∎

In the area of non-plane tree enumerations, there are no explicit formulae but only functional equations implicitly determining the generating functions. However, as we shall see in Section VII. 5 (p. 475), the equations may be used to analyse the dominant singularity of $U(z)$. We shall find that a "universal" law governs the singularities of simple tree generating functions, either plane or non-plane (Figure I.13): the singularities are of the general type $\sqrt{1 - z/\rho}$, which, by singularity analysis, translates into

$$(75) \qquad U_n^{\Omega} \sim \lambda_{\Omega} \frac{(\beta_{\Omega})^n}{\sqrt{n^3}}.$$

Many of these questions have their origin in enumerative combinatorial chemistry, a subject started by Cayley in the nineteenth century [67, Ch. 4]. Pólya re-examined these questions, and, in his important paper [488] published in 1937, he developed at the same time a general theory of combinatorial enumerations under group actions and systematic methods giving rise to estimates such as (75). See the book by Harary and Palmer [319] for more on this topic or Read's edition of Pólya's paper [491].

▷ **I.43.** *Fast determination of the Cayley–Pólya numbers.* Logarithmic differentiation of $H(z)$ provides for the H_n a recurrence by which one computes H_n in time polynomial in n. (Note: a similar technique applies to the partition numbers P_n; see p. 42.) ◁

▷ **I.44.** *Binary non-plane trees.* Unordered binary trees $\mathcal{V}$, with size measured by the number of external nodes, are described by the equation $\mathcal{V} = \mathcal{Z} + \mathrm{MSET}_2(\mathcal{V})$. The functional equation determining $V(z)$ is

$$(76) \qquad V(z) = z + \frac{1}{2}V(z)^2 + \frac{1}{2}V(z^2); \quad V(z) = z + z^2 + z^3 + 2z^4 + 3z^5 + \cdots.$$

The asymptotic analysis of the coefficients (*EIS* **A001190**) was carried out by Otter [466] who established an estimate of type (75). The quantity V_n is also the number of structurally distinct products of n elements under a commutative non-associative binary operation. ◁

▷ **I.45.** *Hierarchies.* Define the class $\mathcal{K}$ of hierarchies to be trees without nodes of outdegree 1 and size determined by the number of external nodes. We have (Cayley 1857, see [67, p.43])

$$\mathcal{K} = \mathcal{Z} + \mathrm{MSET}_{\geq 2}(\mathcal{K}) \implies K(z) = \frac{1}{2}z + \frac{1}{2}\left[\exp\left(K(z) + \frac{1}{2}K(z^2) + \cdots \right) - 1 \right],$$

from which the first values are found (*EIS* **A000669**)

$$K(z) = z + z^2 + 2z^3 + 5z^4 + 12z^5 + 33z^6 + 90z^7 + 261z^8 + 766z^9 + 2312z^{10} + \cdots.$$

These numbers also enumerate hierarchies in statistical classification theory [585]. They are the non-planar analogues of the Hipparchus–Schröder numbers on p. 69. ◁

▷ **I.46.** *Non-plane series–parallel networks.* Consider the class $\mathcal{SP}$ of series–parallel networks as previously considered in relation to the Hipparchus example, p. 69, but ignoring planar embeddings: all parallel arrangements of the (serial) networks $s_1, \ldots, s_k$ are considered equivalent, while the linear arrangement in each serial network matters. For instance, for $n = 2, 3$:

Thus, $SP_2 = 2$ and $SP_3 = 5$. This is modelled by the grammar:

$$\mathcal{S} = \mathcal{Z} + \mathrm{SEQ}_{\geq 2}(\mathcal{P}), \qquad \mathcal{P} = \mathcal{Z} + \mathrm{MSET}_{\geq 2}(\mathcal{S}),$$

and, avoiding to count networks of one element twice,

$$SP(z) = S(z) + P(z) - z = z + 2z^2 + 5z^3 + 15z^4 + 48z^5 + 167z^6 + 602z^7 + 2256z^8 + \cdots,$$

(*EIS* **A003430**). These objects are usually described as networks of electric resistors. ◁

I. 5.3. Related constructions. Trees underlie recursive structures of all sorts. A first illustration is provided by the fact that the Catalan numbers, $C_n = \frac{1}{n+1}\binom{2n}{n}$ count general trees ($\mathcal{G}$) of size $n + 1$, binary trees ($\mathcal{B}$) of size n (if size is defined as the number of internal nodes), as well as triangulations ($\mathcal{T}$) comprised of n triangles. The combinatorialist John Riordan even coined the name *Catalan domain* for the area within combinatorics that deals with objects enumerated by Catalan numbers, and Stanley's book contains an exercise [554, Ex. 6.19] whose statement alone spans ten full pages, with a list of 66 types of object(!) belonging to the Catalan domain. We shall illustrate the importance of Catalan numbers by describing a few fundamental correspondences (combinatorial isomorphisms, bijections) that explain the occurrence of Catalan numbers in several areas of combinatorics.

Rotation of trees. The combinatorial isomorphism relating $\mathcal{G}$ and $\mathcal{B}$ (albeit with a shift in size) coincides with a classical technique of computer science [377, §2.3.2]. To wit, a general tree can be represented in such a way that every node has two types of links, one pointing to the left-most child, the other to the next sibling in left-to-right order. Under this representation, if the root of the general tree is put aside, then every node is linked to two other (possibly empty) subtrees. In other words, general trees with n nodes are equinumerous with pruned binary trees with $n - 1$ nodes:

$$\mathcal{G}_n \cong \mathcal{B}_{n-1}.$$

Graphically, this is illustrated as follows:

The right-most tree is a binary tree drawn in a conventional manner, following a 45° tilt. This justifies the name of "rotation correspondence" often given to this transformation.

Tree decomposition of triangulations. The relation between binary trees $\mathcal{B}$ and triangulations $\mathcal{T}$ is equally simple: draw a triangulation; define the root triangle as the one that contains the edge connecting two designated vertices (for instance, the vertices numbered 0 and 1); associate to the root triangle the root of a binary tree; next, associate recursively to the subtriangulation on the left of the root triangle a left subtree; do similarly for the right subtriangulation giving rise to a right subtree.

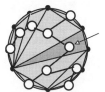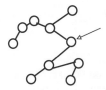

Under this correspondence, tree nodes correspond to triangle faces, while edges connect adjacent triangles. What this correspondence proves is the combinatorial isomorphism

$$\mathcal{T}_n \cong \mathcal{B}_n.$$

We turn next to another type of objects that are in correspondence with trees. These can be interpreted as words encoding tree traversals and, geometrically, as paths in the discrete plane $\mathbb{Z} \times \mathbb{Z}$.

Tree codes and Łukasiewicz words. Any plane tree can be traversed starting from the root, proceeding depth-first and left-to-right, and backtracking upwards once a subtree has been completely traversed. For instance, in the tree

(77) $\qquad\qquad \tau =$

the first visits to nodes take place in the following order

$$a, \quad b, \quad d, \quad h, \quad e, \quad f, \quad c, \quad g, \quad i, \quad j.$$

(Note: the tags $a, b, \ldots$, added for convenience in order to distinguish between nodes, have no special meaning; only the abstract tree shape matters here.) This order is known as *preorder* or *prefix order* since a node is preferentially visited before its children.

Given a tree, the listing of the outdegrees of nodes in prefix order is called the *preorder degree sequence*. For the tree of (77), this is

$$\sigma = (2, 3, 1, 0, 0, 0, 1, 2, 0, 0).$$

It is a fact that the degree sequence determines the tree unambiguously. Indeed, given the degree sequence, the tree is reconstructed step by step, adding nodes one after the other at the left-most available place. For σ, the first steps are then

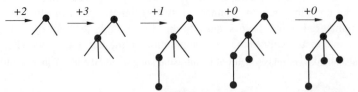

Next, if one represents degree j by a "symbol" f_j, then the degree sequence becomes a *word* over the infinite alphabet $\mathcal{F} = \{f_0, f_1, \ldots\}$, for instance,

$$\sigma \rightsquigarrow f_2 f_3 f_1 f_0 f_0 f_0 f_1 f_2 f_0 f_0.$$

This can be interpreted in the language of logic as a denotation for a functional term built out of symbols from $\mathcal{F}$, where f_j represents a function of degree (or "arity")

j. The correspondence even becomes obvious if superfluous parentheses are added at appropriate places to delimit scope:

$$\sigma \rightsquigarrow f_2(f_3(f_1(f_0), f_0, f_0), f_1(f_2(f_0, f_0))).$$

Such codes are known as Łukasiewicz codes[14], in recognition of the work of the Polish logician with that name. Jan Łukasiewicz (1878–1956) introduced them in order to completely specify the *syntax* of terms in various logical calculi; they prove nowadays basic in the development of parsers and compilers in computer science.

Finally, a tree code can be rendered as a walk over the discrete lattice $\mathbb{Z} \times \mathbb{Z}$. Associate to any f_j (i.e., any node of outdegree j) the displacement $(1, j-1) \in \mathbb{Z} \times \mathbb{Z}$, and plot the sequence of moves starting from the origin. In our example we find:

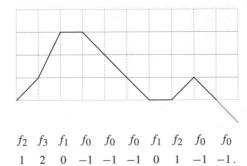

$$f_2 \quad f_3 \quad f_1 \quad f_0 \quad f_0 \quad f_0 \quad f_1 \quad f_2 \quad f_0 \quad f_0$$
$$1 \quad\ 2 \quad\ 0 \quad -1 \quad -1 \quad -1 \quad\ 0 \quad\ 1 \quad -1 \quad -1.$$

There, the last line represents the vertical displacements. The resulting paths are known as Łukasiewicz paths. Such a walk is then characterized by two conditions: the vertical displacements are in the set $\{-1, 0, 1, 2, \ldots\}$; all its points, except for the very last step, lie in the upper half-plane.

By this correspondence, the number of Łukasiewicz paths with n steps is the shifted Catalan number, $\frac{1}{n}\binom{2n-2}{n-1}$.

▷ **I.47.** *Conjugacy principle and cycle lemma.* Let $\mathcal{L}$ be the class of all Łukasiewicz paths. Define a "relaxed" path as one that starts at level 0, ends at level -1 but is otherwise allowed to include arbitrary negative points; let $\mathcal{M}$ be the corresponding class. Then, each relaxed path can be cut-and-pasted uniquely after its left-most minimum as described here:

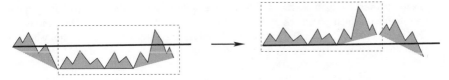

This associates to every relaxed path of length ν a unique standard path. A bit of combinatorial reasoning shows that correspondence is 1-to-ν (each element of $\mathcal{L}$ has *exactly* ν preimages.) One thus has $M_\nu = \nu L_\nu$. This correspondence preserves the number of steps of each type $(f_0, f_1, \ldots)$, so that the number of Łukasiewicz paths with ν_j steps of type f_j is

$$\frac{1}{\nu}[x^{-1}u_0^{\nu_0}u_1^{\nu_1}\cdots]\left(x^{-1}u_0 + u_1 + xu_2 + x^2u_3 + \cdots\right)^\nu = \frac{1}{\nu}\binom{\nu}{\nu_0, \nu_1, \ldots},$$

---

[14]A less dignified name is "Polish prefix notation". The "reverse Polish notation" is a variant based on postorder that has been used in some calculators since the 1970s.

under the necessary condition $(-1)v_0 + 0v_1 + 1v_2 + 2v_3 + \cdots = -1$. This combinatorial way of obtaining refined Catalan statistics is known as the *conjugacy principle* [503] or the *cycle lemma* [129, 155, 184]. It is logically equivalent to the Lagrange Inversion Theorem, as shown by Raney [503]. Dvoretzky & Motzkin [184] have employed this technique to solve a number of counting problems related to circular arrangements.   ◁

Example I.16.   *Binary tree codes and Dyck paths.* Walks associated with binary trees have a very special form since the vertical displacements can only be $+1$ or -1. The paths resulting from the Łukasiewicz correspondence are then equivalently characterized as sequences of numbers $x = (x_0, x_1, \ldots, x_{2n}, x_{2n+1})$ satisfying the conditions

(78) $\qquad x_0 = 0; \qquad x_j \geq 0 \quad \text{for } 1 \leq j \leq 2n; \qquad |x_{j+1} - x_j| = 1; \qquad x_{2n+1} = -1.$

These coincide with "gambler ruin sequences", a familiar object from probability theory: a player plays head and tails. He starts with no capital ($x_0 = 0$) at time 0; his total gain is x_j at time j; he is allowed no credit ($x_j \geq 0$) and loses at the very end of the game $x_{2n+1} = -1$; his gains are ± 1 depending on the outcome of the coin tosses ($|x_{j+1} - x_j| = 1$).

It is customary to drop the final step and consider "excursions' that take place in the upper half-plane. The resulting objects defined as sequences $(x_0 = 0, x_1, \ldots, x_{2n-1}, x_{2n} = 0)$ satisfying the first three conditions of (78) are known in combinatorics as *Dyck paths*[15]. By construction, Dyck paths of length $2n$ correspond bijectively to binary trees with n internal nodes and are consequently enumerated by Catalan numbers. Let $\mathcal{D}$ be the combinatorial class of Dyck paths, with size defined as length. This property can also be checked directly: the quadratic decomposition

(79)

$$\mathcal{D} \quad = \quad \{\epsilon\} + (\nearrow \mathcal{D} \searrow) \times \mathcal{D}$$

$$\implies \quad D(z) \quad = \quad 1 + (zD(z)z)\, D(z).$$

From this OGF, the Catalan numbers are found (as expected): $D_{2n} = \frac{1}{n+1}\binom{2n}{n}$. The decomposition (79) is known as the "first passage" decomposition as it is based on the first time the accumulated gain in the coin-tossing game passes through the value zero.

Dyck paths also arise in connection will well-parenthesized expressions. These are recognized by keeping a counter that records at each stage the excess of the number of opening brackets "(" over closing brackets ")". Finally, one of the origins of the Dyck path is the famous *ballot problem*, which goes back to the nineteenth century [423]: there are two candidates A and B that stand for election, $2n$ voters, and the election eventually results in a tie; what is the probability that A is always ahead of or tied with B when the ballots are counted? The answer is

$$\frac{D_{2n}}{\binom{2n}{n}} = \frac{1}{n+1},$$

since there are $\binom{2n}{n}$ possibilities in total, of which the number of favourable cases is D_{2n}, a Catalan number. The central rôle of Dyck paths and Catalan numbers in problems coming from such diverse areas is quite remarkable. Section V. 4, p. 318 presents refined counting results regarding lattice paths (e.g., the analysis of height) and Subsection VII. 8.1, p. 506 introduces exact and asymptotic results in the harder case of an arbitrary finite collection of step types (not just ± 1). ∎

---

[15]Dyck paths are closely associated with free groups on one generator and are named after the German mathematician Walther (von) Dyck (1856–1934) who introduced free groups around 1880.

▷ **I.48.** *Dyck paths, parenthesis systems, and general trees.* The class of Dyck paths admits an alternative sequence decomposition

(80)

$$\mathcal{D} = \text{SEQ}(\mathcal{Z} \times \mathcal{D} \times \mathcal{Z}),$$

which again leads to the Catalan GF. The decomposition (80) is known as the "arch decomposition" (see Subsection V. 4.1, p. 319, for more). It can also be directly related to traversal sequences of general trees, but with the directions of *edge* traversals being recorded (instead of traversals based on node degrees): for a general tree τ, define its encoding $\kappa(\tau)$ over the binary alphabet $\{\nearrow, \searrow\}$ recursively by the rules:

$$\kappa(\tau) = \epsilon, \qquad \kappa(\bullet(\tau_1, \ldots, \tau_r)) = \nearrow \kappa(\tau_1) \cdots \kappa(\tau_r) \searrow.$$

This is the classical representation of trees by a parenthesis system (interpret "$\nearrow$" and "$\searrow$" as "(" and ")", respectively), which associates to a tree of n nodes a path of length $2n - 2$. ◁

▷ **I.49.** *Random generation of Dyck paths.* Dyck paths of length $2n$ can be generated uniformly at random in time linear in n. (Hint: By Note I.47, it suffices to generate uniformly a sequence of n as and $(n + 1)$ bs, then reorganize it according to the conjugacy principle.) ◁

▷ **I.50.** *Excursions, bridges, and meanders.* Adapting a terminology from probability theory, one sets the following definitions: (i) a *meander* $(\mathcal{M})$ is a word over $\{-1, +1\}$, such that the sum of the values of any of its prefixes is always a non-negative integer; (ii) a *bridge* $(\mathcal{B})$ is a word whose values of letters sum to 0. Thus a meander represents a walk that wanders in the first quadrant; a bridge, regarded as a walk, may wander above and below the horizontal line, but its final altitude is constrained to be 0; an excursion is both a meander and a bridge. Simple decompositions provide

$$M(z) = \frac{D(z)}{1 - zD(z)}, \qquad B(z) = \frac{1}{1 - 2z^2 D(z)},$$

implying $M_n = \binom{n}{\lfloor n/2 \rfloor}$ [*EIS* **A001405**] and $B_{2n} = \binom{2n}{n}$ [*EIS* **A000984**]. ◁

▷ **I.51.** *Motzkin paths and unary–binary trees.* Motzkin paths are defined by changing the third condition of (78) defining Dyck paths into $|x_{j+1} - x_j| \le 1$. They appear as codes for unary–binary trees and are enumerated by the Motzkin numbers of Note I.39, p. 68. ◁

***Example* I.17.** *The complexity of boolean functions.* Complexity theory provides many surprising applications of enumerative combinatorics and asymptotic estimates. In general, one starts with a finite set of abstract mathematical objects Ω and a combinatorial class $\mathcal{D}$ of concrete *descriptions*. By assumption, to every element of $\delta \in \mathcal{D}$ is associated an object $\mu(\delta) \in \Omega$, its "meaning"; conversely any object of Ω admits at least one description in $\mathcal{D}$ (that is, the function μ is surjective). It is then of interest to quantify properties of the shortest description function defined for $\omega \in \Omega$ as

$$\sigma(\omega) := \min \{ |\delta|_{\mathcal{D}} \mid \mu(\delta) = \omega \},$$

and called the *complexity* of the element $\omega \in \Omega$ (with respect to $\mathcal{D}$).

We take here Ω to be the class of all boolean functions on m variables. Their number is $\|\Omega\| = 2^{2^m}$. As descriptions, we adopt the class of logical expressions involving the logical connectives $\lor, \land$ and pure or negated variables. Equivalently, $\mathcal{D}$ is the class of binary trees, where internal nodes are tagged by a logical disjunction ("$\lor$") or a conjunction ("$\land$"), and each external node is tagged by either a boolean variable of $\{x_1, \ldots, x_m\}$ or a negated variable of

$\{\neg x_1, \ldots, \neg x_m\}$. Define the size of a tree description as the number of internal nodes; that is, the number of logical operators. Then, one has

$$(81) \qquad D_n = \left(\frac{1}{n+1} \binom{2n}{n} \right) \cdot 2^n \cdot (2m)^{n+1},$$

as seen by counting tree shapes and possibilities for internal as well as external node tags.

The crux of the matter is that if the inequality

$$(82) \qquad \sum_{j=0}^{\nu} D_j < \|\Omega\|,$$

holds, then there are not enough descriptions of size $\leq \nu$ to exhaust Ω. (This is analogous to the coding argument of Note I.23, p. 53.) In other terms, there must exist at least one object in Ω whose complexity exceeds ν. If the left side of (82) is much smaller than the right side, then it must even be the case that "most" Ω–objects have a complexity that exceeds ν.

In the case of boolean functions and tree descriptions, the asymptotic form (33) is available. From (81) it can be seen that, for n, ν getting large, one has

$$D_n = O(16^n m^n n^{-3/2}), \qquad \sum_{j=0}^{\nu} D_j = O(16^\nu m^\nu \nu^{-3/2}).$$

Choose ν such that the second expression is $o(\|\Omega\|)$, which is ensured for instance by taking for ν the value

$$\nu(m) := \frac{2^m}{4 + \log_2 m}.$$

With this choice, one has the following suggestive statement:

A fraction tending to 1 (as $m \to \infty$) of boolean functions in m variables have tree complexity at least $2^m / (4 + \log_2 m)$.

Regarding upper bounds on boolean function complexity, a function always has a tree complexity that is at most $2^{m+1} - 3$. To see this, note that for $m = 1$, the four functions are

$$0 \equiv (x_1 \wedge \neg x_1), \quad 1 \equiv (x_1 \vee \neg x_1), \quad x_1, \quad \neg x_1.$$

Next, a function of m variables is representable by a technique known as the binary decision tree (BDT),

$$f(x_1, \ldots, x_{m-1}, x_m) = \left(\neg x_m \wedge f(x_1, \ldots, x_{m-1}, 0) \right) \vee \left(x_m \wedge f(x_1, \ldots, x_{m-1}, 1) \right),$$

which provides the basis of the induction as it reduces the representation of an m–ary function to the representation of two $(m - 1)$–ary functions, consuming on the way three logical connectives.

Altogether, basic counting arguments have shown that "most" boolean functions have a tree-complexity $(2^m / \log m)$ that is fairly close to the maximum possible, namely, $O(2^m)$. A similar result has been established by Shannon for the measure called circuit complexity: circuits are more powerful than trees, but Shannon's result states that *almost all boolean functions of m variables have circuit complexity $O(2^m / m)$*. See the chapter by Li and Vitányi in [591] and Gardy's survey [283] on random boolean expressions for a discussion of such counting techniques within the framework of complexity theory and logic. We resume this thread in Example VII.17, p. 487, where we quantify the probability that a large random boolean expression computes a fixed function. .... ∎

I. 5.4. Context-free specifications and languages. Many of the combinatorial examples encountered so far in this section can be organized into a common framework, which is fundamental in formal linguistics and theoretical computer science.

Definition I.13. *A class $\mathcal{C}$ is said to be* context-free *if it coincides with the first component* $(\mathcal{T} = \mathcal{S}_1)$ *of a system of equations*

$$
(83) \qquad
\begin{cases}
\mathcal{S}_1 & = & \mathfrak{F}_1(\mathcal{Z}, \mathcal{S}_1, \ldots, \mathcal{S}_r) \\
\vdots & \vdots & \vdots \\
\mathcal{S}_r & = & \mathfrak{F}_r(\mathcal{Z}, \mathcal{S}_1, \ldots, \mathcal{S}_r),
\end{cases}
$$

where each $\mathfrak{F}_j$ is a constructor that only involves the operations of combinatorial sum $(+)$ and cartesian product $(\times)$, as well as the neutral class, $\mathcal{E} = \{\epsilon\}$.

A language $\mathcal{L}$ is said to be an unambiguous context-free *language if it is combinatorially isomorphic to a context-free class of trees: $\mathcal{C} \cong \mathcal{T}$.*

The classes of general trees $(\mathcal{G})$ and binary trees $(\mathcal{B})$ are context-free, since they are specifiable as

$$
\begin{cases}
\mathcal{G} & = & \mathcal{Z} \times \mathcal{F} \\
\mathcal{F} & = & \{\epsilon\} + (\mathcal{G} \times \mathcal{F}), \qquad & \mathcal{B} = \mathcal{Z} + (\mathcal{B} \times \mathcal{B});
\end{cases}
$$

here $\mathcal{F}$ designates ordered forests of general trees. Context-free specifications may be used to describe all sorts of combinatorial objects. For instance, the class $\mathcal{U} = \mathcal{T} \setminus \mathcal{T}_0$ of non-empty triangulations of convex polygons (Note 10, p. 36) is specified symbolically by

$$
(84) \qquad \mathcal{U} = \nabla + (\nabla \times \mathcal{U}) + (\mathcal{U} \times \nabla) + (\mathcal{U} \times \nabla \times \mathcal{U}),
$$

where $\nabla \cong \mathcal{Z}$ represents a generic triangle. The Łukasiewicz language and the set of Dyck paths are context-free classes since they are bijectively equivalent to $\mathcal{G}$ and $\mathcal{U}$.

The term "context-free" comes from linguistics: it stresses the fact that objects can be "freely" generated by the rules of (83), this without any constraints imposed by an outside context[16]. There, one classically defines a context-free language as the language formed with words that are obtained as sequences of leaf tags (read in left-to-right order) of a context-free variety of trees. In formal linguistics, the one-to-one mapping between trees and words is not generally imposed; when it is satisfied, the context-free language is said to be *unambiguous*; in such cases, words and trees determine each other uniquely, cf Note I.54 below.

An immediate consequence of the admissibility theorems is the following proposition first encountered by Chomsky and Schützenberger [119] in the course of their research relating formal languages and formal power series.

---

[16]Formal language theory also defines context-sensitive grammars where each rule (called a production) is applied only if it is enabled by some external context. Context-sensitive grammars have greater expressive power than context-free ones, but they depart significantly from decomposability and are surrounded by strong undecidability properties. Accordingly, context-sensitive grammars cannot be associated with any global generating function formalism.

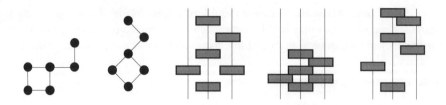

Figure I.16. A directed animal, its tilted version, (after a $+\pi/4$ rotation), and three of its equivalent representations as a heap of dimers.

Proposition I.7. *A combinatorial class C that is* context-free *admits an OGF that is* an *algebraic function. In other words, there exists a (non-null) bivariate polynomial $P(z, y) \in \mathbb{C}[z, y]$ such that*

$$P(z, C(z)) = 0.$$

Proof. By the basic sum and product rules, the context-free system (83) translates into a system of OGF equations,

$$\begin{cases} S_1(z) &= \Phi_1(z, S_1(z), \dots, S_r(z)) \\ \vdots \quad \vdots & \qquad \vdots \\ S_r(z) &= \Phi_r(z, S_1(z), \dots, S_r(z)), \end{cases}$$

where the Φ_j are the polynomials translating the constructions $\mathfrak{F}_j$.

It is then well known that algebraic elimination is possible in polynomial systems. Here, it is possible to eliminate the auxiliary variables $S_2, \dots, S_r$, one by one, preserving the polynomial character of the system at each stage. The end result is then a single polynomial equation satisfied by $C(z) \equiv S_1(z)$. (Methods for effectively performing polynomial elimination include a repeated use of resultants as well as Gröbner basis algorithms; see Appendix B.1: *Algebraic elimination*, p. 739 for a brief discussion and references.) ∎

Proposition I.7 is a counterpart of Proposition I.3 (p. 57) according to which rational generating functions arise from finite state devices, and it justifies the importance of algebraic functions in enumeration theory. We shall encounter applications of such algebraic generating functions to planar non-crossing configurations (p. 485) walks (p. 506) and planar maps (p. 513), when we develop a general asymptotic theory of their coefficients in Chapter VII, based on singularity theory. The example below shows the way certain lattice configurations can be modelled by a context-free specification.

Example I.18. *Directed animals.* Consider the square lattice $\mathbb{Z}^2$. A *directed animal with a compact source of size k* is a finite set of points α of the lattice such that: (i) for $0 \le i < k$, the points $(-i, i)$, called source points, belong to α; (ii) all other points in α can be reached from one of the source points by a path made of North and East steps and having all its vertices in α. (The animal in Figure I.16 has one source.) Such lattice configurations have been introduced by statistical physicists Dhar *et al.* [162], since they provide a tractable model of 2-dimensional percolation. Our discussion follows Bousquet-Mélou's insightful presentation in [84], itself based on Viennot's elegant theory of *heaps of pieces* [597].

The best way to visualize an animal is as follows (Figure I.16): rotate the lattice by $+\pi/4$ and associate to each vertex of the animal a horizontal piece, also called a dimer. The length of a piece is taken to be slightly less than the diagonal of a mesh of the original lattice. Pieces are allowed to slide vertically (up or down) in their column, but *not* to jump over each other. One can then think of an animal as being a heap of pieces, where pieces take their places naturally, under the effect of gravity, and each one stops as soon as it is blocked by a piece immediately below. (The heap associated to an animal satisfies the additional property that no two pieces in a column can be immediately adjacent to one another.)

Define a *pyramid* to be a one-source animal and a *half-pyramid* to be a pyramid that has no vertex strictly to the left of its source point, in the tilted representation. Let $\mathcal{P}$ and $\mathcal{H}$ be respectively the class of pyramids and half-pyramids, viewed as heaps. By a corner decomposition (Note I.52), pyramids and half-pyramids can be constructed as suggested by the following diagram:

(85)

The pictorial description (85) is equivalent to a context-free specification:

$$
\begin{cases}
\mathcal{P} &= \mathcal{H} + \mathcal{P} \times \mathcal{H} \\
\mathcal{H} &= \mathcal{Z} + \mathcal{Z} \times \mathcal{H} + \mathcal{Z} \times \mathcal{H} \times \mathcal{P}
\end{cases}
\implies
\begin{cases}
P = H + PH \\
H = z + zH + zH^2,
\end{cases}
$$

in which the second equation, a quadratic, is readily solved to provide H, which in turn gives P, by the first equation. One finds:

(86)
$$
\begin{cases}
P(z) &= \dfrac{1}{2}\left(\sqrt{\dfrac{1+z}{1-3z}} - 1\right) &= z + 2z^2 + 5z^3 + 13z^4 + 35z^5 + \cdots \\[3mm]
H(z) &= \dfrac{1 - z - \sqrt{(1+z)(1-3z)}}{2z} &= z + z^2 + 2z^3 + 4z^4 + 9z^5 + \cdots,
\end{cases}
$$

corresponding respectively to *EIS* **A005773** and *EIS* **A001006** (Motzkin numbers, cf Notes I.39, p. 68 and I.51, p. 77). See Example VI.3 and Note VI.11, p. 396, for relevant asymptotics.

Similar constructions permit us to decompose compact-source directed animals, whose class we denote by $\mathcal{A}$. For instance:

Compact-source animals with k sources are then specified by $\mathcal{P} \times \mathrm{SEQ}_{k-1}(\mathcal{H})$, and we have

(87)
$$
\mathcal{A} \cong \mathcal{P} \times \mathrm{SEQ}(\mathcal{H}) \implies A(z) = \frac{P(z)}{1 - H(z)} = \frac{z}{1 - 3z},
$$

where the last form results from basic algebraic simplifications. A consequence of (87) is the surpringly simple (but non-trivial) result that there are 3^{n-1} compact-source animals of size n. The papers [61, 87] develop further aspects of the rich counting theory of animals. .... ∎

▷ **I.52.** *Understanding animals.* In the first equation of (85), a pyramid π that is not a half-pyramid has a unique dimer which is of lowest altitude and immediately to the left of the source. Take that dimer and push it upwards, in the direction of imaginary infinity; it will then carry with it a group of dimers that constitute, by construction, a pyramid ω. What remains has no dimer to the left of its source, and hence forms a half-pyramid χ. The following diagram illustrates the decomposition, with the dimers of ω equipped with an upward-pointing arrow:

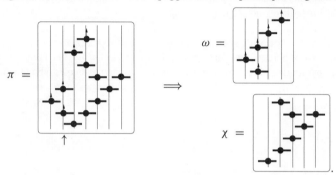

Conversely, given a pair $(\omega, \chi) \in \mathcal{P} \times \mathcal{H}$, attach first χ to the base; then, let ω fall down from imaginary infinity. The dimers of ω will take their place above the dimers of χ, blocked in various manners on their way down, the whole set eventually forming a pyramid. A moment of reflection convinces one that the original pyramid π is recovered in this way; that is, the transformation $\pi \to (\omega, \chi)$ is bijective.   ◁

▷ **I.53.** *"Tree-like" structures.* A context-free specification can always be regarded as defining a class of trees. Indeed, if the jth term in the construction $\mathfrak{F}_i$ of (83) is "coloured" with the pair (i, j), it is seen that a context-free system yields a class of trees whose nodes are tagged by pairs (i, j) in a way consistent with the system's rules. However, despite this correspondence, it is often convenient to preserve the possibility of operating directly with objects when the tree aspect may be unnatural. (Some authors have developed a parallel notion of "object grammars"; see for instance [183], itself inspired by techniques of polyomino surgery in [150].) By a terminology borrowed from the theory of syntax analysis in computer science, such trees are referred to as "parse trees" or "syntax trees".   ◁

▷ **I.54.** *Context-free languages.*   Let $\mathcal{A}$ be a fixed finite alphabet whose elements are called letters. A *grammar* G is a collection of equations

$$
\text{(88)} \qquad\qquad G \quad : \quad \begin{cases} \mathcal{L}_1 & = & \mathfrak{F}_1(\mathbf{a}, \mathcal{L}_1, \dots, \mathcal{L}_m) \\ \vdots & & \vdots \\ \mathcal{L}_m & = & \mathfrak{F}_m(\mathbf{a}, \mathcal{L}_1, \dots, \mathcal{L}_m), \end{cases}
$$

where each $\mathfrak{F}_j$ involves only the operations of union ($\cup$) and concatenation product ($\cdot$) with $\mathbf{a}$ the vector of letters in $\mathcal{A}$. For instance,

$$\mathfrak{F}_1(\mathbf{a}, \mathcal{L}_1, \mathcal{L}_2, \mathcal{L}_3) = a_2 \cdot \mathcal{L}_2 \cdot \mathcal{L}_3 \cup a_3 \cup \mathcal{L}_3 \cdot a_2 \cdot \mathcal{L}_1.$$

A solution to (88) is an m–tuple of languages over the alphabet $\mathcal{A}$ that satisfies the system. By convention, one declares that the grammar G defines the first component, $\mathcal{L}_1$.

   To each grammar (88), one can associate a context-free specification (60) by transforming unions into disjoint union, "$\cup \mapsto +$", and catenation into cartesian products, "$\cdot \mapsto \times$". Let $\widehat{G}$ be the specification associated in this way to the grammar G. The objects described by $\widehat{G}$ appear in this perspective to be trees (see the discussion above regarding parse trees). Let h

be the transformation from trees of $\widehat{G}$ to languages of G that lists letters in infix (i.e., left-to-right) order: we call such an h the erasing transformation since it "forgets" all the structural information contained in the parse tree and only preserves the succession of letters. Clearly, application of h to the combinatorial specifications determined by $\widehat{G}$ yields languages that obey the grammar G. For a grammar G and a word $w \in \mathcal{A}^\star$, the number of parse trees $t \in \widehat{G}$ such that $h(t) = w$ is called the *ambiguity coefficient* of w with respect to the grammar G.

A grammar G is *unambiguous* if all the corresponding ambiguity coefficients are either 0 or 1. This means that there is a bijection between parse trees of $\widehat{G}$ and words of the language described by G: each word generated is uniquely "parsable" according to the grammar. One has, from Proposition I.7: *The OGF of an unambiguous context-free language satisfies a polynomial system of the form* (61), *and is consequently an algebraic function.* ◁

▷ **I.55.** *Extended context-free specifications.* If $\mathcal{A}, \mathcal{B}$ are context-free specifications then: (*i*) the sequence class $\mathcal{C} = \text{SEQ}(\mathcal{A})$ is context-free; (*ii*) the substitution class $\mathcal{D} = \mathcal{A}[b \mapsto \mathcal{B}]$, formally defined in the next section, is also context-free. ◁

I. 6. Additional constructions

This section is devoted to the constructions of sequences, sets, and cycles in the presence of restrictions on the number of components as well as to mechanisms that enrich the framework of core constructions; namely, pointing, substitution, and the use of implicit combinatorial definitions.

I. 6.1. Restricted constructions.

An immediate formula for OGFs is that of the *diagonal* Δ of a cartesian product $\mathcal{B} \times \mathcal{B}$ defined as

$$\mathcal{A} \equiv \Delta(\mathcal{B} \times \mathcal{B}) := \{(\beta, \beta) \mid \beta \in \mathcal{B}\}.$$

Then, one has the relation $A(z) = B(z^2)$, as shown by the combinatorial derivation

$$A(z) = \sum_{(\beta, \beta)} z^{2|\beta|} = B(z^2),$$

or by the equally obvious observation that $A_{2n} = B_n$.

The diagonal construction permits us to access the class of all unordered pairs of (distinct) elements of $\mathcal{B}$, which is $\mathcal{A} = \text{PSET}_2(\mathcal{B})$. A direct argument then runs as follows: the unordered pair $\{\alpha, \beta\}$ is associated to the two ordered pairs (α, β) and (β, α) except when $\alpha = \beta$, where an element of the diagonal is obtained. In other words, one has the combinatorial isomorphism,

$$\text{PSET}_2(\mathcal{B}) + \text{PSET}_2(\mathcal{B}) + \Delta(\mathcal{B} \times \mathcal{B}) \cong \mathcal{B} \times \mathcal{B},$$

meaning that

$$2A(z) + B(z^2) = B(z)^2.$$

This gives the translation of PSET_2, and, by a similar argument for MSET_2 and CYC_2 (observe also that $\text{CYC}_2 \cong \text{MSET}_2$), one has:

$$
\begin{aligned}
\mathcal{A} = \text{PSET}_2(\mathcal{B}) \qquad &\Longrightarrow \qquad A(z) = \tfrac{1}{2}B(z)^2 - \tfrac{1}{2}B(z^2) \\
\mathcal{A} = \text{MSET}_2(\mathcal{B}) \qquad &\Longrightarrow \qquad A(z) = \tfrac{1}{2}B(z)^2 + \tfrac{1}{2}B(z^2) \\
\mathcal{A} = \text{CYC}_2(\mathcal{B}) \qquad &\Longrightarrow \qquad A(z) = \tfrac{1}{2}B(z)^2 + \tfrac{1}{2}B(z^2).
\end{aligned}
$$

This type of direct reasoning could in principle be extended to treat triples, and so on, but the computations easily grow out of control. The classical treatment of these questions relies on what is known as *Pólya theory*, of which we offer a glimpse in Notes I.58–I.60. We follow instead here an easier global approach, based on multivariate generating functions, that suffices to generate *simultaneously* all cardinality-restricted constructions of our standard collection.

Theorem I.3 (Component-restricted constructions). *The OGF of sequences with k components* $\mathcal{A} = \text{SEQ}_k(\mathcal{B})$ *satisfies*

$$A(z) = B(z)^k.$$

The OGF of sets, $\mathcal{A} = \text{PSET}_k(\mathcal{B})$, *is a polynomial in the quantities* $B(z), \ldots, B(z^k)$,

$$A(z) = [u^k] \exp\left(\frac{u}{1}B(z) - \frac{u^2}{2}B(z^2) + \frac{u^3}{3}B(z^3) - \cdots\right).$$

The OGF of multisets, $\mathcal{A} = \text{MSET}_k(\mathcal{B})$, *is*

$$A(z) = [u^k] \exp\left(\frac{u}{1}B(z) + \frac{u^2}{2}B(z^2) + \frac{u^3}{3}B(z^3) + \cdots\right).$$

The OGF of cycles, $\mathcal{A} = \text{CYC}_k(\mathcal{B})$, *is, with* φ *the Euler totient function (p. 721)*

$$A(z) = [u^k] \sum_{\ell=1}^{\infty} \frac{\varphi(\ell)}{\ell} \log \frac{1}{1 - u^\ell B(z^\ell)}.$$

The explicit forms for small values of k are summarized in Figure I.18, p. 93.

Proof. The result for sequences is obvious since $\text{SEQ}_k(\mathcal{B})$ means $\mathcal{B} \times \cdots \times \mathcal{B}$ (k times). For the other constructions, the proof makes use of the techniques of Theorem I.1, p. 27, but it is best based on bivariate generating functions that are otherwise developed fully in Chapter III to which we refer for details (p. 171). The idea consists in describing all composite objects and introducing a supplementary marking variable to keep track of the number of components.

Take $\mathfrak{K}$ to be a construction among SEQ, CYC, MSET, PSET. Consider the relation $\mathcal{A} = \mathfrak{K}(\mathcal{B})$, and let $\chi(\alpha)$ for $\alpha \in \mathcal{A}$ be the parameter "number of $\mathcal{B}$–components". Define the multivariate quantities

$$A_{n,k} := \text{card}\left\{\alpha \in \mathcal{A} \mid |\alpha| = n, \; \chi(\alpha) = k\right\}$$

$$A(z, u) := \sum_{n,k} A_{n,k} u^k z^n = \sum_{\alpha \in \mathcal{A}} z^{|\alpha|} u^{\chi(\alpha)}.$$

For instance, a direct calculation shows that, for sequences,

$$A(z, u) = \sum_{k \geq 0} u^k B(z)^k = \frac{1}{1 - uB(z)}.$$

For multisets and powersets, a simple adaptation of the already seen argument gives $A(z, u)$ as

$$A(z, u) = \prod_n (1 - uz^n)^{-B_n}, \qquad A(z, u) = \prod_n (1 + uz^n)^{B_n},$$

respectively. The result follows from here by the exp–log transformation upon extracting $[u^k]A(z, u)$. The case of cycles results from the bivariate generating function derived in Appendix A.4: *Cycle construction*, p. 729 (alternatively use Note I.60). ∎

▷ **I.56.** *Aperiodic words.* An aperiodic word is a primitive sequence of letters (in the sense of Appendix A.4: *Cycle construction*, p. 729); that is, the word w is aperiodic provided it is *not* obtained by repetition of a proper factor: $w \neq u \cdots u$. The number of aperiodic words of length n over an m–ary alphabet is (with $\mu(k)$ the Möbius function, p. 721)

$$PW_n^{(m)} = \sum_{d \,|\, n} \mu(d)m^{n/d}.$$

For $m = 2$, the sequence starts as 2, 2, 6, 12, 30, 54, 126, 240, 504, 990 (*EIS* **A027375**). ◁

▷ **I.57.** *Around the cycle construction.* A calculation with arithmetical functions (APPENDIX A, p. 721) yields the OGFs of *multisets of cycles* and *multisets of aperiodic cycles* as

$$\prod_{k \geq 1} \frac{1}{1 - A(z^k)} \qquad \text{and} \qquad \frac{1}{1 - A(z)},$$

respectively [144]. (The latter fact corresponds to the combinatorial property that any word can be written as a decreasing product of Lyndon words; notably, it serves to construct bases of free Lie algebras [413, Ch. 5].) ◁

▷ **I.58.** *Pólya theory I: the cycle indicator.* Consider a finite set $\mathcal{M}$ of cardinality m and a group G of permutations of $\mathcal{M}$. Whenever convenient, the set $\mathcal{M}$ can be identified with the interval $[1 \mathinner{.\,.} m]$. The *cycle indicator* ("*Zyklenzeiger*") of G is, by definition, the multivariate polynomial

$$Z(G) \equiv Z(G; x_1, \ldots, x_m) = \frac{1}{\mathrm{card}(G)} \sum_{g \in G} x_1^{j_1(g)} \cdots x_m^{j_m(g)},$$

where $j_k(g)$ is the number of cycles of length k in the permutation g. For instance, if $\mathfrak{I}_m = \{\mathrm{Id}\}$ is the group reduced to the identity permutation, $\mathfrak{S}_m$ is the group of all permutations of size m, and $\mathfrak{R}_m$ is the group consisting of the identity permutation and the "mirror-reflection" permutation $\left(\begin{smallmatrix} 1 \, \cdots \, m \\ m \, \cdots \, 1 \end{smallmatrix}\right)$, then

(89)
$$Z(\mathfrak{I}_m) = x_1^m; \qquad Z(\mathfrak{S}_m) = \sum_{j_1, \ldots, j_m \geq 0} \frac{x_1^{j_1} \cdots x_m^{j_m}}{j_1! \, 1^{j_1} \cdots j_m! \, m^{j_m}};$$

$$Z(\mathfrak{R}_m) = \begin{cases} \frac{1}{2} x_2^\nu + \frac{1}{2} x_1^{2\nu} & \text{if } m = 2\nu \text{ is even} \\ \frac{1}{2} x_1 x_2^\nu + \frac{1}{2} x_1^{2\nu+1} & \text{if } m = 2\nu + 1 \text{ is odd.} \end{cases}$$

(For the case of $\mathfrak{S}_m$, see Equation (40), Chapter III, p. 188.) ◁

▷ **I.59.** *Pólya theory II: the fundamental theorem.* Let $\mathcal{B}$ be a combinatorial class and $\mathcal{M}$ a finite set on which the group G acts. Consider the set $\mathcal{B}^\mathcal{M}$ of all mappings from $\mathcal{M}$ into $\mathcal{B}$. Two mappings $\phi_1, \phi_2 \in \mathcal{B}^\mathcal{M}$ are declared to be equivalent if there exists a $g \in G$ such that $\phi_1 \circ g = \phi_2$, and we let $(\mathcal{B}^\mathcal{M}/G)$ be the set of equivalence classes. The problem is to enumerate $(\mathcal{B}^\mathcal{M}/G)$, given the data $\mathcal{B}$, $\mathcal{M}$, and the "symmetry group" G.

Let w be a weight function that assigns to any $\beta \in \mathcal{B}$ a weight $w(\beta)$; the weight is extended *multiplicatively* to any $\phi \in \mathcal{B}^\mathcal{M}$, hence to $(\mathcal{B}^\mathcal{M}/G)$, by $w(\phi) := \prod_{k \in \mathcal{M}} w(\phi(k))$. The *Pólya–Redfield Theorem* expresses the identity

(90)
$$\sum_{\phi \in (\mathcal{B}^\mathcal{M}/G)} w(\phi) = Z\left(G; \sum_{\beta \in \mathcal{B}} w(\beta), \ldots, \sum_{\beta \in \mathcal{B}} w(\beta)^m\right).$$

In particular, we can choose $w(\beta) = z^{|\beta|}$ with z a formal parameter; the Pólya–Redfield Theorem (90) then provides the OGF of objects of $\mathcal{B}^{\mathcal{M}}$ up to symmetries by G:

$$(91) \qquad \sum_{\phi \in (\mathcal{B}^{\mathcal{M}}/G)} z^{|\phi|} = Z\left(G; B(z), \dots, B(z^m)\right).$$

(There are many excellent presentations of this classic theory, starting with Pólya himself [488, 491]; see for instance Comtet [129, §6.6], De Bruijn [142], and Harary–Palmer [319, Ch. 2]. The proof relies on orbit counting and Burnside's lemma.)   ◁

▷ **I.60.** *Pólya theory III: basic constructions.* Say we want to obtain the OGF of $\mathcal{A} = $ MSET$_3(\mathcal{B})$. We view A as the set of triples $\mathcal{B}^{\mathcal{M}}$, with $\mathcal{M} = [1..3]$, taken up to $\mathfrak{S}_3$, the set of all permutations of three elements. The cycle indicator is given by (89), from which the translation of MSET$_3$ results (see Figure I.18, p. 93, for the outcome); the calculation extends to all MSET$_m$, providing an alternative approach to Theorem I.3. The translation of the CYC$_m$ construction can be obtained in this way via the cycle index of the group $\mathfrak{C}_m$ of all cyclic permutations; namely,

$$Z(\mathfrak{C}_m) = \frac{1}{m} \sum_{d \mid m} \varphi(d) x_d^{n/d},$$

where $\varphi(k)$ is the Euler totient function. The use of the groups $\mathfrak{R}_m$ gives rise to the *undirected sequence* construction,

$$\mathcal{A} = \text{USEQ}(\mathcal{B}) \qquad \Longrightarrow \qquad A(z) = \frac{1}{2}\frac{1}{1 - B(z)} + \frac{1}{2}\frac{1 + B(z)}{1 - B(z^2)},$$

where a sequence and its mirror image are identified. Similar principles give rise to the *undirected cycle* construction UCYC, generated by cyclic permutations *and* mirror reflection. (The approach taken in the text can be seen, in the perspective of Pólya theory, as a direct determination of $\sum_{m \geq 0} Z(\mathfrak{G}_m)$, for an entire family of symmetry groups $\{G_m\}$, where $G_m = \mathfrak{C}_m, \mathfrak{S}_m, \dots$)   ◁

▷ **I.61.** *Sets with distinct component sizes.* Let $\mathcal{A}$ be the class of the finite sets of elements from $\mathcal{B}$, with the additional constraint that no two elements in a set have the same size. One has

$$A(z) = \prod_{n=1}^{\infty} (1 + B_n z^n).$$

Similar identities serve in the analysis of polynomial factorization algorithms [236].   ◁

▷ **I.62.** *Sequences without repeated components.* The generating function is formally

$$\int_0^{\infty} \exp\left(\sum_{j \geq 1} (-1)^{j-1} \frac{u^j}{j} B(z^j)\right) e^{-u} \, du.$$

(This representation is based on the Eulerian integral: $k! = \int_0^{\infty} e^{-u} u^k \, du$.)   ◁

I.6.2. Pointing and substitution. Two more constructions, namely pointing and substitution, translate agreeably into generating functions. Combinatorial structures are viewed as always as formed of atoms (letters, nodes, etc), which determine their sizes. Pointing means "pointing at a distinguished atom"; substitution, written $\mathcal{B} \circ \mathcal{C}$ or $\mathcal{B}[\mathcal{C}]$, means "substitute elements of $\mathcal{C}$ for atoms of $\mathcal{B}$".

Definition I.14. *Let $\{\epsilon_1, \epsilon_2, \dots\}$ be a fixed collection of distinct neutral objects of size 0. The* pointing *of a class $\mathcal{B}$, denoted $\mathcal{A} = \Theta\mathcal{B}$, is formally defined as*

$$\Theta\mathcal{B} := \sum_{n \geq 0} \mathcal{B}_n \times \{\epsilon_1, \dots, \epsilon_n\}.$$

The substitution *of C into B (also known as composition of B and C), noted $B \circ C$ or $B[C]$, is formally defined as*

$$B \circ C \equiv B[C] := \sum_{k \geq 0} B_k \times \mathrm{SEQ}_k(C).$$

With B_n the number of B structures of size n, the quantity $n B_n$ can be interpreted as counting pointed structures where *one* of the n atoms composing a B–structure has been distinguished (here by a special "pointer" of size 0 attached to it). Elements of $B \circ C$ may also be viewed as obtained by selecting in all possible ways an element $\beta \in B$ and replacing each of its atoms by an arbitrary element of C, while preserving the underlying structure of β.

The interpretations above rely (silently) on the fact that atoms in an object can be eventually distinguished from each other. This can be obtained by "canonicalizing"[17] the representations of objects: first define inductively the lexicographic ordering for products and sequences; next represent powersets and multisets as increasing sequences with the induced lexicographic ordering (more complicated rules can also canonicalize cycles). In this way, any constructible object admits a unique "rigid" representation in which each particular atom is determined by its place. Such a canonicalization thus reconciles the abstract definitions of Definition I.14 with the intuitive interpretation of pointing and substitution.

Theorem I.4 (Pointing and substitution). *The constructions of pointing and substitution are admissible*[18]:

$$A = \Theta B \quad \Longrightarrow \quad A(z) = z \partial_z B(z) \quad \partial_z := \frac{d}{dz}$$

$$A = B \circ C \quad \Longrightarrow \quad A(z) = B(C(z))$$

Proof. By the definition of pointing, one has

$$A_n = n \cdot B_n, \qquad \text{so that} \qquad A(z) = z \partial_z B(z).$$

The definition of substitution implies, by the sum and product rules,

$$A(z) = \sum_{k \geq 0} B_k \cdot (C(z))^k = B(C(z)),$$

and the proof is completed. ∎

---

[17] Such canonicalization techniques also serve to develop fast algorithms for the exhaustive listing of objects of a given size as well as for the range of problems known as "ranking" and "unranking", with implications in fast random generation. See, for instance, [430, 456, 607] for the general theory as well as [500, 623] for particular cases such as necklaces and trees.

[18] In this book, we borrow from differential algebra the convenient notation $\partial_z := \frac{d}{dz}$ to represent derivatives.

Permutations as pointed objects. As an example of pointing, consider the class $\mathcal{P}$ of all permutations written as words over integers starting from 1. One can go from a permutation of size $n - 1$ to a permutation of size n by selecting a "gap" and inserting the value n. When this is done in all possible ways, it gives rise to the combinatorial relation

$$\mathcal{P} = \mathcal{E} + \Theta(\mathcal{Z} \times \mathcal{P}), \qquad \mathcal{E} = \{\epsilon\}, \qquad \Longrightarrow \qquad P(z) = 1 + z\frac{d}{dz}(zP(z)).$$

The OGF satisfies an ordinary differential equation whose formal solution is $P(z) = \sum_{n \geq 0} n! z^n$, since it is equivalent to the recurrence $P_n = n P_{n-1}$.

Unary–binary trees as substituted objects. As an example of substitution, consider the class $\mathcal{B}$ of (plane–rooted) binary trees, where all nodes contribute to size. If at each node a linear chain of nodes (linked by edges placed on top of the node) is substituted, one forms an element of the class $\mathcal{M}$ of unary–binary trees; in symbols:

$$\mathcal{M} = \mathcal{B} \circ \text{SEQ}_{\geq 1}(\mathcal{Z}) \qquad \Longrightarrow \qquad M(z) = B\left(\frac{z}{1-z}\right).$$

Thus from the known OGF, $B(z) = (1 - \sqrt{1 - 4z^2})/(2z)$, one derives

$$M(z) = \frac{1 - \sqrt{1 - 4z^2(1-z)^{-2}}}{2z(1-z)^{-1}} = \frac{1 - z - \sqrt{1 - 2z - 3z^2}}{2z},$$

which matches the direct derivation on p. 68 (Motzkin numbers).

▷ **I.63.** *Combinatorics of derivatives.* The combinatorial operation **D** of "erasing–pointing" points to an atom in an object and replaces it by a neutral object, otherwise preserving the overall structure of the object. The translation of **D** on OGFs is then simply $\partial := \partial_z$. Classical identities of analysis then receive transparent combinatorial interpretations: for instance,

$$\partial(A \times B) = (A \times \partial B) + (\partial A \times B)$$

as well as Leibniz's identity, $\partial^m(f \cdot g) = \sum_j \binom{m}{j}(\partial^j f) \cdot (\partial^{m-j} g)$, also follow from basic logic. Similarly, for the "chain rule" $\partial(f \circ g) = ((\partial f) \circ g) \cdot \partial g$. (Example VII.25, p. 529, illustrates the use of these methods for analytically solving many *urn processes*.)   ◁

▷ **I.64.** *The combinatorics of Newton–Raphson iteration.* Given a real function f, the iteration scheme of Newton–Raphson finds (conditionally) a root of the equation $f(y) = 0$ by repeated use of the transformation $\alpha^\star = \alpha - f(\alpha)/f'(\alpha)$, starting for instance from $\alpha = 0$. (For sufficiently smooth functions, this scheme is quadratically convergent.) The application of Newton–Raphson iteration to the equation $y = z\phi(y)$ associated with a simple variety of trees in the sense of Proposition I.5, p. 66, leads to the scheme:

$$\alpha_{m+1} = \alpha_m + \frac{z\phi(\alpha_m) - \alpha_m}{1 - z\phi'(\alpha_m)}; \qquad \alpha_0 = 0.$$

It can be seen, analytically *and* combinatorially, that α_m has a contact of order at least $2^m - 1$ with $y(z)$. The interesting combinatorics is due to Décoste, Labelle, and Leroux [147]; it involves a notion of "heavy" trees (such that at least one of the root subtrees is large enough, in a suitable sense); see [50, §3.3] and [485] for further developments.   ◁

I. 6.3. Implicit structures.

There are many cases where a combinatorial class $\mathcal{X}$ is determined by a relation $\mathcal{A} = \mathcal{B} + \mathcal{X}$, where $\mathcal{A}$ and $\mathcal{B}$ are known. (An instance of this is the equational technique of Subsection I. 4.2, p. 56 for enumerating words that *do* contain a given pattern $\mathfrak{p}$.) Less trivial examples involve inverting cartesian products as well as sequences and multisets (examples below).

Theorem I.5 (Implicit specifications). *The generating functions associated to the implicit equations with unknown $\mathcal{X}$*

$$\mathcal{A} = \mathcal{B} + \mathcal{X}, \qquad \mathcal{A} = \mathcal{B} \times \mathcal{X}, \qquad \mathcal{A} = \text{SEQ}(\mathcal{X}),$$

are, respectively,

$$X(z) = A(z) - B(z), \qquad X(z) = \frac{A(z)}{B(z)}, \qquad X(z) = 1 - \frac{1}{A(z)}.$$

For the implicit construction $\mathcal{A} = \text{MSET}(\mathcal{X})$, one has

$$X(z) = \sum_{k \geq 1} \frac{\mu(k)}{k} \log A(z^k),$$

where $\mu(k)$ is the Möbius function[19].

Proof. The first two cases result from kindergarten algebra, since in terms of OGFs one has $A = B + X$ and $A = BX$, respectively. For sequences, the relation $A(z) = (1 - X(z))^{-1}$ is readily inverted as stated. For multisets, start from the fundamental relation of Theorem I.1 (p. 27) and take logarithms:

$$\log(A(z)) = \sum_{k=1}^{\infty} \frac{1}{k} X(z^k).$$

Let $L = \log A$ and $L_n = [z^n] L(z)$. One has

$$n L_n = \sum_{d \mid n} (d X_d),$$

to which it suffices to apply Möbius inversion (p. 721). ∎

Example I.19. *Indecomposable permutations.* A permutation $\sigma = \sigma_1 \cdots \sigma_n$ (written here as a word of distinct letters) is said to be *decomposable* if, for some $k < n$, $\sigma_1 \cdots \sigma_k$ is a permutation of $\{1, \ldots, k\}$; i.e., a strict prefix of the permutation (in word form) is itself a permutation. Any permutation decomposes uniquely as a concatenation of indecomposable permutations, as shown in Figure I.17.

As a consequence of our definitions, the class $\mathcal{P}$ of all permutations and the class $\mathcal{I}$ of indecomposable ones are related by

$$\mathcal{P} = \text{SEQ}(\mathcal{I}).$$

This determines $I(z)$ implicitly, and Theorem I.5 gives

$$I(z) = 1 - \frac{1}{P(z)} \quad \text{where} \quad P(z) = \sum_{n \geq 0} n! \, z^n.$$

This example illustrates the utility of implicit constructions, and at the same time the possibility of *bona fide* algebraic calculations with power series even in cases where they are divergent (Appendix A.5: *Formal power series*, p. 730). One finds

$$I(z) = z + z^2 + 3 z^3 + 13 z^4 + 71 z^5 + 461 z^6 + 3447 z^7 + \cdots,$$

---

[19]The Möbius function $\mu(n)$ is $\mu(n) = (-1)^r$ if n is the product of r distinct primes and $\mu(n) = 0$ otherwise (Appendix A.1: *Arithmetical functions*, p. 721).

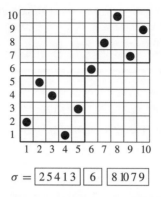

$$\sigma = \boxed{2\,5\,4\,1\,3} \ \boxed{6} \ \boxed{8\,10\,7\,9}$$

Figure I.17. The decomposition of a permutation (σ).

where the coefficients (*EIS* **A003319**) are

$$I_n = n! - \sum_{\substack{n_1+n_2=n \\ n_1,n_2 \geq 1}} (n_1!n_2!) + \sum_{\substack{n_1+n_2+n_3=n \\ n_1,n_2,n_3 \geq 1}} (n_1!n_2!n_3!) - \cdots .$$

From this, simple majorizations of the terms imply that $I_n \sim n!$, so that *almost all permutations are indecomposable* [129, p. 262]. .... ∎

▷ **I.65.** *Two-dimensional wanderings.* A drunkard starts from the origin in the $\mathbb{Z} \times \mathbb{Z}$ plane and, at each second, he makes a step in either one of the four directions, NW, NE, SW, SE. The steps are thus ↖, ↗, ↙, ↘. Consider the class $\mathcal{L}$ of "primitive loops" defined as walks that start and end at the origin, but do not otherwise touch the origin. The GF of $\mathcal{L}$ is (*EIS* **A002894**)

$$L(z) = 1 - \frac{1}{\sum_{n=0}^{\infty} \binom{2n}{n}^2 z^{2n}} = 4z^2 + 20z^4 + 176z^6 + 1876z^8 + \cdots .$$

(Hint: a walk is determined by its projections on the horizontal and vertical axes; one-dimensional walks that return to the origin in $2n$ steps are enumerated by $\binom{2n}{n}$.) In particular $[z^{2n}]L(z/4)$ is the probability that the random walk first returns to the origin in $2n$ steps.

Such problems largely originate with Pólya and implicit constructions were well-mastered by him [490]; see also [85] for certain multidimensional extensions. The first-return problem is analysed asymptotically in Chapter VI, p. 425, based on singularity theory and Hadamard closure properties. ◁

Example I.20. *Irreducible polynomials over finite fields.* Objects not obviously of a combinatorial nature can sometimes be enumerated by symbolic methods. Here is an indirect construction relative to polynomials over finite fields. We fix a prime number p and consider the base field $\mathbb{F}_p$ of integers taken modulo p. The polynomial ring $\mathbb{F}_p[X]$ is the ring of polynomials in X with coefficients in $\mathbb{F}_p$.

For all practical purposes, one may restrict attention to polynomials that are *monic*; that is, ones whose leading coefficient is 1. We regard the set $\mathcal{P}$ of monic polynomials in $\mathbb{F}_p[X]$ as a combinatorial class, with the size of a polynomial being identified to its degree. Since a polynomial is specified by the sequence of its coefficients, one has, with $\mathcal{A}$ the "alphabet" of coefficients, $\mathcal{A} = \mathbb{F}_p$ treated as a collection of atomic objects,

(92) $$\mathcal{P} = \text{SEQ}(\mathcal{A}) \quad \Longrightarrow \quad P(z) = \frac{1}{1 - pz},$$

in agreement with the fact that there are p^n monic polynomials of degree n.

Polynomials are a *unique factorization domain*, since they can be subjected to Euclidean division. A polynomial that has no proper non-constant divisor is termed *irreducible*—irreducibles are thus the analogues of the primes in the integer realm. For instance, over $\mathbb{F}_3$, one has

$$X^{10} + X^8 + 1 = (X+1)^2(X+2)^2(X^6 + 2X^2 + 1).$$

Let $\mathcal{I}$ be the set of monic irreducible polynomials. The unique factorization property implies that the collection of all polynomials is combinatorially isomorphic to the multiset class (there may be repeated factors) of the collection of irreducibles:

(93) $\qquad \mathcal{P} \cong \mathrm{MSET}(\mathcal{I}) \qquad \Longrightarrow \qquad P(z) = \exp\left(I(z) + \frac{1}{2}I(z^2) + \frac{1}{3}I(z^3) + \cdots \right).$

The irreducibles are thus determined *implicitly* from the class of all polynomials whose OGF is known by (92). Theorem I.5 then implies the identity

(94) $\qquad I(z) = \sum_{k \geq 1} \frac{\mu(k)}{k} \log \frac{1}{1 - pz^k} \qquad$ and $\qquad I_n = \frac{1}{n} \sum_{k \mid n} \mu(k) p^{n/k}.$

In particular, I_n is asymptotic to p^n/n. This estimate constitutes the density theorem for irreducible polynomials, a result already known to Gauss (see the scholarly notes of von zur Gathen and Gerhard in [599, p. 396]):

The fraction of irreducible polynomials among all polynomials of degree n over the finite field $\mathbb{F}_p$ is asymptotic to $\frac{1}{n}$.

This property is analogous to the Prime Number Theorem (which however lies *much* deeper, see [22, 138]), according to which the proportion of prime numbers in the interval $[1, n]$ is asymptotic to $1/\log n$. Indeed, a polynomial of degree n appears to be roughly comparable to a number written in base p having n digits. (On the basis of such properties, Knopfmacher has further developed in [370] an abstract theory of statistical properties of arithmetical semigroups.) We pursue this thread further in the book: we shall prove that the number of factors in a random polynomial of degree n is on average $\sim \log n$ (Example VII.4, p. 449) and that the corresponding distribution is asymptotically Gaussian (Example IX.21, p. 672). . . . . ∎

▷ **I.66.** *Square-free polynomials.* Let Q be the class of monic square-free polynomials (i.e., polynomials not divisible by the square of a polynomial). One has by "Vallée's identity" (p. 30) $Q(z) = P(z)/P(z^2)$, hence

$$Q(z) = \frac{1 - pz^2}{1 - pz} \qquad \text{and} \qquad Q_n = p^n - p^{n-1} \quad (n \geq 2).$$

Berlekamp's book [51] discusses such facts together with relations to error correcting codes. ◁

▷ **I.67.** *Balanced trees.* The class $\mathcal{E}$ of balanced 2-3 trees contains all the (rooted planar) trees whose internal nodes have degree 2 or 3 and such that all leaves are at the same distance from the root. Only leaves contribute to size. Such trees, which are particular cases of B–trees, are a useful data structure for implementing dynamic dictionaries [378, 537]. Balanced trees satisfy an implicit equation based on combinatorial substitution:

$$\mathcal{E} = \mathcal{Z} + \mathcal{E} \circ [(\mathcal{Z} \times \mathcal{Z}) + (\mathcal{Z} \times \mathcal{Z} \times \mathcal{Z})] \qquad \Longrightarrow \qquad E(z) = z + E(z^2 + z^3).$$

The expansion starts as (*EIS* **A014535**)

$$E(z) = z + z^2 + z^3 + z^4 + 2z^5 + 2z^6 + 3z^7 + 4z^8 + 5z^9 + 8z^{10} + \cdots.$$

Odlyzko [459] has determined the growth of E_n to be roughly as φ^n/n, where $\varphi = (1 + \sqrt{5})/2$ is the golden ratio. See Subsection IV.7.2, p. 280 for an analysis. ◁

I. 7. Perspective

This chapter and the next amount to a survey of elementary combinatorial enumerations, organized in a coherent manner and summarized in Figure I.18, in the case of the unlabelled universe that is considered here. We refer to the process of specifying combinatorial classes using these constructions and then automatically having access to the corresponding generating functions as the *symbolic method*. The symbolic method is the "combinatorics" in analytic combinatorics: it allows us to structure classical results in combinatorics with a unifying overall approach, to derive new results that generalize and extend classical problems, and to address new classes of problems that are arising in computer science, computational biology, statistical physics, and other scientific disciplines.

More importantly, the symbolic method leaves us with generating functions that we can handle with the "analytic" part of analytic combinatorics. A full treatment of this feature of the approach is premature, but a brief discussion may help place the rest of the book in context.

For a given family of problems, the symbolic method typically leads to a natural class of functions in which the corresponding generating functions lie. Even though the symbolic method is completely formal, we can often successfully proceed by using classical techniques from complex and asymptotic analysis. For example, denumerants with a finite set of coin denominations always lead to rational generating functions with poles on the unit circle. Such an observation is useful as a common strategy for coefficient extraction can then be applied (partial fraction expansion, in the case of denumerants with fixed coin denominations). In the same vein, run statistics constitute a particular case of the general theorem of Chomsky and Schützenberger to the effect that the generating function of a regular language is necessarily a rational function. Similarly, context-free structures are attached to generating functions that are invariably algebraic. Theorems of this sort establish a bridge between combinatorial analysis and special functions.

Not all applications of the symbolic method are automatic (although that is certainly one goal underlying the approach). The example of counting set partitions shows that application of the symbolic method may require finding an adequate presentation of the combinatorial structures to be counted. In this way, bijective combinatorics enters the game in a non-trivial fashion.

Our introductory examples of compositions and partitions correspond to classes of combinatorial structures with *explicit* "iterative" definitions, a fact leading in turn to explicit generating function expressions. The tree examples then introduce *recursively defined* structures. In that case, the recursive definition translates into a *functional equation* that only determines the generating function implicitly. In simpler situations (such as binary or general trees), the generating function equations can be solved and explicit counting results often follow. In other cases (such as non-plane trees) one can usually conduct an analysis of singularities directly from the functional equations and obtain very precise *asymptotic estimates*: Chapters IV–VIII of Part B offer an abundance of illustrations of this paradigm. The further development on a

1. The main constructions of disjoint union (combinatorial sum), product, sequence, powerset, multiset, and cycle and their translation into generating functions (Theorem I.1).

Construction		OGF
Union	$\mathcal{A} = \mathcal{B} + \mathcal{C}$	$A(z) = B(z) + C(z)$
Product	$\mathcal{A} = \mathcal{B} \times \mathcal{C}$	$A(z) = B(z) \cdot C(z)$
Sequence	$\mathcal{A} = \text{SEQ}(\mathcal{B})$	$A(z) = \dfrac{1}{1 - B(z)}$
Powerset	$\mathcal{A} = \text{PSET}(\mathcal{B})$	$A(z) = \exp\left(B(z) - \dfrac{1}{2}B(z^2) + \cdots \right)$
Multiset	$\mathcal{A} = \text{MSET}(\mathcal{B})$	$A(z) = \exp\left(B(z) + \dfrac{1}{2}B(z^2) + \cdots \right)$
Cycle	$\mathcal{A} = \text{CYC}(\mathcal{B})$	$A(z) = \log\dfrac{1}{1 - B(z)} + \dfrac{1}{2}\log\dfrac{1}{1 - B(z^2)} + \cdots$

2. The translation for sequences, powersets, multisets, and cycles constrained by the number of components (Theorem I.3, p. 84).

$$\text{SEQ}_k(\mathcal{B}) : \quad B(z)^k$$

$$\text{PSET}_2(\mathcal{B}) : \quad \frac{B(z)^2}{2} - \frac{B(z^2)}{2}$$
$$\text{MSET}_2(\mathcal{B}) : \quad \frac{B(z)^2}{2} + \frac{B(z^2)}{2}$$
$$\text{CYC}_2(\mathcal{B}) : \quad \frac{B(z)^2}{2} + \frac{B(z^2)}{2}$$

$$\text{PSET}_3(\mathcal{B}) : \quad \frac{B(z)^3}{6} - \frac{B(z)\, B(z^2)}{2} + \frac{B(z^3)}{3}$$
$$\text{MSET}_3(\mathcal{B}) : \quad \frac{B(z)^3}{6} + \frac{B(z)\, B(z^2)}{2} + \frac{B(z^3)}{3}$$
$$\text{CYC}_3(\mathcal{B}) : \quad \frac{B(z)^3}{3} + \frac{2B(z^3)}{3}$$

$$\text{PSET}_4(\mathcal{B}) : \quad \frac{B(z)^4}{24} - \frac{B(z)^2 B(z^2)}{4} + \frac{B(z)B(z^3)}{3} + \frac{B(z^2)^2}{8} - \frac{B(z^4)}{4}$$
$$\text{MSET}_4(\mathcal{B}) : \quad \frac{B(z)^4}{24} + \frac{B(z)^2 B(z^2)}{4} + \frac{B(z)B(z^3)}{3} + \frac{B(z^2)^2}{8} + \frac{B(z^4)}{4}$$
$$\text{CYC}_4(\mathcal{B}) : \quad \frac{B(z)^4}{4} + \frac{B(z^2)^2}{4} + \frac{B(z^4)}{2}.$$

3. The additional constructions of pointing and substitution (Section I. 6).

Construction		OGF
Pointing	$\mathcal{A} = \Theta\mathcal{B}$	$A(z) = z\dfrac{d}{dz}B(z)$
Substitution	$\mathcal{A} = \mathcal{B} \circ \mathcal{C}$	$A(z) = B(C(z))$

Figure I.18. A dictionary of constructions applicable to *unlabelled* structures, together with their translation into ordinary generating functions (OGFs). (The labelled counterpart of this table appears in Figure II.18, p. 148.)

suitable perturbative theory will then lead us to systematic ways of quantifying parameters (not just counting sequences) of large combinatorial structures—this is the subject of Chapter IX, in Part C of this book.

Bibliographic notes. Modern presentations of combinatorial analysis appear in the books of Comtet [129] (a beautiful book largely example-driven), Stanley [552, 554] (a rich set with an algebraic orientation), Wilf [608] (generating functions oriented), and Lando [400] (a neat modern introduction). An elementary but insightful presentation of the basic techniques appears in Graham, Knuth, and Patashnik's classic [307], a popular book with a highly original design. An encyclopaedic reference is the book of Goulden & Jackson [303] whose descriptive approach very much parallels ours.

The sources of the modern approaches to combinatorial analysis are hard to trace since they are usually based on earlier traditions and informally stated mechanisms that were well-mastered by practicing combinatorial analysts. (See for instance MacMahon's book [428] *Combinatory Analysis* first published in 1917, the introduction of denumerant generating functions by Pólya as presented in [489, 493], or the "domino theory" in [307, Sec. 7.1].) One source in recent times is the Chomsky–Schützenberger theory of formal languages and enumerations [119]. Rota [518] and Stanley [550, 554] developed an approach which is largely based on partially ordered sets. Bender and Goldman developed a theory of "prefabs" [42] whose purposes are similar to the theory developed here. Joyal [359] proposed an especially elegant framework, the "theory of species", that addresses foundational issues in combinatorial theory and constitutes the starting point of the superb exposition by Bergeron, Labelle, and Leroux [50]. Parallel (but largely independent) developments by the "Russian School" are nicely synthesized in the books by Sachkov [525, 526].

One of the reasons for the revival of interest in combinatorial enumerations and properties of random structures is the analysis of algorithms (a subject founded in modern times by Knuth [381]), in which the goal is to model the performance of computer algorithms and programs. The symbolic ideas expounded here have been applied to the analysis of algorithms in surveys [221, 598], with elements presented in our book [538]. Further implications of the symbolic method in the area of the random generation of combinatorial structures appear in [177, 228, 264, 456].

> *[...] une propriété qui se traduit par une égalité $|A| = |B|$ est mieux explicitée lorsque l'on construit une bijection entre deux ensembles A et B, plutôt qu'en calculant les coefficients d'un polynôme dont les variables n'ont pas de significations particulières. La méthode des fonctions génératrices, qui a exercé ses ravages pendant un siècle, est tombée en désuétude*
> *pour cette raison.*
>
> *("[...] a property, which is translated by an equality $|A| = |B|$, is understood better, when one constructs a bijection between the two sets A and B, than when one calculates the coefficients of a polynomial whose variables have no particular meaning. The method of generating functions, which has had devastating effects for a century, has fallen into obsolescence, for this reason.")*
>
> —CLAUDE BERGE [48, p. 10]

Labelled Structures and Exponential Generating Functions

Cette approche évacue pratiquement tous les calculs[1].

— DOMINIQUE FOATA &
MARCO SCHÜTZENBERGER [267]

Many objects of classical combinatorics present themselves naturally as *labelled structures*, where atoms of an object (typically nodes in a graph or a tree) are distinguishable from one another by the fact that they bear distinct *labels*. Without loss of generality, we may take the set from which labels are drawn to be the set of integers. For instance, a permutation can be viewed as a linear arrangement of distinct integers, and the classical cycle decomposition represents it as an unordered collection of circular digraphs, whose vertices are themselves integers.

Operations on labelled structures are based on a special product: the *labelled product* that distributes labels between components. This operation is a natural analogue of the cartesian product for plain unlabelled objects. The labelled product in turn leads to labelled analogues of the sequence, set, and cycle constructions.

Labelled constructions translate over *exponential generating functions*—the translation schemes turn out to be even simpler than in the unlabelled case. At the same time, these constructions enable us to take into account structures that are in some ways combinatorially richer than their unlabelled counterparts of Chapter I, in particular with regard to order properties. Labelled constructions constitute the second pillar of the symbolic method for combinatorial enumeration.

In this chapter, we examine some of the most important classes of labelled objects, including surjections, set partitions, permutations, as well as labelled graphs, trees, and mappings from a finite set into itself. Certain aspects of words can also be treated

---

[1] *"This approach eliminates virtually all calculations."* Foata and Schützenberger refer here to a "geometric" approach to combinatorics, much akin to ours, that permits one to relate combinatorial properties and special function identities.

by this theory, a fact which has important consequences not only in combinatorics itself but also in probability and statistics. In particular, labelled constructions of words provide an elegant solution to two classical problems, the birthday problem and the coupon collector problem, as well as several of their variants that have numerous applications in other fields, including the analysis of hashing algorithms in computer science.

II.1. Labelled classes

Throughout this chapter, we consider combinatorial classes in the sense of Definition I.1, p. 16: we deal exclusively with finite objects; a combinatorial class $\mathcal{A}$ is a set of objects, with a notion of size attached, so that the number of objects of each size in $\mathcal{A}$ is finite. To these basic concepts, we now add that the objects are *labelled*, by which we mean that each atom carries with it a distinctive colour, or equivalently an integer label, in such a way that all the labels occurring in an object are distinct. Precisely:

Definition II.1. *A* weakly labelled object *of size n is a graph whose set of vertices is a subset of the integers. Equivalently, we say that the vertices bear labels, with the implied condition that labels are distinct integers from* $\mathbb{Z}$. *An object of size n is said to be* well-labelled, *or, simply,* labelled, *if it is weakly labelled and, in addition, its collection of labels is the complete integer interval* $[1 \mathinner{.\,.} n]$. *A labelled class is a combinatorial class comprised of well-labelled objects.*

The graphs considered may be directed or undirected. In fact, when the need arises, we shall take "object" in a broad sense to mean any kind of discrete structure enriched by integer labels. Virtually all labelled classes considered in this book can eventually be encoded as graphs of sorts, so that this extended use of the notion of a labelled class is a harmless convenience. (See Section II.7, p. 147 for a brief discussion of alternative but logically equivalent frameworks for the notion of a labelled class.)

Example **II.1.** *Labelled graphs.* By definition, a *labelled graph* is an undirected graph such that distinct integer labels forming an interval of the form $\{1, 2, \ldots, n\}$ are supported by vertices. A particular labelled graph of size 4 is for instance

$$g = \begin{array}{c} 1 \rule[0.5ex]{1em}{0.4pt} 3 \\ | \quad\ | \\ 4 \rule[0.5ex]{1em}{0.4pt} 2 \end{array},$$

which represents a graph whose vertices bear the labels $\{1, 2, 3, 4\}$ and whose set of edges is

$$\{\{1, 3\}, \{2, 3\}, \{2, 4\}, \{1, 4\}\}.$$

Only the graph structure (as defined by its adjacency structure, i.e., its set of edges) counts, so that this is the same abstract graph as in the alternative physical representations

$$g = \begin{array}{c} 1 \rule[0.5ex]{1em}{0.4pt} 4 \\ | \quad\ | \\ 3 \rule[0.5ex]{1em}{0.4pt} 2 \end{array}, \quad \begin{array}{c} 3 \rule[0.5ex]{1em}{0.4pt} 2 \\ | \quad\ | \\ 1 \rule[0.5ex]{1em}{0.4pt} 4 \end{array}.$$

However, this graph is different from either of

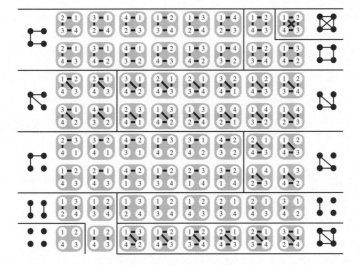

There are altogether $G_4 = \mathbf{64} = 2^6$ labelled graphs of size 4, i.e., comprising 4 nodes, in agreement with the general formula (see p. 105 for details): $G_n = 2^{n(n-1)/2}$. The labelled graphs can be grouped into equivalence classes up to arbitrary permutation of the labels, which determines the $\widehat{G}_4 = \mathbf{11}$ unlabelled graphs of size 4. Each unlabelled graph corresponds to a variable number of labelled graphs: for instance, the totally disconnected graph (bottom, left) and the complete graph (top right) correspond to 1 labelling only, while the line graph (top left) admits $\frac{1}{2} 4! = 12$ possible labellings.

Figure II.1. Labelled versus unlabelled graphs for size $n = 4$.

$$h = \begin{matrix} 4 - 1 \\ | \quad | \\ 3 - 2 \end{matrix}, \qquad j = \begin{matrix} 3 - 1 \\ | \quad | \\ 4 - 2 \end{matrix},$$

since, for instance, 1 and 2 are adjacent in h and j, but not in g. Altogether, there are 3 different labelled graphs (namely, g, h, j), that have the same "shape", corresponding to the single unlabelled quadrangle graph

$$Q = \begin{matrix} \bullet - \bullet \\ | \quad | \\ \bullet - \bullet \end{matrix}.$$

Figure II.1 lists all the 64 labelled graphs of size 4 as well as their 11 unlabelled counterparts viewed as equivalence classes of labelled graphs when labels are ignored. .... ∎

In order to count labelled objects, we appeal to exponential generating functions.

Definition II.2. *The* exponential generating function *(EGF) of a sequence* (A_n) *is the formal power series*

(1)
$$A(z) = \sum_{n \geq 0} A_n \frac{z^n}{n!}.$$

The exponential generating function *(EGF) of a class* A *is the exponential generating function of the numbers* $A_n = \text{card}(A_n)$. *Equivalently, the EGF of class* A *is*

$$A(z) = \sum_{n \geq 0} A_n \frac{z^n}{n!} = \sum_{\alpha \in A} \frac{z^{|\alpha|}}{|\alpha|!}.$$

It is also said that the variable z *marks* size *in the generating function.*

With the standard notation for coefficients of series, the coefficient A_n in an exponential generating function is then recovered by[2]

$$A_n = n! \cdot [z^n] A(z),$$

since $[z^n]A(z) = A_n/n!$ by the definition of EGFs and in accordance with the coefficient extractor notation, Equation (9), p. 19, in Chapter I.

Note that, as in the previous chapter, we adhere to a systematic *naming convention* for generating functions of combinatorial structures. A labelled class A, its counting sequence (A_n) (or (a_n)), and its exponential generating function $A(z)$ (or $a(z)$) are all denoted by the same group of letters. As usual, combinatorially isomorphic classes (Definition I.3, p. 19) are freely identified.

Neutral and atomic classes. As in the unlabelled universe (p. 24), it proves useful to introduce a neutral (empty, null) object ϵ that has size 0 and bears no label at all, and consider it as a special labelled object; a *neutral class* $\mathcal{E}$ is then by definition $\mathcal{E} = \{\epsilon\}$ and is also denoted by boldface **1**. The (labelled) *atomic class* $\mathcal{Z} = \{\textcircled{1}\}$ is formed of a unique object of size 1 that, being well-labelled, bears the integer label $\textcircled{1}$. The EGFs of the neutral class and the atomic class are, respectively,

$$E(z) = 1, \qquad Z(z) = z.$$

Permutations, urns, and circular graphs. These structures, described in Examples II.2–II.4, are undoubtedly the most fundamental ones for labelled enumeration.

Example II.2. Permutations. The class $\mathcal{P}$ of all permutations is prototypical of labelled classes. Under the linear representation of permutations, where

$$\sigma = \begin{pmatrix} 1 & 2 & \cdots & n \\ \sigma_1 & \sigma_2 & \cdots & \sigma_n \end{pmatrix}$$

is represented as the sequence $(\sigma_1, \sigma_2, \ldots, \sigma_n)$, the class $\mathcal{P}$ is schematically

$$\mathcal{P} = \left\{ \epsilon, \; \textcircled{1}, \; \begin{matrix} \textcircled{1}-\textcircled{2} \\ \textcircled{2}-\textcircled{1} \end{matrix}, \; \begin{matrix} \textcircled{1}-\textcircled{2}-\textcircled{3} \\ \textcircled{2}-\textcircled{3}-\textcircled{1} \\ \textcircled{3}-\textcircled{1}-\textcircled{2} \\ \textcircled{2}-\textcircled{1}-\textcircled{3} \\ \textcircled{1}-\textcircled{3}-\textcircled{2} \\ \textcircled{3}-\textcircled{2}-\textcircled{1} \end{matrix}, \; \ldots \right\},$$

so that $P_0 = 1$, $P_1 = 1$, $P_2 = 2$, $P_3 = 6$, etc. There, by definition, all the possible orderings of the distinct labels are taken into account, so that the class $\mathcal{P}$ can be equivalently viewed as the class of all labelled linear digraphs (with an implicit direction, from left to right, say, in the representation). Accordingly, the class $\mathcal{P}$ of permutations has the counting sequence $P_n = n!$

---

[2] Some authors prefer the notation $[\frac{z^n}{n!}]A(z)$ to $n![z^n]A(z)$, which we avoid in this book. Indeed, Knuth [376] argues convincingly that the variant notation is not consistent with many desirable properties of a "good" coefficient operator (e.g., bilinearity).

(argument: there are n choices of where to place the element 1, then $(n - 1)$ possible positions for 2, and so on). Thus the EGF of $\mathcal{P}$ is

$$P(z) = \sum_{n \geq 0} n! \frac{z^n}{n!} = \sum_{n \geq 0} z^n = \frac{1}{1 - z}.$$

Permutations, as they contain information relative to the ordering of their elements are essential in many applications related to order statistics. . . . . ∎

Example II.3. *Urns.* The class $\mathcal{U}$ of totally disconnected graphs starts as

The ordering between the labelled atoms does *not* matter, so that for each n, there is only *one* possible arrangement and $U_n = 1$. The class $\mathcal{U}$ can be regarded as the class of *urns*, where an urn of size n contains n distinguishable balls in an unspecified (and irrelevant) order. The corresponding EGF is

$$U(z) = \sum_{n \geq 0} 1 \frac{z^n}{n!} = \exp(z) = e^z.$$

(The fact that the EGF of the constant sequence $(1)_{n \geq 0}$ is the exponential function explains the term "exponential generating function".) It also proves convenient, in several applications, to represent elements of an urn in a sorted sequence, which leads to an equivalent representation of urns as *increasing linear graphs*; for instance,

$$①\text{--}②\text{--}③\text{--}④\text{--}⑤$$

may be equivalently used to represent the urn of size 5. Though urns look trivial at first glance, they are of particular importance as building blocks of complex labelled structures (e.g., allocations of various sorts), as we shall see shortly. . . . . ∎

Example II.4. *Circular graphs.* Finally, the class of *circular graphs*, in which cycles are oriented in some conventional manner (say, positively here) is

Circular graphs correspond bijectively to *cyclic permutations*. One has $C_n = (n - 1)!$ (argument: a directed cycle is determined by the succession of elements that "follow" 1, hence by a permutation of $n - 1$ elements). Thus, one has

$$C(z) = \sum_{n \geq 1} (n - 1)! \frac{z^n}{n!} = \sum_{n \geq 1} \frac{z^n}{n} = \log \frac{1}{1 - z}.$$

As we shall see in the next section, the logarithm is characteristic of circular arrangements of labelled objects. . . . . ∎

▷ **II.1.** *Labelled trees.* Let U_n now be the number of labelled graphs with n vertices that are connected and acyclic; equivalently, U_n is the number of labelled unrooted non-plane trees. Let T_n be the number of labelled rooted non-plane trees. The identity $T_n = nU_n$ is elementary, since all vertices in a labelled tree are distinguished by their labels and a root can be chosen in n ways. In Section II. 5, p. 125, we shall prove that $U_n = n^{n-2}$ and $T_n = n^{n-1}$. ◁

II. 2. Admissible labelled constructions

We now describe a toolkit of constructions that make it possible to build complex labelled classes from simpler ones. Combinatorial sum, also known as disjoint union is taken in the sense of Chapter I, p. 25: it is the union of disjoint copies. Next, in order to define a product adapted to labelled structures, we cannot rely on the cartesian product, since a pair of two labelled objects is not well-labelled (for instance the label 1 would invariably appear repeated twice). Instead, we define a new operation, the *labelled product*, which translates naturally into exponential generating functions. From here, simple translation rules follow for labelled sequences, sets, and cycles.

Binomial convolutions. As a preparation to the translation of labelled constructions, we first briefly review the effect of products over EGFs. Let $a(z), b(z), c(z)$ be EGFs, with $a(z) = \sum_n a_n z^n / n!$, and so on. The *binomial convolution* formula is:

$$(2) \qquad \text{if} \quad a(z) = b(z) \cdot c(z), \quad \text{then} \quad a_n = \sum_{k=0}^{n} \binom{n}{k} b_k c_{n-k},$$

where $\binom{n}{k} = n!/(k!\,(n-k)!)$ represents, as usual, a binomial coefficient. This formula results from the usual product of formal power series,

$$\frac{a_n}{n!} = \sum_{k=0}^{n} \frac{b_k}{k!} \cdot \frac{c_{n-k}}{(n-k)!} \quad \text{and} \quad \binom{n}{k} = \frac{n!}{k!\,(n-k)!}.$$

In the same vein, if $a(z) = b^{(1)}(z)\, b^{(2)}(z) \cdots b^{(r)}(z)$, then

$$(3) \qquad a_n = \sum_{n_1+n_2+\cdots+n_r=n} \binom{n}{n_1, n_2, \ldots, n_r} b^{(1)}_{n_1} b^{(2)}_{n_2} \cdots b^{(r)}_{n_r}.$$

In Equation (3) there occurs the multinomial coefficient

$$\binom{n}{n_1, n_2, \ldots, n_r} = \frac{n!}{n_1!\, n_2! \cdots n_r!},$$

which counts the number of ways of splitting n elements into r distinguishable classes of cardinalities $n_1, \ldots, n_r$. This property lies at the very heart of enumerative applications of binomial convolutions and EGFs.

II. 2.1. Labelled constructions.
A labelled object may be relabelled. *We only consider* consistent relabellings *defined by the fact that they preserve the order relations among labels.* Then two dual modes of relabellings prove important:

— *Reduction*: For a weakly labelled structure of size n, this operation reduces its labels to the standard interval $[1 \mathinner{..} n]$ while preserving the relative order of labels. For instance, the sequence $\langle 7, 3, 9, 2 \rangle$ reduces to $\langle 3, 2, 4, 1 \rangle$. We use $\rho(\alpha)$ to denote this canonical reduction of the structure α.

— *Expansion*: This operation is defined relative to a relabelling function $e :$ $[1 \mathinner{..} n] \mapsto \mathbb{Z}$ that is assumed to be strictly increasing. To a well-labelled object α of size n, it associates a weakly labelled object $\widetilde{\alpha}$, in which label j of α is replaced by labelled $e(j)$. For instance, $\langle 3, 2, 4, 1 \rangle$ may expand as

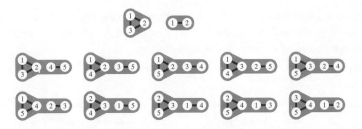

Figure II.2. The $10 \equiv \binom{5}{2}$ elements in the labelled product of a triangle and a segment.

$\langle 33, 22, 44, 11 \rangle$, $\langle 7, 3, 9, 2 \rangle$, and so on. We use $e(\alpha)$ to denote the result of relabelling α by e.

These notions enable us to devise a product well suited to labelled objects, which was originally formalized under the name of "partitional product" by Foata [265]. The idea is simply to relabel objects, so as to avoid duplicate labels.

Given two labelled objects $\beta \in \mathcal{B}$ and $\gamma \in \mathcal{C}$, their *labelled product*, or simply *product*, denoted by $\beta \star \gamma$, is a set comprised of the collection of well-labelled ordered pairs (β', γ') that reduce to (β, γ):

(4) $\beta \star \gamma := \{ (\beta', \gamma') \mid (\beta', \gamma') \text{ is well-labelled, } \rho(\beta') = \beta, \ \rho(\gamma') = \gamma \}.$

An equivalent form, via expansion of labels, is

(5) $\beta \star \gamma = \{ (e(\beta), f(\gamma) \mid \mathrm{Im}(e) \cap \mathrm{Im}(f) = \emptyset, \ \mathrm{Im}(e) \cup \mathrm{Im}(f) = [1 \,..\, |\beta| + |\gamma|] \},$

where e, f are relabelling functions with ranges $\mathrm{Im}(e)$, $\mathrm{Im}(f)$, respectively.

Note that elements of a labelled product are, by construction, well-labelled. The labelled product $(\beta \star \gamma)$ of two elements β, γ of respective sizes n_1, n_2 is a set whose cardinality is, with $n = n_1 + n_2$, expressed as

$$\binom{n_1 + n_2}{n_1, n_2} \equiv \binom{n}{n_1},$$

since this quantity is the number of legal relabellings by expansion of the pair (β, γ). (Figure II.2 displays the $\binom{5}{2} = 10$ elements of the labelled product of a particular object of size 3 with another object of size 2.) The labelled product of classes is then defined by the natural extension of operations to sets.

Definition II.3. *The* labelled product *of $\mathcal{B}$ and $\mathcal{C}$, denoted $\mathcal{B} \star \mathcal{C}$, is obtained by forming ordered pairs from $\mathcal{B} \times \mathcal{C}$ and performing all possible order-consistent relabellings. In symbols:*

(6) $\displaystyle \mathcal{B} \star \mathcal{C} = \bigcup_{\beta \in \mathcal{B}, \ \gamma \in \mathcal{C}} (\beta \star \gamma).$

Equipped with this notion, we can build sequences, sets, and cycles, in a way much similar to the unlabelled case. We proceed to do so and, at the same time, establish *admissibility*[3] of the constructions.

Labelled product. When $\mathcal{A} = \mathcal{B} \star \mathcal{C}$, the corresponding counting sequences satisfy the relation,

$$(7) \qquad A_n = \sum_{|\beta|+|\gamma|=n} \binom{|\beta|+|\gamma|}{|\beta|,|\gamma|} B_{|\beta|} C_{|\gamma|} = \sum_{n_1+n_2=n} \binom{n}{n_1,n_2} B_{n_1} C_{n_2}.$$

The product $B_{n_1} C_{n_2}$ keeps track of all the possibilities for the $\mathcal{B}$ and $\mathcal{C}$ components and the binomial coefficient accounts for the number of possible relabellings, in accordance with our earlier discussion. The binomial convolution property (7) then implies admissibility

$$\mathcal{A} = \mathcal{B} \star \mathcal{C} \qquad \Longrightarrow \qquad A(z) = B(z) \cdot C(z),$$

with the labelled product simply translating into the product operation on EGFs.

▷ **II.2.** *Multiple labelled products.* The (binary) labelled product satisfies the associativity property,

$$\mathcal{B} \star (\mathcal{C} \star \mathcal{D}) \cong (\mathcal{B} \star \mathcal{C}) \star \mathcal{D},$$

which serves to define $\mathcal{B} \star \mathcal{C} \star \mathcal{D}$. The corresponding EGF is the product $B(z) \cdot C(z) \cdot D(z)$. This rule generalizes to r factors with coefficients given by a multinomial convolution (3). ◁

k–sequences and sequences. The kth (labelled) *power* of $\mathcal{B}$ is defined as $(\mathcal{B} \star \mathcal{B} \cdots \mathcal{B})$, with k factors equal to $\mathcal{B}$. It is denoted $\text{SEQ}_k(\mathcal{B})$ as it corresponds to forming k–sequences and performing all consistent relabellings. The (labelled) *sequence* class of $\mathcal{B}$ is denoted by $\text{SEQ}(\mathcal{B})$ and is defined by

$$\text{SEQ}(\mathcal{B}) := \{\epsilon\} + \mathcal{B} + (\mathcal{B} \star \mathcal{B}) + (\mathcal{B} \star \mathcal{B} \star \mathcal{B}) + \cdots = \bigcup_{k \geq 0} \text{SEQ}_k(\mathcal{B}).$$

The product relation for EGFs extends to arbitrary products (Note II.2 above), so that

$$\begin{cases} \mathcal{A} = \text{SEQ}_k(\mathcal{B}) & \Longrightarrow \quad A(z) = B(z)^k \\ \mathcal{A} = \text{SEQ}(\mathcal{B}) & \Longrightarrow \quad A(z) = \sum_{k=0}^{\infty} B(z)^k = \dfrac{1}{1 - B(z)}, \end{cases}$$

where the last equation requires $\mathcal{B}_0 = \emptyset$.

k–sets and sets. We denote by $\text{SET}_k(\mathcal{B})$ the class of k–sets formed from $\mathcal{B}$. The set class is defined formally, as in the case of the unlabelled multiset: it is the quotient $\text{SET}_k(\mathcal{B}) := \text{SEQ}_k(\mathcal{B})/\mathbf{R}$, where the equivalence relation $\mathbf{R}$ identifies two sequences when the components of one are a permutation of the components of the other (p. 26). A "set" is like a sequence, but the order between components is immaterial. The (labelled) *set* construction applied to $\mathcal{B}$, denoted $\text{SET}(\mathcal{B})$, is then defined by

$$\text{SET}(\mathcal{B}) := \{\epsilon\} + \mathcal{B} + \text{SET}_2(\mathcal{B}) + \cdots = \bigcup_{k \geq 0} \text{SET}_k(\mathcal{B}).$$

---

[3]We recall that a construction is admissible (Definition I.5, p. 22) if the counting sequence of the result only depends on the counting sequences of the operands. An admissible construction therefore induces a well-defined transformation over exponential generating functions.

A labelled k–set is associated with exactly $k!$ different sequences, since all its components are distinguishable by their labels. Precisely, one may choose to identify each component in a labelled set or sequence by its "*leader*"; that is, the value of its smallest label. There is then a uniform $k!$–to–one correspondence between k–sequences and k–sets, as illustrated in a particular case ($k = 3$) by the diagram below:

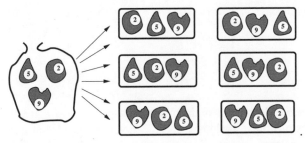

In figurative terms: the contents of a bag containing k different items can be laid on a table in $k!$ ways. Thus in terms of EGFs, one has, assuming $\mathcal{B}_0 = \emptyset$,

$$
\begin{cases}
\mathcal{A} = \text{SET}_k(\mathcal{B}) & \implies & A(z) = \dfrac{1}{k!} B(z)^k \\[2mm]
\mathcal{A} = \text{SET}(\mathcal{B}) & \implies & A(z) = \displaystyle\sum_{k=0}^{\infty} \dfrac{1}{k!} B(z)^k = \exp(B(z)).
\end{cases}
$$

In the unlabelled case, formulae are more complex, since components in multisets are not necessarily different. Note also that the distinction between multisets and powersets, which is meaningful for unlabelled structures is here immaterial, and we have the unlabelled-to-labelled analogy: MSET, PSET $\rightsquigarrow$ SET.

k–cycles and cycles. We also introduce the class of k–cycles, $\text{CYC}_k(\mathcal{B})$ and the cycle class. The cycle class is defined formally, as in the unlabelled case, to be the quotient $\text{CYC}_k(\mathcal{B}) := \text{SEQ}_k(\mathcal{B})/\text{S}$, where the equivalence relation **S** identifies two sequences when the components of one are a cyclic permutation of the components of the other (p. 26). A cycle is like a sequence whose components can be cyclically shifted, so that there is now a uniform k–to–one correspondence between k–sequences and k–cycles. In terms of EGFs, we have (assuming $\mathcal{B}_0 = \emptyset$ and $k \geq 1$)

$$
\begin{cases}
\mathcal{A} = \text{CYC}_k(\mathcal{B}) & \implies & A(z) = \dfrac{1}{k} B(z)^k \\[2mm]
\mathcal{A} = \text{CYC}(\mathcal{B}) & \implies & A(z) = \displaystyle\sum_{k=1}^{\infty} \dfrac{1}{k} B(z)^k = \log \dfrac{1}{1 - B(z)},
\end{cases}
$$

since each cycle admits exactly k representations as a sequence. In summary:

Theorem II.1 (Basic admissibility, labelled universe). *The constructions of combinatorial sum, labelled product, sequence, set, and cycle are all admissible. Associated operators on EGFs are:*

Sum:	$\mathcal{A} = \mathcal{B} + \mathcal{C}$	$\implies$	$A(z) = B(z) + C(z),$
Product:	$\mathcal{A} = \mathcal{B} \star \mathcal{C}$	$\implies$	$A(z) = B(z) \cdot C(z),$

Sequence: $\mathcal{A} = \text{SEQ}(\mathcal{B})$ $\Longrightarrow$ $A(z) = \dfrac{1}{1 - B(z)}$,

— k components: $\mathcal{A} = \text{SEQ}_k(\mathcal{B}) \equiv (\mathcal{B})^{\star k}$ $\Longrightarrow$ $A(z) = B(z)^k$,

Set: $\mathcal{A} = \text{SET}(\mathcal{B})$ $\Longrightarrow$ $A(z) = \exp(B(z))$,

— k components: $\mathcal{A} = \text{SET}_k(\mathcal{B})$ $\Longrightarrow$ $A(z) = \dfrac{1}{k!} B(z)^k$,

Cycle: $\mathcal{A} = \text{CYC}(\mathcal{B})$ $\Longrightarrow$ $A(z) = \log \dfrac{1}{1 - B(z)}$,

— k components: $\mathcal{A} = \text{CYC}_k(\mathcal{B})$ $\Longrightarrow$ $A(z) = \dfrac{1}{k} B(z)^k$.

Constructible classes. As in the previous chapter, we say that a class of labelled objects is constructible if it admits a specification in terms of sums (disjoint unions), the labelled constructions of product, sequence, set, cycle, and the initial classes defined by the neutral structure of size 0 and the atomic class $\mathcal{Z} = \{\textcircled{1}\}$. Regarding the elementary classes discussed in Section II. 1, it is immediately recognized that

$$\mathcal{P} = \text{SEQ}(\mathcal{Z}), \quad \mathcal{U} = \text{SET}(\mathcal{Z}), \quad \mathcal{C} = \text{CYC}(\mathcal{Z}),$$

specify permutations, urns, and circular graphs, respectively. These classes are basic building blocks out of which more complex objects can be constructed. In particular, as we shall explain shortly (Section II. 3 and Section II. 4), set partitions ($\mathcal{S}$), surjections ($\mathcal{R}$), permutations under their cycle decomposition ($\mathcal{P}$), and alignments ($\mathcal{O}$) are constructible classes corresponding to

Surjections: $\mathcal{R} \cong \text{SEQ}(\text{SET}_{\geq 1}(\mathcal{Z}))$ (sequences-of-sets);

Set partitions: $\mathcal{S} \cong \text{SET}(\text{SET}_{\geq 1}(\mathcal{Z}))$ (sets-of-sets);

Alignments: $\mathcal{O} \cong \text{SEQ}(\text{CYC}(\mathcal{Z}))$ (sequences-of-cycles);

Permutations: $\mathcal{P} \cong \text{SET}(\text{CYC}(\mathcal{Z}))$,   (sets-of-cycles).

An immediate consequence of Theorem II.1 is the fact that a functional equation for the EGF of a constructible labelled class can be computed automatically.

Theorem II.2 (Symbolic method, labelled universe). *The exponential generating function of a constructible class of labelled objects is a component of a system of generating function equations whose terms are built from 1 and z using the operators*

$$+, \quad \times, \quad Q(f) = \frac{1}{1 - f}, \quad E(f) = e^f \quad L(f) = \log \frac{1}{1 - f}.$$

When we further allow restrictions in composite constructions, the operators f^k (for SEQ_k), $f^k/k!$ (for SET_k), and f^k/k (for CYC_k) are to be added to the list.

II. 2.2. Labelled versus unlabelled enumeration. Any labelled class $\mathcal{A}$ has an unlabelled counterpart $\widehat{\mathcal{A}}$: objects in $\widehat{\mathcal{A}}$ are obtained from objects of $\mathcal{A}$ by ignoring the labels. This idea is formalized by identifying two labelled objects if there is an *arbitrary* relabelling (not just an order-consistent one, as has been used so far) that transforms one into the other. For an object of size n, each equivalence class contains *a priori* between 1 and $n!$ elements. Thus:

Proposition II.1. *The counts of a labelled class $\mathcal{A}$ and its unlabelled counterpart $\widehat{\mathcal{A}}$ are related by*

$$(8) \qquad \widehat{A}_n \leq A_n \leq n!\,\widehat{A}_n \quad \text{or equivalently} \quad 1 \leq \frac{A_n}{\widehat{A}_n} \leq n!.$$

***Example* II.5.** *Labelled and unlabelled graphs.* This phenomenon has been already encountered in our discussion of graphs (Figure II.1, p. 97). Let in general G_n and $\widehat{G}_n$ be the number of graphs of size n in the labelled and unlabelled case, respectively. One finds for $n = 1 \mathinner{.\,.} 15$:

$\widehat{G}_n$ (unlabelled)	G_n (labelled)
1	1
2	2
4	8
11	64
34	1024
156	32768
1044	2097152
12346	268435456
274668	68719476736
12005168	35184372088832
1018997864	36028797018963968
165091172592	73786976294838206464

The sequence $(\widehat{G}_n)$ constitutes *EIS* **A000088**, which can be obtained by an extension of methods of Chapter I, p. 85, specifically by Pólya theory [319, Ch. 4]. The sequence (G_n) is determined directly by the fact that a graph of n vertices can have each of the $\binom{n}{2}$ possible edges either present or not, so that

$$G_n = 2^{\binom{n}{2}} = 2^{n(n-1)/2}.$$

The sequence of labelled counts obviously grows much faster than its unlabelled counterpart. We may then verify the inequality (8) in this particular case. The normalized ratios,

$$\rho_n := G_n/\widehat{G}_n, \quad \sigma_n := G_n/(n!\widehat{G}_n),$$

are observed to be

n	$\rho_n = G_n/\widehat{G}_n$	$\sigma_n = G_n/(n!\widehat{G}_n)$
1	1.000000000	1.0000000000
2	1.000000000	0.5000000000
3	2.000000000	0.3333333333
4	5.818181818	0.2424242424
6	210.0512821	0.2917378918
8	21742.70663	0.5392536367
12	446946830.2	0.9330800361
16	$0.2076885783 \cdot 10^{14}$	0.9926428522

From these data, it is natural to conjecture that σ_n tends rapidly to 1 as n tends to infinity. This is indeed a non-trivial fact originally established by Pólya (see Chapter 9 of Harary and Palmer's book [319] dedicated to asymptotics of graph enumerations):

$$\widehat{G}_n \sim \frac{1}{n!} 2^{\binom{n}{2}} = \frac{G_n}{n!}.$$

In other words, "almost all" graphs of size n should admit a number of labellings close to $n!$. (Combinatorially, this corresponds to the fact that in a random unlabelled graph, with high probability, all of the nodes can be distinguished via the adjacency structure of the graph; in such a case, the graph has no non-trivial automorphism and the number of distinct labellings is $n!$ exactly.) .... ■

In contrast with the case of all graphs, where $\widehat{G}_n \sim G_n/n!$, urns (totally disconnected graphs) illustrate the other extreme situation where

$$\widehat{U}_n = U_n = 1.$$

These examples indicate that, beyond the general bounds of Proposition II.1, there is no automatic way to translate between labelled and unlabelled enumerations. But at least, if the class $\mathcal{A}$ is constructible, its unlabelled counterpart $\widehat{\mathcal{A}}$ can be obtained by interpreting all the intervening constructions as unlabelled ones in the sense of Chapter I (with SET $\mapsto$ MSET); both generating functions are computable, and their coefficients can then be compared.

▷ **II.3.** *Permutations and their unlabelled counterparts.* The labelled class of permutations can be specified by $\mathcal{P} = \text{SEQ}(\mathcal{Z})$; the unlabelled counterpart is the set $\widehat{\mathcal{P}}$ of integers in unary notation, and $\widehat{P}_n \equiv 1$, so that $P_n = n!\widehat{P}_n$ exactly. The specification $\mathcal{P}' = \text{SET}(\text{CYC}(\mathcal{Z}))$ describes sets of cycles and, in the labelled universe, one has $\mathcal{P}' \cong \mathcal{P}$; however, the unlabelled counterpart of $\mathcal{P}'$ is the class $\widehat{\mathcal{P}'} \neq \widehat{\mathcal{P}}$ of integer partitions examined in Chapter I. [In the unlabelled universe, there are special combinatorial isomorphisms such as $\text{SEQ}_{\geq 1}(\mathcal{Z}) \cong \text{MSET}_{\geq 1}(\mathcal{Z}) \cong \text{CYC}(\mathcal{Z})$. In the labelled universe, the identity $\text{SET} \circ \text{CYC} \equiv \text{SEQ}$ holds.]   ◁

II.3. Surjections, set partitions, and words

This section and the next are devoted to what could be termed level-two non-recursive structures defined by the fact that they combine two constructions. In this section, we discuss surjections and set partitions (Subsection II.3.1), which constitute labelled analogues of integer compositions and integer partitions in the unlabelled universe. The symbolic method then extends naturally to words over a finite alphabet, where it opens access to an analysis of the frequencies of letters composing words. This in turn has useful consequences for the study of classical random allocation problems, of which the birthday paradox and the coupon collector problem stand out (Subsection II.3.2). Figure II.3 summarizes some of the main enumeration results derived in this section.

II.3.1. Surjections and set partitions. We examine classes

$$\mathcal{R} = \text{SEQ}(\text{SET}_{\geq 1}(\mathcal{Z})) \qquad \text{and} \qquad \mathcal{S} = \text{SET}(\text{SET}_{\geq 1}(\mathcal{Z})),$$

corresponding to sequences-of-sets ($\mathcal{R}$) and sets-of-sets ($\mathcal{S}$), or equivalently, sequences of urns and sets of urns, respectively. Such abstract specifications model basic objects of discrete mathematics, namely surjections ($\mathcal{R}$) and set partitions ($\mathcal{S}$)

	Specification	EGF	coefficient	
Surjections:	$\mathcal{R} = \mathrm{SEQ}(\mathrm{SET}_{\geq 1}(\mathcal{Z}))$	$\dfrac{1}{2 - e^z}$	$\sim \dfrac{n!}{2(\log 2)^{n+1}}$	(pp. 109, 259)
— r images	$\mathcal{R}^{(r)} = \mathrm{SEQ}_r(\mathrm{SET}_{\geq 1}(\mathcal{Z}))$	$(e^z - 1)^r$	$r! \begin{Bmatrix} n \\ r \end{Bmatrix}$	(p. 107)
Set partitions:	$\mathcal{S} = \mathrm{SET}(\mathrm{SET}_{\geq 1}(\mathcal{Z}))$	$e^{e^z - 1}$	$\approx \dfrac{n!}{(\log n)^n}$	(pp. 109, 560)
— r blocks	$\mathcal{S}^{(r)} = \mathrm{SET}_r(\mathrm{SET}_{\geq 1}(\mathcal{Z}))$	$\dfrac{1}{r!}(e^z - 1)^r$	$\begin{Bmatrix} n \\ r \end{Bmatrix}$	(p. 108)
— blocks $\leq b$	$\mathcal{S} = \mathrm{SET}(\mathrm{SET}_{1 \ldots b}(\mathcal{Z}))$	$e^{e_b(z) - 1}$	$\approx n^{n(1 - 1/b)}$	pp. 111, 568
Words:	$\mathcal{W} = \mathrm{SEQ}_r(\mathrm{SET}(\mathcal{Z}))$	e^{rz}	r^n	(p. 112)

Figure II.3. Major enumeration results relative to surjections, set partitions, and words.

Surjections with r images. In elementary mathematics, a surjection from a set A to a set B is a function from A to B that assumes each value *at least once* (an onto mapping). Fix some integer $r \geq 1$ and let $\mathcal{R}_n^{(r)}$ denote the class of all surjections from the set $[1 \ldots n]$ *onto* $[1 \ldots r]$ whose elements are also called r–surjections. A particular object $\phi \in \mathcal{R}_9^{(5)}$ is depicted in Figure II.4.

We set $\mathcal{R}^{(r)} = \bigcup_n \mathcal{R}_n^{(r)}$ and proceed to compute the corresponding EGF, $R^{(r)}(z)$. First, let us observe that an r–surjection $\phi \in \mathcal{R}_n^{(r)}$ is determined by the *ordered r–tuple* formed with the collection of all preimage sets, $\big(\phi^{-1}(1), \phi^{-1}(2), \ldots, \phi^{-1}(r)\big)$, themselves disjoint non-empty sets of integers that cover the interval $[1 \ldots n]$. In the case of the surjection ϕ of Figure II.4, this alternative representation is

$$\phi : \qquad [\,\{2\},\ \{1, 3\},\ \{4, 6, 8\},\ \{9\},\ \{5, 7\}\,]\,.$$

One has the combinatorial specification and EGF relation:

$$(9) \quad \mathcal{R}^{(r)} = \mathrm{SEQ}_r(\mathcal{V}), \quad \mathcal{V} = \mathrm{SET}_{\geq 1}(\mathcal{Z}) \quad \Longrightarrow \quad R^{(r)}(z) = (e^z - 1)^r.$$

Here $\mathcal{V} \cong \mathcal{U} \setminus \{\epsilon\}$ designates the class of urns ($\mathcal{U}$) that are non-empty, with EGF $V(z) = e^z - 1$. In words: "a surjection is a sequence of non-empty sets". (Figure II.4).

Expression (9) does solve the counting problem for surjections. For small r, one finds

$$R^{(2)}(z) = e^{2z} - 2e^z + 1, \qquad R^{(3)}(z) = e^{3z} - 3e^{2z} + 3e^z - 1,$$

whence, by expanding,

$$R_n^{(2)} = 2^n - 2, \qquad R_n^{(3)} = 3^n - 3 \cdot 2^n + 3\,.$$

The general formula follows similarly from expanding the rth power in (9) by the binomial theorem, and then extracting coefficients:

$$(10) \quad R_n^{(r)} = n!\,[z^n] \sum_{j=0}^{r} \binom{r}{j}(-1)^j e^{(r-j)z} = \sum_{j=0}^{r} \binom{r}{j}(-1)^j (r - j)^n.$$

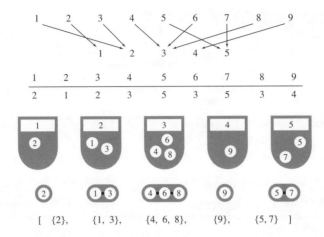

[{2},   {1, 3},   {4, 6, 8},   {9},   {5, 7}]

Figure II.4. The decomposition of surjections as sequences-of-sets: a surjection ϕ given by its graph (top), its table (second line), and its sequence of preimages (bottom lines).

▷ **II.4.** *A direct derivation of the surjection EGF.* One can verify the result provided by the symbolic method by returning to first principles. The preimage of value j by a surjection is a non-empty set of some cardinality $n_j \geq 1$, so that

$$(11) \qquad R_n^{(r)} = \sum_{(n_1, n_2, \ldots, n_r)} \binom{n}{n_1, n_2, \ldots, n_r},$$

the sum being over $n_j \geq 1$, $n_1 + n_2 + \cdots + n_r = n$. Introduce the numbers $V_n := [\![n \geq 1]\!]$, where $[\![P]\!]$ is Iverson's bracket (p. 58). The formula (11) then assumes the simple form

$$(12) \qquad R_n^{(r)} \equiv \sum_{n_1, n_2, \ldots, n_r} \binom{n}{n_1, n_2, \ldots, n_r} V_{n_1} V_{n_2} \cdots V_{n_r},$$

where the summation now extends to *all* tuples $(n_1, n_2, \ldots, n_r)$. The EGF of the V_n is $V(z) = \sum V_n z^n / n! = e^z - 1$. Thus the convolution relation (12) leads again to (9). ◁

Set partitions into r blocks. Let $S_n^{(r)}$ denote the number of ways of partitioning the set $[1 \mathbin{.\,.} n]$ into r disjoint and non-empty equivalence classes also known as *blocks*. We set $\mathcal{S}^{(r)} = \bigcup_n \mathcal{S}_n^{(r)}$; the corresponding objects are called *set partitions* (the latter not to be confused with integer partitions examined in Section I. 3). The enumeration problem for set partitions is closely related to that of surjections. Symbolically, a partition is determined as a labelled *set* of classes (blocks), each of which is a non-empty urn. Thus, one has

$$(13) \quad \mathcal{S}^{(r)} = \text{SET}_r(\mathcal{V}), \ \mathcal{V} = \text{SET}_{\geq 1}(\mathcal{Z}) \quad \Longrightarrow \quad S^{(r)}(z) = \frac{1}{r!} \left(e^z - 1 \right)^r.$$

The basic formula connecting the two counting sequences $R_n^{(r)}$ and $S_n^{(r)}$ is

$$S_n^{(r)} = \frac{1}{r!} R_n^{(r)},$$

in accordance with (9) and (13). This can also be interpreted directly: an r–partition is associated with a group of exactly $r!$ distinct r–surjections, two surjections belonging to the same group iff one is obtained from the other by permuting the range values, $[1 . . r]$.

The numbers $S_n^{(r)} = n![z^n]S^{(r)}(z)$ are known as the *Stirling numbers of the second kind*, or better, the *Stirling partition numbers*. They were already encountered in connection with encodings by words (Chapter I, p. 62). Knuth, following Karamata, advocated for the $S_n^{(r)}$ the notation $\left\{{n \atop r}\right\}$. From (10), an explicit form also exists:

$$(14) \qquad S_n^{(r)} \equiv \left\{{n \atop r}\right\} = \frac{1}{r!} \sum_{j=0}^{r} \binom{r}{j} (-1)^j (r-j)^n.$$

The books by Graham, Knuth, and Patashnik [307] and Comtet [129] contain a thorough discussion of these numbers; see also Appendix A.8: *Stirling numbers*, p. 735.

All surjections and set partitions. Define now the collection of all surjections and all set partitions by

$$\mathcal{R} = \bigcup_r \mathcal{R}^{(r)}, \qquad \mathcal{S} = \bigcup_r \mathcal{S}^{(r)}.$$

Thus $\mathcal{R}_n$ is the class of all surjections of $[1 . . n]$ onto *any* initial segment of the integers, and $\mathcal{S}_n$ is the class of all partitions of the set $[1 . . n]$ into *any* number of blocks (Figure II.5). Symbolically, one has

$$(15) \qquad \begin{aligned} \mathcal{R} &= \text{SEQ}(\text{SET}_{\geq 1}(\mathcal{Z})) &\implies& \quad R(z) = \frac{1}{2 - e^z} \\ \mathcal{S} &= \text{SET}(\text{SET}_{\geq 1}(\mathcal{Z})) &\implies& \quad S(z) = e^{e^z - 1}. \end{aligned}$$

The numbers $R_n = n! \, [z^n]R(z)$ are called *surjection numbers* (also, "preferential arrangements", *EIS* **A000670**). The numbers S_n are the *Bell numbers* (*EIS* **A000110**). These numbers are easily determined by expanding the EGFs:

$$\begin{aligned} R(z) &= 1 + z + 3\frac{z^2}{2!} + 13\frac{z^3}{3!} + 75\frac{z^4}{4!} + 541\frac{z^5}{5!} + 4683\frac{z^6}{6!} + 47293\frac{z^7}{7!} + \cdots \\ S(z) &= 1 + z + 2\frac{z^2}{2!} + 5\frac{z^3}{3!} + 15\frac{z^4}{4!} + 52\frac{z^5}{5!} + 203\frac{z^6}{6!} + 877\frac{z^7}{7!} + \cdots . \end{aligned}$$

Explicit expressions as finite double sums result from summing Stirling numbers,

$$R_n = \sum_{r \geq 0} r! \left\{{n \atop r}\right\}, \qquad \text{and} \qquad S_n = \sum_{r \geq 0} \left\{{n \atop r}\right\},$$

where each Stirling number is itself a sum given by (14). Alternatively, single (though infinite) sums arise from the expansions

$$\left\{ \begin{aligned} R(z) &= \frac{1}{2}\frac{1}{1 - \frac{1}{2}e^z} \\ &= \sum_{\ell=0}^{\infty} \frac{1}{2^{\ell+1}}e^{\ell z} \end{aligned} \right. \qquad \text{and} \qquad \left\{ \begin{aligned} S(z) &= e^{e^z - 1} = \frac{1}{e}e^{e^z} \\ &= \frac{1}{e}\sum_{\ell=0}^{\infty}\frac{1}{\ell!}e^{\ell z}, \end{aligned} \right.$$

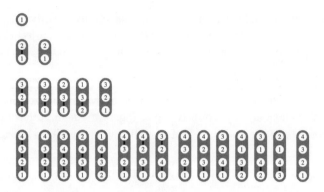

Figure II.5. A complete listing of all set partitions for sizes $n = 1, 2, 3, 4$. The corresponding sequence $1, 1, 2, 5, 15, \ldots$ is formed of Bell numbers, *EIS* **A000110**.

from which coefficient extraction yields

$$R_n = \frac{1}{2} \sum_{\ell=0}^{\infty} \frac{\ell^n}{2^\ell} \quad \text{and} \quad S_n = \frac{1}{e} \sum_{\ell=0}^{\infty} \frac{\ell^n}{\ell!}.$$

The formula for Bell numbers was found by Dobinski in 1877.

The asymptotic analysis of the surjection numbers (R_n) will be performed in Example IV.7 (p. 259), as one of the very first illustrations of complex asymptotic methods (the meromorphic case); that of Bell's partition numbers is best done by means of the saddle-point method (Example VIII.6, p. 560). The asymptotic forms found are

(16) $\qquad R_n \sim \frac{n!}{2} \frac{1}{(\log 2)^{n+1}} \quad \text{and} \quad S_n \sim n! \frac{e^{e^r-1}}{r^n \sqrt{2\pi r(r+1)e^r}},$

where $r \equiv r(n)$ is the positive root of the equation $re^r = n + 1$. One has $r(n) \sim \log n - \log\log n$, so that

$$\log S_n = n \left(\log n - \log\log n - 1 + o(1)\right).$$

Elementary derivations (i.e., based solely on real analysis) of these asymptotic forms are also possible, a fact discussed briefly in Appendix B.6: *Laplace's method*, p. 755.

The line of reasoning adopted for enumerating surjections viewed as sequences-of-sets and partitions viewed as sets-of-sets yields a general result that is applicable to a wide variety of constrained objects.

Proposition II.2. *The class* $\mathcal{R}^{(A,B)}$ *of surjections, where the cardinalities of the preimages lie in* $A \subseteq \mathbb{Z}_{\geq 1}$ *and the cardinality of the range belongs to* B, *has EGF*

$$R^{(A,B)}(z) = \beta(\alpha(z)) \quad \text{where} \quad \alpha(z) = \sum_{a \in A} \frac{z^a}{a!}, \quad \beta(z) = \sum_{b \in B} z^b.$$

The class $\mathcal{S}^{(A,B)}$ of set partitions with block sizes in $A \subseteq \mathbb{Z}_{\geq 1}$ and with a number of blocks that belongs to B has EGF

$$S^{(A,B)}(z) = \beta(\alpha(z)) \qquad where \qquad \alpha(z) = \sum_{a \in A} \frac{z^a}{a!}, \quad \beta(z) = \sum_{b \in B} \frac{z^b}{b!}.$$

Proof. One has $\mathcal{R}^{(A,B)} = \mathrm{SEQ}_B(\mathrm{SET}_A(\mathcal{Z}))$ and $\mathcal{S}^{(A,B)} = \mathrm{SET}_B(\mathrm{SET}_A(\mathcal{Z}))$, where, in accordance with our general convention of p. 30, the notation $\mathfrak{K}_\Omega$ specifies a construction $\mathfrak{K}$ with a number of components restricted to set Ω. ∎

Example II.6. *Smallest and largest blocks in set partitions.* Let $e_b(z)$ denote the truncated exponential function,

$$e_b(z) := 1 + \frac{z}{1!} + \frac{z^2}{2!} + \cdots + \frac{z^b}{b!}.$$

The EGFs $S^{(\leq b)}(z) = \exp(e_b(z) - 1)$ and $S^{(>b)}(z) = \exp(e^z - e_b(z))$ correspond to partitions with all blocks of size $\leq b$ and all blocks of size $> b$, respectively. .... ∎

▷ **II.5.** *No singletons.* The EGF of partitions without singleton parts is $e^{e^z - 1 - z}$. The EGF of "double surjections" (each preimage contains at least two elements) is $(2 + z - e^z)^{-1}$. ◁

Example II.7. *Comtet's square.* An exercise in Comtet's book [129, Ex. 13, p. 225] serves beautifully to illustrate the power of the symbolic method. The question is to enumerate set partitions such that a parity constraint is satisfied by the number of blocks and/or the number of elements in each block. Then, the EGFs are tabulated as follows:

Set partitions:	Any # of blocks	Odd # of blocks	Even # of blocks
any block sizes	$e^{e^z - 1}$	$\sinh(e^z - 1)$	$\cosh(e^z - 1)$
odd block sizes	$e^{\sinh z}$	$\sinh(\sinh z)$	$\cosh(\sinh z)$
even block sizes	$e^{\cosh z - 1}$	$\sinh(\cosh z - 1)$	$\cosh(\cosh z - 1)$

The proof is a direct application of Proposition II.2, upon noting that e^z, $\sinh z$, $\cosh z$ are the characteristic EGFs of $\mathbb{Z}_{\geq 0}$, $2\mathbb{Z}_{\geq 0} + 1$, and $2\mathbb{Z}_{\geq 0}$ respectively. The sought EGFs are then obtained by forming the compositions

$$\left\{ \begin{array}{c} \exp \\ \sinh \\ \cosh \end{array} \right\} \circ \left\{ \begin{array}{c} -1 + \exp \\ \sinh \\ -1 + \cosh \end{array} \right\},$$

in accordance with general principles. .... ∎

II. 3.2. Applications to words and random allocations.

Numerous enumeration problems present themselves when analysing statistics on letters in words. They find applications in the study of *random allocations* [388] and the design of *hashing algorithms* in computer science [378, 538]. Fix an alphabet

$$\mathcal{X} = \{a_1, a_2, \ldots, a_r\}$$

of cardinality r, and let $\mathcal{W}$ be the class of all words over the alphabet $\mathcal{X}$, the size of a word being its length. A word $w \in \mathcal{W}_n$ of length n can be viewed as a function from $[1 \, .. \, n]$ to $[1 \, .. \, r]$, namely the function associating to each position the value of the corresponding letter (canonically numbered from 1 to r) in the word. For instance, let $\mathcal{X} = \{\mathsf{a}, \mathsf{b}, \mathsf{c}, \mathsf{d}, \mathsf{p}, \mathsf{q}, \mathsf{r}\}$ and take the letters of $\mathcal{X}$ canonically numbered as $a_1 = \mathsf{a}, \ldots, a_7 = \mathsf{r}$; for the word $w = $ "abracadabra", the table giving the position-to-letter mapping is

$$\begin{pmatrix} \mathsf{a} & \mathsf{b} & \mathsf{r} & \mathsf{a} & \mathsf{c} & \mathsf{a} & \mathsf{d} & \mathsf{a} & \mathsf{b} & \mathsf{r} & \mathsf{a} \\ \hline 1 & 2 & 3 & 4 & 5 & 6 & 7 & 8 & 9 & 10 & 11 \\ 1 & 2 & 7 & 1 & 3 & 1 & 4 & 1 & 2 & 7 & 1 \end{pmatrix},$$

which is itself determined by its sequence of preimages:

$$\overbrace{\{1, 4, 6, 8, 11\}}^{\mathsf{a}=a_1}, \quad \overbrace{\{2, 9\}}^{\mathsf{b}=a_2}, \quad \overbrace{\{5\}}^{\mathsf{c}=a_3}, \quad \overbrace{\{7\}}^{\mathsf{d}=a_4}, \quad \overbrace{\{\,\}}^{\mathsf{p}=a_5}, \quad \overbrace{\{\,\}}^{\mathsf{q}=a_6}, \quad \overbrace{\{3, 10\}}^{\mathsf{r}=a_7}.$$

This decomposition is the same as the one used for surjections; only, it is no longer imposed that all preimages should be non-empty.

The decomposition based on preimages then gives, with $\mathcal{U}$ the class of all urns

$$(17) \qquad \mathcal{W} \cong \mathcal{U}^r \equiv \mathrm{SEQ}_r(\mathcal{U}) \qquad \Longrightarrow \qquad W(z) = (e^z)^r = e^{rz},$$

which yields back $W_n = r^n$, as was to be expected. In summary: words over an r–ary alphabet are equivalent to functions into a set of cardinality r and are described by an r-fold labelled product.

For the situation where restrictions are imposed on the number of occurrences of letters, the decomposition (17) generalizes as follows.

Proposition II.3. *Let $\mathcal{W}^{(A)}$ denote the family of words over an alphabet of cardinality r, such that the number of occurrences of each letter lies in a set A. Then*

$$(18) \qquad W^{(A)}(z) = \alpha(z)^r \qquad \text{where} \qquad \alpha(z) = \sum_{a \in A} \frac{z^a}{a!}.$$

The proof is a one-liner: $\mathcal{W}^{(A)} \cong \mathrm{SEQ}_r(\mathrm{SET}_A(\mathcal{Z}))$. Although this result is technically a shallow consequence of the symbolic method, it has several important applications in discrete probability, as we see next.

Example II.8. *Restricted words.* The EGF of words containing each letter *at most b* times, and that of words containing each letter *more* than b times are

$$(19) \qquad \mathcal{W}^{\langle \le b \rangle}(z) = e_b(z)^r, \qquad \mathcal{W}^{\langle > b \rangle}(z) = \left(e^z - e_b(z)\right)^r,$$

respectively. (Observe the analogy with Example II.6, p. 111.) Taking $b = 1$ in the first formula gives the number of n-arrangements of r elements (i.e., of ordered combinations of n elements among r possibilities),

$$(20) \qquad n! [z^n](1 + z)^r = n! \binom{r}{n} = r(r - 1) \cdots (r - n + 1),$$

as anticipated; taking $b = 0$, but now in the second formula, gives back the number of r–surjections. For general b, the generating functions of (19) contain valuable information on the least frequent and most frequent letter in random words. . . . . ∎

Example II.9. *Random allocations (balls-in-bins model).* Throw at random n distinguishable balls into m distinguishable bins. A particular realization is described by a word of length n (balls are distinguishable, say, as numbers from 1 to n) over an alphabet of cardinality m (representing the bins chosen). Let Min and Max represent the size of the least filled and most filled bins, respectively. Then[4],

(21)
$$\mathbb{P}\{\text{Max} \le b\} = n!\,[z^n]e_b\left(\frac{z}{m}\right)^m$$
$$\mathbb{P}\{\text{Min} > b\} = n!\,[z^n]\left(e^{z/m} - e_b\left(\frac{z}{m}\right)\right)^m.$$

The justification of this formula relies on the easy identity

(22)
$$\frac{1}{m^n}\,[z^n]f(z) \equiv [z^n]f\left(\frac{z}{m}\right),$$

and on the fact that a probability is determined as the ratio between the number of favorable cases (given by (19)) and the total number of cases (m^n). The formulae of (21) lend themselves to evaluation using symbolic manipulations systems; for instance, with $m = 100$ and $n = 200$, one finds, for $\mathbb{P}(\text{Max} = k)$:

k	2	4	5	6	7	8	9	12	15	20
$\mathbb{P}(\text{Max} = k)$	10^{-55}	$1.4 \cdot 10^{-3}$	0.17	0.46	0.26	0.07	0.01	$9 \cdot 10^{-5}$	$2 \cdot 10^{-7}$	$4 \cdot 10^{-10}$

The values $k = 5, 6, 7, 8$ concentrate about 99% of the probability mass.

An especially interesting case is when m and n are asymptotically proportional, that is, $n/m = \alpha$ and α lies in a compact subinterval of $(0, +\infty)$. In that case, with probability tending to 1 as n tends to infinity, one has

$$\text{Min} = 0, \qquad \text{Max} \sim \frac{\log n}{\log\log n}.$$

In other words, there are, almost surely, empty urns (in fact many of them, see Example III.10, p. 177) and the most filled urn grows logarithmically in size (Example VIII.14, p. 598). Such probabilistic properties are best established by complex analytic methods, whose starting point is exact generating function representations such as (19) and (21). They form the core of the reference book [388] by Kolchin, Sevastyanov, and Chistyakov. The resulting estimates are in turn invaluable in the analysis of hashing algorithms [301, 378, 538] to which the balls-in-bins model has been recognized to apply with great accuracy [425]. .... ■

▷ **II.6.** *Number of different letters in words.* The probability that a random word of length n over an alphabet of cardinality r contains k different letters is (with $\left\{{n \atop k}\right\}$ a Stirling number)

$$p_{n,k}^{(r)} := \frac{1}{r^n}\binom{r}{k}\left\{{n \atop k}\right\}k!$$

(Choose k letters among r, then split the n positions into k distinguished non-empty classes.) The quantity $p_{n,k}^{(r)}$ is also the probability that a random mapping from $[1 \mathrel{..} n]$ to $[1 \mathrel{..} r]$ has an image of cardinality k. ◁

▷ **II.7.** *Arrangements.* An *arrangement* of size n is an ordered combination of (some) elements of $[1 \mathrel{..} n]$. Let $\mathcal{A}$ be the class of all arrangements. Grouping together into an urn all the elements *not* present in the arrangement shows that a specification and its companion EGF are [129, p. 75]

$$\mathcal{A} \cong \mathcal{U} \star \mathcal{P}, \quad \mathcal{U} = \text{SET}(\mathcal{Z}), \quad \mathcal{P} = \text{SEQ}(\mathcal{Z}) \quad \Longrightarrow \quad A(z) = \frac{e^z}{1-z}.$$

---

[4]We let $\mathbb{P}(E)$ represent the probability of an event E and $\mathbb{E}(X)$ the expectation of the random variable X; cf Appendix A.3: *Combinatorial probability*, p. 727 and Appendix C.2: *Random variables*, p. 771.

The counting sequence $A_n = \sum_{k=0}^n \frac{n!}{k!}$ starts as 1, 2, 5, 16, 65, 326, 1957 (*EIS* **A000522**). $\lhd$

Birthday paradox and coupon collector problem. The next two examples show applications of EGFs to two classical problems of probability theory, the *birthday paradox* and the *coupon collector problem*. They constitute a neat illustration of the fact that the symbolic method may be used to analyse discrete probabilistic models— this theme is explored systematically in Chapter III, as regards exact results, and Chapter IX, which is dedicated to asymptotic laws.

Assume that there is a very long line of persons ready to enter a very large room one by one. Each person is let in and declares her birthday upon entering the room. How many people must enter in order to find two that have the same birthday? The birthday paradox is the counterintuitive fact that on average a birthday collision is likely to take place as early as at time $n \doteq 24$. Dually, the coupon collector problem asks for the average number of persons that must enter in order to exhaust all the possible days in the year as birthdates. In this case, the average is the rather large number $n' \doteq 2364$. (The term "coupon collection" refers to the situation where images or coupons of various sorts are inserted in sales items and some premium is given to those who succeed in gathering a complete collection.) The birthday problem and the coupon collector problem are relative to a potentially infinite sequence of events; however, the fact that the first birthday collision or the first complete collection occurs at any fixed time n only involves finite events. The following diagram illustrates the events of interest:

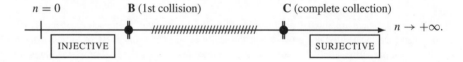

In other words, we seek the time at which injectivity *ceases* to hold (the first birthday collision, B) and the time at which surjectivity *begins* to be satisfied (a complete collection, C). In what follows, we consider a year with r days (readers from Earth may take $r = 365$) and let $\mathcal{X}$ represent an alphabet with r letters (the days in the year).

Example II.10. *Birthday paradox.* Let B be the time of the first collision, which is a random variable ranging between 2 and $r + 1$ (where the upper bound is derived from the pigeonhole principle). A collision has not yet occurred at time n, if the sequence of birthdates $\beta_1, \ldots, \beta_n$ has no repetition. In other words, the function β from $[1 .. n]$ to $\mathcal{X}$ must be injective; equivalently, $\beta_1, \ldots, \beta_n$ is an n-arrangement of r objects. Thus, we have the fundamental relation

$$
\begin{aligned}
\mathbb{P}\{B > n\} &= \frac{r(r-1)\cdots(r-n+1)}{r^n} \\
&= \frac{n!}{r^n}[z^n](1+z)^r \\
&= n!\,[z^n]\left(1 + \frac{z}{r}\right)^r,
\end{aligned}
$$

(23)

where the second line repeats (20) and the third results from the series transformation (22).

The expectation of the random variable B is elementarily

$$(24) \qquad \mathbb{E}(B) = \sum_{n=0}^{\infty} \mathbb{P}\{B > n\},$$

this by virtue of a general formula valid for all discrete random variables (Appendix C.2: *Random variables*, p. 771). From (23), line 1, this gives us a sum expressing the expectation: namely,

$$(25) \qquad \mathbb{E}(B) = 1 + \sum_{n=1}^{r} \frac{r(r-1)\cdots(r-n+1)}{r^n}.$$

For instance, with $r = 365$, one finds that the expectation is the rational number,

$$\mathbb{E}(B) = \frac{12681\cdots06674}{5151\cdots0625} \doteq 24.61658,$$

where the denominator comprises as much as 864 digits.

An alternative form of the expectation is derived from the generating function involved in (23), line 3. Let $f(z) = \sum_n f_n z^n$ be an entire function with non-negative coefficients. Then the formula

$$(26) \qquad \sum_{n=0}^{\infty} f_n n! = \int_0^{\infty} e^{-t} f(t)\, dt,$$

a particular case of the Laplace transform, is valid provided either the sum or the integral on the right converges. The proof is a direct consequence of the usual Eulerian representation of factorials,

$$n! = \int_0^{\infty} e^{-t} t^n \, dt.$$

Applying this principle to (24) with the probabilities given by (23) [third line], one finds

$$(27) \qquad \mathbb{E}(B) = \int_0^{\infty} e^{-t} \left(1 + \frac{t}{r}\right)^r dt.$$

Asymptotic analysis can take up from here. The Laplace method[5] can be applied either in its version for discrete sums to (25) or in its version for integrals to (27); see Appendix B.6: *Laplace's method*, p. 755. Either way provides the estimate

$$(28) \qquad \mathbb{E}(B) = \sqrt{\frac{\pi r}{2}} + \frac{2}{3} + O(r^{-1/2}),$$

as r tends to infinity. In particular, the approximation provided by the first two terms of (28), for $r = 365$, is 24.61*119*, which only represents a relative error of $2 \cdot 10^{-4}$. See also a sample realization in Figure II.6, corresponding to $r = 20$. The quantity $\mathbb{E}(B)$ is related to Ramanujan's Q-function (see Equation (50), p. 130) by $\mathbb{E}(B) = 1 + Q(r)$, and we shall examine a global way to deal with an entire class of related sums in Example VI.13, p. 416.

The interest of such integral representations based on generating functions is that they are *robust*: they adjust naturally to many kinds of combinatorial conditions. For instance, the same calculations applied to (21) prove the following: *the expected time necessary for the*

---

[5]Knuth [377, Sec. 1.2.11.3] uses this calculation as a pilot example for (real) asymptotic analysis.

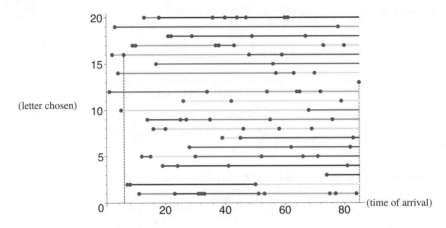

(letter chosen)

(time of arrival)

Figure II.6. A sample realization of the "birthday paradox" and "coupon collection" with an alphabet of $r = 20$ letters. The first collision occurs at time $B = 6$ and the collection becomes complete at time $C = 87$.

first occurrence of the event "b persons have the same birthday" has expectation given by the integral

$$(29) \qquad\qquad I(r, b) := \int_0^\infty e^{-t} e_{b-1}\left(\frac{t}{r}\right)^r dt.$$

(The basic birthday paradox corresponds to $b = 2$.) The formula (29) was first derived by Klamkin and Newman in 1967; their paper [366] shows in addition that

$$I(r, b) \underset{r \to \infty}{\sim} \sqrt[b]{b!}\, \Gamma\left(1 + \frac{1}{b}\right) r^{1-1/b},$$

once more a consequence of Laplace's method. The asymptotic form evaluates to 82.87, for $r = 365$ and $b = 3$, and the exact value of the expectation is 88.73891. Thus three-way collisions also tend to occur much sooner than one might think, after about 89 persons on average. Globally, such developments illustrate the versatility of the symbolic approach and its applicability to many basic probabilistic problems (see also Subsection III. 6.1, p. 189). ... ∎

▷ **II.8.** *The probability distribution of time till a birthday collision.* Elementary approximations show that, for large r, and in the "central" regime $n = t\sqrt{r}$, one has

$$\mathbb{P}(B > t\sqrt{r}) \sim e^{-t^2/2}, \qquad \mathbb{P}(B = t\sqrt{r}) \sim \frac{1}{\sqrt{r}} t e^{-t^2/2}.$$

The continuous probability distribution with density $te^{-t^2/2}$ is called a *Rayleigh distribution*. Saddle-point methods (Chapter VIII) may be used to show that for the first occurrence of a b-fold birthday collision: $\mathbb{P}(B > tr^{1-1/b}) \sim e^{-t^b/b!}$.   ◁

***Example* II.11.** *Coupon collector problem.* This problem is dual to the birthday paradox. We ask for the first time C when $\beta_1, \ldots, \beta_C$ contains all the elements of $\mathcal{X}$: that is, all the possible birthdates have been "collected". In other words, the event $\{C \le n\}$ means the equality between sets, $\{\beta_1, \ldots, \beta_n\} = \mathcal{X}$. Thus, the probabilities satisfy

$$
\begin{aligned}
\mathbb{P}\{C \leq n\} &= \frac{R_n^{(r)}}{r^n} = \frac{r!\{^n_r\}}{r^n} \\
&= \frac{n!}{r^n}\, [z^n]\,(e^z - 1)^r \\
&= n![z^n]\left(e^{z/r} - 1\right)^r,
\end{aligned}
$$
(30)

by our earlier enumeration of surjections. The complementary probabilities are then

$$
\mathbb{P}\{C > n\} = 1 - \mathbb{P}\{C \leq n\} = n![z^n]\left(e^z - \left(e^{z/r} - 1\right)^r\right).
$$

An application of the Eulerian integral trick of (27) then provides a representation of the expectation of the time needed for a full collection as

$$
\mathbb{E}(C) = \int_0^\infty \left(1 - (1 - e^{-t/r})^r\right)\, dt.
$$
(31)

A simple calculation (expand by the binomial theorem and integrate termwise) shows that

$$
\mathbb{E}(C) = r \sum_{j=1}^r \binom{r}{j} \frac{(-1)^{j-1}}{j},
$$

which constitutes a first answer to the coupon collector problem in the form of an alternating sum. Alternatively, in (31), perform the change of variables $v = 1 - e^{-t/r}$, then expand and integrate termwise; this process provides the more tractable form

$$
\mathbb{E}(C) = r\, \mathrm{H}_r,
$$
(32)

where H_r is the harmonic number:

$$
\mathrm{H}_r = 1 + \frac{1}{2} + \frac{1}{3} + \cdots + \frac{1}{r}.
$$
(33)

Formula (32) is by the way easy to interpret directly[6]: one needs on average $1 = r/r$ trials to get the first day, then $r/(r-1)$ to get a different day, etc.

Regarding (32), one has available the well-known formula (by comparing sums with integrals or by Euler–Maclaurin summation),

$$
\mathrm{H}_r = \log r + \gamma + \frac{1}{2r} + O(r^{-2}), \quad \gamma \doteq 0.57721\,56649,
$$

where γ is known as Euler's constant. Thus, the expected time for a full collection satisfies

$$
\mathbb{E}(C) = r \log r + \gamma r + \frac{1}{2} + O(r^{-1}).
$$
(34)

Here the "surprise" lies in the nonlinear growth of the expected time for a full collection. For a year on Earth, $r = 365$, the exact expected value is $\doteq 2364.64602$ whereas the approximation provided by the first three terms of (34) yields 2364.64625, representing a relative error of only one in ten million.

As usual, the symbolic treatment adapts to a variety of situations, for instance, to multiple collections. One finds: *the expected time till each item (birthday or coupon) is obtained b times is*

$$
J(r, b) = \int_0^\infty \left(1 - \left(1 - e_{b-1}(t/r)e^{-t/r}\right)^r\right)\, dt.
$$

---

[6]Such elementary derivations are very much problem specific: contrary to the symbolic method, they do not usually generalize to more complex situations.

This expression vastly generalizes the standard case (31), which corresponds to $b = 1$. From it, one finds [454]

$$J(r, b) = r \left(\log r + (b - 1) \log \log r + \gamma - \log(b - 1)! + o(1) \right),$$

so that only a few more trials are needed in order to obtain additional collections. .... ∎

▷ **II.9.** *The little sister.* The coupon collector has a little sister to whom he gives his duplicates. Foata, Lass, and Han [266] show that the little sister misses on average H_r coupons when her big brother first obtains a complete collection.   ◁

▷ **II.10.** *The probability distribution of time till a complete collection.* The saddle-point method (Chapter VIII) may be used to prove that, in the regime $n = r \log r + tr$, we have

$$\lim_{t \to \infty} \mathbb{P}(C \leq r \log r + tr) = e^{-e^{-t}}.$$

This continuous probability distribution is known as a *double exponential distribution.* For the time $C^{(b)}$ till a collection of multiplicity b, one has

$$\lim_{t \to \infty} P(C^{(b)} < r \log r + (b - 1)r \log \log r + tr) = \exp(-e^{-t}/(b - 1)!),$$

a property known as the Erdős–Rényi law, which finds application in the study of random graphs [195].   ◁

Words as both labelled and unlabelled objects. What distinguishes a labelled structure from an unlabelled one? There is nothing intrinsic there, and everything is in the eye of the beholder—or rather in the type of construction adopted when modelling a specific problem. Take the class of words $\mathcal{W}$ over an alphabet of cardinality r. The two generating functions (an OGF and an EGF respectively),

$$\widehat{W}(z) \equiv \sum_n W_n z^n = \frac{1}{1 - rz} \qquad \text{and} \qquad W(z) \equiv \sum_n W_n \frac{z^n}{n!} = e^{rz},$$

leading in both cases to $W_n = r^n$, correspond to two different ways of constructing words: the first one directly as an unlabelled sequence, the other as a labelled power of letter positions. A similar situation arises for r–partitions, for which we find as OGF and EGF,

$$\widehat{S}^{(r)}(z) = \frac{z^r}{(1 - z)(1 - 2z) \cdots (1 - rz)} \qquad \text{and} \qquad S^{(r)}(z) = \frac{(e^z - 1)^r}{r!},$$

by viewing these either as unlabelled structures (an encoding via words of a regular language in Section I. 4.3, p. 62) or directly as labelled structures (this chapter, p. 108).

▷ **II.11.** *Balls switching chambers: the Ehrenfest² model.* Consider a system of two chambers A and B (also classically called "urns"). There are N distinguishable balls, and, initially, chamber A contains them all. At any instant $\frac{1}{2}, \frac{3}{2}, \ldots$, one ball is allowed to change from one chamber to the other. Let $E_n^{[\ell]}$ be the number of possible evolutions that lead to chamber A containing ℓ balls at instant n and $E^{[\ell]}(z)$ the corresponding EGF. Then

$$E^{[\ell]}(z) = \binom{N}{\ell} (\cosh z)^\ell (\sinh z)^{N-\ell}, \qquad E^{[N]}(z) = (\cosh z)^N \equiv 2^{-N} (e^z + e^{-z})^N.$$

[Hint: the EGF $E^{[N]}$ enumerates mappings where each preimage has an even cardinality.] In particular the probability that urn A is again full at time $2n$ is

$$\frac{1}{2^N N^{2n}} \sum_{k=0}^{N} \binom{N}{k} (N - 2k)^{2n}.$$

This famous model was introduced by Paul and Tatiana Ehrenfest [188] in 1907, as a simplified model of heat transfer. It helped resolve the apparent contradiction between irreversibility in thermodynamics (the case $N \to \infty$) and recurrence of systems undergoing ergodic transformations (the case $N < \infty$). See especially Mark Kac's discussion [361]. The analysis can also be carried out by combinatorial methods akin to those of weighted lattice paths: see Note V.25, p. 336 and [304]. ◁

II. 4. Alignments, permutations, and related structures

In this section, we start by considering specifications built by piling up two constructions, sequences-of-cycles and sets-of-cycles respectively. They define a new class of objects, alignments, while serving to specify permutations in a novel way. (These specifications otherwise parallel surjections and set partitions.) In this context, permutations are examined under their cycle decomposition, the corresponding enumeration results being the most important ones combinatorially (Subsection II. 4.1 and Figure II.8, p. 123). In Subsection II. 4.2, we recapitulate the meaning of classes that can be defined iteratively by a combination of any two nested labelled constructions.

II. 4.1. Alignments and permutations. The two specifications under consideration now are

$$(35) \qquad \mathcal{O} = \text{SEQ}(\text{CYC}(\mathcal{Z})), \qquad \text{and} \qquad \mathcal{P} = \text{SET}(\text{CYC}(\mathcal{Z})),$$

specifying new objects called *alignments* ($\mathcal{O}$) as well as an important decomposition of *permutations* ($\mathcal{P}$).

Alignments. An alignment is a well-labelled sequence of cycles. Let $\mathcal{O}$ be the class of all alignments. Schematically, one can visualize an alignment as a collection of directed cycles arranged in a linear order, somewhat like slices of a sausage fastened on a skewer:

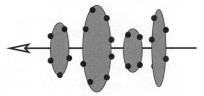

The symbolic method provides,

$$\mathcal{O} = \text{SEQ}(\text{CYC}(\mathcal{Z})) \qquad \Longrightarrow \qquad O(z) = \frac{1}{1 - \log(1 - z)^{-1}},$$

and the expansion starts as

$$O(z) = 1 + z + 3\frac{z^2}{2!} + 14\frac{z^3}{3!} + 88\frac{z^4}{4!} + 694\frac{z^5}{5!} + \cdots,$$

but the coefficients (see *EIS* **A007840**: "ordered factorizations of permutations into cycles") appear to admit no simple form.

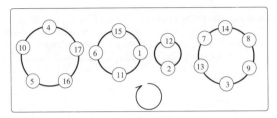

A permutation may be viewed as a *set* of cycles that are labelled circular digraphs. The diagram shows the decomposition of the permutation

$$\sigma = \begin{pmatrix} 1 & 2 & 3 & 4 & 5 & 6 & 7 & 8 & 9 & 10 & 11 & 12 & 13 & 14 & 15 & 16 & 17 \\ 11 & 12 & 13 & 17 & 10 & 15 & 14 & 9 & 3 & 4 & 6 & 2 & 7 & 8 & 1 & 5 & 16 \end{pmatrix}.$$

(Cycles here read clockwise and i is connected to σ_i by an edge in the graph.)

Figure II.7.   The cycle decomposition of permutations.

Permutations and cycles. From elementary mathematics, it is known that a permutation admits a unique decomposition into cycles. Let σ be a permutation. Start with any element, say 1, and draw a directed edge from 1 to $\sigma(1)$, then continue connecting to $\sigma^2(1), \sigma^3(1)$, and so on; a cycle containing 1 is obtained after at most n steps. If one repeats the construction, taking at each stage an element not yet connected to earlier ones, the cycle decomposition of the permutation σ is obtained; see Figure II.7. This argument shows that the class of sets-of-cycles (corresponding to $\mathcal{P}$ in (35)) is isomorphic to the class of permutations as defined in Example II.2, p. 98:

$$(36) \qquad\qquad \mathcal{P} \cong \mathrm{SET}(\mathrm{CYC}(\mathcal{Z})) \cong \mathrm{SEQ}(\mathcal{Z}).$$

This combinatorial isomorphism is reflected by the obvious series identity

$$P(z) = \exp\left(\log \frac{1}{1-z}\right) = \frac{1}{1-z}.$$

The property that exp and log are inverse of one another is nothing but an analytic reflex of the combinatorial fact that permutations uniquely decompose into cycles!

As regards combinatorial applications, what is especially fruitful is the variety of special results derived from the decomposition of permutations into cycles. By a use of restricted construction that entirely parallels Proposition II.2, p. 110, we obtain the following statement.

Proposition II.4. *The class $\mathcal{P}^{(A,B)}$ of permutations with cycle lengths in $A \subseteq \mathbb{Z}_{>0}$ and with cycle number that belongs to $B \subseteq \mathbb{Z}_{\geq 0}$ has EGF*

$$P^{(A,B)}(z) = \beta(\alpha(z)) \qquad \text{where} \qquad \alpha(z) = \sum_{a \in A} \frac{z^a}{a}, \; \beta(z) = \sum_{b \in B} \frac{z^b}{b!}.$$

▷ **II.12.** *What about alignments?* With similar notations, one has for alignments

$$O^{(A,B)}(z) = \beta(\alpha(z)) \qquad \text{where} \qquad \alpha(z) = \sum_{a \in A} \frac{z^a}{a}, \; \beta(z) = \sum_{b \in B} z^b,$$

corresponding to $\mathcal{O}^{(A,B)} = \mathrm{SEQ}_B(\mathrm{CYC}_A(\mathcal{Z}))$.   ◁

***Example* II.12.** *Stirling cycle numbers.* The class $\mathcal{P}^{(r)}$ of permutations that decompose into r cycles, satisfies

$$(37) \qquad \mathcal{P}^{(r)} = \mathrm{SET}_r(\mathrm{CYC}(\mathcal{Z})) \qquad \Longrightarrow \quad P^{(r)}(z) = \frac{1}{r!} \left(\log \frac{1}{1-z} \right)^r.$$

The number of such permutations of size n is then

$$(38) \qquad P_n^{(r)} = \frac{n!}{r!} \, [z^n] \left(\log \frac{1}{1-z} \right)^r.$$

These numbers are fundamental quantities of combinatorial analysis. They are known as the Stirling numbers of the first kind, or better, according to a proposal of Knuth, the *Stirling cycle numbers*. Together with the Stirling partition numbers, the properties of the Stirling cycle numbers are explored in the book by Graham, Knuth, and Patashnik [307] where they are denoted by $\left[{n \atop r} \right]$. See Appendix A.8: *Stirling numbers*, p. 735. (Note that the number of alignments formed with r cycles is $r!\left[{n \atop r} \right]$.) As we shall see shortly (p. 140) Stirling numbers also surface in the enumeration of permutations by their number of records.

It is also of interest to determine what happens regarding cycles in a random permutation of size n. Clearly, when the uniform distribution is placed over all elements of $\mathcal{P}_n$, each particular permutation has probability exactly $1/n!$. Since the probability of an event is the quotient of the number of favorable cases over the total number of cases, the quantity

$$p_{n,k} := \frac{1}{n!} \left[{n \atop k} \right]$$

is the probability that a random element of P_n has k cycles. This probabilities can be effectively determined for moderate values of n from (38) by means of a computer algebra system. Here are for instance selected values for $n = 100$:

k	1	2	3	4	5	6	7	8	9	10
$p_{n,k}$	0.01	0.05	0.12	0.19	0.21	0.17	0.11	0.06	0.03	0.01

For this value $n = 100$, we expect in a vast majority of cases the number of cycles to be in the interval $[1, 10]$. (The residual probability is only about 0.005.) Under this probabilistic model, the mean is found to be about 5.18. Thus: *A random permutation of size 100 has on average a little more than 5 cycles; it rarely has more than 10 cycles.*

Such procedures demonstrate a direct exploitation of symbolic methods. They do not however tell us how the number of cycles could depend on n, as n increases unboundedly. Such questions are to be investigated systematically in Chapters III and IX. Here, we shall content ourselves with a brief sketch. First, form the *bivariate generating function*,

$$P(z, u) := \sum_{r=0}^{\infty} P^{(r)}(z) u^r,$$

and observe that

$$P(z, u) = \sum_{r=0}^{\infty} \frac{u^r}{r!} \left(\log \frac{1}{1-z} \right)^r = \exp \left(u \log \frac{1}{1-z} \right) = (1-z)^{-u}.$$

Newton's binomial theorem then provides

$$[z^n](1-z)^{-u} = (-1)^n \binom{-u}{n}.$$

In other words, a simple formula

(39)
$$\sum_{k=0}^{n} \begin{bmatrix} n \\ k \end{bmatrix} u^k = u(u+1)(u+2)\cdots(u+n-1)$$

encodes precisely all the Stirling cycle numbers corresponding to a fixed value of n. From here, the expected number of cycles, $\mu_n := \sum_k k p_{n,k}$ is easily found to be expressed in terms of harmonic numbers (use logarithmic differentiation of (39)):

$$\mu_n = H_n \equiv 1 + \frac{1}{2} + \cdots + \frac{1}{n}.$$

In particular, one has $\mu_{100} \equiv H_{100} \doteq 5.18738$. In general: *The mean number of cycles in a random permutation of size n grows logarithmically with n, $\mu_n \sim \log n$.* .... ∎

Example II.13. *Involutions and permutations without long cycles.* A permutation σ is an *involution* if $\sigma^2 = \mathrm{Id}$, with Id the identity permutation. Clearly, an involution can have only cycles of sizes 1 and 2. The class $\mathcal{I}$ of all involutions thus satisfies

(40)
$$\mathcal{I} = \mathrm{SET}(\mathrm{CYC}_{1,2}(\mathcal{Z})) \quad\Longrightarrow\quad I(z) = \exp\left(z + \frac{z^2}{2}\right).$$

The explicit form of the EGF lends itself to expansion,

$$I_n = \sum_{k=0}^{\lfloor n/2 \rfloor} \frac{n!}{(n-2k)! 2^k k!},$$

which solves the counting problem explicitly. A *pairing* is an involution without a fixed point. In other words, only cycles of length 2 are allowed, so that

$$\mathcal{J} = \mathrm{SET}(\mathrm{CYC}_2(\mathcal{Z})) \quad\Longrightarrow\quad J(z) = e^{z^2/2}, \quad J_{2n} = 1 \cdot 3 \cdot 5 \cdots (2n-1).$$

(The formula for J_n, hence that of I_n, can be checked by a direct reasoning.)

Generally, the EGF of permutations, all of whose cycles (in particular the largest one) have length at most equal to r, satisfies

$$B^{(r)}(z) = \exp\left(\sum_{j=1}^{r} \frac{z^j}{j}\right).$$

The numbers $b_n^{(r)} = [z^n] B^{(r)}(z)$ satisfy the recurrence

$$(n+1) b_{n+1}^{(r)} = (n+1) b_n^{(r)} - b_{n-r}^{(r)},$$

by which they can be computed quickly, while they can be analysed asymptotically by means of the saddle-point method (Chapter VIII, p. 568). This gives access to the statistics of the longest cycle in a permutation. .... ∎

Example II.14. *Derangements and permutations without short cycles.* Classically, a derangement is defined as a permutation without fixed points, i.e., $\sigma_i \neq i$ for all i. Given an integer r, an r–derangement is a permutation all of whose cycles (in particular the shortest one) have length larger than r. Let $\mathcal{D}^{(r)}$ be the class of all r–derangements. A specification is

(41)
$$\mathcal{D}^{(r)} = \mathrm{SET}(\mathrm{CYC}_{>r}(\mathcal{Z})),$$

	Specification	EGF	coefficient	
Permutations:	$\text{SEQ}(\mathcal{Z})$	$\dfrac{1}{1-z}$	$n!$	(p. 104)
r cycles	$\text{SET}_r(\text{CYC}(\mathcal{Z}))$	$\dfrac{1}{r!}\left(\log\dfrac{1}{1-z}\right)^r$	$\begin{bmatrix} n \\ r \end{bmatrix}$	(p. 121)
involutions	$\text{SET}(\text{CYC}_{1..2}(\mathcal{Z}))$	$e^{z+z^2/2}$	$\approx n^{n/2}$	(pp. 122, 558)
all cycles $\leq r$	$\text{SET}(\text{CYC}_{1..r}(\mathcal{Z}))$	$\exp\left(\dfrac{z}{1}+\cdots+\dfrac{z^r}{r}\right)$	$\approx n^{1-1/r}$	(pp. 122, 568)
derangements	$\text{SET}(\text{CYC}_{>1}(\mathcal{Z}))$	$\dfrac{e^{-z}}{1-z}$	$\sim n!e^{-1}$	(pp. 122, 261)
all cycles $> r$	$\text{SET}(\text{CYC}_{>r}(\mathcal{Z}))$	$\dfrac{\exp\left(-\frac{z}{1}-\cdots-\frac{z^r}{r}\right)}{1-z}$	$\sim n!e^{-H_r}$	(pp. 123, 261)

Figure II.8. A summary of permutation enumerations.

the corresponding EGF then being

$$(42) \qquad D^{(r)}(z) = \exp\left(\sum_{j>r}\frac{z^j}{j}\right) = \frac{\exp(-\sum_{j=1}^{r}\frac{z^j}{j})}{1-z}.$$

For instance, when $r = 1$, a direct expansion yields

$$\frac{D_n^{(1)}}{n!} = 1 - \frac{1}{1!} + \frac{1}{2!} - \cdots + \frac{(-1)^n}{n!},$$

a truncation of the series expansion of $\exp(-1)$ that converges rapidly to e^{-1}. Phrased differently, this becomes a famous combinatorial problem with a pleasantly quaint nineteenth-century formulation [129]: "A number n of people go to the opera, leave their hats on hooks in the cloakroom and grab them at random when leaving; the probability that nobody gets back his own hat is asymptotic to $1/e$, which is nearly 37%." The usual proof uses inclusion–exclusion; see Section III. 7, p. 198 for both the classical and symbolic arguments. (It is a sign of changing times that Motwani and Raghavan [451, p. 11] describe the problem as one of sailors that return to their ship in a state of inebriation and choose random cabins to sleep in.)

For the generalized derangement problem, we have, for any fixed r (with H_r a harmonic number, p. 117),

$$(43) \qquad \frac{D_n^{(r)}}{n!} \sim e^{-H_r},$$

which is proved easily by complex asymptotic methods (Chapter IV, p. 261). .... ∎

Similar to several other structures that we have been considering previously, permutation allow for transparent connections between structural constraints and the forms of generating functions. The major counting results encountered in this section are summarized in Figure II.8.

▷ **II.13.** *Permutations such that* $\sigma^f = $ Id. Such permutations are "roots of unity" in the symmetric group. Their EGF is

$$\exp\left(\sum_{d \mid f} \frac{z^d}{d}\right),$$

where the sum extends to all divisors d of f.   ◁

▷ **II.14.** *Parity constraints in permutations.* The EGFs of permutations having only even-size cycles or odd-size cycles ($O(z)$) are, respectively,

$$E(z) = \exp\left(\frac{1}{2}\log\frac{1}{1-z^2}\right) = \frac{1}{\sqrt{1-z^2}}, \qquad O(z) = \exp\left(\frac{1}{2}\log\frac{1+z}{1-z}\right) = \sqrt{\frac{1+z}{1-z}}.$$

One finds $E_{2n} = (1 \cdot 3 \cdot 5 \cdots (2n-1))^2$ and $O_{2n} = E_{2n}$, $O_{2n+1} = (2n+1)E_{2n}$.

The EGFs of permutations having an even number of cycles ($E^*(z)$) and an odd number of cycles ($O^*(z)$) are, respectively,

$$E^*(z) = \cosh\left(\log\frac{1}{1-z}\right) = \frac{1}{2}\frac{1}{1-z} + \frac{1-z}{2}, \quad O^*(z) = \sinh\left(\log\frac{1}{1-z}\right) = \frac{1}{2}\frac{1}{1-z} + \frac{z-1}{2},$$

so that parity of the number of cycles is evenly distributed among permutations of size n as soon as $n \geq 2$. The generating functions obtained in this way are analogous to the ones appearing in the discussion of "Comtet's square", p. 111.   ◁

▷ **II.15.** *A hundred prisoners I.* This puzzle originates with a paper of Gál and Miltersen [275, 612]. A hundred prisoners, each uniquely identified by a number between 1 and 100, have been sentenced to death. The director of the prison gives them a last chance. He has a cabinet with 100 drawers (numbered 1 to 100). In each, he'll place at random a card with a prisoner's number (all numbers different). Prisoners will be allowed to enter the room one after the other and open, then close again, 50 drawers of their own choosing, but will not in any way be allowed to communicate with one another afterwards. The goal of each prisoner is to locate the drawer that contains his own number. If *all* prisoners succeed, then they will all be spared; if at least one fails, they will all be executed.

There are two mathematicians among the prisoners. The first one, a pessimist, declares that their overall chances of success are only of the order of $1/2^{100} \doteq 8 \cdot 10^{-31}$. The second one, a combinatorialist, claims he has a strategy for the prisoners, which has a greater than 30% chance of success. Who is right? [Note III.10, p. 176 provides a solution, but our gentle reader is advised to reflect on the problem for a few moments, before she jumps there.]   ◁

II. 4.2. Second-level structures.

Consider the three basic constructors of labelled sequences (SEQ), sets (SET), and cycles (CYC). We can play the formal game of examining what the various combinations produce as combinatorial objects. Restricting attention to superpositions of two constructors (an external one applied to an internal one) gives nine possibilities summarized by the table of Figure II.9.

The classes of surjections, alignments, set partitions, and permutations appear naturally as SEQ ∘ SET, SEQ ∘ CYC, SET ∘ SET, and SET ∘ CYC (top right corner). The others represent essentially non-classical objects. The case of the class $\mathcal{L} = $ SEQ(SEQ$_{\geq 1}(\mathcal{Z})$) describes objects that are (ordered) sequences of linear graphs; this can be interpreted as permutations with separators inserted, e.g, 53|264|1, or alternatively as integer compositions with a labelling superimposed, so that $L_n = n! \, 2^{n-1}$. The class $\mathcal{F} = $ SET(SEQ$_{\geq 1}(\mathcal{Z})$) corresponds to unordered collections of permutations; in other words, "fragments" are obtained by breaking a permutation into pieces

ext.\int.	SEQ$_{\geq 1}$	SET$_{\geq 1}$	CYC
	Labelled compositions ($\mathcal{L}$)	Surjections ($\mathcal{R}$)	Alignments ($\mathcal{O}$)
SEQ	SEQ ∘ SEQ $$\frac{1-z}{1-2z}$$	SEQ ∘ SET $$\frac{1}{2-e^z}$$	SEQ ∘ CYC $$\frac{1}{1-\log(1-z)^{-1}}$$
	Fragmented permutations ($\mathcal{F}$)	Set partitions ($\mathcal{S}$)	Permutations ($\mathcal{P}$)
SET	SET ∘ SEQ $$e^{z/(1-z)}$$	SET ∘ SET $$e^{e^z-1}$$	SET ∘ CYC $$\frac{1}{1-z}$$
	Supernecklaces ($\mathcal{S}^I$)	Supernecklaces ($\mathcal{S}^{II}$)	Supernecklaces ($\mathcal{S}^{III}$)
CYC	CYC ∘ SEQ $$\log\frac{1-z}{1-2z}$$	CYC ∘ SET $$\log(2-e^z)^{-1}$$	CYC ∘ CYC $$\log\frac{1}{1-\log(1-z)^{-1}}$$

Figure II.9. The nine second-level structures.

(pieces must be non-empty for definiteness). The interesting EGF is

$$F(z) = e^{z/(1-z)} = 1 + z + 3\frac{z^2}{2!} + 13\frac{z^3}{3!} + 73\frac{z^4}{4!} + \cdots,$$

(*EIS* **A000262**: "sets of lists"). The corresponding asymptotic analysis serves to illustrate an important aspect of the saddle-point method in Chapter VIII (p. 562). What we termed "supernecklaces" in the last row represents cyclic arrangements of composite objects existing in three brands.

All sorts of refinements, of which Figures II.8 and II.9 may give an idea, are clearly possible. We leave to the reader's imagination the task of determining which among the level 3 structures may be of combinatorial interest. . .

▷ **II.16.** *A meta-exercise: Counting specifications of level n.* The algebra of constructions satisfies the combinatorial isomorphism SET(CYC($\mathcal{X}$)) $\cong$ SEQ($\mathcal{X}$) for all $\mathcal{X}$. How many different terms involving n constructions can be built from three symbols CYC, SET, SEQ satisfying a semi-group law ("∘") together with the relation SET ∘ CYC = SEQ? This determines the number of specifications of level n. [Hint: the OGF is rational as normal forms correspond to words with an excluded pattern.] ◁

II. 5. Labelled trees, mappings, and graphs

In this section, we consider *labelled trees* as well as other important structures that are naturally associated with them. As in the unlabelled case considered in Section I. 6, p. 83, the corresponding combinatorial classes are inherently recursive, since a tree is obtained by appending a root to a collection (set or sequence) of subtrees. From here, it is possible to build the "functional graphs" associated to *mappings* from a finite set to itself—these decompose as sets of connected components that are cycles of trees. Variations of these construction finally open up the way to the enumeration of *graphs* having a fixed excess of the number of edges over the number of vertices.

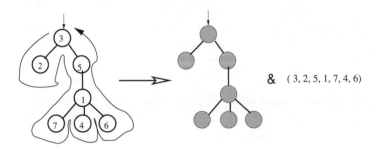

Figure II.10. A labelled plane tree is determined by an unlabelled tree (the "shape") and a permutation of the labels $1, \ldots, n$.

II.5.1. Trees. The trees to be studied here are labelled, meaning that nodes bear distinct integer labels. Unless otherwise specified, they are rooted, meaning as usual that one node is distinguished as the root. Labelled trees, like their unlabelled counterparts, exist in two varieties: (i) plane trees where an embedding in the plane is understood (or, equivalently, subtrees dangling from a node are ordered, say, from left to right); (ii) non-plane trees where no such embedding is imposed (such trees are then nothing but connected undirected acyclic graphs with a distinguished root). Trees may be further restricted by the additional constraint that the nodes' outdegrees should belong to a fixed set $\Omega \subseteq \mathbb{Z}_{\geq 0}$ where $\Omega \ni 0$.

Plane labelled trees. We first dispose of the plane variety of labelled trees. Let $\mathcal{A}$ be the set of (rooted labelled) plane trees constrained by Ω. This family is

$$\mathcal{A} = \mathcal{Z} \star \mathrm{SEQ}_{\Omega}(\mathcal{A}),$$

where $\mathcal{Z}$ represents the atomic class consisting of a single labelled node: $\mathcal{Z} = \{1\}$. The sequence construction appearing here reflects the planar embedding of trees, as subtrees stemming from a common root are ordered between themselves. Accordingly, the EGF $A(z)$ satisfies

$$A(z) = z\phi(A(z)) \qquad \text{where} \quad \phi(u) = \sum_{\omega \in \Omega} u^{\omega}.$$

This is exactly the same equation as the one satisfied by the *ordinary* GF of Ω–restricted *unlabelled* plane trees (see Proposition I.5, p. 66). Thus, $\frac{1}{n!} A_n$ is the number of unlabelled trees. In other words: *in the plane rooted case, the number of labelled trees equals $n!$ times the corresponding number of unlabelled trees.* As illustrated by Figure II.10, this is easily understood combinatorially: each labelled tree can be defined by its "shape" that is an unlabelled tree and by the sequence of node labels where nodes are traversed in some fixed order (preorder, say). In a way similar to Proposition I.5, p. 66, one has, by Lagrange inversion (Appendix A.6: *Lagrange Inversion*, p. 732):

$$A_n = n![z^n]A(z) = (n-1)![u^{n-1}]\phi(u)^n.$$

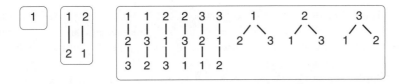

Figure II.11. There are $T_1 = 1$, $T_2 = 2$, $T_3 = 9$, and in general $T_n = n^{n-1}$ Cayley trees of size n.

This simple analytic–combinatorial relation enables us to transpose all of the enumeration results of Subsection I. 5.1, p. 65, to plane labelled trees, upon multiplying the evaluations by $n!$, of course. In particular, the total number of "general" plane labelled trees (with no degree restriction imposed, i.e., $\Omega = \mathbb{Z}_{\geq 0}$) is

$$n! \times \frac{1}{n}\binom{2n-2}{n-1} = \frac{(2n-2)!}{(n-1)!} = 2^{n-1}\left(1 \cdot 3 \cdots (2n-3)\right).$$

The corresponding sequence starts as $1, 2, 12, 120, 1680$ and is *EIS* **A001813**.

Non-plane labelled trees. We next turn to non-plane labelled trees (Figure II.11) to which the rest of this section will be devoted. The class $\mathcal{T}$ of all such trees is definable by a symbolic equation, which provides an implicit equation satisfied by the EGF:

$$(44) \qquad \mathcal{T} = \mathcal{Z} \star \text{SET}(\mathcal{T}) \qquad \Longrightarrow \qquad T(z) = ze^{T(z)}.$$

There the set construction translates the fact that subtrees stemming from the root are not ordered between themselves. From the specification (44), the EGF $T(z)$ is defined implicitly by the "functional equation"

$$(45) \qquad T(z) = ze^{T(z)}.$$

The first few values are easily found, for instance by the method of indeterminate coefficients:

$$T(z) = z + 2\frac{z^2}{2!} + 9\frac{z^3}{3!} + 64\frac{z^4}{4!} + 625\frac{z^5}{5!} + \cdots.$$

As suggested by the first few coefficients($9 = 3^2$, $64 = 4^3$, $625 = 5^4$), the general formula is

$$(46) \qquad T_n = n^{n-1}$$

which is established (as in the case of plane unlabelled trees) by Lagrange inversion:

$$(47) \qquad T_n = n!\,[z^n]T(z) = n!\left(\frac{1}{n}[u^{n-1}](e^u)^n\right) = n^{n-1}.$$

The enumeration result $T_n = n^{n-1}$ is a famous one, attributed to the prolific British mathematician Arthur Cayley (1821–1895) who had keen interest in combinatorial mathematics and published altogether over 900 papers and notes. Consequently, formula (46) given by Cayley in 1889 is often referred to as "Cayley's formula" and unrestricted non-plane labelled trees are often called "Cayley trees". See [67, p. 51] for a historical discussion. The function $T(z)$ is also known as the

(Cayley) "tree function"; it is a close relative of the W-function [131] defined implicitly by $We^W = z$, which was introduced by the Swiss mathematician Johann Lambert (1728–1777) otherwise famous for first proving the irrationality of the number π.

A similar process gives the number of (non-plane rooted) trees where all outdegrees of nodes are restricted to lie in a set Ω. This corresponds to the specification

$$\mathcal{T}^{(\Omega)} = \mathcal{Z} \star \mathrm{SET}_\Omega(\mathcal{T}^{(\Omega)}) \quad \Longrightarrow \quad T^{(\Omega)}(z) = z\overline{\phi}(T^{(\Omega)}(z)), \quad \overline{\phi}(u) := \sum_{\omega \in \Omega} \frac{u^\omega}{\omega!}.$$

What the last formula involves is the "exponential characteristic" of the degree sequence (as opposed to the ordinary characteristic, in the planar case). It is once more amenable to Lagrange inversion. In summary:

Proposition II.5. *The number of rooted non-plane trees, where all nodes have outdegree in Ω, is*

$$T_n^{(\Omega)} = (n-1))^n \quad \text{where} \quad \overline{\phi}(u) = \sum_{\omega \in \Omega} \frac{u^\omega}{\omega!}.$$

In particular, when all node degrees are allowed, i.e., when $\Omega \equiv \mathbb{Z}_{\geq 0}$, the number of trees is $T_n = n^{n-1}$ and its EGF is the Cayley tree function satisfying $T(z) = ze^{T(z)}$.

As in the unlabelled case (p. 66), we refer to a class of labelled trees defined by degree restrictions as a *simple variety of trees*: its EGF satisfies an equation of the form $y = z\phi(y)$.

▷ **II.17.** *Prüfer's bijective proofs of Cayley's formula.* The simplicity of Cayley's formula calls for a combinatorial explanation. The most famous one is due to Prüfer (in 1918). It establishes as follows a bijective correspondence between unrooted Cayley trees whose number is n^{n-2} for size n and sequences $(a_1, \ldots, a_{n-2})$ with $1 \leq a_j \leq n$ for each j. Given an unrooted tree τ, remove the endnode (and its incident edge) with the smallest label; let a_1 denote the label of the node that was joined to the removed node. Continue with the pruned tree τ' to get a_2 in a similar way. Repeat the construction of the sequence until the tree obtained only consists of a single edge. For instance:

$$\longrightarrow \quad (4, 8, 4, 8, 8, 4).$$

It can be checked that the correspondence is bijective; see [67, p. 53] or [445, p. 5].   ◁

▷ **II.18.** *Forests.* The number of unordered k–forests (i.e., k–sets of trees) is

$$F_n^{(k)} = n![z^n]\frac{T(z)^k}{k!} = \frac{(n-1)!}{(k-1)!}[u^{n-k}](e^u)^n = \binom{n-1}{k-1}n^{n-k},$$

as follows from Bürmann's form of Lagrange inversion, relative to powers (p. 66).   ◁

▷ **II.19.** *Labelled hierarchies.* The class $\mathcal{L}$ of *labelled hierarchies* is formed of trees whose internal nodes are unlabelled and are constrained to have outdegree larger than 1, while their leaves have labels attached to them. As for other labelled structures, size is the number of labels (internal nodes do not contribute). Hierarchies satisfy the specification (compare with p. 72)

$$\mathcal{L} = \mathcal{Z} + \mathrm{SET}_{\geq 2}(\mathcal{L}), \quad \Longrightarrow \quad L = z + e^L - 1 - L.$$

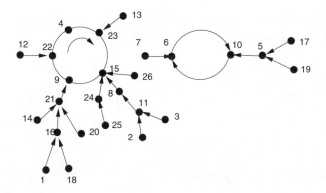

Figure II.12. A functional graph of size $n = 26$ associated to the mapping φ such that $\varphi(1) = 16$, $\varphi(2) = \varphi(3) = 11$, $\varphi(4) = 23$, and so on.

This happens to be solvable in terms of the Cayley function: $L(z) = T(\frac{1}{2}e^{z/2-1/2}) + \frac{z}{2} - \frac{1}{2}$. The first few values are 0, 1, 4, 26, 236, 2752 (*EIS* **A000311**): these numbers count phylogenetic trees, used to describe the evolution of a genetically-related group of organisms, and they correspond to Schröder's "fourth problem" [129, p. 224]. The asymptotic analysis is done in Example VII.12, p. 472.

The class of binary (labelled) hierarchies defined by the additional fact that internal nodes can have degree 2 only is expressed by

$$\mathcal{M} = \mathcal{Z} + \text{SET}_2(\mathcal{M}) \qquad \Longrightarrow \qquad M(z) = 1 - \sqrt{1 - 2z} \quad \text{and} \quad M_n = 1 \cdot 3 \cdots (2n - 3),$$

where the counting numbers are now, surprisingly perhaps, the odd factorials. $\triangleleft$

II. 5.2. Mappings and functional graphs. Let $\mathcal{F}$ be the class of mappings (or "functions") from $[1 .. n]$ to itself. A mapping $f \in [1 .. n] \mapsto [1 .. n]$ can be represented by a directed graph over the set of vertices $[1 .. n]$ with an edge connecting x to $f(x)$, for all $x \in [1 .. n]$. The graphs so obtained are called *functional graphs* and they have the characteristic property that the outdegree of each vertex is exactly equal to 1.

Mappings and associated graphs. Given a mapping (or function) f, upon starting from any point x_0, the succession of (directed) edges in the graph traverses the vertices corresponding to iterated values of the mapping,

$$x_0, \quad f(x_0), \quad f(f(x_0)), \ldots .$$

Since the domain is finite, each such sequence must eventually loop back on itself. When the operation is repeated, starting each time from an element not previously hit, the vertices group themselves into (weakly connected) components. This leads to a valuable characterization of functional graphs (Figure II.12): *a functional graph is a set of connected functional graphs; a connected functional graph is a collection of rooted trees arranged in a cycle.* (This decomposition is seen to extend the decomposition of permutations into cycles, p. 120.)

Thus, with $\mathcal{T}$ being as before the class of all Cayley trees, and with $\mathcal{K}$ the class of all connected functional graphs, we have the specification:

$$
(48) \quad
\begin{cases}
\mathcal{F} = \text{SET}(\mathcal{K}) \\
\mathcal{K} = \text{CYC}(\mathcal{T}) \\
\mathcal{T} = \mathcal{Z} \star \text{SET}(\mathcal{T})
\end{cases}
\implies
\begin{cases}
F(z) = e^{K(z)} \\
K(z) = \log \dfrac{1}{1 - T(z)} \\
T(z) = z e^{T(z)}.
\end{cases}
$$

What is especially interesting here is a specification binding three types of related structures. From (48), the EGF $F(z)$ is found to satisfy $F = (1 - T)^{-1}$. It can be checked from this, by Lagrange inversion once again (p. 733), that we have

$$(49) \qquad\qquad\qquad F_n = n^n,$$

as was to be expected (!) from the origin of the problem. More interestingly, Lagrange inversion also gives the number of connected functional graphs (expand $\log(1 - T)^{-1}$ and recover coefficients by Bürmann's form, p. 66):

$$(50) \quad K_n = n^{n-1} Q(n) \quad \text{where} \quad Q(n) := 1 + \frac{n-1}{n} + \frac{(n-1)(n-2)}{n^2} + \cdots.$$

The quantity $Q(n)$ that appears in (50) is a famous one that surfaces in many problems of discrete mathematics (including the birthday paradox, Equation (27), p. 115). Knuth has proposed naming it "Ramanujan's Q-function" as it already appears in the first letter of Ramanujan to Hardy in 1913. The asymptotic analysis is elementary and involves developing a continuous approximation of the general term and approximating the resulting Riemann sum by an integral: this is an instance of the Laplace method for sums briefly explained in Appendix B.6: *Laplace's method*, p. 755 (see also [377, Sec. 1.2.11.3] and [538, Sec. 4.7]). In fact, very precise estimates come out naturally from an analysis of the singularities of the EGF $K(z)$, as we shall see in Chapters VI (p. 416) and VII (p. 449). The net result is

$$K_n \sim n^n \sqrt{\frac{\pi}{2n}},$$

so that a fraction about $1/\sqrt{n}$ of all the graphs consist of a single component.

Constrained mappings. As is customary with the symbolic method, basic constructions open the way to a large number of related counting results (Figure II.13). First, by an adaptation of (48), the mappings without fixed points, $(\forall x : f(x) \neq x)$ and those without 1, 2–cycles, (additionally, $\forall x : f(f(x)) \neq x$), have EGFs, respectively,

$$\frac{e^{-T(z)}}{1 - T(z)}, \qquad \frac{e^{-T(z) - T^2(z)/2}}{1 - T(z)}.$$

The first term is consistent with what a direct count yields, namely $(n - 1)^n$, which is asymptotic to $e^{-1} n^n$, so that the fraction of mappings without fixed point is asymptotic to e^{-1}. The second one lends itself easily to complex asymptotic methods that give

$$n! [z^n] \frac{e^{-T - T^2/2}}{1 - T} \sim e^{-3/2} n^n,$$

	EGF	coefficient	
Mappings:	$\dfrac{1}{1-T}$	n^n	(p. 130)
connected	$\log \dfrac{1}{1-T}$	$\sim n^n \sqrt{\dfrac{\pi}{2n}}$	(pp. 130, 449)
no fixed-point	$\dfrac{e^{-T}}{1-T}$	$\sim e^{-1} n^n$	(p. 130)
idempotent	e^{ze^z}	$\approx \dfrac{n^n}{(\log n)^n}$	(pp. 131, 571)
partial	$\dfrac{e^T}{1-T}$	$\sim e\, n^n$	(p. 132)

Figure II.13. A summary of various counting results relative to mappings, with $T \equiv T(z)$ the Cayley tree function. (Bijections, surjections, involutions, and injections are covered by previous constructions.)

and the proportion is asymptotic to $e^{-3/2}$. These two particular estimates are of the same form as that found for permutations (the generalized derangements, Equation (43)). Such facts are not quite obvious by elementary probabilistic arguments, but they are neatly explained by the singular theory of combinatorial schemas developed in Part B of this book.

Next, idempotent mappings, i.e., ones satisfying $f(f(x)) = f(x)$ for all x, correspond to $\mathcal{I} \cong \mathrm{SET}(\mathcal{Z} \star \mathrm{SET}(\mathcal{Z}))$, so that

$$I(z) = e^{ze^z} \qquad \text{and} \qquad I_n = \sum_{k=0}^{n} \binom{n}{k} k^{n-k}.$$

(The specification translates the fact that idempotent mappings can have only cycles of length 1 on which are grafted sets of direct antecedents.) The latter sequence is *EIS* **A000248**, which starts as 1,1,3,10,41,196,1057. An asymptotic estimate can be derived either from the Laplace method or, better, from the saddle-point method expounded in Chapter VIII (p. 571).

Several analyses of this type are of relevance to cryptography and the study of random number generators. For instance, the fact that a random mapping over $[1 \mathinner{.\,.} n]$ tends to reach a cycle in $O(\sqrt{n})$ steps (Subsection VII. 3.3, p. 462) led Pollard to design a surprising Monte Carlo integer factorization algorithm; see [378, p. 371] and [538, Sec 8.8], as well as our discussion in Example VII.11, p. 465. This algorithm, once suitably optimized, first led to the factorization of the Fermat number $F_8 = 2^{2^8} + 1$ obtained by Brent in 1980.

▷ **II.20.** *Binary mappings.* The class $\mathcal{BF}$ of binary mappings, where each point has either 0 or 2 preimages, is specified by

$$\mathcal{BF} = \mathrm{SET}(\mathcal{K}), \ \mathcal{K} = \mathrm{CYC}(\mathcal{P}), \ \mathcal{P} = \mathcal{Z} \star \mathcal{B}, \ \mathcal{B} = \mathcal{Z} \star \mathrm{SET}_{0,2}(\mathcal{B})$$

(planted trees $\mathcal{P}$ and binary trees $\mathcal{B}$ are needed), so that

$$BF(z) = \frac{1}{\sqrt{1 - 2z^2}}, \qquad BF_{2n} = \frac{((2n)!)^2}{2^n (n!)^2}.$$

The class $\mathcal{BF}$ is an approximate model of the behaviour of (modular) quadratic functions under iteration. See [18, 247] for a general enumerative theory of random mappings including degree-restricted ones.   ◁

▷ **II.21. *Partial mappings.*** A partial mapping may be undefined at some points, and at those we consider it takes a special value, ⊥. The iterated preimages of ⊥ form a forest, while the remaining values organize themselves into a standard mapping. The class $\mathcal{PF}$ of partial mappings is thus specified by $\mathcal{PF} = \text{SET}(\mathcal{T}) \star \mathcal{F}$, so that

$$
PF(z) = \frac{e^{T(z)}}{1 - T(z)} \qquad \text{and} \qquad PF_n = (n+1)^n.
$$

This construction lends itself to all sorts of variations. For instance, the class PFI of *injective* partial maps is described as sets of chains of linear and circular graphs, $PFI = \text{SET}(\text{CYC}(\mathcal{Z}) + \text{SEQ}_{\geq 1}(\mathcal{Z}))$, so that

$$
PFI(z) = \frac{1}{1-z} e^{z/(1-z)}, \qquad PFI_n = \sum_{i=0}^{n} i! \binom{n}{i}^2.
$$

(This is a symbolic rewriting of part of the paper [78]; see Example VIII.13, p. 596, for asymptotics.)   ◁

II.5.3. Labelled graphs.

Random graphs form a major chapter of the theory of random discrete structures [76, 355]. We examine here enumerative results concerning graphs of low "complexity", that is, graphs which are very nearly trees. (Such graphs for instance play an essential rôle in the analysis of early stages of the evolution of a random graph, when edges are successively added, as shown in [241, 354].)

Unrooted trees and acyclic graphs. The simplest of all connected graphs are certainly the ones that are acyclic. These are trees, but contrary to the case of Cayley trees, no root is specified. Let $\mathcal{U}$ be the class of all *unrooted* trees. Since a rooted tree (rooted trees are, as we know, counted by $T_n = n^{n-1}$) is an unrooted tree combined with a choice of a distinguished node (there are n such possible choices for trees of size n), one has

$$
T_n = n U_n \qquad \text{implying} \qquad U_n = n^{n-2}.
$$

At generating function level, this combinatorial equality translates into

$$
U(z) = \int_0^z T(w) \frac{dw}{w},
$$

which integrates to give (take T as the independent variable)

$$
U(z) = T(z) - \frac{1}{2} T(z)^2.
$$

Since $U(z)$ is the EGF of acyclic connected graphs, the quantity

$$
A(z) = e^{U(z)} = e^{T(z) - T(z)^2/2}
$$

is the EGF of all acyclic graphs. (Equivalently, these are unordered forests of unrooted trees; the sequence is *EIS* **A001858**: 1, 1, 2, 7, 38, 291, ...) Singularity analysis methods (Note VI.14, p. 406) imply the estimate $A_n \sim e^{1/2} n^{n-2}$. Surprisingly, perhaps, there are barely more acyclic graphs than unrooted trees—such phenomena are easily explained by singularity analysis.

Unicyclic graphs. The *excess* of a graph is defined as the difference between the number of edges and the number of vertices. For a connected graph, this quantity must be at least -1, this minimal value being precisely attained by unrooted trees. The class $\mathcal{W}_k$ is the class of connected graphs of excess equal to k; in particular $\mathcal{U} = \mathcal{W}_{-1}$. The successive classes $\mathcal{W}_{-1}, \mathcal{W}_0, \mathcal{W}_1, \ldots$, may be viewed as describing connected graphs of increasing complexity.

The class $\mathcal{W}_0$ comprises all connected graphs with the number of edges equal to the number of vertices. Equivalently, a graph in $\mathcal{W}_0$ is a connected graph with exactly one cycle (a sort of "eye"), and for that reason, elements of $\mathcal{W}_0$ are sometimes referred to as "unicyclic components" or "unicycles". In a way, such a graph looks very much like an undirected version of a connected functional graph. In precise terms, a graph of $\mathcal{W}_0$ consists of a cycle of length at least 3 (by definition, graphs have neither loops nor multiple edges) that is undirected (the orientation present in the usual cycle construction is killed by identifying cycles isomorphic up to reflection) and on which are grafted trees (these are implicitly rooted by the point at which they are attached to the cycle). With UCYC representing the (new) undirected cycle construction, one thus has

$$\mathcal{W}_0 \cong \text{UCYC}_{\geq 3}(\mathcal{T}).$$

We claim that this construction is reflected by the EGF equation

$$(51) \qquad W_0(z) = \frac{1}{2} \log \frac{1}{1 - T(z)} - \frac{1}{2} T(z) - \frac{1}{4} T(z)^2.$$

Indeed one has the isomorphism

$$\mathcal{W}_0 + \mathcal{W}_0 \cong \text{CYC}_{\geq 3}(\mathcal{T}),$$

since we may regard the two disjoint copies on the left as instantiating two possible orientations of the undirected cycle. The result of (51) then follows from the usual translation of the cycle construction—it is originally due to the Hungarian probabilist Rényi in 1959. Asymptotically, one finds (using methods of Chapter VI, p. 406):

$$(52) \qquad n![z^n] W_0 \sim \frac{1}{4} \sqrt{2\pi} n^{n-1/2}.$$

(The sequence starts as 0, 0, 1, 15, 222, 3660, 68295 and is *EIS* **A057500**.)

Finally, the number of graphs made only of trees and unicyclic components has EGF

$$e^{W_{-1}(z) + W_0(z)} = \frac{e^{T/2 - 3T^2/4}}{\sqrt{1 - T}},$$

which asymptotically yields $n![z^n] e^{W_{-1} + W_0} \sim \Gamma(3/4)(2e)^{-1/4} \pi^{-1/2} n^{n-1/4}$. Such graphs stand just next to acyclic graphs in order of structural complexity. They are the undirected counterparts of functional graphs encountered in the previous subsection.

▷ **II.22.** *2–Regular graphs.* This is based on Comtet's account [129, Sec. 7.3]. A *2-regular graph* is an undirected graph in which each vertex has degree exactly 2. Connected 2–regular graphs are thus undirected cycles of length $n \geq 3$, so that their class $\mathcal{R}$ satisfies

$$(53) \qquad \mathcal{R} = \text{SET}(\text{UCYC}_{\geq 3}(\mathcal{Z})) \qquad \Longrightarrow \qquad R(z) = \frac{e^{-z/2 - z^2/4}}{\sqrt{1 - z}}.$$

	EGF	coefficient
Graphs:		$2^{n(n-1)/2}$
acyclic, connected	$U \equiv W_{-1} = T - T^2/2$	n^{n-2}
acyclic (forest)	$A = e^{T - T^2/2}$	$\sim e^{1/2} n^{n-2}$
unicycle	$W_0 = \dfrac{1}{2} \log \dfrac{1}{1-T} - \dfrac{T}{2} - \dfrac{T^2}{4}$	$\sim \frac{1}{4}\sqrt{2\pi}\, n^{n-1/2}$
set of trees & unicycles	$B = \dfrac{e^{T/2 - 3T^2/4}}{\sqrt{1-T}}$	$\sim \Gamma(3/4)\dfrac{(2e)^{-1/4}}{\sqrt{\pi}} n^{n-1/4}$
connected, excess k	$W_k = \dfrac{P_k(T)}{(1-T)^{3k}}$	$\sim \dfrac{P_k(1)\sqrt{2\pi}}{2^{3k/2}\Gamma(3k/2)} n^{n+(3k-1)/2}$

Figure II.14. A summary of major enumeration results relative to labelled graphs. The asymptotic estimates result from singularity analysis (Note VI.14, p. 406).

Given n straight lines in general position in the plane, a *cloud* is defined to be a set of n intersection points, no three being collinear. Clouds and 2–regular graphs are equinumerous. [Hint: Use duality.] The asymptotic analysis will serve as a prime example of the singularity analysis process (Examples VI.1, p. 379 and VI.2, p. 395).

The general enumeration of r–regular graphs becomes somewhat more difficult as soon as $r > 2$. Algebraic aspects are discussed in [289, 303] while Bender and Canfield [39] have determined the asymptotic formula (for rn even)

$$(54) \qquad R_n^{(r)} \sim \sqrt{2} e^{(r^2-1)/4} \frac{r^{r/2}}{e^{r/2} r!} n^{rn/2},$$

for the number of r–regular graphs of size n. (See also Example VIII.9, p. 583, for regular multigraphs.) ◁

Graphs of fixed excess. The previous discussion suggests considering more generally the enumeration of connected graphs according to excess. E. M. Wright made important contributions in this area [620, 621, 622] that are revisited in the famous "giant paper on the giant component" by Janson, Knuth, Łuczak, and Pittel [354]. Wright's result are summarized by the following proposition.

Proposition II.6. *The EGF $W_k(z)$ of connected graphs with excess (of edges over vertices) equal to k is, for $k \geq 1$, of the form*

$$(55) \qquad W_k(z) = \frac{P_k(T)}{(1-T)^{3k}}, \quad T \equiv T(z),$$

where P_k is a polynomial of degree $3k + 2$. For any fixed k, as $n \to \infty$, one has

$$(56) \qquad W_{k,n} = n![z^n]W_k(z) = \frac{P_k(1)\sqrt{2\pi}}{2^{3k/2}\Gamma(3k/2)} n^{n+(3k-1)/2} \left(1 + O(n^{-1/2})\right).$$

The combinatorial part of the proof (see Note II.23 below) is an interesting exercise in *graph surgery* and symbolic methods. The analytic part of the statement follows straightforwardly from singularity analysis. The polynomials $P(T)$ and the constants $P_k(1)$ are determined by an explicit nonlinear recurrence; one finds for instance:

$$W_1 = \frac{1}{24} \frac{T^4(6-T)}{(1-T)^3}, \qquad W_2 = \frac{1}{48} \frac{T^4(2+28T-23T^2+9T^3-T^4)}{(1-T)^6}.$$

▷ **II.23.** *Wright's surgery.* The full proof of Proposition II.6 by symbolic methods requires the notion of *pointing* in conjunction with multivariate generating function techniques of Chapter III. It is convenient to define $w_k(z, y) := y^k W_k(zy)$, which is a bivariate generating function with y marking the number of edges. Pick up an edge in a connected graph of excess $k + 1$, then remove it. This results either in a connected graph of excess k with two pointed vertices (and no edge in between) or in two connected components of respective excess h and $k - h$, each with a pointed vertex. Graphically (with connected components in grey):

This translates into the differential recurrence on the w_k ($\partial_x := \frac{\partial}{\partial x}$),

$$2\partial_y w_{k+1} = \left(z^2 \partial_z^2 w_k - 2y\partial_y w_k\right) + \sum_{h=-1}^{k+1} (z\partial_z w_h) \cdot (z\partial_z w_{k-h}),$$

and similarly for $W_k(z) = w_k(z, 1)$. From here, it can be verified by induction that each W_k is a rational function of $T \equiv W_{-1}$. (See Wright's original papers [620, 621, 622] or [354] for details; constants related to the $P_k(1)$ occur in Subsection VII. 10.1, p. 532.) ◁

As explained in the giant paper [354], such results combined with complex analytic techniques provide, with great detail, information about a random graph $\Gamma(n, m)$ with n nodes and m edges. In the sparse case where m is of the order of n, one finds the following properties to hold "*with high probability*" (w.h.p.)[7]; that is, with probability tending to 1 as $n \to \infty$.

- For $m = \mu n$, with $\mu < \frac{1}{2}$, the random graph $\Gamma(m, n)$ has w.h.p. only tree and unicycle components; the largest component is w.h.p. of size $O(\log n)$.
- For $m = \frac{1}{2}n + O(n^{2/3})$, w.h.p. there appear one or several semi-giant components that have size $O(n^{2/3})$.
- For $m = \mu n$, with $\mu > \frac{1}{2}$, there is w.h.p. a unique giant component of size proportional to n.

In each case, refined estimates follow from a detailed analysis of corresponding generating functions, which is a main theme of [241] and especially [354]. Raw forms of these results were first obtained by Erdős and Rényi who launched the subject in a famous series of papers dating from 1959–60; see the books [76, 355] for a probabilistic context and the paper [40] for the finest counting estimates available. In contrast, the enumeration of *all* connected graphs (irrespective of the number of edges, that is, without excess being taken into account) is a relatively easy problem treated in the

---

[7]Synonymous expressions are "asymptotically almost surely" (a.a.s) and "in probability". The term "almost surely" is sometimes used, though it lends itself to confusion with properties of continuous measures.

next section. Many other classical aspects of the enumerative theory of graphs are covered in the book *Graphical Enumeration* by Harary and Palmer [319].

▷ **II.24.** *Graphs are not specifiable.* The class of *all* graphs does not admit a specification that starts from single atoms and involves only sums, products, sets and cycles. Indeed, the growth of G_n is such that the EGF $G(z)$ has radius of convergence 0, whereas EGFs of constructible classes must have a non-zero radius of convergence. (Section IV. 4, p. 249, provides a detailed proof of this fact for iterative structures; for recursively specified classes, this is a consequence of the analysis of inverse functions, p. 402, and systems, p. 489, with suitable adaptations based on the technique of majorant series. p. 250.)   ◁

II. 6. Additional constructions

As in the unlabelled case, pointing and substitution are available in the world of labelled structures (Subsection II. 6.1), and implicit definitions enlarge the scope of the symbolic method (Subsection II. 6.2). The inversion process needed to enumerate implicit structures is even simpler, since in the labelled universe sets and cycles have more concise translations as operators over EGF. Finally, and this departs significantly from Chapter I, the fact that integer labels are naturally ordered makes it possible to take into account certain order properties of combinatorial structures (Subsection II. 6.3).

II. 6.1. Pointing and substitution. The pointing operation is entirely similar to its unlabelled counterpart since it consists in distinguishing one atom among all the ones that compose an object of size n. The definition of composition for labelled structures is however a bit more subtle as it requires singling out "leaders" in components.

Pointing. The *pointing* of a class $\mathcal{B}$ is defined by

$$\mathcal{A} = \Theta\mathcal{B} \qquad \text{iff} \qquad \mathcal{A}_n = [1 . . n] \times \mathcal{B}_n.$$

In other words, in order to generate an element of $\mathcal{A}$, select one of the n labels and point at it. Clearly

$$A_n = n \cdot B_n \implies A(z) = z\frac{d}{dz}B(z).$$

Substitution (composition). Composition or *substitution* can be introduced so that it corresponds *a priori* to composition of generating functions. It is formally defined as

$$\mathcal{B} \circ \mathcal{C} = \sum_{k=0}^{\infty} \mathcal{B}_k \times \text{SET}_k(\mathcal{C}),$$

so that its EGF is

$$\sum_{k=0}^{\infty} B_k \frac{(C(z))^k}{k!} = B(C(z)).$$

A combinatorial way of realizing this definition and forming an arbitrary object of $\mathcal{B} \circ \mathcal{C}$, is as follows. First select an element of $\beta \in \mathcal{B}$ called the "base" and let $k = |\beta|$ be its size; then pick up a k–set of elements of $\mathcal{C}$; the elements of the k–set are naturally ordered by the value of their "leader" (the *leader* of an object being by convention the value of its smallest label); the element with leader of rank r is then substituted to the node labelled by value r of β. Gathering the above, we obtain:

Theorem II.3. *The combinatorial constructions of pointing and substitution are admissible*

$$\mathcal{A} = \Theta\mathcal{B} \quad \Longrightarrow \quad A(z) = z\partial_z B(z), \qquad \partial_z \equiv \frac{d}{dz}$$

$$\mathcal{A} = \mathcal{B} \circ \mathcal{C} \quad \Longrightarrow \quad A(z) = B(C(z)).$$

For instance, the EGF of (relabelled) pairings of elements drawn from $\mathcal{C}$ is

$$e^{C(z)^2/2},$$

since the EGF of involutions without fixed points is $e^{z^2/2}$.

▷ **II.25.** *Standard constructions based on substitutions.* The sequence class of $\mathcal{A}$ may be defined by composition as $\mathcal{P} \circ \mathcal{A}$ where $\mathcal{P}$ is the set of all permutations. The set class of $\mathcal{A}$ may be defined as $\mathcal{U} \circ \mathcal{A}$ where $\mathcal{U}$ is the class of all urns. Similarly, cycles are obtained by substitution into circular graphs. Thus,

$$\text{SEQ}(\mathcal{A}) \cong \mathcal{P} \circ \mathcal{A}, \qquad \text{SET}(\mathcal{A}) \cong \mathcal{U} \circ \mathcal{A}, \qquad \text{CYC}(\mathcal{A}) \cong \mathcal{C} \circ \mathcal{A}.$$

In this way, permutation, urns and circle graphs appear as archetypal classes in a development of combinatorial analysis based on composition. (Joyal's "theory of species" [359] and the book by Bergeron, Labelle, and Leroux [50] show that a far-reaching theory of combinatorial enumeration can be based on the concept of substitution.) ◁

▷ **II.26.** *Distinct component sizes.* The EGFs of permutations with cycles of distinct lengths and of set partitions with parts of distinct sizes are

$$\prod_{n=1}^{\infty}\left(1 + \frac{z^n}{n}\right), \qquad \prod_{n=1}^{\infty}\left(1 + \frac{z^n}{n!}\right).$$

The probability that a permutation of $\mathcal{P}_n$ has distinct cycle sizes tends to $e^{-\gamma}$; see [309, Sec. 4.1.6] for a Tauberian argument and [495] for precise asymptotics. The corresponding analysis for set partitions is treated in the seven-author paper [368]. ◁

II. 6.2. Implicit structures. Let $\mathcal{X}$ be a labelled class implicitly characterized by either of the combinatorial equations

$$\mathcal{A} = \mathcal{B} + \mathcal{X}, \qquad \mathcal{A} = \mathcal{B} \star \mathcal{X}.$$

Then, solving the corresponding EGF equations leads to

$$X(z) = A(z) - B(z), \qquad X(z) = \frac{A(z)}{B(z)},$$

respectively. For the composite labelled constructions SEQ, SET, CYC, the algebra is equally easy.

Theorem II.4 (Implicit specifications). *The generating functions associated with the implicit equations in $\mathcal{X}$*

$$\mathcal{A} = \text{SEQ}(\mathcal{X}), \qquad \mathcal{A} = \text{SET}(\mathcal{X}), \qquad \mathcal{A} = \text{CYC}(\mathcal{X}),$$

are, respectively,

$$X(z) = 1 - \frac{1}{A(z)}, \qquad X(z) = \log A(z), \qquad X(z) = 1 - e^{-A(z)}.$$

***Example* II.15.**   *Connected graphs.* In the context of graphical enumerations, the labelled set construction takes the form of an enumerative formula relating a class $\mathcal{G}$ of graphs and the subclass $\mathcal{K} \subset \mathcal{G}$ of its connected graphs:

$$\mathcal{G} = \text{SET}(\mathcal{K}) \qquad \Longrightarrow \qquad G(z) = e^{K(z)}.$$

This basic formula is known in graph theory [319] as the *exponential formula*.

Consider the class $\mathcal{G}$ of all (undirected) labelled graphs, the size of a graph being the number of its nodes. Since a graph is determined by the choice of its set of edges, there are $\binom{n}{2}$ potential edges each of which may be taken in or out, so that $G_n = 2^{\binom{n}{2}}$. Let $\mathcal{K} \subset \mathcal{G}$ be the subclass of all connected graphs. The exponential formula determines $K(z)$ implicitly:

$$
\begin{aligned}
K(z) \;&=\; \log\left(1 + \sum_{n \geq 1} 2^{\binom{n}{2}} \frac{z^n}{n!}\right) \\[2mm]
&=\; z + \frac{z^2}{2!} + 4\frac{z^3}{3!} + 38\frac{z^4}{4!} + 728\frac{z^5}{5!} + \cdots ,
\end{aligned}
$$

(57)

where the sequence is *EIS* **A001187**. The series is divergent, that is, it has radius of convergence 0. It can nonetheless be manipulated as a formal series (Appendix A.5: *Formal power series*, p. 730). Expanding by means of $\log(1 + u) = u - u^2/2 + \cdots$, yields a complicated convolution expression for K_n:

$$K_n = 2^{\binom{n}{2}} - \frac{1}{2}\sum\binom{n}{n_1, n_2} 2^{\binom{n_1}{2}+\binom{n_2}{2}} + \frac{1}{3}\sum\binom{n}{n_1, n_2, n_3} 2^{\binom{n_1}{2}+\binom{n_2}{2}+\binom{n_3}{2}} - \cdots .$$

(The kth term is a sum over $n_1 + \cdots + n_k = n$, with $0 < n_j < n$.) Given the very fast increase of G_n with n, for instance

$$2^{\binom{n+1}{2}} = 2^n\, 2^{\binom{n}{2}},$$

a detailed analysis of the various terms of the expression of K_n shows predominance of the first sum, and, in that sum itself, the extreme terms corresponding to $n_1 = n - 1$ or $n_2 = n - 1$ predominate, so that

(58) $$K_n = 2^{\binom{n}{2}}\left(1 - 2n2^{-n} + o(2^{-n})\right).$$

Thus: *almost all labelled graphs of size n are connected.* In addition, the error term decreases very quickly: for instance, for $n = 18$, an exact computation based on the generating function formula reveals that a proportion only 0.0001373291074 of all the graphs are not connected—this is *extremely* close to the value 0.0001373291016 predicted by the main terms in the asymptotic formula (58). Notice that good use could be made here of a purely divergent generating function for asymptotic enumeration purposes. . . . . ■

$\triangleright$ **II.27.** *Bipartite graphs.* A plane bipartite graph is a pair (G, ω) where G is a labelled graph, $\omega = (\omega_W, \omega_E)$ is a bipartition of the nodes (into *West* and *East* categories), and the edges are such that they only connect nodes from ω_W to nodes of ω_E. A direct count shows that the EGF of plane bipartite graphs is

$$\Gamma(z) = \sum_n \gamma_n \frac{z^n}{n!} \text{ with } \gamma_n = \sum_k \binom{n}{k} 2^{k(n-k)}.$$

The EGF of plane bipartite graphs that are connected is $\log \Gamma(z)$.

A bipartite graph is a labelled graph whose nodes can be partitioned into two groups so that edges only connect nodes of different groups. The EGF of bipartite graphs is

$$\exp\left(\frac{1}{2}\log\Gamma(z)\right) = \sqrt{\Gamma(z)}.$$

[Hint. The EGF of a connected bipartite graph is $\frac{1}{2} \log \Gamma(z)$, since a factor of $\frac{1}{2}$ kills the East–West orientation present in a connected plane bipartite graph. See Wilf's book [608, p. 78] for details.] ◁

▷ **II.28.** *Do two permutations generate the symmetric group?* To two permutations σ, τ of the same size, associate a graph $\Gamma_{\sigma,\tau}$ whose set vertices is $V = [1 .. n]$, if $n = |\sigma| = |\tau|$, and set of edges is formed of all the pairs $(x, \sigma(x))$, $(x, \tau(x))$, for $x \in V$. The probability that a random $\Gamma_{\sigma,\tau}$ is connected is

$$\pi_n = \frac{1}{n!} [z^n] \log \left(\sum_{n \geq 0} n! z^n \right).$$

This represents the probability that two permutations generate a transitive group (that is for all $x, y \in [0 .. n]$, there exists a composition of $\sigma, \sigma^{-1}, \tau, \tau^{-1}$ that maps x to y). One has

$$\text{(59)} \qquad \pi_n \sim 1 - \frac{1}{n} - \frac{1}{n^2} - \frac{4}{n^3} - \frac{23}{n^4} - \frac{171}{n^5} - \frac{1542}{n^6} - \cdots,$$

Surprisingly, the coefficients $1, 1, 4, 23, \ldots$ (*EIS* **A084357**) in the asymptotic formula (59) enumerate a "third-level" structure (Subsection II. 4.2, p. 124 and Note VIII.15, p. 571), namely: $\text{SET}(\text{SET}_{\geq 1}(\text{SEQ}_{\geq 1}(\mathcal{Z})))$. In addition, one has $n!^2 \pi_n = (n-1)! I_n$, where I_{n+1} is the number of indecomposable permutations (Example I.19, p. 89).

Let $\pi_n^\star$ be the probability that two random permutations generate the whole symmetric group. Then, by a result of Babai based on the classification of groups, the quantity $\pi_n - \pi_n^\star$ is exponentially small, so that (59) also applies to $\pi_n^\star$; see Dixon [167]. ◁

II. 6.3. Order constraints.

A construction well-suited to dealing with many of the order properties of combinatorial structures is the modified labelled product:

$$\mathcal{A} = (\mathcal{B}^\square \star \mathcal{C}).$$

This denotes the subset of the product $\mathcal{B} \star \mathcal{C}$ formed with elements such that the smallest label is constrained to lie in the $\mathcal{B}$ component. (To make this definition consistent, it must be assumed that $B_0 = 0$.) We call this binary operation on structures the *boxed product*.

Theorem II.5. *The boxed product is admissible:*

$$\text{(60)} \qquad \mathcal{A} = (\mathcal{B}^\square \star \mathcal{C}) \quad \Longrightarrow \quad A(z) = \int_0^z (\partial_t B(t)) \cdot C(t) \, dt, \qquad \partial_t \equiv \frac{d}{dt}.$$

Proof. The definition of boxed products implies the coefficient relation

$$A_n = \sum_{k=1}^{n} \binom{n-1}{k-1} B_k C_{n-k}.$$

The binomial coefficient that appears in the standard convolution, Equation (2), p. 100, is to be modified since only $n-1$ labels need to be distributed between the two components: $k-1$ go to the $\mathcal{B}$ component (that is already constrained to contain the label 1) and $n-k$ to the $\mathcal{C}$ component. From the equivalent form

$$A_n = \frac{1}{n} \sum_{k=0}^{n} \binom{n}{k} (k B_k) C_{n-k},$$

the result follows by taking EGFs, via $A(z) = (\partial_z B(z)) \cdot C(z)$. ∎

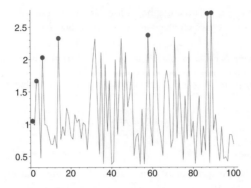

Figure II.15.   A numerical sequence of size 100 with records marked by circles:
there are 7 records that occur at times 1, 3, 5, 11, 60, 86, 88.

A useful special case is the min-rooting operation,

$$A = \left(\mathcal{Z}^{\square} \star \mathcal{C} \right),$$

for which a variant definition goes as follows: take in all possible ways elements
$\gamma \in \mathcal{C}$, prepend an atom with a label, for instance 0, smaller than the labels of γ, and
relabel in the canonical way over $[1 .. (n+1)]$ by shifting all label values by 1. Clearly
$A_{n+1} = C_n$, which yields

$$A(z) = \int_0^z C(t)\, dt,$$

a result which is also consistent with the general formula (60) of boxed products.

For some applications, it is convenient to impose constraints on the *maximal* label
rather than the minimum. The max-boxed product written

$$A = (\mathcal{B}^{\blacksquare} \star \mathcal{C}),$$

is then defined by the fact the maximum is constrained to lie in the $\mathcal{B}$–component of
the labelled product. Naturally, translation by an integral in (60) remains valid for this
trivially modified boxed product.

▷ **II.29.** *Combinatorics of integration.* In the perspective of this book, integration by parts has
an immediate interpretation. Indeed, the equality

$$\int_0^z A'(t) \cdot B(t)\, dt + \int_0^z A(t) \cdot B'(t)\, dt = A(z) \cdot B(z)$$

reads: *"The smallest label in an ordered pair appears either on the left or on the right."*   ◁

***Example* II.16.**   *Records in permutations.* Given a sequence of numbers $x = (x_1, \dots, x_n)$,
assumed all distinct, a *record* is defined to be an element x_j such that $x_k < x_j$ for all $k < j$. (A
record is an element "better" than its predecessors!) Figure II.15 displays a numerical sequence
of length $n = 100$ that has 7 records. Confronted by such data, a statistician will typically
want to determine whether the data obey purely random fluctuations or if there could be some
indications of a "trend" or of a "bias" [139, Ch. 10]. (Think of the data as reflecting share prices
or athletic records, say.) In particular, if the x_j are independently drawn from a continuous
distribution, then the number of records obeys the same laws as in a random permutation of

[1 .. *n*]. This statistical preamble then invites the question: *How many permutations of n have k records?*

First, we start with a special brand of permutations, the ones that have their *maximum* at the beginning. Such permutations are defined as ("■" indicates the boxed product based on the maximum label)

$$\mathcal{Q} = (Z^{\blacksquare} \star \mathcal{P}),$$

where $\mathcal{P}$ is the class of all permutations. Observe that this gives the EGF

$$Q(z) = \int_0^z \left(\frac{d}{dt}t\right) \cdot \frac{1}{1-t} \, dt = \log\frac{1}{1-z},$$

implying the obvious result $Q_n = (n-1)!$ for all $n \geq 1$. These are exactly the permutations with *one* record. Next, consider the class

$$\mathcal{P}^{(k)} = \text{SET}_k(\mathcal{Q}).$$

The elements of $\mathcal{P}^{(k)}$ are unordered sets of cardinality k with elements of type $\mathcal{Q}$. Define the max–leader ("el lider máximo") of any component of $\mathcal{P}^{(k)}$ as the value of its maximal element. Then, if we place the components in sequence, ordered by increasing values of their leaders, then read off the whole sequence, we obtain a permutation with exactly k records. The correspondence[8] is clearly revertible. Here is an illustration, with leaders underlined:

$$\{(\underline{7}, 2, 6, 1), \ (\underline{4}, 3), \ (\underline{9}, 8, 5)\} \ \cong \ [(\underline{4}, 3), \ (\underline{7}, 2, 6, 1), \ (\underline{9}, 8, 5))]$$

$$\cong \ \underline{4}, 3, \underline{7}, 2, 6, 1, \underline{9}, 8, 5.$$

Thus, the number of permutations with k records is determined by

$$P^{(k)}(z) = \frac{1}{k!}\left(\log\frac{1}{1-z}\right)^k, \qquad P_n^{(k)} = \begin{bmatrix} n \\ k \end{bmatrix},$$

where we recognize Stirling cycle numbers from Example II.12, p. 121. In other words:

> *The number of permutations of size n having k records is counted by the Stirling "cycle" number* $\begin{bmatrix} n \\ k \end{bmatrix}$.

Returning to our statistical problem, the treatment of Example II.12 p. 121 (to be revisited in Chapter III, p. 189) shows that the expected number of records in a random permutation of size n equals H_n, the harmonic number. One has $H_{100} \doteq 5.18$, so that for 100 data items, a little more than 5 records are expected on average. The probability of observing 7 records or more is still about 23%, an altogether not especially rare event. In contrast, observing twice as many records as we did, namely 14, would be a fairly strong indication of a bias—on random data, the event has probability very close to 10^{-4}. Altogether, the present discussion is consistent with the hypothesis for the data of Figure II.15 to have been generated independently at random (and indeed they were). .... ■

---

[8]This correspondence can also be viewed as a transformation on permutations that maps the number of records to the number of cycles—it is known as Foata's fundamental correspondence [413, Sec. 10.2].

It is possible to base a fair part of the theory of labelled constructions on sums and products in conjunction with the boxed product. In effect, consider the three relations

$$\mathcal{F} = \text{SEQ}(\mathcal{G}) \quad \Longrightarrow \quad f(z) = \frac{1}{1 - g(z)}, \qquad f = 1 + gf$$

$$\mathcal{F} = \text{SET}(\mathcal{G}) \quad \Longrightarrow \quad f(z) = e^{g(z)}, \qquad f = 1 + \int g' f$$

$$\mathcal{F} = \text{CYC}(\mathcal{G}) \quad \Longrightarrow \quad f(z) = \log \frac{1}{1 - g(z)}, \qquad f = \int g' \frac{1}{1 - g}.$$

The last column is easily checked, by standard calculus, to provide an alternative form of the standard operator corresponding to sequences, sets, and cycles. Each case can in fact be deduced directly from Theorem II.5 and the labelled product rule as follows.

(*i*) *Sequences*: they obey the recursive definition

$$\mathcal{F} = \text{SEQ}(\mathcal{G}) \quad \Longrightarrow \quad \mathcal{F} \cong \{\epsilon\} + (\mathcal{G} \star \mathcal{F}).$$

(*ii*) *Sets*: we have

$$\mathcal{F} = \text{SET}(\mathcal{G}) \quad \Longrightarrow \quad \mathcal{F} \cong \{\epsilon\} + (\mathcal{G}^{\blacksquare} \star \mathcal{F}),$$

which means that, in a set, one can always single out the component with the largest label, the rest of the components forming a set. In other words, when this construction is repeated, the elements of a set can be canonically arranged according to increasing values of their largest labels, the "leaders". (We recognize here a generalization of the construction used for records in permutations.)

(*iii*) *Cycles*: The element of a cycle that contains the largest label can be taken canonically as the cycle "starter", which is then followed by an arbitrary sequence of elements upon traversing the cycle in cyclic order. Thus

$$\mathcal{F} = \text{CYC}(\mathcal{G}) \quad \Longrightarrow \quad \mathcal{F} \cong (\mathcal{G}^{\blacksquare} \star \text{SEQ}(\mathcal{G})).$$

Greene [308] has developed a complete framework of labelled grammars based on standard and boxed labelled products. In its basic form, its expressive power is essentially equivalent to ours, because of the above relations. More complicated order constraints, dealing simultaneously with a collection of larger and smaller elements, can be furthermore taken into account within this framework.

▷ **II.30.** *Higher order constraints, after Greene.* Let the symbols □, ⊡, ■ represent smallest, second smallest, and largest labels, respectively. One has the correspondences (with $\partial_z = \frac{d}{dz}$)

$$A = \left(B^{\square} \star C^{\blacksquare}\right) \qquad \partial_z^2 A(z) = (\partial_z B(z)) \cdot (\partial_z C(z))$$

$$A = \left(B^{\square \blacksquare} \star C\right) \qquad \partial_z^2 A(z) = \left(\partial_z^2 B(z)\right) \cdot C(z)$$

$$A = \left(B^{\square} \star C^{\boxdot} \star D^{\blacksquare}\right) \qquad \partial_z^3 A(z) = (\partial_z B(z)) \cdot (\partial_z C(z)) \cdot (\partial_z D(z)),$$

and so on. These can be transformed into (iterated) integral representations. (See [308] for more.)   ◁

The next three examples demonstrate the utility of min/max-rooting used in conjunction with recursion. Examples II.17 and II.18 introduce two important classes of

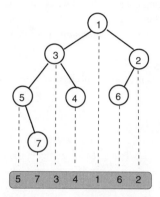

Figure II.16. A permutation of size 7 and its increasing binary tree lifting.

trees that are tightly linked to permutations. Example II.19 provides a simple symbolic solution to a famous parking problem, on which many analyses can be built.

Example II.17. *Increasing binary trees and alternating permutations.* To each permutation, one can associate bijectively a binary tree of a special type called an *increasing binary tree* and sometimes a heap-ordered tree or a tournament tree. This is a plane rooted binary tree in which internal nodes bear labels in the usual way, but with the additional constraint that node labels increase along any branch stemming from the root. Such trees are closely related to many classical data structures of computer science, such as heaps and binomial queues.

The correspondence (Figure II.16) is as follows: Given a permutation written as a word, $\sigma = \sigma_1 \sigma_2 \ldots \sigma_n$, factor it into the form $\sigma = \sigma_L \cdot \min(\sigma) \cdot \sigma_R$, with $\min(\sigma)$ the smallest label value in the permutation, and σ_L, σ_R the factors left and right of $\min(\sigma)$. Then the binary tree $\beta(\sigma)$ is defined recursively in the format $\langle \text{root, left, right} \rangle$ by

$$\beta(\sigma) = \langle \min(\sigma), \beta(\sigma_L), \beta(\sigma_R) \rangle, \qquad \beta(\epsilon) = \epsilon.$$

The empty tree (consisting of a unique external node of size 0) goes with the empty permutation ϵ. Conversely, reading the labels of the tree in symmetric (infix) order gives back the original permutation. (The correspondence is described for instance in Stanley's book [552, p. 23–25] who says that "it has been primarily developed by the French", pointing at [267].)

Thus, the family $\mathcal{I}$ of binary increasing trees satisfies the recursive definition

(61) $$\mathcal{I} = \{\epsilon\} + \left(\mathcal{Z}^\square \star \mathcal{I} \star \mathcal{I} \right),$$

which implies the nonlinear integral equation for the EGF

$$I(z) = 1 + \int_0^z I(t)^2 \, dt.$$

This equation reduces to $I'(z) = I(z)^2$ and, under the initial condition $I(0) = 1$, it admits the solution $I(z) = (1 - z)^{-1}$. Thus $I_n = n!$, which is consistent with the fact that there are as many increasing binary trees as there are permutations.

The construction of increasing trees is instrumental in deriving EGFs relative to various local order patterns in permutations. We illustrate its use here by counting the number of *up-and-down* (or *zig-zag*) permutations, also known as *alternating* permutations. The result, already mentioned in our *Invitation* chapter (p. 2) was first derived by Désiré André in 1881 by means of a direct recurrence argument.

A permutation $\sigma = \sigma_1 \sigma_2 \cdots \sigma_n$ is an alternating permutation if

(62) $$\sigma_1 > \sigma_2 < \sigma_3 > \sigma_4 < \cdots ,$$

so that pairs of consecutive elements form a succession of ups and downs; for instance,

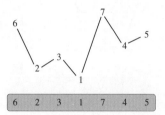

Consider first the case of an alternating permutation of *odd* size. It can be checked that the corresponding increasing trees have no one-way branching nodes, so that they consist solely of binary nodes and leaves. Thus, the corresponding specification is

$$\mathcal{J} = \mathcal{Z} + \left(\mathcal{Z}^{\square} \star \mathcal{J} \star \mathcal{J}\right),$$

so that

$$J(z) = z + \int_0^z J(t)^2 \, dt \qquad \text{and} \qquad \frac{d}{dz} J(z) = 1 + J(z)^2.$$

The equation admits separation of variables, which implies, since $J(0) = 0$, that $\arctan(J(z)) = z$, hence:

$$J(z) = \tan(z) = z + 2\frac{z^3}{3!} + 16\frac{z^5}{5!} + 272\frac{z^7}{7!} + \cdots .$$

The coefficients J_{2n+1} are known as the *tangent numbers* or the *Euler numbers* of odd index (*EIS* **A000182**).

Alternating permutations of *even* size defined by the constraint (62) and denoted by $\mathcal{K}$ can be determined from

$$\mathcal{K} = \{\epsilon\} + \left(\mathcal{Z}^{\square} \star \mathcal{J} \star \mathcal{K}\right),$$

since now all internal nodes of the tree representation are binary, except for the right-most one that only branches on the left. Thus, $K'(z) = \tan(z)K(z)$, and the EGF is

$$K(z) = \frac{1}{\cos(z)} = 1 + 1\frac{z^2}{2!} + 5\frac{z^4}{4!} + 61\frac{z^6}{6!} + 1385\frac{z^8}{8!} + \cdots ,$$

where the coefficients K_{2n} are the *secant numbers* also known as Euler numbers of even index (*EIS* **A000364**).

Use will be made later in this book (Chapter III, p. 202) of this important tree representation of permutations as it opens access to parameters such as the number of descents, runs, and (once more!) records in permutations. Analyses of increasing trees also inform us of crucial performance issues regarding binary search trees, quicksort, and heap-like priority queue structures [429, 538, 598, 600]. .... ∎

▷ **II.31.** *Combinatorics of trigonometrics.* Interpret $\tan\frac{z}{1-z}$, $\tan\tan z$, $\tan(e^z - 1)$ as EGFs of combinatorial classes. ◁

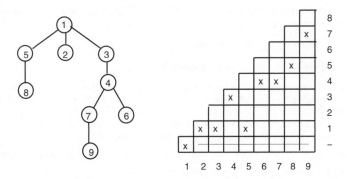

Figure II.17. An increasing Cayley tree (left) and its associated regressive mapping (right).

***Example* II.18.** *Increasing Cayley trees and regressive mappings.* An increasing Cayley tree is a Cayley tree (i.e., it is labelled, non-plane, and rooted) whose labels along any branch stemming from the root form an increasing sequence. In particular, the minimum must occur at the root, and no plane embedding is implied. Let $\mathcal{L}$ be the class of such trees. The recursive specification is now

$$\mathcal{L} = \left(\mathcal{Z}^\square \star \text{SET}(\mathcal{L})\right).$$

The generating function thus satisfies the functional relations

$$L(z) = \int_0^z e^{L(t)}\, dt, \qquad L'(z) = e^{L(z)},$$

with $L(0) = 0$. Integration of $L'e^{-L} = 1$ shows that $e^{-L} = 1 - z$, hence

$$L(z) = \log \frac{1}{1-z} \qquad \text{and} \qquad L_n = (n-1)!.$$

Thus the number of increasing Cayley trees is $(n-1)!$, which is also the number of permutations of size $n-1$. These trees have been studied by Meir and Moon [435] under the name of "recursive trees", a terminology that we do not, however, retain here.

The simplicity of the formula $L_n = (n-1)!$ certainly calls for a combinatorial interpretation. In fact, an increasing Cayley tree is fully determined by its child–parent relationship (Figure II.17). In other words, to each increasing Cayley tree τ, we associate a partial map $\phi = \phi_\tau$ such that $\phi(i) = j$ iff the label of the parent of i is j. Since the root of tree is an orphan, the value of $\phi(1)$ is undefined, $\phi(1) = \bot$; since the tree is increasing, one has $\phi(i) < i$ for all $i \geq 2$. A function satisfying these last two conditions is called a *regressive mapping*. The correspondence between trees and regressive mappings is then easily seen to be bijective.

Thus regressive mappings on the domain $[1 .. n]$ and increasing Cayley trees are equinumerous, so that we may as well use $\mathcal{L}$ to denote the class of regressive mappings. Now, a regressive mapping of size n is evidently determined by a single choice for $\phi(2)$ (since $\phi(2) = 1$), two possible choices for $\phi(3)$ (either of $1, 2$), and so on. Hence the formula

$$L_n = 1 \times 2 \times 3 \times \cdots \times (n-1)$$

receives a natural interpretation. .... ∎

▷ **II.32.** *Regressive mappings and permutations.* Regressive mappings can be related directly to permutations. The construction that associates a regressive mapping to a permutation is called the "inversion table" construction; see [378, 538]. Given a permutation $\sigma = \sigma_1, \ldots, \sigma_n$, associate to it a function $\psi = \psi_\sigma$ from $[1 \ldots n]$ to $[0 \ldots n - 1]$ by the rule

$$\psi(j) = \operatorname{card} \left\{ k < j \mid \sigma_k > \sigma_j \right\}.$$

The function ψ is a trivial variant of a regressive mapping.   ◁

▷ **II.33.** *Rotations and increasing trees.* An increasing Cayley tree can be canonically drawn by ordering descendants of each node from left to right according to their label values. The rotation correspondence (p. 73) then gives rise to a binary increasing tree. Hence, increasing Cayley trees and increasing binary trees are also directly related. Summarizing this note and the previous one, we have a quadruple combinatorial connection,

Increasing Cayley trees ≅ Regressive mappings ≅ Permutations ≅ Increasing binary trees,

which opens the way to yet more permutation enumerations.   ◁

Example II.19. *A parking problem.* Here is Knuth's introduction to the problem, dating back from 1973 (see [378, p. 545]), which nowadays might be regarded by some as politically incorrect:

> "A certain one-way street has m parking spaces in a row numbered 1 to m. A man and his dozing wife drive by, and suddenly, she wakes up and orders him to park immediately. He dutifully parks at the first available space [...]."

Consider $n = m - 1$ cars and condition by the fact that everybody eventually finds a parking space *and* the last space remains empty. There are $m^n = (n + 1)^n$ possible sequences of "wishes", among which only a certain number F_n satisfy the condition—this number is to be determined. (An important motivation for this problem is the analysis of hashing algorithms examined in Note III.11, p. 178, under the "linear probing" strategy.)

A sequence satisfying the condition called an *almost-full* allocation, its size n being the number of cars involved. Let $\mathcal{F}$ represent the class of almost-full allocations. We claim the decomposition:

(63) $$\mathcal{F} = \left[(\Theta\mathcal{F} + \mathcal{F}) \star \mathcal{Z}^\blacksquare \star \mathcal{F} \right].$$

Indeed, consider the car that arrived last, before it will eventually land in some position $k + 1$ from the left. Then, there are two islands, which are themselves almost-full allocations (of respective sizes k and $n - k - 1$). This last car's intended parking wish must have been either one of the first k occupied cells on the left (the factor $\Theta\mathcal{F}$ in (63)) or the last empty cell of the first island (the term $\mathcal{F}$ in the left factor); the right island is not affected (the factor $\mathcal{F}$ on the right). Finally, the last car is inserted into the street (the factor $\mathcal{Z}^\blacksquare$). Pictorially, we have a sort of binary tree decomposition of almost-full allocations:

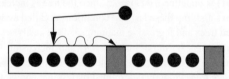

Analytically, the translation of (63) into EGF is

(64) $$F(z) = \int_0^z (w F'(w) + F(w)) F(w) \, dw,$$

which, through differentiation gives

(65) $$F'(z) = (z F(z))' \cdot F(z).$$

Simple manipulations do the rest: we have $F'/F = (zF)'$, which by integration gives $\log F = (zF)$ and $F = e^{zF}$. Thus $F(z)$ satisfies a functional equation strangely similar to that of the Cayley tree function $T(z)$; indeed, it is not hard to see that one has

$$(66) \qquad\qquad F(z) = \frac{1}{z}T(z) \qquad \text{and} \qquad F_n = (n+1)^{n-1},$$

which solves the original counting problem. The derivation above is based on articles by Flajolet, Poblete, Viola, and Knuth [249, 380], who show that probabilistic properties of parking allocations can be precisely analysed (for instance, total displacement, examined in Note VII.54, p. 534, is found to be governed by an Airy distribution). . . . . ■

II. 7. Perspective

Together with the previous chapter and Figure I.18, this chapter and Figure II.18 provide the basis for the symbolic method that is at the core of analytic combinatorics. The translations of the basic constructions for labelled classes to EGFs could hardly be simpler, but, as we have seen, they are sufficiently powerful to embrace numerous classical results in combinatorics, ranging from the birthday and coupon collector problems to tree and graph enumeration.

The examples that we have considered for second-level structures, trees, mappings, and graphs lead to EGFs that are simple to express and natural to generalize. (Often, the simple form is misleading—direct derivations of many of these EGFs that do not appeal to the symbolic method can be rather intricate.) Indeed, the symbolic method provides a framework that allows us to understand the nature of many of these combinatorial classes. From here, numerous seemingly scattered counting problems can be organized into broad structural categories and solved in an almost mechanical manner.

Again, the symbolic method is only half of the story (the "combinatorics" in analytic combinatorics), leading to EGFs for the counting sequences of numerous interesting combinatorial classes. While some of these EGFs lead immediately to explicit counting results, others require classical techniques in complex analysis and asymptotic analysis that are covered in Part B (the "analytic" part of analytic combinatorics) to deliver asymptotic estimates. Together with these techniques, the basic constructions, translations, and applications that we have discussed in this chapter reinforce the overall message that the symbolic method is a systematic approach that is successful for addressing classical and new problems in combinatorics, generalizations, and applications.

We have been focusing on *enumeration problems*—counting the number of objects of a given size in a combinatorial class. In the next chapter, we shall consider how to extend the symbolic method to help analyse other properties of combinatorial classes.

Bibliographic notes. The labelled set construction and the exponential formula were recognized early by researchers working in the area of graphical enumerations [319]. Foata [265] proposed a detailed formalization in 1974 of labelled constructions, especially sequences and sets, under the names of partitional complex; a brief account is also given by Stanley in his survey [550]. This is parallel to the concept of "prefab" due to Bender and Goldman [42]. The

1. The main constructions of union, and product, sequence, set, and cycle for labelled structures together with their translation into exponential generating functions.

Construction		EGF
Union	$\mathcal{A} = \mathcal{B} + \mathcal{C}$	$A(z) = B(z) + C(z)$
Product	$\mathcal{A} = \mathcal{B} \star \mathcal{C}$	$A(z) = B(z) \cdot C(z)$
Sequence	$\mathcal{A} = \text{SEQ}(\mathcal{B})$	$A(z) = \dfrac{1}{1 - B(z)}$
Set	$\mathcal{A} = \text{SET}(\mathcal{B})$	$A(z) = \exp(B(z))$
Cycle	$\mathcal{A} = \text{CYC}(\mathcal{B})$	$A(z) = \log \dfrac{1}{1 - B(z)}$

2. Sets, multisets, and cycles of fixed cardinality.

Construction		EGF
Sequence	$\mathcal{A} = \text{SEQ}_k(\mathcal{B})$	$A(z) = B(z)^k$
Set	$\mathcal{A} = \text{SET}_k(\mathcal{B})$	$A(z) = \dfrac{1}{k!} B(z)^k$
Cycle	$\mathcal{A} = \text{CYC}_k(\mathcal{B})$	$A(z) = \dfrac{1}{k} B(z)^k$

3. The additional constructions of pointing and substitution.

Construction		EGF
Pointing	$\mathcal{A} = \Theta \mathcal{B}$	$A(z) = z \frac{d}{dz} B(z)$
Substitution	$\mathcal{A} = \mathcal{B} \circ \mathcal{C}$	$A(z) = B(C(z))$

4. The "boxed" product.

$$\mathcal{A} = (\mathcal{B}^{\square} \star \mathcal{C}) \implies A(z) = \int_0^z \left(\frac{d}{dt} B(t) \right) \cdot C(t)\, dt.$$

Figure II.18. A "dictionary" of *labelled* constructions together with their translation into *exponential* generating functions (EGFs). The first constructions are counterparts of the unlabelled constructions of the previous chapter (the multiset construction is not meaningful here). Translation for composite constructions of bounded cardinality appears to be simple. Finally, the boxed product is specific to labelled structures. (Compare with the unlabelled counterpart, Figure I.18, p. 18.)

books by Comtet [129], Wilf [608], Stanley [552], or Goulden and Jackson [303] have many examples of the use of labelled constructions in combinatorial analysis.

Greene [308] has introduced in his 1983 dissertation a general framework of "labelled grammars" largely based on the boxed product with implications for the random generation of combinatorial structures. Joyal's theory of species dating from 1981 (see [359] for the original article and the book by Bergeron, Labelle, and Leroux [50] for a rich exposition) is based on category theory; it presents the advantage of uniting in a common framework the unlabelled and the labelled worlds.

Flajolet, Salvy, and Zimmermann have developed a specification language closely related to the system expounded here. They show in [255] how to compile automatically specifications into generating functions; this is complemented by a calculus that produces fast random generation algorithms [264].

> *I can see looming ahead one of those terrible exercises in probability where six men have white hats and six men have black hats and you have to work it out by mathematics how likely it is that the hats will get mixed up and in what proportion. If you start thinking about things like that, you would go round the bend. Let me assure you of that!*
>
> —AGATHA CHRISTIE
> (*The Mirror Crack'd.* Toronto, Bantam Books, 1962.)

III

Combinatorial Parameters and Multivariate Generating Functions

> *Generating functions find averages, etc.*
>
> — HERBERT WILF [608]

Many scientific endeavours demand precise quantitative information on probabilistic properties of *parameters* of combinatorial objects. For instance, when designing, analysing, and optimizing a sorting algorithm, it is of interest to determine the typical disorder of data obeying a given model of randomness, and to do so in the mean, or even in distribution, either exactly or asymptotically. Similar situations arise in a broad variety of fields, including probability theory and statistics, computer science, information theory, statistical physics, and computational biology. The exact problem is then a refined counting problem with two parameters, namely, size and an additional characteristic: this is the subject addressed in this chapter and treated by a natural extension of the generating function framework. The asymptotic problem can be viewed as one of characterizing in the limit a family of probability laws indexed by the values of the possible sizes: this is a topic to be discussed in Chapter IX. As demonstrated here, the symbolic methods initially developed for counting combinatorial objects adapt gracefully to the analysis of various sorts of parameters of constructible classes, unlabelled and labelled alike.

Multivariate generating functions (MGFs)—ordinary or exponential—can keep track of a collection of parameters defined over combinatorial objects. From the knowledge of such generating functions, there result either explicit probability distributions or, at least, mean and variance evaluations. For *inherited* parameters, all the combinatorial classes discussed so far are amenable to such a treatment. Technically, the translation schemes that relate combinatorial constructions and multivariate generating functions present no major difficulty—they appear to be natural (notational, even) refinements of the paradigm developed in Chapters I and II for the univariate case. Typical applications from classical combinatorics are the number of summands in a composition, the number of blocks in a set partition, the number of cycles in a permutation, the root degree or path length of a tree, the number of fixed points in a permutation, the number of singleton blocks in a set partition, the number of leaves in trees of various sorts, and so on.

Beyond its technical aspects anchored in symbolic methods, this chapter also serves as a first encounter with the general area of random combinatorial structures. The general question is: *What does a random object of large size look like?* Multivariate generating functions first provide an easy access to *moments* of combinatorial parameters—typically the mean and variance. In addition, when combined with basic probabilistic inequalities, moment estimates often lead to precise characterizations of properties of large random structures that hold with high probability. For instance, a large integer partition conforms with high probability to a deterministic profile, a large random permutation almost surely has at least one long cycle and a few short ones, and so on. Such a highly constrained behaviour of large objects may in turn serve to design dedicated algorithms and optimize data structures; or it may serve to build statistical tests—when does one depart from randomness and detect a "signal" in large sets of observed data? Randomness forms a recurrent theme of the book: it will be developed much further in Chapter IX, where the complex asymptotic methods of Part B are grafted on the exact modelling by multivariate generating functions presented in this chapter.

This chapter is organized as follows. First a few pragmatic developments related to bivariate generating functions are presented in Section III. 1. Next, Section III. 2 presents the notion of bivariate enumeration and its relation to discrete probabilistic models, including the determination of moments, since the language of elementary probability theory does indeed provide an intuitively appealing way to conceive of bivariate counting data. The symbolic method *per se*, declined in its general multivariate version, is centrally developed in Sections III. 3 and III. 4: with suitable multi-index notations, the extension of the symbolic method to the multivariate case is almost immediate. Recursive parameters that often arise in particular from tree statistics form the subject of Section III. 5, while complete generating functions and associated combinatorial models are discussed in Section III. 6. Additional constructions such as pointing, substitution, and order constraints lead to interesting developments, in particular, an original treatment of the inclusion–exclusion principle in Section III. 7. The chapter concludes, in Section III. 8, with a brief abstract discussion of extremal parameters like height in trees or smallest and largest components in composite structures—such parameters are best treated via families of univariate generating functions.

III. 1. An introduction to bivariate generating functions (BGFs)

We have seen in Chapters I and II that a number sequence (f_n) can be encoded by means of a generating function in one variable, either ordinary or exponential:

$$(f_n) \quad \rightsquigarrow \quad f(z) = \begin{cases} \displaystyle\sum_n f_n z^n & \text{(ordinary GF)} \\[2ex] \displaystyle\sum_n f_n \frac{z^n}{n!} & \text{(exponential GF).} \end{cases}$$

$$
\begin{array}{lll}
f_{00} & & & \longrightarrow & f_0(u) \\
f_{10} & f_{11} & & \longrightarrow & f_1(u) \\
f_{20} & f_{21} & f_{22} & \longrightarrow & f_2(u) \\
\vdots & \vdots & \vdots \\
\downarrow & \downarrow & \downarrow \\
f^{\langle 0 \rangle}(z) & f^{\langle 1 \rangle}(z) & f^{\langle 2 \rangle}(z)
\end{array}
$$

Figure III.1. An array of numbers and its associated horizontal and vertical GFs.

This encoding is powerful, since many combinatorial constructions admit a translation as operations over such generating functions. In this way, one gains access to many useful counting formulae.

Similarly, consider a sequence of numbers $(f_{n,k})$ depending on two integer-valued indices, n and k. Usually, in this book, $(f_{n,k})$ will be an array of numbers (often a triangular array), where $f_{n,k}$ is the number of objects φ in some class $\mathcal{F}$, such that $|\varphi| = n$ and some parameter $\chi(\varphi)$ is equal to k. We can encode this sequence by means of a *bivariate generating function (BGF)* involving two variables: a primary variable z attached to n and a secondary u attached to k.

Definition III.1. *The* bivariate generating functions (BGFs), *either* ordinary *or* exponential, *of an array* $(f_{n,k})$ *are the formal power series in two variables defined by*

$$
f(z, u) = \begin{cases}
\displaystyle\sum_{n,k} f_{n,k} z^n u^k & \text{(ordinary BGF)} \\[2ex]
\displaystyle\sum_{n,k} f_{n,k} \frac{z^n}{n!} u^k & \text{(exponential BGF)}.
\end{cases}
$$

(The "double exponential" GF corresponding to $\frac{z^n}{n!} \frac{u^k}{k!}$ is not used in the book.)

As we shall see shortly, parameters of constructible classes become accessible through such BGFs. According to the point of view adopted for the moment, one starts with an array of numbers and forms a BGF by a double summation process. We present here two examples related to binomial coefficients and Stirling cycle numbers illustrating how such BGFs can be determined, then manipulated. In what follows it is convenient to refer to the *horizontal* and *vertical* generating functions (Figure III.1) that are each a one-parameter family of GFs in a single variable defined by

$$
\begin{array}{lll}
\text{horizontal GF:} & f_n(u) & := & \sum_k f_{n,k} u^k; \\[1ex]
\text{vertical GF:} & f^{\langle k \rangle}(z) & := & \sum_n f_{n,k} z^n & \text{(ordinary case)} \\[1ex]
& f^{\langle k \rangle}(z) & := & \sum_n f_{n,k} \frac{z^n}{n!} & \text{(exponential case)}.
\end{array}
$$

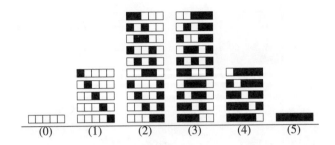

(0)   (1)   (2)   (3)   (4)   (5)

Figure III.2. The set $\mathcal{W}_5$ of the 32 binary words over the alphabet $\{\square, \blacksquare\}$ enumerated according to the number of occurrences of the letter '$\blacksquare$' gives rise to the bivariate counting sequence $\{W_{5,j}\} = 1, 5, 10, 10, 5, 1$.

The terminology is transparently explained if the elements $(f_{n,k})$ are arranged as an infinite matrix, with $f_{n,k}$ placed in row n and column k, since the horizontal and vertical GFs appear as the GFs of the rows and columns respectively. Naturally, one has

$$f(z, u) = \sum_k u^k f^{\langle k \rangle}(z) = \begin{cases} \displaystyle\sum_n f_n(u) z^n & \text{(ordinary BGF)} \\[2ex] \displaystyle\sum_n f_n(u) \frac{z^n}{n!} & \text{(exponential BGF)}. \end{cases}$$

Example III.1. *The ordinary BGF of binomial coefficients.* The binomial coefficient $\binom{n}{k}$ counts binary words of length n having k occurrences of a designated letter; see Figure III.2. In order to compose the bivariate GF, start from the simplest case of Newton's binomial theorem and directly form the horizontal GFs corresponding to a fixed n:

$$(1) \qquad W_n(u) := \sum_{k=0}^{n} \binom{n}{k} u^k = (1 + u)^n,$$

Then a summation over all values of n gives the ordinary BGF

$$(2) \qquad W(z, u) = \sum_{k,n \geq 0} \binom{n}{k} u^k z^n = \sum_{n \geq 0} (1 + u)^n z^n = \frac{1}{1 - z(1 + u)}.$$

Such calculations are typical of BGF manipulations. What we have done amounts to starting from a sequence of numbers, $W_{n,k}$, determining the horizontal GFs $W_n(u)$ in (1), then the bivariate GF $W(z, u)$ in (2), according to the scheme:

$$W_{n,k} \quad \leadsto \quad W_n(u) \quad \leadsto \quad W(z, u).$$

The BGF in (2) reduces to the OGF $(1 - 2z)^{-1}$ of all words, as it should, upon setting $u = 1$.

In addition, one can deduce from (2) the vertical GFs of the binomial coefficients corresponding to a fixed value of k

$$W^{\langle k \rangle}(z) = \sum_{n \geq 0} \binom{n}{k} z^n = \frac{z^k}{(1 - z)^{k+1}},$$

from an expansion of the BGF with respect to u

$$(3) \qquad W(z, u) = \frac{1}{1 - z} \frac{1}{1 - u\frac{z}{1-z}} = \sum_{k \geq 0} u^k \frac{z^k}{(1 - z)^{k+1}},$$

and the result naturally matches what a direct calculation would give. .... ■

▷ **III.1.** *The exponential BGF of binomial coefficients.* This is

$$(4) \qquad \widetilde{W}(z, u) = \sum_{k,n} \binom{n}{k} u^k \frac{z^n}{n!} = \sum (1 + u)^n \frac{z^n}{n!} = e^{z(1+u)}.$$

The vertical GFs are $e^z z^k / k!$. The horizontal GFs are $(1 + u)^n$, as in the ordinary case. ◁

Example III.2. *The exponential BGF of Stirling cycle numbers.* As seen Example II.12, p. 121, the number $P_{n,k}$ of permutations of size n having k cycles equals the Stirling cycle number $\begin{bmatrix} n \\ k \end{bmatrix}$, a vertical EGF being

$$P^{\langle k \rangle}(z) := \sum_n \begin{bmatrix} n \\ k \end{bmatrix} \frac{z^n}{n!} = \frac{L(z)^k}{k!}, \qquad L(z) := \log \frac{1}{1 - z}.$$

From this, the exponential BGF is formed as follows (this revisits the calculations on p. 121):

$$(5) \qquad P(z, u) := \sum_k P^{\langle k \rangle}(z) u^k = \sum_k \frac{u^k}{k!} L(z)^k = e^{uL(z)} = (1 - z)^{-u}.$$

The simplification is quite remarkable but altogether quite typical, as we shall see shortly, in the context of a labelled set construction. The starting point is thus a collection of vertical EGFs and the scheme is now

$$P_n^{\langle k \rangle} \quad \rightsquigarrow \quad P^{\langle k \rangle}(z) \quad \rightsquigarrow \quad P(z, u).$$

The BGF in (5) reduces to the EGF $(1 - z)^{-1}$ of all permutations, upon setting $u = 1$.

Furthermore, an expansion of the BGF in terms of the variable z provides useful information; namely, the horizontal GF is obtained by Newton's binomial theorem:

$$(6) \qquad \begin{aligned} P(z, u) &= \sum_{n \geq 0} \binom{n + u - 1}{n} z^n &= \sum_{n \geq 0} P_n(u) \frac{z^n}{n!}, \\ \text{where} \quad P_n(u) &= u(u + 1) \cdots (u + n - 1). \end{aligned}$$

This last polynomial is called the *Stirling cycle polynomial* of index n and it describes completely the distribution of the number of cycles in all permutations of size n. In addition, the relation

$$P_n(u) = P_{n-1}(u)(u + (n - 1)),$$

is equivalent to the recurrence

$$\begin{bmatrix} n \\ k \end{bmatrix} = (n - 1) \begin{bmatrix} n - 1 \\ k \end{bmatrix} + \begin{bmatrix} n - 1 \\ k - 1 \end{bmatrix},$$

by which Stirling numbers are often defined and easily evaluated numerically; see also Appendix A.8: *Stirling numbers*, p. 735. (The recurrence is susceptible to a direct combinatorial interpretation—add n either to an existing cycle or as a "new" singleton.) .... ■

Numbers	Horizontal GFs
$\dbinom{n}{k}$	$(1+u)^n$
Vertical OGFs	Ordinary BGF
$\dfrac{z^k}{(1-z)^{k+1}}$	$\dfrac{1}{1-z(1+u)}$

Numbers	Horizontal GFs
$\begin{bmatrix} n \\ k \end{bmatrix}$	$u(u+1)\cdots(u+n-1)$
Vertical EGFs	Exponential BGF
$\dfrac{1}{k!}\left(\log\dfrac{1}{1-z}\right)^k$	$(1-z)^{-u}$

Figure III.3.   The various GFs associated with binomial coefficients (left) and Stirling cycle numbers (right).

Concise expressions for BGFs, like (2), (3), (5), or (18), are summarized in Figure III.3; they are invaluable for deriving moments, variance, and even finer characteristics of distributions, as we see next. The determination of such BGFs can be covered by a simple extension of the symbolic method, as will be detailed in Sections III. 3 and III. 4.

III. 2. Bivariate generating functions and probability distributions

Our purpose in this book is to analyse characteristics of a broad range of combinatorial types. The eventual goal of multivariate enumeration is the quantification of properties present with high regularity in large random structures.

We shall be principally interested in enumeration according to size *and* an auxiliary parameter, the corresponding problems being naturally treated by means of BGFs. In order to avoid redundant definitions, it proves convenient to introduce the sequence of *fundamental factors* $(\omega_n)_{n\geq 0}$, defined by

(7) $\omega_n = 1$ for ordinary GFs, $\omega_n = n!$ for exponential GFs.

Then, the OGF and EGF of a sequence (f_n) are jointly represented as

$$f(z) = \sum f_n \frac{z^n}{\omega_n} \quad \text{and} \quad f_n = \omega_n [z^n] f(z).$$

Definition III.2. *Given a combinatorial class $\mathcal{A}$, a (scalar) parameter is a function from $\mathcal{A}$ to $\mathbb{Z}_{\geq 0}$ that associates to any object $\alpha \in \mathcal{A}$ an integer value $\chi(\alpha)$. The sequence*

$$A_{n,k} = \operatorname{card}\left(\{\alpha \in \mathcal{A} \mid |\alpha| = n, \ \chi(\alpha) = k\}\right),$$

is called the counting sequence *of the pair $\mathcal{A}, \chi$. The* bivariate generating function *(BGF) of $\mathcal{A}, \chi$ is defined as*

$$A(z, u) := \sum_{n,k \geq 0} A_{n,k} \frac{z^n}{\omega_n} u^k,$$

and is ordinary *if $\omega_n \equiv 1$ and* exponential *if $\omega_n \equiv n!$. One says that the variable z* marks size *and the variable u* marks the parameter χ.

Naturally $A(z, 1)$ reduces to the usual counting generating function $A(z)$ associated with $\mathcal{A}$, and the cardinality of $\mathcal{A}_n$ is expressible as

$$A_n = \omega_n [z^n] A(z, 1).$$

III. 2.1. Distributions and moments. Within this subsection, we examine the relationship between probabilistic models needed to interpret bivariate counting sequences and bivariate generating functions. The elementary notions needed are recalled in Appendix A.3: *Combinatorial probability*, p. 727.

Consider a combinatorial class $\mathcal{A}$. The *uniform probability distribution* over $\mathcal{A}_n$ assigns to any $\alpha \in \mathcal{A}_n$ a probability equal to $1/A_n$. We shall use the symbol $\mathbb{P}$ to denote probability and occasionally subscript it with an indication of the probabilistic model used, whenever this model needs to be stressed: we shall then write $\mathbb{P}_{\mathcal{A}_n}$ (or simply $\mathbb{P}_n$ if $\mathcal{A}$ is understood) to indicate probability relative to the uniform distribution over $\mathcal{A}_n$.

Probability generating functions. Consider a parameter χ. It determines over each $\mathcal{A}_n$ a discrete *random variable* defined over the discrete probability space $\mathcal{A}_n$:

$$(8) \qquad \mathbb{P}_{\mathcal{A}_n}(\chi = k) = \frac{A_{n,k}}{A_n} = \frac{A_{n,k}}{\sum_k A_{n,k}}.$$

Given a discrete random variable X, typically, a parameter χ taken over a subclass $\mathcal{A}_n$, we recall that its *probability generating function* (PGF) is by definition the quantity

$$(9) \qquad p(u) = \sum_k \mathbb{P}(X = k) u^k.$$

From (8) and (9), one has immediately:

Proposition III.1 (PGFs from BGFs). *Let $A(z, u)$ be the bivariate generating function of a parameter χ defined over a combinatorial class $\mathcal{A}$. The probability generating function of χ over $\mathcal{A}_n$ is given by*

$$\sum_k \mathbb{P}_{\mathcal{A}_n}(\chi = k) u^k = \frac{[z^n] A(z, u)}{[z^n] A(z, 1)},$$

and is thus a normalized version of a horizontal generating function.

The translation into the language of probability enables us to make use of whichever intuition might be available in any particular case, while allowing for a natural interpretation of data (Figure III.4). Indeed, instead of noting that the quantity 381922055502195 represents the number of permutations of size 20 that have 10 cycles, it is perhaps more informative to state the probability of the event, which is 0.00015, i.e., about 1.5 per 10 000. Discrete distributions are conveniently represented by *histograms* or "bar charts", where the height of the bar at abscissa k indicates the value of $\mathbb{P}\{X = k\}$. Figure III.4 displays two classical combinatorial distributions in this way. Given the uniform probabilistic model that we have been adopting, such histograms are eventually nothing but a condensed form of the "stacks" corresponding to exhaustive listings, like the one displayed in Figure III.2.

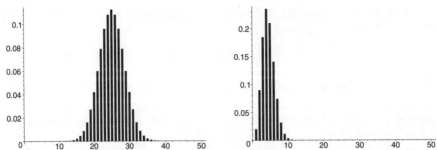

Figure III.4. Histograms of two combinatorial distributions. Left: the number of occurrences of a designated letter in a random binary word of length 50 (binomial distribution). Right: the number of cycles in a random permutation of size 50 (Stirling cycle distribution).

Moments. Important information is conveyed by *moments*. Given a discrete random variable X, the *expectation* of $f(X)$ is by definition the linear functional

$$\mathbb{E}(f(X)) := \sum_k \mathbb{P}\{X = k\} \cdot f(k).$$

The (power) *moments* are

$$\mathbb{E}(X^r) := \sum_k \mathbb{P}\{X = k\} \cdot k^r.$$

Then the expectation (or average, mean) of X, its variance, and its standard deviation, respectively, are expressed as

$$\mathbb{E}(X), \qquad \mathbb{V}(X) = \mathbb{E}(X^2) - \mathbb{E}(X)^2, \qquad \sigma(X) = \sqrt{\mathbb{V}(X)}.$$

The expectation corresponds to what is typically seen when forming the arithmetic mean value of a large number of observations: this property is the *weak law of large numbers* [205, Ch X]. The standard deviation then measures the dispersion of values observed from the expectation and it does so in a mean-quadratic sense.

The *factorial moment* defined for order r as

(10) $$E\left(X(X - 1) \cdots (X - r + 1)\right)$$

is also of interest for computational purposes, since it is obtained plainly by differentiation of PGFs (Appendix A.3: *Combinatorial probability*, p. 727). Power moments are then easily recovered as linear combinations of factorial moments, see Note III.9 of Appendix A. In summary:

Proposition III.2 (Moments from BGFs). *The factorial moment of order r of a parameter χ is determined from the BGF $A(z, u)$ by r-fold differentiation followed by evaluation at 1:*

$$\mathbb{E}_{\mathcal{A}_n}\left(\chi(\chi - 1) \cdots (\chi - r + 1)\right) = \frac{[z^n]\partial_u^r A(z, u)\big|_{u=1}}{[z^n]A(z, 1)}.$$

In particular, the first two moments satisfy

$$\mathbb{E}_{\mathcal{A}_n}(\chi) = \frac{[z^n]\partial_u A(z, u)|_{u=1}}{[z^n]A(z, 1)}$$

$$\mathbb{E}_{\mathcal{A}_n}(\chi^2) = \frac{[z^n]\partial_u^2 A(z, u)|_{u=1}}{[z^n]A(z, 1)} + \frac{[z^n]\partial_u A(z, u)|_{u=1}}{[z^n]A(z, 1)},$$

the variance and standard deviation being determined by

$$\mathbb{V}(\chi) = \sigma(\chi)^2 = \mathbb{E}(\chi^2) - \mathbb{E}(\chi)^2.$$

Proof. The PGF $p_n(u)$ of χ over $\mathcal{A}_n$ is given by Proposition III.1. On the other hand, factorial moments are on general grounds obtained by differentiation and evaluation at $u = 1$. The result follows. ∎

In other words, the quantities

$$\Omega_n^{(k)} := \omega_n \cdot \left([z^n]\, \partial_u^k A(z, u)\Big|_{u=1}\right)$$

give, after a simple normalization (by $\omega_n \cdot [z^n]A(z, 1)$), the factorial moments:

$$\mathbb{E}\left(\chi(\chi - 1)\cdots(\chi - k + 1)\right) = \frac{1}{A_n}\Omega_n^{(k)}.$$

Most notably, $\Omega_n^{(1)}$ is the *cumulated value* of χ over all objects of $\mathcal{A}_n$:

$$\Omega_n^{(1)} \equiv \omega_n \cdot [z^n]\, \partial_u A(z, u)|_{u=1} = \sum_{\alpha \in \mathcal{A}_n} \chi(\alpha) \equiv A_n \cdot \mathbb{E}_{\mathcal{A}_n}(\chi).$$

Accordingly, the GF (ordinary or exponential) of the $\Omega_n^{(1)}$ is sometimes named the *cumulative* generating function. It can be viewed as an unnormalized generating function of the sequence of expected values. These considerations explain Wilf's suggestive motto quoted on p. 151: "*Generating functions find averages, etc*". (The "*etc*" can be interpreted as a token for higher moments and probability distributions.)
▷ **III.2.** *A combinatorial form of cumulative GFs.* One has

$$\Omega^{(1)}(z) \equiv \sum_n \mathbb{E}_{\mathcal{A}_n}(\chi)A_n\frac{z^n}{\omega_n} = \sum_{\alpha \in \mathcal{A}} \chi(\alpha)\frac{z^{|\alpha|}}{\omega_{|\alpha|}},$$

where $\omega_n = 1$ (ordinary case) or $\omega_n = n!$ (exponential case). ◁

Example III.3. *Moments of the binomial distribution.* The binomial distribution of index n can be defined as the distribution of the number of as in a random word of length n over the binary alphabet $\{a, b\}$. The determination of moments results easily from the ordinary BGF,

$$W(z, u) = \frac{1}{1 - z - zu}.$$

By differentiation, one finds

$$\frac{\partial^r}{\partial u^r}W(z, u)\Big|_{u=1} = \frac{r!z^r}{(1 - 2z)^{r+1}}.$$

Coefficient extraction then gives the form of the factorial moments of orders $1, 2, 3, \ldots, r$ as

$$\frac{n}{2}, \quad \frac{n(n - 1)}{4}, \quad \frac{n(n - 1)(n - 2)}{8}, \ldots, \quad \frac{r!}{2^r}\binom{n}{r}.$$

In particular, the mean and the variance are $\frac{1}{2}n$ and $\frac{1}{4}n$. The standard deviation is thus $\frac{1}{2}\sqrt{n}$ which is of a smaller order than the mean: this indicates that the distribution is somehow concentrated around its mean value, as suggested by Figure III.4. .... ∎

▷ **III.3.** *De Moivre's approximation of the binomial coefficients.* The fact that the mean and the standard deviation of the binomial distribution are respectively $\frac{1}{2}n$ and $\frac{1}{2}\sqrt{n}$ suggests we examine what goes on at a distance of x standard deviations from the mean. Consider for simplicity the case of $n = 2\nu$ even. From the ratio

$$r(\nu, \ell) := \frac{\binom{2\nu}{\nu+\ell}}{\binom{2\nu}{\nu}} = \frac{(1 - \frac{1}{\nu})(1 - \frac{2}{\nu}) \cdots (1 - \frac{k-1}{\nu})}{(1 + \frac{1}{\nu})(1 + \frac{2}{\nu}) \cdots (1 + \frac{k}{\nu})},$$

the approximation $\log(1 + x) = x + O(x^2)$ shows that, for any fixed $y \in \mathbb{R}$,

$$\lim_{n \to \infty, \, \ell = \nu + y\sqrt{\nu/2}} \frac{\binom{2\nu}{\nu+\ell}}{\binom{2\nu}{\nu}} = e^{-y^2/2}.$$

(Alternatively, Stirling's formula can be employed.) This Gaussian approximation for the binomial distribution was discovered by Abraham de Moivre (1667–1754), a close friend of Newton. General methods for establishing such approximations are developed in Chapter IX.   ◁

***Example* III.4.** *Moments of the Stirling cycle distribution.* Let us return to the example of cycles in permutations which is of interest in connection with certain sorting algorithms like bubble sort or insertion sort, maximum finding, and *in situ* rearrangement [374].

We are dealing with labelled objects, hence exponential generating functions. As seen earlier on p. 155, the BGF of permutations counted according to cycles is

$$P(z, u) = (1 - z)^{-u}.$$

By differentiating the BGF with respect to u, then setting $u = 1$, we next get the expected number of cycles in a random permutation of size n as a Taylor coefficient:

$$(11) \qquad \mathbb{E}_n(\chi) = [z^n] \frac{1}{1 - z} \log \frac{1}{1 - z} = 1 + \frac{1}{2} + \cdots + \frac{1}{n},$$

which is the harmonic number H_n. Thus, on average, a random permutation of size n has about $\log n + \gamma$ cycles, a well-known fact of discrete probability theory, derived on p. 122 by means of horizontal generating functions.

For the variance, a further differentiation of the bivariate EGF gives

$$(12) \qquad \sum_{n \geq 0} \mathbb{E}_n(\chi(\chi - 1))z^n = \frac{1}{1 - z} \left(\log \frac{1}{1 - z} \right)^2.$$

From this expression and Note III.4 (or directly from the Stirling cycle polynomials of p. 155), a calculation shows that

$$(13) \qquad \sigma_n^2 = \left(\sum_{k=1}^{n} \frac{1}{k} \right) - \left(\sum_{k=1}^{n} \frac{1}{k^2} \right) = \log n + \gamma - \frac{\pi^2}{6} + O\left(\frac{1}{n} \right).$$

Thus, asymptotically,

$$\sigma_n \sim \sqrt{\log n}.$$

The standard deviation is of an order smaller than the mean, and therefore large deviations from the mean have an asymptotically negligible probability of occurrence (see below the discussion of moment inequalities). Furthermore, the distribution is asymptotically Gaussian, as we shall see in Chapter IX, p. 644. .... ∎

▷ **III.4.** *Stirling cycle numbers and harmonic numbers.* By the "exp–log trick" of Chapter I, p. 29, the PGF of the Stirling cycle distribution satisfies

$$\frac{1}{n!}u(u+1)\cdots(u+n-1) = \exp\left(v\,H_n - \frac{v^2}{2}\,H_n^{(2)} + \frac{v^3}{3}\,H_n^{(3)} + \cdots\right), \qquad u = 1 + v,$$

where $H_n^{(r)}$ is the generalized harmonic number $\sum_{j=1}^{n} j^{-r}$. Consequently, any moment of the distribution is a polynomial in generalized harmonic numbers; compare (11) and (13). Furthermore, the kth moment satisfies $\mathbb{E}_{\mathcal{P}_n}(\chi^k) \sim (\log n)^k$. (The same technique expresses the Stirling cycle number $\left[{n \atop k}\right]$ as a polynomial in generalized harmonic numbers $H_{n-1}^{(r)}$.)

Alternatively, start from the expansion of $(1 - z)^{-\alpha}$ and differentiate repeatedly with respect to α; for instance, one has

$$(1-z)^{-\alpha}\log\frac{1}{1-z} = \sum_{n\geq 0}\left(\frac{1}{\alpha} + \frac{1}{\alpha+1} + \cdots + \frac{1}{n-1+\alpha}\right)\binom{n+\alpha-1}{n}z^n,$$

which provides (11) upon setting $\alpha = 1$, while the next differentiation gives (13). ◁

The situation encountered with cycles in permutations is typical of iterative (non-recursive) structures. In many other cases, especially when dealing with recursive structures, the bivariate GF may satisfy complicated functional equations in two variables (see the example of path length in trees, Section III. 5 below), which means we do not know them explicitly. However, asymptotic laws can be determined in a large number of cases (Chapter IX). In all cases, the BGFs are the central tool in obtaining mean and variance estimates, since their derivatives evaluated at $u = 1$ become univariate GFs that usually satisfy much simpler relations than the BGFs themselves.

III. 2.2. Moment inequalities and concentration of distributions.

Qualitatively speaking, families of distributions can be classified into two categories: (i) distributions that are *spread*, i.e., the standard deviation is of order at least as large as the mean (e.g.the uniform distributions over $[0 . . n]$, which have totally flat histograms); (ii) distributions for which the standard deviation is of an asymptotic order smaller than the mean (e.g., the Stirling cycle distribution, Figure III.4, and the binomial distribution, Figure III.5.) Such informal observations are indeed supported by the Markov–Chebyshev inequalities, which take advantage of information provided by the first two moments. (A proof is found in Appendix A.3: *Combinatorial probability*, p. 727.)

Markov–Chebyshev inequalities. *Let X be a non-negative random variable and Y an arbitrary real variable. One has for any $t > 0$:*

$$\mathbb{P}\{X \geq t\mathbb{E}(X)\} \qquad\qquad \leq \quad \frac{1}{t} \qquad \textit{(Markov inequality)}$$

$$\mathbb{P}\{|Y - \mathbb{E}(Y)| \geq t\sigma(Y)\} \quad \leq \quad \frac{1}{t^2} \qquad \textit{(Chebyshev inequality)}.$$

This result informs us that the probability of being much larger than the mean must decay (Markov) and that an upper bound on the decay is measured in units given by the standard deviation (Chebyshev).

The next proposition formalizes a concentration property of distributions. It applies to a *family* of distributions indexed by the integers.

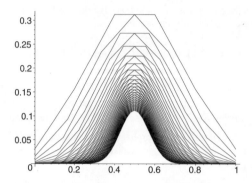

Figure III.5. Plots of the binomial distributions for $n = 5, \ldots, 50$. The horizontal axis is normalized (by a factor of $1/n$) and rescaled to 1, so that the curves display $\left\{ \mathbb{P}(\frac{X_n}{n} = x) \right\}$, for $x = 0, \frac{1}{n}, \frac{2}{n}, \ldots$.

Proposition III.3 (Concentration of distribution). *Consider a family of random variables X_n, typically, a scalar parameter χ on the subclass $\mathcal{A}_n$. Assume that the means $\mu_n = \mathbb{E}(X_n)$ and the standard deviations $\sigma_n = \sigma(X_n)$ satisfy the condition*

$$\lim_{n \to +\infty} \frac{\sigma_n}{\mu_n} = 0.$$

Then the distribution of X_n is concentrated *in the sense that, for any $\epsilon > 0$, there holds*

$$(14) \qquad \lim_{n \to +\infty} \mathbb{P} \left\{ 1 - \epsilon \leq \frac{X_n}{\mu_n} \leq 1 + \epsilon \right\} = 1.$$

Proof. The result is a direct consequence of Chebyshev's inequality. ∎

The concentration property (14) expresses the fact that values of X_n tend to become closer and closer (in relative terms) to the mean μ_n as n increases. Another figurative way of describing concentration, much used in random combinatorics, is to say that "X_n/μ_n *tends to 1 in probability*"; in symbols:

$$\frac{X_n}{\mu_n} \xrightarrow{P} 1.$$

When this property is satisfied, the expected value is in a strong sense a typical value—this fact is an extension of the *weak law of large numbers* of probability theory.

Concentration properties of the binomial and Stirling cycle distributions. The binomial distribution *is* concentrated, since the mean of the distribution is $n/2$ and the standard deviation is $\sqrt{n/4}$, a much smaller quantity. Figure III.5 illustrates concentration by displaying the graphs (as polygonal lines) associated to the binomial distributions for $n = 5, \ldots, 50$. Concentration is also quite perceptible on simulations as n gets large: the table below describes the results of batches of ten (sorted) simulations from the binomial distribution $\left\{ \frac{1}{2^n} \binom{n}{k} \right\}_{k=0}^n$:

$n = 100$	$39, 42, 43, 49, 50, 52, 54, 55, 55, 57$
$n = 1000$	$487, 492, 494, 494, 506, 508, 512, 516, 527, 545$
$n = 10\,000$	$4972, 4988, 5000, 5004, 5012, 5017, 5023, 5025, 5034, 5065$
$n = 100\,000$	$49798, 49873, 49968, 49980, 49999, 50017, 50029, 50080, 50101, 50284;$

the maximal deviations from the mean observed on such samples are 22% ($n = 10^2$), 9% ($n = 10^3$), 1.3% ($n = 10^4$), and 0.6% ($n = 10^5$). Similarly, the mean and variance computations of (11) and (13) imply that the number of cycles in a random permutation of large size is concentrated.

Finer estimates on distributions form the subject of our Chapter IX dedicated to limit laws. The reader may get a feeling of some of the phenomena at stake when examining Figure III.5 and Note III.3, p. 160: the visible emergence of a continuous curve (the bell-shaped curve) corresponds to a common asymptotic shape for the whole family of distributions—the *Gaussian law*.

III. 3. Inherited parameters and ordinary MGFs

In this section and the next, we address the question of determining BGFs directly from combinatorial specifications. The answer is provided by a simple extension of the symbolic method, which is formulated in terms of *multivariate generating functions* (MGFs). Such generating functions have the capability of taking into account a finite collection (equivalently, a vector) of combinatorial parameters. Bivariate generating functions discussed earlier appear as a special case.

III. 3.1. Multivariate generating functions (MGFs). The theory is best developed in full generality for the joint analysis of a fixed finite collection of parameters.

Definition III.3. *Consider a combinatorial class $\mathcal{A}$. A (multidimensional) parameter $\chi = (\chi_1, \ldots, \chi_d)$ on the class is a function from $\mathcal{A}$ to the set $\mathbb{Z}_{\geq 0}^d$ of d–tuples of natural numbers. The counting sequence of $\mathcal{A}$ with respect to size and the parameter χ is then defined by*

$$A_{n, k_1, \ldots, k_d} = \text{card}\left\{ \alpha \mid |\alpha| = n, \ \chi_1(\alpha) = k_1, \ldots, \chi_d(\alpha) = k_d \right\}.$$

We sometimes refer to such a parameter as a "multiparameter" when $d > 1$, and a "simple" or "scalar" parameter otherwise. For instance, one may take the class $\mathcal{P}$ of all permutations σ, and for χ_j ($j = 1, 2, 3$) the number of cycles of length j in σ. Alternatively, we may consider the class $\mathcal{W}$ of all words w over an alphabet with four letters, $\{\alpha_1, \ldots, \alpha_4\}$ and take for χ_j ($j = 1, \ldots, 4$) the number of occurrences of the letter α_j in w, and so on.

The *multi-index convention* employed in various branches of mathematics greatly simplifies notations: let $\mathbf{x} = (x_1, \ldots, x_d)$ be a vector of d formal variables and $\mathbf{k} = (k_1, \ldots, k_d)$ be a vector of integers of the same dimension; then, the multipower $\mathbf{x}^{\mathbf{k}}$ is defined as the monomial

(15) $$\mathbf{x}^{\mathbf{k}} := x_1^{k_1} x_2^{k_2} \cdots x_d^{k_d}.$$

With this notation, we have:

Definition III.4. *Let $A_{n,k}$ be a multi-index sequence of numbers, where $\mathbf{k} \in \mathbb{N}^d$. The multivariate generating function (MGF) of the sequence of either ordinary or exponential type is defined as the formal power series*

(16)
$$A(z, \mathbf{u}) = \sum_{n,\mathbf{k}} A_{n,\mathbf{k}} \mathbf{u}^{\mathbf{k}} z^n \quad \text{(ordinary MGF)}$$

$$A(z, \mathbf{u}) = \sum_{n,\mathbf{k}} A_{n,\mathbf{k}} \mathbf{u}^{\mathbf{k}} \frac{z^n}{n!} \quad \text{(exponential MGF)}.$$

Given a class $\mathcal{A}$ and a parameter χ, the MGF of the pair $\langle \mathcal{A}, \chi \rangle$ is the MGF of the corresponding counting sequence. In particular, one has the combinatorial forms*:*

(17)
$$A(z, \mathbf{u}) = \sum_{\alpha \in \mathcal{A}} \mathbf{u}^{\chi(\alpha)} z^{|\alpha|} \quad \text{(ordinary MGF; unlabelled case)}$$

$$A(z, \mathbf{u}) = \sum_{\alpha \in \mathcal{A}} \mathbf{u}^{\chi(\alpha)} \frac{z^{|\alpha|}}{|\alpha|!} \quad \text{(exponential MGF; labelled case)}.$$

One also says that $A(z, \mathbf{u})$ is the MGF of the combinatorial class with the formal variable u_j marking the parameter χ_j and z marking size.

From the very definition, with $\mathbf{1}$ a vector of all 1's, the quantity $A(z, \mathbf{1})$ coincides with the generating function of $\mathcal{A}$, either ordinary or exponential as the case may be. One can then view an MGF as a deformation of a univariate GF by way of a vector $\mathbf{u}$, with the property that the multivariate GF reduces to the univariate GF at $\mathbf{u} = \mathbf{1}$. If all but one of the u_j are set to 1, then a BGF results; in this way, the symbolic calculus that we are going to develop gives full access to BGFs (and, from here, to moments).

▷ **III.5.** *Special cases of MGFs.* The exponential MGF of permutations with u_1, u_2 marking the number of 1–cycles and 2–cycles respectively is

(18)
$$P(z, u_1, u_2) = \frac{\exp\left((u_1 - 1)z + (u_2 - 1)\frac{z^2}{2}\right)}{1 - z}.$$

(This will be proved later in this chapter, p. 187.) The formula is checked to be consistent with three already known special cases derived in Chapter II: (*i*) setting $u_1 = u_2 = 1$ gives back the counting of *all* permutations, $P(z, 1, 1) = (1 - z)^{-1}$, as it should; (*ii*) setting $u_1 = 0$ and $u_2 = 1$ gives back the EGF of derangements, namely $e^{-z}/(1 - z)$; (*iii*) setting $u_1 = u_2 = 0$ gives back the EGF of permutations with cycles all of length greater than 2, $P(z, 0, 0) = e^{-z-z^2/2}/(1 - z)$, a generalized derangement GF. In addition, the particular BGF

$$P(z, u, 1) = \frac{e^{(u-1)z}}{1 - z},$$

enumerates permutations according to singleton cycles. This last BGF interpolates between the EGF of derangements ($u = 0$) and the EGF of all permutations ($u = 1$). ◁

III. 3.2. Inheritance and MGFs.

Parameters that are *inherited* from substructures (definition below) can be taken into account by a direct extension of the symbolic method. With a suitable use of the multi-index conventions, it is even the case that the translation rules previously established in Chapters I and II can be copied verbatim. This approach provides a large quantity of multivariate enumeration results that follow automatically by the symbolic method.

Definition III.5. *Let* $\langle \mathcal{A}, \chi \rangle$, $\langle \mathcal{B}, \xi \rangle$, $\langle \mathcal{C}, \zeta \rangle$ *be three combinatorial classes endowed with parameters of the same dimension d. The parameter* χ *is said to be* inherited *in the following cases.*

- *Disjoint union: when* $\mathcal{A} = \mathcal{B} + \mathcal{C}$, *the parameter* χ *is inherited from* ξ, ζ *iff its value is determined by cases from* ξ, ζ:

$$
\chi(\omega) = \begin{cases} \xi(\omega) & \textit{if } \omega \in \mathcal{B} \\ \zeta(\omega) & \textit{if } \omega \in \mathcal{C}. \end{cases}
$$

- *Cartesian product: when* $\mathcal{A} = \mathcal{B} \times \mathcal{C}$, *the parameter* χ *is inherited from* ξ, ζ *iff its value is obtained additively from the values of* ξ, ζ:

$$
\chi(\beta, \gamma) = \xi(\beta) + \zeta(\gamma).
$$

- *Composite constructions: when* $\mathcal{A} = \mathfrak{K}\{B\}$, *where* $\mathfrak{K}$ *is a metasymbol representing any of* SEQ, MSET, PSET, CYC, *the parameter* χ *is inherited from* ξ *iff its value is obtained additively from the values of* ξ *on components; for instance, for sequences:*

$$
\chi(\beta_1, \ldots, \beta_r) = \xi(\beta_1) + \cdots + \xi(\beta_r).
$$

With a natural extension of the notation used for constructions, we shall write

$$
\langle \mathcal{A}, \chi \rangle = \langle \mathcal{B}, \xi \rangle + \langle \mathcal{C}, \zeta \rangle, \quad \langle \mathcal{A}, \chi \rangle = \langle \mathcal{B}, \xi \rangle \times \langle \mathcal{C}, \zeta \rangle, \quad \langle \mathcal{A}, \chi \rangle = \mathfrak{K}\{\langle \mathcal{B}, \xi \rangle\}.
$$

This definition of inheritance is seen to be a natural extension of the axioms that size itself has to satisfy (Chapter I): size of a disjoint union is defined by cases; size of a pair, and similarly of a composite construction, is obtained by addition.

Next, we need a bit of formality. Consider a pair $\langle \mathcal{A}, \chi \rangle$, where $\mathcal{A}$ is a combinatorial class endowed with its usual size function $|\cdot|$ and $\chi = (\chi_1, \ldots, \chi_d)$ is a d-dimensional (multi)parameter. Write χ_0 for size and z_0 for the variable marking size (previously denoted by z). The key point is to define an *extended multiparameter* $\overline{\chi} = (\chi_0, \chi_1, \ldots, \chi_d)$; that is, *we treat size and parameters on an equal opportunity basis*. Then the ordinary MGF in (16) assumes an extremely simple and symmetrical form:

$$
(19) \qquad A(\mathbf{z}) = \sum_{\mathbf{k}} A_{\mathbf{k}} \mathbf{z}^{\mathbf{k}} = \sum_{\alpha \in \mathcal{A}} \mathbf{z}^{\overline{\chi}(\alpha)}.
$$

Here, the indeterminates are the vector $\mathbf{z} = (z_0, z_1, \ldots, z_d)$, the indices are $\mathbf{k} = (k_0, k_1, \ldots, k_d)$, where k_0 indexes size (previously denoted by n) and the usual multi-index convention introduced in (15) is in force:

$$
(20) \qquad \mathbf{z}^{\mathbf{k}} := z_0^{k_0} z_1^{k_1} \cdots z_d^{k_d},
$$

but it is now applied to $(d + 1)$-dimensional vectors. With this convention, we have:

Theorem III.1 (Inherited parameters and ordinary MGFs). *Let* $\mathcal{A}$ *be a combinatorial class constructed from* $\mathcal{B}, \mathcal{C}$, *and let* χ *be a parameter inherited from* ξ *defined on* $\mathcal{B}$ *and (as the case may be) from* ζ *on* $\mathcal{C}$. *Then the translation rules of admissible constructions stated in Theorem I.1, p. 27, are applicable, provided the multi-index*

convention (19) *is used. The associated operators on ordinary MGFs are then* ($\varphi(k)$
is the Euler totient function, defined on p. 721):

Union: $\mathcal{A} = \mathcal{B} + \mathcal{C}$ $\implies$ $A(\mathbf{z}) = B(\mathbf{z}) + C(\mathbf{z})$,

Product: $\mathcal{A} = \mathcal{B} \times \mathcal{C}$ $\implies$ $A(\mathbf{z}) = B(\mathbf{z}) \cdot C(\mathbf{z})$,

Sequence: $\mathcal{A} = \text{SEQ}(\mathcal{B})$ $\implies$ $A(\mathbf{z}) = \dfrac{1}{1 - B(\mathbf{z})}$,

Powerset: $\mathcal{A} = \text{PSET}(\mathcal{B})$ $\implies$ $A(\mathbf{z}) = \exp\left(\displaystyle\sum_{\ell=1}^{\infty} \dfrac{(-1)^{\ell-1}}{\ell} B(\mathbf{z}^{\ell}) \right).$

Multiset: $\mathcal{A} = \text{MSET}(\mathcal{B})$ $\implies$ $A(\mathbf{z}) = \exp\left(\displaystyle\sum_{\ell=1}^{\infty} \dfrac{1}{\ell} B(\mathbf{z}^{\ell}) \right),$

Cycle: $\mathcal{A} = \text{CYC}(\mathcal{B})$ $\implies$ $A(\mathbf{z}) = \displaystyle\sum_{\ell=1}^{\infty} \dfrac{\varphi(\ell)}{\ell} \log \dfrac{1}{1 - B(\mathbf{z}^{\ell})},$

Proof. For disjoint unions, one has

$$A(\mathbf{z}) = \sum_{\alpha \in \mathcal{A}} \mathbf{z}^{\overline{\chi}(\alpha)} = \sum_{\beta \in \mathcal{B}} \mathbf{z}^{\overline{\xi}(\beta)} + \sum_{\gamma \in \mathcal{C}} \mathbf{z}^{\overline{\zeta}(\gamma)},$$

since inheritance is defined by cases on unions. For cartesian products, one has

$$A(\mathbf{z}) = \sum_{\alpha \in \mathcal{A}} \mathbf{z}^{\overline{\chi}(\alpha)} = \sum_{\beta \in \mathcal{B}} \mathbf{z}^{\overline{\xi}(\beta)} \times \sum_{\gamma \in \mathcal{C}} \mathbf{z}^{\overline{\zeta}(\gamma)},$$

since inheritance corresponds to additivity on products.

The translation of composite constructions in the case of sequences, powersets, and multisets is then built up from the union and product schemes, in exactly the same manner as in the proof of Theorem I.1. Cycles are dealt with by the methods of Appendix A.4: *Cycle construction*, p. 729.   ∎

The multi-index notation is a crucial ingredient for developing the general theory of multivariate enumerations. When we work with only a small number of parameters, typically one or two, we will however often find it convenient to return to vectors of variables like (z, u) or (z, u, v). In this way, unnecessary subscripts are avoided.

The reader is especially encouraged to study the treatment of integer compositions in Examples III.5 and III.6 below carefully, since it illustrates the power of the multivariate symbolic method, in its bare bones version.

***Example* III.5.** *Integer compositions and MGFs I.* The class $\mathcal{C}$ of all integer compositions (Chapter I) is specified by

$$\mathcal{C} = \text{SEQ}(\mathcal{I}), \qquad \mathcal{I} = \text{SEQ}_{\geq 1}(\mathcal{Z}),$$

where $\mathcal{I}$ is the set of all positive numbers. The corresponding OGFS are

$$C(z) = \frac{1}{1 - I(z)}, \qquad I(z) = \frac{z}{1 - z},$$

so that $C_n = 2^{n-1}$ ($n \geq 1$). Say we want to enumerate compositions according to the number χ of summands. One way to proceed, in accordance with the formal definition of inheritance, is as follows. Let ξ be the parameter that takes the constant value 1 on all elements of $\mathcal{I}$. The parameter χ on compositions is inherited from the (almost trivial) parameter $\xi \equiv 1$ defined on summands. The ordinary MGF of $\langle \mathcal{I}, \xi \rangle$ is

$$I(z, u) = zu + z^2 u + z^3 u + \cdots = \frac{zu}{1 - z}.$$

Let $C(z, u)$ be the BGF of $\langle \mathcal{C}, \chi \rangle$. By Theorem III.1, the schemes translating admissible constructions in the univariate case carry over to the multivariate case, so that

$$(21) \qquad C(z, u) = \frac{1}{1 - I(z, u)} = \frac{1}{1 - u\frac{z}{1-z}} = \frac{1 - z}{1 - z(u + 1)}.$$

Et voilà! .... ∎

   Markers. There is an alternative way of arriving at MGFs, as in (21), which is important and will be of much use thoughout this book. A *marker* (or *mark*) in a specification Σ is a neutral object (i.e., an object of size 0) attached to a construction or an atom by a product. Such a marker does not modify size, so that the univariate counting sequence associated to Σ remains unaffected. On the other hand, the total number of markers that an object contains determines by design an inherited parameter, so that Theorem III.1 is automatically applicable. In this way, one may decorate specifications so as to keep track of "interesting" substructures and get BGFs automatically. The insertion of several markers similarly gives MGFs.

   For instance, say we are interested in the number of summands in compositions, as in Example III.5 above. Then, one has an enriched specification, and its translation into MGF,

$$(22) \qquad \mathcal{C} = \text{SEQ}\left(\mu\,\text{SEQ}_{\geq 1}(\mathcal{Z})\right) \quad \Longrightarrow \quad C(z, u) = \frac{1}{1 - uI(z)},$$

based on the correspondence: $\mathcal{Z} \mapsto z, \mu \mapsto u$.

Example **III.6.** *Integer compositions and MGFs II.* Consider the double parameter $\chi = (\chi_1, \chi_2)$ where χ_1 is the number of parts equal to 1 and χ_2 the number of parts equal to 2. One can write down an extended specification, with μ_1 a combinatorial mark for summands equal to 1 and μ_2 for summands equal to 2,

$$(23) \qquad \mathcal{C} = \text{SEQ}\left(\mu_1 \mathcal{Z} + \mu_2 \mathcal{Z}^2 + \text{SEQ}_{\geq 3}(\mathcal{Z})\right)$$
$$\Longrightarrow \qquad C(z, u_1, u_2) = \frac{1}{1 - (u_1 z + u_2 z^2 + z^3(1 - z)^{-1})},$$

where u_j $(j = 1, 2)$ records the number of marks of type μ_j.

   Similarly, let μ mark each summand and μ_1 mark summands equal to 1. Then, one has,

$$(24) \quad \mathcal{C} = \text{SEQ}\left(\mu\mu_1 \mathcal{Z} + \mu\,\text{SEQ}_{\geq 2}(\mathcal{Z})\right) \quad \Longrightarrow \quad C(z, u_1, u) = \frac{1}{1 - (uu_1 z + uz^2(1 - z)^{-1})},$$

where u keeps track of the total number of summands and u_1 records the number of summands equal to 1.

MGFs obtained in this way via the multivariate extension of the symbolic method can then provide explicit counts, after suitable series expansions. For instance, the number of compositions of n with k parts is, by (21),

$$[z^n u^k]\frac{1-z}{1-(1+u)z} = \binom{n}{k} - \binom{n-1}{k} = \binom{n-1}{k-1},$$

a result otherwise obtained in Chapter I by direct combinatorial reasoning (the balls-and-bars model). The number of compositions of n containing k parts equal to 1 is obtained from the special case $u_2 = 1$ in (23),

$$[z^n u^k]\frac{1}{1-uz-\frac{z^2}{(1-z)}} = [z^{n-k}]\frac{(1-z)^{k+1}}{(1-z-z^2)^{k+1}},$$

where the last OGF closely resembles a power of the OGF of Fibonacci numbers.

Following the discussion of Section III.2, such MGFs also carry complete information about moments. In particular, the cumulated value of the number of parts in all compositions of n has OGF

$$\partial_u C(z,u)|_{u=1} = \frac{z(1-z)}{(1-2z)^2},$$

since cumulated values are obtained via differentiation of a BGF. Therefore, the expected number of parts in a random composition of n is exactly (for $n \geq 1$)

$$\frac{1}{2^{n-1}}[z^n]\frac{z(1-z)}{(1-2z)^2} = \frac{1}{2}(n+1).$$

One further differentiation will give rise to the variance. The standard deviation is found to be $\frac{1}{2}\sqrt{n-1}$, which is of an order (much) smaller than the mean. Thus, the distribution of the number of summands in a random composition satisfies the concentration property as $n \to \infty$.

In the same vein, the number of parts equal to a fixed number r in compositions is determined by

$$\mathcal{C} = \mathrm{SEQ}\left(\mu \mathcal{Z}^r + \mathrm{SEQ}_{\neq r}(\mathcal{Z})\right) \quad \Longrightarrow \quad C(z,u) = \left(1 - \left(\frac{z}{1-z} + (u-1)z^r\right)\right)^{-1}.$$

It is then easy to pull out the expected number of r-summands in a random composition of size n. The differentiated form

$$\partial_u C(z,u)|_{u=1} = \frac{z^r(1-z)^2}{(1-2z)^2}$$

gives, by partial fraction expansion,

$$\partial_u C(z,u)|_{u=1} = \frac{2^{-r-2}}{(1-2z)^2} + \frac{2^{-r-1} - r2^{-r-2}}{1-2z} + q(z),$$

for a polynomial $q(z)$ that we do not need to make explicit. Extracting the nth coefficient of the cumulative GF $\partial_u C(z,1)$ and dividing by 2^{n-1} yields the mean number of r–parts in a random composition. Another differentiation gives access to the second moment. One obtains the following proposition.

Proposition III.4 (Summands in integer compositions). *The total number of summands in a random composition of size n has mean $\frac{1}{2}(n+1)$ and a distribution that is concentrated around the mean. The number of r summands in a composition of size n has mean*

$$\frac{n}{2^{r+1}} + O(1);$$

and a standard deviation of order $\sqrt{n}$, which also ensures concentration of distribution.

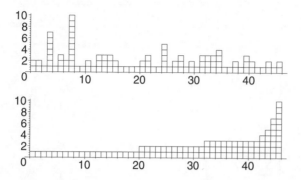

Figure III.6. A random composition of $n = 100$ represented as a ragged landscape (top); its associated profile $1^{20}2^{12}3^{10}4^{1}5^{1}7^{1}10^{1}$, defined as the partition obtained by sorting the summands (bottom).

Results of a simulation illustrating the proposition are displayed in Figure III.6 to which Note III.6 below adds further comments. .... ∎

▷ **III.6.** *The profile of integer compositions.* From the point of view of random structures, Proposition III.4 shows that random compositions of large size tend to conform to a global "profile". With high probability, a composition of size n should have about $n/4$ parts equal to 1, $n/8$ parts equal to 2, and so on. Naturally, there are statistically unavoidable fluctuations, and for any finite n, the regularity of this law cannot be perfect: it tends to fade away, especially with regard to largest summands that are $\log_2(n) + O(1)$ with high probability. (In this region mean and standard deviation both become of the same order and are $O(1)$, so that concentration no longer holds.) However, such observations *do* tell us a great deal about what a typical random composition must (probably) look like—it should conform to a "geometric profile",

$$1^{n/4} 2^{n/8} 3^{n/16} 4^{n/32} \cdots .$$

Here are for instance the profiles of two compositions of size $n = 1024$ drawn uniformly at random:

$$1^{250} 2^{138} 3^{70} 4^{29} 5^{15} 6^{10} 7^{4} 8^{0}, 9^{1} \quad \text{and} \quad 1^{253} 2^{136} 3^{68} 4^{31} 5^{13} 6^{8} 7^{3} 8^{1} 9^{1} 10^{2}.$$

These are to be compared with the "ideal" profile

$$1^{256} 2^{128} 3^{64} 4^{32} 5^{16} 6^{8} 7^{4} 8^{2} 9^{1}.$$

It is a striking fact that samples of a very few elements or even just *one* element (this would be ridiculous by the usual standards of statistics) are often sufficient to illustrate asymptotic properties of large random structures. The reason is once more to be attributed to concentration of distributions whose effect is manifest here. Profiles of a similar nature present themselves among objects defined by the sequence construction, as we shall see throughout this book. (Establishing such general laws is usually not difficult but it requires the full power of complex analytic methods developed in Chapters IV–VIII.) ◁

▷ **III.7.** *Largest summands in compositions.* For any $\epsilon > 0$, with probability tending to 1 as $n \to \infty$, the largest summand in a random integer composition of size n is in the interval $[(1 - \epsilon) \log_2 n, (1 + \epsilon) \log_2 n]$. (Hint: use the first and second moment methods. More precise estimates are obtained by the methods of Example V.4, p. 308.) ◁

$\mathfrak{K}$	BGF $(A(z,u))$	cumulative GF $(\Omega(z))$
SEQ :	$\dfrac{1}{1-uB(z)}$	$A(z)^2 \cdot B(z) = \dfrac{B(z)}{(1-B(z))^2}$
PSET :	$\begin{cases} \exp\left(\displaystyle\sum_{k=1}^{\infty}(-1)^{k-1}\dfrac{u^k}{k}B(z^k)\right) \\[1em] \displaystyle\prod_{n=1}^{\infty}(1+uz^n)^{B_n} \end{cases}$	$A(z)\cdot\displaystyle\sum_{k=1}^{\infty}(-1)^{k-1}B(z^k)$
MSET :	$\begin{cases} \exp\left(\displaystyle\sum_{k=1}^{\infty}\dfrac{u^k}{k}B(z^k)\right) \\[1em] \displaystyle\prod_{n=1}^{\infty}(1-uz^n)^{-B_n} \end{cases}$	$A(z)\cdot\displaystyle\sum_{k=1}^{\infty}B(z^k)$
CYC :	$\displaystyle\sum_{k=1}^{\infty}\dfrac{\varphi(k)}{k}\log\dfrac{1}{1-u^kB(z^k)}$	$\displaystyle\sum_{k=1}^{\infty}\varphi(k)\dfrac{B(z^k)}{1-B(z^k)}.$

Figure III.7. Ordinary GFs relative to the number of components in $\mathcal{A} = \mathfrak{K}(\mathcal{B})$.

Simplified notation for markers. It proves highly convenient to simplify notations, much in the spirit of our current practice, where the atom $\mathcal{Z}$ is reflected by
the name of the variable z in GFs. The following convention will be systematically
adopted: *the same symbol (usually u, v, u_1, u_2 ...) is freely employed to designate a
combinatorial marker (of size 0) and the corresponding marking variable in MGFs.*

For instance, we can write directly, for compositions,

$$\mathcal{C} = \text{SEQ}(u\,\text{SEQ}_{\geq 1}\,\mathcal{Z})), \qquad \mathcal{C} = \text{SEQ}(uu_1\mathcal{Z} + u\,\text{SEQ}_{\geq 2}\,\mathcal{Z})),$$

where u marks all summands and u_1 marks summands equal to 1, giving rise to (22)
and (24) above. The symbolic scheme of Theorem III.1 invariably applies to enumeration according to the number of markers.

III.3.3. Number of components in abstract unlabelled schemas. Consider a
construction $\mathcal{A} = \mathfrak{K}(\mathcal{B})$, where the metasymbol $\mathfrak{K}$ designates any standard *unlabelled*
constructor among SEQ, MSET, PSET, CYC. What is sought is the BGF $A(z,u)$ of
class $\mathcal{A}$, with u marking each component. The specification is then of the form

$$\mathcal{A} = \mathfrak{K}(u\mathcal{B}), \qquad \mathfrak{K} = \text{SEQ, MSET, PSET, CYC}.$$

Theorem III.1 applies and yields immediately the BGF $A(z,u)$. In addition, differentiating with respect to u then setting $u = 1$ provides the GF of cumulated values
(hence, in a non-normalized form, the OGF of the sequence of mean values of the
number of components):

$$\Omega(z) = \frac{\partial}{\partial u}A(z,u)\bigg|_{u=1}.$$

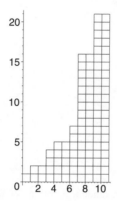

Figure III.8. A random partition of size $n = 100$ has an aspect rather different from the profile of a random composition of the same size (Figure III.6).

In summary:

Proposition III.5 (Components in unlabelled schemas). *Given a construction, $\mathcal{A} = \mathfrak{K}(\mathcal{B})$, the BGF $A(z, u)$ and the cumulated GF $\Omega(z)$ associated to the number of components are given by the table of Figure III.7.*

Mean values are then recovered with the usual formula,

$$\mathbb{E}_{\mathcal{A}_n}(\# \text{ components}) = \frac{[z^n]\Omega(z)}{[z^n]A(z)}.$$

▷ **III.8.** *r–Components in abstract unlabelled schemas.* Consider unlabelled structures. The BGF of the number of r–components in $\mathcal{A} = \mathfrak{K}\{\mathcal{B}\}$ is given by

$$A(z, u) = \left(1 - B(z) - (u - 1)B_r z^r\right)^{-1}, \qquad A(z, u) = A(z) \cdot \left(\frac{1 - z^r}{1 - uz^r}\right)^{B_r},$$

in the case of sequences ($\mathfrak{K} = \text{SEQ}$) and multisets ($\mathfrak{K} = \text{MSET}$), respectively. Similar formulae hold for the other basic constructions and for cumulative GFs. ◁

▷ **III.9.** *Number of distinct components in a multiset.* The specification and the BGF are

$$\prod_{\beta \in B} \left(1 + u \, \text{SEQ}_{\geq 1}(\beta)\right) \qquad \Longrightarrow \qquad \prod_{n \geq 1} \left(1 + \frac{uz^n}{1 - z^n}\right)^{B_n},$$

as follows from first principles. ◁

As an illustration of Proposition III.5, we discuss the profile of random partitions (Figure III.8).

***Example* III.7.** *The profile of partitions.* Let $\mathcal{P} = \text{MSET}(\mathcal{I})$ be the class of all integer partitions, where $\mathcal{I} = \text{SEQ}_{\geq 1}(\mathcal{Z})$ represents integers in unary notation. The BGF of $\mathcal{P}$ with u marking the number χ of parts (or summands) is obtained from the specification

$$\mathcal{P} = \text{MSET}(u\mathcal{I}) \qquad \Longrightarrow \qquad P(z, u) = \exp\left(\sum_{k=1}^{\infty} \frac{u^k}{k} \frac{z^k}{1 - z^k}\right).$$

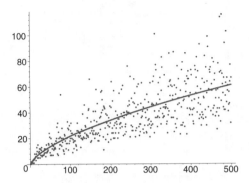

Figure III.9. The number of parts in random partitions of size $1, \ldots, 500$: exact values of the mean and simulations (circles, one for each value of n).

Equivalently, from first principles,

$$\mathcal{P} \cong \prod_{n=1}^{\infty} \text{SEQ}(u\mathcal{I}_n) \quad \Longrightarrow \quad \prod_{n=1}^{\infty} \frac{1}{1 - uz^n}.$$

The OGF of cumulated values then results from the second form of the BGF by logarithmic differentiation:

$$(25) \qquad\qquad\qquad \Omega(z) = P(z) \cdot \sum_{k=1}^{\infty} \frac{z^k}{1 - z^k}.$$

Now, the factor on the right in (25) can be expanded as

$$\sum_{k=1}^{\infty} \frac{z^k}{1 - z^k} = \sum_{n=1}^{\infty} d(n) z^n,$$

with $d(n)$ the number of divisors of n. Thus, the mean value of χ is

$$(26) \qquad\qquad\qquad \mathbb{E}_n(\chi) = \frac{1}{P_n} \sum_{j=1}^{n} d(j) P_{n-j}.$$

The same technique applies to the number of parts equal to r. The form of the BGF is

$$\widetilde{\mathcal{P}} \cong \text{SEQ}(u\mathcal{I}_r) \times \prod_{n \neq r} \text{SEQ}(\mathcal{I}_n) \quad \Longrightarrow \quad \widetilde{P}(z, u) = \frac{1 - z^r}{1 - uz^r} \cdot P(z),$$

which implies that the mean value of the number $\widetilde{\chi}$ of r–parts satisfies

$$\mathbb{E}_n(\widetilde{\chi}) = \frac{1}{P_n} [z^n] \left(P(z) \cdot \frac{z^r}{1 - z^r} \right) = \frac{1}{P_n} \left(P_{n-r} + P_{n-2r} + P_{n-3r} + \cdots \right).$$

From these formulae and a decent symbolic manipulation package, the means are calculated easily up to values of n well into the range of several thousand. . . . . ∎

The comparison between Figures III.6 and III.8 shows that different combinatorial models may well lead to rather different types of probabilistic behaviours. Figure III.9 displays the exact value of the mean number of parts in random partitions of size $n = 1, \ldots, 500$, (as calculated from (26)) accompanied with the observed values of one

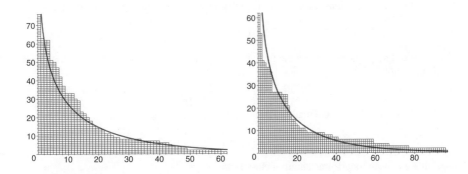

Figure III.10. Two partitions of $\mathcal{P}_{1000}$ drawn at random, compared to the limiting shape $\Psi(x)$ defined by (27).

random sample for each value of n in the range. The mean number of parts is known to be asymptotic to
$$\frac{\sqrt{n}\log n}{\pi\sqrt{2/3}},$$
and the distribution, though it admits a comparatively large standard deviation $O(\sqrt{n})$, is still concentrated, in the technical sense of the term. We shall prove some of these assertions in Chapter VIII, p. 581.

In recent years, Vershik and his collaborators [152, 595] have shown that most integer partitions tend to conform to a definite profile given (after normalization by $\sqrt{n}$) by the continuous plane curve $y = \Psi(x)$ defined implicitly by

$$(27) \qquad y = \Psi(x) \qquad \text{iff} \qquad e^{-\alpha x} + e^{-\alpha y} = 1, \qquad \alpha = \frac{\pi}{\sqrt{6}}.$$

This is illustrated in Figure III.10 by two randomly drawn elements of $\mathcal{P}_{1000}$ represented together with the "most likely" limit shape. The theoretical result explains the huge differences that are manifest on simulations between integer compositions and integer partitions.

The last example of this section demonstrates the application of BGFs to estimates regarding the root degree of a tree drawn uniformly at random among the class $\mathcal{G}_n$ of general Catalan trees of size n. Tree parameters such as number of leaves and path length that are more global in nature and need a recursive definition will be discussed in Section III. 5 below.

Example III.8. *Root degree in general Catalan trees.* Consider the parameter χ equal to the degree of the root in a tree, and take the class $\mathcal{G}$ of all plane unlabelled trees, i.e., general Catalan trees. The specification is obtained by first defining trees $(\mathcal{G})$, then defining trees with a mark for subtrees $(\mathcal{G}^\circ)$ dangling from the root:

$$\begin{cases} \mathcal{G} = \mathcal{Z} \times \mathrm{SEQ}(\mathcal{G}) \\ \mathcal{G}^\circ = \mathcal{Z} \times \mathrm{SEQ}(u\mathcal{G}) \end{cases} \implies \begin{cases} G(z) = \dfrac{z}{1-G(z)} \\ G(z,u) = \dfrac{z}{1-uG(z)}. \end{cases}$$

This set of equations reveals that the probability that the root degree equals r is

$$\mathbb{P}_n\{\chi = r\} = \frac{1}{G_n}[z^{n-1}]G(z)^r = \frac{r}{n-1}\binom{2n-3-r}{n-2} \sim \frac{r}{2^{r+1}},$$

this by Lagrange inversion and elementary asymptotics. Furthermore, the cumulative GF is found to be

$$\Omega(z) = \frac{zG(z)}{(1-G(z))^2}.$$

The relation satisfied by G entails a further simplification,

$$\Omega(z) = \frac{1}{z}G(z)^3 = \left(\frac{1}{z} - 1\right)G(z) - 1,$$

so that the mean root degree admits a closed form,

$$\mathbb{E}_n(\chi) = \frac{1}{G_n}\left(G_{n+1} - G_n\right) = 3\frac{n-1}{n+1},$$

a quantity clearly asymptotic to 3.

A random plane tree is thus usually composed of a small number of root subtrees, at least one of which should accordingly be fairly large. .... ∎

III. 4. Inherited parameters and exponential MGFs

The theory of inheritance developed in the last section applies almost *verbatim* to labelled objects. The only difference is that the variable marking size must carry a factorial coefficient dictated by the needs of relabellings. Once more, with a suitable use of multi-index conventions, the translation mechanisms developed in the univariate case (Chapter II) remain in force, this in a way that parallels the unlabelled case.

Let us consider a pair $\langle \mathcal{A}, \chi \rangle$, where $\mathcal{A}$ is a labelled combinatorial class endowed with its size function $|\cdot|$ and $\chi = (\chi_1, \ldots, \chi_d)$ is a d-dimensional parameter. As before, the parameter χ is extended into $\overline{\chi}$ by inserting size as zeroth coordinate and a vector $\mathbf{z} = (z_0, \ldots, z_d)$ of $d+1$ indeterminates is introduced, with z_0 marking size and z_j marking χ_j. Once the multi-index convention of (20) defining $\mathbf{z}^\mathbf{k}$ has been brought into play, the exponential MGF of $\langle \mathcal{A}, \chi \rangle$ (see Definition III.4, p. 164) can be rephrased as

$$(28) \qquad A(\mathbf{z}) = \sum_{\mathbf{k}} A_\mathbf{k}\frac{\mathbf{z}^\mathbf{k}}{k_0!} = \sum_{\alpha \in \mathcal{A}} \frac{\mathbf{z}^{\overline{\chi}(\alpha)}}{|\alpha|!}.$$

This MGF is exponential in z (alias z_0) but ordinary in the other variables; only the factorial $k_0!$ is needed to take into account relabelling induced by labelled products.

We *a priori* restrict attention to parameters that do not depend on the absolute values of labels (but may well depend on the relative order of labels): a parameter is said to be *compatible* if, for any α, it assumes the same value on any labelled object α and all the order-consistent relabellings of α. A parameter is said to be *inherited* if it is compatible and it is defined by cases on disjoint unions and determined additively on labelled products—this is Definition III.5 (p. 165) with labelled products replacing cartesian products. In particular, for a compatible parameter, *inheritance signifies additivity on components of labelled sequences, sets, and cycles*. We can then cut-and-paste (with minor adjustments) the statement of Theorem III.1, p. 165:

Theorem III.2 (Inherited parameters and exponential MGFs). *Let $\mathcal{A}$ be a labelled combinatorial class constructed from $\mathcal{B}, \mathcal{C}$, and let χ be a parameter inherited from ξ defined on $\mathcal{B}$ and (as the case may be) from ζ on $\mathcal{C}$. Then the translation rules of admissible constructions stated in Theorem II.1, p. 103, are applicable, provided the multi-index convention (28) is used. The associated operators on exponential MGFs are then:*

$$
\begin{aligned}
&\text{Union:} && \mathcal{A} = \mathcal{B} + \mathcal{C} && \Longrightarrow && A(\mathbf{z}) = B(\mathbf{z}) + C(\mathbf{z}) \\
&\text{Product:} && \mathcal{A} = \mathcal{B} \star \mathcal{C} && \Longrightarrow && A(\mathbf{z}) = B(\mathbf{z}) \cdot C(\mathbf{z}) \\
&\text{Sequence:} && \mathcal{A} = \text{SEQ}(\mathcal{B}) && \Longrightarrow && A(\mathbf{z}) = \frac{1}{1 - B(\mathbf{z})} \\
&\text{Cycle:} && \mathcal{A} = \text{CYC}(\mathcal{B}) && \Longrightarrow && A(\mathbf{z}) = \log \frac{1}{1 - B(\mathbf{z})}. \\
&\text{Set:} && \mathcal{A} = \text{SET}(\mathcal{B}) && \Longrightarrow && A(\mathbf{z}) = \exp\left(B(\mathbf{z}) \right).
\end{aligned}
$$

Proof. Disjoint unions are treated in a similar manner to the unlabelled multivariate case. Labelled products result from

$$
A(\mathbf{z}) = \sum_{\alpha \in \mathcal{A}} \frac{\mathbf{z}^{\overline{\chi}(\alpha)}}{|\alpha|!} = \sum_{\beta \in \mathcal{B}, \gamma \in \mathcal{C}} \binom{|\beta| + |\gamma|}{|\beta|, |\gamma|} \frac{\mathbf{z}^{\overline{\xi}(\beta)} \mathbf{z}^{\overline{\zeta}(\gamma)}}{(|\beta| + |\gamma|)!},
$$

and the usual translation of binomial convolutions that reflect labellings by means of products of exponential generating functions (like in the univariate case detailed in Chapter II). The translation for composite constructions is then immediate. ∎

This theorem can be exploited to determine moments, in a way that entirely parallels its unlabelled counterpart.

Example III.9. *The profile of permutations.* Let $\mathcal{P}$ be the class of all permutations and χ the number of components. Using the concept of marking, the specification and the exponential BGF are

$$
\mathcal{P} = \text{SET}\left(u \, \text{CYC}(\mathcal{Z}) \right) \qquad \Longrightarrow \qquad P(z, u) = \exp\left(u \log \frac{1}{1 - z} \right) = (1 - z)^{-u},
$$

as was already obtained by an *ad hoc* calculation in (5). We also know (p. 160) that the mean number of cycles is the harmonic number H_n and that the distribution is concentrated, since the standard deviation is much smaller than the mean.

Regarding the number $\overline{\chi}$ of cycles of length r, the specification and the exponential BGF are now

$$
\mathcal{P} = \text{SET}\left(\text{CYC}_{\neq r}(\mathcal{Z}) + u \, \text{CYC}_{=r}(\mathcal{Z}) \right)
$$

(29)

$$
\Longrightarrow \quad P(z, u) = \exp\left(\log \frac{1}{1 - z} + (u - 1) \frac{z^r}{r} \right) = \frac{e^{(u-1)z^r / r}}{1 - z}.
$$

The EGF of cumulated values is then

(30)
$$
\Omega(z) = \frac{z^r}{r} \frac{1}{1 - z}.
$$

The result is a remarkably simple one: *In a random permutation of size n, the mean number of r–cycles is equal to $1/r$ for any $r \leq n$.*

Thus, the profile of a random permutation, where profile is defined as the ordered sequence of cycle lengths, departs significantly from what has been encountered for integer compositions

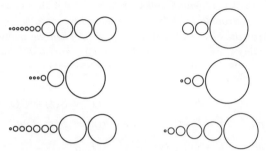

Figure III.11. The profile of permutations: a rendering of the cycle structure of six random permutations of size 500, where circle areas are drawn in proportion to cycle lengths. Permutations tend to have a few small cycles (of size $O(1)$), a few large ones (of size $\Theta(n)$), and altogether have $H_n \sim \log n$ cycles on average.

and partitions. Formula (30) also sheds a new light on the harmonic number formula for the mean number of cycles—each term $1/r$ in the harmonic number expresses the mean number of r–cycles.

As formulae are so simple, one can extract more information. By (29) one has

$$\mathbb{P}\{\overline{\chi} = k\} = \frac{1}{k!\,r^k}[z^{n-kr}]\frac{e^{-z^r/r}}{1-z},$$

where the last factor counts permutations without cycles of length r. From this (and the asymptotics of generalized derangement numbers in Note IV.9, p. 261), one proves easily that the asymptotic law of the number of r–cycles is Poisson[1] of rate $1/r$; in particular it is not concentrated. (This interesting property to be established in later chapters constitutes the starting point of an important study by Shepp and Lloyd [540].)

Furthermore, the mean number of cycles whose size is between $n/2$ and n is $H_n - H_{\lfloor n/2 \rfloor}$, a quantity that equals the probability of *existence* of such a long cycle and is approximately $\log 2 \doteq 0.69314$. In other words, we expect a random permutation of size n to have one or a few large cycles. (See the article of Shepp and Lloyd [540] for the original discussion of largest and smallest cycles.) .... ∎

▷ **III.10.** *A hundred prisoners II.* This is the solution to the prisoners problem of Note II.15, p. 124 The better strategy goes as follows. Each prisoner will first open the drawer which corresponds to his number. If his number is not there, he'll use the number he just found to access another drawer, then find a number there that points him to a third drawer, and so on, hoping to return to his original drawer in at most 50 trials. (The last opened drawer will then contain his number.) This strategy globally succeeds provided the initial permutation σ defined by σ_i (the number contained in drawer i) has *all* its cycles of length at most 50. The probability of the event is

$$p = [z^{100}]\exp\left(\frac{z}{1} + \frac{z^2}{2} + \cdots + \frac{z^{50}}{50}\right) = 1 - \sum_{j=51}^{100} \frac{1}{j} \doteq 0.31182\,78206.$$

---

[1] The Poisson distribution of rate $\lambda > 0$ has the non-negative integers as support and is determined by

$$\mathbb{P}\{k\} = e^{-\lambda}\frac{\lambda^k}{k!}.$$

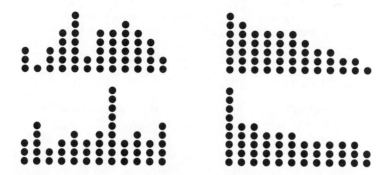

Figure III.12. Two random allocations with $m = 12$, $n = 48$, corresponding to $\lambda \equiv n/m = 4$ (left). The right-most diagrams display the bins sorted by decreasing order of occupancy.

Do the prisoners stand a chance against a malicious director who would not place the numbers in drawers at random? For instance, the director might organize the numbers in a cyclic permutation. [Hint: randomize the problem by renumbering the drawers according to a randomly chosen permutation.] ◁

Example III.10. *Allocations, balls-in-bins models, and the Poisson law.* Random allocations and the balls-in-bins model were introduced in Chapter II in connection with the birthday paradox and the coupon collector problem. Under this model, there are n balls thrown into m bins in all possible ways, the total number of allocations being thus m^n. By the labelled construction of words, the bivariate EGF with z marking the number of balls and u marking the number $\chi^{(s)}$ of bins that contain s balls (s a fixed parameter) is given by

$$\mathcal{A} = \text{SEQ}_m\left(\text{SET}_{\neq s}(\mathcal{Z}) + u\,\text{SET}_{=s}(\mathcal{Z})\right) \implies A^{(s)}(z, u) = \left(e^z + (u - 1)\frac{z^s}{s!}\right)^m.$$

In particular, the distribution of the number of empty bins ($\chi^{(0)}$) is expressible in terms of Stirling partition numbers:

$$\mathbb{P}_{m,n}(\chi^{(0)} = k) \equiv \frac{n!}{m^n}[u^k z^n]A^{(0)}(z, u) = \frac{(m - k)!}{m^n}\binom{m}{k}\begin{Bmatrix} n \\ m - k \end{Bmatrix}.$$

By differentiating the BGF, we get an exact expression for the mean (any $s \geq 0$):

(31) $$\frac{1}{m}\mathbb{E}_{m,n}(\chi^{(s)}) = \frac{1}{s!}\left(1 - \frac{1}{m}\right)^{n-s}\frac{n(n - 1)\cdots(n - s + 1)}{m^s}.$$

Let m and n tend to infinity in such a way that $n/m = \lambda$ is a fixed constant. This regime is extremely important in many applications, some of which are listed below. The average proportion of bins containing s elements is $\frac{1}{m}\mathbb{E}_{m,n}(\chi^{(s)})$, and from (31), one obtains by straightforward calculations the asymptotic limit estimate,

(32) $$\lim_{n/m=\lambda,\ n\to\infty}\frac{1}{m}\mathbb{E}_{m,n}(\chi^{(s)}) = e^{-\lambda}\frac{\lambda^s}{s!}.$$

(See Figure III.12 for two simulations corresponding to $\lambda = 4$.) In other words, a Poisson formula describes the average proportion of bins of a given size in a large random allocation. (Equivalently, the occupancy of a random bin in a random allocation satisfies a Poisson law in the limit.)

$\mathfrak{K}$	exponential BGF $(A(z, u))$	cumulative GF $(\Omega(z))$
SEQ :	$\dfrac{1}{1 - uB(z)}$	$A(z)^2 \cdot B(z) = \dfrac{B(z)}{(1 - B(z))^2}$
SET :	$\exp(uB(z))$	$A(z) \cdot B(z) = B(z)e^{B(z)}$
CYC :	$\log \dfrac{1}{1 - uB(z)}$	$\dfrac{B(z)}{1 - B(z)}.$

Figure III.13. Exponential GFs relative to the number of components in $\mathcal{A} = \mathfrak{K}(\mathcal{B})$.

The variance of each $\chi^{(s)}$ (with fixed s) is estimated similarly via a second derivative and one finds:

$$\mathbb{V}_{m,n}(\chi^{(s)}) \sim me^{-2\lambda}\frac{\lambda^s}{s!}E(\lambda), \quad E(\lambda) := \left(e^\lambda - \frac{s\lambda^{s-1}}{(s-1)!} - (1 - 2s)\frac{\lambda^s}{s!} - \frac{\lambda^{s+1}}{s!}\right).$$

As a consequence, one has the convergence in probability,

$$\frac{1}{m}\chi^{(s)} \xrightarrow{P} e^{-\lambda}\frac{\lambda^s}{s!},$$

valid for any *fixed* $s \geq 0$. See Example VIII.14, p. 598 for an analysis of the most filled urn. ∎

▷ **III.11.** *Hashing and random allocations.* Random allocations of balls into bins are central in the understanding of a class of important algorithms of computer science known as *hashing* [378, 537, 538, 598]: given a universe $\mathcal{U}$ of data, set up a function (called a hashing function) $h : \mathcal{U} \longrightarrow [1 .. m]$ and arrange for an array of m bins; an element $x \in \mathcal{U}$ is placed in bin number $h(x)$. If the hash function scrambles the data in a way that is suitably (pseudo)uniform, then the process of hashing a file of n records (keys, data items) into m bins is adequately modelled by a random allocation scheme. If $\lambda = n/m$, representing the "load", is kept reasonably bounded (say, $\lambda \leq 10$), the previous analysis implies that hashing allows for an almost direct access to data. (See also Example II.19, p. 146 for a strategy that folds colliding items into a table.) ◁

Number of components in abstract labelled schemas. As in the unlabelled universe, a general formula gives the distribution of the number of components for the basic constructions.

Proposition III.6. *Consider labelled structures and the parameter χ equal to the number of components in a construction $\mathcal{A} = \mathfrak{K}\{\mathcal{B}\}$, where $\mathfrak{K}$ is one of* SEQ, SET CYC. *The exponential BGF $A(z, u)$ and the exponential GF $\Omega(z)$ of cumulated values are given by the table of Figure III.13.*

Mean values are then easily recovered, and one finds

$$\mathbb{E}_n(\chi) = \frac{\Omega_n}{A_n} = \frac{[z^n]\Omega(z)}{[z^n]A(z)},$$

by the same formula as in the unlabelled case.

▷ **III.12.** *r–Components in abstract labelled schemas.* The BGF $A(z, u)$ and the cumulative EGF $\Omega(z)$ are given by the following table,

$$
\text{SEQ}: \quad \frac{1}{1 - \left(B(z) + (u - 1)\frac{B_r z^r}{r!} \right)} \qquad\qquad \frac{1}{(1 - B(z))^2} \cdot \frac{B_r z^r}{r!}
$$

$$
\text{SET}: \quad \exp\left(B(z) + (u - 1)\frac{B_r z^r}{r!} \right) \qquad\qquad e^{B(z)} \cdot \frac{B_r z^r}{r!}
$$

$$
\text{CYC}: \quad \log \frac{1}{1 - \left(B(z) + (u - 1)\frac{B_r z^r}{r!} \right)} \qquad\qquad \frac{1}{(1 - B(z))} \cdot \frac{B_r z^r}{r!},
$$

in the labelled case.   ◁

Example III.11. *Set partitions.* Set partitions $\mathcal{S}$ are sets of blocks, themselves non-empty sets of elements. The enumeration of set partitions according to the number of blocks is then given by

$$
\mathcal{S} = \text{SET}(u\,\text{SET}_{\geq 1}(\mathcal{Z})) \quad\Longrightarrow\quad S(z, u) = e^{u(e^z - 1)}.
$$

Since set partitions are otherwise known to be enumerated by the Stirling partition numbers, one has the BGF and the vertical EGFs as a corollary,

$$
\sum_{n,k} \begin{Bmatrix} n \\ k \end{Bmatrix} u^k \frac{z^n}{n!} = e^{u(e^z - 1)}, \qquad \sum_{n} \begin{Bmatrix} n \\ k \end{Bmatrix} \frac{z^n}{n!} = \frac{1}{k!}(e^z - 1)^k,
$$

which is consistent with earlier calculations of Chapter II.

The EGF of cumulated values, $\Omega(z)$ is then almost a derivative of $S(z)$:

$$
\Omega(z) = (e^z - 1)e^{e^z - 1} = \frac{d}{dz}S(z) - S(z).
$$

Thus, the mean number of blocks in a random partition of size n equals

$$
\frac{\Omega_n}{S_n} = \frac{S_{n+1}}{S_n} - 1,
$$

a quantity directly expressible in terms of Bell numbers. A delicate computation based on the asymptotic expansion of the Bell numbers reveals that the expected value and the standard deviation are asymptotic to

$$
\frac{n}{\log n}, \qquad \frac{\sqrt{n}}{\log n},
$$

respectively (Chapter VIII, p. 595). Similarly the exponential BGF of the number of blocks of size k is

$$
\mathcal{S} = \text{SET}(u\,\text{SET}_{=k}(\mathcal{Z}) + \text{SET}_{\neq 0,k}(\mathcal{Z})) \quad\Longrightarrow\quad S(z, u) = e^{e^z - 1 + (u-1)z^k/k!},
$$

out of which mean and variance can also be derived. . . . . ∎

Example III.12. *Root degree in Cayley trees.* Consider the class $\mathcal{T}$ of Cayley trees (non-plane labelled trees) and the parameter "root-degree". The basic specifications are

$$
\begin{cases} \mathcal{T} &= \mathcal{Z} \star \text{SET}(\mathcal{T}) \\ \mathcal{T}^\circ &= \mathcal{Z} \star \text{SET}(u\mathcal{T}) \end{cases} \quad\Longrightarrow\quad \begin{cases} T(z) &= ze^{T(z)} \\ T(z, u) &= ze^{uT(z)}. \end{cases}
$$

The set construction reflects the non-planar character of Cayley trees and the specification $\mathcal{T}^\circ$ is enriched by a mark associated to subtrees dangling from the root. Lagrange inversion provides the fraction of trees with root degree k,

$$\frac{1}{(k-1)!} \frac{n!}{(n-1-k)!} \frac{(n-1)^{n-2-k}}{n^{n-1}} \sim \frac{e^{-1}}{(k-1)!}, \qquad k \geq 1.$$

Similarly, the cumulative GF is found to be $\Omega(z) = T(z)^2$, so that the mean root degree satisfies

$$\mathbb{E}_{\mathcal{T}_n}(\text{root degree}) = 2\left(1 - \frac{1}{n}\right) \sim 2.$$

Thus the law of root degree is asymptotically a Poisson law of rate 1, shifted by 1. Probabilistic phenomena qualitatively similar to those encountered in plane trees are observed here, since the mean root degree is asymptotic to a constant. However a Poisson law eventually reflecting the non-planarity condition replaces the modified geometric law (known as a negative binomial law) present in plane trees. .... ■

▷ **III.13.** *Numbers of components in alignments.* Alignments ($\mathcal{O}$) are sequences of cycles (Chapter II, p. 119). The expected number of components in a random alignment of $\mathcal{O}_n$ is

$$\frac{[z^n]\log(1-z)^{-1}(1-\log(1-z)^{-1})^{-2}}{[z^n](1-\log(1-z)^{-1})^{-1}}.$$

Methods of Chapter V imply that the number of components in a random alignment has expectation $\sim n/(e-1)$ and standard deviation $\Theta(\sqrt{n})$. ◁

▷ **III.14.** *Image cardinality of a random surjection.* The expected cardinality of the image of a random surjection in $\mathcal{R}_n$ (Chapter II, p. 106) is

$$\frac{[z^n]e^z(2-e^z)^{-2}}{[z^n](2-e^z)^{-1}}.$$

The number of values whose preimages have cardinality k is obtained upon replacing the factor e^z by $z^k/k!$. By the methods of Chapters IV (p. 259) and V (p. 296), the image cardinality of a random surjection has expectation $n/(2\log 2)$ and standard deviation $\Theta(\sqrt{n})$. ◁

▷ **III.15.** *Distinct component sizes in set partitions.* Take the number of *distinct* block sizes and cycle sizes in set partitions and permutations. The bivariate EGFs are

$$\prod_{n=1}^{\infty}\left(1 - u + ue^{z^n/n!}\right), \qquad \prod_{n=1}^{\infty}\left(1 - u + ue^{z^n/n}\right),$$

as follows from first principles. ◁

Postscript: Towards a theory of schemas. Let us look back and recapitulate some of the information gathered in pages 167–180 regarding the number of components in composite structures. The classes considered in Figure III.14 are compositions of two constructions, either in the unlabelled or the labelled universe. Each entry contains the BGF for the number of components (e.g., cycles in permutations, parts in integer partitions, and so on), and the asymptotic orders of the mean and standard deviation of the number of components for objects of size n.

Some obvious facts stand out from the data and call for explanation. First the outer construction appears to play the essential rôle: outer *sequence* constructs (compare integer compositions, surjections and alignments) tend to dictate a number of

Unlabelled structures	
Integer partitions, MSET ∘ SEQ	Integer compositions, SEQ ∘ SEQ
$$\exp\left(u\frac{z}{1-z} + \frac{u^2}{2}\frac{z^2}{1-z^2} + \cdots\right)$$	$$\left(1 - u\frac{z}{1-z}\right)^{-1}$$
$\sim \dfrac{\sqrt{n}\log n}{\pi\sqrt{2/3}}, \quad \Theta(\sqrt{n})$	$\sim \dfrac{n}{2}, \quad \Theta(\sqrt{n})$

Labelled structures	
Set partitions, SET ∘ SET	Surjections, SEQ ∘ SET
$$\exp\left(u\left(e^z - 1\right)\right)$$	$$\left(1 - u\left(e^z - 1\right)\right)^{-1}$$
$\sim \dfrac{n}{\log n} \quad \sim \dfrac{\sqrt{n}}{\log n}$	$\sim \dfrac{n}{2\log 2}, \quad \Theta(\sqrt{n})$
Permutations, SET ∘ CYC	Alignments, SEQ ∘ CYC
$$\exp\left(u\log(1-z)^{-1}\right)$$	$$\left(1 - u\log(1-z)^{-1}\right)^{-1}$$
$\sim \log n, \quad \sim \sqrt{\log n}$	$\sim \dfrac{n}{e-1}, \quad \Theta(\sqrt{n})$

Figure III.14. Major properties of the number of components in six level-two structures. For each class, from top to bottom: (*i*) specification type; (*ii*) BGF; (*iii*) mean and standard deviation of the number of components.

components that is $\Theta(n)$ on average, while outer *set* constructs (compare integer partitions, set partitions, and permutations) are associated with a greater variety of asymptotic regimes. Eventually, such facts can be organized into broad *analytic schemas*, as will be seen in Chapters V–IX.

▷ **III.16.** *Singularity and probability.* The differences in behaviour are to be assigned to the rather different types of singularity involved (Chapters IV–VIII): on the one hand sets corresponding algebraically to an $\exp(\cdot)$ operator induce an exponential blow-up of singularities; on the other hand sequences expressed algebraically by quasi-inverses $(1 - \cdot)^{-1}$ are likely to induce polar singularities. Recursive structures such as trees lead to yet other types of phenomena with a number of components, e.g., the root degree, that is bounded in probability. ◁

III. 5. Recursive parameters

In this section, we adapt the general methodology of previous sections in order to treat parameters that are defined by recursive rules over structures that are themselves recursively specified. Typical applications concern trees and tree-like structures.

Regarding the number of leaves, or more generally, the number of nodes of some fixed degree, in a tree, the method of placing marks applies, as in the non-recursive case. It suffices to distinguish elements of interest and mark them by an auxiliary variable. For instance, in order to mark composite objects made of r components, where r is an integer and $\mathfrak{K}$ designates any of SEQ, SET (or MSET, PSET), CYC, one should split a construction $\mathfrak{K}(\mathcal{C})$ as follows:

$$\mathfrak{K}(\mathcal{C}) = u\mathfrak{K}_{=r}(\mathcal{C}) + \mathfrak{K}_{\neq r}(\mathcal{C}) = (u-1)\mathfrak{K}_r(\mathcal{C}) + \mathfrak{K}(\mathcal{C}).$$

This technique gives rise to specifications decorated by marks to which Theorems III.1 and III.2 apply. For a recursively-defined structure, the outcome is a functional equation defining the BGF recursively. The situation is illustrated by Examples III.13 and III.14 below in the case of Catalan trees and the parameter number of leaves.

Example III.13. *Leaves in general Catalan trees.* How many leaves does a random tree of some variety have? Can different varieties of trees be somehow distinguished by the proportion of their leaves? Beyond the botany of combinatorics, such considerations are for instance relevant to the analysis of algorithms since tree leaves, having no descendants, can be stored more economically; see [377, Sec. 2.3] for an algorithmic motivation for such questions.

Consider once more the class $\mathcal{G}$ of plane unlabelled trees, $\mathcal{G} = \mathcal{Z} \times \text{SEQ}(\mathcal{G})$, enumerated by the Catalan numbers: $G_n = \frac{1}{n}\binom{2n-2}{n-1}$. The class $\mathcal{G}^\circ$ where each leaf is marked is

$$\mathcal{G}^\circ = \mathcal{Z}u + \mathcal{Z} \times \text{SEQ}_{\geq 1}(\mathcal{G}^\circ) \qquad \Longrightarrow \qquad G(z,u) = zu + \frac{zG(z,u)}{1 - G(z,u)}.$$

The induced quadratic equation can be solved explicitly

$$G(z,u) = \frac{1}{2}\left(1 + (u-1)z - \sqrt{1 - 2(u+1)z + (u-1)^2z^2}\right).$$

It is however simpler to expand using the Lagrange inversion theorem which yields

$$\begin{aligned}
G_{n,k} &= [u^k]\left([z^n]G(z,u)\right) = [u^k]\left(\frac{1}{n}[y^{n-1}]\left(u + \frac{y}{1-y}\right)^n\right) \\
&= \frac{1}{n}\binom{n}{k}[y^{n-1}]\frac{y^{n-k}}{(1-y)^{n-k}} = \frac{1}{n}\binom{n}{k}\binom{n-2}{k-1}.
\end{aligned}$$

These numbers are known as Narayana numbers, see *EIS* **A001263**, and they surface repeatedly in connection with ballot problems. The mean number of leaves is derived from the cumulative GF, which is

$$\Omega(z) = \partial_u G(z,u)|_{u=1} = \frac{1}{2}z + \frac{1}{2}\frac{z}{\sqrt{1-4z}},$$

so that the mean is $n/2$ exactly for $n \geq 2$. The distribution is concentrated since the standard deviation is easily calculated to be $O(\sqrt{n})$. .... ∎

Example III.14. *Leaves and node types in binary trees.* The class $\mathcal{B}$ of binary plane trees, also enumerated by Catalan numbers $(B_n = \frac{1}{n+1}\binom{2n}{n})$ can be specified as

(33) $$\mathcal{B} = \mathcal{Z} + (\mathcal{B} \times \mathcal{Z}) + (\mathcal{Z} \times \mathcal{B}) + (\mathcal{B} \times \mathcal{Z} \times \mathcal{B}),$$

which stresses the distinction between four types of nodes: leaves, left branching, right branching, and binary. Let u_0, u_1, u_2 be variables that mark nodes of degree $0, 1, 2$, respectively. Then the root decomposition (33) yields, for the MGF $B = B(z, u_0, u_1, u_2)$, the functional equation

$$B = zu_0 + 2zu_1 B + zu_2 B^2,$$

which, by Lagrange inversion, gives

$$B_{n,k_0,k_1,k_2} = \frac{2^{k_1}}{n}\binom{n}{k_0, k_1, k_2},$$

subject to the natural conditions: $k_0 + k_1 + k_2 = n$ and $k_0 = k_2 + 1$. Moments can be easily calculated using this approach [499]. In particular, the mean number of nodes of each type is asymptotically:

$$\text{leaves:} \sim \frac{n}{4}, \qquad \text{1-nodes:} \sim \frac{n}{2}, \qquad \text{2-nodes:} \sim \frac{n}{4}.$$

There is an equal asymptotic proportion of leaves, double nodes, left branching, and right branching nodes. Furthermore, the standard deviation is in each case $O(\sqrt{n})$, so that all the corresponding distributions are concentrated. . . . . ∎

▷ **III.17.** *Leaves and node-degree profile in Cayley trees.* For Cayley trees, the bivariate EGF with u marking the number of leaves is the solution to

$$T(z, u) = uz + z(e^{T(z,u)} - 1).$$

(By Lagrange inversion, the distribution is expressible in terms of Stirling partition numbers.) The mean number of leaves in a random Cayley tree is asymptotic to ne^{-1}. More generally, the mean number of nodes of outdegree k in a random Cayley tree of size n is asymptotic to

$$n \cdot e^{-1} \frac{1}{k!}.$$

Degrees are thus approximately described by a Poisson law of rate 1. ◁

▷ **III.18.** *Node-degree profile in simple varieties of trees.* For a family of trees generated by $T(z) = z\phi(T(z))$ with ϕ a power series, the BGF of the number of nodes of degree k satisfies

$$T(z, u) = z \left(\phi(T(z, u)) + \phi_k(u - 1)T(z, u)^k \right),$$

where $\phi_k = [u^k]\phi(u)$. The cumulative GF is

$$\Omega(z) = z \frac{\phi_k T(z)^k}{1 - z\phi'(T(z))} = \phi_k z^2 T(z)^{k-1} T'(z),$$

from which expectations can be determined. ◁

▷ **III.19.** *Marking in functional graphs.* Consider the class $\mathcal{F}$ of finite mappings discussed in Chapter II:

$$\mathcal{F} = \text{SET}(\mathcal{K}), \qquad \mathcal{K} = \text{CYC}(\mathcal{T}), \qquad \mathcal{T} = \mathcal{Z} \star \text{SET}(\mathcal{T}).$$

The translation into EGFs is

$$F(z) = e^{K(z)}, \qquad K(z) = \log \frac{1}{1 - T(z)}, \qquad T(z) = ze^{T(z)}.$$

Here are the bivariate EGFs for (i) the number of components, (ii) the number of maximal trees, (iii) the number of leaves:

$$(i) \; e^{uK(z)}, \qquad (ii) \; \frac{1}{1 - uT(z)},$$

$$(iii) \; \frac{1}{1 - T(z, u)} \quad \text{with} \quad T(z, u) = (u - 1)z + ze^{T(z,u)}.$$

The trivariate EGF $F(u_1, u_2, z)$ of functional graphs with u_1 marking components and u_2 marking trees is

$$F(z, u_1, u_2) = \exp(u_1 \log(1 - u_2 T(z))^{-1}) = \frac{1}{(1 - u_2 T(z))^{u_1}}.$$

An explicit expression for the coefficients involves the Stirling cycle numbers. ◁

We shall now stop supplying examples that could be multiplied *ad libitum*, since such calculations greatly simplify when interpreted in the light of asymptotic analysis, as developed in Part B. The phenomena observed asymptotically are, for good reasons, especially close to what the classical theory of branching processes provides (see the books by Athreya–Ney [21] and Harris [324], as well as our discussion in the context of "complete" GFs on p. 196).

Linear transformations on parameters and path length in trees. We have so far been dealing with a parameter defined directly by recursion. Next, we turn to other parameters such as path length. As a preamble, one needs a simple linear transformation on combinatorial parameters. Let $\mathcal{A}$ be a class equipped with two scalar parameters, χ and ξ, related by

$$\chi(\alpha) = |\alpha| + \xi(\alpha).$$

Then, the combinatorial form of BGFs yields

$$\sum_{\alpha \in \mathcal{A}} z^{|\alpha|} u^{\chi(\alpha)} = \sum_{\alpha \in \mathcal{A}} z^{|\alpha|} u^{|\alpha| + \xi(\alpha)} = \sum_{\alpha \in \mathcal{A}} (zu)^{|\alpha|} u^{\xi(\alpha)} \, ;$$

that is,

(34) $A_\chi(z, u) = A_\xi(zu, u).$

This is clearly a general mechanism:

Linear transformations and MGFs: *A linear transformation on parameters induces a monomial substitution on the corresponding marking variables in MGFs.*

We now put this mechanism to use in the recursive analysis of path length in trees.

Example **III.15.** *Path length in trees.* The path length of a tree is defined as the sum of distances of all nodes to the root of the tree, where distances are measured by the number of edges on the minimal connecting path of a node to the root. Path length is an important characteristic of trees. For instance, when a tree is used as a data structure with nodes containing additional information, path length represents the total cost of accessing all data items when a search is started from the root. For this reason, path length surfaces, under various models, in the analysis of algorithms, in particular, in the area of algorithms and data structures for searching and sorting (e.g., tree-sort, quicksort, radix-sort [377, 538]).

The formal definition of path length of a tree is

(35) $\lambda(\tau) := \sum_{\nu \in \tau} \text{dist}(\nu, \text{root}(\tau)),$

where the sum is over all nodes of the tree and the distance between two nodes is measured by the number of connecting edges. The definition implies an inductive rule

(36) $\lambda(\tau) = \sum_{\upsilon \prec \tau} (\lambda(\upsilon) + |\upsilon|),$

in which $\upsilon \prec \tau$ indicates a summation over all the root subtrees υ of τ. (To verify the equivalence of (35) and (36), observe that path length also equals the sum of all subtree sizes.)

From this point on, we focus the discussion on general Catalan trees (see Note III.20 for other cases): $\mathcal{G} = \mathcal{Z} \times \text{SEQ}(\mathcal{G})$. Introduce momentarily the parameter $\mu(\tau) = |\tau| + \lambda(\tau)$. Then, one has from the inductive definition (36) and the general transformation rule (34):

$$(37) \qquad G_\lambda(z, u) = \frac{z}{1 - G_\mu(z, u)} \qquad \text{and} \qquad G_\mu(z, u) = G_\lambda(zu, u).$$

In other words, $G(z, u) \equiv G_\lambda(z, u)$ satisfies a nonlinear functional equation of the difference type:

$$G(z, u) = \frac{z}{1 - G(uz, u)}.$$

(This functional equation will be revisited in connection with area under Dyck paths in Chapter V, p. 330.) The generating function $\Omega(z)$ of cumulated values of λ is then obtained by differentiation with respect to u, then setting $u = 1$. We find in this way that the cumulative GF $\Omega(z) := \partial_u G(z, u)|_{u=1}$ satisfies

$$\Omega(z) = \frac{z}{(1 - G(z))^2} \left(zG'(z) + \Omega(z) \right),$$

which is a linear equation that solves to

$$\Omega(z) = z^2 \frac{G'(z)}{(1 - G(z))^2 - z} = \frac{z}{2(1 - 4z)} - \frac{z}{2\sqrt{1 - 4z}}.$$

Consequently, one has ($n \geq 1$)

$$\Omega_n = 2^{2n-3} - \frac{1}{2} \binom{2n - 2}{n - 1},$$

where the sequence starting $1, 5, 22, 93, 386$ for $n \geq 2$ constitutes *EIS* **A000346**. By elementary asymptotic analysis, we get:

> *The mean path length of a random Catalan tree of size n is asymptotic to $\frac{1}{2}\sqrt{\pi n^3}$; in short: a branch from the root to a random node in a random Catalan tree of size n has expected length of the order of $\sqrt{n}$.*

Random Catalan trees thus tend to be somewhat imbalanced—by comparison, a fully balanced binary tree has all paths of length at most $\log_2 n + O(1)$. .... ∎

The imbalance in random Catalan trees is a general phenomenon—it holds for binary Catalan and more generally for all simple varieties of trees. Note III.20 below and Example VII.9 (p. 461) imply that path length is invariably of order $n\sqrt{n}$ on average in such cases. Height is of typical order $\sqrt{n}$ as shown by Rényi and Szekeres [507], de Bruijn, Knuth, and Rice [145], Kolchin [386], as well as Flajolet and Odlyzko [246]: see Subsection VII. 10.2, p. 535 for the outline of a proof. Figure III.15 borrowed from [538] illustrates this on a simulation. (The contour of the histogram of nodes by levels, once normalized, has been proved to converge to the process known as Brownian excursion.)

▷ **III.20.** *Path length in simple varieties of trees.* The BGF of path length in a variety of trees generated by $T(z) = z\phi(T(z))$ satisfies

$$T(z, u) = z\phi(T(zu, u)).$$

In particular, the cumulative GF is

$$\Omega(z) \equiv \partial_u (T(z, u))_{u=1} = \frac{\phi'(T(z))}{\phi(T(z))} (zT'(z))^2,$$

from which coefficients can be extracted. ◁

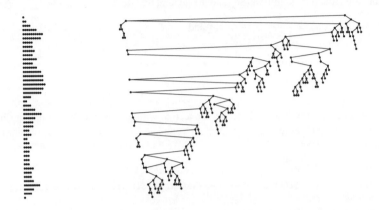

Figure III.15. A random pruned binary tree of size 256 and its associated level profile: the histogram on the left displays the number of nodes at each level in the tree.

III. 6. Complete generating functions and discrete models

By a *complete* generating function, we mean, loosely speaking, a generating function in a (possibly large, and even infinite in the limit) number of variables that mark a homogeneous collection of characteristics of a combinatorial class[2]. For instance one might be interested in the joint distribution of *all* the different letters composing words, the number of cycles of *all* lengths in permutations, and so on. A complete MGF naturally entails detailed knowledge on the enumerative properties of structures to which it is relative. Complete generating functions, given their expressive power, also make weighted models amenable to calculation, a situation that covers in particular Bernoulli trials (p. 190) and branching processes from classical probability theory (p. 196).

Complete GFs for words. As a basic example, consider the class of all words $\mathcal{W} = \text{SEQ}\{\mathcal{A}\}$ over some finite alphabet $\mathcal{A} = \{a_1, \ldots, a_r\}$. Let $\chi = (\chi_1, \ldots, \chi_r)$, where $\chi_j(w)$ is the number of occurrences of the letter a_j in word w. The MGF of $\mathcal{A}$ with respect to χ is

$$\mathcal{A} = u_1 a_1 + u_2 a_2 + \cdots u_r a_r \qquad \Longrightarrow \qquad A(z, \mathbf{u}) = z u_1 + z u_2 + \cdots + z u_r,$$

and χ on $\mathcal{W}$ is clearly inherited from χ on $\mathcal{A}$. Thus, by the sequence rule, one has

$$(38) \qquad \mathcal{W} = \text{SEQ}(\mathcal{A}) \qquad \Longrightarrow \qquad W(z, \mathbf{u}) = \frac{1}{1 - z(u_1 + u_2 + \cdots + u_r)},$$

which describes all words according to their compositions into letters. In particular, the number of words with n_j occurrences of letter a_j and with $n = \sum n_j$ is in this

---

[2]Complete GFs are *not* new objects. They are simply an avatar of multivariate GFs. Thus the term is only meant to be suggestive of a particular usage of MGFs, and essentially no new theory is needed in order to cope with them.

framework obtained as

$$[u_1^{n_1} u_2^{n_2} \cdots u_r^{n_r}] (u_1 + u_2 + \cdots + u_r)^n = \binom{n}{n_1, n_2, \ldots, n_r} = \frac{n!}{n_1! n_2! \cdots n_r}.$$

We are back to the usual multinomial coefficients.

▷ **III.21.** *After Bhaskara Acharya (circa 1150*AD*).* Consider all the numbers formed in decimal with digit 1 used once, with digit 2 used twice,..., with digit 9 used nine times. Such numbers all have 45 digits. Compute their sum S and discover, much to your amazement that S equals

458755596000061532190847692863999999999999999999954124440399993846780915230713600000.

This number has a long run of nines (and further nines are hidden!). Is there a simple explanation? This exercise is inspired by the Indian mathematician Bhaskara Acharya who discovered multinomial coefficients near 1150AD; see [377, pp. 23–24] for a brief historical note. ◁

Complete GFs for permutations and set partitions. Consider permutations and the various lengths of their cycles. The MGF where u_k marks cycles of length k for $k = 1, 2, \ldots$ can be written as an MGF in *infinitely many* variables:

$$(39) \qquad P(z, \mathbf{u}) = \exp\left(u_1 \frac{z}{1} + u_2 \frac{z^2}{2} + u_3 \frac{z^3}{3} + \cdots\right).$$

This MGF expression has the neat feature that, upon restricting all but a finite number of u_j to 1, we derive all the particular cases of interest with respect to any finite collection of cycles lengths. Observe also that one can calculate in the usual way any coefficient $[z^n]P$ as it only involves the variables $u_1, \ldots, u_n$.

▷ **III.22.** *The theory of formal power series in infinitely many variables.* (This note is for formalists.) Mathematically, an object like P in (39) is perfectly well defined. Let $U = \{u_1, u_2, \ldots\}$ be an infinite collection of indeterminates. First, the ring of polynomials $R = \mathbb{C}[U]$ is well defined and a given element of R involves only finitely many indeterminates. Then, from R, one can define the ring of formal power series in z, namely $R[[z]]$. (Note that, if $f \in R[[z]]$, then each $[z^n]f$ involves only finitely many of the variables u_j.) The basic operations and the notion of convergence, as described in Appendix A.5: *Formal power series*, p. 730, apply in a standard way.

For instance, in the case of (39), the complete GF $P(z, \mathbf{u})$ is obtainable as the formal limit

$$P(z, \mathbf{u}) = \lim_{k \to \infty} \exp\left(u_1 \frac{z}{1} + \cdots + u_k \frac{z^k}{k} + \frac{z^{k+1}}{k+1} + \cdots\right)$$

in $R[[z]]$ equipped with the formal topology. (In contrast, the quantity evocative of a generating function of words over an infinite alphabet

$$W \overset{!}{=} \left(1 - z \sum_{j=1}^{\infty} u_j\right)^{-1}$$

cannot be soundly defined as an element of the formal domain $R[[z]]$.) ◁

Henceforth, we shall keep in mind that verifications of formal correctness regarding power series in infinitely many indeterminates are always possible by returning to basic definitions.

Complete generating functions are often surprisingly simple to expand. For instance, the equivalent form of (39)

$$P(z, \mathbf{u}) = e^{u_1 z/1} \cdot e^{u_2 z^2/2} \cdot e^{u_3 z^3/3} \cdots$$

implies immediately that the number of permutations with k_1 cycles of size 1, k_2 of size 2, and so on, is

(40)
$$\frac{n!}{k_1!\,k_2!\,\cdots k_n!\,1^{k_1}\,2^{k_2}\cdots n^{k_n}},$$

provided $\sum jk_j = n$. This is a result originally due to Cauchy. Similarly, the EGF of set partitions with u_j marking the number of blocks of size j is

$$S(z, \mathbf{u}) = \exp\left(u_1\frac{z}{1!} + u_2\frac{z^2}{2!} + u_3\frac{z^3}{3!} + \cdots\right).$$

A formula analogous to (40) follows: the number of partitions with k_1 blocks of size 1, k_2 of size 2, and so on, is

$$\frac{n!}{k_1!\,k_2!\,\cdots k_n!\,1!^{k_1}\,2!^{k_2}\cdots n!^{k_n}}.$$

Several examples of such complete generating functions are presented in Comtet's book; see [129], pages 225 and 233.

▷ **III.23.** *Complete GFs for compositions and surjections.*   The complete GFs of integer compositions and surjections with u_j marking the number of components of size j are

$$\frac{1}{1 - \sum_{j=1}^{\infty} u_j z^j}, \qquad \frac{1}{1 - \sum_{j=1}^{\infty} u_j \frac{z^j}{j!}}.$$

The associated counts with $n = \sum_j jk_j$ are given by

$$\binom{k_1 + k_2 + \cdots}{k_1, k_2, \ldots}, \qquad \frac{n!}{1!^{k_1} 2!^{k_2}\cdots}\binom{k_1 + k_2 + \cdots}{k_1, k_2, \ldots}.$$

These factored forms follow directly from the multinomial expansion. The symbolic form of the multinomial expansion of powers of a generating function is sometimes expressed in terms of Bell polynomials, themselves nothing but a rephrasing of the multinomial expansion; see Comtet's book [129, Sec. 3.3] for a fair treatment of such polynomials.   ◁

▷ **III.24.** *Faà di Bruno's formula.* The formulae for the successive derivatives of a functional composition $h(z) = f(g(z))$

$$\partial_z h(z) = f'(g(z))g'(z), \quad \partial_z^2 h(z) = f''(g(z))g'(z)^2 + f'(z)g''(z), \ldots,$$

are clearly equivalent to the expansion of a formal power series composition. Indeed, assume without loss of generality that $z = 0$ and $g(0) = 0$; set $f_n := \partial_z^n f(0)$, and similarly for g, h. Then

$$h(z) \equiv \sum_n h_n\frac{z^n}{n!} = \sum_k \frac{f_k}{k!}\left(g_1 z + \frac{g_2}{2!}z^2 + \cdots\right)^k.$$

Thus in one direct application of the multinomial expansion, one finds

$$\frac{h_n}{n!} = \sum_k \frac{f_k}{k!}\sum_{\mathcal{C}}\binom{k}{\ell_1, \ell_2, \ldots, \ell_k}\left(\frac{g_1}{1!}\right)^{\ell_1}\left(\frac{g_2}{2!}\right)^{\ell_2}\cdots\left(\frac{g_k}{k!}\right)^{\ell_k},$$

where the summation condition $\mathcal{C}$ is: $1\ell_1 + 2\ell_2 + \cdots + k\ell_k = n$, $\ell_1 + \ell_2 + \cdots + \ell_k = k$. This shallow identity is known as Faà di Bruno's formula [129, p. 137]. (Faà di Bruno (1825–1888) was canonized by the Catholic Church in 1988, presumably for reasons unrelated to his formula.)   ◁

▷ **III.25.** *Relations between symmetric functions.* Symmetric functions may be manipulated by mechanisms that are often reminiscent of the set and multiset construction. They appear in many areas of combinatorial enumeration. Let $X = \{x_i\}_{i=1}^r$ be a collection of formal variables. Define the symmetric functions

$$\prod_i (1 + x_i z) = \sum_n a_n z^n, \quad \prod_i \frac{1}{1 - x_i z} = \sum_n b_n z^n, \quad \sum_i \frac{x_i z}{1 - x_i z} = \sum_n c_n z^n.$$

The a_n, b_n, c_n, called, respectively, elementary, monomial, and power symmetric functions, are expressible as

$$a_n = \sum_{i_1 < i_2 < \cdots < i_r} x_{i_1} x_{i_2} \cdots x_{i_r}, \quad b_n = \sum_{i_1 \leq i_2 \leq \cdots \leq i_r} x_{i_1} x_{i_2} \cdots x_{i_r}, \quad c_n = \sum_{i=1}^r x_i^r.$$

The following relations hold for the OGFs $A(z), B(z), C(z)$ of a_n, b_n, c_n:

$$\begin{aligned} B(z) &= \frac{1}{A(-z)}, & A(z) &= \frac{1}{B(-z)}, \\ C(z) &= z \frac{d}{dz} \log B(z), & B(z) &= \exp \int_0^z C(t) \frac{dt}{t}. \end{aligned}$$

Consequently, each of a_n, b_n, c_n is polynomially expressible in terms of any of the other quantities. (The connection coefficients, as in Note III.24, involve multinomials.) ◁

▷ **III.26.** *Regular graphs.* A graph is r–regular iff each node has degree exactly equal to r. The number of r–regular graphs of size n is

$$[x_1^r x_2^r \cdots x_n^r] \prod_{1 \leq i < j \leq n} (1 + x_i x_j).$$

[Gessel [289] has shown how to extract explicit expressions from such huge symmetric functions; see Appendix B.4: *Holonomic functions*, p. 748.] ◁

III. 6.1. Word models.

The enumeration of words constitutes a rich chapter of combinatorial analysis, and complete GFs serve to generalize many results to the case of non-uniform letter probabilities, such as the coupon collector problem and the birthday paradox considered in Chapter II. Applications are to be found in classical probability theory and statistics [139] (the so-called Bernoulli trial models), as well as in computer science [564] and mathematical models of biology [603].

Example III.16. *Words and records.* Fix an alphabet $\mathcal{A} = \{a_1, \ldots, a_r\}$ and let $\mathcal{W} = \text{SEQ}\{\mathcal{A}\}$ be the class of all words over $\mathcal{A}$, where $\mathcal{A}$ is naturally ordered by $a_1 < a_2 < \cdots < a_r$. Given a word $w = w_1 \cdots w_n$, a (strict) record is an element w_j that is larger than all preceding elements: $w_j > w_i$ for all $i < j$. (Refer to Figure III.15 of Chapter II for a graphical rendering of records in the case of permutations.)

Consider first the subset of $\mathcal{W}$ comprising all words that have the letters $a_{i_1}, \ldots, a_{i_k}$ as successive records, where $i_1 < \cdots < i_k$. The symbolic description of this set is in the form of a product of k terms

(41) $$\left(a_{i_1} \text{ SEQ}(a_1 + \cdots + a_{i_1})\right) \quad \cdots \quad \left(a_{i_k} \text{ SEQ}(a_1 + \cdots + a_{i_k})\right).$$

Consider now MGFs of words where z marks length, v marks the number of records, and each u_j marks the number of occurrences of letter a_j. The MGF associated to the subset described in (41) is then

$$\left(zvu_{i_1}(1 - z(u_1 + \cdots + u_{i_1}))^{-1}\right) \quad \cdots \quad \left(zvu_{i_k}(1 - z(u_1 + \cdots + u_{i_k}))^{-1}\right).$$

Summing over all values of k and of $i_1 < \cdots < i_k$ gives

$$(42) \qquad W(z, v, \mathbf{u}) = \prod_{s=1}^{r} \left(1 + zvu_s \left(1 - z(u_1 + \cdots + u_s) \right)^{-1} \right),$$

the rationale being that, for arbitrary quantities y_s, one has by distributivity:

$$\sum_{k=0}^{r} \sum_{1 \le i_1 < \cdots < i_k \le r} y_{i_1} y_{i_2} \cdots y_{i_k} = \prod_{s=1}^{r} (1 + y_s).$$

We shall encounter more applications of (42) below. For the time being let us simply examine the mean number of records in a word of length n over the alphabet $\mathcal{A}$, when all such words are taken equally likely. One should set $u_j \mapsto 1$ (the composition into specific letters is forgotten), so that W assumes the simpler form

$$W(z, v) = \prod_{j=1}^{r} \left(1 + \frac{vz}{1 - jz} \right).$$

Logarithmic differentiation then gives access to the generating function of cumulated values,

$$\Omega(z) \equiv \left. \frac{\partial}{\partial v} W(z, v) \right|_{v=1} = \frac{z}{1 - rz} \sum_{j=1}^{r} \frac{1}{1 - (j-1)z}.$$

Thus, by partial fraction expansion, the mean number of records in $\mathcal{W}_n$ (whose cardinality is r^n) has the exact value

$$(43) \qquad \mathbb{E}_{\mathcal{W}_n} (\# \text{ records}) = H_r - \sum_{j=1}^{r-1} \frac{(j/r)^n}{r - j}.$$

There appears the harmonic number H_r, as in the permutation case, but now with a negative correction term which, for fixed r, vanishes exponentially with n. .... ∎

Example III.17. *Weighted word models and Bernoulli trials.* Let $\mathcal{A} = \{a_1, \ldots, a_r\}$ be an alphabet of cardinality r, and let $\Lambda = \{\lambda_1, \ldots, \lambda_r\}$ be a system of numbers called *weights*, where weight λ_j is viewed as attached to letter a_j. Weights may be extended from letters to words multiplicatively by defining the weight $\pi(w)$ of word w as

$$\begin{aligned} \pi(w) &= \lambda_{i_1} \lambda_{i_2} \cdots \lambda_{i_n} \qquad \text{if} \quad w = a_{i_1} a_{i_2} \cdots a_{i_n} \\ &= \prod_{j=1}^{r} \lambda_j^{\chi_j(w)}, \end{aligned}$$

where $\chi_j(w)$ is the number of occurrences of letter a_j in w. Finally, the weight of a set is by definition the *sum* of the weights of its elements.

Combinatorially, weights of sets are immediately obtained once the corresponding generating function is known. Indeed, let $\mathcal{S} \subseteq \mathcal{W} = \text{SEQ}\{\mathcal{A}\}$ have the complete GF

$$S(z, u_1, \ldots, u_r) = \sum_{w \in \mathcal{S}} z^{|w|} u_1^{\chi_1(w)} \cdots u_r^{\chi_r(w)},$$

where $\chi_j(w)$ is the number of occurrences of letter a_j in w. Then one has

$$S(z, \lambda_1, \ldots, \lambda_r) = \sum_{w \in \mathcal{S}} z^{|w|} \pi(w),$$

so that extracting the coefficient of z^n gives the total weight of $S_n = S \cap W_n$ under the weight system Λ. In other words, *the GF of a weighted set is obtained by substitution of the numerical values of the weights inside the associated complete MGF.*

In probability theory, Bernoulli trials refer to sequences of independent draws from a fixed distribution with finitely many possible values. One may think of the succession of flippings of a coin or castings of a die. If any trial has r possible outcomes, then the various possibilities can be described by letters of the r–ary alphabet $\mathcal{A}$. If the probability of the jth outcome is taken to be λ_j, then the Λ-weighted models on words becomes the usual probabilistic model of independent trials. (In this situation, the λ_j are often written as p_j.) Observe that, in the probabilistic situation, one must have $\lambda_1 + \cdots + \lambda_r = 1$ with each λ_j satisfying $0 \le \lambda_j \le 1$. The equiprobable case, where each outcome has probability $1/r$ can be obtained by setting $\lambda_j = 1/r$, leaving us with the usual enumerative model. In terms of GFs, the coefficient $[z^n]S(z, \lambda_1, \ldots, \lambda_r)$ then represents the probability that a random word of W_n belongs to S. Multivariate generating functions and cumulative generating functions then obey properties similar to their usual (ordinary, exponential) counterparts.

As an illustration, assume one has a biased coin with probability p for heads (H) and $q = 1 - p$ for tails (T). Consider the event: "*in n tosses of the coin, there never appear ℓ contiguous heads*". The alphabet is $\mathcal{A} = \{H, T\}$. The construction describing the events of interest is, as seen in Subsection I. 4.1 (p. 51),

$$S = \text{SEQ}_{<\ell}\{H\}\,\text{SEQ}\{T\,\text{SEQ}_{<\ell}\{H\}\}.$$

Its GF, with u marking heads and v marking tails, is then

$$W(z, u, v) = \frac{1 - z^\ell u^\ell}{1 - zu}\left(1 - zv\frac{1 - z^\ell u^\ell}{1 - zu}\right)^{-1}.$$

Thus, the probability of the absence of ℓ–runs among a sequence of n random coin tosses is obtained after the substitution $u \to p$, $v \to q$ in the MGF,

$$[z^n]\frac{1 - p^\ell z^\ell}{1 - z + qp^\ell z^{\ell+1}},$$

leading to an expression which is amenable to numerical or asymptotic analysis. For instance, Feller's book [206, p. 322–326] offers a classical discussion of the problem. .... ∎

Example III.18. *Records in Bernoulli trials.* We pursue the discussion of probabilistic models on words and come back to the analysis of records. Assume now that the alphabet $\mathcal{A} = \{a_1, \ldots, a_r\}$ has in all generality the probability p_j associated with the letter a_j. The mean number of records is analysed by a process entirely parallel to the derivation of (43): one finds by logarithmic differentiation of (42)

$$(44) \quad \mathbb{E}_{W_n}(\text{\# records}) = [z^n]\Omega(z) \quad \text{where} \quad \Omega(z) = \frac{z}{1 - z}\sum_{j=1}^{r}\frac{p_j}{1 - z(p_1 + \cdots + p_{j-1})}.$$

The cumulative GF $\Omega(z)$ in (44) has simple poles at the points $1, 1/P_{r-1}, 1/P_{r-2}$, and so on, where $P_s = p_1 + \cdots + p_s$. For asymptotic purposes, only the dominant pole at $z = 1$ counts (see Chapter IV for a systematic discussion), near which

$$\Omega(z) \underset{z \to 1}{\sim} \frac{1}{1 - z}\sum_{j=1}^{r}\frac{p_j}{1 - P_{j-1}}.$$

Consequently, one has an elegant asymptotic formula, generalizing the case of permutations:

The mean number of records in a random word of length n with non-uniform letter probabilities p_j satisfies asymptotically ($n \to +\infty$)

$$\mathbb{E}_{\mathcal{W}_n}(\# \text{ records}) \sim \sum_{j=1}^{r} \frac{p_j}{p_j + p_{j+1} + \cdots + p_r}.$$

This relation and similar ones were obtained by Burge [97]; analogous ideas may serve to analyse the sorting algorithm *Quicksort* under equal keys [536] as well as the hybrid data structures of Bentley and Sedgewick; see [47, 124]. . . . . ∎

Coupon collector problem and birthday paradox. Similar considerations apply to weighted EGFs of words, as considered in Chapter II. For instance, the probability of having a complete coupon collection at time n in the case a company issues coupon j with probability p_j, for $1 \le j \le r$, is (coupon collector problem, p. 114)

$$\mathbb{P}(C \le n) = n! [z^n] \prod_{j=1}^{r} \left(e^{p_j z} - 1 \right).$$

The probability that all coupons are different at time n is (birthday paradox, p. 114)

$$\mathbb{P}(B > n) = n! [z^n] \prod_{j=1}^{r} \left(1 + p_j z \right),$$

which corresponds to the birthday problem in the case of non-uniform mating periods. Integral representations comparable those of Chapter II are also available:

$$\mathbb{E}(C) = \int_{0}^{\infty} \left(1 - \prod_{j=1}^{r} (1 - e^{-p_j t}) \right) dt, \qquad \mathbb{E}(B) = \int_{0}^{\infty} \prod_{j=1}^{r} \left(1 + p_j t \right) e^{-t} dt.$$

See the study by Flajolet, Gardy, and Thimonier [231] for variations on this theme.
▷ **III.27.** *Birthday paradox with leap years.* Assume that the 29th of February exists precisely once every fourth year. Estimate the effect on the expectation of the first birthday collision. ◁

Example III.19. *Rises in Bernoulli trials: Simon Newcomb's problem.* Simon Newcomb (1835–1909), otherwise famous for his astronomical work, was reportedly fond of playing the following patience game: one draws from a deck of 52 playing cards, stacking them in piles in such a way that one new pile is started each time a card appears whose number is smaller than its predecessor. What is the probability of obtaining t piles? A solution to this famous problem is found in MacMahon's book [428] and a concise account by Andrews appears in [14, §4.4].

Simon Newcomb's problem can be rephrased in terms of rises. Given a word $w = w_1 \cdots w_n$ over the alphabet $\mathcal{A}$ ordered by $a_1 < a_2 < \cdots$, a *weak rise* is a position $j < n$ such that $w_j \le w_{j+1}$. (The numbers of piles in Newcomb's problem is the number of cards minus 1 minus the number of weak rises.) Let $W \equiv W(z, v, \mathbf{u})$ be the MGF of all words where z marks length, v marks the number of weak rises, and u_j marks the number of occurrences of letter j. Set $z_j = z u_j$ and let $W_j \equiv W_j(z, v, \mathbf{u})$ be the MGF relative to those non-empty words that start with letter a_j, so that

$$W = 1 + (W_1 + \cdots + W_r).$$

The W_j satisfy the set of equations ($j = 1, \ldots, r$),

$$(45) \qquad W_j = z_j + z_j \left(W_1 + \cdots + W_{j-1} \right) + v z_j \left(W_j + \cdots + W_r \right),$$

as seen by considering the first letter of each word. The linear system (45) is easily solved upon setting $W_j = z_j X_j$. Indeed, by differencing, one finds that

$$(46) \qquad X_{j+1} - X_j = z_j X_j (1 - v), \qquad X_{j+1} = X_j (1 + z_j(1 - v)).$$

In this way, each X_j can be determined in terms of X_1. Then transporting the resulting expressions into the relation (45) taken with $j = 1$, and solving for X_1 leads to an expression for X_1, hence for all the X_j and finally for W itself:

$$(47) \qquad W = \frac{v-1}{v - P^{-1}}, \qquad P := \prod_{j=1}^{r} (1 + (1 - v)z_j).$$

Goulden and Jackson obtain a similar expressions in [303] (pp. 72 and 236).

The result of (47) gives access to moments (e.g., mean and variance) of the number of rises in a Bernoulli sequence as well as to counting results, once coefficients of the MGF are extracted. (See also [289, 303] for an approach based on the theory of symmetric functions.) The OGF (47) can alternatively be derived by an inclusion–exclusion argument: refer to the particular case of rises in permutations and Eulerian numbers, p. 210. ∎

▷ **III.28.** *The final solution to Simon Newcomb's problem.* Consider a deck of cards with a suits and r distinct card values. Set $N = ra$. (The original problem has $r = 13$, $a = 4$, $N = 52$.) One has from (47): $W = (v - 1)P/(1 - vP)$. The expansion of $(1 - y)^{-1}$ and the collection of coefficients yields

$$[z_1^a \cdots z_r^a]W = (1 - v) \sum_{k \geq 1} v^{k-1}[z_1^a \cdots z_r^a]P^k = (1 - v)^{N+1} \sum_{k \geq 1} \binom{k}{a}^r v^{k-1},$$

so that $[z_1^a \cdots z_r^a v^t]W = \sum_{k=0}^{t+1} (-1)^{t+1-k} \binom{N+1}{t+1-k} \binom{k}{a}^r.$ ◁

III. 6.2. Tree models. We examine here two important GFs associated with tree models; these provide valuable information concerning the *degree profile* and the *level profile* of trees, while being tightly coupled with an important class of stochastic processes, namely *branching processes*.

The major classes of trees that we have encountered so far are the unlabelled plane trees and the labelled non-plane trees, prototypes being general Catalan trees (Chapter I) and Cayley trees (Chapter II). In both cases, the counting GFs satisfy a relation of the form

$$(48) \qquad Y(z) = z\phi(Y(z)),$$

where the GF is either ordinary (plane unlabelled trees) or exponential (non-plane labelled trees). Corresponding to the two cases, the function ϕ is determined, respectively, by

$$(49) \qquad \phi(w) = \sum_{\omega \in \Omega} w^\omega, \qquad \phi(w) = \sum_{\omega \in \Omega} \frac{w^\omega}{\omega!},$$

where $\Omega \subseteq \mathbb{N}$ is the set of allowed node degrees. Meir and Moon in an important paper [435] have described some common properties of tree families that are determined by the Axiom (48). (For instance mean path length is invariably of order $n\sqrt{n}$, see Chapter VII, and height is $O(\sqrt{n})$.) Following these authors, we call a *simple variety of trees* any class whose counting GF is defined by an equation of type (48). For each of the two cases of (49), we write

$$(50) \qquad \phi(w) = \sum_{j=0}^{\infty} \phi_j w^j.$$

Degree profile of trees. First we examine the *degree profile* of trees. Such a profile is determined by the collection of parameters χ_j, where $\chi_j(\tau)$ is the number of nodes of outdegree j in τ. The variable u_j will be used to mark χ_j, that is, nodes of outdegree j. The discussion already conducted regarding recursive parameters shows that the GF $Y(z, \mathbf{u})$ satisfies the equation

$$Y(z, \mathbf{u}) = z\Phi(Y(z, \mathbf{u})) \qquad \text{where} \quad \Phi(w) = u_0\phi_0 + u_1\phi_1 w + u_2\phi_2 w^2 + \cdots.$$

Formal Lagrange inversion can then be applied to $Y(z, \mathbf{u})$, to the effect that its coefficients are given by the coefficients of the powers of Φ.

Proposition III.7 (Degree profile of trees). *The number of trees of size n and degree profile $(n_0, n_1, n_2, \ldots)$ in a simple variety of trees defined by the "generator" (50) is*

$$(51) \qquad Y_{n;n_0,n_1,n_2,\ldots} = \omega_n \cdot \frac{1}{n} \binom{n}{n_0, n_1, n_2, \ldots} \phi_0^{n_0} \phi_1^{n_1} \phi_2^{n_2} \cdots.$$

There, $\omega_n = 1$ in the unlabelled case, whereas $\omega_n = n!$ in the labelled case. The values of the n_j are assumed to satisfy the two consistency conditions: $\sum_j n_j = n$ and $\sum_j j n_j = n - 1$.

Proof. The consistency conditions translate the fact that the total number of nodes should be n while the total number of edges should equal $n - 1$ (each node of degree j is the originator of j edges). The result follows from Lagrange inversion

$$Y_{n;n_0,n_1,n_2,\ldots} = \omega_n \cdot [u_0^{n_0} u_1^{n_1} u_2^{n_2} \cdots] \left(\frac{1}{n}[w^{n-1}]\Phi(w)^n \right),$$

to which a standard multinomial expansion applies, yielding (51).

For instance, for general Catalan trees ($\phi_j = 1$) and for Cayley trees ($\phi_j = 1/j!$) these formulae become

$$\frac{1}{n} \binom{n}{n_0, n_1, n_2, \ldots} \qquad \text{and} \qquad \frac{(n-1)!}{0!^{n_0} 1!^{n_1} 2!^{n_2} \cdots} \binom{n}{n_0, n_1, n_2, \ldots}.$$

∎

The proof above also reveals the logical equivalence between the general tree counting result of Proposition III.7 and the most general case of Lagrange inversion. (This equivalence is due to the fact that any fixed series is a special case of Φ.) Put another way, any direct proof of (51) provides a combinatorial proof of the Lagrange inversion theorem. Such direct derivations have been proposed by Raney [503] and are based on simple but cunning surgery performed on lattice path representations of trees (the "conjugation principle" of which a particular case is the "cycle lemma" of Dvoretzky–Motzkin [184]; see Note I.47, p. 75).

Level profile of trees. The next example demonstrates the utility of complete GFs for investigating the level profile of trees.

Example III.20. *Trees and level profile.* Given a rooted tree τ, its *level profile* is defined as the vector $(n_0, n_1, n_2, \ldots)$ where n_j is the number of nodes present at level j (i.e., at distance j from the root) in tree τ. Continuing within the framework of a simple variety of trees, we now define the quantity $Y_{n;n_0,n_1,\ldots}$ to be the number of trees with size n and level profile given by the n_j. The corresponding complete GF $Y(z, \mathbf{u})$ with z marking size and u_j marking nodes at level j is expressible in terms of the fundamental "generator" ϕ:

$$(52) \qquad Y(z, \mathbf{u}) = zu_0\phi\,(zu_1\phi\,(zu_2\phi\,(zu_3\phi(\cdots)))) .$$

We may call this a "continued ϕ-form". For instance, general Catalan trees have generator $\phi(w) = (1 - w)^{-1}$, so that in this case the complete GF is the continued fraction:

$$(53) \qquad Y(z, \mathbf{u}) = \cfrac{u_0 z}{1 - \cfrac{u_1 z}{1 - \cfrac{u_2 z}{1 - \cfrac{u_3 z}{\ddots}}}} .$$

(See Section V. 4, p. 318, for complementary aspects.) In contrast, Cayley trees are generated by $\phi(w) = e^w$, so that

$$Y(z, \mathbf{u}) = zu_0 e^{zu_1 e^{zu_2 e^{zu_3 e^{\cdot^{\cdot^{\cdot}}}}}} ,$$

which is a "continued exponential"; that is, a tower of exponentials. Expanding such generating functions with respect to $u_0, u_1, \ldots$, in order gives the following proposition straightforwardly.

Proposition III.8 (Level profile of trees). *The number of trees of size n, having $(n_0, n_1, n_2, \ldots)$ as level profile, in a simple variety of trees with generator $\phi(w)$ is*

$$Y_{n;n_0,n_1,n_2,\ldots} = \omega_{n-1} \cdot \phi_{n_1}^{(n_0)} \phi_{n_2}^{(n_1)} \phi_{n_3}^{(n_2)} \cdots \qquad \text{where} \quad \phi_v^{(\mu)} := [w^v]\phi(w)^\mu.$$

There, the consistency conditions are $n_0 = 1$ and $\sum_j n_j = n$. In particular, the counts for general Catalan trees and for Cayley trees are, respectively,

$$\binom{n_0 + n_1 - 1}{n_1}\binom{n_1 + n_2 - 1}{n_2}\binom{n_2 + n_3 - 1}{n_3}\cdots , \qquad \frac{(n-1)!}{n_0!n_1!n_2!\cdots}n_0^{n_1} n_1^{n_2} n_2^{n_3}\cdots .$$

(Note that one must always have $n_0 = 1$ for a single tree; the general formula with $n_0 \neq 1$ and ω_{n-1} replaced by ω_{n-n_0} gives the level profile of forests.) The first of these enumerative results is due to Flajolet [214] and it places itself within a general combinatorial theory of continued fractions (Section V. 4, p. 318); the second one is due to Rényi and Szekeres [507] , who developed such a formula in the course of a deep study relative to the distribution of height in random Cayley trees (Chapter VII, p. 537). .... ∎

▷ **III.29.** *Continued forms for path length.* The BGF of path length is obtained from the level profile MGF by means of the substitution $u_j \mapsto q^j$. For general Catalan trees and Cayley trees, this gives

$$
(54) \qquad G(z, q) = \cfrac{z}{1 - \cfrac{zq}{1 - \cfrac{zq^2}{\ddots}}}, \qquad T(z, q) = ze^{zq}e^{zq^2}e^{\cdot^{\cdot^{\cdot}}},
$$

where q marks path length. The MGFs are ordinary and exponential. (Combined with differentiation, such MGFs represent an attractive option for mean value analysis.)   ◁

Trees and processes. The next example is an especially important application of complete GFs, as these GFs provide a bridge between combinatorial models and a major class of stochastic processes, the *branching processes* of probability theory.

Example III.21. *Weighted tree models and branching processes.* Consider the family $\mathcal{G}$ of all general plane trees. Let $\Lambda = (\lambda_0, \lambda_1, \ldots)$ be a system of numeric weights. The weight of a node of outdegree j is taken to be λ_j and the weight of a tree is the product of the individual weights of its nodes:

$$
(55) \qquad \pi(\tau) = \prod_{j=0}^{\infty} \lambda_j^{\chi_j(\tau)},
$$

with $\chi_j(\tau)$ the number of nodes of degree j in τ. One can view the weighted model of trees as a model in which a tree receives a probability proportional to $\pi(\tau)$. Precisely, the probability of selecting a particular tree τ under this model is, for a fixed size n,

$$
(56) \qquad \mathbb{P}_{\mathcal{G}_n, \Lambda}(\tau) = \frac{\pi(\tau)}{\sum_{|T|=n} \pi(T)}.
$$

This defines a probability measure over the set $\mathcal{G}_n$ and one can consider events and random variables under this weighted model.

The weighted model defined by (55) and (56) covers any simple variety of trees: just replace each λ_j by the quantity ϕ_j given by the "generator' (50) of the model. For instance, plane unlabelled unary–binary trees are obtained by $\Lambda = (1, 1, 1, 0, 0, \ldots)$, while Cayley trees correspond to $\lambda_j = 1/j!$. Two *equivalence-preserving transformations* are then especially important in this context:

(i) Let Λ^* be defined by $\lambda_j^* = c\lambda_j$ for some non-zero constant c. Then the weight corresponding to Λ^* satisfies $\pi^*(\tau) = c^{|\tau|}\pi(\tau)$. Consequently, the models associated to Λ and Λ^* are equivalent as regards (56).

(ii) Let Λ° be defined by $\lambda_j^\circ = \theta^j \lambda_j$ for some non-zero constant θ. Then the weight corresponding to Λ° satisfies $\pi^\circ(\tau) = \theta^{|\tau|-1}\pi(\tau)$, since $\sum_j j\chi_j(\tau) = |\tau| - 1$ for any tree τ. Thus the models Λ° and Λ are again equivalent.

Each transformation has a simple effect on the generator ϕ, namely:

$$
(57) \qquad \phi(w) \mapsto \phi^*(w) = c\phi(w) \qquad \text{and} \qquad \phi(w) \mapsto \phi^\circ(w) = \phi(\theta w).
$$

Once equipped with such equivalence transformations, it becomes possible to describe probabilistically the process that generates trees according to a weighted model. Assume that $\lambda_j \geq 0$ and that the λ_j are summable. Then the normalized quantities

$$
p_j = \frac{\lambda_j}{\sum_j \lambda_j}
$$

form a probability distribution over $\mathbb{N}$. By the first equivalence-preserving transformation the model induced by the weights p_j is the same as the original model induced by the λ_j. (By the second equivalence transformation, one can furthermore assume that the generator ϕ is the probability generating function of the p_j.)

Such a model defined by non-negative weights $\{p_j\}$ summing to 1 is nothing but the classical model of *branching processes* (also known as Galton–Watson processes); see [21, 324]. In effect, a realization T of the branching process is classically defined by the two rules: (*i*) produce a root node of degree j with probability p_j; (*ii*) if $j \geq 1$, attach to the root node a collection $T_1, \ldots, T_j$ of independent realizations of the process. This may be viewed as the development of a "family" stemming from a common ancestor where any individual has probability p_j of giving birth to j children. Clearly, the probability of obtaining a particular finite tree τ has probability $\pi(\tau)$, where π is given by (55) and the weights are $\lambda_j = p_j$. The generator

$$\phi(w) = \sum_{j=0}^{\infty} p_j w^j$$

is then nothing but the probability generating function of (one-generation) offspring, with the quantity $\mu = \phi'(1)$ being its mean size.

For the record, we recall that branching processes can be classified into three categories depending on the values of μ.

> *Subcriticality*: when $\mu < 1$, the random tree produced is finite with probability 1 and its expected size is also finite.
> *Criticality*: when $\mu = 1$, the random tree produced is finite with probability 1 but its expected size is infinite.
> *Supercriticality*: when $\mu > 1$, the random tree produced is finite with probability strictly less than 1.

From the discussion of equivalence transformations (57), it is furthermore true that, regarding trees of a *fixed size* n, there is complete equivalence between all branching processes with generators of the form

$$\phi_\theta(w) = \frac{\phi(\theta w)}{\phi(\theta)}.$$

Such families of related functions are known as "exponential families" in probability theory. In this way, one may always regard at will the random tree produced by a weighted model of some fixed size n as originating from a branching process (of subcritical, critical, or supercritical type) conditioned upon the size of the total progeny.

Finally, take a set $\mathcal{S} \subseteq \mathcal{G}$ for which the complete generating function of $\mathcal{S}$ with respect to the degree profile is available,

$$S(z, u_0, u_1, \ldots) = \sum_{\tau \in \mathcal{S}} z^{|\tau|} \left(u_0^{\chi_0(\tau)} u_1^{\chi_1(\tau)} \cdots \right).$$

Then, for a system of weights Λ, one has

$$S(z, \lambda_0, \lambda_1, \ldots) = \sum_{\tau \in \mathcal{S}} \pi(\tau) z^{|\tau|}.$$

Thus, we can find the probability that a weighted tree of size n belongs to $\mathcal{S}$, by extracting the coefficient of z^n. This applies *a fortiori* to branching processes as well. In summary, *the analysis of parameters of trees of size n under either weighted models or branching process models follows from substituting weights or probability values in the corresponding* complete *generating functions.* .... ■

The reduction of combinatorial tree models to branching processes was pursued early, most notably by the "Russian School": see especially the books by Kolchin [386, 387] and references therein. (For asymptotic purposes, the equivalence between combinatorial models and critical branching processes often turns out to be most fruitful.) Conversely, symbolic-combinatorial methods may be viewed as a systematic way of obtaining equations relative to characteristics of branching processes. We do not elaborate further along these lines as this would take us outside of the scope of the present book.

$\triangleright$ **III.30.** *Catalan trees, Cayley trees, and branching processes.* Catalan trees of size n are defined by the weighted model in which $\lambda_j \equiv 1$, but also equivalently by $\widehat{\lambda}_j = c\theta^j$, for any $c > 0$ and $\theta \le 1$. In particular they coincide with the random tree produced by the critical branching process whose offspring probabilities are geometric: $p_j = 1/2^{j+1}$.

Cayley trees are *a priori* defined by $\lambda_j = 1/j!$. They can be generated by the critical branching process with Poisson probabilities, $p_j = e^{-1}/j!$, and more generally with an arbitrary Poisson distribution $p_j = e^{-\lambda}\lambda^j/j!$. $\lhd$

III. 7. Additional constructions

We discuss here additional constructions already examined in earlier chapters; namely pointing and substitution (Section III. 7.1), order constraints (Section III. 7.2), and implicit structures (Section III. 7.3). Given that basic translation mechanisms can be directly adapted to the multivariate realm, such extensions involve basically no new concept, and the methods of Chapters I and II can be easily recycled. In Section III. 7.4, we revisit the classical principle of inclusion–exclusion under a generating function perspective. In this light, the principle appears as a typically multivariate device well suited to enumerating objects according the number of occurrences of subconfigurations.

III. 7.1. Pointing and substitution. Let $\langle \mathcal{F}, \chi \rangle$ be a class–parameter pair, where χ is multivariate of dimension $r \ge 1$, and let $F(\mathbf{z})$ be the MGF associated to it in the notations of (19) and (28). In particular $z_0 \equiv z$ marks size, and z_k marks the component k of the multiparameter χ. If z marks size, then, as in the univariate case, $\theta_z \equiv z\partial_z$ translates the fact of distinguishing one atom. Generally, pick up a variable $x \equiv z_j$ for some j with $0 \le j \le r$. Then since

$$x\partial_x(s^a t^b x^f) = f \cdot (s^a t^b x^f),$$

the interpretation of the operator $\theta_x \equiv x\partial_x$ is immediate; it means "pick up in all possible ways in objects of $\mathcal{F}$ a configuration marked by x and point to it". For instance, if $F(z, u)$ is the BGF of trees where z marks size and u marks leaves, then $\theta_u F(z, u) = u\partial_u F(z, u)$ enumerates trees with one distinguished leaf.

Similarly, the substitution $x \mapsto S(\mathbf{z})$ in a GF F, where $S(\mathbf{z})$ is the MGF of a class S, means attaching an object of type S to configurations marked by the variable x in $\mathcal{F}$. The process is better understood by practice than by long formal developments. Justification in each particular case can be easily obtained by returning to the combinatorial representation of generating functions as images of combinatorial classes.

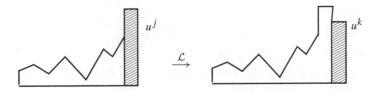

Figure III.16. The technique of "adding a slice" for constrained compositions.

Example III.22. *Constrained integer compositions and "slicing".* This example illustrates variations around the substitution scheme. Consider compositions of integers where successive summands have sizes that are constrained to belong to a fixed set $\mathcal{R} \subseteq \mathbb{N}^2$. For instance, the relations

$$\mathcal{R}_1 = \{(x, y) \mid 1 \le x \le y\}, \quad \mathcal{R}_2 = \{(x, y) \mid 1 \le y \le 2x\},$$

correspond to weakly increasing summands in the case of $\mathcal{R}_1$ and to summands that can at most double at each stage in the case of $\mathcal{R}_2$. In the "ragged landscape" representation of compositions, this means considering diagrams of unit cells aligned in columns along the horizontal axis, with successive columns obeying the constraint imposed by $\mathcal{R}$.

Let $F(z, u)$ be the BGF of such $\mathcal{R}$–restricted compositions, where z marks total sum and u marks the value of the last summand; that is, the height of the last column. The function $F(z, u)$ satisfies a functional equation of the form

$$(58) \qquad F(z, u) = f(zu) + (\mathcal{L}[F(z, u)])_{u \mapsto zu},$$

where $f(z)$ is the generating function of the one-column objects and $\mathcal{L}$ is a linear operator over formal series in u given by

$$(59) \qquad \mathcal{L}[u^j] := \sum_{(j,k) \in \mathcal{R}} u^k.$$

In effect, Equation (58) describes inductively objects as comprising either one column ($f(zu)$) or else as being formed by adding a new column to an existing one; see Figure III.16. The process of appending a slice of size j to one of size k, with $(j, k) \in \mathcal{R}$, is precisely what (59) expresses; the functional equation (58) is obtained by effecting the final substitution $u \mapsto zu$, in order to take into account the k atoms contributed by the new slice. The special case $F(z, 1)$ gives the enumeration of $\mathcal{F}$–objects irrespective of the size of the last column.

For a rule $\mathcal{R}$ that is "simple", the basic equation (58) will often involve a substitution. Let us first rederive in this way the enumeration of partitions. We take $\mathcal{R} = \mathcal{R}_1$ and assume that the first column can have any positive size. Compositions into increasing summands are clearly the same as partitions. Since

$$\mathcal{L}[u^j] = u^j + u^{j+1} + u^{j+2} + \cdots = \frac{u^j}{1 - u},$$

the function $F(z, u)$ satisfies a functional equation involving a substitution,

$$(60) \qquad F(z, u) = \frac{zu}{1 - zu} + \frac{1}{1 - zu} F(z, zu).$$

This relation iterates: *any linear functional equation of the substitution type*

$$\phi(u) = \alpha(u) + \beta(u)\phi(\sigma(u))$$

is solved formally by

$$\phi(u) = \alpha(u) + \beta(u)\alpha(\sigma(u)) + \beta(u)\beta(\sigma(u))\alpha(\sigma^{(2)}(u)) + \cdots, \tag{61}$$

where $\sigma^{(j)}(u)$ designates the jth iterate of u.

We can now return to partitions. The turnkey solution (61) gives, upon iterating on the second argument and treating the first argument as a parameter,

$$F(z, u) = \frac{zu}{1 - zu} + \frac{z^2 u}{(1 - zu)(1 - z^2 u)} + \frac{z^3 u}{(1 - zu)(1 - z^2 u)(1 - z^3 u)} + \cdots. \tag{62}$$

Equivalence with the alternative form

$$F(z, u) = \frac{zu}{1 - z} + \frac{z^2 u^2}{(1 - z)(1 - z^2)} + \frac{z^3 u^3}{(1 - z)(1 - z^2)(1 - z^3)} + \cdots \tag{63}$$

is then easily verified from (60) by expanding $F(z, u)$ as a series in u and applying the method of indeterminate coefficients to the form $(1 - zu)F(z, u) = zu + F(z, zu)$. (The representation (63) is furthermore consistent with the treatment of partitions given in Chapter I since the quantity $[u^k]F(z, u)$ clearly represents the OGF of non-empty partitions whose largest summand is k. In passing, the equality between (62) and (63) is a shallow but curious identity that is quite typical of the area of q–analogues.)

This same method has been applied in [250] to compositions satisfying condition $\mathcal{R}_2$ above. In this case, successive summands are allowed to double at most at each stage. The associated linear operator is

$$\mathcal{L}[u^j] = u + \cdots + u^{2j} = u \frac{1 - u^{2j}}{1 - u}.$$

For simplicity, it is assumed that the first column has size 1. Thus, F satisfies a functional equation of the substitution type:

$$F(z, u) = zu + \frac{zu}{1 - zu}\left(F(z, 1) - F(z, z^2 u^2)\right).$$

This can be solved by means of the general iteration mechanism (61), treating for the moment $F(z, 1)$ as a known quantity: with $a(u) := zu + F(z, 1)/(1 - zu)$, one has

$$F(z, u) = a(u) - \frac{zu}{1 - zu}a(z^2 u^2) + \frac{zu}{1 - zu}\frac{z^2 u^2}{1 - z^2 u^2}a(z^6 u^4) - \cdots.$$

Then, the substitution $u = 1$ in the solution becomes permissible. Upon solving for $F(z, 1)$, one eventually gets the somewhat curious GF for compositions satisfying $\mathcal{R}_2$:

$$F(z, 1) = \frac{\sum_{j \geq 1}(-1)^{j-1}z^{2^{j+1}-j-2}/Q_{j-1}(z)}{\sum_{j \geq 0}(-1)^j z^{2^{j+1}-j-2}/Q_j(z)} \tag{64}$$

$$\text{where} \quad Q_j(z) = (1 - z)(1 - z^3)(1 - z^7)\cdots(1 - z^{2^j-1}).$$

The sequence of coefficients starts as $1, 1, 2, 3, 5, 9, 16, 28, 50$ and is *EIS* **A002572**: it represents, for instance, the number of possible level profiles of binary trees, or equivalently the number of partitions of 1 into summands of the form $1, \frac{1}{2}, \frac{1}{4}, \frac{1}{8}, \ldots$ (this is related to the number of solutions to Kraft's inequality). See [250] for details, including precise asymptotic estimates, and Tangora's paper [571] for relations to algebraic topology. .... ■

The reason for presenting the slicing method[3] in some detail is that it is very general. It has been particularly employed to derive a number of original enumerations of polyominoes by area, a topic of interest in some branches of statistical mechanics: for instance, the book by Janse van Rensburg [592] discusses many applications of such lattice models to polymers and vesicles. Bousquet-Mélou's review paper [82] offers a methodological perspective. Some of the origins of the method point to Pólya in the 1930s, see [490], and independently to Temperley [574, pp. 65–67].

▷ **III.31.** *Pointing–erasing and the combinatorics of Taylor's formula.* The derivative operator ∂_x corresponds combinatorially to a "pointing–erasing" operation: select in all possible ways an atom marked by x and make it transparent to x-marking (e.g., by replacing it by a neutral object). The operator $\frac{1}{k!}\partial_x^k f(x)$, then corresponds to picking up in all possible way a *subset* (order does not count) of k configurations marked by x. The identity (Taylor's formula)

$$f(x + y) = \sum_{k \geq 0} \left(\frac{1}{k!}\partial_x^k f(x)\right) y^k$$

can then receive a simple combinatorial interpretation: Given a population of individuals ($\mathcal{F}$ enumerated by f), form the bicoloured population of individuals enumerated by $f(x + y)$, where each atom of each object can be repainted either in x-colour or y-colour; the process is equivalent to deciding *a priori* for each individual to repaint k of its atoms from x to y, this for all possible values of $k \geq 0$. Conclusion: *seen from combinatorics, Taylor's formula merely expresses the logical equivalence between two ways of counting.* ◁

▷ **III.32.** *Carlitz compositions I.* Let $\mathcal{K}$ be the class of compositions such that all pairs of adjacent summands are formed of distinct values. These can be generated by the operator $\mathcal{L}[u^j] = \frac{uz}{1-uz} - u^j z^j$, so that $L[f(u)] = \frac{uz}{1-uz} f(1) - f(uz)$. The BGF $K(z, u)$, with u marking the value of the last summand, then satisfies a functional equation,

$$K(z, u) = \frac{uz}{1 - uz} + \frac{uz}{1 - uz} K(z, 1) - K(z, zu),$$

giving eventually $K(z) \equiv K(z, 1)$ under the form

(65)
$$\begin{aligned}
K(z) &= \left(1 + \sum_{j \geq 1} \frac{(-z)^j}{1 - z^j}\right)^{-1} \\
&= 1 + z + z^2 + 3z^3 + 4z^4 + 7z^5 + 14z^6 + 23z^7 + 39z^8 + \cdots.
\end{aligned}$$

The sequence of coefficients constitutes *EIS* **A003242**. Such compositions were introduced by Carlitz in 1976; the derivation above is from a paper by Knopfmacher and Prodinger [369] who provide early references and asymptotic properties. (We resume this thread in Note III.35, p. 206, then in Chapter IV, p. 263, with regard to asymptotics.) ◁

III. 7.2. Order constraints. We refer in this subsection to the discussion of order constraints in labelled products that has been given in Subsection II. 6.3 (p. 139). We recall that the modified labelled product

$$\mathcal{A} = (\mathcal{B}^\square \star \mathcal{C})$$

only includes the elements of $(\mathcal{B} \star \mathcal{C})$ such that the minimal label lies in the $\mathcal{A}$ component. Once more the univariate rules generalize verbatim for parameters that are

---

[3]For other applications, see Examples V.20, p. 365 (horizontally convex polyominoes) and IX.14, p. 660 (parallelogram polyominoes), as well as Subsection VII. 8.1, p. 506 (walks and the kernel method).

peak:	$\sigma_{i-1} < \sigma_i > \sigma_{i+1}$	leaf node (u_0)
double rise:	$\sigma_{i-1} < \sigma_i < \sigma_{i+1}$	unary right-branching (u_1)
double fall:	$\sigma_{i-1} > \sigma_i > \sigma_{i+1}$	unary left-branching (u'_1)
valley:	$\sigma_{i-1} > \sigma_i < \sigma_{i+1}$	binary node (u_2)

Figure III.17. Local order patterns in a permutation and the four types of nodes in the corresponding increasing binary tree.

inherited and the corresponding exponential MGFs are related by

$$A(z, \mathbf{u}) = \int_0^z (\partial_t B(t, \mathbf{u})) \cdot C(t, \mathbf{u}) \, dt.$$

To illustrate this multivariate extension, we shall consider a quadrivariate statistic on permutations.

Example III.23.   *Local order patterns in permutations.*   An element σ_i of a permutation written $\sigma = \sigma_1, \ldots, \sigma_n$ when compared to its immediate neighbours can be categorized into one of four types[4] summarized in the first two columns of Figure III.17. The correspondence with binary increasing trees described in Example II.17 and Figure II.16 (p. 143) then shows the following: peaks and valleys correspond to leaves and binary nodes, respectively, while double rises and double falls are associated with right-branching and left-branching unary nodes. Consider the class $\widehat{\mathcal{I}}$ of *non-empty* increasing binary trees (so that $\widehat{\mathcal{I}} = \mathcal{I} \setminus \{\epsilon\}$ in the notations of p. 143) and let u_0, u_1, u'_1, u_2 be markers for the number of nodes of each type, as summarized in Figure III.17. Then the exponential MGF of non-empty increasing trees under this statistic is given by

$$\widehat{\mathcal{I}} = u_0 \mathcal{Z} + u_1(\mathcal{Z}^\square \star \widehat{\mathcal{I}}) + u'_1(\widehat{\mathcal{I}} \star \mathcal{Z}^\square) + u_2(\widehat{\mathcal{I}} \star \mathcal{Z}^\square \star \widehat{\mathcal{I}})$$
$$\implies \quad \widehat{I}(z) = u_0 z + \int_0^z \left((u_1 + u'_1)\widehat{I}(w) + u_2 \widehat{I}(w)^2 \right) dw,$$

which gives rise to the differential equation:

$$\frac{\partial}{\partial z} \widehat{I}(z, \mathbf{u}) = u_0 + (u_1 + u'_1)\widehat{I}(z, \mathbf{u}) + u_2 \widehat{I}(z, \mathbf{u})^2.$$

This is solved by separation of variables as

(66) $$\widehat{I}(z, \mathbf{u}) = \frac{\delta}{u_2} \frac{v_1 + \delta \tan(z\delta)}{\delta - v_1 \tan(z\delta)} - \frac{v_1}{u_2},$$

where the following abbreviations are used:

$$v_1 = \frac{1}{2}(u_1 + u'_1), \qquad \delta = \sqrt{u_0 u_2 - v_1^2}.$$

One finds

$$\widehat{I} = u_0 z + u_0(u_1 + u'_1)\frac{z^2}{2!} + u_0((u_1 + u'_1)^2 + 2u_0 u_2)\frac{z^3}{3!} + \cdots,$$

----

[4]Here, for $|\sigma| = n$, we regard σ as *bordered* by $(-\infty, -\infty)$, i.e., we set $\sigma_0 = \sigma_{n+1} = -\infty$ and let the index i in Figure III.17 vary in $[1 \ldots n]$. Alternative bordering conventions prove occasionally useful.

Figure III.18. The level profile of a random increasing binary tree of size 256. (Compare with Figure III.15, p. 186, for binary trees drawn under the uniform Catalan statistics.)

which agrees with the small cases. This calculation is consistent with what has been found in Chapter II regarding the EGF of all non-empty permutations and of alternating permutations,

$$\frac{z}{1-z}, \quad \tan(z),$$

that follow from the substitutions $\{u_0 = u_1 = u_1' = u_2 = 1\}$ and $\{u_0 = u_2 = 1, u_1 = u_1' = 0\}$, respectively. The substitution $\{u_0 = u_1 = u, u_1' = u_2 = 1\}$ gives a simple variant (without the empty permutation) of the BGF of Eulerian numbers (75) on p. 209.

From the quadrivariate GF, there results that, in a tree of size n the mean number of nodes of nullary, unary, or binary type is asymptotic to $n/3$, with a variance that is $O(n)$, thereby ensuring concentration of distribution. .... ∎

A similar analysis yields path length. It is found that a random increasing binary tree of size n has mean path length

$$2n \log n + O(n).$$

Contrary to what the uniform combinatorial model gives, such trees tend to be rather well balanced, and a typical branch is only about 38.6% longer than in a perfect binary tree (since $2/\log 2 \doteq 1.386$): see Figure III.18 for an illustration. This fact applies to binary search trees (Note III.33) and it justifies the fact that the performance of such trees is quite good, when they are applied to random data [378, 429, 538] or subjected to randomization [451, 520]. See Subsection VI. 10.3 (p. 427) dedicated to tree recurrences for a general analysis of additive functionals on such trees and Example IX.28, p. 684, for a distributional analysis of depth.

▷ **III.33.** *Binary search trees (BSTs).* Given a permutation τ, one defines inductively a tree BST(τ) by

$$\text{BST}(\epsilon) = \emptyset; \qquad \text{BST}(\tau) = \langle \tau_1, \text{BST}(\tau|_{<\tau_1}), \text{BST}(\tau|_{>\tau_1}) \rangle.$$

(Here, $\tau|_P$ represents the subword of τ consisting of those elements that satisfy predicate P.) Let IBT(σ) be the increasing binary tree canonically associated to σ. Then one has the fundamental *Equivalence Principle*,

$$\text{IBT}(\sigma) \stackrel{\text{shape}}{\equiv} \text{BST}(\sigma^{-1}),$$

where $A \stackrel{\text{shape}}{\equiv} B$ means that A and B have identical tree shapes. (Hint: relate the trees to the cartesian representation of permutations [538, 600], as in Example II.17, p. 143.) ◁

III. 7.3. Implicit structures. For implicit structures defined by a relation of the form $\mathcal{A} = \mathfrak{K}[\mathcal{X}]$, we note that equations involving sums and products, either labelled or not, are easily solved just as in the univariate case. The same remark applies for sequence and set constructions: refer to the corresponding sections of Chapters I (p. 88) and II (p. 137). Again, the process is best understood by examples.

Suppose for instance one wants to enumerate connected labelled graphs by the number of nodes (marked by z) and the number of edges (marked by u). The class $\mathcal{K}$ of connected graphs and the class $\mathcal{G}$ of all graphs are related by the set construction,

$$\mathcal{G} = \text{SET}(\mathcal{K}),$$

meaning that every graph decomposes uniquely into connected components. The corresponding exponential BGFs then satisfy

$$G(z, u) = e^{K(z,u)} \qquad \text{implying} \qquad K(z, u) = \log G(z, u),$$

since the number of edges in a graph is inherited (additively) from the corresponding numbers in connected components. Now, the number of graphs of size n having k edges is $\binom{n(n-1)/2}{k}$, so that

$$(67) \qquad\qquad K(z, u) = \log\left(1 + \sum_{n=1}^{\infty}(1 + u)^{n(n-1)/2}\frac{z^n}{n!}\right).$$

This formula, which appears as a refinement of the univariate formula of Chapter II (p. 138), then simply reads: *connected graphs are obtained as components (the* log *operator) of general graphs, where a general graph is determined by the presence or absence of an edge (corresponding to* $(1+u)$*) between any pair of nodes (the exponent* $n(n-1)/2$*).*

To pull information out of the formula (67) is, however, not obvious due to the alternation of signs in the expansion of $\log(1 + w)$ and due to the strongly divergent character of the involved series. As an aside, we note here that the quantity

$$\widehat{K}(z, u) = K\left(\frac{z}{u}, u\right)$$

enumerates connected graphs according to size (marked by z) and excess (marked by u) of the number of edges over the number of nodes. This means that the results of Note II.23 (p. 135), obtained by Wright's decomposition, can be rephrased as the expansion (within $\mathbb{C}(u)[\![z]\!]$):

$$
\begin{aligned}
(68) \qquad \log\left(1 + \sum_{n=1}^{\infty}(1 + u)^{n(n-1)/2}\frac{z^n u^{-n}}{n!}\right) &= \frac{1}{u}W_{-1}(z) + W_0(z) + \cdots \\
&= \frac{1}{u}\left(T - \frac{1}{2}T^2\right) + \left(\frac{1}{2}\log\frac{1}{1 - T} - \frac{1}{2}T - \frac{1}{4}T^2\right) + \cdots,
\end{aligned}
$$

with $T \equiv T(z)$. See Temperley's early works [573, 574] as well as the "giant paper on the giant component" [354] and the paper [254] for direct derivations that eventually constitute analytic alternatives to Wright's combinatorial approach.

Example III.24. *Smirnov words.* Following the treatment of Goulden and Jackson [303], we define a Smirnov word to be any word that has no consecutive equal letters. Let $\mathcal{W} = \text{SEQ}(\mathcal{A})$ be the set of words over the alphabet $\mathcal{A} = \{a_1, \ldots, a_r\}$ of cardinality r, and $\mathcal{S}$ be the set of

Smirnov words. Let also v_j mark the number of occurrences of the jth letter in a word. One has[5]

$$W(v_1, \ldots, v_r) = \frac{1}{1 - (v_1 + \cdots + v_r)}$$

a_j that appears in it an arbitrary non-
lone at all places of a Smirnov word,
y word can be associated to a unique
groups of contiguous equal letters. In
ords by a simultaneous substitution:

$\mapsto \text{SEQ}_{\geq 1}\{a_r\}].$

$$\ldots, \frac{v_r}{1 - v_r}\Big).$$

tly. Now, since the inverse function of

$$= \Big(1 - \sum_{j=1}^{r} \frac{v_j}{1 + v_j}\Big)^{-1}.$$

composition of the words into letters,
to length as

$$\sum r(r-1)^{n-1} z^n.$$

a Smirnov word of length n is deter-
owed by a sequence of $n-1$ choices
ing to $r-1$ possibilities for each po-
e Bernoulli model where letters may
torial argument does not appear to be
this case: see Example IV.10, p. 262

GF of words that never contain more
n (70) the substitution $v_j \mapsto v_j +$
quivalently, the case where letters are

$$\frac{,m+1}{-1)z^{m+1}}.$$

runs and double runs in binary words
that was performed in Subsection I.4.1, p. 51. Naturally, the present approach applies equally well to non-uniform letter probabilities and to a collection of run-length upper-bounds and lower-bounds dependent on each particular letter. This topic is in particular pursued by different methods in several works of Karlin and coauthors (see, e.g., [446]), themselves motivated by applications to life sciences. .... ■

---

[5]The variable z marking length, being redundant, is best omitted in this calculation.

▷ **III.34.** *Enumeration in free groups.*   Consider the composite alphabet $\mathcal{B} = \mathcal{A} \cup \overline{\mathcal{A}}$, where $\mathcal{A} = \{a_1, \ldots, a_r\}$ and $\overline{\mathcal{A}} = \{\overline{a_1}, \ldots, \overline{a_r}\}$. A word over alphabet $\mathcal{B}$ is said to be *reduced* if it arises from a word over $\mathcal{B}$ by a maximal application of the reductions $a_j \overline{a_j} \mapsto \epsilon$ and $\overline{a_j} a_j \mapsto \epsilon$ (with ϵ the empty word). A reduced word thus has no factor of the form $a_j \overline{a_j}$ or $\overline{a_j} a_j$. Such a reduced word serves as a canonical representation of an element in the free group $\mathbf{F}_r$ generated by $\mathcal{A}$, upon identifying $\overline{a_j} = a_j^{-1}$. The GF of the class $\mathcal{R}$ of reduced words, with u_j and $\overline{u}_j$ marking the number of occurrences of letter a_j and $\overline{a_j}$, respectively, is

$$R(u_1, \ldots, u_r, \overline{u_1}, \ldots, \overline{u_r}) = S\left(\frac{u_1}{1 - u_1} + \frac{\overline{u_1}}{1 - \overline{u_1}}, \ \ldots, \ \frac{u_r}{1 - u_r} + \frac{\overline{u_r}}{1 - \overline{u_r}} \right),$$

where S is the GF of Smirnov words, as in (70). In particular this gives the OGF of reduced words with z marking length as $R(z) = (1+z)/(1-(2r-1)z)$; this implies $R_n = 2r(2r-1)^n$, which matches the result given by elementary combinatorics.

The Abelian image $\lambda(w)$ of an element w of the free group $\mathbf{F}_k$ is obtained by letting all letters commute and applying the reductions $a_j \cdot a_j^{-1} = 1$. It can then be put under the form $a_1^{m_1} \cdots a_r^{m_r}$, with each m_j in $\mathbb{Z}$, so that it can be identified with an element of $\mathbb{Z}^r$. Let $\mathbf{x} = (x_1, \ldots, x_r)$ be a vector of indeterminates and define $\mathbf{x}^{\lambda(w)}$ to be the monomial $x_1^{m_1} \cdots x_r^{m_r}$. Of interest in certain group-theoretic investigations is the MGF of reduced words

$$Q(z; \mathbf{x}) := \sum_{w \in \mathcal{R}} z^{|w|} \mathbf{x}^{\lambda(w)} = S\left(\frac{zx_1}{1 - zx_1} + \frac{zx_1^{-1}}{1 - zx_1^{-1}}, \ \ldots, \ \frac{zx_r}{1 - zx_r} + \frac{zx_r^{-1}}{1 - zx_r^{-1}} \right),$$

which is found to simplify to

$$Q(z; \mathbf{x}) = \frac{1 - z^2}{1 - z\sum_{j=1}^{r}(x_j + x_j^{-1}) + (2r - 1)z^2}.$$

This last form appears in a paper of Rivin [514], where it is obtained by matrix techniques. Methods developed in Chapter IX can then be used to establish central and local limit laws for the asymptotic distribution of $\lambda(w)$ over $\mathcal{R}_n$, providing an alternative to the methods of Rivin [514] and Sharp [539]. (This note is based on an unpublished memo of Flajolet, Noy, and Ventura, 2006.)   ◁

▷ **III.35.** *Carlitz compositions II.*   Here is an alternative derivation of the OGF of Carlitz compositions (Note III.32, p. 201). Carlitz compositions with largest summand $\leq r$ are obtained from the OGF of Smirnov words by the substitution $v_j \mapsto z^j$:

$$(71) \qquad\qquad K^{[r]}(z) = \left(1 - \sum_{j=1}^{r} \frac{z^j}{1 + z^j} \right)^{-1},$$

The OGF of all Carlitz compositions then results from letting $r \to \infty$:

$$(72) \qquad\qquad K(z) = \left(1 - \sum_{j=1}^{\infty} \frac{z^j}{1 + z^j} \right)^{-1}.$$

The asymptotic form of the coefficients is derived in Chapter IV, p. 263.   ◁

III. 7.4. Inclusion–exclusion.
Inclusion–exclusion is a familiar type of reasoning rooted in elementary mathematics. Its principle, in order to count *exactly*, consists in grossly *overcounting*, then performing a simple correction of the overcounting, then correcting the correction, and so on. Characteristically, enumerative results provided by inclusion exclusion involve an alternating sum. We revisit this process here in the

perspective of multivariate generating functions, where it essentially reduces to a combined use of substitution and implicit definitions. Our approach follows Goulden and Jackson's encyclopaedic treatise [303].

Let $\mathcal{E}$ be a set endowed with a real- or complex-valued measure $|\cdot|$ in such a way that, for $A, B \subset \mathcal{E}$, there holds

$$|A \cup B| = |A| + |B| \qquad \text{whenever} \qquad A \cap B = \emptyset.$$

Thus, $|\cdot|$ is an additive measure, typically taken as set cardinality (i.e., $|e| = 1$ for $e \in E$) or a discrete probability measure on $\mathcal{E}$ (i.e., $|e| = p_e$ for $e \in E$). The general formula

$$|A \cup B| = |A| + |B| - |AB| \qquad \text{where} \qquad AB := A \cap B,$$

follows immediately from basic set-theoretic principles:

$$\sum_{c \in A \cup B} |c| = \sum_{a \in A} |a| + \sum_{b \in B} |b| - \sum_{i \in A \cap B} |i|.$$

What is called the *inclusion–exclusion principle* or *sieve formula* is the following multivariate generalization, for an arbitrary family $A_1, \ldots, A_r \subset \mathcal{E}$:

(73)
$$|A_1 \cup \cdots \cup A_r| \equiv \left| \mathcal{E} \setminus (\overline{A}_1 \overline{A}_2 \cdots \overline{A}_r) \right|$$
$$= \sum_{1 \le i \le r} |A_i| - \sum_{1 \le i_1 < i_2 \le r} |A_{i_1} A_{i_2}| + \cdots + (-1)^{r-1} |A_1 A_2 \cdots A_r|,$$

where $\overline{A} := \mathcal{E} \setminus A$ denotes complement. (The easy proof by induction results from elementary properties of the boolean algebra formed by the subsets of $\mathcal{E}$; see, e.g., [129, Ch. IV].) An alternative formulation results from setting $B_j = \overline{A}_j$, $\overline{B}_j = A_j$:

(74) $\displaystyle |B_1 B_2 \cdots B_r| = |\mathcal{E}| - \sum_{1 \le i \le r} |\overline{B}_i| + \sum_{1 \le i_1 < i_2 \le r} |\overline{B}_{i_1} \overline{B}_{i_2}| - \cdots + (-1)^r |\overline{B}_1 \overline{B}_2 \cdots \overline{B}_r|.$

In terms of measure, this equality quantifies the set of objects satisfying *exactly* a collection of *simultaneous* conditions (all the B_j) in terms of those that violate *at least some* of the conditions (the $\overline{B}_j$).

Derangements. Here is a textbook example of an inclusion–exclusion argument, namely, the enumeration of *derangements*. Recall that a derangement (p. 122) is a permutation σ such that $\sigma_i \ne i$, for all i. Fix $\mathcal{E}$ as the set of all permutations of $[1, n]$, take the measure $|\cdot|$ to be set cardinality, and let B_i be the subset of permutations in $\mathcal{E}$ associated to the property $\sigma_i \ne i$. (There are consequently $r = n$ conditions.) Thus, B_i means having no fixed point at i, while $\overline{B}_i$ means having a fixed point at the *distinguished* value i. Then, the left-hand side of (74) gives the number of permutations that are derangements; that is, D_n. As regards the right-hand side, the kth sum comprises itself $\binom{n}{k}$ terms counting possibilities attached to the choices of indices $i_1 < \cdots < i_k$; each such choice is associated to a factor $\overline{B}_{i_1} \cdots \overline{B}_{i_k}$ that describes all permutations with fixed points at the distinguished points $i_1, \ldots, i_k$ (i.e., $\sigma(i_1) = i_1, \ldots, \sigma_{i_k} = i_k$). Clearly, $|\overline{B}_{i_1} \cdots \overline{B}_{i_k}| = (n - k)!$. Therefore one has

$$D_n = n! - \binom{n}{1}(n - 1)! + \binom{n}{2}(n - 2)! - \cdots + (-1)^n \binom{n}{n} 0!,$$

which rewrites into the more familiar form

$$\frac{D_n}{n!} = 1 - \frac{1}{1!} + \frac{1}{2!} - \cdots + \frac{(-1)^n}{n!}.$$

This gives an elementary derivation of the derangement numbers already encountered in Chapter II and obtained there by means of the labelled set and cycle constructions.

Symbolic inclusion–exclusion. The derivation above is perfectly fine but complex examples may represent somewhat of a challenge. In contrast, as we now explain, there exists a *symbolic* alternative based on multivariate generating functions, which is technically easy and has great versatility.

Let us now re-examine derangements in a generating function perspective. Consider the set $\mathcal{P}$ of all permutations and build a superset $\mathcal{Q}$ as follows. The set $\mathcal{Q}$ is comprised of permutations in which an arbitrary number of fixed points—some, possibly none, possibly all—have been *distinguished*. (This corresponds to arbitrary products of the $\overline{B}_j$ in the argument above.) For instance $\mathcal{Q}$ contains elements like

$$\underline{1}, 3, 2, \quad 1, 3, 2, \quad \underline{1}, 2, 3, \quad 1, \underline{2}, \underline{3}, \quad 1, \underline{2}, 3, \quad \underline{1}, \underline{2}, \underline{3},$$

where distinguished fixed points are underlined. Clearly, if one removes the distinguished elements of a $\gamma \in \mathcal{Q}$, what is left constitutes an arbitrary permutation of the remaining elements. One has

$$\mathcal{Q} \cong \mathcal{U} \star \mathcal{P},$$

where $\mathcal{U}$ denotes the class of urns that are sets of atoms. In particular, the EGF of $\mathcal{Q}$ is $Q(z) = e^z/(1-z)$. (What we have just done is to enumerate the quantities that appear in (74), but with the signs "wrong", i.e., all pluses.)

Introduce now the variable v to mark the distinguished fixed points in objects of $\mathcal{Q}$. The exponential BGF is then, by the general principles of this chapter,

$$Q(z, v) = e^{vz} \frac{1}{1-z}.$$

Let now $P(z, u)$ be the BGF of permutations where u marks the number of fixed points. Permutations with *some* fixed points distinguished are generated by the substitution $u \mapsto 1 + v$ inside $P(z, u)$. In other words one has the fundamental relation

$$Q(z, v) = P(z, 1 + v).$$

This is then immediately solved to give

$$P(z, u) = Q(z, u - 1),$$

so that knowledge of (the easy) Q gives (the harder) P. For the case at hand, this yields

$$P(z, u) = \frac{e^{(u-1)z}}{1-z}, \qquad P(z, 0) = D(z) = \frac{e^{-z}}{1-z},$$

and, in particular, the EGF of derangements has been retrieved. Note that the desired quantity $P(z, 0)$ comes out as $Q(z, -1)$, so that signs corresponding to the sieve formula (74) have now been put "right", i.e., alternating.

The process employed for derangements is clearly very general: counting objects that contain an exact number of "patterns" is reduced to counting objects that contain the pattern at distinguished places—the latter is usually a simpler problem. The generating function analogue of inclusion–exclusion is then simply the substitution $v \mapsto u - 1$, if a bivariate GF is sought, or $v \mapsto -1$ in the univariate case, when patterns are altogether to be excluded.

Rises in permutations and patterns in words. The book by Goulden and Jackson [303, pp. 45–48] describes a useful formalization of the inclusion process operating on MGFs. Conceptually, it combines substitution and implicit definitions, just as in the case of derangements above. Again, the *modus operandi* is best grasped through examples, two of which are detailed now.

Example III.25. *Rises and ascending runs in permutations.* A *rise* (also called an *ascent*) in a permutation $\sigma = \sigma_1 \cdots \sigma_n$ is a pair of consecutive elements $\sigma_i \sigma_{i+1}$ satisfying $\sigma_i < \sigma_{i+1}$ (with $1 \leq i < n$). The problem is to determine the number $A_{n,k}$ of permutations of size having exactly k rises, together with the exponential BGF $A(z, u)$. By symmetry, we are also enumerating descents (defined by $\sigma_i > \sigma_{i+1}$) as well as ascending runs that are each terminated by a descent.

Guided by the inclusion–exclusion principle, we tackle the easier problem of enumerating permutations with *distinguished* rises, of which the set is denoted by $\mathcal{B}$. For instance, $\mathcal{B}$ contains elements such as

$$2\,6\,1\,\boxed{3\nearrow4\nearrow8\nearrow9\nearrow11}\,15\,12\,\boxed{5\nearrow10}\,13\,7\,14,$$

where those rises that are distinguished are represented by arrows. (Note that some rises may *not* be distinguished.) Maximal sequences of adjacent distinguished rises (boxed in the representation) will be called *clusters*. Then, $\mathcal{B}$ can be specified by the sequence construction applied to atoms ($\mathcal{Z}$) and clusters ($\mathcal{C}$) as

$$\mathcal{B} = \text{SEQ}(\mathcal{Z} + \mathcal{C}), \qquad \text{where} \quad \mathcal{C} = (\mathcal{Z} \nearrow \mathcal{Z}) + (\mathcal{Z} \nearrow \mathcal{Z} \nearrow \mathcal{Z}) + \cdots = \text{SET}_{\geq 2}(\mathcal{Z}).$$

since a cluster is an ordered sequence, or equivalently a set, furthermore having at least two elements. This gives the EGF of $\mathcal{B}$ as

$$B(z) = \frac{1}{1 - (z + (e^z - 1 - z))} = \frac{1}{2 - e^z},$$

which happens to coincide with the EGF of surjections.

For inclusion–exclusion purposes, we need the BGF of $\mathcal{B}$ with v marking the number of distinguished rises. A cluster of size k contains $k - 1$ rises, so that

$$B(z, v) = \frac{1}{1 - (z + (e^{zv} - 1 - zv)/v)} = \frac{v}{v + 1 - e^{zv}}.$$

Now, the usual argument applies: the BGF $A(z, u)$ satisfies $B(z, v) = A(z, 1 + v)$, so that $A(z, u) = B(z, u - 1)$, which yields the particularly simple form

(75) $$A(z, u) = \frac{u - 1}{u - e^{z(u-1)}}.$$

In particular, this GF expands as

$$A(z, u) = 1 + z + (u + 1)\frac{z^2}{2!} + (u^2 + 4u + 1)\frac{z^3}{3!} + (u^3 + 11u^2 + 11u + 1)\frac{z^4}{4!} + \cdots.$$

The coefficients $A_{n,k}$ are known as the *Eulerian numbers* (*Invitation*, p. 9). In combinatorial analysis, these numbers are almost as classic as the Stirling numbers; a detailed discussion of their properties is to be found in classical treatises such as Comtet [129] or Graham *et al.* [307].

Moments derive easily from an expansion of (75) at $u = 1$, which gives

$$A(z, u) = \frac{1}{1-z} + \frac{1}{2}\frac{z^2}{(1-z)^2}(u-1) + \frac{1}{12}\frac{z^3(2+z)}{(1-z)^3}(u-1)^2 + \cdots.$$

In particular: *the mean of the number of rises in a random permutation of size n is $\frac{1}{2}(n-1)$ and the variance is $\sim \frac{1}{12}n$, ensuring concentration of distribution.*

The same method applies to the enumeration of *ascending runs*: for a fixed parameter ℓ, an ascending run of length ℓ is a sequence of consecutive elements $\sigma_i \sigma_{i+1} \cdots \sigma_{i+\ell}$ such that $\sigma_i < \sigma_{i+1} < \cdots < \sigma_{i+\ell}$. (Thus, a rise is an ascending run of length 1.) We define a cluster as a sequence of distinguished runs which overlap in the sense that they share some of the elements of the permutation. The exponential BGF of permutations with distinguished ascending runs is then

$$B(z, v) = \frac{1}{1 - z - \widehat{I}(z, v)}, \qquad \text{where} \quad \widehat{I}(z, v) = \sum_{n,k} I_{n,k} v^k \frac{z^n}{n!},$$

and $I_{n,k}$ is the number of ways of covering the segment $[1, n]$ with k distinct intervals of length ℓ that are contained in $[1, n]$ and have integral end points. The numbers $I_{n,k}$ themselves result from elementary combinatorics (see also the case of patterns in words below) and one has for the OGF corresponding to $\widehat{I}$:

$$I(z, v) = \frac{z^{\ell+1} v}{1 - v(z + z^2 + \cdots + z^\ell)}.$$

(Proof: The first segment in the covering must be placed on the left, the others appear in succession, each shifted right by 1 to ℓ positions from the previous one.) The last two equations finally determine the exponential BGF of permutations with size marked by z and ascending runs of length $\ell + 1$ marked by u,

(76) $$A(z, u) = B(z, u - 1),$$

given the inclusion–exclusion principle.

The resulting formulae generalize the case of rises ($\ell = 1$). They can be made explicit by first expanding the OGF $I(z, v)$ into partial fractions, then applying the transformation $(1 - \omega z)^{-1} \mapsto e^{\omega z}$ in order to translate $I(z, v)$ into $\widehat{I}(z, v)$. The net result is

$$A(z, u) = \frac{1}{1 - z - \widehat{I}(z, u-1)}, \qquad \text{where} \quad \widehat{I}(z, v) = (1-z)(v+1) + \sum_{j=1}^{\ell} c_j(v) e^{\omega_j(v)z}$$

involves a sum of exponentials. In this last equation, the $\omega_j(v)$ are the roots of the characteristic equation $\omega^\ell = v(1 + \cdots + \omega^{\ell-1})$ and the $c_j(v)$ are the corresponding coefficients in the partial fraction decomposition of $I(z, v)$. These expressions were first published by Elizalde and Noy [190] who obtained them by means of tree decompositions.

The BGF (76) can be exploited in order to determine quantitative information on long runs in permutations. First, an expansion at $u = 1$ (also, by a direct reasoning: see the discussion of hidden words in Chapter I) shows that the mean number of ascending runs of length $\ell - 1$ is $(n - \ell + 1)/\ell!$ exactly, as soon as $n \geq \ell$. This entails that, if $n = o(\ell!)$, the probability of finding an ascending run of length $\ell - 1$ tends to 0 as $n \to \infty$. What is used in passing in this argument is the general fact that for a discrete variable X with values in $0, 1, 2, \ldots$, one has (with Iverson's notation),

$$\mathbb{P}(X \geq 1) = \mathbb{E}(\llbracket X \geq 1 \rrbracket) = \mathbb{E}(\min(X, 1)) \leq \mathbb{E}(X).$$

An inequality in the converse direction can be obtained from the second moment method. In effect, the variance of the number of ascending runs of length $\ell - 1$ is found to be of the exact form $\alpha_\ell n + \beta_\ell$, in which α_ℓ is essentially $1/\ell!$ and β_ℓ is of comparable order (details omitted). Then, by Chebyshev's inequalities, concentration of distribution holds as long as ℓ is such that $(\ell + 1)! = o(n)$. In this case, with high probability (i.e., with probability tending to 1 as n tends to ∞), there is at least one ascending run of length $\ell - 1$ (in fact, many). In particular:

> Let L_n be the length of the longest ascending run in a random permutation of n elements. Let $\ell_0(n)$ be the smallest integer such that $\ell! \geq n$. Then the distribution of L_n is concentrated: $L_n/\ell_0(n)$ converges in probability to 1 (in the sense of Equation (14), p. 162).

What has been found here is a fairly sharp threshold phenomenon. .... ∎

▷ **III.36.** *Permutations without ℓ–ascending runs.* The EGF of permutations without 1–, 2– and 3–ascending runs are respectively

$$\left(\sum_{i \geq 0} \frac{x^{2i}}{(2i)!} - \frac{x^{2i+1}}{(2i+1)!} \right)^{-1}, \quad \left(\sum_{i \geq 0} \frac{x^{3i}}{(3i)!} - \frac{x^{3i+1}}{(3i+1)!} \right)^{-1}, \quad \left(\sum_{i \geq 0} \frac{x^{4i}}{(4i)!} - \frac{x^{4i+1}}{(4i+1)!} \right)^{-1},$$

and so on. (See Carlitz's review [103] as well as Elizalde and Noy's article [190] for interesting results involving several types of order patterns in permutations.) ◁

Many variations on the theme of rises and ascending runs are clearly possible. Local order patterns in permutations have been intensely researched, notably by Carlitz in the 1970s. Goulden and Jackson [303, Sec. 4.3] offer a general theory of patterns in sequences and permutations. Special permutations patterns associated with binary increasing trees are also studied by Flajolet, Gourdon, and Martínez [235] (by combinatorial methods) and Devroye [159] (by probabilistic arguments). On another register, the longest ascending run has been found above to be of order $(\log n)/\log\log n$ in probability. The superficially resembling problem of analysing the length of the *longest increasing sequence* in random permutations (elements must be in ascending order but need not be adjacent) has attracted a lot of attention, but is considerably harder. This quantity is $\sim 2\sqrt{n}$ on average and in probability, as shown by a penetrating analysis of the shape of random Young tableaux due to Logan and Shepp [411] and Vershik and Kerov [596]. Solving a problem that had been open for over 20 years, Baik, Deift, and Johansson [24] have eventually determined its limiting distribution. The undemanding survey by Aldous and Diaconis [10] discusses some of the background of this problem, while Chapter VIII (p. 596) shows how to derive bounds that are of the right order of magnitude, using saddle-point methods.

Example III.26. *Patterns in words.* Take the set of all words $W = \text{SEQ}\{\mathcal{A}\}$ over a finite alphabet $\mathcal{A} = \{a_1, \ldots, a_r\}$. A pattern $\mathfrak{p} = p_1 p_2 \cdots p_k$, which is a particular word of length k has been fixed. What is sought is the BGF $W(z, u)$ of W, where u marks the number of occurrences of pattern $\mathfrak{p}$ inside a word of W. The results of Chapter I already give access to $W(z, 0)$, which is the OGF of words not containing the pattern.

In accordance with the inclusion–exclusion principle, one should introduce the class $\mathcal{X}$ of words augmented by distinguishing an arbitrary number of occurrences of $\mathfrak{p}$. Define a *cluster* as a maximal collection of distinguished occurrences that have an overlap. For instance, if $\mathfrak{p} = aaaaa$, a particular word may give rise to the particular cluster:

$$a\,b\,\underline{a\,a\,a\,a\,a\,a\,a\,a\,a\,a\,a\,a\,a\,a\,a}\,b\,\underline{a\,a\,a\,a\,a\,a\,a\,a}\,b\,b$$
$$\qquad\quad a\,a\,a\,a\,a$$
$$\qquad\qquad\quad a\,a\,a\,a\,a$$
$$\qquad\qquad\qquad\quad a\,a\,a\,a\,a$$

Then objects of $\mathcal{X}$ decompose as sequences of either arbitrary letters from $\mathcal{A}$ or clusters:

$$\mathcal{X} = \text{SEQ}\,(\mathcal{A} + \mathcal{C}),$$

with $\mathcal{C}$ the class of all clusters.

Clusters are themselves obtained by repeatedly sliding the pattern, but with the constraint that it should constantly overlap partly with itself. Let $c(z)$ be the autocorrelation polynomial of $\mathfrak{p}$ as defined in Chapter I (p. 61), and set $\widehat{c}(z) = c(z) - 1$. A moment's reflection should convince the reader that $z^k \widehat{c}(z)^{s-1}$ when expanded describes all the possibilities for forming clusters of s overlapping occurrences. On the example above, one has $\widehat{c}(z) = z + z^2 + z^3 + z^4$, and a particular cluster of 3 overlapping occurrences corresponds to one of the terms in $z^k \widehat{c}(z)^2$ as follows:

$$\overbrace{a\;a\;a\;a\;a}^{z^5}$$
$$\qquad\overbrace{a\;a\;a}^{z^2}\!\overset{\frown}{a\;a}$$
$$\qquad\qquad\underbrace{a\;a\;a\;a\;a}_{z^4}$$

$$z^5$$
$$\times\,(z + \underline{z^2} + z^3 + z^4)$$
$$\times\,(z + z^2 + z^3 + \underline{z^4}).$$

The OGF of clusters is consequently $C(z) = z^k/(1 - \widehat{c}(z))$ since this quantity describes all the ways to write the pattern (z^k) and then slide it so that it should overlap with itself (this is given by $(1 - \widehat{c}(z))^{-1}$).

By a similar reasoning, the BGF of clusters is $vz^k/(1 - v\widehat{c}(z))$, and the BGF of $\mathcal{X}$ with the supplementary variable v marking the number of distinguished occurrences is

$$X(z, v) = \frac{1}{1 - rz - vz^k/(1 - v\widehat{c}(z))}.$$

Finally, the usual inclusion–exclusion argument (change v to $u - 1$) yields $W(z, u) = X(z, u - 1)$. As a result:

> *For a pattern $\mathfrak{p}$ with correlation polynomial $c(z)$ and length k, the BGF of words over an alphabet of cardinality r, where u marks the number of occurrences of $\mathfrak{p}$, is*

$$(77) \qquad W(z, u) = \frac{(u - 1)c(z) - u}{(1 - rz)((u - 1)c(z) - u) + (u - 1)z^k}.$$

The specialization $u = 0$ gives back the formula already found in Chapter I, p. 61. The same principles clearly apply to weighted models corresponding to unequal letter probabilities, provided a suitably weighted version of the correlation polynomial is introduced (see Note III.39 below). . . . . ∎

There are a very large number of formulae related to patterns in strings. For instance, BGFs are known for occurrences of one or several patterns under either Bernoulli or Markov models; see Note III.39 below. We refer to Szpankowski's book [564] and Lothaire's chapter [347], where such questions are treated systematically in great detail. Bourdon and Vallée [81] have succeeded in extending this approach to *dynamical sources* of information, thereby uniting a large number of previously known results. Their approach even makes it possible to analyse the occurrence of patterns in continued fraction representations of real numbers.

▷ **III.37.** *Moments of number of occurrences.* The derivatives of $X(z, v)$ at $v = 0$ give access to the factorial moments of the number of occurrences of a pattern. In this way or directly, one determines

$$W(z, u) = \frac{1}{1 - rz} + \frac{z^k}{(1 - rz)^2}(u - 1) + 2\frac{z^k((1 - rz)(c(z) - 1) + z^k)}{(1 - rz)^3}\frac{(u - 1)^2}{2!} + \cdots.$$

The mean number of occurrences is r^{-n} times the coefficient of z^n in the coefficient of $(u - 1)$ and is $(n - k + 1)r^{-k}$, as anticipated. The coefficient of $(u - 1)^2/2!$ is of the form

$$\frac{2r^{-2k}}{(1 - rz)^3} + \frac{2r^{-k}(1 + 2kr^{-k} - c(1/r))}{(1 - rz)^2} + \frac{P(z)}{1 - rz},$$

with P a polynomial. This shows that the variance of the number of occurrences is of the form

$$\alpha n + \beta, \qquad \alpha = r^{-k}(2c(1/r) - 1 + r^{-k}(1 - 2k)).$$

Consequently, the distribution is concentrated around its mean. (See also the discussion of "Borges' Theorem" in Chapter I, p. 61.)   ◁

▷ **III.38.** *Words with fixed repetitions.* Let $W^{\langle s \rangle}(z) = [u^s]W(z, u)$ be the OGF of words containing a pattern exactly s times. One has, for $s > 0$ and $s = 0$, respectively,

$$W^{\langle s \rangle}(z) = \frac{z^k N(z)^{s-1}}{D(z)^{s+1}}, \qquad W^{\langle 0 \rangle}(z) = \frac{c(z)}{D(z)},$$

with $N(z)$ and $D(z)$ given by

$$N(z) = (1 - rz)(c(z) - 1) + z^k, \qquad D(z) = (1 - rz)c(z) + z^k.$$

The expression of $W^{\langle 0 \rangle}$ is in agreement with Chapter I, Equation (62), p. 61.   ◁

▷ **III.39.** *Patterns in Bernoulli sequences.* Let $\mathcal{A}$ be an alphabet where letter α has probability π_α and consider the Bernoulli model where letters in words are chosen independently. Fix a pattern $\mathfrak{p} = p_1 \cdots p_k$ and define the finite language of *protrusions* as

$$\Gamma = \bigcup_{i \,:\, c_i \neq 0} \{p_{i+1}p_{i+2} \cdots p_k\},$$

where the union is over all correlation positions of the pattern. Define now the correlation polynomial $\gamma(z)$ (relative to $\mathfrak{p}$ and the π_α) as the generating polynomial of the finite language of protrusions weighted by (π_α). For instance, $\mathfrak{p} = ababa$ gives rise to $\Gamma = \{\epsilon, ba, baba\}$ and

$$\gamma(z) = 1 + \pi_a\pi_b z^2 + \pi_a^2\pi_b^2 z^4.$$

The BGF of words with z marking length and u marking the number of occurrences of $\mathfrak{p}$ is

$$W(z, u) = \frac{(u - 1)\gamma(z) - u}{(1 - z)((u - 1)\gamma(z) - u) + (u - 1)\pi[\mathfrak{p}]z^k},$$

where $\pi[\mathfrak{p}]$ is the product of the probabilities of letters of $\mathfrak{p}$.   ◁

▷ **III.40.** *Patterns in trees I.* Consider the class $\mathcal{B}$ of pruned binary trees. An occurrence of pattern t in a tree τ is defined by a node of τ whose dangling subtree is isomorphic to t. We seek the BGF $B(z, u)$ of class $\mathcal{B}$ where u marks the number of occurrences of t.

The OGF of $\mathcal{B}$ is $B(z) = (1 - \sqrt{1 - 4z})/(2z)$. The quantity $vB(zv)$ is the BGF of $\mathcal{B}$ with v marking external nodes. By virtue of the pointing operation, the quantity

$$U_k := \left(\frac{1}{k!}\partial_v^k (vB(zv))\right)_{v=1},$$

describes trees with k distinct external nodes distinguished (pointed). Let $m = |t|$. The quantity

$$V := \sum U_k u^k (z^m)^k \quad \text{satisfies} \quad V = (vB(zv))_{v=1+uz^m},$$

by virtue of Taylor's formula. It is also the BGF of trees with distinguished occurrences of t marked by v. Setting $v \mapsto u - 1$ in V then gives $B(z, u)$ as

$$(78) \qquad\qquad B(z, u) = \frac{1}{2z} \left(1 - \sqrt{1 - 4z - 4(u - 1)z^{m+1}} \right).$$

In particular $B(z, 0) = \frac{1}{2z} \left(1 - \sqrt{1 - 4z + 4z^{m+1}} \right)$ represents the OGF of trees *not* containing pattern t. The method generalizes to any simple variety of trees. It can be used to prove that the factored representation (as a directed acyclic graph) of a random tree of size n has expected size $O(n/\sqrt{\log n})$. (These results appear in [257]; see also Example IX.26, p. 680, for a related Gaussian law.)   ◁

▷ **III.41.** *Patterns in trees II.* Here follows an alternative derivation of (78) that is based on the root decomposition of trees. A pattern t occurs either in the left root subtree τ_0, or in the right root subtree τ_1, or at the root iself in the case in which t coincides with τ. Thus the number $\omega[\tau]$ of occurrences of t in τ satisfies the recursive definition

$$\omega[\tau] = \omega[\tau_0] + \omega[\tau_1] + [\![\tau = t]\!], \qquad \omega[\emptyset] = 0.$$

The function $u^{\omega[\tau]}$ is almost multiplicative, and

$$u^{\omega[\tau]} = u^{[\![\tau=t]\!]} u^{\omega[\tau_0]} u^{\omega[\tau_1]} = u^{\omega[\tau_0]} u^{\omega[\tau_1]} + [\![\tau = t]\!] \cdot (u - 1).$$

Thus, the bivariate generating function $B(z, u) := \sum_t z^{|t|} u^{\omega[t]}$ satisfies the quadratic equation,

$$B(z, u) = 1 + (u - 1)z^m + zB(z, u)^2,$$

which, when solved, yields (78).   ◁

III. 8. Extremal parameters

Apart from additively inherited parameters already examined at length in this chapter, another important category is that of parameters defined by a maximum rule. Two major cases are the largest component in a combinatorial structure (for instance, the largest cycle of a permutation) and the maximum degree of nesting of constructions in a recursive structure (typically, the height of a tree). In this case, bivariate generating functions are of little help, because of the nonlinear character of the max-function. The standard technique consists in introducing *a collection of univariate generating functions* defined by imposing a bound on the parameter of interest. Such GFs can then be constructed by the symbolic method in its univariate version.

III. 8.1. Largest components. Consider a construction $\mathcal{B} = \Phi[\mathcal{A}]$, where Φ may involve an arbitrary combination of basic constructions, and assume here for simplicity that the construction for $\mathcal{B}$ is a non-recursive one. This corresponds to a relation between generating functions

$$B(z) = \Psi[A(z)],$$

where Ψ is the functional that is the "image" of the combinatorial construction Φ. Elements of $\mathcal{A}$ thus appear as components in an object $\beta \in \mathcal{B}$. Let $\mathcal{B}^{\langle b \rangle}$ denote the subclass of $\mathcal{B}$ formed with objects whose $\mathcal{A}$–components all have a size at most b. The

GF of $\mathcal{B}^{\langle b \rangle}$ is obtained by the same process as that of $\mathcal{B}$ itself, save that $A(z)$ should be replaced by the GF of elements of size at most b. Thus,

$$B^{\langle b \rangle}(z) = \Psi[\mathbf{T}_b A(z)],$$

where the *truncation operator* is defined on series by

$$\mathbf{T}_b f(z) = \sum_{n=0}^{b} f_n z^n \qquad \left(f(z) = \sum_{n=0}^{\infty} f_n z^n \right).$$

Example III.27. *A pot-pourri of largest components.* Several instances of largest components have already been analysed in Chapters I and II. For instance, the cycle decomposition of permutations translated by

$$\mathcal{P} = \text{SET}(\text{CYC}(\mathcal{Z})) \qquad \Longrightarrow \qquad P(z) = \exp\left(\log \frac{1}{1-z} \right)$$

gives more generally the EGF of permutations with longest cycle $\leq b$,

$$P^{\langle b \rangle}(z) = \exp\left(\frac{z}{1} + \frac{z^2}{2} + \cdots + \frac{z^b}{b} \right),$$

which involves the truncated logarithm.

The labelled specification of words over an m–ary alphabet

$$\mathcal{W} = \text{SET}_m(\text{SET}(\mathcal{Z})) \qquad \Longrightarrow \qquad W(z) = \left(e^z\right)^m$$

leads to the EGF of words such that each letter occurs at most b times:

$$W^{\langle b \rangle}(z) = \left(1 + \frac{z}{1!} + \frac{z^2}{2!} + \cdots + \frac{z^b}{b!} \right)^m,$$

which now involves the truncated exponential. Similarly, the EGF of set partitions with largest block of size at most b is

$$S^{\langle b \rangle}(z) = \exp\left(\frac{z}{1!} + \frac{z^2}{2!} + \cdots + \frac{z^b}{b!} \right).$$

A slightly less direct example is that of the longest run in a binary string (p. 51), which we now revisit. The collection $\mathcal{W}$ of binary words over the alphabet $\{a, b\}$ admits the unlabelled specification

$$\mathcal{W} = \text{SEQ}(a) \cdot \text{SEQ}(b\ \text{SEQ}(a)),$$

corresponding to a "scansion" dictated by the occurrences of the letter b. The corresponding OGF then appears under the form

$$W(z) = Y(z) \cdot \frac{1}{1 - zY(z)}, \qquad \text{where} \quad Y(z) = \frac{1}{1-z}$$

corresponds to $\mathcal{Y} = \text{SEQ}(a)$. Thus, the OGF of strings with at most $k - 1$ consecutive occurrences of the letter a obtains upon replacing $Y(z)$ by its truncation:

$$W^{\langle k \rangle}(z) = Y^{\langle k \rangle}(z) \frac{1}{1 - zY^{\langle k \rangle}(z)}, \qquad \text{where} \quad Y^{\langle k \rangle}(z) = 1 + z + z^2 + \cdots + z^{k-1},$$

so that

$$W^{\langle k \rangle}(z) = \frac{1 - z^k}{1 - 2z + z^{k+1}}.$$

An asymptotic analysis is given in Example V.4, p. 308. .... ∎

Generating functions for largest components are thus easy to derive. The asymptotic analysis of their coefficients is however often hard when compared to additive parameters, owing to the need to rely on complex analytic properties of the truncation operator. The bases of a general asymptotic theory have been laid by Gourdon [305].

▷ **III.42.** *Smallest components.* The EGF of permutations with smallest cycle of size $> b$ is

$$\frac{1}{1-z} \exp\left(-\frac{z}{1} - \frac{z^2}{2} - \cdots - \frac{z^b}{b}\right).$$

A symbolic theory of *smallest* components in combinatorial structures is easily developed as regards formal GFs. Elements of the corresponding asymptotic theory are provided by Panario and Richmond in [470].   ◁

III. 8.2. Height. The degree of nesting of a recursive construction is a generalization of the notion of height in the simpler case of trees. Consider for instance a recursively defined class

$$\mathcal{B} = \Phi[\mathcal{B}],$$

where Φ is a construction. Let $\mathcal{B}^{[h]}$ denote the subclass of $\mathcal{B}$ composed solely of elements whose construction involves at most h applications of Φ. We have by definition

$$\mathcal{B}^{[h+1]} = \Phi\{\mathcal{B}^{[h]}\}.$$

Thus, with Ψ the image functional of construction Φ, the corresponding GFs are defined by a *recurrence*,

$$B^{[h+1]} = \Psi[B^{[h]}].$$

(This discussion is related to the semantics of recursion, p. 33.)

Example **III.28.** *Generating functions for tree height.* Consider first general plane trees:

$$\mathcal{G} = \mathcal{Z} \times \text{SEQ}(\mathcal{G}) \qquad \Longrightarrow \qquad G(z) = \frac{z}{1 - G(z)}.$$

Define the height of a tree as the number of edges on its longest branch. Then the set of trees of height $\leq h$ satisfies the recurrence

$$\mathcal{G}^{[0]} = \mathcal{Z}, \qquad \mathcal{G}^{[h+1]} = \mathcal{Z} \times \text{SEQ}(\mathcal{G}^{[h]}).$$

Accordingly, the OGF of trees of bounded height satisfies

$$G^{[0]}(z) = z, \qquad G^{[h+1]}(z) = \frac{z}{1 - G^{[h]}(z)}.$$

The recurrence unwinds and one finds

$$(79) \qquad G^{[h]}(z) = \cfrac{z}{1 - \cfrac{z}{1 - \cfrac{z}{\ddots \atop {1 - z}}}},$$

where the number of stages in the fraction equals b. This is the finite form (technically known as a "convergent") of a *continued fraction* expansion. From implied linear recurrences and an analysis based on Mellin transforms, de Bruijn, Knuth, and Rice [145] have determined the average height of a general plane tree to be $\sim \sqrt{\pi n}$. We provide a proof of this fact in Chapter V (p. 329) dedicated to applications of rational and meromorphic asymptotics.

For plane binary trees defined by

$$\mathcal{B} = \mathcal{Z} + \mathcal{B} \times \mathcal{B} \qquad \text{so that} \qquad B(z) = z + (B(z))^2,$$

(size here is the number of external nodes), the recurrence is

$$B^{[0]}(z) = z, \quad B^{[h+1]}(z) = z + (B^{[h]}(z))^2.$$

In this case, the $B^{[h]}$ are the approximants to a "continuous quadratic form", namely

$$B^{[h]}(z) = z + (z + (z + (\cdots)^2)^2)^2.$$

These are polynomials of degree 2^h for which no closed form expression is known, nor even likely to exist[6]. However, using complex asymptotic methods and singularity analysis, Flajolet and Odlyzko [246] have shown that the average height of a binary plane tree is $\sim 2\sqrt{\pi n}$. See Subsection VII. 10.2, p. 535 for the sketch of a proof.

For Cayley trees, finally, the defining equation is

$$\mathcal{T} = \mathcal{Z} \star \text{SET}(\mathcal{T}) \qquad \Longrightarrow \qquad T(z) = ze^{T(z)}.$$

The EGF of trees of bounded height satisfy the recurrence

$$T^{[0]}(z) = z, \qquad T^{[h+1]}(z) = ze^{T^{[h]}(z)}.$$

We are now confronted with a "continuous exponential",

$$T^{[h]}(z) = ze^{ze^{ze^{\cdot^{\cdot^{\cdot ze^z}}}}}.$$

The average height was found by Rényi and Szekeres who appealed again to complex analytic methods and found it to be $\sim \sqrt{2\pi n}$. .... ∎

These examples show that height statistics are closely related to iteration theory. Except in a few cases like general plane trees, normally no algebra is available and one has to resort to complex analytic methods as expounded in forthcoming chapters.

III. 8.3. Averages and moments.

For extremal parameters, the GFs of mean values obey a general pattern. Let $\mathcal{F}$ be some combinatorial class with GF $f(z)$. Consider for instance an extremal parameter χ such that $f^{[h]}(z)$ is the GF of objects with χ-parameter *at most h*. The GF of objects for which $\chi = h$ *exactly* is equal to

$$f^{[h]}(z) - f^{[h-1]}(z).$$

Thus differencing gives access to the probability distribution of height over $\mathcal{F}$. The generating function of cumulated values (providing mean values after normalization) is then

$$\begin{aligned}
\Xi(z) &= \sum_{h=0}^{\infty} h \left[f^{[h]}(z) - f^{[h-1]}(z) \right] \\
&= \sum_{h=0}^{\infty} \left[f(z) - f^{[h]}(z) \right],
\end{aligned}$$

as is readily checked by rearranging the second sum, or equivalently using summation by parts.

---

[6]These polynomials are exactly the much-studied Mandelbrot polynomials whose behaviour in the complex plane gives rise to extraordinary graphics (Figure VII.23, p. 536).

For the largest components, the formulae involve truncated Taylor series. For height, analysis involves in all generality the differences between the fixed point of a functional Φ (the GF $f(z)$) and the approximations to the fixed point ($f^{[h]}(z)$) provided by iteration. This is a common scheme in extremal statistics.

▷ **III.43.** *The height of increasing binary trees.* Given the specification of increasing binary trees in Equation (61), p. 143, the EGF of trees of height at most h is given by the recurrence

$$I^{[0]}(z) = 1, \qquad I^{[h+1]}(z) = 1 + \int_0^z I^{[h]}(w)^2 \, dw.$$

Devroye [157, 158] showed in 1986 that the expected height of a tree of size n is asymptotic to $c \log n$ where $c \doteq 4.31107$ is a solution of $c \log((2e)/c) = 1$. ◁

▷ **III.44.** *Hierarchical partitions.* Let $\varepsilon(z) = e^z - 1$. The generating function

$$\varepsilon(\varepsilon(\cdots(\varepsilon(z)))) \qquad (h \text{ times}).$$

can be interpreted as the EGF of certain hierarchical partitions. (Such structures show up in statistical classification theory [585, 586].) ◁

▷ **III.45.** *Balanced trees.* Balanced structures lead to counting GFs close to the ones obtained for height statistics. The OGF of balanced 2–3 trees of height h counted by the number of leaves satisfies the recurrence

$$Z^{[h+1]}(z) = Z^{[h]}(z^2 + z^3) = (Z^{[h]}(z))^2 + (Z^{[h]}(z))^3,$$

which can be expressed in terms of the iterates of $\sigma(z) = z^2 + z^3$ (see Note I.67, p. 91, as well as Chapter IV, p. 281, for asymptotics). It is possible to express the OGF of cumulated values of the number of internal nodes in such trees in terms of the iterates of σ. ◁

▷ **III.46.** *Extremal statistics in random mappings.* One can express the EGFs relative to the largest cycle, longest branch, and diameter of functional graphs. Similarly for the largest tree, largest component. [Hint: see [247] for details.] ◁

▷ **III.47.** *Deep nodes in trees.* The BGF giving the number of nodes at maximal depth in a general plane tree or a Cayley tree can be expressed in terms of a continued fraction or a continuous exponential. ◁

III. 9. Perspective

The message of this chapter is that we can use the symbolic method not just to count combinatorial objects but also to quantify their properties. The relative ease with which we are able to do so is testimony to the power of the method as a major organizing principle of analytic combinatorics.

The global framework of the symbolic method leads us to a natural structural categorization of parameters of combinatorial objects. First, the concept of *inherited parameters* permits a direct extension of the already seen formal translation mechanisms from combinatorial structures to GFs, for both labelled and unlabelled objects—this leads to MGFs useful for solving a broad variety of classical combinatorial problems. Second, the adaptation of the theory to *recursive parameters* provides information about trees and similar structures, this even in the absence of explicit representations of the associated MGFs. Third, *extremal parameters*, which are defined by a maximum rule (rather than an additive rule), can be studied by analysing families of univariate GFs. Yet another illustration of the power of the symbolic method is found in the notion of *complete GF*, which in particular enables us to study Bernoulli trials and branching processes.

As we shall see starting with Chapter IV, these approaches become especially powerful since they serve as the basis for the *asymptotic analysis of properties of structures*. Not only does the symbolic method provide precise information about particular parameters, but it also paves the way for the discovery of general *schemas* and theorems that tell us what to expect about a broad variety of combinatorial types.

Bibliographic notes. Multivariate generating functions are a common tool from classical combinatorial analysis. Comtet's book [129] is once more an excellent source of examples. A systematization of multivariate generating functions for inherited parameters is given in the book by Goulden and Jackson [303].

In contrast generating functions for cumulated values of parameters (related to averages) seemed to have received relatively little attention until the advent of digital computers and the analysis of algorithms. Many important techniques are implicit in Knuth's treatises, especially [377, 378]. Wilf discusses related issues in his book [608] and the paper [606]. Early systems specialized to tree algorithms were proposed by Flajolet and Steyaert in the 1980s [215, 261, 262, 560]; see also Berstel and Reutenauer's work [56]. Some of the ideas developed there initially drew their inspiration from the well-established treatment of formal power series in non-commutative indeterminates; see the books by Eilenberg [189] and Salomaa and Soittola [527] as well as the proceedings edited by Berstel [54]. Several computations in this area can nowadays even be automated with the help of computer algebra systems [255, 528, 628].

Je n'ai jamais été assez loin pour bien sentir l'application de l'algèbre à la géométrie. Je n'aimais point cette manière d'opérer sans voir ce qu'on fait, et il me sembloit que résoudre un problème de géométrie par les équations, c'étoit jouer un air en tournant une manivelle.

("*I never went far enough to get a good feel for the application of algebra to geometry. I was not pleased with this method of operating according to the rules without seeing what one does; solving geometrical problems by means of equations seemed like playing a tune by turning a crank.*")

— JEAN-JACQUES ROUSSEAU, *Les Confessions*, Livre VI

Part B

COMPLEX ASYMPTOTICS

IV

Complex Analysis, Rational and Meromorphic Asymptotics

Entre deux vérités du domaine réel, le chemin le plus facile et le plus court passe bien souvent par le domaine complexe.

PAUL PAINLEVÉ [467, p. 2]

It has been written that the shortest and best way between two truths of the real domain often passes through the imaginary one[1].

— JACQUES HADAMARD [316, p. 123]

Generating functions are a central concept of combinatorial theory. In Part A, we have treated them as formal objects; that is, as formal power series. Indeed, the major theme of Chapters I–III has been to demonstrate how the algebraic structure of generating functions directly reflects the structure of combinatorial classes. From now on, we examine generating functions in the light of *analysis*. This point of view involves assigning *values* to the variables that appear in generating functions.

Comparatively little benefit results from assigning only real values to the variable z that figures in a univariate generating function. In contrast, assigning *complex* values turns out to have serendipitous consequences. When we do so, a generating function becomes a geometric transformation of the complex plane. This transformation is very regular near the origin—one says that it is *analytic* (or *holomorphic*). In other words, near 0, it only effects a smooth distortion of the complex plane. Farther away from the origin, some cracks start appearing in the picture. These cracks—the dignified name is *singularities*—correspond to the disappearance of smoothness. It turns out that a function's singularities provide a wealth of information regarding the function's coefficients, and especially their asymptotic rate of growth. Adopting a geometric point of view for generating functions has a large pay-off.

---

[1]Hadamard's quotation (1945) is a free rendering of the original one due to Painlevé (1900); namely, *"The shortest and easiest path between two truths of the real domain most often passes through the complex domain."*

By focusing on singularities, analytic combinatorics treads in the steps of many respectable older areas of mathematics. For instance, Euler recognized that for the Riemann zeta function $\zeta(s)$ to become infinite (hence have a singularity) at 1 implies the existence of infinitely many prime numbers; Riemann, Hadamard, and de la Vallée-Poussin later uncovered deep connections between quantitative properties of prime numbers and singularities of $1/\zeta(s)$.

The purpose of this chapter is largely to serve as an accessible introduction or a refresher of basic notions regarding analytic functions. We start by recalling the elementary theory of functions and their singularities in a style tuned to the needs of analytic combinatorics. Cauchy's integral formula expresses coefficients of analytic functions as contour integrals. Suitable uses of Cauchy's integral formula then make it possible to estimate such coefficients by suitably selecting an appropriate contour of integration. For the common case of functions that have singularities at a finite distance, the exponential growth formula relates the *location* of the singularities closest to the origin—these are also known as *dominant* singularities—to the *exponential order of growth* of coefficients. The *nature* of these singularities then dictates the fine structure of the asymptotics of the function's coefficients, especially the *subexponential factors* involved.

As regards generating functions, combinatorial enumeration problems can be broadly categorized according to a hierarchy of increasing structural complexity. At the most basic level, we encounter scattered classes, which are simple enough, so that the associated generating function and coefficients can be made explicit. (Examples of Part A include binary and general plane trees, Cayley trees, derangements, mappings, and set partitions). In that case, elementary real-analysis techniques usually suffice to estimate asymptotically counting sequences. At the next, intermediate, level, the generating function is still explicit, but its form is such that no simple expression is available for coefficients. This is where the theory developed in this and the next chapters comes into play. It usually suffices to have an expression for a generating function, but *not* necessarily its coefficients, so as to be able to deduce precise asymptotic estimates of its coefficients. (Surjections, generalized derangements, unary–binary trees are easily subjected to this method. A striking example, that of trains, is detailed in Section IV. 4.) Properties of analytic functions then make this analysis depend only on *local properties* of the generating function at a few points, its dominant singularities. The third, highest, level, within the perspective of analytic combinatorics, comprises generating functions that can no longer be made explicit, but are only determined by a *functional equation*. This covers structures defined recursively or implicitly by means of the basic constructors of Part A. The analytic approach even applies to a large number of such cases. (Examples include simple families of trees, balanced trees, and the enumeration of certain molecules treated at the end of this chapter. Another characteristic example is that of non-plane unlabelled trees treated in Chapter VII.)

As we shall see throughout this book, the analytic methodology applies to almost all the combinatorial classes studied in Part A, which are provided by the symbolic method. In the present chapter we carry out this programme for *rational functions* and *meromorphic functions* (i.e., functions whose singularities are *poles*).

IV. 1. Generating functions as analytic objects

Generating functions, considered in Part A as purely *formal* objects subject to al-gebraic operations, are now going to be interpreted as *analytic* objects. In so doing one gains easy access to the asymptotic form of their coefficients. This informal section offers a glimpse of themes that form the basis of Chapters IV–VII.

In order to introduce the subject, let us start with two simple generating functions, one, $f(z)$, being the OGF of the Catalan numbers (cf $G(z)$, p. 35), the other, $g(z)$, being the EGF of derangements (cf $D^{(1)}(z)$, p. 123):

$$(1) \qquad f(z) = \frac{1}{2}\left(1 - \sqrt{1 - 4z}\right), \qquad g(z) = \frac{\exp(-z)}{1 - z}.$$

At this stage, the forms above are merely compact descriptions of formal power series built from the elementary series

$$(1-y)^{-1} \;=\; 1 + y + y^2 + \cdots, \qquad\qquad (1-y)^{1/2} \;=\; 1 - \frac{1}{2}y - \frac{1}{8}y^2 - \cdots,$$

$$\exp(y) \;\;=\; 1 + \frac{1}{1!}y + \frac{1}{2!}y^2 + \cdots,$$

by standard composition rules. Accordingly, the coefficients of both GFs are known in explicit form:

$$f_n := [z^n] f(z) = \frac{1}{n}\binom{2n-2}{n-1}, \qquad g_n := [z^n] g(z) = \left(\frac{1}{0!} - \frac{1}{1!} + \cdots + \frac{(-1)^n}{n!}\right).$$

Stirling's formula and the comparison with the alternating series giving $\exp(-1)$ pro-vide, respectively,

$$(2) \qquad\qquad f_n \underset{n\to\infty}{\sim} \frac{4^{n-1}}{\sqrt{\pi n^3}}, \qquad g_n = \underset{n\to\infty}{\sim} e^{-1} \doteq 0.36787.$$

Our purpose now is to provide intuition on how such approximations could be derived without appealing to explicit forms. We thus examine, heuristically for the moment, the direct relationship between the asymptotic forms (2) and the structure of the corresponding generating functions in (1).

Granted the growth estimates available for f_n and g_n, it is legitimate to substitute in the power series expansions of the GFs $f(z)$ and $g(z)$ any real or complex value of a small enough modulus, the upper bounds on modulus being $\rho_f = 1/4$ (for f) and $\rho_g = 1$ (for g). Figure IV.1 represents the graph of the resulting functions when such *real* values are assigned to z. The graphs are smooth, representing functions that are differentiable any number of times for z interior to the interval $(-\rho, +\rho)$. However, at the right boundary point, smoothness stops: $g(z)$ become infinite at $z = 1$, and so it even ceases to be finitely defined; $f(z)$ does tend to the limit $\frac{1}{2}$ as $z \to (\frac{1}{4})^-$, but its derivative becomes infinite there. Such special points at which smoothness stops are called *singularities*, a term that will acquire a precise meaning in the next sections.

Observe also that, in spite of the series expressions being divergent outside the specified intervals, the functions $f(z)$ and $g(z)$ can be *continued* in certain regions: it

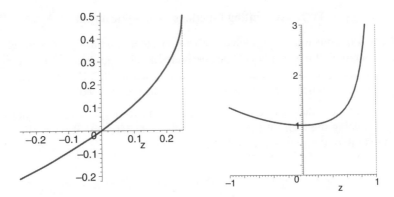

Figure IV.1. Left: the graph of the Catalan OGF, $f(z)$, for $z \in (-\frac{1}{4}, +\frac{1}{4})$; right: the graph of the derangement EGF, $g(z)$, for $z \in (-1, +1)$.

suffices to make use of the global expressions of Equation (1), with exp and $\sqrt{}$ being assigned their usual real-analytic interpretation. For instance:

$$f(-1) = \frac{1}{2}\left(1 - \sqrt{5}\right), \qquad g(-2) = \frac{e^2}{3}.$$

Such continuation properties, most notably to the *complex* realm, will prove essential in developing efficient methods for coefficient asymptotics.

One may proceed similarly with complex numbers, starting with numbers whose modulus is less than the radius of convergence of the series defining the GF. Figure IV.2 displays the images of regular grids by f and g, as given by (1). This illustrates the fact that a regular grid is transformed into an orthogonal network of curves and more precisely that f and g preserve angles—this property corresponds to complex differentiability and is equivalent to analyticity to be introduced shortly. The singularity of f is clearly perceptible on the right of its diagram, since, at $z = 1/4$ (corresponding to $f(z) = 1/2$), the function f folds lines and divides angles by a factor of 2. The singularity of g at $z = 1$ is indirectly perceptible from the fact that $g(z) \to \infty$ as $z \to 1$ (the square grid had to be truncated at $z = 0.75$, since this book can only accommodate finite graphs).

Let us now turn to coefficient asymptotics. As is expressed by (2), the coefficients f_n and g_n each belong to a general asymptotic type for coefficients of a function F, namely,

$$(3) \qquad\qquad [z^n]F(z) = A^n \theta(n),$$

corresponding to an *exponential growth* factor A^n modulated by a tame factor $\theta(n)$, which is *subexponential*. Here, one has $A = 4$ for f_n and $A = 1$ for g_n; also, $\theta(n) \sim \frac{1}{4}(\sqrt{\pi n^3})^{-1}$ for f_n and $\theta(n) \sim e^{-1}$ for g_n. Clearly, A should be related to the radius of convergence of the series. We shall see that, invariably, for combinatorial generating functions, the exponential rate of growth is given by $A = 1/\rho$, where ρ is the first singularity encountered along the positive real axis (Theorem IV.6,

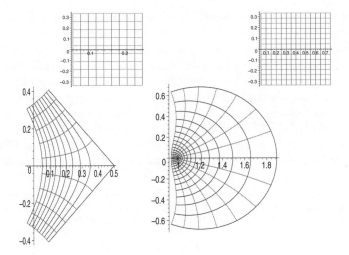

Figure IV.2. The images of regular grids by $f(z)$ (left) and $g(z)$ (right).

---

p. 240). In addition, under general complex analytic conditions, it will be established that $\theta(n) = O(1)$ is systematically associated to a simple pole of the generating function (Theorem IV.10, p. 258), while $\theta(n) = O(n^{-3/2})$ systematically arises from a singularity that is of the square-root type (Chapters VI and VII). We enunciate:

> **First Principle of Coefficient Asymptotics.** *The* location *of a function's singularities dictates the* exponential growth (A^n) *of its coefficients.*
>
> **Second Principle of Coefficient Asymptotics.** *The* nature *of a function's singularities determines the associate* subexponential factor $(\theta(n))$.

Observe that the rescaling rule,

$$[z^n]F(z) = \rho^{-n}[z^n]F(\rho z),$$

enables one to normalize functions so that they are singular at 1. Then, various theorems, starting with Theorems IV.9 and IV.10, provide sufficient conditions under which the following fundamental implication is valid,

$$(4) \qquad h(z) \sim \sigma(z) \quad \Longrightarrow \quad [z^n]h(z) \sim [z^n]\sigma(z).$$

There $h(z)$, whose coefficients are to be estimated, is a function singular at 1 and $\sigma(z)$ is a local approximation near the singularity; usually σ is a much simpler function, typically like $(1-z)^\alpha \log^\beta(1-z)$ whose coefficients are comparatively easy to estimate (Chapter VI). The relation (4) expresses *a mapping between asymptotic scales of functions near singularities and asymptotics scales of coefficients.* Under suitable conditions, it then suffices to estimate a function locally at a few special points (singularities), in order to estimate its coefficients asymptotically.

A succinct roadmap. Here is what now awaits the reader. Section IV. 2 serves to introduce basic notions of complex function theory. Singularities and exponential growth of coefficients are examined in Section IV. 3, which justifies the First Principle. Next, in Section IV. 4, we establish the computability of exponential growth rates for all the non-recursive structures that are specifiable. Section IV. 5 presents two important theorems that deal with rational and meromorphic functions and illustrate the Second Principle, in its simplest version (the subexponential factors are merely polynomials). Then, Section IV. 6 examines constructively ways to locate singularities and treats in detail the case of patterns in words. Finally, Section IV. 7 shows how functions only known through a functional equation may be accessible to complex asymptotic methods.

▷ **IV.1.** *Euler, the discrete, and the continuous.* Eulers's proof of the existence of infinitely many prime numbers illustrates in a striking manner the way analysis of generating functions can inform us on the discrete realm. Define, for real $s > 1$ the function

$$\zeta(s) := \sum_{n=1}^{\infty} \frac{1}{n^s},$$

known as the Riemann zeta function. The decomposition (p ranges over the prime numbers $2, 3, 5, \ldots$)

$$\zeta(s) = \left(1 + \frac{1}{2^s} + \frac{1}{2^{2s}} + \cdots\right)\left(1 + \frac{1}{3^s} + \frac{1}{3^{2s}} + \cdots\right)\left(1 + \frac{1}{5^s} + \frac{1}{5^{2s}} + \cdots\right)\cdots$$

(5)

$$= \prod_{p}\left(1 - \frac{1}{p^s}\right)^{-1}$$

expresses precisely the fact that each integer has a unique decomposition as a product of primes. Analytically, the identity (5) is easily checked to be valid for all $s > 1$. Now suppose that there were only finitely many primes. Let s tend to 1^+ in (5). Then, the left-hand side becomes infinite, while the right-hand side tends to the finite limit $\prod_p (1 - 1/p)^{-1}$: a contradiction has been reached.   ◁

▷ **IV.2.** *Elementary transfers.* Elementary series manipulation yield the following general result: *Let $h(z)$ be a power series with radius of convergence > 1 and assume that $h(1) \neq 0$; then one has*

$$[z^n]\frac{h(z)}{1-z} \sim h(1), \qquad [z^n]h(z)\sqrt{1-z} \sim -\frac{h(1)}{2\sqrt{\pi n^3}}, \qquad [z^n]h(z)\log\frac{1}{1-z} \sim \frac{h(1)}{n}.$$

See our discussion on p. 434 and Bender's survey [36] for many similar statements, of which this chapter and Chapter VI provide many far-reaching extensions.   ◁

▷ **IV.3.** *Asymptotics of generalized derangements.* The EGF of permutations without cycles of length 1 and 2 satisfies (p. 123)

$$j(z) = \frac{e^{-z-z^2/2}}{1-z} \qquad \text{with} \qquad j(z) \underset{z \to 1}{\sim} \frac{e^{-3/2}}{1-z}.$$

Analogy with derangements suggests that $[z^n]j(z) \underset{n \to \infty}{\sim} e^{-3/2}$. [For a proof, use Note IV.2 or refer to Example IV.9 below, p. 261.] Here is a table of exact values of $[z^n]j(z)$ (with relative error of the approximation by $e^{-3/2}$ in parentheses):

	$n = 5$	$n = 10$	$n = 20$	$n = 50$
j_n :	0.2	0.22317	0.2231301600	0.2231301601484298289332804707640122
error :	(10^{-1})	$(2 \cdot 10^{-4})$	$(3 \cdot 10^{-10})$	(10^{-33})

The quality of the asymptotic approximation is extremely good, such a property being, as we shall see, invariably attached to polar singularities.   ◁

IV. 2. Analytic functions and meromorphic functions

Analytic functions are a primary mathematical concept of asymptotic theory. They can be characterized in two essentially equivalent ways (see Subsection IV. 2.1): by means of convergent series expansions (à la Cauchy and Weierstrass) and by differentiability properties (à la Riemann). The first aspect is directly related to the use of generating functions for enumeration; the second one allows for a powerful abstract discussion of closure properties that usually requires little computation.

Integral calculus with analytic functions (see Subsection IV. 2.2) assumes a shape radically different from that which prevails in the real domain: integrals become quintessentially independent of details of the integration contour—certainly the prime example of this fact is Cauchy's famous residue theorem. Conceptually, this independence makes it possible to relate properties of a function at a point (e.g., the coefficients of its expansion at 0) to its properties at another far-away point (e.g., its residue at a pole).

The presentation in this section and the next one constitutes an informal review of basic properties of analytic functions tuned to the needs of asymptotic analysis of counting sequences. The entry in Appendix B.2: *Equivalent definitions of analyticity*, p. 741, provides further information, in particular a proof of the Basic Equivalence Theorem, Theorem IV.1 below. For a detailed treatment, we refer the reader to one of the many excellent treatises on the subject, such as the books by Dieudonné [165], Henrici [329], Hille [334], Knopp [373], Titchmarsh [577], or Whittaker and Watson [604]. The reader previously unfamiliar with the theory of analytic functions should essentially be able to adopt Theorems IV.1 and IV.2 as "*axioms*" and start from here using basic definitions and a fair knowledge of elementary calculus. Figure IV.19 at the end of this chapter (p. 287) recapitulates the main results of relevance to *Analytic Combinatorics*.

IV. 2.1. Basics. We shall consider functions defined in certain *regions* of the complex domain ℂ. By a region is meant an *open* subset Ω of the complex plane that is *connected*. Here are some examples:

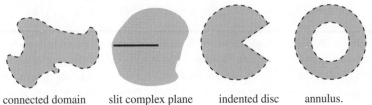

simply connected domain   slit complex plane   indented disc   annulus.

Classical treatises teach us how to extend to the complex domain the standard functions of real analysis: polynomials are immediately extended as soon as complex addition and multiplication have been defined, while the exponential is definable by means of Euler's formula. One has for instance

$$z^2 = (x^2 - y^2) + 2ixy, \qquad e^z = e^x \cos y + i e^x \sin y,$$

if $z = x + iy$, that is, $x = \Re(z)$ and $y = \Im(z)$ are the real and imaginary parts of z. Both functions are consequently defined over the whole complex plane $\mathbb{C}$.

The square-root and logarithm functions are conveniently described in polar co-ordinates:

(6)
$$\sqrt{z} = \sqrt{\rho} e^{i\theta/2}, \qquad \log z = \log \rho + i\theta,$$

if $z = \rho e^{i\theta}$. One can take the domain of validity of (6) to be the complex plane slit along the axis from 0 to $-\infty$, that is, restrict θ to the open interval $(-\pi, +\pi)$, in which case the definitions above specify what is known as the *principal determination*. There is no way for instance to extend by continuity the definition of $\sqrt{z}$ in any domain containing 0 in its interior since, for $a > 0$ and $z \to -a$, one has $\sqrt{z} \to i\sqrt{a}$ as $z \to -a$ from above, whereas $\sqrt{z} \to -i\sqrt{a}$ as $z \to -a$ from below. This situation is depicted here:

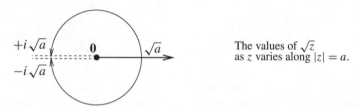

The values of $\sqrt{z}$ as z varies along $|z| = a$.

The point $z = 0$, where several determinations "meet", is accordingly known as a *branch point*.

Analytic functions. First comes the main notion of an *analytic function* that arises from convergent series expansions and is of obvious relevance to generating-functionology.

Definition IV.1. *A function $f(z)$ defined over a region Ω is analytic at a point $z_0 \in \Omega$ if, for z in some open disc centred at z_0 and contained in Ω, it is representable by a convergent power series expansion*

(7)
$$f(z) = \sum_{n \geq 0} c_n (z - z_0)^n.$$

A function is analytic in a region Ω iff it is analytic at every point of Ω.

As derived from an elementary property of power series (Note IV.4), given a function f that is analytic at a point z_0, there exists a disc (of possibly infinite radius) with the property that the series representing $f(z)$ is convergent for z inside the disc and divergent for z outside the disc. The disc is called the *disc of convergence* and its radius is the *radius of convergence* of $f(z)$ at $z = z_0$, which will be denoted by $R_{\mathrm{conv}}(f; z_0)$. The radius of convergence of a power series conveys basic information regarding the rate at which its coefficients grow; see Subsection IV. 3.2 below for developments. It is also easy to prove by simple series rearrangement that if a function is analytic at z_0, it is then analytic at all points interior to its disc of convergence (see Appendix B.2: *Equivalent definitions of analyticity*, p. 741).

▷ **IV.4.** *The disc of convergence of a power series.* Let $f(z) = \sum f_n z^n$ be a power series. Define R as the supremum of all values of $x \geq 0$ such that $\{f_n x^n\}$ is bounded. Then, for

$|z| < R$, the sequence $f_n z^n$ tends geometrically to 0; hence $f(z)$ is convergent. For $|z| > R$, the sequence $f_n z^n$ is unbounded; hence $f(z)$ is divergent. In short: *a power series converges in the interior of a disc; it diverges in its exterior.*   ◁

Consider for instance the function $f(z) = 1/(1 - z)$ defined over $\mathbb{C} \setminus \{1\}$ in the usual way via complex division. It is analytic at 0 by virtue of the geometric series sum,

$$\frac{1}{1 - z} = \sum_{n \geq 0} 1 \cdot z^n,$$

which converges in the disc $|z| < 1$. At a point $z_0 \neq 1$, we may write

$$
\begin{aligned}
(8) \qquad \frac{1}{1 - z} &= \frac{1}{1 - z_0 - (z - z_0)} = \frac{1}{1 - z_0} \frac{1}{1 - \frac{z - z_0}{1 - z_0}} \\
&= \sum_{n \geq 0} \left(\frac{1}{1 - z_0} \right)^{n+1} (z - z_0)^n.
\end{aligned}
$$

The last equation shows that $f(z)$ is analytic in the disc centred at z_0 with radius $|1 - z_0|$, that is, the interior of the circle centred at z_0 and passing through the point 1. In particular $R_{\text{conv}}(f, z_0) = |1 - z_0|$ and $f(z)$ is globally analytic in the punctured plane $\mathbb{C} \setminus \{1\}$.

The example of $(1 - z)^{-1}$ illustrates the definition of analyticity. However, the series rearrangement approach that it uses might be difficult to carry out for more complicated functions. In other words, a more manageable approach to analyticity is called for. The differentiability properties developed now provide such an approach.

Differentiable (holomorphic) functions. The next important notion is a geometric one based on differentiability.

Definition IV.2. *A function $f(z)$ defined over a region Ω is called* complex-differentiable *(also* holomorphic*) at z_0 if the limit, for complex δ,*

$$\lim_{\delta \to 0} \frac{f(z_0 + \delta) - f(z_0)}{\delta}$$

exists. (In particular, the limit is independent of the way δ tends to 0 in $\mathbb{C}$.) This limit is denoted as usual by $f'(z_0)$, or $\frac{d}{dz} f(z) \big|_{z_0}$, or $\partial_z f(z_0)$. A function is complex-differentiable in Ω iff it is complex-differentiable at every $z_0 \in \Omega$.

From the definition, if $f(z)$ is complex-differentiable at z_0 and $f'(z_0) \neq 0$, it acts locally as a linear transformation:

$$f(z) - f(z_0) = f'(z_0)(z - z_0) + o(z - z_0) \qquad (z \to z_0).$$

Then, $f(z)$ behaves in small regions almost like a similarity transformation (composed of a translation, a rotation, and a scaling). In particular, it preserves angles[2] and infinitesimal squares get transformed into infinitesimal squares; see Figure IV.3 for a rendering. Further aspects of the local shape of an analytic function will be examined in Section VIII. 1, p. 543, in relation with the saddle-point method.

---

[2] A mapping of the plane that locally preserves angles is also called a *conformal* map. Section VIII. 1 (p. 543) presents further properties of the local "shape" of an analytic function.

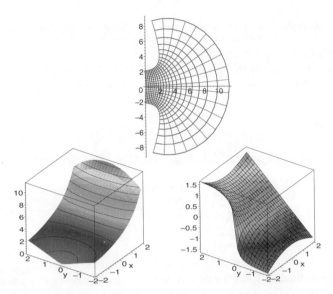

Figure IV.3. Multiple views of an analytic function. The image of the domain $\Omega = \{z \mid |\Re(z)| < 2, |\Im(z)| < 2\}$ by $f(z) = \exp(z) + z + 2$: [top] transformation of a square grid in Ω by f; [bottom] the modulus and argument of $f(z)$.

For instance the function $\sqrt{z}$, defined by (6) in the complex plane slit along the ray $(-\infty, 0)$, is complex-differentiable at any z_0 of the slit plane since

$$(9) \qquad \lim_{\delta \to 0} \frac{\sqrt{z_0 + \delta} - \sqrt{z_0}}{\delta} = \lim_{\delta \to 0} \sqrt{z_0} \frac{\sqrt{1 + \delta/z_0} - 1}{\delta} = \frac{1}{2\sqrt{z_0}},$$

which extends the customary proof of real analysis. Similarly, $\sqrt{1 - z}$ is complex-differentiable in the complex plane slit along the ray $(1, +\infty)$. More generally, the usual proofs from real analysis carry over almost verbatim to the complex realm, to the effect that

$$(f+g)' = f' + g', \quad (fg)' = f'g + fg', \quad \left(\frac{1}{f}\right)' = -\frac{f'}{f^2}, \quad (f \circ g)' = (f' \circ g)g'.$$

The notion of complex differentiability is thus much more manageable than the notion of analyticity.

It follows from a well known theorem of Riemann (see for instance [329, vol. 1, p 143] and Appendix B.2: *Equivalent definitions of analyticity*, p. 741) that analyticity and complex differentiability are equivalent notions.

Theorem IV.1 (Basic Equivalence Theorem). *A function is analytic in a region Ω if and only if it is complex-differentiable in Ω.*

The following are known facts (see p. 236 and Appendix B): (*i*) if a function is analytic (equivalently complex-differentiable) in Ω, it admits (complex) derivatives of any order there—this property markedly differs from real analysis: complex-differentiable, equivalently analytic, functions are all smooth; (*ii*) derivatives of a function may be obtained through term-by-term differentiation of the series representation of the function.

Meromorphic functions. We finally introduce *meromorphic*[3] functions that are mild extensions of the concept of analyticity (or holomorphy) and are essential to the theory. The quotient of two analytic functions $f(z)/g(z)$ ceases to be analytic at a point a where $g(a) = 0$; however, a simple structure for quotients of analytic functions prevails.

Definition IV.3. *A function $h(z)$ is* meromorphic *at z_0 iff, for z in a neighbourhood of z_0 with $z \neq z_0$, it can be represented as $f(z)/g(z)$, with $f(z)$ and $g(z)$ being analytic at z_0. In that case, it admits near z_0 an expansion of the form*

$$(10) \qquad h(z) = \sum_{n \geq -M} h_n(z - z_0)^n.$$

If $h_{-M} \neq 0$ and $M \geq 1$, then $h(z)$ is said to have a pole *of order M at $z = z_0$. The coefficient h_{-1} is called the* residue *of $h(z)$ at $z = z_0$ and is written as*

$$\mathrm{Res}[h(z); z = z_0].$$

A function is meromorphic in a region iff it is meromorphic at every point of the region.

IV. 2.2. Integrals and residues. A path in a region Ω is described by its parameterization, which is a continuous function γ mapping $[0, 1]$ into Ω. Two paths γ, γ' in Ω that have the same end points are said to be *homotopic* (in Ω) if one can be continuously deformed into the other while staying within Ω as in the following examples:

homotopic paths:

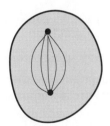

A closed path is defined by the fact that its end points coincide: $\gamma(0) = \gamma(1)$, and a path is *simple* if the mapping γ is one-to-one. A closed path is said to be a *loop* of Ω if it can be continuously deformed *within* Ω to a single point; in this case one also says that the path is homotopic to 0. In what follows paths are taken to be piecewise continuously differentiable and, by default, *loops are oriented positively*.

Integrals along curves in the complex plane are defined in the usual way as curvilinear integrals of complex-valued functions. Explicitly: let $f(x + iy)$ be a function

---

[3]"Holomorphic" and "meromorphic" are words coming from Greek, meaning, respectively, "of complete form" and "of partial form".

and γ be a path; then,

$$
\begin{aligned}
\int_{\gamma} f(z)\, dz \ &:= \ \int_{0}^{1} f(\gamma(t))\gamma'(t)\, dt \\
&= \ \int_{0}^{1} [AC - BD]\, dt + i \int_{0}^{1} [AD + BC]\, dt,
\end{aligned}
$$

where $f \circ \gamma = A + iB$ and $\gamma' = C + iD$. However, integral calculus in the complex plane greatly differs from its form on the real line—in many ways, it is much *simpler* and much more *powerful*. One has:

Theorem IV.2 (Null Integral Property). *Let f be analytic in Ω and let λ be a simple loop of Ω. Then, one has $\int_{\lambda} f = 0$.*

Equivalently, integrals are largely independent of details of contours: for f analytic in Ω, one has

(11)
$$
\int_{\gamma} f = \int_{\gamma'} f,
$$

provided γ and γ' are homotopic (not necessarily closed) paths in Ω. A proof of Theorem IV.2 is sketched in Appendix B.2: *Equivalent definitions of analyticity*, p. 741.

　　Residues. The important *Residue Theorem* due to Cauchy relates *global* properties of a meromorphic function (its integral along closed curves) to purely *local* characteristics at designated points (its residues at poles).

Theorem IV.3 (Cauchy's residue theorem). *Let $h(z)$ be meromorphic in the region Ω and let λ be a positively oriented simple loop in Ω along which the function is analytic. Then*

$$
\frac{1}{2i\pi} \int_{\lambda} h(z)\, dz = \sum_{s} \mathrm{Res}[h(z); z = s],
$$

where the sum is extended to all poles s of $h(z)$ enclosed by λ.

Proof. (Sketch) To see it in the representative case where $h(z)$ has only a pole at $z = 0$, observe by appealing to primitive functions that

$$
\int_{\lambda} h(z)\, dz = \sum_{\substack{n \geq -M \\ n \neq -1}} h_n \left[\frac{z^{n+1}}{n + 1} \right]_{\lambda} + h_{-1} \int_{\lambda} \frac{dz}{z},
$$

where the bracket notation $\left[u(z) \right]_{\lambda}$ designates the variation of the function $u(z)$ along the contour λ. This expression reduces to its last term, itself equal to $2i\pi h_{-1}$, as is checked by using integration along a circle (set $z = re^{i\theta}$). The computation extends by translation to the case of a unique pole at $z = a$.

　　Next, in the case of multiple poles, we observe that the simple loop can only enclose finitely many poles (by compactness). The proof then follows from a simple decomposition of the interior domain of λ into cells, each containing only one pole. Here is an illustration in the case of three poles.

(Contributions from internal edges cancel.) ∎

Global (integral) to local (residues) connections. Here is a textbook example of a reduction from global to local properties of analytic functions. Define the integrals

$$I_m := \int_{-\infty}^{\infty} \frac{dx}{1 + x^{2m}},$$

and consider specifically I_1. Elementary calculus teaches us that $I_1 = \pi$ since the antiderivative of the integrand is an arc tangent:

$$I_1 = \int_{-\infty}^{\infty} \frac{dx}{1 + x^2} = [\arctan x]_{-\infty}^{+\infty} = \pi.$$

Here is an alternative, and in many ways more fruitful, derivation. In the light of the residue theorem, we consider the integral over the whole line as the limit of integrals over large intervals of the form $[-R, +R]$, then complete the contour of integration by means of a large semi-circle in the upper half-plane, as shown below:

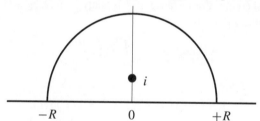

Let γ be the contour comprised of the interval and the semi-circle. Inside γ, the integrand has a pole at $x = i$, where

$$\frac{1}{1 + x^2} \equiv \frac{1}{(x + i)(x - i)} = -\frac{i}{2}\frac{1}{x - i} + \cdots,$$

so that its residue there is $-i/2$. By the residue theorem, the integral taken over γ is equal to $2i\pi$ times the residue of the integrand at i. As $R \to \infty$, the integral along the semi-circle vanishes (it is less than $\pi R/(R^2 - 1)$ in modulus), while the integral along the real segment gives I_1 in the limit. There results the relation giving I_1:

$$I_1 = 2i\pi \operatorname{Res}\left(\frac{1}{1 + x^2}; x = i\right) = (2i\pi)\left(-\frac{i}{2}\right) = \pi.$$

The evaluation of the integral in the framework of complex analysis rests solely upon the local expansion of the integrand at special points (here, the point i). This is a remarkable feature of the theory, one that confers it much simplicity, when compared with real analysis.

▷ **IV.5.** *The general integral I_m.* Let $\alpha = \exp(\frac{i\pi}{2m})$ so that $\alpha^{2m} = -1$. Contour integration of the type used for I_1 yields

$$I_m = 2i\pi \sum_{j=1}^{m} \mathrm{Res}\left(\frac{1}{1 + x^{2m}}; x = \alpha^{2j-1}\right),$$

while, for any $\beta = \alpha^{2j-1}$ with $1 \le j \le m$, one has

$$\frac{1}{1 + x^{2m}} \underset{x \to \beta}{\sim} \frac{1}{2m\beta^{2m-1}} \frac{1}{x - \beta} \equiv -\frac{\beta}{2m} \frac{1}{x - \beta}.$$

As a consequence,

$$I_{2m} = -\frac{i\pi}{m}\left(\alpha + \alpha^3 + \cdots + \alpha^{2m-1}\right) = \frac{\pi}{m \sin \frac{\pi}{2m}}.$$

In particular, $I_2 = \pi/\sqrt{2}$, $I_3 = 2\pi/3$, $I_4 = \frac{\pi}{4}\sqrt{2}\sqrt{2 + \sqrt{2}}$, and $\frac{1}{\pi}I_5$, $\frac{1}{\pi}I_6$ are expressible by radicals, but $\frac{1}{\pi}I_7$, $\frac{1}{\pi}I_9$ are not. The special cases $\frac{1}{\pi}I_{17}$, $\frac{1}{\pi}I_{257}$ are expressible by radicals. ◁

▷ **IV.6.** *Integrals of rational fractions.* Generally, all integrals of rational functions taken over the whole real line are computable by residues. In particular,

$$J_m = \int_{-\infty}^{+\infty} \frac{dx}{(1 + x^2)^m}, \qquad K_m = \int_{-\infty}^{+\infty} \frac{dx}{(1^2 + x^2)(2^2 + x^2) \cdots (m^2 + x^2)}$$

can be explicitly evaluated. ◁

Cauchy's coefficient formula. Many function-theoretic consequences are derived from the residue theorem. For instance, if f is analytic in Ω, $z_0 \in \Omega$, and λ is a simple loop of Ω encircling z_0, one has

$$(12) \qquad\qquad f(z_0) = \frac{1}{2i\pi} \int_\lambda f(\zeta) \frac{d\zeta}{\zeta - z_0}.$$

This follows directly since

$$\mathrm{Res}\left[f(\zeta)/(\zeta - z_0); \zeta = z_0\right] = f(z_0).$$

Then, by differentiation with respect to z_0 under the integral sign, one has similarly

$$(13) \qquad\qquad \frac{1}{k!} f^{(k)}(z_0) = \frac{1}{2i\pi} \int_\lambda f(\zeta) \frac{d\zeta}{(\zeta - z_0)^{k+1}}.$$

The values of a function and its derivatives at a point can thus be obtained as values of integrals of the function away from that point. The world of analytic functions is a very friendly one in which to live: contrary to real analysis, a function is differentiable *any number of times* as soon as it is differentiable *once*. Also, Taylor's formula invariably holds: as soon as $f(z)$ is analytic at z_0, one has

$$(14) \qquad f(z) = f(z_0) + f'(z_0)(z - z_0) + \frac{1}{2!} f''(z_0)(z - z_0)^2 + \cdots,$$

with the representation being convergent in a disc centred at z_0. [Proof: a verification from (12) and (13), or a series rearrangement as in Appendix B, p. 742.]

A very important application of the residue theorem concerns coefficients of analytic functions.

Theorem IV.4 (Cauchy's Coefficient Formula). *Let $f(z)$ be analytic in a region Ω containing 0 and let λ be a simple loop around 0 in Ω that is positively oriented. Then, the coefficient $[z^n]f(z)$ admits the integral representation*

$$f_n \equiv [z^n]f(z) = \frac{1}{2i\pi} \int_\lambda f(z) \frac{dz}{z^{n+1}}.$$

Proof. This formula follows directly from the equalities

$$\frac{1}{2i\pi} \int_\lambda f(z) \frac{dz}{z^{n+1}} = \operatorname{Res}\left[f(z)z^{-n-1}; z = 0\right] = [z^n]f(z),$$

of which the first one follows from the residue theorem, and the second one from the identification of the residue at 0 as a coefficient. ∎

Analytically, the coefficient formula allows us to deduce information about the coefficients from the values of the function itself, using adequately chosen contours of integration. It thus opens the possibility of estimating the coefficients $[z^n]f(z)$ in the expansion of $f(z)$ near 0 by using information on $f(z)$ *away* from 0. The rest of this chapter will precisely illustrate this process in the case of rational and meromorphic functions. Observe also that the residue theorem provides the simplest proof of the Lagrange inversion theorem (see Appendix A.6: *Lagrange Inversion*, p. 732) whose rôle is central to tree enumerations, as we saw in Chapters I and II. The notes below explore some independent consequences of the residue theorem and the coefficient formula.

▷ **IV.7.** *Liouville's Theorem.* If a function $f(z)$ is analytic in the whole of $\mathbb{C}$ and is of modulus bounded by an absolute constant, $|f(z)| \leq B$, then it must be a constant. [By trivial bounds, upon integrating on a large circle, it is found that the Taylor coefficients at the origin of index ≥ 1 are all equal to 0.] Similarly, if $f(z)$ is of at most polynomial growth, $|f(z)| \leq B(|z|+1)^r$, over the whole of $\mathbb{C}$, then it must be a polynomial. ◁

▷ **IV.8.** *Lindelöf integrals.* Let $a(s)$ be analytic in $\Re(s) > \frac{1}{4}$ where it is assumed to satisfy $a(s) = O(\exp((\pi - \delta)|s|))$ for some δ with $0 < \delta < \pi$. Then, one has for $|\arg(z)| < \delta$,

$$\sum_{k=1}^\infty a(k)(-z)^k = -\frac{1}{2i\pi} \int_{1/2-i\infty}^{1/2+i\infty} a(s)z^s \frac{\pi}{\sin \pi s} ds,$$

in the sense that the integral exists and provides the analytic continuation of the sum in $|\arg(z)| < \delta$. [Close the integration contour by a large semi-circle on the right and evaluate by residues.] Such integrals, sometimes called Lindelöf integrals, provide representations for many functions whose Taylor coefficients are given by an explicit rule [268, 408]. ◁

▷ **IV.9.** *Continuation of polylogarithms.* As a consequence of Lindelöf's representation, the generalized *polylogarithm* functions,

$$\operatorname{Li}_{\alpha,k}(z) = \sum_{n \geq 1} n^{-\alpha}(\log n)^k z^n \qquad (\alpha \in \mathbb{R}, \quad k \in \mathbb{Z}_{\geq 0}),$$

are analytic in the complex plane $\mathbb{C}$ slit along $(1+, \infty)$. (More properties are presented in Section VI. 8, p. 408; see also [223, 268].) For instance, one obtains in this way

$$\text{``}\sum_{n=1}^\infty (-1)^n \log n\text{''} = -\frac{1}{4} \int_{-\infty}^{+\infty} \frac{\log(\frac{1}{4} + t^2)}{\cosh(\pi t)} dt = 0.22579\cdots = \log\sqrt{\frac{\pi}{2}},$$

when the divergent series on the left is interpreted as $\operatorname{Li}_{0,1}(-1) = \lim_{z \to -1+} \operatorname{Li}_{0,1}(z)$. ◁

▷ **IV.10.** *Magic duality.* Let ϕ be a function initially defined over the non-negative integers but admitting a meromorphic extension over the whole of $\mathbb{C}$. Under growth conditions in the style of Note IV.8, the function

$$F(z) := \sum_{n \geq 1} \phi(n)(-z)^n,$$

which is analytic at the origin, is such that, near positive infinity,

$$F(z) \underset{z \to +\infty}{\sim} E(z) - \sum_{n \geq 1} \phi(-n)(-z)^{-n},$$

for some elementary function $E(z)$, which is a linear combination of terms of the form $z^\alpha (\log z)^k$. [Starting from the representation of Note IV.8, close the contour of integration by a large semi-circle to the left.] In such cases, the function is said to satisfy the principle of *magic duality*—its expansion at 0 and ∞ are given by one and the same rule. Functions

$$\frac{1}{1+z}, \quad \log(1+z), \quad \exp(-z), \quad \mathrm{Li}_2(-z), \quad \mathrm{Li}_3(-z),$$

satisfy a form of magic duality. Ramanujan [52] made a great use of this principle, which applies to a wide class of functions including hypergeometric ones; see Hardy's insightful discussion [321, Ch XI].   ◁

▷ **IV.11.** *Euler–Maclaurin and Abel–Plana summations.* Under simple conditions on the analytic function f, one has Plana's (also known as Abel's) complex variables version of the Euler–Maclaurin summation formula:

$$\sum_{n=0}^{\infty} f(n) = \frac{1}{2}f(0) + \int_0^\infty f(x)\,dx + \int_0^\infty \frac{f(iy) - f(-iy)}{e^{2i\pi y} - 1}\,dy.$$

(See [330, p. 274] for a proof and validity conditions.)   ◁

▷ **IV.12.** *Nörlund–Rice integrals.* Let $a(z)$ be analytic for $\Re(z) > k_0 - \frac{1}{2}$ and of at most polynomial growth in this right half-plane. Then, with γ a simple loop around the interval $[k_0, n]$, one has

$$\sum_{k=k_0}^{n} \binom{n}{k}(-1)^{n-k} a(k) = \frac{1}{2i\pi}\int_\gamma a(s)\,\frac{n!\,ds}{s(s-1)(s-2)\cdots(s-n)}.$$

If $a(z)$ is meromorphic and suitably small in a larger region, then the integral can be estimated by residues. For instance, with

$$S_n = \sum_{k=1}^{n} \binom{n}{k}\frac{(-1)^k}{k}, \qquad T_n = \sum_{k=1}^{n} \binom{n}{k}\frac{(-1)^k}{k^2+1},$$

it is found that $S_n = -H_n$ (a harmonic number), while T_n oscillates boundedly as $n \to +\infty$. [This technique is a classical one in the calculus of finite differences, going back to Nörlund [458]. In computer science it is known as the method of Rice's integrals [256] and is used in the analysis of many algorithms and data structures including digital trees and radix sort [378, 564].]   ◁

IV. 3. Singularities and exponential growth of coefficients

For a given function, a singularity can be informally defined as a point where the function ceases to be analytic. (Poles are the simplest type of singularity.) Singularities are, as we have stressed repeatedly, essential to coefficient asymptotics. This section presents the bases of a discussion within the framework of analytic function theory.

IV. 3.1. Singularities. Let $f(z)$ be an analytic function defined over the interior region determined by a simple closed curve γ, and let z_0 be a point of the bounding curve γ. If there exists an analytic function $f^\star(z)$ defined over some open set $\Omega^\star$ containing z_0 and such that $f^\star(z) = f(z)$ in $\Omega^\star \cap \Omega$, one says that f is *analytically continuable* at z_0 and that $f^\star$ is an *immediate analytic continuation* of f. Pictorially:

Analytic continuation: 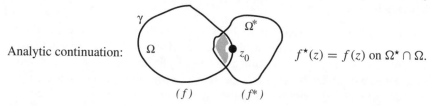 $f^\star(z) = f(z)$ on $\Omega^\star \cap \Omega$.

$$(f) \qquad (f^*)$$

Consider for instance the quasi-inverse function, $f(z) = 1/(1 - z)$. Its power series representation $f(z) = \sum_{n\geq 0} z^n$ initially converges in $|z| < 1$. However, the calculation of (8), p. 231, shows that it is representable locally by a convergent series near any point $z_0 \neq 1$. In particular, it is continuable at any point of the unit disc except 1. (Alternatively, one may appeal to complex-differentiability to verify directly that $f(z)$, which is given by a "global" expression, is holomorphic, hence analytic, in the punctured plane $\mathbb{C} \setminus \{1\}$.)

In sharp contrast with real analysis, where a smooth function admits of uncountably many extensions, analytic continuation is essentially *unique*: if $f^\star$ (in $\Omega^\star$) and $f^{\star\star}$ (in $\Omega^{\star\star}$) continue f at z_0, then one must have $f^\star(z) = f^{\star\star}(z)$ in the intersection $\Omega^\star \cap \Omega^{\star\star}$, which in particular includes a small disc around z_0. Thus, the notion of immediate analytic continuation at a boundary point is intrinsic. The process can be iterated and we say that g is an *analytic continuation*[4] of f along a path, even if the domains of definition of f and g do not overlap, provided a finite chain of intermediate function elements connects f and g. This notion is once more intrinsic—this is known as the principle of *unicity of analytic continuation* (Rudin [523, Ch. 16] provides a thorough discussion). An analytic function is then much like a hologram: as soon as it is specified in any tiny region, it is rigidly determined in any wider region to which it can be continued.

Definition IV.4. *Given a function f defined in the region interior to the simple closed curve γ, a point z_0 on the boundary (γ) of the region is a* singular point *or a* singularity[5] *if f is not analytically continuable at z_0.*

Granted the intrinsic character of analytic continuation, we can usually dispense with a detailed description of the original domain Ω and the curve γ. In simple terms, a function is singular at z_0 if it cannot be continued as an analytic function beyond z_0. A point at which a function is analytic is also called by contrast a *regular point*.

The two functions $f(z) = 1/(1 - z)$ and $g(z) = \sqrt{1 - z}$ may be taken as initially defined over the open unit disc by their power series representation. Then, as we already know, they can be analytically continued to larger regions, the punctured plane

---

[4]The collection of all function elements continuing a given function gives rise to the notion of *Riemann surface*, for which many good books exist, e.g., [201, 549]. We shall not need to appeal to this theory.

[5]For a detailed discussion, see [165, p. 229], [373, vol. 1, p. 82], or [577].

$\Omega = \mathbb{C} \setminus \{1\}$ for f [e.g., by the calculation of (8), p. 231] and the complex plane slit along $(1, +\infty)$ for g [e.g., by virtue of continuity and differentiability as in (9), p. 232]. But both are singular at 1: for f, this results (say) from the fact that $f(z) \to \infty$ as $z \to 1$; for g this is due to the branching character of the square-root. Figure IV.4 displays a few types of singularities that are traceable by the way they deform a regular grid near a boundary point.

A converging power series is analytic inside its disc of convergence; in other words, it can have no singularity inside this disc. However, it *must* have at least one singularity on the boundary of the disc, as asserted by the theorem below. In addition, a classical theorem, called Pringsheim's theorem, provides a refinement of this property in the case of functions with non-negative coefficients, which happens to include all counting generating functions.

Theorem IV.5 (Boundary singularities). *A function $f(z)$ analytic at the origin, whose expansion at the origin has a finite radius of convergence R, necessarily has a singularity on the boundary of its disc of convergence, $|z| = R$.*

Proof. Consider the expansion

$$(15) \qquad f(z) = \sum_{n \geq 0} f_n z^n,$$

assumed to have radius of convergence exactly R. We already know that there can be no singularity of f within the disc $|z| < R$. To prove that there is a singularity on $|z| = R$, suppose *a contrario* that $f(z)$ is analytic in the disc $|z| < \rho$ for some ρ satisfying $\rho > R$. By Cauchy's coefficient formula (Theorem IV.4, p. 237), upon integrating along the circle of radius $r = (R + \rho)/2$, and by trivial bounds, it is seen that the coefficient $[z^n] f(z)$ is $O(r^{-n})$. But then, the series expansion of f would have to converge in the disc of radius $r > R$, a contradiction. ∎

Pringsheim's Theorem stated and proved now is a refinement of Theorem IV.5 that applies to *all* series having non-negative coefficients, in particular, generating functions. It is central to asymptotic enumeration, as the remainder of this section will amply demonstrate.

Theorem IV.6 (Pringsheim's Theorem). *If $f(z)$ is representable at the origin by a series expansion that has non-negative coefficients and radius of convergence R, then the point $z = R$ is a singularity of $f(z)$.*

▷ **IV.13.** *Proof of Pringsheim's Theorem.* (See also [577, Sec. 7.21].) In a nutshell, the idea of the proof is that if f has positive coefficients and is analytic at R, then its expansion slightly to the left of R has positive coefficients. Then, the power series of f would converge in a disc larger than the postulated disc of convergence—a clear contradiction.

Suppose then *a contrario* that $f(z)$ is analytic at R, implying that it is analytic in a disc of radius r centred at R. We choose a number h such that $0 < h < \frac{1}{3}r$ and consider the expansion of $f(z)$ around $z_0 = R - h$:

$$(16) \qquad f(z) = \sum_{m \geq 0} g_m (z - z_0)^m.$$

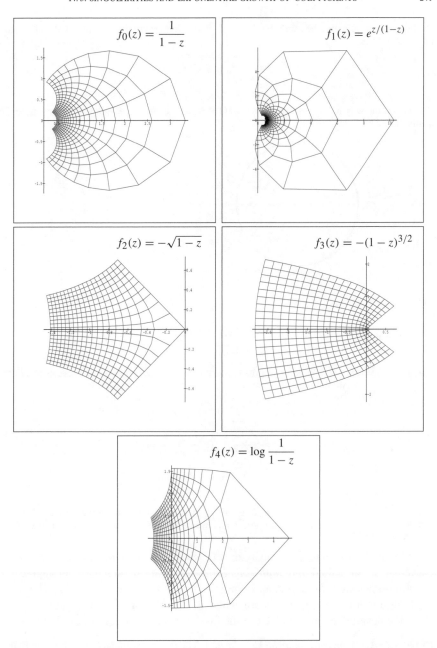

Figure IV.4. The images of a grid on the unit square (with corners $\pm 1 \pm i$) by various functions singular at $z = 1$ reflect the nature of the singularities involved. Singularities are apparent near the right of each diagram where small grid squares get folded or unfolded in various ways. (In the case of functions f_0, f_1, f_4 that become infinite at $z = 1$, the grid has been slightly truncated to the right.)

By Taylor's formula and the representability of $f(z)$ together with its derivatives at z_0 by means of (15), we have

$$g_m = \sum_{n \geq 0} \binom{n}{m} f_n z_0^{n-m},$$

and in particular, $g_m \geq 0$.

Given the way h was chosen, the series (16) converges at $z = R + h$ (so that $z - z_0 = 2h$) as illustrated by the following diagram:

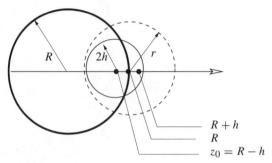

Consequently, one has

$$f(R+h) = \sum_{m \geq 0} \left(\sum_{n \geq 0} \binom{n}{m} f_n z_0^{m-n} \right) (2h)^m.$$

This is a converging double sum of positive terms, so that the sum can be reorganized in any way we like. In particular, one has convergence of all the series involved in

$$
\begin{aligned}
f(R+h) &= \sum_{m,n \geq 0} \binom{n}{m} f_n (R-h)^{m-n} (2h)^m \\
&= \sum_{n \geq 0} f_n \left[(R-h) + (2h) \right]^n \\
&= \sum_{n \geq 0} f_n (R+h)^n.
\end{aligned}
$$

This establishes the fact that $f_n = o((R+h)^{-n})$, thereby reaching a contradiction with the assumption that the series representation of f has radius of convergence exactly R. Pringsheim's theorem is proved. ◁

Singularities of a function analytic at 0, which lie on the boundary of the disc of convergence, are called *dominant singularities*. Pringsheim's theorem appreciably simplifies the search for dominant singularities of combinatorial generating functions since these have non-negative coefficients—it is sufficient to investigate analyticity along the positive real line and detect the first place at which it ceases to hold.

***Example* IV.1.** *Some combinatorial singularities.* The derangement and the surjection EGFs,

$$D(z) = \frac{e^{-z}}{1-z}, \qquad R(z) = (2 - e^z)^{-1}$$

are analytic, except for a simple pole at $z = 1$ in the case of $D(z)$, and for points $\chi_k = \log 2 + 2ik\pi$ that are simple poles in the case of $R(z)$. Thus the dominant singularities for derangements and surjections are at 1 and $\log 2$, respectively.

It is known that $\sqrt{Z}$ cannot be unambiguously defined as an analytic function in a neighbourhood of $Z = 0$. As a consequence, the function

$$G(z) = \frac{1 - \sqrt{1 - 4z}}{2},$$

which is the generating function of general Catalan trees, is an analytic function in regions that must exclude $1/4$; for instance, one may take the complex plane slit along the ray $(1/4, +\infty)$. The OGF of Catalan numbers $C(z) = G(z)/z$ is, as $G(z)$, *a priori* analytic in the slit plane, except perhaps at $z = 0$, where it has the indeterminate form $0/0$. However, after $C(z)$ is extended by continuity to $C(0) = 1$, it becomes an analytic function at 0, where its Taylor series converges in $|z| < \frac{1}{4}$. In this case, we say that that $C(z)$ has an *apparent* or *removable* singularity at 0. (See also Morera's Theorem, Note B.6, p. 743.)

Similarly, the EGF of cyclic permutations

$$L(z) = \log \frac{1}{1 - z}$$

is analytic in the complex plane slit along $(1, +\infty)$.

A function having no singularity at a finite distance is called *entire*; its Taylor series then converges everywhere in the complex plane. The EGFs,

$$e^{z+z^2/2} \qquad \text{and} \qquad e^{e^z-1},$$

associated, respectively, with involutions and set partitions, are entire. .... ∎

IV. 3.2. The Exponential Growth Formula.

We say that a number sequence $\{a_n\}$ is of *exponential order* K^n, which we abbreviate as (the symbol $\bowtie$ is a "bowtie")

$$a_n \bowtie K^n \qquad \text{iff} \qquad \limsup |a_n|^{1/n} = K.$$

The relation "$a_n \bowtie K^n$" reads as "a_n is of exponential order K^n". It expresses both an upper bound and a lower bound, and one has, for any $\epsilon > 0$:

   (i) $|a_n| >_{\text{i.o}} (K - \epsilon)^n$; that is to say, $|a_n|$ exceeds $(K - \epsilon)^n$ *infinitely often* (for infinitely many values of n);

   (ii) $|a_n| <_{\text{a.e.}} (K + \epsilon)^n$; that is to say, $|a_n|$ is dominated by $(K + \epsilon)^n$ *almost everywhere* (except for possibly finitely many values of n).

This relation can be rephrased as $a_n = K^n\theta(n)$, where θ is a *subexponential factor* :

$$\limsup |\theta(n)|^{1/n} = 1;$$

such a factor's modulus is thus bounded from above almost everywhere by any increasing exponential (of the form $(1 + \epsilon)^n$) and bounded from below infinitely often by any decaying exponential (of the form $(1 - \epsilon)^n$). Typical subexponential factors are

$$1, \; n^3, \; (\log n)^2, \; \sqrt{n}, \; \frac{1}{\sqrt[3]{\log n}}, \; n^{-3/2}, \; (-1)^n, \; \log\log n.$$

(Functions such as $e^{\sqrt{n}}$ and $\exp(\log^2 n)$ are also to be treated as subexponential factors for the purpose of this discussion.) The lim sup definition also allows in principle for factors that are infinitely often very small or 0, such as $n^2 \sin n\frac{\pi}{2}$, $\log n \cos \sqrt{n}\frac{\pi}{2}$, and so on. In this and the next chapters, we shall develop systematic methods that enable one to extract such subexponential factors from generating functions.

It is an elementary observation that the radius of convergence of the series representation of $f(z)$ at 0 is related to the exponential growth rate of the coefficients $f_n = [z^n] f(z)$. To wit, if $R_{\mathrm{conv}}(f; 0) = R$, then we claim that

$$(17) \qquad f_n \bowtie \left(\frac{1}{R}\right)^n, \qquad \text{i.e.,} \quad f_n = R^{-n} \theta(n) \quad \text{with } \limsup |\theta(n)|^{1/n} = 1.$$

▷ **IV.14.** *Radius of convergence and exponential growth.* This only requires the basic definition of a power series. (*i*) By definition of the radius of convergence, we have for any small $\epsilon > 0$, $f_n(R - \epsilon)^n \to 0$. In particular, $|f_n|(R - \epsilon)^n < 1$ for all sufficiently large n, so that $|f_n|^{1/n} < (R - \epsilon)^{-1}$ "almost everywhere". (*ii*) In the other direction, for any $\epsilon > 0$, $|f_n|(R + \epsilon)^n$ cannot be a bounded sequence, since otherwise, $\sum_n |f_n|(R + \epsilon/2)^n$ would be a convergent series. Thus, $|f_n|^{1/n} > (R + \epsilon)^{-1}$ "infinitely often". ◁

A global approach to the determination of growth rates is desirable. This is made possible by Theorem IV.5, p. 240, as shown by the following statement.

Theorem IV.7 (Exponential Growth Formula). *If $f(z)$ is analytic at 0 and R is the modulus of a singularity nearest to the origin in the sense that[6]*

$$R := \sup \left\{ r \geq 0 \mid f \text{ is analytic in } |z| < r \right\},$$

then the coefficient $f_n = [z^n] f(z)$ satisfies

$$f_n \bowtie \left(\frac{1}{R}\right)^n.$$

For functions with non-negative coefficients, including all combinatorial generating functions, one can also adopt

$$R := \sup \left\{ r \geq 0 \mid f \text{ is analytic at all points of } 0 \leq z < r \right\}.$$

Proof. Let R be as stated. We cannot have $R < R_{\mathrm{conv}}(f; 0)$ since a function is analytic everywhere in the interior of the disc of convergence of its series representation. We cannot have $R > R_{\mathrm{conv}}(f; 0)$ by the Boundary Singularity Theorem. Thus $R = R_{\mathrm{conv}}(f; 0)$. The statement then follows from (17). The adaptation to non-negative coefficients results from Pringsheim's theorem. ∎

The exponential growth formula thus directly relates the exponential growth of coefficients of a function to the *location* of its singularities nearest to the origin. This is precisely expressed by the *First Principle of Coefficient Asymptotics* (p. 227), which, given its importance, we repeat here:

> **First Principle of Coefficient Asymptotics.** *The location of a function's singularities dictates the exponential growth (A^n) of its coefficient.*

Example IV.2. *Exponential growth and combinatorial enumeration.* Here are a few immediate applications of exponential bounds.

> *Surjections.* The function

$$R(z) = (2 - e^z)^{-1}$$

---

[6] One should think of the process defining R as follows: take discs of increasing radii r and stop as soon as a singularity is encountered on the boundary. (The dual process that would start from a large disc and restrict its radius is in general ill-defined—think of $\sqrt{1 - z}$.)

n	$\frac{1}{n}\log r_n$	$\frac{1}{n}\log r_n^*$
10	0.33385	−0.22508
20	0.35018	−0.18144
50	0.35998	−0.154449
100	0.36325	−0.145447
∞	0.36651	−0.13644
	$(\log 1/\rho)$	$(\log(1/\rho^*))$

Figure IV.5. The growth rate of simple and double surjections.

is the EGF of surjections. The denominator is an entire function, so that singularities may only arise from its zeros, to be found at the points $\chi_k = \log 2 + 2ik\pi$, $k \in \mathbb{Z}$. The dominant singularity of R is then at $\rho = \chi_0 = \log 2$. Thus, with $r_n = [z^n]R(z)$,

$$r_n \bowtie \left(\frac{1}{\log 2}\right)^n.$$

Similarly, if "double" surjections are considered (each value in the range of the surjection is taken at least twice), the corresponding EGF is

$$R^*(z) = \frac{1}{2 + z - e^z},$$

with the counts starting as $1,0,1,1,7,21,141$ (*EIS* **A032032**). The dominant singularity is at ρ^* defined as the positive root of equation $e^{\rho^*} - \rho^* = 2$, and the coefficient r_n^* satisfies: $r_n^* \bowtie (1/\rho^*)^n$ Numerically, this gives

$$r_n \bowtie 1.44269^n \qquad \text{and} \qquad r_n^* \bowtie 0.87245^n,$$

with the actual figures for the corresponding logarithms being given in Figure IV.5.

These estimates constitute a weak form of a more precise result to be established later in this chapter (p. 260): If random surjections of size n are considered equally likely, the probability of a surjection being a double surjection is exponentially small.

Derangements. For the cases $d_{1,n} = [z^n]e^{-z}(1-z)^{-1}$ and $d_{2,n} = [z^n]e^{-z-z^2/2}(1-z)^{-1}$, we have, from the poles at $z = 1$,

$$d_{1,n} \bowtie 1^n \qquad \text{and} \qquad d_{2,n} \bowtie 1^n.$$

The implied upper bound is combinatorially trivial. The lower bound expresses that the probability for a random permutation to be a derangement is *not* exponentially small. For $d_{1,n}$, we have already proved (p. 225) by an elementary argument the stronger result $d_{1,n} \to e^{-1}$; in the case of $d_{2,n}$, we shall establish later (p. 261) the precise asymptotic estimate $d_{2,n} \to e^{-3/2}$.

Unary–binary trees. The expression

$$U(z) = \frac{1 - z - \sqrt{1 - 2z - 3z^2}}{2z} = z + z^2 + 2z^3 + 4z^4 + 9z^5 + \cdots,$$

represents the OGF of (plane unlabelled) unary–binary trees. From the equivalent form,

$$U(z) = \frac{1 - z - \sqrt{(1 - 3z)(1 + z)}}{2z},$$

it follows that $U(z)$ is analytic in the complex plane slit along $(\frac{1}{3}, +\infty)$ and $(-\infty, -1)$ and is singular at $z = -1$ and $z = 1/3$ where it has branch points. The closest singularity to the origin being at $\frac{1}{3}$, one has

$$U_n \bowtie 3^n.$$

In this case, the stronger upper bound $U_n \leq 3^n$ results directly from the possibility of encoding such trees by words over a ternary alphabet using Łukasiewicz codes (Chapter I, p. 74). A complete asymptotic expansion will be obtained, as one of the first applications of singularity analysis, in Chapter VI (p. 396). .... ∎

▷ **IV.15. Coding theory bounds and singularities.** Let C be a combinatorial class. We say that it *can be encoded with* $f(n)$ *bits* if, for all sufficiently large values of n, elements of C_n can be encoded as words of $f(n)$ bits. (An interesting example occurs in Note I.23, p. 53.) Assume that C has OGF $C(z)$ with radius of convergence R satisfying $0 < R < 1$. Then, for any ϵ, C can be encoded with $(1 + \epsilon)\kappa n$ bits where $\kappa = -\log_2 R$, but C cannot be encoded with $(1 - \epsilon)\kappa n$ bits.

Similarly, if C has EGF $\widehat{C}(z)$ with radius of convergence R satisfying $0 < R < \infty$, then C can be encoded with $n\log(n/e) + (1+\epsilon)\kappa n$ bits where $\kappa = -\log_2 R$, but C cannot be encoded with $n\log(n/e) + (1 - \epsilon)\kappa n$ bits. Since the radius of convergence is determined by the distance to singularities nearest to the origin, we have the following interesting fact: *singularities convey information on optimal codes.*   ◁

Saddle-point bounds. The exponential growth formula (Theorem IV.7, p. 244) can be supplemented by effective upper bounds which are very easy to derive and often turn out to be surprisingly accurate. We state:

Proposition IV.1 (Saddle-point bounds). *Let $f(z)$ be analytic in the disc $|z| < R$ with $0 < R \leq \infty$. Define $M(f; r)$ for $r \in (0, R)$ by $M(f; r) := \sup_{|z|=r} |f(z)|$. Then, one has, for any r in $(0, R)$, the family of* saddle-point upper bounds

$$(18) \quad [z^n]f(z) \leq \frac{M(f; r)}{r^n} \quad implying \quad [z^n]f(z) \leq \inf_{r \in (0,R)} \frac{M(f; r)}{r^n}.$$

If in addition $f(z)$ has non-negative coefficients at 0, then

$$(19) \quad [z^n]f(z) \leq \frac{f(r)}{r^n} \quad implying \quad [z^n]f(z) \leq \inf_{r \in (0,R)} \frac{f(r)}{r^n}.$$

Proof. In the general case of (18), the first inequality results from trivial bounds applied to the Cauchy coefficient formula, when integration is performed along a circle:

$$[z^n]f(z) = \frac{1}{2i\pi} \int_{|z|=r} f(z) \frac{dz}{z^{n+1}}.$$

It is consequently valid for any r smaller than the radius of convergence of f at 0. The second inequality in (18) plainly represents the best possible bound of this type.

In the positive case of (19), the bounds can be viewed as a direct specialization of (18). (Alternatively, they can be obtained in a straightforward manner, since

$$f_n \leq \frac{f_0}{r^n} + \cdots + \frac{f_{n-1}}{r} + f_n + \frac{f_{n+1}}{r^{n+1}} + \cdots,$$

whenever the f_k are non-negative.)   ∎

Note that the value s that provides the best bound in (19) can be determined by setting a derivative to zero,

$$(20) \qquad s\frac{f'(s)}{f(s)} = n.$$

Thanks to the universal character of the first bound, *any* approximate solution of this last equation will in fact provide a valid upper bound.

We shall see in Chapter VIII another way to conceive of these bounds as a first step in an important method of asymptotic analysis; namely, the *saddle-point method*, which explains where the term "saddle-point bound" originates from (Theorem VIII.2, p. 547). For reasons that are well developed there, the bounds usually capture the actual asymptotic behaviour up to a polynomial factor. A typical instance is the weak form of Stirling's formula,

$$\frac{1}{n!} \equiv [z^n]e^z \le \frac{e^n}{n^n},$$

which only overestimates the true asymptotic value by a factor of $\sqrt{2\pi n}$.

▷ **IV.16.** *A suboptimal but easy saddle-point bound.* Let $f(z)$ be analytic in $|z| < 1$ with non-negative coefficients. Assume that $f(x) \le (1-x)^{-\beta}$ for some $\beta \ge 0$ and all $x \in (0,1)$. Then

$$[z^n]f(z) = O(n^\beta).$$

(Better bounds of the form $O(n^{\beta-1})$ are usually obtained by the method of singularity analysis expounded in Chapter VI.)   ◁

Example IV.3. *Combinatorial examples of saddle-point bounds.* Here are applications to fragmented permutations, set partitions (Bell numbers), involutions, and integer partitions.

Fragmented permutations. First, fragmented permutations (Chapter II, p. 125) are labelled structures defined by $\mathcal{F} = \text{SET}(\text{SEQ}_{\ge 1}(\mathcal{Z}))$. The EGF is $e^{z/(1-z)}$; we claim that

$$(21) \qquad \frac{1}{n!}F_n \equiv [z^n]e^{z/(1-z)} \le e^{2\sqrt{n}-\frac{1}{2}+O(n^{-1/2})}.$$

Indeed, the minimizing radius of the saddle-point bound (19) is s such that

$$0 = \frac{d}{ds}\left(\frac{s}{1-s} - n\log s\right) = \frac{1}{(1-s)^2} - \frac{n}{s}.$$

The equation is solved by $s = (2n+1-\sqrt{4n+1})/(2n)$. One can either use this exact value and compute an asymptotic approximation of $f(s)/s^n$, or adopt right away the approximate value $s_1 = 1 - 1/\sqrt{n}$, which leads to simpler calculations. The estimate (21) results. It is off from the actual asymptotic value only by a factor of order $n^{-3/4}$ (cf Example VIII.7, p. 562).

Bell numbers and set partitions. Another immediate application is an upper bound on Bell numbers enumerating set partitions, $\mathcal{S} = \text{SET}(\text{SET}_{\ge 1}(\mathcal{Z}))$, with EGF e^{e^z-1}. According to (20), the best saddle-point bound is obtained for s such that $se^s = n$. Thus,

$$(22) \qquad \frac{1}{n!}S_n \le e^{e^s-1-n\log s} \qquad \text{where} \quad s \, : \, se^s = n;$$

additionally, one has $s = \log n - \log\log n + o(\log\log n)$. See Chapter VIII, p. 561 for the complete saddle-point analysis.

n	$\widetilde{I}_n$	I_n
100	$0.106579 \cdot 10^{85}$	$0.240533 \cdot 10^{83}$
200	$0.231809 \cdot 10^{195}$	$0.367247 \cdot 10^{193}$
300	$0.383502 \cdot 10^{316}$	$0.494575 \cdot 10^{314}$
400	$0.869362 \cdot 10^{444}$	$0.968454 \cdot 10^{442}$
500	$0.425391 \cdot 10^{578}$	$0.423108 \cdot 10^{576}$

Figure IV.6. A comparison of the exact number of involutions I_n to its approximation $\widetilde{I}_n = n! e^{\sqrt{n}+n/2} n^{-n/2}$: [left] a table; [right] a plot of $\log_{10}(I_n/\widetilde{I}_n)$ against $\log_{10} n$ suggesting that the ratio satisfies $I_n/\widetilde{I}_n \sim K \cdot n^{-1/2}$, the slope of the curve being $\approx -\frac{1}{2}$.

Involutions. Involutions are specified by $\mathcal{I} = \text{SET}(\text{CYC}_{1,2}(\mathcal{Z}))$ and have EGF $I(z) = \exp(z + \frac{1}{2}z^2)$. One determines, by choosing $s = \sqrt{n}$ as an approximate solution to (20):

$$(23) \qquad \frac{1}{n!} I_n \le \frac{e^{\sqrt{n}+n/2}}{n^{n/2}}.$$

(See Figure IV.6 for numerical data and Example VIII.5, p. 558 for a full analysis.) Similar bounds hold for permutations with all cycle lengths $\le k$ and permutations σ such that $\sigma^k = Id$.

Integer partitions. The function

$$(24) \qquad P(z) = \prod_{k=1}^{\infty} \frac{1}{1-z^k} = \exp\left(\sum_{\ell=1}^{\infty} \frac{1}{\ell} \frac{z^\ell}{1-z^\ell}\right)$$

is the OGF of integer partitions, an unlabelled analogue of set partitions. Its radius of convergence is *a priori* bounded from above by 1, since the set $\mathcal{P}$ is infinite and the second form of $P(z)$ shows that it is exactly equal to 1. Therefore $P_n \bowtie 1^n$. A finer upper bound results from the estimate (see also p. 576)

$$(25) \qquad L(t) := \log P(e^{-t}) \sim \frac{\pi^2}{6t} + \log\sqrt{\frac{t}{2\pi}} - \frac{1}{24}t + O(t^2),$$

which is obtained from Euler–Maclaurin summation or, better, from a Mellin analysis following Appendix B.7: *Mellin transform*, p. 762. Indeed, the Mellin transform of L is, by the harmonic sum rule,

$$L^\star(s) = \zeta(s)\zeta(s+1)\Gamma(s), \qquad s \in \langle 1, +\infty \rangle,$$

and the successive left-most poles at $s = 1$ (simple pole), $s = 0$ (double pole), and $s = -1$ (simple pole) translate into the asymptotic expansion (25). When $z \to 1^-$, we have

$$(26) \qquad P(z) \sim \frac{e^{-\pi^2/12}}{\sqrt{2\pi}} \sqrt{1-z} \exp\left(\frac{\pi^2}{6(1-z)}\right),$$

from which we derive (choose $s = D\sqrt{n}$ as an approximate solution to (20))

$$P_n \le C n^{-1/4} e^{\pi\sqrt{2n/3}},$$

for some $C > 0$. This last bound is once more only off by a polynomial factor, as we shall prove when studying the saddle-point method (Proposition VIII.6, p. 578). . . . . ∎

▷ **IV.17.** *A natural boundary.* One has $P(re^{i\theta}) \to \infty$ as $r \to 1^-$, for any angle θ that is a rational multiple of 2π. The points $e^{2i\pi p/q}$ being dense on the unit circle, the function $P(z)$ admits the unit circle as a *natural boundary*; that is, it cannot be analytically continued beyond this circle. ◁

IV. 4. Closure properties and computable bounds

Analytic functions are robust: they satisfy a rich set of closure properties. This fact makes possible the determination of exponential growth constants for coefficients of a wide range of classes of functions. Theorem IV.8 below expresses computability of growth rate for all specifications associated with iterative specifications. It is the first result of several that relate symbolic methods of Part A with analytic methods developed here.

Closure properties of analytic functions. The functions analytic at a point $z = a$ are closed under sum and product, and hence form a ring. If $f(z)$ and $g(z)$ are analytic at $z = a$, then so is their quotient $f(z)/g(z)$ provided $g(a) \neq 0$. Meromorphic functions are furthermore closed under quotient and hence form a field. Such properties are proved most easily using complex-differentiability and extending the usual relations from real analysis, for instance, $(f + g)' = f' + g'$, $(fg)' = fg' + f'g$.

Analytic functions are also closed under composition: if $f(z)$ is analytic at $z = a$ and $g(w)$ is analytic at $b = f(a)$, then $g \circ f(z)$ is analytic at $z = a$. Graphically:

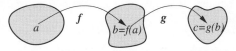

The proof based on complex-differentiability closely mimicks the real case. Inverse functions exist conditionally: if $f'(a) \neq 0$, then $f(z)$ is locally linear near a, hence invertible, so that there exists a g satisfying $f \circ g = g \circ f = Id$, where Id is the identity function, $Id(z) \equiv z$. The inverse function is itself locally linear, hence complex-differentiable, hence analytic. In short: *the inverse of an analytic function f at a place where the derivative does not vanish is an analytic function.* We shall return to this important property later in this chapter (Subsection IV. 7.1, p. 275), then put it to full use in Chapter VI (p. 402) and VII (p. 452) in order to derive strong asymptotic properties of simple varieties of trees.

▷ **IV.18.** *A Mean Value Theorem for analytic functions.* Let f be analytic in Ω and assume the existence of $M := \sup_{z \in \Omega} |f'(z)|$. Then, for all a, b in Ω, one has

$$|f(b) - f(a)| \leq 2M|b - a|.$$

(Hint: a simple consequence of the Mean Value Theorem applied to $\Re(f), \Im(f)$.) ◁

▷ **IV.19.** *The analytic inversion lemma.* Let f be analytic on $\Omega \ni z_0$ and satisfy $f'(z_0) \neq 0$. Then, there exists a small region $\Omega_1 \subseteq \Omega$ containing z_0 and a $C > 0$ such that $|f(z) - f(z')| > C|z - z'|$, for all $z, z' \in \Omega_1, z \neq z'$. Consequently, f maps bijectively Ω_1 on $f(\Omega_1)$. (See also Subsection IV. 6.2, p. 269, for a proof based on integration.) ◁

One way to establish closure properties, as suggested above, is to deduce analyticity criteria from complex differentiability by way of the Basic Equivalence Theorem (Theorem IV.1, p. 232). An alternative approach, closer to the original notion of analyticity, can be based on a two-step process: (*i*) closure properties are shown to hold true for formal power series; (ii) the resulting formal power series are proved to be locally convergent by means of suitable majorizations on their coefficients. This is the basis of the classical method of *majorant series* originating with Cauchy.

▷ **IV.20.** *The majorant series technique.* Given two power series, define $f(z) \preceq g(z)$ if $\left|[z^n]f(z)\right| \le [z^n]g(z)$ for all $n \ge 0$. The following two conditions are equivalent: (i) $f(z)$ is analytic in the disc $|z| < \rho$; (ii) for any $r > \rho^{-1}$ there exists a c such that

$$ f(z) \preceq \frac{c}{1 - rz}. $$

If f, g are majorized by $c/(1-rz), d/(1-rz)$, respectively, then $f + g$ and $f \cdot g$ are majorized,

$$ f(z) + g(z) \preceq \frac{c+d}{1-rz}, \qquad f(z) \cdot g(z) \preceq \frac{e}{1-sz}, $$

for any $s > r$ and for some e dependent on s. Similarly, the composition $f \circ g$ is majorized:

$$ f \circ g(z) \preceq \frac{c}{1 - r(1+d)z}. $$

Constructions for $1/f$ and for the functional inverse of f can be similarly developed. See Cartan's book [104] and van der Hoeven's study [587] for a systematic treatment.   ◁

As a consequence of closure properties, for functions defined by analytic expressions, singularities can be determined inductively in an intuitively transparent manner. If $\mathrm{Sing}(f)$ and $\mathrm{Zero}(f)$ are, respectively, the set of singularities and zeros of the function f, then, due to closure properties of analytic functions, the following informally stated guidelines apply.

$$ \left\{ \begin{array}{rcl} \mathrm{Sing}(f \pm g) & \subseteq & \mathrm{Sing}(f) \cup \mathrm{Sing}(g) \\ \mathrm{Sing}(f \times g) & \subseteq & \mathrm{Sing}(f) \cup \mathrm{Sing}(g) \\ \mathrm{Sing}(f/g) & \subseteq & \mathrm{Sing}(f) \cup \mathrm{Sing}(g) \cup \mathrm{Zero}(g) \\ \mathrm{Sing}(f \circ g) & \subseteq & \mathrm{Sing}(g) \cup g^{(-1)}(\mathrm{Sing}(f)) \\ \mathrm{Sing}(\sqrt{f}) & \subseteq & \mathrm{Sing}(f) \cup \mathrm{Zero}(f) \\ \mathrm{Sing}(\log(f)) & \subseteq & \mathrm{Sing}(f) \cup \mathrm{Zero}(f) \\ \mathrm{Sing}(f^{(-1)}) & \subseteq & f(\mathrm{Sing}(f)) \cup f(\mathrm{Zero}(f')). \end{array} \right. $$

A mathematically rigorous treatment would require considering multivalued functions and Riemann surfaces, so that we do not state detailed validity conditions and keep for these formulae the status of useful heuristics. In fact, because of Pringsheim's theorem, the search of dominant singularities of combinatorial generating function can normally avoid considering the complete multivalued structure of functions, since only some initial segment of the positive real half-line needs to be considered. This in turn implies a powerful and easy way of determining the exponential order of coefficients of a wide variety of generating functions, as we explain next.

Computability of exponential growth constants. As defined in Chapters I and II, a combinatorial class is *constructible* or *specifiable* if it can be specified by a finite set of equations involving only the basic constructors. A specification is *iterative* or *non-recursive* if in addition the dependency graph (p. 33) of the specification is acyclic. In that case, no recursion is involved and a single functional term (written with sums, products, sequences, sets, and cycles) describes the specification.

Our interest here is in effective computability issues. We recall that a real number α is computable iff there exists a program Π_α, which, on input m, outputs a rational number α_m guaranteed to be within $\pm 10^{-m}$ of α. We state:

Theorem IV.8 (Computability of growth). *Let $\mathcal{C}$ be a* constructible *unlabelled class that admits an* iterative *specification in terms of* (SEQ, PSET, MSET, CYC; $+$, $\times$) *starting with* $(1, \mathcal{Z})$. *Then, the radius of convergence $\rho_\mathcal{C}$ of the OGF $C(z)$ of $\mathcal{C}$ is either $+\infty$ or a (strictly) positive computable real number.*

Let $\mathcal{D}$ be a constructible *labelled class that admits an* iterative *specification in terms of* (SEQ, SET, CYC; $+$, $\star$) *starting with* $(1, \mathcal{Z})$. *Then, the radius of convergence $\rho_\mathcal{D}$ of the EGF $D(z)$ of $\mathcal{D}$ is either $+\infty$ or a (strictly) positive computable real number.*

Accordingly, if finite, the constants $\rho_\mathcal{C}$, $\rho_\mathcal{D}$ in the exponential growth estimates,

$$[z^n]C(z) \equiv C_n \bowtie \left(\frac{1}{\rho_\mathcal{C}}\right)^n, \qquad [z^n]D(z) \equiv \frac{1}{n!}D_n \bowtie \left(\frac{1}{\rho_\mathcal{D}}\right)^n,$$

are computable numbers.

Proof. In both cases, the proof proceeds by induction on the structural specification of the class. For each class $\mathcal{F}$, with generating function $F(z)$, we associate a *signature*, which is an ordered pair $\langle \rho_F, \tau_F \rangle$, where ρ_F is the radius of convergence of F and τ_F is the value of F at ρ_F, precisely,

$$\tau_F := \lim_{x \to \rho_F^-} F(x).$$

(The value τ_F is well defined as an element of $\mathbb{R} \cup \{+\infty\}$ since F, being a counting generating function, is necessarily increasing on $(0, \rho_F)$.)

Unlabelled case. An unlabelled class $\mathcal{G}$ is either finite, in which case its OGF $G(z)$ is a polynomial, or infinite, in which case it diverges at $z = 1$, so that $\rho_G \leq 1$. It is clearly decidable, given the specification, whether a class is finite or not: a necessary and sufficient condition for a class to be infinite is that one of the unary constructors (SEQ, MSET, CYC) intervenes in the specification. We prove (by induction) the assertion of the theorem together with the stronger property that $\tau_F = \infty$ as soon as the class is infinite.

First, the signatures of the neutral class 1 and the atomic class $\mathcal{Z}$, with OGF 1 and z, are $\langle +\infty, 1 \rangle$ and $\langle +\infty, +\infty \rangle$. Any non-constant polynomial which is the OGF of a finite set has the signature $\langle +\infty, +\infty \rangle$. The assertion is thus easily verified in these cases.

Next, let $\mathcal{F} = \text{SEQ}(\mathcal{G})$. The OGF $G(z)$ must be non-constant and satisfy $G(0) = 0$, in order for the sequence construction to be properly defined. Thus, by the induction hypothesis, one has $0 < \rho_G \leq +\infty$ and $\tau_G = +\infty$. Now, the function G being increasing and continuous along the positive axis, there must exist a value β such that $0 < \beta < \rho_G$ with $G(\beta) = 1$. For $z \in (0, \beta)$, the quasi-inverse $F(z) = (1 - G(z))^{-1}$ is well defined and analytic; as z approaches β from the left, $F(z)$ increases unboundedly. Thus, the smallest singularity of F along the positive axis is at β, and by Pringsheim's theorem, one has $\rho_F = \beta$. The argument shows at the same time that $\tau_F = +\infty$. There only remains to check that β is computable. The coefficients of

G form a computable sequence of integers, so that $G(x)$, which can be well approximated via a truncated Taylor series, is an effectively computable number[7] if x is itself a positive computable number less than ρ_G. Then, binary search provides an effective procedure for determining β.

Next, we consider the multiset construction, $\mathcal{F} = \text{MSET}(\mathcal{G})$, whose translation into OGFs necessitates the Pólya exponential of Chapter I (p. 34):

$$F(z) = \text{Exp}(G(z)) \quad \text{where} \quad \text{Exp}(h(z)) := \exp\left(h(z) + \frac{1}{2}h(z^2) + \frac{1}{3}h(z^3) + \cdots\right).$$

Once more, the induction hypothesis is assumed for G. If G is a polynomial, then F is a rational function with poles at roots of unity only. Thus, $\rho_F = 1$ and $\tau_F = \infty$ in that particular case. In the general case of $\mathcal{F} = \text{MSET}(\mathcal{G})$ with $\mathcal{G}$ infinite, we start by fixing arbitrarily a number r such that $0 < r < \rho_G \leq 1$ and examine $F(z)$ for $z \in (0, r)$. The expression for F rewrites as

$$\text{Exp}(G(z)) = e^{G(z)} \cdot \exp\left(\frac{1}{2}G(z^2) + \frac{1}{3}G(z^3) + \cdots\right).$$

The first factor is analytic for z on $(0, \rho_G)$ since, the exponential function being entire, e^G has the singularities of G. As to the second factor, one has $G(0) = 0$ (in order for the set construction to be well-defined), while $G(x)$ is convex for $x \in [0, r]$ (since its second derivative is positive). Thus, there exists a positive constant K such that $G(x) \leq Kx$ when $x \in [0, r]$. Then, the series $\frac{1}{2}G(z^2) + \frac{1}{3}G(z^3) + \cdots$ has its terms dominated by those of the convergent series

$$\frac{K}{2}r^2 + \frac{K}{3}r^3 + \cdots = K\log(1-r)^{-1} - Kr.$$

By a well-known theorem of analytic function theory, a uniformly convergent sum of analytic functions is itself analytic; consequently, $\frac{1}{2}G(z^2) + \frac{1}{3}G(z^3) + \cdots$ is analytic at all z of $(0, r)$. Analyticity is then preserved by the exponential, so that $F(z)$, being analytic at $z \in (0, r)$ for any $r < \rho_G$ has a radius of convergence that satisfies $\rho_F \geq \rho_G$. On the other hand, since $F(z)$ dominates termwise $G(z)$, one has $\rho_F \leq \rho_G$. Thus finally one has $\rho_F = \rho_G$. Also, $\tau_G = +\infty$ implies $\tau_F = +\infty$.

A parallel discussion covers the case of the powerset construction (PSET) whose associated functional $\overline{\text{Exp}}$ is a minor modification of the Pólya exponential Exp. The cycle construction can be treated by similar arguments based on consideration of "Pólya's logarithm" as $\mathcal{F} = \text{CYC}(\mathcal{G})$ corresponds to

$$F(z) = \text{Log}\,\frac{1}{1-G(z)}, \quad \text{where} \quad \text{Log}\,h(z) = \log h(z) + \frac{1}{2}\log h(z^2) + \cdots.$$

In order to conclude with the unlabelled case, it only remains to discuss the binary constructors $+, \times$, which give rise to $F = G + H$, $F = G \cdot H$. It is easily verified that

---

[7] The present argument only establishes non-constructively the *existence* of a program, based on the fact that truncated Taylor series converge geometrically fast at an interior point of their disc of convergence. Making explicit this program and the involved parameters from the specification itself however represents a much harder problem (that of "uniformity" with respect to specifications) that is not addressed here.

$\rho_F = \min(\rho_G, \rho_H)$. Computability is granted since the minimum of two computable numbers is computable. That $\tau_F = +\infty$ in each case is immediate.

Labelled case. The labelled case is covered by the same type of argument as above, the discussion being even simpler, since the ordinary exponential and logarithm replace the Pólya operators Exp and Log. It is still a fact that all the EGFs of infinite non-recursive classes are infinite at their dominant positive singularity, though the radii of convergence can now be of any magnitude (compared to 1). ∎

▷ **IV.21.** *Restricted constructions.* This is an exercise in induction. Theorem IV.8 is stated for specifications involving the basic constructors. Show that the conclusion still holds if the corresponding restricted constructions ($\mathfrak{K}_{=r}$, $\mathfrak{K}_{<r}$, $\mathfrak{K}_{>r}$, with $\mathfrak{K}$ being any of the basic constructors) are also allowed. ◁

▷ **IV.22.** *Syntactically decidable properties.* For unlabelled classes $\mathcal{F}$, the property $\rho_F = 1$ is decidable. For labelled and unlabelled classes, the property $\rho_F = +\infty$ is decidable. ◁

▷ **IV.23.** *Pólya–Carlson and a curious property of OGFs.* Here is a statement first conjectured by Pólya, then proved by Carlson in 1921 (see [164, p. 323]): *If a function is represented by a power series with integer coefficients that converges inside the unit disc, then either it is a rational function or it admits the unit circle as a natural boundary.* This theorem applies in particular to the OGF of any combinatorial class. ◁

▷ **IV.24.** *Trees are recursive structures only!* General and binary trees cannot receive an iterative specification since their OGFs assume a finite value at their Pringsheim singularity. [The same is true of most simple families of trees; cf Proposition VI.6, p. 404]. ◁

▷ **IV.25.** *Non-constructibility of permutations and graphs.* The class $\mathcal{P}$ of all permutations cannot be specified as a constructible unlabelled class since the OGF $P(z) = \sum_n n! z^n$ has radius of convergence 0. (It is of course constructible as a labelled class.) Graphs, whether labelled or unlabelled, are too numerous to form a constructible class. ◁

Theorem IV.8 establishes a link between analytic combinatorics, computability theory, and symbolic manipulation systems. It is based on an article of Flajolet, Salvy, and Zimmermann [255] devoted to such computability issues in exact and asymptotic enumeration. Recursive specifications are not discussed now since they tend to give rise to branch points, themselves amenable to singularity analysis techniques to be fully developed in Chapters VI and VII. The inductive process, implied by the proof of Theorem IV.8, that decorates a specification with the radius of convergence of each of its subexpressions, provides a practical basis for determining the exponential growth rate of counts associated to a non-recursive specification.

Example IV.4. *Combinatorial trains.* This purposely artificial example from [219] (see Figure IV.7) serves to illustrate the scope of Theorem IV.8 and demonstrate its inner mechanisms at work. Define the class of all *labelled trains* by the following specification,

(27)
$$\begin{cases} \mathcal{T}r &= \mathcal{W}a \star \text{SEQ}(\mathcal{W}a \star \text{SET}(\mathcal{P}a)), \\ \mathcal{W}a &= \text{SEQ}_{\geq 1}(\mathcal{P}\ell), \\ \mathcal{P}\ell &= \mathcal{Z} \star \mathcal{Z} \star (1 + \text{CYC}(\mathcal{Z})), \\ \mathcal{P}a &= \text{CYC}(\mathcal{Z}) \star \text{CYC}(\mathcal{Z}). \end{cases}$$

In figurative terms, a train ($\mathcal{T}r$) is composed of a first wagon ($\mathcal{W}a$) to which is appended a sequence of passenger wagons, each of the latter capable of containing a set of passengers ($\mathcal{P}a$). A wagon is itself composed of "planks" ($\mathcal{P}\ell$) conventionally identified by their two end points ($\mathcal{Z} \star \mathcal{Z}$) and to which a circular wheel ($\text{CYC}(\mathcal{Z})$) may optionally be attached. A passenger is

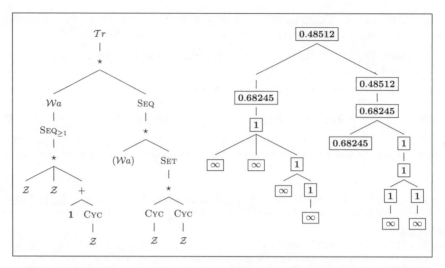

Figure IV.7. The inductive determination of the radius of convergence of the EGF of trains: (left) a hierarchical view of the specification of Tr; (right) the corresponding radii of convergence for each subspecification.

composed of a head and a belly that are each circular arrangements of atoms. Here is a depiction of a random train:

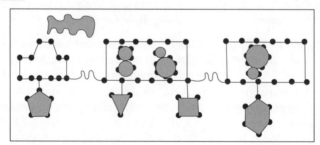

The translation into a set of EGF equations is immediate and a symbolic manipulation system readily provides the form of the EGF of trains as

$$Tr(z) = \frac{z^2\left(1+\log((1-z)^{-1})\right)}{\left(1-z^2\left(1+\log((1-z)^{-1})\right)\right)}\left(1 - \frac{z^2\left(1+\log((1-z)^{-1})\right)e^{\left(\log((1-z)^{-1})\right)^2}}{1-z^2\left(1+\log((1-z)^{-1})\right)}\right)^{-1},$$

together with the expansion

$$Tr(z) = 2\frac{z^2}{2!} + 6\frac{z^3}{3!} + 60\frac{z^4}{4!} + 520\frac{z^5}{5!} + 6660\frac{z^6}{6!} + 93408\frac{z^7}{7!} + \cdots .$$

The specification (27) has a hierarchical structure, as suggested by the top representation of Figure IV.7, and this structure is itself directly reflected by the form of the expression tree of the GF $Tr(z)$. Then, each node in the expression tree of $Tr(z)$ can be tagged with the corresponding value of the radius of convergence. This is done according to the principles of Theorem IV.8;

see the right diagram of Figure IV.7. For instance, the quantity 0.68245 associated to $Wa(z)$ is given by the sequence rule and is determined as the smallest positive solution of the equation

$$z^2 \left(1 - \log(1 - z)^{-1} \right) = 1.$$

The tagging process works upwards till the root of the tree is reached; here the radius of convergence of Tr is determined to be $\rho \doteq 0.48512 \cdots$, a quantity that happens to coincide with the ratio $[z^{49}]Tr(z)/[z^{50}]Tr(z)$ to more than 15 decimal places. .... ∎

IV. 5. Rational and meromorphic functions

The last section has fully justified the *First Principle of Coefficient Asymptotics* leading to the exponential growth formula $f_n \bowtie A^n$ for the coefficients of an analytic function $f(z)$. Indeed, as we saw, one has $A = 1/\rho$, where ρ equals both the radius of convergence of the series representing f and the distance of the origin to the dominant, i.e., closest, singularities. We are going to start examining here the *Second Principle*, already given on p. 227 and relative to the form

$$f_n = A^n \theta(n),$$

with $\theta(n)$ the subexponential factor:

Second Principle of Coefficient Asymptotics. *The* nature *of a function's* singularities determines the associate *subexponential factor ($\theta(n)$).*

In this section, we develop a complete theory in the case of rational functions (that is, quotients of polynomials) and, more generally, meromorphic functions. The net result is that, for such functions, the subexponential factors are essentially polynomials:

Polar singularities ⤳ subexponential factors $\theta(n)$ of *polynomial growth.*

A distinguishing feature is the extremely good quality of the asymptotic approximations obtained; for naturally occurring combinatorial problems, 15 digits of accuracy is not uncommon in coefficients of index as low as 50 (see Figure IV.8, p. 260 below for a striking example).

IV. 5.1. Rational functions. A function $f(z)$ is a *rational function* iff it is of the form $f(z) = N(z)/D(z)$, with $N(z)$ and $D(z)$ being polynomials, which we may, without loss of generality, assume to be relatively prime. For rational functions that are analytic at the origin (e.g., generating functions), we have $D(0) \neq 0$.

Sequences $\{f_n\}_{n \geq 0}$ that are coefficients of rational functions satisfy linear recurrence relations with constant coefficients. This fact is easy to establish: compute $[z^n]f(z) \cdot D(z)$; then, with $D(z) = d_0 + d_1 z + \cdots + d_m z^m$, one has, for all $n > \deg(N(z))$,

$$\sum_{j=0}^{m} d_j f_{n-j} = 0.$$

The main theorem we prove now provides an *exact* finite expression for coefficients of $f(z)$ in terms of the poles of $f(z)$. Individual terms in these expressions are sometimes called *exponential–polynomials.*

Theorem IV.9 (Expansion of rational functions). *If $f(z)$ is a rational function that is analytic at zero and has poles at points $\alpha_1, \alpha_2, \ldots, \alpha_m$, then its coefficients are a sum of exponential–polynomials: there exist m polynomials $\{\Pi_j(x)\}_{j=1}^m$ such that, for n larger than some fixed n_0,*

$$(28) \qquad\qquad f_n \equiv [z^n] f(z) = \sum_{j=1}^m \Pi_j(n) \alpha_j^{-n}.$$

Furthermore the degree of Π_j is equal to the order of the pole of f at α_j minus one.

Proof. Since $f(z)$ is rational it admits a partial fraction expansion. To wit:

$$f(z) = Q(z) + \sum_{(\alpha, r)} \frac{c_{\alpha, r}}{(z - \alpha)^r},$$

where $Q(z)$ is a polynomial of degree $n_0 := \deg(N) - \deg(D)$ if $f = N/D$. Here α ranges over the poles of $f(z)$ and r is bounded from above by the multiplicity of α as a pole of f. Coefficient extraction in this expression results from Newton's expansion,

$$[z^n] \frac{1}{(z - \alpha)^r} = \frac{(-1)^r}{\alpha^r} [z^n] \frac{1}{\left(1 - \frac{z}{\alpha}\right)^r} = \frac{(-1)^r}{\alpha^r} \binom{n + r - 1}{r - 1} \alpha^{-n}.$$

The binomial coefficient is a polynomial of degree $r - 1$ in n, and collecting terms associated with a given α yields the statement of the theorem.   ∎

Notice that the expansion (28) is also an asymptotic expansion in disguise: when grouping terms according to the α's of increasing modulus, each group appears to be *exponentially smaller* than the previous one. In particular, if there is a unique dominant pole, $|\alpha_1| < |\alpha_2| \le |\alpha_3| \le \cdots$, then

$$f_n \sim \alpha_1^{-n} \Pi_1(n),$$

and the error term is exponentially small as it is $O(\alpha_2^{-n} n^r)$ for some r. A classical instance is the OGF of Fibonacci numbers,

$$F(z) = \frac{z}{1 - z - z^2},$$

with poles at $\dfrac{-1 + \sqrt{5}}{2} \doteq 0.61803$ and $\dfrac{-1 - \sqrt{5}}{2} \doteq -1.61803$, so that

$$[z^n] F(z) \equiv F_n = \frac{1}{\sqrt{5}} \varphi^n - \frac{1}{\sqrt{5}} \bar{\varphi}^n = \frac{\varphi^n}{\sqrt{5}} + O\left(\frac{1}{\varphi^n}\right),$$

with $\varphi = (1 + \sqrt{5})/2$ the golden ratio, and $\bar{\varphi}$ its conjugate.

▷ **IV.26.** *A simple exercise.* Let $f(z)$ be as in Theorem IV.9, assuming additionally a single dominant pole α_1, with multiplicity r. Then, by inspection of the proof of Theorem IV.9:

$$f_n = \frac{C}{(r - 1)!} \alpha_1^{-n + r} n^{r - 1} \left(1 + O\left(\frac{1}{n}\right)\right) \qquad \text{with} \quad C = \lim_{z \to \alpha_1} (z - \alpha_1)^r f(z).$$

This is certainly the most direct illustration of the Second Principle: under the assumptions, a one-term asymptotic expansion of the function at its dominant singularity suffices to determine the asymptotic form of the coefficients.   ◁

Example IV.5. *Qualitative analysis of a rational function.* This is an artificial example designed to demonstrate that all the details of the full decomposition are usually not required. The rational function

$$f(z) = \frac{1}{(1 - z^3)^2(1 - z^2)^3(1 - \frac{z^2}{2})}$$

has a pole of order 5 at $z = 1$, poles of order 2 at $z = \omega, \omega^2$ ($\omega = e^{2i\pi/3}$ a cubic root of unity), a pole of order 3 at $z = -1$, and simple poles at $z = \pm\sqrt{2}$. Therefore,

$$f_n = P_1(n) + P_2(n)\omega^{-n} + P_3(n)\omega^{-2n} + P_4(n)(-1)^n +$$
$$+ P_5(n)2^{-n/2} + P_6(n)(-1)^n 2^{-n/2}$$

where the degrees of $P_1, \ldots, P_6$ are $4, 1, 1, 2, 0, 0$. For an asymptotic equivalent of f_n, only the poles at roots of unity need to be considered since they correspond to the fastest exponential growth; in addition, only $z = 1$ needs to be considered for first-order asymptotics; finally, at $z = 1$, only the term of fastest growth needs to be taken into account. In this way, we find the correspondence

$$f(z) \sim \frac{1}{3^2 \cdot 2^3 \cdot (\frac{1}{2})} \frac{1}{(1 - z)^5} \implies f_n \sim \frac{1}{3^2 \cdot 2^3 \cdot (\frac{1}{2})} \binom{n+4}{4} \sim \frac{n^4}{864}.$$

The way the analysis can be developed *without computing details* of partial fraction expansion is typical. .... ∎

Theorem IV.9 applies to any specification leading to a GF that is a rational function[8]. Combined with the qualitative approach to rational coefficient asymptotics, it gives access to a large number of effective asymptotic estimates for combinatorial counting sequences.

Example IV.6. *Asymptotics of denumerants.* Denumerants are integer partitions with summands restricted to be from a *fixed* finite set (Chapter I, p. 43). We let $\mathcal{P}^\mathcal{T}$ be the class relative to set $\mathcal{T} \subset \mathbb{Z}_{>0}$, with the known OGF,

$$P^\mathcal{T}(z) = \prod_{\omega \in \mathcal{T}} \frac{1}{1 - z^\omega}.$$

Without loss of generality, we assume that $\gcd(\mathcal{T}) = 1$; that is, the coin denomination are *not* all multiples of a number $d > 1$.

A particular case is the one of integer partitions whose summands are in $\{1, 2, \ldots, r\}$,

$$P^{\{1,\ldots,r\}}(z) = \prod_{m=1}^{r} \frac{1}{1 - z^m}.$$

The GF has all its poles being roots of unity. At $z = 1$, the order of the pole is r, and one has

$$P^{\{1,\ldots,r\}}(z) \sim \frac{1}{r!} \frac{1}{(1 - z)^r},$$

as $z \to 1$. Other poles have strictly smaller multiplicity. For instance the multiplicity of $z = -1$ is equal to the number of factors $(1 - z^{2j})^{-1}$ in $P^{\{1,\ldots,r\}}$, which is the same as the number of coin denominations that are even; this last number is at most $r - 1$ since, by the gcd assumption $\gcd(\mathcal{T}) = 1$, at least one is odd. Similarly, a primitive qth root of unity is found to have

---

[8] In Part A, we have been occasionally led to discuss coefficients of some simple enough rational functions, thereby anticipating the statement of the theorem: see for instance the discussion of parts in compositions (p. 168) and of records in sequences (p. 190).

multiplicity at most $r - 1$. It follows that the pole $z = 1$ contributes a term of the form n^{r-1} to the coefficient of index n, while each of the other poles contributes a term of order at most n^{r-2}. We thus find

$$P_n^{\{1,\dots,r\}} \sim c_r n^{r-1} \qquad \text{with} \quad c_r = \frac{1}{r!(r-1)!}.$$

The same argument provides the asymptotic form of $P_n^{\mathcal{T}}$, since, to first order asymptotics, only the pole at $z = 1$ counts.

Proposition IV.2. *Let $\mathcal{T}$ be a finite set of integers without a common divisor ($\gcd(\mathcal{T}) = 1$). The number of partitions with summands restricted to $\mathcal{T}$ satisfies*

$$P_n^{\mathcal{T}} \sim \frac{1}{\tau} \frac{n^{r-1}}{(r-1)!}, \qquad \text{with} \quad \tau := \prod_{\omega \in \mathcal{T}} \omega, \quad r := \text{card}(\mathcal{T}).$$

For instance, in a strange country that would have pennies (1 cent), nickels (5 cents), dimes (10 cents), and quarters (25 cents), the number of ways to make change for a total of n cents is

$$[z^n] \frac{1}{(1 - z)(1 - z^5)(1 - z^{10})(1 - z^{25})} \sim \frac{1}{1 \cdot 5 \cdot 10 \cdot 25} \frac{n^3}{3!} \equiv \frac{n^3}{7500},$$

asymptotically. $\dotfill$ $\blacksquare$

IV. 5.2. Meromorphic functions.

An expansion similar to that of Theorem IV.9 (p. 256) holds true for coefficients of a much larger class; namely, meromorphic functions.

Theorem IV.10 (Expansion of meromorphic functions). *Let $f(z)$ be a function meromorphic at all points of the closed disc $|z| \leq R$, with poles at points $\alpha_1, \alpha_2, \dots, \alpha_m$. Assume that $f(z)$ is analytic at all points of $|z| = R$ and at $z = 0$. Then there exist m polynomials $\{\Pi_j(x)\}_{j=1}^m$ such that:*

$$(29) \qquad f_n \equiv [z^n] f(z) = \sum_{j=1}^m \Pi_j(n) \alpha_j^{-n} + O(R^{-n}).$$

Furthermore the degree of Π_j is equal to the order of the pole of f at α_j minus one.

Proof. We offer two different proofs, one based on subtracted singularities, the other one based on contour integration.

(i) Subtracted singularities. Around any pole α, $f(z)$ can be expanded locally:

$$(30) \qquad f(z) = \sum_{k \geq -M} c_{\alpha,k} (z - \alpha)^k$$

$$(31) \qquad = S_\alpha(z) + H_\alpha(z)$$

where the "singular part" $S_\alpha(z)$ is obtained by collecting all the terms with index in $[-M .. -1]$ (that is, forming $S_\alpha(z) = N_\alpha(z)/(z - \alpha)^M$ with $N_\alpha(z)$ a polynomial of degree less than M) and $H_\alpha(z)$ is analytic at α. Thus setting $S(z) := \sum_j S_{\alpha_j}(z)$, we observe that $f(z) - S(z)$ is analytic for $|z| \leq R$. In other words, by collecting the singular parts of the expansions and subtracting them, we have "removed" the singularities of $f(z)$, whence the name of *method of subtracted singularities* sometimes given to the method [329, vol. 2, p. 448].

Taking coefficients, we get:

$$[z^n]f(z) = [z^n]S(z) + [z^n](f(z) - S(z)).$$

The coefficient of $[z^n]$ in the rational function $S(z)$ is obtained from Theorem IV.9. It suffices to prove that the coefficient of z^n in $f(z) - S(z)$, a function analytic for $|z| \le R$, is $O(R^{-n})$. This fact follows from trivial bounds applied to Cauchy's integral formula with the contour of integration being $\lambda = \{z : |z| = R\}$, as in the proof of Proposition IV.1, p 246 (saddle-point bounds):

$$\left| [z^n](f(z) - S(z)) \right| = \frac{1}{2\pi} \left| \int_{|z|=R} (f(z) - S(z)) \frac{dz}{z^{n+1}} \right| \le \frac{1}{2\pi} \frac{O(1)}{R^{n+1}} 2\pi R.$$

(ii) *Contour integration.* There is another line of proof for Theorem IV.10 which we briefly sketch as it provides an insight which is useful for applications to other types of singularities treated in Chapter VI. It consists in using Cauchy's coefficient formula and "pushing" the contour of integration past singularities. In other words, one computes directly the integral

$$I_n = \frac{1}{2i\pi} \int_{|z|=R} f(z) \frac{dz}{z^{n+1}}$$

by residues. There is a pole at $z = 0$ with residue f_n and poles at the α_j with residues corresponding to the terms in the expansion stated in Theorem IV.10; for instance, if $f(z) \sim c/(z - a)$ as $z \to a$, then

$$\text{Res}(f(z)z^{-n-1}; z = a) = \text{Res}\left(\frac{c}{(z-a)} z^{-n-1}; z = a \right) = \frac{c}{a^{n+1}}.$$

Finally, by the same trivial bounds as before, I_n is $O(R^{-n})$. ∎

▷ **IV.27.** *Effective error bounds.* The error term $O(R^{-n})$ in (29), call it ε_n, satisfies

$$|\varepsilon_n| \le R^{-n} \cdot \sup_{|z|=R} |f(z)|.$$

This results immediately from the second proof. This bound may be useful, even in the case of rational functions to which it is clearly applicable. ◁

As a consequence of Theorem IV.10, all GFs whose dominant singularities are poles can be easily analysed. Prime candidates from Part A are specifications that are "driven" by a sequence construction, since the translation of sequences involves a quasi-inverse, itself conducive to polar singularities. This covers in particular surjections, alignments, derangements, and constrained compositions, which we treat now.

Example IV.7. *Surjections.* These are defined as sequences of sets ($\mathcal{R} = \text{SEQ}(\text{SET}_{\ge 1}(\mathcal{Z}))$) with EGF $R(z) = (2 - e^z)^{-1}$ (see p. 106). We have already determined the poles in Example IV.2 (p. 244), the one of smallest modulus being at $\log 2 \doteq 0.69314$. At this dominant pole, one finds $R(z) \sim -\frac{1}{2}(z - \log 2)^{-1}$. This implies an approximation for the number of surjections:

$$R_n \equiv n![z^n]R(z) \sim \xi(n), \qquad \text{with} \quad \xi(n) := \frac{n!}{2} \cdot \left(\frac{1}{\log 2} \right)^{n+1}.$$

$$
\begin{array}{r|l}
3 & 3 \\
75 & 75 \\
4683 & 4683 \\
545835 & 545835 \\
102247563 & 102247563 \\
28091567595 & 28091567595 \\
10641342970443 & 10641342970443 \\
5315654681981355 & 5315654681981355 \\
3385534663256845323 & 3385534663256845326 \\
2677687796244384203115 & 2677687796244384203088 \\
2574844419803190384544203 & 2574844419803190384544450 \\
29582791210741454726650648875 & 295827912107414547265064 6597 \\
4002225759844168492486127539083 & 4002225759844168492486127555859 \\
6297562064950066033518373935334635 & 6297562064950066033518373935416161 \\
1140356879401188048374246419618490196 3 & 1140356879401188048374246419617 4527074 \\
2354515408573489664918449063714485547639 5 & 2354515408573489664918449063714 5314147690 \\
\end{array}
$$

Figure IV.8. The surjection numbers pyramid: for $n = 2, 4, \ldots, 32$, the exact values of the numbers R_n (left) compared to the approximation $\lceil \xi(n) \rfloor$ with discrepant digits in boldface (right).

Figure IV.8 gives, for $n = 2, 4, \ldots, 32$, a table of the values of the surjection numbers (left) compared with the asymptotic approximation rounded[9] to the nearest integer, $\lceil \xi(n) \rfloor$: It is piquant to see that $\lceil \xi(n) \rfloor$ provides the exact value of R_n for all values of $n = 1, \ldots, 15$, and it starts losing one digit for $n = 17$, after which point a few "wrong" digits gradually appear, but in very limited number; see Figure IV.8. (A similar situation prevails for tangent numbers discussed in our *Invitation*, p. 5.) The explanation of such a faithful asymptotic representation owes to the fact that the error terms provided by meromorphic asymptotics are exponentially small. In effect, there is no other pole in $|z| \leq 6$, the next ones being at $\log 2 \pm 2i\pi$ with modulus of about 6.32. Thus, for $r_n = [z^n]R(z)$, there holds

$$
(32) \qquad \frac{R_n}{n!} \sim \frac{1}{2} \cdot \left(\frac{1}{\log 2} \right)^{n+1} + O(6^{-n}).
$$

For the double surjection problem, $R^*(z) = (2 + z - e^z)$, we get similarly

$$
[z^n]R^*(z) \sim \frac{1}{e^{\rho^*} - 1} (\rho^*)^{-n-1},
$$

with $\rho^* = 1.14619$ the smallest positive root of $e^{\rho^*} - \rho^* = 2$. .... ∎

It is worth reflecting on this example as it is representative of a "production chain" based on the two successive implications which are characteristic of Part A and Part B of the book:

$$
\left\{
\begin{array}{lll}
\mathcal{R} = \mathrm{SEQ}(\mathrm{SET}_{\geq 1}(\mathcal{Z})) & \Longrightarrow & R(z) = \dfrac{1}{2 - e^z} \\[3mm]
R(z) \underset{z \to \log 2}{\sim} -\dfrac{1}{2} \dfrac{1}{(z - \log 2)} & \longrightarrow & \dfrac{1}{n!} R_n \sim \dfrac{1}{2}(\log 2)^{-n-1}.
\end{array}
\right.
$$

[9]The notation $\lceil x \rfloor$ represents x rounded to the nearest integer: $\lceil x \rfloor := \lfloor x + \frac{1}{2} \rfloor$.

The first implication (written "$\Longrightarrow$", as usual) is provided *automatically* by the symbolic method. The second one (written here "$\longrightarrow$") is a direct translation from the expansion of the GF at its dominant singularity to the asymptotic form of coefficients; it is valid *conditionally* upon complex analytic conditions, here those of Theorem IV.10.

Example IV.8. *Alignments.* These are sequences of cycles ($\mathcal{O} = \text{SEQ}(\text{CYC}(\mathcal{Z}))$, p. 119) with EGF

$$O(z) = \frac{1}{1 - \log \frac{1}{1-z}}.$$

There is a singularity when $\log(1 - z)^{-1} = 1$, which is at $\rho = 1 - e^{-1}$ and which arises before $z = 1$, where the logarithm becomes singular. Then, the computation of the asymptotic form of $[z^n]O(z)$ only requires a local expansion near ρ,

$$O(z) \sim \frac{-e^{-1}}{z - 1 + e^{-1}} \qquad \longrightarrow \qquad [z^n]O(z) \sim \frac{e^{-1}}{(1 - e^{-1})^{n+1}},$$

and the coefficient estimates result from Theorem IV.10. .... ∎

▷ **IV.28.** *Some "supernecklaces".* One estimates

$$[z^n] \log \left(\frac{1}{1 - \log \frac{1}{1-z}} \right) \sim \frac{1}{n}(1 - e^{-1})^{-n},$$

where the EGF enumerates labelled cycles of cycles (supernecklaces, p. 125). [Hint: Take derivatives.] ◁

Example IV.9. *Generalized derangements.* The probability that the shortest cycle in a random permutation of size n has length larger than k is

$$[z^n]D^{(k)}(z), \qquad \text{where} \quad D^{(k)}(z) = \frac{1}{1-z}e^{-\frac{z}{1} - \frac{z^2}{2} - \cdots - \frac{z^k}{k}},$$

as results from the specification $\mathcal{D}^{(k)} = \text{SET}(\text{CYC}_{>k}(\mathcal{Z}))$. For any *fixed* k, one has (easily) $D^{(k)}(z) \sim e^{-H_k}/(1 - z)$ as $z \to 1$, with 1 being a simple pole. Accordingly the coefficients $[z^n]D^{(k)}(z)$ tend to e^{-H_k} as $n \to \infty$. In summary, due to meromorphy, we have the characteristic implication

$$D^{(k)}(z) \sim \frac{e^{-H_k}}{1 - z} \qquad \longrightarrow \qquad [z^n]D^{(k)}(z) \sim e^{-H_k}.$$

Since there is no other singularity at a finite distance, the error in the approximation is (at least) exponentially small,

(33) $$[z^n]\frac{1}{1-z}e^{-\frac{z}{1} - \frac{z^2}{2} - \cdots - \frac{z^k}{k}} = e^{-H_k} + O(R^{-n}),$$

for *any* $R > 1$. The cases $k = 1, 2$ in particular justify the estimates mentioned at the beginning of this chapter, on p. 228. .... ∎

This example is also worth reflecting upon. In prohibiting cycles of length $< k$, we modify the EGF of all permutations, $(1 - z)^{-1}$ by a factor $e^{-z/1 - \cdots - z^k/k}$. The resulting EGF is meromorphic at 1; thus only the value of the modifying factor at $z = 1$ matters, so that this value, namely e^{-H_k}, provides the asymptotic proportion of k–derangements. We shall encounter more and more shortcuts of this sort as we progress into the book.

▷ **IV.29.** *Shortest cycles of permutations are not too long.* Let S_n be the random variable denoting the length of the shortest cycle in a random permutation of size n. Using the circle $|z| = 2$ to estimate the error in the approximation e^{-H_k} above, one finds that, for $k \leq \log n$,

$$\left| \mathbb{P}(S_n > k) - e^{-H_k} \right| \leq \frac{1}{2^n} e^{2^{k+1}},$$

which is exponentially small in this range of k-values. Thus, the approximation e^{-H_k} remains usable when k is allowed to tend sufficiently slowly to ∞ with n. One can also explore the possibility of better bounds and larger regions of validity of the main approximation. (See Panario and Richmond's study [470] for a general theory of smallest components in sets.) ◁

▷ **IV.30.** *Expected length of the shortest cycle.* The classical approximation of the harmonic numbers, $H_k \approx \log k + \gamma$, suggests $e^{-\gamma}/k$ as a possible approximation to (33) for *both* large n and large k in suitable regions. In agreement with this heuristic argument, the expected length of the shortest cycle in a random permutation of size n is effectively asymptotic to

$$\sum_{k=1}^{n} \frac{e^{-\gamma}}{k} \sim e^{-\gamma} \log n,$$

a property first discovered by Shepp and Lloyd [540]. ◁

The next example illustrates the analysis of a collection of rational generating functions (Smirnov words) paralleling nicely the enumeration of a special type of integer composition (Carlitz compositions), which belongs to meromorphic asymptotics.

Example IV.10. *Smirnov words and Carlitz compositions.* Bernoulli trials have been discussed in Chapter III (p. 204), in relation to weighted word models. Take the class $\mathcal{W}$ of all words over an r–ary alphabet, where letter j is assigned probability p_j and letters of words are drawn independently. With this weighting, the GF of all words is $W(z) = 1/(1 - \sum p_j z) = (1 - z)^{-1}$. Consider the problem of determining the probability that a random word of length n is of Smirnov type, that is, all blocks of length 2 are formed with unequal letters. In order to avoid degeneracies, we impose $r \geq 3$ (since for $r = 2$, the only Smirnov words are ababa...and babab...).

By our discussion in Example III.24 (p. 204), the GF of Smirnov words (again with the probabilistic weighting) is

$$S(z) = \frac{1}{1 - \sum \frac{p_j z}{1 + p_j z}}.$$

By monotonicity of the denominator, this rational function has a dominant singularity at the unique positive solution of the equation

(34)
$$\sum_{j=1}^{r} \frac{p_j \rho}{1 + p_j \rho} = 1,$$

and the point ρ is a simple pole. Consequently, ρ is a well-characterized algebraic number defined implicitly by a polynomial equation of degree $\leq r$. One can furthermore check, by studying the variations of the denominator, that the other roots are all real and negative; thus, ρ is the unique dominant singularity. (Alternatively, appeal to the Perron–Frobenius argument of Example V.11, p. 349) It follows that the probability for a word to be Smirnov is, not too surprisingly, exponentially small, the precise formula being

$$[z^n]S(z) \sim C \cdot \rho^{-n}, \qquad C = \left(\sum_{j=1}^{r} \frac{p_j \rho}{(1 + p_j \rho)^2} \right)^{-1}.$$

A similar analysis, using bivariate generating functions, shows that in a random word of length n conditioned to be Smirnov, the letter j appears with asymptotic frequency

$$(35) \qquad q_j = \frac{1}{Q} \frac{p_j}{(1 + p_j \rho)^2}, \qquad Q := \sum_{j=1}^{r} \frac{p_j}{(1 + p_j \rho)^2},$$

in the sense that the mean number of occurrences of letter j is asymptotic to $q_j n$. All these results are seen to be consistent with the equiprobable letter case $p_j = 1/r$, for which $\rho = r/(r-1)$.

Carlitz compositions illustrate a limit situation, in which the alphabet is infinite, while letters have different sizes. Recall that a Carlitz composition of the integer n is a composition of n such that no two adjacent summands have equal value. By Note III.32, p. 201, such compositions can be obtained by substitution from Smirnov words, to the effect that

$$(36) \qquad K(z) = \left(1 - \sum_{j=1}^{\infty} \frac{z^j}{1 + z^j} \right)^{-1}.$$

The asymptotic form of the coefficients then results from an analysis of dominant poles. The OGF has a simple pole at ρ, which is the smallest positive root of the equation

$$(37) \qquad \sum_{j=1}^{\infty} \frac{\rho^j}{1 + \rho^j} = 1.$$

(Note the analogy with (34) due to commonality of the combinatorial argument.) Thus:

$$K_n \sim C \cdot \beta^n, \qquad C \doteq 0.4563634740, \qquad \beta \doteq 1.7502412917.$$

There, $\beta = 1/\rho$ with ρ as in (37). In a way analogous to Smirnov words, the asymptotic frequency of summand k appears to be proportional to $k\rho^k/(1 + \rho^k)^2$; see [369, 421] for further properties. .... ∎

IV. 6. Localization of singularities

There are situations where a function possesses several dominant singularities, that is, several singularities are present on the boundary of the disc of convergence. We examine here the induced effect on coefficients and discuss ways to locate such dominant singularities.

IV. 6.1. Multiple singularities. In the case when there exists more than one dominant singularity, several geometric terms of the form β^n sharing the same modulus (and each carrying its own subexponential factor) must be combined. In simpler situations, such terms globally induce a pure periodic behaviour for coefficients that is easy to describe. In the general case, irregular fluctuations of a somewhat arithmetic nature may prevail.

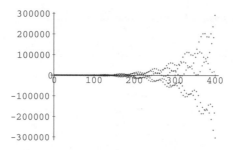

Figure IV.9. The coefficients $[z^n]f(z)$ of the rational function $f(z) = \left(1 + 1.02z^4\right)^{-3}\left(1 - 1.05z^5\right)^{-1}$ illustrate a periodic superposition of regimes, depending on the residue class of n modulo 40.

Pure periodicities. When several dominant singularities of $f(z)$ have the same modulus and are regularly spaced on the boundary of the disc of convergence, they may induce complete cancellations of the main exponential terms in the asymptotic expansion of the coefficient f_n. In that case, different regimes will be present in the coefficients f_n based on congruence properties of n. For instance, the functions

$$\frac{1}{1+z^2} = 1 - z^2 + z^4 - z^6 + z^8 - \cdots, \qquad \frac{1}{1-z^3} = 1 + z^3 + z^6 + z^9 + \cdots,$$

exhibit patterns of periods 4 and 3, respectively, this corresponding to poles that are roots of unity or order 4 ($\pm i$), and 3 ($\omega : \omega^3 = 1$). Then, the function

$$\phi(z) = \frac{1}{1+z^2} + \frac{1}{1-z^3} = \frac{2 - z^2 + z^3 + z^4 + z^8 + z^9 - z^{10}}{1 - z^{12}}$$

has coefficients that obey a pattern of period 12 (for example, the coefficients ϕ_n such that $n \equiv 1, 5, 6, 7, 11$ modulo 12 are zero). Accordingly, the coefficients of

$$[z^n]\psi(z) \qquad \text{where} \qquad \psi(z) = \phi(z) + \frac{1}{1 - z/2},$$

manifest a different exponential growth when n is congruent to $1, 5, 6, 7, 11$ mod 12. See Figure IV.9 for such a superposition of pure periodicities. In many combinatorial applications, generating functions involving periodicities can be decomposed at sight, and the corresponding asymptotic subproblems generated are then solved separately.

▷ **IV.31.** *Decidability of polynomial properties.* Given a polynomial $p(z) \in \mathbb{Q}[z]$, the following properties are decidable: (i) whether one of the zeros of p is a root of unity; (ii) whether one of the zeros of p has an argument that is commensurate with π. [One can use resultants. An algorithmic discussion of this and related issues is given in [306].]   ◁

Nonperiodic fluctuations. As a representative example, consider the polynomial $D(z) = 1 - \frac{6}{5}z + z^2$, whose roots are

$$\alpha = \frac{3}{5} + i\frac{4}{5}, \qquad \bar{\alpha} = \frac{3}{5} - i\frac{4}{5},$$

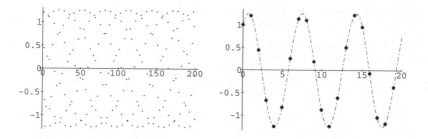

Figure IV.10. The coefficients of $f(z) = 1/(1 - \frac{6}{5}z + z^2)$ exhibit an apparently chaotic behaviour (left) which in fact corresponds to a discrete sampling of a sine function (right), reflecting the presence of two conjugate complex poles.

---

both of modulus 1 (the numbers 3, 4, 5 form a Pythagorean triple), with argument $\pm\theta_0$ where $\theta_0 = \arctan(\frac{4}{3}) \doteq 0.92729$. The expansion of the function $f(z) = 1/D(z)$ starts as

$$\frac{1}{1 - \frac{6}{5}z + z^2} = 1 + \frac{6}{5}z + \frac{11}{25}z^2 - \frac{84}{125}z^3 - \frac{779}{625}z^4 - \frac{2574}{3125}z^5 + \cdots,$$

the sign sequence being

$$+++---++++---+++----+++----+++---,$$

which indicates a somewhat irregular oscillating behaviour, where blocks of three or four pluses follow blocks of three or four minuses.

The exact form of the coefficients of f results from a partial fraction expansion:

$$f(z) = \frac{a}{1 - z/\alpha} + \frac{b}{1 - z/\bar{\alpha}} \quad \text{with} \quad a = \frac{1}{2} + \frac{3}{8}i, \quad b = \frac{1}{2} - \frac{3}{8}i,$$

where $\alpha = e^{i\theta_0}, \bar{\alpha} = e^{-i\theta_0}$ Accordingly,

$$(38) \qquad\qquad f_n = ae^{-in\theta_0} + be^{in\theta_0} = \frac{\sin((n+1)\theta_0)}{\sin(\theta_0)}.$$

This explains the sign changes observed. Since the angle θ_0 is not commensurate with π, the coefficients fluctuate but, unlike in our earlier examples, no exact periodicity is present in the sign patterns. See Figure IV.10 for a rendering and Figure V.3 (p. 299) for a meromorphic case linked to compositions into prime summands.

Complicated problems of an arithmetical nature may occur if several such singularities with non-commensurate arguments combine, and some open problem remain even in the analysis of linear recurring sequences. (For instance no decision procedure is known to determine whether such a sequence ever vanishes [200].) Fortunately, such problems occur infrequently in combinatorial applications, where dominant poles of rational functions (as well as many other functions) tend to have a simple geometry as we explain next.

▷ **IV.32.** *Irregular fluctuations and Pythagorean triples.* The quantity θ_0/π is an irrational number, so that the sign fluctuations of (38) are "irregular" (i.e., non-purely periodic). [Proof: *a contrario*. Indeed, otherwise, $\alpha = (3 + 4i)/5$ would be a root of unity. But then the minimal polynomial of α would be a cyclotomic polynomial with non-integral coefficients, a contradiction; see [401, VIII.3] for the latter property.]   ◁

▷ **IV.33.** *Skolem-Mahler-Lech Theorem.* Let f_n be the sequence of coefficients of a rational function, $f(z) = A(z)/B(z)$, where $A, B \in \mathbb{Q}[z]$. The set of all n such that $f_n = 0$ is the union of a finite (possibly empty) set and a finite number (possibly zero) of infinite arithmetic progressions. (The proof is based on p-adic analysis, but the argument is intrinsically non-constructive; see [452] for an attractive introduction to the subject and references.)   ◁

Periodicity conditions for positive generating functions. By the previous discussion, it is of interest to locate dominant singularities of combinatorial generating functions, and, in particular, determine whether their arguments (the "dominant directions") are commensurate to 2π. In the latter case, different asymptotic regimes of the coefficients manifest themselves, depending on the congruence properties of n.

Definition IV.5. *For a sequence (f_n) with GF $f(z)$, the support of f, denoted $\mathrm{Supp}(f)$, is the set of all n such that $f_n \neq 0$. The sequence (f_n), as well as its GF $f(z)$, is said to admit a span d if for some r, there holds*

$$\mathrm{Supp}(f) \subseteq r + d\mathbb{Z}_{\geq 0} \equiv \{r, \ r+d, \ r+2d, \ldots\}.$$

The largest span, p, is the period, *all other spans being divisors of p. If the period is equal to 1, the sequence and its GF are said to be* aperiodic.

If f is analytic at 0, with span d, there exists a function g analytic at 0 such that $f(z) = z^r g(z^d)$, for some $r \in \mathbb{Z}_{\geq 0}$. With $E := \mathrm{Supp}(f)$, the maximal span [the period] is determined as $p = \gcd(E - E)$ (pairwise differences) as well as $p = \gcd(E - \{r\})$ where $r := \min(E)$. For instance $\sin(z)$ has period 2, $\cos(z) + \cosh(z)$ has period 4, $z^3 e^{z^5}$ has period 5, and so on.

In the context of periodicities, a basic property is expressed by what we have chosen to name figuratively the "Daffodil Lemma". By virtue of this lemma, the span of a function f with non-negative coefficients is related to the behaviour of $|f(z)|$ as z varies along circles centred at the origin (Figure IV.11).

Lemma IV.1 ("Daffodil Lemma"). *Let $f(z)$ be analytic in $|z| < \rho$ and have non-negative coefficients at 0. Assume that f does not reduce to a monomial and that for some non-zero non-positive z satisfying $|z| < \rho$, one has*

$$|f(z)| = f(|z|).$$

Then, the following hold: (i) the argument of z must be commensurate to 2π, i.e., $z = Re^{i\theta}$ with $\theta/(2\pi) = \frac{r}{p} \in \mathbb{Q}$ (an irreducible fraction) and $0 < r < p$; (ii) f admits p as a span.

Proof. This classical lemma is a simple consequence of the strong triangle inequality. Indeed, for Part (i) of the statement, with $z = Re^{i\theta}$, the equality $|f(z)| = f(|z|)$ implies that the complex numbers $f_n R^n e^{in\theta}$, for $n \in \mathrm{Supp}(f)$, all lie on the same ray (a half-line emanating from 0). This is impossible if $\theta/(2\pi)$ is irrational, since, by assumption, the expansion of f contains at least two monomials (one cannot have $n_1\theta \equiv n_2\theta \pmod{2\pi}$). Thus, $\theta/(2\pi) = r/p$ is a rational number. Regarding Part (ii), consider two distinct indices n_1 and n_2 in $\mathrm{Supp}(f)$ and let $\theta/(2\pi) = r/p$. Then, by the strong triangle inequality again, one must have $(n_1 - n_2)\theta \equiv 0 \pmod{2\pi}$; that

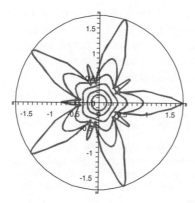

Figure IV.11. Illustration of the "Daffodil Lemma": the images of circles $z = Re^{i\theta}$ ($R = 0.4 \ldots 0.8$) rendered by a polar plot of $|f(z)|$ in the case of $f(z) = z^7 e^{z^{25}} + z^2/(1 - z^{10})$), which has span 5.

is, $(n_i - n_j)r/p = (k_1 - k_2)$, for some $k_1, k_2 \in \mathbb{Z} \geq 0$. This is only possible if p divides $n_1 - n_2$. Hence, p is a span. ∎

Berstel [53] first realized that rational generating functions arising from regular languages can only have dominant singularities of the form $\rho\omega^j$, where ω is a certain root of unity. This property in fact extends to many non-recursive specifications, as shown by Flajolet, Salvy, and Zimmermann in [255].

Proposition IV.3 (Commensurability of dominant directions). *Let S be a constructible labelled class that is non-recursive, in the sense of Theorem IV.8. Assume that the EGF $S(z)$ has a finite radius of convergence ρ. Then there exists a computable integer $d \geq 1$ such that the set of dominant singularities of $S(z)$ is contained in the set $\{\rho\omega^j\}$, where $\omega^d = 1$.*

Proof. (Sketch; see [53, 255]) By definition, a non-recursive class S is obtained from $\mathbf{1}$ and $\mathcal{Z}$ by means of a finite number of union, product, sequence, set, and cycle constructions. We have seen earlier, in Section IV. 4 (p. 249), an inductive algorithm that determines radii of convergence. It is then easy to enrich that algorithm and determine simultaneously (by induction on the specification) the period of its GF and the set of dominant directions.

The period is determined by simple rules. For instance, if $S = T \star U$ ($S = T \cdot U$) and T, U are infinite series with respective periods p, q, one has the implication

$$\text{Supp}(T) \subseteq a + p\mathbb{Z}, \quad \text{Supp}(U) \subseteq b + q\mathbb{Z} \implies \text{Supp}(S) \subseteq a + b + \xi\mathbb{Z},$$

with $\xi = \gcd(p, q)$. Similarly, for $S = \text{SEQ}(T)$,

$$\text{Supp}(T) \subseteq a + p\mathbb{Z} \implies \text{Supp}(S) \subseteq \delta\mathbb{Z},$$

where now $\delta = \gcd(a, p)$.

Regarding dominant singularities, the case of a sequence construction is typical. It corresponds to $g(z) = (1 - f(z))^{-1}$. Assume that $f(z) = z^a h(z^p)$, with p the maximal period, and let $\rho > 0$ be such that $f(\rho) = 1$. The equations determining any dominant singularity ζ are $f(\zeta) = 1$, $|\zeta| = \rho$. In particular, the equations imply $|f(\zeta)| = f(|\zeta|)$, so that, by the Daffodil Lemma, the argument of ζ must be of the form $2\pi r/s$. An easy refinement of the argument shows that, for $\delta = \gcd(a, p)$, all the dominant directions coincide with the multiples of $2\pi/\delta$. The discussion of cycles is entirely similar since $\log(1 - f)^{-1}$ has the same dominant singularities as $(1 - f)^{-1}$. Finally, for exponentials, it suffices to observe that e^f does not modify the singularity pattern of f, since $\exp(z)$ is an entire function.   ∎

▷ **IV.34.** *Daffodil lemma and unlabelled classes.* Proposition IV.3 applies to any unlabelled class S that admits a non-recursive specification, provided its radius of convergence ρ satisfies $\rho < 1$. (When $\rho = 1$, there is a possibility of having the unit circle as a natural boundary, a property that is otherwise decidable from the specification.) The case of regular specifications will be investigated in detail in Section V. 3, p. 300.   ◁

Exact formulae. The error terms appearing in the asymptotic expansion of coefficients of meromorphic functions are already exponentially small. By peeling off the singularities of a meromorphic function layer by layer, in order of increasing modulus, one is led to extremely precise, sometimes even exact, expansions for the coefficients. Such exact representations are found for Bernoulli numbers B_n, surjection numbers R_n, as well as Secant numbers E_{2n} and Tangent numbers E_{2n+1}, defined by

$$(39) \quad \begin{cases} \displaystyle\sum_{n=0}^{\infty} B_n \frac{z^n}{n!} & = \quad \dfrac{z}{e^z - 1} & \text{(Bernoulli numbers)} \\[2em] \displaystyle\sum_{n=0}^{\infty} R_n \frac{z^n}{n!} & = \quad \dfrac{1}{2 - e^z} & \text{(Surjection numbers)} \\[2em] \displaystyle\sum_{n=0}^{\infty} E_{2n} \frac{z^{2n}}{(2n)!} & = \quad \dfrac{1}{\cos(z)} & \text{(Secant numbers)} \\[2em] \displaystyle\sum_{n=0}^{\infty} E_{2n+1} \frac{z^{2n+1}}{(2n+1)!} & = \quad \tan(z) & \text{(Tangent numbers).} \end{cases}$$

Bernoulli numbers. These numbers traditionally written B_n can be defined by their EGF $B(z) = z/(e^z - 1)$, and they are central to Euler–Maclaurin expansions (p. 726). The function $B(z)$ has poles at the points $\chi_k = 2ik\pi$, with $k \in \mathbb{Z} \setminus \{0\}$, and the residue at χ_k is equal to χ_k,

$$\frac{z}{e^z - 1} \sim \frac{\chi_k}{z - \chi_k} \qquad (z \to \chi_k).$$

The expansion theorem for meromorphic functions is applicable here: start with the Cauchy integral formula, and proceed as in the proof of Theorem IV.10, using as external contours a large circle of radius R that passes half-way between poles. As R tends to infinity, the integrand tends to 0 (as soon as $n \geq 2$) because the Cauchy kernel z^{-n-1} decreases as an inverse power of R while the EGF remains $O(R)$. In the limit, corresponding to an infinitely large contour, the coefficient integral becomes equal to the sum of all residues of the meromorphic function over the whole of the complex plane.

From this argument, we get the representation $B_n = -n! \sum_{k \in \mathbb{Z} \setminus \{0\}} \chi_k^{-n}$. This verifies that $B_n = 0$ if n is odd and $n \geq 3$. If n is even, then grouping terms two by two, we get the exact representation (which also serves as an asymptotic expansion):

$$(40) \qquad \frac{B_{2n}}{(2n)!} = (-1)^{n-1} 2^{1-2n} \pi^{-2n} \sum_{k=1}^{\infty} \frac{1}{k^{2n}}.$$

Reverting the equality, we have also established that

$$\zeta(2n) = (-1)^{n-1} 2^{2n-1} \pi^{2n} \frac{B_{2n}}{(2n)!}, \qquad \text{with} \quad \zeta(s) = \sum_{k=1}^{\infty} \frac{1}{k^s}, \qquad B_n = n! [z^n] \frac{z}{e^z - 1},$$

a well-known identity that provides values of the Riemann zeta function $\zeta(s)$ at even integers as rational multiples of powers of π.

Surjection numbers. In the same vein, the surjection numbers have EGF $R(z) = (2 - e^z)^{-1}$ with simple poles at

$$\chi_k = \log 2 + 2ik\pi \qquad \text{where} \qquad R(z) \sim \frac{1}{2} \frac{1}{\chi_k - z}.$$

Since $R(z)$ stays bounded on circles passing half-way in between poles, we find the exact formula, $R_n = \frac{1}{2} n! \sum_{k \in \mathbb{Z}} \chi_k^{-n-1}$. An equivalent real formulation is

$$(41) \qquad \frac{R_n}{n!} = \frac{1}{2} \left(\frac{1}{\log 2} \right)^{n+1} + \sum_{k=1}^{\infty} \frac{\cos((n+1)\theta_k)}{(\log^2 2 + 4k^2 \pi^2)^{(n+1)/2}}, \qquad \theta_k := \arctan(\frac{2k\pi}{\log 2}),$$

which exhibits infinitely many harmonics of fast decaying amplitude.

▷ **IV.35.** *Alternating permutations, tangent and secant numbers.* The relation (40) also provides a representation of the *tangent numbers* since $E_{2n-1} = (-1)^{n-1} B_{2n} 4^n (4^n - 1)/(2n)$. The secant numbers E_{2n} satisfy

$$\sum_{k=1}^{\infty} \frac{(-1)^k}{(2k+1)^{2n+1}} = \frac{(\pi/2)^{2n+1}}{2 (2n)!} E_{2n},$$

which can be read either as providing an asymptotic expansion of E_{2n} or as an evaluation of the sums on the left (the values of a Dirichlet L-function) in terms of π. The asymptotic number of alternating permutations (pp. 5 and 143) is consequently known to great accuracy. ◁

▷ **IV.36.** *Solutions to the equation* $\tan(x) = x$. Let x_n be the nth positive root of the equation $\tan(x) = x$. For any integer $r \geq 1$, the sum $S(r) := \sum_n x_n^{-2r}$ is a computable rational number. For instance: $S_2 = 1/10$, $S_4 = 1/350$, $S_6 = 1/7875$. [From mathematical folklore.] ◁

IV. 6.2. Localization of zeros and poles.

We gather here a few results that often prove useful in determining the location of zeros of analytic functions, and hence of poles of meromorphic functions. A detailed treatment of this topic may be found in Henrici's book [329, §4.10].

Let $f(z)$ be an analytic function in a region Ω and let γ be a simple closed curve interior to Ω, and on which f is assumed to have no zeros. We claim that the quantity

$$(42) \qquad N(f; \gamma) = \frac{1}{2i\pi} \int_{\gamma} \frac{f'(z)}{f(z)} dz$$

exactly equals the number of zeros of f inside γ counted with multiplicity. [Proof: the function f'/f has its poles exactly at the zeros of f, and the residue at each pole α equals the multiplicity of α as a root of f; the assertion then results from the residue theorem.]

Since a primitive function (antiderivative) of f'/f is $\log f$, the integral also represents the variation of $\log f$ along γ, which is written $[\log f]_\gamma$. This variation itself reduces to $2i\pi$ times the variation of the argument of f along γ, since $\log(re^{i\theta}) = \log r + i\theta$ and the modulus r has variation equal to 0 along a closed contour ($[\log r]_\gamma = 0$). The quantity $[\theta]_\gamma$ is, by its definition, 2π multiplied by the number of times the transformed contour $f(\gamma)$ winds about the origin, a number known as the *winding number*. This observation is known as the *Argument Principle*:

> **Argument Principle.** *The number of zeros of $f(z)$ (counted with multiplicities) inside the simple loop γ equals the winding number of the transformed contour $f(\gamma)$ around the origin.*

By the same argument, if f is meromorphic in $\Omega \ni \gamma$, then $N(f; \gamma)$ equals the difference between the number of zeros and the number of poles of f inside γ, multiplicities being taken into account. Figure IV.12 exemplifies the use of the argument principle in localizing zeros of a polynomial.

By similar devices, we get Rouché's theorem:

> **Rouché's theorem.** *Let the functions $f(z)$ and $g(z)$ be analytic in a region containing in its interior the closed simple curve γ. Assume that f and g satisfy $|g(z)| < |f(z)|$ on the curve γ. Then $f(z)$ and $f(z) + g(z)$ have the same number of zeros inside the interior domain delimited by γ.*

An intuitive way to visualize Rouché's Theorem is as follows: since $|g| < |f|$, then $f(\gamma)$ and $(f + g)(\gamma)$ must have the same winding number.

▷ **IV.37.** *Proof of Rouché's theorem.* Under the hypothesis of Rouché's theorem, for $0 \leq t \leq 1$, the function $h(z) = f(z) + tg(z)$ is such that $N(h; \gamma)$ is both an integer and an analytic, hence continuous, function of t in the given range. The conclusion of the theorem follows. ◁

▷ **IV.38.** *The Fundamental Theorem of Algebra.* Every complex polynomial $p(z)$ of degree n has exactly n roots. A proof follows by Rouché's theorem from the fact that, for large enough $|z| = R$, the polynomial assumed to be monic is a "perturbation" of its leading term, z^n. [Other proofs can be based on Liouville's Theorem (Note IV.7, p. 237) or on the Maximum Modulus Principle (Theorem VIII.1, p. 545).] ◁

▷ **IV.39.** *Symmetric function of the zeros.* Let $S_k(f; \gamma)$ be the sum of the kth powers of the roots of equation $f(z) = 0$ inside γ. One has

$$S_k(f; \gamma) = \frac{1}{2i\pi} \int \frac{f'(z)}{f(z)} z^k \, dz,$$

by a variant of the proof of the Argument Principle. ◁

These principles form the basis of numerical algorithms for locating zeros of analytic functions, in particular the ones closest to the origin, which are of most interest to us. One can start from an initially large domain and recursively subdivide it until roots have been isolated with enough precision—the number of roots in a subdomain being at each stage determined by numerical integration; see Figure IV.12 and refer for instance to [151] for a discussion. Such algorithms even acquire the status of full

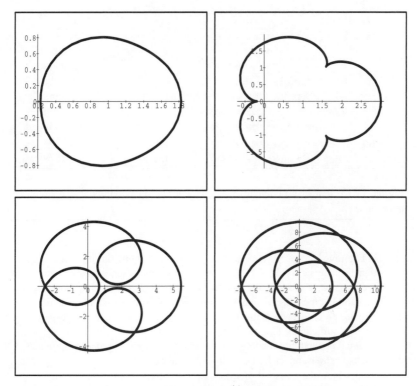

Figure IV.12. The transforms of $\gamma_j = \{|z| = \frac{4j}{10}\}$ by $P_4(z) = 1 - 2z + z^4$, for $j =$ 1, 2, 3, 4, demonstrate, via winding numbers, that $P_4(z)$ has no zero inside $|z| < 0.4$, one zero inside $|z| < 0.8$, two zeros inside $|z| < 1.2$ and four zeros inside $|z| < 1.6$. The actual zeros are at $\rho_4 = 0.54368, 1$ and $1.11514 \pm 0.77184i$.

proofs if one operates with guaranteed precision routines (using, for instance, careful implementations of interval arithmetics).

IV. 6.3. Patterns in words: a case study. Analysing the coefficients of a single generating function that is rational is a simple task, often even bordering on the trivial, granted the exponential–polynomial formula for coefficients (Theorem IV.9, p. 256). However, in analytic combinatorics, we are often confronted with problems that involve an *infinite family* of functions. In that case, Rouché's Theorem and the Argument Principle provide decisive tools for localizing poles, while Theorems IV.3 (Residue Theorem, p. 234) and IV.10 (Expansion of meromorphic functions, p. 258) serve to determine effective error terms. An illustration of this situation is the analysis of patterns in words for which GFs have been derived in Chapters I (p. 60) and III (p. 212).

Example IV.11. *Patterns in words: asymptotics.* All patterns are not born equal. Surprisingly, in a random sequence of coin tossings, the pattern HTT is likely to occur much sooner (after 8 tosses on average) than the pattern HHH (needing 14 tosses on average); see the preliminary

Length (k)	types	$c(z)$	ρ
$k = 3$	aab, abb, bba, baa	1	0.61803
	aba, bab	$1 + z^2$	0.56984
	aaa, bbb	$1 + z + z^2$	0.54368
$k = 4$	aaab, aabb, abbb,		
	bbba, bbaa, baaa	1	0.54368
	aaba, abba, abaa,		
	bbab, baab, babb	$1 + z^3$	0.53568
	abab, baba	$1 + z^2$	0.53101
	aaaa, bbbb	$1 + z + z^2 + z^3$	0.51879

Figure IV.13. Patterns of length $3, 4$: autocorrelation polynomial and dominant poles of $S(z)$.

---

discussion in Example I.12 (p. 59). Questions of this sort are of obvious interest in the statistical analysis of genetic sequences [414, 603]. Say you discover that a sequence of length 100,000 on the four letters A, G, C, T contains the pattern TACTAC twice. Can this be assigned to chance or is this likely to be a meaningful signal of some yet unknown structure? The difficulty here lies in quantifying precisely where the asymptotic regime starts, since, by Borges's Theorem (Note I.35, p. 61), sufficiently long texts will almost certainly contain any fixed pattern. The analysis of rational generating functions supplemented by Rouché's theorem provides definite answers to such questions, under Bernoulli models at least.

We consider here the class $\mathcal{W}$ of words over an alphabet $\mathcal{A}$ of cardinality $m \geq 2$. A pattern $\mathfrak{p}$ of some length k is given. As seen in Chapters I and III, its autocorrelation polynomial is central to enumeration. This polynomial is defined as $c(z) = \sum_{j=0}^{k-1} c_j z^j$, where c_j is 1 if $\mathfrak{p}$ coincides with its jth shifted version and 0 otherwise. We consider here the enumeration of words containing the pattern $\mathfrak{p}$ at least once, and dually of words excluding the pattern $\mathfrak{p}$. In other words, we look at problems such as: What is the probability that a random text of length n does (or does not) contain your name as a block of consecutive letters?

The OGF of the class of words excluding $\mathfrak{p}$ is, we recall,

$$(43) \qquad\qquad S(z) = \frac{c(z)}{z^k + (1 - mz)c(z)}.$$

(Proposition I.4, p. 61), and we shall start with the case $m = 2$ of a binary alphabet. The function $S(z)$ is simply a rational function, but the location and nature of its poles is yet unknown. We only know *a priori* that it should have a pole in the positive interval somewhere between $\frac{1}{2}$ and 1 (by Pringsheim's Theorem and since its coefficients are in the interval $[1, 2^n]$, for n large enough). Figure IV.13 gives a small list, for patterns of length $k = 3, 4$, of the pole ρ of $S(z)$ that is nearest to the origin. Inspection of the figure suggests ρ to be close to $\frac{1}{2}$ as soon as the pattern is long enough. We are going to prove this fact, based on Rouché's Theorem applied to the denominator of (43).

As regards termwise domination of coefficients, the autocorrelation polynomial lies between 1 (for less correlated patterns like aaa...ab) and $1 + z + \cdots + z^{k-1}$ (for the special case aaa...aa). We set aside the special case of $\mathfrak{p}$ having only equal letters, i.e., a "maximal" autocorrelation polynomial—this case is discussed at length in the next chapter. Thus, in this scenario, the autocorrelation polynomial starts as $1 + z^\ell + \cdots$ for some $\ell \geq 2$. Fix the

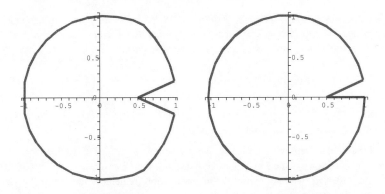

Figure IV.14. Complex zeros of $z^{31} + (1 - 2z)c(z)$ represented as joined by a polygonal line: (left) correlated pattern $a(ba)^{15}$; (right) uncorrelated pattern $a(ab)^{15}$.

number $A = 0.6$, which proves suitable for our subsequent analysis. On $|z| = A$, we have

$$(44) \qquad |c(z)| \geq \left|1 - (A^2 + A^3 + \cdots)\right| = \left|1 - \frac{A^2}{1 - A}\right| = \frac{1}{10}.$$

In addition, the quantity $(1 - 2z)$ ranges over the circle of diameter $[-0.2, 1.2]$ as z varies along $|z| = A$, so that $|1 - 2z| \geq 0.2$. All in all, we have found that, for $|z| = A$,

$$|(1 - 2z)c(z)| \geq 0.02.$$

On the other hand, for $k > 7$, we have $|z^k| < 0.017$ on the circle $|z| = A$. Then, among the two terms composing the denominator of (43), the first is strictly dominated by the second along $|z| = A$. By virtue of Rouché's Theorem, the number of roots of the denominator inside $|z| \leq A$ is then same as the number of roots of $(1 - 2z)c(z)$. The latter number is 1 (due to the root $\frac{1}{2}$) since $c(z)$ cannot be 0 by the argument of (44). Figure IV.14 exemplifies the extremely well-behaved characters of the complex zeros.

In summary, we have found that for all patterns with at least two different letters ($\ell \geq 2$) and length $k \geq 8$, the denominator has a unique root in $|z| \leq A = 0.6$. The same property for lengths k satisfying $4 \leq k \leq 7$ is then easily verified directly. The case $\ell = 1$ where we are dealing with long runs of identical letters can be subjected to an entirely similar argument (see also Example V.4, p. 308, for details). Therefore, unicity of a simple pole ρ of $S(z)$ in the interval $(0.5, 0.6)$ is granted, for a binary alphabet.

It is then a simple matter to determine the local expansion of $S(z)$ near $z = \rho$,

$$S(z) \underset{z \to \rho}{\sim} \frac{\widetilde{\Lambda}}{\rho - z}, \qquad \widetilde{\Lambda} := \frac{c(\rho)}{2c(\rho) - (1 - 2\rho)c'(\rho) - k\rho^{k-1}},$$

from which a precise estimate for coefficients results from Theorems IV.9 (p. 256) and IV.10 (p. 258).

The computation finally extends almost verbatim to non-binary alphabets, with ρ being now close to $1/m$. It suffices to use the disc of radius $A = 1.2/m$. The Rouché part of the argument grants us unicity of the dominant pole in the interval $(1/m, A)$ for $k \geq 5$ when $m = 3$, and for $k \geq 4$ and any $m \geq 4$. (The remaining cases are easily checked individually.)

Proposition IV.4. *Consider an m–ary alphabet. Let $\mathfrak{p}$ be a fixed pattern of length $k \geq 4$, with autocorrelation polynomial $c(z)$. Then the probability that a random word of length n does not contain $\mathfrak{p}$ as a pattern (a block of consecutive letters) satisfies*

$$(45) \qquad \mathbb{P}_{\mathcal{W}_n}(\mathfrak{p} \text{ does not occur}) = \Lambda_{\mathfrak{p}}(m\rho)^{-n-1} + O\left(\left(\frac{5}{6}\right)^n\right),$$

where $\rho \equiv \rho_{\mathfrak{p}}$ is the unique root in $(\frac{1}{m}, \frac{6}{5m})$ of the equation $z^k + (1 - mz)c(z) = 0$ and $\Lambda_{\mathfrak{p}} := mc(\rho)/(mc(\rho) - c'(\rho)(1 - m\rho) - k\rho^{k-1})$.

Despite their austere appearance, these formulae have indeed a fairly concrete content. First, the equation satisfied by ρ can be put under the form $mz = 1 + z^k/c(z)$, and, since ρ is close to $1/m$, we may expect the approximation (remember the use of "$\approx$" as meaning "numerically approximately equal", but *not* implying strict asymptotic equivalence)

$$m\rho \approx 1 + \frac{1}{\gamma m^k},$$

where $\gamma := c(m^{-1})$ satisfies $1 \leq \gamma < m/(m - 1)$. By similar principles, the probabilities in (45) are approximately

$$\mathbb{P}_{\mathcal{W}_n}(\mathfrak{p} \text{ does not occur}) \approx \left(1 + \frac{1}{\gamma m^k}\right)^{-n} \approx e^{-n/(\gamma m^k)}.$$

For a binary alphabet, this tells us that the occurrence of a pattern of length k starts becoming likely when n is of the order of 2^k, that is, when k is of the order of $\log_2 n$. The more precise moment when this happens must depend (via γ) on the autocorrelation of the pattern, with strongly correlated patterns having a tendency to occur a little late. (This vastly generalizes our empirical observations of Chapter I.) However, the mean number of occurrences of a pattern in a text of length n does not depend on the shape of the pattern. The apparent paradox is easily resolved, as we already observed in Chapter I: correlated patterns tend to occur late, while being prone to appear in clusters. For instance, the "late" pattern aaa, when it occurs, still has probability $\frac{1}{2}$ to occur at the next position as well and cash in another occurrence; in contrast no such possibility is available to the "early" uncorrelated pattern aab, whose occurrences must be somewhat spread out.

Such analyses are important as they can be used to develop a precise understanding of the behaviour of data compression algorithms (the Lempel–Ziv scheme); see Julien Fayolle's contribution [204] for details. .... ■

▷ **IV.40.** *Multiple pattern occurrences.* A similar analysis applies to the generating function $S^{\langle s \rangle}(z)$ of words containing a fixed number s of occurrences of a pattern $\mathfrak{p}$. The OGF is obtained by expanding (with respect to u) the BGF $W(z, u)$ obtained in Chapter III, p. 212, by means of an inclusion–exclusion argument. For $s \geq 1$, one finds

$$S^{\langle s \rangle}(z) = z^k \frac{N(z)^{s-1}}{D(z)^{s+1}}, \qquad D(z) = z^k + (1 - mz)c(z), \qquad N(z) = z^k + (1 - mz)(c(z) - 1)),$$

which now has a pole of multiplicity $s + 1$ at $z = \rho$.   ◁

▷ **IV.41.** *Patterns in Bernoulli sequences—asymptotics.* Similar results hold when letters are assigned non-uniform probabilities, $p_j = \mathbb{P}(a_j)$, for $a_j \in \mathcal{A}$. The weighted autocorrelation polynomial is then defined by protrusions, as in Note III.39 (p. 213). Multiple pattern occurrences can be also analysed.   ◁

IV. 7. Singularities and functional equations

In the various combinatorial examples discussed so far in this chapter, we have been dealing with functions that are given by explicit expressions. Such situations essentially cover non-recursive structures as well as the very simplest recursive ones, such as Catalan or Motzkin trees, whose generating functions are expressible in terms of radicals. In fact, as we shall see extensively in this book, complex analytic methods are instrumental in analysing coefficients of functions *implicitly* specified by functional equations. In other words: *the nature of a functional equation can often provide information regarding the singularities of its solution.* Chapter V will illustrate this philosophy in the case of rational functions defined by systems of positive equations; a very large number of examples will then be given in Chapters VI and VII, where singularities that are much more general than poles are treated.

In this section, we discuss three representative functional equations,

$$f(z) = ze^{f(z)}, \qquad f(z) = z + f(z^2 + z^3), \qquad f(z) = \frac{1}{1 - zf(z^2)},$$

associated, respectively, to Cayley trees, balanced 2–3 trees, and Pólya's alcohols. These illustrate the use of fundamental inversion or iteration properties for locating dominant singularities and derive exponential growth estimates of coefficients.

IV. 7.1. Inverse functions. We start with a generic problem already introduced on p. 249: given a function ψ analytic at a point y_0 with $z_0 = \psi(y_0)$ what can be said about its inverse, namely the solution(s) to the equation $\psi(y) = z$ when z is near z_0 and y near y_0?

Let us examine what happens when $\psi'(y_0) \neq 0$, first without paying attention to analytic rigour. One has locally ("$\approx$" means as usual "approximately equal")

$$(46) \qquad \psi(y) \approx \psi(y_0) + \psi'(y_0)(y - y_0),$$

so that the equation $\psi(y) = z$ should admit, for z near z_0, a solution satisfying

$$(47) \qquad y \approx y_0 + \frac{1}{\psi'(y_0)}(z - z_0).$$

If this is granted, the solution being locally linear, it is differentiable, hence analytic. The Analytic Inversion Lemma[10] provides a firm foundation for such calculations.

Lemma IV.2 (Analytic Inversion). *Let $\psi(z)$ be analytic at y_0, with $\psi(y_0) = z_0$. Assume that $\psi'(y_0) \neq 0$. Then, for z in some small neighbourhood Ω_0 of z_0, there exists an analytic function $y(z)$ that solves the equation $\psi(y) = z$ and is such that $y(z_0) = y_0$.*

Proof. (Sketch) The proof involves ideas analogous to those used to establish Rouché's Theorem and the Argument Principle (see especially the argument justifying Equation (42), p. 269). As a preliminary step, define the integrals ($j \in \mathbb{Z}_{\geq 0}$)

$$(48) \qquad \sigma_j(z) := \frac{1}{2i\pi} \int_\gamma \frac{\psi'(y)}{\psi(y) - z} y^j \, dy,$$

---

[10]A more general statement and several proof techniques are also discussed in Appendix B.5: *Implicit Function Theorem*, p. 753.

where γ is a small enough circle centred at y_0 in the y-plane.

First consider σ_0. This function satisfies $\sigma_0(z_0) = 1$ [by the Residue Theorem] and is a continuous function of z whose value can only be an integer, this value being the number of roots of the equation $\psi(y) = z$. Thus, for z close enough to z_0, one must have $\sigma_0(z) \equiv 1$. In other words, the equation $\psi(y) = z$ has exactly one solution, the function ψ is locally invertible and a solution $y = y(z)$ that satisfies $y(z_0) = y_0$ is well-defined.

Next examine σ_1. By the Residue Theorem once more, the integral defining $\sigma_1(z)$ is the sum of the roots of the equation $\psi(y) = z$ that lie inside γ, that is, in our case, the value of $y(z)$ itself. (This is also a particular case of Note IV.39, p. 270.) Thus, one has $\sigma_1(z) \equiv y(z)$. Since the integral defining $\sigma_1(z)$ depends analytically on z for z close enough to z_0, analyticity of $y(z)$ results. ∎

▷ **IV.42. Details.** Let ψ be analytic in an open disc D centred at y_0. Then, there exists a small circle γ centred at y_0 and contained in D such that $\psi(y) \neq y_0$ on γ. [Zeros of analytic functions are isolated, a fact that results from the definition of an analytic expansion]. The integrals $\sigma_j(z)$ are thus well defined for z restricted to be close enough to z_0, which ensures that there exists a $\delta > 0$ such that $|\psi(y) - z| > \delta$ for all $y \in \gamma$. One can then expand the integrand as a power series in $(z - z_0)$, integrate the expansion termwise, and form in this way the analytic expansions of σ_0, σ_1 at z_0. (This line of proof follows [334, I, §9.4].) ◁

▷ **IV.43. Inversion and majorant series.** The process corresponding to (46) and (47) can be transformed into a sound proof: first derive a formal power series solution, then verify that the formal solution is locally convergent using the method of majorant series (p. 250). ◁

The Analytic Inversion Lemma states the following: *An analytic function locally admits an analytic inverse near any point where its first derivative is non-zero.* However, as we see next, a function cannot be analytically inverted in a neighbourhood of a point where its first derivative vanishes.

Consider now a function $\psi(y)$ such that $\psi'(y_0) = 0$ but $\psi''(y_0) \neq 0$, then, by the Taylor expansion of ψ, one expects

$$(49) \qquad \psi(y) \approx \psi(y_0) + \frac{1}{2}(y - y_0)^2 \psi''(y_0).$$

Solving formally for y now indicates a *locally quadratic* dependency

$$(y - y_0)^2 \approx \frac{2}{\psi''(y_0)}(z - z_0),$$

and the inversion problem admits *two* solutions satisfying

$$(50) \qquad y \approx y_0 \pm \sqrt{\frac{2}{\psi''(y_0)}} \sqrt{z - z_0}.$$

What this informal argument suggests is that the solutions have a singularity at z_0, and, in order for them to be suitably specified, one must somehow restrict their domain of definition: the case of $\sqrt{z}$ (the root(s) of $y^2 - z = 0$) discussed on p. 230 is typical.

Given some point z_0 and a neighbourhood Ω of z_0, the *slit neighbourhood* along direction θ is the set

$$\Omega^{\backslash \theta} := \left\{ z \in \Omega \mid \arg(z - z_0) \neq \theta \bmod 2\pi, z \neq z_0 \right\}.$$

We state:

Lemma IV.3 (Singular Inversion). *Let $\psi(y)$ be analytic at y_0, with $\psi(y_0) = z_0$. Assume that $\psi'(y_0) = 0$ and $\psi''(y_0) \neq 0$. There exists a small neighbourhood Ω_0 of z_0 such that the following holds: for any fixed direction θ, there exist two functions, $y_1(z)$ and $y_2(z)$ defined on $\Omega_0^{\backslash\theta}$ that satisfy $\psi(y(z)) = z$; each is analytic in $\Omega_0^{\backslash\theta}$, has a singularity at the point z_0, and satisfies $\lim_{z \to z_0} y(z) = y_0$.*

Proof. (Sketch) Define the functions $\sigma_j(z)$ as in the proof of the previous lemma, Equation (48). One now has $\sigma_0(z) = 2$, that is, the equation $\psi(y) = z$ possesses *two* roots near y_0, when z is near z_0. In other words ψ effects a double covering of a small neighbourhood Ω of y_0 onto the image neighbourhood $\Omega_0 = \psi(\Omega) \ni z_0$. By possibly restricting Ω, we may furthermore assume that $\psi'(y)$ only vanishes at y_0 in Ω (zeros of analytic functions are isolated) and that Ω is simply connected.

Fix any direction θ and consider the slit neighbourhood $\Omega_0^{\backslash\theta}$. Fix a point ζ in this slit domain; it has two preimages, $\eta_1, \eta_2 \in \Omega$. Pick up the one named η_1. Since $\psi'(\eta_1)$ is non-zero, the Analytic Inversion Lemma applies: there is a local analytic inverse $y_1(z)$ of ψ. This $y_1(z)$ can then be uniquely continued[11] to the whole of $\Omega_0^{\backslash\theta}$, and similarly for $y_2(z)$. We have thus obtained two *distinct* analytic inverses.

Assume *a contrario* that $y_1(z)$ can be analytically continued at z_0. It would then admit a local expansion

$$y_1(z) = \sum_{n \geq 0} c_n (z - z_0)^n,$$

while satisfying $\psi(y_1(z)) = z$. But then, composing the expansions of ψ and y would entail

$$\psi(y_1(z)) = z_0 + O\left((z - z_0)^2\right) \qquad (z \to z_0),$$

which cannot coincide with the identity function (z). A contradiction has been reached. The point z_0 is thus a singular point for y_1 (as well as for y_2). $\blacksquare$

▷ **IV.44.** *Singular inversion and majorant series.* In a way that parallels Note IV.43, the process summarized by Equations (49) and (50) can be justified by the method of majorant series, which leads to an alternative proof of the Singular Inversion Lemma. ◁

▷ **IV.45.** *Higher order branch points.* If all derivatives of ψ till order $r - 1$ inclusive vanish at y_0, there are r inverses, $y_1(z), \ldots, y_r(z)$, defined over a slit neighbourhood of z_0. ◁

Tree enumeration. We can now consider the problem of obtaining information on the coefficients of a function $y(z)$ defined by an implicit equation

$$(51) \qquad y(z) = z\phi(y(z)),$$

when $\phi(u)$ is analytic at $u = 0$. In order for the problem to be well-posed (i.e., algebraically, in terms of formal power series, as well as analytically, near the origin, there should be a unique solution for $y(z)$), we assume that $\phi(0) \neq 0$. Equation (51) may then be rephrased as

$$(52) \qquad \psi(y(z)) = z \qquad \text{where} \qquad \psi(u) = \frac{u}{\phi(u)},$$

---

[11]The fact of slitting Ω_0 makes the resulting domain simply connected, so that analytic continuation becomes uniquely defined. In contrast, the punctured domain $\Omega_0 \setminus \{z_0\}$ is *not* simply connected, so that the argument cannot be applied to it. As a matter of fact, $y_1(z)$ gets continued to $y_2(z)$, when the ray of angle θ is crossed: the point z_0 where two determinations meet is a *branch point*.

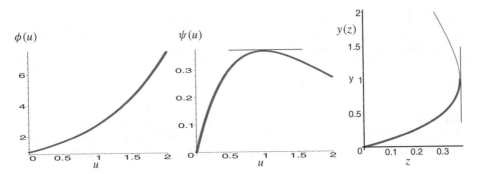

Figure IV.15. Singularities of inverse functions: $\phi(u) = e^u$ (left); $\psi(u) = u/\phi(u)$ (centre); $y = \mathrm{Inv}(\psi)$ (right).

so that it is in fact an instance of the inversion problem for analytic functions.

Equation (51) occurs in the counting of various types of trees, as seen in Subsections I. 5.1 (p. 65), II. 5.1 (p. 126), and III. 6.2 (p. 193). A typical case is $\phi(u) = e^u$, which corresponds to labelled non-plane trees (Cayley trees). The function $\phi(u) = (1+u)^2$ is associated to unlabelled plane binary trees and $\phi(u) = 1+u+u^2$ to unary–binary trees (Motzkin trees). A full analysis was developed by Meir and Moon [435], themselves elaborating on earlier ideas of Pólya [488, 491] and Otter [466]. In all these cases, the exponential growth rate of the number of trees can be automatically determined.

Proposition IV.5. *Let ϕ be a function analytic at 0, having non-negative Taylor co-efficients, and such that $\phi(0) \neq 0$. Let $R \leq +\infty$ be the radius of convergence of the series representing ϕ at 0. Under the condition,*

$$(53) \qquad \lim_{x \to R^-} \frac{x\phi'(x)}{\phi(x)} > 1,$$

there exists a unique solution $\tau \in (0, R)$ of the characteristic equation,

$$(54) \qquad \frac{\tau\phi'(\tau)}{\phi(\tau)} = 1.$$

Then, the formal solution $y(z)$ of the equation $y(z) = z\phi(y(z))$ is analytic at 0 and its coefficients satisfy the exponential growth formula:

$$[z^n]\, y(z) \bowtie \left(\frac{1}{\rho}\right)^n \qquad where \quad \rho = \frac{\tau}{\phi(\tau)} = \frac{1}{\phi'(\tau)}.$$

Note that condition (53) is automatically realized as soon as $\phi(R^-) = +\infty$, which covers our earlier examples as well as all the cases where ϕ is an entire function (e.g., a polynomial). Figure IV.15 displays graphs of functions on the real line associated to a typical inversion problem, that of Cayley trees, where $\phi(u) = e^u$.

Proof. By Note IV.46 below, the function $x\phi'(x)/\phi(x)$ is an increasing function of x for $x \in (0, R)$. Condition (53) thus guarantees the existence and unicity of a solution

Type	$\phi(u)$	(R)	τ	ρ	$y_n \bowtie \rho^{-n}$	
binary tree	$(1+u)^2$	(∞)	1	$\frac{1}{4}$	$y_n \bowtie 4^n$	(p. 67)
Motzkin tree	$1 + u + u^2$	(∞)	1	$\frac{1}{3}$	$y_n \bowtie 3^n$	(p. 68)
gen. Catalan tree	$\dfrac{1}{1-u}$	(1)	$\frac{1}{2}$	$\frac{1}{4}$	$y_n \bowtie 4^n$	(p. 65)
Cayley tree	e^u	(∞)	1	e^{-1}	$y_n \bowtie e^n$	(p. 128)

Figure IV.16. Exponential growth for classical tree families.

of the characteristic equation. (Alternatively, rewrite the characteristic equation as $\phi_0 = \phi_2 \tau^2 + 2\phi_3 \tau^3 + \cdots$, where the right side is clearly an increasing function.)

Next, we observe that the equation $y = z\phi(y)$ admits a unique formal power series solution, which furthermore has non-negative coefficients. (This solution can for instance be built by the method of indeterminate coefficients.) The Analytic Inversion Lemma (Lemma IV.2) then implies that this formal solution represents a function, $y(z)$, that is analytic at 0, where it satisfies $y(0) = 0$.

Now comes the hunt for singularities and, by Pringsheim's Theorem, one may restrict attention to the positive real axis. Let $r \le +\infty$ be the radius of convergence of $y(z)$ at 0 and set $y(r) := \lim_{x \to r^-} y(x)$, which is well defined (although possibly infinite), given positivity of coefficients. Our goal is to prove that $y(r) = \tau$.

— Assume *a contrario* that $y(r) < \tau$. One would then have $\psi'(y(r)) \neq 0$. By the Analytic Inversion Lemma, $y(z)$ would be analytic at r, a contradiction.
— Assume *a contrario* that $y(r) > \tau$. There would then exist $r^* \in (0, r)$ such that $\psi'(y(r^*)) = 0$. But then y would be singular at r^*, by the Singular Inversion Lemma, also a contradiction.

Thus, one has $y(r) = \tau$, which is finite. Finally, since y and ψ are inverse functions, one must have

$$r = \psi(\tau) = \tau/\phi(\tau) = \rho,$$

by continuity as $x \to r^-$, which completes the proof. ∎

Proposition IV.5 thus yields an *algorithm* that produces the exponential growth rate associated to tree functions. This rate is itself invariably a computable number as soon as ϕ is computable (i.e., its sequence of coefficients is computable). This computability result complements Theorem IV.8 (p. 251), which is relative to non-recursive structures only.

As an example of application of Proposition IV.5, general Catalan trees correspond to $\phi(y) = (1 - y)^{-1}$, whose radius of convergence is $R = 1$. The characteristic equation is $\tau/(1 - \tau) = 1$, which implies $\tau = 1/2$ and $\rho = 1/4$. We obtain (not a surprise!) $y_n \bowtie 4^n$, a weak asymptotic formula for the Catalan numbers. Similarly, for Cayley trees, $\phi(u) = e^u$ and $R = +\infty$. The characteristic equation reduces to $(\tau - 1)e^\tau = 0$, so that $\tau = 1$ and $\rho = e^{-1}$, giving a weak form of Stirling's formula: $[z^n]y(z) = n^{n-1}/n! \bowtie e^n$. Figure IV.16 summarizes the application of the method to a few already encountered tree families.

As our previous discussion suggests, the dominant singularity of tree generating functions is, under mild conditions, of the square-root type. Such a singular behaviour can then be analysed by the methods of Chapter VI: the coefficients admit an asymptotic form

$$[z^n]\, y(z) \sim C \cdot \rho^{-n} n^{-3/2},$$

with a subexponential factor of the form $n^{-3/2}$; see Section VI. 7, p. 402.

▷ **IV.46.** *Convexity of GFs, Boltzmann models, and the Variance Lemma.* Let $\phi(z)$ be a non-constant analytic function with non-negative coefficients and a non-zero radius of convergence R, such that $\phi(0) \neq 0$. For $x \in (0, R)$ a parameter, define the *Boltzmann random variable* Ξ (of parameter x) by the property

$$(55) \qquad\qquad \mathbb{P}(\Xi = n) = \frac{\phi_n x^n}{\phi(x)}, \qquad \text{with} \quad \mathbb{E}(s^\Xi) = \frac{\phi(sx)}{\phi(x)}$$

the probability generating function of Ξ. By differentiation, the first two moments of Ξ are

$$\mathbb{E}(\Xi) = \frac{x\phi'(x)}{\phi(x)}, \qquad \mathbb{E}(\Xi^2) = \frac{x^2\phi''(x)}{\phi(x)} + \frac{x\phi'(x)}{\phi(x)}.$$

There results, for any non-constant GF ϕ, the general convexity inequality valid for $0 < x < R$:

$$(56) \qquad\qquad \frac{d}{dx}\left(\frac{x\phi'(x)}{\phi(x)}\right) > 0,$$

due to the fact that the variance of a non-degenerate random variable is always positive. Equivalently, the function $\log(\phi(e^t))$ is convex for $t \in (-\infty, \log R)$. (In statistical physics, a Boltzmann model (of parameter x) corresponds to a class Φ (with OGF ϕ) from which elements are drawn according to the size distribution (55). An alternative derivation of (56) is given in Note VIII.4, p. 550.)   ◁

▷ **IV.47.** *A variant form of the inversion problem.* Consider the equation $y = z + \phi(y)$, where ϕ is assumed to have non-negative coefficients and be entire, with $\phi(u) = O(u^2)$ at $u = 0$. This corresponds to a simple variety of trees in which trees are counted by the number of their leaves only. For instance, we have already encountered labelled hierarchies (phylogenetic trees in Section II. 5, p. 128) corresponding to $\phi(u) = e^u - 1 - u$, which gives rise to one of "Schröder's problems". Let τ be the root of $\phi'(\tau) = 1$ and set $\rho = \tau - \phi(\tau)$. Then, $[z^n]y(z) \bowtie \rho^{-n}$. For the EGF L of labelled hierarchies ($L = z + e^L - 1 - L$), this gives $L_n/n! \bowtie (2\log 2 - 1)^{-n}$. (Observe that Lagrange inversion also provides $[z^n]y(z) = \frac{1}{n}[w^{n-1}](1 - y^{-1}\phi(y))^{-n}$.)   ◁

IV. 7.2. Iteration.

The study of iteration of analytic functions was launched by Fatou and Julia in the first half of the twentieth century. Our reader is certainly aware of the beautiful images associated with the name of Mandelbrot whose works have triggered renewed interest in these questions, now classified as belonging to the field of "complex dynamics" [31, 156, 443, 473]. In particular, the sets that appear in this context are often of a fractal nature. Mathematical objects of this sort are occasionally encountered in analytic combinatorics. We present here the first steps of a classic analysis of balanced trees published by Odlyzko [459] in 1982.

Example IV.12. *Balanced trees.* Consider the class $\mathcal{E}$ of *balanced 2–3 trees* defined as trees whose node degrees are restricted to the set $\{0, 2, 3\}$, with the additional property that all leaves are at the same distance from the root (Note I.67, p. 91). We adopt as notion of size the number

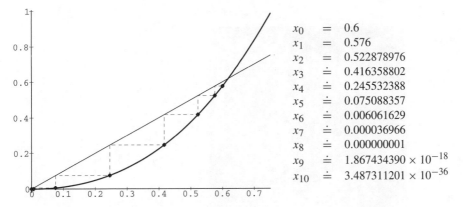

$$
\begin{aligned}
x_0 &= 0.6 \\
x_1 &= 0.576 \\
x_2 &= 0.522878976 \\
x_3 &\doteq 0.416358802 \\
x_4 &\doteq 0.245532388 \\
x_5 &\doteq 0.075088357 \\
x_6 &\doteq 0.006061629 \\
x_7 &\doteq 0.000036966 \\
x_8 &\doteq 0.000000001 \\
x_9 &\doteq 1.867434390 \times 10^{-18} \\
x_{10} &\doteq 3.487311201 \times 10^{-36}
\end{aligned}
$$

Figure IV.17. The iterates of a point $x_0 \in (0, \frac{1}{\varphi})$, here $x_0 = 0.6$, by $\sigma(z) = z^2 + z^3$ converge fast to 0.

of leaves (also called external nodes), the list of all 4 trees of size 8 being:

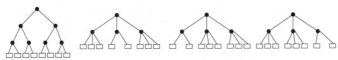

Given an existing tree, a new tree is obtained by substituting in all possible ways to each external node ($\square$) either a pair ($\square, \square$) or a triple ($\square, \square, \square$), and symbolically, one has

$$
\mathcal{E}[\square] = \square + \mathcal{E}\Big[\square \to (\square\square + \square\square\square)\Big].
$$

In accordance with the specification, the OGF of $\mathcal{E}$ satisfies the functional equation

(57) $$E(z) = z + E(z^2 + z^3),$$

corresponding to the seemingly innocuous recurrence

$$
E_n = \sum_{k=0}^{n} \binom{k}{n-2k} E_k \quad \text{with} \quad E_0 = 0,\ E_1 = 1.
$$

Let $\sigma(z) = z^2 + z^3$. Equation (57) can be expanded by iteration in the ring of formal power series,

(58) $$E(z) = z + \sigma(z) + \sigma^{[2]}(z) + \sigma^{[3]}(z) + \cdots,$$

where $\sigma^{[j]}(z)$ denotes the jth iterate of the polynomial σ: $\sigma^{[0]}(z) = z$, $\sigma^{[h+1]}(z) = \sigma^{[h]}(\sigma(z)) = \sigma(\sigma^{[h]}(z))$. Thus, $E(z)$ is nothing but the sum of all iterates of σ. The problem is to determine the radius of convergence of $E(z)$, and, by Pringsheim's theorem, the quest for dominant singularities can be limited to the positive real line.

For $z > 0$, the polynomial $\sigma(z)$ has a unique fixed point, $\rho = \sigma(\rho)$, at

$$
\rho = \frac{1}{\varphi} \quad \text{where} \quad \varphi = \frac{1 + \sqrt{5}}{2}
$$

is the golden ratio. Also, for any positive x satisfying $x < \rho$, the iterates $\sigma^{[j]}(x)$ do converge to 0; see Figure IV.17. Furthermore, since $\sigma(z) \sim z^2$ near 0, these iterates converge to 0 doubly

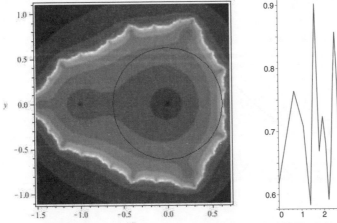
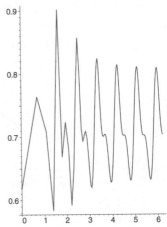

Figure IV.18. Left: the fractal domain of analyticity of $E(z)$ (inner domain in white and gray, with lighter areas representing slower convergence of the iterates of σ) and its circle of convergence. Right: the ratio $E_n/(\varphi^n n^{-1})$ plotted against $\log n$ for $n = 1 \ldots 500$ confirms that $E_n \bowtie \varphi^n$ and illustrates the periodic fluctuations of (60).

exponentially fast (Note IV.48). By the triangle inequality, we have $|\sigma(z)| \leq \sigma(|z|)$, so that the sum in (58) is a normally converging sum of analytic functions, and is thus itself analytic for $|z| < \rho$. Consequently, $E(z)$ is analytic in the whole of the open disc $|z| < \rho$.

It remains to prove that the radius of convergence of $E(z)$ is exactly equal to ρ. To that purpose it suffices to observe that $E(z)$, as given by (58), satisfies

$$E(x) \to +\infty \qquad \text{as} \qquad x \to \rho^-.$$

Let N be an arbitrarily large but fixed integer. It is possible to select a positive x_N sufficiently close to ρ with $x_N < \rho$, such that the Nth iterate $\sigma^{[N]}(x_N)$ is larger than $\frac{1}{2}$ (the function $\sigma^{[N]}(x)$ admits ρ as a fixed point and it is continuous and increasing at ρ). Given the sum expression (58), this entails the lower bound $E(x_N) > \frac{N}{2}$ for such an $x_N < \rho$. Thus $E(x)$ is unbounded as $x \to \rho^-$ and ρ is a singularity.

The dominant positive real singularity of $E(z)$ is thus $\rho = \varphi^{-1}$, and the Exponential Growth Formula gives the following estimate.

Proposition IV.6. *The number of balanced 2–3 trees satisfies:*

$$(59) \qquad\qquad [z^n] E(z) \bowtie \left(\frac{1 + \sqrt{5}}{2} \right)^n.$$

It is notable that this estimate could be established so simply by a purely qualitative examination of the basic functional equation and of a fixed point of the associated iteration scheme.

The complete asymptotic analysis of the E_n requires the full power of singularity analysis methods to be developed in Chapter VI. Equation (60) below states the end result, which involves fluctuations that are clearly visible on Figure IV.18 (right). There is overconvergence of the representation (58), that is, convergence in certain domains beyond the disc of convergence of $E(z)$. Figure IV.18 (left) displays the domain of analyticity of $E(z)$ and reveals its fractal nature (compare with Figure VII.23, p. 536). .... ∎

▷ **IV.48.** *Quadratic convergence.* First, for $x \in [0, \frac{1}{2}]$, one has $\sigma(x) \leq \frac{3}{2}x^2$, so that $\sigma^{[j]}(x) \leq (3/2)^{2^j-1} x^{2^j}$. Second, for $x \in [0, A]$, where A is any number $< \rho$, there is a number k_A such that $\sigma^{[k_A]}(x) < \frac{1}{2}$, so that $\sigma^{[k]}(x) \leq (3/2)(3/4)^{2^{k-k_A}}$. Thus, for any $A < \rho$, the series of iterates of σ is quadratically convergent when $z \in [0, A]$. ◁

▷ **IV.49.** *The asymptotic number of 2–3 trees.* This analysis is from [459, 461]. The number of 2–3 trees satisfies asymptotically

$$(60) \qquad E_n = \frac{\varphi^n}{n}\Omega(\log n) + O\left(\frac{\varphi^n}{n^2}\right),$$

where Ω is a periodic function with mean value $(\varphi \log(4-\varphi))^{-1} \doteq 0.71208$ and period $\log(4-\varphi) \doteq 0.86792$. Thus oscillations are inherent in E_n; see Figure IV.18 (right). ◁

IV. 7.3. Complete asymptotics of a functional equation.

George Pólya (1887–1985) is mostly remembered by combinatorialists for being at the origin of Pólya theory, a branch of combinatorics that deals with the enumeration of objects invariant under symmetry groups. However, in his classic article [488, 491] which founded this theory, Pólya discovered at the same time a number of startling applications of complex analysis to asymptotic enumeration[12]. We detail one of these now.

Example IV.13. *Pólya's alcohols.* The combinatorial problem of interest here is the determination of the number M_n of chemical isomeres of alcohols $C_n H_{2n+1} OH$ without asymmetric carbon atoms. The OGF $M(z) = \sum_n M_n z^n$ that starts as (*EIS* **A000621**)

$$(61) \qquad M(z) = 1 + z + z^2 + 2z^3 + 3z^4 + 5z^5 + 8z^6 + 14z^7 + 23z^8 + 39z^9 + \cdots,$$

is accessible through a functional equation,

$$(62) \qquad M(z) = \frac{1}{1 - zM(z^2)}.$$

which we adopt as our starting point. Iteration of the functional equation leads to a continued fraction representation,

$$M(z) = \cfrac{1}{1 - \cfrac{z}{1 - \cfrac{z^2}{1 - \cfrac{z^4}{1 - \ddots}}}},$$

from which Pólya found:

Proposition IV.7. *Let* $M(z)$ *be the solution analytic around 0 of the functional equation*

$$M(z) = \frac{1}{1 - zM(z^2)}.$$

Then, there exist constants K, β, *and* $B > 1$, *such that*

$$M_n = K \cdot \beta^n \left(1 + O(B^{-n})\right), \qquad \beta \doteq 1.68136\,75244, \qquad K \doteq 0.36071\,40971.$$

---

[12]In many ways, Pólya can be regarded as the grandfather of the field of analytic combinatorics.

We offer two proofs. The first one is based on direct consideration of the functional equation and is of a fair degree of applicability. The second one, following Pólya, makes explicit a special linear structure present in the problem. As suggested by the main estimate, the dominant singularity of $M(z)$ is a simple pole.

First proof. By positivity of the functional equation, $M(z)$ dominates coefficientwise any GF $(1 - zM^{<m}(z^2))^{-1}$, where $M^{<m}(z) := \sum_{0 \le j < m} M_n z^n$ is the mth truncation of $M(z)$. In particular, one has the domination relation (use $M^{<2}(z) = 1 + z$)

$$M(z) \succeq \frac{1}{1 - z - z^3}.$$

Since the rational fraction has its dominant pole at $z \doteq 0.68232$, this implies that the radius ρ of convergence of $M(z)$ satisfies $\rho < 0.69$. In the other direction, since $M(z^2) < M(z)$ for $z \in (0, \rho)$, then, one has the numerical inequality

$$M(z) \le \frac{1}{1 - zM(z)}, \qquad 0 \le z < \rho.$$

This can be used to show (Note IV.50) that the Catalan generating function $C(z) = (1 - \sqrt{1 - 4z})/(2z)$ is a majorant of $M(z)$ on the interval $(0, \frac{1}{4})$, which implies that $M(z)$ is well defined and analytic for $z \in (0, \frac{1}{4})$. In other words, one has $\frac{1}{4} \le \rho < 0.69$. Altogether, the radius of convergence of M lies strictly between 0 and 1.

▷ **IV.50.** *Alcohols, trees, and bootstrapping.* Since $M(z)$ starts as $1 + z + z^2 + \cdots$ while $C(z)$ starts as $1 + z + 2z^2 + \cdots$, there is a small interval $(0, \epsilon)$ such that $M(z) \le C(z)$. By the functional equation of $M(z)$, one has $M(z) \le C(z)$ for z in the larger interval $(0, \sqrt{\epsilon})$. Bootstrapping then shows that $M(z) \le C(z)$ for $z \in (0, \frac{1}{4})$. ◁

Next, as $z \to \rho^-$, one must have $zM(z^2) \to 1$. (Indeed, if this was not the case, we would have $zM(z^2) < A < 1$ for some A. But then, since $\rho^2 < \rho$, the quantity $(1 - zM(z^2))^{-1}$ would be analytic at $z = \rho$, a clear contradiction.) Thus, ρ is determined implicitly by the equation

$$\rho M(\rho^2) = 1, \qquad 0 < \rho < 1.$$

One can then estimate ρ numerically (Note IV.51), and the stated value of $\beta = 1/\rho$ follows. (Pólya determined ρ to five decimals by hand!)

The previous discussion also implies that ρ is a pole of $M(z)$, which must be simple (since $\partial_z (zM(z^2))\big|_{z=\rho} > 0$). Thus

(63) $$M(z) \underset{z \to \rho}{\sim} K \frac{1}{1 - z/\rho}, \qquad K := \frac{1}{\rho M(\rho^2) + 2\rho^3 M'(\rho^2)}.$$

The argument shows at the same time that $M(z)$ is meromorphic in $|z| < \sqrt{\rho} \doteq 0.77$. That ρ is the only pole of $M(z)$ on $|z| = \rho$ results from the fact that $zM(z^2) = z + z^3 + \cdots$ can be subjected to the type of argument encountered in the context of the Daffodil Lemma (see the discussion of quasi-inverses in the proof of Proposition IV.3, p. 267). The translation of the singular expansion (63) then yields the statement.

▷ **IV.51.** *The growth constant of molecules.* The quantity ρ can be obtained as the limit of the ρ_m satisfying $\sum_{n=0}^m M_n \rho_m^{2n+1} = 1$, together with $\rho \in [\frac{1}{4}, 0.69]$. In each case, only a few of the M_n (provided by the functional equation) are needed. One obtains: $\rho_{10} \doteq 0.595$, $\rho_{20} \doteq 0.594756$, $\rho_{30} \doteq 0.59475397$, $\rho_{40} \doteq 0.594753964$. This algorithms constitutes a geometrically convergent scheme with limit $\rho \doteq 0.59475\,39639$. ◁

Second proof. First, a sequence of formal approximants follows from (62) starting with

$$1, \quad \frac{1}{1-z}, \quad \frac{1}{1-\dfrac{z}{1-z^2}} = \frac{1-z^2}{1-z-z^2}, \quad \frac{1}{1-\dfrac{z}{1-\dfrac{z^2}{1-z^4}}} = \frac{1-z^2-z^4}{1-z-z^2-z^4+z^5},$$

which permits us to compute any number of terms of the series $M(z)$. Closer examination of (62) suggests to set

$$M(z) = \frac{\psi(z^2)}{\psi(z)},$$

where $\psi(z) = 1 - z - z^2 - z^4 + z^5 - z^8 + z^9 + z^{10} - z^{16} + \cdots$. Back substitution into (62) yields

$$\frac{\psi(z^2)}{\psi(z)} = \frac{1}{1 - z\dfrac{\psi(z^4)}{\psi(z^2)}} \quad \text{or} \quad \frac{\psi(z^2)}{\psi(z)} = \frac{\psi(z^2)}{\psi(z^2) - z\psi(z^4)},$$

which shows $\psi(z)$ to be a solution of the functional equation

$$\psi(z) = \psi(z^2) - z\psi(z^4), \qquad \psi(0) = 1.$$

The coefficients of ψ satisfy the recurrence

$$\psi_{4n} = \psi_{2n}, \qquad \psi_{4n+1} = -\psi_n, \qquad \psi_{4n+2} = \psi_{2n+1}, \qquad \psi_{4n+3} = 0,$$

which implies that their values are all contained in the set $\{0, -1, +1\}$.

Thus, $M(z)$ appears to be the quotient of two function, $\psi(z^2)/\psi(z)$, each analytic in the unit disc, and $M(z)$ is meromorphic in the unit disc. A numerical evaluation then shows that $\psi(z)$ has its smallest positive real zero at $\rho \doteq 0.59475$, which is a simple root. The quantity ρ is thus a pole of $M(z)$ (since, numerically, $\psi(\rho^2) \neq 0$). Thus

$$M(z) \sim \frac{\psi(\rho^2)}{(z-\rho)\psi'(\rho)} \implies M_n \sim -\frac{\psi(\rho^2)}{\rho\psi'(\rho)} \left(\frac{1}{\rho}\right)^n.$$

Numerical computations then yield Pólya's estimate. Et voilà! . . . . ∎

The example of Pólya's alcohols is exemplary, both from a historical point of view and from a methodological perspective. As the first proof of Proposition IV.7 demonstrates, quite a lot of information can be pulled out of a functional equation without solving it. (A similar situation will be encountered in relation to coin fountains, Example V.9, p. 330.) Here, we have made great use of the fact that if $f(z)$ is analytic in $|z| < r$ and some *a priori* bounds imply the strict inequalities $0 < r < 1$, then one can regard functions like $f(z^2)$, $f(z^3)$, and so on, as "known" since they are analytic in the disc of convergence of f and even beyond, a situation also evocative of our earlier discussion of Pólya operators in Section IV. 4, p. 249. Globally, the lesson is that functional equations, even complicated ones, can be used to bootstrap the local singular behaviour of solutions, and one can often do so even in the absence of any explicit generating function solution. The transition from singularities to coefficient asymptotics is then a simple jump.

▷ **IV.52.** *An arithmetic exercise.* The coefficients $\psi_n = [z^n]\psi(z)$ can be characterized simply in terms of the binary representation of n. Find the asymptotic proportion of the ψ_n for $n \in [1 .. 2^N]$ that assume each of the values 0, $+1$, and -1. ◁

IV. 8. Perspective

In this chapter, we have started examining generating functions under a new light. Instead of being merely *formal algebraic* objects—power series—that encode *exactly* counting sequences, generating functions can be regarded as *analytic* objects—transformations of the complex plane—whose singularities provide a wealth of information concerning *asymptotic* properties of structures.

Singularities provide a royal road to coefficient asymptotics. We could treat here, with a relatively simple apparatus, singularities that are poles. In this perspective, the two main statements of this chapter are the theorems relative to the expansion of rational and meromorphic functions, (Theorems IV.9, p. 256, and IV.10, p. 258). These are classical results of analysis. Issai Schur (1875–1941) is to be counted among the very first mathematicians who recognized their rôle in combinatorial enumerations (denumerants, Example IV.6, p. 257). The complex analytic thread was developed much further by George Pólya in his famous paper of 1937 (see [488, 491]), which Read in [491, p. 96] describes as a "landmark in the history of combinatorial analysis". There, Pólya laid the groundwork of combinatorial chemistry, the enumeration of objects under group actions, and, last but not least, the complex asymptotic theory of graphs and trees. Thanks to complex analytic methods, many combinatorial classes amenable to symbolic descriptions can be thoroughly analysed, with regard to their asymptotic properties, by means of a selected collection of basic theorems of complex analysis. The case of structures such as balanced trees and molecules, where only a functional equation of sorts is available, is exemplary.

The present chapter then serves as the foundation stone of a rich theory to be developed in future chapters. Chapter V will elaborate on the analysis of rational and meromorphic functions, and present a coherent theory of paths in graphs, automata, and transfer matrices in the perspective of analytic combinatorics. Next, the method of singularity analysis developed in Chapter VI considerably extends the range of applicability of the Second Principle to functions having singularities appreciably more complicated that poles (e.g., those involving fractional powers, logarithms, iterated logarithms, and so on). Applications will be given to recursive structures, including many types of trees, in Chapter VII. Chapter VIII, dedicated to saddle-point methods will then complete the picture of univariate asymptotics by providing a unified treatment of counting GFs that are either entire functions (hence, have no singularity at a finite distance) or manifest a violent growth at their singularities (hence, fall outside of the scope of meromorphic or singularity-analysis asymptotics). Finally, in Chapter IX, the corresponding perturbative methods will be put to use in order to distil limit laws for parameters of combinatorial structures.

Bibliographic notes. This chapter has been designed to serve as a refresher of basic complex analysis, with special emphasis on methods relevant for analytic combinatorics. See Figure IV.19 for a concise summary of results. References most useful for the discussion given here include the books of Titchmarsh [577] (oriented towards classical analysis), Whittaker and Watson [604] (stressing special functions), Dieudonné [165], Hille [334], and Knopp [373]. Henrici [329] presents complex analysis under the perspective of constructive and numerical methods, a highly valuable point of view for this book.

Basics. The theory of analytic functions benefits from the equivalence between two notions, analyticity and differentiability. It is the basis of a powerful integral calculus, much different from its real variable counterpart. The following two results can serve as "axioms" of the theory.

THEOREM IV.1 [Basic Equivalence Theorem] (p. 232): Two fundamental notions are equivalent, namely, analyticity (defined by convergent power series) and holomorphy (defined by differentiability). Combinatorial generating functions, *a priori* determined by their expansions at 0 thus satisfy the rich set of properties associated with these two equivalent notions.

THEOREM IV.2 [Null Integral Property] (p. 234): The integral of an analytic function along a simple loop (closed path that can be contracted to a single point) is 0. Consequently, integrals are largely independent of particular details of the integration contour.

Residues. For meromorphic functions (functions with poles), residues are essential. Coefficients of a function can be evaluated by means of integrals. The following two theorems provide connections between local properties of a function (e.g., coefficients at one point) and global properties of the function elsewhere (e.g., an integral along a distant curve).

THEOREM IV.3 [Cauchy's residue theorem] (p. 234): In the realm of meromorphic functions, integrals of a function can be evaluated based on local properties of the function at a few specific points, its poles.

THEOREM IV.4 [Cauchy's Coefficient Formula] (p. 237): This is an almost immediate consequence of Cauchy's residue theorem: The coefficients of an analytic function admit of a representation by a contour integral. Coefficients can then be evaluated or estimated using properties of the function at points away from the origin.

Singularities and growth. Singularities (places where analyticity stops), provide essential information on the growth rate of a function's coefficients. The "First Principle" relates the exponential growth rate of coefficients to the location of singularities.

THEOREM IV.5 [Boundary singularities] (p. 240): A function (given by its series expansion at 0) always has a singularity on the boundary of its disc of convergence.

THEOREM IV.6 [Pringsheim's Theorem] (p. 240): This theorem refines the previous one for functions with non-negative coefficients. It implies that, in the case of combinatorial generating functions, the search for a dominant singularity can be restricted to the positive real axis.

THEOREM IV.7 [Exponential Growth Formula] (p. 244): The exponential growth rate of coefficients is dictated by the *location* of the singularities nearest to the origin—the *dominant* singularities.

THEOREM IV.8 [Computability of growth] (p. 251): For any combinatorial class that is non-recursive (iterative), the exponential growth rate of coefficients is invariably a computable number. This statement can be regarded as the first general theorem of analytic combinatorics.

Coefficient asymptotics. The "Second Principle" relates subexponential factors of coefficients to the nature of singularities. For rational and meromorphic functions, everything is simple.

THEOREM IV.9 [Expansion of rational functions] (p. 256): Coefficients of rational functions are explicitly expressible in terms of the poles, given their location (values) and nature (multiplicity).

THEOREM IV.10 [Expansion of meromorphic functions] (p. 258): Coefficients of meromorphic functions admit of a precise asymptotic form with exponentially small error terms, given the location and nature of the dominant poles.

Figure IV.19. A summary of the main results of Chapter IV.

De Bruijn's classic booklet [143] is a wonderfully concrete introduction to effective asymptotic theory, and it contains many examples from discrete mathematics thoroughly worked out using a complex analytic approach. The use of such analytic methods in combinatorics was pioneered in modern times by Bender and Odlyzko, whose first publications in this area go back to the 1970s. The state of affairs in 1995 regarding analytic methods in combinatorial enumeration is superbly summarized in Odlyzko's scholarly chapter [461]. Wilf devotes Chapter 5 of his *Generatingfunctionology* [608] to this question. The books by Hofri [335], Mahmoud [429], and Szpankowski [564] contain useful accounts in the perspective of analysis of algorithms. See also our book [538] for a light introduction and the chapter by Vitter and Flajolet [598] for more on this specific topic.

> *Despite all appearances they [generating functions] belong to algebra and not to analysis.*
>
> *Combinatorialists use recurrence, generating functions, and such transformations as the Vandermonde convolution; others to my horror, use contour integrals, differential equations, and other resources of mathematical analysis.*
>
> — JOHN RIORDAN [513, p. viii] and [512, Pref.]

V

Applications of Rational and Meromorphic Asymptotics

Analytic methods are extremely powerful and when they apply,
they often yield estimates of unparalleled precision.

— ANDREW ODLYZKO [461]

The primary goal of this chapter is to provide combinatorial illustrations of the power of complex analytic methods, and specifically of the rational–meromorphic framework developed in the previous chapter. At the same time, we shift gears and envisage counting problems at a *new level of generality*. Precisely, we organize combinatorial problems into wide *families* of combinatorial types amenable to a common treatment and associated with a common collection of asymptotic properties. Without attempting a formal definition, we call *schema* any such family determined by combinatorial and analytic conditions that covers an infinity of combinatorial classes.

First, we discuss a general schema of analytic combinatorics known as the *supercritical sequence* schema, which provides a neat illustration of the power of meromorphic asymptotics (Theorem IV.10, p. 258), while being of wide applicability. This schema unifies the analysis of compositions, surjections, and alignments; it applies to any class which is defined as a sequence, provided components satisfy a simple analytic condition ("supercriticality"). For instance, one can predict very precisely (and easily) the number of ways in which an integer can be decomposed additively as a sum of primes (or twin primes), this even though many details of the distribution of primes are still surrounded in mystery.

The next schema comprises *regular specifications* and languages, which *a priori* lead to rational generating functions and are thus systematically amenable to Theorem IV.9 (p. 256), to the effect that coefficients are described as exponential polynomials. In the case of regular specifications, much additional structure is present, especially positivity. Accordingly, counting sequences are of a simple exponential–polynomial form and fluctuations can be systematically circumvented. Applications presented in this chapter include the analysis of longest runs, attached to maximal sequences of good (or bad) luck in games of chance, pure birth processes, and the occurrence of hidden patterns (subsequences) in random texts.

We then consider an important subset of regular specifications, corresponding to *nested sequences*, that combinatorially describe a variety of lattice paths. Such nested sequences naturally lead to nested quasi-inverses, which are none other than *continued fractions*. A wealth of combinatorial, algebraic, and analytic properties then surround such constructions. A prime illustration is the complete analysis of height in Dyck paths and general Catalan trees; other interesting applications relate to coin fountain and interconnection networks.

Finally, the last two sections examine *positive linear systems of generating functions*, starting with the simplest case of finite graphs and automata, and concluding with the general framework of transfer matrices. Although the resulting generating functions are once more bound to be rational, there is benefit in examining them as defined implicitly (rather than solving explicitly) and work out singularities directly. The spectrum of matrices (the set of eigenvalues) then plays a central rôle. An important case is the *irreducible linear system* schema, which is closely related to the Perron–Frobenius theory of non-negative matrices, whose importance has been long recognized in the theory of finite Markov chains. A general discussion of singularities can then be conducted, leading to valuable consequences on a variety of models— paths in graphs, finite automata, and transfer matrices. The last example discussed in this chapter treats locally constrained permutations, where rational functions combined with inclusion–exclusion provide an entry to the world of value-constrained permutations.

In the various combinatorial examples encountered in this chapter, the generating functions are meromorphic in some domain extending beyond their disc of convergence at 0. As a consequence, the asymptotic estimates of coefficients involve main terms that are explicit exponential–polynomials and error terms that are exponentially smaller. This is a situation well summarized by Odlyzko's aphorism quoted on p. 289: "*Analytic methods [...] often yield estimates of unparalleled precision*".

V. 1. A roadmap to rational and meromorphic asymptotics

The key character in this chapter is the combinatorial sequence construction SEQ. Since its translation into generating functions involves a *quasi-inverse*, $(1 - f)^{-1}$, the construction should in many cases be expected to induce polar singularities. Also, linear systems of equations, of which the simplest case is $X = 1 + AX$, are solvable by means of inverses: the solution is $X = (1 - A)^{-1}$ in the scalar case, and it is otherwise expressible as a quotient of determinants (by Cramer's rule) in the matrix case. Consequently, linear systems of equations are also conducive to polar singularities.

This chapter accordingly develops along two main lines. First, we study non-recursive families of combinatorial problems that are, in a suitable sense, driven by a sequence construction (Sections V. 2–V. 4). Second, we examine families of recursive problems that are naturally described by linear systems of equations (Sections V. 5–V. 6). Clearly, the general theorems giving the asymptotic forms of coefficients of rational and meromorphic functions apply. As we shall see, the additional positivity structure arising from combinatorics entails notable simplifications in the asymptotic form of counting sequences.

The supercritical sequence schema. This schema, fully described in Section V. 2 (p. 293) corresponds to the general form $\mathcal{F} = \text{SEQ}(\mathcal{G})$, together with a simple analytic condition, *"supercriticality"*, attached to the generating function $G(z)$ of $\mathcal{G}$. Under this condition, the sequence (F_n) happens to be predictable and an asymptotic estimate,

$$(1) \qquad F_n = c\beta^n + O(B^n), \qquad 0 \le B < \beta, \quad c \in \mathbb{R}_{>0},$$

applies with β such that $G(1/\beta) = 1$. Integer compositions, surjections, and alignments presented in Chapters I and II can then be treated in a unified manner. The supercritical sequence schema even covers situations where $\mathcal{G}$ is *not* necessarily constructible: this includes compositions into summands that are prime numbers or twin primes. Parameters, like the number of components and more generally profiles, are under these circumstances governed by laws that hold with a high probability.

Regular specification and languages. This topic is treated in Section V. 3 (p. 300). Regular specifications are non-recursive specifications that only involve the constructions $(+, \times, \text{SEQ})$. In the unlabelled case, they can always be interpreted as describing a regular language in the sense of Section I. 4, p. 49. The main result here is the following: given a regular specification $\mathcal{R}$, it is possible to determine constructively a number D, so that an asymptotic estimate of the form

$$(2) \qquad R_n = P(n)\beta^n + O(B^n), \qquad 0 \le B < \beta, \quad P \text{ a polynomial},$$

holds, once the index n is restricted to a fixed congruence class modulo D. (Naturally, the quantities P, β, B may depend on the particular congruence class considered.) In other words, a *"pure" exponential polynomial form* holds for each of the D "sections" [subsequences defined on p. 302] of the counting sequence $(R_n)_{n \ge 0}$. In particular, irregular fluctuations, which might otherwise arise from the existence of several dominant poles sharing the same modulus but having incommensurable arguments (see the discussion in Subsection IV. 6.1, p. 263 dedicated to multiple singularities), are simply not present in regular specifications and languages. Similar estimates hold for *profiles* of regular specifications, where the profile of an object is understood to be the number of times any fixed construction is employed.

Nested sequences, lattice paths, and continued fractions. The material considered in Section V. 4 (p. 318) could be termed the $\text{SEQ} \circ \cdots \circ \text{SEQ}$ schema, corresponding to nested sequences. The associated GFs are chains of quasi-inverses; that is, continued fractions. Although the general theory of regular specifications applies, the additional structure resulting from nested sequences implies, in essence, uniqueness and simplicity of the dominant pole, resulting directly in an estimate of the form

$$(3) \qquad S_n = c\beta^n + O(B^n), \qquad 0 \le B < \beta, \quad c \in \mathbb{R}_{>0},$$

for objects enumerated by nested sequences. This schema covers lattice paths of bounded height, their weighted versions, as well as several other bijectively equivalent classes, like interconnection networks. In each case, profiles can be fully characterized, the estimates being of a simple form.

Paths in graphs and automata. The framework of paths in directed graphs expounded in Section V.5 (p. 336) is of considerable generality. In particular, it covers the case of finite automata introduced in Subsection I.4.2, p. 56. Although, in the abstract, the descriptive power of this framework is formally equivalent to the one of regular specifications (Appendix A.7: *Regular languages*, p. 733), there is great advantage in considering directly problems whose natural formulation is recursive and phrased in terms of graphs or automata. (The reduction of automata to regular expressions is non-trivial so that it does not tend to preserve the original combinatorial structure.) The algebraic theory is that of matrices of the form $(I - zT)^{-1}$, where T is a matrix with non-negative entries. The analytic theory behind the scene is now that of positive matrices and the companion *Perron–Frobenius theory*. Uniqueness and simplicity of dominant poles of generating functions can be guaranteed under easily testable structural conditions—principally, the condition of *irreducibility* that corresponds to a strong connectedness of the system. Then a pure exponential polynomial form holds,

$$(4) \qquad C_n \sim c\lambda_1^n + O(\Lambda^n), \qquad 0 \le \Lambda < \lambda_1, \quad c \in \mathbb{R}_{>0},$$

where λ_1 is the (unique) dominant eigenvalue of the transition matrix T. Applications include walks over various types of graphs (the interval graph, the devil's staircase) and words excluding one or several patterns (walks on the De Bruijn graph).

Transfer matrices. This framework, whose origins lie in statistical physics, is an extension of automata and paths in graphs. What is retained is the notion of a finite state system, but transitions can now take place at different speeds. Algebraically, one is dealing with matrices of the form $(I - T(z))^{-1}$, where T is a matrix whose entries are polynomials (in z) with non-negative coefficients. Perron–Frobenius theory can be adapted to cover such cases, that, to a probabilist, look like a mixture of Markov chain and renewal theory. The consequence, for this category of models, is once more an estimate of the type (4), under irreducibility conditions; namely

$$(5) \qquad D_n \sim c\mu_1^n + O(M^n), \qquad 0 \le M < \mu_1, \quad c \in \mathbb{R}_{>0},$$

where $\mu_1 = 1/\sigma$ and σ is the smallest positive value of z such that $T(z)$ has dominant eigenvalue 1. A striking application of transfer matrices is a study, with an experimental mathematics flavour, of self-avoiding walks and polygons in the plane: it turns out to be possible to predict, with a high degree of confidence (but no mathematical certainty, yet), what the number of polygons is and which distribution of area is to be expected. A combination of the transfer matrix approach with a suitable use of inclusion–exclusion (Subsection V.6.4, p. 367) finally provides a solution to the classic *ménage problem* of combinatorial theory as well as to many related questions regarding value-constrained permutations.

Browsing notes. We, authors, recommend that our gentle reader first gets a bird's eye view of this chapter, by skimming through sections, before descending to ground level and studying examples in detail—some of the latter are indeed somewhat technically advanced (e.g., they make use of Mellin transforms and/or develop limit laws). The contents of this chapter are not needed for Chapters VI–VIII, so that the reader who is impatient to penetrate further the logic of analytic combinatorics can at any time have a peek at Chapters VI–VIII. We shall see in Chapter IX (specifically, Section IX. 6, p. 650) that all the schemas considered here are, under simple non-degeneracy conditions, associated to Gaussian limit laws.

Sections V. 2 to V. 6 are organized following a common pattern: first, we discuss "combinatorial aspects", then "analytic aspects", and finally "applications". Each of Sections V. 2 to V. 5 is furthermore centred around two analytic–combinatorial theorems, one describing *asymptotic enumeration*, the other quantifying the *asymptotic profiles* of combinatorial structures. We examine in this way the supercritical sequence schema (Section V. 2), general regular specifications (Section V. 3), nested sequences (Section V. 4), and path-in-graphs models (Section V. 5). The last section (Section V. 6) departs slightly from this general pattern, since transfer matrices are reducible rather simply to the framework of paths in graphs and automata, so that we do not need specifically new statements.

V. 2. The supercritical sequence schema

This schema is combinatorially the simplest treated in this chapter, since it plainly deals with the sequence construction. An auxiliary analytic condition, named "supercriticality" ensures that meromorphic asymptotics applies and entails strong statistical regularities. The paradigm of supercritical sequences unifies the asymptotic properties of a number of seemingly different combinatorial types, including integer compositions, surjections, and alignments.

V. 2.1. Combinatorial aspects. We consider a sequence construction, which may be taken in either the unlabelled or the labelled universe. In either case, we have

$$\mathcal{F} = \mathrm{SEQ}(\mathcal{G}) \quad \Longrightarrow \quad F(z) = \frac{1}{1 - G(z)},$$

with $G(0) = 0$. It will prove convenient to set

$$f_n = [z^n]F(z), \qquad g_n = [z^n]G(z),$$

so that the number of $\mathcal{F}_n$ structures is f_n in the unlabelled case and $n! f_n$ otherwise.

From Chapter III, the BGF of $\mathcal{F}$–structures with u marking the number of $\mathcal{G}$–components is

(6) $$\mathcal{F} = \mathrm{SEQ}(u\mathcal{G}) \quad \Longrightarrow \quad F(z, u) = \frac{1}{1 - uG(z)}.$$

We also have access to the BGF of $\mathcal{F}$ with u marking the number of $\mathcal{G}_k$–components:

(7) $$\mathcal{F}^{\langle k \rangle} = \mathrm{SEQ}\left(u\mathcal{G}_k + (\mathcal{G} \setminus \mathcal{G}_k)\right) \quad \Longrightarrow \quad F^{\langle k \rangle}(z, u) = \frac{1}{1 - \left(G(z) + (u - 1)g_k z^k\right)}.$$

V. 2.2. Analytic aspects. We restrict attention to the case where the radius of convergence ρ of $G(z)$ is non-zero, in which case, the radius of convergence of $F(z)$ is also non-zero by virtue of closure properties of analytic functions. Here is the basic concept of this section.

Definition V.1. *Let F, G be generating functions with non-negative coefficients that are analytic at 0, with $G(0) = 0$. The analytic relation $F(z) = (1 - G(z))^{-1}$ is said to be* supercritical *if $G(\rho) > 1$, where $\rho = \rho_G$ is the radius of convergence of G. A combinatorial schema $\mathcal{F} = \text{SEQ}(\mathcal{G})$ is said to be supercritical if the relation $F(z) = (1 - G(z))^{-1}$ between the corresponding generating functions is supercritical.*

Note that $G(\rho)$ is well defined in $\mathbb{R} \cup \{+\infty\}$ as the limit $\lim_{x \to \rho^-} G(x)$ since $G(x)$ increases along the positive real axis, for $x \in (0, \rho)$. (The value $G(\rho)$ corresponds to what has been denoted earlier by τ_G when discussing "signatures" in Section IV. 4, p. 249.) From now on we assume that $G(z)$ is *strongly aperiodic* in the sense that there does not exist an integer $d \geq 2$ such that $G(z) = h(z^d)$ for some h analytic at 0. (Put otherwise, the span of $1 + G(z)$, as defined on p. 266, is equal to 1.) This condition entails no loss of analytic generality.

Theorem V.1 (Asymptotics of supercritical sequence). *Let the schema $\mathcal{F} = \text{SEQ}(\mathcal{G})$ be supercritical and assume that $G(z)$ is strongly aperiodic. Then, one has*

$$[z^n] F(z) = \frac{1}{\sigma G'(\sigma)} \cdot \sigma^{-n} \left(1 + O(A^n)\right),$$

where σ is the root in $(0, \rho_G)$ of $G(\sigma) = 1$ and A is a number less than 1. The number X of $\mathcal{G}$–components in a random $\mathcal{F}$–structure of size n has mean and variance satisfying

$$\mathbb{E}_n(X) = \frac{1}{\sigma G'(\sigma)} \cdot (n + 1) - 1 + \frac{G''(\sigma)}{G'(\sigma)^2} + O(A^n)$$

$$\mathbb{V}_n(X) = \frac{\sigma G''(\sigma) + G'(\sigma) - \sigma G'(\sigma)^2}{\sigma^2 G'(\sigma)^3} \cdot n + O(1).$$

In particular, the distribution of X on $\mathcal{F}_n$ is concentrated.

Proof. See also [260, 547]. The basic observation is that G increases continuously from $G(0) = 0$ to $G(\rho_G) = \tau_G$ (with $\tau_G > 1$ by assumption) when x increases from 0 to ρ_G. Therefore, the positive number σ, which satisfies $G(\sigma) = 1$ is well defined. Then, F is analytic at all points of the interval $(0, \sigma)$. The function G being analytic at σ, satisfies, in a neighbourhood of σ

$$G(z) = 1 + G'(\sigma)(z - \sigma) + \frac{1}{2!} G''(\sigma)(z - \sigma)^2 + \cdots .$$

so that $F(z)$ has a pole at $z = \sigma$; also, this pole is simple since $G'(\sigma) > 0$, by positivity of the coefficients of G. Thus, we have

$$F(z) \underset{z \to \rho}{\sim} - \frac{1}{G'(\sigma)(z - \sigma)} \equiv \frac{1}{\sigma G'(\sigma)} \frac{1}{1 - z/\sigma}.$$

Pringsheim's theorem (Theorem IV.6, p. 240) then implies that the radius of convergence of F must coincide with σ.

There remains to show that $F(z)$ is meromorphic in a disc of some radius $R > \sigma$ with the point σ as the only singularity inside the disc. This results from the assumption that G is strongly aperiodic. In effect, as a consequence of the Daffodil Lemma (Lemma IV.3, p. 267), one has $G(\sigma e^{i\theta}) \neq 1$, for all $\theta \not\equiv 0 \pmod{2\pi}$. Thus, by compactness, there exists a closed disc of radius $R > \sigma$ in which F is analytic except for a unique pole at σ. We can now apply the main theorem of meromorphic function asymptotics (Theorem IV.10, p. 258) to deduce the stated formula with $A = \sigma/R$.

Next, the number of $\mathcal{G}$–components in a random $\mathcal{F}$ structure of size n has BGF given by (6), and by differentiation, we get

$$\mathbb{E}_n(X) = \frac{1}{f_n} [z^n] \left. \frac{\partial}{\partial u} \frac{1}{1 - uG(z)} \right|_{u=1} = \frac{1}{f_n} [z^n] \frac{G(z)}{(1 - G(z))^2}.$$

The problem is now reduced to extracting coefficients in a univariate generating function with a double pole at $z = \sigma$, and it suffices to expand the GF locally at σ:

$$\frac{G(z)}{(1 - G(z))^2} \underset{z \to \rho}{\sim} \frac{1}{G'(\sigma)^2(z - \sigma)^2} \equiv \frac{1}{\sigma^2 G'(\sigma)^2} \frac{1}{(1 - z/\sigma)^2}.$$

The variance calculation is similar, with a triple pole being involved. ∎

When a sequence construction is supercritical, the number of components is in the mean of order n while its standard deviation is $O(\sqrt{n})$. Thus, the distribution is concentrated (in the sense of Section III. 2.2, p. 161). In fact, there results from a general theorem of Bender [35] that the distribution of the number of components is asymptotically Gaussian, a property to be established in Section IX. 6, p. 650.

Profiles of supercritical sequences. We have seen in Chapter III that integer compositions and integer partitions, when sampled at random, tend to assume rather different aspects. Given a sequence construction, $\mathcal{F} = \text{SEQ}(\mathcal{G})$, the profile of an element $\alpha \in \mathcal{F}$ is the vector $(X^{\langle 1 \rangle}, X^{\langle 2 \rangle}, \ldots)$ where $X^{\langle j \rangle}(\alpha)$ is the number of $\mathcal{G}$–components in α that have size j. In the case of (unrestricted) integer compositions, it could be proved elementarily (Example III.6, p. 167) that, on average, for size n, the number of 1-summands is $\sim n/2$, the number of 2-summands is $\sim n/4$, and so on. Now that meromorphic asymptotics is available, such a property can be placed in a much wider perspective.

Theorem V.2 (Profiles of supercritical sequences). *Consider a supercritical sequence construction, $\mathcal{F} = \text{SEQ}(\mathcal{G})$, with $G(z)$ strongly aperiodic, as in Theorem V.1. The number of $\mathcal{G}$–components of any fixed size k in a random $\mathcal{F}$–object of size n satisfies*

$$(8) \qquad \mathbb{E}_n(X^{\langle k \rangle}) = \frac{g_k \sigma^k}{\sigma G'(\sigma)} n + O(1), \quad \mathbb{V}_n(X^{\langle k \rangle}) = O(n),$$

where σ in $(0, \sigma_G)$ is such that $G(\sigma) = 1$, and $g_k = [z^k]G(z)$.

Proof. The BGF with u marking the number of $\mathcal{G}$–components of size k is given in (7). The mean value is then obtained as a quotient,

$$\mathbb{E}_n(X^{\langle k \rangle}) = \frac{1}{f_n} [z^n] \left. \frac{\partial}{\partial u} F(z, u) \right|_{u=1} = \frac{1}{f_n} [z^n] \frac{g_k z^k}{(1 - G(z))^2}.$$

The GF of cumulated values has a double pole at $z = \sigma$, and the estimate of the mean value follows. The variance is estimated similarly, after two successive differentiations and the analysis of a triple pole. ∎

The total number of components X satisfies $X = \sum X^{\langle k \rangle}$, and, by Theorem V.1, its mean is asymptotic to $n/(\sigma G'(\sigma))$. Thus, Equation (8) indicates that, at least in some average-value sense, the "proportion" of components of size k among all components is given by $g_k \sigma^k$.

▷ **V.1.** *Proportion of k–components and convergence in probability.* For any fixed k, the random variable $X_n^{\langle k \rangle}/X_n$ converges in probability to the value $g_k \sigma^k$,

$$\frac{X_n^{\langle k \rangle}}{X_n} \xrightarrow{P} g_k \sigma^k, \quad \text{i.e.,} \quad \lim_{n \to \infty} \mathbb{P}\left\{ g_k \sigma^k (1 - \epsilon) \leq \frac{X_n^{\langle k \rangle}}{X_n} \leq g_k \sigma^k (1 + \epsilon) \right\} = 1,$$

for any $\epsilon > 0$. The proof is an easy consequence of the Chebyshev inequalities (the distributions of X_n and $X_n^{\langle k \rangle}$ are both concentrated).   ◁

V.2.3. Applications.

We examine here two types of applications of the super-critical sequence schema. Example V.1 makes explicit the asymptotic enumeration and the analysis of profiles of compositions, surjections and alignments. What stands out is the way the mean profile of a structure reflects the underlying inner construction $\mathfrak{K}$ in schemas of the form $\text{SEQ}(\mathfrak{K}(\mathcal{Z}))$. Example V.2 discusses compositions into restricted summands, including the striking case of compositions into primes.

Example V.1. *Compositions, surjections, and alignments.* The three classes of interest here are integer compositions ($\mathcal{C}$), surjections ($\mathcal{R}$) and alignments ($\mathcal{O}$), which are specified as

$$\mathcal{C} = \text{SEQ}(\text{SEQ}_{\geq 1}(\mathcal{Z})), \qquad \mathcal{R} = \text{SEQ}(\text{SET}_{\geq 1}(\mathcal{Z})), \qquad \mathcal{O} = \text{SEQ}(\text{CYC}(\mathcal{Z}))$$

and belong to either the labelled universe ($\mathcal{C}$) or to the labelled universe ($\mathcal{R}$ and $\mathcal{O}$). The generating functions (of type OGF, EGF, and EGF, respectively) are

$$C(z) = \frac{1}{1 - \frac{z}{1-z}}, \qquad R(z) = \frac{1}{1 - (e^z - 1)}, \qquad O(z) = \frac{1}{1 - \log(1 - z)^{-1}}.$$

A direct application of Theorem V.1 (p. 294) gives us back the known results

$$C_n = 2^{n-1}, \qquad \frac{1}{n!} R_n \sim \frac{1}{2} (\log 2)^{-n-1}, \qquad \frac{1}{n!} O_n = e^{-1}(1 - e^{-1})^{-n-1},$$

corresponding to σ equal to $\frac{1}{2}$, $\log 2$, and $1 - e^{-1}$, respectively.

Similarly, the expected number of summands in a random composition of the integer n is $\sim n/2$; the expected cardinality of the range of a random surjection whose domain has cardinality n is asymptotic to βn with $\beta = 1/(2 \log 2)$; the expected number of components in a random alignment of size n is asymptotic to $n/(e - 1)$.

Theorem V.2 also applies, providing the mean number of components of size k in each case. The following table summarizes the conclusions.

Structures	*specification*	*law* $(g_k \sigma^k)$	*type*	σ
Compositions	$\text{SEQ}(\text{SEQ}_{\geq 1}(\mathcal{Z}))$	$\dfrac{1}{2^k}$	Geometric	$\dfrac{1}{2}$
Surjections	$\text{SEQ}(\text{SET}_{\geq 1}(\mathcal{Z}))$	$\dfrac{1}{k!}(\log 2)^k$	Poisson	$\log 2$
Alignments	$\text{SEQ}(\text{CYC}(\mathcal{Z}))$	$\dfrac{1}{k}(1 - e^{-1})^k$	Logarithmic	$1 - e^{-1}$

Note that the stated laws necessitate $k \geq 1$. The geometric and Poisson law are classical; the *logarithmic distribution* (also called "logarithmic-series distribution") of a parameter $\lambda > 0$ is

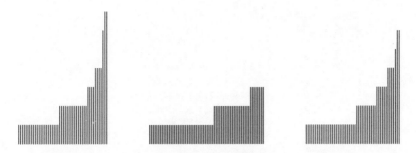

Figure V.1. Profile of structures drawn at random represented by the sizes of their components in sorted order: (from left to right) a random composition, surjection, and alignment of size $n = 100$.

by definition the law of a discrete random variable Y such that

$$\mathbb{P}(Y = k) = \frac{1}{\log(1 - \lambda)^{-1}} \frac{\lambda^k}{k}, \qquad k \geq 1.$$

The way the internal construction $\mathfrak{K}$ in the schema $\mathrm{SEQ}(\mathfrak{K}(\mathcal{Z}))$ determines the asymptotic proportion of component of each size,

Sequence $\mapsto$ Geometric;   Set $\mapsto$ Poisson;   Cycle $\mapsto$ Logarithmic,

stands out. Figure V.1 exemplifies the phenomenon by displaying components sorted by size and represented by vertical segments of corresponding lengths for three randomly drawn objects of size $n = 100$. .... ∎

Example V.2.   *Compositions with restricted summands, compositions into primes.* Unrestricted integer compositions are well understood as regards enumeration: their number is exactly $C_n = 2^{n-1}$, their OGF is $C(z) = (1 - z)/(1 - 2z)$, and compositions with k summands are enumerated by binomial coefficients. Such simple exact formulae disappear when restricted compositions are considered, but, as we now show, asymptotics is much more robust to changes in specifications.

Let S be a subset of the integers $\mathbb{Z}_{\geq 1}$ such that $\gcd(S) = 1$, i.e., not all members of S are multiples of a common divisor $d \geq 2$. In order to avoid trivialities, we also assume that S has at least two elements. The class $\mathcal{C}^S$ of compositions with summands constrained to the set S then satisfies:

$$\mathcal{C}^S = \mathrm{SEQ}(\mathrm{SEQ}_S(\mathcal{Z})) \qquad \Longrightarrow \qquad C^S(z) = \frac{1}{1 - S(z)}, \quad S(z) = \sum_{s \in S} z^s.$$

By assumption, $S(z)$ is strongly aperiodic, so that Theorem V.1 (p. 294) applies directly. There is a well-defined number σ such that

$$S(\sigma) = 1, \qquad 0 < \sigma < 1,$$

and the number of S–restricted compositions satisfies

(9) $$C_n^S := [z^n] C^S(z) = \frac{1}{\sigma S'(\sigma)} \cdot \sigma^{-n} \left(1 + O(A^n) \right).$$

Among the already discussed cases, $S = \{1, 2\}$ gives rise to Fibonacci numbers F_n and, more generally, $S = \{1, \ldots, r\}$ corresponds to partitions with summands at most r. In this case, the

10	16	15
20	732	73 **4**
30	36039	360 **57**
40	1772207	17722 **61**
50	87109263	871092 **48**
60	4281550047	42815 **49331**
70	210444532770	21044453 **0095**
80	10343662267187	1034366226 **5182**
90	508406414757253	5084064147 **81706**
100	24988932929490838	24988932929 **612479**

Figure V.2. The pyramid relative to compositions into prime summands for $n = 10 \ldots 100$: (left: exact values; right: asymptotic formula rounded).

OGF,

$$C^{\{1,\ldots,r\}}(z) = \frac{1}{1 - z\frac{1-z^r}{1-z}} = \frac{1-z}{1 - 2z + z^{r+1}}$$

is a simple variant of the OGF associated to longest runs in strings, which is studied at length in Example V.4, p. 308. The treatment of the latter can be copied almost verbatim to the effect that the largest component in a random composition of n is found to be $\log_2 n + O(1)$, both on average and with high probability.

Compositions into primes. Here is a surprising application of the general theory. Consider the case where S is taken to be the set of prime numbers, Prime $= \{2, 3, 5, 7, 11, \ldots\}$, thereby defining the class of *compositions into prime summands*. The sequence starts as

$$1, 0, 1, 1, 1, 3, 2, 6, 6, 10, 16, 20, 35, 46, 72, 105,$$

corresponding to $G(z) = z^2 + z^3 + z^5 + \cdots$, and is *EIS* **A023360** in Sloane's *Encyclopedia*. The formula (9) provides the asymptotic shape of the number of such compositions (Figure V.2). It is also worth noting that the constants appearing in (9) are easily determined to great accuracy, as we now explain.

By (9) and the preceding equation, the dominant singularity of the OGF of compositions into primes is the positive root $\sigma < 1$ of the characteristic equation

$$S(z) \equiv \sum_{p \text{ Prime}} z^p = 1.$$

Fix a threshold value m_0 (for instance $m_0 = 10$ or 100) and introduce the two series

$$S^-(z) := \sum_{s \in S, \ s < m_0} z^s, \quad S^+(z) := \left(\sum_{s \in S, \ s < m_0} z^s \right) + \frac{z^{m_0}}{1-z}.$$

Clearly, for $x \in (0,1)$, one has $S^-(x) < S(x) < S^+(x)$. Define then two constants σ^-, σ^+ by the conditions

$$S^-(\sigma^-) = 1, \quad S^+(\sigma^+) = 1, \quad 0 < \sigma^-, \sigma^+ < 1.$$

These constants are algebraic numbers that are accessible to computation. At the same time, they satisfy $\sigma^+ < \sigma < \sigma^-$. As the order of truncation, m_0, increases, the values of σ^+, σ^- provide better and better approximations to σ, together with an interval in which σ provably lies. For instance, $m_0 = 10$ is enough to determine that $0.66 < \sigma < 0.69$, and the choice

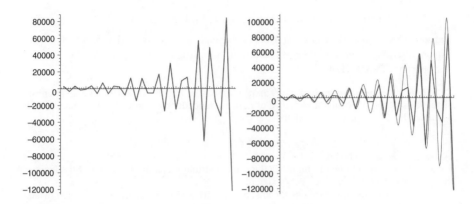

Figure V.3. Errors in the approximation of the number of compositions into primes for $n = 70 . . 100$: left, the values of $C_n^{\text{Prime}} - g(n)$; right, the correction arising from the next two poles, which are complex conjugate, and its continuous extrapolation $g_2(n)$, for $n \in [70, 100]$.

$m_0 = 100$ gives σ to 15 guaranteed digits of accuracy, namely, $\sigma \doteq 0.67740\,17761\,30660$. Then, the asymptotic formula (9) instantiates as

$$(10) \quad C_n^{\text{Prime}} \sim g(n), \qquad g(n) := \lambda \cdot \beta^n, \quad \lambda \doteq 0.30365\,52633, \quad \beta \doteq 1.47622\,87836.$$

(The constant $\beta \equiv \sigma^{-1} \doteq 1.47622$ is akin to the family of Backhouse constants described in [211].)

Once more, the asymptotic approximation is very good, as is exemplified by the "pyramid" of Figure V.2. The difference between C_n^{Prime} and its approximation $g(n)$ from Equation (10) is plotted on the left-hand part of Figure V.3. The seemingly haphazard oscillations that manifest themselves are well explained by the principles discussed in Section IV. 6.1 (p. 263). It appears that the next poles of the OGF are complex conjugate and lie near $-0.76 \pm 0.44i$, having modulus about 0.88. The corresponding residues then jointly contribute a quantity of the form

$$g_2(n) = c \cdot A^n \sin(\omega n + \omega_0), \qquad A \doteq 1.13290,$$

for some constants c, ω, ω_0. Comparing the left-hand and right-hand parts of Figure V.3, we see that this next layer of poles explains quite well the residual error $C_n^{\text{Prime}} - g(n)$.

Here is finally a variant of compositions into primes that demonstrates in a striking way the scope of the method. Define the set Prime_2 of "twinned primes" as the set of primes that belong to a twin prime pair, that is, $p \in \text{Prime}_2$ if one of $p - 2, p + 2$ is prime. The set Prime_2 starts as $3, 5, 7, 11, 13, 17, 19, 29, 31, \ldots$ (prime numbers like 23 or 37 are thus excluded). The asymptotic formula for the number of compositions of the integer n into summands that are twinned primes is

$$C_n^{\text{Prime}_2} \sim 0.18937 \cdot 1.29799^n,$$

where the constants are found by methods analogous to the case of all primes. It is quite remarkable that the constants involved are still computable real numbers (and of low complexity, even), this despite the fact that it is not known whether the set of twinned primes is finite or infinite. Incidentally, a sequence that starts like C_n^{Prime2},

$$1, 0, 0, 1, 0, 1, 1, 1, 2, 1, 3, 4, 3, 7, 7, 8, 14, 15, 21, 28, 33, 47, 58, \ldots$$

and coincides till index 22 included (!), but not beyond, was encountered by MacMahon[1], as the authors discovered, much to their astonishment, from scanning Sloane's *Encyclopedia*, where it appears as *EIS* **A002124**. .... ∎

▷ **V.2.** *Random generation of supercritical sequences.* Let $\mathcal{F} = \text{SEQ}(\mathcal{G})$ be a supercritical sequence scheme. Consider a sequence of i.i.d. (independently identically distributed) random variables $Y_1, Y_2, \ldots$ each of them obeying the discrete law

$$\mathbb{P}(Y = k) = g_k \sigma^k, \qquad k \geq 1.$$

A sequence is said to be hitting n if $Y_1 + \cdots + Y_r = n$ for some $r \geq 1$. The vector $(Y_1, \ldots, Y_r)$ for a sequence conditioned to hit n has the same distribution as the sequence of the lengths of components in a random $\mathcal{F}$–object of size n.

For probabilists, this explains the shape of the formulae in Theorem V.1, which resemble renewal relations [205, Sec. XIII.10]. It also implies that, given a uniform random generator for $\mathcal{G}$–objects, one can generate a random $\mathcal{F}$–object of size n in $O(n)$ steps on average [177]. This applies to surjections, alignments, and compositions in particular. ◁

▷ **V.3.** *Largest components in supercritical sequences.* Let $\mathcal{F} = \text{SEQ}(\mathcal{G})$ be a supercritical sequence. Assume that $g_k = [z^k]G(z)$ satisfies the asymptotic "smoothness" condition

$$g_k \underset{k \to \infty}{\sim} c\rho^{-k}k^\beta, \qquad c, \rho \in \mathbb{R}_{>0}, \ \beta \in \mathbb{R}.$$

Then the size L of the largest $\mathcal{G}$ component in a random $\mathcal{F}$–object satisfies, for size n,

$$\mathbb{E}_{\mathcal{F}_n}(L) = \frac{1}{\log(\rho/\sigma)} (\log n + \beta \log \log n) + o(\log \log n).$$

This covers integer compositions ($\rho = 1$, $\beta = 0$) and alignments ($\rho = 1$, $\beta = -1$). [The analysis generalizes the case of longest runs in Example V.4 (p. 308) and is based on similar principles. The GF of $\mathcal{F}$ objects with $L \leq m$ is $F^{\langle m \rangle}(z) = \left(1 - \sum_{k \leq m} g_k z^k\right)^{-1}$, according to Section III.7. For m large enough, this has a dominant singularity which is a simple pole at σ_m such that $\sigma_m - \sigma \sim c_1(\sigma/\rho)^m m^\beta$. There follows a double-exponential approximation

$$\mathbb{P}_{\mathcal{F}_n}(L \leq m) \approx \exp\left(-c_2 n m^\beta (\sigma/\rho)^m\right)$$

in the "central" region. See Example V.4 (p. 308) for a particular instance and Gourdon's study [305] for a general theory.] ◁

V. 3. Regular specifications and languages

The purpose of this section is the general study of the $(+, \times, \text{SEQ})$ schema, which covers all regular specifications. As we show now, "pure" exponential–polynomial forms (ones with a single dominating exponential) can always be extracted. Theorems V.3 and V.4 below provide a universal framework for the asymptotic analysis of regular classes. Additional structural conditions to be introduced in later sections (nested sequences, irreducibility of the dependency graph and of transfer matrices) will then be seen to induce further simplifications in asymptotic formulae.

---

[1]See "Properties of prime numbers deduced from the calculus of symmetric functions", *Proc. London Math. Soc.*, 23 (1923), 290-316). MacMahon's sequence corresponds to compositions into arbitrary *odd* primes, and 23 is the first such prime that is not twinned.

V. 3.1. Combinatorial aspects. For convenience and without loss of analytic generality, we consider here unlabelled structures. According to Chapter I (Definition I.10, p. 51, and the companion Proposition I.2, p. 52), a combinatorial specification is *regular* if it is non-recursive ("iterative") and it involves only the constructions of Atom, Union, Product, and Sequence. A language $\mathcal{L}$ is S–regular if it is combinatorially isomorphic to a class $\mathcal{M}$ described by a regular specification. Alternatively, a language is S–regular if all the operations involved in its description (unions, catenation products and star operations) are unambiguous. The dictionary translating constructions into OGFs is

$$(11) \quad \mathcal{F} + \mathcal{G} \mapsto F + G, \qquad \mathcal{F} \times \mathcal{G} \mapsto F \times G, \qquad \text{SEQ}(\mathcal{F}) \mapsto (1 - F)^{-1},$$

and for languages, under the essential condition of *non-ambiguity* (Appendix A.7: *Regular languages*, p. 733),

$$(12) \quad \mathcal{L} \cup \mathcal{M} \mapsto L + M, \qquad \mathcal{L} \cdot \mathcal{M} \mapsto L \times M, \qquad \mathcal{L}^{\star} \mapsto (1 - L)^{-1}.$$

The rules (11) and (12) then give rise to generating functions that are invariably *rational functions*. Consequently, given a regular class $\mathcal{C}$, the exponential–polynomial form of coefficients expressed by Theorem IV.9 (p. 256) systematically applies, and one has

$$(13) \qquad C_n \equiv [z^n]C(z) = \sum_{j=1}^{m} \Pi_j(n)\alpha_j^{-n},$$

for a family of algebraic numbers α_j (the poles of $C(z)$) and a family of polynomials Π_j.

As we know from the discussion of periodicities in Section IV. 6.1 (p. 263), the collective behaviour of the sum in (13) depends on whether or not a single α dominates. In the case where several dominant singularities coexist, fluctuations of sorts (either periodic or irregular) may manifest themselves. In contrast, if a single α dominates, then the exponential–polynomial formula acquires a transparent asymptotic meaning. Accordingly, we set:

Definition V.2. *An exponential–polynomial form $\sum_{j=1}^{m} \Pi_j(n)\alpha_j^{-n}$ is said to be* pure *if* $|\alpha_1| < |\alpha_j|$, *for all $j \geq 2$. In that case, a single exponential dominates asymptotically all the other ones.*

As we see next for regular languages and specifications, the corresponding counting coefficients can always be described by a *finite collection* of pure exponential–polynomial forms. The fundamental reason is that we are dealing with a special subset of rational functions, one that enjoys strong positivity properties.

▷ **V.4.** *Positive rational functions.* Define the class Rat$^+$ of *positive rational functions* as the smallest class containing polynomials with positive coefficients ($\mathbb{R}_{\geq 0}[z]$) and closed under sum, product, and quasi-inverse, where $Q(f) = (1 - f)^{-1}$ is applied to elements f such that $f(0) = 0$. The OGF of any regular class with positive weights attached to neutral structures and atoms is in Rat$^+$. Conversely, any function in Rat$^+$ is the OGF of a positively weighted regular class. The notion of a Rat$^+$ function is for instance relevant to the analysis of weighted word models and Bernoulli trials (Section III. 6.1, p. 189). ◁

V.3.2. Analytic aspects. First we need the notion of sections of a sequence.

Definition V.3. *Let (f_n) be a sequence of numbers. Its* section *of parameters D, r, where $D \in \mathbb{Z}_{>0}$ and $r \in \mathbb{Z}_{\geq 0}$ is the subsequence (f_{nD+r}). The numbers D and r are referred to as the* modulus *and the* base, *respectively.*

The main theorem describing the asymptotic behaviour of regular classes is a consequence of Proposition IV.3 (p. 267) and is originally due to Berstel. (See Soittola's article [546] as well as the books by Eilenberg [189, Ch VII] and Berstel–Reutenauer [56] for context.)

Theorem V.3 (*Asymptotics of regular classes*). *Let S be a class described by a regular specification. Then there exists an integer D such that each section of modulus D of S_n that is not eventually 0 admits a pure exponential–polynomial form: for n larger than some n_0, and any such section of base r, one has*

$$S_n = \Pi(n)\beta^n + \sum_{j=1}^{m} P_j(n)\beta_j^n \qquad n \equiv r \bmod D,$$

where the quantities β, β_j, with $\beta > |\beta_j|$, and the polynomials Π, P_j, with $\Pi(x) \not\equiv 0$, depend on the base r.

Proof. (Sketch.) Let α_1 be the dominant pole of $S(z)$ that is positive. Proposition IV.3 (p. 267) asserts that any dominant pole, α is such that $\alpha/|\alpha|$ is a root of unity. Let D_0 be such that the dominant singularities are all contained in the set $\{\alpha_1 \omega^{j-1}\}_{j=1}^{D_0}$, where $\omega = \exp(2i\pi/D_0)$. By collecting all contributions arising from dominant poles in the general expansion (13) and by restricting n to a fixed congruence class modulo D_0, namely $n = \nu D_0 + r$ with $0 \leq r < D_0$, one gets

(14) $$S_{\nu D_0 + r} = \Pi^{[r]}(n)\alpha_1^{-D_0\nu} + O(A^{-n}).$$

There $\Pi^{[r]}$ is a polynomial depending on r and the remainder term represents an exponential polynomial with growth at most $O(A^{-n})$ for some $A > \alpha_1$.

The sections with modulus D_0 that are not eventually 0 can then be categorized into two classes.

— Let $\mathcal{R}_{\neq 0}$ be the set of those values of r such that $\Pi^{[r]}$ is not identically 0. The set $\mathcal{R}_{\neq 0}$ is non-empty (else the radius of convergence of $S(z)$ would be larger than α_1.) For any base $r \in \mathcal{R}_{\neq 0}$, the assertion of the theorem is then established with $\beta = 1/\alpha_1$.

— Let $\mathcal{R}_0$ be the set of those values of r such that $\Pi^{[r]}(x) \equiv 0$, with $\Pi^{[r]}$ as given by (14). Then one needs to examine the next layer of poles of $S(z)$, as detailed below.

Consider a number r such that $r \in \mathcal{R}_0$, so that the polynomial $\Pi^{[r]}$ is identically 0. First, we isolate in the expansion of $S(z)$ those indices that are congruent to r modulo D_0. This is achieved by means of a Hadamard product, which, given two power series $a(z) = \sum a_n z^n$ and $b(z) = \sum b_n z^n$, is defined as the series $c(z) = \sum c_n z^n$ such that

$c_n = a_n b_n$ and is written $c = a \odot b$. In symbols:

(15)
$$\left(\sum_{n\geq0} a_n z^n\right) \odot \left(\sum_{n\geq0} b_n z^n\right) = \sum_{n\geq0} a_n b_n z^n.$$

We have:

(16)
$$g(z) = S(z) \odot \left(\frac{z^r}{1 - z^{D_0}}\right).$$

A classical theorem [57, 189] from the theory of positive rational functions (in the sense of Note V.4) asserts that such functions are closed under Hadamard product. (A dedicated construction for (16) is also possible and is left as an exercise to the reader.) Then the resulting function $G(z)$ is of the form

$$g(z) = z^r \gamma(z^{D_0}),$$

with the rational function $\gamma(z)$ being analytic at 0. Note that we have $[z^\nu]\gamma(z) = S_{\nu D_0 + r}$, so that γ is exactly the generating function of the section of base r of $S(z)$. One verifies next that $\gamma(z)$, which is obtained by the substitution $z \mapsto z^{1/D_0}$ in $g(z)z^{-r}$, is itself a positive rational function. Then, by a fresh application of Berstel's Theorem (Proposition IV.3, p. 267), this function, if not a polynomial, has a radius of convergence ρ with all its dominant poles σ being such that σ/ρ is a root of unity of order D_1, for some $D_1 \geq 1$. The argument originally applied to $S(z)$ can thus be repeated, with $\gamma(z)$ replacing $S(z)$. In particular, one finds at least one section (of modulus D_1) of the coefficients of $\gamma(z)$ that admits a pure exponential–polynomial form. The other sections of modulus D_1 can themselves be further refined, and so on

In other words, successive refinements of the sectioning process provide at each stage at least one pure exponential–polynomial form, possibly leaving a few congruence classes open for further refinements. Define the *layer index* of a rational function f as the integer $\kappa(f)$, such that

$$\kappa(f) = \text{card}\left\{|\zeta| \mid f(\zeta) = \infty\right\}.$$

(This index is thus the number of different moduli of poles of f.) It is seen that each successive refinement step decreases by at least 1 the layer index of the rational function involved, thereby ensuring *termination* of the whole refinement process. Finally, the collection of the iterated sectionings obtained can be reduced to a single sectioning according to a common modulus D, which is the least common multiple of the collection of all the finite products $D_0 D_1 \cdots$ that are generated by the algorithm. ∎

For instance the coefficients (Figure V.4) of the function

(17)
$$L(z) = \frac{1}{(1 - z)(1 - z^2 - z^4)} + \frac{z}{1 - 3z^3},$$

associated to the regular language $a^\star(bb + cccc)^\star + d(ddd + eee + fff)^\star$, exhibit an apparently irregular behaviour, with the expansion of $L(z)$ starting as

$$1 + 2z + 2z^2 + 2z^3 + 7z^4 + 4z^5 + 7z^6 + 16z^7 + 12z^8 + 12z^9 + 47z^{10} + 20z^{11} + \cdots.$$

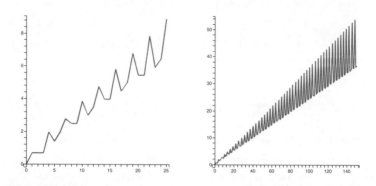

Figure V.4. Plots of $\log F_n$ with $F_n = [z^n]F(z)$ and $F(z)$ as in (17) display fluctuations that disappear as soon as sections of modulus 6 are considered.

The first term in (17) has a periodicity modulo 2, while the second one has an obvious periodicity modulo 3. In accordance with the theorem, the sections modulo 6 each admit a pure exponential–polynomial form and, consequently, they become easy to describe (Note V.5).

▷ **V.5.** *Sections and asymptotic regimes.* For the function $L(z)$ of (17), one finds, with $\varphi :=$ $(1 + \sqrt{5})/2$ and $c_1, c_2 \in \mathbb{R}_{>0}$,

$$
\begin{aligned}
L_n &= 3^{-1/3} \cdot 3^{n/3} + O(\varphi^{n/2}) &\quad (n \equiv 1, 4 \bmod 6), \\
L_n &= c_1 \varphi^{n/2} + O(1) &\quad (n \equiv 0, 2 \bmod 6), \\
L_n &= c_2 \varphi^{n/2} + O(1) &\quad (n \equiv 3, 5 \bmod 6),
\end{aligned}
$$

in accordance with the general form predicted by Theorem V.3.   ◁

▷ **V.6.** *Extension to* Rat$^+$ *functions.* The conclusions of Theorem V.3 hold for any function in Rat$^+$ in the sense of Note V.4.   ◁

▷ **V.7.** *Soittola's Theorem.* This is a converse to Theorem V.3 proved in [546]. Assume that coefficients of an *arbitrary* rational function $f(z)$ are non-negative and that there exists a sectioning such that each section admits a pure exponential–polynomial form. Then $f(z)$ is in Rat$^+$ in the sense of Note V.4; in particular, f is the OGF of a (weighted) regular class.   ◁

Theorem V.3 is useful for interpreting the enumeration of regular classes and languages. It serves a similar purpose with regards to structural parameters of regular classes. Indeed, consider a regular specification $\mathcal{C}$ augmented with a mark u that is, as usual, a neutral object of size 0 (see Chapter III). We let $C(z, u)$ be the corresponding BGF of $\mathcal{C}$, so that $C_{n,k} = [z^n u^k]C(z, u)$ is the number of $\mathcal{C}$–objects of size n that bear k marks. A suitable placement of marks makes it possible to record the number of times any given construction enters an object. For instance, in the augmented specification of binary words,

$$
\mathcal{C} = (\text{SEQ}_{<r}(b) + u\,\text{SEQ}_{\geq r}(b))\,\text{SEQ}(a(\text{SEQ}_{<r}(b) + u\,\text{SEQ}_{\geq r}(b))),
$$

all maximal runs of b having length at least r are marked by a u. There results the following BGF for the corresponding parameter "number of runs of bs of length $\geq r$",

(18) $$C(z, u) = \left(\frac{1 - z^r}{1 - z} + \frac{uz^r}{1 - z}\right) \cdot \frac{1}{1 - z\left(\frac{1-z^r}{1-z} + \frac{uz^r}{1-z}\right)},$$

from which mean and variance can be determined. In general, marks make it possible to analyse profile, with respect to constructions entering the specification, of a random object.

Theorem V.4 (Profile of regular classes). *Consider a regular specification of a class $\mathcal{C}$, augmented with a mark and let χ be the parameter corresponding to the number of occurrences of that mark. There exists a sectioning index d such that for any fixed section of (C_n) of modulus d, the following holds: the moment of integral order $s \geq 1$ of χ satisfies an asymptotic formula*

(19) $$\mathbb{E}_{C_n}[\chi^s] = Q(n)\beta^n + O(G^n),$$

where the quantities β, Q, G depend on the particular section considered, with $0 < \beta \leq 1$, $Q(n)$ a rational fraction, and $G < \beta$.

(Only sections that are not eventually 0 are to be considered.)

Proof. The case of expectations suffices to indicate the lines of a general proof. One possible approach[2] is to build a derived specification $\mathcal{E}$ such that

$$\mathbb{E}_{C_n}[\chi] = \frac{E_n}{C_n},$$

which is also a regular specification. To this purpose, define a transformation on specifications defined inductively by the rules

$$\partial(A + B) = \partial A + \partial B, \qquad\qquad \partial(A \times B) = \partial A \times B + A \times \partial B,$$
$$\partial \,\mathrm{SEQ}(A) = \mathrm{SEQ}(A) \times \partial A \times \mathrm{SEQ}(A),$$

together with the initial conditions $\partial u = \mathbf{1}$ and $\partial \mathcal{Z} = \emptyset$. This is a form of combinatorial differentiation: an object $\gamma \in \mathcal{C}$ corresponds to $\chi(\gamma)$ objects in $\mathcal{E}$, namely, one for each choice of an occurrence of the mark.

As a consequence, E_n is the cumulated value of χ over C_n, so that $E_n/C_n = \mathbb{E}_{C_n}[\chi]$. On the other hand, $\mathcal{E}$ is a regular specification to which Theorem V.3 applies. The result follows upon considering (if necessary) a sectioning that refines the sectionings of both $\mathcal{C}$ and $\mathcal{E}$. The argument extends easily to higher moments. ∎

▷ **V.8.** *A rational mean.* Consider the regular language $\mathcal{C} = a^\star (b + c)^\star d (b + c)^\star$. Let χ be the length of the initial run of a's. Then one finds

$$C(z) = \frac{z}{(1 - z)(1 - 2z)^2}, \qquad E(z) = \frac{z^2}{(1 - z)^2(1 - 2z)^2}.$$

Thus the mean of χ satisfies

$$\mathbb{E}_{C_n}[\chi] = \frac{E_n}{C_n} = \frac{(n - 3)2^n + (n + 3)}{(n - 1)2^n + 1} = \frac{n - 3}{n - 1} + O\left(\left(\frac{3}{4}\right)^n\right).$$

---

[2]Equivalently, one may operate at generating function level and observe that the derivative of a Rat$^+$ function is Rat$^+$; cf Notes V.4 and V.6.

Class	Asymptotics	
Integer compositions	2^{n-1}	
— k summands	$\sim \frac{n^{k-1}}{(k-1)!}$	(§I. 3.1, p. 44)
— summands $\leq r$	$\sim c\beta_r^n$	(§I. 3.1, p. 42)
Integer partitions		
— k summands	$\sim \frac{n^{k-1}}{k!\,(k-1)!}$	(§I. 3.1, p. 44)
— summands $\leq r$	$\sim \frac{n^{r-1}}{r!\,(r-1)!}$	(§I. 3.1, p. 43)
Set partitions, k classes	$\sim \frac{k^n}{k!}$	(§I. 4.3, p. 62)
Words excluding a pattern p	$\sim c\beta_{\mathrm{p}}^n$	(§IV. 6.3, p. 271)

Figure V.5. A pot-pourri of regular classes and their asymptotics.

Generally, in the statement of Theorem V.4, let $Q(n) = A(n)/B(n)$ with A, B polynomials and $a = \deg(A)$, $b = \deg(B)$. The following combinations prove to be possible (for first moments): $\beta = 1$ and (a, b) any pair such that $0 \leq a \leq b + 1$; also, $\beta < 1$ and (a, b) any pair of elements ≥ 0. ◁

▷ **V.9.** *Shuffle products.* Let $\mathcal{L}$, $\mathcal{M}$ be two languages over two disjoint alphabets. Then, the shuffle product $\mathcal{S}$ of $\mathcal{L}$ and $\mathcal{M}$ is such that $\widehat{S}(z) = \widehat{L}(z) \cdot \widehat{M}(z)$, where $\widehat{S}, \widehat{L}, \widehat{M}$ are the exponential generating functions of $\mathcal{S}, \mathcal{L}, \mathcal{M}$. Accordingly, if the OGF $L(z)$ and $M(z)$ are rational then the OGF $S(z)$ is also rational. (This technique may be used to analyse generalized birthday paradox and coupon collector problems; see [231].) ◁

V. 3.3. Applications.

V. 3.3. Applications. This subsection details several examples that illustrate the explicit determination of exponential–polynomial forms in regular specifications, in accordance with Theorems V.3 and V.4. We start by recapitulating a collection, a "pot-pourri", of combinatorial problems already encountered in Part A, where rational generating functions have been used *en passant*. We then examine longest runs in words, walks of the pure-birth type, and subsequence (hidden pattern) statistics.

Example V.3. *A pot-pourri of regular specifications.* A few combinatorial problems, to be found scattered across Chapters I–IV, are reducible to regular specifications: see Figure V.5 for a summary.

Compositions of integers (Section I. 3, p. 39) are specified by $\mathcal{C} = \text{SEQ}(\text{SEQ}_{\geq 1}(\mathcal{Z}))$, whence the OGF $(1 - z)/(1 - 2z)$ and the closed form $C_n = 2^{n-1}$, an especially transparent exponential–polynomial form. Polar singularities are also present for compositions into k summands that are described by $\text{SEQ}_k(\text{SEQ}_{\geq 1}(\mathcal{Z}))$ and for compositions whose summands are restricted to the interval $[1 .. r]$ (i.e., $\text{SEQ}(\text{SEQ}_{1 .. r}(\mathcal{Z}))$), with corresponding generating functions

$$\frac{z^k}{(1 - z)^k}, \qquad \frac{1 - z}{1 - 2z + z^{r+1}}.$$

In the first case, there is an explicit form for the coefficients, $\binom{n-1}{k-1}$, which constitutes a particular exponential–polynomial form (with the basis of the exponential being 1). The second case requires a dedicated analysis of the dominant polar singularity, which is recognizably a variant of Example V.4 (p. 308 below) dedicated to longest runs in random binary words.

Integer partitions involve the multiset construction. However, when summands are restricted to the interval $[1 \ldots r]$, the specification and the OGF are given by

$$\text{MSET}(\text{SEQ}_{1 \ldots r}(\mathcal{Z})) \cong \text{SEQ}(\mathcal{Z}) \times \text{SEQ}(\mathcal{Z}^2) \times \cdots \text{SEQ}(\mathcal{Z}^r) \quad \Longrightarrow \quad \prod_{j=1}^{r} \frac{1}{1 - z^j}.$$

This case, introduced in Section I. 3 (p. 39) also served as a leading example in our discussion of denumerants in Example IV.6 (p. 257): the analysis of the pole at 1 furnishes the dominant asymptotic behaviour, $n^{r-1}/(r!(r-1)!)$, for such special partitions. The enumeration of partitions by number of parts then follows, by duality, from the staircase representation.

Set partitions are typically labelled objects. However, when suitably constrained, they can be encoded by regular expressions; see Section I. 4.3 (p. 62) for partitions into k classes, where the OGF found is

$$S^{(k)}(z) = \frac{z^k}{(1 - z)(1 - 2z) \cdots (1 - kz)} \qquad \text{implying} \qquad S_n^{(k)} \sim \frac{k^n}{k!},$$

and the asymptotic estimate results from the partial fraction decomposition and the dominant pole at $1/k$.

Words lead to many problems that are prototypical of the regular specification framework. In Section I. 4 (p. 49), we saw that one could give a regular expression describing the set of words containing the pattern abb, from which the exact and asymptotic forms of counting coefficients derive. For a general pattern $\mathfrak{p}$, the generating functions of words constrained to include (or dually exclude) $\mathfrak{p}$ are rational. The corresponding asymptotic analysis has been given in Section IV. 6.3 (p. 271).

Words can also be analysed under the Bernoulli model, where letter i is selected with probability p_i; cf Section III. 6.1, p. 189, for a general discussion including the analysis of records in random words (p. 190). . . . . ∎

▷ **V.10.** *Partially commutative monoids.* Let $\mathcal{W} = \mathcal{A}^\star$ be the set of all words over a finite alphabet $\mathcal{A}$. Consider a collection C of commutation rules between pairs of elements of A. For instance, if $\mathcal{A} = \{a, b, c\}$, then $C = \{ab = ba, ac = ca\}$ means that a commutes with both b and c, but bc is not a commuting pair: $bc \neq cb$. Let $\mathcal{M} = \mathcal{W}/[C]$ be the set of equivalent classes of words (monomials) under the rules induced by C. The set $\mathcal{M}$ is said to be a *partially commutative monoid* or a *trace monoid* [105].

If $A = \{a, b\}$, then the two possibilities for C are $C = \emptyset$ and $C := \{ab = ba\}$. Normal forms for $\mathcal{M}$ are given by the regular expressions $(a + b)^\star$ and $a^\star b^\star$ corresponding to the OGFs

$$\frac{1}{1 - a - b}, \qquad \frac{1}{1 - a - b + ab}.$$

If $\mathcal{A} = \{a, b, c\}$, the possibilities for C, the corresponding normal forms, and the OGFs M are as follows. If $C = \emptyset$, then $\mathcal{M} \cong (a + b + c)^\star$ with OGF $(1 - a - b - c)^{-1}$; the other cases are

$ab = ba$	$ab = ba, \; ac = ca$	$ab = ba, \; ac = ca, \; bc = cb$
$(a^\star b^\star c)^\star a^\star b^\star$	$a^\star(b + c)^\star$	$a^\star b^\star c^\star$
$\dfrac{1}{1 - a - b - c + ab}$	$\dfrac{1}{1 - a - b - c + ab + ac}$	$\dfrac{1}{1 - a - b - c + ab + ac + bc - abc}$

Cartier and Foata [105] have discovered the general form (based on extended Möbius inversion),

$$M = \left(\sum_F (-1)^{|F|} F \right)^{-1},$$

where the sum is over all monomials F composed of distinct letters that all commute pairwise.

Viennot [597] has discovered an attractive geometric presentation of partially commutative monoids in terms of *heaps of pieces*, which has startling applications to several areas of combinatorial theory. (Example I.18, p. 80, relative to animals provides an example.) Goldwurm and Santini [298] have shown that $[z^n]M(z) \sim K \cdot \alpha^n$ for $K, \alpha > 0$. ◁

Longest runs. It is possible to develop a complete analysis of runs of consecutive equal letters in random sequences: this is in theory a special case of the analysis of patterns in random texts (Section IV.6.3, p. 271), but the particular nature of the patterns makes it possible to derive much more explicit results, including asymptotic distributions.

Example V.4. *Longest runs in words* Longest runs in words, introduced in Section I.4.1 (p. 51), provide an illustration of the technique of localizing dominant singularities in rational functions and of the corresponding coefficient extraction process. The probabilistic problem is a famous one, discussed by Feller in [205]: it represents a basic question in the analysis of runs of good (or bad) luck in a succession of independent events. Our presentation closely follows an insightful note of Knuth [375] whose motivation was the analysis of carry propagation in certain binary adders.

Start from the class $\mathcal{W}$ of all binary words over the alphabet $\{a, b\}$. Our interest lies in the length L of the longest consecutive block of a's in a word. For the property $L < k$, the specification and the corresponding OGF are

$$\mathcal{W}^{\langle k \rangle} = \text{SEQ}_{<k}(a)\, \text{SEQ}(b\, \text{SEQ}_{<k}(a)) \qquad \Longrightarrow \qquad W^{\langle k \rangle}(z) = \frac{1 - z^k}{1 - z} \cdot \frac{1}{1 - z\frac{1 - z^k}{1 - z}};$$

that is,

$$(20) \qquad\qquad W^{\langle k \rangle}(z) = \frac{1 - z^k}{1 - 2z + z^{k+1}}.$$

This represents a collection of OGFs indexed by k, which contain all the information relative to the distribution of longest runs in random words. We propose to prove:

Proposition V.1. *The longest run parameter L taken over the set of binary words of length n (endowed with the uniform distribution) satisfies the uniform estimate*[3]

$$(21) \qquad \mathbb{P}_n\left(L < \lfloor \lg n \rfloor + h\right) = e^{-\alpha(n)2^{-h-1}} + O\left(\frac{\log n}{\sqrt{n}}\right), \qquad \alpha(n) := 2^{\{\lg n\}}.$$

In particular, the mean satisfies

$$\mathbb{E}_n(L) = \lg n + \frac{\gamma}{\log 2} - \frac{3}{2} + P(\lg n) + O\left(\frac{\log^2 n}{\sqrt{n}}\right),$$

where P is a continuous periodic function whose Fourier expansion is given by (29). The variance satisfies $\mathbb{V}_n(L) = O(1)$ and the distribution is concentrated around its mean.

The probability distributions appearing in (21) are known as *double exponential distributions* (Figure V.6, p. 311). The formula (21) does not represent a single limit distribution in the usual sense of Chapter IX, but rather a whole *family of distributions* indexed by the fractional part of $\lg n$, thus dictated by the way n places itself with respect to powers of 2.

---

[3]The symbol $\lg x$ denotes the binary logarithm, $\lg x = \log_2 x$, and $\{x\}$ is the fractional part function $(\{\pi\} = 0.14159 \cdots,$.

Proof. The proof consists of the following steps: locate the dominant pole; estimate the corresponding contribution; separate the dominant pole from the other poles in order to derive constructive error terms; finally approximate the main quantities of interest.

(*i*) *Location of the dominant pole.* The OGF $W^{\langle k \rangle}$ has, by the first form of (20), a dominant pole ρ_k, which is a root of the equation $1 = s(\rho_k)$, where $s(z) = z(1-z^k)/(1-z)$. We consider $k \geq 2$. Since $s(z)$ is an increasing polynomial and $s(0) = 0$, $s(1/2) < 1$, $s(1) = k$, the root ρ_k must lie in the open interval $(1/2, 1)$. In fact, as one easily verifies, the condition $k \geq 2$ guarantees that $s(0.6) > 1$, hence the first estimate

$$(22) \qquad \frac{1}{2} < \rho_k < \frac{3}{5} \qquad (k \geq 2).$$

It now becomes possible to derive precise estimates by bootstrapping. (This technique is a form of iteration for approaching a fixed point—its use in the context of asymptotic expansions is detailed in De Bruijn's book [143].) Writing the defining equation for ρ_k as a fixed point equation,

$$z = \frac{1}{2}(1 + z^{k+1}),$$

and making use of the rough estimates (22) yields next

$$(23) \qquad \frac{1}{2}\left(1 + \left(\frac{1}{2}\right)^{k+1}\right) < \rho_k < \frac{1}{2}\left(1 + \left(\frac{3}{5}\right)^{k+1}\right).$$

Thus, ρ_k is exponentially close to $\frac{1}{2}$, and further iteration from (23) shows

$$(24) \qquad \rho_k = \frac{1}{2} + \frac{1}{2^{k+2}} + O\left(\frac{k}{2^{2k}}\right),$$

(*ii*) *Contribution from the dominant pole.* A straightforward calculation provides the value of the residue,

$$(25) \qquad R_{n,k} := -\operatorname{Res}\left[W^{\langle k \rangle}(z)z^{-n-1}; z = \rho_k\right] = \frac{1 - \rho_k^k}{2 - (k+1)\rho_k^k}\rho_k^{-n-1},$$

which is expected to provide the main approximation to the coefficients of $W^{\langle k \rangle}$ as $n \to \infty$. The quantity in (25) is of the rough form $2^n e^{-n/2^{k+1}}$; we shall return to such approximations shortly.

(*iii*) *Separation of the subdominant poles.* Consider the circle $|z| = 3/4$ and take the second form of the denominator of $W^{\langle k \rangle}$, namely, that of (20):

$$1 - 2z + z^{k+1}.$$

In view of Rouché's theorem (p. 270), we may regard this polynomial as the sum $f(z) + g(z)$, where $f(z) = 1 - 2z$ and $g(z) = z^{k+1}$. The term $f(z)$ has on the circle $|z| = 3/4$ a modulus that varies between $1/2$ and $5/2$; the term $g(z)$ is at most $27/64$ for any $k \geq 2$. Thus, on the circle $|z| = 3/4$, one has $|g(z)| < |f(z)|$, so that $f(z)$ and $f(z) + g(z)$ have the same number of zeros inside the circle. Since $f(z)$ admits $z = 1/2$ as only zero there, the denominator must also have a unique root in $|z| \leq 3/4$, and that root must coincide with ρ_k.

Similar arguments also give bounds on the error term when the number of words w satisfying $L(w) < k$ is estimated by the residue (25) at the dominant pole. On the circle $|z| = 3/4$, the denominator of $W^{\langle k \rangle}$ stays bounded away from 0 (its modulus is at least $5/64$ when $k \geq 2$, by previous considerations). Thus, the modulus of the remainder integral is $O((4/3)^n)$, and in fact bounded from above by $35(4/3)^n$. In summary, letting $q_{n,k}$ represent the probability that the longest run in a random word of length n is less than k, one obtains the main estimate ($k \geq 2$)

$$(26) \qquad q_{n,k} := \mathbb{P}_n(L < k) = \frac{1 - \rho_k^k}{1 - (k+1)\rho_k^k/2} \left(\frac{1}{2\rho_k}\right)^{n+1} + O\left(\left(\frac{2}{3}\right)^n\right),$$

which holds *uniformly* with respect to k. Here is a table of the numerical values of the quantities appearing in the approximation of $q_{n,k}$, written under the form $c_k \cdot (2\rho_k)^{-n}$:

k	$c_k \cdot (2\rho_k)^{-n}$
2	$1.17082 \cdot 0.80901^n$
3	$1.13745 \cdot 0.91964^n$
4	$1.09166 \cdot 0.96378^n$
5	$1.05753 \cdot 0.98297^n$
10	$1.00394 \cdot 0.99950^n$

(*iv*) *Final approximations.* There only remains to transform the main estimate (26) into the limit form asserted in the statement. First, the "tail inequalities" (with $\lg x \equiv \log_2 x$)

$$(27) \qquad \mathbb{P}_n\left(L < \frac{3}{4}\lg n\right) = O\left(e^{-\frac{1}{2}\sqrt[4]{n}}\right), \qquad \mathbb{P}_n(L \geq 2\lg n + y) = O\left(\frac{e^{-2y}}{n}\right)$$

describe the tail of the probability distribution of L_n. They are derived from simple bounding techniques applied to the main approximation (26) using (24). Thus, for asymptotic purposes, only a relatively small region around $\lg n$ needs to be considered.

Regarding the central regime, for $k = \lg n + x$ and x in $[-\frac{1}{4}\lg n, \lg n]$, the approximation (24) of ρ_k and related quantities applies, and one finds

$$(2\rho_k)^{-n} = \exp\left(-\frac{n}{2^{k+1}} + O(kn2^{-2k})\right) = e^{-n/2^{k+1}}\left(1 + O\left(\frac{\log n}{\sqrt{n}}\right)\right).$$

(This results from standard expansions of the form $(1-a)^n = e^{-na}\exp(O(na^2))$.) At the same time, the coefficient in (26) of the quantity $(2\rho_k)^{-n}$ is

$$1 + O(k\rho_k^k) = 1 + O\left(\frac{\log n}{\sqrt{n}}\right).$$

Thus a double exponential approximation holds (Figure V.6): for $k = \lg n + x$ with x in $[-\frac{1}{4}\lg n, \lg n]$, one has (uniformly)

$$(28) \qquad q_{n,k} = e^{-n/2^{k+1}}\left(1 + O\left(\frac{\log n}{\sqrt{n}}\right)\right).$$

In particular, upon setting $k = \lfloor \lg n \rfloor + h$ and making use of the tail inequalities (27), the first part of the statement, namely Equation (21), follows. (The floor function takes into account the fact that k must be an integer.)

The mean and variance estimates are derived from the fact that the distribution quickly decays at values away from $\lg n$ (by (27)) while it satisfies Equation (28) in the central region. The mean satisfies

$$\mathbb{E}_n(L) := \sum_{h \geq 1} [1 - \mathbb{P}_n(L < h)] = \Phi\left(\frac{n}{2}\right) - 1 + O\left(\frac{\log^2 n}{n}\right), \qquad \Phi(x) := \sum_{h \geq 0}\left[1 - e^{-x/2^h}\right].$$

Consider the three cases $h < h_0$, $h \in [h_0, h_1]$, and $h > h_1$ with $h_0 = \lg x - \log\log x$ and $h_1 = \lg x + \log\log x$, where the general term is (respectively) close to 1, between 0 and 1, and

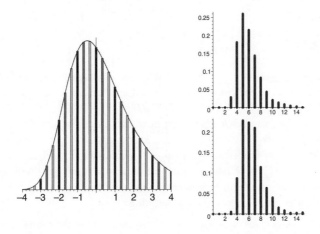

Figure V.6. The double exponential laws: Left, histograms for n at 2^p (black), $2^{p+1/3}$ (dark gray), and $2^{p+2/3}$ (light gray), where $x = k - \lg n$. Right, empirical histograms for 1000 simulations with $n = 100$ (top) and $n = 140$ (bottom).

close to 0. By summing, one finds elementarily $\Phi(x) = \lg x + O(\log\log x)$ as $x \to \infty$. (An elementary way of catching the next $O(1)$ term is discussed for instance in [538, p. 403].)

The method of choice for precise asymptotics is to treat $\Phi(x)$ as a harmonic sum and apply Mellin transform techniques (Appendix B.7: *Mellin transforms*, p. 762). The Mellin transform of $\Phi(x)$ is

$$\Phi^\star(s) := \int_0^\infty \Phi(x) x^{s-1}\, dx = \frac{\Gamma(s)}{1 - 2^s} \qquad \Re(s) \in (-1, 0).$$

The double pole of $\Phi^\star$ at 0 and the simple poles at $s = \frac{2ik\pi}{\log 2}$ are reflected by an asymptotic expansion that involves a Fourier series:
(29)
$$\Phi(x) = \lg x + \frac{\gamma}{\log 2} + \frac{1}{2} + P(\lg x) + O(x^{-1}), \qquad P(w) := -\frac{1}{\log 2} \sum_{k \in \mathbb{Z}\setminus\{0\}} \Gamma\left(\frac{2ik\pi}{\log 2}\right) e^{-2ik\pi w}.$$

The oscillating function $P(w)$ is found to have tiny fluctuations, of the order of 10^{-6}; for instance, the first Fourier coefficient has amplitude: $|\Gamma(2i\pi/\log 2)|/\log 2 \doteq 7.86 \cdot 10^{-7}$. (See also [234, 311, 375, 564] for more on this topic.) The variance is similarly analysed. This concludes the proof of Proposition V.1. ∎

The double exponential approximation in (21) is typical of extremal statistics. What is striking here is the existence of a family of distributions indexed by the fractional part of $\lg n$. This fact is then reflected by the presence of oscillating functions in moments of the random variable L. .... ∎

▷ **V.11.** *Longest runs in Bernoulli sequences.* Consider an alphabet $\mathcal{A} = \{a_j\}$ with letter a_j independently chosen with probability $\{p_j\}$. The OGF of words where each run of equal letters has length at most k is derived from the construction of Smirnov words (pp. 204 and 262), and it is found to be

$$W^{[k]}(z) = \left(1 - \sum_i p_i z \frac{1 - (p_i z)^k}{1 - (p_i z)^{k+1}}\right)^{-1}.$$

Let p_{max} be the largest of the p_j. Then the expected length of the longest run of any letter is $\log n / \log p_{max} + O(1)$, and precise quantitative information can be derived from the OGFs by methods akin to Example IV.10 (Smirnov words and Carlitz compositions, p. 262). ◁

Walks of the pure-birth type. The next two examples develop the analysis of walks in a special type of graphs. These examples serve two purposes: they illustrate further cases of modelling by means of regular specifications, and they provide a bridge to the analysis of lattice paths in the next section. Furthermore, some specific walks of the pure-birth type turn out to have applications to the analysis of a probabilistic algorithm (Approximate Counting).

Example V.5. *Walks of the pure-birth type.* Consider a walk on the non-negative integers that starts at 0 and is only allowed either to stay at the same place or move by an increment of $+1$. Our goal is to enumerate the walks that start from 0 and reach point m in n steps. A step from j to $j + 1$ will be encoded by a letter a_j; a step from j to j will be encoded by c_j, in accordance with the following state diagram:

(30)

The language encoding all legal walks from state 0 to state m can be described by a regular expression:

$$\mathcal{H}_{0,m} = \text{SEQ}(c_0)a_0 \, \text{SEQ}(c_1)a_1 \cdots \text{SEQ}(c_{m-1})a_{m-1} \, \text{SEQ}(c_m).$$

Symbolicly using letters as variables, the corresponding ordinary multivariate generating function is then (with $\mathbf{a} = (a_0, \ldots)$ and $\mathbf{c} = (c_0, \ldots)$)

$$H_{0,m}(\mathbf{a}, \mathbf{c}) = \frac{a_0 a_1 \cdots a_{m-1}}{(1 - c_0)(1 - c_1) \cdots (1 - c_m)}.$$

Assume now that the steps are assigned weights, with α_j corresponding to a_j and γ_j to c_j. Weights of letters are extended multiplicatively to words in the usual way (cf Section III.6.1, p. 189). In addition, upon taking $\gamma_j = 1 - \alpha_j$, one obtains a probabilistic weighting: the walker starts from position 0, and, if at j, at each clock tick, she either stays at the same place with probability $1 - \alpha_j$ or moves to the right with probability α_j. The OGF of such weighted walks then becomes

(31) $$H_{0,m}(z) = \frac{\alpha_0 \alpha_1 \cdots \alpha_{m-1} z^m}{(1 - (1 - \alpha_0)z)(1 - (1 - \alpha_1)z) \cdots (1 - (1 - \alpha_m)z)},$$

and $[z^n]H_{0,m}$ is the probability for the walker to be found at position m at (discrete) time n. This walk process can be alternatively interpreted as a (discrete-time) *pure-birth process*[4] in the usual sense of probability theory: There is a population of individuals and, at each discrete epoch, a new birth may take place, the probability of a birth being α_j when the population is of size j.

---

[4]The theory of pure-birth processes is discussed under a calculational and non measure-theoretic angle in the book by Bharucha-Reid [62]. See also the *Course* by Karlin and Taylor [363] for a concrete presentation.

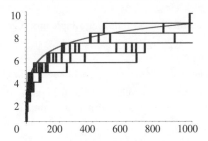

Figure V.7. A simulation of 10 trajectories of the pure-birth process till $n = 1024$, with geometric probabilities corresponding to $q = 1/2$, compared to the curve $\log_2 x$.

The form (31) readily lends itself to a partial fraction decomposition. Assume for simplicity that the α_j are all distinct. The poles of $H_{0,m}$ are at the points $(1 - \alpha_j)^{-1}$ and one finds as $z \to (1 - \alpha_j)^{-1}$:

$$H_{0,m}(z) \sim \frac{r_{j,m}}{1 - z(1 - \alpha_j)} \quad \text{where} \quad r_{j,m} := \frac{\alpha_0 \alpha_1 \cdots \alpha_{m-1}}{\prod_{k \in [0,m], \, k \neq j} (\alpha_k - \alpha_j)}.$$

Thus, the probability of being in state m at time n is given by a sum:

$$(32) \qquad [z^n] H_{0,m}(z) = \sum_{j=0}^{m} r_{j,m} (1 - \alpha_j)^n.$$

An especially interesting case of the pure-birth walk is when the quantities α_k are geometric: $\alpha_k = q^k$ for some q with $0 < q < 1$. In that case, the probability of being in state m after n transitions becomes (cf (32))

$$(33) \qquad \sum_{j=0}^{m} \frac{(-1)^j q^{\binom{j}{2}}}{(q)_j (q)_{m-j}} (1 - q^{m-j})^n, \quad (q)_j := (1 - q)(1 - q^2) \cdots (1 - q^j).$$

This corresponds to a stochastic progression in a medium with exponentially increasing hardness or, equivalently, to the growth of a population whose size adversely affects fertility in an exponential manner. On intuitive grounds, we expect an evolution of the process to stay reasonably close to the curve $y = \log_{1/q} x$; see Figure V.7 for a simulation confirming this fact, which can be justified by means of formula (33). This particular analysis is borrowed from [218], where it was initially developed in connection with the "approximate counting" algorithm to be studied next. .... ∎

Example V.6. *Approximate Counting.* Assume you need to keep a counter that is able to record the number of certain events (say impulses) and should have the capability of keeping counts till a certain maximal value N. A standard information-theoretic argument (with ℓ bits, one can only keep track of 2^ℓ possibilities) implies that one needs $\lceil \log_2(N+1) \rceil$ bits to perform the task—a standard binary counter will indeed do the job. However, in 1977, Robert Morris has proposed a way to maintain counters that only requires of the order of $\log \log N$ bits. What's the catch?

Morris' elegant idea consists in relaxing the constraint of exactness in the counting process and, by playing with probabilities, tolerate a small error on the counts obtained. Precisely, his solution maintains a random quantity Q which is initialized by $Q = 0$. Upon receiving an impulse, one updates Q according to the following simple procedure (with $q \in (0, 1)$ a design parameter):

> procedure Update(Q);
>   with probability q^Q do $Q := Q + 1$ (else keep Q unchanged).

When asked the number of impulses (number of times the update procedure was called) at any moment, simply use the following procedure to return an estimate:

> procedure Answer(Q);
>   output $X = \dfrac{q^{-Q} - 1}{1 - q}$.

Let Q_n be the value of the random quantity Q after n executions of the update procedure and X_n the corresponding estimate output by the algorithm. It is easy to verify (by recurrence or by generating functions; see Note V.12 below for higher moments) that, for $n \geq 1$,

$$(34) \qquad \mathbb{E}(q^{-Q_n}) = n(1 - q) + 1, \qquad \text{so that} \quad \mathbb{E}(X_n) = n.$$

Thus the answer provided at any instant is an *unbiased estimator* (in a mean value sense) of the actual count n. On the other hand, the analysis of the geometric pure-birth process in the previous example applies. In particular, the exponential approximation $(1 - \alpha)^n \approx e^{-n\alpha}$ in conjunction with the basic formula (33) shows that for large n and m sufficiently near to $\log_{1/q} n$, one has (asymptotically) the *geometric-birth distribution*

$$(35) \qquad \mathbb{P}(Q_n = m) = \sum_{j=0}^{\infty} \frac{(-1)^j q^{\binom{j}{2}}}{(q)_j (q)_\infty} \exp(-q^{x-j}) + o(1), \qquad x \equiv m - \log_{1/q} n.$$

(We refer to [218] for details.) Such calculations imply that Q_n is with high probability (w.h.p.) close to $\log_{1/q} n$. Thus, if $n \leq N$, the value of Q_n will be w.h.p. bounded from above by $(1 + \epsilon) \log_{1/q} N$, with ϵ a small constant. But this means that the integer Q, which can itself be represented in binary, will only require

$$(36) \qquad \qquad \log_2 \log n + O(1)$$

bits for storage, for fixed q.

A closer examination of the formulae reveals that the accuracy of the estimate improves considerably when q becomes close to 1. The *standard error* is defined as $\frac{1}{n}\sqrt{\mathbb{V}(X_n)}$ and it measures, in a mean-quadratic sense, the relative error likely to be made. The variance of Q_n is, as for the mean, determined by recurrence or generating functions, and one finds

$$(37) \qquad \mathbb{V}(q^{-Q_n+1}) = \binom{n}{2}\frac{(1 - q)^3}{q}, \qquad \frac{1}{n}\sqrt{\mathbb{V}(X_n)} \sim \sqrt{\frac{1 - q}{2q}}.$$

(see also Note V.12 below). This means that accuracy increases as q approaches 1 and, by suitably dimensioning q, one can make it asymptotically as small as desired. In summary, (34), (37), and (36) express the following property: *Approximate counting makes it possible to count till N using only about $\log \log N$ bits of storage, while achieving a standard error that is asymptotically a constant and can be set to any prescribed small value.* Morris' trick is now fully understood.

For instance, with $q = 2^{-1/16}$, it proves possible to count up to $2^{16} = 65536$ using only 8 bits (instead of 16), with an error likely not to exceed 20%. Naturally, there's not too much reason to appeal to the algorithm when a *single* counter needs to be managed (everybody can afford a few bits!): Approximate Counting turns out to be useful when a very large number of counts need to be kept *simultaneously*. It constitutes one of the early examples of a probabilistic algorithm in the extraction of information from large volumes of data, an area also known as *data mining*; see [224] for a review of connections with analytic combinatorics and references.

Functions akin to those of (35) also surface in other areas of probability theory. Guillemin, Robert, and Zwart [314] have detected them in processes that combine an additive increase and a multiplicative decrease (AIMD processes), in a context motivated by the adaptive transmission of "windows" of varying sizes in large communication networks (the TCP protocol of the internet). Biane, Bertoin, and Yor [58] encountered a function identical to (35) in their study of exponential functionals of Poisson processes. .... ∎

▷ **V.12.** *Moments of* q^{-Q_n}. It is a perhaps surprising fact that any integral moment of q^{-Q_n} is a polynomial in n, q, and q^{-1}, as in (34), (37). To see it, define

$$\Phi(w) \equiv \Phi(w, \xi, q) := \sum_{m \geq 0} q^{m(m+1)/2} \frac{\xi^m w^m}{(1 + \xi q)(1 + \xi q^2) \cdots (1 + \xi q^{m+1})}.$$

By (31), one has

$$\sum_{m \geq 0} H_{0,m}(z) w^m = \frac{1}{1 - z} \Phi\left(w; \frac{z}{1 - z}, q\right).$$

On the other hand, Φ satisfies $\Phi(w) = 1 - q\xi(1 - w)\Phi(qw)$, hence the q–identity,

$$\Phi(w) = \sum_{j \geq 0} (-q\xi)^j \left[(1 - w)(1 - qw) \cdots (1 - q^{j-1}w)\right],$$

which belongs to the area of q–calculus[5]. Thus $\Phi(q^{-r}; \xi, q)$ is a polynomial for any $r \in \mathbb{Z}_{\geq 0}$, as the expansion terminates. See Prodinger's study [498] for connections with basic hypergeometric functions and Heine's transformation. ◁

Hidden patterns: regular expression modelling and moments. We return here to the analysis of the number of occurrences of a pattern $\mathfrak{p}$ as a *subsequence* in a random text. The mean number of occurrences can be obtained by enumerating contexts of occurrences: in a sense we are then enumerating the language of all words by means of a dedicated regular expression where the ambiguity coefficient (the multiplicity) of a word is precisely equal to the number of occurrences of the pattern. This technique, which gives an easy access to expectations, also works for higher moments. It supplements the fact that there is no easy way to get a BGF in such cases, and it appears to be sufficient to derive a concentration of distribution property.

Example V.7. *Occurrences of "hidden" patterns in Bernoulli texts.* Fix an alphabet $\mathcal{A} = \{a_1, \ldots, a_r\}$ of cardinality r and assume a probability distribution on $\mathcal{A}$ to be given, with p_j the probability of letter a_j. We consider the Bernoulli model on $\mathcal{W} = \text{SEQ}(\mathcal{A})$, where the probability of a word is the product of the probabilities of its letters (cf Subsection III. 6.1, p. 189). A word $\mathfrak{p} = y_1 \cdots y_k$ called the pattern is fixed. The problem is to gather information on the random variable X representing the number of occurrences of $\mathfrak{p}$ in the set $\mathcal{W}_n$, where occurrences as a *"hidden pattern"*, i.e., as a *subsequence*, are counted (see Example I.11, p. 54, for the case of equiprobable letters).

---

[5]By q–calculus is roughly meant the collection of special function identities relating power series of the form $\sum a_n(q)z^n$, where $a_n(q)$ is a rational fraction whose degree is quadratic in n. See [15, Ch. 10] for basics and [284] for more advanced (q–hypergeometric) material.

Mean value analysis. The generating function associated to $\mathcal{W}$ endowed with its probabilistic weighting is

$$W(z) = \frac{1}{1 - \sum p_j z} = \frac{1}{1 - z}.$$

The regular specification

(38) $\mathcal{O} = \text{SEQ}(\mathcal{A})y_1 \, \text{SEQ}(\mathcal{A}) \cdots \text{SEQ}(\mathcal{A})y_{k-1} \, \text{SEQ}(\mathcal{A})y_k \, \text{SEQ}(\mathcal{A})$

describes all *contexts of occurrences* of $\mathfrak{p}$ as a subsequence in all words. Graphically, this may be rendered as follows, for a pattern of length 3 such as $\mathfrak{p} = y_1 y_2 y_3$:

(39)

There the boxes indicate distinguished positions where letters of the pattern appear and the horizontal lines represent arbitrary separating words ($\text{SEQ}(\mathcal{A})$). The corresponding OGF

(40) $O(z) = \dfrac{\pi(\mathfrak{p})z^k}{(1 - z)^{k+1}},$ $\pi(\mathfrak{p}) := p_{y_1} \cdots p_{y_{k-1}} p_{y_k}$

counts elements of $\mathcal{W}$ with *multiplicity*[6], where the multiplicity coefficient $\lambda(w)$ of a word $w \in \mathcal{W}$ is precisely equal to the number of occurrences of $\mathfrak{p}$ as a subsequence in w:

$$O(z) \equiv \sum_{w \in \mathcal{A}^\star} \lambda(w)\pi(w)z^{|w|}.$$

This shows that the mean value of the number X of hidden occurrences of $\mathfrak{p}$ in a random word of length n satisfies

(41) $\mathbb{E}_{\mathcal{W}_n}(X) = [z^n]O(z) = \pi(\mathfrak{p})\dbinom{n}{k},$

which is consistent with what a direct probabilistic reasoning would give.

Variance analysis. In order to determine the variance of X over $\mathcal{W}_n$, we need contexts in which *pairs* of occurrences appear. Let $\mathcal{Q}$ denote the set of all words in $\mathcal{W}$ with *two* occurrences (i.e., an ordered pair of occurrences) of $\mathfrak{p}$ as a subsequence being distinguished. Then clearly $[z^n]Q(z)$ represents $\mathbb{E}_{\mathcal{W}_n}(X^2)$. There are several cases to be considered. Graphically, a pair of occurrences may share no common position, like in what follows:

(42)

But they may also have one or several overlapping positions, like in

(43)

(44)

(This last situation necessitates $y_2 = y_3$, typical patterns being *abb* and *aaa*.)

---

[6] In language-theoretic terms, we make use of the regular expression $\mathcal{O} = \mathcal{A}^\star y_1 \mathcal{A}^\star \cdots y_{k-1} \mathcal{A}^\star y_k \mathcal{A}^\star$ that describes a subset of $\mathcal{A}^\star$ in an ambiguous manner and takes into account the *ambiguity coefficients*.

In the first case corresponding to (42), where there are no overlapping positions, the configurations of interest have OGF

(45)
$$Q^{[0]}(z) = \binom{2k}{k} \frac{\pi(\mathfrak{p})^2 z^{2k}}{(1-z)^{2k+1}}.$$

There, the binomial coefficient $\binom{2k}{k}$ counts the total number of ways of freely interleaving two copies of $\mathfrak{p}$; the quantity $\pi(\mathfrak{p})^2 z^{2k}$ takes into account the $2k$ distinct positions where the letters of the two copies appear; the factor $(1-z)^{-2k-1}$ corresponds to all the possible $2k+1$ fillings of the gaps between letters.

In the second case, let us start by considering pairs where exactly one position is overlapping, like in (43). Say this position corresponds to the rth and sth letters of $\mathfrak{p}$ (r and s may be unequal). Obviously, we need $y_r = y_s$ for this to be possible. The OGF of the configurations is now

$$\binom{r+s-2}{r-1}\binom{2k-r-s}{k-r} \frac{\pi(\mathfrak{p})^2 (p_{y_r})^{-1} z^{2k-1}}{(1-z)^{2k}}.$$

There, the first binomial coefficient $\binom{r+s-2}{r-1}$ counts the total number of ways of interleaving $y_1 \cdots y_{r-1}$ and $y_1 \cdots y_{s-1}$; the second binomial $\binom{2k-r-s}{k-r}$ is similarly associated to the interleavings of $y_{r+1} \cdots y_k$ and $y_{s+1} \cdots y_k$; the numerator takes into account the fact that $2k - 1$ positions are now occupied by predetermined letters; finally the factor $(1-z)^{-2k}$ corresponds to all the $2k$ fillings of the gaps between letters. Summing over all possibilities for r, s gives the OGF of pairs with one overlapping position as

(46)
$$Q^{[1]}(z) = \left(\sum_{1 \le r,s \le k} \binom{r+s-2}{r-1}\binom{2k-r-s}{k-r} \frac{[\![y_r = y_s]\!]}{p_{y_r}} \right) \frac{\pi(\mathfrak{p})^2 z^{2k-1}}{(1-z)^{2k}}.$$

Similar arguments show that the OGF of pairs of occurrences with at least *two* shared positions (see, e.g., (44)) is of the form, with P a polynomial,

(47)
$$Q^{[\ge 2]}(z) = \frac{P(z)}{(1-z)^{2k-1}},$$

for the essential reason that, in the finitely many remaining situations, there are at most $(2k-1)$ possible gaps.

We can now examine (45), (46), (47) in the light of singularities. The coefficient $[z^n]Q^{[0]}(z)$ is seen to cancel to first asymptotic order with the square of the mean as given in (41). The contribution of the coefficient $[z^n]Q^{[\ge 2]}(z)$ appears to be negligible as it is $O(n^{2k-2})$. The coefficient $[z^n]Q^{[1]}(z)$, which is $O(n^{2k-1})$, is seen to contribute to the asymptotic growth of the variance. In summary, after a trite calculation, we obtain:

Proposition V.2. *The number X of occurrences of a hidden pattern* $\mathfrak{p}$ *in a random text of size n obeying a Bernoulli model satisfies*

$$\mathbb{E}_{\mathcal{W}_n}(X) = \pi(\mathfrak{p})\binom{n}{k} \sim \frac{\pi(\mathfrak{p})}{k!} n^k, \qquad \mathbb{V}_{\mathcal{W}_n}(X) = \frac{\pi(\mathfrak{p})^2 \kappa(\mathfrak{p})^2}{(2k-1)!} n^{2k-1} \left(1 + O(\tfrac{1}{n})\right),$$

where the "correlation coefficient" $\kappa(\mathfrak{p})^2$ *is given by*

$$\kappa(\mathfrak{p})^2 = \sum_{1 \le r,s \le k} \binom{r+s-2}{r-1}\binom{2k-r-s}{k-r} \left(\frac{[\![y_r = y_s]\!]}{p_{y_r}} - 1 \right).$$

In particular, the distribution of X is concentrated around its mean.

This example is based on an article by Flajolet, Szpankowski, and Vallée [263]. There the authors show further that the asymptotic behaviour of moments of higher order can be worked out. By the Moment Convergence Theorem (Theorem C.2, p. 778), this calculation entails that *the distribution of X over W_n is asymptotically normal*. The method also extends to a much more general notion of "hidden" pattern; e.g., distances between letters of $\mathfrak{p}$ can be constrained in various ways so as to determine a valid occurrence in the text [263]. It also extends to the very general framework of dynamical sources [81], which include Markov models as a special case. The two references [81, 263] thus provide a set of analyses that interpolate between the two extreme notions of pattern occurrence—as a block of consecutive symbols or as a subsequence ("hidden pattern"). Such studies demonstrate that hidden patterns are with high probability bound to occur an extremely large number of times in a long enough text—this might cast some doubts on numerological interpretations encountered in various cultures: see in particular the critical discussion of the "Bible Codes" by McKay *et al.* in [433]. .... ∎

▷ **V.13.** *Hidden patterns and shuffle relations.* To each pairs u, v of words over $\mathcal{A}$ associate the weighted-shuffle polynomial in the indeterminates $\mathcal{A}$ denoted by $\left(\!\!\left(\begin{smallmatrix} u \\ v \end{smallmatrix}\right)\!\!\right)_t$ and defined by the properties

$$\begin{cases} \left(\!\!\left(\begin{matrix} xu \\ yv \end{matrix}\right)\!\!\right)_t = x \left(\!\!\left(\begin{matrix} u \\ yv \end{matrix}\right)\!\!\right)_t + y \left(\!\!\left(\begin{matrix} xu \\ v \end{matrix}\right)\!\!\right)_t + t[\![x = y]\!]x \left(\!\!\left(\begin{matrix} u \\ v \end{matrix}\right)\!\!\right)_t \\ \left(\!\!\left(\begin{matrix} \mathbf{1} \\ u \end{matrix}\right)\!\!\right)_t = \left(\!\!\left(\begin{matrix} u \\ \mathbf{1} \end{matrix}\right)\!\!\right)_t = u \end{cases}$$

where t is a parameter, x, y are elements of $\mathcal{A}$, and $\mathbf{1}$ is the empty word. Then the OGF of $Q(z)$ above is

$$Q(z) = \sigma\left[\left(\!\!\left(\begin{matrix} \mathfrak{p} \\ \mathfrak{p} \end{matrix}\right)\!\!\right)_{(1-z)}\right]\frac{1}{(1-z)^{2k+1}},$$

where σ is the substitution $a_j \mapsto p_j z$. ◁

V. 4. Nested sequences, lattice paths, and continued fractions

This section treats the *nested sequence* schema, corresponding to a cascade of sequences of the rough form SEQ ∘ SEQ ∘ ⋯ ∘ SEQ. Such a schema covers Dyck and Motzkin path, a particular type of Łukasiewicz paths already encountered in Section I. 5.3 (p. 73). Equipped with probabilistic weights, these paths appear as trajectories of birth-and-death processes (the case of pure-birth processes has already been dealt with in Example V.5, p. 312). They also have great descriptive power since, once endowed with integer weights, they can encode a large variety of combinatorial classes, including trees, permutations, set partitions, and surjections.

Since a combinatorial sequence translates into a quasi-inverse, $Q(f) = (1 - f)^{-1}$, a class described by nested sequences has its generating function expressed by a cascade of fractions, that is, a *continued fraction*[7]. Analytically, these GFs have two dominant poles (the Dyck case) or a single pole (the Motzkin case) on their disc of convergence, so that the implementation of the process underlying Theorem V.3 is easy: we encounter a pure polynomial form of the simplest type that describes all counting sequences of interest. The profile of a nested sequence can also be easily characterized.

---

[7] Characteristically, the German term for "continued fraction", is "*Kettenbruch*", literally "*chain-fraction*".

This section starts with a statement of the "Continued Fraction Theorem" (Proposition V.3, p. 321) taken from an old study of Flajolet [214], which provides the general set-up for the rest of the section. It then proceeds with the general analytic treatment of nested sequences. A number of examples from various areas of discrete mathematics are then detailed, including the important analysis of height in Dyck paths and general Catalan trees. Some of these examples make use of structures that are described as infinitely nested sequences, that is, infinite continued fractions, to which the finite theory often extends—the analysis of coin fountains below is typical.

V. 4.1. Combinatorial aspects. We discuss here a special type of lattice paths connecting points of the discrete cartesian plane $\mathbb{Z} \times \mathbb{Z}$.

Definition V.4 (Lattice path). *A* Motzkin path $v = (U_0, U_1, \ldots, U_n)$ *is a sequence of points in the discrete quarter-plane* $\mathbb{Z}_{\geq 0} \times \mathbb{Z}_{\geq 0}$, *such that* $U_j = (j, y_j)$ *and the jump condition* $|y_{j+1} - y_j| \leq 1$ *is satisfied. An edge* $\langle U_j, U_{j+1} \rangle$ *is called an* ascent *if* $y_{j+1} - y_j = +1$, *a* descent *if* $y_{j+1} - y_j = -1$, *and a* level step *if* $y_{j+1} - y_j = 0$. *A path that has no level steps is called a* Dyck path.

The quantity n *is the* length of the path, $\mathrm{ini}(v) := y_0$ *is the* initial altitude, $\mathrm{fin}(v) := y_n$ *is the* final altitude. *A path is called an* excursion *if both its initial and final altitudes are zero. The extremal quantities* $\sup\{v\} := \max_j y_j$ *and* $\inf\{v\} := \min_j y_j$ *are called the* height *and* depth *of the path.*

A path can always be encoded by a word with a, b, c representing ascents, descents, and level steps, respectively. What we call the *standard encoding* is such a word in which each step a, b, c is (redundantly) subscripted by the value of the y-coordinate of its initial point. For instance,

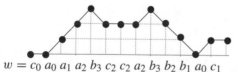

$$w = c_0 \, a_0 \, a_1 \, a_2 \, b_3 \, c_2 \, c_2 \, a_2 \, b_3 \, b_2 \, b_1 \, a_0 \, c_1$$

encodes a path that connects the initial point $(0, 0)$ to the point $(13, 1)$. Such a path can also be regarded as the evolution in discrete time of a walk over the integer line with jumps restricted to $\{-1, 0, +1\}$, or equivalently as a path in the graph:

(48)
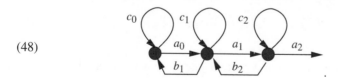

.

Lattice paths can also be interpreted as trajectories of birth-and-death processes, where a population can evolve at any discrete time by a birth or a death. (Compare with the pure-birth case in (30), p. 312.)

As a preparation for later developments, let us examine the description of the class written $\mathcal{H}_{0,0}^{[<1]}$ of Motzkin excursions of height < 1. We have

$$\mathcal{H}_{0,0}^{[<1]} \cong \text{SEQ}(c_0) \qquad \Longrightarrow \qquad H_{0,0}^{[<1]} = \frac{1}{1 - c_0}.$$

The class of excursions of height < 2 is obtained from here by a substitution

$$c_0 \mapsto c_0 + a_0 \, \text{SEQ}(c_1) b_1,$$

to the effect that

$$\mathcal{H}_{0,0}^{[<2]} \cong \text{SEQ}\left(c_0 + a_0 \, \text{SEQ}(c_1) b_1\right)$$

$$\Longrightarrow \qquad H_{0,0}^{[<2]} = \frac{1}{1 - c_0 - \dfrac{a_0 b_1}{1 - c_1}} = \frac{1 - c_1}{1 - c_0 - c_1 + c_0 c_1 - a_0 b_1}.$$

Iteration of this simple mechanism lies at the heart of the calculations performed below. Clearly, generating functions written in this way are nothing but a concise description of usual counting generating functions: for instance if individual weights[8] $\alpha_j, \beta_j, \gamma_j$ are assigned to the letters a_j, b_j, c_j, respectively, then the OGF of multiplicatively weighted paths with z marking length is obtained by setting

(49)
$$a_j = \alpha_j z, \qquad b_j = \beta_j z, \qquad c_j = \gamma_j z.$$

The general class of paths of interest in this subsection is defined by arbitrary combinations of *flooring* (by m) *ceiling* (by h), as well as fixing initial (k) and final (l) altitudes. Accordingly, we define the following subclasses of the class $\mathcal{H}$ of all Motzkin paths:

$$\mathcal{H}_{k,l}^{[m \leq \bullet < h]} \quad := \quad \{w \in \mathcal{H} : \text{ini}(w) = k, \ \text{fin}(w) = l, \ m \leq \inf\{w\}, \ \sup\{w\} < h\}.$$

We shall also need the special cases:

$$\mathcal{H}_{k,l}^{[<h]} = \mathcal{H}_{k,l}^{[0 \leq \bullet < h]}, \qquad \mathcal{H}_{k,l}^{[\geq m]} = \mathcal{H}_{k,l}^{[m \leq \bullet < \infty]}, \qquad \mathcal{H}_{k,l} = \mathcal{H}_{k,l}^{[0 \leq \bullet < \infty]}.$$

(Thus, the supercript indicates the condition that is to be satisfied by *all abscissae* of vertices of the path.) Three simple combinatorial decompositions of paths (Figure V.8) then suffice to derive all the basic formulae.

(*i*) *Arch decomposition*: An excursion from and to level 0 consists of a sequence of "arches", each made of either a c_0 or an $a_0 \mathcal{H}_{1,1}^{[\geq 1]} b_1$, so that

(50)
$$\mathcal{H}_{0,0} = \text{SEQ}\left(c_0 \cup a_0 \mathcal{H}_{1,1}^{[\geq 1]} b_1\right),$$

which relativizes to height $< h$.

(*ii*) *Last passages decomposition*. Recording the times at which each level $0, \ldots, k$ is last traversed gives

(51)
$$\mathcal{H}_{0,k} = \mathcal{H}_{0,0}^{[\geq 0]} a_0 \mathcal{H}_{1,1}^{[\geq 1]} a_1 \cdots a_{k-1} \mathcal{H}_{k,k}^{[\geq k]}.$$

---

[8]Throughout this chapter, all weights are assumed to be *non-negative*.

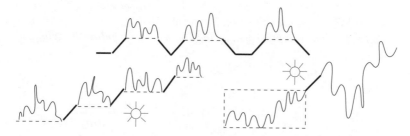

Figure V.8. The three major decompositions of lattice paths: the arch decomposition (top), the last passages decomposition (bottom left), and the first passage decomposition (bottom right).

(iii) First passage decomposition. The quantities $H_{k,l}$ with $k \le l$ are implicitly determined by the first passage through k in a path connecting level 0 to l, so that

$$(52) \qquad \mathcal{H}_{0,l} = \mathcal{H}_{0,k-1}^{[<k]} a_{k-1} \mathcal{H}_{k,l} \quad (k \le l),$$

(A dual decomposition holds when $k \ge l$.)

The basic results of the theory express the generating functions in terms of a fundamental continued fraction and its associated convergent polynomials. They involve the "numerator" and "denominator" polynomials, denoted by P_h and Q_h that are defined as solutions to the second-order (or "three-term") linear recurrence equation

$$(53) \qquad Y_{h+1} = (1 - c_h)Y_h - a_{h-1}b_h Y_{h-1}, \ h \ge 0,$$

together with the initial conditions $(P_{-1}, Q_{-1}) = (-1, 0)$, $(P_0, Q_0) = (0, 1)$, and with the convention $a_{-1}b_0 = 1$. In other words, setting $C_j = 1 - c_j$ and $A_j = a_{j-1}b_j$, we have:

(54)

$$
\begin{aligned}
&P_0 = 0, \quad P_1 = 1, \quad P_2 = C_1, \qquad\qquad P_3 = C_1 C_2 - A_2 \\
&Q_0 = 1, \quad Q_1 = C_0, \quad Q_2 = C_0 C_1 - A_1, \quad Q_3 = C_0 C_1 C_2 - C_2 A_1 - C_0 A_2.
\end{aligned}
$$

These polynomials are also known as continuant polynomials [379, 601].

▷ **V.14. Combinatorics of continuant polynomials.** The polynomial Q_h is obtained by the following process: start with the product $\Pi := C_0 C_1 \cdots C_{h-1}$; then cross out in all possible ways pairs of adjacent elements $C_{j-1}C_j$, replacing each such crossed pair by $-A_j$. For instance, Q_4 is obtained as

$$
C_0 C_1 C_2 C_3 + \overbrace{\cancel{C_0}\cancel{C_1}}^{-A_1} C_2 C_3 + C_0 \overbrace{\cancel{C_1}\cancel{C_2}}^{-A_2} C_3 + C_0 C_1 \overbrace{\cancel{C_2}\cancel{C_3}}^{-A_3} + \overbrace{\cancel{C_0}\cancel{C_1}}^{-A_1} \overbrace{\cancel{C_2}\cancel{C_3}}^{-A_3}.
$$

The polynomials P_h are obtained similarly after a shift of indices. (These observations are due to Euler; see [307, §6.7].) ◁

Proposition V.3 (Continued Fraction Theorem [214]). *(i) The generating function $H_{0,0}$ of all excursions is represented by the fundamental continued fraction:*

$$(55) \qquad H_{0,0} = \cfrac{1}{1 - c_0 - \cfrac{a_0 b_1}{1 - c_1 - \cfrac{a_1 b_2}{1 - c_2 - \cfrac{a_2 b_3}{\ddots}}}}.$$

(ii) The generating function of ceiled excursion $H_{0,0}^{[<h]}$ is given by a convergent of the fundamental continued fraction (55), with P_h, Q_h as in Equation (53):

$$(56) \qquad H_{0,0}^{[<h]} = \cfrac{1}{1 - c_0 - \cfrac{a_0 b_1}{1 - c_1 - \cfrac{a_1 b_2}{\cfrac{\ddots}{1 - c_{h-1}}}}} = \frac{P_h}{Q_h}.$$

(iii) The generating function of floored excursions is given by a truncation of the fundamental fraction:

$$(57) \qquad H_{h,h}^{[\geq h]} = \cfrac{1}{1 - c_h - \cfrac{a_h b_{h+1}}{1 - c_{h+1} - \cfrac{a_{h+1} b_{h+2}}{\ddots}}}$$

$$(58) \qquad = \frac{1}{a_{h-1} b_h} \frac{Q_h H_{0,0} - P_h}{Q_{h-1} H_{0,0} - P_{h-1}},$$

Proof. Repeated use of the arch decomposition (50) provides a form of $H_{0,0}^{[<h]}$ with nested quasi-inverses $(1 - f)^{-1}$ that is the finite fraction representation (56); for instance,

$$\mathcal{H}_{00}^{[<1]} \cong \mathrm{SEQ}(c_0), \qquad \mathcal{H}_{00}^{[<2]} \cong \mathrm{SEQ}(c_0 + a_0\, \mathrm{SEQ}(c_1)b_1),$$
$$\mathcal{H}_{00}^{[<3]} \cong \mathrm{SEQ}(c_0 + a_0\, \mathrm{SEQ}(c_1 + a_1\, \mathrm{SEQ}(c_2)b_2)b_1).$$

The continued fraction representation for basic paths without height constraints (namely $H_{0,0}$) is then obtained by taking the limit $h \to \infty$ in (56). Finally, the continued fraction form (57) for ceiled excursions is nothing but the fundamental form (55), when the indices are shifted. The three continued fraction expansions (55), (56), (57) are hence established.

Finding explicit expressions for the fractions $H_{0,0}^{[<h]}$ and $H_{h,h}^{[\geq h]}$ next requires determining the polynomials that appear in the convergents of the basic fraction (55). By definition, the convergent polynomials P_h and Q_h are the numerator and denominator of the fraction $H_{0,0}^{[<h]}$. For the computation of $H_{0,0}^{[<h]}$ and P_h, Q_h, one classically

introduces the linear fractional transformations

$$g_j(y) = \frac{1}{1 - c_j - a_j b_{j+1} y},$$

so that

(59) $H_{0,0}^{[<h]} = g_0 \circ g_1 \circ g_2 \circ \cdots \circ g_{h-1}(0)$ and $H_{0,0} = g_0 \circ g_1 \circ g_2 \circ \cdots ,$.

Now, linear fractional transformations are representable by 2×2 matrices

(60) $$\frac{ay + b}{cy + d} \mapsto \left(\begin{array}{cc} a & b \\ c & d \end{array} \right),$$

in such a way that the composition corresponds to matrix product. By induction on the compositions that build up $H_{0,0}^{[<h]}$, there follows the equality

(61) $$g_0 \circ g_1 \circ g_2 \circ \cdots \circ g_{h-1}(y) = \frac{P_h - P_{h-1} a_{h-1} b_h y}{Q_h - Q_{h-1} a_{h-1} b_h y},$$

where P_h and Q_h are seen to satisfy the recurrence (53). Setting $y = 0$ in (61) proves (56).

Finally, $H_{h,h}^{[\geq h]}$ is determined implicitly as the root y of the equation $g_0 \circ \cdots \circ g_{h-1}(y) = H_{0,0}$, an equation that, when solved using (61), yields the form (58). ∎

A large number of generating functions can be derived by similar techniques. We refer to the article [214], where this theory was first systematically developed and to the synthesis given in [303, Chapter 5]. Our presentation also draws upon [238] where the theory was put to use in order to develop a formal algebraic theory of general birth-and-death processes in continuous time.

▷ **V.15.** *Transitions and crossings.* The lattice paths $\mathcal{H}_{0,l}$ corresponding to the transitions from altitude 0 to l and $\mathcal{H}_{k,0}$ (from k to 0) have OGFs

$$H_{0,l} = \frac{1}{\mathcal{B}_l} \left(Q_l H_{0,0} - P_l \right), \qquad H_{k,0} = \frac{1}{\mathcal{A}_k} (Q_k H_{0,0} - P_k).$$

The crossings $\mathcal{H}_{0,h-1}^{[<h]}$ and $\mathcal{H}_{h-1,0}^{[<h]}$ have OGFs,

$$H_{0,h-1}^{[<h]} = \frac{\mathcal{A}_{h-1}}{Q_h}, \qquad H_{h-1,0}^{[<h]} = \frac{\mathcal{B}_{h-1}}{Q_h}.$$

(Abbreviations used here are: $\mathcal{A}_m = a_0 \cdots a_{m-1}$, $\mathcal{B}_m = b_1 \cdots b_m$.) These extensions provide combinatorial interpretations for fractions of the form $1/Q$. They result from the basic decompositions combined with Proposition V.3; see [214, 238] for details.   ◁

▷ **V.16.** *Denominator polynomials and orthogonality.* Let $H_n = [z^n] H_{0,0}(z)$ represent the number of all excursions of length n equipped with *non-negative* weights. Define a linear functional $\mathcal{L}$ on the space $\mathbb{C}(z)$ of polynomials by $\mathcal{L}[z^n] = H_n$. Introduce the reciprocal polynomials: $\overline{Q}_h(z) = z^h Q(1/z)$. The fact, deducible from Note V.15, that $Q_l H_{0,0} - P_l = O(z^{2l})$ corresponds to the property $\mathcal{L}[z^j \overline{Q}_l] = 0$ for all $0 \leq j < l$. In other words, the polynomials $\overline{Q}_l$ are orthogonal with respect to the special scalar product $\langle f, g \rangle := \mathcal{L}[fg]$. (Historically, the theory of orthogonal polynomials evolved from the theory of continued fractions, before living a life of its own; see [118, 343, 563] for its many facets.)   ◁

▷ **V.17.** *Discrete time birth-and-death processes.* Assume that, at discrete times $n = 0, 1, 2, \ldots,$ a population of size j can grow by one element [a birth] with probability α_j, decrease by one element [a death] with probability β_j, and stay the same with probability $\gamma_j = 1 - \alpha_j - \beta_j$. Let ω_n be the probability that an initially empty population is again empty at time n. Then the GF of the sequence (ω_n) is

$$\sum_{n \geq 0} \omega_n z^n = \cfrac{1}{1 - \gamma_0 z - \cfrac{\alpha_0 \beta_1 z^2}{1 - \gamma_1 z - \cfrac{\alpha_1 \beta_2 z^2}{\cdots}}}.$$

This result was found by I. J. Good in 1958: see [302].   ◁

▷ **V.18.** *Continuous time birth-and-death processes.* Consider a continuous time birth-and-death process, where a transition from state j to $j + 1$ takes place according to an exponential distribution of rate λ_j and a transition from j to $j - 1$ has rate μ_j. Let $\varpi(t)$ be the probability to be in state 0 at time t starting from state 0 at time 0. One has

$$\int_0^\infty e^{-st} \varpi(t)\, dt = \cfrac{1}{s + \lambda_0 - \cfrac{\lambda_0 \mu_1}{s + \lambda_1 + \mu_1 - \cfrac{\lambda_1 \mu_2}{\cdots}}} = \cfrac{1}{s + \cfrac{\lambda_0}{1 + \cfrac{\mu_1}{s + \cfrac{\lambda_1}{\cdots}}}}.$$

Thus, continued fractions and orthogonal polynomials may be used to analyse birth-and-death processes. (This fact was originally discovered by Karlin and McGregor [362], with later additions due to Jones and Magnus [358]. See [238] for a systematic discussion in relation to combinatorial theory.)   ◁

V.4.2. Analytic aspects. We now consider the general asymptotic properties of lattice paths of height bounded from above by a fixed integer $h \geq 1$. Letters denoting elementary steps are weighted, as previously indicated, with

$$a_j = \alpha_j z, \qquad b_j = \beta_j z, \qquad c_j = \gamma_j z,$$

the weights being invariably non-negative. We shall limit the discussion to excursions, which are often the most interesting objects from the combinatorial point of view.

As a preamble, in the Dyck case, where all γ_j are 0 (level steps are disallowed), the GF $H^{[<h]}$ is a function of z^2 only, since it takes an even number of steps to return to altitude 0 when starting from altitude 0. In such a case, we shall systematically assume that, when considering $[z^n] H^{[<h]}$, the index $n = 2\nu$ is even. In order to avoid trivialities, we also assume that none of the coefficients attached to ascents and descents are 0.

Theorem V.5 (Asymptotics of nested sequences). *Consider the class $\mathcal{H}_{0,0}^{[<h]}$ of weighted Motzkin excursions of height $< h$. In the non-Dyck case (at least one $\gamma_j \neq 0$), their number satisfies a pure exponential–polynomial formula,*

$$H_{0,0,n}^{[<h]} = cB^n + O(C^n),$$

where $B > 0$ and $0 \leq C < B$. In the Dyck case, the formula holds, assuming furthermore that $n \equiv 0 \pmod 2$.

Proof. The proof proceeds by induction according to the depth of nesting of the sequence constructions, starting with the innermost construction. (The present discussion is similar to the analysis of the supercritical sequence schema in Section V. 2, p. 293.) Write

$$f_j(z) := H_{h-j-1,h-j-1}^{[h-j-1\leq\bullet<h]}(z),$$

and let ρ_j denote the dominant singularity of f_j that is positive (existence is guaranteed by Pringsheim's Theorem).

For ease of discussion, we first examine the case where all γ_j are non-zero. The function $f_0(z)$ is

$$f_0(z) = \frac{1}{1 - \gamma_{h-1}z},$$

and one has $\rho_0 = 1/\gamma_{h-1}$. The function f_1 is given by

$$f_1(z) = \frac{1}{1 - \gamma_{h-2}z - \alpha_{h-2}\beta_{h-1}z^2 f_0(z)}.$$

The quantity $\gamma_{h-2}z + \alpha_{h-2}\beta_{h-1}z^2 f_0(z)$ in its denominator increases continuously from 0 to $+\infty$ as z increases from 0 to ρ_0; consequently, it crosses the value 1 at some point which must be ρ_1. In particular, one must have $\rho_1 < \rho_0$. Our assumption that all the γ_j are non-zero implies the absence of periodicities, so that ρ_1 is the unique dominant singularity. The argument can be repeated, implying that the sequence of radii is decreasing $\rho_0 > \rho_1 > \rho_2 > \cdots$, the corresponding poles are all simple, and they are uniquely dominating. The statement is thus established in the case that all the γ_j are non-zero.

Dually, in the Dyck case where all the γ_j are zero, one can reason in a similar manner, operating with the collection of "condensed" series $f_j(\sqrt{z})$, which are seen to have a unique dominant singularity. This implies that $f_j(z)$ itself has exactly two dominant singularities, namely ρ_h and $-\rho_h$, both being simple poles.

In the mixed case, the f_j are initially of the Dyck type, until a certain $\gamma_{h-1-j_0} \neq 0$ is encountered. In that case the function f_{j_0} is aperiodic (its span in the sense of Definition IV.5, p. 266, is equal to 1). The reasoning then continues in a similar manner to the Motzkin case, with all the subsequent f_j (for $j \geq j_0$) including $f_{h-1}(z) \equiv H_{0,0}^{[<h]}(z)$ having a unique dominant singularity. ∎

Similar devices yield a characterization of the profile of a random path, that is, the number of times a given step appears in a random excursion.

Theorem V.6 (Profile of nested sequences). *Let X_n be the random variable representing the number of times a given step (of type a_j, b_j, or c_j) with non-zero weight appears in a random excursion of length n and height $< h$. The moments of X_n satisfy*

$$\mathbb{E}(X_n) = c_1 n + d_1 + O(D^n), \qquad \mathbb{V}(X_n) = c_2 n + d_2 + O(D^n),$$

for constants c_1, c_2, d_1, d_2, D, with $c_1, c_2 > 0$ and $0 \leq D < 1$. In particular the distribution of X_n is concentrated.

Proof. Introduce an auxiliary variable u marking the number of designated steps, and form the corresponding BGF $H(z, u)$. We only detail the case of expectations. The function H is a linear fractional transformation in u of the form

$$H(z, u) = A(z) + \frac{1}{C(z) + uD(z)}.$$

(The coefficients A, B, C are *a priori* in $\mathbb{C}(z)$; they are in fact computable from Proposition V.3.) Then, one has

$$\frac{\partial}{\partial u} H(z, u) \Big|_{u=1} = -\frac{D(z)}{(C(z) + D(z))^2}.$$

This function resembles $H(z, 1)^2$. An application of the chain rule permits us to verify that indeed

$$\frac{\partial}{\partial u} H(z, u) \Big|_{u=1} = E(z) H(z, 1)^2,$$

where $E(z)$ is analytic in a disc larger than the disc of analyticity of $H(z, 1)$. The analysis of the dominant double pole then yields the result. (The determination of the second moment follows along similar lines: a triple pole is involved.) ∎

▷ **V.19.** *All poles are real.* Assume again $\alpha_j \beta_{j+1} > 0$ and $\gamma_j \geq 0$. By Note V.16, the denominator polynomials Q_h are reciprocals of a family of polynomials $\overline{Q}_h$ that are formally orthogonal with respect to a scalar product. Thus the zeros of any of the $\overline{Q}_h$ are all real, and so are the zeros of Q_h. Consequently: *The poles of the OGF of ceiled excursions* $H_{0,0}^{[<h]}$ *are all real.* (See for instance [563, §3.3] for the basic argument.) ◁

V. 4.3. Applications.

Lattice paths have quite a wide range of descriptive power, especially when weights are allowed. We illustrate this fact by three types of examples.

Example V.8 provides a complete analysis of height in Dyck paths and general plane rooted trees, as regards moments as well as distribution. This is the simplest case of a continued fraction (one with constant coefficients) attached to the OGF of Catalan numbers and involving Fibonacci-Chebyshev polynomials. Example V.9 discusses coin fountains. There, we are dealing with an infinite continued fraction to which the techniques of the previous subsection can be extended. (The developments take us close to the realm of q–calculus and to the analysis of alcohols seen in Chapter IV.) Example V.10 constitutes a typical application of the possibility of encoding combinatorial structures—here, interconnection networks—by means of lattice paths weighted by integers. The enumeration involves Hermite polynomials. (Other examples related to set partitions and permutations are described in the accompanying notes.)

Example V.8. *Height of Dyck paths and plane rooted trees.* In order to count lattice paths of the Dyck (D) or Motzkin (M) type, it suffices to effect one of the substitutions,

$$\sigma_M : a_j \mapsto z, \; b_j \mapsto z, c_j \mapsto z; \qquad \sigma_D : a_j \mapsto z, \; b_j \mapsto z, c_j \mapsto 0.$$

We henceforth restrict attention to the case of Dyck paths. See Figure V.9 for three simulations suggesting that the distribution of height is somewhat spread. Given the parenthesis system representation (Note I.48, p.77), the height of a Dyck path automatically translates into as height of the corresponding plane rooted tree.

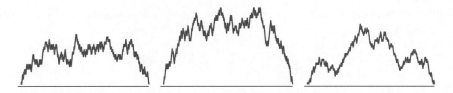

Figure V.9. Three random Dyck paths of length $2n = 500$ have heights, respectively, 20, 31, 24: the distribution is spread, see Proposition V.4.

Expressions of GFs. The continued fraction expressing $H_{0,0}$ results immediately from Proposition V.3 and is in this case periodic (here, in the sense that its stages are all alike); it represents a quadratic function,

$$H_{0,0}(z) = \cfrac{1}{1 - \cfrac{z^2}{1 - \cfrac{z^2}{1 - \cdots}}} = \frac{1}{2z^2}\left(1 - \sqrt{1 - 4z^2}\right),$$

since $H_{0,0}$ satisfies $y = (1 - z^2 y)^{-1}$. The families of polynomials P_h, Q_h are in this case determined by a recurrence with constant coefficients. Define classically the Fibonacci polynomials by the recurrence

(62) $$F_{h+2}(z) = F_{h+1}(z) - z F_h(z), \quad F_0(z) = 0, \quad F_1(z) = 1.$$

One finds $Q_h = F_{h+1}(z^2)$ and $P_h = F_h(z^2)$. (The Fibonacci polynomials are reciprocals of Chebyshev polynomials; see Note V.20, p. 329.) By Proposition V.3, the GF of paths of height $< h$ is then

$$H_{00}^{[<h]}(z) = \frac{F_h(z^2)}{F_{h+1}(z^2)}.$$

(We get more and, for instance, the number of ways of crossing a strip of width $h - 1$ is $H_{0,h-1}^{[<h]}(z) = z^{h-1}/F_{h+1}(z^2)$.) The Fibonacci polynomials have an explicit form,

$$F_h(z) = \sum_{k=0}^{\lfloor (h-1)/2 \rfloor} \binom{h-1-k}{k}(-z)^k,$$

as follows from the generating function expression: $\sum_h F_h(z) y^h = y/(1 - y + z y^2)$.

The equivalence between Dyck paths and (general) plane tree traversals discussed in Chapter I (p. 73) implies that trees of height at most h and size $n + 1$ are equinumerous with Dyck paths of length $2n$ and height at most h. Set for convenience

$$G^{[h]}(z) = z H_{00}^{[<h+1]}(z^{1/2}) = z\frac{F_{h+1}(z)}{F_{h+2}(z)},$$

which is precisely the OGF of general plane trees having height $\leq h$. (This is otherwise in agreement with the continued fraction forms obtained directly in Chapter III: cf (53), p. 195 and (79), p. 216.) It is possible to go much further as first shown by De Bruijn, Knuth, and Rice in a landmark paper [145], which also constitutes a historic application of Mellin transforms in analytic combinatorics. (We refer to this paper for historical context and references.)

First, solving the linear recurrence (62) with z treated as a parameter yields the alternative closed form expression

$$(63) \qquad F_h(z) = \frac{G^h - \overline{G}^h}{G - \overline{G}}, \qquad G = \frac{1 - \sqrt{1 - 4z}}{2}, \qquad \overline{G} = \frac{1 + \sqrt{1 - 4z}}{2}.$$

There, $G(z)$ is the OGF of all trees, and an equivalent form of $G^{[h]}$ is provided by

$$(64) \qquad G - G^{[h-2]} = \sqrt{1 - 4z} \frac{u^h}{1 - u^h}, \qquad \text{where} \quad u = \frac{1 - \sqrt{1 - 4z}}{1 + \sqrt{1 - 4z}} = \frac{G^2}{z},$$

as is easily verified. Thus $G^{[h]}$ can be expressed in terms of $G(z)$ and z:

$$G - G^{[h-2]} = \sqrt{1 - 4z} \sum_{j \geq 1} z^{-jh} G(z)^{2jh}.$$

The Lagrange–Bürmann inversion theorem (p. 732) then gives after a simple calculation

$$(65) \qquad G_{n+1} - G_{n+1}^{[h-2]} = \sum_{j \geq 1} \Delta^2 \binom{2n}{n - jh},$$

where

$$\Delta^2 \binom{2n}{n - m} := \binom{2n}{n + 1 - m} - 2 \binom{2n}{n - m} + \binom{2n}{n - 1 - m}.$$

Consequently, the number of trees of height $\geq h - 1$ admits a closed form: it is a "sampled" sum, by steps of h, of the $2n$th line of Pascal's triangle (upon taking second-order differences).

Probability distribution of height. The relation (65) leads easily to the asymptotic distribution of height in random trees of size n. Stirling's formula yields the Gaussian approximation of binomial numbers: for $k = o(n^{3/4})$ and with $w = k/\sqrt{n}$, one finds

$$(66) \qquad \frac{\binom{2n}{n-k}}{\binom{2n}{n}} \sim e^{-w^2} \left(1 - \frac{w^4 - 3w^2}{6n} + \frac{5w^8 - 54w^6 + 135w^4 - 60w^2}{360n^2} + \cdots \right).$$

The use of the Gaussian approximation (66) inside the exact formula (65) then implies: *The probability that a tree of size $n + 1$ has height at least $h - 1$ satisfies uniformly for $h \in [\alpha\sqrt{n}, \beta\sqrt{n}]$ (for any α, β such that $0 < \alpha < \beta < \infty$) the estimate*

$$(67) \qquad \frac{G_{n+1} - G_{n+1}^{[h-2]}}{G_{n+1}} = \Theta\left(\frac{h}{\sqrt{n}}\right) + O\left(\frac{1}{n}\right), \qquad \Theta(x) := \sum_{j \geq 1} e^{-j^2 x^2} (4j^2 x^2 - 2).$$

The function $\Theta(x)$ is a "theta function" which classically arises in the theory of elliptic functions [604]. Since binomial coefficients decay rapidly, away from the centre, simple bounds also show that the probability of the height being at least $n^{1/2+\epsilon}$ decays as $\exp(-n^{2\epsilon})$, so that it is exponentially small. Note also that the probability distribution of height H itself admits an exact expression obtained by differencing (65), which is reflected asymptotically by differentiation of the estimate of (67):

$$(68)$$

$$\mathbb{P}_{\mathcal{G}_{n+1}}[H = \lfloor x\sqrt{n} \rfloor] = -\frac{1}{\sqrt{n}} \Theta'(x) + O\left(\frac{1}{n}\right), \qquad \Theta'(x) := \sum_{j \geq 1} e^{-j^2 x^2} (12j^2 x - 8j^4 x^3).$$

The forms (67) and (68) also give access to moments of the distribution of height. We find

$$\mathbb{E}_{\mathcal{G}_{n+1}}[H^r] \sim \frac{1}{\sqrt{n}} S_r\left(\frac{1}{\sqrt{n}}\right), \qquad \text{where} \quad S_r(y) := -\sum_{h \geq 1} h^r \Theta'(hy).$$

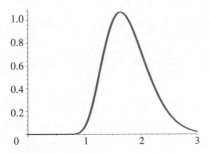

Figure V.10. The limit density of the distribution of height $-\Theta'(x)$.

The quantity $y^{r+1}S_r(y)$ is a Riemann sum relative to the function $-x^r\Theta'(x)$, and the step $y = n^{-1/2}$ decreases to 0 as $n \to \infty$. Approximating the sum by the integral, one gets:

$$\mathbb{E}_{\mathcal{G}_{n+1}}[H^r] \sim n^{r/2}\mu_r \qquad \text{where} \quad \mu_r := -\int_0^\infty x^r\Theta'(x)\,dx.$$

The integral giving μ_r is a Mellin transform in disguise (set $s = r + 1$) to which the treatment of harmonic sums applies. We then get upon replacing $n + 1$ by n:

Proposition V.4. *The expected height of a random plane rooted tree comprising $n + 1$ nodes is*

$$(69) \qquad\qquad \sqrt{\pi n} - \frac{3}{2} + o(1).$$

More generally, the moment of order r of height is asymptotic to

$$(70) \qquad\qquad \mu_r n^{r/2} \qquad \text{where} \quad \mu_r = r(r-1)\Gamma(r/2)\zeta(r).$$

The random variable $H/\sqrt{n}$ obeys asymptotically a Theta distribution, in the sense of both the "central" estimate (67) and the "local" estimate (68). The same asymptotic estimates hold for height of Dyck paths having length $2n$.

　　The improved estimate of the mean (69) is from [145]. The general form of moments in (70) is in fact valid for any real r (not just integers). An alternative formula for the Theta function appears in Note V.20 below. Figure V.10 plots the limit density $-\Theta'(x)$, which surfaces again in the height of binary and other simple trees (Example VII.27, p. 535). .... ∎

▷ **V.20.** *Height and Fibonacci–Chebyshev polynomials.* The reciprocal polynomials $\overline{F}_h(z) = F_{h-1}(z) = z^{h-1}F_h(1/z^2)$ are related to the classical Chebyshev polynomials by $\overline{F}_h(2z) = U_h(z)$, where $U_h(\cos(\theta)) = \sin((h+1)\theta)/\sin(\theta)$. (This is readily verified from the recurrence (62) and elementary trigonometry.) Then, the roots of $F_h(z)$ are $(4\cos^2 j\pi/(h+1))^{-1}$ and the partial fraction expansion of $G^{[h]}(z)$ can be worked out explicitly [145]. Thus, for $n \geq 1$,

$$(71) \qquad\qquad G_{n+1}^{[h-2]} = \frac{4^{n+1}}{h}\sum_{1\leq j<h/2}\sin^2\frac{j\pi}{h}\cos^{2n}\frac{j\pi}{h},$$

which provides in particular an asymptotic form for any fixed h. (This formula can also be found directly from the sampled sum (65) by multisection of series.) Asymptotic analysis of this last expression when $h = x\sqrt{n}$ yields the alternative expression

$$\lim_{n\to\infty}\mathbb{P}_{\mathcal{G}_{n+1}}[H \leq x\sqrt{n}] = 4\pi^{5/2}x^{-3}\sum_{j\geq 0}j^2 e^{-j^2\pi^2/x^2} \qquad (\equiv 1 - \Theta(x)),$$

which, when compared with (67), reflects an important transformation formula of elliptic functions [604]. See the study by Biane, Pitman, and Yor [64] for fascinating connections with Brownian motion and the functional equation of the Riemann zeta function. Height in simple varieties of trees also obeys a Theta law, but the proofs (Example VII.27, p. 535) require the full power of singularity analysis.   ◁

▷ **V.21.** *Motzkin paths.* The OGF of Motzkin paths of height $< h$ is $\frac{1}{1-z} \cdot {}^D H_{0,0}^{[<h]}\left(\frac{z}{1-z}\right)$, where ${}^D H_{0,0}^{[<h]}$ refers to Dyck paths. Therefore, such paths can be enumerated exactly by formulae derived from Equations(65) to (71). Accordingly, the mean height is $\sim \sqrt{3\pi n}$.   ◁

Example V.9. *Area under Dyck path and coin fountains.*   Consider Dyck paths and the area parameter: *area* under a lattice path is taken here as the sum of the indices (i.e., the starting altitudes) of all the variables that enter the standard encoding of the path. Thus, the BGF $D(z, q)$ of Dyck paths with z marking half-length and q marking area is obtained by the substitution

$$a_j \mapsto q^j z, \quad b_j \mapsto q^j, \quad c_j \mapsto 0$$

inside the fundamental continued fraction (55). (We rederive here Equation (54) of Chapter III, p. 196.) It proves convenient to operate with the continued fraction

$$(72) \qquad\qquad F(z, q) = \cfrac{1}{1 - \cfrac{zq}{1 - \cfrac{zq^2}{\ddots}}},$$

so that $D(z, q) = F(q^{-1}z, q^2)$. Since F satisfies a difference equation,

$$(73) \qquad\qquad F(z, q) = \frac{1}{1 - zq F(qz, q)},$$

moments of area can be determined by differentiating and setting $q = 1$ (see Chapter III, p. 184, for a direct approach).

   A general trick from q–calculus is effective for deriving an alternative form of F. Express the continued fraction F of (72) as a quotient $F(z, q) = A(z)/B(z)$. Then, the relation (73) implies

$$\frac{A(z)}{B(z)} = \frac{1}{1 - qz\frac{A(qz)}{B(qz)}},$$

and, by identifying numerators and denominators, we get

$$A(z) = B(qz), \quad B(z) = B(qz) - qzB(q^2 z),$$

with q treated as a parameter. The difference equation satisfied by $B(z)$ is then readily solved by indeterminate coefficients. (This classical technique was introduced in the theory of integer partitions by Euler.) With $B(z) = \sum b_n z^n$, the coefficients satisfy the recurrence

$$b_0 = 1, \quad b_n = q^n b_n - q^{2n-1} b_{n-1}.$$

This is a first-order recurrence on b_n that unwinds to give

$$b_n = (-1)^n \frac{q^{n^2}}{(1-q)(1-q^2)\cdots(1-q^n)}.$$

In other words, introducing the "q–exponential function",

(74) $$E(z,q) = \sum_{n=0}^{\infty} \frac{(-z)^n q^{n^2}}{(q)_n}, \quad \text{where} \quad (q)_n = (1-q)(1-q^2)\cdots(1-q^n),$$

one finds

(75) $$F(z,q) = \frac{E(qz,q)}{E(z,q)}.$$

The exact distribution of area in Dyck paths can then be regarded as known, in the sense that it is fully characterized by (74) and (75). (Example VII.26, p. 533, presents an analysis of the corresponding limit distribution, based on "moment pumping", to the effect that an Airy law prevails.)

Given the importance of the functions under discussion in various branches of mathematics, we cannot resist a quick digression. The name of the q–exponential comes form the obvious property that $E(z(1-q), q)$ reduces to e^{-z} as $q \to 1^-$. The explicit form (74) constitutes in fact the "easy half" of the proof of the celebrated Rogers–Ramanujan identities, namely,

(76)
$$E(-1,q) = \sum_{n=0}^{\infty} \frac{q^{n^2}}{(q)_n} \quad = \quad \prod_{n=0}^{\infty} (1-q^{5n+1})^{-1}(1-q^{5n+4})^{-1}$$
$$E(-q,q) = \sum_{n=0}^{\infty} \frac{q^{n(n+1)}}{(q)_n} \quad = \quad \prod_{n=0}^{\infty} (1-q^{5n+2})^{-1}(1-q^{5n+3})^{-1},$$

that relate the q–exponential to modular forms. See Andrews' book [14, Ch. 7] for context.

Coin fountains. Here is finally a cute application of these ideas to the asymptotic enumeration of some special polyominoes. Odlyzko and Wilf define in [461, 464] an (n, m) coin fountain as an arrangement of n coins in rows in such a way that there are m coins in the bottom row, and that each coin in a higher row touches exactly two coins in the next lower row. Let $C_{n,m}$ be the number of (n, m) fountains and $C(z, q)$ be the corresponding BGF with q marking n and z marking m. Set $C(q) = C(1, q)$. The question is to determine the total number of coin fountains of area n, $[q^n]C(q)$. The series starts as (this is *EIS* **A005169**)

$$C(q) = 1 + q + q^2 + 2q^3 + 3q^4 + 5q^5 + 9q^6 + 15q^7 + 26q^8 + \cdots,$$

as results from inspection of the first few cases.

There is a clear bijection with Dyck paths (do a 135° scan) that takes area into account: a coin fountain of size n with m coins on its base is equivalent to a Dyck path of length $2m$ and area $2n - m$ (with our earlier definition of area of Dyck paths). From this bijection, one has $C(z, q) = F(z, q)$ (with F as defined earlier) and, in particular, $C(q) = F(1, q)$. Consequently, by (72) and (75), we find

$$C(q) = \cfrac{1}{1 - \cfrac{q}{1 - \cfrac{q^2}{1 - \cfrac{q^3}{\ddots}}}} = \frac{E(q,q)}{E(1,q)}.$$

Objects	weights $(\alpha_j, \beta_j\gamma_j)$	counting	orthogonal pol.
Simple paths	$1, 1, 0$	Catalan #	Chebyshev
Permutations	$j+1, j, 2j+1$	Factorial #	Laguerre
Alternating perm.	$j+1, j, 0$	Secant #	Meixner
Involutions	$1, j, 0$	Odd factorial #	Hermite
Set partition	$1, j, j+1$	Bell #	Poisson–Charlier
Non-overlap. set part.	$1, 1, j+1$	Bessel #	Lommel

Figure V.11. Some special families of combinatorial objects together with corresponding weights, counting sequences, and orthogonal polynomials. (See also Notes V.23— 25.)

The rest of the discussion is analogous to Section IV. 7.3 (p. 283) relative to alcohols. The function $C(q)$ is *a priori* meromorphic in $|q| < 1$. An exponential lower bound of the form 1.6^n holds for $[q^n]C(q)$, since $(1 - q)/(1 - q - q^2)$ is dominated by $C(q)$ for $q > 0$. At the same time, the number $[q^n]C(q)$ is majorized by the number of compositions, which is 2^{n-1}. Thus, the radius of convergence of $C(q)$ has to lie somewhere between 0.5 and 0.61803 . . . . It is then easy to check by numerical analysis the existence of a simple zero of the denominator, $E(1, q)$, near $\rho \doteq 0.57614$. Routine computations based on Rouché's theorem then make it possible to verify formally that ρ is the only pole in $|q| \le 3/5$ and that this pole is simple (the process is detailed in [461]). Thus, singularity analysis of meromorphic functions applies.

Proposition V.5. *The number of coin fountains made of n coins satisfies asymptotically*

$$[q^n]C(q) = cA^n + O((5/3)^n), \quad c \doteq 0.31236, \quad A = \rho^{-1} \doteq 1.73566.$$

This example illustrates the power of modelling by continued fractions as well as the smooth articulation with meromorphic function asymptotics. . . . . ∎

Lattice path encodings of classical structures. The systematic theory of lattice path enumerations and continued fractions was developed initially because of the need to count weighted lattice paths, notably in the context of the analysis of dynamic data structures in computer science [226]. In this framework, a system of multiplicative weights $\alpha_j, \beta_j, \gamma_j$ is associated with the steps a_j, b_j, c_j, each weight being an integer that represents a number of "possibilities" for the corresponding step type. A system of weighted lattice paths has counting generating functions given by the usual substitution from the corresponding multivariate expressions; namely,

(77) $a_j \mapsto \alpha_j z, \quad b_j \mapsto \beta_j z, \quad c_j \mapsto \gamma_j z,$

where z marks the length of paths. One can then attempt to solve an enumeration problem expressible in this way by reverse-engineering the known collection of continued fractions as found in reference books such as those by Perron [479], Wall [601], and Lorentzen–Waadeland [412]. Next, for general reasons, the polynomials P, Q are always elementary variants of a family of orthogonal polynomials that is determined by the weights (see Note V.16, p. 323, and [118, 563]). When the multiplicities have enough structural regularity, the weighted lattice paths are likely to correspond to classical combinatorial objects and to classical families of orthogonal polynomials; see [214, 226, 295, 303] and Figure V.11 for an outline. We illustrate this by a simple

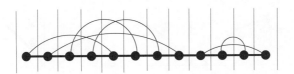

Figure V.12. An interconnection network on $2n = 12$ points.

example due to Lagarias, Odlyzko, and Zagier [394], which is relative to involutions without fixed points.

***Example* V.10.** *Interconnection networks and involutions.* The problem treated here is the following [394]. *There are 2n points on a line, with n point-to-point connections between pairs of points. What is the probable behaviour of the* width *of such an interconnection network?* Imagine the points to be $1, \ldots, 2n$, the connections as circular arcs between points, and let a vertical line sweep from left to right; width is defined as the maximum number of arcs met by such a line. One may freely imagine a tunnel of fixed capacity (this corresponds to the width) inside which wires can be placed to connect points pairwise (Figure V.12).

Let $\mathcal{J}_{2n}$ be the class of all interconnection networks on $2n$ points, which is precisely the collection of ways of grouping $2n$ elements into n pairs, or, equivalently, the class of all involutions without fixed points, i.e., permutations with cycles of length 2 only. The number J_{2n} equals the "odd factorial",

$$J_{2n} = 1 \cdot 3 \cdot 5 \cdots (2n-1),$$

whose EGF is $e^{z^2/2}$ (see Chapter II, p. 122). The problem calls for determining the quantity $J_{2n}^{[h]}$ that is the number of networks having width $\leq h$.

The relation to lattice paths is as follows. First, when sweeping a vertical line across a network, define an active arc at an abscissa as one that straddles that abscissa. Then build the sequence of active arc counts at half-integer positions $\frac{1}{2}, \frac{3}{2}, \ldots, 2n - \frac{1}{2}, 2n + \frac{1}{2}$. This constitutes a sequence of integers in which each member is ± 1 the previous one; that is, a lattice path without level steps. In other words, there is an ascent in the lattice path for each element that is smaller in its cycle and a descent otherwise. One may view ascents as associated to situations where a node "opens" a new cycle, while descents correspond to "closing" a cycle.

Involutions are much more numerous than lattice paths, so that the correspondence from involutions to lattice paths has to be many-to-one. However, one can easily enrich lattice paths, so that the enriched objects are in one-to-one correspondence with involutions. Consider again a scanning position at a half-integer where the vertical line crosses ℓ (active) arcs. If the next node is of the closing type, there are ℓ possibilities to choose from. If the next node is of the opening type, then there is only one possibility, namely, to start a new cycle. A complete encoding of a network is accordingly obtained by recording additionally the sequence of the n possible choices corresponding to descents in the lattice path (some canonical order is fixed, for instance, oldest first). If we write these choices as superscripts, this means that the set of all enriched encodings of networks is obtained from the set of standard lattice path encodings by effecting the substitutions

$$b_j \mapsto \sum_{k=1}^{j} b_j^{(k)}.$$

Figure V.13. Three simulations of random networks with $2n = 1000$ illustrate the tendency of the profile to conform to a parabola with height close to $n/2 = 250$.

The OGF of all involutions is obtained from the generic continued fraction of Proposition V.3 by the substitution

$$a_j \mapsto z, \quad b_j \mapsto j \cdot z,$$

where z records the number of steps in the enriched lattice path, or equivalently, the number of nodes in the network. In other words, we have obtained combinatorially a *formal* continued fraction representation,

$$\sum_{n=0}^{\infty}(1 \cdot 3 \cdots (2n-1))z^{2n} = \cfrac{1}{1 - \cfrac{1 \cdot z^2}{1 - \cfrac{2 \cdot z^2}{1 - \cfrac{3 \cdot z^2}{\ddots}}}},$$

which was originally discovered by Gauss [601]. Proposition V.3 also gives immediately the OGF of involutions of width at most h as a quotient of polynomials. Define

$$J^{[h]}(z) := \sum_{n\geq 0} J^{[h]}_{2n} z^{2n}.$$

One has

$$J^{[h]}(z) = \cfrac{1}{1 - \cfrac{1 \cdot z^2}{1 - \cfrac{2 \cdot z^2}{\ddots}}} = \frac{P_{h+1}(z)}{Q_{h+1}(z)}$$

$$\overline{1 - h \cdot z^2}$$

where P_h and Q_h satisfy the recurrence

$$Y_{h+1} = Y_h - hz^2 Y_{h-1}.$$

The polynomials are readily determined by their generating functions that satisfies a first-order linear differential equation reflecting the recurrence. In this way, the denominator polynomials are identified to be reciprocals of the Hermite polynomials,

$$\mathrm{He}_h(z) = (2z)^h Q_h\left(\frac{1}{z\sqrt{2}}\right),$$

themselves defined classically [3, Ch. 22] as orthogonal with respect to the measure $e^{-x^2} dx$ on $(-\infty, \infty)$ and expressible via

$$\mathrm{He}_m(x) = \sum_{m=0}^{\lfloor m/2 \rfloor} \frac{(-1)^j m!}{j!(m-2j)!}(2x)^{m-2j}, \qquad \sum_{m \geq 0} \mathrm{He}_m(x)\frac{t^m}{m!} = e^{2xt - t^2}.$$

In particular, one finds

$$J^{[0]} = 1, \quad J^{[1]} = \frac{1}{1 - z^2}, \quad J^{[2]} = \frac{1 - 2z^2}{1 - 3z^2}, \quad J^{[3]} = \frac{1 - 5z^2}{1 - 6z^2 + 3z^4}, \quad \&c.$$

The interesting analysis of the dominant poles of the rational GFs, for any fixed h, is discussed in the paper [394]. Furthermore, simulations strongly suggest that the width of a random interconnection network on $2n$ nodes is tightly concentrated around $n/2$; see Figure V.13. Louchard [418] (see also Janson's study [353]) succeeded in proving this fact and a good deal more. With high probability, the altitude (the altitude is defined here as the number of active arcs as time evolves) of a random network conforms asymptotically to a deterministic parabola $2nx(1 - x)$ (with $x \in [0, 1]$) to which are superimposed random fluctuations of a smaller amplitude, $O(\sqrt{n})$, well-characterized by a Gaussian process. In particular, *the width of a random network of $2n$ nodes converges in probability to $n/2$*. .... ∎

▷ **V.22. Bell numbers and continued fractions.** With $S_n = n![z^n]e^{e^z - 1}$ a Bell number:

$$\sum_{n \geq 0} S_n z^n = \cfrac{1}{1 - 1z - \cfrac{1z^2}{1 - 2z - \cfrac{2z^2}{\cdots}}}.$$

[Hint: Define an encoding like for networks, with level steps representing intermediate elements of blocks [214].] Refinements include Stirling partition numbers and involution numbers. ◁

▷ **V.23. Factorial numbers and continued fractions.** One has

$$\sum_{n \geq 0} n! z^n = \cfrac{1}{1 - 1z - \cfrac{1^2 z^2}{1 - 3z - \cfrac{2^2 z^2}{\cdots}}}.$$

Refinements include tangent and secant numbers, as well as Stirling cycle numbers and Eulerian numbers. (This continued fraction goes back to Euler [198]; see [214] for a proof based on a bijection of Françon–Viennot [269] and Biane [63] for an alternative bijection.) ◁

▷ **V.24. Surjection numbers and continued fractions.** Let $R_n = n![z^n](2 - e^z)^{-1}$. Then

$$\sum_{n=0}^{\infty} R_n z^n = \cfrac{1}{1 - 1z - \cfrac{2 \cdot 1^2 z^2}{1 - 4z - \cfrac{2 \cdot 2^2 z^2}{1 - 7z - \cdots}}}.$$

This continued fraction is due to Flajolet [216]. ◁

▷ **V.25.** *The Ehrenfest^2 two-chambers model.* (See Note II.11, p. 118 for context.) The OGF of the number of evolutions that lead to chamber A full satisfies

$$\sum_{n \geq 0} E_n^{[N]} z^n = \cfrac{1}{1 - \cfrac{1Nz^2}{1 - \cfrac{2(N-1)z^2}{\cdots}}} = \frac{1}{2^N} \sum_{k=0}^{N} \frac{\binom{N}{k}}{1 - (N-2k)z}.$$

This results from the EGF of Note II.11 (p. 118), the Continued Fraction Theorem, and basic properties of the Laplace transform. (This continued fraction expansion is originally due to Stieltjes [562] and Rogers [516]. See also [304] for additional formulae.)   ◁

V. 5. Paths in graphs and automata

In this section, we develop the framework of *paths in graphs*: given a graph, a source node, and a destination node, the problem is to enumerate all paths from the source to the destination in the graph. Non-negative weights acting multiplicatively (probabilities, multiplicities) may be attached to edges. Applications include the analysis of walks in various types of graphs as well as languages described by finite automata. Under a fundamental structural condition, known as *irreducibility* and corresponding to strong-connectedness of the graph, generating functions of paths all have the same dominant singularity, which is a *simple pole*. This essential property implies simple exponential forms for the asymptotics of coefficients (possibly tempered by explicit congruence conditions in the periodic case). The corresponding results can equivalently be formulated in terms of the set of *eigenvalues* (the spectrum) of the corresponding adjacency matrix and are related to the classical Perron–Frobenius theory of non-negative matrices—under irreducibility, only the largest positive eigenvalue matters asymptotically.

V. 5.1. Combinatorial aspects. A *directed graph* or *digraph* Γ is determined by the pair (V, E) of its vertex set V and its edge set $E \subseteq V \times V$. Here, self-loops corresponding to edges of the form (v, v) are allowed. Given an edge, $e = (a, b)$, we denote its origin by orig$(e) := a$ and its destination by destin$(e) := b$. For Γ a digraph with vertex set identified to the set $\{1, \ldots, m\}$, we allow each edge (a, b) to be weighted by a quantity $g_{a,b}$, which we may take as a formal indeterminate for which we allow the possibility of substituting positive weight values; the matrix $\mathbf{G}$ such that

(78) $\mathbf{G}_{a,b} = g_{a,b}$ if the edge $(a, b) \in \Gamma$, $\mathbf{G}_{a,b} = 0$ otherwise,

is called the *weighted adjacency matrix* of the (weighted) graph Γ (Figure V.14). The usual adjacency matrix of Γ is obtained by the substitution $g_{a,b} \mapsto 1$.

A *path* is a sequence of edges, $\varpi = (e_1, \ldots, e_n)$, such that, for all j with $1 \leq j < n$, one has destin$(e_j) = $ orig(e_{j+1}). The parameter n is called the length of the path and we define: orig$(\varpi) := $ orig(e_1), destin$(\varpi) := $ destin(e_n). A *circuit* is a path whose origin and destination are the same vertex. Note that, with our definition, a circuit has its origin that is distinguished. We *do not* identify here two circuits such that one is obtained by circular permutation from the other: the circuits that we consider, with such a distinguished root, are *rooted circuits*.

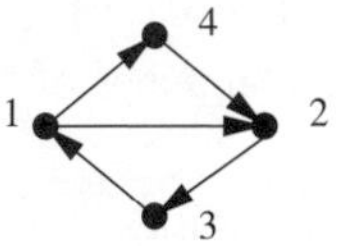

$$\Gamma = \begin{matrix} & 4 \\ & \\ 1 & \quad 2 \\ & \\ & 3 \end{matrix} \quad , \qquad G = \begin{pmatrix} 0 & g_{1,2} & 0 & g_{1,4} \\ 0 & 0 & g_{2,3} & 0 \\ g_{3,1} & 0 & 0 & 0 \\ 0 & g_{4,2} & 0 & 0 \end{pmatrix},$$

$$F^{\langle 1,1 \rangle}(z) = 1 + g_{1,2}g_{2,3}g_{3,1}z^3 + g_{1,4}g_{4,2}g_{2,3}g_{3,1}z^4 + \cdots .$$

Figure V.14. A graph Γ, its formal adjacency matrix G, and the generating function $F^{\langle 1,1 \rangle}(z)$ of paths from 1 to 1.

From the standard definition of matrix products, the powers G^n have elements that are *path polynomials*. More precisely, one has the simple but essential relation,

$$(79) \qquad (G)^n_{i,j} = \sum_{w \in \mathcal{F}_n^{\langle i,j \rangle}} w,$$

where $\mathcal{F}_n^{\langle i,j \rangle}$ is the set of paths in Γ that connect i to j and have length n, and a path w is identified with the monomial in indeterminates $\{g_{i,j}\}$ that represents multiplicatively the succession of its edges; for instance:

$$(G)^3_{i,j} = \sum_{v_1=i,v_2,v_3,v_4=j} g_{v_1,v_2}g_{v_2,v_3}g_{v_3,v_4}.$$

In other words: *powers of the matrix associated to a graph generate all paths in graph*, the weight of a path being the *product* of the weights of the individual edges it comprises. (This fact probably constitutes the most basic result of algebraic graph theory [66, p. 9].) One may then treat simultaneously all lengths of paths (and all powers of matrices) by introducing the variable z to record length.

Proposition V.6. (*i*) *Let* Γ *be a digraph and let* G *be the formal adjacency matrix of* Γ *as given by* (78). *The OGF* $F^{\langle i,j \rangle}(z)$ *of the set of all paths from* i *to* j *in* Γ, *with* z *marking length and* $g_{a,b}$ *the weight associated to edge* (a,b), *is the entry* i, j *of the matrix* $(I - zG)^{-1}$; *namely*

$$(80) \qquad F^{\langle i,j \rangle}(z) = \left((I - zG)^{-1} \right)_{i,j} = (-1)^{i+j} \frac{\Delta^{\langle i,j \rangle}(z)}{\Delta(z)},$$

where $\Delta(z) = \det(I - zG)$ *is the reciprocal polynomial of the characteristic polynomial of* G *and* $\Delta^{\langle j,i \rangle}(z)$ *is the determinant of the minor of index* j, i *of* $I - zG$.

(*ii*) *The generating function of (rooted) circuits is expressible in terms of a logarithmic derivative:*

$$(81) \qquad \sum_i (F^{\langle i,i \rangle}(z) - 1) = -z \frac{\Delta'(z)}{\Delta(z)}.$$

In this algebraic statement, if one takes the $\{g_{a,b}\}$ as formal indeterminates, then $F^{\langle i,j \rangle}(z)$ is a multivariate GF of paths in z with the variable $\{g_{a,b}\}$ marking the number of occurrences of edge (a,b). The result applies, in particular, to the case where the $g_{a,b}$ are assigned numerical values, in which case $[z^n]F^{\langle i,j \rangle}(z)$ becomes the total weight of paths of length n, which we also refer to as "number of paths" in the weighted graph.

Proof. For the proof, it is convenient to assume that the quantities $g_{a,b}$ are assigned arbitrary real numbers, so that usual matrix operations (triangularization, diagonalization, and so on) can be easily applied. As the properties expressed by the statement are ultimately equivalent to a collection of multivariate polynomial identities, their general validity is implied by the fact that they hold for all real assignments of values.

Part (i) is a consequence of the fundamental equivalence between paths and matrix products (79), which implies

$$F^{\langle i,j \rangle}(z) = \sum_{n=0}^{\infty} z^n \left(\mathbf{G}^n\right)_{i,j} = \left((I - z\mathbf{G})^{-1}\right)_{i,j},$$

and from the cofactor formula of matrix inversion.

Part (ii) results from elementary properties of the matrix trace[9] functional. With m the dimension of $\mathbf{G}$ and $\{\lambda_1, \ldots, \lambda_m\}$ the multiset of its eigenvalues, we have

$$(82) \qquad \sum_{i=1}^{m} F_n^{\langle i,i \rangle} = \mathrm{Tr}\,\mathbf{G}^n = \sum_{j=1}^{m} \lambda_j^n,$$

where $F_n^{\langle i,j \rangle} = [z^n]F^{\langle i,j \rangle}(z)$. Upon taking a generating function, there results that

$$(83) \qquad \sum_{i=1}^{m} \sum_{n=1}^{\infty} F_n^{\langle i,i \rangle} z^n = \sum_{j=1}^{m} \frac{\lambda_j z}{1 - \lambda_j z},$$

which, up to a factor of $-z$, is none other than the logarithmic derivative of $\Delta(z)$. ∎

▷ **V.26.** *Positivity of inverses of characteristic polynomials.* Let $\mathbf{G}$ have non-negative coefficients. Then, the rational function $Z_{\mathbf{G}}(z) := 1/\det(I - z\mathbf{G})$ has non-negative Taylor coefficients. More generally, if $\mathbf{G} = (g_{a,b})$ is a matrix in the formal indeterminates $g_{a,b}$, then $[z^n]Z_{\mathbf{G}}(z)$ is a polynomial in the $g_{a,b}$ with non-negative coefficients. (Hint: The proof proceeds by integration from (81): we have, for $1/\Delta(z)$, the equivalent expressions

$$\frac{1}{\Delta(z)} \equiv \exp\left(-\int_0^z \frac{\Delta'(t)}{\Delta(t)}\,dt\right) = \exp\left(\int_0^z \sum_{i=1}^{m}(F^{\langle i,i \rangle}(t) - 1)\,\frac{dt}{t}\right) = \exp\left(\sum_{n \ge 1} \frac{z^n}{n}\,\mathrm{Tr}\,\mathbf{G}^n\right),$$

which ensure positivity of the coefficients of $Z_{\mathbf{G}}$.)   ◁

▷ **V.27.** *MacMahon's Master Theorem.* Let J be the determinant

$$J(z_1, \ldots, z_m) := \begin{vmatrix} 1 - z_1 g_{11} & -z_2 g_{12} & \cdots & -z_m g_{1m} \\ -z_1 g_{21} & 1 - z_2 g_{22} & \cdots & -z_m g_{2m} \\ \vdots & \vdots & \ddots & \vdots \\ -z_m g_{m1} & -z_2 g_{2m} & \cdots & 1 - z_m g_{mm} \end{vmatrix}.$$

MacMahon's "Master Theorem" asserts the identity of coefficients,

$$[z_1^{\alpha_1} \cdots z_m^{\alpha_m}]\frac{1}{J(z_1, \ldots, z_m)} = [z_1^{\alpha_1} \cdots z_m^{\alpha_m}]Y_1^{\alpha_1} \cdots Y_m^{\alpha_m}, \quad \text{where} \quad Y_j = \sum_i g_{ij} z_j.$$

---

[9]If $\mathbf{H}$ is an $m \times m$ matrix with multiset of eigenvalues $\{\mu_1, \ldots, \mu_m\}$, the trace is defined by $\mathrm{Tr}\,\mathbf{H} := \sum_{i=1}^{m}(\mathbf{H})_{ii}$ and, by triangularization (Jordan form), it satisfies $\mathrm{Tr}\,\mathbf{H} = \sum_{j=1}^{m} \mu_j$.

This result can be obtained by a simple change of variables in a multivariate Cauchy integral and is related to multivariate Lagrange inversion [303, pp. 21–23]. Cartier and Foata [105] provide a general combinatorial interpretation related to trace monoids of Note V.10, p. 307. ◁

▷ **V.28.** *The Jacobi trace formula.* this trace formula [303, p. 11] for square matrices is

$$
\text{(84)} \qquad \det \circ \exp(M) = \exp \circ \text{Tr}(M);
$$

equivalently, with due care paid to determinations: $\log \circ \det(M) = \text{Tr} \circ \log(M)$. It generalizes the scalar identities $e^a e^b = e^{a+b}$ and $\log ab = \log a + \log b$. (Hint: recycle the computations of Note V.26.) ◁

▷ **V.29.** *Fast computation of the characteristic polynomial.* The following algorithm is due to Leverrier (1811–1877), the astronomer and mathematician who, together with Adams, first predicted the position of the planet Neptune. Since, by (82) and (83), one has

$$
\sum_{n \geq 1} z^n \, \text{Tr} \, \mathbf{G}^n = \sum_{j=1}^{m} \frac{\lambda_j z}{1 - \lambda_j z},
$$

it is possible to deduce an algorithm that determines the characteristic polynomial of a matrix of dimension m in $O(m^4)$ arithmetic operations. [Hint: computing the quantities $\text{Tr} \, \mathbf{G}^j$ for $j = 1, \ldots, m$ is sufficient and requires precisely m matrix multiplications.] ◁

▷ **V.30.** *The Matrix Tree Theorem.* Let Γ be a directed graph without loops and associated matrix $\mathbf{G}$, with $g_{a,b}$ the weight of edge (a, b). The Laplacian matrix $\mathbf{L}[\mathbf{G}]$ is defined by

$$
\mathbf{L}[\mathbf{G}]_{i,j} = -g_{i,j} + [\![i = j]\!]\delta_i, \qquad \text{where} \quad \delta_i := \sum_k g_{i,k}.
$$

Let $\mathbf{L}_1[\mathbf{G}]$ be the matrix obtained by deleting the first row and first column of $\mathbf{L}[\mathbf{G}]$. Then, the "tree polynomial"

$$
T_1[\mathbf{G}] := \det \mathbf{L}_1[\mathbf{G}]
$$

enumerates all (oriented) spanning trees of Γ rooted at node 1. (This classic result belongs to a circle of ideas initiated by Kirchhoff, Sylvester, Borchardt and others in the nineteenth century. See, for instance, the discussions by Knuth [377, p. 582–583] and Moon [445].) ◁

Weighted graphs, word models, and finite automata. The numeric substitution $\sigma : g_{a,b} \mapsto 1$ transforms the formal adjacency matrix $\mathbf{G}$ of Γ into the usual *adjacency matrix*. In particular, the number of paths of length n is obtained, under this substitution, as $[z^n](1 - z\mathbf{G})^{-1}$. As already noted, it is possible to consider weighted graphs, where the $g_{a,b}$ are assigned *positive real-valued weights*; with the weight of a path being defined by the product of its edge weights. One finds that $[z^n](I - z\mathbf{G})^{-1}$ equals the total weight of all paths of length n. If furthermore the assignment is made in such a way that $\sum_b g_{a,b} = 1$, for all a, then the matrix $\mathbf{G}$, which is called a *stochastic matrix*, can be interpreted as the transition matrix of a Markov chain. Naturally, the formulae of Proposition V.6 continue to hold in all these cases.

Word problems corresponding to regular languages can be treated by the theory of regular specifications whenever they have enough structure and an unambiguous regular expression description is of tractable form. (This is the main theme of Subsection I.4.1, p. 51, further pursued in Sections V.3 and V.4.) The dual point of view of automata theory introduced in Subsection I.4.2 (p. 56) proves useful whenever no such direct description is in sight. Finite automata can be reduced to the theory of paths in graphs, so that Proposition V.6 is applicable to them. Indeed, the language $\mathcal{L}$

accepted by a finite automaton A, with set of states Q, initial state q_0, and Q_f the set of final states, decomposes as

$$\mathcal{L} = \sum_{q \in Q_f} \mathcal{F}^{\langle q_0, q \rangle},$$

where $\mathcal{F}^{\langle q_0, q \rangle}$ is the set of paths from the initial state q_0 to the final state, q. (The corresponding graph Γ is obtained from A by collapsing multiple edges between any two vertices, i and j, into a single edge equipped with a weight that is the *sum* of the weights of all the letters leading from i to j.) Proposition V.6 is then clearly applicable.

Profiles. The term "profile of a set of paths", as used here, means the collection of the m^2 statistics $N = (N_{1,1}, \ldots, N_{m,m})$ where $N_{i,j}$ is the number of times the edge $(i \longrightarrow j)$ is traversed. This notion is, for instance, consistent with the notion of profile given earlier for lattice paths in Section V. 4. It also contains the information regarding the letter composition of words in a regular language and is thus compatible with the notion of profile introduced in Section V. 3.

Let Γ be a graph with edge (a, b) weighted by $\gamma_{a,b}$. Then, the BGF of paths with u marking the number of times a particular edge (c, d) is traversed is in matrix form

$$(I - z\widetilde{\mathbf{G}})^{-1}, \qquad \text{with} \quad \widetilde{\mathbf{G}} = \mathbf{G}\left[g_{a,b} \mapsto g_{a,b}u^{[\![(a,b)=(c,d)]\!]}\right].$$

The entry (i, j) in this matrix gives the BGF of paths with origin i and destination j. The GF of cumulated values (moments of order 1) is then obtained in the usual way, by differentiation followed by the substitution $u = 1$. Higher moments are similarly attainable by successive differentiations.

V. 5.2. Analytic aspects. In full generality, the components of a linear system of equations may exhibit the whole variety of behaviours obtained for the OGFs of regular languages in Section V. 3, p. 300. However, positivity coupled with some simple ancillary conditions (irreducibility and aperiodicity defined below) entails that the GFs of interest closely resemble the extremely simple rational function,

$$\frac{1}{1 - z/\rho} \equiv \frac{1}{1 - \lambda_1 z},$$

where ρ is the dominant positive singularity and $\lambda_1 = 1/\rho$ is a well-characterized eigenvalue of T. Accordingly, the asymptotic phenomena associated with such systems are highly predictable and coefficients involve the pure exponential form $c \cdot \rho^{-n}$. We propose first to expound the general theory, then treat classical applications to statistics of paths in graphs and languages recognized by finite automata.

Irreducibility and aperiodicity of matrices and graphs. From this point on, we only consider matrices with non-negative entries. Two notions are essential, irreducibility and aperiodicity (the terms are borrowed from Markov chain theory and matrix theory).

For A a scalar matrix of dimension $m \times m$ (with non-negative entries), a crucial rôle is played by the *dependency graph* (p. 33); this is the (directed) graph with vertex set $V = \{1 . . m\}$ and edge set containing the directed edge $(a \to b)$ iff $A_{a,b} \neq 0$.

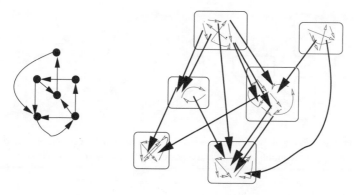

Figure V.15. Irreducibility conditions. Left: a strongly connected digraph. Right: a weakly connected digraph that is not strongly connected decomposes as a collection of strongly connected components linked by a directed acyclic graph.

The reason for this terminology is the following: Let A represent the linear transformation $\left\{ y_i^\star = \sum_j A_{i,j} y_j \right\}_i$; then, the fact that an entry $A_{i,j}$ is non-zero means that $y_i^\star$ depends effectively on y_j and is translated by the directed edge $(i \to j)$ in the dependency graph.

Definition V.5. *The non-negative matrix A is called* irreducible *if its dependency graph is strongly connected (i.e., any two vertices are connected by a directed path).*

By considering only simple paths, it is then seen that irreducibility is equivalent to the condition that $(I + A)^m$ has all its entries that are strictly positive. See Figure V.15 for a graphical rendering of irreducibility and for the general structure of a (weakly connected) digraph.

Definition V.6. *A strongly connected digraph Γ is said to be* periodic *with parameter d iff the vertex set V can be partitioned into d classes, $V = V_0 \cup \cdots \cup V_{d-1}$, in such a way that any edge whose source is an element of a V_j has its destination in $V_{j+1 \bmod d}$.*

The largest possible d is called the period. *If no decomposition exists with $d \geq 2$, so that the period has the trivial value 1, then the graph and all the matrices that admit it as their dependency graph are called* aperiodic.

For instance, a directed 10–cycle is periodic with parameters $d = 1, 2, 5, 10$ and the period is 10. Figure V.16 illustrates the notion. Periodicity implies that the existence of paths of length n between any two given nodes i, j is constrained by the congruence class $n \bmod d$. Conversely, aperiodicity entails the existence, for all n sufficiently large, of paths of length n connecting i, j.

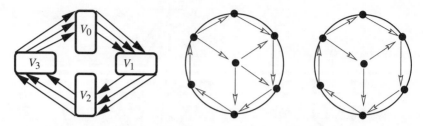

Figure V.16. Periodicity notions: the overall structure of a periodic graph with $d = 4$ (left), an aperiodic graph (middle) and a periodic graph of period 2 (right).

From the definition, a matrix A with period d has, up to simultaneous permutation of its rows and columns, a cyclic block structure

$$\begin{pmatrix} 0 & \boxed{A_{0,1}} & 0 & \cdots & 0 \\ 0 & 0 & \boxed{A_{1,2}} & \cdots & 0 \\ \vdots & \vdots & \vdots & \ddots & \vdots \\ 0 & 0 & 0 & \cdots & \boxed{A_{d-2,d-1}} \\ \boxed{A_{d-1,0}} & 0 & 0 & \cdots & 0 \end{pmatrix}$$

where the blocks $A_{i,i+1}$ reflect the connectivity between V_i and V_{i+1}. In the case of a period d, the matrix A^d admits a diagonal square block decomposition where each of its diagonal block is aperiodic (and of a smaller dimension than the original matrix). Then, the matrices $A^{\nu d}$ can be analysed block by block, and the analysis reduces to the aperiodic case. Similarly for powers $A^{\nu d+r}$ for any fixed r as ν varies. In other words, *the irreducible periodic case with period $d \geq 2$ can always be reduced to a collection of d irreducible aperiodic subproblems.* For this reason, we usually postulate in our statements both an irreducibility condition *and* an aperiodicity condition.

▷ **V.31.** *Sufficient conditions for aperiodicity.* Any one of the following conditions suffices to guarantee aperiodicity of the non-negative matrix T:

     (*i*) T has (strictly) positive entries;
    (*ii*) some power T^s has (strictly) positive entries;
  (*iii*) T is irreducible and at least one diagonal element of T is non-zero;
  (*iv*) T is irreducible and the dependency graph of T is such that there exist two circuits (closed paths) that are of relatively prime lengths.

(Any such condition implies in turn the existence of a unique dominant eigenvalue of T, which is simple, according to Theorem V.7 and Note V.34 below.)　　　◁

▷ **V.32.** *Computability of the period.* There exists a polynomial time algorithm that determines the period of a matrix. (Hint: in order to verify that Γ is periodic with parameter d, develop a breadth-first search tree, label nodes by their level, and check that edges have endpoints satisfying suitable congruence conditions modulo d.)　　　◁

Paths in strongly connected graphs. For analytic combinatorics, the importance of irreducibility and aperiodicity conditions stems from the fact that they guarantee uniqueness and simplicity of a dominant pole of path generating functions.

Theorem V.7 (Asymptotics of paths in graphs). *Consider the matrix*

$$F(z) = (I - zT)^{-1},$$

where T is a scalar non-negative matrix, in particular, the adjacency matrix of a graph Γ equipped with positive weights. Assume that T is irreducible. Then all entries $F^{\langle i,j \rangle}(z)$ of $F(z)$ have the same radius of convergence ρ, which can be defined in two equivalent ways:

   (i) as $\rho = \lambda_1^{-1}$ with λ_1 the largest positive eigenvalue of T;
   (ii) as the smallest positive root of the determinantal equation: $\det(I - zT) = 0$.

Furthermore, the point $\rho = \lambda_1^{-1}$ is a simple pole of each $F^{\langle i,j \rangle}(z)$.

If T is irreducible and aperiodic, then $\rho = \lambda_1^{-1}$ is the unique dominant singularity of each $F^{\langle i,j \rangle}(z)$, and

$$[z^n]F^{\langle i,j \rangle}(z) = \varphi_{i,j}\lambda_1^n + O(\Lambda^n), \qquad 0 \le \Lambda < \lambda_1,$$

for computable constants $\varphi_{i,j} > 0$.

Proof. The proof proceeds by stages, building up properties of the $F^{\langle i,j \rangle}$ by means of the relations that bind them, with suitable exploitation of Proposition V.6, p. 337 in conjunction with *Pringsheim's Theorem* (p. 240). In parts (i)–(v), we *assume that the matrix T is aperiodic*. Periodicity is finally examined in part (vi).

   (i) *All $F^{\langle i,j \rangle}$ have the same radius of convergence.* Simple upper and lower bounds show that each $F^{\langle i,j \rangle}$ has a finite non-zero radius of convergence $\rho_{i,j}$. By Pringsheim's Theorem, this $\rho_{i,j}$ is necessarily a singularity of the function $F^{\langle i,j \rangle}$. Since each $F^{\langle i,j \rangle}$ is a rational function, it then has a pole at $\rho_{i,j}$, hence becomes infinite as $z \to \rho_{i,j}$. Now, the matrix F satisfies the identities

(85) $$F = I + zTF, \qquad \text{and} \qquad F = I + zFT.$$

Thus, given that T is irreducible, each $F^{\langle i,j \rangle}$ is positively (linearly) related to any other $F^{\langle k,\ell \rangle}$. Then, the $F^{\langle i,j \rangle}$ must all become infinite as soon as one of them does. Consequently, all the $\rho_{i,j}$ are equal—we let ρ denote their common value.

   (ii) *All poles are of the same multiplicity.* By a similar argument, we see that all the $F^{\langle i,j \rangle}$ must have the same multiplicity κ of their common pole ρ, since otherwise, one function would be of slower growth, and a contradiction would result with the linear relations stemming from (85). We thus have, for some $\varphi_{i,j} > 0$ and $\kappa \ge 1$:

$$F^{\langle i,j \rangle}(z) \underset{z \to \rho}{\sim} \frac{\varphi_{i,j}}{(1 - z/\rho)^\kappa}.$$

   (iii) *The common multiplicity of poles is $\kappa = 1$.* This property results from the expression of the GF of all rooted circuits (Proposition V.6, Part (ii)) in terms of a logarithmic derivative, which has by construction only simple poles. Hence, a positive linear combination of some of the $F^{\langle i,j \rangle}$ has only a simple pole, so that $\kappa = 1$ and

(86) $$F^{\langle i,j \rangle}(z) \underset{z \to \rho}{\sim} \frac{\varphi_{i,j}}{1 - z/\rho}.$$

Another consequence is that we have $\rho = 1/\lambda_1$, where λ_1 is an eigenvalue of matrix T, which then satisfies the property that $\lambda_1 \geq |\lambda|$ for any eigenvalue λ of T: in matrix theory terminology, such an eigenvalue is called *dominant*[10].

(*iv*) *There are positive dominant eigenvectors.* From the relations (85) satisfied by the $\mathcal{F}^{(i,j)}(z)$ with j fixed and from (86), one finds as $z \to \rho$

$$(87) \qquad \frac{\varphi_{i,j}}{1 - z/\rho} \sim \rho \sum_k \frac{t_{i,k}\varphi_{k,j}}{1 - z/\rho}, \qquad \text{where} \quad T = (T_{i,j}).$$

This expresses the fact that the column vector $(\varphi_{1,j}, \ldots, \varphi_{m,j})^t$ is a right eigenvector corresponding to the eigenvalue $\lambda_1 = \rho^{-1}$. Similarly, for each fixed i, the row vector $(\varphi_{i,1}, \ldots, \varphi_{i,m})$ is found to be a left eigenvector. By part (*ii*), these eigenvector have all their components strictly positive.

(*v*) *The eigenvalue λ_1 is simple.* This property is needed in order to identify the $\varphi_{i,j}$ coefficients. We base our proof on the Jordan normal form and simple inequalities.

Assume first that there are two different Jordan blocks corresponding to the eigenvalue λ_1. Then there exist two vectors, $v = (v_1, \ldots, v_m)^t$ and $w = (w_1, \ldots, w_m)^t$, such that

$$Tv = \lambda_1 v, \qquad Tw = \lambda_1 w,$$

where we may assume that the eigenvector v has positive coordinates, given part (*iv*). Let j_0 be an index such that

$$\frac{|w_{j_0}|}{v_{j_0}} = \max_{j=1..m} \frac{|w_j|}{v_j}.$$

By possibly changing w to $-w$ and by rescaling, we may freely assume that $w_{j_0} = v_{j_0}$. Also, since v and w are not collinear, there must exist j_1 such that $|w_{j_1}| < v_{j_1}$. In summary:

$$(88) \qquad w_{j_0} = v_{j_0}, \qquad |w_{j_1}| < v_{j_1}, \qquad \forall j : |w_j| \leq v_j.$$

Consider finally the two relations $T^m v = \lambda_1^m v$ and $T^m w = \lambda_1^m w$, and examine consequences for the j_0 components. One has

$$(89) \qquad v_{j_0} = \sum_{k=1}^m U_{j_0,k} v_k, \qquad w_{j_0} = \sum_{k=1}^m U_{j_0,k} w_k,$$

where each $U_{j,k}$, the (j,k) entry of T^m, is positive, by the irreducibility and aperiodicity assumptions. But then, by the triangle inequality, there is a contradiction between (89) and (88). Thus, there cannot be two distinct Jordan blocks corresponding to λ_1.

It only remains to exclude the existence of a Jordan block of dimension ≥ 2 associated with λ_1. If such a Jordan block were present, there would exists a vector w

---

[10]In matrix theory, a dominant eigenvalue (λ_1) is one that is *largest* in modulus, while, for an analytic function, a dominant singularity (ρ) is one that is *smallest* in modulus. The two notions are reconciled by the fact that the singularities of generating functions are *inverses* of eigenvalues of matrices ($\rho = 1/\lambda_1$).

such that

$$(90) \quad \begin{cases} Tv = \lambda_1 w \\ Tw = \lambda_1 w + v \end{cases} \quad \text{implying} \quad \begin{cases} T^{vm} v = \lambda_1^{vm} w, \\ T^{vm} w = \lambda_1^{vm} w + vm\lambda_1^{vm-1} v. \end{cases}$$

By simple bounds obtained from comparing w to v componentwise, it is found that the vector $T^{vm} w$ must have all its coordinates that are $O(\lambda_1^{vm})$. Upon taking $v \to \infty$, a contradiction is reached with the last relation of (90), where the growth of these coordinates is of the form $v\lambda_1^{vm}$. Thus, a Jordan block of dimension ≥ 2 is also excluded, and the eigenvalue λ_1 is simple.

(vi) Aperiodicity of T is equivalent to the existence of a unique dominant eigenvalue. If λ_1 uniquely dominates, meaning that $\lambda_1 > |\lambda|$ for all eigenvalues $\lambda \neq \lambda_1$, then each $F^{\langle i,j \rangle}$ has a simple pole at ρ that is its unique dominant singularity. Hence the coefficients $[z^n]F^{\langle i,j \rangle}(z)$ are non-zero for n large enough, since they are asymptotic to $\varphi_{i,j}\rho^{-n}$ by (86). This last property ensures aperiodicity.

Conversely, if T is aperiodic, then λ_1 uniquely dominates. Indeed, suppose that μ is an eigenvalue of T such that $|\mu| = \lambda_1$, with w a corresponding eigenvector. We would have $T^m v = \lambda_1^m v$ and $T^m w = \mu^m w$. But then, by an argument similar to the one used in part (v), upon making use of inequalities (88), we would need to have w and v collinear, which is absurd.

We leave it as an exercise to the reader to verify the stronger property that identifies the period with the number of dominant eigenvalues: see Note V.33. ∎

Several of these arguments will inspire the discussion, in Chapter VII, of the harder problem of analysing coefficients of algebraic functions defined by positive polynomial systems (Subsection VII. 6.3, p. 488).

▷ **V.33.** *Periodicities.* If T has period d, then the support of each $F^{\langle j,j \rangle}(z)$ is included in $d\mathbb{Z}$, hence there are at least d conjugate singularities, corresponding to eigenvalues of the form $\lambda_1 e^{2ik\pi/d}$. There are no other eigenvalues since T^d is built out of irreducible blocks, each with the unique dominant eigenvalue λ_1^d. ◁

▷ **V.34.** *The classical Perron–Frobenius Theorem.* The proof of Theorem V.7 immediately gives the following famous statement.

Theorem (Perron–Frobenius Theorem). *Let A be a matrix with non-negative elements that is assumed to be irreducible. The eigenvalues of A can be ordered in such a way that*

$$\lambda_1 = |\lambda_2| = \cdots = |\lambda_d| > |\lambda_{d+1}| \geq |\lambda_{d+2}| \geq \cdots,$$

and all the eigenvalues of largest modulus are simple. Furthermore, the quantity d is precisely equal to the period of the dependency graph. In particular, in the aperiodic case $d = 1$, there is unicity of the dominant eigenvalue. In the periodic case $d \geq 2$, the whole spectrum has a rotational symmetry: it is invariant under the set of transformations

$$\lambda \mapsto \lambda e^{2ij\pi/d}, \qquad j = 0, 1, \ldots, d-1.$$

The properties of positive and of non-negative matrices have been superbly elicited by Perron [478] in 1907 and by Frobenius [271] in 1908–1912. The corresponding theory has far-reaching implications: it lies at the basis of the theory of finite Markov chains and it extends to positive operators in infinite-dimensional spaces [390]. Excellent treatments of Perron–Frobenius theory are to be found in the books of Bellman [34, Ch. 16], Gantmacher [276, Ch. 13], as well as Karlin and Taylor [363, p. 536–551]. ◁

▷ **V.35.** *Unrooted circuits.* Consider a strongly connected weighted graph Γ with adjacency matrix $\mathbf{G} = (g_{i,j})$. Let $\mathcal{RC}$ be the class of all *rooted* circuits and $\mathcal{PRC}$ the subclass of those that are primitive (i.e., they differ from all their cyclic shifts). Let also $\mathcal{UC}$ be the class of all *unrooted* circuits (no origin distinguished) and $\mathcal{PUC}$ the subclass of those that are primitive. Define the adjacency matrix $\mathbf{G}^{\odot s} := ((g_{i,j})^s)$ obtained by raising each entry of $\mathbf{G}$ to the sth power. Set finally $\Delta_{\mathbf{G}}(z) := \det(I - z\mathbf{G})$. We find

$$
\begin{cases}
RC(z, \mathbf{G}) &= \displaystyle\sum_{k \geq 1} PRC(z^k, \mathbf{G}^{\odot k}), \qquad PUC(z, \mathbf{G}) &= \displaystyle\int_0^z PRC(t, \mathbf{G})\, \frac{dt}{t}, \\
UC(z, \mathbf{G}) &= \displaystyle\sum_{k \geq 1} PUC(z^k, \mathbf{G}^{\odot k}),
\end{cases}
$$

upon mimicking the reasoning of Appendix A.4: *Cycle construction*, p. 729. This results in

$$
UC(z) = \sum_{k \geq 1} \frac{\varphi(k)}{k} \log\left(1/\Delta_{\mathbf{G}^{\odot k}}(z)\right),
$$

$$
[z^n] UC(z) = \frac{\lambda_1^n}{n} + O(\Lambda^n), \qquad [z^n] PUC(z) = \frac{\lambda_1^n}{n} + O(\Lambda^n),
$$

where the two asymptotic estimates hold under irreducibility and aperiodicity conditions. These estimates can be regarded as a Prime Number Theorem for walks in graphs. (See [555] for related facts and zeta functions of graphs.) ◁

Profiles. The proof of Theorem V.7 additionally provides the form of a certain "residue matrix", from which several probabilistic properties of paths follow.

Lemma V.1 (Iteration of irreducible matrices). *Let the non-negative matrix T be irreducible and aperiodic, with λ_1 its dominant eigenvalue. Then the residue matrix Φ such that*

$$
(91) \qquad\qquad (I - zT)^{-1} = \frac{\Phi}{1 - z\lambda_1} + O(1) \qquad (z \to \lambda_1^{-1})
$$

has entries given by ($\langle x, y \rangle$ *represents the scalar product* $\sum_i x_i y_i$)

$$
\varphi_{i,j} = \frac{r_i \ell_j}{\langle r, \ell \rangle},
$$

where r and ℓ are, respectively, right and left eigenvectors of T corresponding to the eigenvalue λ_1.

Proof. We have seen that the matrix $\Phi = (\varphi_{i,j})$ has its rows and columns proportional, respectively, to right and left eigenvectors belonging to the eigenvalue λ_1. Thus, we have

$$
\frac{\varphi_{i,j}}{\varphi_{1,j}} = \frac{\varphi_{i,1}}{\varphi_{1,1}},
$$

while the $\varphi_{1,j}$ (respectively, $\varphi_{i,1}$) are the coordinates of a left (respectively, right) eigenvector. There results that there exists a normalization constant ξ such that

$$
\varphi_{i,j} = \xi r_i \ell_j.
$$

That normalization constant is then determined by the fact that the GF of circuits has residue equal to $\rho = \lambda_1^{-1}$ at $z = \rho$, so that $\sum_i \varphi_{j,j} = 1$, leading to

$$
1 = \xi \sum_j r_j \ell_j,
$$

which implies the statement.   ∎

Equipped with the lemma, we can now state:

Theorem V.8 (Profiles of paths in graphs). *Let* $\mathbf{G}$ *be a non-negative matrix associated to a weighted digraph* Γ, *assumed to be irreducible and aperiodic. Let* ℓ, r *be, respectively, the left and right eigenvectors corresponding to the dominant (Perron–Frobenius) eigenvalue* λ_1. *Consider the collection* $\mathcal{F}^{\langle a,b\rangle}$ *of (weighted) paths in* Γ *with fixed origin* a *and final destination* b. *Then, the number of traversals of edge* (s, t) *in a random element of* $\mathcal{F}_n^{\langle a,b\rangle}$ *has mean*

$$(92) \qquad\qquad \tau_{s,t} n + O(1) \qquad \text{where} \quad \tau_{s,t} := \frac{\ell_s g_{s,t} r_t}{\langle \ell, r\rangle}.$$

In other words, a long random path tends to spend asymptotically a fixed (non-zero) fraction of its time traversing any given edge. Accordingly, the number of visits to vertex s is also proportional to n and obtained by summing the expression of (92) over all the possible values of t.

Proof. First, the total weight ("number") of paths in $\mathcal{F}_{a,b}$ satisfies

$$(93) \qquad\qquad [z^n]\left[(I - z\mathbf{G})^{-1}\right]_{a,b} \sim \frac{r_a \ell_b}{\langle \ell, r\rangle} \lambda_1^n,$$

as follows from Lemma V.1. Next, introduce the modified matrix $\mathbf{H} = (h_{i,j})$ defined by

$$h_{i,j} = g_{i,j}\, u^{\llbracket i=s \wedge j=t \rrbracket}.$$

In other words, we mark each traversal of edge i, j by the variable u. Then, the quantity

$$(94) \qquad\qquad [z^n]\left[\frac{\partial}{\partial u}(I - z\mathbf{H})^{-1}\Big|_{u=1}\right]_{a,b}$$

represents the total number of traversals of edge (s, t), with weights taken into account. Simple algebra[11] shows that

$$(95) \qquad\qquad \frac{\partial}{\partial u}(I - z\mathbf{H})^{-1}\Big|_{u=1} = (I - z\mathbf{G})^{-1}\,(z\mathbf{H}')\,(I - z\mathbf{G}),$$

where $\mathbf{H}' := (\partial_u \mathbf{H})_{u=1}$ has all its entries equal to 0, except for the s, t entry whose value is $g_{s,t}$. By the calculation of the residue matrix in Lemma V.1, the coefficient of (94) is then asymptotic to

$$(96) \qquad [z^n]\frac{\varphi_{a,s}}{1 - \lambda_1 z} g_{s,t} z \frac{\varphi_{t,b}}{1 - \lambda_1 z} \sim \upsilon n \lambda_1^n, \qquad \upsilon := \frac{r_a \ell_s g_{s,t} r_t \ell_b}{\langle \ell, r\rangle^2}.$$

Comparison of (96) and (93) finally yields the result since the relative error terms are $O(n^{-1})$ in each case.   ∎

---

[11] If A is an operator depending on u, one has $\partial_u(A^{-1}) = -A^{-1}(\partial_u A)A^{-1}$, which is a non-commutative generalization of the usual differentiation rule for inverses.

Another consequence of this last proof and Equation (93) is that the numbers of paths starting at a and ending at either b or c satisfy

$$(97) \qquad \lim_{n \to \infty} \frac{F_n^{\langle a,b \rangle}}{F_n^{\langle a,c \rangle}} = \frac{\ell_b}{\ell_c}.$$

In other words, the quantity

$$\frac{\ell_b}{\sum_j \ell_j}$$

is the asymptotic probability that a random path with origin fixed at some point a but otherwise unconstrained will end up at point b after a large number of steps. Such properties are strongly evocative of Markov chain theory discussed below in Example V.13, p. 352.

▷ **V.36. *Residues and projections.*** Let $\mathcal{E} = \mathbb{C}^m$ be the ambient space, where m is the dimension of T, assumed to be irreducible and aperiodic. There exists a direct sum decomposition $\mathcal{E} = \mathcal{F}_1 \oplus \mathcal{F}_2$ where $\mathcal{F}_1$ is the one-dimensional eigenspace generated by the eigenvector (r) corresponding to eigenvalue λ_1 and $\mathcal{F}_2$ is the supplementary space which is the direct sum of characteristic spaces corresponding to the other eigenvalues $\lambda_2, \ldots$. (For the purposes of the present discussion, one may freely think of the matrix as diagonalizable, with $\mathcal{F}_2$ the union of eigenspaces associated to $\lambda_2, \ldots$.) Then T as a linear operator acting on $\mathcal{F}$ admits the decomposition

$$T = \lambda_1 P + S,$$

where P is the projector on $\mathcal{F}_1$ and S acts on $\mathcal{F}_2$ with spectral radius $|\lambda_2|$, as illustrated by the diagram:

(98)

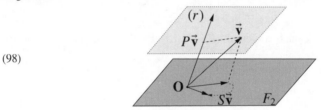

By standard properties of projections, $P^2 = P$ and $PS = SP = 0$ so that $T^n = \lambda_1^n P + S^n$. Consequently, there holds,

$$(99) \qquad (I - zT)^{-1} = \sum_{n \geq 0} \left(z^n \lambda_1^n P + z^n S^n \right) = \frac{P}{1 - \lambda_1 z} + (I - zS)^{-1}.$$

Thus, the residue matrix Φ coincides with the projector P.

From this, one finds also

$$(100) \qquad (I - zT)^{-1} = \frac{\Phi}{1 - \lambda_1 z} + \sum_{k \geq 0} R_k \left(z - \lambda_1^{-1} \right)^k, \qquad R_k := S^k (I - \lambda_1^{-1} S)^{-k-1},$$

which provides a full expansion. ◁

▷ **V.37. *Algebraicity of the residues.*** One only needs to solve one polynomial equation in order to determine λ_1. Then the entries of Φ and the R_k in (100) are all obtained by rational operations in the field generated by the entries of T extended by the algebraic quantity λ_1: for instance, in order to get an eigenvector, it suffices to replace one of the equations of the system $Tr = \lambda_1 r$ by a normalization condition, like $r_1 + \cdots + r_m = 1$. (Numerical procedures are likely to be used instead for large matrices.) ◁

Automata and words. By proposition V.6 (p. 337), the OGF of the language defined by a deterministic finite automaton is expressible in terms of the *quasi-inverse* $(1 - zT)^{-1}$, where the matrix T is a direct encoding of the automaton's transitions. Corollary V.7 and Lemma V.1 have been precisely custom-tailored for this situation. We shall allow weights on letters of the alphabet, corresponding to a Bernoulli model on words. We say that an automaton is irreducible (respectively, aperiodic) if the underlying graph and the associated matrix are irreducible (respectively, aperiodic).

Proposition V.7 (Random words and automata). *Let $\mathcal{L}$ be a language recognized by a deterministic finite automaton A whose graph is irreducible and aperiodic. The number of words of $\mathcal{L}$ satisfies*

$$L_n \sim c\lambda_1^n + O(\Lambda^n),$$

where λ_1 is the dominant (Perron–Frobenius) eigenvalue of the transition matrix of A and c, Λ are real constants with $c > 0$ and $0 \le \Lambda < \lambda_1$.

In a random word of $\mathcal{L}_n$, the number of traversals of a designated vertex or edge has a mean that is asymptotically linear in n, as given by Theorem V.8.

▷ **V.38.** *Unambiguous automata.* A non-deterministic finite state automaton is said to be unambiguous if the set of accepting paths for any given words comprises at most one element. The translation into a generating function as described above also applies to such automata, even though they are non-deterministic. ◁

▷ **V.39.** *Concentration of distribution for the number of passages.* Under the conditions of the theorem, the standard deviation of the number of traversals of a designated node or edge is $O(\sqrt{n})$. Thus in a random long path, the distribution of the number of such traversals is concentrated. [Compared to (95), the calculation of the second moment requires taking a further derivative, which leads to a triple pole. The second moment and the square of the mean, which are each $O(n^2)$, are then found to cancel to main asymptotic order.] ◁

V. 5.3. Applications. We now provide a few application of Theorems V.7 and V.8. First, Example V.11 studies briefly the case of words that are locally constrained in the sense that certain transitions between letters are forbidden; Example V.12 revisits walks on an interval and develops an alternative matrix view of a problem otherwise amenable to continued fraction theory. Next, Example V.13 makes explicit the way the fundamental theorem of finite Markov chain theory can be derived effortlessly as a consequence of the more general Theorem V.8, and Example V.14 compares on a simple problem, the devil's staircase, the combinatorial and the Markovian approaches. Example V.15 comes back to words and develops simple consequences of an important combinatorial construction, that of De Bruijn graphs. This graph is invaluable in predicting in many cases the *shape* of the asymptotic results that are to be expected when confronted with word problems; Finally, Example V.16 concludes this section with a brief discussion of the special case of words with excluded patterns, thereby leading to a quantitative version of Borges' Theorem (Note I.35, p. 61).

In all these cases, the counting estimates are of the form $c\lambda_1^n$, whereas the expectations of parameters of interest have a linear growth.

Example V.11. *Locally constrained words.* Consider a fixed alphabet $\mathcal{A} = \{a_1, \ldots, a_m\}$ and a set $\mathcal{F} \subseteq \mathcal{A}^2$ of forbidden transitions between consecutive letters. The set of words over $\mathcal{A}$ with no such forbidden transition is denoted by $\mathcal{L}$ and is called a *locally constrained language*. (The

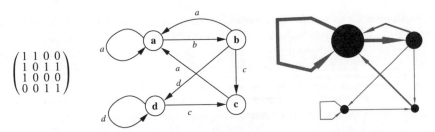

$$\begin{pmatrix} 1 & 1 & 0 & 0 \\ 1 & 0 & 1 & 1 \\ 1 & 0 & 0 & 0 \\ 0 & 0 & 1 & 1 \end{pmatrix}$$

Figure V.17. Locally constrained words: The transition matrix (T) associated to the forbidden pairs $F = \{ac, ad, bb, cb, cc, cd, da, db\}$, the corresponding automaton, and the graph with widths of vertices and edges drawn in proportion to their asymptotic frequencies.

particular case where exactly all pairs of equal letters are forbidden corresponds to Smirnov words and has been discussed on p. 262.)

Clearly, the words of $\mathcal{L}$ are recognized by an automaton whose state space is isomorphic to $\mathcal{A}$: state q simply memorizes the fact that the last letter read was a q. The graph of the automaton is then obtained by the collection of allowed transitions $(q, r) \mapsto a$, with $(q, r) \notin F$. (In other words, the graph of the automaton is the complete graph in which all edges that correspond to forbidden transitions are deleted.) Consequently, the OGF of any locally constrained language is a rational function. Its OGF is given by

$$(1, 1, \ldots, 1)(I - zT)^{-1}(1, 1, \ldots, 1)^t,$$

where T_{ij} is 0 if $(a_i, a_j) \in F$ and 1 otherwise. If each letter can occur later than any other letter in an accepted word, the automaton is irreducible. Also, the graph is aperiodic except in a few degenerate cases (e.g., in the case where the allowed transitions would be $a \to b, c; b \to d; c \to d; d \to a$). Under irreducibility and aperiodicity, the number of words must be $\sim c\lambda_1^n$ and each letter has on average an asymptotically constant frequency. (See (34) and (35) of Chapter IV, p. 262, for the case of Smirnov words.)

For the example of Figure V.17, the alphabet is $\mathcal{A} = \{a, b, c, d\}$. There are eight forbidden transitions and the characteristic polynomial $\chi_G(\lambda) := \det(\lambda I - G)$ is found to be $\lambda^3(\lambda - 2)$. Thus, one has $\lambda_1 = 2$. The right and left eigenvectors are found to be

$$r = (2, 2, 1, 1)^t, \qquad \ell = (2, 1, 1, 1).$$

Then, the matrix τ, where $\tau_{s,t}$ represents the asymptotic frequency of transitions from letter s to letter t, is found in accordance with Theorem V.8:

$$\tau = \begin{pmatrix} \frac{1}{4} & \frac{1}{4} & 0 & 0 \\ \frac{1}{8} & 0 & \frac{1}{16} & \frac{1}{16} \\ \frac{1}{8} & 0 & 0 & 0 \\ 0 & 0 & \frac{1}{16} & \frac{1}{16} \end{pmatrix}.$$

This means that a random path spends a proportion equal to 1/4 of its time on a transition between an a and a b, but much less (1/16) on transitions between pairs of letters bc, bd, cc, ca. The letter frequencies in a random word of $\mathcal{L}$ are (1/2, 1/4, 1/8, 1/8), so that an a is four times more frequent than a c or a d, and so on. See Figure V.17 (right) for a rendering.

Various specializations, including multivariate GFs and non-uniform letter models, are readily treated by this method. Bertoni *et al.* [59] develop related variance and distribution calculations for the number of occurrences of a symbol in an arbitrary regular language. ... ∎

Example V.12. *Walks on the interval.* As a direct illustration, consider the walks associated to the graph $\Gamma(5)$ with vertex set $1, \ldots, 5$ and edges being formed of all pairs (i, j) such that $|i - j| \leq 1$. The graph $\Gamma(5)$ and its incidence matrix $G(5)$ are

$$
\Gamma(5) = \quad\underset{1\quad\;2\quad\;3\quad\;4\quad\;5}{\text{\raisebox{0pt}{}}}\quad, \qquad
G(5) = \begin{pmatrix} 1 & 1 & 0 & 0 & 0 \\ 1 & 1 & 1 & 0 & 0 \\ 0 & 1 & 1 & 1 & 0 \\ 0 & 0 & 1 & 1 & 1 \\ 0 & 0 & 0 & 1 & 1 \end{pmatrix}.
$$

The characteristic polynomial $\chi_{G(5)}(z) := \det(zI - G(5))$ factorizes as

$$
\chi_{G(5)}(z) = z(z - 1)(z - 2)(z^2 - 2z - 2),
$$

and its dominant root is $\lambda_1 = 1 + \sqrt{3}$. From here, one finds a left eigenvector (which is also a right eigenvector since the matrix is symmetric):

$$
r = \ell^t = (1, \sqrt{3}, 2, \sqrt{3}, 1).
$$

Thus a random path (with the uniform distribution over all paths corresponding to the weights being equal to 1) visits nodes $1, \ldots, 5$ with frequencies proportional to

$$
1, \quad 1.732, \quad 2, \quad 1.732, \quad 1,
$$

implying that the non-extremal nodes are visited more often—such nodes have higher degrees of freedom, so that there tend to be more paths that traverse them.

In fact, this example has structure. For instance, the graph $\Gamma(11)$ defined by an interval of length 10, leads to a matrix with a highly factorable characteristic polynomial

$$
\chi_{G(11)} = z\,(z - 1)\,(z - 2)\left(z^2 - 2z - 2\right)\left(z^2 - 2z - 1\right)\left(z^4 - 4z^3 + 2z^2 + 4z - 2\right).
$$

The reader may have recognized here a particular case of lattice paths, which is covered by the theory presented in Section V. 4, p. 318. Indeed, according to Proposition V.3, the OGF of paths from vertex 1 to vertex 1 in the graph $\Gamma(k)$ with vertex set $\{1, \ldots, k\}$ is given by the continued fraction

$$
\cfrac{1}{1 - z - \cfrac{z^2}{1 - z - \cfrac{z^2}{\ddots \cfrac{}{1 - z - \cfrac{z^2}{1 - z}}}}}.
$$

(The number of fraction bars is k.) From this it can be shown that the characteristic polynomial of G is an elementary variant of the Fibonacci–Chebyshev polynomial of Example V.8, p. 326. The analysis based on Theorem V.8 is simpler, albeit more rudimentary, as it only provides a first-order asymptotic solution to the problem.

This example is typical: whenever combinatorial problems have the appropriate amount of regularity, all the resources of linear algebra are available, including the vast body of knowledge gathered over years on calculations of structured determinants, which is well summarized in Krattenthaler's survey [391] and the book by Vein and Dale [594]. .... ∎

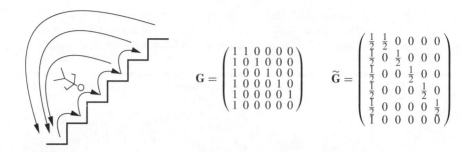

$$G = \begin{pmatrix} 1 & 1 & 0 & 0 & 0 & 0 \\ 1 & 0 & 1 & 0 & 0 & 0 \\ 1 & 0 & 0 & 1 & 0 & 0 \\ 1 & 0 & 0 & 0 & 1 & 0 \\ 1 & 0 & 0 & 0 & 0 & 1 \\ 1 & 0 & 0 & 0 & 0 & 0 \end{pmatrix} \qquad \widetilde{G} = \begin{pmatrix} \frac{1}{2} & \frac{1}{2} & 0 & 0 & 0 & 0 \\ \frac{1}{2} & 0 & \frac{1}{2} & 0 & 0 & 0 \\ \frac{1}{2} & 0 & 0 & \frac{1}{2} & 0 & 0 \\ \frac{1}{2} & 0 & 0 & 0 & \frac{1}{2} & 0 \\ \frac{1}{2} & 0 & 0 & 0 & 0 & \frac{1}{2} \\ 1 & 0 & 0 & 0 & 0 & 0 \end{pmatrix}$$

Figure V.18. The devil's staircase ($m = 6$) and the two matrices that can model it.

Example **V.13.** *Elementary theory of finite Markov chains.* Consider the case where the row sums of matrix **G** are all equal to 1, that is, $\sum_j g_{i,j} = 1$. Such a matrix is called a *stochastic matrix*. The quantity $g_{i,j}$ can then be interpreted as the probability of leaving state i for state j, assuming one is in state i. Assume that the matrix **G** is irreducible and aperiodic. Clearly, the matrix **G** admits the column vector $r = (1, 1, \ldots, 1)^t$ as a right eigenvector corresponding to the dominant eigenvalue $\lambda_1 = 1$. The left eigenvector ℓ normalized so that its elements sum to 1 is called the (row) vector of stationary probabilities. It must be calculated by linear algebra and its determination involves finding an element of the kernel of matrix $I - G$, which can be done in a standard way.

Theorem V.8 and Equation (93) immediately imply the following:

Proposition V.8 (Stationary probabilities of Markov chains). *Consider a weighted graph corresponding to a stochastic matrix **G** which is irreducible and aperiodic. Let ℓ be the normalized left eigenvector corresponding to the eigenvalue 1. A random (weighted) path of length n with fixed origin and destination visits node s a mean number of times asymptotic to $\ell_s n$ and traverses edge (s, t) a mean number of times asymptotic to $\ell_s g_{s,t} n$. A random path of length n with fixed origin ends at vertex s with probability asymptotic to ℓ_s.*

The vector ℓ is also known as the vector of *stationary probabilities*. The first-order asymptotic property expressed by Proposition V.8 certainly constitutes the most fundamental result in the theory of finite Markov chains. .... ∎

Example **V.14.** *The devil's staircase.* This example illustrates an elementary technique often employed in calculations of eigenvalues and eigenvectors. It presupposes that the matrix to be analysed can be reduced to a sparse form and has a sufficiently regular structure.

You live in a house that has a staircase with m steps. You come back home a bit loaded and at each second, you can either succeed in climbing a step or fall back all the way down. On the last step, you always stumble and fall back down (Figure V.18). Where are you likely to be found at time n?

Precisely, two slightly different models correspond to this informally stated problem. The probabilistic model views it as a Markov chain with equally likely possibilities at each step and is reflected by matrix $\widetilde{G}$ in Figure V.18. The combinatorial model just assumes all possible evolutions ("histories") of the system as equally likely and it corresponds to matrix **G**. We opt here for the latter, keeping in mind that the same method basically applies to both cases.

We first write down the constraints expressing the joint properties of an eigenvalue λ and its right eigenvector $x = (x_1, \ldots, x_m)^t$. The equations corresponding to $(\lambda I - G)x = 0$ are formed of a first batch of $m - 1$ relations,

(101) $\qquad (\lambda - 1)x_1 - x_2 = 0, \quad -x_1 + \lambda x_2 - x_3 = 0, \quad \cdots, -x_1 + \lambda x_{m-1} - x_m = 0,$

together with the additional relation (one cannot go higher than the last step):

(102) $\qquad\qquad\qquad\qquad -x_1 + \lambda x_m = 0.$

The solution to (101) is readily found by pulling out successively $x_2, \ldots, x_m$ as functions of x_1:

(103) $\quad x_2 = (\lambda - 1)x_1, \quad x_3 = (\lambda^2 - \lambda - 1)x_1, \quad \cdots, \quad x_m = (\lambda^{m-1} - \lambda^{m-2} - \cdots - 1)x_1.$

Combined with the special relation (102), this last relation shows that λ must satisfy the equation

(104) $\qquad\qquad\qquad\qquad 1 - 2\lambda^m + \lambda^{m+1} = 0.$

Let λ_1 be the largest positive root of this equation, existence and dominance being guaranteed by Perron–Frobenius properties. Note that the quantity $\rho := 1/\lambda_1$ satisfies the characteristic equation

$$1 - 2\rho + \rho^{m+1} = 0,$$

already encountered when discussing longest runs in words; the discussion of Example V.4 then grants us the existence of an isolated ρ near $\frac{1}{2}$, hence the fact that λ_1 is slightly less than 2.

Similar devices yield the left eigenvector $y = (y_1, \ldots, y_m)$. It is found easily that y_j must be proportional to λ_1^{-j}. We thus obtain from Theorem V.8 and Equation (97): *The probability of being in state j (i.e., being on step j of the stair) at time n tends to the limit*

$$\varpi_j = \gamma \lambda_1^{-j}$$

where λ_1 is the root near 2 of the polynomial (104) and the normalization constant γ is determined by $\sum_j \varpi_j = 1$. In other words, the distribution of the altitude at time n is a truncated geometric distribution with parameter $1/\lambda_1$. For instance, $m = 6$ leads to $\lambda_1 = 1.98358$, and the asymptotic probabilities of being in states $1, \ldots, 6$ are

(105) $\qquad 0.50413, \quad 0.25415, \quad 0.12812, \quad 0.06459, \quad 0.03256, \quad 0.01641,$

exhibiting a clear geometric decay. Here is the simulation of a random trajectory for $n = 100$:

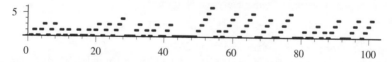

In this case, the frequencies observed are 0.44, 0.26, 0.17, 0.08, 0.04, 0.01, pretty much in agreement with what is expected.

Finally, the similarity with the longest run problem in words is easily explained. Let u and d be letters representing steps upwards and downwards, respectively. The set of paths from state 1 to state 1 is described by the regular expression

$$\mathcal{P}_{1,1} = \left(d + ud + \cdots + u^{m-1}d\right)^\star,$$

corresponding to the generating function

$$P_{1,1}(z) = \frac{1}{1 - z - z^2 - \cdots - z^m},$$

a variant of the OGF of words without m–runs of the letter u, which also corresponds to the enumeration of compositions with summands $\leq m$. (The case of the probabilistic transition matrix $\widetilde{\mathbf{G}}$ is left as an exercise to the reader.) .... ∎

Example V.15.   *De Bruijn graphs.*   Two thieves want to break into a house whose entrance is protected by a digital lock with an unknown four-digit code. As soon as the four digits of the code are typed consecutively, the gate opens. The first thief proposes to try in order all the four-digit sequences, resulting in as much as 40 000 key strokes in the worst-case. The second thief, who is a mathematician, says he can try *all* four-digit combinations with only 10 003 key strokes. What is the mathematician's trade secret?

Clearly certain optimizations are possible: for instance, for an alphabet of cardinality 2 and codes of two letters, the sequence 00110 is better than the naïve one, 00 01 10 11, which is redundant; a few more attempts will lead to an optimal solution for three-digit codes that has length 10 (rather than 24), for instance,

$$0001110100.$$

The general question is then: How far can one go and how to construct such sequences?

Fix an alphabet of cardinality m. A sequence that contains as factors (contiguous blocks) *all* the k letter words is called a *de Bruijn sequence*. Clearly, its length must be at least $\delta(m, k) = m^k + k - 1$, as it must have at least m^k positions at distance at least $k - 1$ from the end. A sequence of smallest possible length $\delta(m, k)$ is called a *minimal* de Bruijn sequence. Such sequences were discovered by N. G. de Bruijn [140] in 1946, in response to a question coming from electrical engineering, where all possible reactions of a device presented as a black box must be tested at minimal cost. We shall treat here the case of a binary alphabet, $m = 2$, the generalization to $m > 2$ being obvious.

Let $\ell = k - 1$ and consider the automaton $\mathcal{B}_\ell$ that memorizes the last block of length ℓ read when scanning the input text from left to right. A state is thus assimilated to a string of length ℓ and the total number of states is 2^ℓ. The transitions are easily calculated: let $q \in \{0, 1\}^\ell$ be a state and let $\sigma(w)$ be the function that shifts all letters of a word w one position to the left, dropping the first letter of w in the process (thus σ maps $\{0, 1\}^\ell$ to $\{0, 1\}^{\ell-1}$); the transitions are

$$q \stackrel{0}{\mapsto} \sigma(q)0, \qquad q \stackrel{1}{\mapsto} \sigma(q)1.$$

If one further interprets a state q as the integer in the interval $[0 . . 2^\ell - 1]$ that it represents, then the transition matrix assumes a remarkably simple form:

$$T_{i,j} = [\![(j \equiv 2i \bmod 2^\ell) \text{ or } (j \equiv 2i + 1 \bmod 2^\ell)]\!].$$

See Figure V.19 for a rendering borrowed from [263].

Combinatorially, the de Bruijn graph is such that each node has indegree 2 and outdegree 2. By a well known theorem going back to Euler: *A necessary and sufficient condition for an undirected connected graph to have an Eulerian circuit (that is, a closed path that traverses each vertex exactly once) is that every node has even degree. For a strongly connected digraph, the condition is that each node has an outdegree equal to its indegree.* This last condition is obviously satisfied here. Take an Eulerian circuit starting and ending at node 0^ℓ; its length is $2^{\ell+1} = 2^k$. Then, clearly, the sequence of edge labels encountered when prefixed with the word $0^{k-1} = 0^\ell$ constitutes a minimal de Bruijn sequence. In general, the argument gives a de Bruijn sequence with minimal length $m^k + k - 1$. Et voilà! The trade secret of the thief-mathematician is exposed.

Back to enumeration. The de Bruijn matrix is irreducible since a path labelled by sufficiently many zeros always leads any state to the state 0^ℓ, while a path ending with the letters of $w \in \{0, 1\}^\ell$ leads to state w. The matrix is aperiodic since it has a loop on states 0^ℓ and 1^ℓ. Thus, by Perron–Frobenius properties, it has a unique dominant eigenvalue, and it is not hard to check that its value is $\lambda_1 = 2$, corresponding to the right eigenvector $(1, 1, \ldots, 1)^t$. If one fixes

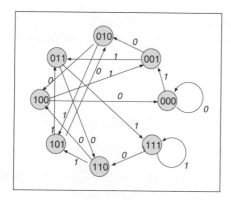
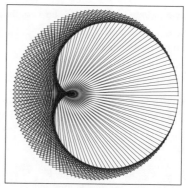

Figure V.19. The de Bruijn graph: (left) $\ell = 3$; (right) $\ell = 7$.

a *pattern* $w \in \{0, 1\}^\ell$, Theorem V.8 yields back the known fact that a random word contains on average $\sim \frac{n}{2^\ell}$ occurrences of pattern w, while Note V.39, p. 349, further implies that the distribution of the number of occurrences is concentrated around the mean, as the variance is $O(n)$. The de Bruijn graph may be used to quantify many properties of occurrences of patterns in random words: see for instance [43, 240, 263]. . . . . ∎

Example V.16. *Words with excluded patterns.* Fix a finite set of patterns $\Omega = \{w_1, \ldots, w_r\}$, where each w_j is a word of $\mathcal{A}^\star$. The language $\mathcal{E} \equiv \mathcal{E}^\Omega$ of words that contain no factor in Ω is described by the extended regular expression

$$\mathcal{E} = \mathcal{A}^\star \setminus \bigcup_{j=1}^{r} (\mathcal{A}^\star w_j \mathcal{A}^\star),$$

which constitutes a concise but highly ambiguous description. By closure properties of regular languages, $\mathcal{E}$ is itself regular and there must exist a deterministic automaton that recognizes it.

An automaton recognizing $\mathcal{E}$ can be constructed starting from the de Bruijn automaton of index $k = -1 + \max |w_j|$ and deleting all the vertices and edges that correspond to a word of Ω. Precisely, vertex q is deleted whenever q contains a factor in Ω; the transition (edge) from q associated with letter α gets deleted whenever the word $q\alpha$ contains a factor in Ω. The pruned de Bruijn automaton, call it $\mathcal{B}_k^\circ$, accepts all words of $0^k \mathcal{E}$, when it is equipped with the initial state 0^k and all states are final. Thus, the OGF $E(z)$ is in all cases a rational function.

The matrix of $\mathcal{B}_k^\circ$ is the matrix of the de Bruijn graph $\mathcal{B}_k$ with some non-zero entries replaced by 0. Assume that $\mathcal{B}_k^\circ$ is irreducible. This assumption only eliminates a few pathological cases (e.g., $\Omega = \{01\}$ on the alphabet $\{0, 1\}$). Then, the matrix of $\mathcal{B}_k^\circ$ admits a simple Perron–Frobenius eigenvalue λ_1. By domination properties ($\Omega \neq \emptyset$), we must have $\lambda_1 < m$, where m is the cardinality of the alphabet. Aperiodicity is automatically granted. We then get by a purely qualitative argument: *The number of words of length n excluding patterns from the finite set Ω is, under the assumption of irreducibility, asymptotic to $c(\lambda_1/m)^n$, for some $c > 0$ and $\lambda_1 < m$.* This gives us in a simple manner a strong version of what has been earlier nicknamed "Borges's Theorem" (Note V.35, p. 61): *Almost every sufficiently long text contains all patterns of some predetermined length ℓ.*

The construction of a pruned automaton is clearly a generalization of the case of words obeying local constraints in Example V.11 above. . . . . ∎

Transfer matrix method. Let $\mathcal{C}$ be a combinatorial class to be enumerated.

(i) Determine a collection $\mathcal{C}_1, \mathcal{C}_2, \ldots, \mathcal{C}_m$ of classes, with $\mathcal{C}_1 \cong \mathcal{C}$ such that the following system of equation holds:

$$(106) \qquad \mathcal{C}_j = \sum_{k \in \{1,2,\ldots,m\}} \Omega_{j,k} \mathcal{C}_k + \mathcal{I}_j, \qquad j = 1, 2, \ldots, m,$$

where each $\Omega_{j,k}$ and each $\mathcal{I}_j$ is a finite class.

(ii) The OGF $C(z) = C_1(z)$ is then given by the solution of the linear system

$$(107) \qquad C_j(z) = \sum_j \Omega_{j,k}(z) C_k(z) + I_j(z), \qquad j = 1, \ldots, m,$$

where $\Omega_{j,k}(z)$ and $I_j(z)$ are the generating polynomials of $\Omega_{j,k}$ and $\mathcal{I}_j$, respectively. Accordingly, $C(z)$ is a $\mathbb{C}[z]$–linear combination of entries of the quasi-inverse matrix $(I - \Omega(z))^{-1}$.

Figure V.20. A summary of the basic transfer matrix method.

▷ **V.40.** *Walks on undirected graphs.* Consider an undirected graph Γ, where one moves by following at each step a random edge of the graph, uniformly at random from the current position. Then, the transition matrix $P = (p_{ij})$ of the associated Markov chain is: $p_{i,j} = 1/\deg(i)$ if (i, j) is an edge, where $\deg(i)$ is the degree of vertex i. The stationary distribution is given by $\pi_i = (\deg(i))/(2\|E\|)$, where $\|E\|$ is the number of edges of Γ. In particular, if the graph is *regular*, the stationary distribution is uniform. (See Aldous and Fill's forthcoming book [11] for (much) more.)   ◁

▷ **V.41.** *Words with excluded patterns and digital trees.* Let S be a finite set of words. An automaton recognizing S, considered as a finite language, can be constructed as a tree. The tree obtained is akin to the classical *digital tree* or *trie* that serves as a data structure for maintaining dictionaries [378]. A modification of the construction yields an automaton of size linear in the total number of characters that appear in words of S. [Hint. The construction can be based on the Aho–Corasick automaton [5, 538]).   ◁

V. 6. Transfer matrix models

There exists a cluster of applications of rational functions to problems that are naturally described as paths in digraphs, but with edges that may be of different *sizes*. In physics, such models lie at the heart of what is known as the "transfer matrix method". Technically, the theory is a simple extension of the standard case of paths in graphs developed in Section V. 5. Its main interest lies in its expressiveness as regards a number of combinatorial problems, including trees of bounded width, partial models of self-avoiding walks, and certain constrained permutation problems.

V. 6.1. Combinatorial aspects. The transfer matrix method constitutes a variant of the modelling by deterministic automata and by paths in standard graphs. The general framework is summarized in Figure V.20. The idea is to set up a system of linear equations that relate a cleverly crafted collection of classes ("states") $\mathcal{C}_j$, which are of the same nature as the original class $\mathcal{C}$ that is to be enumerated. The combinatorial system (106) in Figure V.20 can then be visualized as a graph, with the objects of the $\Omega_{j,k}$ classes attached to edges ("transitions between states") being generally of different sizes.

Definition V.7. *Given a directed multigraph Γ with vertex set V and edge set E, a size function on Γ is any function $\sigma : E \to \mathbb{Z}_{\geq 1}$. A sized graph is a pair (G, σ), where σ is a size function.*

Paths are defined in the same way as in Section V. 5. The *length* of a path is, as usual, the number of edges it comprises; the *size* of a path is defined to be the sum of the sizes of its edges. As in the basic case treated in the previous section, we also allow edges to carry positive weights (multiplicities, probability coefficients), the *weight* of a path being the product of the weights of its edges.

Definition V.8. *A matrix $T(z)$ is a transfer matrix if each of its entries is a polynomial in z with non-negative coefficients. A transfer matrix $T(z)$ is said to be proper if $T(0)$ is nilpotent, that is, $T(0)^r = 0$ for some $r \geq 1$.*

Examples of transfer matrices are

$$z \begin{pmatrix} \frac{1}{4} & \frac{3}{4} \\ \frac{1}{2} & \frac{1}{2} \end{pmatrix}, \quad \begin{pmatrix} 0 & 1 \\ z^3 & z + z^2 \end{pmatrix},$$

and both are proper. For the graphs and automata considered in Section V. 5, all edges were taken to be of unit size. In that case, the associated (weighted) adjacency matrices are invariably of the form $T(z) = zS$, with S a scalar matrix having non-negative entries, and thus are very particular cases of proper transfer matrices.

Given a sized graph Γ equipped with weight function $w : E \to \mathbb{R}_{>0}$ (with $w(e) \equiv 1$ in the pure enumerative case), we can associate to it a transfer matrix $T(z)$ as follows:

$$(108) \qquad\qquad T_{a,b}(z) = \sum_{e \in \text{Edge}(a,b)} w(e) z^{|e|}.$$

There, $\text{Edge}(a, b)$ represents the set of all edges connecting a to b; $w(e)$ and $|e| \equiv \sigma(e)$ represent, respectively, the weight and the size of edge e. The matrix $T(z)$ whose (a, b)–entry is the polynomial $T_{a,b}(z)$, as given in (108), is called the *transfer matrix* of the (weighted, sized) graph. By Definition V.7, the transfer matrix of a sized graph is always proper. Since $T(z)^m$ describes all paths in the graph with z marking size, the proof techniques of Proposition V.6 (p. 337) immediately provide:

Proposition V.9. *Given a sized graph with associated transfer matrix $T(z)$, the OGF $F^{\langle i,j \rangle}(z)$ of the set of paths from i to j, where z marks size, is the entry i, j of the matrix $(I - T(z))^{-1}$:*

$$F^{\langle i,j \rangle}(z) = \left((I - T(z))^{-1} \right)_{i,j}.$$

V. 6.2. Analytic aspects. In order to apply the general results from Section V. 5 to transfer matrices, we must first take note of an easy reduction of transfer matrices to the standard case of paths in graphs where all edges have size 1.

Given a sized graph Γ, one can build as follows a standard graph $\widehat{G}$ where all edges of $\widehat{G}$ have unit size. The set of vertices of $\widehat{G}$ is the set of vertices of Γ augmented by additional vertices called *relay nodes*. For each edge e of size $\sigma(e) = m$ in Γ, introduce $m - 1$ additional relay nodes and connect these in $\widehat{G}$ by a simple path from *a* to *b*, with edges all of size 1. Here is for instance the transcription of an edge of length 4 in Γ by means of three relay nodes in $\widehat{G}$:

Clearly, the vertices of Γ are a subset of the vertices of $\widehat{G}$ and all paths of Γ correspond to paths of $\widehat{G}$. Let $\widehat{T}$ be the (scalar) adjacency matrix of Γ. Then, the quasi-inverse $(I - z\widehat{T})^{-1}$ describes all the paths in Γ, with size taken into account, in the sense that the entry of index (i, j) in this quasi-inverse is the OGF of paths from node numbered i to node numbered j in the sized graph Γ.

This construction permits us to apply the main results of Section V.5 to transfer matrices and sized graphs. Let us say that the sized graph Γ and its transfer matrix $T(z)$ are *irreducible* (respectively, *aperiodic*) if $\widehat{G}$ and $\widehat{T}$ are irreducible (respectively, aperiodic). We can then immediately transcribe Theorems V.7 and V.8 as follows.

Corollary V.1. (*i*) *Consider a sized graph* Γ *that is irreducible and aperiodic. Then, there exist a computable constant* λ_1 *and numbers* $\varphi_{i,j}$ *such that the OGF of paths from i to j in* Γ *satisfies*

$$(109) \qquad [z^n]F^{\langle i,j\rangle}(z) = \varphi_{i,j}\lambda_1^n + O(\Lambda^n), \qquad 0 \le \Lambda < \lambda_1.$$

(*ii*) *In a random path from a to b of large size, the number of occurrences of a designated edge* (s, t) *is asymptotically*

$$(110) \qquad\qquad \varpi_{s,t}n + O(1),$$

for a computable constant $\varpi_{s,t}$.

Thus, on general grounds, the behaviour of paths is predictable. The notes below explore some further properties that make it possible to operate directly with the transfer matrix and the sized graph, without necessitating the explicit construction of $\widehat{T}$ and $\widehat{G}$.

▷ **V.42.** *Irreducibility for sized graphs.* The sized graph Γ is irreducible if and only if the graph G_1 where all edges of Γ are taken to be of size 1 is strongly connected. The transfer matrix $T(z)$ of Γ is irreducible (in the sense above) if and only if $T(1)$ is irreducible in the usual sense of scalar transfer matrices.   ◁

▷ **V.43.** *Aperiodicity for sized graphs.* A polynomial $p(z) = \sum_j c_j z^{e_j}$, with every $c_j \neq 0$, is said to be primitive if the quantity $\delta = \gcd(\{e_j\})$ is equal to 1; it is imprimitive otherwise. Equivalently, $p(z)$ is imprimitive iff $p(z) = q(z^\delta)$ for some *bona fide* polynomial q and some $\delta > 1$. An irreducible sized graph is aperiodic (in the sense above) if and only if at least one diagonal entry of some power $T(z)^e$ is a primitive polynomial. Equivalently: there exist two circuits of the same length, whose sizes, s_1, s_2, satisfy $\gcd(s_1, s_2) = 1$.   ◁

▷ **V.44.** *Direct determination of the asymptotic growth constant.* Let Γ be a sized graph assumed to be irreducible and aperiodic. Then, one has $\lambda_1 = 1/\rho$, where ρ is the smallest positive root of $\det(I - T(z)) = 0$, with $T(z)$ the transfer matrix of Γ.   ◁

V.6.3. Applications.

The quantitative properties summarized by (109) and (110) apply with full strength to classes that are amenable to the transfer matrix method. We shall first illustrate the situation by the width of trees following an early article by Odlyzko and Wilf [463], then continue with an example that draws its inspiration from the insightful exposition of domino tilings and generating functions in the book

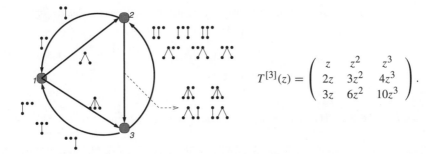

Figure V.21. The sized graph corresponding to general plane trees of width at most 3 and its transfer matrix. (For readability, the transitions from a node to itself are omitted.)

of Graham, Knuth, and Patashnik [307], and conclude with an exactly solvable polyomino model.

***Example* V.17.** *Width of trees.* The width of a tree is defined as the maximal number of nodes that can appear on any layer at a fixed distance from the root. If a tree is drawn in the discrete plane, then width and height can be seen as the horizontal and vertical dimensions of a bounding rectangle. Also, width is an indicator of the complexity of traversing the tree in breadth-first search (by a queue), while height is associated to depth-first search (by a stack).

Transfer matrices are ideally suited to the problem of analysing the number of trees of fixed width. Consider a simple variety of trees $\mathcal{Y}$ corresponding to the equation $Y(z) = z\phi(Y(z))$, where the "generator" ϕ describes the basic formation of trees (Proposition I.5, p. 66). Let $\mathcal{C} := \mathcal{Y}^{[w]}$ be the subclass of trees of width at most w. Such trees are easily built layer by layer. Indeed, with reference to our general description of the transfer matrix method at the beginning of the section, let us introduce a collection of classes $\mathcal{C}_k$, where each $\mathcal{C}_k$ ($k = 1, \ldots, w$) comprises all trees of width $\le w$ having exactly k nodes at the deepest level. We then have $\mathcal{C} = \sum_{k=1}^{w} \mathcal{C}_k$ (this is a trivial variant of the case considered in our general description). Thus the states of the transfer matrix model, equivalently the nodes of the sized graph, correspond to the number of nodes on the deepest layer of the tree. The transition between configurations $\mathcal{C}_j$ corresponding to state j and configurations $\mathcal{C}_k$ corresponding to state k is effected by grafting in all possible ways a forest of j trees, of total height equal to 1, having k leaves. See Figure V.21 for the case of width $w = 3$.

The number of j–forests of depth 1 having k leaves is the quantity

$$t_{j,k} = [u^k]\phi(y)^j.$$

Let T be the $w \times w$ matrix with entry $T_{j,k} = z^k t_{j,k}$. Then, clearly, the quantity $z^i (T^h)_{i,j}$ (with $1 \le i, j \le w$) is the number of i–forests of height h and width at most w, having j nodes on level h. Thus, the GF of $\mathcal{Y}$–trees having width at most w is

$$Y^{[w]}(z) = (z, 0, 0, \ldots)(I - T)^{-1}(1, 1, 1, \ldots)^t.$$

For instance, in the case of general Catalan trees, the matrix T has the shape,

$$T^{[w]}(z) = \begin{pmatrix} z\binom{1}{0} & z^2\binom{2}{0} & z^3\binom{3}{0} & z^4\binom{4}{0} \\ z\binom{2}{1} & z^2\binom{3}{1} & z^3\binom{4}{1} & z^4\binom{5}{1} \\ z\binom{3}{2} & z^2\binom{4}{2} & z^3\binom{5}{2} & z^4\binom{6}{2} \\ z\binom{4}{3} & z^2\binom{5}{3} & z^3\binom{6}{3} & z^4\binom{7}{3} \end{pmatrix},$$

for width 4. The analysis of dominant poles provides asymptotic formulae for $[z^n]Y^{[w]}(z)$:

$w = 2$	$w = 3$	$w = 4$	$w = 5$	$w = 6$
$0.0085 \cdot 2.1701^n$	$0.0026 \cdot 2.8050^n$	$0.0012 \cdot 3.1638^n$	$0.0006 \cdot 3.3829^n$	$0.0004 \cdot 3.5259^n$.

Irreducibility is granted since all entries in the transfer matrix are non-zero. Aperiodicity derives from aperiodicity of the generator ϕ, as verified by a simple argument (e.g., using Note V.43).

Proposition V.10. *The number of trees of width at most w in a simple family of trees satisfies an asymptotic estimate of the form*

$$Y_n^{[w]} = c_w \rho_w^{-n} + O(n),$$

for some computable positive constants c_w, ρ_w.

In addition, the exact distribution of height in trees of size n becomes computable in polynomial time.

The character of these generating functions has not been investigated in detail since the original work [463], so that, at the moment, complex analysis does not lead us any further. Fortunately, probability theory takes over. Chassaing and Marckert [111] have shown, for Cayley trees, that the width satisfies

$$\mathbb{E}_n(W) = \sqrt{\frac{\pi n}{2}} + O\left(n^{1/4}\sqrt{\log n}\right), \qquad \mathbb{P}_n(\sqrt{2}W \le x) \to 1 - \Theta(x),$$

where $\Theta(x)$ is the Theta function defined in (67), p. 328. This answers very precisely an open question of Odlyzko and Wilf [463]. The distributional results of [111] extend to trees in any simple variety (under mild and natural analytic assumptions on the generator ϕ): see the paper by Chassaing, Marckert, and Yor [112], which builds upon earlier results of Drmota and Gittenberger [173]. In essence, the conclusion of these works is that the breadth first search traversal of a large tree in a simple variety gives rise to a queue whose size fluctuates asymptotically like a Brownian excursion, and is thus, in a strong sense, of a complexity comparable to depth-first search: trees taken uniformly don't have much of a preference as to the way they may be traversed. . . . . ∎

▷ **V.45.** *A question on width polynomials.* It is unknown whether the following assertion is true. The smallest positive root ρ_k of the denominator of $Y^{[k]}(z)$ satisfies

$$\rho_k = \rho + \frac{c}{k^2} + o(k^{-2}),$$

for some $c > 0$. If such an estimate were established, together with suitable companion bounds, it would yield a purely analytic proof of the fact that the expected width of n–trees is $\Theta(\sqrt{n})$, as well as detailed probability estimates. (The classical theory of Fredholm equations may be useful in this context.) ◁

Example V.18. *Monomer-dimer tilings of a rectangle.* Suppose one is given pieces that may be one of the three forms: monomers (m) that are 1×1 squares, and dimers that are dominoes, either vertically (v) oriented 1×2, or horizontally (h) oriented 2×1. In how many ways can an $n \times 3$ rectangle be covered completely and without overlap ('tiled') by such pieces?

The pieces are thus of the following types,

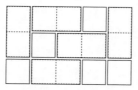

$$m = \square, \quad h = \square\square, \quad v = \square,$$

and here is a particular tiling of a 5×3 rectangle:

In order to approach this counting problem, one first defines a suitable collection, generically denoted by $\mathcal{C}$, of combinatorial classes called configurations, in accordance with the strategy summarized in Figure V.20, p. 356. A configuration relative to an $n \times k$ rectangle is a partial tiling, such that all the first $n - 1$ columns are entirely covered by dominoes while between zero and three unit cells of the last column are covered. Here are for instance, configurations corresponding to the example above.

These diagrams suggest the way configurations can be built by successive addition of dominoes. Starting with the empty rectangle 0×3, one adds at each stage a collection of at most three dominoes in such a way that there is no overlap. This creates a configuration where, like in the example above, the dominoes may not be aligned in a flush-right manner. Continue to add successively dominoes whose left border is at abscissa $1, 2, 3$, etc, in a way that creates no internal "holes".

Depending on the state of filling of their last column, configuration can thus be classified into 8 classes that we may index in binary as $\mathcal{C}_{000}, \dots, \mathcal{C}_{111}$. For instance $\mathcal{C}_{001}$ represent configurations such that the first two cells (from top to bottom, by convention) are free, while the third one is occupied. Then, a set of rules describes the new type of configuration obtained, when the sweep line is moved one position to the right and dominoes are added. For instance, we have

$$\mathcal{C}_{010} \quad \odot \quad \begin{array}{c}\square\square \\ \square\square\end{array} \quad \Longrightarrow \quad \mathcal{C}_{101}.$$

In this way, one can set up a system of linear equations (resembling a grammar or a deterministic finite automaton) that expresses all the possible constructions of longer rectangles from shorter ones according to the last layer added. The system contains equations like

$$
\begin{aligned}
\mathcal{C}_{000} \;=\; & \epsilon + \underline{mmm}\mathcal{C}_{000} + \underline{mv}\mathcal{C}_{000} + \underline{vm}\mathcal{C}_{000} \\
& + \underline{\cdot mm}\mathcal{C}_{100} + \underline{m \cdot m}\mathcal{C}_{010} + \underline{mm \cdot}\mathcal{C}_{001} + \underline{v \cdot}\mathcal{C}_{001} + \underline{\cdot v}\mathcal{C}_{100} \\
& + \underline{m \cdot \cdot}\mathcal{C}_{011} + \underline{\cdot m \cdot}\mathcal{C}_{101} + \underline{\cdot \cdot m}\mathcal{C}_{110} + \underline{\cdot \cdot \cdot}\mathcal{C}_{111} .
\end{aligned}
$$

Here, a "letter" like $\underline{mv}$ represent the addition of dominoes, in top to bottom order, of types m, v, respectively; the letter $\underline{m \cdot m}$ means adding two m-dominoes on the top and on the bottom, etc.

The system transforms into a linear system of equations with polynomial coefficients, upon performing the substitutions

$$m \mapsto z, \quad h \mapsto z^2, \quad v \mapsto z^2.$$

Solving it gives the generating functions of configurations with z marking the area covered:

$$C_{000}(z) = \frac{(1 - 2z^3 - z^6)(1 + z^3 - z^6)}{(1 + z^3)(1 - 5z^3 - 9z^6 + 9z^9 + z^{12} - z^{15})}.$$

In particular, the coefficient $[z^{3n}]C_{000}(z)$ is the number of tilings of an $n \times 3$ rectangle:

$$C_{000}(z) = 1 + 3z^3 + 22z^6 + 131z^9 + 823z^{12} + 5096z^{15} + \cdots.$$

The sequence grows like $c\,\alpha^n$ (for $n \equiv 0 \pmod 3$) where $\alpha \doteq 1.83828$ (α is the cube root of an algebraic number of degree 5). (See [109] for a computer algebra session.) On average, for large n, there is a fixed proportion of monomers and the distribution of monomers in a random tiling of a large rectangle is asymptotically normally distributed, a result that follows from the developments of Section IX. 6, p. 650. . . . . ∎

The tiling example is a typical illustration of the transfer matrix method as described in Figure V.20, p. 356. One seeks to enumerate a "special" set of configurations: in the example above, this is C_{000} representing complete rectangle coverings. One determines an extended set of configurations $\mathcal{C}$ (the partial coverings, in the example) such that: (i) $\mathcal{C}$ is partitioned into finitely many classes; (ii) there is a finite set of "actions" that operate on the classes; (iii) size is affected in a well-defined additive way by the actions. The similarity with finite automata is apparent: classes play the rôle of states and actions the rôle of letters.

Often, the method of transfer matrices is used to approximate a hard combinatorial problem that is not known to decompose, the approximation being by means of a family of models of increasing "widths". For instance, the enumeration of the number T_n of tilings of an $n \times n$ square by monomers and dimers remains a famous unsolved problem of statistical physics. Here, transfer matrix methods may be used to solve the $n \times w$ version of the monomer–dimer coverings, in principle at least, for any fixed width w: the result will always be a rational function, although its degree, dictated by the dimension of the transfer matrix, will grow exponentially with w. (The "diagonal" sequence of the $n \times w$ rectangular models corresponds to the square model.) It has been at least determined by computer search that the diagonal sequence T_n starts as (this is *EIS* **A028420**):

$$1, \ 7, \ 131, \ 10012, \ 2810694, \ 2989126727, \ 11945257052321, \ldots.$$

From this and other numerical data, one estimates numerically that $(T_n)^{1/n^2}$ tends to a constant, $1.94021\ldots$, for which no expression is known to exist. The difficulty of coping with the finite-width models is that their complexity (as measured, e.g., by the number of states) increases exponentially with w—such models are best treated by computer algebra; see [627]—but no law allowing to take a diagonal is visible. At least, the finite-width models have the merit of providing provable upper and lower bounds on the exponential growth rate of the hard "diagonal problem".

In contrast, for coverings by dimers only, a strong algebraic structure is available and the number of covers of an $n \times n$ square by horizontal and vertical dimers satisfies a beautiful formula originally discovered by Kasteleyn (n even):

$$(111) \qquad U_n = 2^{n^2/2} \prod_{j=1}^{n/2} \prod_{k=1}^{n/2} \left(\cos^2 \frac{j\pi}{n+1} + \cos^2 \frac{k\pi}{n+1} \right).$$

This sequence is *EIS* **A004003**,

1, 2, 36, 6728, 12988816, 258584046368, 53060477521960000, ...

It is elementary to prove from (111) that

$$\lim_{n \to +\infty} (U_n)^{1/n^2} = \exp \left(\frac{1}{\pi} \sum_{n=0}^{\infty} \frac{(-1)^n}{(2n+1)^2} \right) = e^{G/\pi} \doteq 1.33851 \ldots,$$

where G is Catalan's constant. This means in substance that each cell has a number of degrees of freedom equivalent to 1.33851. See Percus' monograph [477] for proofs of this famous result and Finch's book [211, Sec. 5.23] for context and references.

▷ **V.46.** *Powers of Fibonacci numbers.* Consider the OGFs

$$G(z) := \frac{1}{1 - z - z^2} = \sum_{n \geq 0} F_{n+1} \, z^n, \qquad G^{[k]}(z) := \sum_{n \geq 0} \left(F_{n+1} \right)^k z^n,$$

where F_n is a Fibonacci number. The OGF of monomer–dimer placements on a $k \times n$ board when only monomers (m) and horizontal dimers (h) are allowed is obviously $G^{[k]}(z)$. On the other hand, it is possible to set up a transfer matrix model with state i ($0 \leq i \leq k$) corresponding to i positions of the current column occupied by a previous domino. Consequently,

$$G^{[k]}(z) = \mathrm{coeff}_{k,k} \left((I - zT)^{-1} \right), \qquad \text{where} \quad T_{i,j} = \binom{i}{i+j-k},$$

for $0 \leq i, j \leq k$. (The denominator of $G^{[k]}(z)$ is known exactly: see [377, Ex. 1.2.8.30].)   ◁

▷ **V.47.** *Tours on chessboards.* The OGF of Hamiltonian tours on an $n \times w$ rectangle is rational (one is allowed to move from any cell to any other vertically or horizontally adjacent cell). The same holds for king's tours and knight's tours.   ◁

▷ **V.48.** *Cover time of graphs.* Given a fixed digraph Γ assumed to be strongly connected, and a designated start vertex, one travels at random, moving at each time to any neighbour of the current vertex, making choices with equal likelihood. The expectation of the time to visit all the vertices is a rational number that is effectively (though perhaps not efficiently!) computable. [Hint: set up a transfer matrix, a state of which is a subset of vertices representing those vertices that have been already visited. For an interval $[0, \ldots m]$, this can be treated by the dedicated theory of walks on the integer interval, as in Section V. 4; for the complete graph, this is equivalent to the coupon collector problem. Most other cases are "hard" to solve analytically and one has to turn to probabilistic approximations; see Aldous and Fill's forthcoming book [11] for a probabilistic approach.]   ◁

Example V.19. *Self-avoiding walks and polygons.* A long-standing open problem, shared by statistical physics, combinatorics, and probability theory alike, is that of quantifying properties of self-avoiding configurations on the square lattice (Figure V.22). Here we consider objects that, starting from the origin (the "root"), follow a path, and are solely composed of horizontal and vertical steps of amplitude ± 1. The *self-avoiding walk* or *SAW* can wander but is subject to the condition that it never crosses nor touches itself. The *self-avoiding polygons* or *SAPs*, whose class is denoted by $\mathcal{P}$, are self-avoiding walks, with only an exception at the end, where the endpoint must coincide with the origin. We shall focus here on polygons. It proves convenient also to consider *unrooted polygons* (also called simply-connected *polyominoes*), which are polygons

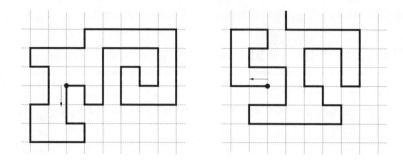

Figure V.22. A self-avoiding polygon or SAP (left) and a self-avoiding walk or SAW (right).

in which the origin is discarded, so that they plainly represent the possible shapes of SAPs up to translation. For length $2n$, the number p_n of unrooted polygons satisfies $p_n = P_n/(4n)$ since the origin ($2n$ possibilities) and the starting vertex (2 possibilities) of the corresponding SAPs are disregarded in that case. Here is a table, for small values of n, listing polyominoes and the corresponding counting sequences p_n, P_n.

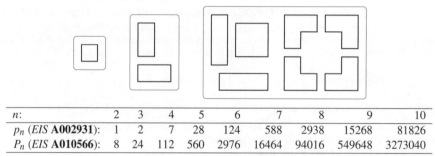

n:	2	3	4	5	6	7	8	9	10
p_n (EIS A002931):	1	2	7	28	124	588	2938	15268	81826
P_n (EIS A010566):	8	24	112	560	2976	16464	94016	549648	3273040

Take the (widely open) problem of determining *exactly* the number P_n of SAPs of perimeter $2n$. This (intractable) problem can be approached as a limit of the (tractable) problem[12] that consists in enumerating the collection $\mathcal{P}^{[w]}$ of SAPs of width w, for increasing values of w. The latter problem is amenable to the transfer matrix method, as first discovered by Enting in 1980; see [192]. Indeed, take a polygon and consider a vertical sweepline, that moves from left to right. Once width is fixed, there are at most 2^{2w+2} possibilities for the ways such a line may intersect the polygon's edges at half integer abscissae. (There are $w + 1$ edges and for each of these, one should "remember" whether they connect with the upper or lower boundary.) The transitions are then themselves finitely described. In this way, it becomes possible to set up a transfer matrix for any fixed width w. For fixed n, by computing values of $P_n^{[w]}$ with increasing w, one finally determines (in principle) the exact value of any P_n.

The program suggested above has been carried out to record values by the "Melbourne School" under the impulse of Tony Guttmann. For instance, Jensen [356] found in 2003 that the number of unrooted polygons of perimeter 100 is

$$p_{50} = 75456497744850697064688603335686 2162.$$

---

[12]We limit ourselves here to a succinct description and refer to the original papers [192, 356] for details.

Attaining such record values necessitates algorithms that are much more sophisticated than the naïve approach we have just described, as well as a number of highly ingenious programming optimizations.

It is an equally open problem to estimate *asymptotically* the number of SAPs of perimeter n. Given the exact values up to perimeter 100 or more, a battery of fitting tests for asymptotic formulae can be applied, leading to highly *convincing* (though still heuristic) formulae. Thanks to several workers in this area, we can regard the final answer as "known". From the works of Jensen and his predecessors, it results that a reliable empirical estimate is of the form

$$\begin{cases} p_n = B\mu^{2n}(2n)^{-\beta}(1 + o(1)), \\ \mu \doteq 2.63815\,85303, \qquad \beta = -\frac{5}{2} \pm 3 \cdot 10^{-7}, \qquad B \doteq 0.5623013. \end{cases}$$

Thus, the answer is almost certainly of the form $p_n \asymp \mu^{2n} n^{-5/2}$ for unrooted polygons and $P_n \asymp \mu^{2n} n^{-3/2}$ for rooted polygons. It is believed that the same connective constant μ dictates the exponential growth rate of self-avoiding walks. See Finch's book [211, Sec. 5.10] for a perspective and numerous references.

There is also great interest in the number $p_{m,n}$ of polyominoes with perimeter $2n$ and area m, with area defined as the number of square cells composing the polyomino. Studies conducted by the Melbourne school yield numerical data that are consistent to an amazing degree (e.g., moments up to order ten and small-n corrections are considered) with the following assumption: *The distribution of area in a fixed-perimeter polyomino obeys in the asymptotic limit an "Airy area distribution"*. This distribution is defined as the limit distribution of the area under Dyck paths, a problem that was introduced on p. 330 and to which we propose to return in Chapter VII (p. 535) and IX (p. 706). See [356, 509, 510] and references therein for a specific discussion of polyomino area. It is finally of great interest to note that the interpretation of data was strongly guided by what is already known for exactly solvable models of the type we are repeatedly considering in this book. .... ∎

Example V.20. *Horizontally convex polyominoes.* Pólya [490] and Temperley [574] independently discovered an exactly solvable polyomino model. (See also the text by van Rensburg [592] for more.) Define as usual a polyomino as a collection of unit squares with vertices in $\mathbb{Z}_{\geq 0} \times \mathbb{Z}_{\geq 0}$ that forms a connected set without articulation points. Such a polyomino is said to be *horizontally convex* (H.C.) if its intersection with any horizontal line is either empty or an interval. An H.C. polyomino is thus a stack of a certain number of rows of squares, where each row has a segment of length ≥ 1 in common with the next row up. (We imagine H.C. polyominoes growing from bottom to top.) The enumeration of such polyominoes, following Temperley [574, p. 66] constitutes a nice extension of the transfer matrix method in the case when the set of states is *infinite*.

Let $T^{[k]}$ be the class of polyominoes with exactly k square cells on their top row. The size of a polyomino is its number of cells. We wish to enumerate the class $T := \bigcup_k T^{[k]}$. In order to do so, according to the transfer matrix method, one needs to relate the $T^{[k]}$ to one another. Let z be the variable marking size. The transition from one $T^{[k]}$ to a $T^{[\ell]}$ has a multiplicity equal to $k + \ell - 1$. Thus the generating functions $t_k := T^{[k]}(z)$ satisfy an infinite system of equations, which starts as

$$\begin{array}{rcl} t_1 & = & z + z(t_1 + 2t_2 + 3t_3 + \cdots) \\ t_2 & = & z^2 + z^2(2t_1 + 3t_2 + 4t_3 + \cdots) \\ t_3 & = & z^3 + z^3(3t_1 + 4t_2 + 5t_3 + \cdots). \end{array} \tag{112}$$

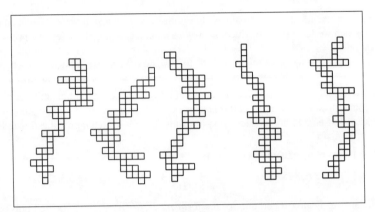

Figure V.23. Five horizontally convex polyominoes of size $n = 50$ drawn uniformly at random.

This corresponds to an infinite transfer matrix which is highly structured:

$$M(z)_{k,\ell} = (k + \ell - 1)z^\ell,$$

and, as shown by Temperley [574, p. 66], the system can be solved by elementary manipulations. We shall however prefer to take another route, more in line with the spirit of this book.

In a case like this, it is well worth trying a bivariate generating function. Define

$$T(z, u) = \sum_{n,k} T^{[k]}(z)u^k.$$

The action of "adding a slice" on the top row of a polyomino is reflected by a linear operator $\mathcal{L}$ that transforms u^k, representing the top row of the polyomino before addition, into a sum of monomials $u^\ell z^\ell$, with the proper multiplicities:

$$\mathcal{L}[u^k] = k(uz)^k + (k+1)(uz)^{k+1} + \cdots = (k-1)\frac{uz}{1 - uz} + \frac{uz}{(1 - uz)^2}.$$

(An earlier instance of the technique of "adding a slice" appears in the context of constrained compositions, Example III.22, p. 199.) A better formula results if one expresses more generally the quantity $\mathcal{L}[f(u)]$:

$$(113) \qquad \mathcal{L}[f(u)] = \frac{uz}{(1 - uz)^2} f(1) + \frac{uz}{1 - uz} \left(f'(1) - f(1) \right).$$

Treat now the BGF $T(z, u)$ as a function of u, keeping z as a parameter, and write for readability $\tau(u) := T(z, u)$. A horizontally convex polyomino is obtained by starting from a bottom row that can have any number of cells and repeatedly adding a slice. This construction is thus reflected by the main functional equation

$$(114) \qquad \begin{aligned} \tau(u) &= \frac{zu}{1 - zu} + \mathcal{L}[\tau(u)] \\ &= \frac{zu}{1 - zu} + \frac{zu}{1 - zu}\tau'(1) + \frac{z^2u^2}{(1 - zu)^2}\tau(1), \end{aligned}$$

upon making use of (113). Instantiating at $u = 1$ provides the first relation

$$(115) \qquad \tau(1) = \frac{z}{1 - z} + \frac{z}{1 - z}\tau'(1) + \frac{z^2}{(1 - z)^2}\tau(1),$$

while differentiation of (114) with respect to u followed by the specialization $u = 1$ provides the second relation

$$(116) \qquad \tau'(1) = \frac{z}{(1-z)^2} + \frac{z}{(1-z)^2}\tau'(1) + 2\frac{z}{(1-z)^3}\tau(1).$$

We now have a linear system of two equations in two unknowns, resulting in an expression of $\tau(1) = T(z) = T(z, 1)$, which enumerates all horizontally convex polyominoes:

$$(117) \qquad T(z) = \frac{z(1-z)^3}{1 - 5z + 7z^2 - 4z^3}.$$

(From (114) to (117), the whole calculation is barely three lines of code under a decent computer algebra system.) Note that, the original system being infinite, it is far from obvious *a priori* that the generating function should be rational—in the present context, rationality devolves from the highly structured character of the transfer matrix.

The counting sequence obtained by expansion,

$$T(z) = z + 2z^2 + 6z^3 + 19z^4 + 61z^5 + 196z^6 + 629z^7 + 2017z^8 + \cdots$$

is *EIS* **A001169** (*"Number of board-pile polyominoes with n cells"*). The asymptotic form is then easily obtained: we find

$$T_n \sim CA^n, \qquad C \doteq 0.18091, \qquad A \doteq 3.20556,$$

with A a cubic irrational.

An alternative derivation, which is more sophisticated, is due to Klarner and is presented in Stanley's book [552, §4.7]. Hickerson [333] has found a direct construction, which explains the rationality of the GF by means of a regular language encoding. (The drawings of Figure V.23 have been obtained by an application of the recursive method [264] to Hickerson's specification.) Louchard [420] has conducted an in-depth study of probabilistic properties of several parameters of H.C. polyominoes, using generating functions. .... ∎

▷ **V.49.** *Height of H.C. polyominoes.* Upon introducing an extra variable v to encode height, one finds that height grows on average linearly with n and the variance is $O(n)$, so that the distribution is concentrated [420]. (This explains the skinny aspects of polyominoes drawn in Figure V.23.) ◁

▷ **V.50.** *A transfer matrix model for lattice paths.* Consider the general context of weighted lattice paths in Section V. 4. Let $\alpha_j, \beta_j, \gamma_j$ be the weights of ascents, descents, and level steps, respectively, when the starting altitude is j. The infinite transfer matrix,

$$T = \begin{pmatrix} \gamma_0 & \alpha_0 & 0 & 0 & 0 & \cdots \\ \beta_1 & \gamma_1 & \alpha_1 & 0 & 0 & \cdots \\ 0 & \beta_2 & \gamma_2 & \alpha_2 & 0 & \cdots \\ \vdots & \vdots & \vdots & \vdots & \vdots & \ddots \end{pmatrix},$$

which has a tridiagonal form, "generates" all lattice paths via the quasi-inverse $(I - zT)^{-1}$. In particular, any exactly solvable weighted lattice path model is equivalent to an explicit structured matrix inversion. ◁

V. 6.4. Value-constrained permutations.
We conclude this chapter with a discussion of a construction that combines transfer matrix methods with an inclusion–exclusion argument. We treat a collection of constrained permutation problems whose origin lies in nineteenth century recreational mathematics. For instance, the *ménage* problem solved and popularized by Édouard Lucas in 1891, see [129], has the following quaint formulation: *What is the number of possible ways one can arrange n*

married couples ("ménages") around a table in such a way that men and women alternate, but no woman sits next to her husband?

The ménage problem is equivalent to a permutation enumeration problem. Sit first conventionally the men at places numbered $1, 2, \ldots, n$ and the wives at positions $\frac{3}{2}, \frac{5}{2}, \ldots, n + \frac{1}{2}$. Let σ_i be such that the ith wife is placed at $\sigma_i + \frac{1}{2}$. Then, a ménage placement imposes the conditions $\sigma_i \neq i$ and $\sigma_i \neq i - 1$ for each i. We consider here a linearly arranged table (see remarks at the end for the other classical formulation that considers a round table), so that the condition $\sigma_i \neq i - 1$ becomes vacuous when $i = 1$. Here is a ménage placement for $n = 6$ and its corresponding permutation

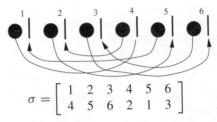

$$\sigma = \begin{bmatrix} 1 & 2 & 3 & 4 & 5 & 6 \\ 4 & 5 & 6 & 2 & 1 & 3 \end{bmatrix}$$

This is a generalization of the derangement problem (for which only the weaker condition $\sigma_i \neq i$ is imposed and the cycle decomposition of permutations suffices to provide a direct solution; see Example II.14, p. 122).

Definition V.9. *Given a permutation $\sigma = \sigma_1 \cdots \sigma_n$, any quantity $\sigma_i - i$ is called an exceedance of σ. Given a finite set of integers $\Omega \subset \mathbb{Z}_{\geq 0}$, a permutation is said to be Ω–avoiding if none of its exceedances lies in Ω.*

The original ménage problem is modelled by $\Omega = \{-1, 0\}$, or, up to a simple transformation, by $\Omega = \{0, 1\}$.

Inclusion–exclusion. The set Ω being fixed, consider first for all j the class of augmented permutations $\mathcal{P}_{n,j}$ that are permutations of size n such that j of the positions are distinguished and the corresponding exceedances lie in Ω, the remaining positions having arbitrary values (but with the permutation property being satisfied). Loosely speaking, the objects in $\mathcal{P}_{n,j}$ can be regarded as permutations with "at least" j exceedances in Ω. For instance, with $\Omega = \{1\}$ and

$$\sigma = \begin{pmatrix} 1 & 2 & 3 & 4 & 5 & 6 & 7 & 8 & 9 \\ 2 & 3 & 4 & 8 & 6 & 7 & 1 & 5 & 9 \end{pmatrix},$$

there are 5 exceedances that lie in Ω (at positions $1, 2, 3, 5, 6$) and with 3 of these distinguished (say by enclosing them in a box), one obtains an element counted by $\mathcal{P}_{9,3}$, such as

$$2 \boxed{3} \boxed{4} 8 6 \boxed{7} 1 5 9.$$

Let $P_{n,j}$ be the cardinality of $\mathcal{P}_{n,j}$. We claim that the number $Q_n = Q_n^\Omega$ of Ω–avoiding permutations of size n satisfies

(118)
$$Q_n = \sum_{j=0}^{n} (-1)^j P_{n,j}.$$

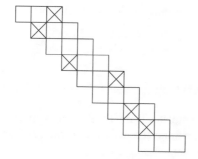

Figure V.24. A graphical rendering of the legal template **20?02?11?** relative to $\Omega = \{0, 1, 2\}$.

Equation (118) is typically an *inclusion–exclusion* relation. To prove it formally[13], define the number $R_{n,k}$ of permutations that have exactly k exceedances in Ω and the generating polynomials

$$P_n(w) = \sum_j P_{n,j} w^j, \qquad R_n(w) = \sum_k R_{n,k} w^k.$$

The GFs are related by

$$P_n(w) = R_n(w+1) \qquad \text{or} \qquad R_n(w) = P_n(w-1)..$$

(The relation $P_n(w) = R_n(w+1)$ simply expresses symbolically the fact that each Ω-exceedance in $\mathcal{R}$ may or may not be taken in when composing an element of $\mathcal{P}$.) In particular, we have $P_n(-1) = R_n(0) = R_{n,0} = Q_n$ as was to be proved.

   Transfer matrix model. The preceding discussion shows that everything relies on the enumeration $P_{n,j}$ of permutations with distinguished exceedances in Ω. Introduce the alphabet $\mathcal{A} = \Omega \cup \{\text{'?'}\}$, where the symbol '?' is called the 'don't-care symbol'. A word on $\mathcal{A}$, an instance with $\Omega = \{0, 1, 2\}$ being 20?02?11?, is called a *template*. To an augmented permutation, one associates a template as follows: each exceedance that is not distinguished is represented by a don't care symbol; each distinguished exceedance (thereby an exceedance with value in Ω) is represented by its value. A template is said to be legal if it arises from an augmented permutation. For instance a template $2\,1\,\cdots$ cannot be legal since the corresponding constraints, namely $\sigma_1 - 1 = 2$, $\sigma_2 - 2 = 1$, are incompatible with the permutation structure (one would have $\sigma_1 = \sigma_2 = 3$). In contrast, the template 20?02?11? is seen to be legal. Figure V.24 is a graphical rendering; there, letters of templates are represented by dominoes, with a cross at the position of a numeric value in Ω, and with the domino being blank in the case of a don't-care symbol.

   Let $T_{n,j}$ be the set of legal templates relative to Ω that have length n and comprise j don't care symbols. Any such legal template is associated to exactly $j!$ permutations, since $n - j$ position-value pairs are fixed in the permutation, while the j remaining

---

[13]See also the discussion in Subsection III. 7.4, p. 206.

positions and values can be taken arbitrarily. There results that

(119) $P_{n,n-j} = j!\, T_{n,j}$ and $Q_n = \sum_{j=0}^{n} (-1)^{n-j}\, j!\, T_{n,j},$

by (118). Thus, the enumeration of avoiding permutations rests entirely on the enumeration of legal templates.

The enumeration of legal templates is finally effected by means of a transfer matrix method, or equivalently, by a finite automaton. If a template $\tau = \tau_1 \cdots \tau_n$ is legal, then the following condition is met,

(120) $\tau_j + j \neq \tau_i + i,$

for all pairs (i, j) such that $i < j$ and neither of τ_i, τ_j is the don't-care symbol. (There are additional conditions to characterize templates fully, but these only concern a few letters at the end of templates and we may ignore them in this discussion.) In other words, a τ_i with a numerical value preempts the value $\tau_i + i$. Figure V.24 exemplifies the situation in the case $\Omega = \{0, 1, 2\}$. The dominoes are shifted one position each time (since it is the value of $\sigma - i$ that is represented) and the compatibility constraint (120) is that no two crosses should be vertically aligned. More precisely the constraints (120) are recognized by a deterministic finite automaton whose states are indexed by subsets of $\{0, \ldots, b-1\}$ where the "span" b is defined as $b = \max_{\omega \in \Omega} \omega$. The initial state is the one associated with the empty set (no constraint is present initially), the transitions are of the form ($j \in \{0, \ldots, b\}$):

$$\begin{cases} (qs, j) \mapsto q_{S'} & \text{where } S' = ((S - 1) \cup \{j - 1\}) \cap \{0, \ldots, b-1\} \\ (qs, ?) \mapsto q_{S'} & \text{where } S' = (S - 1) \cap \{0, \ldots, b-1\}. \end{cases}$$

The initial state (is $q_{\{\}}$ and it is equal to the final state (this translates the fact that no domino can protrude from the right, and is implied by the linear character of the ménage problem under consideration). In essence, the automaton only needs a finite memory since the dominoes slide along the diagonal and, accordingly, constraints older than the span can be forgotten. Notice that the complexity of the automaton, as measured by its number of states, is 2^b.

Here are the automata corresponding to $\Omega = \{0\}$ (derangements) and to $\Omega = \{0, 1\}$ (ménages).

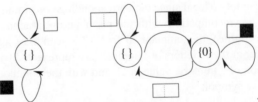

For the ménage problem, there are two states depending on whether or not the currently examined value has been preempted at the preceding step.

From the automaton construction, the bivariate GF $T^{\Omega}(z, u)$ of legal templates, with u marking the position of don't care symbols, is a rational function that can be determined in an automatic fashion from Ω. For the derangement and ménage problems, one finds

$$T^{\{0\}}(z, u) = \frac{1}{1 - z(1 + u)}, \qquad T^{\{0,1\}}(z, u) = \frac{1 - z}{1 - z(2 + u) + z^2}.$$

In general, this gives access to the OGF of the corresponding permutations. Indeed, the OGF of Ω–avoiding permutations is obtained from T^{Ω} by a transformation akin to the Laplace transform: we have

$$(121) \quad z^n u^j \mapsto (-z)^n (-1)^j j!, \quad \text{so that} \quad Q^{\Omega}(z) = \int_0^{\infty} e^{-u} T^{\Omega}(-z, -u) \, du,$$

which transcribes (119) and constitutes a first closed-form solution. In addition, consider the partial expansion of $T^{\Omega}(z, u)$ with respect to u, taken as

$$(122) \qquad\qquad T^{\Omega}(z, u) = \sum_r \frac{c_r(z)}{1 - u u_r(z)},$$

assuming for simplicity only simple poles. There, the sum is finite and it involves algebraic functions c_r and u_r of the variable z. Define next the (divergent) OGF of all permutations,

$$F(y) = \sum_{n=0}^{\infty} n! \, y^n = {}_2 F_0[1, 1; y],$$

in the terminology of hypergeometric functions (Note B.15, p. 751). Then, by (121) and (122), we find

$$(123) \qquad\qquad Q^{\Omega}(z) = \sum_r c_r(-z) F(-u_j(-z)).$$

In other words: *the OGF of Ω–avoiding permutations is expressible both as the Laplace transform of a bivariate rational function (121) and as a composition (123) of the OGF of the factorial series with algebraic functions.*

The expressions (122) simplify much in the case of ménages and derangements where the denominators of T are of degree 1 in u. One finds

$$Q^{\{0\}}(z) = \frac{1}{1 + z} F\left(\frac{z}{1 + z}\right) = 1 + z^2 + 2z^3 + 9z^4 + 44z^5 + 265z^6 + 1854z^7 + \cdots,$$

for derangements, whence a new derivation of the known formula,

$$Q_n^{\{0\}} = \sum_{k=0}^{n} (-1)^k \binom{n}{k} (n - k)!.$$

Similarly, for (linear) ménage placements, one finds

$$Q^{\{0,1\}}(z) = \frac{1}{1 + z} F\left(\frac{z}{(1 + z)^2}\right) = 1 + z^3 + 3z^4 + 16z^5 + 96z^6 + 675z^7 + \cdots,$$

which is *EIS* **A00027** and corresponds to the formula

$$Q_n^{\{0,1\}} = \sum_{k=0}^{n} (-1)^k \binom{2n - k}{k} (n - k)!.$$

Finally, the same techniques adapts to constraints that "wrap around", that is, constraints taken modulo n. (This corresponds to a round table in the ménage problem.) In that case, what should be considered is the circuits in the automaton recognizing templates (see also the discussion on p. 337). One obtains in this way the OGF of the circular (i.e., classical) ménage problem (*EIS* **A000179**),

$$\widehat{Q}^{\{0,1\}}(z) = \frac{1-z}{1+z} F\left(\frac{z}{(1+z)^2}\right) + 2z = 1 + z + z^3 + 2z^4 + 13z^5 + 80z^6 + 579z^7 + \cdots,$$

which yields the classical solution of the (circular) ménage problem,

$$\widehat{Q}_n^{\{0,1\}} = \sum_{k=0}^{n} (-1)^k \frac{2n}{2n-k} \binom{2n-k}{k} (n-k)!.$$

This last formula is due to Touchard; see [129, p. 185] for pointers to the vast classical literature on the subject. The algebraic part of the treatment above is close to the inspiring discussion found in Stanley's book [552]. An application to robustness of interconnections in random graphs is presented in [239].

Asymptotic analysis. For asymptotic analysis purposes, the following property proves useful. *Let F be the OGF of factorial numbers and assume that $y(z)$ is analytic at the origin where it satisfies $y(z) = z - \lambda z^2 + O(z^3)$; then the following estimate holds:*

(124) $$[z^n]F(y(z)) \sim [z^n]F(z(1-\lambda z)) \sim n!e^{-\lambda}.$$

(The proof results from simple manipulations of divergent series in the style of [36, §5].) This gives at sight the estimates

$$Q_n^{\{0\}} \sim n!e^{-1}, \qquad Q_n^{\{0,1\}} \sim n!e^{-2}.$$

Generally, one has:

Proposition V.11. *For any set Ω containing λ elements, the number of permutations without exceedances in Ω satisfies*

$$Q_n^{\{\Omega\}} \sim n!e^{-\lambda}.$$

Furthermore, the number $R_{n,k}^{\Omega}$ of permutations having exactly k occurrences (k fixed) of an exceedance in Ω is asymptotic to

$$R_{n,k}^{\{\Omega\}} \sim n!e^{-\lambda}\frac{\lambda^k}{k!}.$$

That is, the rare event that an exceedance belongs to Ω is asymptotically governed by a Poisson distribution of rate $\lambda = |\Omega|$.

This statement is established by means of elementary combinatorial manipulations in Bender's survey [36, §4.2] and by probabilistic techniques in the book of Barbour, Holst, and Janson [29, Sec. 4.3]. *The relation* (124) *provides a way of arriving at such estimates by purely analytic–combinatorial techniques.*

▷ **V.51.** *Other constrained permutations.* Given a permutation $\sigma = \sigma_1 \cdots \sigma_n$, a *succession gap* is defined as any difference $\sigma_{i+1} - \sigma_i$.

In how many ways can a kangaroo jump through all points of the integer interval $[1, n+1]$ starting at 1 and ending at $n + 1$, while making hops that are restricted to $\{-1, 1, 2\}$? (The OGF is the rational function $1/(1 - z - z^3)$ corresponding to *EIS* **A000930**.)

The number R_n of permutations of size n, such that $\sigma_{i+1} - \sigma_i \neq 1$ has OGF $F(z/(1+z))$, the coefficients being *EIS* **A000255**, with asymptotics $R_n \sim n!e^{-1}$. The number S_n of those, such that $|\sigma_{i+1} - \sigma_i| \neq 1$ has OGF $F(z(1-z)/(1+z))$. Proof (for S_n): Use inclusion–exclusion based on configurations with distinguished sequences of ± 1 successions, like

$$\boxed{8\;7\;6}\;\overset{\longrightarrow}{10\;15}\;\boxed{2\;3\;4}\;5\;9\;1\;13\;\boxed{12\;11}\;14 \;\cong\; \boxed{\bullet\;\bullet\;4}\;6\;10\;\boxed{2\;\bullet\;\bullet}\;3\;5\;1\;8\;\boxed{\bullet\;7}\;9,$$

which leads to the OGF

$$\left[\sum_{m \geq 0} m!\left(z + \frac{2z^2 u}{1 - zu}\right)^m\right]_{u=-1} = \sum_{m \geq 0} m!\left(z\frac{1-z}{1+z}\right)^m$$

$$= 1 + z + 2z^4 + 14z^5 + 90z^6 + 646z^7 + \cdots ;$$

cf *EIS* **A002464** and [4]; this is the number of placements of n kings on a chessboard, one per line, one per column, and in non-attacking position. Asymptotically, one has $S_n \sim n!e^{-2}$, see [572], in accordance with (124). In general, what about the counting of permutations whose succession gaps are constrained to lie outside of a finite set Ω?　　$\lhd$

$\rhd$ **V.52.** *Superménage numbers.* Let T_n be the number of permutations of size n such that $(\sigma_{i+1} - \sigma_i) \notin \{0, 1, 2\}$. The OGF is

$$T(z) = \frac{1}{1 - z^2}\left(-z + F\left(\frac{z(1-z)}{(1+z)(1+z-z^3)}\right)\right) = 1 + z^4 + 5z^5 + 33z^6 + 236z^7 + \cdots ;$$

see [222] and *EIS* **A001887**. Asymptotically: $T_n \sim n!e^{-3}$.　　$\lhd$

V. 7. Perspective

The theorems in this chapter demonstrate the power of the fundamental techniques developed in Chapter IV, which exploit classical theorems in complex analysis to develop coefficient asymptotics. As we start seeing it here, this approach applies to many of the generating functions derived from the formal combinatorial techniques of Part A of this book. By paying careful attention to the types of combinatorial constructions involved, we are able to identify abstract schemas that help us solve whole classes of problems at once. Each schema connects a type of combinatorial construction to a complex asymptotic method. In this way, it becomes possible to discuss properties shared by an infinite collection of combinatorial classes. In this chapter, we have presented the method in detail for classes that involve a sequence construction and classes recursively defined by a linear system of equations (paths in graphs, automata, transfer matrices).

In an ideal world, we might wish to have a direct correspondence between combinatorial constructions and analytic methods—a theory that would carry all the way from combinatorial objects of any description to full analysis of all their properties. The case of paths in graphs and automata, with its strong connectedness condition leading to Perron–Frobenius theory, is an instance of this ideal situation. Reality is however usually a bit more complex: theorems for deriving asymptotic results from combinatorial specifications must often have some sort of analytic side conditions. A typical example is the radius of convergence condition for supercritical sequences. As soon as such side conditions are satisfied, the asymptotic properties of large structures become highly predictable. This is the very essence of analytic combinatorics.

In the next two chapters, we investigate generating functions whose singularities are no longer poles—fractional exponents and logarithmic factors become allowed. This first necessitates investing in general methodology, a task undertaken in Chapter VI where the method known as singularity analysis is developed. Then, a chapter parallel to the present one, Chapter VII, will present a number of new schemas based on the set and cycle constructions, as well as on recursion.

Bibliographic notes. Applications of rational functions in discrete and continuous mathematics are in abundance. Many examples are to be found in Goulden and Jackson's book [303]. Stanley [552] even devotes a full chapter of his book *Enumerative Combinatorics*, vol. I, to rational generating functions. These two books push the theory further than we can do here, but the corresponding asymptotic aspects which we develop lie outside of their scope. The analytic theory of positive rational functions starts with the works of Perron and Frobenius at the beginning of the twentieth century and is explained in books on matrix theory likes those of Bellman [34] and Gantmacher [276]. Its importance has been long recognized in the theory of finite Markov chains, so that the basic theory of positive matrices is well developed in many elementary treatises on probability theory. For such aspects, we refer for instance to the classic presentations by Feller [205] or Karlin and Taylor [363].

The supercritical sequence schema is the first in a list of abstract schemas that neatly exemplify the interplay between combinatorial, analytic, and probabilistic properties of large random structures. The origins of this approach are to be traced to early works of Bender [35, 36] followed by Soria and Flajolet [258, 260, 547].

Turning to more specific topics, we mention in relation to Section V. 4 the first global attempt at a combinatorial theory of continued fractions by Flajolet in [214] together with related works of Jackson of which an exposition is to be found in [303, Ch. 5] and a synthesis in [238], in relation to birth and death processes. Walks on graphs from an algebraic standpoint are well discussed in Godsil's book [295]; for infinite graphs and groups, see Woess [613]. The discussion of local constraints in permutations based on [239] combines some of the combinatorial elements bound in Stanley's book [552] with the general philosophy of analytic combinatorics. Our treatment of words and languages largely draws its inspiration from the line of research started by Schützenberger in the early 1960s and on the subsequent account to be found in Lothaire's book [413]. A nice review of transfer matrix methods (including a discussion of limit distributions) is offered by Bender, Richmond, and Williamson in [46].

Applied mathematics is bad mathematics.

— PAUL HALMOS [317]

Good applied mathematics is like the unicorn:
something we can all recognize but seldom actually see.

— DAVID ALDOUS
(in *Statistical Science*, Vol. 5, No. 4 (Nov., 1990), pp. 446–447)

VI

Singularity Analysis of Generating Functions

Es ist eine Tatsache, daß die genauere Kenntnis
des Verhaltens einer analytischen Funktion
in der Nähe ihrer singulären Stellen
eine Quelle von arithmetischen Sätzen ist[1].

— ERICH HECKE [326, Kap. VIII]

A function's singularities are reflected in the function's coefficients. Chapters IV and V have treated in detail rational fractions and meromorphic functions, where the local analysis of polar singularities provides contributions to coefficients in the form of exponential–polynomials (products of polynomials and exponentials). In this chapter, we present a general approach to the analysis of coefficients of generating functions that is not restricted to polar singularities and extends to a large class of functions that have moderate growth or decay at their dominant singularities. It includes a number of functions coming from combinatorial constructions of Part A. The basic principle behind the extension is the existence of a *general correspondence* between

the asymptotic expansion of a function near its dominant singularities
and
the asymptotic expansion of the function's coefficients.

This mapping preserves orders of growth in the sense that larger functions tend to have have larger coefficients. It extends considerably the analysis of meromorphic functions in Chapters IV–V and further justifies the *Principles of Coefficient Asymptotics* enunciated in Chapter IV, p. 227.

---

[1] *"It is a fact that the precise knowledge of the behaviour of an analytic function in the vicinity of its singular points is a source of arithmetic properties."*

Precisely, the method of *singularity analysis* applies to functions whose singular expansion involves fractional powers and logarithms—one sometimes refers to such singularities as "*algebraic–logarithmic*". It centrally relies on two ingredients.

- (*i*) A *catalogue* of asymptotic expansions for coefficients of the standard functions that occur in such singular expansions.
- (*ii*) *Transfer theorems*, which allow us to extract the asymptotic order of coefficients of error terms in singular expansions.

The developments are based on Cauchy's coefficient formula, used in conjunction with special contours of integration known as *Hankel contours*. The contours come very close to the singularities then steer away: by design, they capture essential asymptotic information contained in the functions' singularities.

The method of singularity analysis is robust: functions amenable to it are closed under a variety of operations, including sum, product, integration, differentiation, and composition. Another important feature of the method is that it only necessitates *local asymptotic properties* of the function to be analysed. In this way, it often proves instrumental in the case of functions that are only indirectly accessible through functional equations.

This chapter is meant to develop the basic technology of singularity analysis and, like Chapter IV, it is largely of a methodological nature. We illustrate the approach with a few combinatorial problems, including simple varieties of trees (e.g, unary–binary trees), combinatorial sums, the supercritical cycle construction, supertrees, Pólya's drunkard walks, and tree recurrences. The next chapter, Chapter VII, will systematically explore combinatorial structures and schemas as well as functional equations that can be asymptotically analysed by means of singularity analysis in a way that parallels the applications of rational and meromorphic asymptotics in Chapter V.

VI. 1. A glimpse of basic singularity analysis theory

Rational and meromorphic functions involve, locally near a singularity ζ, elements of the form $(1 - z/\zeta)^{-r}$, with $r \in \mathbb{Z}_{\geq 1}$. Accordingly their coefficients involve asymptotically exponential–polynomials, that is, finite linear combinations of elements of the type $\zeta^{-n} n^{r-1}$, with r a positive integer. We examine here an approach that takes into account functions whose singularities are of a richer nature than mere poles found in rational and meromorphic functions. Specifically, we consider functions whose expansion at a singularity ζ involves elements of the form

$$\left(1 - \frac{z}{\zeta}\right)^{-\alpha} \left(\log \frac{1}{1 - \frac{z}{\zeta}}\right)^{\beta}.$$

Under suitable conditions to be discussed in detail in this chapter, any such element contributes a term of the form

$$\zeta^{-n} n^{\alpha-1} (\log n)^{\beta}.$$

Here, α and β can be arbitrary complex numbers.

Location of singularities and exponential factors. The exponential factor ζ^{-n} present in earlier expansions is easily accounted for, since the location of the dominant singularities always induces a multiplicative exponential factor for coefficients. Indeed, if $f(z)$ is singular at $z = \zeta$, then $g(z) \equiv f(z\zeta)$ satisfies, by the scaling rule of Taylor expansions,

$$[z^n] f(z) = \zeta^{-n} [z^n] f(z\zeta) = \zeta^{-n} [z^n] g(z),$$

where $g(z)$ now has a singularity at $z = 1$. Consequently, in the discussion that follows, we shall examine functions that are singular at 1, a condition that entails no loss of generality.

Basic scale. Consider the following table of commonly encountered functions that are singular at 1, together with their coefficients:

(1)

	Function	coefficient (exact)		coefficient (asympt.)
(f_1)	$1 - \sqrt{1 - z}$	$\dfrac{2}{n4^n}\dbinom{2n-2}{n-1}$	$\sim$	$\dfrac{1}{2\sqrt{\pi n^3}}$
(f_2)	$\dfrac{1}{\sqrt{1-z}}$	$\dfrac{1}{4^n}\dbinom{2n}{n}$	$\sim$	$\dfrac{1}{\sqrt{\pi n}}$
(f_3)	$\dfrac{1}{1-z}$	1	$\sim$	1
(f_4)	$\dfrac{1}{1-z}\log\dfrac{1}{1-z}$	H_n	$\sim$	$\log n$
(f_5)	$\dfrac{1}{(1-z)^2}$	$n+1$	$\sim$	$n.$

Some structure is apparent in this table: a logarithmic factor in the function is reflected by a similar factor in the coefficients, square-roots somehow induce square-roots, and functions involving larger powers *do* have larger coefficients.

It is easy to come up at least with a partial explanation of these observations. Regarding basic functions such as f_1, f_2, f_3, and f_5, the Newton expansion

$$(1-z)^{-\alpha} = \sum_{n=0}^{\infty} \binom{n+\alpha-1}{n} z^n$$

when specialized to an integer $\alpha = r \in \mathbb{Z}_{\geq 1}$ immediately gives the asymptotic form of the coefficients involved,

(2) $\quad [z^n](1-z)^{-r} \equiv \dfrac{(n+1)(n+2)\cdots(n+r-1)}{(r-1)!} = \dfrac{n^{r-1}}{(r-1)!}\left(1+O\left(\dfrac{1}{n}\right)\right).$

For general α, it is therefore natural to expect

(3) $\quad [z^n](1-z)^{-\alpha} \equiv \binom{n+\alpha-1}{\alpha-1} = \dfrac{n^{\alpha-1}}{(\alpha-1)!}\left(1+O\left(\dfrac{1}{n}\right)\right).$

It turns out that this asymptotic formula remains valid for real or complex α, provided we interpret $(\alpha-1)!$ suitably. We shall prove the estimate

(4) $\quad [z^n](1-z)^{-\alpha} \sim \dfrac{n^{\alpha-1}}{\Gamma(\alpha)}\left(1+\dfrac{\alpha(\alpha-1)}{2n}+\cdots\right),$

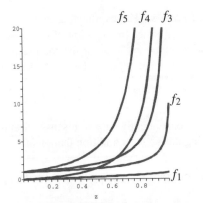 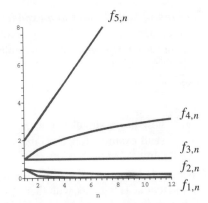

Figure VI.1. The five functions from Equation (1) and a plot of their coefficient sequences illustrate the tendency of coefficient extraction to be consistent with orders of growth of functions.

where $\Gamma(\alpha)$ is the *Euler Gamma function* defined as

(5) $$\Gamma(\alpha) := \int_0^\infty e^{-t} t^{\alpha-1} \, dt,$$

for $\Re(\alpha) > 0$, which coincides with $(\alpha - 1)!$ whenever α is an integer. (Basic properties of this function are recalled in Appendix B.3: *Gamma function*, p. 743.)

We observe from the pair (2)–(3) that functions that are larger at the singularity $z = 1$ have indeed larger coefficients (see Figure VI.1). The correspondence that this observation suggests is general, as we are going to see repeatedly throughout this chapter. A *catalogue* of exact or asymptotic forms for coefficients of standard singular functions is obtained in Section VI. 2 (see Theorem VI.1, p. 381).

Transfer of error terms. An asymptotic expansion of a function $f(z)$ that is singular at $z = 1$ is typically of the form

(6) $f(z) = \sigma(z) + O(\tau(z)),$ where $\tau(z) = o(\sigma(z))$ as $z \to 1,$

with σ and τ belonging to an asymptotic scale of standard functions such as the collection $\{(1 - z)^{-\alpha}\}_{\alpha \in \mathbb{R}}$, in simpler cases. Taking formally Taylor coefficients in the expansion (6), we arrive at

(7) $f_n \equiv [z^n] f(z) = [z^n]\sigma(z) + [z^n]O(\tau(z)).$

The term $[z^n]\sigma(z)$ is described asymptotically by (4). Therefore, in order to extract asymptotic informations on the coefficients of $f(z)$, one needs a way of extracting coefficients of functions known only by their order of growth around the singularity. Such a translation of error terms from functions to coefficients is achieved by *transfer theorems*, which, under conditions of analytic continuation, guarantee that

$$[z^n]O(\tau(z)) = O([z^n]\tau(z));$$

see Section VI. 3 and Theorem VI.3, p. 390. This relation is much more profound than its symbolic form would seem to imply.

In summary, it is the goal of this chapter to expound the (favorable) conditions under which we have available the correspondence

$$(8) \qquad f(z) = \sigma(z) + O(\tau(z)) \qquad \longrightarrow \qquad f_n = \sigma_n + O(\tau_n),$$

which defines the process known as *singularity analysis*: cf Section VI. 4 and Theorem VI.4, p. 393. (This is seen to parallel the analysis of coefficients of rational and meromorphic functions presented in Chapters IV and V.) We develop the method for functions from the scale

$$(1-z)^{-\alpha} \left(\log \frac{1}{1-z} \right)^{\beta} \qquad (z \to 1),$$

whose coefficients have subexponential factors of the form

$$n^{\alpha-1}(\log n)^{\beta}.$$

(The range of singular behaviours taken into account by singularity analysis is even considerably larger: iterated logarithms (log log's) as well as more exotic functions can be encapsulated in the method.)

***Example* VI.1.** *First asymptotics of 2–regular graphs.* As an illustration of the *modus operandi* of singularity analysis, consider the class $\mathcal{R}$ of labelled 2–regular graphs (Note II.22, p. 133):

$$\mathcal{R} = \text{SET}(\text{UCYC}_{\geq 3}(\mathcal{Z})) \qquad \Longrightarrow \qquad R(z) = \exp\left(\frac{1}{2}\left(\log(1-z)^{-1} - z - \frac{z^2}{2} \right) \right),$$

where UCYC is the undirected cycle construction.

Singularity analysis permits us to reason as follows. The function

$$R(z) = \frac{e^{-z/2 - z^2/4}}{\sqrt{1-z}}$$

is only singular at $z = 1$ where it has a branch point. Expanding the numerator around $z = 1$, we have

$$(9) \qquad R(z) = \frac{e^{-3/4}}{\sqrt{1-z}} + O((1-z)^{1/2}).$$

Therefore (see Theorems VI.1 and VI.3, as well as the discussion in Example VI.2 below, p. 395), upon translating formally term by term, one obtains

$$(10) \qquad [z^n]R(z) = e^{-3/4}\binom{n-1/2}{n} + O\binom{n-3/2}{n} = \frac{e^{-3/4}}{\sqrt{\pi n}} + O(n^{-3/2}).$$

Furthermore, a full asymptotic expansion into descending powers of n can be derived in the same way, from a complete expansion of the numerator $e^{-z/2 - z^2/4}$ at $z = 1$. .... ∎

Plan of this chapter. The first part of this chapter, Sections VI. 2–VI. 5, is dedicated to the basic technology of singularity analysis along the lines of our foregoing discussion, and including the case of functions with finitely many singularities on the boundary of their disc of convergence. An "Intermezzo", Section VI. 6, serves a prelude to the second part of the chapter, where we investigate operations on generating functions whose effect on singularities is predictable. The most important of these is inversion, which, under a broad set of conditions, leads to square-root singularity and provides a unified asymptotic theory of simple varieties of trees (Section VI. 7). Polylogarithms are proved to be amenable to singularity analysis in Section VI. 8, a fact that permits us to take into account weights such as $\sqrt{n}$ or $\log n$ in combinatorial sums. Composition of functions is studied in Section VI. 9. Then Section VI. 10 presents several closure properties of functions of singularity analysis class, including differentiation, integration, and Hadamard product. The chapter concludes with a brief discussion of two classical alternatives to singularity analysis: Tauberian theory and Darboux's method (Section VI. 11).

VI. 2. Coefficient asymptotics for the standard scale

This section and the next two present the fundamentals of *singularity analysis*, a theory which was developed by Flajolet and Odlyzko in [248]. Technically the theory relies on a systematic use of Hankel contours in Cauchy coefficient integrals. Such Hankel contours classically serve to express the Gamma function: see Appendix B.3: *Gamma function*, p. 743. Here they are first used to estimate coefficients of a standard scale of functions, and then to prove transfer theorems for error terms (Section VI. 3). With this basic process, an asymptotic expansion of a function near a singularity is directly mapped to a matching asymptotic expansion of its coefficients.

Starting from the binomial expansion, we have for general α,

$$[z^n](1-z)^{-\alpha} = (-1)^n \binom{-\alpha}{n} = \binom{n+\alpha-1}{n} = \frac{\alpha(\alpha+1)\cdots(\alpha+n-1)}{n!}.$$

This quantity is expressible in terms of Gamma factors, and

$$(11) \qquad \binom{n+\alpha-1}{n} = \frac{\Gamma(n+\alpha)}{\Gamma(\alpha)\Gamma(n+1)},$$

provided α is neither 0 nor a negative integer. (When $\alpha \in \{0, -1, \ldots\}$, the coefficients $\binom{n+\alpha-1}{n}$ eventually vanish, so that the asymptotic problem of estimating $[z^n](1-z)^{-\alpha}$ becomes void.) The asymptotic analysis of the coefficients $\binom{n+\alpha-1}{n}$ is straightforward, by means of Stirling's formula and real integral estimates: see Notes VI.1 and VI.2.

A method far more productive than elementary real analysis techniques consists in estimating coefficients of a function $f(z)$ by means of Cauchy's coefficient formula:

$$[z^n]f(z) = \frac{1}{2i\pi}\int_\gamma f(z)\,\frac{dz}{z^{n+1}}.$$

The basic principle is simple: it consists in choosing a contour of integration γ that comes at distance $1/n$ of the singularity $z = 1$. Under the change of variables $z = 1 + t/n$, the kernel z^{-n-1} in the integral transforms (asymptotically) into an

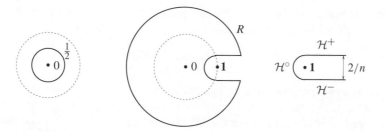

Figure VI.2. The contours $\mathcal{C}_0$, $\mathcal{C}_1$, and $\mathcal{C}_2 \equiv \mathcal{H}(n)$ used for estimating the coefficients of functions from the standard function scale.

exponential, and the function can be expanded locally, with the differential coefficient only introducing a rescaling factor of $1/n$:

(12)
$$
\begin{aligned}
z &\mapsto \left(1 + \frac{t}{n}\right), & dz &\mapsto \frac{1}{n}\, dt \\
\frac{1}{z^{n+1}} &\mapsto e^{-t}, & (1-z)^{-\alpha} &\mapsto n^{\alpha}(-t)^{-\alpha}.
\end{aligned}
$$

This gives us for instance (precise justification below):

$$
[z^n](1-z)^{-\alpha} \sim g_\alpha n^{\alpha-1}, \qquad g_\alpha := \frac{1}{2i\pi}\int e^{-t}(-t)^{-\alpha}\, dt.
$$

The contour and the associated rescaling capture the behaviour of the function near its singularity, thereby enabling coefficient estimation.

Theorem VI.1 (Standard function scale). *Let α be an arbitrary complex number in $\mathbb{C} \setminus \mathbb{Z}_{\leq 0}$. The coefficient of z^n in*

$$
f(z) = (1-z)^{-\alpha}
$$

admits for large n a complete asymptotic expansion in descending powers of n,

$$
[z^n]f(z) \sim \frac{n^{\alpha-1}}{\Gamma(\alpha)}\left(1 + \sum_{k=1}^{\infty} \frac{e_k}{n^k}\right),
$$

where e_k is a polynomial in α of degree $2k$. In particular[2]:

(13)
$$
\begin{aligned}
[z^n]f(z) \sim \frac{n^{\alpha-1}}{\Gamma(\alpha)}\bigg(1 &+ \frac{\alpha(\alpha-1)}{2n} + \frac{\alpha(\alpha-1)(\alpha-2)(3\alpha-1)}{24n^2} \\
&+ \frac{\alpha^2(\alpha-1)^2(\alpha-2)(\alpha-3)}{48n^3} + O\left(\frac{1}{n^4}\right)\bigg).
\end{aligned}
$$

---

[2]The quantity e_k is a polynomial in α that is divisible by $\alpha(\alpha-1)\cdots(\alpha-k)$, in accordance with the fact that the asymptotic expansion terminates when $\alpha \in \mathbb{Z}_{\geq 0}$. The factor $1/\Gamma(\alpha)$ vanishes identically when $\alpha \in \mathbb{Z}_{\leq 0}$, in accordance with the fact that coefficients are asymptotically 0 in that case.

Proof. The first step is to express the coefficient $[z^n](1-z)^{-\alpha}$ as a complex integral by means of Cauchy's coefficient formula,

$$(14) \qquad f_n = \frac{1}{2i\pi} \int_C (1-z)^{-\alpha} \frac{dz}{z^{n+1}},$$

where C is a small enough contour that encircles the origin; see Figure VI.2. We can start with $C \equiv C_0$, where C_0 is the positively oriented circle $C_0 = \{z, |z| = \frac{1}{2}\}$. The second step is to deform C_0 into another simple closed curve C_1 around the origin that does not cross the half-line $\Re(z) \geq 1$: the contour C_1 consists of a large circle of radius $R > 1$ with a notch that comes back near and to the left of $z = 1$. Since the integrand along large circles decreases as $O(R^{-n})$, we can finally let R tend to infinity and are left with an integral representation for f_n where C has been replaced by a contour C_2 that starts from $+\infty$ in the lower half-plane, winds clockwise around 1, and ends at $+\infty$ in the upper half-plane. The latter is a typical case of a *Hankel contour*. A judicious choice of its distance to the half-line $\mathbb{R}_{\geq 1}$ yields the expansion.

To specify precisely the integration path, we particularize C_2 to be the contour $\mathcal{H}(n)$ that passes at a distance $\frac{1}{n}$ from the half line $\mathbb{R}_{\geq 1}$:

$$(15) \qquad \mathcal{H}(n) = \mathcal{H}^-(n) \cup \mathcal{H}^+(n) \cup \mathcal{H}^\circ(n)$$

where

$$(16) \qquad \begin{cases} \mathcal{H}^-(n) &= \{z = w - \frac{i}{n}, \ w \geq 1\} \\ \mathcal{H}^+(n) &= \{z = w + \frac{i}{n}, \ w \geq 1\} \\ \mathcal{H}^\circ(n) &= \{z = 1 - \frac{e^{i\phi}}{n}, \ \phi \in [-\frac{\pi}{2}, \frac{\pi}{2}]\}. \end{cases}$$

Now, a change of variable

$$(17) \qquad z = 1 + \frac{t}{n}$$

in the integral (14) gives the form

$$(18) \qquad f_n = \frac{n^{\alpha-1}}{2i\pi} \int_{\mathcal{H}} (-t)^{-\alpha} \left(1 + \frac{t}{n}\right)^{-n-1} dt.$$

(The Hankel contour $\mathcal{H}$ winds about 0, being at distance 1 from the positive real axis; it is the same as the one in the proof of Theorem B.1, p. 745.)

We have the asymptotic expansion

$$(19)$$
$$\left(1 + \frac{t}{n}\right)^{-n-1} = e^{-(n+1)\log(1+t/n)} = e^{-t}\left[1 + \frac{t^2 - 2t}{2n} + \frac{3t^4 - 20t^3 + 24t^2}{24n^2} + \cdots\right],$$

which tells us that the integrand in (18) converges pointwise (as well as uniformly in any bounded domain of the t plane) to $(-t)^{-\alpha}e^{-t}$. Substitution of the asymptotic form

$$\left(1 + \frac{t}{n}\right)^{-n-1} = e^{-t}\left(1 + O\left(\frac{1}{n}\right)\right),$$

as $n \to \infty$ inside the integral (18) suggests (formally) that

$$
\begin{aligned}
[z^n](1-z)^{-\alpha} &= \frac{n^{\alpha-1}}{2i\pi} \int_{\mathcal{H}} (-t)^{-\alpha} e^{-t} \, dt \left(1 + O\left(\frac{1}{n}\right)\right) \\
&= \frac{n^{\alpha-1}}{\Gamma(\alpha)} \left(1 + O\left(\frac{1}{n}\right)\right),
\end{aligned}
$$

when use is made of Hankel's formula for the Gamma function (p. 745).

To justify this formal argument, we proceed as follows:

(i) Split the contour $\mathcal{H}$ according to $\Re(t) \leq \log^2 n$ and $\Re(t) \geq \log^2 n$, as in the corresponding diagram:

(20)

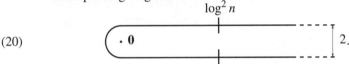

(ii) Verify that the part corresponding to $\Re(t) \geq \log^2 n$ is negligible in the scale of the problem; for instance:

$$
\left(1 + \frac{t}{n}\right)^{-n} = O(\exp(-\log^2 n)) \qquad \text{for } \Re(t) \geq \log^2 n.
$$

(iii) Use a terminating form of (19) to develop an expansion to any predetermined order, with uniform error terms, for the part corresponding to $\Re(t) \leq \log^2 n$. (This is possible because $t/n = O(\log^2 n/n)$ is small.)

These considerations validate term-by-term integration of expansion (19) within the integral of (18), so that the full expansion of f_n is determined as follows: a term of the form t^r/n^s in the expansion (19) induces, by Hankel's formula, a term of the form $n^{-s}/\Gamma(\alpha - r)$. (The expansion so obtained is non-degenerate provided α differs from a negative integer or zero; see also Note VI.3 for details.) Since

$$
\frac{1}{\Gamma(\alpha - k)} = \frac{1}{\Gamma(\alpha)}(\alpha - 1)(\alpha - 2) \cdots (\alpha - k).
$$

the expansion in the statement of the theorem eventually follows.   ∎

The asymptotic approximations obtained from Theorem VI.2 differ from the ones that are associated with meromorphic asymptotics (Chapter IV), where exponentially small error terms could be derived. However, it is not uncommon to obtain results with about 10^{-6} accuracy, already for values of n in the range 10^1–10^2 with just a few terms of the asymptotic expansion. Figure VI.3 exemplifies this situation by displaying the approximations obtained for the Catalan numbers,

$$
C_n = \frac{4^n}{n+1}[z^n](1-z)^{-1/2},
$$

when C_{10}, C_{20}, C_{50} are considered and up to eight asymptotic terms are taken into account.

	$n = 10$	$n = 20$	$n = 50$
$\dfrac{4^n}{\sqrt{\pi n^3}}\big(1$	1 8708	6 935533866	2022877684829178931751713264
$-\dfrac{9}{8}n^{-1}$	16 603	65 45410086	197 7362936920522405787299715
$+\dfrac{145}{128}n^{-2}$	16 815	656 5051735	19782 7955337146062749074971 0
$-\dfrac{1155}{1024}n^{-3}$	1679 4	6564 073885	1978261 300061101426696482732
$+\dfrac{36939}{32768}n^{-4}$	16796	656412 2750	19782616 6491988462935781359 1
$-\dfrac{295911}{262144}n^{-5}$	16796	6564120 303	1978261657 612856326190245636
$+\dfrac{4735445}{4194304}n^{-6}$	16796	656412042 6	19782616 5775 9023715384519184
$-\dfrac{37844235}{33554432}n^{-7}\big)$	16796	6564120420	1978261657561 03402179527600
C_n	16796	6564120420	19782616577561 60653623774456

Figure VI.3. Improved approximations to the Catalan numbers obtained by successive terms of their asymptotic expansion (with exact digits in boldface).

▷ **VI.1.** *Stirling's formula and asymptotics of binomial coefficients.* The Gamma function form (11) of the binomial coefficients yields

$$[z^n](1 - z)^{-\alpha} = \frac{n^{\alpha-1}}{\Gamma(\alpha)}\left(1 + O(\tfrac{1}{n})\right),$$

when Stirling's formula is applied to the Gamma factors.   ◁

▷ **VI.2.** *Beta integrals and asymptotics of binomial coefficients.* A direct way of obtaining the general asymptotic form of $\binom{n+\alpha-1}{n}$ bases itself on the Eulerian Beta integral (see [604, p.254] and Appendix B.3: *Gamma function*, p. 743). Consider the quantity ($\alpha > 0$)

$$\phi(n, \alpha) = \int_0^1 t^{\alpha-1}(1-t)^{n-1}\, dt = \frac{(n-1)!}{\alpha(\alpha+1)\cdots(\alpha+n-1)} \equiv \frac{1}{n\binom{n+\alpha-1}{n}},$$

where the second form results elementarily from successive integrations by parts. The change of variables $t = x/n$ yields

$$\phi(n, \alpha) = \frac{1}{n^\alpha}\int_0^n x^{\alpha-1}(1-x/n)^{n-1}\, dt \underset{n\to\infty}{\sim} \frac{1}{n^\alpha}\int_0^\infty x^{\alpha-1}e^{-x}\, dx \equiv \frac{\Gamma(\alpha)}{n^\alpha},$$

where the asymptotic form results from the standard limit formula of the exponential: $\exp(a) = \lim_{n\to\infty}(1 + a/n)^n$.   ◁

▷ **VI.3.** *Computability of full expansions.* The coefficients e_k of Theorem VI.1 satisfy

$$e_k = \sum_{\ell=k}^{2k} \lambda_{k,\ell}(\alpha - 1)(\alpha - 2)\cdots(\alpha - \ell),$$

where $\lambda_{k,\ell} := [v^k t^\ell]e^t(1 + vt)^{-1-1/v}$.   ◁

▷ **VI.4.** *Oscillations and complex exponents.* Oscillations occur in the case of singular expansions involving complex exponents. From the consideration of $[z^n](1 - z)^{\pm i} \asymp n^{\mp i-1}$, one finds

$$[z^n]\cos\left(\log\frac{1}{1-z}\right) = \frac{P(\log n)}{n} + O(\tfrac{1}{n^2}),$$

where $P(u)$ is a continuous and 1–periodic function. In general, such oscillations are present in $[z^n](1-z)^{-\alpha}$ for any non-real α. ◁

Logarithmic factors. The basic principle underlying the method of proof of Theorem VI.1 (see also the summary Equation (12)) has the advantage of being easily extended to a wide class of singular functions, most notably the ones that involve logarithmic terms.

Theorem VI.2 (Standard function scale, logarithms). *Let α be an arbitrary complex number in $\mathbb{C} \setminus \mathbb{Z}_{\leq 0}$. The coefficient of z^n in the function*[3]

$$f(z) = (1-z)^{-\alpha} \left(\frac{1}{z} \log \frac{1}{1-z} \right)^{\beta}$$

admits for large n a full asymptotic expansion in descending powers of $\log n$,

$$\text{(21)} \qquad f_n \equiv [z^n] f(z) \sim \frac{n^{\alpha-1}}{\Gamma(\alpha)} (\log n)^{\beta} \left[1 + \frac{C_1}{\log n} + \frac{C_2}{\log^2 n} + \cdots \right],$$

where $C_k = \binom{\beta}{k} \Gamma(\alpha) \frac{d^k}{ds^k} \frac{1}{\Gamma(s)} \Big|_{s=\alpha}$.

Proof. The proof is a simple variant of that of Theorem VI.1 (see [248] for details). The basic expansion used is now

$$f\left(1+\frac{t}{n}\right)\left(1+\frac{t}{n}\right)^{-n-1} \sim e^{-t}\left(\frac{-n}{t}\right)^{\alpha}\left(\log\left(\frac{-n}{t}\right)\right)^{\beta}$$

$$\sim e^{-t}(-t)^{-\alpha}n^{\alpha}(\log n)^{\beta}\left(1 - \frac{\log(-t)}{\log n}\right)^{\beta}$$

$$\sim e^{-t}(-t)^{-\alpha}n^{\alpha}(\log n)^{\beta}\left(1 - \beta\frac{\log(-t)}{\log n} + \frac{\beta(\beta-1)}{2!}\left(\frac{\log(-t)}{\log n}\right)^2 + \cdots\right).$$

Again, we are justified in using this expansion inside Cauchy's integral representation of coefficients. What comes out from term-by-term integration is a collection of Hankel integrals of the form

$$\frac{1}{2i\pi}\int_{+\infty}^{(0)}(-t)^{-s}e^{-t}(\log(-t))^k\,dt = (-1)^k\frac{d^k}{ds^k}\left[\frac{1}{2i\pi}\int_{+\infty}^{(0)}(-t)^{-s}e^{-t}\,dt\right]$$

$$= (-1)^k\frac{d^k}{ds^k}\frac{1}{\Gamma(s)},$$

where the reduction to derivatives of $1/\Gamma(s)$ results from differentiation with respect to s under the integral sign. ∎

A typical example of application of Theorem VI.2 is the estimate

$$[z^n]\frac{1}{\sqrt{1-z}}\frac{1}{\frac{1}{z}\log\frac{1}{1-z}} = \frac{1}{\sqrt{\pi n}\,\log n}\left(1 - \frac{\gamma + 2\log 2}{\log n} + O\left(\frac{1}{\log^2 n}\right)\right).$$

---

[3] A coefficient of $1/z$ is introduced in front of the logarithm since $\log(1-z)^{-1} = z + O(z^2)$: in this way, $f(z)$ is a *bona fide* power series in z, even when β is *not* an integer. Such a factor does not affect asymptotic expansions in a logarithmic scale near $z = 1$.

	$\alpha \notin \{0, -1, -2, \ldots\}$	(Eq.)	$\alpha \in \{0, -1, -2, \ldots\}$	(Eq.)
$\beta \notin \mathbb{Z}_{\geq 0}$	$\dfrac{n^{\alpha-1}}{\Gamma(\alpha)}(\log n)^{\beta} \displaystyle\sum_{j=0}^{\infty} \dfrac{C_j}{(\log n)^j}$	(21)	$f_n \sim n^{\alpha-1}(\log n)^{\beta} \displaystyle\sum_{j=1}^{\infty} \dfrac{D_j}{(\log n)^j}$	(24)
$\beta \in \mathbb{Z}_{\geq 0}$	$\dfrac{n^{\alpha-1}}{\Gamma(\alpha)} \displaystyle\sum_{j=0}^{\infty} \dfrac{E_j(\log n)}{n^j}$	(25)	$n^{\alpha-1} \displaystyle\sum_{j=0}^{\infty} \dfrac{F_j(\log n)}{n^j}$	(27)

Figure VI.4. The general and special cases of $f_n \equiv [z^n] f(z)$ when $f(z)$ is as in Theorem VI.2.

(Such singular functions do occur in combinatorics and analysis of algorithms [257].)

▷ **VI.5.** *Singularity analysis of slowly varying functions.* A function $\Lambda(u)$ is said to be *slowly varying* towards infinity (in the complex plane) if there exists a $\phi \in (0, \frac{\pi}{2})$ such that, for any fixed $c > 0$ and all θ satisfying $|\theta| \leq \pi - \phi$, there holds

(22)
$$\lim_{u \to +\infty} \frac{\Lambda(ce^{i\theta}u)}{\Lambda(u)} = 1.$$

(Powers of logarithms and iterated logarithms are typically slowly varying functions.) Under uniformity assumptions on (22), the following estimate holds [248]:

(23)
$$[z^n](1-z)^{-\alpha} \Lambda\left(\frac{1}{1-z}\right) \sim \frac{n^{\alpha-1}}{\Gamma(\alpha)} \Lambda(n).$$

For instance, we have:

$$[z^n] \frac{\exp\left(\sqrt{\frac{1}{z}\log\frac{1}{1-z}}\right)}{\sqrt{1-z}} \sim \frac{\exp\left(\sqrt{\log n}\right)}{\sqrt{\pi n}}.$$

See also the discussion of Tauberian theory, p. 435.　　　　　　　　　　◁

▷ **VI.6.** *Iterated logarithms.* For a general $\alpha \notin \mathbb{Z}_{\leq 0}$, the relation (23) admits as a special case

$$[z^n](1-z)^{-\alpha} \left(\frac{1}{z}\log\frac{1}{1-z}\right)^{\beta} \left(\frac{1}{z}\log\left(\frac{1}{z}\log\frac{1}{1-z}\right)\right)^{\delta} \sim \frac{n^{\alpha-1}}{\Gamma(\alpha)}(\log n)^{\beta}(\log\log n)^{\delta}.$$

A full asymptotic expansion can be derived in this case.　　　　　　　　　　◁

Special cases. The conditions of Theorems VI.1 and VI.2 exclude explicitly the case when α is a negative integer: the formulae actually remain valid in this case, provided one interprets them as limit cases, making use of $1/\Gamma(0) = 1/\Gamma(-1) = \cdots = 0$. Also, when β is a positive integer, the expansion of Theorem VI.2 terminates: in that situation, stronger forms are valid. Such cases are summarized in Figure VI.4 and discussed below.

The case of integral $\alpha \in \mathbb{Z}_{\leq 0}$ and general $\beta \notin \mathbb{Z}_{\geq 0}$. When α is a negative integer, the coefficients of $f(z) = (1-z)^{-\alpha}$ eventually reduce to zero, so that the asymptotic coefficient expansion becomes trivial: this situation is implicitly covered by the statement of Theorem VI.1 since, in that case, $1/\Gamma(\alpha) = 0$. When logarithms are present (with $\alpha \in \mathbb{Z}_{\leq 0}$ still), the expansion of Theorem VI.2 regarding

$$f(z) = (1-z)^{-\alpha} \left(\frac{1}{z}\log\frac{1}{1-z}\right)^{\beta}$$

remains valid provided we again take into account the equality $1/\Gamma(\alpha) = 0$ in formula (21) after effecting simplifications by Gamma factors: it is only the first term of (21) that vanishes, and one has

$$(24) \qquad [z^n]f(z) \sim n^{\alpha-1}(\log n)^\beta \left[\frac{D_1}{\log n} + \frac{D_2}{\log^2 n} + \cdots\right],$$

where D_k is given by $D_k = \binom{\beta}{k}\dfrac{d^k}{ds^k}\dfrac{1}{\Gamma(s)}\bigg|_{s=\alpha}$. For instance, we find

$$[z^n]\frac{z}{\log(1-z)^{-1}} = -\frac{1}{n\log^2 n} + \frac{2\gamma}{n\log^3 n} + O(\frac{1}{n\log^4 n}).$$

The case of general $\alpha \notin \mathbb{Z}_{\leq 0}$ and integral $\beta \in \mathbb{Z}_{\geq 0}$. When $\beta = k$ is a nonnegative integer, the error terms can be further improved with respect to the ones predicted by the general statement of Theorem VI.2. For instance, we have:

$$[z^n]\frac{1}{1-z}\log\frac{1}{1-z} = \log n + \gamma + \frac{1}{2n} - \frac{1}{12n^2} + O(\frac{1}{n^4})$$

$$[z^n]\frac{1}{\sqrt{1-z}}\log\frac{1}{1-z} \sim \frac{1}{\sqrt{\pi n}}\left(\log n + \gamma + 2\log 2 + O(\frac{\log n}{n})\right).$$

(In such a case, the expansion of Theorem VI.2 terminates since only its first $(k+1)$ terms are non-zero.) In fact, in the general case of non-integral α, there exists an expansion of the form

$$(25) \qquad [z^n](1-z)^{-\alpha}\log^k\frac{1}{1-z} \sim \frac{n^{\alpha-1}}{\Gamma(\alpha)}\left[E_0(\log n) + \frac{E_1(\log n)}{n} + \cdots\right],$$

where the E_j are polynomials of degree k, as can be proved by adapting the argument employed for general α (Note VI.8).

The joint case of integral $\alpha \in \mathbb{Z}_{\leq 0}$ and integral $\beta \in \mathbb{Z}_{\geq 0}$. If α is a negative integer, the coefficients appear as finite differences of coefficients of logarithmic powers. Explicit formulae are then available elementarily from the calculus of finite differences when β is a positive integer. For instance, with $\alpha = -m$ for $m \in \mathbb{Z}_{\geq 0}$, one has

$$(26) \qquad [z^n](1-z)^m\log\frac{1}{1-z} = (-1)^m\frac{m!}{n(n-1)\cdots(n-m)}.$$

The case $\alpha = -m$ and $\beta = k$ (with $m, k \in \mathbb{Z}_{\geq 0}$) is covered by (28) in Note VI.7 below: there is a formula analogous to (25),

$$(27) \qquad [z^n](1-z)^m\log^k\frac{1}{1-z} \sim n^{-m-1}\left[F_0(\log n) + \frac{F_1(\log n)}{n} + \cdots\right],$$

but now with $\deg(F_j) = k - 1$.

Figure VI.5 provides the asymptotic form of coefficients of a few standard functions illustrating Theorems VI.1 and VI.2 as well as some of the "special cases".

Function	coefficients
$(1-z)^{3/2}$	$\dfrac{1}{\sqrt{\pi n^5}}(\dfrac{3}{4}+\dfrac{45}{32n}+\dfrac{1155}{512n^2}+O(\dfrac{1}{n^3}))$
$(1-z)$	(0)
$(1-z)^{1/2}$	$-\dfrac{1}{\sqrt{\pi n^3}}(\dfrac{1}{2}+\dfrac{3}{16n}+\dfrac{25}{256n^2}+O(\dfrac{1}{n^3}))$
$(1-z)^{1/2}L(z)$	$-\dfrac{1}{\sqrt{\pi n^3}}(\dfrac{1}{2}\log n+\dfrac{\gamma+2\log 2-2}{2}+O(\dfrac{\log n}{n}))$
$(1-z)^{1/3}$	$-\dfrac{1}{3\Gamma(\frac{2}{3})n^{4/3}}(1+\dfrac{2}{9n}+\dfrac{7}{81n^2}+O(\dfrac{1}{n^3}))$
$z/L(z)$	$\dfrac{1}{n\log^2 n}(-1+\dfrac{2\gamma}{\log n}+\dfrac{\pi^2-6\gamma^2}{2\log^2 n}+O(\dfrac{1}{\log^3 n}))$
1	(0)
$\log(1-z)^{-1}$	$\dfrac{1}{n}$
$\log^2(1-z)^{-1}$	$\dfrac{1}{n}(2\log n+2\gamma-\dfrac{1}{n}-\dfrac{1}{6n^2}+O(\dfrac{1}{n^4}))$
$(1-z)^{-1/3}$	$\dfrac{1}{\Gamma(\frac{1}{3})n^{2/3}}(1+O(\dfrac{1}{n}))$
$(1-z)^{-1/2}$	$\dfrac{1}{\sqrt{\pi n}}(1-\dfrac{1}{8n}+\dfrac{1}{128n^2}+\dfrac{5}{1024n^3}+O(\dfrac{1}{n^4}))$
$(1-z)^{-1/2}L(z)$	$\dfrac{1}{\sqrt{\pi n}}(\log n+\gamma+2\log 2-\dfrac{\log n+\gamma+2\log 2}{8n}+O(\dfrac{\log n}{n^2}))$
$(1-z)^{-1}$	1
$(1-z)^{-1}L(z)$	$\log n+\gamma+\dfrac{1}{2n}-\dfrac{1}{12n^2}+\dfrac{1}{120n^4}+O(\dfrac{1}{n^6}))$
$(1-z)^{-1}L(z)^2$	$\log^2 n+2\gamma\log n+\gamma^2-\dfrac{\pi^2}{6}+O(\dfrac{\log n}{n})$
$(1-z)^{-3/2}$	$\sqrt{\dfrac{n}{\pi}}(2+\dfrac{3}{4n}-\dfrac{7}{64n^2}+O(\dfrac{1}{n^3}))$
$(1-z)^{-3/2}L(z)$	$\sqrt{\dfrac{n}{\pi}}(2\log n+2\gamma+4\log 2-4+\dfrac{3\log n}{4n}+O(\dfrac{1}{n}))$
$(1-z)^{-2}$	$n+1$
$(1-z)^{-2}L(z)$	$n\log n+(\gamma-1)n+\log n+\dfrac{1}{2}+\gamma+O(\dfrac{1}{n})$
$(1-z)^{-2}L(z)^2$	$n(\log^2 n+2(\gamma-1)\log n+\gamma^2-2\gamma+2-\dfrac{\pi^2}{6}+O(\dfrac{\log n}{n}))$
$(1-z)^{-3}$	$\frac{1}{2}n^2+\frac{3}{2}n+1$

Figure VI.5. A table of some commonly encountered functions and the asymptotic forms of their coefficients. The following abbreviation is used:

$$L(z) := \log\frac{1}{1-z}.$$

▷ **VI.7.** *The method of Frobenius and Jungen.* This is an alternative approach to the case $\beta \in \mathbb{Z}_{\geq 0}$ (see [360]). Start from the observation that

$$(1-z)^{-\alpha} \left(\log \frac{1}{1-z} \right)^k = \frac{\partial^k}{\partial \alpha^k} (1-z)^{-\alpha},$$

then let the operators of differentiation ($\partial/\partial \alpha$) and coefficient extraction ($[z^n]$) commute (this can be justified by Cauchy's coefficient formula upon differentiating under the integral sign). This yields

(28) $$[z^n](1-z)^{-\alpha} \left(\log \frac{1}{1-z} \right)^k = \frac{\partial^k}{\partial \alpha^k} \frac{\Gamma(n+\alpha)}{\Gamma(\alpha)\Gamma(n+1)},$$

which leads to an "exact" formula (Note VI.8 below). ◁

▷ **VI.8.** *Shifted harmonic numbers.* Define the α-shifted harmonic number by

$$h_n(\alpha) := \sum_{j=0}^{n-1} \frac{1}{j+\alpha}.$$

With $L(z) := -\log(1-z)$, still, one has

$$[z^n](1-z)^{-\alpha} L(z) = \binom{n+\alpha-1}{n} h_n(\alpha)$$

$$[z^n](1-z)^{-\alpha} L(z)^2 = \binom{n+\alpha-1}{n} \left(h_n'(\alpha) + h_n(\alpha)^2 \right).$$

(Note: $h_n(\alpha) = \psi(\alpha+n) - \psi(\alpha)$, where $\psi(s) := \partial_s \log \Gamma(s)$.) In particular,

$$[z^n] \frac{1}{\sqrt{1-z}} \log \frac{1}{1-z} = \frac{1}{4^n} \binom{2n}{n} [2 H_{2n} - H_n],$$

where $H_n \equiv h_n(1)$ is the usual harmonic number. ◁

VI. 3. Transfers

Our general objective is to translate an *approximation of a function* near a singularity into an asymptotic *approximation of its coefficients*. What is required at this stage is a way to extract coefficients of error terms (known usually in $O(\cdot)$ or $o(\cdot)$ form) in the expansion of a function near a singularity. This task is technically simple as a fairly coarse analysis suffices. As in the previous section, it relies on contour integration by means of Hankel-type paths; see for instance the summary in Equation (12), p. 381, above.

A natural extension of the approach of the previous section is to assume the error terms to be valid in the complex plane slit along the real half line $\mathbb{R}_{\geq 1}$. In fact, weaker conditions suffice: any domain whose boundary makes an acute angle with the half line $\mathbb{R}_{\geq 1}$ appears to be suitable.

Definition VI.1. *Given two numbers ϕ, R with $R > 1$ and $0 < \phi < \frac{\pi}{2}$, the open domain $\Delta(\phi, R)$ is defined as*

$$\Delta(\phi, R) = \{z \mid |z| < R, \ z \neq 1, \ |\arg(z-1)| > \phi\}.$$

A domain is a Δ–domain at 1 if it is a $\Delta(\phi, R)$ for some R and ϕ. For a complex number $\zeta \neq 0$, a Δ–domain at ζ is the image by the mapping $z \mapsto \zeta z$ of a Δ–domain at 1. A function is Δ–analytic if it is analytic in some Δ–domain.

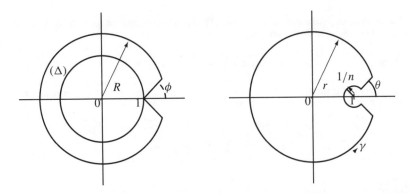

Figure VI.6. A Δ–domain and the contour used to establish Theorem VI.3.

Analyticity in a Δ–domain (Figure VI.6, left) is the basic condition for *transfer* to coefficients of error terms in asymptotic expansions.

Theorem VI.3 (Transfer, Big-Oh and little-oh). *Let α, β be arbitrary real numbers, $\alpha, \beta \in \mathbb{R}$ and let $f(z)$ be a function that is Δ–analytic.*

(i) Assume that $f(z)$ satisfies in the intersection of a neighbourhood of 1 with its Δ–domain the condition

$$f(z) = O\left((1-z)^{-\alpha} (\log \frac{1}{1-z})^{\beta} \right).$$

Then one has: $[z^n] f(z) = O(n^{\alpha-1} (\log n)^{\beta}).$

(ii) Assume that $f(z)$ satisfies in the intersection of a neighbourhood of 1 with its Δ–domain the condition

$$f(z) = o\left((1-z)^{-\alpha} (\log \frac{1}{1-z})^{\beta} \right).$$

Then one has: $[z^n] f(z) = o(n^{\alpha-1} (\log n)^{\beta}).$

Proof. (i) The starting point is Cauchy's coefficient formula,

$$f_n \equiv [z^n] f(z) = \frac{1}{2i\pi} \int_{\gamma} f(z) \frac{dz}{z^{n+1}},$$

where γ is any simple loop around the origin which is internal to the Δ–domain of f. We choose the positively oriented contour (Figure VI.6, right) $\gamma = \gamma_1 \cup \gamma_2 \cup \gamma_3 \cup \gamma_4$, with

$$
\begin{cases}
\gamma_1 & = \left\{ z \mid |z-1| = \dfrac{1}{n}, |\arg(z-1)| \geq \theta] \right\} & \text{(inner circle)} \\[2mm]
\gamma_2 & = \left\{ z \mid \dfrac{1}{n} \leq |z-1|, |z| \leq r, \arg(z-1) = \theta \right\} & \text{(top line segment)} \\[2mm]
\gamma_3 & = \left\{ z \mid |z| = r, |\arg(z-1)| \geq \theta] \right\} & \text{(outer circle)} \\[2mm]
\gamma_4 & = \left\{ z \mid \dfrac{1}{n} \leq |z-1|, |z| \leq r, \arg(z-1) = -\theta \right\} & \text{(bottom line segment)}.
\end{cases}
$$

If the Δ domain of f is $\Delta(\phi, R)$, we assume that $1 < r < R$, and $\phi < \theta < \frac{\pi}{2}$, so that the contour γ lies entirely inside the domain of analyticity of f.

For $j = 1, 2, 3, 4$, let

$$f_n^{(j)} = \frac{1}{2i\pi} \int_{\gamma_j} f(z) \frac{dz}{z^{n+1}}.$$

The analysis proceeds by bounding the absolute value of the integral along each of the four parts. In order to keep notations simple, *we detail the proof in the case where* $\beta = 0$.

(1) *Inner circle* (γ_1). From trivial bounds, the contribution from γ_1 satisfies

$$|f_n^{(1)}| = O(\frac{1}{n}) \cdot O\left(\left(\frac{1}{n}\right)^{-\alpha}\right) = O\left(n^{\alpha-1}\right),$$

as the function is $O(n^\alpha)$ (by assumption on $f(z)$), the contour has length $O(n^{-1})$, and z^{-n-1} remains $O(1)$ on this part of the contour.

(2) *Rectilinear parts* (γ_2, γ_4). Consider the contribution $f_n^{(2)}$ arising from the part γ_2 of the contour. Setting $\omega = e^{i\theta}$, and performing the change of variable $z = 1 + \frac{\omega t}{n}$, we find

$$|f_n^{(2)}| \le \frac{1}{2\pi} \int_1^\infty K \left(\frac{t}{n}\right)^{-\alpha} \left|1 + \frac{\omega t}{n}\right|^{-n-1} dt,$$

for some constant $K > 0$ such that $|f(z)| < K(1-z)^{-\alpha}$ over the Δ–domain, which is granted by the growth assumption on f. From the relation

$$\left|1 + \frac{\omega t}{n}\right| \ge 1 + \Re(\frac{\omega t}{n}) = 1 + \frac{t}{n}\cos\theta,$$

there results the inequality

$$|f_n^{(2)}| \le \frac{K}{2\pi} J_n n^{\alpha-1}, \quad \text{where} \quad J_n = \int_1^\infty t^{-\alpha}\left(1 + \frac{t\cos\theta}{n}\right)^{-n} dt.$$

For a given α, the integrals J_n are all bounded above by some constant since they admit a limit as n tends to infinity:

$$J_n \to \int_1^\infty t^{-\alpha} e^{-t\cos\theta} dt.$$

The condition on θ that $0 < \theta < \pi/2$ precisely ensures convergence of the integral. Thus, globally, on the part γ_2 of the contour, we have

$$|f_n^{(2)}| = O(n^{\alpha-1}).$$

A similar bound holds for $f_n^{(4)}$ relative to γ_4.

(3) *Outer circle* (γ_3). There, $f(z)$ is bounded while z^{-n} is of the order of r^{-n}. Thus, the integral $f_n^{(3)}$ is exponentially small.

In summary, each of the four integrals of the split contour contributes $O(n^{\alpha-1})$. The statement of part (i) of the theorem thus follows, when $\beta = 0$. Entirely similar bounding techniques cover the case of logarithmic factors ($\beta \neq 0$).

(ii) An adaptation of the proof shows that $o(.)$ error terms may be translated similarly. All that is required is a further break-up of the rectilinear part at a distance $\log^2 n/n$ from 1 (see the discussion surrounding Equation (20), p. 383 or [248] for details). ∎

An immediate corollary of Theorem VI.3 is the possibility of transferring *asymptotic equivalence* from singular forms to coefficients:

Corollary VI.1 (sim–transfer). *Assume that $f(z)$ is Δ–analytic and*

$$f(z) \sim (1-z)^{-\alpha}, \qquad as\ z \to 1, \quad z \in \Delta,$$

with $\alpha \notin \{0, -1, -2, \cdots\}$. Then, the coefficients of f satisfy

$$[z^n]f(z) \sim \frac{n^{\alpha-1}}{\Gamma(\alpha)}.$$

Proof. It suffices to observe that, with $g(z) = (1-z)^{-\alpha}$, one has

$$f(z) \sim g(z) \qquad \text{iff} \qquad f(z) = g(z) + o(g(z)),$$

then apply Theorem VI.1 to the first term, and Theorem VI.3 (little-oh transfer) to the remainder. ∎

▷ **VI.9.** *Transfer of nearly polynomial functions.* Let $f(z)$ be Δ–analytic and satisfy the singular expansion $f(z) \sim (1-z)^r$, where $r \in \mathbb{Z}_{\geq 0}$. Then, $f_n = o(n^{-r-1})$. [This is a direct consequence of the little-oh transfer.] ◁

▷ **VI.10.** *Transfer of large negative exponents.* The Δ–analyticity condition can be weakened for functions that are large at their singularity. Assume that $f(z)$ is analytic in the open disc $|z| < 1$, and that in the whole of the open disc it satisfies

$$f(z) = O((1-z)^{-\alpha}).$$

Then, provided $\alpha > 1$, one has

$$[z^n]f(z) = O(n^{\alpha-1}).$$

[Hint. Integrate on the circle of radius $1 - \frac{1}{n}$; see also [248].] ◁

VI. 4. The process of singularity analysis

In Sections VI. 2 and VI. 3, we have developed a collection of statements granting us the existence of *correspondences* between properties of a function $f(z)$ singular at an isolated point ($z = 1$) and the asymptotic behaviour of its coefficients $f_n = [z^n]f(z)$. Using the symbol '$\longrightarrow$' to represent such a correspondence[4], we

---

[4]The symbol "$\Longrightarrow$" represents an *unconditional* logical implication and is accordingly used in this book to represent the systematic correspondence between combinatorial specifications and generating function equations. In contrast, the symbol '$\longrightarrow$' represents a mapping from functions to coefficients, under suitable *analytic conditions*, like those of Theorems VI.1–VI.3.

can summarize some of our results relative to the scale $\{(1-z)^{-\alpha}, \ \alpha \in \mathbb{C} \setminus \mathbb{Z}_{\leq 0}\}$ as follows:

$$
\begin{cases}
f(z) = (1-z)^{-\alpha} & \longrightarrow \quad f_n = \dfrac{n^{\alpha-1}}{\Gamma(\alpha)} + \cdots & \text{(Theorem VI.1)} \\[2ex]
f(z) = O((1-z)^{-\alpha}) & \longrightarrow \quad f_n = O(n^{\alpha-1}) & \text{(Theorem VI.3 (i))} \\[2ex]
f(z) = o((1-z)^{-\alpha}) & \longrightarrow \quad f_n = o(n^{\alpha-1}) & \text{(Theorem VI.3 (ii))} \\[2ex]
f(z) \sim (1-z)^{-\alpha} & \longrightarrow \quad f_n \sim \dfrac{n^{\alpha-1}}{\Gamma(\alpha)} & \text{(Corollary VI.1).}
\end{cases}
$$

The important requirement is that the function should have an isolated singularity (the condition of Δ–analyticity) and that the asymptotic property of the function near its singularity should be valid in an area of the complex plane extending beyond the disc of convergence of the original series, (in a Δ–domain). Extensions to logarithmic powers and special cases like $\alpha \in \mathbb{Z}_{\leq 0}$ are also, as we know, available. We let $\mathcal{S}$ denote the set of such singular functions:

$$
(29) \quad \mathcal{S} = \left\{ (1-z)^{-\alpha} \lambda(z)^{\beta} \ \Big| \ \alpha, \beta \in \mathbb{C} \right\}, \qquad \lambda(z) := \frac{1}{z} \log \frac{1}{1-z} \equiv \frac{1}{z} L(z).
$$

At this stage, we thus have available tools by which, starting from the expansion of a function at its singularity, also called *singular expansion*, one can justify the term-by-term transfer from an approximation of the function to an asymptotic estimate of the coefficients[5]. We state the following theorem.

Theorem VI.4 (Singularity analysis, single singularity). *Let $f(z)$ be function analytic at 0 with a singularity at ζ, such that $f(z)$ can be continued to a domain of the form $\zeta \cdot \Delta_0$, for a Δ–domain Δ_0, where $\zeta \cdot \Delta_0$ is the image of Δ_0 by the mapping $z \mapsto \zeta z$. Assume that there exist two functions σ, τ, where σ is a (finite) linear combination of functions in $\mathcal{S}$ and $\tau \in \mathcal{S}$, so that*

$$
f(z) = \sigma(z/\zeta) + O(\tau(z/\zeta)) \qquad as \quad z \to \zeta \quad in \quad \zeta \cdot \Delta_0.
$$

Then, the coefficients of $f(z)$ satisfy the asymptotic estimate

$$
f_n = \zeta^{-n} \sigma_n + O(\zeta^{-n} \tau_n^{\star}),
$$

where $\sigma_n = [z^n] \sigma(z)$ has its coefficients determined by Theorems VI.1, VI.2 and $\tau_n^{\star} = n^{a-1}(\log n)^b$, if $\tau(z) = (1-z)^{-a} \lambda(z)^b$.

We observe that the statement is equivalent to $\tau_n^{\star} = [z^n] \tau(z)$, except when $a \in \mathbb{Z}_{\leq 0}$, where the $1/\Gamma(a)$ factor should be omitted. Also, generically, we have $\tau_n^{\star} = o(\sigma_n)$, so that orders of growth of functions at singularities are mapped to orders of growth of coefficients.

Proof. The normalized function $g(z) = f(z/\zeta)$ is singular at 1. It is Δ–analytic and satisfies the relation $g(z) = \sigma(z) + O(\tau(z))$ as $z \to 1$ within Δ_0. Theorem VI.3, (i) (the big-Oh transfer) applies to the O-error term. The statement follows finally since $[z^n] f(z) = \zeta^{-n} [z^n] g(z)$. ∎

---

[5]Functions with a singularity of type $(1-z)^{-\alpha}$, possibly with logarithmic factors, are sometimes called *algebraic–logarithmic*.

Let $f(z)$ be a function analytic at 0 whose coefficients are to be asymptotically analysed.

1. *Preparation.* This consists in locating dominant singularities and checking analytic continuation.

   1a. *Locate singularities.* Determine the dominant singularities of $f(z)$ (assumed not to be entire). Check that $f(z)$ has a single singularity ζ on its circle of convergence.

   1b. *Check continuation.* Establish that $f(z)$ is analytic in some domain of the form $\zeta \Delta_0$.

2. *Singular expansion.* Analyse the function $f(z)$ as $z \to \zeta$ in the domain $\zeta \cdot \Delta_0$ and determine in that domain an expansion of the form

$$f(z) \underset{z \to 1}{=} \sigma(z/\zeta) + O(\tau(z/\zeta)) \qquad \text{with} \quad \tau(z) = o(\sigma(z)).$$

For the method to succeed, the functions σ and τ should belong to the standard scale of functions $\mathcal{S} = \{(1-z)^{-\alpha}\lambda(z)^{\beta}\}$, with $\lambda(z) := z^{-1}\log(1-z)^{-1}$.

3. *Transfer* Translate the main term term $\sigma(z)$ using the catalogues provided by Theorems VI.1 and VI.2. Transfer the error term (Theorem VI.3) and conclude that

$$[z^n]f(z) \underset{n \to +\infty}{=} \zeta^{-n}\sigma_n + O\left(\zeta^{-n}\tau_n^\star\right),$$

where $\sigma_n = [z^n]\sigma(z)$ and $\tau_n^\star = [z^n]\tau(z)$ provided the corresponding exponent $\alpha \notin \mathbb{Z}_{\leq 0}$ (otherwise, the factor $1/\Gamma(\alpha) = 0$ should be dropped).

Figure VI.7. A summary of the singularity analysis process (single dominant singularity).

The statement of Theorem VI.4 can be concisely expressed by the correspondence:

$$(30) \quad f(z) \underset{z \to 1}{=} \sigma(z/\zeta) + O\left(\tau(z/\zeta)\right) \qquad \longrightarrow \qquad f_n \underset{n \to \infty}{=} \zeta^{-n}\sigma_n + O(\zeta^{-n}\tau_n^\star).$$

The conditions of analytic continuation and validity of the expansion in a Δ–domain are essential. Similarly, we have

$$(31) \qquad f(z) \underset{z \to 1}{=} \sigma\left(z/\zeta\right)) + o\left(\tau\left(z/\zeta\right)\right) \qquad \longrightarrow \qquad f_n \underset{n \to \infty}{=} \zeta^{-n}\sigma_n + o(\zeta^{-n}\tau_n^\star),$$

as a simple consequence of Theorem VI.3, part (ii) (little-oh transfer). The mappings (30) and (31) supplemented by the accompanying analysis constitute the heart of the *singularity analysis* process summarized in Figure VI.7.

Many of the functions commonly encountered in analysis are found to be Δ–analytic. This fact results from the property of the elementary functions (such as $\sqrt{}$, log, tan) to be continuable to larger regions than what their expansions at 0 imply, as well as to the rich set of composition properties that analytic functions satisfy. Furthermore, asymptotic expansions at a singularity initially determined along the real axis by elementary real analysis often hold in much wider regions of the complex plane. The singularity analysis process is then likely to be applicable to a large number of generating functions that are provided by the symbolic method—most notably the iterative structures described in Section IV. 4 (p. 249). In such cases, singularity analysis greatly refines the exponential growth estimates obtained in Theorem IV.8

(p. 251). The condition is that singular expansions should be of a suitably moderate[6] growth. We illustrate this situation now by treating combinatorial generating functions obtained by the symbolic methods of Chapters I and II, for which explicit expressions are available.

Example VI.2. *Asymptotics of 2–regular graphs.* This example completes the discussion of Example VI.1, p. 379 relative to the EGF

$$R(z) = \frac{e^{-z/2-z^2/4}}{\sqrt{1-z}}.$$

We follow step by step the singularity analysis process, as summarized in Figure VI.7.

1. *Preparation.* The function $R(z)$ being the product of $e^{-z/2-z^2/4}$ (that is entire) and of $(1-z)^{-1/2}$ (that is analytic in the unit disc) is itself analytic in the unit disc. Also, since $(1-z)^{-1/2}$ is Δ–analytic (it is well-defined and analytic in the complex plane slit along $\mathbb{R}_{\geq 1}$), $R(z)$ is itself Δ–analytic, with a singularity at $z = 1$.

2. *Singular expansion.* The asymptotic expansion of $R(z)$ near $z = 1$ is obtained starting from the standard (analytic) expansion of $e^{-z/2-z^2/4}$ at $z = 1$,

$$e^{-z/2-z^2/4} = e^{-3/4} + e^{-3/4}(1-z) + \frac{e^{-3/4}}{4}(1-z)^2 - \frac{e^{-3/4}}{12}(1-z)^3 + \cdots.$$

The factor $(1-z)^{-1/2}$ is its own asymptotic expansion, clearly valid in any Δ–domain. Performing the *multiplication* yields a complete expansion,

$$(32) \quad R(z) \sim \frac{e^{-3/4}}{\sqrt{1-z}} + e^{-3/4}\sqrt{1-z} + \frac{e^{-3/4}}{4}(1-z)^{3/2} - \frac{e^{-3/4}}{12}(1-z)^{5/2} + \cdots,$$

out of which terminating forms, with an O–error term, can be extracted.

3. *Transfer.* Take for instance the expansion of (32) limited to two terms plus an error term. The singularity analysis process allows the transfer of (32) to coefficients, which we can present in tabular form as follows:

$R(z)$	$c_n \equiv [z^n]R(z)$
$e^{-3/4}\dfrac{1}{\sqrt{1-z}}$	$e^{-3/4}\dbinom{n-1/2}{-1/2} \sim \dfrac{e^{-3/4}}{\sqrt{\pi n}}\left[1 - \dfrac{1}{8n} + \dfrac{1}{128n^2} + \cdots\right]$
$+e^{-3/4}\sqrt{1-z}$	$+e^{-3/4}\dbinom{n-3/2}{-3/2} \sim \dfrac{-e^{-3/4}}{2\sqrt{\pi n^3}}\left[1 + \dfrac{3}{8n} + \cdots\right]$
$+ O((1-z)^{3/2})$	$+O\left(\dfrac{1}{n^{5/2}}\right).$

Terms are then collected with expansions suitably truncated to the coarsest error term, so that here a three-term expansion results. In the sequel, we shall no longer need to detail such computations and we shall content ourselves with putting in parallel the function's expansion and the coefficient's expansion, as in the following correspondence:

$$R(z) = \frac{e^{-3/4}}{\sqrt{1-z}} + e^{-3/4}\sqrt{1-z} + O\left((1-z)^{3/2}\right) \quad \longrightarrow \quad c_n = \frac{e^{-3/4}}{\sqrt{\pi n}} - \frac{5e^{-3/4}}{8\sqrt{\pi n^3}} + O\left(\frac{1}{n^{5/2}}\right).$$

---

[6]For functions with fast growth at a singularity, the saddle-point method developed in Chapter VIII becomes effectual.

Here is a numerical check. Set $c_n^{(1)} := e^{-3/4}/\sqrt{\pi n}$ and let $c_n^{(2)}$ represent the sum of the first two terms of the expansion of c_n. One finds:

n	5	50	500
$n! c_n^{(1)}$	14.30212	$1.1462888618 \cdot 10^{63}$	$1.4542120372 \cdot 10^{1132}$
$n! c_n^{(2)}$	12.51435	$1.1319602511 \cdot 10^{63}$	$1.4523942721 \cdot 10^{1132}$
$n! c_n$	12	$1.1319677968 \cdot 10^{63}$	$1.4523943224 \cdot 10^{1132}$

Clearly, a complete asymptotic expansion in descending powers of n can be obtained in this way. .... ∎

Example VI.3. *Asymptotics of unary–binary trees and Motzkin numbers.* Unary–binary trees are unlabelled plane trees that admit the specification and OGF:

$$\mathcal{U} = \mathcal{Z}(1 + \mathcal{U} + \mathcal{U} \times \mathcal{U}) \quad \Longrightarrow \quad U(z) = \frac{1 - z - \sqrt{(1+z)(1-3z)}}{2z}.$$

(See Note I.39 (p. 68) and Subsection V. 4 (p. 318) for the lattice path version.) The GF $U(z)$ is singular at $z = -1$ and $z = 1/3$, the dominant singularity being at $z = 1/3$. By branching properties of the square-root function, $U(z)$ is analytic in a Δ–domain like the one depicted below:

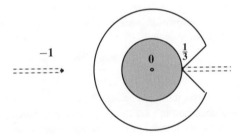

Around the point $1/3$, a singular expansion is obtained by multiplying $(1 - 3z)^{1/2}$ and the analytic expansion of the factor $(1 + z)^{1/2}/(2z)$. The singularity analysis process then applies and yields automatically:

$$U(z) = 1 - 3^{1/2}\sqrt{1 - 3z} + O((1 - 3z)) \quad \longrightarrow \quad U_n = \sqrt{\frac{3}{4\pi n^3}} \, 3^n + O(3^n n^{-2}).$$

Further terms in the singular expansion of $U(z)$ at $z = 1/3$ provide additional terms in the asymptotic expression of the Motzkin numbers U_n; for instance, the form

$$U_n = \sqrt{\frac{3}{4\pi n^3}} \, 3^n \left(1 - \frac{15}{16n} + \frac{505}{512 n^2} - \frac{8085}{8192 n^3} + \frac{505659}{524288 n^4} + O\left(\frac{1}{n^5}\right) \right)$$

results from an expansion of $U(z)$ till $O((1-3z)^{11/2})$. The approximation provided by the first three terms is quite good: for $n = 10$, it estimates $f_{10} = 835$. with an error less than 1. ..... ∎

▷ **VI.11.** *The population of Noah's Ark.* The number of one-source directed lattice animals (pyramids, Example I.18, p. 80) satisfies

$$P_n \equiv [z^n] \frac{1}{2} \left(\sqrt{\frac{1+z}{1-3z}} - 1 \right) = \frac{3^n}{\sqrt{3\pi n}} \left[1 - \frac{1}{16n} + O\left(\frac{1}{n^2}\right) \right].$$

The expected size of the base of a random animal in $\mathcal{A}_n$ is $\sim \sqrt{\frac{4n}{27\pi}}$. What is the asymptotic number of animals with a compact source of size k? ◁

Example VI.4. *Asymptotics of children's rounds.* Stanley [550] has introduced certain combinatorial configurations that he has nicknamed "children's rounds": a round is a labelled set of directed cycles, each of which has a centre attached. The specification and EGF are

$$\mathcal{R} = \mathrm{SET}(\mathcal{Z} \star \mathrm{CYC}(\mathcal{Z})) \quad \Longrightarrow \quad R(z) = \exp\left(z \log \frac{1}{1-z}\right) = (1-z)^{-z}.$$

The function $R(z)$ is analytic in the $\mathbb{C}$-plane slit along $\mathbb{R}_{\geq 1}$, as is seen by elementary properties of the *composition* of analytic functions. The singular expansion at $z = 1$ is then mapped to an expansion for the coefficients:

$$R(z) = \frac{1}{1-z} + \log(1-z) + O((1-z)^{1/2}) \quad \longrightarrow \quad [z^n]R(z) = 1 - \frac{1}{n} + O(n^{-3/2}).$$

A more detailed analysis yields

$$[z^n]R(z) = 1 - \frac{1}{n} - \frac{1}{n^2}(\log n + \gamma - 1) + O\left(\frac{\log^2 n}{n^3}\right),$$

and an expansion to any order can be easily obtained. .... ■

▷ **VI.12.** *The asymptotic shape of the rounds numbers.* A complete asymptotic expansion has the form

$$[z^n]R(z) \sim 1 - \sum_{j\geq 1} \frac{P_j(\log n)}{n^j},$$

where P_j is a polynomial of degree $j - 1$. (The coefficients of P_j are rational combinations of powers of $\gamma, \zeta(2), \ldots, \zeta(j-1)$.) The successive terms in this expansion are easily obtained by a computer algebra program. ◁

Example VI.5. *Asymptotics of coefficients of an elementary function.* Our final example is meant to show the way rather arbitrary compositions of basic functions can be treated by singularity analysis, much in the spirit of Section IV. 4, p. 249. Let $\mathcal{C} = \mathcal{Z} \star \mathrm{SEQ}(\mathcal{C})$ be the class of general labelled plane trees. Consider the labelled class defined by substitution

$$\mathcal{F} = \mathcal{C} \circ \mathrm{CYC}(\mathrm{CYC}(\mathcal{Z})) \quad \Longrightarrow \quad F(z) = C(\mathrm{L}(\mathrm{L}(z))).$$

There, $C(z) = \frac{1}{2}(1 - \sqrt{1 - 4z})$ and $\mathrm{L}(z) = \log\frac{1}{1-z}$. Combinatorially, $\mathcal{F}$ is the class of trees in which nodes are replaced by cycles of cycles, a rather artificial combinatorial object, and

$$F(z) = \frac{1}{2}\left[1 - \sqrt{1 - 4\log\frac{1}{1 - \log\frac{1}{1-z}}}\right].$$

The problem is first to locate the dominant singularity of $F(z)$, then to determine its nature, which can be done inductively on the structure of $F(z)$. The dominant positive singularity ρ of $F(z)$ satisfies $\mathrm{L}(\mathrm{L}(\rho)) = 1/4$ and one has

$$\rho = 1 - e^{e^{-1/4} - 1} \doteq 0.198443,$$

given that $C(z)$ is singular at $1/4$ and $\mathrm{L}(z)$ has positive coefficients. Since $\mathrm{L}(\mathrm{L}(z))$ is analytic at ρ, a local expansion of $F(z)$ is obtained next by *composition* of the singular expansion of $C(z)$ at $1/4$ with the standard Taylor expansion of $\mathrm{L}(\mathrm{L}(z))$ at ρ. We find

$$F(z) = \frac{1}{2} - C_1(\rho - z)^{1/2} + O\left((\rho - z)^{3/2}\right) \quad \longrightarrow \quad [z^n]F(z) = \frac{C_1\rho^{-n+1/2}}{2\sqrt{\pi n^3}}\left[1 + O\left(\frac{1}{n}\right)\right],$$

with $C_1 = e^{\frac{5}{8}-\frac{1}{2}e^{-1/4}} \doteq 1.26566.$ .... ∎

▷ **VI.13.** *The asymptotic number of trains.* Combinatorial trains were introduced in Section IV. 4 (p. 249) as a way to exemplify the power of complex asymptotic methods. One finds that, at its dominant singularity ρ, the EGF $Tr(z)$ is of the form $Tr(z) \sim C/(1-z/\rho)$, and, by singularity analysis,

$$[z^n]Tr(z) \sim 0.11768\,31406\,15497 \cdot 2.06131\,73279\,40138^n.$$

(This asymptotic approximation is good to 15 significant digits for $n = 50$, in accordance with the fact that the dominant singularity is a simple pole.)   ◁

VI. 5. Multiple singularities

The previous section has described in detail the analysis of functions with a single dominant singularity. The extension to functions that have *finitely many* (by necessity isolated) singularities on their circle of convergence follows along entirely similar lines. It parallels the situation of rational and meromorphic functions in Chapter IV (p. 263) and is technically simple, the net result being:

> *In the case of multiple singularities, the separate contributions from each of the singularities, as given by the basic singularity analysis process, are to be added up.*

As in (29), p. 393, we let S be the standard scale of functions singular at 1, namely

$$S = \left\{(1-z)^{-\alpha}\lambda(z)^\beta \mid \alpha, \beta \in \mathbb{C}\right\}, \qquad \lambda(z) := \frac{1}{z}\log\frac{1}{1-z}.$$

Theorem VI.5 (Singularity analysis, multiple singularities). *Let $f(z)$ be analytic in $|z| < \rho$ and have a finite number of singularities on the circle $|z| = \rho$ at points $\zeta_j = \rho e^{i\theta_j}$, for $j = 1..r$. Assume that there exists a Δ–domain Δ_0 such that $f(z)$ is analytic in the indented disc*

$$\mathbf{D} = \bigcap_{j=1}^{r}(\zeta_j \cdot \Delta_0),$$

with $\zeta \cdot \Delta_0$ the image of Δ_0 by the mapping $z \mapsto \zeta z$.

Assume that there exists r functions $\sigma_1, \ldots, \sigma_r$, each a linear combination of elements from the scale S, and a function $\tau \in S$ such that

$$f(z) = \sigma_j(z/\zeta_j) + O\left(\tau(z/\zeta_j)\right) \qquad \text{as } z \to \zeta_j \text{ in } \mathbf{D}.$$

Then the coefficients of $f(z)$ satisfy the asymptotic estimate

$$f_n = \sum_{j=1}^{r}\zeta_j^{-n}\sigma_{j,n} + O\left(\rho^{-n}\tau_n^\star\right),$$

where each $\sigma_{j,n} = [z^n]\sigma_j(z)$ has its coefficients determined by Theorems VI.1, VI.2 and $\tau_n^\star = n^{a-1}(\log n)^b$, if $\tau(z) = (1-z)^{-a}\lambda(z)^b$.

A function analytic in a domain like $\mathbf{D}$ is sometimes said to be *star-continuable*, a notion that naturally generalizes Δ–analyticity for functions with several dominant singularities. Furthermore, a similar statement holds with o–error terms replacing Os.

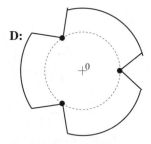 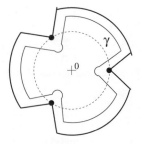

Figure VI.8. Multiple singularities ($r = 3$): analyticity domain (**D**, left) and composite integration contour (γ, right).

Proof. Just as in the case of a single singularity, the proof bases itself on Cauchy's coefficient formula

$$f_n = [z^n] \int_\gamma f(z) \frac{dz}{z^{n+1}},$$

where a composite contour γ depicted on Figure VI.8 is used. Estimates on each part of the contour obey exactly the same principles as in the proofs of Theorems VI.1–VI.3. Let $\gamma^{(j)}$ be the open loop around ζ_j that comes from the outer circle, winds about ζ_j and joins again the outer circle; let r be the radius of the outer circle.

(i) The contribution along the arcs of the outer circle is $O(r^{-n})$, that is, exponentially small.

(ii) The contribution along the loop $\gamma^{(1)}$ (say) separates into

$$\frac{1}{2i\pi} \int_{\gamma^{(1)}} f(z) \frac{dz}{z^{n+1}} = I' + I''$$

$$I' := \frac{1}{2i\pi} \int_{\gamma^{(1)}} \sigma_1(z/\zeta_1) \frac{dz}{z^{n+1}}, \quad I'' := \frac{1}{2i\pi} \int_{\gamma^{(1)}} (f(z) - \sigma_1(z/\zeta_1)) \frac{dz}{z^{n+1}}.$$

The quantity I' is estimated by extending the open loop to infinity by the same method as in the proof of Theorems VI.1 and VI.2: it is found to equal $\zeta_1^{-n} \sigma_{1,n}$ plus an exponentially small term. The quantity I'', corresponding to the error term, is estimated by the same bounding technique as in the proof of Theorem VI.3 and is found to be $O(\rho^{-n} \tau_n^\star)$.

Collecting the various contributions completes the proof of the statement. ∎

Theorem VI.5 expresses that, in the case of multiple singularities, each dominant singularity can be analysed separately; the singular expansions are then each transferred to coefficients, and the corresponding asymptotic contributions are finally collected. Two examples illustrating the process follow.

***Example* VI.6.** *An artificial example.* Let us demonstrate the *modus operandi* on the simple function

(33)
$$g(z) = \frac{e^z}{\sqrt{1 - z^2}}.$$

There are two singularities at $z = +1$ and $z = -1$, with

$$g(z) \sim \frac{e}{\sqrt{2}\sqrt{1-z}} \quad z \to +1 \qquad \text{and} \qquad g(z) \sim \frac{e^{-1}}{\sqrt{2}\sqrt{1+z}} \quad z \to -1.$$

The function is clearly star-continuable with the singular expansions being valid in an indented disc. We have

$$[z^n]\frac{e}{\sqrt{2}\sqrt{1-z}} \sim \frac{e}{\sqrt{2\pi n}} \qquad \text{and} \qquad [z^n]\frac{e^{-1}}{\sqrt{2}\sqrt{1+z}} \sim \frac{e^{-1}(-1)^n}{\sqrt{2\pi n}}.$$

To obtain the coefficient $[z^n]g(z)$, it suffices to add up these two contributions (by Theorem VI.5), so that

$$[z^n]g(z) \sim \frac{1}{\sqrt{2\pi n}}[e + (-1)^n e^{-1}].$$

If expansions at $+1$ (respectively -1) are written with an error term, which is of the form $O((z-1)^{1/2})$ (respectively, $O((z+1)^{1/2})$), there results an estimate of the coefficients $g_n = [z^n]g(z)$, which can be put under the form

$$g_{2n} = \frac{\cosh(1)}{\sqrt{\pi n}} + O\left(n^{-3/2}\right), \qquad g_{2n+1} = \frac{\sinh(1)}{\sqrt{\pi n}} + O\left(n^{-3/2}\right).$$

This makes explicit the dependency of the asymptotic form of g_n on the parity of the index n. Clearly a full asymptotic expansion can be obtained. .... ∎

Example VI.7. *Permutations with cycles of odd length.* Consider the specification and EGF

$$\mathcal{F} = \text{SET}(\text{CYC}_{\text{odd}}(\mathcal{Z})) \quad \Longrightarrow \quad F(z) = \exp\left(\frac{1}{2}\log\frac{1+z}{1-z}\right) = \sqrt{\frac{1+z}{1-z}}.$$

The singularities of f are at $z = +1$ and $z = -1$, the function being obviously star-continuable. By singularity analysis (Theorem VI.5), we have automatically:

$$F(z) = \begin{cases} \dfrac{2^{1/2}}{\sqrt{1-z}} + O\left((1-z)^{1/2}\right) & (z \to 1) \\[2mm] O\left((1+z)^{1/2}\right) & (z \to -1) \end{cases} \quad \longrightarrow \quad [z^n]F(z) = \frac{2^{1/2}}{\sqrt{\pi n}} + O\left(n^{-3/2}\right).$$

For the next asymptotic order, the singular expansions

$$F(z) = \begin{cases} \dfrac{2^{1/2}}{\sqrt{1-z}} - 2^{-3/2}\sqrt{1-z} + O((1-z)^{3/2}) & (z \to 1) \\[2mm] 2^{-1/2}\sqrt{1+z} + O((1+z)^{3/2}) & (z \to -1) \end{cases}$$

yield

$$[z^n]F(z) = \frac{2^{1/2}}{\sqrt{\pi n}} - \frac{(-1)^n 2^{-3/2}}{\sqrt{\pi n^3}} + O(n^{-5/2}).$$

This example illustrates the occurrence of singularities that have different weights, in the sense of being associated with different exponents. .... ∎

The discussion of multiple dominant singularities ties well with the earlier discussion of Subsection IV. 6.1, p. 263. In the periodic case where the dominant singularities are at roots of unity, different regimes manifest themselves cyclically depending on congruence properties of the index n, like in the two examples above. When the dominant singularities have arguments that are not commensurate to π (a comparatively rare situation), irregular fluctuations appear, in which case the situation is similar to what was already discussed, regarding rational and meromorphic functions, in Subsection IV. 6.1.

VI. 6. Intermezzo: functions amenable to singularity analysis

Let us say that a function is *amenable to singularity analysis*, or *SA* for short, if its satisfies the conditions of singularity analysis, as expressed by Theorem VI.4 (single dominant singularity) or Theorem VI.5 (multiple dominant singularities). The property of being of SA is preserved by several basic operations of analysis: we have already seen this feature in passing, when determining singular expansions of functions obtained by sums, products, or compositions in Examples VI.2–VI.5.

As a starting example, it is easily recognized that the assumptions of Δ–analyticity for two functions $f(z)$, $g(z)$ accompanied by the singular expansions

$$f(z) \underset{z \to 1}{\sim} c(1-z)^{-\alpha}, \qquad g(z) \underset{z \to 1}{\sim} d(1-z)^{-\delta},$$

and the condition $\alpha, \delta \notin \mathbb{Z}_{\leq 0}$ imply for the coefficients of the sum

$$[z^n] (f(z) + g(z)) \sim \begin{cases} c\dfrac{n^{\alpha-1}}{\Gamma(\alpha)} & \alpha > \delta \\[2mm] (c+d)\dfrac{n^{\alpha-1}}{\Gamma(\alpha)} & \alpha = \delta, \quad c+d \neq 0 \\[2mm] d\dfrac{n^{\delta-1}}{\Gamma(\delta)} & \alpha < \delta. \end{cases}$$

Similarly, for products, we have

$$[z^n] (f(z)g(z)) \sim cd\frac{n^{\alpha+\delta-1}}{\Gamma(\alpha+\delta)},$$

provided $\alpha + \delta \notin \mathbb{Z}_{\leq 0}$.

The simple considerations above illustrate the robustness of singularity analysis. They also indicate that properties are easy to state in the generic case where no negative integral exponents are present. However, if all cases are to be covered, there can easily be an explosion of the number of particular situations, which may render somewhat clumsy the enunciation of complete statements. Accordingly, in what follows, we shall largely confine ourselves to generic cases, as long as these suffice to develop the important mathematical technique at stake for each particular problem.

In the remainder of this chapter, we proceed to enlarge the class of functions recognized to be of SA, keeping in mind the needs of analytic combinatorics. The following types of functions are treated in later sections.

(*i*) *Inverse functions* (Section VI. 7). The inverse of an analytic function is, under mild conditions, of SA type. In the case of functions attached to simple varieties of trees (corresponding to the inversion of $y/\phi(y)$), the singular expansion invariably has an exponent of $\frac{1}{2}$ attached to it (a square-root singularity). This applies in particular to the Cayley tree function, in terms of which many combinatorial structures and parameters can be analysed.

(ii) *Polylogarithms* (Section VI. 8). These functions are the generating functions of simple arithmetic sequences such as (n^θ) for an arbitrary $\theta \in \mathbb{C}$. The fact that polylogarithms are SA opens the possibility of estimating a large number of sums, which involve both combinatorial terms (e.g., binomial coefficients) and elements like $\sqrt{n}$ and $\log n$. Such sums appear recurrently in the analysis of cost functionals of combinatorial structures and algorithms.

(iii) *Composition* (Section VI. 9). The composition of SA functions often proves to be itself SA This fact has implications for the analysis of composition schemas and makes possible a broad extension of the supercritical sequence schema treated in Section V. 2, (p. 293).

(iv) *Differentiation, integration, and Hadamard products* (Section VI. 10). These are three operations on analytic functions that preserve the property for a function to be SA. Applications are given to tree recurrences and to multidimensional walk problems.

A main theme of this book is that elementary combinatorial classes tend to have generating functions whose singularity structure is strongly constrained—in most cases, singularities are isolated. The singularity analysis process is then a prime technique for extracting asymptotic information from such generating functions.

VI. 7. Inverse functions

Recursively defined structures lead to functional equations whose solutions may often be analysed locally near singularities. An important case is the one of functions defined by inversion. It includes the Cayley tree function as well as all generating functions associated to simple varieties of trees (Subsections I. 5.1 (p. 65), II. 5.1 (p. 126), and III. 6.2 (p. 193)). A common pattern in this context is the appearance of singularities of the square-root type, which proves to be universal among a broad class of problems involving trees and tree-like structures. Accordingly, by singularity analysis, the square-root singularity induces subexponential factors of the asymptotic form $n^{-3/2}$ in expansions of coefficients—we shall further develop this theme in Chapter VII, pp. 452–493.

Inverse functions. Singularities of functions defined by inversion have been located in Subsection IV. 7.1 (p. 275) and our treatment will proceed from there. The goal is to estimate the coefficients of a function defined implicitly by an equation of the form

(34) $y(z) = z\phi(y(z))$ or equivalently $z = \dfrac{y(z)}{\phi(y(z))}.$

The problem of solving (34) is one of functional inversion: we have seen (Lemmas IV.2 and IV.3, pp. 275–277) that *an analytic function admits locally an analytic inverse if and only if its first derivative is non-zero.* We operate here under the following assumptions:

Condition (H$_1$). The function $\phi(u)$ is analytic at $u = 0$ and satisfies

(35) $\phi(0) \neq 0,$ $[u^n]\phi(u) \geq 0,$ $\phi(u) \not\equiv \phi_0 + \phi_1 u.$

(As a consequence, the inversion problem is well defined around 0. The nonlinearity of ϕ only excludes the case $\phi(u) = \phi_0 + \phi_1 u$, corresponding to $y(z) = \phi_0 z/(1 - \phi_1 z)$.)

Condition (H_2). Within the *open* disc of convergence of ϕ at 0, $|z| < R$, there exists a (then necessarily unique) positive solution to the *characteristic equation*:

(36) $$\exists \tau, \; 0 < \tau < R, \quad \phi(\tau) - \tau\phi'(\tau) = 0.$$

(Existence is granted as soon as $\lim x\phi'(x)/\phi(x) > 1$ as $x \to R^-$, with R the radius of convergence of ϕ at 0; see Proposition IV.5, p. 278.)

Then (by Proposition IV.5, p. 278), the radius of convergence of $y(z)$ is the corresponding positive value ρ of z such that $y(\rho) = \tau$, that is to say,

(37) $$\rho = \frac{\tau}{\phi(\tau)} = \frac{1}{\phi'(\tau)}.$$

We start with a calculation indicating in a plain context the occurrence of a square-root singularity.

Example VI.8. *A simple analysis of the Cayley tree function.* The situation corresponding to the function $\phi(u) = e^u$, so that $y(z) = ze^{y(z)}$ (defining the Cayley tree function $T(z)$), is typical of general analytic inversion. From (36), the radius of convergence of $y(z)$ is $\rho = e^{-1}$ corresponding to $\tau = 1$. The image of a circle in the y–plane, centred at the origin and having radius $r < 1$, by the function ye^{-y} is a curve of the z–plane that properly contains the circle $|z| = re^{-r}$ (see Figure VI.9) as $\phi(y) = e^y$, which has non-negative coefficients, satisfies

$$\left| \phi(re^{i\theta}) \right| \leq \phi(r) \quad \text{for all } \theta \in [-\pi, +\pi],$$

the inequality being strict for all $\theta \neq 0$. The following observation is the key to analytic continuation: *Since the first derivative of $y/\phi(y)$ vanishes at 1, the mapping $y \mapsto y/\phi(y)$ is angle-doubling, so that the image of the circle of radius 1 is a curve C that has a cusp at $\rho = e^{-1}$.* (See Figure VI.9; Notes VI.18 and 19 provide interesting generalizations.)

This geometry indicates that the solution of $z = ye^{-y}$ is uniquely defined for z inside C, so that $y(z)$ is Δ–analytic (see the proof of Theorem VI.6 below). A singular expansion for $y(z)$ is then derived from reversion of the power series expansion of $z = ye^{-y}$. We have

(38) $$ye^{-y} = e^{-1} - \boxed{\frac{1}{2e}(y-1)^2} + \frac{1}{3e}(y-1)^3 - \frac{e^{-1}}{8}(y-1)^4 + \cdots .$$

Observe both the absence of a linear term and the presence of a quadratic term (boxed). Then, solving $z = ye^{-y}$ for y gives

$$y - 1 = \sqrt{2}(1 - ez)^{1/2} + \frac{2}{3}(1 - ez) + O((1 - ez)^{3/2}),$$

where the square root arises precisely from inversion of the quadratic term. (A full expansion can furthermore be obtained.) .... ∎

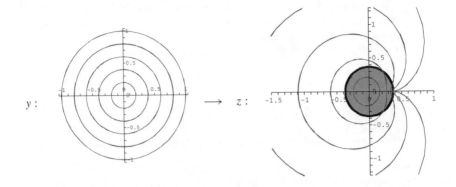

Figure VI.9. The images of concentric circles by the mapping $y \mapsto z = ye^{-y}$. It is seen that $y \mapsto z = ye^{-y}$ is injective on $|y| \le 1$ with an image extending beyond the circle $|z| = e^{-1}$ [in grey], so that the inverse function $y(z)$ is analytically continuable in a Δ–domain around $z = e^{-1}$. Since the direct mapping ye^{-y} is quadratic at 1 (with value e^{-1}, see (38)), the inverse function has a square-root singularity at e^{-1} (with value 1).

Analysis of inverse functions. The calculation of Example VI.8 now needs to be extended to the general case, $y = z\phi(y)$. This involves three steps: (i) all the dominant singularities are to be located; (ii) analyticity of $y(z)$ in a Δ–domain must be established; (iii) the singular expansion, obtained formally so far and involving a square-root singularity, needs to be determined. Step (i) requires a special discussion and is related to periodicities.

A basic example like $\phi(u) = 1 + u^2$ (binary trees), for which

$$y(z) = \frac{1 - \sqrt{1 - 4z^2}}{2z},$$

shows that $y(z)$ may have several dominant singularities—here, two conjugate singularities at $-\frac{1}{2}$ and $+\frac{1}{2}$. The conditions for this to happen are related to our discussion of periodicities in Definition IV.5, p. 266. As a consequence of this definition, $\phi(u)$, which satisfies $\phi(0) \ne 0$, is *p–periodic* if $\phi(u) = g(u^p)$ for some power series g (see p. 266) and $p \ge 2$; it is aperiodic otherwise. An elementary argument developed in Note VI.17, p. 407, shows that the aperiodicity assumption entails no loss of analytic generality (periodicity does not occur for $y(z)$ unless $\phi(u)$ is itself periodic, a case which, in addition, turns out to be reducible to the aperiodic situation).

Theorem VI.6 (Singular Inversion). *Let ϕ be a nonlinear function satisfying the conditions* (**H₁**) *and* (**H₂**) *of Equations (35) and (36), and let $y(z)$ be the solution of $y = z\phi(y)$ satisfying $y(0) = 0$. Then, the quantity $\rho = \tau/\phi(\tau)$ is the radius of convergence of $y(z)$ at 0 (with τ the root of the characteristic equation), and the singular* *expansion of $y(z)$ near ρ is of the form*

$$y(z) = \tau - d_1\sqrt{1 - z/\rho} + \sum_{j\geq 2}(-1)^j d_j(1 - z/\rho)^{j/2}, \qquad d_1 := \sqrt{\frac{2\phi(\tau)}{\phi''(\tau)}},$$

with the d_j being some computable constants.

Assume that, in addition, ϕ is aperiodic[7]. Then, one has

$$[z^n]y(z) \sim \sqrt{\frac{\phi(\tau)}{2\phi''(\tau)}} \frac{\rho^{-n}}{\sqrt{\pi n^3}} \left(1 + \sum_{k=1}^{\infty} \frac{e_k}{n^k}\right),$$

for a family e_k of computable constants.

Proof. Proposition IV.5, p. 278, shows that ρ is indeed the radius of convergence of $y(z)$. The Singular Inversion Lemma (Lemma IV.3, p. 277) also shows that $y(z)$ can be continued to a neighbourhood of ρ slit along the ray $\mathbb{R}_{\geq\rho}$.

The singular expansion at ρ is determined as in Example VI.8. Indeed, the relation between z and y, in the vicinity of $(z, y) = (\rho, \tau)$, may be put under the form

$$(39) \qquad \rho - z = H(y), \qquad \text{where} \quad H(y) := \left(\frac{\tau}{\phi(\tau)} - \frac{y}{\phi(y)}\right),$$

the function $H(y)$ in the right-hand side being such that $H(\tau) = H'(\tau) = 0$. Thus, the dependency between y and z is locally a quadratic one:

$$\rho - z = \frac{1}{2!}H''(\tau)(y - \tau)^2 + \frac{1}{3!}H'''(\tau)(y - \tau)^3 + \cdots.$$

When this relation is locally inverted: a square-root appears:

$$-\sqrt{\rho - z} = \sqrt{\frac{H''(\tau)}{2}}(y - \tau)\left[1 + c_1(y - \tau) + c_2(y - \tau)^2 + \ldots\right].$$

The determination with a $-\sqrt{}$ should be chosen there as $y(z)$ increases to τ^- as $z \to \rho^-$. This implies, by solving with respect to $y - \tau$, the relation

$$y - \tau \sim -d_1^\star(\rho - z)^{1/2} + d_2^\star(\rho - z) - d_3^\star(\rho - z)^{3/2} + \cdots,$$

where $d_1^\star = \sqrt{2/H''(\tau)}$ with $H''(\tau) = \tau\phi''(\tau)/\phi(\tau)^2$. The singular expansion at ρ results.

It now remains to exclude the possibility for $y(z)$ to have singularities other than ρ on the circle $|z| = \rho$, in the aperiodic case. Observe that $y(\rho)$ is well defined (in fact $y(\rho) = \tau$), so that the series representing $y(z)$ converges at ρ as well as on the whole circle (given positivity of the coefficients). If $\phi(z)$ is aperiodic, then so is $y(z)$. Consider any point ζ such that $|\zeta| = \rho$ and $\zeta \neq \rho$ and set $\eta = y(\zeta)$. We then have $|\eta| < \tau$ (by the Daffodil Lemma: Lemma IV.1, p. 266). The function $y(z)$ is analytic

---

[7]If ϕ has maximal period p, then one must restrict n to $n \equiv 1 \bmod p$; in that case, there is an extra factor of p in the estimate of y_n: see Note VI.17 and Equation (40).

Type	$\phi(u)$	singular expansion of $y(z)$	coefficient $[z^n]y(z)$
binary	$(1+u)^2$	$1 - 4\sqrt{\frac{1}{4} - z} + \cdots$	$\dfrac{4^n}{\sqrt{\pi n^3}} + O(n^{-5/2})$
unary–binary	$1 + u + u^2$	$1 - 3\sqrt{\frac{1}{3} - z} + \cdots$	$\dfrac{3^{n+1/2}}{2\sqrt{\pi n^3}} + O(n^{-5/2})$
general	$(1-u)^{-1}$	$\frac{1}{2} - \sqrt{\frac{1}{4} - z}$	$\dfrac{4^{n-1}}{\sqrt{\pi n^3}} + O(n^{-5/2})$
Cayley	e^u	$1 - \sqrt{2e}\sqrt{e^{-1} - z} + \cdots$	$\dfrac{e^n}{\sqrt{2\pi n^3}} + O(n^{-5/2})$

Figure VI.10. Singularity analysis of some simple varieties of trees.

at ζ by virtue of the Analytic Inversion Lemma (Lemma IV.2, p. 275) and the property
that

$$\frac{d}{dy}\frac{y}{\phi(y)}\bigg|_{y=\eta} \neq 0.$$

(This last property is derived from the fact that the numerator of the quantity on the
left,

$$\phi(\eta) - \eta\phi'(\eta) = \phi_0 - \phi_2\eta^2 - 2\phi_3\eta^3 - 3\phi_4\eta^4 - \cdots,$$

cannot vanish, by the triangle inequality since $|\eta| < \tau$.) Thus, under the aperiodicity
assumption, $y(z)$ is analytic on the circle $|z| = \rho$ punctured at ρ. The expansion of
the coefficients then results from basic singularity analysis. ■

Figure VI.10 provides a table of the most basic varieties of simple trees and the
corresponding asymptotic estimates. With Theorem VI.6, we now have available a
powerful method that permits us to analyse not only implicitly defined functions but
also expressions built upon them. This fact will be put to good use in Chapter VII,
when analysing a number of parameters associated to simple varieties of trees.

▷ **VI.14.** *All kinds of graphs.* In relation with the classes of graphs listed in Figure II.14,
p. 134, one has the following correspondence between an EGF $f(z)$ and the asymptotic form
of $n![z^n]f(z)$:

function:	$e^{T-T^2/2}$	$\log\dfrac{1}{1-T}$	$\dfrac{1}{\sqrt{1-T}}$	$\dfrac{1}{(1-T)^m}$
coefficient:	$e^{1/2}n^{n-2}$	$\frac{1}{2}\sqrt{2\pi}n^{n-1/2}$	$C_1 n^{n-1/4}$	$C_2 n^{n+(m-1)/2}$

($m \in \mathbb{Z}_{\geq 1}$; C_1, C_2 represent computable constants). In this way, the estimates of Subsection II.5.3, p. 132, are justifiable by singularity analysis. ◁

▷ **VI.15.** *Computability of singular expansions.* Define

$$h(w) := \sqrt{\frac{\tau/\phi(\tau) - w/\phi(w)}{(\tau - w)^2}},$$

so that $y(z)$ satisfies $\sqrt{\rho - z} = (\tau - y)h(y)$. The singular expansion of y can then be deduced by Lagrange inversion from the expansion of the negative powers of $h(w)$ at $w = \tau$. This technique yields for instance explicit forms for coefficients in the singular expansion of $y = ze^y$.   ◁

▷ **VI.16.** *Stirling's formula via singularity analysis.* The solution to $T = ze^T$ analytic at 0 is the Cayley tree function. It satisfies $[z^n] = n^{n-1}/n!$ (by Lagrange inversion) and, at the same time, its singularity is known from Theorem VI.6 and Example VI.8. As a consequence:

$$\frac{n^{n-1}}{n!} \sim \frac{e^n}{\sqrt{2\pi n^3}} \left(1 - \frac{1}{12n} + \frac{1}{288n^2} + \frac{139}{51840n^3} - \cdots \right).$$

Thus Stirling's formula *also* results from singularity analysis.   ◁

▷ **VI.17.** *Periodicities.* Assume that $\phi(u) = \psi(u^p)$ with ψ analytic at 0 and $p \geq 2$. Let $y = y(z)$ be the root of $y = z\phi(y)$. Set $Z = z^p$ and let $Y(Z)$ be the root of $Y = Z\psi(Y)^p$. One has by construction $y(z) = Y(z^p)^{1/p}$, given that $y^p = z^p\phi(y)^p$. Since $Y(Z) = Y_1 Z + Y_2 Z^2 + \cdots$, we verify that the non-zero coefficients of $y(z)$ are among those of index $1, 1+p, 1+2p, \ldots$.

  If p is chosen maximal, then $\psi(u)^p$ is aperiodic. Then Theorem VI.6 applies to $Y(Z)$: the function $Y(Z)$ is analytically continuable beyond its dominant singularity at $Z = \rho^p$; it has a square root singularity at ρ^p and no other singularity on $|Z| = \rho^p$. Furthermore, since $Y = Z\psi(Y)^p$, the function $Y(Z)$ cannot vanish on $|Z| \leq \rho^p$, $Z \neq 0$. Thus, $Y(Z)^{1/p}$ is analytic in $|Z| \leq \rho^p$, except at ρ^p where it has a $\sqrt{\ }$ branch point. All computations done, we find that

$$(40) \qquad\qquad [z^n]y(z) \sim p \cdot \frac{d_1 \rho^{-n}}{2\sqrt{\pi n^3}} \qquad \text{when} \quad n \equiv 1 \pmod{p}.$$

The argument also shows that $y(z)$ has p conjugate roots on its circle of convergence. (This is a kind of Perron–Frobenius property for periodic tree functions.)   ◁

▷ **VI.18.** *Boundary cases I.* The case when τ lies on the boundary of the disc of convergence of ϕ may lead to asymptotic estimates differing from the usual $\rho^{-n}n^{-3/2}$ prototype. Without loss of generality, take ϕ aperiodic to have radius of convergence equal to 1 and assume that ϕ is of the form

$$(41) \qquad\qquad \phi(u) = u + c(1-u)^\alpha + o((1-u)^\alpha), \qquad \text{with} \quad 1 < \alpha \leq 2,$$

as u tends to 1 within $|u| < 1$. (Thus, continuation of $\phi(u)$ beyond $|u| < 1$ is *not* assumed.) The solution of the characteristic equation $\phi(\tau) - \tau\phi'(\tau) = 0$ is then $\tau = 1$. The function $y(z)$ defined by $y = z\phi(y)$ is Δ–analytic (by a mapping argument similar to the one exemplified by Figure VI.9 and related to the fact that ϕ "multiplies" angles near 1). The singular expansion of $y(z)$ and the coefficients then satisfy

$$(42) \quad y(z) = 1 - c^{-1/\alpha}(1-z)^{1/\alpha} + o\left((1-z)^{1/\alpha}\right) \qquad \longrightarrow \qquad y_n \sim c^{-1/\alpha}\frac{n^{-1/\alpha-1}}{-\Gamma(-1/\alpha)}.$$

[The case $\alpha = 2$ was first observed by Janson [350]. Trees with $\alpha \in (1, 2)$ have been investigated in connection with stable Lévy processes [180]. The singular exponent $\alpha = 3/2$ occurs for instance in planar maps (Subsection VII. 8.2, p. 513), so that GFs with coefficients of the form $\rho^{-n}n^{-5/3}$ would arise, if considering trees whose nodes are themselves maps.]   ◁

▷ **VI.19.** *Boundary cases II.* Let $\phi(u)$ be the probability generating function of a random variable X with mean equal to 1 and such that $\phi_n \sim \lambda n^{-\alpha-1}$, with $1 < \alpha < 2$. Then, by a complex version of an Abelian theorem (see, e.g., [69, §1.7] and [232]), the singular expansion (41) holds when $u \to 1$, $|u| < 1$, within a cone, so that the conclusions of (42) hold in that case. Similarly, if $\phi''(1)$ exists, meaning that X has a second moment, then the estimate (42) holds with $\alpha = 2$, and then coincides with what Theorem VI.6 predicts [350]. (In probabilistic terms, the condition of Theorem VI.6 is equivalent to postulating the existence of exponential moments for the one-generation offspring distribution.)   ◁

VI. 8. Polylogarithms

Generating functions involving sequences such as $(\sqrt{n})$ or $(\log n)$ can be subjected to singularity analysis. The starting point is the definition of the generalized *polylogarithm*, commonly denoted[8] by $\mathrm{Li}_{\alpha,r}$, where α is an arbitrary complex number and r a non-negative integer:

$$\mathrm{Li}_{\alpha,r}(z) := \sum_{n \geq 1} (\log n)^r \frac{z^n}{n^\alpha},$$

The series converges for $|z| < 1$, so that the function $\mathrm{Li}_{\alpha,r}$ is *a priori* analytic in the unit disc. The quantity $\mathrm{Li}_{1,0}(z)$ is the usual logarithm, $\log(1 - z)^{-1}$, hence the established name, polylogarithm, assigned to these functions [406]. In what follows, we make use of the abbreviation $\mathrm{Li}_{\alpha,0}(z) \equiv \mathrm{Li}_\alpha(z)$, so that $\mathrm{Li}_1(z) \equiv \mathrm{Li}_{1,0}(z) \equiv \log(1-z)^{-1}$ is the GF of the sequence $(1/n)$. Similarly, $\mathrm{Li}_{0,1}$ is the GF of the sequence $(\log n)$ and $\mathrm{Li}_{-1/2}(z)$ is the GF of the sequence $(\sqrt{n})$.

Polylogarithms are continuable to the whole of the complex plane slit along the ray $\mathbb{R}_{\geq 1}$, a fact established early in the twentieth century by Ford [268], which results from the integral representation (48), p. 409. They are amenable to singularity analysis [223] and their singular expansions involve the Riemann zeta function defined by

$$\zeta(s) = \sum_{n=1}^{\infty} \frac{1}{n^s},$$

for $\Re(s) > 1$, and by analytic continuation elsewhere [578].

Theorem VI.7 (Singularities of polylogarithms). *For all $\alpha \in \mathbb{Z}$ and $r \in \mathbb{Z}_{\geq 0}$, the function $\mathrm{Li}_{\alpha,r}(z)$ is analytic in the slit plane $\mathbb{C} \setminus \mathbb{R}_{\geq 1}$. For $\alpha \notin \{1, 2, \ldots\}$, there exists an infinite singular expansion (with logarithmic terms when $r > 0$) given by the two rules:*

$$(43) \quad \begin{cases} \mathrm{Li}_\alpha(z) \ \sim \ \Gamma(1 - \alpha)w^{\alpha-1} + \sum_{j \geq 0} \dfrac{(-1)^j}{j!} \zeta(\alpha - j)w^j, \quad w := \sum_{\ell=1}^{\infty} \dfrac{(1 - z)^\ell}{\ell} \\[2mm] \mathrm{Li}_{\alpha,r}(z) = (-1)^r \dfrac{\partial^r}{\partial \alpha^r} \mathrm{Li}_\alpha(z) \qquad (r \geq 0). \end{cases}$$

The expansion of Li_α is conveniently described by the composition of two expansions (Figure VI.11, p. 410): the expansion of $w = \log z$ at $z = 1$, namely, $w = (1 - z) + \frac{1}{2}(1 - z)^2 + \cdots$, is to be substituted inside the formal power series involving powers of w. The exponents of $(1-z)$ involved in the resulting expansion are $\{\alpha - 1, \alpha, \ldots\} \cup \{0, 1, \ldots\}$. For $\alpha < 1$, the main asymptotic term of $\mathrm{Li}_{\alpha,r}$ is, as $z \to 1$,

$$\mathrm{Li}_{\alpha,r}(z) \sim \Gamma(1 - \alpha)(1 - z)^{\alpha-1} L(z)^r, \qquad L(z) := \log \frac{1}{1 - z},$$

---

[8]The notation $\mathrm{Li}_\alpha(z)$ is nowadays well established. It is evocative of the fact that polylogarithms of integer order $m \geq 2$ are expressible by a logarithmic integral:

$$\mathrm{Li}_{m,0}(x) = \frac{(-1)^{m-1}}{(m - 1)!} \int_0^1 \log(1 - xt) \log^{m-2} t \, \frac{dt}{t}$$

(not to be confused with the unrelated "logarithmic integral function" $\mathrm{li}(z) := \int_0^z \frac{dt}{\log t}$; see [3, p. 228]).

while, for $\alpha > 1$, we have $\mathrm{Li}_{\alpha,r}(z) \sim (1-)^r \zeta^{(r)}(\alpha)$, since the sum defining $\mathrm{Li}_{\alpha,r}$ converges at 1.

Proof. The analysis crucially relies on the Mellin transform (see Appendix B.7: *Mellin transforms*, p. 762). We start with the case $r = 0$ and consider several ways in which z may approach the singularity 1. Step (i) below describes the main ingredient needed in *obtaining* the expansion, the subsequent steps being only required for *justifying* it in larger regions of the complex plane.

(i) *When $z \to 1^-$ along the real line.* Set $w = -\log z$ and introduce

$$
(44) \qquad \Lambda(w) := \mathrm{Li}_\alpha(e^{-w}) = \sum_{n \geq 1} \frac{e^{-nw}}{n^\alpha}.
$$

This is a *harmonic sum* in the sense of Mellin transform theory, so that the Mellin transform of Λ satisfies $(\Re(s) > \max(0, 1 - \alpha))$

$$
(45) \qquad \Lambda^\star(s) \equiv \int_0^\infty \Lambda(w) w^{s-1} \, dw = \zeta(s + \alpha) \Gamma(s).
$$

The function $\Lambda(w)$ can be recovered from the inverse Mellin integral,

$$
(46) \qquad \Lambda(w) = \frac{1}{2i\pi} \int_{c-i\infty}^{c+i\infty} \zeta(s + \alpha) \Gamma(s) w^{-s} \, ds,
$$

with c taken in the half-plane in which $\Lambda^\star(s)$ is defined. There are poles at $s = 0, -1, -2, \ldots$ due to the Gamma factor and a pole at $s = 1 - \alpha$ due to the zeta function. Take d to be of the form $-m - \frac{1}{2}$ and smaller than $1 - \alpha$. Then, a standard residue calculation, taking into account poles to the left of c and based on

$$
(47) \qquad
\begin{aligned}
\Lambda(w) &= \sum_{s_0 \in \{0, -1, \ldots, -m\} \cup \{1 - \alpha\}} \mathrm{Res}\left(\zeta(s + \alpha) \Gamma(s) w^{-s}\right)_{s=s_0} \\
&\quad + \frac{1}{2i\pi} \int_{d-i\infty}^{d+i\infty} \zeta(s + \alpha) \Gamma(s) w^{-s} \, ds,
\end{aligned}
$$

then yields a finite form of the estimate (43) of Li_α (as $w \to 0$, corresponding to $z \to 1^-$).

(ii) *When $z \to 1^-$ in a cone of angle less than π inside the unit disc.* In that case, we observe that the identity in (46) remains valid by analytic continuation, since the integral there is still convergent (this property owes to the fast decay of $\Gamma(s)$ towards $\pm i\infty$). Then the residue calculation (47), on which the expansion of $\Lambda(w)$ is based in the real case $w > 0$, still makes sense. The extension of the asymptotic expansion of Li_α within the unit disc is thus granted.

(iii) *When z tends to 1 vertically.* Details of the proof are given in [223]. What is needed is a justification of the validity of expansion (43), when z is allowed to tend to 1 from the exterior of the unit disc. The key to the analysis is a Lindelöf integral representation of the polylogarithm (Notes IV.8 and IV.9, p. 237), which provides analytic continuation; namely,

$$
(48) \qquad \mathrm{Li}_\alpha(-z) = -\frac{1}{2i\pi} \int_{1/2-i\infty}^{1/2+i\infty} \frac{z^s}{s^\alpha} \frac{\pi}{\sin \pi s} \, ds.
$$

$$\mathrm{Li}_{-1/2}(z) = \sum_{n \geq 1} \sqrt{n} z^n = \frac{\sqrt{\pi}}{2(1-z)^{3/2}} - \frac{3\sqrt{\pi}}{8(1-z)^{1/2}} + \zeta(-\tfrac{1}{2}) + O\left((1-z)^{1/2}\right)$$

$$\mathrm{Li}_0(z) = \sum_{n \geq 1} z^n \equiv \frac{1}{1-z} - 1$$

$$\mathrm{Li}_{0,1}(z) = \sum_{n \geq 1} \log n \, z^n = \frac{L(z) - \gamma}{1-z} - \frac{1}{2} L(z) + \frac{\gamma - 1}{2} + \log\sqrt{2\pi} + O\left((1-z)L(z)\right)$$

$$\mathrm{Li}_{1/2}(z) = \sum_{n \geq 1} \frac{z^n}{\sqrt{n}} = \sqrt{\frac{\pi}{1-z}} + \zeta(\tfrac{1}{2}) - \frac{1}{4}\sqrt{\pi}\sqrt{1-z} + O\left((1-z)^{3/2}\right)$$

$$\mathrm{Li}_{1/2,1}(z) = \sum_{n \geq 1} \frac{\log n}{\sqrt{n}} z^n = \sqrt{\pi} \frac{-L(z) - \gamma - 2\log 2}{\sqrt{1-z}} - \zeta(\tfrac{1}{2})\left(\frac{\gamma}{2} + \frac{\pi}{4} + \log\sqrt{8\pi}\right) + \cdots$$

$$\mathrm{Li}_1(z) = \sum_{n \geq 1} \frac{z^n}{n} \equiv L(z)$$

$$\mathrm{Li}_2(z) = \sum_{n \geq 1} \frac{z^n}{n^2} = \frac{\pi^2}{6} - (L(z) + 1)(1 - z) - (\frac{1}{4} + \frac{1}{2}L(z))(1 - z)^2 + \cdots$$

Figure VI.11.   Sample expansions of polylogarithms ($L(z) := \log(1 - z)^{-1}$).

The proof then proceeds with the analysis of the polylogarithm when $z = e^{i(w-\pi)}$ and $s = 1/2 + it$, the integral (48) being estimated asymptotically as a *harmonic integral* (a continuous analogue of harmonic sums [614]) by means of Mellin transforms. The extension to a cone with vertex at 1, having a vertical symmetry and angle less than π, then follows by an analytic continuation argument. By unicity of asymptotic expansions (the horizontal cone of parts (i) and (ii) and the vertical cone have a non-empty intersection), the resulting expansion must coincide with the one calculated explicitly in part (i), above.

To conclude, regarding the general case $r \geq 0$, we may proceed along similar lines, with each $\log n$ factor introducing a derivative of the Riemann zeta function, hence a multiple pole at $s = 1$. It can then be checked that the resulting expansion coincides with what is given by formally differentiating the expansion of Li_α a number of times equal to r. (See also Note VI.20 below.) ∎

Figure VI.11 provides a table of expansions relative to commonly encountered polylogarithms (the function Li_2 is also known as a *dilogarithm*). Example VI.9 illustrates the use of polylogarithms for establishing a class of asymptotic expansions of which Stirling's formula appears as a special case. Further uses of Theorem VI.7 will appear in the following sections.

Example VI.9. *Stirling's formula, polylogarithms, and superfactorials.* One has

$$\sum_{n \geq 1} \log n! \, z^n = (1 - z)^{-1} \mathrm{Li}_{0,1}(z),$$

to which singularity analysis is applicable. Theorem VI.7 then yields the singular expansion

$$\frac{1}{1-z}\mathrm{Li}_{0,1}(z) \sim \frac{L(z)-\gamma}{(1-z)^2} + \frac{1}{2}\frac{1-L(z)+\gamma-1+\log 2\pi}{1-z} + \cdots,$$

from which Stirling's formula reads off:

$$\log n! \sim n\log n - n + \frac{1}{2}\log n + \log\sqrt{2\pi} + \cdots.$$

(Stirling's constant $\log\sqrt{2\pi}$ comes out as neatly $-\zeta'(0)$.) Similarly, define the *superfactorial* function to be $1^1 2^2 \cdots n^n$. One has

$$\sum_{n\geq 1}\log(1^2 2^2 \cdots n^n)z^n = \frac{1}{1-z}\mathrm{Li}_{-1,1}(z),$$

to which singularity analysis is mechanically applicable. The analogue of Stirling's formula then reads:

$$1^1 2^2 \cdots n^n \quad \sim \quad An^{\frac{1}{2}n^2 + \frac{1}{2}n + \frac{1}{12}}e^{-\frac{1}{4}n^2},$$

$$A \quad = \quad \exp\left(\frac{1}{12} - \zeta'(-1)\right) = \exp\left(-\frac{\zeta'(2)}{2\pi^2} + \frac{\log(2\pi)+\gamma}{12}\right).$$

The constant A is known as the Glaisher–Kinkelin constant [211, p. 135]. Higher order factorials can be treated similarly. .... ∎

▷ **VI.20.** *Polylogarithms of integral index and a general formula.* Let $\alpha = m \in \mathbb{Z}_{\geq 1}$. Then:

$$\mathrm{Li}_m(z) = \frac{(-1)^m}{(m-1)!}w^{m-1}(\log w - H_{m-1}) + \sum_{j\geq 0,\, j\neq m-1}\frac{(-1)^j}{j!}\zeta(m-j)w^j,$$

where H_m is the harmonic number and $w = -\log z$. [The line of proof is the same as in Theorem VI.7, only the residue calculation at $s = 1$ differs.] The general formula,

$$\mathrm{Li}_{\alpha,r}(z) \underset{z\to 1}{\sim} (-1)^r \frac{\partial^r}{\partial\alpha^r}\sum_{s\in\mathbb{Z}_{\geq 0}\cup\{1-\alpha\}}\mathrm{Res}\left[\zeta(s+\alpha)\Gamma(s)w^{-s}\right], \qquad w := -\log z,$$

holds for all $\alpha \in \mathbb{C}$ and $r \in \mathbb{Z}_{\geq 0}$ and is amenable to symbolic manipulation.   ◁

VI.9. Functional composition

Let f and g be functions analytic at the origin that have non-negative coefficients. We consider the composition

$$h = f \circ g, \qquad h(z) = f(g(z)),$$

assuming $g(0) = 0$. Let ρ_f, ρ_g, ρ_h be the corresponding radii of convergence, and let $\tau_f = f(\rho_f)$, and so on. We shall assume that f and g are Δ–continuable and that they admit singular expansions in the scale of powers. There are three cases to be distinguished depending on the value of τ_g in comparison with ρ_f.

— *Supercritical case*, when $\tau_g > \rho_f$. In that case, when z increases from 0, there is a value r strictly less than ρ_g such that $g(r)$ attains the value ρ_f, which triggers a singularity of $f \circ g$. In other words $r \equiv \rho_h = g^{(-1)}(\rho_f)$. Around this point, g is analytic and a singular expansion of $f \circ g$ is obtained by combining the singular expansion of f with the regular expansion of g at r. *The singularity type is that of the external function* (f).

— *Subcritical case*, when $\tau_g < \rho_f$. In this dual situation, the singularity of $f \circ g$ is driven by that of the inside function g. We have $\rho_h = \rho_g$, $\tau_h = f(\rho_g)$ and the singular expansion of $f \circ g$ is obtained by combining the regular expansion of f with the singular expansion of g at ρ_g. *The singularity type is that of the internal function* (g).

— *Critical case*, when $\tau_g = \rho_f$. In this boundary case, there is a confluence of singularities. We have $\rho_h = \rho_g$, $\tau_h = \tau_f$, and the singular expansion is obtained by applying the composition rules of the singular expansions involved. *The singularity type is a mix of the types of the internal and external functions* (f, g).

This classification extends the notion of a supercritical sequence schema in Section V. 2, p. 293, for which the external function reduces to $f(z) = (1 - z)^{-1}$, with $\rho_f = 1$. In this chapter, we limit ourselves to discussing examples directly, based on the guidelines above supplemented by the plain algebra of generalized power series expansions. Finer probabilistic properties of composition schemas are studied at several places in Chapter IX starting on p. 629.

Example VI.10. *"Supertrees".* Let $\mathcal{G}$ be the class of general Catalan trees:

$$\mathcal{G} = \mathcal{Z} \times \mathrm{SEQ}(\mathcal{G}) \qquad \Longrightarrow \qquad G(z) = \frac{1}{2}(1 - \sqrt{1 - 4z}).$$

The radius of convergence of $G(z)$ is $1/4$ and the singular value is $G(1/4) = 1/2$. The class $\mathcal{Z}\mathcal{G}$ consists of planted trees, which are such that to the root is attached a stem and an extra node, with OGF equal to $zG(z)$. We then introduce two classes of *supertrees* defined by substitution:

$$\mathcal{H} = \mathcal{G}[\mathcal{Z}\mathcal{G}] \qquad \Longrightarrow \qquad H(z) = G(zG(z))$$
$$\mathcal{K} = \mathcal{G}[(\mathcal{Z} + \mathcal{Z})\mathcal{G}] \quad \Longrightarrow \qquad K(z) = G(2zG(z)).$$

These are "trees of trees": the class $\mathcal{H}$ is formed of trees such that, on each node there is grafted a planted tree (by the combinatorial substitution of Section I. 6, p. 83); the class $\mathcal{K}$ similarly corresponds to the case when the stems can be of any two colours. Incidentally, combinatorial sum expressions are available for the coefficients,

$$H_n = \sum_{k=1}^{\lfloor n/2 \rfloor} \frac{1}{n-k} \binom{2k-2}{k-1} \binom{2n-3k-1}{n-k-1}, \quad K_n = \sum_{k=1}^{\lfloor n/2 \rfloor} \frac{2^k}{n-k} \binom{2k-2}{k-1} \binom{2n-3k-1}{n-k-1},$$

the initial values being given by

$$H(z) = z^2 + z^3 + 3z^4 + 7z^5 + 21z^6 + \cdots, \quad K(z) = 2z^2 + 2z^3 + 8z^4 + 18z^5 + 64z^6 + \cdots.$$

Since $\rho_G = 1/4$ and $\tau_G = 1/2$, the composition scheme is subcritical in the case of $\mathcal{H}$ and critical in the case of $\mathcal{K}$. In the first case, the singularity is of square-root type and one finds easily:

$$H(z) \underset{z \to \frac{1}{4}}{\sim} \frac{2 - \sqrt{2}}{4} - \frac{1}{\sqrt{8}} \sqrt{\frac{1}{4} - z}, \qquad \longrightarrow \qquad H_n \sim \frac{4^n}{8\sqrt{2\pi}n^{3/2}}.$$

In the second case, the two square-roots combine to produce a fourth root:

$$K(z) \underset{z \to \frac{1}{4}}{\sim} \frac{1}{2} - \frac{1}{\sqrt{2}} \left(\frac{1}{4} - z \right)^{1/4} \qquad \longrightarrow \qquad K_n \sim \frac{4^n}{8\Gamma(\frac{3}{4})n^{5/4}}.$$

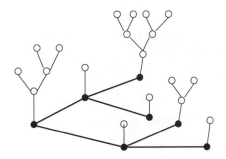

Figure VI.12. A binary supertree is a "tree of trees", with component trees all binary. The number of binary supertrees with $2n$ nodes has the unusual asymptotic form $c4^n n^{-5/4}$.

On a similar register, consider the class $\mathcal{B}$ of complete binary trees:

$$\mathcal{B} = \mathcal{Z} + \mathcal{Z} \times \mathcal{B} \times \mathcal{B} \quad \Longrightarrow \quad B(z) = \frac{1 - \sqrt{1 - 4z^2}}{2z},$$

and define the class of *binary supertrees* (Figure VI.12) by

$$\mathcal{S} = \mathcal{B}(\mathcal{Z} \times \mathcal{B}) \quad \Longrightarrow \quad S(z) = \frac{1 - \sqrt{2\sqrt{1 - 4z^2} - 1 + 4z^2}}{1 - \sqrt{1 - 4z^2}}.$$

The composition is critical since $zB(z) = \frac{1}{2}$ at the dominant singularity $z = \frac{1}{2}$. It is enough to consider the reduced function

$$\overline{S}(z) = S(\sqrt{z}) = z + z^2 + 3z^3 + 8z^4 + 25z^5 + 80z^6 + 267z^7 + 911z^8 + \cdots,$$

whose coefficients constitute *EIS* **A101490** and occur in Bousquet-Mélou's study of integrated superbrownian excursion [83]. We find

$$\overline{S}(z) \sim 1 - \sqrt{2}(1 - 4z)^{1/4} + (1 - 4z)^{1/2} + \cdots \quad \longrightarrow \quad \overline{S}_n = \frac{4^n}{n^{5/4}} \left(\frac{\sqrt{2}}{4\Gamma(\frac{3}{4})} - \frac{1}{2\sqrt{\pi}n^{1/4}} + \cdots \right).$$

For instance, a seven-term expansion yields a relative accuracy better than 10^{-4} for $n \geq 100$, so that such approximations are quite usable in practice.

The occurrence of the exponent $-\frac{5}{4}$ in the enumeration of bicoloured and binary supertrees is noteworthy. Related constructions have been considered by Kemp [364] who obtained more generally exponents of the form $-1 - 2^{-d}$ by iterating the substitution construction (in connection with so-called "multidimensional trees"). It is significant that asymptotic terms of the form $n^{p/q}$ with $q \neq 1, 2$ appear in elementary combinatorics, even in the context of simple algebraic functions. Such exponents tend to be associated with non-standard limit laws, akin to the stable distributions of probability theory: see our discussion in Section IX. 12, p. 715. .... ∎

▷ **VI.21.** *Supersupertrees.* Define supersupertrees by

$$S^{[2]}(z) = B(zB(zB(z))).$$

We find automatically (with the help of B. Salvy's program)

$$[z^{2n+1}]S^{[2]}(z) \sim 2^{-13/4}\Gamma\left(\frac{7}{8}\right)^{-1} 4^n n^{-9/8},$$

and further extensions involving an asymptotic term $n^{-1-2^{-d}}$ are possible [364].   ◁

▷ **VI.22.** *Valuated trees.* Consider the family of (rooted) general plane trees, whose vertices are decorated by integers from $\mathbb{Z}_{\geq 0}$ (called "values") and such that the values of two adjacent vertices differ by ± 1. Size is taken to be the number of edges. Let T_j be the class of valuated trees whose root has value j and $\mathcal{T} = \cup T_j$. The OGFs $T_j(z)$ satisfy the system of equations

$$T_j = 1 + z(T_{j-1} + T_{j+1})T_j,$$

so that $T(z)$ solves $T = 1 + 2zT^2$ and is a simple variant of the Catalan OGF:

$$T(z) = \frac{1 - \sqrt{1 - 8z}}{4z}.$$

Bouttier, Di Francesco, and Guitter [90, 91] found an amazing explicit form for the T_j; namely,

$$T_j = T \frac{(1 - Y^{j+1})(1 - Y^{j+5})}{(1 - Y^{j+2})(1 - Y^{j+4})}, \quad \text{with} \quad Y = z\frac{(1 + Y)^4}{1 + Y^2}.$$

In particular, each T_j is an algebraic function. The function T_0 counts maps (p. 513) that are Eulerian triangulations, or dually bipartite trivalent maps. The coefficients of the T_j as well as the distributions of labels in such trees can be analysed asymptotically: see Bousquet-Mélou's article [83] for a rich set of combinatorial connections.   ◁

Schemas. Singularity analysis also enables us to discuss at a fair level of generality the behaviour of *schemas*, in a way that parallels the discussion of the supercritical sequence schema, based on a meromorphic analysis (Section V. 2, p. 293). We illustrate this point here by means of the *supercritical cycle schema*. Deeper examples relative to recursively defined structures are developed in Chapter VII.

Example VI.11. *Supercritical cycle schema.* The schema $\mathcal{H} = \text{CYC}(\mathcal{G})$ forms labelled cycles from basic components in $\mathcal{G}$:

$$\mathcal{H} = \text{CYC}(\mathcal{G}) \quad \Longrightarrow \quad H(z) = \log \frac{1}{1 - G(z)}.$$

Consider the case where G attains the value 1 before becoming singular, that is, $\tau_G > 1$. This corresponds to a supercritical composition schema, which can be discussed in a way that closely parallels the supercritical sequence schema (Section V. 2, p. 293): a logarithmic singularity replaces a polar singularity.

Let $\sigma := \rho_H$, which is determined by $G(\sigma) = 1$. First, one finds:

$$H(z) \underset{z \to \sigma}{\sim} \log \frac{1}{1 - z/\sigma} - \log(\sigma G'(\sigma)) + A(z),$$

where $A(z)$ is analytic at $z = \sigma$. Thus:

$$[z^n]H(z) \sim \frac{\sigma^{-n}}{n}.$$

(The error term implicit in this estimate is exponentially small).

The BGF $H(z, u) = \log(1 - uG(z))^{-1}$ has the variable u marking the number of components in $\mathcal{H}$–objects. In particular, the mean number of components in a random $\mathcal{H}$–object of size n is $\sim \lambda n$, where $\lambda = 1/(\sigma G'(\sigma))$, and the distribution is concentrated around its mean. Similarly, the mean number of components with size k in a random $\mathcal{H}_n$ object is found to be asymptotic to $\lambda g_k \sigma^k$, where $g_k = [z^k]G(z)$. .... ∎

Weights

(49)

f_k	$\frac{1}{k}$	$\frac{1}{4^k}\binom{2k}{k}$	1	H_k	k	k^2
$f(z)$	$\log\frac{1}{1-z}$	$\frac{1}{\sqrt{1-z}}$	$\frac{1}{1-z}$	$\frac{1}{1-z}\log\frac{1}{1-z}$	$\frac{z}{(1-z)^2}$	$\frac{z+z^2}{(1-z)^3}$.

Triangular arrays

(50)

$g_n^{(k)}$	$\binom{n-1}{k-1}$	$\frac{k^{n-k}}{(n-k)!}$	$\binom{k}{n-k}$	$\frac{k}{n}\binom{2n-k-1}{n-1}$	$\frac{k}{n}\binom{2n}{n-k}$	$k\frac{n^{n-k-1}}{(n-k)!}$
$g(z)$	$\frac{z}{1-z}$	ze^z	$z(1+z)$	$\frac{1-\sqrt{1-4z}}{2}$	$\frac{1-2z-\sqrt{1-4z}}{2z}$	$T(z)$

Figure VI.13. Typical weights (top) and triangular arrays (bottom) illustrating the discussion of combinatorial sums $S_n = \sum_{k=1}^{n} f_k g_n^{(k)}$.

Combinatorial sums. Singularity analysis permits us to discuss the asymptotic behaviour of entire classes of combinatorial sums at a fair level of generality, with asymptotic estimates coming out rather automatically. We examine here combinatorial sums of the form

$$S_n = \sum_{k=0}^{n} f_k g_n^{(k)},$$

where f_k is a sequence of numbers, usually of a simple form and called the *weights*, while the $g_n^{(k)}$ are a triangular array of numbers, for instance Pascal's triangle.

As weights f_k we shall consider sequences such that $f(z)$ is Δ–analytic with a singular expansion involving functions of the standard scale of Theorems VI.1, VI.2, VI.3. Typical examples[9] for $f(z)$ and (f_k) are displayed in Figure VI.13, Equation (49). The triangular arrays discussed here are taken to be coefficients of the *powers* of some fixed function, namely,

$$g_n^{(k)} = [z^n](g(z))^k \qquad \text{where} \qquad g(z) = \sum_{n=1}^{\infty} g_n z^n,$$

with $g(z)$ an analytic function at the origin having non-negative coefficients and satisfying $g(0) = 0$. Examples are given in Figure VI.13, Equation (50). An interesting class of such arrays arises from the Lagrange Inversion Theorem (p. 732). Indeed, if $g(z)$ is implicitly defined by $g(z) = zG(g(z))$, one has $g_{n,k} = \frac{k}{n}[w^{n-k}]G(w)^n$; the last three cases of (50) are obtained in this way (by taking $G(w)$ as $1/(1-w)$, $(1+w)^2$, e^w).

By design, the generating function of the S_n is simply

$$S(z) = \sum_{n=0}^{\infty} S_n z^n = f(g(z)) \qquad \text{with} \qquad f(z) = \sum_{k=0}^{\infty} f_k z^k.$$

Consequently, the asymptotic analysis of S_n results by inspection from the way singularities of $f(z)$ and $g(z)$ get transformed by composition.

---

[9]Weights such as $\log k$ and $\sqrt{k}$, also satisfy these conditions, as seen in Section VI.8.

Example VI.12. *Bernoulli sums.* Let ϕ be a function from $\mathbb{Z}_{\geq 0}$ to $\mathbb{R}$ and write $f_k := \phi(k)$. Consider the sums

$$S_n := \sum_{k=0}^{n} \phi(k) \frac{1}{2^n} \binom{n}{k}.$$

If X_n is a binomial random variable[10], $X_n \in \mathrm{Bin}(n, \frac{1}{2})$, then $S_n = \mathbb{E}(\phi(X_n))$ is exactly the expectation of $\phi(X_n)$. Then, by the binomial theorem, the OGF of the sequence (S_n) is:

$$S(z) = \frac{2}{2-z} f\left(\frac{z}{2-z}\right).$$

Considering weights whose generating function, as in (49), has radius of convergence 1, what we have is a variant of the composition schema, with an additional prefactor. The composition scheme is of the *supercritical type* since the function $g(z) = z/(2-z)$, which has radius of convergence equal to 2, satisfies $\tau_g = \infty$. The singularities of $S(z)$ are then of the same type as those of the weight generating function $f(z)$ and one verifies, in all cases of (49), that, to first asymptotic order, $S_n \sim \phi(n/2)$: this is in agreement with the fact that the binomial distribution is concentrated near its mean $n/2$. Singularity analysis furthermore provides complete asymptotic expansions; for instance,

$$\mathbb{E}\left(\frac{1}{X_n} \mid X_n > 0\right) = \frac{2}{n} + \frac{2}{n^2} + \frac{6}{n^3} + O(n^{-4})$$

$$\mathbb{E}(H_{X_n}) = \log \frac{n}{2} + \gamma + \frac{1}{2n} - \frac{1}{12n^2} + O(n^{-3}).$$

See [208, 223] for more along these lines. .... ■

Example VI.13. *Generalized Knuth–Ramanujan Q-functions.* For reasons motivated by analysis of algorithms, Knuth has encountered repeatedly sums of the form

$$Q_n(\{f_k\}) = f_0 + f_1 \frac{n}{n} + f_2 \frac{n(n-1)}{n^2} + f_3 \frac{n(n-1)(n-2)}{n^3} + \cdots.$$

(See, e.g., [384, pp. 305–307].) There (f_k) is a sequence of coefficients (usually of at most polynomial growth). For instance, the case $f_k \equiv 1$ yields the expected time until the first collision in the birthday paradox problem (Section II.3, p. 114).

A closer examination shows that the analysis of such Q_n is reducible to singularity analysis. Writing

$$Q_n(\{f_k\}) = f_0 + \frac{n!}{n^{n-1}} \sum_{k \geq 1} f_k \frac{n^{n-k-1}}{(n-k)!}$$

reveals the closeness with the last column of (50). Indeed, setting

$$F(z) = \sum_{k \geq 1} \frac{f_k}{k} z^k,$$

one has $(n \geq 1)$

$$Q_n = f_0 + \frac{n!}{n^{n-1}} [z^n] S(z) \qquad \text{where} \qquad S(z) = F(T(z)),$$

and $T(z)$ is the Cayley tree function $(T = ze^T)$.

For weights $f_k = \phi(k)$ of polynomial growth, the schema is *critical*. Then, the singular expansion of S is obtained by composing the singular expansion of f with the expansion of T,

---

[10]A binomial random variable (p. 775) is a sum of Bernoulli variables: $X_n = \sum_{j=1}^{n} Y_j$, where the Y_j are independent and distributed as a Bernoulli variable Y, with $\mathbb{P}(Y = 1) = p$, $\mathbb{P}(Y = 0) = q = 1 - p$.

namely, $T(z) \sim 1 - \sqrt{2}\sqrt{1 - ez}$ as $z \to e^{-1}$. For instance, if $\phi(k) = k^r$ for some integer $r \geq 1$ then $F(z)$ has an rth order pole at $z = 1$. Then, the singularity type of $F(T(z))$ is $Z^{-r/2}$ where $Z = (1 - ez)$, which is reflected by $S_n \asymp e^n n^{r/2-1}$ (we use '$\asymp$' to represent order-of-growth information, disregarding multiplicative constants). After the final normalization, we see that $Q_n \asymp n^{(r+1)/2}$. Globally, for many weights of the form $f_k = \phi(k)$, we expect Q_n to be of the form $\sqrt{n}\phi(\sqrt{n})$, in accordance with the fact that the expectation of the first collision in the birthday problem is on average near $\sqrt{\pi n/2}$. . . . . ∎

▷ **VI.23.** *General Bernoulli sums.* Let $X_n \in \mathrm{Bin}(n; p)$ be a binomial random variable with general parameters p, q:

$$\mathbb{P}(X_n = k) = \binom{n}{k} p^k q^{n-k}, \qquad q = 1 - p.$$

Then with $f_k = \phi(k)$, one has

$$\mathbb{E}(\phi(X_n)) = [z^n] \frac{1}{1 - qz} f\left(\frac{pz}{1 - qz}\right),$$

so that the analysis develops as in the case $\mathrm{Bin}(n; \frac{1}{2})$. ◁

▷ **VI.24.** *Higher moments of the birthday problem.* Take the model where there are n days in the year and let B be the random variable representing the first birthday collision. Then $\mathbb{P}_n(B > k) = k! n^{-k} \binom{n}{k}$, and

$$\mathbb{E}_n(\Phi(B)) = \Phi(1) + Q_n(\{\Delta\Phi(k)\}), \qquad \text{where} \quad \Delta\Phi(k) := \Phi(k+1) - \Phi(k).$$

For instance $\mathbb{E}_n(B) = 1 + Q_n(\langle 1, 1, \ldots \rangle)$. We thus get moments of various functionals (here stated to two asymptotic terms)

$\Phi(x)$	x	$x^2 + x$	$x^3 + x^2$	$x^4 + x^3$
$\mathbb{E}_n(\Phi(B))$	$\sqrt{\frac{\pi n}{2}} + \frac{2}{3}$	$2n + 2$	$3\sqrt{\frac{\pi n^3}{2}} - 2n$	$8n^2 - 7\sqrt{\frac{\pi n^3}{2}}$

via singularity analysis. ◁

▷ **VI.25.** *How to weigh an urn? The "shake-and-paint" algorithm.* You are given an urn containing an unknown number N of identical looking balls. How to estimate this number in much fewer than $O(N)$ operations? A probabilistic solution due to Brassard and Bratley [92] uses a brush and some paint. Shake the urn, pull out a ball, then mark it with paint and replace it into the urn. Repeat until you find an already painted ball. Let X be the number of operations. One has $\mathbb{E}(X) \sim \sqrt{\pi N/2}$. Furthermore the quantity $Y := X^2/2$ constitutes, by the previous note, an asymptotically unbiased estimator of N, in the sense that $\mathbb{E}(Y) \sim N$. In other words, count the time till an already painted ball is first found, and return half of the square of this time. One also has $\sqrt{\mathbb{V}(Y)} \sim N$. By performing the experiment m times (using m different colours of paint) and by taking the arithmetic average of the m estimates, one obtains an unbiased estimator whose typical relative accuracy is $\sqrt{1/m}$. For instance, $m = 16$ gives an accuracy of 25%. (Similar principles are used in the design of data mining algorithms.) ◁

▷ **VI.26.** *Catalan sums.* These are defined by

$$S_n := \sum_{k \geq 0} f_k \binom{2n}{n-k}, \qquad S(z) = \frac{1}{\sqrt{1 - 4z}} f\left(\frac{1 - 2z - \sqrt{1 - 4z}}{2z}\right).$$

The case when $\rho_f = 1$ corresponds to a critical composition, which can be discussed much in the same way as Ramanujan sums. ◁

VI. 10. Closure properties

At this stage[11], we have available composition rules for singular expansions under operations such as $\pm$, $\times$, $\div$: these are induced by corresponding rules for extended formal power series, where generalized exponents and logarithmic factors are allowed. Also, from Section VI. 7, inversion of analytic functions normally gives rise to square-root singularities, and, from Section VI. 9, functions amenable to singularity analysis are essentially closed under composition.

In this section we show that functions amenable to singularity analysis (SA functions) satisfy explicit closure properties under differentiation, integration, and Hadamard product. (The contents are liberally borrowed from an article of Fill, Flajolet, and Kapur [208], to which we refer for details.) In order to keep the developments simple, we shall mostly restrict attention to functions that are Δ–analytic and admit a *simple* singular expansion of the form

$$(51) \qquad f(z) = \sum_{j=0}^{J} c_j (1-z)^{\alpha_j} + O((1-z)^A),$$

or a *simple* singular expansion *with logarithmic terms*

$$(52) \qquad f(z) = \sum_{j=0}^{J} c_j \, (L(z)) \, (1-z)^{\alpha_j} + O((1-z)^A), \qquad L(z) := \log \frac{1}{1-z},$$

where each c_j is a polynomial. These are the cases most frequently occurring in applications (the proof techniques are easily extended to more general situations).

Subsection VI. 10.1 treats differentiation and integration; Subsection VI. 10.2 presents the closure of functions that admit simple expansions under Hadamard product. Finally, Subsection VI. 10.3 concludes with an examination of several interesting classes of tree recurrences, where all the closure properties previously established are put to use in order to quantify precisely the asymptotic behaviour of recurrences that are attached to tree models.

VI. 10.1. Differentiation and integration. Functions that are SA happen to be closed under differentiation, this is in sharp contrast with real analysis. In the simple cases[12] of (51) and (52), closure under integration is also granted. The general principle (Theorems VI.8 and VI.9 below) is the following: *Derivatives and primitives of functions that are amenable to singularity analysis admit singular expansions obtained term by term, via formal differentiation and integration.*

The following statement is a version, tuned to our needs, of well-known differentiability properties of complex asymptotic expansions (see, e.g., Olver's book [465, p. 9]).

---

[11]This section represents supplementary material not needed elsewhere in the book, so that it may be omitted on first reading.

[12]It is possible but unwieldy to treat a larger class, which then needs to include arbitrarily nested logarithms, since, for instance, $\int dx/x = \log x$, $\int dx/(x \log x) = \log \log x$, and so on.

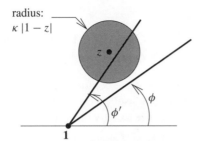

radius:
$\kappa |1 - z|$

Figure VI.14. The geometry of the contour $\gamma(z)$ used in the proof of the differentiation theorem.

Theorem VI.8 (Singular differentiation). *Let $f(z)$ be Δ–analytic with a singular expansion near its singularity of the simple form*

$$f(z) = \sum_{j=0}^{J} c_j (1 - z)^{\alpha_j} + O((1 - z)^A).$$

Then, for each integer $r > 0$, the derivative $\frac{d^r}{dz^r} f(z)$ is Δ–analytic. The expansion of the derivative at the singularity is obtained through term-by-term differentiation:

$$\frac{d^r}{dz^r} f(z) = (-1)^r \sum_{j=0}^{J} c_j \frac{\Gamma(\alpha_j + 1)}{\Gamma(\alpha_j + 1 - r)} (1 - z)^{\alpha_j - r} + O((1 - z)^{A-r}).$$

Proof. All that is required is to establish the effect of differentiation on error terms, which is expressed symbolically as

$$\frac{d}{dz} O((1 - z)^A) = O((1 - z)^{A-1}).$$

By bootstrapping, only the case of a single differentiation ($r = 1$) needs to be considered.

Let $g(z)$ be a function that is regular in a domain $\Delta(\phi, \eta)$ where it is assumed to satisfy $g(z) = O((1 - z)^A)$ for $z \in \Delta$. Choose a subdomain $\Delta' := \Delta(\phi', \eta')$, where $\phi < \phi' < \frac{\pi}{2}$ and $0 < \eta' < \eta$. By elementary geometry, for a sufficiently small $\kappa > 0$, the disc of radius $\kappa |z - 1|$ centred at a value $z \in \Delta'$ lies entirely in Δ; see Figure VI.14. We fix such a small value κ and let $\gamma(z)$ represent the boundary of that disc oriented positively.

The starting point is Cauchy's integral formula

(53) $$g'(z) = \frac{1}{2\pi i} \int_C g(w) \frac{dw}{(w - z)^2},$$

a direct consequence of the residue theorem. Here C should encircle z while lying inside the domain of regularity of g, and we opt for the choice $C \equiv \gamma(z)$. Then trivial bounds applied to (53) give

$$\begin{aligned}
|g'(z)| &= O\left(\|\gamma(z)\| \cdot (1-z)^A |1-z|^{-2}\right) \\
&= O\left(|1-z|^{A-1}\right).
\end{aligned}$$

The estimate involves the length of the contour, $\|\gamma(z)\|$, which is $O(1-z)$ by construction, as well as the bound on g itself, which is $O((1-z)^A)$ since all points of the contour are themselves at a distance exactly of the order of $|1-z|$ from 1. ∎

▷ **VI.27.** *Differentiation and logarithms.* Let $g(z)$ satisfy

$$g(z) = O\left((1-z)^A L(z)^k\right), \qquad L(z) = \log\frac{1}{1-z},$$

for $k \in \mathbb{Z}_{\geq 0}$. Then, one has

$$\frac{d^r}{dz^r} g(z) = O\left((1-z)^{A-r} L(z)^k\right).$$

(The proof is similar to that of Theorem VI.8.) ◁

It is well known that integration of asymptotic expansions is usually easier than differentiation. Here is a statement custom-tailored to our needs.

Theorem VI.9 (Singular integration). *Let $f(z)$ be Δ–analytic and admit an expansion near its singularity of the form*

$$f(z) = \sum_{j=0}^{J} c_j (1-z)^{\alpha_j} + O((1-z)^A).$$

Then $\int_0^z f(t)\, dt$ is Δ–analytic. Assume further that none of the quantities α_j and A equal -1.

(i) If $A < -1$, then the singular expansion of $\int f$ is

$$(54) \qquad \int_0^z f(t)\, dt = -\sum_{j=0}^{J} \frac{c_j}{\alpha_j + 1}(1-z)^{\alpha_j+1} + O\left((1-z)^{A+1}\right).$$

(ii) If $A > -1$, then the singular expansion of $\int f$ is

$$(55) \qquad \int_0^z f(t)\, dt = -\sum_{j=0}^{J} \frac{c_j}{\alpha_j + 1}(1-z)^{\alpha_j+1} + L_0 + O\left((1-z)^{A+1}\right),$$

where the "integration constant" L_0 has the value

$$L_0 := \sum_{\alpha_j < -1} \frac{c_j}{\alpha_j + 1} + \int_0^1 \left[f(t) - \sum_{\alpha_j < -1} c_j(1-t)^{\alpha_j}\right] dt.$$

Proof. The basic technique consists in integrating term by term the singular expansion of f. We let $r(z)$ be the remainder term in the expansion of f, that is,

$$r(z) := f(z) - \sum_{j=0}^{J} c_j(1-z)^{\alpha_j}.$$

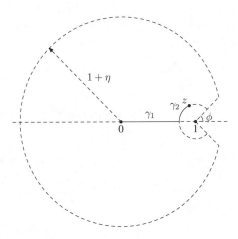

Figure VI.15. The contour used in the proof of the integration theorem.

By assumption, throughout the Δ–domain one has, for some positive constant K,

$$|r(z)| \le K|1 - z|^A.$$

(*i*) *Case $A < -1$.* Straight-line integration between 0 and z, provides (54), as soon as it has been established that

$$\int_0^z r(t)\, dt = O\left(|1 - z|^{A+1}\right).$$

By Cauchy's integral formula, we can choose any path of integration that stays within the region of analyticity of r. We choose the contour $\gamma := \gamma_1 \cup \gamma_2$, shown in Figure VI.15. Then, one has

$$\left| \int_\gamma r(t)\, dt \right| \le \left| \int_{\gamma_1} r(t)\, dt \right| + \left| \int_{\gamma_2} r(t)\, dt \right|$$

$$\le K \int_{\gamma_1} |1 - t|^A \, |dt| + K \int_{\gamma_2} |1 - t|^A |\, |dt|$$

$$= O(|1 - z|^{A+1}),$$

where the symbol $|dt|$ designates the differential line-length element in the corresponding curvilinear integral. Both integrals are $O(|1-z|^{A+1})$: for the integral along γ_1, this results from explicitly carrying out the integration; for the integral along γ_2, this results from the trivial bound $O(\|\gamma_2\|(1 - z)^A)$.

(*ii*) *Case $A > -1$.* We let $f_-(z)$ represent the "divergence part" of f that gives rise to non-integrability:

$$f_-(z) := \sum_{\alpha_j < -1} c_j (1 - z)^{\alpha_j}.$$

Then with the decomposition $f = [f - f_-] + f_-$, integrations can be performed separately. First, one finds

$$\int_0^z f_-(t)\,dt = - \sum_{\alpha_j < -1} \frac{c_j}{\alpha_j + 1}(1 - z)^{\alpha_j + 1} + \sum_{\alpha_j < -1} \frac{c_j}{\alpha_j + 1}.$$

Next, observe that the asymptotic condition guarantees the existence of $\int_0^1$ applied to $[f - f_-]$, so that

$$\int_0^z \left[f(t) - f_-(t) \right] dt = \int_0^1 \left[f(t) - f_-(t) \right] dt + \int_1^z \left[f(t) - f_-(t) \right] dt.$$

The first of these two integrals is a constant that contributes to L_0. As to the second integral, term-by-term integration yields

$$\int_1^z \left[f(t) - f_-(t) \right] dt = - \sum_{\alpha_j > -1} \frac{c_j}{\alpha_j + 1}(1 - z)^{\alpha_j + 1} + \int_1^z r(t)\,dt.$$

The remainder integral is finite, given the growth condition on the remainder term, and, upon carrying out the integration along the rectilinear segment joining 1 to z, trivial bounds show that it is indeed $O(|1 - z|^{A+1})$.   ∎

▷ **VI.28.** *Logarithmic cases.* The case in which either some α_j or A is -1 is easily treated by the additional rules

$$\int_0^z (1 - t)^{-1}\,dt = \mathrm{L}(z), \qquad \int_0^z O((1 - t)^{-1})\,dt = O(\mathrm{L}(z)).$$

that are consistent with elementary integration, and similar rules are easily derived for powers of logarithms. Furthermore, the corresponding O–transfers hold true. (The proofs are simple modifications of the one given above for the basic case.)   ◁

VI. 10.2. Hadamard Products. The *Hadamard product* of two functions $f(z)$ and $g(z)$ analytic at the origin is defined as their term-by-term product,

$$(56) \quad f(z) \odot g(z) = \sum_{n \geq 0} f_n g_n z^n, \quad \text{where} \quad f(z) = \sum_{n \geq 0} f_n z^n, \quad g(z) = \sum_{n \geq 0} g_n z^n.$$

As we are going to see, following Fill, Flajolet, and Kapur [208], functions amenable to singularity analysis are closed under Hadamard product. Establishing such a closure property requires methods for composing functions from the basic scale, namely $(1 - z)^a$, as well as error terms of the form $O((1 - z)^A)$. We address each problem in turn.

Theorem VI.10 (Hadamard Composition). *When neither of a, b, $a + b$ is an integer, the Hadamard product $(1 - z)^a \odot (1 - z)^b$ has an infinite expansion, valid in a Δ–domain, with exponent scale $\{0, 1, 2, \ldots\} \cup \{a + b + 1, a + b + 2, \ldots\}$; namely,*

$$(1 - z)^a \odot (1 - z)^b \sim \sum_{k \geq 0} \lambda_k^{(a,b)} \frac{(1 - z)^k}{k!} + \sum_{k \geq 0} \mu_k^{(a,b)} \frac{(1 - z)^{a+b+1+k}}{k!},$$

where the coefficients λ and μ are given by

$$\lambda_k^{(a,b)} = \frac{\Gamma(1 + a + b)}{\Gamma(1 + a)\Gamma(1 + b)} \frac{(-a)^{\overline{k}}(-b)^{\overline{k}}}{(-a - b)^{\overline{k}}}, \qquad \mu_k^{(a,b)} = \frac{\Gamma(-a - b - 1)}{\Gamma(-a)\Gamma(-b)} \frac{(1 + a)^{\overline{k}}(1 + b)^{\overline{k}}}{(2 + a + b)^{\overline{k}}}.$$

Here $x^{\overline{k}}$ is defined for $k \in \mathbb{Z}_{\geq 0}$ by $x^{\overline{k}} := x(x+1) \cdots (x+k-1)$.

Proof. The expansion around the origin,

$$(57) \qquad (1-z)^a = 1 + \frac{-a}{1}z + \frac{(-a)(-a+1)}{2!}z^2 + \cdots ,$$

gives through term-by-term multiplication

$$(58) \qquad (1-z)^a \odot (1-z)^b = {}_2F_1[-a, -b; 1; z].$$

Here ${}_2F_1$ represents the classical *hypergeometric function* of Gauss (p. 751) defined by

$$(59) \qquad {}_2F_1[\alpha, \beta; \gamma; z] = 1 + \frac{\alpha\beta}{\gamma}\frac{z}{1!} + \frac{\alpha(\alpha+1)\beta(\beta+1)}{\gamma(\gamma+1)}\frac{z^2}{2!} + \cdots .$$

From their transformation theory (see for instance [604, Ch XIV] and Appendix B.4: *Holonomic functions*, p. 748, for proof techniques), hypergeometric functions can generally be expanded in the vicinity of $z = 1$ by means of the $z \mapsto 1 - z$ transformation. Instantiation of this transformation with $\gamma = 1$ yields

$$(60) \quad {}_2F_1[\alpha, \beta; 1; z] = \frac{\Gamma(1-\alpha-\beta)}{\Gamma(1-\alpha)\Gamma(1-\beta)} {}_2F_1[\alpha, \beta; \alpha+\beta; 1-z]$$
$$+ \frac{\Gamma(\alpha+\beta-1)}{\Gamma(\alpha)\Gamma(\beta)}(1-z)^{-\alpha-\beta+1} {}_2F_1[1-\alpha, 1-\beta; 2-\alpha-\beta; 1-z].$$

The statement follows, upon appealing to the definition (59) of hypergeometric functions. ∎

▷ **VI.29.** *Special cases.* The case where either a or b is an integer poses no difficulty, since, for $m \in \mathbb{Z}_{\geq 0}$, the function $(1-z)^m \odot g(z)$ is a polynomial, while $(1-z)^{-m} \odot g(z)$ is reducible to a derivative of g, to which the Singular Differentiation Theorem (p. 419) can be applied.

The case $a + b \in \mathbb{Z}$ needs transformation formulae that extend (60): the principles (based on a Lindelöf integral representation, p. 237, and developed by Barnes) are described in [604, §14.53], and the formulae appear explicitly in [3, pp. 559–560]. ◁

▷ **VI.30.** *Simple expansions with logarithmic terms.* The technique of differentiation with respect to a parameter,

$$\left[(1-z)^a \mathrm{L}(z)\right] \odot (1-z)^b = -\frac{\partial}{\partial a}\left[(1-z)^a \odot (1-z)^b\right],$$

makes it possible to derive explicit composition rules for expansions involving logarithmic terms. ◁

The way Hadamard products preserve Δ–analyticity and compose error terms in singular expansions is summarized by the next statement.

Theorem VI.11 (Hadamard closure). (*i*) *Assume that $f(z)$ and $g(z)$ are analytic in a Δ–domain, $\Delta(\psi_0, \eta)$. Then, the Hadamard product $(f \odot g)(z)$ is analytic in a (possibly smaller) Δ–domain, Δ'.*

(*ii*) *Assume further that*

$$f(z) = O((1-z)^a) \text{ and } g(z) = O((1-z)^b), \quad z \in \Delta(\psi_0, \eta).$$

Then the Hadamard product $(f \odot g)(z)$ admits in Δ' an expansion given by the following rules:

— If $a + b + 1 < 0$, then

$$(f \odot g)(z) = O((1 - z)^{a+b+1}).$$

— If $k < a + b + 1 < k + 1$, for some integer $k \in \mathbb{Z}_{\geq -1}$, then

$$(f \odot g)(z) = \sum_{j=0}^{k} \frac{(-1)^j}{j!} (f \odot g)^{(j)} (1)(1 - z)^j + O\left((1 - z)^{a+b+1}\right).$$

— If $a + b + 1$ is a non-negative integer, then (with $\mathrm{L}(z) = \log(1 - z)^{-1}$)

$$(f \odot g)(z) = \sum_{j=0}^{k} \frac{(-1)^j}{j!} (f \odot g)^{(j)} (1)(1 - z)^j + O\left((1 - z)^{a+b+1} \mathrm{L}(z)\right).$$

Proof. (Sketch) The starting point is an important formula due to Hadamard that expresses Hadamard products as a contour integral:

$$(61) \qquad f(z) \odot g(z) = \frac{1}{2i\pi} \int_\gamma f(w) g\left(\frac{z}{w}\right) \frac{dw}{w}.$$

The contour γ in the w-plane should be chosen such that both factors, $f(w)$ and $g(z/w)$, are analytic. In other words, given the domain Δ in which both f and g are analytic, one should have $\gamma \subset \Delta \cap (z\Delta^{-1})$.

In the first case ($a + b + 1 < 0$), the precise geometry of a feasible contour γ is described in [208], the principles being similar to those employed in the construction of Hankel contours elsewhere in this chapter. The integral giving the value of the Hadamard product is finally estimated trivially, based on the order of growth assumptions on f and g, as $z \to 1$. This approach extends to the case $a + b + 1 = 0$, where a logarithmic factor comes in,

For the remaining cases, the easy identity

$$\vartheta^{c+d}(f \odot g) = \left(\vartheta^c f\right) \odot \left(\vartheta^d g\right), \qquad \text{where} \quad \vartheta \equiv z \frac{d}{dz},$$

reduces the analysis to the situation where $a + b + 1 < 0$. It suffices to differentiate sufficiently many times and finally integrate back, as permitted by the Singular Integration Theorem (p. 420). ∎

Globally, Theorems VI.10 and VI.11 establish the closure under Hadamard products of functions amenable to singularity analysis, which satisfy an expansion (51). In practice, in order to derive the singular expansion of a function at a singularity, one may conveniently appeal to the *Zigzag Algorithm* described in Figure VI.16, whose validity is ensured by the *a priori* knowledge of the *existence* of an expansion guaranteed by Theorems VI.10 and VI.11. (The "zigzag" qualifier reflects the fact that the algorithm proceeds back and forth, by making a repeated use of the correspondences between coefficient asymptotics and singularity asymptotics.) A typical application of this algorithm appears in (64) and (65) below, in the context of Pólya's drunkard problem.

Let $f(z)$ and $g(z)$ be Δ–analytic and admit simple singular expansions of the form (51) or (52). What is sought is the singular expansion of

$$h(z) := f(z) \odot g(z).$$

Step 1. Determine the asymptotic expansions $f_n = [z^n]f(z)$ and $g_n = [z^n]g(z)$ induced by the singular expansions of f and g in accordance with the singularity analysis process. Given finite singular expansions of f and g, the order C of the error in the expansion of h is known *a priori* by Theorem VI.11.

Step 2. Deduce from Step 1 an asymptotic expansion of $h_n = [z^n]h(z)$ by usual multiplication from the expansions of f_n and g_n.

Step 3. Reconstruct by singularity analysis a function $H(z)$ that is singular at 1 and is such that

$$[z^n]H(z) \sim [z^n]h(z).$$

This can be done by using the expansions of basic functions, as provided by Theorems VI.1 and VI.2 in the reverse direction. By construction, $H(z)$ is a sum of functions of the form $(1 - z)^\alpha L(z)^k$, which are all singular at 1.

Step 4. Output the singular expansion of $f \odot g$ as

$$h(z) = H(z) + P(z) + O\left((1 - z)^C\right),$$

where P is a polynomial of degree δ, which is the largest integer $< C$. The polynomial $P(z)$ is needed, since polynomials (and more generally functions analytic at 1) do not leave a trace in asymptotic expansions of coefficients. Since $h(z) - H(z)$ is δ times differentiable at 1, one must take

$$P(z) = \sum_{j=0}^{\delta} \frac{(-1)^j}{j!} \partial_z^j (h(z) - H(z))_{z=1} (1 - z)^j.$$

Figure VI.16.   The Zigzag Algorithm for computing singular expansions of Hadamard products.

Example VI.14. *Pólya's drunkard problem.* (This example is taken from Fill *et al.* [208].) In the d-dimensional lattice $\mathbb{Z}^d$ of points with integer coordinates, the drunkard performs a random walk starting from the origin with steps in $\{-1, +1\}^d$, each taken with equal likelihood. The probability that the drunkard is back at the origin after $2n$ steps is

(62)
$$q_n^{(d)} = \left(\frac{1}{2^{2n}}\binom{2n}{n}\right)^d,$$

since the walk is a product d independent one-dimensional walks. The probability that $2n$ is the epoch of the *first* return to the origin is the quantity $p_n^{(d)}$, which is determined implicitly by

(63)
$$\left(1 - \sum_{n=1}^{\infty} p_n^{(d)} z^n\right)^{-1} = \sum_{n=0}^{\infty} q_n^{(d)} z^n,$$

as results from the decomposition of loops into primitive loops (see also Note I.65, p. 90). In terms of the associated ordinary generating functions P and Q, this relation reads as $(1 - P(z))^{-1} = Q(z)$, implying $P(z) = 1 - 1/Q(z)$.

The asymptotic analysis of the q_n is straightforward; that of the p_n is more involved and is of interest in connection with recurrence and transience of the random walk; see, e.g., [170,

403]. The Hadamard closure theorem provides a direct tool to solve this problem. Define

$$\beta(z) := \sum_{n \geq 0} \frac{1}{2^{2n}} \binom{2n}{n} z^n \equiv \frac{1}{\sqrt{1-z}}.$$

Then, Equations (62) and (63) entail

$$P(z) = 1 - \frac{1}{\beta(z)^{\odot d}}, \qquad \text{where} \quad \beta(z)^{\odot d} := \beta(z) \odot \cdots \odot \beta(z) \quad (d \text{ times}).$$

The singularities of $P(z)$ are found as follows.

Case $d = 1$: No Hadamard product is involved and

$$P(z) = 1 - \sqrt{1-z}, \qquad \text{implying} \quad p_n^{(1)} = \frac{1}{n 2^{2n-1}} \binom{2n-2}{n-1} \sim \frac{1}{2\sqrt{\pi n^3}}.$$

(This agrees with the classical combinatorial solution expressed in terms of Catalan numbers.)

Case $d = 2$: By the Hadamard closure theorem, the function $Q(z) = \beta(z) \odot \beta(z)$ admits _a priori_ a singular expansion at $z = 1$ that is composed solely of elements of the form $(1-z)^{\alpha}$ possibly multiplied by integral powers of the logarithmic function $L(z) = \log(1/(1-z))$. From a computational standpoint (cf the Zigzag Algorithm), it is then best to start from the coefficients themselves,

(64) $$q_n^{(2)} \sim \left(\frac{1}{\sqrt{\pi n}} - \frac{1}{8\sqrt{\pi n^3}} + \cdots \right)^2 \sim \frac{1}{\pi} \left(\frac{1}{n} - \frac{1}{4n^2} + \cdots \right),$$

and reconstruct the only singular expansion that is compatible, namely

(65) $$Q(z) = \frac{1}{\pi} L(z) + K + O((1-z)^{1-\epsilon}),$$

where $\epsilon > 0$ is an arbitrarily small constant and K is fully determined as the limit as $z \to 1$ of $Q(z) - \pi^{-1} L(z)$. Then it can be seen that the function P is Δ–continuable. (Proof: Otherwise, there would be complex poles arising from zeros of the function Q on the unit disc, and this would entail in $p_n^{(2)}$ the presence of terms oscillating around 0, a fact that contradicts the necessary positivity of probabilities.) The singular expansion of $P(z)$ at $z = 1$ results immediately from that of $Q(z)$:

$$P(z) \sim 1 - \frac{\pi}{L(z)} + \frac{\pi^2 K}{L(z)^2} + \cdots .$$

so that, by Theorems VI.2 and VI.3, one has

$$p_n^{(2)} = \frac{\pi}{n \log^2 n} - 2\pi \frac{\gamma + \pi K}{n \log^3 n} + O\left(\frac{1}{n \log^4 n} \right)$$

$$K = 1 + \sum_{n=1}^{\infty} \left(16^{-n} \binom{2n}{n}^2 - \frac{1}{\pi n} \right)$$

$$\doteq 0.88254240061060637358582 57 .$$

(See the study by Louchard _et al._ [422, Sec. 4] for somewhat similar calculations.)

Case $d = 3$: This case is easy since $Q(z)$ remains finite at its singularity $z = 1$ where it admits an expansion in powers of $(1-z)^{1/2}$, with the consequence that

$$q_n^{(3)} \sim \left(\frac{1}{\sqrt{\pi n}} - \frac{1}{8\sqrt{\pi n^3}} + \cdots \right)^3 \sim \frac{1}{\pi^{3/2}} \left(\frac{1}{n^{3/2}} - \frac{3}{8n^{5/2}} + \cdots \right).$$

The function $Q(z)$ is *a priori* Δ–continuable and its singular expansion can be reconstructed from the form of coefficients:

$$Q(z) \underset{z \to 1}{\sim} Q(1) - \frac{2}{\pi}\sqrt{1-z} + O(|1-z|),$$

leading to

$$P(z) = \left(1 - \frac{1}{Q(1)}\right) - \frac{2}{\pi Q(1)^2}\sqrt{1-z} + O(|1-z|).$$

By singularity analysis, the last expansion gives

$$p_n^{(3)} = \frac{1}{\pi^{3/2}Q(1)^2}\frac{1}{n^{3/2}} + O\left(\frac{1}{n^2}\right)$$

$$Q(1) = \frac{\pi}{\Gamma\left(\frac{3}{4}\right)^4} \doteq 1.3932039296856768591842463.$$

A complete asymptotic expansion in powers $n^{-3/2}, n^{-5/2}, \ldots$ can be obtained by the same devices. In particular this improves the error term above to $O(n^{-5/2})$. The explicit form of $Q(1)$ results from its expression as the generalized hypergeometric $_3F_2[\frac{1}{2}, \frac{1}{2}, \frac{1}{2}; 1, 1; 1]$, which evaluates by Clausen's theorem and Kummer's identity to the square of a complete elliptic integral. (See the papers by Larry Glasser for context, for instance [293]; nowadays, several computer algebra systems even provide this value automatically.)

Higher dimensions are treated similarly, with logarithmic terms surfacing in asymptotic expansions for all even dimensions. .... ∎

VI. 10.3. Applications to tree recurrences.

To conclude with singularity analysis theory, we present the general framework of *tree recurrences*, also known as *probabilistic divide-and-conquer recurrences*, which are of the general form

(66)
$$f_n = t_n + \sum_k p_{n,k}(f_k + f_{n-a-k}), \qquad (n \geq n_0).$$

There, (f_n) is the sequence implicitly determined by the recurrence, assuming known initial conditions $f_0, \ldots, f_{n_0-1}$; the sequence (t_n) is known as the sequence of *tolls*; the array $(p_{n,k})$ is a triangular array of numbers that are probabilities in the sense that, for each fixed $n \geq 0$, one has $\sum_k p_{n,k} = 1$; the number a is a small fixed integer (usually 0 or 1).

The interpretation of the recurrence is in the form of a splitting process: a collection of n elements is given; a number a of these is put aside and what remains is partitioned into two subgroups, a "left" subgroup of cardinality K_n and a "right" subgroup of cardinality $n - a - K_n$. The quantity K_n is a random variable with probability distribution

$$\mathbb{P}(K_n = k) = p_{n,k}.$$

The splitting is repeated (recursively) till only groups of size less than the threshold n_0 are obtained. Assuming stochastic independence of all the random variables K involved, it is seen that f_n represents the expectation of the (total) *cost* C_n of a random (recursive) splitting, when a single stage involving n elements incurs a toll equal to t_n. In symbols:

$$f_n = \mathbb{E}(C_n), \qquad C_n = t_n + C_{K_n} + C_{n-a-K_n}.$$

Clearly, a particular realization of the splitting process can be represented by a binary tree. With a suitable choice of probabilities, such processes can be used to analyse cost functional of increasing binary trees, and binary Catalan trees, for instance. A prime motivation is the analysis of divide-and-conquer algorithms in computer science, like quicksort, mergesort, union-find algorithms, and so on [132, 383, 384, 537, 538, 598]. Our treatment once more follows the article [208].

A general approach to the asymptotic solution of a tree recurrence goes as follows. First, introduce generating functions,

$$f(z) = \sum_n f_n \omega_n z^n, \qquad t(z) = \sum_n t_n \omega'_n z^n,$$

for some normalization sequences (ω_n) and (ω'_n) that are problem-specific. (So, $\omega_n \equiv 1$ gives rise to an OGF, $\omega_n \equiv 1/n!$ to an EGF, with other normalizations being also useful.) Then, by linearity of the original recurrence, there exists a linear operator $\mathcal{L}$ on series (and functions), such that

$$f(z) = \mathcal{L}[t(z)].$$

Provided the splitting probabilities $p_{n,k}$ have expressions of a tractable form, it is reasonable to attempt expressing $\mathcal{L}$ in terms of the usual operations of analysis. One may then investigate the way $\mathcal{L}$ affects singularities and deduce the asymptotic form of the cost sequence (f_n) from the singularities of its generating function, $f(z)$. An interesting feature of this approach is to allow for a powerful discussion of the relationship between tolls and induced costs, in a way that parallels composition of singularities in Section VI. 9. Closure properties discussed earlier in this section are a crucial ingredient in the intervening singularity analysis process.

The three examples that we present combine closure properties with the singularity analysis of polylogarithms of Section VI. 8. Example VI.15 is relative to increasing binary trees (defined in Example II.17, p. 143), which model binary search trees of computer science. Example VI.16 discusses additive costs of random binary Catalan trees in the perspective of tree recurrences. Finally, Example VI.17 shows the applicability of singularity analysis to a basic coalescence–fragmentation process.

Example VI.15. *The binary search tree recurrence.* One of the simplest random tree models is defined as follows: a random binary tree of size $n \geq 1$ is obtained by taking a root and appending to it a left subtree of size K_n and a right subtree of size $n - 1 - K_n$, where K_n is uniformly distributed over the set of permissible values $\{0, 1, \ldots, n - 1\}$. (Trees under this model are equivalent to *increasing binary trees* encountered in Example II.17, p. 143, and to *binary search trees* of Note III.33, p. 203.) In the notations of (66), this process corresponds to

$$p_{n,k} \equiv \mathbb{P}(K_n = k) = \frac{1}{n}, \qquad 0 \leq k \leq n - 1.$$

The associated tree recurrence is then

$$f_n = t_n + \frac{2}{n} \sum_{k=0}^{n-1} f_k, \qquad f_0 = t_0,$$

which translates for OGFs,

$$f(z) := \sum_{n \geq 0} f_n z^n, \qquad t(z) = \sum_{n \geq 0} t_n z^n,$$

into a linear integral equation:

$$(67) \qquad f(z) = t(z) + 2 \int_0^z f(w) \frac{dw}{1-w}.$$

Differentiation yields the ordinary differential equation

$$f'(z) = t'(z) + \frac{2}{1-z} f(z), \qquad f(0) = t_0,$$

which is then solved by the variation-of-constants method. In this way, it is found that an integral transform expresses the relation between the GF of tolls and the GF of total costs. Assuming without loss of generality $t_0 = 0$, we have (with $\partial_w \equiv \frac{d}{dw}$)

$$(68) \qquad f(z) = \mathfrak{L}[t(z)], \qquad \text{where} \quad \mathfrak{L}[t(z)] = \frac{1}{(1-z)^2} \int_0^z (\partial_w t(w)) (1-w)^2 \, dw.$$

First, simple toll sequences that admit generating functions of a simple form can be employed to build a *repertoire*[13] that already provides useful indications on the relations between the orders of growth of (t_n) and (f_n). For instance, we find, for the rising-factorial tolls

$$\begin{cases} t_n^{\overline{\alpha}} := \binom{n+\alpha}{\alpha}, & t^{\overline{\alpha}}(z) = (1-z)^{-\alpha-1}, \\ f^{\overline{\alpha}}(z) = \frac{\alpha-1}{\alpha+1} \left[(1-z)^{-\alpha-1} - (1-z)^{-2} \right], & f_n^{\overline{\alpha}} = \frac{\alpha-1}{\alpha+1} \left[\binom{n+\alpha}{\alpha} - n - 1 \right], \end{cases}$$

for $\alpha \neq 1$, while $\alpha = 1$ corresponding to $t_n^{\overline{1}} = n+1$ leads to

$$f^{\overline{1}}(z) = \frac{2}{(1-z)^2} \log \frac{1}{1-z}, \qquad f_n^{\overline{1}} = 2(n+1)(H_{n+1} - 1) = 2n \log n + O(n),$$

with H_n a harmonic number. The emergence of an extra logarithmic factor for $\alpha = 1$ is to be noted: it corresponds to the fact that path length in an increasing binary tree of size n is $\sim 2n \log n$. Such elementary techniques provide the top two entries of Figure VI.17.

Singularity analysis furthermore permits us to develop a complete asymptotic expansion for tolls of the form $\sqrt{n}$, $\log n$, and many others. Consider for instance the toll $t_n^\alpha = n^\alpha$, for which the generating function $t(z)$ is recognized to be a polylogarithm. From Theorem VI.7 (p. 408), the function $t(z)$ admits a singular expansions in terms of elements of the form $(1-z)^\beta$, with the main term corresponding to $\beta = -\alpha - 1$ when $\alpha > -1$. The $\mathfrak{L}$ transformation of (68) reads as a succession of operations, *"differentiate, multiply by $(1-z)^2$, integrate, multiply by $(1-z)^{-2}$"*, which are covered by Theorems VI.8 and VI.9. Consequently, the chain on any particular element starts as

$$c(1-z)^\beta \quad \xrightarrow{\partial} \quad c\beta(1-z)^{\beta-1} \quad \xrightarrow{\times (1-z)^2} \quad c\beta(1-z)^{\beta+1}.$$

At this stage, integration intervenes: according to Theorem VI.9, assuming $\beta \neq -2$ and ignoring integration constants, we find

$$c\beta(1-z)^{\beta+1} \quad \xrightarrow{\int} \quad -c\frac{\beta}{\beta+2}(1-z)^{\beta+2} \quad \xrightarrow{\times (1-z)^{-2}} \quad -c\frac{\beta}{\beta+2}(1-z)^\beta.$$

---

[13]The repertoire approach is developed in an attractive manner by Greene and Knuth in [310].

Tolls (t_n)		costs (f_n)
$t_n = \binom{n+\alpha}{\alpha}$	$(\alpha > 1)$	$\dfrac{\alpha-1}{\alpha+1}\left[\binom{n+\alpha}{\alpha} - n + 1\right] \sim \dfrac{\alpha+1}{\alpha-1}\dfrac{n^\alpha}{\Gamma(\alpha+1)}$
$t_n = \binom{n+\alpha}{\alpha}$	$(\alpha < 1)$	$\dfrac{1-\alpha-1}{1+\alpha}\left[n+1 - \binom{n+\alpha}{\alpha}\right] \sim \dfrac{1+\alpha}{1-\alpha}n$
$t_n = n^\alpha$	$(2 < \alpha)$	$f_n = \dfrac{\alpha+1}{\alpha-1}n^\alpha + O(n^{\alpha-1})$
$t_n = n^\alpha$	$(1 < \alpha < 2)$	$f_n = \dfrac{\alpha+1}{\alpha-1}n^\alpha + O(n)$
$t_n = n^\alpha$	$(0 < \alpha < 1)$	$K_\alpha n + O(n^\alpha)$
$t_n = \log n$		$K_0' n - \log n + O(1)$

Figure VI.17. Tolls and costs for the binary search tree recurrence, with $t_0 = 0$.

Thus, the singular element $(1 - z)^\beta$ corresponds to a contribution

$$-c\frac{\beta}{\beta+2}\binom{n-\beta-1}{-\beta-1},$$

which is of order $O(n^{-\beta-1})$. This chain of operations suffices to determine the leading order of f_n when $t_n = n^\alpha$ and $\alpha > 1$.

The derivation above is representative of the main lines of the analysis, but it has left aside the determination of integration constants, which play a dominant rôle when $t_n = n^\alpha$ and $\alpha < 1$ (because a term of the form $K/(1-z)^2$ then dominates in $f(z)$). Introduce, in accordance with the statement of the Singular Integration Theorem (Theorem VI.9, p. 420) the quantity

$$\mathbf{K}[t] := \int_0^1 \left[t'(w)(1-w)^2 - \left(t'(w)(1-w)^2\right)_-\right] dw,$$

where f_- represents the sum of singular terms of exponent < -1 in the singular expansion of $f(z)$. Then, for $t_n = n^\alpha$ with $0 < \alpha < 1$, taking into account the integration constant (which gets multiplied by $(1-z)^{-2}$, given the shape of $\mathfrak{L}$), we find for $\alpha < 1$:

$$f_n \sim K_\alpha n, \qquad K_\alpha = \mathbf{K}[\mathrm{Li}_{-\alpha}] = 2\sum_{n=1}^\infty \frac{n^\alpha}{(n+1)(n+2)}.$$

Similarly, the toll $t_n = \log n$ gives rise to

$$f_n \sim K_0' n, \qquad K_0' = 2\sum_{n=1}^\infty \frac{\log n}{(n+1)(n+2)} \doteq 1.2035649167.$$

This last estimate quantifies the *entropy* of the distribution of binary search trees, which is studied by Fill in [207], and discussed in the reference book by Cover and Thomas on information theory [134, p. 74-76]. .... ∎

***Example* VI.16.** *The binary tree recurrence.* Consider a procedure that, given a (pruned) binary tree, performs certain calculations (without affecting the tree itself) at a cost of t_n, for size n, then recursively calls itself on the left and right subtrees. If the binary tree to which the

Tolls (t_n)		costs (f_n)
n^α	$(\frac{3}{2} < \alpha)$	$\dfrac{\Gamma(\alpha - \frac{1}{2})}{\Gamma(\alpha)} n^{\alpha+1/2} + O(n^{\alpha-1/2})$
$n^{3/2}$		$\dfrac{2}{\sqrt{\pi}} n^2 + O(n \log n)$
n^α	$(\frac{1}{2} < \alpha < \frac{3}{2})$	$\dfrac{\Gamma(\alpha - \frac{1}{2})}{\Gamma(\alpha)} n^{\alpha+1/2} + O(n)$
$n^{1/2}$		$\dfrac{1}{\sqrt{\pi}} n \log n + O(n)$
n^α	$(0 < \alpha < \frac{1}{2})$	$\overline{K}_\alpha n + O(1)$
$\log n$		$\overline{K}'_0 n + O(\sqrt{n}$

Figure VI.18. Tolls and costs for the binary tree recurrence.

procedure is applied is drawn uniformly among all binary trees of size n the expectation of the total cost of the procedure satisfies the recurrence

$$(69) \qquad f_n = t_n + \sum_{k=0}^{n-1} \frac{C_k C_{n-1-k}}{C_n} (f_k + f_{n-k}) \qquad \text{with} \quad C_n = \frac{1}{n+1} \binom{2n}{n}.$$

Indeed, the quantity

$$p_{n,k} = \frac{C_k C_{n-1-k}}{C_n}$$

represents the probability that a random tree of size n has a left subtree of size k and a right subtree of size $n - k$. It is then natural to introduce the generating functions

$$t(z) = \sum_{n \geq 0} t_n C_n z^n, \qquad f(z) = \sum_{n \geq 0} f_n C_n z^n,$$

and the recurrence (69) translates into a linear equation:

$$f(z) = t(z) + 2zC(z)f(z),$$

with $C(z)$ the OGF of Catalan numbers. Now, given a toll sequence (t_n) with *ordinary* generation function

$$\tau(z) := \sum_{n \geq 0} t_n z^n,$$

the function $t(z)$ is a Hadamard product: $t(z) = \tau(z) \odot C(z)$. Furthermore, $C(z)$ is well known, so that the fundamental relation is

$$(70) \qquad f(z) = \mathcal{L}[\tau(z)], \qquad \text{where} \quad \mathcal{L}[\tau(z)] = \frac{\tau(z) \odot C(z)}{\sqrt{1 - 4z}}, \qquad C(z) = \frac{1 - \sqrt{1 - 4z}}{2z}.$$

This transform relates the ordinary generating function of tolls to the normalized generating function of the total costs via a Hadamard product.

Tolls (t_n)		costs (f_n)
n^α	$(\frac{3}{2} < \alpha)$	$\dfrac{\Gamma(\alpha - \frac{1}{2})}{\sqrt{2}\Gamma(\alpha)} n^{\alpha+1/2} + O(n^{\alpha-1/2})$
$n^{3/2}$		$\sqrt{\dfrac{2}{\pi}} n^2 + O(n \log n)$
n^α	$(\frac{1}{2} < \alpha < \frac{3}{2})$	$\dfrac{\Gamma(\alpha - \frac{1}{2})}{\sqrt{2}\Gamma(\alpha)} n^{\alpha+1/2} + O(n)$
$n^{1/2}$		$\dfrac{1}{\sqrt{2\pi}} n \log n + O(n)$
n^α	$(0 < \alpha < \frac{1}{2})$	$\widehat{K}_\alpha n + O(1)$
$\log n$		$\widehat{K}'_0 n + O(\sqrt{n})$

Figure VI.19. Tolls and costs for the Cayley tree recurrence.

The calculation for simple tolls like n^r with $r \in \mathbb{Z}_{\geq 0}$ can be carried out elementarily. For the tolls $t_n^\alpha = n^\alpha$ what is required is the singular expansion of

$$\tau(z) \odot C\left(\frac{z}{4}\right) = \mathrm{Li}_{-\alpha}(z) \odot C\left(\frac{z}{4}\right) = \sum_{n=1}^{\infty} \frac{n^\alpha}{n+1} \binom{2n}{n} \left(\frac{z}{4}\right)^n.$$

This is precisely covered by Theorems VI.7 (p. 408), VI.10 (p. 422), and VI.11 (p. 423). The results of Figure VI.18 follow, after routine calculations. .... ∎

Example **VI.17.** *The Cayley tree recurrence.* Consider n vertices labelled $1, \ldots, n$. There are $(n-1)! n^{n-2}$ sequences of edges,

$$\langle u_1, v_1, \rangle, \langle u_2, v_2, \rangle, \cdots, \langle u_{n-1}, v_{n-1} \rangle,$$

that give rise to a tree over $\{1, \ldots, n\}$, and the number of such sequences is $(n-1)! n^{n-2}$ since there are n^{n-2} unrooted trees of size n. At each stage k, the edges numbered 1 to k determine a forest. Each addition of an edge connects two trees [that then become rooted] and reduces the number of trees in the forest by 1, so that the forest evolves from the totally disconnected graph (at time 0) to an unrooted tree (at time $n-1$). If we consider each of the sequences to be equally likely, the probability that u_{n-1} and v_{n-1} belong to components of size k and $(n-k)$ is

$$\frac{1}{2(n-1)} \binom{n}{k} \frac{k^{k-1}(n-k)^{n-k-1}}{n^{n-2}}.$$

(The reason is that there are k^{k-1} rooted trees of size k; the last added edge has $n-1$ possibilities and 2 possible orientations.)

Assume that the aggregation of two trees into a tree of size equal to ℓ incurs a toll of t_ℓ. The total cost of the aggregation process for a final tree of size n satisfies the recurrence

$$(71) \quad f_n = t_n + \sum_{0 < k < n} p_{n,k}(f_k + f_{n-k}), \qquad p_{n,k} = \frac{1}{2(n-1)} \binom{n}{k} \frac{k^{k-1}(n-k)^{n-k-1}}{n^{n-2}}.$$

The recurrence (71) has been studied in detail by Knuth and Pittel [383], building upon an earlier analysis of Knuth and Schönhage [384]. A prime motivation of the cited works is the emergence of this recurrence in the study algorithms that dynamically manage equivalence relations (the so-called union-find algorithm [384]).

Given the sequence of tolls (t_n), we introduce the generating function

$$\tau(z) = \sum_{n \geq 1} t_n z^n,$$

and let T be the Cayley tree function ($T = ze^T$). For total costs, the generating function adopted is

$$f(z) = \sum_{n \geq 1} f_n n^{n-1} z^n.$$

The basic recurrence (71) can then be rephrased as a linear ordinary differential equation, which is solved by the variation-of-constant method. This gives rise to an integral transform involving a Hadamard product, namely,

$$(72) \quad f(z) = \mathfrak{L}[\tau(z)], \quad \text{with} \quad \mathfrak{L}[\tau](z) = \frac{1}{2} \frac{T(z)}{1 - T(z)} \int_0^z \partial_w \left(\tau(w) \odot T(w)^2 \right) \frac{dw}{T(w)}.$$

Though the expression of the transform looks formidable at first sight, it is really nothing but a short sequence of basic operations, "Hadamard product, multiplication, differentiation, division, integration, multiplication", each of which has a quantifiable effect on functions of singularity analysis class. (The singularity structure of $T(z)$ is itself determined by the Singular Inversion Theorem, Theorem VI.6, p. 404.)

The net result is that the effect of tolls of the form n^α, $\log n$, and so on, can be analysed: see Figure VI.19 for a listing of estimates. Details of the proof are left as an exercise to our reader and are otherwise found in [208, §5.3]. The analogy of behaviour with the Catalan tree recurrence stands out. This example is also of interest since it furnishes an analytically tractable model of a coalescence-fragmentation process, which is of great interest in several areas of science, for which we refer to Aldous' survey [9]. .... ∎

VI. 11. Tauberian theory and Darboux's method

There are several alternative approaches to the analysis of coefficients of functions that are of moderate growth. Naturally, all such methods must provide estimates compatible with singularity analysis theory (Theorems VI.1, VI.2, and VI.3). Each one requires some sort of "*regularity condition*" either on the part of the function or on the part of the coefficient sequence, the regularity condition of singularity analysis being in essence analytic continuation.

The methods briefly surveyed here fall into three broad categories: (*i*) Elementary real analytic methods; (*ii*) Tauberian theorems; (*iii*) Darboux's method.

Elementary real analytic methods assume some *a priori* smoothness conditions on the coefficient sequence; they are included here for the sake of completeness, though properly speaking they do not belong to the galaxy of complex asymptotic methods. Their scope is mostly limited to the analysis of products while the other methods permit one to approach more general functional composition patterns. Tauberian theorems belong to the category of advanced real analysis methods; they also need some *a priori* regularity on the coefficients, typically positivity or monotonicity. Darboux's method requires some smoothness of the function on the closed unit disc, and, by its techniques and scope, it is the closest to singularity analysis.

We content ourselves with a brief discussion of the main results. For more information, the reader is referred to Odlyzko's excellent survey [461].

Elementary real analytic methods. An asymptotic equivalent of the coefficients of a function can sometimes be worked out elementarily from simple properties of the component functions. The regularity conditions are a smooth asymptotic behaviour of the coefficients of one of the two factors in a product of generating functions. A prime source for these techniques is Bender's survey [36].

Theorem VI.12 (Real analysis asymptotics). *Let $a(z) = \sum a_n z^n$ and $b(z) = \sum b_n z^n$ be two power series with radii of convergence $\alpha > \beta \geq 0$, respectively. Assume that $b(z)$ satisfies the ratio test,*

$$\frac{b_{n-1}}{b_n} \to \beta \qquad as \qquad n \to \infty.$$

Then the coefficients of the product $f(z) = a(z) \cdot b(z)$ satisfy, provided $a(\beta) \neq 0$:

$$[z^n] f(z) \sim a(\beta) b_n \qquad as \qquad n \to \infty.$$

Proof. (Sketch) The basis of the proof is the following chain:

$$
\begin{aligned}
f_n &= a_0 b_n + a_1 b_{n-1} + a_2 b_{n-2} + \cdots + a_n b_0) \\
&= b_n \left(a_0 + a_1 \frac{b_{n-1}}{b_n} + a_2 \frac{b_{n-2}}{b_n} + \cdots + a_n \frac{b_0}{b_n} \right) \\
&= b_n \left(a_0 + a_1 \left(\frac{b_{n-1}}{b_n} \right) + a_2 \left(\frac{b_{n-2}}{b_{n-1}} \right) \left(\frac{b_{n-1}}{b_n} \right) + \cdots \right) \\
&\sim b_n (a_0 + a_1 \beta + a_2 \beta^2 + \cdots).
\end{aligned}
$$

There, only the last line requires a little elementary analysis that is left as an exercise to the reader (see Pólya–Szegő [492], Problem 178, Part I, Volume I). ∎

This theorem applies for instance to the EGF of 2–regular graphs:

$$f(z) = a(z) \cdot b(z) \qquad with \qquad a(z) = e^{-z/2 - z^2/4}, \quad b(z) = \frac{1}{\sqrt{1-z}},$$

for which it gives $f_n \sim e^{-3/4} \binom{n-1/2}{n} \sim \frac{e^{-3/4}}{\sqrt{\pi n}}$, in accordance with Example VI.2 (p. 395). Clearly, a whole collection of lemmas can be stated in the same vein. Singularity analysis usually provides more complete expansions, although Theorem VI.12 does apply to a few situations not covered by it.

Tauberian theory. Tauberian methods apply to functions whose growth is only known along the positive real line. The regularity conditions are in the form of additional assumptions on the coefficients (positivity or monotonicity) known under the name of Tauberian "side conditions". An insightful introduction to the subject may be found in Titchmarsh's book [577], and a detailed exposition in Postnikov's monograph [494] and Korevaar's compendium [389]. We cite the most famous of all Tauberian theorems due to Hardy, Littlewood, and Karamata. For the purpose of this section, a function is said to be *slowly varying* at infinity iff, for any $c > 0$, one has

$\Lambda(cx)/\Lambda(x) \to 1$ as $x \to +\infty$. (Examples of slowly varying functions are provided by powers of logarithms or iterated logarithms.)

Theorem VI.13 (The HLK Tauberian theorem). *Let $f(z)$ be a power series with radius of convergence equal to 1, satisfying*

$$(73) \qquad f(z) \sim \frac{1}{(1-z)^\alpha} \Lambda\left(\frac{1}{1-z}\right),$$

for some $\alpha \geq 0$ with Λ a slowly varying function. Assume that the coefficients $f_n = [z^n]f(z)$ are all non-negative (this is the "side condition"). Then

$$(74) \qquad \sum_{k=0}^{n} f_k \sim \frac{n^\alpha}{\Gamma(\alpha+1)} \Lambda(n).$$

The conclusion (74) is consistent with the result given by singularity analysis: under the conditions, and if in addition analytic continuation is assumed, then

$$(75) \qquad f_n \sim \frac{n^{\alpha-1}}{\Gamma(\alpha)} \Lambda(n),$$

which by summation yields the estimate (74).

It must be noted that a Tauberian theorem requires very little on the part of the function. However, it gives little, since it does *not* include error estimates. Also, the result it provides is valid in the more restrictive sense of mean values, or Cesàro averages. (If further regularity conditions on the f_n are available, for instance monotonicity, then the conclusion of (75) can then be deduced from (74) by purely elementary real analysis.) The method applies only to functions that are large enough at their singularity (the assumption $\alpha \geq 0$), and despite numerous efforts to improve the conclusions, it is the case that Tauberian theorems do not have much to offer in terms of error estimates.

Appeal to a Tauberian theorem may be justified when a function has, apart from the positive half line, a very irregular behaviour near its circle of convergence, for instance when each point of the unit circle is a singularity. (The function is then said to admit the unit circle as a natural boundary.) An interesting example of this situation is discussed by Greene and Knuth [309] who consider the function

$$(76) \qquad f(z) = \prod_{k=1}^{\infty}\left(1 + \frac{z^k}{k}\right),$$

which is the EGF of permutations having cycles all of different lengths. A little computation shows that

$$\log \prod_{k=1}^{\infty}\left(1 + \frac{z^k}{k}\right) = \sum_{k=1}^{\infty} \frac{z^k}{k} - \frac{1}{2}\sum_{k=1}^{\infty}\frac{z^{2k}}{k^2} + \frac{1}{3}\sum_{k=1}^{\infty}\frac{z^{3k}}{k^3} - \cdots$$

$$\sim \log\frac{1}{1-z} - \gamma + o(1).$$

(Only the last line requires some care, see [309].) Thus, we have

$$f(z) \sim \frac{e^{-\gamma}}{1-z} \qquad \longrightarrow \qquad \frac{1}{n}(f_0 + f_1 + \cdots + f_n) \sim e^{-\gamma},$$

by virtue of Theorem VI.12. In fact, Greene and Knuth were able to supplement this argument by a "bootstrapping" technique and show a stronger result, namely

$$f_n \to e^{-\gamma}.$$

▷ **VI.31.** *Fine asymptotics of the Greene–Knuth problem.* With $f(z)$ as in (76), we have

$$
\begin{aligned}
[z^n]f(z) \;=\; & e^{-\gamma} + \frac{e^{-\gamma}}{n} + \frac{e^{-\gamma}}{n^2}(-\log n - 1 - \gamma + \log 2) \\
& + \frac{1}{n^3}\left[e^{-\gamma}\log^2 n + c_1 \log n + c_2 + 2(-1)^n + \Omega(n)\right] + O\left(\frac{1}{n^4}\right),
\end{aligned}
$$

where c_1, c_2 are computable constants and $\Omega(n)$ has period 3. (The paper [227] derives a complete expansion based on a combination of Darboux's method and singularity analysis.) ◁

Darboux's method. The method of Darboux (also known as the Darboux–Pólya method) requires, as regularity condition, that functions be sufficiently differentiable ("smooth") on their circle of convergence. What lies at the heart of the method is a simple relation between the smoothness of a function and the decrease of its Taylor coefficients.

Theorem VI.14 (Darboux's method). *Assume that $f(z)$ is continuous in the closed disc $|z| \le 1$ and is, in addition, k times continuously differentiable ($k \ge 0$) on $|z| = 1$. Then*

$$
(77) \qquad\qquad [z^n]f(z) = o\left(\frac{1}{n^k}\right).
$$

Proof. Start from Cauchy's coefficient formula

$$f_n = \frac{1}{2i\pi}\int_{\mathcal{C}} f(z)\, \frac{dz}{z^{n+1}}.$$

Because of the continuity assumption, one may take as integration contour $\mathcal{C}$ the unit circle. Setting $z = e^{i\theta}$ yields the Fourier version of Cauchy's coefficient formula,

$$
(78) \qquad\qquad f_n = \frac{1}{2\pi}\int_0^{2\pi} f(e^{i\theta})e^{-ni\theta}\, d\theta.
$$

The integrand in (78) is strongly oscillating. The Riemann–Lebesgue lemma of classical analysis [577, p. 403] shows that the integral tends to 0 as $n \to \infty$.

The argument above covers the case $k = 0$. For a general k, successive integrations by parts give

$$
[z^n]f(z) = \frac{1}{2\pi(in)^k}\int_0^{2\pi} f^{(k)}(e^{i\theta})e^{-ni\theta}\, d\theta,
$$

a quantity that is $o(n^k)$, by Riemann–Lebesgue again. ∎

Various consequences of Theorem VI.14 are given in reference texts also under the name of Darboux's method. See for instance [129, 309, 329, 608]. We shall only illustrate the mechanism by rederiving in this framework the analysis of the EGF of 2–regular graphs (Example VI.2, p. 395). We have

$$
(79) \qquad f(z) = \frac{e^{-z/2-z^2/4}}{\sqrt{1-z}} = \frac{e^{-3/4}}{\sqrt{1-z}} + e^{-3/4}\sqrt{1-z} + R(z).
$$

There $R(z)$ is the product of $(1 - z)^{3/2}$ with a function analytic at $z = 1$ that is a remainder in the Taylor expansion of $e^{-z/2 - z^2/4}$. Thus, $R(z)$ is of class $\mathbf{C}^1$, i.e., continuously differentiable once. By Theorem VI.14, we have

$$[z^n]R(z) = o\left(\frac{1}{n}\right),$$

so that

(80) $$[z^n]f(z) = \frac{e^{-3/4}}{\sqrt{\pi n}} + o\left(\frac{1}{n}\right).$$

Darboux's method bears some resemblance to singularity analysis in that the estimates are derived from translating error terms in expansions. However, smoothness conditions, rather than plain order of growth information, are required by it. The method is often applied, in situations similar to (79)–(80), to functions that are products of the type $h(z)(1-z)^\alpha$ with $h(z)$ analytic at 1. In such particular cases, Darboux's method is however subsumed by singularity analysis.

It is inherent in Darboux's method that it cannot be applied to functions whose singular expansion only involves terms that become infinite, while singularity analysis can. A clear example arises in the analysis of the common subexpression problem [257] where there occurs a function with a singular expansion of the form

$$\frac{1}{\sqrt{1-z}} \frac{1}{\sqrt{\log \frac{1}{1-z}}} \left[1 + \frac{c_1}{\log \frac{1}{1-z}} + \cdots \right].$$

▷ **VI.32.** *Darboux versus singularity analysis.* This note provides an instance where Darboux's method applies whereas singularity analysis does not. Let

$$F_r(z) = \sum_{n=0}^{\infty} \frac{z^{2^n}}{(2^n)^r}.$$

The function $F_0(z)$ is singular at every point of the unit circle, and the same property holds for any F_r with $r \in \mathbb{Z}_{\geq 0}$. [Hint: F_0, which satisfies the functional equation $F(z) = z + F(z^2)$, grows unboundedly near 2^nth roots of unity.] Darboux's method can be used to derive

$$[z^n]\frac{1}{\sqrt{1-z}}F_5(z) = \frac{c}{\sqrt{\pi n}} + o\left(\frac{1}{n}\right), \qquad c := \frac{32}{31}.$$

What is the best error term that can be obtained? ◁

VI. 12. Perspective

The method of singularity analysis expands our ability to extract coefficient asymptotics to a far wider class of functions than the meromorphic and rational functions of Chapters IV and V. This ability is the fundamental tool for analysing many of the generating functions provided by the symbolic method of Part A, and it is applicable at a considerable level of generality.

The basic method is straightforward and appealing: we locate singularities, establish analyticity in a domain around them, expand the functions around the singularities, and apply general transfer theorems to take each term in the function expansion to a term in the asymptotic expansion of its coefficients. The method applies directly to a large variety of explicitly given functions, for instance combinations of rational functions, square roots, and logarithms, as well as to functions that are implicitly defined, like generating functions for tree structures, which are obtained by analytic inversion. Functions amenable to singularity analysis also enjoy rich closure properties, and the corresponding operations mirror the natural operations on generating functions implied by the combinatorial constructions of Chapters I–III.

This approach again sets us in the direction of the ideal situation of having a theory where combinatorial constructions and analytic methods fully correspond, but, again, the very essence of analytic combinatorics is that the theorems that provide asymptotic results cannot be so general as to be free of analytic side conditions. In the case of singularity analysis, these side conditions have to do with establishing analyticity in a domain around singularities. Such conditions are automatically satisfied by a large number of functions with moderate (at most polynomial) growth near their dominant singularities, justifying precisely what we need: the term-by-term transfer from the expansion of a generating function at its singularity to an asymptotic form of coefficients, including error terms. The calculations involved in singularity analysis are rather mechanical. (Salvy [528] has indeed succeeded in automating the analysis of a large class of generating functions in this way.)

Again, we can look carefully at specific combinatorial constructions and then apply singularity analysis to general abstract schemas, thereby solving whole classes of combinatorial problems at once. This process, along with several important examples, is the topic of Chapter VII, to come next. After that, we introduce, in Chapter VIII, the saddle-point method, which is appropriate for functions without singularities at a finite distance (entire functions) as well as those whose growth is rapid (exponential) near their singularities. Singularity analysis will surface again in Chapter IX, given its crucial technical rôle in obtaining uniform expansions of multivariate generating functions near singularities.

Bibliographic notes. Excellent surveys of asymptotic methods in enumeration have been given by Bender [36] and more recently Odlyzko [461]. A general reference to asymptotic analysis that has a remarkably concrete approach is De Bruijn's book [143]. Comtet's [129] and Wilf's [608] books each devote a chapter to these questions.

This chapter is largely based on the theory developed by Flajolet and Odlyzko in [248], where the term "singularity analysis" originates. An important early (and unduly neglected) reference is the study by Wong and Wyman [615]. The theory draws its inspiration from classical analytic number theory, for instance the prime number theorem where similar contours are used (see the discussion in [248] for sources). Another area where Hankel contours are used is the inversion theory of integral transforms [168], in particular in the case of algebraic and logarithmic singularities. Closure properties developed here are from the articles [208, 223] by Flajolet, Fill, and Kapur.

Darboux's method can often be employed as an alternative to singularity analysis. Although it is still a widely used technique in the literature, the direct mapping of asymptotic scales afforded by singularity analysis appears to us to be much more transparent. Darboux's method is well explained in the books by Comtet [129], Henrici [329], Olver [465], and Wilf [608]. Tauberian theory is treated in detail in Postnikov's monograph [494] and Korevaar's encyclopaedic treatment [389], with an excellent introduction to be found in Titchmarsh's book [577].

VII

Applications of Singularity Analysis

Mathematics is being lazy. Mathematics is letting the principles do the work for you so that you do not have to do the work for yourself[1].

— GEORGE PÓLYA

I wish to God these calculations had been executed by steam.

— CHARLES BABBAGE (1792–1871)

ऊर्ध्वमूलमधःशाखमश्वत्थं प्राहुरव्ययम् ।
छन्दांसि यस्य पर्णानि यस्तं वेद स वेदवित् ॥

— *The Bhagavad Gita XV.1*[2]

Singularity analysis paves the way to the analysis of a large quantity of generating functions, as provided by the symbolic method expounded in Chapters I–III. In accordance with Pólya's aphorism quoted above, it makes it possible to "be lazy" and "let the principles work for you". In this chapter we illustrate this situation with numerous examples related to languages, permutations, trees, and graphs of various sorts. As in Chapter V, most analyses are organized into broad classes called *schemas*.

First, we develop the general *exp–log schema*, which covers the *set* construction, either labelled or unlabelled, applied to generators whose dominant singularity is of logarithmic type. This typically non-recursive schema parallels in generality the supercritical schema of Chapter V, which is relative to sequences. It permits us to quantify various constructions of permutations, derangements, 2–regular graphs, mappings, and functional graphs, and provides information on factorization properties of polynomials over finite fields.

---

[1]Quoted in M Walter, T O'Brien, Memories of George Pólya, Mathematics Teaching 116 (1986)

[2]*"There is an imperishable tree, it is said, that has its roots upward and its branches down and whose leaves are the Hymns [Vedas]. He who knows it possesses knowledge."*

Next, we deal with *recursively defined structures*, whose study constitutes the main theme of this chapter. In that case, generating functions are accessible by means of equations or systems that implicitly define them. A distinctive feature of many such combinatorial types is that their generating functions have a *square-root singularity*, that is, the singular exponent equals $1/2$. As a consequence, the counting sequences characteristically involve asymptotic terms of the form $A^n n^{-3/2}$, where the latter asymptotic exponent, $-3/2$, precisely reflects the singular exponent $1/2$ in the function's singular expansion, in accordance with the general principles of singularity analysis presented in Chapter VI.

Trees are the prototypical recursively defined combinatorial type. Square-root singularities automatically arise for all varieties of trees constrained by a finite set of allowed node degrees, including binary trees, unary–binary trees, ternary trees, and many more. The counting estimates involve the characteristic $n^{-3/2}$ subexponential factor, a property that holds in the labelled and unlabelled frameworks alike.

Simple varieties of trees have many properties in common, beyond the subexponential growth factor of tree counts. Indeed, in a random tree of some large size n, almost all nodes are found to be at level about $\sqrt{n}$, path length grows on average like $n\sqrt{n}$, and height is of order $\sqrt{n}$, with high probability. These results serve to unify classical tree types—we say that such properties of random trees are *universal*[3] among all simply generated families sharing the square-root singularity property. (This notion of universality, borrowed from physics, is also nowadays finding increasing popularity among probabilists, for reasons much similar to ours.) In this perspective, the motivation for organizing the theory along the lines of major *schemas* fits perfectly with the quest of *universal laws* in analytic combinatorics.

In the context of simple varieties of trees, the square-root singularity arises from general properties of the inverse of an analytic function. Under suitable conditions, this characteristic feature can be extended to functions defined implicitly by a functional equation. Consequences are the general enumeration of non-plane unlabelled trees, including isomers of alkanes in theoretical chemistry, as well as secondary structures of molecular biology.

Much of this chapter is devoted to *context-free specifications and languages*. In that case, *a priori*, generating functions are *algebraic functions*, meaning that they satisfy a system of polynomial equations, itself optionally reducible (by elimination) to a single equation. For solutions of positive polynomial systems, square-root singularities are found to be the rule under a simple technical condition of irreducibility that is evocative of the Perron–Frobenius conditions encountered in Chapter V in relation to finite-state and transfer-matrix models. As an illustration, we show how to develop a

---

[3]The following quotation illustrates well the notion of universality in physics: "[...] *this echoes the notion of universality in statistical physics. Phenomena that appear at first to be unconnected, such as magnetism and the phase changes of liquids and gases, share some identical features. This universal behaviour pays no heed to whether, say, the fluid is argon or carbon dioxide. All that matters are broad-brush characteristics such as whether the system is one-, two- or three-dimensional and whether its component elements interact via long- or short-range forces. Universality says that sometimes the details do not matter.*" [From "Utopia Theory", in *Physics World*, August 2003].

coherent theory of topological configurations in the plane (trees, forests, graphs) that satisfy a non-crossing constraint.

For arbitrary algebraic functions (the ones that are not necessarily associated with positive coefficients and equations, or irreducible positive systems), a richer set of singular behaviours becomes possible: singular expansions involve fractional exponents (not just $1/2$, corresponding to the square-root paradigm above). Singularity analysis is invariably applicable: algebraic functions are viewed as plane algebraic curves, and the famous Newton–Puiseux theorem of elementary algebraic geometry completely describes the types of singularities thay may occur. Algebraic functions also surface as solutions of various types of functional equations: this turns out to be the case for many classes of walks that generalize Dyck and Motzkin paths, via what is known as the kernel method, as well as for many types of planar maps (embedded planar graphs), via the so-called quadratic method. In all these cases, singular exponents of a predictable (rational) form are bound to occur, implying in turn numerous quantitative properties of random discrete structure and universality phenomena..

Differential equations and systems are associated to recursively defined structure, when either pointing constructions or order constraints appear. For counting generating functions, the equations are nonlinear, while the GFs associated to additive parameters lead to linear versions. Differential equations are also central in connection with the *holonomic framework*[4], which intervenes in the enumeration of many classes of "hard" objects, like regular graphs and Latin rectangles. Singularity analysis is once more instrumental in working out precise asymptotic estimates—the appearance of singular exponents that are algebraic (rather than rational) numbers is a characteristic feature of many such estimates. We examine here applications relative to quadtrees and to varieties of increasing trees, some of which are closely related to permutations as well as to algorithms and data structures for sorting and searching.

VII. 1. A roadmap to singularity analysis asymptotics

The singularity analysis theorems of Chapter VI, which may be coarsely summarized by the correspondence

$$
(1) \qquad f(z) \sim (1 - z/\rho)^{-\alpha} \qquad \longrightarrow \qquad f_n \sim \frac{1}{\Gamma(\alpha)} \rho^{-n} n^{\alpha-1},
$$

serve as our main asymptotic engine throughout this chapter. Singularity analysis is instrumental in quantifying properties of non-recursive as well as recursive structures. Our reader might be surprised not to encounter integration contours anymore in this chapter. Indeed, it now suffices to work out the *local analysis of functions at their singularities*, then the general theorems of singularity analysis (Chapter VI) effect the translation to counting sequences and parameters *automatically*.

---

[4]Holonomic functions (Appendix B.4: *Holonomic functions*, p. 748) are defined as solutions of linear differential equations with coefficients that are rational functions.

The exp–log schema. This schema, examined in Section VII. 2, is relative to the labelled *set construction*,

$$(2) \qquad\qquad \mathcal{F} = \text{SET}(\mathcal{G}) \qquad \Longrightarrow \qquad F(z) = \exp\left(G(z)\right),$$

as well as its unlabelled counterparts, MSET and PSET: an $\mathcal{F}$–structure is thus constructed (non-recursively) as an unordered assembly of $\mathcal{G}$–components. In the case where the GF of components is logarithmic at its dominant singularity,

$$(3) \qquad\qquad G(z) \sim \kappa \log \frac{1}{1 - z/\rho} + \lambda,$$

an immediate computation shows that $F(z)$ has a singularity of the power type,

$$F(z) \sim e^{\lambda} \left(1 - z/\rho\right)^{-\kappa},$$

which is clearly in the range of singularity analysis. The construction (2), supplemented by simple technical conditions surrounding (3), defines the *exp–log schema*. Then, for such $\mathcal{F}$–structures that are assemblies of logarithmic components, the asymptotic counting problem is systematically solvable (Theorem VII.1, p. 446): the number of $\mathcal{G}$–components in a large random $\mathcal{F}$–structure is $O(\log n)$, both in the mean and in probability, while more refined estimates describe precisely the likely shape of profiles. This schema has a generality comparable to the supercritical schema examined in Section V. 2, p. 293, but the probabilistic phenomena at stake appear to be in sharp contrast: the number of components is typically small, being logarithmic for exp–log sets, as opposed to a linear growth in the case of supercritical sequences. The schema can be used to analyse properties of permutations, functional graphs, mappings, and polynomial over finite fields.

Recursion and the universality of square-root singularity. A major theme of this chapter is the study of asymptotic properties of recursive structures. In a large number of cases, functions with a square root singularity are encountered, and given the usual correspondence,

$$f(z) \sim -(1 - z)^{1/2} \qquad \longrightarrow \qquad f_n \sim \frac{1}{2\sqrt{\pi n^3}};$$

the corresponding coefficients are of the asymptotic form $C\rho^{-n} n^{-3/2}$. Several schemas can be described to capture this phenomenon; we develop here, in order of increasing structural complexity, the ones corresponding to simple varieties of trees, implicit structures, Pólya operators, and irreducible polynomial systems.

Simple varieties of trees and inverse functions. Our treatment of *recursive combinatorial types* starts with simple varieties of trees, studied in Section VII. 3. In the basic situation, that of plane unlabelled trees, the equation is

$$(4) \qquad\qquad \mathcal{Y} = \mathcal{Z} \times \text{SEQ}_\Omega(\mathcal{Y}) \qquad \Longrightarrow \qquad Y(z) = z\phi(Y(z)),$$

with, as usual, $\phi(w) = \sum_{\omega \in \Omega} w^{\omega}$. Thus, the OGF $Y(z)$ is determined as the inverse of $w/\phi(w)$, where the function ϕ reflects the collection of all allowed node degrees (Ω). From analytic function theory, we know that singularities of the inverse of an analytic function are generically of the square-root type (Subsection IV. 7.1, p. 275 and Section VI. 7, p. 402), and such is the case whenever Ω is a "well-behaved" set of integers, in particular, a finite set. Then, the number of trees invariably satisfies an estimate of the form

$$(5) \qquad\qquad Y_n = [z^n]Y(z) \sim CA^n n^{-3/2}.$$

Square-root singularity is also attached to several universality phenomena, as evoked in the general introduction to this chapter.

Tree-like structures and implicit functions. Functions defined implicitly by an equation of the form

$$(6) \qquad\qquad Y(z) = G(z, Y(z))$$

where G is bivariate analytic, has non-negative coefficients, and satisfies a natural set of conditions also lead to square-root singularity (Section VII. 4 and Theorem VII.3, p. 468)). The schema (6) obviously generalizes (4): simply take $G(z, y) = z\phi(y)$. Again, such functions invariably satisfy an estimate (5).

Trees under symmetries and Pólya operators. The analytic methods mentioned above can be further extended to Pólya operators, which translate unlabelled set and cycle constructions; see Section VII. 5. A typical application is to the class of non-plane unlabelled trees whose OGF satisfies the *infinite* functional equation,

$$H(z) = z \exp\left(\frac{H(z)}{1} + \frac{H(z^2)}{2} + \cdots \right).$$

Singularity analysis applies more generally to varieties of non-plane unlabelled trees (Theorem VII.4, p. 479), which covers the enumeration of various types of interesting molecules in combinatorial chemistry.

Context-free structures and polynomial systems. The generating function of any context-free class or language is known to be a component of a *system* of positive polynomial equations

$$\begin{cases} y_1 &=& P_1(z, y_1, \ldots, y_r) \\ \vdots & \vdots & \vdots \\ y_r &=& P_r(z, y_1, \ldots, y_r). \end{cases}$$

The $n^{-3/2}$ counting law is once more universal among such combinatorial classes under a basic condition of "irreducibility" (Section VII. 6 and Theorem VII.5, p. 483). In that case, the GFs are algebraic functions satisfying a strong positivity constraint; the corresponding analytic statement constitutes the important *Drmota–Lalley–Woods Theorem* (Theorem VII.6, p. 489).

Note that there is a progression in the complexity of the schemas leading to square-root singularity. From the analytic standpoint, this can be roughly rendered by a chain

$$\text{inverse functions} \longrightarrow \text{implicit functions} \longrightarrow \text{systems.}$$

It is, however, often meaningful to treat each combinatorial problem at its minimal level of generality, since expressions tend to become less and less explicit as complexity increases.

General algebraic functions. In essence, the coefficients of *all algebraic functions* can be analysed asymptotically (Section VII. 7). There are only minor limitations arising from the possible presence of several dominant singularities, like in the rational function case. The starting point is the characterization of the local behaviour of an algebraic function at any of its singularities, which is provided by the Newton–Puiseux theorem: if ζ is a singularity, then the branch $Y(z)$ of an algebraic function admits near ζ a representation of the form

$$(7) \qquad Y(z) = Z^{r/s} \left(\sum_{k \geq 0} c_k Z^{k/s} \right), \qquad Z := (1 - z/\zeta),$$

for some $r/s \in \mathbb{Q}$, so that the singular exponent is invariably a *rational number*. Singularity analysis is systematically applicable, so that the nth coefficient of Y is expressible as a finite linear combination of terms, each of the asymptotic form

$$(8) \qquad \zeta^{-n} n^{p/q}, \qquad \frac{p}{q} \in \mathbb{Q} \setminus \{-1, -2, \ldots\};$$

see also Figure VII.1. The various quantities (like ζ, r, s) entering the asymptotic expansion of the coefficients of an algebraic function turn out to be effectively computable.

Beside providing a wide-encompassing conceptual framework of independent interest, the general theory of algebraic coefficient asymptotics is applicable whenever the combinatorial problems considered are not amenable to any of the special schemas previously described. For instance, certain kinds of supertrees (these are defined as trees composed with trees, Example VII.10, p. 412) lead to the singular type $Z^{1/4}$, which is reflected by an unusual subexponential factor of $n^{-5/4}$ present in asymptotic counts. Maps, which are planar graphs drawn in the plane (or on the sphere), satisfy a universality law with a singular exponent equal to $3/2$, which is associated to counting sequences involving an asymptotic $n^{-5/2}$ factor.

Differential equations and systems. When recursion is combined with pointing or with order constraints, enumeration problems translate into integro-differential equations. Section VII. 9 examines the types of singularities that may occur in two important cases: (i) linear differential equations; (ii) nonlinear differential equations.

Linear differential equations arise from the analysis of parameters of splitting processes that extend the framework of tree recurrences (Subsection VI. 10.3, p. 427), and we treat the geometric quadtree structure in this perspective. An especially notable source of linear differential equations is the class of *holonomic functions* (solutions of linear equations with rational coefficients, cf Appendix B.4: *Holonomic functions*, p. 748), which includes GFs of Latin rectangles, regular graphs, permutations constrained by the length of their longest increasing subsequence, Young tableaux and many more structures of combinatorial theory. In an important case, that of a "*regular*" *singularity*, asymptotic forms can be systematically extracted. The singularities that may occur extend the algebraic ones (7), and the corresponding coefficients are then asymptotically composed of elements of the form

$$(9) \qquad \zeta^{-n} n^{\theta} (\log n)^{\ell},$$

Rational	Irred. linear system	ζ^{-n}	Perron–Frob., merom. fns, Ch. V
—	General rational	$\zeta^{-n} n^{\ell}$	meromorphic functions, Ch. V
Algebraic	Irred. positive sys.	$\zeta^{-n} n^{-3/2}$	DLW Th., sing. analysis, *this chapter, §VII. 6, p. 482*
—	General algebraic	$\zeta^{-n} n^{p/q}$	Puiseux, sing. analysis, *this chapter, §VII. 7, p. 493*
Holonomic	Regular sing.	$\zeta^{-n} n^{\theta} \log^{\ell} n$	ODE, sing. analysis, *this chapter, §VII. 9.1, p. 518*
—	Irregular sing.	$\zeta^{-n} e^{P(n^{1/r})} n^{\theta} \log^{\ell} n$	ODE, saddle-point, §VIII. 7, p. 581

Figure VII.1. A telegraphic summary of a hierarchy of special functions by increasing level of generality: asymptotic elements composing coefficients and the coefficient extraction method (with $\ell, r \in \mathbb{Z}_{\geq 0}$, $p/q \in \mathbb{Q}$, ζ and θ algebraic, and P a polynomial).

(θ an algebraic quantity, $\ell \in \mathbb{Z}_{\geq 0}$), a type which is much more general than (8).

Nonlinear differential equations are typically attached to the enumeration of trees satisfying various kinds of order constraints. A global treatment is intrinsically not possible, given the extreme diversity of singular expansions that may occur. Accordingly, we restrict attention to first-order nonlinear equations of the form

$$\frac{d}{dz} Y(z) = \phi(Y(z)),$$

which covers varieties of increasing trees and certain urn processes, including several models closely related to permutations.

Figure VII.1 summarizes three classes of special functions encountered in this book, namely, rational, algebraic, and holonomic. When structural complexity increases, a richer set of asymptotic coefficient behaviours becomes possible. (The complex asymptotic methods employed extend much beyond the range summarized in the figure. For instance, the class of irreducible positive systems of polynomial equations are part of the general square-root singularity paradigm, also encountered with Pólya operators, as well as inverse and implicit functions in non-algebraic cases.)

VII. 2. Sets and the exp–log schema

We begin by examining a schema that is structurally comparable to the supercritical sequence schema of Section V. 2, p. 293, but one that requires singularity analysis for coefficient extraction. The starting point is the construction of permutations ($\mathcal{P}$) as labelled sets of cyclic permutations ($\mathcal{K}$):

(10) $\quad \mathcal{P} = \text{SET}(\mathcal{K}) \quad \Longrightarrow \quad P(z) = \exp(K(z)), \text{ where } K(z) = \log \dfrac{1}{1-z},$

which gives rise to many easy explicit calculations. For instance, the probability that a random permutation consists of a unique cycle is $1/n$ (since it equals K_n/P_n); the number of cycles is asymptotic to $\log n$, both on average (p. 122) and in probability (Example III.4, p. 160); the probability that a random permutation has no singleton cycle is $\sim e^{-1}$ (the derangement problem; see pp. 123 and 228).

Similar properties hold true under surprisingly general conditions. We start with definitions that describe the combinatorial classes of interest.

Definition VII.1. *A function $G(z)$ analytic at 0, having non-negative coefficients and finite radius of convergence ρ is said to be of (κ, λ)-logarithmic type, where $\kappa \neq 0$, if the following conditions hold:*

    (i) the number ρ is the unique singularity of $G(z)$ on $|z| = \rho$;

    (ii) $G(z)$ is continuable to a Δ–domain at ρ;

    (iii) $G(z)$ satisfies

$$(11) \quad G(z) = \kappa \log \frac{1}{1 - z/\rho} + \lambda + O\left(\frac{1}{(\log(1 - z/\rho))^2}\right), \qquad as\ z \to \rho\ in\ \Delta.$$

Definition VII.2. *The labelled construction $\mathcal{F} = \text{SET}(\mathcal{G})$ is said to be a* labelled exp–log schema *("exponential–logarithmic schema") if the exponential generating function $G(z)$ of $\mathcal{G}$ is of logarithmic type. The unlabelled construction $\mathcal{F} = \text{MSET}(\mathcal{G})$ is said to be an* unlabelled exp–log schema *if the ordinary generating function $G(z)$ of $\mathcal{G}$ is of logarithmic type, with $\rho < 1$. In each case, the quantities (κ, λ) of (11) are referred to as the* parameters *of the schema.*

By the fact that $G(z)$ has positive coefficients, we must have $\kappa > 0$, while the sign of λ is arbitrary. The definitions and the main properties to be derived for unlabelled multisets easily extend to the powerset construction: see Notes VII.1 and VII.5 below.

Theorem VII.1 (Exp–log schema). *Consider an exp–log schema with parameters (κ, λ).*

    (i) The counting sequences satisfy

$$\begin{cases} [z^n]G(z) &=& \dfrac{\kappa}{n}\rho^{-n}\left(1 + O\left((\log n)^{-2}\right)\right), \\ [z^n]F(z) &=& \dfrac{e^{\lambda + r_0}}{\Gamma(\kappa)}n^{\kappa-1}\rho^{-n}\left(1 + O\left((\log n)^{-2}\right)\right), \end{cases}$$

where $r_0 = 0$ in the labelled case and $r_0 = \sum_{j\geq 2} G(\rho^j)/j$ in the case of unlabelled multisets.

    (ii) The number X of $\mathcal{G}$–components in a random $\mathcal{F}$–object satisfies

$$\mathbb{E}_{\mathcal{F}_n}(X) \quad = \quad \kappa(\log n - \psi(\kappa)) + \lambda + r_1 + O\left((\log n)^{-1}\right) \quad (\psi(s) \equiv \tfrac{d}{ds}\Gamma(s)),$$

where $r_1 = 0$ in the labelled case and $r_1 = \sum_{j\geq 2} G(\rho^j)$ in the case of unlabelled multisets. The variance satisfies $\mathbb{V}_{\mathcal{F}_n}(X) = O(\log n)$, and, in particular, the distribution[5] of X is concentrated around its mean.

---

[5] We shall see in Subsection IX.7.1 (p. 667) that, in addition, the *asymptotic distribution* of X is invariably *Gaussian* under such exp–log conditions.

Proof. This result is from an article by Flajolet and Soria [258], with a correction to the logarithmic type condition given by Jennie Hansen [318]. We first discuss the *labelled case*, $\mathcal{F} = \mathrm{SET}(\mathcal{G})$, so that $F(z) = \exp G(z)$.

(*i*) The estimate for $[z^n]G(z)$ follows directly from singularity analysis with logarithmic terms (Theorem VI.4, p. 393). Regarding $F(z)$, we find, by exponentiation,

$$(12) \qquad F(z) = \frac{e^\lambda}{(1 - z/\rho)^\kappa}\left[1 + O\left(\frac{1}{(\log(1 - z/\rho))^2}\right)\right].$$

Like G, the function $F = e^G$ has an isolated singularity at ρ, and is continuable to the Δ–domain in which the expansion (11) is valid. The basic transfer theorem then provides the estimate of $[z^n]F(z)$.

(*ii*) Regarding the number of components, the BGF of $\mathcal{F}$ with u marking the number of $\mathcal{G}$-components is $F(z, u) = \exp(uG(z))$, in accordance with the general developments of Chapter III. The function

$$f_1(z) := \left.\frac{\partial}{\partial u}F(z, u)\right|_{u=1} = F(z)G(z),$$

is the EGF of the cumulated values of X. It satisfies near ρ

$$f_1(z) = \frac{e^\lambda}{(1 - z/\rho)^\kappa}\left(\kappa \log \frac{1}{1 - z/\rho} + \lambda\right)\left[1 + O\left(\frac{1}{(\log(1 - z/\rho))^2}\right)\right],$$

whose translation, by singularity analysis theory is immediate:

$$[z^n]f_1(z) \equiv \mathbb{E}_{\mathcal{F}_n}(X) = \frac{e^\lambda}{\Gamma(\kappa)}\rho^{-n}\left(\kappa \log n - \kappa \psi(\kappa) + \lambda + O\left((\log n)^{-1}\right)\right).$$

This provides the mean value estimate of X as $[z^n]f_1(z)/[z^n]F(z)$. The variance analysis is conducted in the same way, using a second derivative.

For the *unlabelled case*, the analysis of $[z^n]G(z)$ can be recycled verbatim. First, given the assumptions, we must have $\rho < 1$ (since otherwise $[z^n]G(z)$ could not be an integer). The classical translation of multisets (Chapter I) rewrites as

$$F(z) = \exp\left(G(z) + R(z)\right), \qquad R(z) := \sum_{j=2}^{\infty} \frac{G(z^j)}{j},$$

where $R(z)$ involves terms of the form $G(z^2), \ldots$, each being analytic in $|z| < \rho^{1/2}$. Thus, $R(z)$ is itself analytic, as a uniformly convergent sum of analytic functions, in $|z| < \rho^{1/2}$. (This follows the usual strategy for treating Pólya operators in asymptotic theory.) Consequently, $F(z)$ is Δ–analytic. As $z \to \rho$, we then find

$$(13) \quad F(z) = \frac{e^{\lambda + r_0}}{(1 - z/\rho)^\kappa}\left[1 + O\left(\frac{1}{(\log(1 - z/\rho))^2}\right)\right], \qquad r_0 \equiv \sum_{j=2}^{\infty}\frac{G(\rho^j)}{j}.$$

The asymptotic expansion of $[z^n]F(z)$ then results from singularity analysis.

The BGF $F(z, u)$ of $\mathcal{F}$, with u marking the number of $\mathcal{G}$-components, is

$$F(z, u) = \exp\left(\frac{uG(z)}{1} + \frac{u^2 G(z^2)}{2} + \cdots\right).$$

$\mathcal{F}$	κ	$n = 100$	$n = 272$	$n = 739$
Permutations	1	5.18737	6.18485	7.18319
Derangements	1	4.19732	5.18852	6.18454
2–regular	$\frac{1}{2}$	2.53439	3.03466	3.53440
Mappings	$\frac{1}{2}$	2.97898	3.46320	3.95312

Figure VII.2. Some exp–log structures ($\mathcal{F}$) and the mean number of $\mathcal{G}$–components for $n = 100, 272 \equiv \lceil 100 \cdot e \rceil, 739 \equiv \lceil 100 \cdot e^2 \rceil$: the columns differ by about κ, as expected.

Consequently,

$$f_1(z) := \left. \frac{\partial}{\partial u} F(z, u) \right|_{u=1} = F(z)\,(G(z) + R_1(z)), \qquad R_1(z) = \sum_{j=2}^{\infty} G(z^j).$$

Again, the singularity type is that of $F(z)$ multiplied by a logarithmic term,

$$(14) \qquad\qquad f_1(z) \underset{z \to \rho}{\sim} F(z)(G(z) + r_1), \qquad r_1 \equiv \sum_{j=2}^{\infty} G(\rho^j).$$

The mean value estimate results. Variance analysis follows similarly. ∎

▷ **VII.1.** *Unlabelled powersets.* For the powerset construction $\mathcal{F} = \text{PSET}(\mathcal{G})$, the statement of Theorem VII.1 holds with

$$r_0 = \sum_{j \geq 2} (-1)^{j-1} \frac{G(\rho^j)}{j},$$

as seen by an easy adaptation of the proof technique of Theorem VII.1. ◁

As we see below, beyond permutations, mappings, unlabelled functional graphs, polynomials over finite fields, 2–regular graphs, and generalized derangements belong to the exp–log schema; see Figure VII.2 for representative numerical data. Furthermore, singularity analysis gives precise information on the decomposition of large $\mathcal{F}$ objects into $\mathcal{G}$ components.

Example VII.1. *Cycles in derangements.* The case of all *permutations*,

$$P(z) = \exp(K(z)), \qquad K(z) = \log \frac{1}{1 - z},$$

is immediately seen to satisfy the conditions of Theorem VII.1: it corresponds to the radius of convergence $\rho = 1$ and parameters $(\kappa, \lambda) = (1, 0)$.

Let Ω be a finite set of integers and consider next the class $\mathcal{D} \equiv \mathcal{D}^{\Omega}$ of permutations *without* any cycle of length in Ω. This includes standard *derangements* (where $\Omega = \{1\}$). The specification is then

$$
\begin{cases}
\mathcal{D} &= \text{SET}(\mathcal{K}) \\
\mathcal{G} &= \text{CYC}_{\mathbb{Z}_{>0} \setminus \Omega}(\mathcal{Z})
\end{cases}
\implies
\begin{cases}
D(z) &= \exp(K(z)) \\
G(z) &= \log \dfrac{1}{1 - z} - \displaystyle\sum_{\omega \in \Omega} \dfrac{z^{\omega}}{\omega}.
\end{cases}
$$

The theorem applies, with $\kappa = 1$, $\lambda := -\sum_{\omega \in \Omega} \omega^{-1}$. In particular, the mean number of cycles in a random generalized derangement of size n is $\log n + O(1)$. .... ∎

***Example* VII.2.** *Connected components in 2–regular graphs.* The class of (undirected) 2– regular graphs is obtained by the set construction applied to components that are themselves *undirected cycles* of length ≥ 3 (see p. 133 and Example VI.2, p. 395). In that case:

$$
\begin{cases}
\mathcal{F} &= \mathrm{SET}(\mathcal{G}) \\
\mathcal{G} &= \mathrm{UCYC}_{\geq 3}(\mathcal{Z})
\end{cases}
\implies
\begin{cases}
F(z) &= \exp(G(z)) \\
G(z) &= \dfrac{1}{2} \log \dfrac{1}{1-z} - \dfrac{z}{2} - \dfrac{z^2}{4}.
\end{cases}
$$

This is an exp–log scheme with $\kappa = 1/2$ and $\lambda = -3/4$. In particular the number of components is asymptotic to $\frac{1}{2} \log n$, both in the mean and in probability. .... ∎

***Example* VII.3.** *Connected components in mappings.* The class $\mathcal{F}$ of *mappings* (functions from a finite set to itself) was introduced in Subsection II. 5.2, p. 129. The associated digraphs are described as labelled sets of connected components ($\mathcal{K}$), themselves (directed) cycles of trees ($\mathcal{T}$), so that the class of all mappings has an EGF given by

$$
F(z) = \exp(K(z)), \qquad K(z) = \log \frac{1}{1 - T(z)}, \qquad T(z) = z e^{T(z)},
$$

with T the Cayley tree function. The analysis of inverse functions (Section VI. 7 and Example VI.8, p. 403) has shown that $T(z)$ is singular at $z = e^{-1}$, where it admits the singular expansion $T(z) \sim 1 - \sqrt{2}\sqrt{1 - ez}$. Thus $G(z)$ is logarithmic with $\kappa = 1/2$ and $\lambda = -\log \sqrt{2}$. As a consequence, the number of connected mappings satisfies

$$
K_n \equiv n![z^n]K(z) = n^n \sqrt{\frac{\pi}{2n}} \left(1 + O(n^{-1/2})\right).
$$

In other words: *the probability for a random mapping of size n to consist of a single component is* $\sim \sqrt{\frac{\pi}{2n}}$. Also, the mean number of components in a random mapping of size n is

$$
\frac{1}{2} \log n + \log \sqrt{2 e^\gamma} + O(n^{-1/2}).
$$

Similar properties hold for mappings without fixed points, which are analogous to derangements and were discussed in Chapter II, p. 130. We shall establish below, p. 480, that *unlabelled functional graphs* also belong to the exp–log schema. .... ∎

***Example* VII.4.** *Factors of polynomials over finite fields.* Factorization properties of random polynomials over finite fields are of importance in various areas of mathematics and have applications to coding theory, symbolic computation, and cryptography [51, 599, 541]. Example I.20, p. 90, offers a preliminary discussion.

Let $\mathbb{F}_p$ be the finite field with p elements and $\mathcal{P} \subset \mathbb{F}_p[X]$ the set of monic polynomials with coefficients in the field. We view these polynomials as (unlabelled) combinatorial objects with size identified to degree. Since a polynomial is specified by the sequence of its coefficients, one has, with $\mathcal{A}$ the "alphabet" of coefficients, $\mathcal{A} = \mathbb{F}_p$ treated as a collection of atomic objects:

$$
(15) \qquad\qquad \mathcal{P} = \mathrm{SEQ}(\mathcal{A}) \implies P(z) = \frac{1}{1 - pz},
$$

On the other hand, the unique factorization property of polynomials entails that the class $\mathcal{I}$ of all monic irreducible polynomials and the class $\mathcal{P}$ of all polynomials are related by $\mathcal{P} = \mathrm{MSET}(\mathcal{I})$.

$$(X+1)\left(X^{10}+X^9+X^8+X^6+X^4+X^3+1\right)\left(X^{14}+X^{11}+X^{10}+X^3+1\right)$$
$$X^3(X+1)\left(X^2+X+1\right)^2\left(X^{17}+X^{16}+X^{15}+X^{11}+X^9+X^6+X^2+X+1\right)$$
$$X^5(X+1)\left(X^5+X^3+X^2+X+1\right)\left(X^{12}+X^8+X^7+X^6+X^5+X^3+X^2+X+1\right)\left(X^2+X+1\right)$$
$$X^2\left(X^2+X+1\right)^2\left(X^3+X^2+1\right)\left(X^8+X^7+X^6+X^4+X^2+X+1\right)\left(X^8+X^7+X^5+X^4+1\right)$$
$$\left(X^7+X^6+X^5+X^3+X^2+X+1\right)\left(X^{18}+X^{17}+X^{13}+X^9+X^8+X^7+X^6+X^4+1\right)$$

Figure VII.3. The factorizations of five random polynomials of degree 25 over $\mathbb{F}_2$. One out of five polynomials in this sample has no root in the base field (the asymptotic probability is $\frac{1}{4}$ by Note VII.4).

As a consequence of Möbius inversion, one then gets (Equation (94) of Chapter I, p. 91):

$$(16)\qquad I(z)=\log\frac{1}{1-z}+R(z),\qquad R(z):=\sum_{k\geq2}\frac{\mu(k)}{k}\log\frac{1}{1-pz^k}.$$

Regarding complex asymptotics, the function $R(z)$ of (16) is analytic in $|z|<p^{-1/2}$. Thus $I(z)$ is of logarithmic type with radius of convergence $1/p$ and parameters

$$\kappa=1,\qquad\lambda=\sum_{k\geq2}\frac{\mu(k)}{k}\log\frac{1}{1-p^{1-k}}.$$

As already noted in Chapter I, a consequence is the asymptotic estimate $I_n\sim p^n/n$, which constitutes a "Prime Number Theorem" for polynomials over finite fields: *a fraction asymptotic to $1/n$ of the polynomials in $\mathbb{F}_p[X]$ are irreducible.* Furthermore, since $I(z)$ is logarithmic and $\mathcal{P}$ is obtained by a multiset construction, we have an unlabelled exp–log scheme, to which Theorem VII.1 applies. As a consequence:

The number of factors of a random polynomial of degree n has mean and variance each asymptotic to $\log n$; its distribution is concentrated.

(See Figure VII.3 for an illustration; the mean value estimate appears in [378, Ex. 4.6.2.5].) We shall revisit this example in Chapter IX, p. 672, and establish a companion Gaussian limit law for the number of irreducible factors in a random polynomial of large degree. This and similar developments lead to a complete analysis of some of the basic algorithms known for factoring polynomials over finite fields; see [236]. .... ∎

▷ **VII.2.** *The divisor function for polynomials.* Let $\delta(\varpi)$ for $\varpi\in\mathcal{P}$ be the total number of monic polynomials (not necessarily irreducible) dividing ϖ: if $\varpi=\iota_1^{e_1}\cdots\iota_k^{e_k}$, where the ι_j are distinct irreducibles, then $\delta(\varpi)=(e_1+1)\cdots(e_k+1)$. One has

$$\mathbb{E}_{\mathcal{P}_n}(\delta)=\frac{[z^n]\prod_{j\geq1}(1+2z^j+3z^{2j}+\cdots)}{[z^n]\prod_{j\geq1}(1+z^j+z^{2j}+\cdots)}=\frac{[z^n]P(z)^2}{[z^n]P(z)},$$

so that the mean value of δ over $\mathcal{P}_n$ is exactly $(n+1)$. This evaluation is relevant to polynomial factorization over $\mathbb{Z}$ since it gives an upper bound on the number of irreducible factor combinations that need to be considered in order to lift a factorization from $\mathbb{F}_p(X)$ to $\mathbb{Z}(X)$; see [379, 599]. ◁

▷ **VII.3.** *The cost of finding irreducible polynomials.* Assume that it takes expected time $t(n)$ to *test* a random polynomial of degree n for irreducibility. Then it takes expected time $\sim nt(n)$ to *find* a random irreducible polynomial of degree n: simply draw a polynomial at random and test it for irreducibility. (Testing for irreducibility can itself be achieved by developing a polynomial factorization algorithm which is stopped as soon as a non-trivial factor is found. See works by Panario *et al.* for detailed analyses of this strategy [468, 469].)   ◁

Profiles of exp–log structures. Under the exp–log conditions, it is also possible to analyse the *profile* of structures, that is, the number of components of size r for each fixed r. The Poisson distribution (Appendix C.4: *Special distributions*, p. 774) of parameter v is the law of a discrete random variable Y such that

$$\mathbb{E}(u^Y) = e^{-v(1-u)}, \qquad \mathbb{P}(Y = k) = e^{-v}\frac{v^k}{k!}.$$

A variable Y is said to be *negative binomial* of parameter (m, α) if its probability generating function and its individual probabilities satisfy:

$$\mathbb{E}(u^Y) = \left(\frac{1-\alpha}{1-\alpha u}\right)^m, \qquad \mathbb{P}(Y = k) = \binom{m+k-1}{k}\alpha^k(1-\alpha)^m.$$

(The quantity $\mathbb{P}(Y = k)$ is the probability that the mth success in a sequence of independent trials with individual success probability α occurs at time $m + k$; see [206, p. 165] and Appendix C.4: *Special distributions*, p. 774.)

Proposition VII.1 (Profiles of exp–log structures). *Assume the conditions of Theorem VII.1 and let $X^{(r)}$ be the number of $\mathcal{G}$–components of size r in an $\mathcal{F}$–object. In the labelled case, $X^{(r)}$ admits a limit distribution of the* Poisson *type: for any fixed k,*

$$(17) \qquad \lim_{n\to\infty} \mathbb{P}_{\mathcal{F}_n}(X^{(r)} = k) = e^{-v}\frac{v^k}{k!}, \qquad v = g_r\rho^r, \qquad g_r \equiv [z^r]G(z).$$

In the unlabelled case, $X^{(r)}$ admits a limit distribution of the negative-binomial *type: for any fixed k,*
(18)

$$\lim_{n\to\infty} \mathbb{P}_{\mathcal{F}_n}(X^{(r)} = k) = \binom{G_r + k - 1}{k}\alpha^k(1-\alpha)^{G_r}, \qquad \alpha = \rho^r, \; G_r \equiv [z^r]G(z).$$

Proof. In the labelled case, the BGF of $\mathcal{F}$ with u marking the number $X^{(r)}$ of r–components is

$$F(z, u) = \exp\big((u - 1)g_r z^r\big) F(z).$$

Extracting the coefficient of u^k leads to

$$\phi_k(z) := [u^k]F(z, u) = \exp\left(-g_r z^r\right)\frac{(g_r z^r)^k}{k!} F(z).$$

The singularity type of $\phi_k(z)$ is that of $F(z)$ since the prefactor (an exponential multiplied by a polynomial) is entire, so that singularity analysis applies directly. As a consequence, one finds

$$[z^n]\phi_k(z) \sim \exp\left(-g_r\rho^r\right)\frac{(g_r\rho^r)^k}{k!} \cdot \big([z^n]F(z)\big),$$

which provides the distribution of $X^{(r)}$ under the form stated in (17).

In the unlabelled case, the starting BGF equation is

$$F(z, u) = \left(\frac{1 - z^r}{1 - u z^r}\right)^{G_r} F(z),$$

and the analytic reasoning is similar to the labelled case.   ∎

Proposition VII.1 will be revisited in Example IX.23, p. 675, when we examine continuity theorems for probability generating functions. Its unlabelled version covers in particular polynomials over finite fields; see [236, 372] for related results.

▷ **VII.4.** *Mean profiles.* The mean value of $X^{(r)}$ satisfies

$$\mathbb{E}_{\mathcal{F}_n}(X^{(r)}) \sim g_r \rho^r, \qquad \mathbb{E}_{\mathcal{F}_n}(X^{(r)}) \sim G_r \frac{\rho^r}{1 - \rho^r},$$

in the labelled and unlabelled (multiset) case, respectively. In particular: *the mean number of roots of a random polynomial over* $\mathbb{F}_p$ *that lie in the base field* $\mathbb{F}_p$ *is asymptotic to* $\frac{p}{p-1}$. Also: *the probability that a polynomial has no root in the base field is asymptotic to* $(1 - 1/p)^p$. (For random polynomials with real coefficients, a famous result of Kac (1943) asserts that the mean number of real roots is $\sim \frac{2}{\pi} \log n$; see [185].)   ◁

▷ **VII.5.** *Profiles of powersets.* In the case of unlabelled powersets $\mathcal{F} = \mathrm{PSET}(\mathcal{G})$ (no repetitions of elements allowed), the distribution of $X^{(r)}$ satisfies

$$\lim_{n \to \infty} \mathbb{P}_{\mathcal{F}_n}(X^{(r)} = k) = \binom{G_r}{k} \alpha^k (1 - \alpha)^{G_r - k}, \qquad \alpha = \frac{\rho^r}{1 + \rho^r};$$

i.e., the limit is a *binomial law* of parameters $(G_r, \rho^r / (1 + \rho^r))$.   ◁

VII. 3. Simple varieties of trees and inverse functions

A unifying theme in this chapter is the enumeration of rooted trees determined by restrictions on the collection of allowed node degrees (Sections I. 5, p. 64 and II. 5, p. 125). Some set $\Omega \subseteq \mathbb{Z}_{\geq 0}$ containing 0 (for leaves) and at least another number $d \geq 2$ (to avoid trivialities) is fixed; in the trees considered, all outdegrees of nodes are constrained to lie in Ω. Corresponding to the four combinations, unlabelled/labelled and plane/non-plane, there are four types of functional equations summarized by Figure VII.4. In three of the four cases, namely,

unlabelled plane, labelled plane, and labelled non-plane,

the generating function (OGF for unlabelled, EGF for labelled) satisfies an equation of the form

(19) $$y(z) = z\phi(y(z)).$$

In accordance with earlier conventions (p. 194), we name *simple variety of trees* any family of trees whose GF satisfies an equation of the form (19). (The functional equation satisfied by the OGF of a degree-restricted variety of unlabelled non-plane trees furthermore involves a Pólya operator Φ, which implies the presence of terms of the form $y(z^2), y(z^3), \ldots$: such cases are discussed below in Section VII. 5.)

The relation $y = z\phi(y)$ has already been examined in Section VI. 7, p. 402, from the point of view of singularity analysis. For convenience, we encapsulate into a definition the conditions of the main theorem of that section, Theorem VI.6, p. 404.

	plane	*non-plane*
Unlabelled (OGF)	$\mathcal{V} = \mathcal{Z} \times \text{SEQ}_\Omega(\mathcal{V})$ $V(z) = z\phi(V(z))$ $\phi(u) := \sum_{\omega \in \Omega} u^\omega$	$\mathcal{V} = \mathcal{Z} \times \text{MSET}_\Omega(\mathcal{V})$ $V(z) = z\Phi(V(z)))$ (Φ a Pólya operator)
Labelled (EGF)	$\mathcal{V} = \mathcal{Z} \star \text{SEQ}_\Omega(\mathcal{V})$ $\widehat{V}(z) = z\phi(\widehat{V}(z))$ $\phi(u) := \sum_{\omega \in \Omega} u^\omega$	$\mathcal{V} = \mathcal{Z} \star \text{SET}_\Omega(\mathcal{V})$ $\widehat{V}(z) = z\phi(\widehat{V}(z))$ $\phi(u) := \sum_{\omega \in \Omega} \frac{u^\omega}{\omega!}$

Figure VII.4. Functional equations satisfied by generating functions (OGF $V(z)$ or EGF $\widehat{V}(z)$) of degree-restricted families of trees.

Definition VII.3. *Let $y(z)$ be a function analytic at 0. It is said to belong to the* smooth inverse-function schema *if there exists a function $\phi(u)$ analytic at 0, such that, in a neighbourhood of 0, one has*

$$y(z) = z\phi(y(z)),$$

and $\phi(u)$ satisfies the following conditions.
   (H_1) *The function $\phi(u)$ is such that*

(20) $$\phi(0) \neq 0, \quad [u^n]\phi(u) \geq 0, \quad \phi(u) \not\equiv \phi_0 + \phi_1 u.$$

   (H_2) *Within the* open *disc of convergence of ϕ at 0, $|z| < R$, there exists a (necessarily unique) positive solution to the* characteristic equation*:*

(21) $$\exists \tau, \ 0 < \tau < R, \quad \phi(\tau) - \tau\phi'(\tau) = 0.$$

A class $\mathcal{Y}$ whose generating function $y(z)$ (either ordinary or exponential) satisfies these conditions is also said to belong to the smooth inverse-function schema.

   The schema is said to be aperiodic *if $\phi(u)$ is an aperiodic function of u (Definition IV.5, p. 266).*

VII. 3.1. Asymptotic counting. As we saw on general grounds in Chapters IV and VI, inversion fails to be analytic when the first derivative of the function to be inverted vanishes. The heart of the matter is that, at the point of failure $y = \tau$, corresponding to $z = \tau/\phi(\tau)$ (the radius of convergence of $y(z)$ at 0), the dependency $y \mapsto z$ becomes quadratic, so that its inverse $z \mapsto y$ gives rise to a square-root singularity (hence the characteristic equation). From here, the typical $n^{-3/2}$ term in coefficient asymptotics results (Theorem VI.6, p. 404). In view of our needs in this chapter, we rephrase Theorem VI.6 as follows.

Theorem VII.2. *Let $y(z)$ belong to the* smooth inverse-function schema *in the aperiodic case. Then, with τ the positive root of the characteristic equation and $\rho =$*

$\tau/\phi(\tau)$, *one has*

$$[z^n]y(z) = \sqrt{\frac{\phi(\tau)}{2\phi''(\tau)}} \frac{\rho^{-n}}{\sqrt{\pi n^3}} \left[1 + O\left(\frac{1}{n}\right)\right].$$

As we also know from Theorem VI.6 (p. 404), a complete (and locally convergent) expansion of $y(z)$ in powers of $\sqrt{1 - z/\rho}$ exists, starting with

(22) $y(z) = \tau - \gamma\sqrt{1 - z/\rho} + O\left(1 - z/\rho\right), \qquad \gamma := \sqrt{\frac{2\phi(\tau)}{\phi''(\tau)}},$

which implies a complete asymptotic expansion for $y_n = [z^n]y(z)$ in odd powers of $1/\sqrt{n}$. (The statement extends to the aperiodic case, with the necessary condition that $n \equiv 1 \mod p$, when ϕ has period p.)

We have seen already that this framework covers binary, unary–binary, general Catalan, as well as Cayley trees (Figure VI.10, p. 406). Here is another typical application.

***Example* VII.5.**   *Mobiles.* A (labelled) *mobile*, as defined by Bergeron, Labelle, and Leroux [50, p. 240], is a (labelled) tree in which subtrees dangling from the root are taken up to cyclic shift:

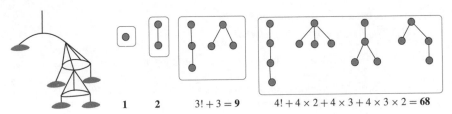

(Think of Alexander Calder's creations.) The specification and EGF equation are

$$\mathcal{M} = \mathcal{Z} \star (1 + \mathrm{CYC}\,\mathcal{M}) \qquad \Longrightarrow \qquad M(z) = z\left(1 + \log \frac{1}{1 - M(z)}\right).$$

(By definition, cycles have at least one components, so that the neutral structure must be added to allow for leaf creation.) The EGF starts as $M(z) = z + 2\frac{z^2}{2!} + 9\frac{z^3}{3!} + 68\frac{z^4}{4!} + 730\frac{z^5}{5!} + \cdots$, whose coefficients constitute *EIS* **A038037**.

The verification of the conditions of the theorem are immediate. We have $\phi(u) = 1 + \log(1 - u)^{-1}$, whose radius of convergence is 1. The characteristic equation reads

$$1 + \log\frac{1}{1 - \tau} - \frac{\tau}{1 - \tau} = 0,$$

which has a unique positive root at $\tau \doteq 0.68215$. (In fact, one has $\tau = 1 - 1/T(e^{-2})$, with T the Cayley tree function.) The radius of convergence is $\rho \equiv 1/\phi'(\tau) = 1 - \tau$. The asymptotic formula for the number of mobiles then results:

$$\frac{1}{n!}M_n \sim C \cdot A^n n^{-3/2}, \qquad \text{where} \quad C \doteq 0.18576, \quad A \doteq 3.14461.$$

(This example is adapted from [50, p. 261], with corrections.) . . . . ∎

▷ **VII.6.** *Trees with node degrees that are prime numbers.* Let $\mathcal{P}$ be the class of all unlabelled plane trees such that the (out)degrees of internal nodes belong to the set of prime numbers, $\{2, 3, 5, \ldots\}$. One has $P(z) = z + z^3 + z^4 + 2z^5 + 6z^6 + 8z^7 + 29z^8 + 50z^9 + \cdots$, and $P_n \sim CA^n n^{-3/2}$, with $A \doteq 2.79256\,84676$. The asymptotic form "forgets" many details of the distribution of primes, so that it can be obtained to great accuracy. (Compare with Example V.2, p. 297 and Note VII.24, p. 480.) ◁

VII. 3.2. Basic tree parameters.

Throughout this subsection, we consider a simple variety of trees $\mathcal{V}$, whose generating function (OGF or EGF, as the case may be) will be denoted by $y(z)$, satisfying the inverse relation $y = z\phi(y)$. In order to place all cases under a single umbrella, we shall write $y_n = [z^n]y(z)$, so that the number of trees of size n is either $V_n = y_n$ (unlabelled case) or $V_n = n!y_n$ (labelled case). We postulate throughout that $y(z)$ *belongs to the smooth inverse-function schema and is aperiodic.*

As already seen on several occasions in Chapter III (Section III. 5, p. 181), additive parameters lead to generating functions that are expressible in terms of the basic tree generating function $y(z)$. Now that singularity analysis is available, such generating functions can be exploited systematically, with a wealth of asymptotic estimates relative to trees of large sizes coming within easy reach. The universality of the square-root singularity among varieties of trees that satisfy the smoothness assumption of Definition VII.3 then implies *universal* behaviour for many tree parameters, which we now list.

(*i*) *Node degrees.* The degree of the root in a large random tree is $O(1)$ on average and with high probability, and its asymptotic distribution can be generally determined (Example VII.6). A similar property holds for the *degree of a random node* in a random tree (Example VII.8).

(*ii*) *Level profiles* can also be determined. The quantity of interest is the mean *number of nodes in the kth layer* from the root in a random tree. It is seen for instance that, near the root, a tree from a simple variety tends to grow linearly (Example VII.7), this in sharp contrast with other random tree models (for instance, increasing trees, Subsection VII. 9.2, p. 526), where the growth is exponential. This property is one of the numerous indications that random trees taken from simple varieties are skinny and far from having a well-balanced shape. A related property is the fact that path length is on average $O(n\sqrt{n})$ (Example VII.9), which means that the typical depth of a random node in a random tree is $O(\sqrt{n})$.

These basic properties are only the tip of an iceberg. Indeed, Meir and Moon, who launched the study of simple varieties of trees (the seminal paper [435] can serve as a good starting point) have worked out literally several dozen analyses of parameters of trees, using a strategy similar to the one presented here[6]. We shall have occasion, in Chapter IX, to return to probabilistic properties of simple varieties of trees satisfying the smooth inverse-function schema—we only indicate here for completeness that

---

[6]The main difference is that Meir and Moon appeal to the Darboux–Pólya method discussed in Section VI. 11 (p. 433) instead of singularity analysis.

Tree	$\phi(w)$	τ, ρ	PGF of root degree	(type)
simple variety	—	—	$u\phi'(\tau u)/\phi'(\tau)$	
binary	$(1+w)^2$	$1, \frac{1}{4}$	$\frac{1}{2}u + \frac{1}{2}u^2$	(Bernoulli)
unary–binary	$1 + w + w^2$	$1, \frac{1}{3}$	$\frac{1}{3}u + \frac{2}{3}u^2$	(Bernoulli)
general	$(1-w)^{-1}$	$\frac{1}{2}, \frac{1}{4}$	$u/(2-u)^2$	(sum of two geometric)
Cayley	e^w	$1, e^{-1}$	ue^{u-1}	(shifted Poisson)

Figure VII.5. The distribution of root degree in simple varieties of trees of the smooth inverse-function schema.

height is known generally to scale as $\sqrt{n}$ and is associated to a limiting theta distribution (see Proposition V.4, p. 329 for the case of Catalan trees and Subsection VII. 10.2, p. 535, for general results), with similar properties holding true for width as shown by Odlyzko–Wilf and Chassaing–Marckert–Yor [112, 463].

Example VII.6. *Root degrees in simple varieties.* Here is an immediate application of singularity analysis, one that exemplifies the synthetic type of reasoning that goes along with the method. Take for notational simplicity a simple family $\mathcal{V}$ that is unlabelled, with OGF $V(z) \equiv y(z)$. Let $\mathcal{V}^{[k]}$ be the subset of $\mathcal{V}$ composed of all trees whose root has degree equal to k. Since a tree in $\mathcal{V}^{[k]}$ is formed by appending a root to a collection of k trees, one has

$$V^{[k]}(z) = \phi_k z y(z)^k, \qquad \phi_k := [w^k]\phi(w).$$

For any *fixed* k, a singular expansion results from raising both members of (22) to the kth power; in particular,

$$(23) \qquad V^{[k]}(z) = \phi_k z \left[\tau^k - k\gamma \tau^{k-1} \sqrt{1 - \frac{z}{\rho}} + O\left(1 - \frac{z}{\rho}\right) \right].$$

This is to be compared with the basic estimate (22): the ratio $V_n^{[k]}/V_n$ is then asymptotic to the ratio of the coefficients of $\sqrt{1 - z/\rho}$ in the corresponding generating functions, $V^{[k]}(z)$ and $V(z) \equiv y(z)$. Thus, for any fixed k, we have found that

$$(24) \qquad \frac{V_n^{[k]}}{V_n} = \rho k \phi_k \tau^{k-1} + O(n^{-1/2}).$$

(The error term can be strengthened to $O(n^{-1})$ by pushing the expansion one step further.)

The ratio $V_n^{[k]}/V_n$ is the probability that the root of a random tree of size n has degree k. Since $\rho = 1/\phi'(\tau)$, one can rephrase (24) as follows: *In a smooth simple variety of trees, the random variable Δ representing root-degree admits a discrete limit distribution given by*

$$(25) \qquad \lim_{n \to \infty} \mathbb{P}_{\mathcal{V}_n}(\Delta = k) = \frac{k \phi_k \tau^{k-1}}{\phi'(\tau)}.$$

(By general principles expounded in Chapter IX, convergence is uniform.) Accordingly, the probability generating function (PGF) of the limit law admits the simple expression

$$\mathbb{E}_{\mathcal{V}_n}\left(u^\Delta\right) = u\phi'(\tau u)/\phi'(\tau).$$

The distribution is thus characterized by the fact that its PGF is a scaled version of the *derivative* of the basic tree constructor $\phi(w)$. Figure VII.5 summarizes this property together with its specialization to our four pilot examples. .... ∎

Additive functionals. Singularity analysis applies to many additive parameters of trees. Consider three tree parameters, ξ, η, σ satisfying the basic relation,

$$(26) \qquad \xi(t) = \eta(t) + \sum_{j=1}^{\deg(t)} \sigma(t_j),$$

which can be taken to define $\xi(t)$ in terms of the simpler parameter $\eta(t)$ (a "toll", cf Subsection VI. 10.3, p. 427) and the sum of values of σ over the root subtrees of t (with $\deg(t)$ the degree of the root and t_j the jth root-subtree of t). In the case of a recursive parameter, $\xi \equiv \sigma$, unwinding the recursion shows that $\xi(t) := \sum_{s \preceq t} \eta(s)$, where the sum is extended to *all* subtrees s of t. As we are interested in average-case analysis, we introduce the cumulative GFs,

$$(27) \qquad \Xi(z) = \sum_t \xi(t) z^{|t|}, \quad H(z) = \sum_t \eta(t) z^{|t|}, \quad \Sigma(z) = \sum_t \sigma(t) z^{|t|},$$

assuming again an unlabelled variety of trees for simplicity.

We first state a simple algebraic result which formalizes several of the calculations of Section III. 5, p. 181, dedicated to recursive tree parameters.

Lemma VII.1 (Iteration lemma for trees). *For tree parameters from a simple variety with GF $y(z)$ that satisfy the additive relation (26), the cumulative generating functions (27), are related by*

$$(28) \qquad \Xi(z) = H(z) + z\phi'(y(z))\Sigma(z).$$

In particular, if ξ is defined recursively in terms of η, that is, $\sigma \equiv \xi$, one has

$$(29) \qquad \Xi(z) = \frac{H(z)}{1 - z\phi'(y(z))} = \frac{zy'(z)}{y(z)} H(z).$$

Proof. We have

$$\Xi(z) = H(z) + \widetilde{\Xi}(z), \qquad \text{where} \quad \widetilde{\Xi}(z) := \sum_{t \in \mathcal{V}} \left(z^{|t|} \sum_{j=1}^{\deg(t)} \sigma(t_j) \right).$$

Spitting the expression of $\widetilde{\Xi}(z)$ according to the values r of root degree, we find

$$
\begin{aligned}
\widetilde{\Xi}(z) &= \sum_{r \geq 0} \phi_r z^{1+|t_1|+\cdots+|t_r|} (\sigma(t_1) + \sigma(t_2) + \cdots + \sigma(t_r)) \\
&= z \sum_{r \geq 0} \phi_r \left(\Sigma(z) y(z)^{r-1} + y(z)\Sigma(z)y(z)^{r-2} + \cdots y(z)^{r-1}\Sigma(z) \right) \\
&= z\Sigma(z) \cdot \sum_{r \geq 0} \left(r\phi_r y(z)^{r-1} \right),
\end{aligned}
$$

which yields the linear relation expressing Ξ in (28).

In the recursive case, the function Ξ is determined by a linear equation, namely $\Xi(z) = H(z) + z\phi'(y(z))\Xi(z)$, which, once solved, provides the first form of (29). Differentiation of the fundamental relation $y = z\phi(y)$ yields the identity

$$y'(1 - z\phi'(y)) = \phi(y) = \frac{y}{z}, \qquad \text{i.e.,} \qquad 1 - z\phi'(y) = \frac{y}{zy'},$$

from which the second form results. ∎

▷ **VII.7.** *A symbolic derivation.* For a recursive parameter, we can view $\Xi(z)$ as the GF of trees with one subtree marked, to which is attached a weight of η. Then (29) can be interpreted as follows: point to an arbitrary node at a tree in $\mathcal{V}$ (the GF is $zy'(z)$), remove the tree attached to this node (a factor of $y(z)^{-1}$), and replace it by the same tree but now weighted by η (the GF is $H(z)$). ◁

▷ **VII.8.** *Labelled varieties.* Formulae (28) and (29) hold verbatim for labelled trees (either of the plane or non-plane type), provided we interpret $y(z)$, $\Xi(z)$, $H(z)$ as EGFs: $\Xi(z) := \sum_{t \in \mathcal{V}} \xi(t)z^{|t|}/|t|!$, and so on. ◁

Example VII.7. *Mean level profile in simple varieties.* The question we address here is that of determining the mean number of nodes at level k (i.e., at distance k from the root) in a random tree of some large size n. (An explicit expression for the joint distribution of nodes at all levels has been developed in Subsection III. 6.2, p. 193, but this multivariate representation is somewhat hard to interpret asymptotically.)

Let $\xi_k(t)$ be the number of nodes at level k in tree t. Define the generating function of cumulated values,

$$X_k(z) := \sum_{t \in \mathcal{V}} \xi_k(t)z^{|t|}.$$

Clearly, $X_0(z) \equiv y(z)$ since each tree has a unique root. Then, since the parameter ξ_k is the sum over subtrees of parameter ξ_{k-1}, we are in a situation exactly covered by (28), with $\eta(t) \equiv 0$. The recurrence $X_k(z) = z\phi'(y(z))\Xi_{k-1}(z)$, is then immediately solved, to the effect that

$$(30) \qquad\qquad X_k(z) = \left(z\phi'(y(z))\right)^k y(z).$$

Making use of the (analytic) expansion of ϕ' at τ, namely, $\phi'(y) \sim \phi'(\tau) + \phi''(\tau)(y - \tau)$ and of $\rho\phi'(\tau) = 1$, one obtains, for any fixed k:

$$X_k(z) \sim \left(1 - k\gamma\rho\phi''(\tau)\sqrt{1 - \frac{z}{\rho}}\right)\left(\tau - \gamma\sqrt{1 - \frac{z}{\rho}}\right) \sim \tau - \gamma(\tau\rho\phi''(\tau)k + 1)\sqrt{1 - \frac{z}{\rho}}.$$

Thus comparing the singular part of $X_k(z)$ to that of $y(z)$, we find: *For* fixed k, *the mean number of nodes at level k in a tree is of the asymptotic form*

$$\mathbb{E}_{\mathcal{V}_n}[\xi_k] \sim Ak + 1, \qquad A := \tau\rho\phi''(\tau).$$

This result was first given by Meir and Moon [435]. The striking fact is that, although the number of nodes at level k can at least double at each level, growth is only linear on average. In figurative terms, the immediate vicinity of the root starts like a "cone", and trees of simple varieties tend to be rather skinny near their base.

When used in conjunction with saddle-point bounds (p. 246), the exact GF expression of (30) additionally provides a probabilistic upper bound on the height of trees of the form $O(n^{1/2+\delta})$ for any $\delta > 0$. Indeed restrict z to the interval $(0, \rho)$ and assume that $k = n^{1/2+\delta}$. Let χ be the height parameter. First, we have

$$(31) \qquad\qquad \mathbb{P}_{\mathcal{V}_n}(\chi \geq k) \equiv \mathbb{E}_{\mathcal{V}_n}(\llbracket \xi_k \geq 1 \rrbracket) \leq \mathbb{E}_{\mathcal{V}_n}(\xi_k).$$

Figure VII.6. Three random 2–3 trees ($\Omega = \{0, 2, 3\}$) of size $n = 500$ have height, respectively, 48, 57, 47, in agreement with the fact that height is typically $O(\sqrt{n})$.

Next by saddle-point bounds, for any legal positive x (that is, $0 < x < R_{\mathrm{conv}}(\phi)$),

$$(32) \qquad \mathbb{E}_{\mathcal{Y}_n}(\xi_k) \le \left(x\phi'(y(x))\right)^k y(x) x^{-n} \le \tau \left(x\phi'(y(x))\right)^k x^{-n}.$$

Fix now $x = \rho - \frac{n^\delta}{n}$. Local expansions then show that

$$(33) \qquad \log\left(\left(x\phi'(y(x))\right)^k x^{-n}\right) \le -Kn^{3\delta/2} + O\left(n^\delta\right),$$

for some positive constant K. Thus, by (31) and (33): *In a* smooth *simple variety of trees, the probability of height exceeding $n^{1/2+\delta}$ is exponentially small, being of the rough form* $\exp(-n^{3\delta/2})$. *Accordingly, the mean height is $O(n^{1/2+\delta})$ for any $\delta > 0$.* The moments of height were characterized in [246]: the mean is asymptotic to $\lambda\sqrt{n}$ and the limit distribution is of the Theta type encountered in Example V.8, p. 326, in the particular case of general Catalan trees, where explicit expressions are available. (Further local limit and large deviation estimates appear in [230]; we shall return to the topic of tree height in Subsection VII. 10.1, p. 532.) Figure VII.6 displays three random trees of size $n = 500$. . . . . ∎

▷ **VII.9.** *The variance of level profiles.* The BGF of trees with u marking nodes at level k has an explicit expression, in accordance with the developments of Chapter III. For instance for $k = 3$, this is $z\phi(z\phi(z\phi(uy(z))))$. Double differentiation followed by singularity analysis shows that

$$\mathbb{V}_{\mathcal{Y}_n}[\xi_k] \sim \frac{1}{2}A^2k^2 - \frac{1}{2}A(3 - 4A)k + \tau A - 1,$$

another result of Meir and Moon [435]. The precise analysis of the mean and variance in the interesting regime where k is proportional to $\sqrt{n}$ is also given in [435], but it requires either the saddle-point method (Chapter VIII) or the adapted singularity analysis techniques of Theorem IX.16, p. 709.   ◁

Example VII.8. *Mean degree profile.* Let $\xi(t) \equiv \xi_k(t)$ be the number of nodes of degree k in random tree of some variety $\mathcal{V}$. The analysis extends that of the root degree seen earlier. The parameter ξ is an additive functional induced by the basic parameter $\eta(t) \equiv \eta_k(t)$ defined by

$\eta_k(t) := [\![\deg(t) = k]\!]$. By the analysis of root degree, we have for the GF of cumulated values associated to η

$$H(z) = \phi_k z y(z)^k, \qquad \phi_k := [w^k]\phi(w),$$

so that, by the fundamental formula (29),

$$X(z) = \phi_k z y(z)^k \frac{z y'(z)}{y(z)} = z^2 \phi_k y(z)^{k-1} y'(z).$$

The singular expansion of $zy'(z)$ can be obtained from that of $y(z)$ by differentiation (Theorem VI.8, p. 419),

$$zy'(z) = \frac{1}{2}\gamma \frac{1}{\sqrt{1 - z/\rho}} + O(1),$$

the corresponding coefficient satisfying $[z^n](zy') = n y_n$. This gives immediately the singularity type of X, which is of the form of an inverse square root. Thus,

$$X(z) \sim \rho \phi_k \tau^{k-1}(zy'(z))$$

implying ($\rho = \tau/\phi(\tau)$)

$$\frac{X_n}{n y_n} \sim \frac{\phi_k \tau^k}{\phi(\tau)}.$$

Consequently, one has:

Proposition VII.2. *In a* smooth *simple variety of trees, the mean number of nodes of degree k is asymptotic to $\lambda_k n$, where $\lambda_k := \phi_k \tau^k/\phi(\tau)$. Equivalently, the probability distribution of the degree $\Delta^\star$ of a random node in a random tree of size n satisfies*

$$\lim_{n\to\infty} \mathbb{P}_n(\Delta^\star) = \lambda_k \equiv \frac{\phi_k \tau^k}{\phi(\tau)}, \qquad \text{with PGF}: \quad \sum_k \lambda_k u^k = \frac{\phi(u\tau)}{\phi(\tau)}.$$

For the usual tree varieties this gives:

Tree	$\phi(w)$	τ, ρ	probability distribution	(type)
binary	$(1+w)^2$	$1, \frac{1}{4}$	PGF: $\frac{1}{4} + \frac{1}{2}u + \frac{1}{4}u^2$	(Bernoulli)
unary–binary	$1 + w + w^2$	$1, \frac{1}{3}$	PGF: $\frac{1}{3} + \frac{1}{3}u + \frac{1}{3}u^2$	(Bernoulli)
general	$(1-w)^{-1}$	$\frac{1}{2}, \frac{1}{4}$	PGF: $1/(2-u)$	(Geometric)
Cayley	e^w	$1, e^{-1}$	PGF: e^{u-1}	(Poisson)

For instance, asymptotically, a general Catalan tree has on average $n/2$ leaves, $n/4$ nodes of degree 1 $n/8$ of degree 2, and so on; a Cayley tree has $\sim n e^{-1}/k!$ nodes of degree k; for binary (Catalan) trees, the four possible types of nodes each appear with asymptotic frequency $1/4$. (These data agree with the fact that a random tree under $\mathcal{V}_n$ is distributed like a branching process tree determined by the PGF $\phi(u\tau)/\phi(\tau)$; see Subsection III.6.2, p. 193.) .... ■

▷ **VII.10.** *Variances.* The variance of the number of k–ary nodes is $\sim \nu n$, so that the distribution of the number of nodes of this type is concentrated, for each fixed k. The starting point is the BGF defined implicitly by

$$Y(z, u) = z\left(\phi(Y(z, u)) + \phi_k(u - 1)Y(z, u)^k\right),$$

upon taking a double derivative with respect to u, setting $u = 1$, and finally performing singularity analysis on the resulting GF.   ◁

▷ **VII.11.** *The mother of a random node.* The discrepancy in distributions between the root degree and the degree of a random node deserves an explanation. Pick up a node distinct from the root at random in a tree and look at the degree of its mother. The PGF of the law is in the limit $u\phi'(u\tau)/\phi'(\tau)$. Thus the degree of the root is asymptotically the same as that of the mother of any non-root node.

More generally, let X have distribution $p_k := \mathbb{P}(X = k)$. Construct a random variable Y such that the probability $q_k := \mathbb{P}(Y = k)$ is proportional both to k and p_k. Then for the associated PGFs, the relation $q(u) = p'(u)/p'(1)$ holds. The law of Y is said to be the *size-biased* version of the law of X. Here, a mother is picked up with an importance proportional to its degree. In this perspective, Eve appears to be just like a random mother.   ◁

***Example* VII.9.** *Path length.* Path length of a tree is the sum of the distances of all nodes to the root. It is defined recursively by

$$\xi(t) = |t| - 1 + \sum_{j=1}^{\deg(t)} \xi(t_j)$$

(Example III.15, p. 184 and Subsection VI. 10.3, p. 427). Within the framework of additive functional of trees (28), we have $\eta(t) = |t| - 1$ corresponding to the GF of cumulated values $H(z) = zy'(z) - y(z)$, and the fundamental relation (29) gives

$$X(z) = (zy'(z) - y(z))\frac{zy'(z)}{y(z)} = \frac{z^2 y'(z)^2}{y(z)} - zy'(z).$$

The type of $y'(z)$ at its singularity is $Z^{-1/2}$, where $Z := (1 - z/\rho)$. The formula for $X(z)$ involves the square of y', so that the singularity of $X(z)$ is of type Z^{-1}, resembling a simple pole. This means that the cumulated value $X_n = [z^n]X(z)$ grows like ρ^{-n}, so that the mean value of ξ over $\mathcal{V}_n$ has growth $n^{3/2}$. Working out the constants, we find

$$X(z) + zy'(z) \sim \frac{\gamma^2}{4\tau} \frac{1}{Z} + O(Z^{-1/2}).$$

As a consequence:

Proposition VII.3. *In a random tree of size n from a smooth simple variety, the expectation of path length satisfies*

(34)
$$\mathbb{E}_{\mathcal{V}_n}(\xi) = \lambda\sqrt{\pi n^3} + O(n), \qquad \lambda := \sqrt{\frac{\phi(\tau)}{2\tau^2 \phi''(\tau)}}.$$

For our classical varieties, the main terms of (34) are then:

Binary	unary–binary	general	Cayley
$\sim \sqrt{\pi n^3}$	$\sim \frac{1}{2}\sqrt{3\pi n^3}$	$\sim \frac{1}{2}\sqrt{\pi n^3}$	$\sim \sqrt{\frac{1}{2}\pi n^3}$.

Observe that the quantity $\frac{1}{n}\mathbb{E}_{\mathcal{V}_n}(\xi)$ represents the expected depth of a random node in a random tree (the model is then $[1 .. n] \times \mathcal{V}_n$), which is thus $\sim \lambda\sqrt{n}$. (This result is consistent with height of a tree being with high probability of order $O(n^{1/2})$.) .... ∎

▷ **VII.12.** *Variance of path length.* Path length can be analysed starting from the bivariate generating function given by a functional equation of the difference type (see Chapter III, p. 185), which allows for the computation of higher moments. The standard deviation is found to be asymptotic to $\Lambda_2 n^{3/2}$ for some computable constant $\Lambda_2 > 0$, so that the distribution is spread. Louchard [416] and Takács [566] have additionally worked out the asymptotic form of all moments, leading to a characterization of the limit law of path length that can be described in terms of the Airy function: see Subsection VII. 10.1, p. 532.   ◁

# components	$\sim \frac{1}{2} \log n$	Tail length (λ)	$\sim \sqrt{\pi n / 8}$
# cyclic nodes	$\sim \sqrt{\pi n / 2}$	Cycle length (μ)	$\sim \sqrt{\pi n / 8}$
# terminal nodes	$\sim n e^{-1}$	Tree size	$\sim n / 3$
# nodes of in-degree k	$\sim n e^{-k} / k!$	Component size	$\sim 2n / 3$

Figure VII.7. Expectations of the main additive parameters of random mappings of size n.

▷ **VII.13.** *Generalizations of path length.* Define the *subtree size index* of order $\alpha \in \mathbb{R}_{\geq 0}$ to be $\xi(t) \equiv \xi_\alpha(t) := \sum_{s \preceq t} |s|^\alpha$, where the sum is extended to all the subtrees s of t. This corresponds to a recursively defined parameter with $\eta(t) = |t|^\alpha$. The results of Section VI. 10 relative to Hadamard products and polylogarithms make it possible to analyse the singularities of $H(z)$ and $X(z)$. It is found that there are three different regimes

$\alpha > \frac{1}{2}$	$\alpha = \frac{1}{2}$	$\alpha < \frac{1}{2}$
$\mathbb{E}_{\mathcal{V}_n}(\xi) \sim K_\alpha n^\alpha$	$\mathbb{E}_{\mathcal{V}_n}(\xi) \sim K_{1/2} n \log n$	$\mathbb{E}_{\mathcal{V}_n}(\xi) \sim K_\alpha n$

where each K_α is a computable constant. (This extends the results of Subsection VI. 10.3, p. 427 to all simple varieties of trees that are smooth.)   ◁

VII. 3.3. Mappings.

The basic construction of mappings (Chapter II, p. 129),

$$
(35) \quad
\begin{cases}
\mathcal{F} & = & \text{SET}(\mathcal{K}) \\
\mathcal{K} & = & \text{CYC}(\mathcal{T}) \\
\mathcal{T} & = & \mathcal{Z} \star \text{SET}(\mathcal{T})
\end{cases}
\implies
\begin{cases}
F & = & \exp(K) \\
K & = & \log \dfrac{1}{1 - T} \\
T & = & z e^T,
\end{cases}
$$

builds maps from Cayley trees, which constitute a smooth simple variety. The construction lends itself to a number of multivariate extensions. For instance, we already know from Example VII.3, p. 449, that the *number of components* is asymptotic to $\frac{1}{2} \log n$, both on average and in probability.

Take next the parameter χ equal to the number of cyclic points, which gives rise to the BGF

$$
F(z, u) = \exp \left(\log \frac{1}{1 - uT} \right) = (1 - uT)^{-1}.
$$

The mean number of a cyclic points, for a random mapping of size n, is accordingly

$$
(36) \quad \mu_n \equiv \mathbb{E}_{\mathcal{F}_n}[\chi] = \frac{n!}{n^n} [z^n] \left(\frac{\partial}{\partial u} F(z, u) \Big|_{u=1} \right) = \frac{n!}{n^n} [z^n] \frac{T}{(1 - T)^2}.
$$

Singularity analysis is immediate, since

$$
\frac{T}{(1 - T)^2} \underset{z \to e^{-1}}{\sim} \frac{1}{2} \frac{1}{1 - ez} \longrightarrow [z^n] \frac{T}{(1 - T)^2} \underset{n \to \infty}{\sim} \frac{1}{2} e^n.
$$

Thus: *The mean number of cyclic points in a random mapping of size n is asymptotic to* $\sqrt{\pi n / 2}$.

Many parameters can be similarly analysed in a systematic manner, thanks to generating function, as shown in the survey [247]: see Figure VII.7 for a summary

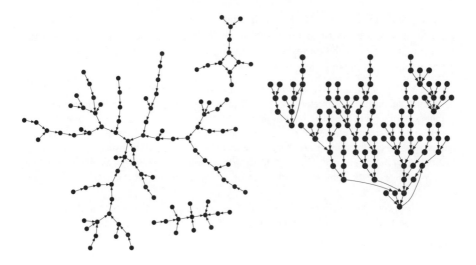

Figure VII.8. Two views of a random mapping of size $n = 100$. The random mapping has three connected components, with cycles of respective size 2, 4, 4; it is made of fairly skinny trees, has a giant component of size 75, and its diameter equals 14.

of results whose proofs we leave as exercises to the reader. The left-most table describes global parameters of mappings; the right-most table is relative to properties of random point in random n-mapping: λ is the *distance to its cycle* of a random point, μ the *length of the cycle* to which the point leads, tree size and component size are, respectively, the size of the largest tree containing the point and the size of its (weakly) connected component. In particular, a random mapping of size n has relatively few components, some of which are expected to be of a large size.

The estimates of Figure VII.7 are in fair agreement with what is observed on the single sample of size $n = 100$ of Figure VII.8: this particular mapping has 3 components (the average is about 2.97), 10 cyclic points (the average, as calculated in (36), is about 12.20), but a fairly large diameter—the maximum value of $\lambda + \mu$, taken over all nodes—equal to 14, and a giant component of size 75. The proportion of nodes of degree $0, 1, 2, 3, 4$ turns out to be, respectively, 39%, 33%, 21%, 7%, 1%, to be compared against the asymptotic values given by a Poisson law of rate 1 (analogous to the degree profile of Cayley trees found in Example VII.8); namely 36.7%, 36.7%, 18.3%, 6.1%, 1.5%.

▷ **VII.14.** *Extremal statistics on mappings.* Let $\lambda^{\max}$, $\mu^{\max}$, and $\rho^{\max}$ be the maximum values of λ, μ, and ρ, taken over all the possible starting points, where $\rho = \lambda + \mu$. Then, the expectations satisfy [247]

$$\mathbb{E}_{\mathcal{F}_n}(\lambda^{\max}) \sim \kappa_1 \sqrt{n}, \quad \mathbb{E}_{\mathcal{F}_n}(\mu^{\max}) \sim \kappa_2 \sqrt{n}, \quad \mathbb{E}_{\mathcal{F}_n}(\rho^{\max}) \sim \kappa_3 \sqrt{n},$$

where $\kappa_1 = \sqrt{2\pi} \log 2 \doteq 1.73746$, $\kappa_2 \doteq 0.78248$ and $\kappa_3 \doteq 2.4149$. (For the estimate relative to $\rho^{\max}$, see also [12].)

The largest tree and the largest components have expectations asymptotic, respectively, to $\delta_1 n$ and $\delta_2 n$, where $\delta_1 \doteq 0.48$ and $\delta_2 \doteq 0.7582$. ◁

The properties outlined above for the class of *all* mappings also prove to be universal for a wide variety of mappings defined by degree restrictions of various sorts: we outline the basis of the corresponding theory in Example VII.10, then show some surprising applications in Example VII.11.

***Example* VII.10.** *Simple varieties of mappings.* Let Ω be a subset of the integers containing 0 and at least another integer greater than 1. Consider mappings $\phi \in \mathcal{F}$ such that the number of preimages of any point is constrained to lie in Ω. Such special mappings may serve to model the behaviour of special classes of functions under iteration, and are accordingly of interest in various areas of computational number theory and cryptography. For instance, the quadratic functions $\phi(x) = x^2 + a$ over $\mathbb{F}_p$ have the property that each element y has either zero, one, or two preimages (depending on whether $y - a$ is a quadratic non-residue, 0, or a quadratic residue).

The basic construction of mappings needs to be amended. Start with the family of trees $\mathcal{T}$ that are the simple variety corresponding to Ω:

$$(37) \qquad\qquad T = z\phi(T), \qquad \phi(w) := \sum_{\omega \in \Omega} \frac{u^{\omega}}{\omega!}.$$

At any vertex on a cycle, one must graft r trees with the constraint that $r + 1 \in \Omega$ (since one edge is coming from the cycle itself). Such legal tuples with a root appended are represented by

$$(38) \qquad\qquad U = z\phi'(T),$$

since ϕ is an exponential generating function and shift $(r \mapsto (r+1))$ corresponds to differentiation. Then connected components and components are formed in the usual way by

$$(39) \qquad\qquad K = \log \frac{1}{1 - U}, \qquad F = \exp(K) = \frac{1}{1 - U}.$$

The three relations (37), (38), (39) fully determine the EGF of Ω–restricted mappings.

The function ϕ is a subseries of the exponential function; hence, it is entire and it satisfies automatically the smoothness conditions of Theorem VII.2, p. 453. With τ the characteristic value, the function $T(z)$ then has a square-root singularity at $\rho = \tau/\phi(\tau)$. The same holds for U, which admits the singular expansion (with γ_1 a constant simply related to γ of equation (22))

$$(40) \qquad\qquad U(z) \sim 1 - \gamma_1 \sqrt{1 - \frac{z}{\rho}},$$

since $U = z\phi'(T)$. Thus, eventually:

$$F(z) \sim \frac{\kappa}{\sqrt{1 - \frac{z}{\rho}}}, \qquad \kappa := \frac{1}{\gamma_1}.$$

There results the *universality of an* $n^{-1/2}$ *counting law* in such constrained mappings:

Proposition VII.4. *Consider mappings with node degrees in a set* $\Omega \subseteq \mathbb{Z}_{\geq 0}$, *such that the corresponding tree family belongs to the smooth implicit function schema and is aperiodic. The number of mappings of size n satisfies*

$$\frac{1}{n!} F_n \sim \frac{\kappa}{\sqrt{\pi n}} \rho^{-n}, \qquad \kappa = \sqrt{\frac{\phi'(\tau)^2}{2\phi(\tau)\phi''(\tau)}}.$$

This statement nicely extends what is known to hold for unrestricted mappings. The analysis of additive functionals can then proceed on lines very similar to the case of standard mappings, to the effect that the estimates of the same form as in Figure VII.7 hold, albeit with different multiplicative factors. The programme just sketched has been carried out in a thorough manner by Arney and Bender [18], whose paper provides a detailed treatment. ...... ∎

Example VII.11. *Applications of random mapping statistics.* There are interesting consequences of the foregoing asymptotic theory of random mappings in several areas of computational mathematics, as we now briefly explain.

Random number generators. Many (pseudo) random number generators operate by iterating a given function φ over a finite domaine $\mathcal{E}$; usually, $\mathcal{E}$ is a large integer interval $[0 .. N - 1]$. Such a scheme produces a pseudo-random sequence $u_0, u_1, u_2, \ldots$, where u_0 is the "seed" and

$$u_{n+1} = \varphi(u_n).$$

Particular strategies are known for the choice of φ, which ensure that the "period" (the maximum of $\rho = \lambda + \mu$, where λ is the distance to cycle and μ is the cycle's length) is of the order of N: this is for instance granted by linear congruential generators and feedback register algorithms; see Knuth's authoritative discussion in [379, Ch. 3]. By contrast, a randomly chosen function φ has expected $O(\sqrt{N})$ cycle time (Figure VII.7, p. 462), so that it is highly likely to give rise to a poor generator. As the popular adage says: "A random *random number generator is bad!*". Accordingly, one can make use of the results of Figure VII.7 and Example VII.10 in order to compare statistical properties of a proposed random number generator to properties of a random function, and discard the former if there is a manifest closeness.

For instance, take φ to be

$$\varphi(x) := x^2 + 1 \mod (10^6 + 3),$$

where the modulus is a prime number. A random mapping of size $(10^6 + 3)$ is expected to cycle on average after about 1250 steps (the expectation of $\rho = \lambda + \mu$ is $\sim \sqrt{\pi N/2}$ by Figure VII.7). From five starting values u_0, we observe the following periods

(41)

u_0 :	3	31	314	3141	31415	314159
$\rho \equiv \lambda + \mu$:	1569	687	985	813	557	932

whose magnitude looks suspiciously like $\sqrt{N}$. Such a random number generator is thus to be discarded. For similar reasons, von Neumann's well-known "middle-square" procedure (start from an ℓ-digit number, then repeatedly square and extract the middle digits) makes for a rather poor random number generator [379, p. 5]. (Related applications to cryptography are presented by Quisquater and Delescaille in [501].)

Floyd's cycle detection. There is a spectacular algorithm due to Floyd [379, Ex. 3.1.6], for *cycle detection*, which is well worth knowing when one needs to experiment with large mappings. Given an initial seed x_0 and a mapping φ, Floyd's algorithm determines, up to a small factor, the value of $\rho(x_0) = \lambda(x_0) + \mu(x_0)$, using only *two registers*. The principle is as follows. Start a tortoise and a hare on u_0 at time 0; then, let the tortoise move at speed 1 along the rho-shaped path and let the hare move at twice the speed. After $\lambda(x_0)$ steps, the tortoise joins the cycle, from which time on, the hare, which is already on the cycle, will catch the tortoise after at most $\mu(x_0)$ steps, since their speed differential on the cycle is one. Pictorially:

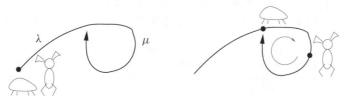

In more dignified terms, setting

$$X_0 = u_0, \quad X_{n+1} = \varphi(X_n), \quad \text{and} \quad Y_0 = u_0, \quad Y_{n+1} = \varphi(\varphi(X_n)),$$

we have the property that the first value v such that $X_v = Y_v \equiv X_{2v}$ must satisfy the inequalities

(42) $\lambda \le v \le \lambda + \mu \le 2v.$

The corresponding algorithm is then extremely short:

> **Algorithm:** *Floyd's Cycle Detector*:
>   tortoise := x_0;   hare := x_0; $v := 0$;
>   **repeat**
>   tortoise :=φ(tortoise);   hare := $\varphi(\varphi$(hare)); $v := v + 1$;
>   **until** tortoise = hare   {v is an estimate of $\lambda + \mu$ in the sense of (42)}.

Pollard's rho method for integer factoring. Pollard [487] had the insight to exploit Floyd's algorithm in order to develop an efficient integer factoring method. Assume *heuristically* that a quadratic function $x \mapsto x^2 + a \bmod p$, with p a prime number, has statistical properties similar to those of a random function (we have verified a particular case by (41) above). It must then tend to cycle after about $\sqrt{p}$ steps. Let N be a (large) number to be factored, and assume for simplicity that $N = pq$, with p and q both prime (but unknown!). Choose a random a and a random initial value x_0, fix

$$\varphi(x) = x^2 + a \quad (\bmod N),$$

and run the hare-and-tortoise algorithm. By the Chinese Remainder Theorem, the value of a number $x \bmod N$ is determined by the pair $(x \bmod p, x \bmod q)$; the tortoise T and the hare H can then be seen as running two simultaneous races, one modulo p, the other modulo q. Say that $p < q$. After about $\sqrt{p}$ steps, one is likely to have

$$H \equiv T \quad (\bmod p),$$

while, most probably, hare and tortoise will be non-congruent mod q. In other words, the greatest common divisor of the difference $(H - T)$ and N will provide p; hence it factors N. The resulting algorithm is also extremely short:

> **Algorithm:** *Pollard's Integer Factoring*:
>   **choose** a, x_0 randomly in $[0 .. N - 1]$;
> $T := x_0$; $H := x_0$;
>   **repeat**
> $T := (T^2 + a) \bmod N$; $H := (H^2 + a)^2 + a \bmod N$;
> $D := \gcd(H - T, N)$;
>   **until** $D \ne 1$ {if $D \ne 0$, a non-trivial divisor has been found}.

The agreement with what the theory of random mappings predicts is excellent: one indeed obtains an algorithm that factors large numbers N in $O(N^{1/4})$ operations with high probability (see for instance the data in [538, p. 470]).

  Although Pollard's algorithm is, for very large N, subsumed by other factoring methods, it is still the best for moderate values of N or for numbers with small divisors, where it proves far superior to trial divisions. Equally importantly, similar ideas serve in many areas of computational number theory; for instance the determination of discrete logarithms. (Proving rigorously what one observes in simulations is another story: it often requires advanced methods of number theory [23, 442].) ....■

▷ **VII.15.** *Probabilities of first-order sentences.* A beautiful theorem of Lynch [426], much in line with the global aims of analytic combinatorics, gives a class of properties of random mappings for which asymptotic probabilities are systematically computable. In mathematical logic, a first-order sentence is built out of variables, equality, boolean connectives ($\lor, \land, \neg$, etc), and quantifiers ($\forall, \exists$). In addition, there is a function symbol φ, representing a generic mapping.

> **Theorem.** *Given a property P expressed by a first-order sentence, let $\mu_n(P)$ be the probability that P is satisfied by a random mapping φ of size n. Then the quantity $\mu_\infty(P) = \lim_{n\to\infty} \mu_n(P)$ exists and its value is given by an expression consisting of integer constants and the operators $+, -, \times, \div,$ and e^x.*

For instance:

P :	φ is perm.	φ without fixed pt.	φ has #leaves ≥ 2
	$\forall x \exists y \varphi(y) = x$	$\forall x \neg \varphi(x) = x$	$\exists x, y\, [x \neq y \land \forall z[\varphi(z) \neq x \land \varphi(z) \neq y]]$
$\mu_\infty(P)$:	0	e^{-1}	1

One can express in this language a property like P_{12} : *"all cycles of length 1 are attached to trees of height at most 2"*, for which the limit probability is $e^{-1+e^{-1+e^{-1}}}$. The proof of the theorem is based on Ehrenfeucht games supplemented by ingenious inclusion–exclusion arguments. (Many cases, like P_{12}, can be directly treated by singularity analysis.) Compton [125, 126, 127] has produced lucid surveys of this area, known as finite model theory.   ◁

VII. 4. Tree-like structures and implicit functions

The aim of this section is to demonstrate the universality of the square-root singularity type for classes of recursively defined structures, which considerably extend the case of (smooth) simple varieties of trees. The starting point is the investigation of recursive classes $\mathcal{Y}$, with associated GF $y(z)$, that correspond to a specification:

$$(43) \qquad \mathcal{Y} = \mathfrak{G}[\mathcal{Z}, \mathcal{Y}] \qquad \Longrightarrow \qquad y(z) = G(z, y(z)).$$

In the labelled case, $y(z)$ is an EGF and $\mathfrak{G}$ may be an arbitrary composition of basic constructors, which is reflected by a bivariate function $G(z, w)$; in the unlabelled case, $y(z)$ is an OGF and $\mathfrak{G}$ may be an arbitrary composition of unions, products, and sequences. (Pólya operators corresponding to unlabelled sets and cycles are discussed in Section VII. 5, p. 475.) This situation covers structures that we have already seen, like Schröder's bracketing systems (Chapter I, p. 69) and hierarchies (Chapter II, p. 128), as well as new ones to be examined here; namely, paths with diagonal steps and trees with variable node sizes or edge lengths.

VII. 4.1. The smooth implicit-function schema. The investigation of (43) necessitates certain analytic conditions to be satisfied by the bivariate function G, which we first encapsulate into the definition of a schema.

Definition VII.4. *Let $y(z)$ be a function analytic at 0, $y(z) = \sum_{n\geq 0} y_n z^n$, with $y_0 = 0$ and $y_n \geq 0$. The function is said to belong to the* smooth implicit-function schema *if there exists a bivariate $G(z, w)$ such that*

$$y(z) = G(z, y(z)),$$

where $G(z, w)$ satisfies the following conditions.

$(\mathbf{I_1})$: $G(z, w) = \sum_{m,n \geq 0} g_{m,n} z^m w^n$ is analytic in a domain $|z| < R$ and $|w| < S$, for some $R, S > 0$.

$(\mathbf{I_2})$: *The coefficients of* G *satisfy*

(44)
$$g_{m,n} \geq 0, \quad g_{0,0} = 0, \quad g_{0,1} \neq 1,$$
$$g_{m,n} > 0 \quad \text{for some } m \text{ and for some } n \geq 2.$$

$(\mathbf{I_3})$: *There exist two numbers* r, s, *such that* $0 < r < R$ *and* $0 < s < S$, *satisfying the system of equations,*

(45) $G(r, s) = s, \quad G_w(r, s) = 1, \qquad \text{with} \quad r < R, \quad s < S,$

which is called the characteristic system.

A class $\mathcal{Y}$ with a generating $y(z)$ satisfying $y(z) = G(z, y(z))$ is also said to belong to the smooth implicit-function schema.

Postulating that $G(z, w)$ is analytic and with non-negative coefficients is a minimal assumption in the context of analytic combinatorics. The problem is assumed to be normalized, so that $y(0) = 0$ and $G(0, 0) = 0$, the condition $g_{0,1} \neq 1$ being imposed to avoid that the implicit equation be of the reducible form $y = y + \cdots$ (first line of (44)). The second condition of (44) means that in $G(z, y)$, the dependency on y is nonlinear (otherwise, the analysis reduces to rational and meromorphic asymptotic methods of Chapter V). The major analytic condition is $(\mathbf{I_3})$, which postulates the existence of positive solutions r, s to the *characteristic system* within the domain of analyticity of G.

The main result[7] due to Meir and Moon [439] expresses universality of the square-root singularity together with its usual consequences regarding asymptotic counting.

Theorem VII.3 (Smooth implicit-function schema). *Let* $y(z)$ *belong to the* smooth implicit-function schema *defined by* $G(z, w)$, *with* (r, s) *the positive solution of the characteristic system. Then,* $y(z)$ *converges at* $z = r$, *where it has a square-root singularity,*

$$y(z) \underset{z \to r}{=} s - \gamma \sqrt{1 - z/r} + O(1 - z/r), \qquad \gamma := \sqrt{\frac{2r G_z(r, s)}{G_{ww}(r, s)}},$$

the expansion being valid in a Δ–domain. *If, in addition,* $y(z)$ *is aperiodic[8], then* r *is the unique dominant singularity of* y *and the coefficients satisfy*

$$[z^n] y(z) \underset{n \to \infty}{=} \frac{\gamma}{2\sqrt{\pi n^3}} r^{-n} \left(1 + O(n^{-1}) \right).$$

---

[7]This theorem has an interesting history. An overly general version of it was first stated by Bender in 1974 (Theorem 5 of [36]). Canfield [102] pointed out ten years later that Bender's conditions were not quite sufficient to grant square-root singularity. A corrected statement was given by Meir and Moon in [439] with a further (minor) erratum in [438]. We follow here the form given in Theorem 10.13 of Odlyzko's survey [461] with the correction of another minor misprint (regarding $g_{0,1}$ which should read $g_{0,1} \neq 1$). A statement concerning a restricted class of functions (either polynomial or entire) already appears in Hille's book [334, vol. I, p. 274].

[8]In the usual sense of Definition IV.5, p. 266. Equivalently, there exist three indices $i < j < k$ such that $y_i y_j y_k \neq 0$ and $\gcd(j - i, k - i) = 1$.

Observe that the statement implies the existence of *exactly one root* of the characteristic system within the part of the positive quadrant where G is analytic, since, obviously, y_n cannot admit two asymptotic expressions with different parameters. A complete expansion exists in powers of $(1 - z/r)^{1/2}$ (for $y(z)$) and in powers of $1/n$ (for y_n), while periodic cases can be treated by a simple extension of the technical apparatus to be developed.

The proof of this theorem first necessitates two lemmas of independent interest: (*i*) Lemma VII.2 is logically equivalent to an analytic version of the classical Implicit Function Theorem found in Appendix B.5: *Implicit Function Theorem*, p. 753. (*ii*) Lemma VII.3 supplements this by describing what happens at a point where the implicit function theorem "fails". These two statements extend the analytic and singular inversion lemmas of Subsection IV. 7.1, p. 275.

Lemma VII.2 (Analytic Implicit Functions). *Let $F(z, w)$ be z bivariate function analytic at $(z, w) = (z_0, w_0)$. Assume that $F(z_0, w_0) = 0$ and $F_w(z_0, w_0) \neq 0$. Then, there exists a unique function $y(z)$ analytic in a neighbourhood of z_0 such that $y(z_0) = w_0$ and $F(z, y(z)) = 0$.*

Proof. This is a restatement of the Analytic Implicit Function Theorem of Appendix B.5: *Implicit Function Theorem*, p. 753, upon effecting a translation $z \mapsto z + z_0$, $w \mapsto w + w_0$. ∎

Lemma VII.3 (Singular Implicit Functions). *Let $F(z, w)$ be a bivariate function analytic at $(z, w) = (z_0, w_0)$. Assume the conditions: $F(z_0, w_0) = 0$, $F_z(z_0, w_0) \neq 0$, $F_w(z_0, w_0) = 0$, and $F_{ww}(z_0, w_0) \neq 0$. Choose an arbitrary ray of angle θ emanating from z_0. Then there exists a neighbourhood Ω of z_0 such that at every point z of Ω with $z \neq z_0$ and z not on the ray, the equation $F(z, y) = 0$ admits two analytic solutions $y_1(z)$ and $y_2(z)$ that satisfy, as $z \to z_0$:*

$$ y_1(z) = y_0 - \gamma\sqrt{1 - z/z_0} + O\left(1 - z/z_0\right), \qquad \gamma := \sqrt{\frac{2z_0 F_z(z_0, w_0)}{F_{ww}(z_0, w_0)}}, $$

and similarly for y_2 whose expansion is obtained by changing $\sqrt{\ }$ to $-\sqrt{\ }$.

Proof. Locally, near (r, s), the function $F(z, w)$ behaves like

$$ \tag{46} F + (w - s)F_w + (z - r)F_z + \frac{1}{2}(w - s)^2 F_{ww}, $$

(plus smaller order terms), where F and its derivatives are evaluated at the point (r, s). Since $F = F_w = 0$, cancelling (46) suggests for the solutions of $F(z, w) = 0$ near $z = r$ the form

$$ w - s = \pm\gamma\sqrt{r - z} + O(z - r), $$

which is consistent with the statement. This informal argument can be justified by the following steps (details omitted): (*a*) establish the existence of a formal solution in powers of $\pm(1 - z/r)^{1/2}$; (*b*) prove, by the method of majorant series, that the formal solutions also converge locally and provide a solution to the equation.

Alternatively, by the Weierstrass Preparation Theorem (Appendix B.5: *Implicit Function Theorem*, p. 753) the two solutions $y_1(z), y_2(z)$ that assume the value s

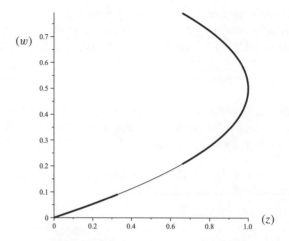

Figure VII.9. The connection problem for the equation $w = \frac{1}{4}z + w^2$ (with explicit forms $w = (1 \pm \sqrt{1-z})/2$): the combinatorial solution $y(z)$ near $z = 0$ and the two analytic solutions $y_1(z)$, $y_2(z)$ near $z = 1$.

at $z = r$ are solutions of a quadratic equation

$$(Y - s)^2 + b(z)(Y - s) + c(z) = 0,$$

where b and c are analytic at $z = r$, with $b(r) = c(r) = 0$. The solutions are then obtained by the usual formula for solving a quadratic equation,

$$Y - s = \frac{1}{2}\left(-b(z) \pm \sqrt{b(z)^2 - 4c(z)}\right),$$

which provides for $y_1(z)$ an expression as the square-root of an analytic function and yields the statement. ∎

It is now possible to return to the proof of our main statement.

Proof. [Theorem VII.3] Given the two lemmas, the general idea of the proof of Theorem VII.3 can be easily grasped. Set $F(z, w) = w - G(z, w)$. There exists a unique analytic function $y(z)$ satisfying $y = G(z, y)$ near $z = 0$, by the analytic lemma. On the other hand, by the singular lemma, near the point $(z, w) = (r, s)$, there exist *two* solutions y_1, y_2, both of which have a square root singularity. Given the positive character of the coefficients of G, it is not hard to see that, of y_1, y_2, the function $y_1(z)$ is increasing as z approaches z_0 from the left (assuming the principal determination of the square root in the definition of γ). A simple picture of the situation regarding the solutions to the equation $y = G(z, y)$ is exemplified by Figure VII.9.

The problem is then to show that a smooth analytic curve (the thin-line curve in Figure VII.9) *does* connect the positive-coefficient solution at 0 to the increasing-branch solution at r. Precisely, one needs to check that $y_1(z)$ (defined near r) is the analytic continuation of $y(z)$ (defined near 0) as z increases along the positive real axis. This is indeed a delicate *connection problem* whose technical proof is discussed in Note VII.16. Once this fact is granted and it has been verified that r is the unique dominant singularity of $y(z)$ (Note VII.17), the statement of Theorem VII.3 follows directly by singularity analysis.   ∎

▷ **VII.16.** *The connection problem for implicit functions.* A proof that $y(z)$ and $y_1(z)$ are well connected is given by Meir and Moon in the study [439], from which our description is adapted.

Let ρ be the radius of convergence of $y(z)$ at 0 and $\tau = y(\rho)$. The point ρ is a singularity of $y(z)$ by Pringsheim's Theorem. The goal is to establish that $\rho = r$ and $\tau = s$. Regarding the curve

$$\mathcal{C} = \{(z, y(z)) \mid 0 \le z \le \rho\},$$

this means that three cases are to be excluded:

(a) $\mathcal{C}$ stays entirely in the interior of the rectangle

$$R := \{(z, y) \mid 0 \le z \le r, \, 0 \le y \le s\}.$$

(b) $\mathcal{C}$ intersects the upper side of the rectangle R at some point of abscissa $r_0 < r$ where $y(r_0) = s$.

(c) $\mathcal{C}$ intersects the right-most side of the rectangle R at the point $(r, y(r))$ with $y(r) < s$.

Graphically, the three cases are depicted in Figure VII.10.

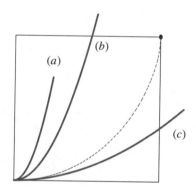

Figure VII.10. The three cases (a), (b), and (c), to be excluded (solid lines).

In the discussion, we make use of the fact that $G(z, w)$, which has non-negative coefficients is an increasing function in each of its argument. Also, the form

(47)
$$y' = \frac{G_z(z, y)}{1 - G_w(z, y)},$$

shows differentiability (hence analyticity) of the solution y as soon as $G_w(z, y) \ne 1$.

Case (a) is excluded. Assume that $0 < \rho < r$ and $0 < \tau < s$. Then, we have $G_w(r, s) = 1$, and by monotonicity properties of G_w, the inequality $G_w(\rho, \tau) < 1$ holds. But then $y(z)$ must be analytic at $z = \rho$, which contradicts the fact that ρ is a singularity.

Case (b) is excluded. Assume that $0 < r_0 < r$ and $y(r_0) = s$. Then there are two distinct points on the implicit curve $y = G(z, y)$ at the same altitude, namely (r_0, s) and (r, s), implying the equalities

$$y(r_0) = G(r_0, y(r_0)) = s = G(r, s),$$

which contradicts the monotonicity properties of G.

Case (c) is excluded. Assume that $y(r) < s$. Let $a < r$ be a point chosen close enough to r. Then above a, there are three branches of the curve $y = G(z, y)$, namely $y(a), y_1(a), y_2(a)$, where the existence of y_1, y_2 results from Lemma VII.3. This means that the function $y \mapsto G(a, y)$ has a graph that intersects the main diagonal at three points, a contradiction with the fact that $G(a, y)$ is a convex function of y.   ◁

▷ **VII.17.** *Unicity of the dominant singularity.* From the previous note, we know that $y(r) = s$, with r the radius of convergence of y. The aperiodicity of y implies that $|y(\zeta)| < y(r)$ for all $|\zeta|$ such that $|\zeta| = r$ and $|\zeta| \neq r$ (see the Daffodil Lemma IV.1, p. 266). One then has for any such ζ the property: $|G_w(\zeta, y(\zeta))| < G(r, s) = 1$, by monotonicity of G_w. But then by (47) above, this implies that $y(\zeta)$ is analytic at ζ. ◁

The solutions to the characteristic system (45) can be regarded as the intersection points of two curves, namely,

$$G(r, s) - s = 0, \qquad G_w(r, s) = 1.$$

Here are plots in the case of two functions G: the first one has non-negative coefficients whereas the second one (corresponding to a counterexample of Canfield [102]) involves negative coefficients. Positivity of coefficients implies convexity properties that avoid pathological situations.

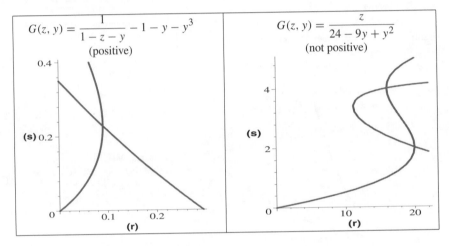

VII. 4.2. Combinatorial applications. Many combinatorial classes, which admit a recursive specification of the form $\mathcal{Y} = \mathfrak{G}(\mathcal{Z}, \mathcal{Y})$, as in (43), p. 467, can be subjected to Theorem VII.3. The resulting structures are, to varying degrees, avatars of tree structures. In what follows, we describe a few instances in which the square-root universality holds.

(*i*) *Hierarchies* are trees enumerated by the number of their leaves (Examples VII.12 and VII.13).

(*ii*) *Trees with variable node sizes* generalize simple families of trees; they occur in particular as mathematical models of secondary structures in biology (Example VII.14).

(*iii*) *Lattice paths with variable edge lengths* are attached to some of the most classical objects of combinatorial theory (Note VII.19).

Example VII.12.   *Labelled hierarchies.* The class $\mathcal{L}$ of labelled hierarchies, as defined in Note II.19, p. 128, satisfies

$$\mathcal{L} = \mathcal{Z} + \text{SET}_{\geq 2}(\mathcal{L}) \qquad \Longrightarrow \qquad L = z + e^L - 1 - L.$$

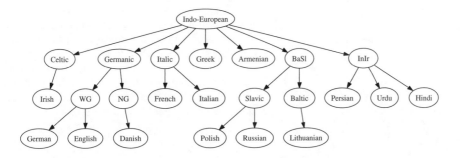

Figure VII.11. A *hierarchy* placed on some of the modern Indo-European languages.

These occur in statistical classification theory: given a collection of n distinguished items, L_n is the number of ways of superimposing a non-trivial classification (cf Figure VII.11). Such abstract classifications usually have no planar structure, hence our modelling by a labelled set construction.

In the notations of Definition VII.4, p. 467, the basic function is $G(z, w) = z + e^w - 1 - w$, which is analytic in $|z| < \infty$, $|w| < \infty$. The characteristic system is

$$r + e^s - 1 - s = s, \qquad e^s - 1 = 1,$$

which has a unique positive solution, $s = \log 2$, $r = 2 \log 2 - 1$, obtained by solving the second equation for s, then propagating the solution to get r. Thus, hierarchies belong to the smooth implicit-function schema, and, by Theorem VII.3, the EGF $L(z)$ has a square-root singularity. One then finds mechanically

$$\frac{1}{n!} L_n \sim \frac{1}{2\sqrt{\pi n^3}} (2 \log 2 - 1)^{-n+1/2}.$$

(The unlabelled counterpart is the object of Note VII.23, p. 479.) .... ∎

▷ **VII.18.** *The degree profile of hierarchies.* Combining BGF techniques and singularity analysis, it is found that a random hierarchy of some large size n has on average about $0.57n$ nodes of degree 2, $0.18n$ nodes of degree 3, $0.04n$ nodes of degree 4, and less than $0.01n$ nodes of degree 5 or higher. ◁

***Example* VII.13.** *Trees enumerated by leaves.* For a (non-empty) set $\Omega \subset \mathbb{Z}_{\geq 0}$ that does not contain 0,1, it makes sense to consider the class of labelled trees,

$$\mathcal{C} = Z + \mathrm{SEQ}_\Omega(\mathcal{C}) \qquad \text{or} \qquad \mathcal{C} = Z + \mathrm{SET}_\Omega(\mathcal{C}).$$

(A similar discussion can be conducted for *unlabelled plane* trees, with OGFs replacing EGFs.) These are rooted trees (plane or non-plane, respectively), with size determined by the number of leaves and with degrees constrained to lie in Ω. The EGF is then of the form

$$C(z) = z + \eta(C(z)).$$

This variety of trees includes the labelled hierarchies, which correspond to $\eta(w) = e^w - 1 - w$.

Assume for simplicity η to be entire (possibly a polynomial). The basic function is $G(z, w) = z + \eta(w)$, and the characteristic system is $s = r + \eta(s)$, $\eta'(s) = 1$. Since $\eta'(0) = 0$ and $\eta'(+\infty) = +\infty$, this system always has a solution:

$$s = \eta^{[-1]}(1), \qquad r = s - \eta(s).$$

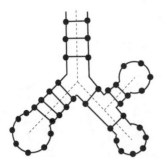

A fragment of RNA is, in first approximation, a tree-like structure with edges corresponding to base pairs and "loops" corresponding to leaves. There are constraints on the sizes of leaves (taken here between 4 and 7) and length of edges (here between 1 and 4 base pairs). We model such an RNA fragment as a planted tree P attached to a binary tree (Y) with equations:

$$\begin{cases} P = AY, \quad Y = AY^2 + B, \\ A = z^2 + z^4 + z^6 + z^8, \quad B = z^4 + z^5 + z^6 + z^7. \end{cases}$$

Figure VII.12. A simplified combinatorial model of RNA structures analogous to those considered by Waterman *et al.*

Thus Theorem VII.3 applies, giving

(48)
$$[z^n]C(z) \sim \frac{\gamma}{2\sqrt{\pi n^3}} r^{-n}, \qquad \gamma = \sqrt{\frac{1}{2} r \eta''(s)},$$

and a complete expansion can be obtained. .... ∎

***Example* VII.14.** *Trees with variable edge lengths and node sizes.* Consider unlabelled plane trees in which nodes can be of different sizes: what is given is a set $\widehat{\Omega}$ of ordered pairs (ω, σ), where a value (ω, σ) means that a node of degree ω and size σ is allowed. Simple varieties in their basic form correspond to $\sigma \equiv 1$; trees enumerated by leaves (including hierarchies) correspond to $\sigma \in \{0, 1\}$ with $\sigma = 1$ iff $\omega = 0$. Figure VII.12 suggests the way such trees can model the self-bonding of single-stranded nucleic acids like RNA, according to Waterman *et al.* [336, 453, 534, 558]. Clearly an extremely large number of variations are possible.

The fundamental equation in the case of a finite $\widehat{\Omega}$ is

$$Y(z) = P(z, Y(z)), \qquad P(z, w) := \sum_{(\omega, \sigma) \in \widehat{\Omega}} z^\sigma w^\omega,$$

with P a polynomial. In the aperiodic case, there is invariably a formula of the form

$$Y_n \sim \kappa \cdot A^n n^{3/2},$$

corresponding to the universal square-root singularity. .... ∎

▷ **VII.19.** *Schröder numbers.* Consider the class $\mathcal{Y}$ of unary–binary trees where unary nodes have size 2, while leaves and binary nodes have the usual size 1. The GF satisfies $Y = z + z^2 Y + z Y^2$, so that

$$Y(z) = z D(z^2), \qquad D(z) = \frac{1 - z - \sqrt{1 - 6z + z^2}}{2z}.$$

We have $D(z) = 1 + 2z + 6z^2 + 22z^3 + 90z^4 + 394z^5 + \cdots$, which is *EIS* **A006318** ("Large Schröder numbers"). By the bijective correspondence between trees and lattice paths, $\mathcal{Y}_{2n+1}$ is in correspondence with excursions of length n made of steps $(1, 1), (2, 0), (1, -1)$. Upon tilting by $45°$, this is equivalent to paths connecting the lower left corner to the upper right corner of an $(n \times n)$ square that are made of horizontal, vertical, and diagonal steps, and never go under the main diagonal. The series $S = \frac{z}{2}(1 + D)$ enumerates Schröder's generalized parenthesis systems (Chapter I, p. 69): $S := z + S^2/(1 - S)$, and the asymptotic formula

$$Y_{2n-1} = S_n = \frac{1}{2}D_{n-1} \sim \frac{1}{4\sqrt{\pi n^3}}\left(3 - 2\sqrt{2}\right)^{-n+1/2}$$

follows straightforwardly.   ◁

VII. 5. Unlabelled non-plane trees and Pólya operators

Essentially all the results obtained earlier for simple varieties of trees can be extended to the case of non-plane unlabelled trees. *Pólya operators* are central, and their treatment is typical of the asymptotic theory of unlabelled objects obeying symmetries (i.e., involving the unlabelled MSET, PSET, CYC constructions), as we have seen repeatedly in this book.

Binary and general trees. We start the discussion by considering the enumeration of two classes of non-plane trees following Pólya [488, 491] and Otter [466], whose articles are important historic sources for the asymptotic theory of non-plane tree enumeration—a brief account also appears in [319]. (These authors used the more traditional method of Darboux instead of singularity analysis, but this distinction is immaterial here, as calculations develop under completely parallel lines under both theories.) The two classes under consideration are those of general and binary non-plane unlabelled trees. In both cases, there is a fairly direct reduction to the enumeration of Cayley trees and of binary trees, which renders explicit several steps of the calculation. The trick is, as usual, to treat values of $f(z^2)$, $f(z^3)$, $\ldots$, arising from Pólya operators, as "known" analytic quantities.

Proposition VII.5 (Special unlabelled non-plane trees). *Consider the two classes of unlabelled non-plane trees*

$$\mathcal{H} = \mathcal{Z} \times \text{MSET}(\mathcal{H}), \quad \mathcal{W} = \mathcal{Z} \times \text{MSET}_{\{0,2\}}(\mathcal{W}),$$

respectively, of the general and binary type. Then, with constants γ_H, A_H and γ_W, A_W given by Notes VII.21 and VII.22, one has

(49) $$H_n \sim \frac{\gamma_H}{2\sqrt{\pi n^3}}A_H^n, \quad W_{2n-1} \sim \frac{\gamma_W}{2\sqrt{\pi n^3}}A_W^n.$$

Proof. (*i*) *General case.* The OGF of non-plane unlabelled trees is the analytic solution to the functional equation

(50) $$H(z) = z\exp\left(\frac{H(z)}{1} + \frac{H(z^2)}{2} + \cdots\right).$$

Let T be the solution to

(51) $$T(z) = ze^{T(z)},$$

that is to say, the Cayley function. The function $H(z)$ has a radius of convergence ρ strictly less than 1 as its coefficients dominate those of $T(z)$, the radius of convergence of the latter being exactly $e^{-1} \doteq 0.367$. The radius ρ cannot be 0 since the number of trees is bounded from above by the number of plane trees whose OGF has radius $1/4$. Thus, one has $1/4 \le \rho \le e^{-1}$.

Rewriting the defining equation of $H(z)$ as

$$H(z) = \zeta e^{H(z)} \qquad \text{with} \qquad \zeta := z \exp\left(\frac{H(z^2)}{2} + \frac{H(z^3)}{3} + \cdots\right),$$

we observe that $\zeta = \zeta(z)$ is analytic for $|z| < \rho^{1/2}$; that is, ζ is analytic in a disc that properly contains the disc of convergence of $H(z)$. We may thus rewrite $H(z)$ as

$$H(z) = T(\zeta(z)).$$

Since $\zeta(z)$ is analytic at $z = \rho$, a singular expansion of $H(z)$ near $z = \rho$ results from composing the singular expansion of T at e^{-1} with the analytic expansion of ζ at ρ. In this way, we get:

$$(52) \qquad H(z) = 1 - \gamma\left(1 - \frac{z}{\rho}\right)^{1/2} + O\left(\left(1 - \frac{z}{\rho}\right)\right), \qquad \gamma = \sqrt{2e\rho\zeta'(\rho)}.$$

Thus,

$$[z^n]H(z) \sim -\frac{\gamma}{2\sqrt{\pi n^3}}\rho^{-n}.$$

(*ii*) *Binary case.* Consider the functional equation

$$(53) \qquad f(z) = z + \frac{1}{2}f(z)^2 + \frac{1}{2}f(z^2).$$

This enumerates non-plane binary trees with size defined as the number of external nodes, so that $W(z) = \frac{1}{z}f(z^2)$. Thus, it suffices to analyse $[z^n]f(z)$, which dispenses us from dealing with periodicity phenomena arising from the parity of n.

The OGF $f(z)$ has a radius of convergence ρ that is at least $1/4$ (since there are fewer non-plane trees than plane ones). It is also at most $1/2$, which is seen from a comparison of f with the solution to the equation $g = z + \frac{1}{2}g^2$. We may then proceed as before: treat the term $\frac{1}{2}f(z^2)$ as a function analytic in $|z| < \rho^{1/2}$, as though it were known, then solve. To this effect, set

$$\zeta(z) := z + \frac{1}{2}f(z^2),$$

which exists in $|z| < \rho^{1/2}$. Then, the equation (53) becomes a plain quadratic equation, $f = \zeta + \frac{1}{2}f^2$, with solution

$$f(z) = 1 - \sqrt{1 - 2\zeta(z)}.$$

The singularity ρ is the smallest positive solution of $\zeta(\rho) = 1/2$. The singular expansion of f is obtained by combining the analytic expansion of ζ at ρ with $\sqrt{1 - 2\zeta}$. The usual square-root singularity results:

$$f(z) \sim 1 - \gamma\sqrt{1 - z/\rho}, \qquad \gamma := \sqrt{2\rho\zeta'(\rho)}.$$

This induces the $\rho^{-n}n^{-3/2}$ form for the coefficients $[z^n]f(z) \equiv [z^{2n-1}]W(z)$.   ∎

The argument used in the proof of the proposition may seem partly non-constructive. However, numerically, the values of ρ and γ can be determined to great accuracy. See the notes below as well as Finch's section on "Otter's tree enumeration constants" [211, Sec. 5.6].

▷ **VII.20.** *Complete asymptotic expansions for* H_n, W_{2n-1}. These can be determined since the OGFs admit complete asymptotic expansions in powers of $\sqrt{1 - z/\rho}$.   ◁

▷ **VII.21.** *Numerical evaluation of constants I.* Here is an unoptimized procedure controlled by a parameter $m \geq 0$ for evaluating the constants γ_H, ρ_H of (49) relative to general unlabelled non-plane trees.

Procedure Get_value_of_ρ(m : integer);
    1. Set up a procedure to compute and memorize the H_n on demand;
    (this can be based on recurrence relations implied by $H'(z)$; see [456])
    2. Define $f^{[m]}(z) := \sum_{j=1}^{m} H_n z^n$;
    3. Define $\zeta^{[m]}(z) := z \exp\left(\sum_{k=2}^{m} \frac{1}{k} f^{[m]}(z^k)\right)$;
    4. Solve numerically $\zeta^{[m]}(x) = e^{-1}$ for $x \in (0, 1)$ to max(m, 10) digits of accuracy;
    5. Return x as an approximation to ρ.

For instance, a conservative estimate of the accuracy attained for $m = 0, 10, \ldots, 50$ (in a few billion machine instructions) is:

$m = 0$	$m = 10$	$m = 20$	$m = 30$	$m = 40$	$m = 50$
$3 \cdot 10^{-2}$	10^{-6}	10^{-11}	10^{-16}	10^{-21}	10^{-26}

Accuracy appears to be a little better than $10^{-m/2}$. This yields to 25D:

$$\rho \doteq 0.3383218568992076951961126, \quad A_H \equiv \rho^{-1} \doteq 2.9557652856519949747148418,$$
$$\gamma_H \doteq 1.5594900203746408855422206.$$

The formula of Proposition VII.5 estimates H_{100} with a relative error of 10^{-3}.   ◁

▷ **VII.22.** *Numerical evaluation of constants II.* The procedure of the previous note adapts easily to binary trees, giving:

$$\rho \doteq 0.4026975036714412909690453, \quad A_W \equiv \rho^{-1} \doteq 2.4832535361726368558562289,$$
$$\gamma_W \doteq 1.1300337163989720071441 37.$$

The formula of Proposition VII.5 estimates $[z^{100}]f(z)$ with a relative error of $7 \cdot 10^{-3}$.   ◁

The results relative to general and binary trees are thus obtained by a modification of the method used for simple varieties of trees, upon treating the Pólya operator part as an analytic variant of the corresponding equations of simple varieties of trees.

Alkanes, alcohols, and degree restrictions. The previous two examples suggest that a general theory is possible for varieties of unlabelled non-plane trees, $\mathcal{T} = \mathcal{Z}\,\mathrm{MSET}_\Omega(\mathcal{T})$, determined by some $\Omega \subset \mathbb{Z}_{\geq 0}$. First, we examine the case of special regular trees defined by $\Omega = \{0, 3\}$, which, when viewed as alkanes and alcohols, are of relevance to combinatorial chemistry (Example VII.15). Indeed, the problem of enumerating isomers of such chemical compounds has been at the origin of Pólya's foundational works [488, 491]. Then, we extend the method to the general situation of trees with degrees constrained to an arbitrary finite set Ω (Proposition VII.5).

Example VII.15. *Non-plane trees and alkanes.* In chemistry, carbon atoms (C) are known to have valency 4 while hydrogen (H) has valency 1. *Alkanes*, also known as paraffins (Figure VII.13), are acyclic molecules formed of carbon and hydrogen atoms according to this rule and without multiple bonds; they are thus of the type $C_n H_{2n+2}$. In combinatorial terms, we are talking of unrooted trees with (total) node degrees in $\{1, 4\}$. The rooted version of these trees are determined by the fact that a root is chosen and (out)degrees of nodes lie in the set $\Omega = \{0, 3\}$; such rooted ternary trees then correspond to alcohols (with the *OH* group marking one of the carbon atoms).

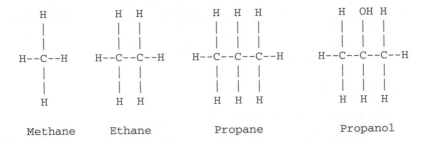

Figure VII.13. A few examples of alkanes (CH_4, C_2H_6, C_3H_8) and an alcohol.

Alcohols ($\mathcal{A}$) are the simplest to enumerate, since they correspond to rooted trees. The OGF starts as (*EIS* **A000598**)

$$A(z) = 1 + z + z^2 + z^3 + 2z^4 + 4z^5 + 8z^6 + 17z^7 + 39z^8 + 89z^9 + \cdots,$$

with size being taken here as the number of internal nodes. The specification is

$$\mathcal{A} = \{\epsilon\} + \mathcal{Z}\,\mathrm{MSET}_3(\mathcal{A}).$$

(Equivalently $\mathcal{A}^+ := \mathcal{A} \setminus \{\epsilon\}$ satisfies $\mathcal{A}^+ = \mathcal{Z}\,\mathrm{MSET}_{0,1,2,3}(\mathcal{A}^+)$.) This implies that $A(z)$ satisfies the functional equation:

$$A(z) = 1 + z\left(\frac{1}{3}A(z^3) + \frac{1}{2}A(z)A(z^2) + \frac{1}{6}A(z)^3\right).$$

In order to apply Theorem VII.3, introduce the function

(54) $$G(z, w) = 1 + z\left(\frac{1}{3}A(z^3) + \frac{1}{2}A(z^2)w + \frac{1}{6}w^3\right),$$

which exists in $|z| < |\rho|^{1/2}$ and $|w| < \infty$, with ρ the (yet unknown) radius of convergence of A. Like before, the Pólya terms $A(z^2)$, $A(z^3)$ are treated as known functions. By methods similar to those earlier in the analysis of binary and general trees, we find that the characteristic system admits a solution,

$$r \doteq 0.3551817423143773928, \quad s \doteq 2.1174207009536310225,$$

so that $\rho = r$ and $y(\rho) = s$. Thus the growth of the number of alcohols is of the form $\kappa\rho^{-n}n^{-3/2}$, with $\rho^{-1} \doteq 2.81546$.

Let $B(z)$ be the OGF of alkanes (*EIS* **A000602**), which are unrooted trees:

$$B(z) = 1 + z + z^2 + z^3 + 2z^4 + 3z^5 + 5z^6 + 9z^7 + 18z^8 35z^9 + 75z^{10} + \cdots.$$

For instance, $B_6 = 5$ because there are five isomers of hexane, C_6H_{14}, for which chemists had to develop a nomenclature system, interestingly enough based on a diameter of the tree:

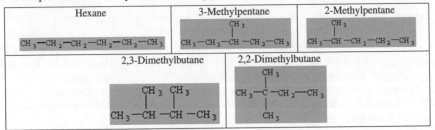

The number of structurally different alkanes can then be found by an adaptation of the dissimilarity formula (Equation (57) below and Note VII.26). This problem has served as a powerful motivation for the enumeration of graphical trees and its fascinating history goes back to Cayley. (See Rains and Sloane's article [502] and [491]). The asymptotic formula of (un-rooted) alkanes is of the global form $\rho^{-n} n^{-5/2}$, which represents roughly a proportion $1/n$ of the number of (rooted) alcohols: see below. .... ∎

The pattern of analysis should by now be clear, and we state:

Theorem VII.4 (Non-plane unlabelled trees). *Let $\Omega \ni 0$ be a finite subset of $\mathbb{Z}_{\geq 0}$ and consider the variety $\mathcal{V}$ of (rooted) unlabelled non-plane trees with outdegrees of nodes in Ω. Assume aperiodicity ($\gcd(\Omega) = 1$) and the condition that Ω contains at least one element larger than 1. Then the number of trees of size n in $\mathcal{V}$ satisfies an asymptotic formula:*

$$V_n \sim C \cdot A^n n^{-3/2}.$$

Proof. The argument given for alcohols is transposed verbatim. Only the existence of a root of the characteristic system needs to be established.

The radius of convergence of $V(z)$ is *a priori* ≤ 1. The fact that ρ is strictly less than 1 is established by means of an exponential lower bound; namely, $V_n > B^n$, for some $B > 1$ and infinitely many values of n. To obtain this "exponential diversity" of the set of trees, first choose an n_0 such that $V_{n_0} > 1$, then build a perfect d–ary tree (for some $d \in \Omega, d \neq 0, 1$) of height h, and finally graft freely subtrees of size n_0 at $n/(4n_0)$ of the leaves of the perfect tree. Choosing d such that $d^h > n/(4n_0)$ yields the lower bound. That the radius of convergence is non-zero results from the upper bound provided by corresponding plane trees whose growth is at most exponential. Thus, one has $0 < \rho < 1$.

By the translation of multisets of bounded cardinality, the function G is polynomial in finitely many of the quantities $\{V(z), V(z^2), \ldots\}$. Thus the function $G(z, w)$ constructed as in the case of alcohols, in Equation (54), converges in $|z| < \rho^{1/2}, |w| < \infty$. As $z \to \rho^{-1}$, we must have $\tau := V(\rho)$ finite, since otherwise, there would be a contradiction in orders of growth in the nonlinear equation $V(z) = \cdots + \cdots V(z)^d \cdots$ as $z \to \rho$. Thus (ρ, τ) satisfies $\tau = G(\rho, \tau)$. For the derivative, one must have $G_w(\rho, \tau) = 1$ since: (*i*) a smaller value would mean that V is analytic at ρ (by the Implicit Function Theorem); (*ii*) a larger value would mean that a singularity has been encountered earlier (by the usual argument on failure of the Implicit Function Theorem). Thus, Theorem VII.3 on positive implicit functions is applicable. ∎

A large number of variations are clearly possible as evidenced by the suggestive title of an article [320] published by Harary, Robinson, and Schwenk in 1975: "Twenty-step algorithm for determining the asymptotic number of trees of various species".

▷ **VII.23.** *Unlabelled hierarchies.* The class $\mathcal{H}$ of unlabelled hierarchies is specified by $\mathcal{H} = \mathcal{Z} + \text{MSET}_{\geq 2}(\mathcal{H})$; see Note I.45, p. 72. One has

$$\widetilde{H}_n \sim \frac{\gamma}{2\sqrt{\pi n^3}} \rho^{-n}, \qquad \rho \doteq 0.29224.$$

(Compare with the labelled case of Example VII.12, p. 472.) What is the asymptotic proportion of internal nodes of degree r, for a fixed $r > 0$? ◁

▷ **VII.24.** *Trees with prime degrees and the BBY theory.* Bell, Burris, and Yeats [33] develop a general theory meant to account for the fact that, in their words, "almost any *family of trees defined by a recursive equation that is nonlinear [. . .] lead[s] to an asymptotic law of the Pólya form $t(n) \sim C\rho^{-n}n^{-3/2}$*". Their most general result [33, Th. 75] implies for instance that the number of unlabelled non-plane trees whose node degrees are restricted to be prime numbers admits such a Pólya form (see also Note VII.6, p. 455). ◁

Unlabelled functional graphs (mapping patterns). Unlabelled functional graphs (named "functions" in [319, pp. 69–70]) are denoted here by $\mathcal{F}$; they correspond to unlabelled digraphs with loops allowed, in which each vertex has outdegree equal to 1. They can be specified as multisets of components ($\mathcal{L}$) that are cycles of non-plane unlabelled trees ($\mathcal{H}$),

$$\mathcal{F} = \mathrm{MSET}(\mathcal{L}); \quad \mathcal{L} = \mathrm{CYC}(\mathcal{H}); \quad \mathcal{H} = \mathcal{Z} \times \mathrm{MSET}(\mathcal{H}),$$

a specification that entirely parallels that of mappings in Equation (35), p. 462. Indeed, an unlabelled functional graph can be used to represent the "shape" of a mapping, as obtained when labels are discarded. That is, functional graphs result when mappings are identified up to a possible permutation of their underlying domain. This explains the alternative term of "mapping pattern" [436] sometimes employed for such graphs. The counting sequence starts as $1, 1, 3, 7, 19, 47, 130, 343, 951$ (*EIS* **A001372**).

The OGF $H(z)$ has a square-root singularity by virtue of (52) above, with additionally $H(\rho) = 1$. The translation of the unlabelled cycle construction,

$$L(z) = \sum_{j \geq 1} \frac{\varphi(j)}{j} \log \frac{1}{1 - H(z^j)},$$

implies that $L(z)$ is logarithmic, and $F(z)$ has a singularity of type $1/\sqrt{Z}$ where $Z := 1 - z/\rho$. Thus, *unlabelled functional graphs constitute an exp–log structure* in the sense of Section VII. 2, p. 445, with $\kappa = 1/2$. The number of unlabelled functional graphs thus grows like $C\rho^{-n}n^{-1/2}$ and the mean number of components in a random functional graph is $\sim \frac{1}{2} \log n$, as for labelled mappings; see [436] for more on this topic.

▷ **VII.25.** *An alternative form of $F(z)$.* Arithmetical simplifications associated with the Euler totient function (APPENDIX A, p. 721) yield:

$$F(z) = \prod_{k=1}^{\infty} \left(1 - H(z^k)\right)^{-1}.$$

A similar form applies generally to multisets of unlabelled cycles (Note I.57, p. 85). ◁

Unrooted trees. All the trees considered so far have been rooted and this version is the one most useful in applications. An *unrooted tree*[9] is by definition a connected acyclic (undirected) graph. In that case, the tree is clearly non-plane and no special root node is distinguished.

The counting of the class $\mathcal{U}$ of *unrooted labelled trees* is easy: there are plainly $U_n = n^{n-2}$ of these, since each node is distinguished by its label, which entails that

---

[9]Unrooted trees are also called sometimes *free trees*.

$nU_n = T_n$, with $T_n = n^{n-1}$ by Cayley's formula. Also, the EGF $U(z)$ satisfies

$$(55) \qquad U(z) = \int_0^z T(y) \, \frac{dy}{y} = T(z) - \frac{1}{2} T(z)^2,$$

as already seen when we discussed labelled graphs in Subsection II. 5.3, p. 132.

For *unrooted unlabelled trees*, symmetries are present and a tree can be rooted in a number of ways that depends on its shape. For instance, a star graph leads to a number of different rooted trees that equals 2 (choose either the centre or one of the peripheral nodes), while a line graph gives rise to $\lceil n/2 \rceil$ structurally different rooted trees. With $\mathcal{H}$ the class of rooted unlabelled trees and $\mathcal{I}$ the class of unrooted trees, we have at this stage only a general inequality of the form

$$I_n \leq H_n \leq n I_n.$$

A table of values of the ratio H_n/I_n suggests that the answer is close to the upper bound:

(56)

n	10	20	30	40	50	60
H_n/I_n	6.78	15.58	23.89	32.15	40.39	48.62

The solution is provided by a famous exact formula due to Otter (Note VII.26):

$$(57) \qquad I(z) = H(z) - \frac{1}{2} \left(H(z)^2 - H(z^2) \right),$$

which gives in particular (*EIS* **A000055**) $I(z) = z + z^2 + z^3 + 2z^4 + 3z^5 + 6z^6 + 11z^7 + 23z^8 + \cdots$. Given (57), it is child's play to determine the singular expansion of I knowing that of H. The radius of convergence of I is the same as that of H, since the term $H(z^2)$ only introduces exponentially small coefficients. Thus, it suffices to analyse $H - \frac{1}{2}H^2$:

$$H(z) - \frac{1}{2}H(z)^2 \sim \frac{1}{2} - \delta_2 Z + \delta_3 Z^{3/2} + O\left(Z^2\right), \qquad Z = \left(1 - \frac{z}{\rho}\right).$$

What is noticeable is the cancellation in coefficients for the term $Z^{1/2}$ (since $1 - x - \frac{1}{2}(1-x)^2 = \frac{1}{2} + O(x^2)$), so that $Z^{3/2}$ is the actual singularity type of I. Clearly, the constant δ_3 is computable from the first four terms in the singular expansion of H at ρ. Then singularity analysis yields: *The number of unrooted trees of size n satisfies the formula*

$$(58) \quad I_n \sim \frac{3\delta_3}{4\sqrt{\pi n^5}} \rho^{-n}, \qquad I_n \sim (0.5349496061\ldots)(2.9955765856\ldots)^n n^{-5/2}.$$

The numerical values are from [211] and the result is Otter's original [466]: an unrooted tree of size n gives rise to about different $0.8n$ rooted trees on average. (The formula (58) corresponds to an error slightly under 10^{-2} for $n = 100$.)

▷ **VII.26.** *Dissimilarity theorem for trees.* Here is how combinatorics justifies (57), following [50, §4.1]. Let $\mathcal{I}^\bullet$ (and $\mathcal{I}^{\bullet-\bullet}$) be the class of unrooted trees with one vertex (respectively, one edge) distinguished. We have $\mathcal{I}^\bullet \cong \mathcal{H}$ (rooted trees) and $\mathcal{I}^{\bullet-\bullet} \cong \mathrm{SET}_2(\mathcal{H})$. The combinatorial isomorphism claimed is

$$(59) \qquad \mathcal{I}^\bullet + \mathcal{I}^{\bullet-\bullet} \cong \mathcal{I} + (\mathcal{I} \times \mathcal{I}).$$

Proof. A *diameter* of an unrooted tree is a simple path of maximal length. If the length of any diameter is even, call "*centre*" its mid-point; otherwise, call "*bicentre*" its mid-edge. (For each tree, there is either *one* centre or *one* bicentre.) The left-hand side of (59) corresponds to trees that are pointed either at a vertex ($\mathcal{I}^\bullet$) or an edge ($\mathcal{I}^{\bullet\bullet}$). The term $\mathcal{I}$ on the right-hand side corresponds to cases where the pointing happens to coincide with the canonical centre or bicentre. If there is not coincidence, then, an ordered pair of trees results from a suitable surgery of the pointed tree. [Hint: cut in some canonical way near the pointed vertex or edge.]   ◁

VII. 6. Irreducible context-free structures

In this section, we discuss an important variety of context-free classes, one that gives rise to the *universal law of square-root singularities*, itself attached to counting sequences that are of the general asymptotic form $A^n n^{-3/2}$. First, we enunciate an abstract structural result (Theorem VII.5, p. 483) that connects *"irreducibility"* of context-free systems to the square-root singularity phenomenon. Before engaging into a proof, we first illustrate its scope by describing applications to non-crossing configurations in the plane (these are richer than triangulations introduced in Chapter I) and to random boolean expressions. Finally, we prove an important complex analytic result, the Drmota–Lalley–Woods Theorem (Theorem VII.6, p. 489), which provides the underlying analytic engine needed to establish Theorem VII.5 and justify the asymptotic properties of irreducible context-free specifications. General algebraic functions are to be treated next, in Section VII. 7, p. 493.

VII. 6.1. Context-free specifications and the irreducibility schema.
We start from the notion of a context-free class already introduced in Subsection I. 5.4, p. 79, which we recall: a class is *context-free* if it is determined as the first component of a system of combinatorial equations

$$(60) \qquad \begin{cases} \mathcal{Y}_1 &= \mathfrak{F}_1(\mathcal{Z}, \mathcal{Y}_1, \ldots, \mathcal{Y}_r) \\ \ \vdots\ \ \vdots & \quad\ \ \vdots \\ \mathcal{Y}_r &= \mathfrak{F}_r(\mathcal{Z}, \mathcal{Y}_1, \ldots, \mathcal{Y}_r), \end{cases}$$

where each $\mathfrak{F}_j$ is a construction that only involves the combinatorial constructions of disjoint union and cartesian product. (This repeats Equation (83) of Chapter I, p. 79.) As seen in Subsection I. 5.4, binary and general trees, triangulations, as well as Dyck and Łukasiewicz languages are typical instances of context-free classes.

As a consequence of the symbolic rules of Chapter I, the OGF of a context-free class $\mathcal{C}$ is the first component ($C(z) \equiv y_1(z)$) of the solution of a polynomial system of equations of the form

$$(61) \qquad \begin{cases} y_1(z) &= \Phi_1(z, y_1(z), \ldots, y_r(z)) \\ \ \vdots\ \ \vdots & \quad\ \ \vdots \\ y_r(z) &= \Phi_r(z, y_1(z), \ldots, y_r(z)), \end{cases}$$

where the Φ_j are polynomials. By elimination (Cf Appendix B.1: *Algebraic elimination*, p. 739), it is always possible to find a bivariate polynomial $P(z, y)$ such that

$$(62) \qquad\qquad\qquad P(z, C(z)) = 0,$$

and $C(z)$ is an *algebraic function*. (Algebraic functions are discussed in all generality in the next section.)

The case of linear systems has been dealt with in Chapter V, when examining the transfer matrix method. Accordingly, we only need to consider here *nonlinear* systems (of equations or specifications) defined by the condition that at least one Φ_j in (61) is a polynomial of degree 2 or more in the y_j, corresponding to the fact that at least one of the constructions $\mathfrak{F}_j$ in (60) involves at least a product $\mathcal{Y}_k \mathcal{Y}_\ell$.

Definition VII.5. *A context-free specification* (60) *is said to belong to the* irreducible context-free *schema if it is nonlinear and its dependency graph (p. 33) is strongly connected. It is said to be aperiodic if all the $y_j(z)$ are aperiodic*[10].

Theorem VII.5 (Irreducible context-free schema). *A class $\mathcal{C}$ that belongs to the irreducible context-free schema has a generating function that has a square-root singularity at its radius of convergence ρ:*

$$C(z) = \tau - \gamma \sqrt{1 - \frac{z}{\rho}} + O\left(1 - \frac{z}{\rho}\right),$$

for computable algebraic numbers ρ, τ, γ. If, in addition, $C(z)$ is aperiodic, then the dominant singularity is unique and the counting sequence satisfies

$$(63) \qquad C_n \sim \frac{\gamma}{2\sqrt{\pi n^3}} \rho^{-n}.$$

This theorem is none other than a transcription, at the combinatorial level, of a remarkable analytic statement, Theorem VII.6, due to Drmota, Lalley, and Woods, which is proved below (p. 489), is slightly stronger, and is of independent interest.

Computability issues. There are two complementary approaches to the calculation of the quantities that appear in (63), one based on the original system (61), the other based on the single equation (62) that results from elimination. We offer at this stage a brief pragmatic discussion of computational aspects, referring the reader to Subsection VII. 6.3, p. 488, and Section VII. 7, p. 493, for context and justifications.

(*a*) *System:* Considering the proof of Theorem VII.6 below, one should solve, in *positive* real numbers, a polynomial system of $m + 1$ equations in the $m + 1$ unknowns $\rho, \tau_1, \ldots, \tau_m$; namely,

$$(64) \qquad \begin{cases} \tau_1 &= \Phi_1(\rho, \tau_1, \ldots, \tau_m) \\ \vdots & \vdots \quad \vdots \\ \tau_m &= \Phi_m(\rho, \tau_1, \ldots, \tau_m) \\ 0 &= J(\rho, \tau_1, \ldots, \tau_m), \end{cases}$$

which one can call the *characteristic system*. There J is the Jacobian determinant:

$$(65) \qquad J(z, y_1, \ldots, y_m) := \det\left(\delta_{i,j} - \frac{\partial}{\partial y_j} \Phi_i(z, y_1, \ldots, y_m)\right),$$

---

[10]An *aperiodic* function is such that the span of the coefficient sequence is equal to 1 (Definition IV.5, p. 266). For an irreducible system, it can be checked that *all* the y_j are aperiodic if and only if *at least* one of the y_j is aperiodic.

with $\delta_{i,j} \equiv [\![i = j]\!]$ being the usual Kronecker symbol. The quantity ρ represents the common radius of convergence of all the $y_j(z)$ and $\tau_j = y_j(\rho)$. (In case several possibilities present themselves for ρ, as in Note VII.28, then one can use either *a priori* combinatorial bounds to filter out the spurious ones[11] or make use of the reduction to a single equation as in point (*b*) below.) The constant $\gamma \equiv \gamma_1$ in Theorem VII.5 is then a component of the solution to a *linear* system of equations (with coefficients in the field generated by ρ, τ_j) and is obtained by the method of undetermined coefficients, since each y_j is of the form

$$(66) \qquad\qquad y_j(z) \sim \tau_j - \gamma_j \sqrt{1 - z/\rho}, \qquad z \to \rho.$$

(*b*) *Equation:* The general techniques are going to be described in Section, §VII. 7, p. 493. They give rise to the following algorithm: (*i*) determine the exceptional set, identify the proper branch of the algebraic curve and the dominant positive singularity; (*ii*) determine the coefficients in the singular (Puiseux) expansion, knowing *a priori* that the singularity is of the square-root type.

In all events, symbolic algebra systems prove invaluable in performing the required algebraic eliminations and isolating the combinatorially relevant roots (see, in particular, Pivoteau *et al.* [485] for a general symbolic–numeric approach). Example VII.16 serves to illustrate some of these computations.

▷ **VII.27.** *Catalan and the Jacobian determinant.* For the Catalan GF, defined by $y = 1 + zy^2$, the characteristic system (64) instantiates to

$$\tau - 1 - \rho\tau^2 = 0, \quad 1 - 2\rho\tau = 0,$$

giving back as expected: $\rho = \frac{1}{4}, \tau = 2$.   ◁

▷ **VII.28.** *Burris' Caveat.* As noted by Stanley Burris (private communication), even some very simple context-free specifications may be such that there exist several positive solutions to the characteristic system (64). Consider

$$(B) : \quad \begin{cases} y_1 &=& z(1 + y_2 + y_1^2) \\ y_2 &=& z(1 + y_1 + y_2^2), \end{cases}$$

which is clearly associated to a redundant way of counting unary–binary trees (via a deterministic 2-colouring). The characteristic system is

$$\left\{ \tau_1 = \rho(1 + \tau_2 + \tau_1^2), \quad \tau_2 = \rho(1 + \tau_1 + \tau_2^2), \quad (1 - 2\rho\tau_1)(1 - 2\rho\tau_2) - \rho^2 = 0 \right\}.$$

The positive solutions are

$$\left\{ \rho = \frac{1}{3}, \quad \tau_1 = \tau_2 = 1 \right\} \quad \cup \quad \left\{ \rho = \frac{1}{7}(2\sqrt{2} - 1), \quad \tau_1 = \tau_2 = \sqrt{2} + 1 \right\}.$$

Only the first solution is combinatorially significant. (A somewhat similar situation, though it relates to a *non-irreducible* context-free specification, arises with supertrees of Example VII.20, p. 503: see Figure VII.19, p. 504.)   ◁

---

[11]This is once more a connection problem, in the sense of p. 470.

VII. 6.2. Combinatorial applications. Lattice animals (Example I.18, p. 80), random walks on free groups [395], directed walks in the plane (see references [27, 392, 395] and p. 506 below), coloured trees [616], and boolean expression trees (reference [115] and Examples VII.17) are only some of the many combinatorial structures belonging to the irreducible context-free schema. Stanley presents in his book [554, Ch. 6] several examples of algebraic GFs, and an inspiring survey is provided by Bousquet-Mélou in [84]. We limit ourselves here to a brief discussion of non-crossing configurations and random boolean expressions.

Example VII.16. *Non-crossing configurations.* Context-free descriptions can model naturally very diverse sorts of objects including particular topological-geometric configurations—we examine here non-crossing planar configurations. The problems considered have their origin in combinatorial musings of the Rev. T.P. Kirkman in 1857 and were revisited in 1974 by Domb and Barett [169] for the purpose of investigating certain perturbative expansions of statistical physics. Our presentation follows closely the synthesis offered by Flajolet and Noy in [245].

Consider, for each value of n, graphs built on vertices that are all the nth complex roots of unity, numbered $0, \ldots, n-1$. A *non-crossing graph* is a graph such that no two of its edges cross. One can also define connected non-crossing graphs, non-crossing forests (acyclic graphs), and non-crossing trees (acyclic connected graphs); see Figure VII.14. Note that the various graphs considered can always be considered as rooted in some canonical way (e.g., at the vertex of smallest index).

Trees. A non-crossing tree is rooted at 0. To the root vertex is attached an ordered collection of vertices, each of which has an end-node v that is the common root of two non-crossing trees, one on the left of the edge $(0, v)$ the other on the right of $(0, v)$. Let $\mathcal{T}$ denote the class of trees and $\mathcal{U}$ denote the class of trees whose root has been severed. With $\bullet \equiv \mathcal{Z}$ denoting a generic node, we have

$$\mathcal{T} = \bullet \times \mathcal{U}, \quad \mathcal{U} = \text{SEQ}(\mathcal{U} \times \bullet \times \mathcal{U}),$$

which corresponds graphically to the "butterfly decomposition":

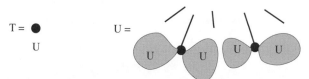

The reduction to a pure context-free form is obtained by noticing that $\mathcal{U} = \text{SEQ}(\mathcal{V})$ is equivalent to $\mathcal{U} = \mathbf{1} + \mathcal{U}\mathcal{V}$: a specification and the associated polynomial system are then

(67) $\{\mathcal{T} = \mathcal{Z}\mathcal{U}, \ \mathcal{U} = \mathbf{1} + \mathcal{U}\mathcal{V}, \ \mathcal{V} = \mathcal{Z}\mathcal{U}\mathcal{U}\} \implies \{T = zU, \ U = 1 + UV, \ V = zU^2\}.$

This system relating U and V is irreducible (then, T is immediately obtained from U), and aperiodicity is obvious from the first few values of the coefficients. The Jacobian (65) of the $\{U, V\}$-system (obtained by $z \to \rho, U \to \upsilon, V \to \beta$), is

$$\begin{vmatrix} 1 - \beta & \upsilon \\ 2\rho\upsilon & 1 \end{vmatrix} = 1 - \beta - 2\rho\upsilon^2.$$

Thus, the characteristic system (64) giving the singularity of U, V is

$$\{\upsilon = 1 + \upsilon\beta, \ \beta = \rho\upsilon^2, \ 1 - \beta - 2\rho\upsilon^2 = 0\},$$

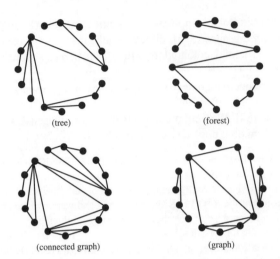

Configuration / OGF	coefficients (exact / asymptotic)
Trees (*EIS*: **A001764**) $$T^3 - zT + z^2 = 0$$	$z + z^2 + 3z^3 + 12z^4 + 55z^5 + \cdots$ $$\frac{1}{2n-1}\binom{3n-3}{n-1}$$ $$\sim \frac{\sqrt{3}}{27\sqrt{\pi n^3}}\left(\frac{27}{4}\right)^n$$
Forests (*EIS*: **A054727**) $$F^3 + (z^2 - z - 3)F^2 + (z+3)F - 1 = 0$$	$1 + z + 2z^2 + 7z^3 + 33z^4 + 181z^5 \cdots$ $$\sum_{j=1}^{n}\frac{1}{2n-j}\binom{n}{j-1}\binom{3n-2j-1}{n-j}$$ $$\sim \frac{0.07465}{\sqrt{\pi n^3}}(8.22469)^n$$
Connected graphs (*EIS*: **A007297**) $$C^3 + C^2 - 3zC + 2z^2 = 0$$	$z + z^2 + 4z^3 + 23z^4 + 156z^5 + \cdots$ $$\frac{1}{n-1}\sum_{j=n-1}^{2n-3}\binom{3n-3}{n+j}\binom{j-1}{j-n+1}$$ $$\sim \frac{2\sqrt{6}-3\sqrt{2}}{18\sqrt{\pi n^3}}\left(6\sqrt{3}\right)^n$$
Graphs (*EIS*: **A054726**) $$G^2 + (2z^2 - 3z - 2)G + 3z + 1 = 0$$	$1 + z + 2z^2 + 8z^3 + 48z^4 + 352z^5 + \cdots$ $$\frac{1}{n}\sum_{j=0}^{n-1}(-1)^j\binom{n}{j}\binom{2n-2-j}{n-1-j}2^{n-1-j}$$ $$\sim \frac{\sqrt{140-99\sqrt{2}}}{4\sqrt{\pi n^3}}\left(6+4\sqrt{2}\right)^n$$

Figure VII.14. (Top) Non-crossing graphs: a tree, a forest, a connected graph, and a general graph. (Bottom) The enumeration of non-crossing configurations by algebraic functions.

whose positive solution is $\rho = \frac{4}{27}$, $\upsilon = \frac{3}{2}$, $\beta = \frac{1}{3}$. The complete asymptotic formula is displayed in Figure VII.14. (In a simple case like this, we have more: T satisfies $T^3 - zT + z^2 = 0$, which, by Lagrange inversion, gives $T_n = \frac{1}{2n-1}\binom{3n-3}{n-1}$.)

Forests. A (non-crossing) forest is a non-crossing graph that is acyclic. In the present context, it is not possible to express forests simply as sequences of trees, because of the geometry of the problem. Starting conventionally from the root vertex 0 and following all connected edges defines a "backbone" tree. To the left of every vertex of the tree, a forest may be placed. There results the decomposition (expressed directly in terms of OGFs)

$$(68) \qquad F = 1 + T[z \mapsto zF],$$

where T is the OGF of trees and F is the OGF of forests. In (68), the term $T[z \mapsto zF]$ denotes a functional composition. A context-free specification in standard form results mechanically from (67) upon replacing z by zF:

$$(69) \qquad \{ F = 1 + T, \quad T = zFU, \quad U = 1 + UV, \quad V = zFU^2 \}.$$

This system is irreducible and aperiodic, so that the asymptotic shape of F_n is *a priori* of the form $\gamma \omega^n n^{-3/2}$ according to Theorem VII.5. The characteristic system is found to have three solutions, of which only one has all its components positive, corresponding to $\rho \doteq 0.12158$, a root of the cubic equation $5\rho^3 - 8\rho^2 - 32\rho + 4 = 0$. (The values of constants are otherwise worked out in Example VII.19, p. 502, by means of the equational approach.)

Graphs. Similar constructions (see [245]) give the OGFs of connected and general graphs, with the results tabulated in Figure VII.14. In summary:

Proposition VII.6. *The number of non-crossing trees, forests, connected graphs, and graphs each satisfy an asymptotic formula of the form*

$$\frac{C}{\sqrt{\pi n^3}} A^n.$$

The common shape of the asymptotic estimates is worthy of note, as is the fact that binomial expressions are available in each particular case (Note VII.34, p. 495, introduces a general framework that "explains" the existence of such binomial expressions). . . . . ∎

***Example* VII.17.** *Random boolean expressions.* We reconsider boolean expressions in the form of and–or trees introduced in Example I.15, p. 69, in connection with Hipparchus of Rhodes and Schröder, and in Example I.17, p. 77. Such an expression is described by a binary tree whose internal nodes can be tagged with "∨" (or-function) or "∧" (and-function); external nodes are formal variables and their negations ("literals"). We fix the number of variables to some number m. The class $\mathcal{E}$ of all such boolean expressions satisfies a symbolic equation of the form

$$\mathcal{E} = {}_{\mathcal{E}}\overset{\vee}{\diagup\diagdown}{}_{\mathcal{E}} + {}_{\mathcal{E}}\overset{\wedge}{\diagup\diagdown}{}_{\mathcal{E}} + \sum_{j=1}^{m}\left(\boxed{x_j} + \boxed{\neg x_j} \right).$$

Size is taken to be the number of internal (binary) nodes; that is, the number of boolean connectives. Each boolean expression given in the form of such an *and–or tree* represents a certain boolean function of m variables, among the 2^{2^m} functions. The corresponding OGF and coefficients are

$$E(z) = \frac{1 - \sqrt{1 - 16mz}}{4z}, \qquad E_n \equiv [z^n]E(z) = 2^n(2m)^{n+1}\frac{1}{n+1}\binom{2n}{n} \sim \frac{2m}{\sqrt{\pi n^3}}(16m)^n,$$

the radius of convergence of $E(z)$ being $\rho = 1/(16m)$.

Our purpose is to establish the following result due to Lefmann and Savický [405], our line of proof following [115].

Proposition VII.7. *Let f be a boolean function of m variables (m fixed). Then the probability that a random and–or formula of size n computes f converges, as n tends to infinity, to a constant value $\varpi(f) \neq 0$.*

Proof. Consider, for each f, the subclass $\mathcal{Y}_f \subset \mathcal{E}$ of expressions that compute f. We thus have 2^{2^m} such classes. It is then immediate to write combinatorial equations describing the $\mathcal{Y}_f$, by considering all the ways in which a function f can arise. Indeed, if f is not a literal, then

$$
\mathcal{Y}_f = \sum_{(g \vee h)=f} \mathcal{Y}_g \overset{\vee}{\diagup} \ \diagdown \mathcal{Y}_h + \sum_{(g \wedge h)=f} \mathcal{Y}_g \overset{\vee}{\diagup} \ \diagdown \mathcal{Y}_h,
$$

while, if $f = x_j$ (say), then

$$
\mathcal{Y}_f = \boxed{x_j} + \sum_{(g \vee h)=f} \mathcal{Y}_g \overset{\vee}{\diagup} \ \diagdown \mathcal{Y}_h + \sum_{(g \wedge h)=f} \mathcal{Y}_g \overset{\vee}{\diagup} \ \diagdown \mathcal{Y}_h.
$$

Thus, at generating function level, we have a system of 2^{2^m} polynomial equations. This system is *irreducible*: given two functions f and g represented by Φ and Γ (say), we can always construct an expression for f involving the expression Γ by building a tree of the form

$$
(\Phi \wedge (\mathrm{True} \vee \Gamma)) = ((\Phi \wedge ((x_1 \vee \neg x_1) \vee \Gamma)).
$$

Thus any $\mathcal{Y}_f$ depends on any other $\mathcal{Y}_g$. Similar arguments, based on the fact that

$$
\mathrm{True} = (\mathrm{True} \wedge \mathrm{True}) = (\mathrm{True} \wedge \mathrm{True} \wedge \mathrm{True}) = \cdots,
$$

with "True" itself representable as $(x_1 \vee \neg x_1) = ((x_1 \wedge x_1) \vee \neg x_1) = \cdots$, guarantee aperiodicity. Thus Theorem VII.5 applies: the $\mathcal{Y}_f$ all have the same radius of convergence, and that radius must be equal to that of $E(z)$ (namely $\rho = 1/(16m)$), since $\mathcal{E} = \sum_f \mathcal{Y}_f$. Thereby the proposition is established. $\blacksquare$

It is an interesting and largely open problem to characterize the relation between the limit probability $\varpi(f)$ of a function f and its structural complexity. At least, the cases $m = 1, 2, 3$ can be solved exactly and numerically: it appears that functions of low complexity tend to occur much more frequently, as shown by the data of [115]. .... $\blacksquare$

VII. 6.3. The analysis of irreducible polynomial systems.

The analytic engine behind Theorem VII.5 is a fundamental result, the "Drmota–Lalley–Woods" (DLW) Theorem, due to independent research by several authors: Drmota [172] developed a version of the theorem in the course of studies relative to limit laws in various families of trees defined by context-free grammars; Woods [616], motivated by questions of boolean complexity and finite model theory, gave a form expressed in terms of colouring rules for trees; finally, Lalley [395] came across a similarly general result when quantifying return probabilities for random walks on groups. Drmota and Lalley show how to pull out limit Gaussian laws for simple parameters (by a perturbative analysis; see Chapter IX); Woods shows how to deduce estimates of coefficients even in some periodic or non-irreducible cases.

In the treatment that follows we start from a polynomial system of equations,

$$
\{y_j = \Phi_j(z, y_1, \ldots, y_m)\}, \quad j = 1, \ldots, m,
$$

in accordance with the notations adopted at the beginning of the section. We only consider *nonlinear systems* defined by the fact that at least one polynomial Φ_j is nonlinear in some of the indeterminates $y_1, \ldots, y_m$. (Linear systems have been discussed extensively in Chapter V.)

For applications to combinatorics, we define four possible attributes of a polynomial system. The first one is a natural positivity condition.

   (*i*) *Algebraic positivity* (or a-positivity). A polynomial system is said to be *a-positive* if all the component polynomials Φ_j have non-negative coefficients.

Next, we want to restrict consideration to systems that determine a unique solution vector $(y_1, \ldots, y_m) \in (\mathbb{C}[[z]])^m$. Define the *z-valuation* $\mathrm{val}(\vec{y})$ of a vector $\vec{y} \in \mathbb{C}[[z]]^m$ as the minimum over all j's of the individual valuations[12] $\mathrm{val}(y_j)$. The distance between two vectors is defined as usual by $d(\vec{u}, \vec{v}) = 2^{-\mathrm{val}(\vec{u}-\vec{v})}$. Then:

   (*ii*) *Algebraic properness* (or a-properness). A polynomial system is said to be *a-proper* if it satisfies a Lipschitz condition

$$d(\Phi(\vec{y}), \Phi(\vec{y}')) < K d(\vec{y}, \vec{y}') \quad \text{for some } K < 1.$$

In that case, the transformation Φ is a contraction on the complete metric space of formal power series and, by the general fixed point theorem, the equation $\vec{y} = \Phi(\vec{y})$ admits a unique solution. This solution may be obtained by the iterative scheme,

$$\vec{y}^{(0)} = (0, \ldots, 0)^t, \quad \vec{y}^{(h+1)} = \Phi(y^{(h)}), \quad \vec{y} = \lim_{h \to \infty} \vec{y}^{(h)}.$$

in accordance with our discussion of the *semantics of recursion*, on p. 31.

The key notion is irreducibility. To a polynomial system, $\vec{y} = \Phi(\vec{y})$, associate its *dependency graph* defined in the usual way as a graph whose vertices are the numbers $1, \ldots, m$ and the edges ending at a vertex j are $k \to j$, if y_j figures in a monomial of Φ_k.

   (*iii*) *Algebraic irreducibility* (or a-irreducibility). A polynomial system is said to be *a-irreducible* if its dependency graph is strongly connected.

(This notion matches that of Definition VII.5, p. 483.)

Finally, one needs the usual technical notion of aperiodicity:

   (*iv*) *Algebraic aperiodicity* (or a-aperiodicity). A proper polynomial system is said to be aperiodic if each of its component solutions y_j is aperiodic in the sense of Definition IV.5, p. 266.

We can now state:

Theorem VII.6 (Irreducible positive polynomial systems, DLW Theorem). *Consider a nonlinear polynomial system $\vec{y} = \Phi(\vec{y})$ that is a-positive, a-proper, and a-irreducible. Then, all component solutions y_j have the same radius of convergence $\rho < \infty$, and there exist functions h_j analytic at the origin such that, in a neighbourhood of ρ:*

$$(70) \qquad\qquad y_j = h_j\left(\sqrt{1 - z/\rho}\right).$$

---

[12]Let $f = \sum_{n=\beta}^{\infty} f_n z^n$ with $f_\beta \neq 0$ and $f_0 = \cdots = f_{\beta-1} = 0$; the valuation of f is by definition $\mathrm{val}(f) = \beta$; see Appendix A.5: *Formal power series*, p. 730.

In addition, all other dominant singularities are of the form $\rho\omega$ with ω a root of unity. If furthermore the system is a-aperiodic, all y_j have ρ as unique dominant singularity. In that case, the coefficients admit a complete asymptotic expansion,

$$(71) \qquad\qquad [z^n] y_j(z) \sim \rho^{-n} \left(\sum_{k \geq 0} d_k n^{-3/2-k} \right),$$

for computable d_k.

Proof. The proof consists in gathering by stages consequences of the assumptions. It is essentially based on a close examination of "failure" of the multivariate implicit function theorem and the way this situation leads to square-root singularities.

(*a*) As a preliminary observation, we note that each component solution y_j is an algebraic function that has a non-zero radius of convergence. This can be checked directly by the method of majorant series (Note IV.20, p. 250), or as a consequence of the multivariate version of the implicit function theorem (Appendix B.5: *Implicit Function Theorem*, p. 753).

(*b*) Properness together with the positivity of the system implies that each $y_j(z)$ has non-negative coefficients in its expansion at 0, since it is a formal limit of approximants that have non-negative coefficients. In particular, by positivity, ρ_j is a singularity of y_j (by virtue of Pringsheim's theorem). From the known nature of singularities of algebraic functions (e.g., the Newton–Puiseux Theorem, p. 498 below), there must exist some order $R \geq 0$ such that each Rth derivative $\partial_z^R y_j(z)$ becomes infinite as $z \to \rho_j^-$.

We establish now that $\rho_1 = \cdots = \rho_m$. In effect, differentiation of the equations composing the system implies that a derivative of arbitrary order r, $\partial_z^r y_j(z)$, is a linear form in other derivatives $\partial_z^r y_j(z)$ of the same order (and a polynomial form in lower order derivatives); also the linear combination and the polynomial form have non-negative coefficients. Assume *a contrario* that the radii were not all equal, say $\rho_1 = \cdots = \rho_s$, with the other radii $\rho_{s+1}, \ldots$ being strictly greater. Consider the system differentiated a sufficiently large number of times, R. Then, as $z \to \rho_1$, we must have $\partial_z^R y_j$ tending to infinity for $j \leq s$. On the other hand, the quantities y_{s+1}, etc., being analytic, their Rth derivatives that are analytic as well must tend to finite limits. In other words, because of the irreducibility assumption (and again positivity), infinity *has to* propagate and we have reached a contradiction. Thus: *all the y_j have the same radius of convergence.* We let ρ denote this common value.

(*c*$_1$) The key step consists in establishing the existence of a square-root singularity at the common singularity ρ. Consider first the scalar case, that is

$$(72) \qquad\qquad y - \phi(z, y) = 0,$$

where ϕ is assumed to be a nonlinear polynomial in y and have non-negative coefficients. This case belongs to the smooth implicit function schema, whose argument we briefly revisit under our present perspective.

Let $y(z)$ be the unique branch of the algebraic function that is analytic at 0. Comparison of the asymptotic orders in y inside the equality $y = \phi(z, y)$ shows that (by nonlinearity) we cannot have $y \to \infty$ when z tends to a finite limit. Let now ρ be the radius of convergence of $y(z)$. Since $y(z)$ is necessarily finite at its singularity ρ, we set $\tau = y(\rho)$ and note that, by continuity, $\tau - \phi(\rho, \tau) = 0$.

By the implicit function theorem, a solution (z_0, y_0) of (72) can be continued analytically as $(z, y_0(z))$ in the vicinity of z_0 as long as the derivative with respect to y (the simplest form of a Jacobian),

$$J(z_0, y_0) := 1 - \phi'_y(z_0, y_0),$$

remains non-zero. The quantity ρ being a singularity, we must thus have $J(\rho, \tau) = 0$. On the other hand, the second derivative $-\phi''_{yy}$ is non-zero at (ρ, τ) (by nonlinearity and positivity). Then, the local expansion of the defining equation (72) at (ρ, τ) binds (z, y) locally by

$$-(z - \rho)\phi'_z(\rho, \tau) - \frac{1}{2}(y - \tau)^2 \phi''_{yy}(\rho, \tau) + \cdots = 0,$$

implying the singular expansion

$$y - \tau = -\gamma(1 - z/\rho)^{1/2} + \cdots .$$

This establishes the first part of the assertion in the scalar case.

(c_2) In the multivariate case, we graft Lalley's ingenious argument [395] that is based on a linearized version of the system to which Perron–Frobenius theory is applicable. First, irreducibility implies that any component solution y_j depends positively and nonlinearly on itself (by possibly iterating Φ), so that a contradiction in asymptotic regimes would result, if we suppose that any y_j tends to infinity. *Each $y_j(z)$ remains finite at the positive dominant singularity ρ.*

Now, the multivariate version of the implicit function theorem (Theorem B.6, p. 755) grants us locally the analytic continuation of any solution $y_1, y_2, \ldots, y_m$ at z_0 provided there is no vanishing of the Jacobian determinant

$$J(z_0, y_1, \ldots, y_m) := \det\left(\delta_{i,j} - \frac{\partial}{\partial y_j}\Phi_i(z_0, y_1, \ldots, y_m)\right)_{i,j=1\ldots m}.$$

Thus, we must have

(73) $$J(\rho, \tau_1, \ldots, \tau_m) = 0 \quad \text{where} \quad \tau_j := y_j(\rho).$$

The next argument uses Perron–Frobenius theory (Subsection V. 5.2 and Note V.34, p. 345) and linear algebra. Consider the Jacobian matrix

$$K(z, y_1, \ldots, y_m) := \left(\frac{\partial}{\partial y_j}\Phi_i(z, y_1, \ldots, y_m)\right)_{i,j=1\ldots m},$$

which represents the "linear part" of Φ. For $z, y_1, \ldots, y_m$ all non-negative, the matrix K has positive entries (by positivity of Φ) so that it is amenable to Perron–Frobenius theory. In particular it has a positive eigenvalue $\lambda(z, y_1, \ldots, y_m)$ that dominates all the other in modulus. The quantity

$$\lambda(z) := \lambda(z, y_1(z), \ldots, y_m(z))$$

is increasing, as it is an increasing function of the matrix entries that themselves increase with z for $z \geq 0$.

We propose to prove that $\lambda(\rho) = 1$, In effect, $\lambda(\rho) < 1$ is excluded since otherwise $(I - K)$ would be invertible at $z = \rho$ and this would imply $J \neq 0$, thereby contradicting the singular character of the $y_j(z)$ at ρ. Assume *a contrario* $\lambda(\rho) > 1$ in order to exclude the other case. Then, by the monotonicity and continuity of $\lambda(z)$, there would exist $\overline{\rho} < \rho$ such that $\lambda(\overline{\rho}) = 1$. Let $\overline{v}$ be a left eigenvector of $K(\overline{\rho}, y_1(\overline{\rho}), \ldots, y_m(\overline{\rho}))$ corresponding to the eigenvalue $\lambda(\overline{\rho})$. Perron–Frobenius theory guarantees that such a vector $\overline{v}$ has all its coefficients that are positive. Then, upon multiplying on the left by $\overline{v}$ the column vectors corresponding to y and $\Phi(y)$ (which are equal), one gets an identity; this derived identity, upon expanding near $\overline{\rho}$, gives

$$(74) \qquad A(z - \overline{\rho}) = -\sum_{i,j} B_{i,j}(y_i(z) - y_i(\overline{\rho}))(y_j(z) - y_j(\overline{\rho})) + \cdots,$$

where $\cdots$ hides lower order terms and the coefficients $A, B_{i,j}$ are non-negative with $A > 0$. There is a contradiction in the orders of growth if each y_i is assumed to be analytic at $\overline{\rho}$, since the left-hand side of (74) is of exact order $(z - \overline{\rho})$ while the right-hand side is at least as small as $(z - \overline{\rho})^2$. Thus, we must have $\lambda(\rho) = 1$ and $\lambda(x) < 1$ for $x \in (0, \rho)$.

A calculation similar to (74) but with $\overline{\rho}$ replaced by ρ shows finally that, if

$$y_i(z) - y_i(\rho) \sim \gamma_i(\rho - z)^\alpha,$$

then consistency of asymptotic expansions implies $2\alpha = 1$, that is $\alpha = \frac{1}{2}$. We have thus proved: *All the component solutions $y_j(z)$ have a square-root singularity at ρ.* (The existence of a complete expansion in powers of $(\rho - z)^{1/2}$ results from a refinement of this argument.) The proof of the general case (70) is thus complete.

(d) In the aperiodic case, we first observe that each $y_j(z)$ cannot assume an infinite value on its circle of convergence $|z| = \rho$, since this would contradict the boundedness of $|y_j(z)|$ in the open disc $|z| < \rho$ (where $y_j(\rho)$ serves as an upper bound). Consequently, by singularity analysis, the Taylor coefficients of any $y_j(z)$ are $O(n^{-1-\eta})$ for some $\eta > 1$ and the series representing y_j at the origin converges on $|z| = \rho$.

For the rest of the argument, we observe that, if $\vec{y} = \Phi(z, \vec{y})$, then $\vec{y} = \Phi^{\langle m \rangle}(z, \vec{y})$ where the superscript denotes iteration of the transformation Φ in the variables $\vec{y} = (y_1, \ldots, y_m)$. By irreducibility, $\Phi^{\langle m \rangle}$ is such that *each* of its component polynomials involves *all* the variables.

Assume *a contrario* the existence of a singularity ρ^* of some $y_j(z)$ on $|z| = \rho$. The triangle inequality yields $|y_j(\rho^*)| \leq y_j(\rho)$, and the stronger form $|y_j(\rho^*)| < y_j(\rho)$ results from the Daffodil Lemma (p. 267). Then, the modified Jacobian matrix $K^{\langle m \rangle}$ of $\Phi^{\langle m \rangle}$ taken at the $y_j(\rho^*)$ has entries dominated strictly by the entries of $K^{\langle m \rangle}$ taken at the $y_j(\rho)$. Therefore, the dominant eigenvalue of $K^{\langle m \rangle}(z, \vec{y}_j(\rho^*))$ must be strictly less than 1. This would imply that $I - K^{\langle m \rangle}(z, \vec{y}_j(\rho^*))$ is invertible so that the $y_j(z)$ would be analytic at ρ^*. A contradiction has been reached: ρ is the sole dominant singularity of each y_j and this concludes the argument.   ∎

Many extensions of the DLW Theorem are possible, as indicated by the notes and references below—the underlying arguments are powerful, versatile, and highly general. Consequences regarding limit distributions, as obtained by Drmota and Lalley, are further explored in Chapter IX (p. 681).

▷ **VII.29.** *Analytic systems.* Drmota [172] has shown that the conclusions of the DLW Theorem regarding universality of the square-root singularity hold more generally for Φ_j that are analytic functions of $\mathbb{C}^{m+1}$ to $\mathbb{C}$, provided there exists a positive solution of the characteristic system within the domain of analyticity of the Φ_j (see the original article [172] and the note [99] for a discussion of precise conditions). This extension then unifies the DLW theorem and Theorem VII.3 relative to the smooth implicit function schema.   ◁

▷ **VII.30.** *Pólya systems.* Woods [616] has shown that several systems built from Pólya operators of the form MSET_k can also be treated by an extension of the DLW Theorem, which then unifies this theorem and Theorem VII.4.   ◁

▷ **VII.31.** *Infinite systems.* Lalley [398] has extended the conclusions of the DLW Theorem to certain infinite systems of generating function equations. This makes it possible to quantify the return probabilities of certain random walks on infinite free products of finite groups.   ◁

The square-root singularity property ceases to be universal when the assumptions of Theorems VII.5 and VII.6, in essence, positivity or irreducibility, fail to be satisfied. For instance, supertrees that are specified by a positive but reducible system have a singularity of the fourth-root type (Example VII.10, p. 412 to be revisited in Example VII.20, p. 503). We discuss next, in Section VII. 7, general methods that apply to *any* algebraic function and are based on the minimal polynomial *equation* (rather than a system) satisfied by the function. Note that the results there do not always subsume the present ones, since structure is not preserved when a system is reduced, by *elimination*, to a single equation. It would at least be desirable to determine directly, from a positive (but *reducible*) system, the type of singular behaviour of the solution, but the systematic research involved in such a programme is yet to be carried out.

VII. 7. The general analysis of algebraic functions

Algebraic series and algebraic functions are simply defined as solutions of a polynomial equation or system. Their singularities are strongly constrained to be *branch points*, with the local expansion at a singularity being a fractional power series known as a Newton–Puiseux expansion (Subsection VII. 7.1). Singularity analysis then turns out to be systematically applicable to algebraic functions, to the effect that their coefficients are asymptotically composed of elements of the form

$$(75) \qquad\qquad C \cdot \omega^n n^{p/q}, \qquad \frac{p}{q} \in \mathbb{Q} \setminus \{-1, -2, \ldots\},$$

see Subsection VII. 7.2. This last form includes as a special case the exponent $p/q = -3/2$, that was encountered repeatedly, when dealing with inverse functions, implicit functions, and irreducible systems. In this section, we develop the basic structural results that lead to the asymptotic forms (75). However, designing effective methods (i.e., decision procedures) to compute the characteristic constants in (75) is not obvious in the algebraic case. Several algorithms will be described in order to locate and analyse singularities (e.g., Newton's polygon method). In particular, the multivalued character of algebraic functions creates a need to solve what are known as *connection problems*.

Basics. We adopt as the starting point of the present discussion the following definition of an algebraic function or series (see also Note VII.32 for a variant).

Definition VII.6. *A function $f(z)$ analytic in a neighbourhood V of a point z_0 is said to be* algebraic *if there exists a (non-zero) polynomial $P(z, y) \in \mathbb{C}[z, y]$, such that*

$$(76) \qquad\qquad P(z, f(z)) = 0, \qquad z \in V.$$

A power series $f \in \mathbb{C}[[z]]$ is said to be an algebraic power series if it coincides with the expansion of an algebraic function at 0.

The *degree* of an algebraic series or function f is by definition the minimal value of $\deg_y P(z, y)$ over all polynomials that are cancelled by f (so that rational series are algebraic of degree 1). One can always assume P to be irreducible over $\mathbb{C}$ (that is $P = QR$ implies that one of Q or R is a scalar) and of minimal degree.

An algebraic function may also be defined by starting with a polynomial system of the form

$$(77) \qquad \begin{cases} P_1(z, y_1, \ldots, y_m) &= 0 \\ \quad\vdots & \vdots \quad \vdots \\ P_m(z, y_1, \ldots, y_m) &= 0, \end{cases}$$

where each P_j is a polynomial. A solution of the system (77) is by definition an m–tuple $(f_1, \ldots, f_m)$ that cancels each P_j; that is, $P_j(z, f_1, \ldots, f_m) = 0$. Any of the f_j is called a component solution. A basic but non-trivial result of *elimination theory* is that any component solution of a non-degenerate polynomial system is an algebraic series (Appendix B.1: *Algebraic elimination*, p. 739). In other words, one can eliminate the auxiliary variables $y_2, \ldots, y_m$ and construct a single bivariate polynomial Q such that $Q(z, y_1) = 0$.

We stress the point that, in the definitions by an equation (76) or a system (77), no positivity of any sort nor irreducibility is assumed. The analysis which is now presented applies to *any* algebraic function, whether or not it comes from combinatorics.

▷ **VII.32.** *Algebraic definition of algebraic series.* It is also customary to define f to be an algebraic series if it satisfies $P(z, f) = 0$ in the sense of formal power series, without *a priori* consideration of convergence issues. Then the technique of majorant series may be used to prove that the coefficients of f grow at most exponentially. Thus, the alternative definition is indeed equivalent to Definition VII.6.   ◁

▷ **VII.33.** *"Alg is in Diag of Rat".* Every algebraic function $F(z)$ over $\mathbb{C}(z)$ is the diagonal of a rational function $G(x, y) = A(x, y)/B(x, y) \in \mathbb{C}(x, y)$. Precisely:

$$F(z) = \sum_{n \geq 0} G_{n,n} z^n, \qquad \text{where} \quad G(x, y) = \sum_{m,n \geq 0} G_{m,n} x^m y^n.$$

This is implied by a theorem of Denef and Lipshitz [154], which is related to the holonomic framework (Appendix B.4: *Holonomic functions*, p. 748).   ◁

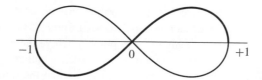

Figure VII.15. The real section of the lemniscate of Bernoulli defined by $P(z, y) = (z^2 + y^2)^2 - (z^2 - y^2) = 0$: the origin is a double point where two analytic branches meet; there are also two real branch points at $z = \pm 1$.

▷ **VII.34.** *Multinomial sums and algebraic coefficients.* Let $F(z)$ be an algebraic function. Then $F_n = [z^n] F(z)$ is a (finite) linear combination of "multinomial forms" defined as

$$S_n(\mathbf{C}; h; c_1, \ldots, c_r) := \sum_{\mathbf{C}} \binom{n_0 + h}{n_1, \ldots, n_r} c_1^{n_1} \cdots c_r^{n_r},$$

where the summation is over all values of $n_0, n_1, \ldots, n_r$ satisfying a collection of linear inequalities $\mathbf{C}$ involving n. [Hint: a consequence of Denef–Lipshitz.] Consequently: *coefficients of any algebraic function over $\mathbb{Q}(z)$ invariably admit combinatorial (i.e., binomial) expressions".* (Eisenstein's lemma, p. 505, can be used to establish algebraicity over $\mathbb{Q}(z)$.) An alternative proof can be based on Note IV.39, p. 270, and Equation (31), p. 753.   ◁

VII. 7.1. Singularities of general algebraic functions.

Let $P(z, y)$ be an irreducible polynomial of $\mathbb{C}[z, y]$,

$$P(z, y) = p_0(z)y^d + p_1(z)y^{d-1} + \cdots + p_d(z).$$

The solutions of the polynomial equation $P(z, y) = 0$ define a locus of points (z, y) in $\mathbb{C} \times \mathbb{C}$ that is known as a *complex algebraic curve*. Let d be the y-degree of P. Then, for each z there are at most d possible values of y. In fact, there exist d values of y "almost always", that is except for a finite number of cases.

— If z_0 is such that $p_0(z_0) = 0$, then there is a reduction in the degree in y and hence a reduction in the number of finite y-solutions for the particular value of $z = z_0$. One can conveniently regard the points that disappear as "points at infinity" (formally, one then operates in the projective plane).

— If z_0 is such that $P(z_0, y)$ has a multiple root, then some of the values of y will coalesce.

Define the *exceptional set* of P as the set (**R** is the resultant of Appendix B.1: *Algebraic elimination*, p. 739):

$$(78) \qquad \Xi[P] := \{z \mid R(z) = 0\}, \quad R(z) := \mathbf{R}(P(z, y), \partial_y P(z, y), y).$$

The quantity $R(z)$ is also known as the *discriminant* of $P(z, y)$, with y as the main variable and z a parameter. If $z \notin \Xi[P]$, then we have a guarantee that there exist d distinct solutions to $P(z, y) = 0$, since $p_0(z) \neq 0$ and $\partial_y P(z, y) \neq 0$. Then, by the Implicit Function Theorem, each of the solutions y_j lifts into a locally analytic function $y_j(z)$. A *branch* of the algebraic curve $P(z, y) = 0$ is the choice of such a $y_j(z)$ together with a simply connected region of the complex plane throughout which this particular $y_j(z)$ is analytic.

Singularities of an algebraic function can thus only occur if z lies in the exceptional set $\Xi[P]$. At a point z_0 such that $p_0(z_0) = 0$, some of the branches escape to infinity, thereby ceasing to be analytic. At a point z_0 where the resultant polynomial $R(z)$ vanishes but $p_0(z) \neq 0$, then two or more branches collide. This can be either a multiple point (two or more branches happen to assume the same value, but each one exists as an analytic function around z_0) or a branch point (some of the branches actually cease to be analytic). An example of an exceptional point that is not a branch point is provided by the classical lemniscate of Bernoulli: at the origin, two branches meet while each one is analytic there (see Figure VII.15).

A partial knowledge of the topology of a complex algebraic curve may be obtained by first looking at its restriction to the reals. Consider for instance the polynomial equation $P(z, y) = 0$, where

$$P(z, y) = y - 1 - zy^2,$$

which defines the OGF of the Catalan numbers. A rendering of the real part of the curve is given in Figure VII.16. The complex aspect of the curve, as given by $\Im(y)$ as a function of z, is also displayed there. In accordance with earlier observations, there are normally two sheets (branches) above each point. The exceptional set is given by the roots of the discriminant,

$$\mathcal{R} = z(1 - 4z),$$

that is, $z = 0, \frac{1}{4}$. For $z = 0$, one of the branches escapes at infinity, while for $z = 1/4$, the two branches meet and there is a branch point: see Figure VII.16.

In summary the exceptional set provides a set of *possible candidates* for the singularities of an algebraic function.

Lemma VII.4 (Location of algebraic singularities). *Let $y(z)$, analytic at the origin, satisfy a polynomial equation $P(z, y) = 0$. Then, $y(z)$ can be analytically continued along any simple path emanating from the origin that does not cross any point of the exceptional set defined in (78).*

Proof. At any z_0 that is not exceptional and for a y_0 satisfying $P(z_0, y_0) = 0$, the fact that the discriminant is non-zero implies that $P(z_0, y)$ has only a simple root at y_0, and we have $P_y(z_0, y_0) \neq 0$. By the Implicit Function Theorem, the algebraic function $y(z)$ is analytic in a neighbourhood of z_0. ∎

Nature of singularities. We start the discussion with an exceptional point that is placed at the origin (by a translation $z \mapsto z + z_0$) and assume that the equation $P(0, y) = 0$ has k equal roots $y_1, \ldots, y_k$ where $y = 0$ is this common value (by a translation $y \mapsto y + y_0$ or an inversion $y \mapsto 1/y$, if points at infinity are considered). Consider a punctured disc $|z| < r$ that does not include any other exceptional point relative to P. In the argument that follows, we let $y_1, (z), \ldots, y_k(z)$ be analytic determinations of the root that tend to 0 as $z \to 0$.

Start at some arbitrary value interior to the real interval $(0, r)$, where the quantity $y_1(z)$ is locally an analytic function of z. By the implicit function theorem, $y_1(z)$ can be continued analytically along a circuit that starts from z and returns to z while simply encircling the origin (and staying within the punctured disc). Then, by permanence of analytic relations, $y_1(z)$ will be taken into another root, say, $y_1^{(1)}(z)$. By repeating the

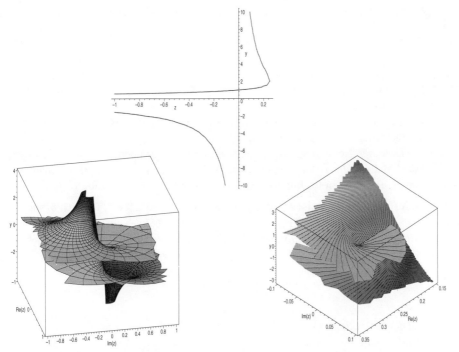

Figure VII.16. The real section of the Catalan curve (top). The complex Catalan curve with a plot of $\Im(y)$ as a function of $z = (\Re(z), \Im(z))$ (bottom left); a blow-up of $\Im(y)$ near the branch point at $z = 1/4$ (bottom right).

process, we see that, after a certain number of times κ with $1 \le \kappa \le k$, we will have obtained a collection of roots $y_1(z) = y_1^{(0)}(z), \ldots, y_1^{(\kappa)}(z) = y_1(z)$ that form a set of κ distinct values. Such roots are said to form a *cycle*. In this case, $y_1(t^\kappa)$ is an analytic function of t except possibly at 0 where it is continuous and has value 0. Thus, by general principles (regarding removable singularities, see Morera's Theorem, p. 743), it is in fact analytic at 0. This in turn implies the existence of a convergent expansion near 0:

$$(79) \qquad\qquad y_1(t^\kappa) = \sum_{n=1}^{\infty} c_n t^n.$$

(The parameter t is known as the *local uniformizing parameter*, as it reduces a multi-valued function to a single-valued one.) This translates back into the world of z: each determination of $z^{1/\kappa}$ yields one of the branches of the multivalued analytic function as

$$(80) \qquad\qquad y_1(z) = \sum_{n=1}^{\infty} c_n z^{n/\kappa}.$$

Alternatively, with $\omega = e^{2i\pi/\kappa}$ a root of unity, the κ determinations are obtained as

$$y_1^{(j)}(z) = \sum_{n=1}^{\infty} c_n \omega^n z^{n/\kappa},$$

each being valid in a sector of opening $< 2\pi$. (The case $\kappa = 1$ corresponds to an analytic branch.)

If $\kappa = k$, then the cycle accounts for all the roots which tend to 0. Otherwise, we repeat the process with another root and, in this fashion, eventually exhaust all roots. Thus, all the k roots that have value 0 at $z = 0$ are grouped into cycles of size $\kappa_1, \ldots, \kappa_\ell$. Finally, values of y at infinity are brought to zero by means of the change of variables $y = 1/u$, then leading to negative exponents in the expansion of y.

Theorem VII.7 (Newton–Puiseux expansions at a singularity). *Let $f(z)$ be a branch of an algebraic function $P(z, f(z)) = 0$. In a circular neighbourhood of a singularity ζ slit along a ray emanating from ζ, $f(z)$ admits a fractional series expansion (Puiseux expansion) that is locally convergent and of the form*

$$f(z) = \sum_{k \geq k_0} c_k (z - \zeta)^{k/\kappa},$$

for a fixed determination of $(z - \zeta)^{1/\kappa}$, where $k_0 \in \mathbb{Z}$ and κ is an integer ≥ 1, called the "branching type"[13].

Newton (1643–1727) discovered the algebraic form of Theorem VII.7 and published it in his famous treatise *De Methodis Serierum et Fluxionum* (completed in 1671). This method was subsequently developed by Victor Puiseux (1820–1883) so that the name of Puiseux series is customarily attached to fractional series expansions. The argument given above is taken from the neat presentation offered by Hille in [334, Ch. 12, vol. II]. It is known as a "monodromy argument", meaning that it consists in following the course of values of an analytic function along paths in the complex plane till it returns to its original value.

Newton polygon. Newton also described a *constructive* approach to the determination of branching types near a point (z_0, y_0), that, by means of the previous discussion, can always be taken to be $(0, 0)$. In order to introduce the discussion, let us examine the Catalan generating function near $z_0 = 1/4$. Elementary algebra gives the explicit form of the two branches

$$y_1(z) = \frac{1}{2z}\left(1 - \sqrt{1 - 4z}\right), \quad y_2(z) = \frac{1}{2z}\left(1 + \sqrt{1 - 4z}\right),$$

whose forms are consistent with what Theorem VII.7 predicts. If however one starts directly with the equation,

$$P(z, y) \equiv y - 1 - zy^2 = 0$$

---

[13]From the general discussion, if $k_0 < 0$, then $\kappa = 1$ is possible (case $f(\zeta) = \infty$, with a polar singularity); if $k_0 \geq 0$, then a singularity only exists if $\kappa \geq 2$ (case of a branch point with $|f(\zeta)| < \infty$).

then, the translation $z = 1/4 - Z$ (the minus sign is a mere notational convenience), $y = 2 + Y$ yields

(81)
$$Q(Z, Y) \equiv -\frac{1}{4}Y^2 + 4Z + 4ZY + ZY^2.$$

Look for solutions of the form $Y = cZ^\alpha(1 + o(1))$ with $c \neq 0$, whose *existence is a priori* granted by Theorem VII.7 (Newton–Puiseux). Each of the monomials in (81) gives rise to a term of a well-determined asymptotic order, respectively, $Z^{2\alpha}$, Z^1, $Z^{\alpha+1}$, $Z^{2\alpha+1}$. If the equation is to be identically satisfied, then the main asymptotic order of $Q(Z, Y)$ should be 0. Since $c \neq 0$, this can only happen if two or more of the exponents in the sequence $(2\alpha, 1, \alpha + 1, 2\alpha + 1)$ coincide *and* the coefficients of the corresponding monomial in $P(Z, Y)$ is zero, a condition that is an algebraic constraint on the constant c. Furthermore, exponents of all the remaining monomials have to be larger since by assumption they represent terms of lower asymptotic order.

Examination of all the possible combinations of exponents leads one to discover that the only possible combination arises from the cancellation of the first two terms of Q, namely $-\frac{1}{4}Y^2 + 4Z$, which corresponds to the set of constraints

$$2\alpha = 1, \quad -\frac{1}{4}c^2 + 4 = 0,$$

with the supplementary conditions $\alpha + 1 > 1$ and $2\alpha + 1 > 1$ being satisfied by this choice $\alpha = 1/2$. We have thus discovered that $Q(Z, Y) = 0$ is consistent asymptotically with

$$Y \sim 4Z^{1/2}, \quad Y \sim -4Z^{1/2}.$$

The process can be iterated upon subtracting dominant terms. It invariably gives rise to complete formal asymptotic expansions that satisfy $Q(Z, Y) = 0$ (in the Catalan example, these are series in $\pm Z^{1/2}$). Furthermore, elementary majorizations establish that such formal asymptotic solutions represent indeed convergent series. Thus, local expansions of branches have indeed been determined.

An algorithmic refinement (also due to Newton) is known as the method of *Newton polygons*. Consider a general polynomial

$$Q(Z, Y) = \sum_{j \in J} Z^{a_j} Y^{b_j},$$

and associate to it the finite set of points (a_j, b_j) in $\mathbb{N} \times \mathbb{N}$, which is called the Newton diagram. It is easily verified that the only asymptotic solutions of the form $Y \propto Z^\tau$ correspond to values of τ that are inverse slopes (i.e., $\Delta x / \Delta y$) of lines connecting two or more points of the Newton diagram (this expresses the cancellation condition between two monomials of Q) *and* such that all other points of the diagram are on this line or to the right of it (as the other monomials must be of smaller order). In other words:

> **Newton's polygon method.** *Any possible exponent τ such that $Y \sim cZ^\tau$ is a solution to a polynomial equation corresponds to one of the inverse slopes of the left-most convex envelope of the Newton diagram. For each viable τ, a polynomial equation constrains the possible values of the corresponding*

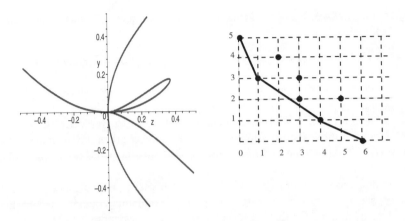

Figure VII.17. The real algebraic curve defined by the equation $P = (y - z^2)(y^2 - z)(y^2 - z^3) - z^3 y^3$ near $(0, 0)$ (left) and the corresponding Newton diagram (right).

coefficient c. Complete expansions are obtained by repeating the process, which means deflating Y from its main term by way of the substitution $Y \mapsto Y - cZ^\tau$.

Figure VII.17 illustrates what goes on in the case of the curve $P = 0$ where

$$
\begin{aligned}
P(z, y) &= (y - z^2)(y^2 - z)(y^2 - z^3) - z^3 y^3 \\
&= y^5 - y^3 z - y^4 z^2 + y^2 z^3 - 2 z^3 y^3 + z^4 y + z^5 y^2 - z^6,
\end{aligned}
$$

considered near the origin. As the factored part suggests, the curve is expected to resemble (locally) the union of two orthogonal parabolas and of a curve $y = \pm z^{3/2}$ having a cusp, i.e., the union of

$$
y = z^2, \quad y = \pm\sqrt{z}, \quad y = \pm z^{3/2},
$$

respectively. It is visible on the Newton diagram that the possible exponents $y \propto z^\tau$ at the origin are the inverse slopes of the segments composing the envelope, that is,

$$
\tau = 2, \quad \tau = \frac{1}{2}, \quad \tau = \frac{3}{2}.
$$

For computational purposes, once determined the branching type κ, the value of k_0 that dictates where the expansion starts, and the first coefficient, the full expansion can be recovered by deflating the function from its first term and repeating the Newton diagram construction. In fact, after a few initial stages of iteration, the method of indeterminate coefficients can always be eventually applied [Bruno Salvy, private communication, August 2000]. Computer algebra systems usually have this routine included as one of the standard packages; see [531].

VII. 7.2. Asymptotic form of coefficients. The Newton–Puiseux theorem describes precisely the local singular structure of an algebraic function. The expansions are valid around a singularity and, in particular, they hold in indented discs of the type required in order to apply the formal translation mechanisms of singularity analysis.

Theorem VII.8 (Algebraic asymptotics). *Let $f(z) = \sum_n f_n z^n$ be the branch of an algebraic function that is analytic at 0. Assume that $f(z)$ has a unique dominant singularity at $z = \alpha_1$ on its circle of convergence. Then, in the* non-polar *case, the coefficient f_n satisfies the asymptotic expansion,*

$$
(82) \qquad f_n \sim \alpha_1^{-n} \left(\sum_{k \geq k_0} d_k n^{-1-k/\kappa} \right),
$$

where $k_0 \in \mathbb{Z}$ and κ is an integer ≥ 2. In the polar *case, $\kappa = 1$ and $k_0 < 0$, the estimate (82) is to be interpreted as a terminating (exponential–polynomial) form.*

If $f(z)$ has several dominant singularities $|\alpha_1| = |\alpha_2| = \cdots = |\alpha_r|$, then there exists an asymptotic decomposition (where ϵ is some small fixed number, $\epsilon > 0$)

$$
(83) \qquad f_n = \sum_{j=1}^{r} \phi^{(j)}(n) + O((|\alpha_1| + \epsilon))^{-n},
$$

where each $\phi^{(j)}(n)$ admits a complete asymptotic expansion,

$$
\phi^{(j)}(n) \sim \alpha_j^{-n} \left(\sum_{k \geq k_0^{(j)}} d_k^{(j)} n^{-1-k/\kappa_j} \right),
$$

with either $k_0^{(j)}$ in $\mathbb{Z}$ and κ_j an integer ≥ 2 or $\kappa_j = 1$ and $k_0 < 0$.

Proof. An early version of this theorem appeared as [220, Th. D, p. 293]. The expansions granted by Theorem VII.7 are of the exact type required by singularity analysis (Theorem VI.4, p. 393). For multiple singularities, Theorem VI.5 (p. 398) based on composite contours is to be used: in that case each $\phi^{(j)}(n)$ is the contribution obtained by transfer of the corresponding local singular element. ∎

In the case of multiple singularities, partial cancellations may occur in some of the dominant terms of (83): consider for instance the case of

$$
\frac{1}{\sqrt{1 - \frac{6}{5}z + z^2}} = 1 + 0.60z + 0.04z^2 - 0.36z^3 - 0.408z^4 - \cdots,
$$

where the function has two complex conjugate singularities with an argument not commensurate to π, and refer to the corresponding discussion of rational coefficients asymptotics (Subsection IV. 6.1, p. 263). Fortunately, such delicate arithmetic situations tend *not* to arise in combinatorial situations.

Example VII.18. *Branches of unary–binary trees.* The generating function of unary–binary trees (Motzkin numbers, pp. 68 and 396) is $f(z)$ defined by $P(z, f(z)) = 0$ where

$$
P(z, y) = y - z - zy - zy^2,
$$

so that

$$
f(z) = \frac{1 - z - \sqrt{1 - 2z - 3z^2}}{2z} = \frac{1 - z - \sqrt{(1 + z)(1 - 3z)}}{2z}.
$$

There exist only two branches: f and its conjugate $\overline{f}$ that form a 2–cycle at $z = 1/3$. The singularities of all branches are at $0, -1, 1/3$ as is apparent from the explicit form of f or from

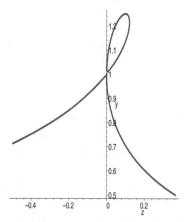

Figure VII.18. The real algebraic curve corresponding to non-crossing forests.

the defining equation. The branch representing $f(z)$ at the origin is analytic there (by a general argument or by the combinatorial origin of the problem). Thus, the dominant singularity of $f(z)$ is at $1/3$ and it is unique in its modulus class. The "easy" case of Theorem VII.8 then applies once $f(z)$ has been expanded near $1/3$. As a rule, the organization of computations is simpler if one makes use of the local uniformizing parameter with a choice of sign in accordance to the direction along which the singularity is approached. In this case, we set $z = 1/3 - \delta^2$ and find

$$f(z) = 1 - 3\delta + \frac{9}{2}\delta^2 - \frac{63}{8}\delta^3 + \frac{27}{2}\delta^4 - \frac{2997}{128}\delta^5 + \cdots, \qquad \delta = \left(\frac{1}{3} - z\right)^{1/2}.$$

This translates immediately into

$$f_n \equiv [z^n] f(z) \sim \frac{3^{n+1/2}}{2\sqrt{\pi n^3}} \left(1 - \frac{15}{16n} + \frac{505}{512n^2} - \frac{8085}{8192n^3} + \cdots\right),$$

which agrees with the direct derivation of Example VI.3, p. 396. .... ∎

▷ **VII.35.** *Meta-asymptotics.* Estimate the growth of the coefficients in the asymptotic expansions of Catalan and Motzkin (unary–binary trees) numbers. ◁

***Example* VII.19.** *Branches of non-crossing forests.* Consider the polynomial equation $P(z, y) = 0$, where

$$P(z, y) = y^3 + (z^2 - z - 3)y^2 + (z + 3)y - 1,$$

(see Figure VII.18 for the real branches) and the combinatorial GF satisfying $P(z, F) = 0$ determined by the initial conditions,

$$F(z) = 1 + 2z + 7z^2 + 33z^3 + 181z^4 + 1083z^5 + \cdots.$$

(*EIS* **A054727**). $F(z)$ is the OGF of non-crossing forests defined in Example VII.16, p. 485.

The exceptional set is mechanically computed: its elements are roots of the discriminant

$$R = -z^3(5z^3 - 8z^2 - 32z + 4).$$

Newton's algorithm shows that two of the branches at 0, say y_0 and y_2, form a cycle of length 2 with $y_0 = 1 - \sqrt{z} + O(z)$, $y_2 = 1 + \sqrt{z} + O(z)$ while it is the "middle branch" $y_1 = 1 + z + O(z^2)$ that corresponds to the combinatorial GF $F(z)$.

The non-zero exceptional points are the roots of the cubic factor of $\mathcal{R}$; namely

$$\Omega \doteq \{-1.93028, \ 0.12158, \ 3.40869\}.$$

Let $\xi \doteq 0.1258$ be the root in $(0, 1)$. By Pringsheim's theorem and the fact that the OGF of an infinite combinatorial class must have a positive dominant singularity in $[0, 1]$, the only possibility for the dominant singularity of $y_1(z)$ is ξ.

For z near ξ, the three branches of the cubic give rise to one branch that is analytic with value approximately 0.67816 and a cycle of two conjugate branches with value near 1.21429 at $z = \xi$. The expansion of the two conjugate branches is of the singular type,

$$\alpha \pm \beta\sqrt{1 - z/\xi},$$

where

$$\alpha = \frac{43}{37} + \frac{18}{37}\xi - \frac{35}{74}\xi^2 \doteq 1.21429, \quad \beta = \frac{1}{37}\sqrt{228 - 981\xi - 5290\xi^2} \doteq 0.14931.$$

The determination with a minus sign must be adopted for representing the combinatorial GF when $z \to \xi^-$ since otherwise one would get negative asymptotic estimates for the non-negative coefficients. Alternatively, one may examine the way the three real branches along $(0, \xi)$ match with one another at 0 and at ξ^-, then conclude accordingly.

Collecting partial results, we finally get by singularity analysis the estimate

$$F_n = \frac{\beta}{2\sqrt{\pi n^3}}\omega^n\left(1 + O(\frac{1}{n})\right), \quad \omega = \frac{1}{\xi} \doteq 8.22469$$

with the cubic algebraic number ξ and the sextic β as above. .... ∎

The example above illustrates several important points in the analysis of coefficients of algebraic functions when there are no simple explicit radical forms. First, a given combinatorial problem determines a unique branch of an algebraic curve at the origin. Next, the dominant singularity has to be identified by "connecting" the combinatorial branch with the branches at every possible singularity of the curve. Finally, computations tend to take place over algebraic numbers and not simply rational numbers.

So far, examples have illustrated the common situation where the function's exponent at its dominant singularity is $1/2$. Our last example shows a case where the exponent assumes a different value, namely $1/4$.

Example VII.20. *Branches of supertrees.* Consider the quartic equation

$$y^4 - 2y^3 + (1 + 2z)y^2 - 2yz + 4z^3 = 0$$

and let K be the branch analytic at 0 determined by the initial conditions:

$$K(z) = 2z^2 + 2z^3 + 8z^4 + 18z^5 + +64z^6 + 188z^7 + \cdots.$$

The OGF K corresponds to bicoloured supertrees of Example VI.10, p. 412; a partial graph is represented in Figure VII.19.

The discriminant is found to be

$$\mathcal{R} = 16z^4\left(16z^2 + 4z - 1\right)(-1 + 4z)^3,$$

with roots at $1/4$ and $(-1 \pm \sqrt{5})/8$. The dominant singularity of the branch of combinatorial interest turns out to be at $z = \frac{1}{4}$ where $K(1/4) = 1/2$. The translation $z = 1/4+Z$, $y = 1/2+Y$

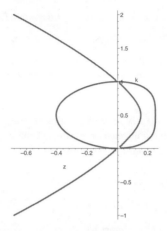

Figure VII.19. The real algebraic curve associated with the generating function of supertrees of type K.

then transforms the basic equation into

$$4\,Y^4 + 8\,ZY^2 + 16\,Z^3 + 12\,Z^2 + Z = 0.$$

According to Newton's polygon method, the main cancellation arises from $4Y^4 + Z = 0$: this corresponds to a segment of inverse slope $1/4$ in the Newton diagram and accordingly to a cycle formed with four conjugate branches, i.e., a fourth-root singularity. Thus, one has

$$K(z) \underset{z \to \frac{1}{4}}{\sim} 1/2 - \frac{1}{\sqrt{2}} \left(\frac{1}{4} - z\right)^{1/4} - \frac{1}{\sqrt{2}} \left(\frac{1}{4} - z\right)^{3/4} + \cdots, \quad [z^n]K(z) \underset{n \to \infty}{\sim} \frac{4^n}{8\Gamma(\frac{3}{4})n^{5/4}},$$

which is consistent with values found earlier (p. 412). .... ∎

Computable coefficient asymptotics. The previous discussion contains the germ of a complete algorithm for deriving an asymptotic expansion of coefficients of any algebraic function. We sketch in Note VII.36 the main principles, while leaving some of the details to the reader. Observe that the problem is a *connection problem*: the "shapes" of the various sheets around each point (including the exceptional points) are known, but it remains to connect them together and see which ones are encountered first when starting with a given branch at the origin.

▷ **VII.36.** *Algebraic Coefficient Asymptotics (ACA).* Here is an outline of the algorithm.

Algorithm ACA:

Input: A polynomial $P(z, y)$ with $d = \deg_y P(z, y)$; a series $Y(z)$ such that $P(z, Y) = 0$ and assumed to be specified by sufficiently many initial terms so as to be distinguished from all other branches.

Output: The asymptotic expansion of $[z^n]Y(z)$ whose existence is granted by Theorem VII.8.

The algorithm consists of three main steps: *Preparation (I)*, *Dominant singularities (II)*, and *Translation (III)*.

I. Preparation: Define the discriminant $R(z) = \mathbf{R}(P, P'_y, y)$.

(P_1) Compute the exceptional set $\Xi = \{z \mid R(z) = 0\}$ and the points of infinity $\Xi_0 = \{z \mid p_0(z) = 0\}$, where $p_0(z)$ is the leading coefficient of $P(z, y)$ considered as a function of y.

(P_2) Determine the Puiseux expansions of all the d branches at each of the points of $\Xi \cup \{0\}$ (by Newton diagrams and/or indeterminate coefficients). This includes the expansion of analytic branches as well. Let $\{y_{\alpha,j}(z)\}_{j=1}^{d}$ be the collection of all such expansions at some $\alpha \in \Xi \cup \{0\}$.

(P_3) Identify the branch at 0 that corresponds to $Y(z)$.

II. Dominant singularities: (Controlled approximate matching of branches). Let $\Xi_1, \Xi_2, \dots$ be a partition of the elements of $\Xi \cup \{0\}$ sorted according to the increasing values of their modulus: it is assumed that the numbering is such that if $\alpha \in \Xi_i$ and $\beta \in \Xi_j$, then $|\alpha| < |\beta|$ is equivalent to $i < j$. Geometrically, the elements of Ξ have been grouped in concentric circles. First, a preparation step is needed.

(D_1) Determine a non-zero lower bound δ on the radius of convergence of any local Puiseux expansion of any branch at any point of Ξ. Such a bound can be constructed from the minimal distance between elements of Ξ and from the degree d of the equation.

The sets Ξ_j are to be examined in sequence until it is detected that one of them contains a singularity. At step j, let $\sigma_1, \sigma_2, \dots, \sigma_s$ be an arbitrary listing of the elements of Ξ_j. The problem is to determine whether any σ_k is a singularity and, in that event, to find the right branch to which it is associated. This part of the algorithm proceeds by controlled numerical approximations of branches and constructive bounds on the minimum separation distance between distinct branches.

(D_2) For each candidate singularity σ_k, with $k \geq 2$, set $\zeta_k = \sigma_k(1 - \delta/2)$. By assumption, each ζ_k is in the domain of convergence of $Y(z)$ and of any $y_{\sigma_k,j}$.

(D_3) Compute a non-zero lower bound η_k on the minimum distance between two roots of $P(\zeta_k, y) = 0$. This separation bound can be obtained from resultant computations.

(D_4) Estimate $Y(\zeta_k)$ and each $y_{\sigma_k,j}(\zeta_k)$ to an accuracy better than $\eta_k/4$. If two elements, $Y(z)$ and $y_{\sigma_k,j}(z)$ are (numerically) found to be at a distance less than η_k for $z = \zeta_k$, then they are matched: σ_k *is a singularity and the corresponding* $y_{\sigma_k,j}$ *is the corresponding singular element.* Otherwise, σ_k is declared to be a regular point for $Y(z)$ and discarded as candidate singularity.

The main loop on j is repeated until a singularity has been detected, when $j = j_0$, say. The radius of convergence ρ is then equal to the common modulus of elements of Ξ_{j_0}; the corresponding singular elements are retained.

III. Coefficient expansion: Collect the singular elements at all the points σ determined to be a dominant singularity at Phase II. Translate termwise using the singularity analysis rule,

$$(\sigma - z)^{p/\kappa} \mapsto \sigma^{p/\kappa - n} \frac{\Gamma(-p/\kappa + n)}{\Gamma(-p/\kappa)\Gamma(n+1)},$$

and reorganize into descending powers of n, if needed. ◁

This algorithm vindicates the following assertion (see also Chabaud's thesis [110]).

Proposition VII.8 (Decidability of algebraic connections.). *The dominant singularities of a branch of an algebraic function can be determined in a finite number of operations by the algorithm* ACA *of Note VII.36.*

▷ **VII.37. Eisenstein's lemma.** Let $y(z)$ be an algebraic function with rational coefficients (for instance a combinatorial generating function) satisfying $\Phi(z, y(z)) = 0$, where the coefficient of the polynomial Φ are in $\mathbb{C}$; then *there exists a polynomial* Ψ *with integer coefficients such that* $\Psi(z, y(z)) = 0$. (Hint [65]. Consider the case where the coefficients of Φ are $\mathbb{Q}$–linear combinations of 1 and an irrational α, and write $\Phi(z, y) = \Phi_1(z, y) + \alpha\Phi_\alpha(z, y)$, where $\Phi_1, \Phi_\alpha \in \mathbb{Q}[z, y]$; extracting $[z^n]\Phi(z, y(z))$ would produce a $\mathbb{Q}$–linear relation between 1 and α, *unless* one of Φ_1, Φ_α is trivial, which must then be the case.) Thus, one can get $\Psi(z, y)$ in $\mathbb{Q}[z, y]$, and by clearing denominators, in $\mathbb{Z}[z, y]$. As a consequence, for algebraic $y(z)$ with rational coefficients, there exists an integer B such that for all n, one has $B^n[z^n]y(z) \in \mathbb{Z}$. Since there are infinitely many primes, the functions e^z, $\log(1+z)$, $\sum z^n/n^2$, $\sum z^n/(n!)^3$, and so on, are transcendental (i.e., not algebraic). $\lhd$

$\rhd$ **VII.38.** *Powers of binomial coefficients.* Define $S_r(z) := \sum_{n \geq 0} \binom{2n}{n}^r z^n$, with $r \in \mathbb{Z}_{>0}$. For even $r = 2\nu$ the function $S_{2\nu}(z)$ is transcendental (not algebraic) since its singular expansion involves a logarithmic term. For odd $r = 2\nu + 1$ and $r \geq 3$, the function $S_{2\nu+1}(z)$ is also transcendental as a consequence of the arithmetic transcendence of the number π; see [220]. These functions intervene in Pólya's drunkard problem (p. 425). In contrast with the "hard" theory of arithmetic transcendence, it is usually "easy" to establish transcendence of functions, by exhibiting a local expansion that contradicts the Newton–Puiseux Theorem (p. 498). $\lhd$

VII. 8. Combinatorial applications of algebraic functions

In this section, we introduce objects whose construction leads to algebraic functions, in a way that extends the basic symbolic method. This includes: walks with a finite number of allowed jumps (Subsection VII. 8.1) and planar maps (Subsection VII. 8.2). In such cases, *bivariate functional equations* reflect the combinatorial decompositions of objects. The common form of these functional equations is

$$(84) \qquad\qquad \Phi(z, u, F(z, u), h_1(z), \ldots, h_r(z)) = 0,$$

where Φ is a known polynomial and the unknown functions are F and $h_1, \ldots, h_r$. Specific methods are needed in order to attain solutions to such functional equations that would seem at first glance to be grossly underdetermined. Walks and excursions lead to a linear version of (84) that is treated by the so-called *kernel method*. Maps lead to nonlinear versions that are solved by means of Tutte's *quadratic method*. In both cases, the strategy consists in binding z and u by forcing them to lie on an algebraic curve (suitably chosen in order to eliminate the dependency on $F(z, u)$), and then pulling out consequences of such a specialization. Asymptotic estimates can then be developed from such algebraic solutions, thanks to the general methods expounded in the previous section.

VII. 8.1. Walks and the kernel method. Start with a set Ω that is a finite subset of $\mathbb{Z}$ and is called the set of *jumps*. A *walk* (relative to Ω) is a sequence $w = (w_0, w_1, \ldots, w_n)$ such that $w_0 = 0$ and $w_{i+1} - w_i \in \Omega$, for all i, $0 \leq i < n$. A *non-negative walk* (also known as a "meander") satisfies $w_i \geq 0$ and an *excursion* is a non-negative walk such that, additionally, $w_n = 0$. A *bridge* is a walk such that $w_n = 0$. The quantity n is called the length of the walk or the excursion. For instance, Dyck paths and Motzkin paths analysed in Section V. 4, p. 318, are excursions that correspond to $\Omega = \{-1, +1\}$ and $\Omega = \{-1, 0, +1\}$, respectively. (Walks and excursions are also somewhat related to paths in graphs in the sense of Section V. 5, p. 336.)

We let $-c$ denote the smallest (negative) value of a jump, and d denote the largest (positive) jump. A fundamental rôle is played in this discussion by the *characteristic polynomial*[14] of the walk,

$$S(y) := \sum_{\omega \in \Omega} y^\omega = \sum_{j=-c}^{d} S_j y^j,$$

which is a Laurent polynomial; that is, it involves negative powers of the variable y. .

Walks. Observe first the rational character of the BGF of walks, with z marking length and u marking final altitude:

(85)
$$W(z, u) = \frac{1}{1 - zS(u)}.$$

Since walks may terminate at a negative altitude, this is a Laurent series in u.

Bridges. The GF of bridges is formally $[u^0]W(z, u)$, since bridges correspond to walks that end at altitude 0. Thus one has

(86)
$$B(z) = \frac{1}{2i\pi} \int_\gamma \frac{1}{1 - zS(u)} \frac{du}{u},$$

upon integrating along a circle γ that separates the small and large branches, as discussed below. The integral can then be evaluated by residues: details are found in [27]; the net result is Equation (97), p. 511.

Excursions and meanders. We propose next to determine the number F_n of excursions of length n and type Ω, via the corresponding OGF

$$F(z) = \sum_{n=0}^{\infty} F_n z^n.$$

In fact, we shall determine the more general BGF

$$F(z, u) := \sum_{n,k} F_{n,k} u^k z^n,$$

where $F_{n,k}$ is the number of non-negative walks (meanders) of length n and final altitude k (*i.e.*, the value of w_n in the definition of a walk is constrained to equal k). In particular, one has $F(z) = F(z, 0)$.

The main result of this subsection can be stated informally as follows (see Propositions VII.9, p. 510 and VII.10, p. 513 for precise versions):

> *For each finite set* $\Omega \in \mathbb{Z}$, *the generating function of excursions is an algebraic function* that is explicitly computable from Ω. The number of excursions of length n satisfies asymptotically *a universal law* of the form
>
> $$CA^n n^{-3/2}.$$

---

[14]If Ω is a set, then the coefficients of S lie in $\{0, 1\}$. The treatment presented here applies in all generality to cases where the coefficients are arbitrary positive real numbers. This accounts for probabilistic situations as well as multisets of jump values.

There are many ways to view this result. The problem is usually treated within probability theory by means of Wiener–Hopf factorizations [515], and Lalley [396] offers an insightful analytic treatment from this angle. On another level, Labelle and Yeh [392] show that an unambiguous context-free specification of excursions can be systematically constructed, a fact that is sufficient to ensure the algebraicity of the GF $F(z)$. (Their approach is implicitly based on the construction of a pushdown automaton itself equivalent, by general principles, to a context-free grammar.) The Labelle–Yeh construction reduces the problem to a large, but somewhat "blind", combinatorial preprocessing. Accordingly, for analysts, it has the disadvantage of not extracting a simpler analytic (but non-combinatorial) structure inherent in the problem: the *shape* of the end result can indeed be predicted by the Drmota–Lalley–Woods Theorem, but the nature of the constants involved is not clearly accessible in this way.

The kernel method. The method described below is often known as the *kernel method*. It takes some of its inspiration from exercises in the 1968 edition of Knuth's book [377] (Ex. 2.2.1.4 and 2.2.1.11), where a new approach was proposed to the enumeration of Catalan and Schröder objects. The technique has since been extended and systematized by several authors; see for instance [26, 27, 86, 202, 203] for relevant combinatorial works. Our presentation below follows that of Lalley [396] and of Banderier and Flajolet [27].

The polynomial $f_n(u) = [z^n]F(z, u)$ is the generating function of non-negative walks of length n, with u recording final altitude. A simple recurrence relates $f_{n+1}(u)$ to $f_n(u)$, namely,

$$(87) \qquad\qquad f_{n+1}(u) = S(u) \cdot f_n(u) - r_n(u),$$

where $r_n(u)$ is a Laurent polynomial consisting of the sum of all the monomials of $S(u) f_n(u)$ that involve negative powers[15] of u:

$$(88) \qquad\qquad r_n(u) := \sum_{j=-c}^{-1} u^j \left([u^j] S(u) f_n(u)\right) = \{u^{<0}\} S(u) f_n(u).$$

The idea behind the formula is to subtract the effect of those steps that would take the walk below the horizontal axis. For instance, one has

$$S(u) = \frac{S_{-1}}{u} + O(1), \qquad \text{so that} \quad r_n(u) = \frac{S_{-1}}{u} f_n(0)$$

$$S(u) = \frac{S_{-2}}{u^2} + \frac{S_{-1}}{u} + O(1), \quad \text{so that} \quad r_n(u) = \left(\frac{S_{-2}}{u^2} + \frac{S_{-1}}{u}\right) f_n(0) + \frac{S_{-2}}{u} f_n'(0).$$

(This technique is similar to that of "adding a slice", p. 199.)

Generally, set

$$(89) \qquad\qquad \lambda_j(u) := \frac{1}{j!} \{u^{<0}\} u^j S(u).$$

---

[15]The convenient notation $\{u^{<0}\}$ denotes the singular part of a Laurent expansion: $\{u^{<0}\} f(z) := \sum_{j<0} \left([u^j] f(u)\right) \cdot u^j$.

Then, from (87) and (88) (multiply by z^{n+1} and sum), the generating function $F(z, u)$ satisfies the fundamental functional equation

$$(90) \qquad F(z, u) = 1 + zS(u)F(z, u) - z\{u^{<0}\} (S(u)F(z, u)) .$$

Thus, one has, explicitly,

$$(91) \qquad F(z, u) = 1 + zS(u)F(z, u) - z \sum_{j=0}^{c-1} \lambda_j(u) \left[\frac{\partial^j}{\partial u^j} F(z, u) \right]_{u=0} ,$$

where the Laurent polynomials $\lambda_j(u)$ depend on $S(u)$ in an effective way by (89).

The main equations (90) and (91) involve one unknown bivariate GF, $F(z, u)$ and c univariate GFs, the partial derivatives of F specialized at $u = 0$. It is true, but not at all obvious, that the single functional equation (91) fully determines the $c + 1$ unknowns. The basic technique is known as "cancelling the kernel" and it relies on strong analyticity properties; see the book by Fayolle *et al.* [203] for deep ramifications in the study of two-dimensional walks. The form of (91) to be employed for this purpose starts by grouping on one side the terms involving $F(z, u)$,

$$(92) \quad F(z, u)(1 - zS(u)) = 1 - z \sum_{j=0}^{c-1} \lambda_j(u)G_j(z), \qquad G_j(z) := \left[\frac{\partial^j}{\partial u^j} F(z, u) \right] .$$

If the right-hand side sum was not present, then the solution would reduce to (85). In the case at hand, from the combinatorial origin of the problem and implied bounds, the quantity $F(z, u)$ is bivariate analytic at $(z, u) = (0, 0)$ (by elementary exponential majorizations on the coefficients). The main principle of the kernel method consists in *coupling* the values of z and u in such a way that $1 - zS(u) = 0$, so that $F(z, u)$ disappears from the picture. A condition is that both z and u should remain small (so that F remains analytic). Relations between the partial derivatives are then obtained from such a specialization, $(z, u) \mapsto (z, u(z))$, which happen to be just in the right number.

Consequently, we consider the "kernel equation",

$$(93) \qquad\qquad\qquad 1 - zS(u) = 0,$$

which is rewritten as

$$u^c = z \cdot (u^c S(u)).$$

Under this form, it is clear that the kernel equation (93) defines $c + d$ branches of an algebraic function. A local analysis shows that, among these $c + d$ branches, there are c branches that tend to 0 as $z \to 0$, whereas the other d tend to infinity as $z \to 0$. (The idea is that, in the equation (93), either one of $zu^{-c} \approx 1$ or $zu^d \approx 1$ predominates; equivalently, a Newton polygon can be constructed.) Let $u_0(z), \ldots, u_{c-1}(z)$ be the c branches that tend to 0, that we call "small" branches. In addition, we single out $u_0(z)$, the "principal" solution, by the reality condition

$$u_0(z) \sim \gamma z^{1/c}, \quad \gamma := (S_c)^{1/c} \in \mathbb{R}_{>0} \qquad (z \to 0^+).$$

By local uniformization (see (79), p. 497), the conjugate branches are given locally by

$$u_\ell(z) = u_0(e^{2i\ell\pi} z) \qquad (z \to 0^+).$$

Coupling z and u by $u = u_\ell(z)$ produces interesting specializations of Equation (92). In that case, (z, u) is close to $(0, 0)$ where F is bivariate analytic so that the substitution is admissible. By substitution, we get

$$(94) \qquad 1 - z \sum_{j=0}^{c-1} \lambda_j(u_\ell(z)) \left[\frac{\partial^j}{\partial u^j} F(z, u) \right]_{u=0}, \qquad \ell = 0..c-1.$$

This is now a linear system of c equations in c unknowns (namely, the partial derivatives) with algebraic coefficients that, in principle, determine $F(z, 0)$.

A convenient approach to the solution of (94) is due to Mireille Bousquet-Mélou. The argument goes as follows. The quantity

$$(95) \qquad M(u) := u^c - zu^c \sum_{j=0}^{c-1} \lambda_j(u) \frac{\partial^j}{\partial u^j} F(z, 0)$$

can be regarded as a polynomial in u. It is monic while it vanishes by construction at the c small branches $u_0, \ldots, u_{c-1}$. Consequently, one has the factorization,

$$(96) \qquad M(u) = \prod_{\ell=0}^{c-1} (u - u_\ell(z)).$$

Now, the constant term of $M(u)$ is otherwise known to equal $-zS_{-c}F(z, 0)$, by the definition (95) of $M(u)$ and by Equation (89) specialized to $\lambda_0(u)$. Thus, the comparison of constant terms between (95) and (96) provides us with an explicit form of the OGF of excursions:

$$F(z, 0) = \frac{(-1)^{c-1}}{S_{-c}z} \prod_{\ell=0}^{c-1} u_\ell(z).$$

One can then finally return to the original functional equation and pull the BGF $F(z, u)$. In summary:

Proposition VII.9. *Let Ω be a finite step of jumps and let $S(u)$ be the characteristic polynomial of Ω. Consider the c small branches of the "kernel" equation,*

$$1 - zS(u) = 0,$$

denoted by $u_0(z), \ldots, u_{c-1}(z)$. The generating function of excursions is given by

$$F(z) = \frac{(-1)^{c-1}}{zS_{-c}} \prod_{\ell=0}^{c-1} u_\ell(z), \qquad \text{where } S_{-c} = [u^{-c}]S(u)$$

is the multiplicity (or weight) of the smallest element $-c \in \Omega$. More generally the bivariate generating function of non-negative walks (meanders) with u marking final altitude is bivariate algebraic and given by

$$F(z, u) = \frac{1}{u^c - zu^c S(u)} \prod_{\ell=0}^{c-1} (u - u_\ell(z)).$$

The OGF of bridges is expressible in terms of the small branches, by

$$(97) \qquad B(z) = z \sum_{j=1}^{c} \frac{u'_j(z)}{u_j(z)} = z \frac{d}{dz} \log \left(u_1(z) \cdots u_c(z) \right).$$

(The proof of (97) is based on a residue evaluation of (86), p. 507.)

Example VII.21. *Trees and Łukasiewicz codes.* A particular class of walks is of special interest; it corresponds to cases where $c = 1$; that is, the largest jump in the negative direction has amplitude 1. Consequently, $\Omega + 1 = \{0, s_1, s_2, \ldots, s_d\}$. In that situation, combinatorial theory teaches us the existence of fundamental isomorphisms between walks defined by steps Ω and trees whose degrees are constrained to lie in $1 + \Omega$. The correspondence is by way of Łukasiewicz codes[16], also known as 'Polish" prefix codes introduced in Chapter I. From this correspondence, we expect to find tree GFs in such cases.

With regard to generating functions, there now exists only *one* small branch, namely the solution $u_0(z)$ to $u_0(z) = z\phi(u_0(z))$ (where $\phi(u) = uS(u)$) that is analytic at the origin. One then has $F(z) = F(z, 0) = \frac{1}{z} u_0(z)$, so that the walk GF is determined by

$$F(z, 0) = \frac{1}{z} u_0(z), \quad u_0(z) = z\phi(u_0(z)), \quad \phi(u) := uS(u).$$

This form is consistent with what is already known regarding the enumeration of simple families of trees. In addition, one finds

$$F(z, u) = \frac{1 - u^{-1} u_0(z)}{1 - zS(u)} = \frac{u - u_0(z)}{u - z\phi(u)}.$$

Classical cases are rederived in this way:

— the Catalan walk (Dyck path), defined by $\Omega = \{-1, +1\}$ and $\phi(u) = 1 + u^2$, has

$$u_0(z) = \frac{1}{2z} \left(1 - \sqrt{1 - 4z^2} \right);$$

— the Motzkin walk, defined by $\Omega = \{-1, 0, +1\}$ and $\phi(u) = 1 + u + u^2$ has

$$u_0(z) = \frac{1}{2z} \left(1 - z - \sqrt{1 - 2z - 3z^2} \right);$$

— the modified Catalan walk, defined by $\Omega = \{-1, 0, 0, +1\}$ (with two steps of type 0) and $\phi(u) = 1 + 2u + u^2$, has

$$u_0(z) = \frac{1}{2z} \left(1 - 2z - \sqrt{1 - 4z} \right);$$

— the d–ary tree walk (the excursions encode d–ary trees) defined by $\Omega = \{-1, d-1\}$, has $u_0(z)$ that is defined implicitly by $u_0(z) = z(1 + u_0(z)^d)$.

The kernel method thus provides a new perspective for the enumeration of Dyck paths and related objects. .... ∎

---

[16]Such a code (p. 74) is obtained by a preorder traversal of the tree, recording a jump of $r - 1$ when a node of outdegree r is encountered. The sequence of jumps gives rise to an excursion followed by an extra -1 jump.

Example VII.22. *Walks with amplitude at most 2.* Take $\Omega = \{-2, -1, 1, 2\}$, so that

$$S(u) = u^{-2} + u^{-1} + u + u^2.$$

Then, $u_0(z), u_1(z)$ are the two branches that vanish as $z \to 0$ of the curve

$$y^2 = z(1 + y + y^3 + y^4).$$

The linear system that determines $F(z, 0)$ and $F'_u(z, 0)$ is

$$\begin{cases} 1 - \left(\dfrac{z}{u_0(z)^2} + \dfrac{z}{u_0(z)} \right) F(z, 0) - \dfrac{z}{u_0(z)} F'_u(z, 0) &= 0 \\[2mm] 1 - \left(\dfrac{z}{u_1(z)^2} + \dfrac{z}{u_1(z)} \right) F(z, 0) - \dfrac{z}{u_1(z)} F'_u(z, 0) &= 0 \end{cases}$$

(derivatives are taken with respect to the second argument) and one finds

$$F(z, 0) = -\frac{1}{z} u_0(z) u_1(z), \qquad F'_u(z, 0) = \frac{1}{z}(u_0(z) + u_1(z) + u_0(z) u_1(z)).$$

This gives the number of walks, through a combination of series expansions,

$$F(z) = 1 + 2z^2 + 2z^3 + 11z^4 + 24z^5 + 93z^6 + 272z^7 + 971z^8 + 3194z^9 + \cdots.$$

A single algebraic equation for $F(z) = F(z, 0)$ is then obtained by elimination (*e.g.*, via Gröbner bases) from the system:

$$\begin{cases} u_0^2 - z(1 + u_0 + u_0^3 + u_0^4) &= 0 \\ u_1^2 - z(1 + u_1 + u_1^3 + u_1^4) &= 0 \\ zF + u_0 u_1 &= 0 \end{cases}$$

Elimination shows that $F(z)$ is a root of the equation

$$z^4 y^4 - z^2(1 + 2z)y^3 + z(2 + 3z)y^2 - (1 + 2z)y + 1 = 0.$$

For $\Omega = \{-2, -1, 0, 1, 2\}$, we find similarly $F(z) = -\frac{1}{z} u_0(z) u_1(z)$, where u_0, u_1 are the small branches of $y^2 = z(1 + y + y^2 + y^3 + y^4)$; the expansion starts as

$$F(z) = 1 + z + 3z^2 + 9z^3 + 32z^4 + 120z^5 + 473z^6 + 1925z^7 + 8034z^8 + \cdots,$$

(*EIS* **A104184**; see also [441]), and $F(z)$ is a root of the equation

$$z^4 y^4 - z^2(1 + z)y^3 + z(2 + z)y^2 - (1 + z)y + 1 = 0.$$

In such cases, the GFs are no longer of the simple tree type. .... ∎

Asymptotic analysis. The singularities of the branches involved in the statement of Proposition VII.9 can be worked out in all generality [27, 396]. The roots of the kernel equation (93) are singular at points z with value u satisfying the simultaneous set of equations,

$$1 - zS(u) = 0, \qquad S'(u) = 0,$$

where the second equation corresponds to a place where the analytic implicit function theorem "fails" to define u as an analytic function of z. The second equation always has a positive root τ, corresponding to a positive value of z, which is $\rho = 1/S(\tau)$. It is then natural to suspect ρ to be radius of convergence of $F(z)$ and the singularity to be of the square-root type ($Z^{1/2}$), this for reasons seen in the proof of Theorem VII.3 (the smooth implicit-function schema). These properties are shown in complete detail in the articles [27, 395, 396], where it is also established that the GF of bridges is of singular type $Z^{-1/2}$, as in the case of Dyck bridges.

Proposition VII.10. *Define the structural constant* τ *by* $S'(\tau) = 0$, $\tau > 0$. *Then assuming aperiodicity, the number of bridges* (B_n) *and the number of excursions* (F_n) *satisfy*

$$B_n \sim \beta_0 \frac{S(\tau)^n}{\sqrt{2\pi n}}, \qquad F_n \sim \epsilon_0 \frac{S(\tau)^n}{2\sqrt{\pi n^3}},$$

where

$$\beta_0 = \frac{1}{\tau}\sqrt{\frac{S(\tau)}{S''(\tau)}}, \qquad \epsilon_0 = \frac{(-1)^{c-1}}{S_{-c}}\sqrt{\frac{2S(\tau)^3}{S''(\tau)}}\prod_{j=1}^{c-1} u_j\left(\frac{1}{S(\tau)}\right).$$

There, the u_j *represent the small branches and* u_0 *is the —principal" branch that is finite and real positive as* $z \to 0$.

Proposition VII.10 expresses a *universal law* of type $n^{-3/2}$ for excursions and $n^{-1/2}$ for bridges, a fact otherwise at least partly accessible to classical probability theory (e.g., via a local limit theorem for bridges and via Brownian motion for excursions). Basic parameters of walks, excursions, bridges, and meanders can then be analysed in a uniform fashion [27].

VII. 8.2. Maps and the quadratic method. A (planar) map is a connected planar graph together with an embedding into the plane. In all generality, loops and multiple edges are allowed. A planar map therefore separates the plane into regions called faces. The maps considered here are in addition rooted, meaning that a face, an incident edge, and an incident vertex are distinguished. In this section, only rooted maps are considered. (Nothing is lost regarding asymptotic properties of random structures when a rooting is imposed. The reason is that a map has, with probability exponentially close to 1, a trivial automorphism group; consequently, almost all maps of m edges can be rooted in $2m$ ways—by choosing an edge, and an orientation of this edge—and there is an almost uniform $2m$-to-1 correspondence between unrooted maps and rooted ones.) When representing rooted maps, we shall agree to draw the root edge with an arrow pointing away from the root node, and to take the root face as that face lying to the left of the directed edge (represented in grey below):

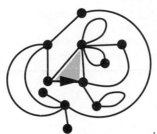

Tutte launched in the 1960s a large census of planar maps, with the intention of attacking the four-colour problem by enumerative techniques[17]; see [96, 579, 580,

---

[17]The four-colour theorem to the effect that every planar graph can be coloured using only four colours was eventually proved by Appel and Haken in 1976, using structural graph theory methods supplemented by extensive computer search.

581, 582]. There is in fact a very large collection of maps defined by various degree or connectivity constraints. In this chapter, we shall limit ourselves to conveying a flavour of this vast theory, with the goal of showing how algebraic functions arise. The presentation takes its inspiration from the book of Goulden and Jackson [303, Sec. 2.9]

The quadratic method. Let $\mathcal{M}$ be the class of all maps where size is taken to be the number of edges. Let $M(z, u)$ be the BGF of maps with u marking the number of edges on the outside face. The basic surgery performed on maps distinguishes two cases based upon the nature of the root edge. A rooted map will be declared to be *isthmic* if the root edge r of map μ is an "isthmus"; that is, an edge whose deletion would disconnect the graph. Clearly, one has

$$(98) \qquad\qquad \mathcal{M} = o + \mathcal{M}^{(i)} + \mathcal{M}^{(n)},$$

where $\mathcal{M}^{(i)}$ (resp. $\mathcal{M}^{(n)}$) represent the class of isthmic (resp. non-isthmic) maps and 'o' is the graph consisting of a single vertex and no edge. There are accordingly two ways to build maps from smaller ones by adding a new edge.

(i) The class of all isthmic maps is constructed by taking two arbitrary maps and joining them together by a new root edge, as shown below:

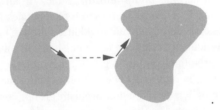

The effect is to increase the number of edges by 1 (the new root edge) and have the root face degree become 2 (the two sides of the new root edge) plus the sum of the root face degrees of the component maps. The construction is clearly revertible. In other words, the BGF of $\mathcal{M}^{(i)}$ is

$$(99) \qquad\qquad M^{(i)}(z, u) = zu^2 M(z, u)^2.$$

(ii) The class of non-isthmic maps is obtained by taking an already existing map and adding an edge that preserves its root node and "cuts across" its root face in some unambiguous fashion (so that the construction should be revertible). This operation will therefore result in a new map with an essentially smaller root-face degree. For instance, there are five ways to cut across a root face of degree 4; namely,

This corresponds to the linear transformation

$$u^4 \quad\mapsto\quad zu^5 \quad+\quad zu^4 \quad+\quad zu^3 \quad+\quad zu^2 \quad+\quad zu^1.$$

In general the effect on a map with root face of degree k is described by the transformation $u^k \mapsto zu(1 - u^{k+1})/(1 - u)$; equivalently, each monomial $g(u) = u^k$ is transformed into $zu(g(1) - ug(u))/(1 - u)$. Thus, the OGF of $\mathcal{M}^{(n)}$ involves a discrete difference operator:

$$(100) \qquad M^{(n)}(z, u) = zu\frac{M(z, 1) - uM(z, u)}{1 - u}.$$

Collecting the contributions from (99) and (100) in (98) then yields the basic functional equation,

$$(101) \qquad M(z, u) = 1 + u^2zM(z, u)^2 + uz\frac{M(z, 1) - uM(z, u)}{1 - u}.$$

The functional equation (101) binds two unknown functions, $M(z, u)$ and $M(z, 1)$. Similar to the case of walks, it would seem to be underdetermined. Now, a method due to Tutte and known as the *quadratic method* provides solutions. Following Tutte and the account in [303, p. 138], we consider momentarily the more general equation

$$(102) \qquad (g_1 F(z, u) + g_2)^2 = g_3,$$

where $g_j = G_j(z, u, h(z))$ and the G_j are explicit functions—here the unknown functions are $F(z, u)$ and $h(z)$ (cf $M(z, u)$ and $M(z, 1)$ in (101)). Bind u and z in such a way that the left side of (102) vanishes; that is, substitute $u = u(z)$ (a yet unknown function) so that $g_1 F + g_2 = 0$. Since the left-hand side of (102) now has a double root in u, so must the right-hand side, which implies

$$(103) \qquad g_3 = 0, \qquad \left.\frac{\partial g_3}{\partial u}\right|_{u=u(z)} = 0.$$

The original equation has become a system of two equations in two unknowns that determines implicitly $h(z)$ and $u(z)$. From this system, elimination provides individual equations for $u(z)$ and for $h(z)$. (If needed, $F(z, u)$ can then be recovered by solving a quadratic equation.) It will be recognized that, if the quantities g_1, g_2, g_3 are polynomials, then the process invariably yields solutions that are algebraic functions.

We now carry out this programme in the case of maps and Equation (101). First, isolate $M(z, u)$ by completing the square, giving

$$(104) \qquad \left(M(z, u) - \frac{1}{2}\frac{1 - u + u^2z}{u^2z(1 - u)}\right)^2 = Q(z, u) + \frac{M(z, 1)}{u(1 - u)},$$

where

$$Q(z, u) = \frac{z^2u^4 - 2zu^2(u - 1)(2u - 1) + (1 - u^2)}{4u^4z^2(1 - u)^2}.$$

Next, the condition expressing the existence of a double root is

$$Q(z, u) + \frac{1}{u(1 - u)}M(z, 1) = 0, \qquad Q_u'(z, u) + \frac{2u - 1}{u^2(1 - u)^2}M(z, 1) = 0.$$

It is now easy to eliminate $M(z, 1)$, since the dependency in M is linear, and a straight-forward calculation shows that $u = u(z)$ should satisfy

$$\left(u^2 z + (u - 1)\right)\left(u^2 z + (u - 1)(2u - 3)\right) = 0.$$

The first parameterization would lead to $M(z, 1) = 1/z$ which is not acceptable. Thus, $u(z)$ is to be taken as the root of the second factor, with $M(z, 1)$ being defined parametrically by

(105) $$z = \frac{(1 - u)(2u - 3)}{u^2}, \quad M(z, 1) = -u\frac{3u - 4}{(2u - 3)^2}.$$

Asymptotic analysis. In principle, the problem of enumerating maps is solved by (105), albeit in a parameterized form. We can then eliminate u (for instance, by resultants) and get an explicit equation for $M \equiv M(z, 1)$:

$$27z^2 M^2 - 18zM + M + 16z - 1 = 0.$$

This quadratic equation is explicitly solvable

$$M(z, 1) = -\frac{1}{54\,z^2}\left(1 - 18z - (1 - 12z)^{3/2}\right),$$

and its singular type is $Z^{3/2}$ (with $Z = (1 - 12z)$). Summarizing, we obtain one of the very first results in the enumerative theory of maps.

Proposition VII.11. *The OGF of maps admits the explicit form*

(106) $$M(z) \equiv M(z, 1) = -\frac{1}{54\,z^2}\left(1 - 18z - (1 - 12z)^{3/2}\right).$$

The number of maps with n edges, $M_n = [z^n]M(z, 1)$, satisfies

(107) $$M_n = 2\frac{(2n)!3^n}{n!(n + 2)!} \sim \frac{2}{\sqrt{\pi n^5}}12^n.$$

The sequence of coefficients is *EIS* **A000168**:

(108) $M(z, 1) = 1 + 2z + 9z^2 + 54z^3 + 378z^4 + 2916z^5 + 24057z^6 + 208494z^7 + \cdots.$

We refer to [303, Sec. 2.9] for detailed calculations (that are nowadays routinely performed with the assistance of a computer algebra system). Currently, there exist many applications of the quadratic method to maps satisfying all sorts of combinatorial constraints, in particular multiconnectivity; see [533] for a panorama. Interestingly enough, the singular exponent of maps is *universally* 3/2, a fact further reflected by the $n^{-5/2}$ factor in the asymptotic form of coefficients. Accordingly, randomness properties of maps are appreciably different from what is observed in trees and many commonly encountered context-free objects (e.g., irreducible ones).

▷ **VII.39.** *Lagrangean parametrization of general maps.* The change of parameter $u = 1 - 1/w$ reduces (105) to the "Lagrangean form",

(109) $$z = \frac{w}{1 - 3w}, \quad M(z, 1) = \frac{1 - 4w}{(1 - 3w)^2},$$

to which the Lagrange Inversion Theorem can be applied, giving back (107).   ◁

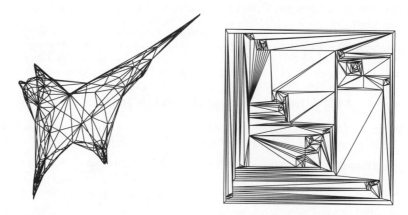

Figure VII.20. The "kitten": a random irreducible triangulation with a quadrangular outer face built out of 69 vertices and 200 edges. Left: a projection of a three-dimensional view (imagine the map drawn on a surface in $\mathbb{R}^3$). Right: a straight-line orthogonal rendering based on Fusy's algorithm [274].

▷ **VII.40**. *Distances in maps*. Chassaing and Schaeffer [113] have shown that the distance between two random vertices of a random planar map with n faces scales as $n^{1/4}$, when $n \to \infty$. Le Gall [404] has proved that a rescaled planar triangulation converges to a random "*continuum planar map*" that has a spherical topology. See Figure VII.20 for some aspects of a random map. (Physicists study similar random planar structures under the name of *2-dimensional quantum gravity*; see also Note VI.22, p. 414, for related material.)　　　　◁

▷ **VII.41**. *Matrix integrals and maps*. Consider an $N \times N$ Hermitian matrix H, such that

$$\Re(H_{i,j}) = \Re(H_{j,i}) = x_{i,j} \qquad \text{and} \qquad \Im(H_{i,j}) = -\Im(H_{j,i}) = y_{i,j},$$

and define the Gaussian measure of parameter λ on the set of Hermitian matrices as (Tr is the matrix trace):

$$d\mu_N(H; \lambda) := \left(\frac{2\pi}{\lambda}\right)^{-N^2/2} e^{-\lambda \operatorname{Tr}(H^2)/2} \prod_{i=1}^{N} dx_{i,i} \prod_{i<j} dx_{i,j} dy_{i,j}.$$

Let $M(t, \mathbf{v})$ be the multivariate generating function of rooted planar maps, where t marks the number of edges, $\mathbf{v}$ represents the vector of indeterminates $(v_1, v_2, \ldots)$, and v_j marks the number of vertices of degree j. One has

$$M(t, \mathbf{v}) = t \frac{d}{dt} \left[\lim_{N \to \infty} \frac{1}{N^2} \log \int \exp\left(N \sum_{m=1}^{\infty} v_m \frac{H^m}{m}\right) d\mu_N(H; N/t) \right].$$

(For this rich theory largely originating with Bessis, Brézin, Itzykson, Parisi and Zuber [60, 94], see Zvonkin's gentle introduction [630], Bouttier's thesis [88], as well as [89] and references therein.)　　　　◁

▷ **VII.42**. *The number of planar graphs*. The asymptotic number of labelled planar graphs with n vertices was determined by Giménez and Noy [290] to be of the form

$$G_n \sim g \cdot \gamma^n n^{-7/2} n!, \qquad g \doteq 0.4970\,04399, \quad \gamma \doteq 27.2268\,77685.$$

This spectacular result, which settled a long standing open question, is obtained by a succession of combinatorial and analytic steps based on: (*i*) the enumeration of 3–connected maps (these are the same as graphs, due to unique embeddability), which can be performed by the quadratic method; (*ii*) the enumeration of 2–connected graphs by Bender, Gao, and Wormald [41]; (*iii*) the integro-differential relations that relate the GFs of 2–connected and 1–connected graphs. The authors of [290] also show that a random planar graph is connected with probability asymptotic to $e^{-\nu} \doteq 0.96325$ and the mean number of connected components is asymptotic to $1 + \nu \doteq 1.03743$. See also the rich survey [291] for much more.   ◁

VII. 9. Ordinary differential equations and systems

In Part A of this book relative to *Symbolic Methods*, we have encountered differential relations attached to several combinatorial constructions.

— *Pointing:* the operation of pointing a specific atom in an object of a combinatorial class $\mathcal{C}$ produces a pointed class $\mathcal{D} = \Theta\mathcal{C}$. If the generating function of $\mathcal{C}$ is $C(z)$ (an OGF in the unlabelled case, an EGF in the labelled case), then one has

(110) $$\mathcal{D} = \Theta\mathcal{C} \quad \Longrightarrow \quad D(z) = z\frac{d}{dz}C(z).$$

See Subsections I. 6.2 (p. 86) and II. 6.1 (p. 136).

— *Order constraints:* in Subsection II. 6.3 (p. 139), we have defined the boxed product $\mathcal{A} = (\mathcal{B}^\square \star \mathcal{C})$ to be the modified labelled product comprised of pairs of elements such that the smallest label is constrained to lie in the $\mathcal{B}$ component. The translation over OGFs is

(111) $$\mathcal{A} = (\mathcal{B}^\square \star \mathcal{C}) \quad \Longrightarrow \quad A(z) = \int_0^z (\partial_t B(t)) \cdot C(t)\, dt.$$

Thus pointing and order constraints systematically lead to integro-differential relation, which can be transformed into *ordinary differential equations* (*ODEs*) and systems. Another rich source of differential equations in combinatorics is provided by the holonomic framework (Appendix B.4: *Holonomic functions*, p. 748). We summarize below some of the major methods that can be used to analyse the corresponding GFs. On the side of differential equations, our analytic arguments largely follow the accessible introductions found in the books by Henrici [329] and Wasow [602]. Linear ODEs are examined in Subsection VII. 9.1, some simple nonlinear ODEs in Subsection VII. 9.2. The main applications discussed here are relative to trees associated to ordered structures—quadtrees and increasing trees principally.

VII. 9.1. Singularity analysis of linear differential equations.
Linear differential equations with analytic coefficients have solutions that, near a reasonably well-behaved singularity ζ, are of the form

$$Z^\theta (\log Z)^k H(Z), \qquad Z := z - \zeta,$$

with $\theta \in \mathbb{C}$ an algebraic number, $k \in \mathbb{Z}_{\geq 0}$, and H a locally analytic function. The coefficients of such equations are composed of elements that are asymptotically of the form

$$n^\beta (\log n)^k, \qquad \beta = -\theta - 1,$$

in accordance with the general correspondence provided by singularity analysis. For instance, a naturally occurring combinatorial structure, the quadtree, gives rise to a number sequence that, surprisingly, turns out to be asymptotically proportional to $n^{(\sqrt{17}-3)/2}$.

Regular singularities. Our starting point is a *linear ordinary differential equation* (linear ODE), which we take to be of the form

$$(112) \qquad c_0(z)\partial^r Y(z) + c_1(z)\partial^{r-1}Y(z) + \cdots + c_r Y(z) = 0, \qquad \partial \equiv \frac{d}{dz}.$$

The integer r is the *order*. We assume the existence of a simply connected domain Ω in which the coefficients $c_j \equiv c_j(z)$ are analytic. At a point z_0 where $c_0(z_0) \neq 0$, a classical existence theorem (Note VII.43 and [602, p. 3]) guarantees that, in a neighbourhood of z_0, there exist r linearly independent *analytic* solutions of (112). Thus, *singularities* can only occur at points ζ that are roots of the leading coefficient $c_0(z)$.

▷ **VII.43.** *Analytic solutions.* Consider the ODE (112) near $z_0 = 0$ and assume $c_0(0) \neq 0$. Then, a formal solution $Y(z)$ can be determined, given any set of initial conditions $Y^{(j)}(0) = w_j$, by the method of indeterminate coefficients. The coefficients can be constructed recurrently, and simple bounds show that they are of at most exponential growth. ◁

To proceed, we rewrite Equation (112) as

$$(113) \qquad \partial^r Y(z) + d_1(z)\partial^{r-1}Y(z) + \cdots + d_r(z)Y(z) = 0,$$

where $d_j = c_j/c_0$. Under our assumptions, the functions $d_j(z)$ are now meromorphic in Ω. Given a meromorphic function $f(z)$, we define $\omega_\zeta(f)$ to be the order of the pole of f at ζ, and $\omega_\zeta(f) = 0$ means that $f(z)$ is analytic at ζ.

Definition VII.7. *The differential equations* (112) *and* (113) *are said to have a singularity at ζ if at least one of the $\omega_\zeta(d_j)$ is positive. The point ζ is said to be a* regular singularity[18] *if*

$$\omega_\zeta(d_1) \leq 1, \quad \omega_\zeta(d_2) \leq 2, \quad \ldots, \quad \omega_\zeta(d_r) \leq r,$$

an irregular singularity *otherwise.*

For instance, the second-order ODE

$$(114) \qquad Y'' + z^{-1}\sin(z)Y' - z^{-2}\cos(z)Y = 0,$$

has a regular singular point at $z = 0$, since the orders are $0, 2$, respectively. It is a notable fact that, even though we do not know how to solve explicitly the equation in terms of the usual special functions of analysis, the asymptotic form of its solutions can be precisely determined.

Let ζ be a regular singular point, and say we attempt to solve (112) by trying a solution of the form $Z^\theta + \cdots$, where $Z := z - \zeta$. For instance, proceeding somewhat optimistically with (114) at $\zeta = 0$, we may expect the left-hand side of the equation to be of the form

$$\left[\theta(\theta - 1)z^{\theta-2} + \cdots\right] + \left[\theta z^{\theta-1} + \cdots\right] - \left[z^{\theta-2} + \cdots\right] = 0.$$

In order to obtain cancellation to main asymptotic order ($z^{\theta-2}$), we must then assume that the coefficient of $z^{\theta-2}$ vanishes; then, θ solves an algebraic equation of degree 2, namely, $\theta(\theta - 1) - 1 = 0$, which suggests the possibility of two solutions of the form

---

[18]For "irregular" singularities, see Section VIII. 7, p. 581.

z^θ near 0, with $\theta = (1 \pm \sqrt{5})/2$. This informal discussion motivates the following definition.

Definition VII.8. *Given an equation of the form* (113) *and a regular singular point* ζ, *the indicial polynomial* $I(\theta)$ *at* ζ *is defined to be*

$$I(\theta) = \theta^{\underline{r}} + \delta_1 \theta^{\underline{r-1}} + \cdots + \delta_r, \qquad \theta^{\underline{\ell}} := \theta(\theta - 1) \cdots (\theta - \ell + 1),$$

where $\delta_j := \lim_{z \to \zeta} (z - \zeta)^j d_j(z)$. *The* indicial equation *(at* ζ*) is the algebraic equation* $I(\theta) = 0$.

If we let $\mathcal{L}$ denote the differential operator corresponding to the left-hand side of (113), we have formally, at a regular singular point,

$$\mathcal{L}[Z^\theta] = I(\theta) Z^{\theta - r} + O\left(Z^{\theta - r - 1}\right), \qquad Z = (z - \zeta),$$

which justifies the rôle of the indicial polynomial. (The process used to determine the solutions by restricting attention to dominant asymptotic terms is analogous to the Newton polygon construction for algebraic equations.) An important structure theorem describes the possible types of solutions of a meromorphic ODE at a regular singularity.

Theorem VII.9 (Regular singularities of ODEs). *Consider a meromorphic differential equation* (113) *and a regular singular point* ζ. *Assume that the indicial equation at* ζ, $I(\theta) = 0$, *is such that no two roots differ by an integer (in particular, all roots are distinct). Then, in a slit neighbourhood of* ζ, *there exists a linear basis of all the solutions that is comprised of functions of the form*

(115) $(z - \zeta)^{\theta_j} H_j(z - \zeta),$

where $\theta_1, \ldots, \theta_r$ *are the roots of the indicial polynomial and each* H_j *is analytic at* 0. *In the case of roots differing by an integer (or multiple roots), the solutions* (115) *may include additional logarithmic terms involving non-negative powers of* $\log(z - \zeta)$.

A description of the logarithmic cases is best based on a matrix treatment of the first-order linear *system* that is equivalent to the ODE [329, 602]. Note VII.44 describes the main lines of a proof of Theorem VII.9; Note VII.45 discusses the representative case of Euler systems, which is explicitly solvable.

▷ **VII.44.** *Singular solutions.* In the first case of Theorem VII.9 (no two roots differing by an integer), it suffices to work out the modified differential equation satisfied by $Z^{-\theta_j} Y(z)$ and verify that one of its solutions is analytic at ζ: the coefficients of H_j satisfy a recurrence, as in the non-singular case, from which their growth is verified to be at most exponential.   ◁

▷ **VII.45.** *Euler equations and systems.* An equation of the form,

$$\partial^r Y + e_1 Z^{-1} \partial^{r-1} Y + \cdots + e_r Z^{-r} Y = 0, \qquad e_j \in \mathbb{C}, \quad Z := (z - \zeta),$$

is known as an *Euler equation.* In the case where all roots of the indicial equation are simple, a basis of solutions is exactly of the form Z^{θ_j}. When θ is a root of multiplicity m, the set of solutions includes $Z^\theta (\log Z)^p$, for $p = 0, \ldots, m - 1$. (Euler equations appear for instance in the median-of-three quicksort algorithm [378, 538]. See [117] for several applications to random tree models and the analysis of algorithms.) *Euler systems* are first-order systems of the form

$$\frac{d}{dz} Y(z) = \frac{\mathbf{A}}{z - \zeta} Y(z),$$

where $\mathbf{A} \in \mathbb{C}^{r \times r}$ is a scalar matrix and $Y = (Y_1, \ldots, Y_r)^{\mathsf{T}}$ is a vector of functions. A formal solution is provided by

$$(z - \zeta)^{\mathbf{A}} = \exp\left(\mathbf{A}\log(z - \zeta)\right),$$

which indicates that the Jordan block decomposition of A plays a rôle in the occurrence of logarithmic factors of solutions. ◁

Theorem VII.10 (Coefficient asymptotics for meromorphic ODEs). *Let $f(z)$ be analytic at 0 and satisfy a linear differential equation*

$$\frac{d^r}{dz^r}f(z) + c_1(z)\frac{d^{r-1}}{dz^{r-1}}f(z) + \cdots + c_r(z)f(z) = 0,$$

where the coefficients $c_j(z)$ are analytic in $|z| < \rho_1$, except for possibly a pole at some ζ satisfying $|\zeta| < \rho_1, \zeta \neq 0$. Assume that ζ is a regular singular point and no two roots of the indicial equation at ζ differ by an integer. Then, there exist scalar constants $\lambda_1, \ldots, \lambda_r \in \mathbb{C}$ such that for any ρ_0 with $|\zeta| < \rho_0 < \rho_1$, one has

(116)
$$[z^n]f(z) = \sum_{j=1}^r \lambda_j \Delta_j(n) + O\left(\rho_0^{-n}\right),$$

where the $\Delta_j(n)$ are of the asymptotic form

(117)
$$\Delta_j(n) \sim \frac{n^{-\theta_j-1}}{\Gamma(-\theta_j)}\zeta^{-n}\left[1 + \sum_{k=1}^{\infty}\frac{s_{i,j}}{n^i}\right],$$

and the θ_j are the roots of the indicial equation at ζ.

Proof. The coefficients λ_j relate the particular solution $f(z)$ to the basis of solutions (115). The rest, by singularity analysis, is nothing but a direct transcription to coefficients of the solutions provided by the structure theorem, Theorem VII.9, with $\Delta_j(n) = [z^n](z - \zeta)^{\theta_j}H_j(z - \zeta)$. ∎

Taking into account multiple roots (as in Note VII.45) and roots differing by an integer, we see that solutions to meromorphic linear ODEs, in the regular case at least, are only composed of linear combinations of asymptotic elements of the form[19]

(118)
$$\zeta^{-n}n^{\beta}(\log n)^{\ell},$$

where ζ is determined as root of a (possibly transcendental) equation, $c_0(\zeta) = 0$, the number β is an algebraic quantity (over the field of constants δ_j) determined by the polynomial equation $I(-\beta - 1) = 0$, and ℓ is an integer.

The coefficients λ_j serve to "connect" the particular function of interest, $f(z)$ to the local basis of singular solutions (115). Their determination thus represents a *connection* problem (see pp. 470 and 505 for the easier algebraic case). However, contrary to what happens for algebraic equations, the determination of the λ_j can only be approached in all generality by numerical methods [252]. (Even when the coefficients $d_j(z) \in \mathbb{Q}(z)$ are rational fractions, no effective procedure is available to decide, from

---

[19] The forms (118) are appreciably more general than the corresponding ones arising in algebraic coefficient asymptotics (Theorem VII.8, p. 501), in which no logarithmic term can be present and the exponents are constrained to be rational numbers only.

an $f(z) \in \mathbb{Q}[[z]]$ determined by initial conditions at 0, which of the connection coefficients λ_j may vanish.) In many combinatorial applications the calculations can be carried out explicitly, in which case the forms (118) serve as a beacon of what to expect asymptotically. (Once existence of such forms is granted, e.g., by Theorems VII.9 and VII.10, it is often possible to identify coefficients and/or exponents in asymptotic expansions directly.) Similar considerations apply to functions defined by *systems* of linear differential equations (Note VII.48 below).

▷ **VII.46.** *Multiple singularities.* In the case of several singularities $\zeta_1, \ldots, \zeta_s$, a sum of s terms, each of the form (117) with $\zeta \to \zeta_i$, expresses $[z^n] f(z)$. [The structure theorem applies at each ζ_i and singularity analysis is known to adapt to multiple singularities; cf Section VI. 5, p. 398.]   ◁

▷ **VII.47.** *A relaxation.* In Theorem VII.10, one may allow the equation to have a singularity of any kind at 0. [Only properties of the basis of solutions near ζ are used.]   ◁

▷ **VII.48.** *Equivalence between equations and systems.* A (first-order) linear differential system is by definition

$$\frac{d}{dz} Y(z) = \mathbf{A}(z) Y(z),$$

where $Y = (Y_1, \ldots, Y_m)^{\mathrm{T}}$ is an m-dimensional column vector and $\mathbf{A}$ is an $m \times m$ coefficient matrix. *A differential equation of order m can always be reduced to a system of dimension m, and conversely.* Only rational operations and derivatives are involved in each of the conversions: technically, coefficient manipulations take place in a differential field $\mathbb{K}$ that contains coefficients of recurrences and systems. (For instance, the set of rational functions $\mathbb{C}(z)$ and the set of meromorphic functions in an open set Ω are differential fields.)

The proofs are simple extensions of the case $m = 2$. Starting from the equation $y'' + by' + cy = 0$, one sets $Y_1 = y$, $Y_2 = y'$ to get the system

$$\{\partial Y_1 = Y_2, \qquad \partial Y_2 = -cY_1 - bY_2\}.$$

Conversely, given the system

$$\{\partial Y_1 = a_{11} Y_1 + a_{12} Y_2, \quad \partial Y_2 = a_{21} Y_1 + a_{22} Y_2\},$$

let $\mathcal{E} = \mathrm{VS}[Y_1, Y_2]$ be the vector space over $\mathbb{K}$ spanned by Y_1, Y_2, which is of dimension ≤ 2. Differentiation of the relation $\partial Y_1 = a_{11} Y_1 + a_{12} Y_2$ shows that $\partial^2 Y_1$ can be expressed as combination of Y_1, Y_2,

$$\partial^2 Y_1 = a'_{11} Y_1 + a'_{12} Y_2 + a_{11}(a_{11} Y_1 + a_{12} Y_2) + a_{12}(a_{21} Y_1 + a_{22} Y_2),$$

hence $\partial^2 Y_1$ lies in $\mathcal{E}$. Thus, the system $\{Y_1, \partial Y_1, \partial Y_1^2\}$ is bound, which corresponds to a differential equation of order 2 being satisfied by Y_1. (In the case where the coefficient matrix $\mathbf{A}$ has a simple pole at ζ, singularities of solutions can be studied by matrix methods akin to those of Note VII.45.)   ◁

Combinatorial applications. The *quadtree* is a structure, discovered by Finkel and Bentley [212], that can be superimposed on any sequence of points in Euclidean space $\mathbb{R}^d$. In computer science, it forms the basis of several algorithms for maintaining and searching dynamically varying geometric objects [532], and it constitutes a natural extension of binary search trees. Quadtrees are associated to differential equations, whose order equals the dimension of the underlying space. Some of their major characteristics can be determined via singularity analysis of these equations [233, 242].

Example VII.23. *The plain quadtree.* Start from the unit square $\mathcal{Q} = [0, 1]^2$ and let $\mathfrak{p} = (P_1, \ldots, P_n)$ be a sequence of n points drawn uniformly and independently from $\mathcal{Q}$, with $P_j =$

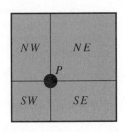 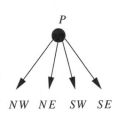 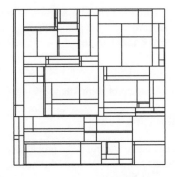

Figure VII.21. The quadtree splitting process (left, center); a hierarchical partition associated to $n = 50$ random points (right).

(x_j, y_j). A quaternary tree, called the *quadtree* and noted QT($\mathfrak{p}$), is built recursively from $\mathfrak{p}$ as follows:

— if $\mathfrak{p}$ is the empty sequence ($n = 0$), then QT($\mathfrak{p}$) = $\emptyset$ is the empty tree;
— otherwise, let $\mathfrak{p}_{NW}, \mathfrak{p}_{NE}, \mathfrak{p}_{SW}, \mathfrak{p}_{SE}$ be the four subsequences of points of $\mathfrak{p}$ that lie, respectively, North-West, North-East, South-West, South-East of P_1. For instance $\mathfrak{p}_{SW}$ is $\mathfrak{p}_{SW} = (P_{j_1}, P_{j_2}, \ldots, P_{j_k})$, where $1 < j_1 < j_2, \cdots < j_k \leq n$, and the $P_{j_\ell} = (x_{j_\ell}, y_{j_\ell})$ are those of the points that satisfy the predicate $x_{j_\ell} < x_1$ and $y_{j_\ell} < y_1$. Then QT($\mathfrak{p}$) is

$$QT(\mathfrak{p}) = \langle P_1; QT(\mathfrak{p}_{NW}), QT(\mathfrak{p}_{NE}), QT(\mathfrak{p}_{SW}), QT(\mathfrak{p}_{SE}) \rangle.$$

In other words, the sequence of points induces a hierarchical partition of the space QT; see Figure VII.21. (For simplicity, the tree is only defined here for points having different x and y coordinates, an event that has probability 1.)

Quadtrees are used for searching in two related ways: (*i*) given a point $P_0 = (x_0, y_0)$, *exact search* aims at determining whether P_0 occurs in $\mathfrak{p}$; (*ii*) given a coordinate $x_0 \in [0, 1]$, a *partial-match query* asks for the set of points $P = (x, y)$ occurring in $\mathfrak{p}$ such that $x = x_0$ (irrespective of the values of y). Both types are accommodated by the quadtree structure: an exact search corresponds to descending in the tree, following a branch guided by the coordinates of the point P_0 that is sought; partial match is implemented by recursive descents into two subtrees (either the pair NW, SW or NE, SE) based on the way x_0 compares with the x coordinate of the root point.

In an ideal world (for computers), trees are perfectly balanced, in which case the search costs satisfy the approximate recurrences,

(119) $$f_n = 1 + f_{n/4}, \qquad g_n = 1 + 2g_{n/4},$$

for exact search and partial match, respectively. The solutions of these recurrences are $\approx \log_4 n$ and $\approx \sqrt{n}$, respectively. To what extent do randomly grown quadtrees differ from the perfect shape, and what is the growth of the cost functions on average? The answer lies in the singularities of certain linear differential equations.

Exact search. Our purpose is to set up recurrences[20] in the spirit of Subsection VI. 10.3, p. 427. We need the probability $\pi_{n,k}$ that a quadtree of size n gives rise to a NW root-subtree of size k and claim that

$$(120) \qquad \pi_{n,k} = \frac{1}{n}\,(H_n - H_k)\,, \qquad H_n = 1 + \frac{1}{2} + \cdots + \frac{1}{n}.$$

Indeed, the probability that ℓ elements are West of the root and k are North-West is

$$(121) \qquad \varpi_{n,\ell,k} = \binom{n-1}{k,\,\ell-k,\,n-1-\ell} \int_0^1 \int_0^1 (xy)^k (x(1-y))^{\ell-k}(1-x)^{n-1-\ell}dx\,dy.$$

(The double integral is the probability that the first k elements fall NW, the next $\ell - k$ fall SW, the rest fall either NE or SE; the integrand corresponds to a conditioning upon the coordinates (x, y) of the root; the multinomial coefficient takes into account the possible shufflings.) The Eulerian Beta integral (p. 747) simplifies the integrals to $\varpi_{n,\ell,k} = 1/(n(\ell+1))$, from which the claimed (120) follows by summation over ℓ.

Given (120), the recurrence

$$(122) \qquad P_n = n + 4 \sum_{k=0}^{n-1} \pi_{n,k} P_k, \qquad P_0 = 0,$$

with $\pi_{n,k}$ as in (120), determines the sequence of expected value of path length. This recurrence translates into the integral equation,

$$(123) \qquad P(z) = \frac{z}{(1-z)^2} + 4 \int_0^z \frac{dt}{t(1-t)} \int_0^t P(u) \frac{du}{1-u},$$

itself equivalent to the linear differential equation of order 2:

$$z(1-z)^4 P''(z) + (1-2z)(1-z)^3 P'(z) - 4(1-z)^2 P(z) = 1 + 3z.$$

The homogeneous equation has a regular singularity at $z = 1$. In such a simple case, it is not difficult to guess the "right" solution, which can then be verified by substitution:

$$P(z) = \frac{1}{3}\frac{1+2z}{(1-z)^2} \log \frac{1}{1-z} + \frac{1}{6}\frac{4z+z^2}{(1-z)^2}, \qquad P_n = \left(n + \frac{1}{3}\right) H_n - \frac{n+1}{6n}.$$

The ratio P_n/n represents the mean level of a random node in a randomly grown quadtree, a quantity which is thus $\log n + O(1)$. Accordingly, quadtrees are on average fairly balanced, the expected level being within a factor $\log 4 \doteq 1.38$ of the corresponding quantity in a perfect tree.

Partial match. The analysis of partial match reveals a curious consequence of the imbalance of quadtrees, where the order of growth differs from that which the perfect tree model (119) predicts. The recurrence satisfied by the expected cost of a partial match query is determined by methods similar to path length [233]. One finds, by a computation similar to (121),

$$(124) \qquad Q_n = 1 + \frac{4}{n(n+1)} \sum_{k=0}^{n-1} (n-k)Q_k, \qquad Q_0 = 0,$$

corresponding, for the GF $Q(z) = \sum Q_n z^n$, to the *inhomogeneous* differential equation, $\mathfrak{L}[Q(z)] = 2/(1-z)$, where the differential operator $\mathfrak{L}$ is

$$(125) \qquad \mathfrak{L}[f] = z(1-z)^2 \partial^2 f + 2(1-z)^2 \partial f - 4f.$$

---

[20]It is also possible, although less convenient, to develop equations starting from basic principles of the symbolic method.

A particular solution of the inhomogeneous equation is $-1/(1-z)$, so that $y(z) := Q(z) + 1/(1-z)$ satisfies the homogeneous equation $\mathcal{L}[y] = 0$.

The differential equation $\mathcal{L}[y] = 0$ is singular at $z = 0, 1, +\infty$ and it has a regular singularity at $z = 1$. Since one has $y_n = O(n)$, by the origin of the problem, the singularity at $z = 1$ is the one that matters. The indicial polynomial can be computed from its definition or, equivalently, by simply substituting $y = (z-1)^\theta$ in the definition of $\mathcal{L}$ and discarding lower order terms. One finds, with $Z = z - 1$:

$$\mathcal{L}[Z^\theta] = \theta(\theta - 1)Z^\theta - 4Z^\theta + O\left(Z^{\theta-1}\right).$$

The roots of the indicial equations are then

$$\theta_1 = \frac{1}{2}\left(1 - \sqrt{17}\right), \qquad \theta_2 = \frac{1}{2}\left(1 + \sqrt{17}\right).$$

Theorem VII.9 guarantees that $y(z)$ admits, near $z = 1$ a representation of the form

(126) $$y(z) = \lambda_1(1-z)^{\theta_1} H_1(z-1) + (1-z)^{\theta_2} H_2(z-1),$$

with H_1, H_2 analytic at 0.

In order to complete the analysis, we still have to verify that the coefficient λ_1, which multiplies the singular element that dominates as $z \to 1$ is non-zero. Indeed, if we had $\lambda_1 = 0$, then, one would have $y(z) \to 0$ as $z \to 1$, which contradicts the fact that $y_n \geq 1$. In other words, here: *the connection problem is solved by means of bounds that are available from the combinatorial origin of the problem*. Singularity analysis then yields the asymptotic form of y_n, hence of Q_n. Summarizing , we have:

Proposition VII.12. *Path length in a randomly grown quadtree of size n is on average $n \log n + O(n)$. The expected cost of a partial match query satisfies, for some positive κ:*

(127) $$Q_n \sim \kappa \cdot n^{\alpha-1}, \qquad \alpha = \frac{\sqrt{17}-1}{2} \doteq 1.56155.$$

The analysis extends to quadtrees of higher dimensions [233]. In general dimension d, path length is on average $\frac{2}{d} n \log n + O(n)$. The cost of a partial match query is of the order of n^β, where β is an algebraic number of degree d. The cost of a random (fully specified) search admits a limit Gaussian distribution, as we prove in Example IX.29, p. 687. . . . . ∎

▷ **VII.49.** *Quadtrees and hypergeometric functions.* For the plain quadtree ($d = 2$), the change of variables $y = (1-z)^{-\theta} \eta(z)$ reduces the differential equation $\mathcal{L}[y] = 0$ to hypergeometric form. The constant κ in (127) is then found to satisfy

$$\kappa = \frac{1}{2}\frac{\Gamma(2\alpha)}{\Gamma(\alpha)^3}, \qquad \alpha = \frac{\sqrt{17}-1}{2}.$$

Hypergeometric solutions (Note B.15, p. 751) are available for $d \geq 2$; see [116, 233, 242]. ◁

▷ **VII.50.** *Closed meanders.* A *closed meander* of size n is a topological configuration describing the way a circuit can cross a river $2n$ times. The sequence starts as $1, 1, 2, 8, 42, 262$ (*EIS* **A005315**). For instance, here is a meander of size 5:

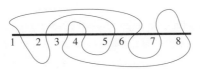

There are good reasons to *believe* that the number M_n of meanders satisfies

$$M_n \sim CA^n n^{-\beta}, \qquad \text{with} \quad \beta = \frac{29 + \sqrt{145}}{12},$$

based on analogies with well-established models of statistical physics [163].   ◁

VII. 9.2. Nonlinear differential equations.
Solutions to nonlinear equations do not necessarily have singularities that arise from the equation itself (as in the linear case). Even the simplest nonlinear equation,

$$Y'(z) = Y(z)^2, \qquad Y(0) = a,$$

has a solution $Y(z) = 1/(a - z)$ whose singularity depends on the initial condition and is not visible on the equation itself. The problem of determining the *location* of singularities is non-obvious in the case of a nonlinear ODE. Furthermore, the problem of determining the *nature* of singularities for nonlinear equations defies classification in general (Note VII.51). In this section, we thus limit ourselves to examining a few examples where enough structure is present in the combinatorics, so that fairly explicit solutions are available, which are then amenable to singularity analysis.

▷ **VII.51.** *A universal differential equation.* Following ideas of Rubel [521, 522], Duffin [178] proved the following: *The differential equation*

$$(D) \qquad 2y''''y'^2 - 5y'''y''y' + 3y''^3 = 0$$

is universal in the sense that any continuous function $\varphi(x)$ on $\mathbb{R}$ can be approximated with arbitrary accuracy by a solution of the equation. Thus, real solutions of nonlinear differential equations cannot be "classified" in general. [Proof: (*i*) construct a third-order differential equation (E) satisfied by the class of functions $g_{a,b,c}(x) = a \cos^4(bx + c)$ for $-\pi/2 \le bx + c \le \pi/2$; (*ii*) verify that any function $G(x)$ that is a juxtaposition of g functions over disjoint intervals and is smooth enough satisfies (E); (*iii*) prove that such a $G(x)$ can be taken so that $\int G$ approximates a continuous $\varphi(x)$ to any predetermined accuracy, and determine (D).]   ◁

Example VII.24. *Varieties of increasing trees.* Consider a labelled class defined by either of

$$(128) \qquad \mathcal{Y} = Z^\square \star \mathrm{SEQ}_\Omega(\mathcal{Y}), \qquad \mathcal{Y} = Z^\square \star \mathrm{SET}_\Omega(\mathcal{Y}),$$

where a set of integers $\Omega \subseteq \mathbb{Z}_{\ge 0}$ has been fixed. This defines trees that are either plane (SEQ) or non-plane (SET) and *increasing*, in the sense that labels go in increasing order along any branch stemming from the root. Such trees have been encountered in Subsection II. 6.3 (p. 139) in relation to alternating permutations, general permutations, and regressive mappings.

Enumeration of trees. By the symbolic translation of the boxed product, the EGF of $\mathcal{Y}$ satisfies a nonlinear differential equation

$$(129) \qquad Y(z) = \int_0^z \phi(Y(w))\, dw,$$

where the structure function ϕ is

$$\phi(y) = \sum_{\omega \in \Omega} y^\omega \quad \text{(case SEQ)}, \qquad \phi(y) = \sum_{\omega \in \Omega} \frac{y^\omega}{\omega!} \quad \text{(case SET)}.$$

The integral equation (129) is our starting point; in order to unify both cases, we set $\phi_\omega := [y^\omega]\phi(y)$. The discussion below is excerpted from the paper of Bergeron, Flajolet, and Salvy [49].

First note that (129) is equivalent to the nonlinear differential equation

$$(130) \qquad Y'(z) = \phi(Y(z)), \qquad Y(0) = 0,$$

	Differential eq.	EGF	ρ	sing. type	coefficient
A :	$Y' = (1+Y)^2$	$\dfrac{z}{1-z}$	1	Z^{-1}	$Y_n = n!$
B :	$Y' = 1 + Y^2$	$\tan z$	$\dfrac{\pi}{2}$	Z^{-1}	$\dfrac{Y_{2n+1}}{(2n+1)!} \asymp (\dfrac{2}{\pi})^{2n+1}$
C :	$Y' = e^Y$	$\log[(1-z)^{-1}]$	1	$\log Z$	$Y_n = (n-1)!$
D :	$Y' = \dfrac{1}{1-Y}$	$1 - \sqrt{1-2z}$	$\dfrac{1}{2}$	$Z^{1/2}$	$Y_n = (2n-3)!!$

Figure VII.22. Some classical varieties of increasing trees: (A) plane binary; (B) strict plane binary; (C) increasing Cayley; (D) increasing plane.

which implies that $Y'/\phi(Y) = 1$ and, upon integrating back,

$$(131) \qquad \int_0^{Y(z)} \frac{d\eta}{\phi(\eta)} = z, \quad \text{i.e.,} \quad K(Y(z)) = z, \quad K(y) := \int_0^y \frac{d\eta}{\phi(\eta)}.$$

Thus, *the EGF $Y(z)$ is the compositional inverse of the integral of the multiplicative inverse of the structure function.* We can visualize this chain of transformation as follows:

$$(132) \qquad Y = \text{Inv} \; \circ \int \circ \; \frac{1}{(\cdot)} \circ \; \phi.$$

In simpler situations, the integration defining $K(y)$ in (131) can be carried out explicitly, so that explicit expressions may become available for $Y(z)$. Figure VII.22 displays data relative to four such classes, the first three of which were already encountered in Chapter II. In each case, there is listed: the differential equation (from which the definition of the trees and the form of ϕ are apparent), the dominant positive singularity, the singularity type, and the corresponding form of coefficients. The general analytic expressions of (131) contain much more: they allow for a general discussion of singularity types and permit us to analyse asymptotically classes that do not admit of an explicit GF.

Assume for simplicity ϕ to be an aperiodic entire function (possibly a polynomial). Let ρ be the radius of convergence of $Y(z)$, which is a singular point (by Pringsheim's Theorem). Consider the limiting value $Y(\rho)$. One cannot have $Y(\rho) < \infty$ since then $K(z)$ being analytic at $Y(\rho)$ would be analytically invertible (by the Implicit Function Theorem). Thus, one must have $Y(\rho) = +\infty$ and, since Y and K are inverses of each other, we get $K(+\infty) = \rho$. The radius of convergence of $Y(z)$ is accordingly

$$(133) \qquad \rho = \int_0^\infty \frac{d\eta}{\phi(\eta)}.$$

The singularity type of $Y(z)$ is then systematically determined by the rules (132). For a general polynomial of degree $d \geq 2$, we have (ignoring coefficients)

$$K(+\infty) - K(y) \approx \int_y^\infty \frac{d\eta}{\eta^d} \approx y^{-d+1}, \qquad Y(z) \approx Z^{-1/(d-1)}, \quad \text{with } Z := (\rho - z).$$

This back-of-the-envelope calculation shows that

$$(134) \qquad \text{for } \phi \text{ a polynomial of degree } d : \qquad Y_n \sim Cn!n^f, \quad \text{with } f = \frac{2-d}{1-d}.$$

In the same vein, the logarithmic singularity of the EGF of increasing Cayley trees (Case C of Figure VII.22) appears as eventually reflecting the inverse of the exponential singularity of $\phi(y) = e^y$. Such a singularity type must then be systematically present when considering increasing non-plane trees (increasing Cayley trees) with a finite collection of node degrees excluded—in other words, whenever the SET constructor is used in (128) and Ω is a cofinite set. This observation "explains" and extends an analysis of [437].

Additive parameters. Consider next an additive parameter of trees[21] defined by a recurrence,

$$(135) \qquad\qquad s(\tau) = t_{|\tau|} + \sum_{\upsilon \propto \tau} s(\upsilon),$$

where (t_n) is a numeric sequence of "tolls" with $t_0 = 0$, and the summation $\upsilon \propto \tau$ is carried out over all root subtrees υ of τ. Introduce the two functions (of cumulated values)

$$S(z) = \sum_{\tau \in \mathcal{Y}} s(\tau) \frac{z^{|\tau|}}{|\tau|!}, \qquad T(z) = \sum_{n \geq 0} t_n Y_n \frac{z^n}{n!},$$

so that the ratio $\frac{[z^n]S(z)}{[z^n]Y(z)}$ equals the mean value of parameter s taken over all increasing trees of size n. By simple algebra similar to Lemma VII.1 (p. 457), it is found that the GF $S(z)$ is

$$(136) \qquad\qquad S(z) = Y'(z) \int_0^z \frac{T'(w)}{Y'(w)} \, dw.$$

The relation (128) defines an integral transform $T \mapsto S$, which can be viewed as a *singularity transformer*. Thanks to the methods of Subsection VI. 10.3, p. 427, its systematic study can be done, once the singularity type of $Y(z)$ is known.

The discussion of path length ($t_n = n$ corresponding to $T(z) = zY'(z)$) is conducted in the present perspective as follows. For polynomial varieties of increasing trees, we have $Y(z) \approx Z^{-\delta}$ with $\delta = 1/(d-1)$, so that

$$T \approx Y' \approx Z^{-\delta - 1}, \; T' \approx Z^{-\delta - 2}, \; \frac{T'}{Y'} \approx Z^{-1}, \; \int \frac{T'}{Y'} \approx \int \frac{1}{Z} \approx \log Z.$$

Thus, the relation between Y and S is of the simplified form $S \approx Y' \log Z$. Singularity analysis, then implies that average path length is of order $n \log n$. Working out the constants involved gives the following proposition.

Proposition VII.13. *Let $\mathcal{Y}$ be an increasing variety of trees defined by a function ϕ that is an aperiodic polynomial of degree $d \geq 2$ and let $\delta = 1/(d-1)$. The number of trees of size n satisfies*

$$Y_n \sim \frac{n!}{\Gamma(\delta)} \left(\frac{\delta}{\rho \phi_d} \right)^\delta \rho^{-n} n^{-1+\delta}, \quad \rho := \int_0^\infty \frac{d\eta}{\phi(\eta)}, \quad \phi_d = [y^d]\phi(y).$$

The expected value of path length on a tree of $\mathcal{Y}_n$ is $(\delta + 1)n \log n + O(n)$.

For naturally occurring models like those of Figure VII.22 and more, many parameters of increasing tree varieties can be analysed in a synthetic way (e.g., the degree profile, the level profile [49]). What stands out is the type of conceptual reasoning afforded by singularity analysis, which provides a direct path to the right order of magnitude of both combinatorial counts and basic parameters of structures. After this, it is only a matter of doing the bookkeeping and getting the constants right! .... ∎

---

[21] Such parameters have been investigated in Subsection VI. 10.3 (p. 427): the binary search tree recurrence there corresponds exacty to the case $\phi(w) = (1+w)^2$ here.

***Example* VII.25.** *Pólya urn processes.* An interesting example of the joint use of nonlinear ODEs and singularity analysis is provided by *urn processes* of probability theory. There, an urn may contain balls of different colours. A fixed set of replacement rules is given (one for each colour). At any discrete instant, a ball is chosen uniformly at random, its colour is inspected, and the corresponding replacement rule is applied. The problem is to determine the evolution of the urn at a large instant n. (The book by Johnson and Kotz [357] can serve as an elementary introduction to the field; Janson otherwise develops a comprehensive probabilistic approach in [349, 351].) In the case of two colours and urns called balanced, it is shown in [130, 225] that the generating function of urn histories is determined by a nonlinear first-order autonomous system, from which many characteristics of the urn can be effectively analysed.

In accordance with the informal description above, an urn model with two colours is determined by a 2×2 matrix with integer entries:

$$(137) \qquad M = \begin{pmatrix} \alpha & \beta \\ \gamma & \delta \end{pmatrix}, \qquad \alpha, \delta \in \mathbb{Z}, \; \beta, \gamma \in \mathbb{Z}_{\geq 0}.$$

At any instant, if a ball of the first colour is drawn, then it is placed back into the urn together with α balls of the first colour and β balls of the second colour; similarly, when a ball of the second colour is drawn, with γ balls of the first colour and δ balls of the second colour. Negative diagonal entries mean that balls are taken out of the urn (rather than added to it). We restrict attention to *balanced* urns, which are such that there exists σ, called the balance:

$$(138) \qquad \sigma = \alpha + \beta = \gamma + \delta.$$

Given an urn initialized with a_0 balls of the first colour and b_0 balls of the second colour, what is sought is the multivariate generating function $H(x, y, z)$ (of exponential type), such that $n![z^n x^a y^b] H(x, y, z)$ is the number of possible evolutions of the urn leading at time n to an urn with colour composition (a, b). For $\sigma \geq 1$, the *total* number of evolutions is clearly

$$(a_0 + b_0)(a_0 + b_0 + \sigma) \cdots (a_0 + b_0 + (n-1)\sigma), \quad \text{so that} \quad H(1, 1, z) = \frac{1}{(1 - \sigma z)^{a_0 + b_0}}.$$

We have the following proposition.

Proposition VII.14. *The exponential MGF of a balanced urn with matrix* (137), *balance* σ, *and initial composition* (a_0, b_0) *satisfies for* $|x_0|, |y_0| \leq 1$, $x_0 y_0 \neq 0$, *and* $|z| < 1/\sigma$

$$H(x_0, y_0, z) = X(z \mid x_0, y_0)^{a_0} \, Y(z \mid x_0, y_0)^{b_0},$$

where $X(t) \equiv X(t \mid x_0, y_0)$ *and* $Y(t) \equiv Y(t \mid x_0, y_0)$ *are the solutions of the* associated *differential system:*

$$(139) \qquad \Sigma : \qquad \begin{cases} \dfrac{d}{dt} X(t) &= X(t)^{\alpha+1} Y(t)^{\beta} \\ \dfrac{d}{dt} Y(t) &= X(t)^{\gamma} Y(t)^{\delta+1} \end{cases}, \qquad X(0) = x_0, \quad Y(0) = y_0.$$

Proof. The proof is an interesting illustration of the modelling of combinatorial structures by differential operators (Note I.63, p. 88). As a starting point, we observe that the obvious rule $\partial_x[x^n] = nx^{n-1}$ of calculus can be interpreted as

$$\partial_x[xx \cdots x] = (\not{x}x \cdots x) + (x\not{x} \cdots) + \cdots + (xx \cdots \not{x}),$$

meaning: *"pick up in all possible ways a single occurrence of the formal variable and delete it"*. Similarly, $x\partial_x$ means: *"pick up an occurrence without deleting it* (this is the pointing operation of Subsection I. 6.2, p. 86).

Guided by this principle, we associate to an urn the linear *partial differential operator*

(140) $$\mathfrak{D} := x^{\alpha+1} y^\beta \partial_x + x^\gamma y^{\delta+1} \partial_y.$$

If $\mathfrak{m} = x^a y^b$ represents an urn with composition (a, b), then it is easily verified that $\mathfrak{D}[\mathfrak{m}]$ generates all the possible evolutions of the urn in one step; similarly $\mathfrak{D}^n[\mathfrak{m}]$ is the generating polynomial of the urn's composition after n steps. This gives us a symbolic form of the exponential MGF H as

(141) $$H(x, y, z) = \sum_{n \geq 0} \mathfrak{D}^n [x^{a_0} y^{b_0}] \frac{z^n}{n!} = e^{z\mathfrak{D}} [x^{a_0} y^{b_0}].$$

Now comes the crucial (and easy) observation that for a solution $X(t)$, $Y(t)$ of the associated differential system (139), one has:

$$
\begin{aligned}
\partial_t (X^a Y^b) &= a X^{a-1} X' Y^b + b X^a Y^{b-1} Y' && \text{(by usual differentiation rules)}\\
&= a X^{a+\alpha} Y^{b+\beta} + b X^{a+\gamma} Y^{b+\delta} && \text{(by system } \Sigma)\\
&= \mathfrak{D} \left[x^a y^b \right]_{\substack{x \to X \\ y \to Y}} && \text{(by definition of } \mathfrak{D}).
\end{aligned}
$$

Induction then provides

(142) $$\partial_t^n (X^a Y^b) = \mathfrak{D}^n \left[x^a y^b \right]_{\substack{x \to X \\ y \to Y}}.$$

In other words: *the evolution of the urn is mimicked by the effect of standard differentiation applied to solutions of the associated system.*

We can now conclude. We have formally, from (141) and the correspondence $\mathfrak{D}^n \leftrightarrow \partial_t^n$,

$$H(X(t), Y(t), z) = \sum_{n \geq 0} \partial_t^n [X(t)^{a_0} Y(y)^{b_0}] \frac{z^n}{n!} = X(t+z)^{a_0} Y(t+z)^{b_0}$$

(the last form plainly expresses Taylor's formula). Setting $t = 0$ yields the statement.   ∎

As a simple illustration, the Ehrenfest urn (Notes II.11, p. 118 and V.25, p. 336) whose matrix is $\left(\begin{smallmatrix} -1 & 1 \\ 1 & -1 \end{smallmatrix} \right)$, with balance $\sigma = 0$, only requires solving the associated system

$$X'(t) = Y(t), \quad Y'(t) = X(t), \quad X(0) = x_0, \quad Y'(0) = y_0,$$

which provides the explicit form

$$H(x, y, z) = (x \cosh z + y \sinh z)^{a_0} (x \sinh z + y \cosh z)^{b_0}.$$

We only discuss one more example, which is typical of the algebraic solution methods and the corresponding singularity analysis. Consider the urn with matrix $\left(\begin{smallmatrix} -1 & 2 \\ 2 & -1 \end{smallmatrix} \right)$, which describes the parity of levels in binary increasing trees [130]. Say we start the urn with one ball of the first colour and seek the probability that, at time n, all balls are of the second colour. We thus need $[z^n] H(0, 1, z)$. The associated system is

$$X' = Y^2, \quad Y' = X^2, \quad X(0) = 0, \quad Y(0) = 1.$$

The system can be solved by a sequence of manipulations (this is general [225]): starting with

$$X'' = 2YY' = 2\sqrt{X'} X^2, \qquad \text{implying} \qquad X'' \sqrt{X'} = 2X' X^2,$$

we can integrate the last form, so that

$$X' = (X^3 + 1)^{2/3}, \qquad \text{i.e.,} \qquad \int_0^X \frac{d\zeta}{(1 + \zeta^3)^{2/3}} = t,$$

meaning that $X(t)$ is implicitly determined as the inverse of the integral of an algebraic function. In this case, it could be verified that the function $X(t)$ is an *elliptic function* (see [225, 471] for other elliptic models), but its dominant singularity can be directly determined by the methods of Example VII.24. The function $X(t)$ is found to become infinite at

$$\rho := \int_0^\infty \frac{d\zeta}{(1+\zeta^3)^{2/3}} = \frac{1}{2\pi\sqrt{3}} \Gamma\left(\frac{1}{3}\right)^3,$$

by an argument similar to (133), p. 527. A local analysis of the integral combined with inversion then reveals that $X(t)$ has a simple pole at ρ. In addition, we have elementarily $X(\omega t) = \omega X(t)$ for $\omega^3 = 1$, which entails the existence of three conjugate singularities at ρ, $\rho e^{2i\pi/3}$, and $\rho e^{-2i\pi/3}$. With the initial conditions $(a_0, b_0) = (1, 0)$, the probability that all balls be of the second colour at time n is then non-zero only if $n \equiv 1 \pmod 3$ and it is found to be *exponentially small*: for some computable $c > 0$, there holds

$$[z^n]X(z) \sim c\rho^{-n}, \qquad n \equiv 1 \pmod 3.$$

In [225, 229] it is shown that one can develop along these lines a complete treatment of 2×2 balanced urns and fully characterize the limit distributions involved. .... ∎

▷ **VII.52.** *Diagrams and combinatorial modelling via differential operators.* Define the linear differential operator

$$\mathfrak{D} := x\partial_x^2.$$

Its meaning, when applied to a monomial x^n, is to pick up two occurrences of x, replace them by unity, and then create a new occurrence of x (this is analogous to a one-colour urn model). It can thus be represented by a "gate" with two "inputs" and one "output". The effect of applying $\mathfrak{D}^n$ to x^{n+1} is then to build all the binary trees, whose external nodes are the occurrences of the original x-variables and whose internal nodes (the gates) are characterized by their order of arrival. Indeed, each particular expansion results in a binary decreasing tree (node labels are decreasing from the root; such a tree is clearly isomorphic to an increasing binary tree) with distinguished external nodes as in the following example relative to $n = 4$,

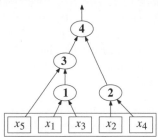

(In this particular expansion, the first application of $\mathfrak{D}$ is to the first (x_1) and third (x_3) occurrence of x in *xxxxx*, corresponding to the first gate (labelled **1**), and it creates one new occurrence of x (the output link of gate 1). The second application is to x_2, x_4 (gate **2**). The third application is to x_5 and to the x produced by gate **1**; and so on.)

Consequently:

$$\mathfrak{D}^n\left[x^{n+1}\right] = n!(n+1)!x, \qquad \text{equivalently,} \qquad \frac{1}{n!}\mathfrak{D}^n\left[\frac{x^{n+1}}{(n+1)!}\right] = 1.$$

Thus, one obtains the EGF of decreasing trees, i.e., permutations, via the coefficient of x in

$$e^{z\mathfrak{D}}\left[e^x\right] = 1 + x\frac{1}{1-z} + \frac{x^2}{2!}\frac{1}{(1-z)^2} + \cdots.$$

Other operators that may be considered include

$$\mathfrak{D} = x + \partial, \ x\partial, \ x^2 + \partial^2, \ x\partial^3, \ x\partial^2 + x\partial, \ldots.$$

It is fascinating to try and model as many classical combinatorial structures as possible in this way, via differential operators and systems of gates. (This exercise was suggested by works of Błasiak, Horzela, Penson, Duchamp, and Solomon [73, 74], themselves motivated by the "boson normal ordering problem" of quantum physics.)   ◁

To conclude this section, it is of interest to compare the properties of increasing trees (Example VII.24) and of simple varieties of trees (Subsection VII. 3.2, p. 455). The conclusion is that simple trees are of the "square-root" type, in the sense that the typical depth of a node and the expected height are of order $\sqrt{n}$. By contrast, increasing trees, which are strongly bound by an order constraint, have logarithmic depth and height [157, 158, 160]—they belong to a "logarithmic" type. From a singular perspective, simple trees are associated to the universal $Z^{1/2}$ law, while increasing trees exhibit a divergence behaviour ($Z^{-1/(d-1)}$ in the polynomial case). Tolls then affect singularities of GFs in rather different ways: through a factor $Z^{-1/2}$ for simple trees, through a factor $\log Z$ in the case of increasing trees. Such abstract observations are typical of the spirit of analytic combinatorics.

A spectacular result in the general area of random discrete structures and nonlinear differential equations is the discovery by Baik, Deift, and Johansson (Note VIII.46, p. 598) of the law governing the longest increasing subsequence in a random permutation. There, the solutions of the nonlinear Painlevé equation $u''(x) = 2u(x)^3 + xu(x)$ play a central rôle.

VII. 10. Singularity analysis and probability distributions

Singularity analysis can often be used to extract information about the probability distribution of a combinatorial parameter. In the central sections of Chapter IX (pp. 650–666), we shall develop perturbation methods grafted on singularity analysis, which are applicable given a bivariate generating function $F(z, u)$, provided it can be continued when u lies in a complex neighbourhood of 1. However, such conditions are not always satisfied. First, it may be the case that $F(z, u)$ is defined for no other value than $z = 0$ (it diverges), as soon as $u > 1$. Second, it may be the case that a parameter is accessible via a collection of univariate GFs rather than a BGF (see typically our discussion of extremal parameters in Section III. 8, p. 214). We briefly indicate in this section ways to deal with such situations.

VII. 10.1. Moment pumping. Our reader should have no difficulty in recognizing as familiar at least the first two steps of the following procedure, nicknamed "*moment pumping*" in [249], which serve to extract moments from bivariate generating functions.

Procedure: *Moment Pumping*
Input: A bivariate generating function $F(z, u)$ determined by a functional equation.
Output: The limit law corresponding to the array of coefficients $[z^n u^k]F(z, u)$; that is, the asymptotic probability distribution of a parameter χ on a class $\mathcal{F}_n$.
   Step 1. Elucidate the singular structure of $F(z, 1)$ corresponding to the counting problem $[z^n]F(z, 1)$. (Tools of Chapters IV–VII are well-suited for this task, the functional equation satisfied by $F(z, 1)$ being usually simpler than that of $F(z, u)$.)
   Step 2. Work out the singular structure (main terms) of each of the partial derivatives

$$\mu_r(z) := \left.\frac{\partial^r}{\partial u^r} F(z, u)\right|_{u=1}$$

for $r = 1, 2, \ldots$, and use meromorphic methods or singularity analysis to conclude as to $[z^n] \mu_r(z)$. If, as it is most often the case, the combinatorial parameter marked by u is of polynomial growth in the size n, then the radius of convergence of each μ_r is *a priori* the same as that of $F(z, 1)$. Furthermore, in many cases, the singular structure of the $\mu_r(z)$ is of the same broad type as that of $\mu_0(z) \equiv F(z, 1)$.

Step 3. From the moments, as given by Step 2, attempt to reconstruct the limit distribution using the *Moment Convergence Theorem* (Theorem C.2, p. 778).

In order for the procedure to succeed[22], we typically need the standard deviation of χ to be of the same order as the mean, which necessitates that the distribution is spread in the sense of Chapter III, p. 161. (Otherwise, there are larger and larger cancellations in moments of the centred and scaled variant of χ, so that the analysis requires an unbounded number of terms in the singular expansions of the GFs $\mu_r(z)$; see also Pittel's study [484] for an insightful discussion of related problems.)

Example VII.26. *The area under Dyck excursions.* We now examine the coefficients in the BGF, which is a solution of the functional equation

(143) $$F(z, q) = \frac{1}{1 - zF(qz, q)}, \quad \text{i.e.,} \quad F(z, q) = 1 + zF(z, q)F(qz, q).$$

It is such that $[z^n q^k] F(z, q)$ represents the number of Dyck excursions of length $2n$ and area $k - n$ (p. 330). Thus we are aiming at characterizing the distribution of area in Dyck paths. We set $\mu_r(z) := \partial_q^r F(z, q)\big|_{q=1}$, which is, up to normalization, the GF of the rth factorial moments.

Clearly, μ_0 satisfies the relation $\mu_0 = 1 + z\mu_0^2$, and $\mu_0 = \frac{1}{2z}\left(1 - \sqrt{1 - 4z}\right)$, as anticipated.

Application of the moment pumping procedure leads to a collection of equations,

$$\begin{aligned}
\mu_1 &= 2z\mu_0\mu_1 + z^2\mu_0\mu_0' \\
\mu_2 &= 2z\mu_0\mu_2 + 2z\mu_1^2 + 2z^2\mu_1\mu_0' + 2z^2\mu_0\mu_1' + z^3\mu_0\mu_0'',
\end{aligned}$$

and so on. Precisely, the shape of the equation giving μ_r, for $r \geq 1$, is

(144) $$\mu_r = z \sum_{j=0}^{r} \binom{r}{j} \mu_{r-j} \sum_{k=0}^{j} \binom{j}{k} z^k \partial_z^k \mu_{j-k},$$

as results, upon setting $q = 1$, from Leibniz's product rule and a computation of the derivatives $\partial_q^j F(qz, q)$. In particular, each μ_r can be expressed from the previous μ and their derivatives, since the equation relative to μ_r is of the linear form $\mu_r = 2z\mu_0\mu_r + \cdots$, so that $\mu_r(z)$ is a rational form in z and $\delta := \sqrt{1 - 4z}$. An examination of the initial values of the μ then suggests that, in terms of dominant singular asymptotics, as $z \to \frac{1}{4}$, there holds

(145) $$\mu_r(z) = \frac{K_r}{(1 - 4z)^{(3r-1)/2}} + O\left((1 - 4z)^{-(3r-2)/2}\right), \quad r \geq 1,$$

a property that is readily verified by induction. (In such situations, the closure of functions of singularity analysis class under differentiation, p. 419, proves handy.) In particular, by singularity analysis, the mean and standard deviation of χ on $\mathcal{F}_n$ are each of order $n^{3/2}$.

Now, equipped with (145), we can trace back the main singular contributions in (144), noting that the "weight", as measured by the exponent of $(1 - 4z)^{-1}$, of the term in (144)

---

[22]The important *Gaussian* case, which is mostly excluded by moment pumping, tends to yield agreeably to the *perturbation methods* of Chapter IX, so that the univariate methods discussed here and those of Chapter IX are indeed complementary.

corresponding to generic indices j, k is $(3r - k - 2)/2$. Then, by identifying the corresponding coefficients, we come up with the recurrence valid for $r \geq 2$

(146)
$$\Lambda_r = \frac{1}{4} \sum_{j=1}^{r-1} \binom{r}{j} \Lambda_{r-j} \Lambda_j + \frac{r(3r-1)}{4} \Lambda_{r-1}$$

(the linear term arises from $j = r, k = 1$) and from (145) and (146), the shape of factorial moments, hence that of the usual power moments, results by plain singularity analysis:

(147)
$$\mathbb{E}_n \left(\chi^r \right) \sim M_r n^{3r/2}, \qquad M_r := \frac{\sqrt{\pi} \Lambda_r}{\Gamma((3r-1)/2)}.$$

It can then be verified [568] that the moment M_r uniquely characterize a probability distribution (Appendix C.5: *Convergence in law*, p. 776).

Proposition VII.15. *The distribution of area χ in Dyck excursions, scaled by $n^{-3/2}$, converges to a limit, known as the* Airy[23] *distribution of the area type, which is determined by its moments M_r, as specified by (146) and (147). In other terms, there exists a distribution function $H(x)$ supported by $\mathbb{R}_{>0}$ such that $\lim_{n \to \infty} \mathbb{P}_n(\chi < xn^{3/2}) = H(x)$.*

Due to the exact correspondence between Dyck excursions and trees, the same limit distribution occurs for *path length* in general Catalan trees. Proposition VII.15 is originally due to Louchard [415, 416], who developed connections with Brownian motion—the limit distribution is indeed up to normalization that of *Brownian excursion area*. (The approach presented here also has the merit of providing finite n corrections.) Our moment pumping approach largely follows the lines of Takács' treatment [568]. The recurrence relation (144) can furthermore be solved by generating functions, to the effect that the Λ_r entertain intimate relations with the Airy function: for surveys, see [244, 352]. Curiously, the Wright constants arising in the enumeration of labelled graphs of fixed excess (the $P_k(1)$ of p. 134) appear to be closely related to the moments M_r: this fact can be explained combinatorially by means of breadth-first search of graphs, as noted by Spencer [548]. . . . . ∎

▷ **VII.53.** *Path length in simple varieties of trees.* Under the usual conditions on ϕ, the limit distribution is an Airy distribution of the area type, as shown by Takács [566]. ◁

▷ **VII.54.** *A parking problem II.* This continues Example II.19, p. 146. Consider m cars and condition by the fact that everybody eventually finds a parking space and the last space remains empty. Define *total displacement* as the sum of the distances (over all cars) between the initially intended parking location and the first available space. The analysis reduces to the difference-differential equation [249, 380], which generalizes (65), p. 146,

$$\frac{\partial}{\partial z} F(z, q) = F(z, q) \cdot \frac{F(z, q) - q F(qz, q)}{1 - q}.$$

Moment pumping is applicable [249]: the limit distribution is once more an Airy (of area type). This problem arises in the analysis of the *linear probing hashing* algorithm [380, §6.4] and is of relevance as a discrete version of important coalescence models. It is also shown in [249] based on [285] that the number of inversions in a Cayley tree is asymptotically Airy. ◁

----

[23] The Airy function Ai(z) is of hypergeometric type and is closely related to Bessel functions of order $\pm 1/3$. It is defined as the solution of $y'' - zy = 0$ satisfying Ai$(0) = 3^{-2/3}/\Gamma(2/3)$ and Ai$'(0) = -3^{-1/3}/\Gamma(1/3)$; see [3, 604] for basic properties. The Λ_r intervene in the expansion of $\log$Ai(z) at infinity [244, 352]. After Louchard and Takács, the distribution function $H(x)$ can be expressed in terms of confluent hypergeometric functions and zeros of the Airy function.

▷ **VII.55.** *The Wiener index and other functionals of trees.* The Wiener index, a structural index of interest to chemists, is defined as the sum of the distances between all pairs of nodes in a tree. For simple families, as shown by Janson [348], it admits a limit distribution. (Similar properties hold for many additive functionals of combinatorial tree families [210]. As regards moment pumping, the methods are also related to those of Subsection VI. 10.3, p. 427, dedicated to tree recurrences.)   ◁

▷ **VII.56.** *Difference equations, polyominoes, and limit laws.* Many of the q–difference equations that are defined by a polynomial relation between $F(z, q)$, $F(qz, q)$, ... (and even systems) may be analysed, as shown by Richard [509, 510]. This covers several models of polyominoes, including the staircase, the horizontally-vertically convex, and the column convex ones. Area (for fixed perimeter) is asymptotically Airy distributed. It is from these and similar results, supplemented by extensive computations based on transfer-matrix methods, that Guttmann and the Melbourne school have been led to conjecturing that the limit area of self-avoiding polygons (closed walks) in the plane is Airy (see our comments on p. 365).   ◁

▷ **VII.57.** *Path length in increasing trees.* For binary increasing trees, the analysis of path length reduces to that of the functional equation,

$$F(z, q) = 1 + \int_0^z F(qt, q)^2 \, dt.$$

There exists a limit law, as first shown by Hennequin [328] using moment pumping, with alternative approaches due to Régnier [505] and Rösler [517]. This law is important in computer science, since it describes the number of comparisons used by the *Quicksort* algorithm and involved in the construction of a binary search tree. The mean is $2n \log n + O(n)$, the variance is $\sim (7 - 4\zeta(2))n^2$, and the moment of order r of the limit law is a polynomial form in zeta values $\zeta(2), \ldots, \zeta(r)$. See [209] for recent news and references.   ◁

VII. 10.2. Families of generating functions.

There is no logical obstacle to applying singularity analysis to a whole family of functions. In a way, this is similar to what was done in Chapter V when analysing longest runs in words (p. 308) and the height of general Catalan trees (p. 326), in the simpler case of meromorphic coefficient asymptotics. One then needs to develop suitable singular expansions together with companion error terms, a task that may be technically demanding when GFs are given by nonlinear functional relations or recurrences. We illustrate below the situation by an *aperçu* of the analysis of height in simple varieties of trees.

***Example* VII.27.** *Height in simple varieties of trees.* The recurrence

(148) $$y_0(z) = 0, \qquad y_{h+1}(z) = 1 + zy_h(z)^2$$

is such that $y_h(z)$ is the OGF of binary trees of height less than h, with size measured by the number of binary nodes (Example III.28, p. 216). Each $y_h(z)$ is a polynomial, with $\deg(y_h) = 2^{h-1} - 1$. Some technical difficulties are to be expected since the y_h have no singularity at a finite distance, whereas their formal limit $y(z)$ is the OGF of Catalan number,

$$y(z) = \frac{1}{2z} \left(1 - \sqrt{1 - 4z}\right),$$

which has a square-root singularity at $z = 1/4$. As a matter of fact, the sequence $w_h = zy_h$ satisfies the recurrence $w_{h+1} = z + w_h^2$, which was made famous by Mandelbrot's studies and gives rise to amazing graphics [473]; see Figure VII.23 for a poor man's version.

When $|z| \leq r < 1/4$, simple majorant series considerations show that the *convergence* $y_h(z) \to y(z)$ is *uniformly geometric*. When $z \geq s > 1/4$, it can be checked that the $y_h(z)$ grow *doubly exponentially*. What happens in-between, in a Δ–domain, needs to be quantified. We do so following Flajolet, Gao, Odlyzko, and Richmond [230, 246].

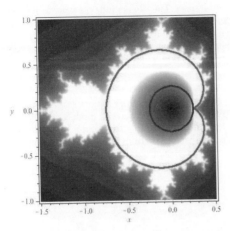

The grey level relative to a point $z = x + iy$ in the diagram indicates the number of iterations necessary for the GFs $y_h(z)$ either to diverge to infinity (the outer, darker region) or to the finite limit $y(z)$ (the inner region, corresponding to the Mandelbrot set, with the darker area around 0 corresponding to faster convergence). The cardioid-shaped region defined by $|1 - \varepsilon(z)| \le 1$ is a guaranteed region of convergence, beyond the circle $|z| = 1/4$. The determination of height reduces to finding what goes on near the cusp $z = 1/4$ of the cardioid.

Figure VII.23. The GFs of binary trees of bounded height: speed of convergence.

Starting from the basic recurrence (148), we have

$$y - y_{h+1} = z(y^2 - y_h^2) = z(y - y_h)(2y - (y - y_h)),$$

which rewrites as

(149) $$e_{h+1} = (2zy)e_h(1 - e_h), \quad \text{where} \quad e_h(z) = \frac{1}{2zy(z)}y(z) - y_h(z)$$

is proportional to the OGF of trees having height at least h. (The function $x \mapsto \lambda x(1 - x)$, which is at the basis of the recurrence (149), is also known as the logistic map; its iterates, for real parameter values λ, give rise to a rich diversity of patterns.)

First, let us examine what happens right at the singularity $1/4$ and consider $e_h \equiv e_h(\frac{1}{4})$. The induced recurrence is

(150) $$e_{h+1} = e_h(1 - e_h), \quad \text{with} \quad e_0 = \tfrac{1}{2},$$

whose solution decreases monotonically to 0 (argument: otherwise, there would need to be a fixed point in $(0, 1)$). This form resembles the familiar recurrence associated with the solution by iteration of a fixed-point equation $\ell = f(\ell)$, but here it corresponds to an "indifferent" fixed-point, $f'(\ell) = 1$, which precludes the usual geometric convergence. A classical trick of iteration theory, found in de Bruijn's book [143, §8.4], neatly solves the problem. Consider instead the quantities $f_h := 1/e_h$, which satisfy the induced recurrence

(151) $$f_{h+1} = \frac{f_h}{1 - f_h^{-1}} \equiv f_h + 1 + \frac{1}{f_h} + \frac{1}{f_h^2} \cdots, \quad \text{with} \quad f_0 = 2.$$

This suggests that $f_h \sim h$. Indeed, by a terminating form of (151),

(152) $$f_{h+1} = f_h + 1 + \frac{1}{f_h} + \frac{f_h^{-2}}{1 - f_h^{-1}}, \quad \text{i.e.,} \quad f_{h+1} = h + 2 + \sum_{j=0}^{h} f_j^{-1} + \sum_{j=0}^{h} \frac{f_j^{-2}}{1 - f_j^{-1}},$$

one can derive properties of the sequence (f_h) by "bootstrapping": the fact that $f_h > h$ implies that the first sum in (152) is $O(\log h)$, while the second one is $O(1)$; then, another round serves to refine the estimates, so that, for some C:

$$f_h = h + \log h + C + O\left(\frac{\log h}{h}\right),$$

and the behaviour of $e_h = 1/f_h$ is now well quantified.

The analysis for $z \neq 1/4$ proceeds along similar lines. We set $\varepsilon \equiv \varepsilon(z) := \sqrt{1 - 4z}$ and again abbreviate $e_h(z)$ as e_h. Upon considering

$$f_h = \frac{e_h}{(1 - \varepsilon)^h}$$

and taking inverses, we obtain

(153) $$f_{h+1} = f_h + (1 - \varepsilon)^h + \frac{f_h e_h^2}{1 - e_h}.$$

Proceeding as before leads to the general approximation

(154) $$e_h(z) \sim \frac{\varepsilon(z)(1 - \varepsilon(z))^h}{1 - (1 - \varepsilon(z))^h}, \qquad \varepsilon(z) := \sqrt{1 - 4z},$$

proved to be valid for any fixed $z \in (0, 1/4)$, as $h \to \infty$. This approximation is compatible both with $e_h(1/4) \sim 1/h$ (derived earlier) and with the geometric convergence of $y_h(z)$ to $y(z)$ valid for $0 < z < 1/4$. With some additional work, it can be proved that (154) remains valid as $z \to \frac{1}{4}$ in a Δ–domain *and* as $h \to \infty$; see Figure VII.23. Obtaining the detailed conditions on (z, h), together with a uniform error term for (154), is the crux of the analysis in [247].

From this point on, we content ourselves with brief indications on subsequent developments. Given (154), one deduces[24] that the *GF of cumulated height* satisfies

$$H(z) := 2y(z) \sum_{h \geq 0} e_h(z) \sim 4 \sum_{h \geq 1} \frac{\varepsilon(1 - \varepsilon)^h}{1 - (1 - \varepsilon)^h} = 4 \log \frac{1}{\varepsilon} + O(1),$$

as $z \to \frac{1}{4}$. Thus, by singularity analysis, one has

$$H(z) \sim 2 \log \frac{1}{1 - 4z} \qquad \longrightarrow \qquad [z^n] H(z) \sim 2 \cdot 4^n / n,$$

which gives the expected height $[z^n]H(z)/[z^n]y(z)$ of a binary tree of size n as $\sim 2\sqrt{\pi n}$. Moments of higher order can be similarly analysed.

It is of interest to note that the GFs that surface explicitly in the analysis of height in general Catalan trees (eventually due to the continued fraction structure and the implied *linear recurrences*) appear here as *analytic approximations* in suitable regions of the complex plane. A precise form of the approximation (154) can also be subjected to singularity analysis, to the effect that the same Theta law expresses in the asymptotic limit the distribution of height in binary trees. Finally, the technique can be extended to all simple varieties of trees satisfying the smooth inverse-function schema (Theorem VII.2, p. 453). In summary, we have the following proposition [230, 246].

Proposition VII.16. *Let $\mathcal{Y}$ be a simple variety of trees satisfying the conditions of Theorem VII.2, with ϕ the basic tree constructor and τ the root of the characteristic equation $\phi(\tau) - \tau\phi(\tau) = 0$. Let χ denote tree height. Then the rth moment of height satisfies*

$$\mathbb{E}_{\mathcal{Y}_n}[\chi^r] \sim r(r - 1)\Gamma(r/2)\zeta(r)\xi^r n^{r/2}, \qquad \xi := \frac{2\phi'(\tau)^2}{\phi(\tau)\phi''(\tau)}.$$

---

[24]In order to obtain the logarithmic approximation of $H(z)$, one can for instance appeal to Mellin transform techniques in a way parallel to the analysis of general Catalan trees (p. 326): set $1 - \varepsilon(z) = e^{-t}$.

The normalized height $\chi/\sqrt{\xi n}$ converges to a Theta law, both in distribution *and in the sense of a* local limit law.

(The Theta distribution is defined in (67), p. 328; Chapter IX develops the notions of convergence in law and of local limits much further.) In particular the expected height in general Catalan trees [145], binary trees, unary–binary trees, pruned t–ary trees, and Cayley trees [507], is found to be, respectively, asymptotic to

$$\sqrt{\pi n}, \quad 2\sqrt{\pi n}, \quad \sqrt{3\pi n}, \quad \sqrt{2\pi t/(t-1)}, \quad \sqrt{2\pi n},$$

and a pleasant *universality* phenomenon manifests itself in the height of simple trees.

A somewhat related analysis of a polynomial iteration in the vicinity of a singularity yields the asymptotic number of balanced trees (Note IV.49, p. 283). .... ∎

VII. 11. Perspective

The theorems in this chapter demonstrate the central rôle of the singularity analysis theory developed in Chapter VI, this in a way that parallels what Chapter V did for Chapter IV with meromorphic function analysis. Exploiting properties of complex functions to develop coefficient asymptotics for abstract schemas helps us solve whole collections of combinatorial constructions at once.

Within the context of analytic combinatorics, the results in this chapter have broad reach, and bring us closer to our ideal of a theory covering full analysis of combinatorial objects of any "reasonable" description. Analytic side conditions defining schemas often play a significant rôle. Adding in this chapter the mathematical support for handling set constructions (with the exp–log schema) and context-free constructions (with coefficient asymptotics of algebraic functions) to the support developed in Chapter V to handle the sequence construction (with the supercritical sequence schema) and regular constructions (with coefficient asymptotics of rational functions) gives us general methods encompassing a broad swathe of combinatorial analysis, with a great many applications (Figure VII.24).

Together, the methods covered in Chapter V, this chapter, and, next, Chapter VIII (relative to the saddle-point method) apply to virtually all of the generating functions derived in Part A of this book by means of the symbolic techniques defined there. The SEQ construction and regular specifications lead to poles; the SET construction leads to algebraic singularities (in the case of logarithmic generators discussed here) or to essential singularities (in most of the remaining cases discussed in Chapter VIII); recursive (context-free) constructions lead to square-root singularities. The surprising end result is that the asymptotic counting sequences from all of these generating functions have one of just a few functional forms. This universality means that comparisons of methods, finding optimal values of parameters, and many other outgrowths of analysis can be very effective in practical situations. Indeed, because of the nature of the asymptotic forms, the results are often extremely accurate, as we have seen repeatedly in this book.

The general theory of coefficient asymptotics based on singularities has many applications outside of analytic combinatorics (see the notes below). The broad reach of the theory provides strong indications that universal laws hold for many combinatorial constructions and schemas yet to be discovered.

Combinatorial Type	coeff. asymptotics (subexp. term)	
Rooted maps	$n^{-5/2}$	§VII. 8.2
Unrooted trees	$n^{-5/2}$	§VII. 5
Rooted trees	$n^{-3/2}$	§VII. 3, §VII. 4
Excursions	$n^{-3/2}$	§VII. 8.1
Bridges	$n^{-1/2}$	§VII. 8.1
Mappings	$n^{-1/2}$	§VII. 3.3
Exp-log sets	$n^{\kappa-1}$	§VII. 2
Increasing d–ary trees	$n^{-(d-2)/(d-1)}$	§VII. 9.2

Analytic form	singularity type	coeff. asymptotics	
Positive irred. (polynomial syst.)	$Z^{1/2}$	$\zeta^{-n}n^{-3/2}$	§VII. 6
General algebraic	$Z^{p/q}$	$\zeta^{-n}n^{-p/q-1}$	§VII. 7
Regular singularity (ODE)	$Z^{\theta}(\log Z)^{\ell}$	$\zeta^{-n}n^{-\theta-1}(\log n)^{\ell}$	§VII. 9.1

Figure VII.24. A collection of *universality laws* summarized by the subexponential factors involved in the asymptotics of counting sequences (top). A summary of the main singularity types and asymptotic coefficient forms of this chapter (bottom).

Bibliographic notes. The exp–log schema, like its companion, the supercritical-sequence schema, illustrates the level of generality that can be attained by singularity analysis techniques. Refinements of the results we have given can be found in the book by Arratia, Barbour, and Tavaré [20], which develops a stochastic process approach to these questions; see also [19] by the same authors for an accessible introduction.

The rest of the chapter deals in an essential manner with recursively defined structures. As noted repeatedly in the course of this chapter, recursion is conducive to square-root singularity and universal behaviours of the form $n^{-3/2}$. Simple varieties of trees have been introduced in an important paper of Meir and Moon [435], that bases itself on methods developed earlier by Pólya [488, 491] and Otter [466]. One of the merits of [435] is to demonstrate that a high level of generality is attainable when discussing properties of trees. A similar treatment can be inflicted more generally to recursively defined structures when their generating functions satisfy an implicit equation. In this way, non-plane unlabelled trees are shown to exhibit properties very similar to their plane counterparts. It is of interest to note that some of the enumerative questions in this area had been initially motivated by problems of theoretical chemistry: see the colourful account of Cayley and Sylvester's works in [67], the reference books by Harary and Palmer [319] and Finch [211], as well as Pólya's original studies [488, 491].

Algebraic functions are the modern counterpart of the study of curves by classical Greek mathematicians. They are either approached by algebraic methods (this is the core of algebraic geometry) or by transcendental methods. For our purposes, however, only rudiments of the theory of curves are needed. For this, there exist several excellent introductory books, of which we recommend the ones by Abhyankar [2], Fulton [273], and Kirwan [365]. On the algebraic side, we have aimed at providing an introduction to algebraic functions that requires minimal apparatus. At the same time the emphasis has been put somewhat on algorithmic aspects, since most algebraic models are nowadays likely to be treated with the help of computer algebra. As regards symbolic computational aspects, we recommend the treatise by von zur Gathen and Gerhard [599] for background, while polynomial systems are excellently reviewed in the book by Cox, Little, and O'Shea [135].

In the combinatorial domain, algebraic functions have been used early: in Euler and Segner's enumeration of triangulations (1753) as well as in Schröder's famous "*Vier combinatorische Probleme*" described by Stanley in [554, p. 177]. A major advance was the realization by Chomsky and Schützenberger that algebraic functions are the "exact" counterpart of context-free grammars and languages (see their historic paper [119]). A masterful summary of the early theory appears in the proceedings edited by Berstel [54] while a modern and precise presentation forms the subject of Chapter 6 of Stanley's book [554]. On the analytic asymptotic side, many researchers have long been aware of the power of Puiseux expansions in conjunction with some version of singularity analysis (often in the form of the Darboux–Pólya method: see [491] based on Pólya's classic paper [488] of 1937). However, there appeared to be difficulties in coping with the fully general problem of algebraic coefficient asymptotics [102, 440]. We believe that Section VII. 7 sketches the first complete theory (though most ingredients are of folklore knowledge). In the case of positive systems, the "Drmota–Lalley–Woods" theorem is the key to most problems encountered in practice—its importance should be clear from the developments of Section VII. 6.

The applicability of algebraic function theory to context-free languages has been known for some time (e.g., [220]). Our presentation of one-dimensional walks of a general type follows articles by Lalley [396] and Banderier and Flajolet [27], which can be regarded as the analytic pendant of algebraic studies by Gessel [286, 287]. The kernel method has its origins in problems of queueing theory and random walks [202, 203] and is further explored in an article by Bousquet-Mélou and Petkovšek [86]. The algebraic treatment of random maps by the quadratic method is due to brilliant studies of Tutte in the 1960s: see for instance his census [579] and the account in the book by Jackson and Goulden [303]. A combinatorial–analytic treatment of multiconnectivity issues is given in [28], where the possibility of treating in a unified manner about a dozen families of maps appears clearly.

Regarding differential equations, an early (and at the time surprising) occurrence in an asymptotic expansion of terms of the form n^α, with α an algebraic number, is found in the study [252], dedicated to multidimensional search trees. The asymptotic analysis of coefficients of solutions to linear differential equations can also, in principle, be approached from the recurrences that these coefficients satisfy. Wimp and Zeilberger [611] propose an interesting approach based on results by George Birkhoff and his school (e.g., [70]), which are relative to difference equations in the complex plane. There are, however, some doubts among specialists regarding the completeness of Birkhoff's programme (see our discussion in Section VIII. 7, p. 581). By contrast, the (easier) singularity theory of linear ODEs is well established, and, as we showed in this chapter, it is possible—in the regular singular case at least—to base a sound method for asymptotic coefficient extraction on it.

VIII

Saddle-point Asymptotics

Like a lazy hiker, the path crosses the ridge at a low point;
but unlike the hiker, the best path takes the steepest ascent to the ridge.
[· · ·] The integral will then be concentrated in a small interval.

— DANIEL GREENE AND DONALD KNUTH [310, sec. 4.3.3]

A *saddle-point* of a surface is a point reminiscent of the inner part of a saddle or of a geographical pass between two mountains. If the surface represents the modulus of an analytic function, saddle-points are simply determined as the *zeros of the derivative* of the function.

In order to estimate *complex integrals* of an analytic function, it is often a good strategy to adopt as contour of integration a curve that "crosses" one or several of the saddle-points of the integrand. When applied to integrals depending on a large parameter, this strategy provides in many cases accurate asymptotic information. In this book, we are primarily concerned with Cauchy integrals expressing coefficients of large index of generating functions. The implementation of the method is then fairly simple, since integration can be performed along a circle centred at the origin.

Precisely, the principle of the *saddle-point method* for the estimation of contour integrals is to choose a path crossing a saddle-point, then estimate the integrand locally near this saddle-point (where the modulus of the integrand achieves its maximum on the contour), and deduce, by local approximations and termwise integration, an asymptotic expansion of the integral itself. Some sort of "localization" or "concentration" property is required to ensure that the contribution near the saddle-point captures the essential part of the integral. A simplified form of the method provides what are known as *saddle-point bounds*—these useful and technically simple upper bounds are obtained by applying trivial bounds to an integral relative to a saddle-point path. In many cases, the saddle-point method can furthermore provide complete asymptotic expansions.

In the context of analytic combinatorics, the method is applicable to Cauchy coefficient integrals, in the case of rapidly varying functions: typical instances are *entire functions* as well as functions with singularities at a finite distance that exhibit some form of *exponential growth*. Saddle-point analysis then complements singularity analysis whose scope is essentially the category of functions having only moderate (i.e., polynomial) growth at their singularities. The saddle-point method is also a method of choice for the analysis of coefficients of *large powers* of some fixed function and, in this context, it paves the way to the study of multivariate asymptotics and limiting Gaussian distributions developed in the next chapter.

Applications are given here to Stirling's formula, as well as the asymptotics of the central binomial coefficients, the involution numbers and the Bell numbers associated to set partitions. The asymptotic enumeration of integer partitions is one of the jewels of classical analysis and we provide an introduction to this rich topic where saddle-points lead to effective estimates of an amazingly good quality. Other combinatorial applications include balls-in-bins models and capacity, the number of increasing subsequences in permutations, and blocks in set partitions. The counting of acyclic graphs (equivalently forests of unrooted trees), finally takes us beyond the basic paradigm of simple saddle-points by making use of multiple saddle-points, also known as "*monkey saddles*".

Plan of this chapter. First, we examine the surface determined by the modulus of an analytic function and give, in Section VIII. 1, a classification of points into three kinds: ordinary points, zeros, and saddle-points. Next we develop general purpose saddle-point bounds in Section VIII. 2, which also serves to discuss the properties of saddle-point crossing paths. The saddle-point method *per se* is presented in Section VIII. 3, both in its most general form and in the way it specializes to Cauchy coefficient integrals. Section VIII. 4 then discusses three examples, involutions, set partitions, and fragmented permutations, which help us get further familiarized with the method. We next jump to a new level of generality and introduce in Section VIII. 5 the abstract concept of *admissibility*—this approach has the merit of providing easily testable conditions, while opening the possibility of determining broad classes of functions to which the saddle-point method is applicable. In particular, many combinatorial types whose leading construction is a SET operation are seen to be "automatically" amenable to saddle-point analysis. The case of integer partitions, which is technically more advanced, is treated in a separate section, Section VIII. 6. The saddle-method is also instrumental in analysing coefficients of many generating functions implicitly defined by differential equations, including holonomic functions: see Section VIII. 7. Next, the framework of "large powers", developed in Section VIII. 8 constitutes a combinatorial counterpart of the central limit theorem of probability theory, and as such it provides a bridge to the study of limit distributions to be treated systematically in Chapter IX. Other applications to discrete probability distributions are examined in Section VIII. 9. Finally, Section VIII. 10 serves as a brief introduction to the rich subject of multiple saddle-points and coalescence.

VIII. 1. Landscapes of analytic functions and saddle-points

This section introduces a well-known classification of points on the surface representing the modulus of an analytic function. In particular, as we are going to see, saddle-points, which are determined by roots of the function's derivative, are associated with a simple geometric property that gives them their name.

Consider any function $f(z)$ analytic for $z \in \Omega$, where Ω is some domain of $\mathbb{C}$. Its modulus $|f(x+iy)|$ can be regarded as a function of the two real quantities, $x = \Re(z)$ and $y = \Im(z)$. As such, it can be represented as a surface in three-dimensional space. This surface is smooth (analytic functions are infinitely differentiable), but far from being arbitrary.

Let z_0 be an interior point of Ω. The local shape of the surface $|f(z)|$ for z near z_0 depends on which of the initial elements in the sequence $f(z_0), f'(z_0), f''(z_0), \ldots$, vanish. As we are going to see, its points can be of only one of three types: ordinary points (the generic case), zeros, and saddle-points; see Figure VIII.1. The classification of points is conveniently obtained by considering polar coordinates, writing $z = z_0 + re^{i\theta}$, with r small.

An *ordinary point* is such that $f(z_0) \neq 0$, $f'(z_0) \neq 0$. This is the generic situation as analytic functions have only isolated zeros. In that case, one has, for small $r > 0$,

$$(1) \quad |f(z)| = \left| f(z_0) + re^{i\theta} f'(z_0) + O(r^2) \right| = |f(z_0)| \left| 1 + \lambda re^{i(\theta+\phi)} + O(r^2) \right|,$$

where we have set $f'(z_0)/f(z_0) = \lambda e^{i\phi}$, with $\lambda > 0$. The modulus then satisfies

$$|f(z)| = |f(z_0)| \left(1 + \lambda r \cos(\theta + \phi) + O(r^2) \right).$$

Thus, for r kept small enough and fixed, as θ varies, $|f(z)|$ is maximum when $\theta = -\phi$ (where it is $\sim |f(z_0)|(1 + \lambda r)$), and minimum when $\theta = -\phi + \pi$ (where it is $\sim |f(z_0)|(1 - \lambda r)$). When $\theta = -\phi \pm \frac{\pi}{2}$, one has $|f(z)| = |f(z_0)| + o(r)$, which means that $|f(z)|$ is essentially constant. This is easily interpreted: the line $\theta \equiv -\phi$ (mod π) is (locally) a *steepest descent line*; the perpendicular line $\theta \equiv -\phi + \frac{\pi}{2}$ (mod π) is locally a *level line*. In particular, near an ordinary point, the surface $|f(z)|$ has neither a minimum nor a maximum. In figurative terms, this is like standing on the flank of a mountain.

A *zero* is by definition a point such that $f(z_0) = 0$. In this case, the function $|f(z)|$ attains its minimum value 0 at z_0. Locally, to first order, one has $|f(z)| \sim |f'(z_0)|r$ for a simple zero and $|f(z)| = O(r^m)$ or a zero of order m. A zero is thus like a sink or the bottom of a lake, save that, in the landscape of an analytic function, all lakes are at sea level.

A *saddle-point* is a point such that $f(z_0) \neq 0$, $f'(z_0) = 0$; it thus corresponds to a *zero of the derivative*, when the function itself does not vanish. It is said to be a *simple saddle-point* if furthermore $f''(z_0) \neq 0$. In that case, a calculation similar to (1),

$$(2)$$
$$|f(z)| = \left| f(z_0) + \frac{1}{2}r^2 e^{2i\theta} f''(z_0) + O(r^3) \right| = |f(z_0)| \left| 1 + \lambda r^2 e^{i(2\theta+\phi)} + O(r^3) \right|,$$

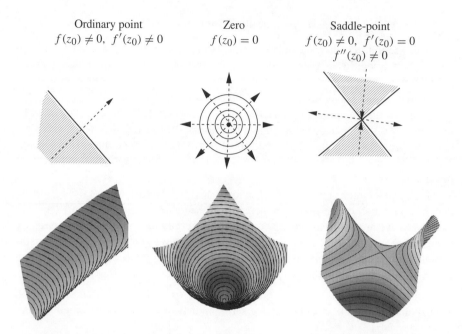

Figure VIII.1. The different types of points on a surface $|f(z)|$: an ordinary point, a zero, a simple saddle-point. Top: a diagram showing the local structure of level curves (in solid lines), steepest descent lines (dashed with arrows pointing towards the direction of increase) and regions (hashed) where the surface lies below the reference value $|f(z_0)|$. Bottom: the function $f(z) = \cosh z$ and the local shape of $|f(z)|$ near an ordinary point ($i\pi/4$), a zero ($i\pi/2$), and a saddle-point (0), with level lines shown on the surfaces.

where we have set $\frac{1}{2} f''(z_0)/f(z_0) = \lambda e^{i\phi}$, shows that the modulus satisfies

$$|f(z)| = |f(z_0)| \left(1 + \lambda r^2 \cos(2\theta + \phi) + O(r^3)\right).$$

Thus, starting at the direction $\theta = -\phi/2$ and turning around z_0, the following sequence of events regarding the modulus $|f(z)| = |f(re^{i\theta})|$ is observed: it is maximal ($\theta = -\phi/2$), stationary ($\theta = -\phi/2 + \pi/4$), minimal ($\theta = -\phi/2 + \pi/2$), stationary, ($\theta = -\phi/2 + 3\pi/4$), maximal again ($\theta = -\phi/2 + \pi$), and so on. The pattern, symbolically "$+ = - =$", repeats itself twice. This is superficially similar to an ordinary point, save for the important fact that changes are observed at twice the angular speed. Accordingly, the shape of the surface looks quite different; it is like the central part of a saddle. Two level curves cross at a right angle: one steepest descent line (away from the saddle-point) is perpendicular to another steepest descent line (towards the saddle-point). In a mountain landscape, this is thus much like a pass between two mountains. The two regions on each side corresponding to points with an altitude *below* a simple saddle-point are often referred to as "valleys".

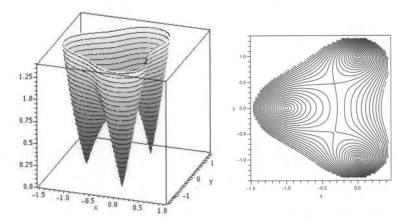

Figure VIII.2. The "tripod": two views of $|1+z+z^2+z^3|$ as function of $x = \Re(z)$, $y = \Im(z)$: (left) the modulus as a surface in $\mathbb{R}^3$; (right) the projection of level lines on the z-plane.

Generally, a *multiple saddle-point* has multiplicity p if $f(z_0) \neq 0$ and all derivatives $f'(z_0), \ldots, f^{(p)}(z_0)$ are equal to zero while $f^{(p+1)}(z_0) \neq 0$. In that case, the basic pattern "$+ = - =$" repeats itself $p + 1$ times. For instance, from a double saddle-point ($p = 2$), three roads go down to three different valleys separated by the flanks of three mountains. A double saddle-point is also called a "monkey saddle" since it can be visualized as a saddle having places for the legs and the tail: see Figure VIII.12 (p. 602) and Figure VIII.14 (p. 605).

Theorem VIII.1 (Classification of points on modulus surfaces). *A surface $|f(z)|$ attached to the modulus of a function analytic over an open set Ω has points of only three possible types: (i) ordinary points, (ii) zeros, (iii) saddle-points. Under projection on the complex plane, a simple saddle-point is locally the common apex of two curvilinear sectors with angle $\pi/2$, referred to as "valleys", where the modulus of the function is smaller than at the saddle-point.*

As a consequence, the surface defined by the modulus of an analytic function has no maximum: this property is known as the *Maximum Modulus Principle*. It has no minimum either, apart from zeros. It is therefore a peakless landscape, in de Bruijn's words [143]. Accordingly, for a meromorphic function, peaks are at ∞ and minima are at 0, the other points being either ordinary points or isolated saddle-points.

Example VIII.1. *The tripod: a cubic polynomial.* An idea of the typical shape of the surface representing the modulus of an analytic function can be obtained by examining Figure VIII.2 relative to the third degree polynomial $f(z) = 1+z+z^2+z^3$. Since $f(z) = (1-z^4)/(1-z)$, the zeros are at

$$-1, \quad i, \quad -i.$$

There are saddle-points at the zeros of the derivative $f'(z) = 1+2z+3z^2$, that is, at the points

$$\zeta := -\frac{1}{3} + \frac{i}{3}\sqrt{2}, \qquad \zeta' := -\frac{1}{3} - \frac{i}{3}\sqrt{2}.$$

The diagram below summarizes the position of these "interesting" points:

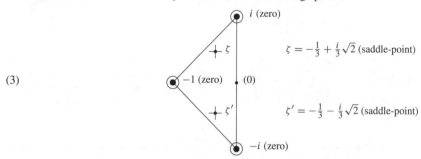

(3)

The three zeros are especially noticeable on Figure VIII.2 (left), where they appear at the end of the three "legs". The two saddle-points are visible on Figure VIII.2 (right) as intersection points of level curves. .... ∎

▷ **VIII.1.** *The Fundamental Theorem of Algebra.* This theorem asserts that a non-constant polynomial has at least one root, hence n roots if its degree is n (Note IV.38, p. 270). Let $P(z) = 1 + a_1 z + \cdots a_n z^n$ be a polynomial of degree n. Consider $f(z) = 1/P(z)$. By basic analysis, one can take R sufficiently large, so that on $|z| = R$, one has $|f(z)| < \frac{1}{2}$. Assume *a contrario* that $P(z)$ has no zero. Then, $f(z)$ which is analytic in $|z| \le R$ should attain its maximum at an interior point (since $f(0) = 1$), so that a contradiction has been reached. ◁

▷ **VIII.2.** *Saddle-points of polynomials and the convex hull of zeros.* Let P be a polynomial and $\mathcal{H}$ the convex hull of its zeros. Then any root of $P'(z)$ lies in $\mathcal{H}$. (Proof: assume distinct zeros and consider

$$\phi(z) := \frac{P'(z)}{P(z)} = \sum_{\alpha \, : \, P(\alpha)=0} \frac{1}{z - \alpha}.$$

If z lies outside $\mathcal{H}$, then z "sees" all zeros α in a half-plane, this by elementary geometry. By projection on the normal to the half-plane boundary, it is found that, for some θ, one has $\Re(e^{i\theta}\phi(z)) < 0$, so that $P'(z) \ne 0$.) ◁

VIII. 2. Saddle-point bounds

Saddle-point analysis is a general method suited to the estimation of integrals of analytic functions $F(z)$,

$$(4) \qquad\qquad I = \int_A^B F(z)\,dz,$$

where $F(z) \equiv F_n(z)$ involves some large parameter n. The method is instrumental when the integrand F is subject to rather violent variations, typically when there occurs in it some exponential or some fixed function raised to a large power $n \to +\infty$. In this section, we discuss some of the *global* properties of saddle-point contours, then particularize the discussion to Cauchy coefficient integrals. General *saddle-point bounds*, which are easy to derive, result from simple geometric considerations (a preliminary discussion appears in Chapter IV, p. 246.).

Starting from the general form (4), we let $\mathcal{C}$ be a contour joining A and B and taken in a domain of the complex plane where $F(z)$ is analytic. By standard inequalities, we have

$$(5) \qquad |I| \leq \|\mathcal{C}\| \cdot \sup_{z \in \mathcal{C}} |F(z)|,$$

with $\|\mathcal{C}\|$ representing the length of $\mathcal{C}$. This is the common *trivial bound* from integration theory applied to a fixed contour $\mathcal{C}$.

For an analytic integrand F with A and B inside the domain of analyticity, there is an infinite class **P** of acceptable paths to choose from, all in the analyticity domain of F. Thus, by optimizing the bound (5), we may write

$$(6) \qquad |I| \leq \inf_{\mathcal{C} \in \mathbf{P}} \left[\|\mathcal{C}\| \cdot \sup_{z \in \mathcal{C}} |F(z)| \right],$$

where the infimum is taken over all paths $\mathcal{C} \in \mathbf{P}$. Broadly speaking, a bound of this type is called a *saddle-point bound*[1].

The length factor $\|\mathcal{C}\|$ usually turns out to be unimportant for asymptotic bounding purposes—this is, for instance, the case when paths remain in finite regions of the complex plane. If there happens to be a path $\mathcal{C}$ from A to B such that no point is at an altitude higher than $\sup(|F(A)|, |F(B)|)$, then a simple bound results, namely, $|I| \leq \|\mathcal{C}\| \cdot \sup(|F(A)|, |F(B)|)$: this is in a sense the uninteresting case. The common situation, typical of Cauchy coefficient integrals of combinatorics, is that paths have to go at some higher altitude than the end points. A path $\mathcal{C}$ that traverses a saddle-point by connecting two points at a lower altitude on the surface $|F(z)|$ and by following two steepest descent lines across the saddle-point is clearly a local minimum for the path functional

$$\Phi(\mathcal{C}) = \sup_{z \in \mathcal{C}} |F(z)|,$$

as neighbouring paths must possess a higher maximum. Such a path is called a *saddle-point path* or *steepest descent path*. Then, the search for a path minimizing

$$\inf_{\mathcal{C}} \left[\sup_{z \in \mathcal{C}} |F(z)| \right]$$

(a simplification of (6) to its essential feature) naturally suggests considering saddle-points and saddle-point paths. This leads to the variant of (6),

$$(7) \qquad |I| \leq \|\mathcal{C}_0\| \cdot \sup_{z \in \mathcal{C}_0} |F(z)|, \qquad \mathcal{C}_0 \text{ minimizes } \sup_{z \in \mathcal{C}} |F(z)|,$$

also referred to as a *saddle-point bound*.

We can summarize this stage of the discussion by a simple generic statement.

Theorem VIII.2 (General saddle-point bounds). *Let $F(z)$ be a function analytic in a domain Ω. Consider the class of integral $\int_\gamma F(z)\, dz$ where the contour γ connects*

---

[1] Notice additionally that the optimization problem need not be solved exactly, as any approximate solution to (6) still furnishes a valid upper bound because of the universal character of the trivial bound (5).

two points A, B and is constrained to a class $\mathbf{P}$ of allowable paths in Ω (e.g., those that encircle 0). Then one has the saddle-point bound[2]*:*

$$
(8) \qquad \left| \int_{\gamma} F(z)\,dz \right| \leq \|C_0\| \cdot \sup_{z \in C_0} |F(z)|,
$$

where C_0 is any path that minimizes $\sup_{z \in C} |F(z)|$.

If A and B lie in opposite valleys of a saddle-point z_0, then the minimization problem is solved by saddle-point paths C_0 made of arcs connecting A to B through z_0. In that case, one has

$$
\left| \int_A^B F(z)\,dz \right| \leq \|C_0\| \cdot |F(z_0)|, \qquad F'(z_0) = 0.
$$

Borrowing a metaphor of de Bruijn [143], the situation may be described as follows. Estimating a path integral is like estimating the difference of altitude between two villages in a mountain range. If the two villages are in different valleys, the best strategy (this is what road networks often do) consists in following paths that cross boundaries between valleys at passes, *i.e.*, through saddle-points.

The statement of Theorem VIII.2 does no fix all details of the contour, when there are several saddle-points "separating" A and B—the problem is like finding the most economical route across a whole mountain range. But at least it suggests the construction of a composite contour made of connected arcs crossing saddle-points from valley to valley. Furthermore, in cases of combinatorial interest, some strong positivity is present and the selection of the suitable saddle-point contour is normally greatly simplified, as we explain next.

▷ **VIII.3.** *An integral of powers.* Consider the polynomial $P(z) = 1 + z + z^2 + z^3$ of Example VIII.1. Define the line integral

$$
I_n = \int_{-1}^{+i} P(z)^n\,dz.
$$

On the segment connecting the end points, the maximum of $|P(z)|$ is 0.63831, giving the weak trivial bound $I_n = O(0.63831^n)$. In contrast, there is a saddle-point at $\zeta = -\frac{1}{3} + \frac{i}{3}\sqrt{2}$ where $|P(\zeta)| = \frac{1}{3}$, resulting in the bound

$$
|I_n| \leq \lambda \left(\frac{1}{3} \right)^n, \qquad \lambda := |\zeta + 1| + |i - \zeta| \doteq 1.44141,
$$

as follows from adopting a contour made of two segments connecting -1 to i through ζ. Discuss further the bounds on $\int_\alpha^{\alpha'}$, when (α, α') ranges over all pairs of roots of P.   ◁

Saddle-point bounds for Cauchy coefficient integrals. Saddle-point bounds can be applied to Cauchy coefficient integrals,

$$
(9) \qquad g_n \equiv [z^n]G(z) = \frac{1}{2i\pi} \oint G(z)\,\frac{dz}{z^{n+1}},
$$

---

[2]The form given by (8) is in principle weaker than the form (6), since it does not take into account the length of the contour itself, but the difference is immaterial in all our asymptotic problems.

for which we can avail ourselves of the previous discussion, with $F(z) = G(z)z^{-n-1}$. In (9) the symbol $\oint$ indicates that the allowable paths are constrained to encircle the origin (the domain of definition of the integrand is a subset of $\mathbb{C} \setminus \{0\}$; the points A, B can then be seen as coinciding and taken somewhere along the negative real line; equivalently, one may take $A = -ae^{i\epsilon}$ and $B = -ae^{-i\epsilon}$, for $a > 0$ and $\epsilon \to 0$).

In the particular case where $G(z)$ is a function with non-negative coefficients, a simple condition guarantees the existence of a saddle-point on the positive real axis. Indeed, assume that $G(z)$, which has radius of convergence R with $0 < R \le +\infty$, satisfies $G(z) \to +\infty$ as $z \to R^-$ along the real axis and $G(z)$ *not* a polynomial. Then the integrand $F(z) = G(z)z^{-n-1}$ satisfies $F(0^+) = F(R^-) = +\infty$. This means that there exists at least one local minimum of F over $(0, R)$, hence, at least one value $\zeta \in (0, R)$ where the derivative F' vanishes. (Actually, there can be only one such point; see Note VIII.4, p. 550.) Since ζ corresponds to a local minimum of F, we have additionally $F''(\zeta) > 0$, so that the saddle-point is crossed transversally by a circle of radius ζ. Thus, the saddle-point bound, specialized to circles centred at the origin, yields the following corollary.

Corollary VIII.1 (Saddle-point bounds for generating functions). *Let $G(z)$, not a polynomial, be analytic at 0 with non-negative coefficients and radius of convergence $R \le +\infty$. Assume that $G(R^-) = +\infty$. Then one has*

(10) $\quad [z^n]G(z) \le \dfrac{G(\zeta)}{\zeta^n}, \quad$ *with $\zeta \in (0, R)$ the unique root of $\zeta \dfrac{G'(\zeta)}{G(\zeta)} = n + 1$.*

Proof. The saddle-point is the point where the derivative of the integrand is 0. Therefore, we consider $(G(z)z^{-n-1})' = 0$, or $G'(z)z^{-n-1} - (n+1)G(z)z^{-n-2} = 0$, or

$$z\frac{G'(z)}{G(z)} = n + 1.$$

We refer to this as the *saddle-point equation* and use ζ to denote its positive root. The perimeter of the circle is $2\pi\zeta$, so that the inequality $[z^n]G(z) \le G(\zeta)/\zeta^n$ follows. ∎

Corollary VIII.1 is equivalent to Proposition IV.1, p. 246, on which it sheds a new light, while paving the way to the full saddle-point method to be developed in the next section.

We examine below two particular cases related to the central binomial and the inverse factorial. The corresponding landscapes of Figure VIII.3, which bear a surprising resemblance to one another, are, by the previous discussion, instances of a general pattern for functions with non-negative coefficients. It is seen on these two examples that the saddle-point bounds already catch the proper exponential growths, being off only by a factor of $O(n^{-1/2})$.

Example VIII.2. *Saddle-point bounds for central binomials and inverse factorials.* Consider the two contour integrals around the origin

(11) $$J_n = \frac{1}{2i\pi} \oint (1+z)^{2n} \frac{dz}{z^{n+1}}, \qquad K_n = \frac{1}{2i\pi} \oint e^z \frac{dz}{z^{n+1}},$$

whose values are otherwise known, by virtue of Cauchy's coefficient formula, to be $J_n = \binom{2n}{n}$ and $K_n = 1/n!$. In that case, one can think of the end points A and B as coinciding and taken

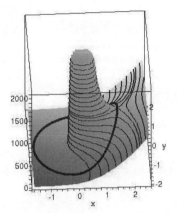 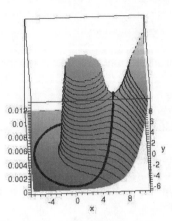

Figure VIII.3. The modulus of the integrands of J_n (central binomials) and K_n (inverse factorials) for $n = 5$ and the corresponding saddle-point contours.

---

somewhat arbitrarily on the negative real axis, while the contour has to encircle the origin once and counter-clockwise.

The saddle-point equations are, respectively,

$$\frac{2n}{1+z} - \frac{n+1}{z} = 0, \qquad 1 - \frac{n+1}{z} = 0,$$

the corresponding saddle-points being $\zeta = \dfrac{n+1}{n-1}$ and $\zeta' = n+1$. This provides the upper bounds

(12) $$J_n = \binom{2n}{n} \le \left(\frac{4n^2}{n^2-1}\right)^n \le \frac{16}{9} 4^n, \qquad K_n = \frac{1}{n!} \le \frac{e^{n+1}}{(n+1)^n},$$

which are valid for all values $n \ge 2$. .... ∎

▷ **VIII.4.** *Upward convexity of* $G(x)x^{-n}$. For $G(z)$ having non-negative coefficients at the origin, the quantity $G(x)x^{-n}$ is upward convex for $x > 0$, so that the saddle-point equation for ζ can have at most one root. Indeed, the second derivative

(13) $$\frac{d^2}{dx^2} \frac{G(x)}{x^n} = \frac{x^2 G''(x) - 2nx G'(x) + n(n+1)G(x)}{x^{n+2}},$$

is positive for $x > 0$ since its numerator,

$$\sum_{k \ge 0} (n+1-k)(n-k)g_k x^k, \qquad g_k := [z^k]G(z),$$

has only non-negative coefficients. (See Note IV.46, p. 280, for an alternative derivation.) ◁

▷ **VIII.5.** *A minor optimization.* The bounds of Equation (6), p. 547, which take the length of the contour into account, lead to estimates that closely resemble (10). Indeed, we have

$$[z^n]G(z) \le \frac{G(\widehat{\zeta})}{\widehat{\zeta}^n}, \qquad \widehat{\zeta} \text{ root of } z\frac{G'(z)}{G(z)} = n,$$

when optimization is carried out over circles centred at the origin. ◁

VIII. 3. Overview of the saddle-point method

Given a complex integral with a contour traversing a *single* saddle-point, the saddle-point corresponds locally to a maximum of the integrand along the path. It is then natural to expect that *a small neighbourhood of the saddle-point may provide the dominant contribution to the integral*. The saddle-point method is applicable precisely when this is the case *and* when this dominant contribution can be estimated by means of local expansions. The method then constitutes the complex analytic counterpart of the method of Laplace (Appendix B.6: *Laplace's method*, p. 755) for the evaluation of real integrals depending on a large parameter, and we can regard it as being

Saddle-point method = Choice of contour + Laplace's method.

Similar to its real-variable counterpart, the saddle-point method is a general strategy rather than a completely deterministic algorithm, since many choices are left open in the implementation of the method concerning details of the contour and choices of its splitting into pieces.

To proceed, it is convenient to set $F(z) = e^{f(z)}$ and consider

$$(14) \qquad\qquad I = \int_A^B e^{f(z)}\, dz,$$

where $f(z) \equiv f_n(z)$, as $F(z) \equiv F_n(z)$ in the previous section, involves some large parameter n. Following possibly some preparation based on Cauchy's theorem, we may assume that the contour $\mathcal{C}$ connects two end points A and B lying in opposite valleys of the saddle-point ζ. The saddle-point equation is $F'(\zeta) = 0$, or equivalently since $F = e^f$:

$$f'(\zeta) = 0.$$

The saddle-point method, of which a summary is given in Figure VIII.4, is based on a fundamental splitting of the integration contour. We decompose $\mathcal{C} = \mathcal{C}^{(0)} \cup \mathcal{C}^{(1)}$, where $\mathcal{C}^{(0)}$ called the "central part" contains ζ (or passes very near to it) and $\mathcal{C}^{(1)}$ is formed of the two remaining "tails". This splitting has to be determined in each case in accordance with the growth of the integrand. The basic principle rests on two major conditions: the contributions of the two tails should be asymptotically negligible (condition $\mathbf{SP_1}$); in the central region, the quantity $f(z)$ in the integrand should be asymptotically well approximated by a quadratic function (condition $\mathbf{SP_2}$). Under these conditions, the integral is asymptotically equivalent to an incomplete Gaussian integral. It then suffices to verify—this is condition $\mathbf{SP_3}$, usually a minor *a posteriori* technical verification—that tails can be completed back, introducing only negligible error terms. By this sequence of steps, the original integral is asymptotically reduced to a complete Gaussian integral, which evaluates in closed form.

Specifically, the three steps of the saddle-point method involve checking conditions expressed by Equations (15), (16), and (18) below.

Goal: Estimate $\int_A^B F(z)\,dz$, setting $F = e^f$; here, $F \equiv F_n$ and $f \equiv f_n$ depend on a large parameter n.

— The end points A, B are assumed to lie in opposite valleys of the saddle-point.
— A contour $\mathcal{C}$ through (or near) a simple saddle-point ζ, so that $f'(\zeta) = 0$, has been chosen.
— The contour is split as $\mathcal{C} = \mathcal{C}^{(0)} \cup \mathcal{C}^{(1)}$.

The following conditions are to be verified.

SP$_1$: *Tails pruning*. On the contour $\mathcal{C}^{(1)}$, the tails integral $\int_{\mathcal{C}^{(1)}}$ is negligible:

$$\int_{\mathcal{C}^{(1)}} F(z)\,dz = o\left(\int_{\mathcal{C}} F(z)\,dz\right).$$

SP$_2$: *Central approximation*. Along $\mathcal{C}^{(0)}$, a quadratic expansion,

$$f(z) = f(\zeta) + \frac{1}{2}f''(\zeta)(z-\zeta)^2 + O(\eta_n),$$

is valid, with $\eta_n \to 0$ as $n \to \infty$, *uniformly* with respect to $z \in \mathcal{C}^{(0)}$.

SP$_3$: *Tails completion*. The incomplete Gaussian integral resulting from **SP$_2$**, taken over the central range, is asymptotically equivalent to a *complete* Gaussian integral (with $f''(\zeta) = e^{i\phi}|f''(\zeta)|$ and $\varepsilon = \pm 1$ depending on orientation):

$$\int_{\mathcal{C}^{(0)}} e^{\frac{1}{2}f''(\zeta)(z-\zeta)^2}\,dz \sim \varepsilon i e^{-i\phi/2} \int_{-\infty}^{\infty} e^{-|f''(\zeta)|x^2/2}\,dx \equiv \varepsilon i e^{-i\phi/2}\sqrt{\frac{2\pi}{|f''(\zeta)|}}.$$

Result: Assuming **SP$_1$**, **SP$_2$**, and **SP$_3$**, one has, with $\varepsilon = \pm 1$ and $\arg(f''(\zeta)) = \phi$:

$$\frac{1}{2i\pi}\int_A^B e^{f(z)}\,dz \sim \varepsilon e^{-i\phi/2}\frac{e^{f(\zeta)}}{\sqrt{2\pi|f''(\zeta)|}} = \pm\frac{e^{f(\zeta)}}{\sqrt{2\pi f''(\zeta)}}.$$

Figure VIII.4. A summary of the basic saddle-point method.

SP$_1$: *Tails pruning*. On the contour $\mathcal{C}^{(1)}$, the tail integral $\int_{\mathcal{C}^{(1)}}$ is negligible:

$$(15) \qquad \int_{\mathcal{C}^{(1)}} F(z)\,dz = o\left(\int_{\mathcal{C}} F(z)\,dz\right).$$

This condition is usually established by proving that $F(z)$ remains small enough (e.g., exponentially small in the scale of the problem) away from ζ, for $z \in \mathcal{C}^{(1)}$.

SP$_2$: *Central approximation*. Along $\mathcal{C}^{(0)}$, a quadratic expansion,

$$(16) \qquad f(z) = f(\zeta) + \frac{1}{2}f''(\zeta)(z-\zeta)^2 + O(\eta_n),$$

is valid, with $\eta_n \to 0$ as $n \to \infty$, uniformly for $z \in \mathcal{C}^{(0)}$. This guarantees that $\int e^f$ is well-approximated by an incomplete Gaussian integral:

$$(17) \qquad \int_{\mathcal{C}^{(0)}} e^{f(z)}\,dz \sim e^{f(\zeta)} \int_{\mathcal{C}^{(0)}} e^{\frac{1}{2}f''(\zeta)(z-\zeta)^2}\,dz.$$

SP$_3$: *Tails completion.* The tails can be completed back, at the expense of asymptotically negligible terms, meaning that the incomplete Gaussian integral is asymptotically equivalent to a complete one (itself given by (12), p. 744),

$$(18) \quad \int_{\mathcal{C}^{(0)}} e^{\frac{1}{2}f''(\zeta)(z-\zeta)^2} \, dz \sim \varepsilon i e^{-i\phi/2} \int_{-\infty}^{\infty} e^{-|f''(\zeta)|x^2/2} \, dx \equiv \varepsilon i e^{-i\phi/2} \sqrt{\frac{2\pi}{|f''(\zeta)|}}.$$

where $\varepsilon = \pm 1$ is determined by the orientation of the original contour $\mathcal{C}$, and $f''(\zeta) = e^{i\phi}|f''(\zeta)|$. This last step deserves a word of explanation. Along a steepest descent curve across ζ, the quantity $f''(\zeta)(z - \zeta)^2$ is real and negative, as we saw when discussing saddle-point landscapes (p. 543). Indeed, with $f''(\zeta) = e^{i\phi}|f''(\zeta)|$, one has $\arg(z-\zeta) \equiv -\phi/2 + \frac{\pi}{2}$ (mod π). Thus, the change of variables $x = \pm i(z-\zeta)e^{-i\phi/2}$ reduces the left side of (18) to an integral taken along (or close to) the real line[3]. The condition (18) then demands that this integral can be completed to a complete Gaussian integral, which itself evaluates in closed form.

If these conditions are granted, one has the chain

$$\int_{\mathcal{C}} e^f \, dz \sim \int_{\mathcal{C}^{(0)}} e^f \, dz \sim e^{f(\zeta)} \int_{\mathcal{C}^{(0)}} e^{\frac{1}{2}f''(\zeta)(z-\zeta)^2} \, dz \sim \pm i e^{-i\phi/2} e^{f(\zeta)} \sqrt{\frac{2\pi}{|f''(\zeta)|}},$$

by virtue of Equations (15), (17), (18). In summary:

Theorem VIII.3 (Saddle-point Algorithm). *Consider an integral $\int_A^B F(z)\,dz$, where the integrand $F = e^f$ is an analytic function depending on a large parameter and A, B lie in opposite valleys across a saddle-point ζ, which is a root of the saddle-point equation*

$$f'(\zeta) = 0$$

(or, equivalently, $F'(\zeta) = 0$). Assume that the contour $\mathcal{C}$ connecting A to B can be split into $\mathcal{C} = \mathcal{C}^{(0)} \cup \mathcal{C}^{(1)}$ in such a way that the following conditions are satisfied:

   (i) tails are negligible, in the sense of Equation (15) of **SP$_1$**,

   (ii) a central approximation hold, in the sense of Equation (16) of **SP$_2$**,

   (iii) tails can be completed back, in the sense of Equation (18) of **SP$_3$**.

Then one has, with $\varepsilon = \pm 1$ reflecting orientation and $\phi = \arg(f''(\zeta))$:

$$(19) \quad \frac{1}{2i\pi} \int_A^B e^{f(z)} \, dz \sim \varepsilon e^{-i\phi/2} \frac{e^{f(\zeta)}}{\sqrt{2\pi|f''(\zeta)|}} = \pm \frac{e^{f(\zeta)}}{\sqrt{2\pi f''(\zeta)}}.$$

It can be verified at once that a blind application of the formula to the two integrals of Example VIII.2 produces the expected asymptotic estimates

$$(20) \quad J_n \equiv \binom{2n}{n} \sim \frac{4^n}{\sqrt{\pi n}} \quad \text{and} \quad K_n \equiv \frac{1}{n!} \sim \frac{1}{n^n e^{-n}\sqrt{2\pi n}}.$$

The complete justification in the case of K_n is given in Example VIII.3 below. The case of J_n is covered by the general theory of "large powers" of Section VIII. 8, p. 585.

---

[3]The sign in (18) is naturally well-defined, once the data A, B, and f are fixed: one possibility is to adopt the determination of $\phi/2$ (mod π) such that A and B are sent close to the negative and the positive real axis, respectively, after the final change of variables $x = i(z - \zeta)e^{-i\phi/2}$.

In order for the saddle-point method to work, conflicting requirements regarding the dimensioning of $\mathcal{C}^{(0)}$ and $\mathcal{C}^{(1)}$ must be satisfied. The tails pruning and tails completion conditions, $\mathbf{SP_1}$ and $\mathbf{SP_3}$, force $\mathcal{C}^{(0)}$ to be chosen large enough, so as to capture the main contribution to the integral; the central approximation condition $\mathbf{SP_2}$ requires $\mathcal{C}^{(0)}$ to be small enough, to the effect that $f(z)$ can be suitably reduced to its quadratic expansion. Usually, one has to take $\|\mathcal{C}^{(0)}\|/\|\mathcal{C}\| \to 0$, and the following observation may help make the right choices. The error in the two-term expansion being likely given by the next term, which involves a third derivative, it is a good guess to dimension $\mathcal{C}^{(0)}$ to be of length $\delta \equiv \delta(n)$ chosen in such a way that

$$(21) \qquad\qquad f''(\zeta)\delta^2 \to \infty, \qquad f'''(\zeta)\delta^3 \to 0,$$

so that both tail and central approximation conditions can be satisfied. We call this choice the *saddle-point dimensioning heuristic*.

On another register, it often proves convenient to adopt integration paths that come close enough to the saddle-point but need not pass exactly through it. In the same vein, a steepest descent curve may be followed only approximately. Such choices will still lead to valid conclusions, as long as the conditions of Theorem VIII.3 are verified. (Note carefully that these conditions *neither* impose that the contour should pass strictly through the saddle-point, *nor* that a steepest descent curve should be exactly followed.)

Saddle-point method for Cauchy coefficient integrals. For the purposes of analytic combinatorics, the general saddle-point method specializes. We are given a generating function $G(z)$, assumed to be analytic at the origin and with non-negative coefficients, and seek an asymptotic form of the coefficients, given in integral form by

$$[z^n]G(z) = \frac{1}{2i\pi} \int_{\mathcal{C}} G(z) \frac{dz}{z^{n+1}}.$$

There, $\mathcal{C}$ encircles the origin, lies within the domain where G is analytic, and is positively oriented. This is a particular case of the general integral (14) considered earlier, with the integrand being $F(z) = G(z)/z^{n+1}$.

The geometry of the problem is now simple, and, for reasons seen in the previous section, it suffices to consider as integration contour a circle centred at the origin and passing through (or very near) a saddle-point present on the positive real line. It is then natural to make use of polar coordinates and set

$$z = re^{i\theta},$$

where the radius r of the circle will be chosen equal to (or close to) the positive saddle-point value. We thus need to estimate

$$(22) \qquad [z^n]G(z) = \frac{1}{2i\pi} \oint G(z) \frac{dz}{z^{n+1}} = \frac{r^{-n}}{2\pi} \int_{-\pi}^{+\pi} G(re^{i\theta})e^{-ni\theta}\,d\theta.$$

Under the circumstances, the basic split of the contour $\mathcal{C} = \mathcal{C}^{(0)} \cup \mathcal{C}^{(1)}$ involves a central part $\mathcal{C}^{(0)}$, which is an arc of the circle of radius r determined by $|\theta| \le \theta_0$ for some suitably chosen θ_0. On $\mathcal{C}^{(0)}$, a quadratic approximation should hold, according to **SP$_2$** [central approximation]. Set

$$(23) \qquad\qquad f(z) := \log G(z) - n \log z.$$

A natural possibility is to adopt for r the value that cancels $f'(r)$,

$$(24) \qquad\qquad r \frac{G'(r)}{G(r)} = n,$$

which is a version of the *saddle-point equation*[4] relative to polar coordinates. This grants us *locally*, a quadratic approximation without linear terms, with $\beta(r)$ a computable quantity (in terms of $f(r)$, $f'(r)$, $f''(r)$), we have

$$(25) \qquad\qquad f(re^{i\theta}) - f(r) = -\frac{1}{2}\beta(r)\theta^2 + o(\theta^3),$$

which is valid at least for fixed r (i.e., for fixed n), as $\theta \to 0$

The cutoff angle θ_0 is to be chosen as a function of n (or, equivalently, r) in accordance with the saddle-point heuristic (21). It then suffices to carry out a verification of the validity of the three conditions of the saddle-point method, **SP$_1$**, **SP$_2$** (for which a suitably *uniform* version of (25) needs to be developed), and **SP$_3$** of Theorem VIII.3, p. 553, adjusted to take into account polar coordinate notations.

The example below details the main steps of the saddle-point analysis of the generating function of inverse factorials, based on the foregoing principles.

***Example* VIII.3.** *Saddle-point analysis of the exponential and the inverse factorial I.* The goal is to estimate $\frac{1}{n!} = [z^n]e^z$, the starting point being

$$K_n = \frac{1}{2i\pi}\int_{|z|=r} e^z \frac{dz}{z^{n+1}},$$

where integration will be performed along a circle centred at the origin. The landscape of the modulus of the integrand has been already displayed in Figure VIII.3, p. 550—there is a saddle-point of $G(z)z^{-n-1}$ at $\zeta = n+1$ with an axis perpendicular to the real line. We thus expect an asymptotic estimate to derive from adopting a circle passing through the saddle-point, or about.

We switch to polar coordinates, fix the choice of the radius $r = n$ in accordance with (24), and set $z = ne^{i\theta}$. The original integral becomes, in polar coordinates,

$$(26) \qquad\qquad K_n = \frac{e^n}{n^n} \cdot \frac{1}{2\pi}\int_{-\pi}^{+\pi} e^{n(e^{i\theta}-1-i\theta)}\,d\theta,$$

where, for readability, we have taken out the factor $G(r)/r^n \equiv e^n/n^n$. Set $h(\theta) = e^{i\theta} - 1 - i\theta$. The function $|e^{h(\theta)}| = e^{\cos\theta - 1}$ is unimodal with its peak at $\theta = 0$ and the same property holds for $|e^{nh(\theta)}|$, representing the modulus of the integrand in (26), which gets more and more strongly peaked at $\theta = 0$, as $n \to +\infty$; see Figure VIII.5.

---

[4]Equation (24) is almost the same as $\zeta G'(\zeta)/G(\zeta) = n+1$ of (10), which defines the saddle-point in z-coordinates. The (minor) difference is accounted for by the fact that saddle-points are sensitive to changes of variables in integrals. In practice, it proves workable to integrate along a circle of radius either r or ζ, or even a suitably close approximation of r, ζ, the choice being often suggested by computational convenience.

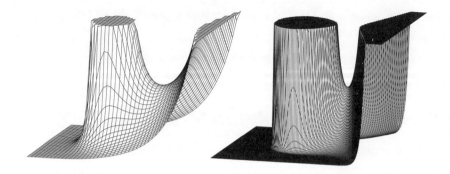

Figure VIII.5. Plots of $|e^z z^{-n-1}|$ for $n = 3$ and $n = 30$ (scaled according to the value of the saddle-point) illustrate the essential concentration condition as higher values of n produce steeper saddle-point paths.

In agreement with the saddle-point strategy, the estimation of K_n proceeds by isolating a small portion of the contour, corresponding to z near the real axis. We thus introduce

$$K_n^{(0)} = \int_{-\theta_0}^{+\theta_0} e^{nh(\theta)} \, d\theta, \qquad K_n^{(1)} = \int_{\theta_0}^{2\pi - \theta_0} e^{nh(\theta)} \, d\theta,$$

and choose θ_0 in accordance with the general heuristic of (21), which corresponds to the two conditions: $n\theta_0^2 \to \infty$ (informally: $\theta_0 \gg n^{-1/2}$) and $n\theta_0^3 \to 0$, (informally: $\theta_0 \ll n^{-1/3}$). One way of realizing the compromise is to adopt $\theta_0 = n^a$, where a is any number between $-1/2$ and $-1/3$. To be specific, we fix $a = -2/5$, so

$$(27) \qquad\qquad \theta_0 \equiv \theta_0(n) = n^{-2/5}.$$

In particular, the angle of the central region tends to zero.

  (*i*) *Tails pruning.* For $z = ne^{i\theta}$ one has $\left|e^z\right| = e^{n\cos\theta}$, and, by unimodality properties of the cosine, the tail integral $K^{(1)}$ satisfies

$$(28) \qquad\qquad \left|K_n^{(1)}\right| = O\left(e^{-n(\cos\theta_0 - 1)}\right) = O\left(\exp\left(-Cn^{1/5}\right)\right),$$

for some $C > 0$. The tail integral is thus is exponentially small.

  (*ii*) *Central approximation.* Near $\theta = 0$, one has $h(\theta) \equiv e^{i\theta} - 1 - i\theta = -\frac{1}{2}\theta^2 + O(\theta^3)$, so that, for $|\theta| \le \theta_0$,

$$e^{nh(\theta)} = e^{-n\theta^2/2 + O(n\theta^3)} = e^{-n\theta^2/2}\left(1 + O(n\theta_0^3)\right).$$

Since $\theta_0 = n^{-2/5}$, we have

$$(29) \qquad\qquad K_n^{(0)} = \int_{-n^{-2/5}}^{+n^{-2/5}} e^{-n\theta^2/2} \, d\theta \left(1 + O(n^{-1/5})\right),$$

which, by the change of variables $t = \theta\sqrt{n}$, becomes

$$(30) \qquad\qquad K_n^{(0)} = \frac{1}{\sqrt{n}} \int_{-n^{1/10}}^{+n^{1/10}} e^{-t^2/2} \, dt \left(1 + O(n^{-1/5})\right).$$

The central integral is thus asymptotic to an incomplete Gaussian integral.

(*iii*) *Tails completion.* Given (30), the task is now easy. We have, elementarily, for $c > 0$,

$$\int_c^{+\infty} e^{-t^2/2}\, dt = O\left(e^{-c^2/2}\right), \tag{31}$$

which expresses the exponential smallness of Gaussian tails. As a consequence,

$$K_n^{(0)} \sim \frac{1}{\sqrt{n}} \int_{-\infty}^{+\infty} e^{-t^2/2}\, dt \equiv \sqrt{\frac{2\pi}{n}}. \tag{32}$$

Assembling (28) and (32), we obtain

$$K_n^{(0)} + K_n^{(1)} \sim \sqrt{\frac{2\pi}{n}}, \qquad \text{i.e.,} \qquad K_n = \frac{1}{2\pi}\frac{e^n}{n^n}\left(K_n^{(0)} + K_n^{(1)}\right) \sim \frac{e^n}{n^n \sqrt{2\pi n}}.$$

The proof also provides a relative error term of $O(n^{-1/5})$. Stirling's formula is thus seen to be (*inter alia!*) a consequence of the saddle-point method. . . . . ∎

Complete asymptotic expansions. Just like Laplace's method, the saddle-point method can often be made to provide *complete asymptotic expansions*. The idea is still to localize the main contribution in the central region, but now take into account *corrections terms to the quadratic approximation*. As an illustration of these general principles, we make explicit here the calculations relative to the inverse factorial.

Example VIII.4. *Saddle-point analysis of the exponential and the inverse factorial II.* For a complete expansion of $[z^n]e^z$, we only need to revisit the estimation of $K^{(0)}$ in the previous example, since $K^{(1)}$ is exponentially small anyhow. One first rewrites

$$
\begin{aligned}
K_n^{(0)} &= \int_{-\theta_0}^{\theta_0} e^{-n\theta^2/2} e^{n(\cos\theta - 1 + \frac{1}{2}\theta^2)}\, d\theta \\
&= \frac{1}{\sqrt{n}} \int_{-\theta_0\sqrt{n}}^{\theta_0\sqrt{n}} e^{-w^2/2} e^{n\xi(w/\sqrt{n})}\, dw, \qquad \xi(\theta) := \cos\theta - 1 + \frac{1}{2}\theta^2.
\end{aligned}
$$

The calculation proceeds exactly in the same way as for the Laplace method (Appendix B.6: *Laplace's method*, p. 755). It suffices to expand $h(\theta)$ to any fixed order, which is legitimate in the central region. In this way, a representation of the form,

$$K_n^{(0)} = \frac{1}{\sqrt{n}} \int_{-\theta_0\sqrt{n}}^{\theta_0\sqrt{n}} e^{-w^2/2}\left(1 + \sum_{k=1}^{M-1} \frac{E_k(w)}{n^{k/2}} + O\left(\frac{1 + w^{3M}}{n^{M/2}}\right)\right) dw,$$

is obtained, where the $E_k(w)$ are computable polynomials of degree $3k$. Distributing the integral operator over terms in the asymptotic expansion and completing the tails yields an expansion of the form

$$K_n^{(0)} \sim \frac{1}{\sqrt{n}}\left(\sum_{k=0}^{M-1} \frac{d_k}{n^{k/2}} + O(n^{-M/2})\right),$$

where $d_0 = \sqrt{2\pi}$ and $d_k := \int_{-\infty}^{+\infty} e^{-w^2/2} E_k(w)\, dw$. All odd terms disappear by parity. The net result is then the following.

Proposition VIII.1 (Stirling's formula). *The factorial numbers satisfy*

$$\frac{1}{n!} \sim \frac{e^n n^{-n}}{\sqrt{2\pi n}}\left(1 - \frac{1}{12n} + \frac{1}{288\,n^2} + \frac{139}{51840\,n^3} - \frac{571}{2488320\,n^4} + \cdots\right).$$

Notice the amazing similarity with the form obtained directly for $n!$ in Appendix B.6: *Laplace's method*, p. 755. .... ∎

▷ **VIII.6.** *A factorial surprise.* Why is it that the expansion of $n!$ and $1/n!$ involve the same set of coefficients, up to sign?   ◁

VIII. 4. Three combinatorial examples

The saddle-point method permits us to solve a number of asymptotic problems coming from analytic combinatorics. In this section, we illustrate its use by treating in some detail three combinatorial examples[5]:

> *Involutions ($\mathcal{I}$), Set partitions ($\mathcal{S}$), Fragmented permutations ($\mathcal{F}$).*

These are all labelled structures introduced in Chapter II. Their specifications and EGFs are

$$(33) \quad \begin{cases} \text{Involutions :} & \mathcal{I} = \text{SET}(\text{SET}_{1,2}(\mathcal{Z})) \implies I(z) = e^{z+z^2/2} \\ \text{Set Partition :} & \mathcal{S} = \text{SET}(\text{SET}_{\geq 1}(\mathcal{Z})) \implies S(z) = e^{e^z - 1} \\ \text{Fragmented perms :} & \mathcal{F} = \text{SET}(\text{SEQ}_{\geq 1}(\mathcal{Z})) \implies F(z) = e^{z/(1-z)}. \end{cases}$$

The first two are entire functions (i.e., they only have a singularity at ∞), while the last one has a singularity at $z = 1$. Each of these functions exhibits a fairly violent growth—of an exponential type—near its positive singularity, at either a finite or infinite distance. As the reader will have noticed, all three combinatorial types are structurally characterized by a set construction applied to some simpler structure.

Each example is treated, starting from the easier saddle-point bounds and proceeding with the saddle-point method. The example of involutions deals with a problem that is only a little more complicated than inverse factorials. The case of set partitions (Bell numbers) illustrates the need in general of a good asymptotic technology for implicitly defined saddle-points. Finally, fragmented permutations, with their singularity at a finite distance, pave the way for the (harder) analysis of integer partitions in Section VIII. 6. We recapitulate the main features of the saddle-point analyses of these three structures, together with the case of inverse factorials (urns), in Figure VIII.6.

***Example* VIII.5.** *Involutions.* An *involution* is a permutation τ such that τ^2 is the identity permutation (p. 122). The corresponding EGF is $I(z) = e^{z+z^2/2}$. We have in the notation of (23)

$$f(z) = z + \frac{z^2}{2} - n \log z,$$

and the saddle-point equation in polar coordinates is

$$r(1+r) = n, \quad \text{implying} \quad r = -\frac{1}{2} + \frac{1}{2}\sqrt{4n+1} \sim \sqrt{n} - \frac{1}{2} + \frac{1}{8\sqrt{n}} + O(n^{-3/2}).$$

---

[5]The purpose of these examples is to become further familiarized with the practice of the saddle-point method in analytic combinatorics. The impatient reader can jump directly to the next section, where she will find a general theory that covers these and many more cases.

Class	EGF	radius (r)	angle (θ_0)	coeff $[z^n]$ in EGF
urns				
SET($\mathcal{Z}$)	e^z	n	$n^{-2/5}$	$\sim \dfrac{e^n n^{-n}}{\sqrt{2\pi n}}$
(Ex. VIII.3, p. 555)				
involutions				
SET(CYC$_{1,2}(\mathcal{Z})$)	$e^{z+z^2/2}$	$\sim \sqrt{n} - \frac{1}{2}$	$n^{-2/5}$	$\sim \dfrac{e^{n/2-1/4}n^{-n/2}}{2\sqrt{\pi n}}e^{\sqrt{n}}$
(Ex. VIII.5, p. 558)				
set partitions				
SET(SET$_{\geq 1}(\mathcal{Z})$)	e^{e^z-1}	$\sim \log n - \log\log n$	$e^{-2r/5}/r$	$\sim \dfrac{e^{e^r}-1}{r^n\sqrt{2\pi r(r+1)e^r}}$
(Ex. VIII.6, p. 560)				
fragmented perms				
SET(SEQ$_{\geq 1}(\mathcal{Z})$)	$e^{z/(1-z)}$	$\sim 1 - \frac{1}{\sqrt{n}}$	$n^{-7/10}$	$\sim \dfrac{e^{-1/2+2\sqrt{n}}}{2\sqrt{\pi}n^{3/4}}$
(Ex. VIII.7, p. 562)				

Figure VIII.6. A summary of some major saddle-point analyses in combinatorics.

The use of the saddle-point bound then gives mechanically

$$(34) \qquad \frac{I_n}{n!} \leq e^{-1/4}\frac{e^{n/2+\sqrt{n}}}{n^{n/2}}(1+o(1)), \quad I_n \leq e^{-1/4}\sqrt{2\pi n}\, e^{-n/2+\sqrt{n}}n^{n/2}(1+o(1)).$$

(Notice that if we use instead the approximate saddle-point value, $\sqrt{n}$, we only lose a factor $e^{-1/4} \doteq 0.77880$.)

The cutoff point between the central and non-central regions is determined, in agreement with (21), by the fact that the length δ of the contour (in z coordinates) should satisfy $f''(r)\delta^2 \to \infty$ and $f'''(r)\delta^3 \to 0$. In terms of angles, this means that we should choose a cutoff angle θ_0 that satisfies

$$r^2 f''(r)\theta_0^2 \to \infty, \qquad r^3 f'''(r)\theta_0^3 \to 0.$$

Here, we have $f''(r) = O(1)$ and $f'''(r) = O(n^{-1/2})$. Thus, θ_0 must be of an order somewhere in between $n^{-1/2}$ and $n^{-1/3}$, and we fix

$$\theta_0 = n^{-2/5}.$$

(i) *Tails pruning.* First, some general considerations are to be made, regarding the behaviour of $|I(z)|$ along large circles, $z = re^{i\theta}$. One has

$$\log|I(re^{i\theta})| = r\cos\theta + \frac{r^2}{2}\cos 2\theta.$$

As a function of θ, this function decreases on $(0, \frac{\pi}{2})$, since it is the sum of two decreasing functions. Thus, $|I(z)|$ attains its maximum ($e^{r+r^2/2}$) at r and its minimum ($e^{-r^2/2}$) at $z = ri$. In the left half-plane, first for $\theta \in (\frac{\pi}{2}, \frac{3\pi}{4})$, the modulus $|I(z)|$ is at most e^r since $\cos 2\theta < 0$. Finally, for $\theta \in (\frac{3\pi}{4}, \pi)$ smallness is granted by the fact that $\cos\theta < -1/\sqrt{2}$ resulting in the bound $|I(z)| \leq e^{r^2/2-r/\sqrt{2}}$. The same argument applies to the lower half plane $\Im(z) < 0$.

As a consequence of these bounds, $I(z)/I(\sqrt{n})$ is strongly peaked at $z = r$; in particular, it is exponentially small away from the positive real axis, in the sense that

$$(35) \qquad \frac{I(re^{i\theta})}{I(r)} = O\left(\frac{I(re^{i\theta_0})}{I(r)}\right) = O\left(\exp(-n^\alpha)\right), \qquad \theta \notin [-\theta_0, \theta_0],$$

for some $\alpha > 0$.

   (ii) *Central approximation.* We then proceed and consider the central integral

$$J_n^{(0)} = \frac{e^{f(r)}}{2\pi} \int_{-\theta_0}^{+\theta_0} \exp\left(f(re^{i\theta}) - f(r)\right) d\theta.$$

What is required is a Taylor expansion with remainder near the point $r \sim \sqrt{n}$. In the central region, the relations $f'(r) = 0$ $f''(r) = 2 + O(1/n)$, and $f'''(z) = O(n^{-1/2})$ yield

$$f(re^{i\theta}) - f(r) = \frac{r^2}{2} f''(r)(e^{i\theta} - 1)^2 + O\left(n^{-1/2}r^3\theta_0^3\right) = -r^2\theta^2 + O(n^{-1/5}).$$

This is enough to guarantee that

$$(36) \qquad J_n^{(0)} = \frac{e^{f(r)}}{2\pi} \int_{-\theta_0}^{+\theta_0} e^{-r^2\theta^2} d\theta \left(1 + O(n^{-1/5})\right).$$

   (iii) *Tails completion.* Since $r \sim \sqrt{n}$ and $\theta_0 = n^{-2/5}$, we have

$$(37) \qquad \int_{-\theta_0}^{+\theta_0} e^{-r^2\theta^2} d\theta = \frac{1}{r} \int_{-\theta_0 r}^{+\theta_0 r} e^{-t^2} dt = \frac{1}{r} \left(\int_{-\infty}^{+\infty} e^{-t^2} dt + O\left(e^{-n^{1/5}}\right)\right).$$

Finally, Equations (35), (36), and (37) give:

Proposition VIII.2. *The number I_n of involutions satisfies*

$$(38) \qquad \frac{I_n}{n!} = \frac{e^{-1/4}}{2\sqrt{\pi n}} n^{-n/2} e^{n/2+\sqrt{n}} \left(1 + O\left(\frac{1}{n^{1/5}}\right)\right).$$

   Comparing the saddle-point bound (34) to the true asymptotic form (38), we see that the former is only off by a factor of $O(n^{1/2})$. Here is a table further comparing the asymptotic estimate $I_n^\star$ provided by the right side of (38) to the exact value of I_n:

n	10	100	1000
I_n	9496	$2.40533 \cdot 10^{82}$	$2.14392 \cdot 10^{1296}$
$I_n^\star$	8839	$2.34149 \cdot 10^{82}$	$2.12473 \cdot 10^{1296}$.

The relative error is empirically close to $0.3/\sqrt{n}$, a fact that could be proved by developing a complete asymptotic expansion along the lines expounded in the previous section, p. 557.

   The estimate (38) of I_n is given by Knuth in [378]: his derivation is carried out by means of the Laplace method applied to the explicit binomial sum that expresses I_n. Our complex analytic derivation follows Moser and Wyman's in [448]. . . . . ∎

Example VIII.6.   *Set partitions and Bell numbers.* The number of partitions of a set of n elements defines the Bell number S_n (p. 109) and one has

$$S_n = n! e^{-1} [z^n] G(z) \qquad \text{where} \qquad G(z) = e^{e^z}.$$

The saddle-point equation relative to $G(z)z^{-n-1}$ (in z-coordinates) is

$$\zeta e^\zeta = n + 1.$$

This famous equation admits an asymptotic solution obtained by iteration (or "bootstrapping"): it suffices to write $\zeta = \log(n+1) - \log \zeta$, and iterate (say, starting from $\zeta = 1$), which provides the solution,

$$(39) \qquad \zeta \equiv \zeta(n) = \log n - \log \log n + \frac{\log \log n}{\log n} + O\left(\frac{\log^2 \log n}{\log^2 n}\right)$$

(see [143, p. 26] for a detailed discussion). The corresponding saddle-point bound reads

$$S_n \leq n! \frac{e^{e^\zeta - 1}}{\zeta^n}.$$

The approximate solution $\widehat{\zeta} = \log n$ yields in particular the simplified upper bound

$$S_n \leq n! \frac{e^{n-1}}{(\log n)^n}.$$

which is enough to check that there are much fewer set partitions than permutations, the ratio being bounded from above by a quantity $e^{-n \log \log n + O(n)}$.

In order to implement the saddle-point strategy, integration will be carried out over a circle of radius $r \equiv \zeta$. We then set

$$f(z) = \log\left(\frac{G(z)}{z^{n+1}}\right) = e^z - (n+1)\log z,$$

and proceed to estimate the integral,

$$J_n = \frac{1}{2i\pi} \int_{\mathcal{C}} G(z) \frac{dz}{z^{n+1}},$$

along the circle $\mathcal{C}$ of radius r. The usual saddle-point heuristic suggests that the range of the saddle-point is determined by a quantity $\theta_0 \equiv \theta_0(n)$ such that the quadratic terms in the expansion of f at r tend to infinity, while the cubic terms tend to zero. In order to carry out the calculations, it is convenient to express all quantities in terms of r alone, which is possible since n can be disposed of by means of the relation $n + 1 = re^r$. We find:

$$f''(r) = e^r(1+r^{-1}), \qquad f'''(r) = e^r(1 - 2r^2).$$

Thus, θ_0 should be chosen such that $r^2 e^r \theta_0^2 \to \infty$, $r^3 e^r \theta_0^3 \to 0$, and the choice $r\theta_0 = e^{-2r/5}$ is suitable.

(*i*) *Tails pruning.* First, observe that the function $G(z)$ is strongly concentrated near the real axis since, with $z = re^{i\theta}$, there holds

$$(40) \qquad \left|e^z\right| = e^{r\cos\theta}, \qquad \left|e^{e^z}\right| \leq e^{e^r \cos\theta}.$$

In particular $G(re^{i\theta})$ is exponentially smaller than $G(r)$ for any fixed $\theta \neq 0$, when r gets large.

(*ii*) *Central approximation.* One then considers the central contribution,

$$J_n^{(0)} := \frac{1}{2i\pi} \int_{\mathcal{C}^{(0)}} G(z) \frac{dz}{z^{n+1}},$$

where $\mathcal{C}^{(0)}$ is the part of the circle $z = re^{i\theta}$ such that $|\theta| \leq \theta_0 \equiv e^{-2r/5} r^{-1}$. Since on $\mathcal{C}^{(0)}$, the third derivative is uniformly $O(e^r)$, one has there

$$f(re^{i\theta}) = f(r) - \frac{1}{2} r^2 \theta^2 f''(r) + O(r^3 \theta^3 e^r).$$

This approximation can then be transported into the integral $J_n^{(0)}$.

(*iii*) *Tails completion.* Tails can be completed in the usual way. The net effect is the estimate

$$[z^n]G(z) = \frac{e^{f(r)}}{\sqrt{2\pi f''(r)}}\left(1 + O\left(r^3\theta^3 e^r\right)\right),$$

which, upon making the error term explicit rephrases, as follows.

Proposition VIII.3. *The number S_n of set partitions of size n satisfies*

$$(41) \qquad\qquad S_n = n!\,\frac{e^{e^r-1}}{r^n\sqrt{2\pi r(r+1)e^r}}\left(1 + O(e^{-r/5})\right),$$

where r is defined implicitly by $re^r = n + 1$, so that $r = \log n - \log\log n + o(1)$.

Here is a numerical table of the exact values S_n compared to the main term $S_n^\star$ of the approximation (41):

n	10	100	1000
S_n	115975	$4.75853 \cdot 10^{115}$	$2.98990 \cdot 10^{1927}$
$S_n^\star$	114204	$4.75537 \cdot 10^{115}$	$2.99012 \cdot 10^{1927}$

The error is about 1.5% for $n = 10$, less than 10^{-3} and 10^{-4} for $n = 100$ and $n = 1000$.

The asymptotic form in terms of r itself is the proper one as no back substitution of an asymptotic expansion of r (in terms of n and $\log n$) can provide an asymptotic expansion for S_n solely in terms of n. Regarding explicit representations in terms of n, it is only $\log S_n$ that can be expanded as

$$\frac{1}{n}\log S_n = \log n - \log\log n - 1 + \frac{\log\log n}{\log n} + \frac{1}{\log n} + O\left(\left(\frac{\log\log n}{\log n}\right)^2\right).$$

(Saddle-point estimates of coefficient integrals often involve such implicitly defined quantities.)

This example probably constitutes the most famous application of saddle-point techniques to combinatorial enumeration. The first correct treatment by means of the saddle-point method is due to Moser and Wyman [447]. It is used for instance by de Bruijn in [143, pp. 104–108] as a lead example of the method. .... ∎

Example VIII.7. *Fragmented permutations.* These correspond to $F(z) = \exp(z/(1 - z))$. The example now illustrates the case of a singularity at a finite distance. We set as usual

$$f(z) = \frac{z}{1 - z} - (n + 1)\log z,$$

and start with saddle-point bounds. The saddle-point equation is

$$(42) \qquad\qquad \frac{\zeta}{(1-\zeta)^2} = n + 1,$$

so that ζ comes close to the singularity at 1 as n gets large:

$$\zeta = \frac{2n + 3 - \sqrt{4n + 5}}{2n + 2} = 1 - \frac{1}{\sqrt{n}} + \frac{1}{2n} + O(n^{-3/2}).$$

Here, the approximation $\widehat{\zeta}(n) = 1 - 1/\sqrt{n}$, leads to

$$(43) \qquad\qquad [z^n]F(z) \le e^{-1/2}e^{2\sqrt{n}}(1 + o(1)).$$

The saddle-point method is then applied with integration along a circle of radius $r \equiv \zeta$. The saddle-point heuristic suggests to restrict the integral to a small sector of angle $2\theta_0$, and, since $f''(r) = O(n^{3/2})$ while $f'''(r) = O(n^2)$, this means taking θ_0 such that $n^{3/4}\theta_0 \to \infty$

and $n^{2/3}\theta_0 \to 0$. For instance, the choice $\theta_0 = n^{-7/10}$ is suitable. Concentration is easily verified: we have

$$\left| e^{1/(1-z)} \right|_{z=re^{i\theta}} = e \cdot \exp\left(\frac{1 - r\cos\theta}{1 - 2r\cos\theta + r^2} \right),$$

which is a unimodal function of θ for $\theta \in (-\pi, \pi)$. (The maximum of this function of θ is of order $\exp((1-r)^{-1})$ and is attained at $\theta = 0$; the minimum is $O(1)$, attained at $\theta = \pi$.) In particular, along the non-central part $|\theta| \geq \theta_0$ of the saddle-point circle, one has

(44)
$$\left| e^{1/(1-z)} \right|_{z=re^{i\theta}} = O\left(\exp\left(\sqrt{n} - n^{1/10}\right)\right),$$

so that tails are exponentially small. Local expansions then enable us to justify the use of the general saddle-point formula in this case. The net result is the following.

Proposition VIII.4. *The number of fragmented permutations, $F_n = n![z^n]F(z)$, satisfies*

(45)
$$\frac{F_n}{n!} \sim \frac{e^{-1/2}e^{2\sqrt{n}}}{2\sqrt{\pi}n^{3/4}}.$$

Quite characteristically, the corresponding saddle-point bound (43) turns out to be off the asymptotic estimate (45) only by a factor of order $n^{3/4}$. The relative error of the approximation (45) is about 4%, 1%, 0.3% for $n = 10, 100, 1000$, respectively.

The expansion above has been extended by E. Maitland Wright [618, 619] to several classes of functions with a singularity whose type is an exponential of a function of the form $(1-z)^{-\rho}$; see Note VIII.7. (For the case of (45), Wright [618] refers to an earlier article of Perron published in 1914.) His interest was due, at least partly, to applications to generalized partition asymptotics, of which the basic cases are discussed in Section VIII. 6, p. 574. .... ∎

▷ **VIII.7.** *Wright's expansions.* Consider the function

$$F(z) = (1-z)^{-\beta} \exp\left(\frac{A}{(1-z)^\rho} \right), \qquad A > 0, \quad \rho > 0.$$

Then, a saddle-point analysis yields, when $\rho < 1$:

$$[z^n]F(z) \sim \frac{N^{\beta-1-\rho/2} \exp\left(A(\rho+1)N^\rho\right)}{\sqrt{2\pi A\rho(\rho+1)}}, \qquad N := \left(\frac{n}{A\rho}\right)^{\frac{1}{\rho+1}}.$$

(The case $\rho \geq 1$ involves more terms of the asymptotic expansion of the saddle-point.) The method generalizes to analytic and logarithmic multipliers, as well as to a sum of terms of the form $A(1-z)^{-\rho}$ inside the exponential. See [618, 619] for details. ◁

▷ **VIII.8.** *Some oscillating coefficients.* Define the function

$$s(z) = \sin\left(\frac{z}{1-z} \right).$$

The coefficients $s_n = [z^n]s(z)$ are seen to change sign at $n = 6, 21, 46, 81, 125, 180, \dots$. Do signs change infinitely many times? (Hint: Yes. there are two complex conjugate saddle-points and the associated asymptotic forms combine a growth of the type $n^a e^{b\sqrt{n}}$ with an oscillating factor similar to $\sin\sqrt{n}$.) The sum

$$U_n = \sum_{k=0}^{n} \binom{n}{k} \frac{(-1)^k}{k!}$$

exhibits similar fluctuations. ◁

VIII. 5. Admissibility

The saddle-point method is a versatile approach to the analysis of coefficients of fast-growing generating functions, but one which is often cumbersome to apply step-by-step. Fortunately, it proves possible to encapsulate the conditions repeatedly encountered in our previous examples into a general framework. This leads to the notion of an *admissible function* presented in Subsection VIII. 5.1. By design, saddle-point analysis applies to such functions and asymptotic forms for their coefficients can be systematically determined: this follows an approach initiated by Hayman in 1956. A great merit of abstraction in this context is that admissible functions satisfy useful closure properties, so that an *infinite class* of admissible functions of relevance to combinatorial applications can be determined—we develop this theme in Subsection VIII. 5.2, relative to enumeration. Finally, Subsection VIII. 5.3 presents an approach to the probabilistic problem known as depoissonization, which is much akin to admissibility.

VIII. 5.1. Admissibility theory. The notion of admissibility is in essence an axiomatization of the conditions underlying Theorem VIII.3 particularized to the case of Cauchy coefficient integrals. In this section, we base our discussion on *H–admissibility*, the prefix *H* being a token of Hayman's original contribution [325]. A crisp account of the theory is given in Section II.7 of Wong's book [614] and in Odlyzko's authoritative survey [461, Sec. 12].

We consider here a function $G(z)$ that is analytic at the origin and whose coefficients $[z^n]G(z)$ are to be estimated by

$$g_n \equiv [z^n]G(z) = \frac{1}{2i\pi} \int_C G(z) \frac{dz}{z^{n+1}}.$$

The switch to polar coordinates is natural, so that the expansion of $G(re^{i\theta})$ for small θ plays a central rôle: with r a positive real number lying within the disc of analyticity of $G(z)$, the fundamental expansion is

$$(46) \qquad \log G(re^{i\theta}) = \log G(r) + \sum_{\nu=1}^{\infty} \alpha_\nu(r) \frac{(i\theta)^\nu}{\nu!}.$$

Not surprisingly, the most important quantities are the first two terms, and once $G(z)$ has been put into exponential form, $G(z) = e^{h(z)}$, a simple computation yields

$$(47) \quad \begin{cases} a(r) & := & \alpha_1(r) & = & rh'(r) \\ b(r) & := & \alpha_2(r) & = & r^2h''(r) + rh'(r), \end{cases} \quad \text{with} \quad h(z) := \log G(z).$$

In terms of G, itself, one thus has

$$(48) \qquad a(r) = r\frac{G'(r)}{G(r)}, \qquad b(r) = r\frac{G'(r)}{G(r)} + r^2\frac{G''(r)}{G(r)} - r^2\left(\frac{G'(r)}{G(r)}\right)^2.$$

Whenever $G(z)$ has non-negative Taylor coefficients at the origin, $b(r)$ is positive for $r > 0$ and $a(r)$ increases as $r \to \rho$, with ρ the radius of convergence of G. (This follows from the argument developed in Note VIII.4, p. 550.)

Definition VIII.1 (Hayman–admissibility). *Let $G(z)$ have radius of convergence ρ with $0 < \rho \le +\infty$ and be always positive on some subinterval (R_0, ρ) of $(0, \rho)$. The function $G(z)$ is said to be H–admissible (Hayman admissible) if, with $a(r)$ and $b(r)$ as defined in (47), it satisfies the following three conditions:*

$\mathbf{H_1}$. *[Capture condition]* $\lim\limits_{r \to \rho} a(r) = +\infty$ *and* $\lim\limits_{r \to \rho} b(r) = +\infty$.

$\mathbf{H_2}$. *[Locality condition] For some function $\theta_0(r)$ defined over (R_0, ρ) and satisfying $0 < \theta_0 < \pi$, one has*

$$G(re^{i\theta}) \sim G(r)e^{i\theta a(r) - \theta^2 b(r)/2} \qquad \text{as } r \to \rho,$$

uniformly in $|\theta| \le \theta_0(r)$.

$\mathbf{H_3}$. *[Decay condition] Uniformly in $\theta_0(r) \le |\theta| < \pi$*

$$G(re^{i\theta}) = o\left(\frac{G(r)}{\sqrt{b(r)}} \right).$$

Note that the conditions in the definition are *intrinsic* to the function: they only make reference to *the function's values along circles*, no parameter n being involved yet. It can be easily verified, from the previous examples, that the functions e^z, $e^{e^z - 1}$, and $e^{z + z^2/2}$ are admissible with $\rho = +\infty$, and that the function $e^{z/(1-z)}$ is admissible with $\rho = 1$ (refer in each case to the discussion of the behaviour of the modulus of $G(re^{i\theta})$, as θ varies). By contrast, functions such as e^{z^2} and $e^{z^2} + e^z$ are *not* admissible since they attain values that are too large when $\arg(z)$ is near π.

Coefficients of H–admissible functions can be systematically analysed to first asymptotic order, as expressed by the following theorem:

Theorem VIII.4 (Coefficients of admissible functions). *Let $G(z)$ be an H–admissible function and $\zeta \equiv \zeta(n)$ be the unique solution in the interval (R_0, ρ) of the equation*

$$(49) \qquad \zeta \frac{G'(\zeta)}{G(\zeta)} = n.$$

The Taylor coefficients of $G(z)$ satisfy, as $n \to \infty$:

$$(50) \quad g_n \equiv [z^n]G(z) \sim \frac{G(\zeta)}{\zeta^n \sqrt{2\pi b(\zeta)}}, \qquad b(z) := z^2 \frac{d^2}{dz^2} \log G(z) + z \frac{d}{dz} \log G(z).$$

Proof. The proof simply amounts to transcribing the definition of admissibility into the conditions of Theorem VIII.3. Integration is carried out over a circle centred at the origin, of some radius r to be specified shortly. Under the change of variable $z = re^{i\theta}$, the Cauchy coefficient formula becomes

$$(51) \qquad g_n \equiv [z^n]G(z) = \frac{r^{-n}}{2\pi} \int_{-\pi}^{+\pi} G(re^{i\theta}) e^{-ni\theta} \, d\theta.$$

In order to obtain a quadratic approximation *without* a linear term, one chooses the radius of the circle as the positive solution ζ of the equation $a(\zeta) = n$, that is, a solution of Equation (49). (Thus ζ is a saddle-point of $G(z)z^{-n}$.) By the capture condition $\mathbf{H_1}$, we have $\zeta \to \rho^-$ as $n \to +\infty$. Following the general saddle-point strategy, we decompose the integration domain and set, with θ_0 as specified in conditions $\mathbf{H_2}$ and $\mathbf{H_3}$:

$$J^{(0)} = \int_{-\theta_0}^{+\theta_0} G(\zeta e^{i\theta}) e^{-ni\theta} \, d\theta, \qquad J^{(1)} = \int_{\theta_0}^{2\pi - \theta_0} G(\zeta e^{i\theta}) e^{-ni\theta} \, d\theta.$$

(*i*) *Tails pruning.* By the decay condition $\mathbf{H_3}$, we have a trivial bound, which suffices for our purposes:

$$(52) \qquad J^{(1)} = o\left(\frac{G(\zeta)}{\sqrt{b(\zeta)}}\right).$$

(*ii*) *Central approximation.* The uniformity of the locality condition $\mathbf{H_2}$ implies

$$(53) \qquad J^{(0)} \sim G(\zeta) \int_{-\theta_0}^{+\theta_0} e^{-\theta^2 b(\zeta)/2} \, d\theta.$$

(*iii*) *Tails completion.* A combination of the locality condition $\mathbf{H_2}$ and the decay condition $\mathbf{H_3}$ instantiated at $\theta = \theta_0$, shows that $b(\zeta)\theta^2 \to +\infty$ as $n \to +\infty$. There results that tails can be completed back, and

$$(54) \quad \int_{-\theta_0}^{+\theta_0} e^{-b(r)\theta^2/2} \, d\theta \sim \frac{1}{\sqrt{b(r)}} \int_{-\theta_0/\sqrt{b(\zeta)}}^{+\theta_0/\sqrt{b(\zeta)}} e^{-t^2/2} \, dt \sim \frac{1}{\sqrt{b(r)}} \int_{-\infty}^{+\infty} e^{-t^2/2} \, dt.$$

From (52), (53), and (54) (or equivalently via an application of Theorem VIII.3), the conclusion of the theorem follows. ∎

The usual comments regarding the choice of the function $\theta_0(r)$ apply. Considering the expansion (46), we must have $\alpha_2(r)\theta_0^2 \to \infty$ and $\alpha_3(r)\theta_0^3 \to 0$. Thus, in order to succeed, the method necessitates *a priori* $\alpha_3(r)^2/\alpha_2(r)^3 \to 0$. Then, θ_0 should be taken according to the *saddle-point dimensioning heuristic*, which can be figuratively summarized as[6]

$$(55) \qquad \frac{1}{\alpha_2(r)^{1/2}} \ll \theta_0 \ll \frac{1}{\alpha_3(r)^{1/3}},$$

a possible choice being the geometric mean of the two bounds $\theta_0 = \alpha_2^{-1/4}\alpha_3^{-1/6}$.

The original proof by Hayman [325] contains in addition a general result that describes the *shape of the individual terms* $g_n r^n$ in the Taylor expansion of $G(r)$ as r gets closer to its limit value ρ: these appear to exhibit a bell-shaped profile. Precisely, for G with non-negative coefficients, define a family of discrete random variables $X(r)$ indexed by $r \in (0, R)$ as follows:

$$\mathbb{P}(X(r) = n) = \frac{g_n r^n}{G(r)}.$$

The model in which a random $\mathcal{F}$ structure with GF $G(z)$ is drawn with its size being the random value $X(r)$ is known as a *Boltzmann model*. Then:

---

[6]We occasionally write $A \ll B$, equivalently, $B \gg A$, if $A = o(B)$.

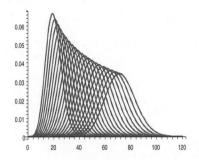 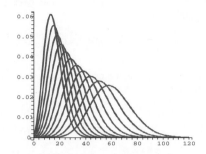

Figure VIII.7. The families of Boltzmann distributions associated with involutions, $G(z) = e^{z+z^2/2}$ with $r = 4..8$, and set partitions, $G(z) = e^{e^z-1}$ with $r = 2..3$, obey an approximate Gaussian profile.

Proposition VIII.5. *The Boltzmann probabilities associated to an admissible function* $G(z)$ *satisfy, as* $r \to \rho^-$*, a "local" Gaussian estimate; namely,*

$$(56) \qquad \frac{g_n r^n}{G(r)} = \frac{1}{\sqrt{2\pi b(r)}} \left[\exp\left(-\frac{(a(r) - n)^2}{2b(r)} \right) + \epsilon_n \right],$$

where the error term satisfies $\epsilon_n = o(1)$ *as* $r \to \rho$ *uniformly with respect to integers* n*; that is,* $\lim_{r \to \rho} \sup_n |\epsilon_n| = 0$*.*

The proof is entirely similar to that of Theorem VIII.4; see Note VIII.9 and Figure VIII.7 for a suggestive illustration.

$\triangleright$ **VIII.9.** *Admissibility and Boltzmann models.* The Boltzmann distribution is accessible from

$$g_n r^n = \frac{1}{2\pi} \int_{-\theta_0}^{2\pi - \theta_0} G(re^{i\theta}) e^{-in\theta} \, d\theta.$$

The estimation of this integral is once more based on a fundamental split

$$g_n r^n = J^{(0)} + J^{(1)} \qquad \text{where} \qquad J^{(0)} = \frac{1}{2\pi} \int_{-\theta_0}^{+\theta_0}, \quad J^{(1)} = \frac{1}{2\pi} \int_{+\theta_0}^{2\pi - \theta_0},$$

and $\theta_0 = \theta_0(n)$ is as specified by the admissibility definition. Only the central approximation and tails completion deserves adjustments. The "locality" condition $\mathbf{H_2}$ gives uniformly in n,

$$(57) \qquad \begin{aligned} J^{(0)} &= \frac{G(r)}{2\pi} \int_{-\theta_0}^{+\theta_0} e^{i(a(r)-n)\theta - \frac{1}{2}b(r)\theta^2} (1 + o(1)) \, d\theta \\ &= \frac{G(r)}{2\pi} \left[\int_{-\theta_0}^{+\theta_0} e^{i(a(r)-n)\theta - \frac{1}{2}b(r)\theta^2} \, d\theta + o\left(\int_{-\infty}^{+\infty} e^{-\frac{1}{2}b(r)\theta^2} \right) d\theta \right]. \end{aligned}$$

and setting $(a(r) - n)(2/b(r))^{1/2} = c$, we obtain

$$(58) \qquad J^{(0)} = \frac{G(r)}{\pi \sqrt{2b(r)}} \left[\int_{-\theta_0\sqrt{b(r)/2}}^{+\theta_0\sqrt{b(r)/2}} e^{-t^2+ict} \, dt + o(1) \right].$$

The integral in (58) can then be routinely extended to a complete Gaussian integral, introducing only $o(1)$ error terms,

$$(59) \qquad J^{(0)} = \frac{G(r)}{\pi\sqrt{2b(r)}} \left[\int_{-\infty}^{+\infty} e^{-t^2 + ict} \, dt + o(1) \right].$$

Finally, the Gaussian integral evaluates to $\sqrt{\pi} e^{-c^2/4}$, as is seen by *completing the square* and *shifting vertically the integration line*.   ◁

▷ **VIII.10.** *Hayman's original.* The condition $\mathbf{H_1}$ of Theorem VIII.4 can be replaced by

  $\mathbf{H'_1}$. [Capture condition] $\lim_{r \to \rho} b(r) = +\infty$.

That is, $a(r) \to +\infty$ is a consequence of $\mathbf{H'_1}, \mathbf{H_2}$, and $\mathbf{H_3}$. (See [325, §5].)   ◁

▷ **VIII.11.** *Non-admissible functions.* Singularity analysis and H–admissibility conditions are in a sense complementary. Indeed, the function $G(z) = (1 - z)^{-1}$ fails to be be admissible as the asymptotic form that Theorem VIII.4 would imply is the erroneous $[z^n] \dfrac{1}{1 - z} \overset{!!}{\sim} \dfrac{e}{\sqrt{2\pi}}$, corresponding to a saddle-point near $1 - n^{-1}$. The explanation of the discrepancy is as follows: Expansion (46) has $\alpha_\nu(r)$ of the order of $(1 - r)^{-\nu}$, so that the locality condition and the decay condition cannot be simultaneously satisfied.

Singularity analysis salvages the situation by using a larger contour and by normalizing to a global Hankel Gamma integral instead of a more "local" Gaussian integral. This is also in accordance with the fact that the saddle-point formula gives, in the case of $[z^n](1 - z)^{-1}$, an estimate, which is within a constant factor of the true value 1. (More generally, functions of the form $(1 - z)^{-\beta}$ are typical instances with too slow a growth to be admissible.)   ◁

Closure properties. An important aspect of Hayman's work is that it leads to general theorems, which guarantee that large classes of functions are admissible.

Theorem VIII.5 (Closure of H–admissible functions). *Let $G(z)$ and $H(z)$ be admissible functions and let $P(z)$ be a polynomial with real coefficients. Then:*

  (i) The product $G(z)H(z)$ and the exponential $e^{G(z)}$ are admissible functions.

  (ii) The sum $G(z) + P(z)$ is admissible. If the leading coefficient of $P(z)$ is positive then $G(z)P(z)$ and $P(G(z))$ are admissible.

  (iii) If the Taylor coefficients of $e^{P(z)}$ are eventually positive, then $e^{P(z)}$ is admissible.

Proof. (Sketch) The easy proofs essentially reduce to making an inspired guess for the choice of the θ_0 function, which may be guided by Equation (55) in the usual way, and then routinely checking the conditions of the admissibility definition. For instance, in the case of the exponential, $K(z) = e^{G(z)}$, the conditions $\mathbf{H_1}, \mathbf{H_2}, \mathbf{H_3}$ of Definition VIII.1 are satisfied if one takes $\theta_0(r) = (G(r))^{-2/5}$. We refer to Hayman's original paper [325] for details.   ∎

Exponentials of polynomials. The closure theorem also implies as a very special case that any GF of the form $e^{P(z)}$ with $P(z)$ a polynomial with positive coefficients can be subjected to saddle-point analysis, a fact first noted by Moser and Wyman [449, 450].

Corollary VIII.2 (Exponentials of polynomials). *Let $P(z) = \sum_{j=1}^{m} a_j z^j$ have non-negative coefficients and be aperiodic in the sense that $\gcd\{j \mid a_j \neq 0\} = 1$. Let*

$f(z) = e^{P(z)}$. *Then, one has*

$$f_n \equiv [z^n] f(z) \sim \frac{1}{\sqrt{2\pi\lambda}} \frac{e^{P(r)}}{r^n}, \qquad \text{where} \quad \lambda = \left(r \frac{d}{dr}\right)^2 P(r),$$

and r is a function of n given implicitly by $r\frac{d}{dr}P(r) = n$.

The computations are in this case purely mechanical, since they only involve the asymptotic expansion (with respect to n) of an algebraic equation.

Granted the basic admissibility theorem and closures properties, many functions are immediately seen to be admissible, including

$$e^z, \qquad e^{e^z - 1}, \qquad e^{z + z^2/2},$$

which have previously served as lead examples for illustrating the saddle-point method. Corollary VIII.2 also covers involutions, permutations of a fixed order in the symmetric group, permutations with cycles of bounded length, as well as set partitions with bounded block sizes: see Note VIII.12 below. More generally, Corollary VIII.2 applies to any labelled set construction, $\mathcal{F} = \text{SET}(\mathcal{G})$, when the sizes of $\mathcal{G}$–components are restricted to a finite set, in which case one has

$$\mathcal{F}^{[m]} = \text{SET}\left(\cup_{j=1}^r \mathcal{G}_j\right), \qquad \Longrightarrow \qquad F^{[m]}(z) = \exp\left(\sum_{j=1}^m G_j \frac{z^j}{j!}\right).$$

This covers all sorts of graphs (plain or functional) whose connected components are of bounded size.

▷ **VIII.12.** *Applications of "exponentials of polynomials".* Corollary VIII.2 applies to the following combinatorial situations:

Permutations of order p ($\sigma^p = 1$)	$f(z) = \exp\left(\sum_{j \mid p} \frac{z^j}{j}\right)$
Permutations with longest cycle $\leq p$	$f(z) = \exp\left(\sum_{j=1}^p \frac{z^j}{j}\right)$
Partitions of sets with largest block $\leq p$	$f(z) = \exp\left(\sum_{j=1}^p \frac{z^j}{j!}\right).$

For instance, the number of solutions of $\sigma^p = 1$ in the symmetric group is asymptotic to

$$\left(\frac{n}{e}\right)^{n(1-1/p)} p^{-1/2} \exp(n^{1/p}),$$

for any fixed prime $p \geq 3$ (Moser and Wyman [449, 450]). ◁

Complete asymptotic expansions. Harris and Schoenfeld have introduced in [323] a technical condition of admissibility that is stronger than Hayman admissibility and is called HS–admissibility. Under such HS–admissibility, a complete asymptotic expansion can be obtained. We omit the definition here due to its technical character but refer instead to the original paper [323] and to Odlyzko's survey [461]. Odlyzko and Richmond [462] later showed that, if $g(z)$ is H–admissible, then $f(z) = e^{g(z)}$ is HS–admissible. Thus, taking H–admissibility to mean at least exponential growth, *full asymptotic expansions are to be systematically expected at double exponential growth and beyond.* The principles of developing full asymptotic expansions are essentially the same as the ones explained on p. 557—only the discussion of the asymptotic scales involved becomes a bit intricate, at this level of generality.

VIII. 5.2. Higher-level structures and admissibility. The concept of admissibility and its surrounding properties (Theorems VIII.4 and VIII.5, Corollary VIII.2) afford a neat discussion of which combinatorial classes should lead to counting sequences that are amenable to the saddle-point method. For simplicity, we restrict ourselves here to the labelled universe.

Start from the *first-level structures*, namely

$$\text{SEQ}(\mathcal{Z}), \qquad \text{CYC}(\mathcal{Z}), \qquad \text{SET}(\mathcal{Z}),$$

corresponding, respectively, to permutations, circular graphs, and urns, with EGFs

$$\frac{1}{1-z}, \qquad \log\frac{1}{1-z}, \qquad e^z.$$

The first two are of singularity analysis class; the last is, as we saw, within the reach of the saddle-point method and is H–admissible.

Next consider *second-level structures* defined by arbitrary composition of two constructions taken among SEQ, CYC, SET; see Subsection II. 4.2, p. 124 for a preliminary discussion (In the case of the internal construction, it is understood that, for definiteness, the number of components is constrained to be ≥ 1.) There are three structures whose external construction is of the sequence type, namely,

$$\text{SEQ} \circ \text{SEQ}, \qquad \text{SEQ} \circ \text{CYC}, \qquad \text{SEQ} \circ \text{SET},$$

corresponding, respectively, to labelled compositions, alignments, and surjections. All three have a dominant singularity that is a pole; hence they are amenable to meromorphic coefficient asymptotics (Chapters IV and V), or, with weaker remainder estimates, to singularity analysis (Chapters VI and VII).

Similarly there are three structures whose external construction is of the cycle type, namely,

$$\text{CYC} \circ \text{SEQ}, \qquad \text{CYC} \circ \text{CYC}, \qquad \text{CYC} \circ \text{SET},$$

corresponding to cyclic versions of the previous ones. In that case, the EGFs have a logarithmic singularity; hence they are amenable to singularity analysis, or, after differentiation, to meromorphic coefficient asymptotics again.

The case of an external set construction is of interest. It gives rise to

$$\text{SET} \circ \text{SEQ}, \qquad \text{SET} \circ \text{CYC}, \qquad \text{SET} \circ \text{SET},$$

corresponding, respectively, to fragmented permutations, the class of all permutations, and set partitions. The composition $\text{SET} \circ \text{CYC}$ appears to be special, because of the general isomorphism, valid for any class $\mathcal{C}$,

$$\text{SET}(\text{CYC}(\mathcal{C})) \cong \text{SEQ}(\mathcal{C}),$$

corresponding to the unicity of the decomposition of a permutation of $\mathcal{C}$–objects into cycles. Accordingly, for generating functions, an exponential singularity "simplifies", when combined with a logarithmic singularity, giving rise to an algebraic (here polar) singularity. The remaining two cases, namely, fragmented permutations and set partitions, characteristically come under the saddle-point method and admissibility, as we have seen already.

Closure properties then make it possible to consider structures defined by an arbitrary nesting of the constructions in {SEQ, CYC, SET}. For instance, "superpartitions" defined by

$$S = \text{SET}(\text{SET}_{\geq 1}(\text{SET}_{\geq 1}(\mathcal{Z}))), \qquad \Longrightarrow \qquad S(z) = e^{e^{e^z - 1} - 1},$$

are *third-level structures*. They can be subjected to admissibility theory and saddle-point estimates apply *a priori*. Notes VIII.14 and VIII.15 further examine such third-level structures.

▷ **VIII.13.** *Idempotent mappings.* Consider functions from a finite set to itself ("mappings" or "functional graphs" in the terminology of Chapter II) that are idempotent, i.e., $\phi \circ \phi = \phi$. The EGF is $I(z) = \exp(ze^z)$ since cycles are constrained to have length 1 exactly. The function $I(z)$ is admissible and

$$I_n \sim \frac{n!}{\sqrt{2\pi n\zeta}} \zeta^{-n} e^{(n+1)/(\zeta+1)},$$

where ζ is the positive solution of $\zeta(\zeta + 1)e^\zeta = n + 1$. This example is discussed by Harris and Schoenfeld in [323]. ◁

▷ **VIII.14.** *The number of societies.* A society on n distinguished individuals is defined by Sloane and Wieder [545] as follows: first partition the n individuals into non-empty subsets and then form an ordered set partition [preferential arrangement] into each subset. The class of societies is thus a third-level (labelled) structure, with specification and EGF

$$S = \text{SET}\left(\text{SEQ}_{\geq 1}(\text{SET}_{\geq 1}(\mathcal{Z}))\right) \qquad \Longrightarrow \qquad S(z) = \exp\left(\frac{1}{2 - e^z} - 1\right).$$

The counting sequence starts as 1, 1, 4, 23, 173, 1602 (*EIS* **75729**); asymptotically

$$S_n \sim C \frac{e^{\sqrt{2n/\log 2}}}{n^{3/4}(\log 2)^{n+1/4}} n!, \qquad C := \frac{1}{4\sqrt{\pi}} \left(\frac{2}{e}\right)^{3/4} e^{1/(4\log 2)}.$$

(The singularity is of the type "exponential-of-pole" at $z = \log 2$.) ◁

▷ **VIII.15.** *Third-level classes.* Consider labelled classes defined from atoms ($\mathcal{Z}$) by *three* nested constructions, each either a sequence or a set. All cases can be analysed, either by saddle-point and admissibility or by singularity analysis. Here is a table recapitulating structures, together with their EGF and radius of convergence (ρ):

Saddle-point:	$\text{SET}(\text{SET}_{\geq 1}(\text{SET}_{\geq 1}(\mathcal{Z})))$	$e^{e^{e^z - 1} - 1}$	$\rho = \infty$
	$\text{SET}(\text{SET}_{\geq 1}(\text{SEQ}_{\geq 1}(\mathcal{Z})))$	$e^{e^{z/(1-z)} - 1}$	$\rho = 1$
	$\text{SET}(\text{SEQ}_{\geq 1}(\text{SET}_{\geq 1}(\mathcal{Z})))$	$\exp(\frac{e^z - 1}{2 - e^z})$	$\rho = \log 2$
	$\text{SET}(\text{SEQ}_{\geq 1}(\text{SEQ}_{\geq 1}(\mathcal{Z})))$	$e^{z/(1-2z)}$	$\rho = \frac{1}{2}$;
Singularity analysis:	$\text{SEQ}(\text{SET}_{\geq 1}(\text{SET}_{\geq 1}(\mathcal{Z})))$	$\dfrac{1}{2 - e^{e^z - 1}}$	$\rho = \log\log(2e)$
	$\text{SEQ}(\text{SET}_{\geq 1}(\text{SEQ}_{\geq 1}(\mathcal{Z})))$	$\dfrac{1}{2 - e^{z/(1-z)}}$	$\rho = \dfrac{\log 2}{1+\log 2}$
	$\text{SEQ}(\text{SEQ}_{\geq 1}(\text{SET}_{\geq 1}(\mathcal{Z})))$	$\dfrac{2 - e^z}{3 - 2e^z}$	$\rho = \log \frac{3}{2}$
	$\text{SEQ}(\text{SEQ}_{\geq 1}(\text{SEQ}_{\geq 1}(\mathcal{Z})))$	$\dfrac{1 - 2z}{1 - 3z}$	$\rho = \frac{1}{3}$.

The outermost construction dictates the analytic type and precise asymptotic equivalents can be developed in all cases. ◁

▷ **VIII.16.** *A Multiple Choice Questionnaire.* Classify all the 27 third-level structures built out of {SEQ, CYC, SET}, according to whether they are of type SA (singularity analysis) or SP (saddle-point). ◁

▷ **VIII.17.** *A meta-MCQ.* Among the 3^n specifications of level n, what is the asymptotic proportion of those that are of type SP? ◁

VIII. 5.3. Analytic depoissonization. We conclude this section on methodology with a sketch of an approach to the analysis of exponential generating functions, which has been termed *analytic depoissonization*, by its proponents, Jacquet and Szpankowski [346, 564]. This approach, which is based on the saddle-point method, has affinities with admissibility theory and it plays a rôle in the investigation of several important models of discrete mathematics.

The *Poisson generating function* of a sequence (a_n) is defined as

$$\alpha(z) = \sum_{n \geq 0} a_n e^{-z} \frac{z^n}{n!}.$$

It is thus a simple variant of the EGF (multiply by e^{-z}) and, when z assumes a non-negative real value λ, it can be viewed as a sum of the a_n, weighted by the Poisson probabilities $\{e^{-\lambda}\lambda^n/n!\}$. Since the Poisson distribution is concentrated around its mean value λ, it is reasonable to expect an approximation

(60) $$\alpha(\lambda) \sim a_{\lfloor \lambda \rfloor} \qquad (\lambda \to \infty)$$

to be valid, provided a_n, assumed to be known, varies sufficiently "regularly". A statement granting us the correctness of (60), based on *a priori* knowledge of the a_n, is an Abelian theorem, in the usual sense of analysis (see Section VI. 11, p. 433, and e.g., [69, §1.7]); it is easily established using the Laplace method for sums (p. 755), upon appealing to a Gaussian approximation of Poisson laws of large rate λ (Note IX.19, p. 643).

What is of interest here is the converse (Tauberian) problem: we seek ways of translating information on the Poisson generating function $\alpha(z)$ into an asymptotic expansion of the coefficients (a_n). Beyond being fully in the spirit of the book (especially, Chapters VI and VII), this situation is of interest, since it is encountered in many probabilistic contexts where a Poisson model intervenes. In this subsection, we stand on the shoulders of Jacquet and Szpankowski [346, 564], who developed a whole theory.

A *sector* S_ϕ, with $\phi \in \mathbb{R}$, is defined to be $S_\phi = \{z : |\arg(z)| \leq \phi\}$. A function $f(z)$ is said to be *small, away from the positive real axis,* if, for some $A > 0$ and $\phi \in (0, \pi/2)$, one has

$$\left| e^z f(z) \right| = O\left(e^{-A|z|} \right), \qquad \text{as} \quad |z| \to \infty, \quad z \notin S_\phi.$$

We have [564, Th. 10.6]:

Theorem VIII.6 (Analytic depoissonization). *Let the Poisson generating function $\alpha(z)$ be small, away from the positive real axis, with sector S_ϕ. Then one has the following* *correspondence between properties of the individual terms in the expansion of $\alpha(z)$ within S_ϕ and asymptotic terms in the expansion of the coefficient a_n:*

$\alpha(z)$	a_n
$O\big(\lvert z\rvert^B \lvert \log(z)\rvert^C\big) \;\longrightarrow$	$O\left(n^B (\log n)^C\right)$
$z^b \qquad\qquad \longrightarrow$	$\sim n^b\left[1 - \dfrac{b(b-1)}{2n} + \dfrac{b(b-1)(b-2)(3b-1)}{24n^2} - \cdots\right]$
$z^b (\log z)^r \qquad \longrightarrow$	$\sim \dfrac{\partial^r}{\partial b^r}\left(n^b\left[1 - \dfrac{b(b-1)}{2n} + \cdots\right]\right).$

Proof. (Sketch) Given the assumptions, we regard $e^z \alpha(z)$ as a variant of the exponential function, to which the saddle-point method is known to be applicable: see the derivation of Example VIII.3 (p. 555), which we closely follow. Accordingly, we appeal to Cauchy's formula,

$$a_n = \frac{n!}{2i\pi} \int_{\lvert z\rvert = n} e^z \alpha(z) \frac{dz}{z^{n+1}},$$

and integrate along the circle $\lvert z\rvert = n$. The smallness condition on $\alpha(z)$ ensures that the integral outside of S_ϕ is exponentially negligible. Setting $z = ne^{i\theta}$, we see that, inside S_ϕ, we can neglect the part corresponding to $\lvert \theta\rvert \geq \theta_0(n) \equiv n^{-2/5}$, since it is again exponentially small. Then, for the central part of the contour,

$$a_n^{(0)} := \frac{n!\, n^{-n} e^n}{2\pi \sqrt{n}} \int_{-\theta_0}^{\theta_0} e^{-n\theta^2/2} \exp\left(n\left[e^{i\theta} - 1 - i\theta + \frac{1}{2}\theta^2\right]\right) \alpha(ne^{i\theta})\, d\theta,$$

it suffices to perform the change of variables $t = \theta\sqrt{n}$, make careful use of the assumed asymptotic approximation of $\alpha(z)$ in each of the three cases, and finally conclude. ∎

The estimates of Theorem VIII.6 are thus considerable refinements of (60). (To some probabilists, it may come as a surprise that one can depoissonize by making use of Poisson laws of *complex rate*!) Analytic depoissonization parallels the philosophy underlying singularity analysis as well as admissibility theory. Its merit is to be well-suited to solving a large number of problems arising in word statistics, the analysis of digital trees and distributed algorithms, as well as data compression: see Szpankowski's book [564, Ch. 10] and the fundamental study [346] for applications and advanced results.

▷ **VIII.18.** *The "Jasz" expansion.* Jacquet and Szpankowski prove more generally that

$$a_n \sim \alpha(n) + \sum_{k=1}^{\infty} \sum_{i=1}^{k} c_{i,k+1} n^i \left(\partial_z^{k+i} \alpha(z)\right)_{z \mapsto n},$$

where $c_{i,j} = [x^i y^j]\exp(x\log(1+y) - xy)$, under suitable conditions on $\alpha(z)$. ◁

▷ **VIII.19.** *The converse "Jasz" expansion.* Jacquet and Szpankowski also give an Abelian result:

$$\alpha(z) \sim g(n) + \sum_{k=1}^{\infty} \sum_{j=1}^{k} d_{i,k+i} z^i \partial_z^{k+i} g(z),$$

where $d_{i,j} = [x^i y^j] \exp(x(e^y - 1) - xy$, the function $g(z)$ extrapolates a_n (i.e., $a_n = g(n)$) to $\mathbb{C}$, and suitable smoothness conditions on g are imposed.   ◁

VIII. 6. Integer partitions

We now examine the asymptotic enumeration of integer partitions, where the saddle-point method serves as the main asymptotic engine. The corresponding generating function enjoys rich properties, and the analysis, which goes back to Hardy and Ramanujan in 1917, constitutes, as pointed out in the introduction of this chapter, a jewel of classical analysis.

Integer partitions represent additive decompositions of integers, when the order of summands is *not* taken into account. When all summands are allowed, the specification and ordinary generating function are (Section I. 3, p. 39)

$$(61) \qquad \mathcal{P} = \text{MSET}(\text{SEQ}_{\geq 1}(\mathcal{Z})) \qquad \Longrightarrow \qquad P(z) = \prod_{m=1}^{\infty} \frac{1}{1 - z^m},$$

which, by the exp–log transformation, admits the equivalent form

$$(62) \qquad \begin{aligned} P(z) &= \exp \sum_{m=1}^{\infty} \log(1 - z^m)^{-1} \\ &= \exp\left(\frac{z}{1 - z} + \frac{1}{2}\frac{z^2}{1 - z^2} + \frac{1}{3}\frac{z^3}{1 - z^3} \cdots \right). \end{aligned}$$

From either of these two forms, it can be seen that the unit circle is a natural boundary, beyond which the function cannot be continued. The second form, which involves the quantity $\exp(z/(1 - z))$ is reminiscent of the EGF of fragmented permutations, examined in Example VIII.7, p. 562, to which the saddle-point method could be successfully applied.

In what follows, we show (Example VIII.8 below) that the saddle-point method is applicable, although the analysis of $P(z)$ near the unit circle is delicate (and pregnant with deep properties). The accompanying notes point to similar methods being applicable to a variety of similar-looking generating functions, including those relative to partitions into primes, squares, and distinct summands, as well as plane partitions: see Figure VIII.8 for a summary of some of the asymptotic results known.

***Example* VIII.8.** *Integer partitions.* We are dealing here with a famous chapter of both asymptotic combinatorics and additive number theory. A problem similar to that of asymptotically enumerating partitions was first raised by Ramanujan in a letter to Hardy in 1913, and subsequently developed in a famous joint work of Hardy and Ramanujan (see the account in Hardy's *Lectures* [321]). The Hardy–Ramanujan expansion was later perfected by Rademacher [22] who, in a sense, gave an "exact" formula for the partition numbers P_n.

A complete derivation with all details would consume more space than we can devote to this questions. We outline here the proof strategy in such a way that, hopefully, the reader can supply the missing details by herself. (The cited references provide a complete treatment).

As before, we start with simple saddle-point bounds, already briefly discussed on p. 248. Let P_n denote the number of integer partitions of n, with OGF as stated in (61). A form

Summands	specification	asymptotics	
all, $\mathbb{Z}_{\geq 1}$	$\text{MSET}(\text{SEQ}_{\geq 1}(\mathcal{Z}))$	$\dfrac{1}{4n\sqrt{3}} e^{\pi\sqrt{2n/3}}$	Ex. VIII.8, p. 574
all distinct, $\mathbb{Z}_{\geq 1}$	$\text{PSET}(\text{SEQ}_{\geq 1}(\mathcal{Z}))$	$\dfrac{1}{4 \cdot 3^{1/4} n^{3/4}} e^{\pi\sqrt{n/3}}$	Note VIII.24, p. 579
squares, 1, 4, 9, 16, $\cdots$		$C n^{-7/6} e^{K n^{1/3}}$	Note VIII.24, p. 579
primes, 2, 3, 5, 7, $\dots$		$\log P_n^{(\Pi)} \sim c\sqrt{\dfrac{n}{\log n}}$	Note VIII.26, p. 580
powers of two, 1, 2, 4, $\dots$		$\log M_{2n} \sim \dfrac{(\log n)^2}{2\log 2}$	Note VIII.27, p. 581
plane	$\left(\displaystyle\prod_m (1 - z^m)^{-m}\right)$	$c_1 n^{-25/36} e^{c_2 n^{2/3}}$	Note VIII.25, p. 580

Figure VIII.8. Asymptotic enumeration of various types of partitions.

amenable to bounds is derived from the exp–log reorganization (62), which yields

$$P(z) = \exp\left(\left(\frac{1}{1-z}\right)\cdot\left(\frac{z}{1} + \frac{z^2}{2(1+z)} + \frac{z^3}{3(1+z+z^2)} + \cdots\right)\right).$$

The denominator of the general term in the exponential satisfies, for $x \in (0, 1)$, the inequalities $mx^{m-1} < (1 + x + \cdots + x^{m-1}) < m$, so that

$$(63) \qquad \frac{1}{1-x}\sum_{m\geq 1}\frac{x}{m^2} > \log P(x) > \frac{1}{1-x}\sum_{m\geq 1}\frac{x^m}{m^2}.$$

This proves for real $x \to 1^-$ that

$$(64) \qquad P(x) = \exp\left(\frac{\pi^2}{6(1-x)}(1+o(1))\right),$$

given the elementary identity $\sum m^{-2} = \pi^2/6$. The singularity type at $z = 1$ resembles that of fragmented permutations (p. 562), and at least the growth along the real axis is similar. An approximate saddle-point is then

$$(65) \qquad \widehat{\zeta}(n) = 1 - \frac{\pi}{\sqrt{6n}},$$

which gives a saddle-point bound

$$(66) \qquad P_n \leq \exp\left(\pi\sqrt{2n/3}(1+o(1))\right).$$

Proceeding further involves transforming the saddle-point bounds into a complete saddle-point analysis. Based on previous experience, we shall integrate along a circle of radius $r = \widehat{\zeta}(n)$. To do so, two ingredients are needed: (i) an approximation in the central range; (ii) bounds establishing that the function $P(z)$ is small away from the central range so that tails can be first neglected, then completed back. Assuming the expansion (62) to lift to an area of the complex plane near the real axis, the range of the saddle-point should be analogous to that already found for $\exp(z/(1-z))$, so that $\theta_0 = n^{-7/10}$ will be adopted. Accordingly, we choose to integrate along a circle of radius $r = \widehat{\zeta}(n)$ given by (65) and define the central region by $\theta_0 = n^{-7/10}$.

Under these conditions, the central region is seen under an angle that is $O(n^{-1/5})$ from the
point $z = 1$.

(i) *Central approximation.* This requires a refinement of (64) till $o(1)$ terms as well as an
argument establishing a lifting to a region near the real axis. We set $z = e^{-t}$ and start with
$t > 0$. The function

$$L(t) := \log P(e^{-t}) = \sum_{m \geq 1} \frac{e^{-mt}}{m(1 - e^{-mt})}$$

is a harmonic sum which is amenable to Mellin transform techniques (as described in Appendix B.7: *Mellin transforms*, p. 762; see also p. 248). The base function is $e^{-t}/(1 - e^{-t})$, the
amplitudes are the coefficients $1/m$ and the frequencies are the quantities m figuring in the exponents. The Mellin transform of the base function, as given in Appendix B (p. 763), is $\Gamma(s)\zeta(s)$.
The Dirichlet series associated to the amplitude frequency pairs is $\sum m^{-1} m^{-s} = \zeta(s + 1)$, so
that

$$L^\star(s) = \zeta(s)\zeta(s + 1)\Gamma(s).$$

Thus $L(t)$ is amenable to Mellin asymptotics and one finds

(67) $$L(t) = \frac{\pi^2}{6t} + \frac{1}{2}\log t - \log\sqrt{2\pi} - \frac{1}{24}t + O(t^2), \qquad t \to 0^+,$$

from the poles of $L^\star(s)$ at $s = 1, 0, -1$. This corresponds to an improved form of (64):

(68) $$\log P(z) = \frac{\pi^2}{6(1 - z)} + \frac{1}{2}\log(1 - z) - \frac{\pi^2}{12} - \log\sqrt{2\pi} + O(1 - z).$$

At this stage, we make a crucial observation: *The precise estimate (67) extends when t lies
in any sector symmetric about the real axis, situated in the half-plane* $\Re(t) > 0$, *and with an
opening angle of the form* $\pi - \delta$ *for an arbitrary* $\delta > 0$. This is derived from the fact that
the Mellin inversion integral and the companion residue calculations giving rise to (67) extend
to the complex realm as long as $|\arg(t)| < \frac{\pi}{2} - \frac{1}{2}\delta$. (See Appendix B.7: *Mellin transforms*,
p. 762 or the article [234].) Thus, the expansion (68) holds throughout the central region given
our choice of the angle θ_0. The analysis in the central region is then practically isomorphic to
that of $\exp(z/(1 - z))$ in the previous example, and it presents no special difficulty.

(ii) *Bounds in the non-central region.* This is here a non-trivial task since half of the
factors entering the product form (61) of $P(z)$ are infinite at $z = -1$, one third are infinite at
$z = e^{\pm 2i\pi/3}$, and so on. Accordingly, the landscape of $|P(z)|$ along a circle of radius r that
tends to 1 is quite chaotic: see Figure VIII.9 for a rendering. It is possible to extend the analysis
of $\log P(z)$ near the real axis by way of the Mellin transform to the case $z = e^{-t-i\phi}$ as $t \to 0$
and $\phi = 2\pi \frac{p}{q}$ is commensurate to 2π. In that case, one must operate with

$$L_\phi(t) = \sum_{m \geq 1} \frac{1}{m} \frac{e^{-m(t+i\phi)}}{1 - e^{-m(t+i\phi)}} = \sum_{m \geq 1} \sum_{k \geq 1} \frac{1}{m} e^{-mk(t+i\phi)},$$

which is yet another harmonic sum. The net result is that when $|z|$ tends radially towards $e^{2\pi i \frac{p}{q}}$,
then $P(z)$ behaves roughly like

(69) $$\exp\left(\frac{\pi^2}{6q^2(1 - |z|)}\right),$$

which is a power $1/q^2$ of the exponential growth as $z \to 1^-$. This analysis extends next to
a small arc. Finally, consider a complete covering of the circle by arcs whose centres are of
argument $2\pi j/N$, $j = 1, \ldots, N - 1$, with N chosen large enough. A uniform version of the

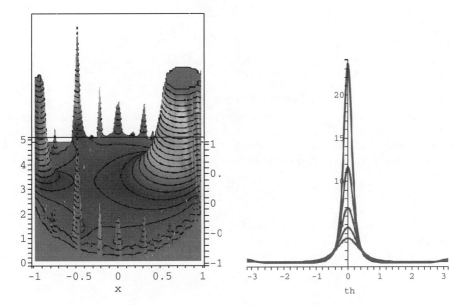

Figure VIII.9. Integer partitions. Left: the surface $|P(z)|$ with $P(z)$ the OGF of integer partitions. The plot shows the major singularity at $z = 1$ and smaller peaks corresponding to singularities at $z = -1$, $e^{\pm 2i\pi/3}$ and other roots of unity. Right: a plot of $P(re^{i\theta})$ as a function of θ, for various values $r = 0.5, \ldots, 0.75$, illustrates the increasing concentration property of $P(z)$ near the real axis.

bound (69) makes it possible to bound the contribution of the non-central region and prove it to be exponentially small. There are several technical details to be filled in order to justify this approach, so that we switch to a more synthetic one based on transformation properties of $P(z)$, following [14, 17, 22, 321]. (Such properties also enter the Hardy–Ramanujan–Rademacher formula for P_n in an essential way.)

The fundamental identity satisfied by $P(z)$ reads

$$(70) \qquad P(e^{-2\pi\tau}) = \sqrt{\tau} \exp\left(\frac{\pi}{12} \left(\frac{1}{\tau} - \tau \right) \right) P(e^{-2\pi/\tau}),$$

which is valid when $\Re(\tau) > 0$. The proof is a simple rephrasing of a transformation formula of Dedekind's η (eta) function, summarized in Note VIII.20 below.

▷ **VIII.20.** *Modular transformation for the Dedekind eta function.* Consider

$$\eta(\tau) := q^{1/24} \prod_{m=1}^{\infty} (1 - q^m), \qquad q = e^{2\pi i \tau},$$

with $\Im(\tau) > 0$. Then $\eta(\tau)$ satisfies the "modular transformation" formula,

$$(71) \qquad \eta\left(-\frac{1}{\tau} \right) = \sqrt{\frac{\tau}{i}} \eta(\tau).$$

This transformation property is first proved when τ is purely imaginary, i.e., $\tau = it$, then extended by analytic continuation. Its logarithmic form results from a residue evaluation of the integral

$$\frac{1}{2\pi i} \int_\gamma \cot \pi s \cot \pi \frac{s}{\tau} \frac{ds}{s},$$

with γ a large contour avoiding poles. (This elementary derivation is due to C. L. Siegel. The function $\eta(\tau)$ satisfies transformation formulae under $S : \tau \mapsto \tau+1$ and $T : \tau \mapsto -1/\tau$, which generate the group of modular (in fact "unimodular") transformations $\tau \mapsto (a\tau + b)/(c\tau + d)$ with $ad - bc = 1$. Such functions are called *modular forms*.)   ◁

Given (70), the behaviour of $P(z)$ away from the positive real axis and near the unit circle can now be quantified. Here, we content ourselves with a representative special case, the situation when $z \to -1$. Consider thus $P(z)$ with $z = e^{-2\pi t + i\pi}$, where, for our purposes, we may take $t = 1/\sqrt{24n}$. Then, Equation (70) relates $P(z)$ to $P(z')$, with $\tau = t - i/2$ and

$$z' = e^{-2\pi/\tau} = \exp\left(-\frac{2\pi t}{t^2 + \frac{1}{4}}\right) e^{i\phi}, \qquad \phi = -\frac{\pi}{t^2 + \frac{1}{4}}.$$

Thus $|z'| \to 1$ as $t \to 0$ with the important condition that $|z'| - 1 = O\big((|z| - 1)^{1/4}\big)$. In other words, z' has moved *away* from the unit circle. Thus, since $|P(z')| < P(|z'|)$, we may apply the estimate (68) to $P(|z'|)$ to the effect that

$$\log |P(z)| \le \frac{\pi}{24(1 - |z|)}(1 + o(1)), \qquad (z \to -1^+).$$

This is an instance of what was announced in (69) and is in agreement with the surface plot of Figure VIII.9. The extension to an arbitrary angle presents no major difficulty.

The two properties developed in (i) and (ii) above guarantee that the approximation (68) can be used and that tails can be completed. We find accordingly that

$$P_n \sim [z^n] e^{-\pi^2/12} \sqrt{1 - z} \exp\left(\frac{\pi^2}{6(1 - z)}\right).$$

All computations done, this provides:

Proposition VIII.6. *The number p_n of partitions of integer n satisfies*

$$(72) \qquad p_n \equiv [z^n] \prod_{k=1}^\infty \frac{1}{1 - z^k} \sim \frac{1}{4n\sqrt{3}} e^{\pi \sqrt{2n/3}}$$

The singular behaviour along and near the real line is comparable to that of $\exp((1-z)^{-1})$, *which explains a growth of the form* $e^{\sqrt{n}}$. . . . . ∎

The asymptotic formula (72) is only the first term of a complete expansion involving decreasing exponentials that was discovered by Hardy and Ramanujan in 1917 and later perfected by Rademacher (see Note VIII.22 below). Whereas the full Hardy–Ramanujan expansion necessitates considering infinitely many saddle-points near the unit circle and require the modular transformation of Note VIII.20, the main term of (72) only requires the asymptotic expansion of the partition generating function near $z = 1$.

The principles underlying the partition example have been made into a general method by Meinardus [434] in 1954. Meinardus' method abstracts the essential features of the proof and singles out sufficient conditions under which the analysis of an infinite product generating function can be achieved. The conditions, in agreement with the Mellin treatment of harmonic sums, require analytic continuation of the

Dirichlet series involved in $\log P(z)$ (or its analogue), as well as smallness towards infinity of that same Dirichlet series. A summary of Meinardus' method constitutes Chapter 6 of Andrews treatise on partitions [14] to which the reader is referred. The method applies to many cases where the summands and their multiplicities have a regular enough arithmetic structure.

▷ **VIII.21.** *A simple yet powerful formula.* Define (cf [321, p. 118])

$$P_n^{\star} = \frac{1}{2\pi\sqrt{2}} \frac{d}{dn}\left(\frac{e^{K\lambda_n}}{\lambda_n}\right), \qquad K = \pi\sqrt{\frac{2}{3}}, \quad \lambda_n := \sqrt{n - \frac{1}{24}}.$$

Then $P_n^{\star}$ approximates P_n with a relative precision of order $e^{-c\sqrt{n}}$ for some $c > 0$. For instance, the error is less than $3 \cdot 10^{-8}$ for $n = 1000$. [Hint: The transformation formula makes it possible to evaluate the central part of the integral giving P_n very precisely.] ◁

▷ **VIII.22.** *The Hardy–Ramanujan–Rademacher expansion.* The number of integer partitions satisfies the *exact* formula

$$P_n = \frac{1}{\pi\sqrt{2}} \sum_{k=1}^{\infty} A_k(n)\sqrt{k}\frac{d}{dn}\frac{\sinh(\frac{\pi}{k}\sqrt{\frac{2}{3}(n - \frac{1}{24})})}{\sqrt{n - \frac{1}{24}}},$$

$$\text{where} \quad A_k(n) = \sum_{h \bmod k, \gcd(h,k)=1} \omega_{h,k}e^{-2i\pi h/k},$$

$\omega_{h,k}$ is a 24th root of unity, $\omega_{h,k} = \exp(\pi i s(h,k))$, and $s_{h,k} = \sum_{\mu=1}^{k-1}\{\{\frac{\mu}{k}\}\}\{\{\frac{h\mu}{k}\}\}$ is known as a

Dedekind sum, with $\{\{x\}\} = x - \lfloor x \rfloor - \frac{1}{2}$. Proofs are found in [14, 17, 22, 321]. ◁

▷ **VIII.23.** *Meinardus' theorem.* Consider the infinite product $(a_n \geq 0)$

$$f(z) = \prod_{n=1}^{\infty}(1 - z^n)^{-a_n}.$$

The associated Dirichlet series is $\alpha(s) = \sum_{n\geq 1}\frac{a_n}{n^s}$. Assume that $\alpha(s)$ is continuable into a meromorphic function to $\Re(s) \geq -C_0$ for some $C_0 > 0$, with only a simple pole at some $\rho > 0$ and corresponding residue A; assume also that $\alpha(s)$ is of moderate growth in the half-plane, namely, $\alpha(s) = O(|s|^{C_1})$, for some $C_1 > 0$ (as $|s| \to \infty$ in $\Re(s) \geq -C_0$). Let $g(z) = \sum_{n\geq 1}a_n z^n$ and assume a concentration condition of the form

$$\Re g(e^{-t-2i\pi y}) - g(e^{-t}) \leq -C_2 y^{-\epsilon}.$$

Then the coefficient $f_n = [z^n]f(z)$ satisfies

$$f_n = Cn^{\kappa}\exp\left(Kn^{\rho/(\rho+1)}\right), \qquad K = (1 + \rho^{-1})[A\Gamma(\rho+1)\zeta(\rho+1)]^{1/(\rho+1)}.$$

The constants C, κ are:

$$C = e^{\alpha'(0)}(2\pi(1+\rho))^{-1/2}[A\Gamma(\rho+1)\zeta(\rho+1)]^{(1-2\alpha(0))/(2\rho+2)}, \quad \kappa = \frac{\alpha(0) - 1 - \frac{1}{2}\rho}{1 + \rho}.$$

Details of the concentration condition, and error terms are found in [14, Ch 6]. ◁

▷ **VIII.24.** *Various types of partitions.* The number of partitions into distinct odd summands, squares, cubes, triangular numbers, are essentially cases of application of Meinardus' method.

For instance the method provides, for the number Q_n of partitions into *distinct* summands, the asymptotic form

$$Q_n \equiv \prod_{m \geq 1} (1 + z^m) \sim \frac{e^{\pi \sqrt{n/3}}}{4 \cdot 3^{1/4} n^{3/4}}.$$

The central approximation is obtained by a Mellin analysis from

$$L(t) := \log Q(e^{-t}) = \sum_{m=1}^{\infty} \frac{(-1)^{m-1}}{m} \frac{e^{-mt}}{1 - e^{-mt}}, \quad L^{\star}(s) = \Gamma(s)\zeta(s)\zeta(s+1)(1 - 2^{-s}),$$

$$L(t) \sim \frac{\pi^2}{12t} - \log \sqrt{2} + \frac{1}{24}t..$$

(See the already cited references [14, 17, 22, 321].) ◁

▷ **VIII.25.** *Plane partitions.* A plane partition of a given number n is a two-dimensional array of integers $n_{i,j}$ that are non-increasing both from left to right and top to bottom and that add up to n. The first few terms (*EIS* **A000219**) are 1, 1, 3, 6, 13, 24, 48, 86, 160, 282, 500, 859 and P. A. MacMahon proved that the OGF is

$$R(z) = \prod_{m=1}^{\infty} (1 - z^m)^{-m}.$$

Meinardus' method applies to give

$$R_n \sim (\zeta(3)2^{-11})^{1/36} n^{-25/36} \exp\left(3 \cdot 2^{-2/3} \zeta(3)^{1/3} n^{2/3} + 2c\right),$$
where $c = -\frac{e}{4\pi^2}(\log(2\pi) + \gamma - 1)$.

(See [14, p. 199] for this result due to Wright [617] in 1931.) ◁

▷ **VIII.26.** *Partitions into primes.* Let $P_n^{(\Pi)}$ be the number of partitions of n into summands that are all prime numbers,

$$P^{(\Pi)}(z) = \prod_{m=1}^{\infty} \frac{1}{1 - z^{p_m}},$$

where p_m is the mth prime ($p_1 = 2$, $p_2 = 3$, ...). The sequence starts as (*EIS* **A000607**):

$$1, 0, 1, 1, 1, 2, 2, 3, 3, 4, 5, 6, 7, 9, 10, 12, 14, 17, 19, 23, 26, 30, 35, 40.$$

Then

(73) $$\log P_n^{(\Pi)} \sim 2\pi \sqrt{\frac{n}{3 \log n}}.$$

An upper bound of a form consistent with (73) can be derived elementarily as a saddle-point bound based on the property

$$\sum_{n \geq 1} e^{-tp_n} \sim \frac{t}{\log t}, \qquad t \to 0.$$

This last fact results either from the Prime Number Theorem or from a Mellin analysis based on the fact that $\Pi(s) := \sum p_n^{-s}$ satisfies, with $\mu(m)$ the Möbius function,

$$\Pi(s) = \sum_{m=1}^{\infty} \mu(m) \log \zeta(ms).$$

(See Roth and Szekeres' study [519] as well as the articles by Yang [625] and Vaughan [593] for relevant references and recent technology.) The present situation is in sharp contrast with that of compositions into primes (see Chapter V, p. 297), for which the analysis turned out to be especially easy. ◁

▷ **VIII.27.** *Partitions into powers of* 2. Let M_n be the number of partitions of integer n into summands that are powers of 2. Thus $M(z) = \prod_{m \geq 0}(1 - z^{2^m})^{-1}$. The sequence (M_n) starts as $1, 1, 2, 2, 4, 4, 6, 6, 10$ (*EIS* **A018819**). One has

$$\log M_{2n} = \frac{1}{2\log 2}\left(\log\frac{n}{\log n}\right)^2 + \left(\frac{1}{2} + \frac{1}{\log 2} + \frac{\log\log 2}{\log 2}\right)\log n + O(\log\log n).$$

De Bruijn [141] determined the precise asymptotic form of M_{2n}. (See also [179] for related problems.) ◁

Averages and moments. Based on the foregoing analysis, it is possible to perform the analysis of several parameters of integer partitions (see also our general discussion of moments in Subsection VIII. 9.1, p. 594). In particular, it becomes possible to justify the empirical observations regarding the profile of partitions made in the course of Example III.7, p. 171.

▷ **VIII.28.** *Mean number of parts in integer partitions.* The mean number of parts (or summands) in a random integer partition of size n is

$$\frac{1}{K}\sqrt{n}\log n + O(n^{1/2}), \qquad K = \pi\sqrt{\frac{2}{3}}.$$

For a partition into distinct parts, the mean number of parts is

$$\frac{2\sqrt{3}\log 2}{\pi}\sqrt{n} + o(n^{1/2}).$$

The complex analytic proof starts from the BGFs of Subsection III. 3.3, p. 170 and, analytically, it only requires the central estimates of $\log P(e^{-t})$ and $\log Q(e^{-t})$, given the concentration properties, as well as the estimates

$$\sum_{m \geq 1}\frac{e^{-mt}}{1 - e^{-mt}} \sim \frac{-\log t + \gamma}{t} + \frac{1}{4}, \qquad \sum_{m \geq 1}(-1)^{m-1}\frac{e^{-mt}}{1 - e^{-mt}} \sim \frac{\log 2}{t} - \frac{1}{4},$$

which result from a standard Mellin analysis, the respective transforms being

$$\Gamma(s)\zeta(s)^2, \qquad \Gamma(s)(1 - 2^{1-s})\zeta(s)^2.$$

Full asymptotic expansions of the mean and of moments of any order can be determined. In addition, the distributions are concentrated around their mean. (The first-order estimates are due to Erdős and Lehner [194] who gave an elementary derivation and also obtained the limit distribution of the number of summands in both cases: they are a double exponential (for P) and a Gaussian (for Q).) ◁

VIII. 7. Saddle-points and linear differential equations.

The purpose of this section is to complete the *classification of singularities* of *linear ordinary differential equations* (see Subsection VII. 9.1, p. 518 for the so-called "regular" case) and briefly point to potentially useful saddle-point connections. What is given is, once more, a linear differential equation (linear ODE) of the form

$$(74) \qquad \partial^r Y(z) + d_1(z)\partial^{r-1}Y(z) + \cdots + d_r Y(z) = 0, \qquad \partial \equiv \frac{d}{dz}$$

(cf Equation (114), p. 519) and a simply connected open domain Ω where the coefficients $d_j(z)$ are meromorphic. It is assumed that the coefficients $d_j(z)$ have a pole at a single point $\zeta \in \Omega$ and are analytic elsewhere. As we know, it is only at such a point ζ that singularities of solutions may arise.

Consider for instance the ODE

(75) $(1 - z)^2 Y'(z) - (2 - z)Y(z) = 0,$

in a neighbourhood of $\zeta = 1$. The method of trying to match an approximate solution of the form $(z - 1)^\theta$ for some $\theta \in \mathbb{C}$ does not succeed: there is no way to find a value of θ for which there is a cancellation between two terms in the main asymptotic order. Accordingly, the conditions of Definition VII.7, p. 519, relative to regular singularities fail to be satisfied: in such cases, we say that the point ζ is an *irregular singularity* of the linear ODE. In fact, the solution of (75), together with $y(0) = 1$, is explicit (see also Example VIII.13 and Note VIII.43, p. 597): $T(z) = 1/(1 - z) \exp(z/(1 - z))$. Thus, we encounter an exponential-of-pole singularity rather than the plain algebraic–logarithmic singularity that prevails in the regular case. The general case is hardly more complicated to state[7].

Theorem VIII.7 (Structure theorem for irregular singularities). *Let there be given a differential equation of the form* (74), *a singular point* ζ, *and a sector S with vertex at* ζ. *Then, for z in a sufficiently small sector S' of S and for* $|z - \zeta|$ *sufficiently small, there exists a basis of d linearly independent solutions of* (74), *such that any solution Y in that basis admits, as* $z \to \zeta$ *in S', an asymptotic expansion*

(76) $Y(z) \sim \exp(P(Z^{-1/r})) \, Z^a \sum Q_j(\log Z) Z^{js},$ $Z := (z - \zeta),$

where P is a polynomial, r an integer of $\mathbb{Z}_{\geq 0}$, *a is a complex number, s is a rational number of* $\mathbb{Q}_{\geq 0}$, *and the* Q_j *are a family of polynomials of uniformly bounded degree.*

Proof. The proof [602, p. 11] starts by constructing a basis of formal solutions, each of the form (76), by the method of indeterminate coefficients and exponents. It continues by appealing to a summation mechanism that transforms such formal solutions into actual analytic ones. (The restriction of the statement to sectors is inherent: it is related to what is known as the "Stokes phenomenon"[8] of ODE theory [602, §15].) ∎

In particular, if the polynomial P that intervenes in the expansion (76) has a positive leading coefficient and the sector is large enough, then the intervening quantities are Hayman admissible. In this way, up to (possibly difficult) connection problems, the coefficients of solutions to meromorphic ODEs can *in principle* be analysed, whether the singularities be of the regular or irregular type. Indeed, proceeding at least formally (see the analysis of fragmented permutations in Example VIII.7, p. 562 and Note VIII.7, p. 563 for similar computations) suggests that the coefficients of a solution to a linear ODE with meromorphic coefficients are finite linear combinations of asymptotic elements of the form

(77) $\zeta^{-n} \exp(R(n^{1/\rho})) n^\alpha \sum S_j(\log n) n^{j\sigma},$

where R is a polynomial, ρ an integer of $\mathbb{Z}_{\geq 0}$, α is a complex number, σ is a rational number of $\mathbb{Q}_{\geq 0}$, and the S_j are a family of polynomials of uniformly bounded degree.

---

[7]Singularities at infinity can be transformed into singularities at 0 via $Z := 1/z$.

[8]The Stokes phenomenon is roughly the fact that solutions of an ODE with irregular singular points may involve certain *discontinuities* in asymptotic expansions, relatively to different sectors.

(The case of entire functions with an irregular singularity at infinity further introduces multipliers in the form of fractional powers of $n!$.)

The fact that expansions of the type (77) hold in all generality is probably true, but far from being accepted as a theorem by experts. Odlyzko [461, p. 1135–1138], Wimp [610, p. 64], and Wimp–Zeilberger [611] offer a lucid (and prudent) discussion of these questions. The result (77) was claimed by G.D. Birkhoff and Trjitzinsky [70, 71], based directly on their general theory of analytic *difference equations*, but in Wimp's words (footnote on p. 64 of [610]):

> "Some now believe that the Birkhoff–Trjitzinsky theory has disabling gaps, see [342]. The alleged deficiencies are difficult to discern by a casual inspection of the papers [70, 71] since they are extremely long and their arguments are very laborious. My policy is not to use the theory unless its results can be substantiated by other arguments."

A sound strategy consists in basing an analysis of linear ODEs with an irregular singularity on the well-established Theorem VIII.7 and accordingly work out local singular expansions. Then determine a suitable integration contour for the Cauchy coefficient formula that wanders from valley to valley, and estimate the local contribution of each singularity that has an exponential growth by means of the saddle-point method—for regular singularities, use a Hankel contour, as in Subsection VII. 9.1, p. 518. (As already noted, this may involve delicate *connection problems* as well as difficulties related to the Stokes phenomenon.) The positivity attached to combinatorial problems can often be used to restrict attention to asymptotically dominant solutions. Estimates involving asymptotic elements of the form (77) must eventually result, whenever the strategy is successful. This is in particular applicable to *holonomic sequences and functions* in the sense of Appendix B.4: *Holonomic functions*, p. 748.

Example VIII.9. *Symmetric matrices with constant row sums.* Let $\mathcal{Y}_{k,n}$ be the class of $n \times n$ symmetric matrices with non-negative integer entries and all row sums (hence also column sums) equal to k. The problem is to determine the cardinalities $Y_{k,n}$ for small values of k. It is equivalent to determining the number of (regular, undirected) multigraphs, where all vertices have degree exactly k. We let $Y_k(z)$ represent the corresponding EGF.

For all k, the EGF $Y_k(z)$ is holonomic; that is, it satisfies a linear ODE with polynomial coefficients. This results from Gessel's theory of holonomic symmetric functions (p. 748). We follow here Chyzak, Mishna, and Salvy [122], who developed an original class of effective algorithms, which *inter alia* provide a means of computing the Y_k. The cases $k = 1$ and $k = 2$ succumb to elementary combinatorics, but the problem becomes non-trivial as soon as $k \geq 3$. We consider here $k = 1, 2, 3$.

Case $k = 1$. A matrix of $\mathcal{Y}_{1,n}$ is none other than a symmetric permutation matrix, which is bijectively associated with an involution, so that $Y_1(z) = e^{z+z^2/2}$. In that case, the saddle-point method applied to the entire function $Y_1(z)$ yields (Example VIII.5, p. 558):

$$(78) \qquad Y_{1,n} \sim \frac{1}{(8e\pi)^{1/4}} n!^{1/2} \frac{e^{\sqrt{n}}}{n^{1/4}}.$$

Case $k = 2$. This one is a classic of combinatorial theory [554, pp. 16–19]. A matrix of $\mathcal{Y}_{2,n}$ is the incidence matrix of a multigraph in which all vertices have degree exactly equal to 2. A bit of combinatorial reasoning (compare with 2–regular graphs in Note II.22, p. 133) shows that connected components can be only one of four types:

single nodes   undirected segments   2–cycles   undirected cycles of length ≥ 3

$$z \qquad \frac{1}{2}\frac{z^2}{1-z} \qquad \frac{z^2}{2} \qquad \frac{1}{2}\log\frac{1}{1-z}-\frac{z}{2}-\frac{z^2}{4}.$$

(The corresponding EGFs are given by the last line; their sum provides $\log Y_2(z)$.) Thus, after simplifications, we obtain

$$(79) \qquad Y_2(z) = \frac{1}{\sqrt{1-z}}\exp\left(\frac{z^2}{4}+\frac{1}{2}\frac{z}{1-z}\right).$$

The sequence $Y_{2,n}$ starts as 1, 1, 3, 11, 56, 348 (*EIS* **AA000985**). An asymptotic estimate results from an analysis entirely similar to that of fragmented permutations (Example VIII.7, p. 562), since the singularity is of an "exponential-of-pole type", only modulated by a function of moderate growth $(1-z)^{-1/2}$. We find:

$$(80) \qquad Y_{2,n} \sim n!\frac{e^{\sqrt{2n}}}{2\sqrt{\pi n}}.$$

Case $k = 3$. Chyzak, Mishna, and Salvy determined that $Y \equiv Y_3$ satisfies the linear ODE

$$\phi_2(z)\partial_z^2 Y(z) + \phi_1(z)\partial_z Y(z) + \phi_0 Y(z) = 0,$$

where the coefficients are as in the following table:

$$\begin{aligned}
\phi_0(z) &= z^{11}+z^{10}-6z^9-4z^8+11z^7-15z^6+8z^5-2z^3+12z^2-24z-24\\
\phi_1(z) &= -3z(z^{10}-2z^8+2z^6-6z^5+8z^4+2z^3+8z^2+16z-8)\\
\phi_2(z) &= 9z^3(z^4-z^2+z-2).
\end{aligned}$$

The first values of $Y_{3,n}$ are 1, 1, 4, 23, 214, 2698. Based on analogy with (78) and (80) supplemented by rough combinatorial bounds, we expect the sequence $Y_{3,n}$ to have a growth comparable to $n!^{3/2}$; that is, the EGF $Y_3(z)$ has radius 0. The authors of [122] then opt to introduce a modified GF, obtained by a Hadamard product,

$$\widehat{Y_3}(z) = Y_3(z) \odot \left(\sum_{n\geq 0}\frac{z^{2n}}{2\cdot 4\cdots 2n}+\sum_{n\geq 0}\frac{z^{2n+1}}{1\cdot 3\cdots(2n+1)}\right),$$

whose radius of convergence is finite and non-zero. Thanks to dedicated symbolic computation algorithms and programs, they determine that $\widehat{Y} \equiv \widehat{Y_3}$ satisfies a linear ODE order 29,

$$z^{27}(3z^2-4)^2\partial_z^{29}\widehat{Y}(z) + \sum_{j=0}^{28}\widehat{\phi}_j(z)\partial_z^j\widehat{Y}(z) = 0,$$

with coefficients $\widehat{\phi}_j(z)$ of degree 37(!). This corresponds to a dominant singularity at $\zeta = 2/\sqrt{3}$, while the square factor $(3z^2-4)^2$ betrays an *irregular singularity*. A local analysis of the ODE then reveals the existence of exactly one singular solution at ζ (up to a multiplicative constant),

$$\sigma(z) \sim \exp\left(\frac{3}{4Z}\right)Z^{-1/2}\left(1-\frac{145}{144}Z-\frac{8591}{41472}Z^2+\cdots\right), \qquad Z := 1-z/\zeta,$$

whose form is in general agreement with Theorem VIII.7. We must then have $\widehat{Y}_3(z) \sim \lambda \sigma(z)$ as $z \to \zeta$, for some constant $\lambda > 0$, and a similar analysis applies to the conjugate root $\zeta' = -2/\sqrt{3}$. The form obtained for $\widehat{Y}_3(z)$ is of the exponential-of-pole type, hence amenable to a saddle-point analysis. Omitting intermediate computations, one finds eventually

$$
(81) \qquad Y_{3,n} \sim C_3 n!^{3/2} \left(\frac{\sqrt{3}}{2} \right)^n \frac{\exp(\sqrt{3n})}{n^{3/4}},
$$

for a connection constant C_3 that is determined numerically: $C_3 \doteq 0.37720.$ .... ∎

▷ **VIII.29.** *An asymptotic pattern.* Based on (78), (79), (81), and further (heavier) computations at $k = 4$, Chyzak *et al.* [122] observe the general asymptotic pattern:

$$
Y_{n,k} \sim C_k n!^{k/2} \left(\frac{k^{k/2}}{k!} \right)^n \frac{\exp(\sqrt{kn})}{n^{k/4}}, \qquad C_k = \frac{1}{\sqrt{2}} \frac{e^{k(k-2)/4}}{(2\pi)^{k/4}}.
$$

This asymptotic formula is indeed valid for each fixed k: it results from estimates of Bender and Canfield [39]. Although it is here limited to small values of k, the method of Chyzak *et al.* still has two advantages: (i) the exact values of the counting sequence are computable in a linear number of arithmetic operations; (ii) complete asymptotic expansions can be obtained comparatively easily. ◁

▷ **VIII.30.** *The number of regular matrices.* The asymptotic enumeration of regular (non-symmetric) matrices is treated by Békéssy, Békéssy, and Kómlos in [32] and by Bender in [37]. Combining their results with estimates of Bender and Canfield [39] yields the following table of asymptotic values for the number of regular matrices with row and column sums equal to k:

	$(0, 1)$–*entries*	*non-negative entries*	
Symmetric	$e^{-(k-1)^2/4} \cdot \dfrac{I_{kn}}{(k!)^n}$	$\left[\dfrac{1}{\sqrt{2}} \dfrac{e^{k(k-2)/4}}{(2\pi)^{k/4}} \right] \cdot n!^{k/2} \left(\dfrac{k^{k/2}}{k!} \right)^n$	$\dfrac{\exp(\sqrt{kn})}{n^{k/4}}$
Non-sym.	$e^{-(k-1)^2/2} \cdot (nk)! \, (k!)^{-2n}$	$e^{(k-1)^2/2} \cdot (nk)! \, (k!)^{-2n}$	

(There, I_n is the number of involutions of size n; see Proposition VIII.2, p. 560.) Thus the number of regular graphs, either directed or undirected, and with or without multiple edges, is asymptotically known. ◁

▷ **VIII.31.** *Multidimensional integral representations.* It is of interest to observe the multidimensional contour integral representation

$$
Y_{k,n} = \frac{1}{(2i\pi)^n} \int \cdots \int \prod_{i<j} \left(\frac{1}{1 - x_i x_j} \right) \prod_i \left(\frac{1}{1 - x_i} \right) \frac{dx_1 \cdots dx_n}{x_1^{k+1} \cdots x_n^{k+1}},
$$

in connection with the advanced saddle-point methods methods of McKay and his coauthors [296, 432]. Find similar integral representations for all the cases of Note VIII.30 above. ◁

VIII. 8. Large powers

The extraction of coefficients in powers of a fixed function and more generally in functions of the form $A(z)B(z)^n$ constitutes a prototypical and easy application of the saddle-point method. We will accordingly be concerned here with the problem of estimating

$$
(82) \qquad [z^N] A(z) \cdot B(z)^n = \frac{1}{2i\pi} \oint A(z) B(z)^n \frac{dz}{z^{N+1}},
$$

as both n and N get large. This situation generalizes directly the example of the exponential and its inverse factorial coefficients, where we have dealt with a coefficient extraction equivalent to $[z^n](e^z)^n$ (see pp. 549 and 555), as well as the case of the central binomial coefficients (p. 549), corresponding to $[z^n](1 + z)^{2n}$. General estimates relative to (82) are derived in Subsections VIII. 8.1 (bounds) and VIII. 8.2 (asymptotics). We finally discuss perturbations of the basic saddle-point paradigm in the case of large powers (Subsection VIII. 8.3): *Gaussian approximations* are obtained in a way that generalizes "local" versions of the Central Limit Theorem for sums of discrete random variables. This last subsection paves the way for the analysis of limit laws in the next chapter, where the rich framework of *"quasi-powers"* will be shown to play a central rôle in so many combinatorial applications.

VIII. 8.1. Large powers: saddle-point bounds.
We consider throughout this section two fixed functions, $A(z)$ and $B(z)$ satisfying the following conditions.

$\mathbf{L_1}$: The functions $A(z) = \sum_{j \geq 0} a_j z^j$ and $B(z) = \sum_{j \geq 0} b_j z^j$ are analytic at 0 and have non-negative coefficients; furthermore it is assumed (without loss of generality) that $B(0) \neq 0$.

$\mathbf{L_2}$: The function $B(z)$ is *aperiodic* in the sense that $\gcd\{j \mid b_j > 0\} = 1$. (Thus $B(z)$ is not a function of the form $\beta(z^p)$ for some integer $p \geq 2$ and some β analytic at 0.)

$\mathbf{L_3}$: Let $R \leq \infty$ be the radius of convergence of $B(z)$; the radius of convergence of $A(z)$ is at least as large as R.

Define the quantity T called the *spread*:

$$(83) \qquad\qquad\qquad T := \lim_{x \to R^-} \frac{x B'(x)}{B(x)}.$$

Our purpose is to analyse the coefficients

$$[z^N] A(z) \cdot B(z)^n,$$

when N and n are linearly related. The condition $N < Tn$ will be imposed: it is both technically needed in our proof and inherent in the nature of the problem. (For B a polynomial of degree d, the spread is $T = d$; for a function B whose derivative at its dominant positive singularity remains bounded, the spread is finite; for $B(z) = e^z$ and more generally for (non-polynomial) entire functions, the spread is $T = \infty$.)

Saddle-point bounds result almost immediately from the previous assumptions.

Proposition VIII.7 (Saddle-point bounds for large powers). *Consider functions $A(z)$ and $B(z)$ satisfying the conditions $\mathbf{L_1}, \mathbf{L_2}, \mathbf{L_3}$ above. Let λ be a positive number with $0 < \lambda < T$ and let ζ be the unique positive root of the equation*

$$\zeta \frac{B'(\zeta)}{B(\zeta)} = \lambda.$$

Then, for $N = \lambda n$ an integer, one has

$$[z^N] A(z) \cdot B(z)^n \leq A(\zeta) B(\zeta)^n \zeta^{-N}.$$

Proof. The existence and unicity of ζ is guaranteed by an argument already encountered several times (Note VIII.46, p. 280, and Note VIII.4, p. 550). The conclusion then follows by an application of general saddle-point bounds (Corollary VIII.1, p. 549). ∎

Example VIII.10. *Entropy bounds for binomial coefficients.* Consider the problem of estimating the binomial coefficients $\binom{n}{\lambda n}$ for some λ with $0 < \lambda < 1$ and $N = \lambda n$. Proposition VIII.7 provides

$$\binom{n}{\lambda n} = [z^N](1+z)^n \leq (1+\zeta)^n \zeta^{-N},$$

where $\frac{\zeta}{1+\zeta} = \lambda$, i.e., $\zeta = \frac{\lambda}{1-\lambda}$. A simple computation then shows that

$$\binom{n}{\lambda n} \leq \exp(nH(\lambda)), \qquad \text{where} \qquad H(\lambda) = -\lambda \log \lambda - (1-\lambda) \log(1-\lambda)$$

is the *entropy function*. Thus, for $\lambda \neq 1/2$, the binomial coefficients $\binom{n}{\lambda n}$ are exponentially smaller than the central coefficient $\binom{n}{n/2}$, and the entropy function precisely quantifies this exponential gap. . . . . ∎

▷ **VIII.32.** *Anomalous dice games.* The probability of a score equal to λn in n casts of an unbiased die is bounded from above by a quantity of the form e^{-nK} where

$$K = -\log 6 + \log \left(\frac{1-\zeta^6}{1-\zeta} \right) - (\lambda - 1) \log \zeta,$$

and ζ is an algebraic function of λ determined by $\sum_{j=0}^{5} (\lambda - j)\zeta^j = 0$. ◁

▷ **VIII.33.** *Large deviation bounds for sums of random variables.* Let $g(u) = \mathbb{E}(u^X)$ be the probability generating function of a discrete random variable $X \geq 0$ and let $\mu = g'(1)$ be the corresponding mean (assume $\mu < \infty$). Set $N = \lambda n$ and let ζ be the root of $\zeta g'(\zeta)/g(\zeta) = \lambda$ assumed to exist within the domain of analyticity of g. Then, for $\lambda < \mu$, one has

$$\sum_{k \leq N} [u^k]g(u)^n \leq \frac{1}{1-\zeta} g(\zeta)^n \zeta^{-N}.$$

Dually, for $\lambda > \mu$, one finds

$$\sum_{k \geq N} [u^k]g(u)^n \leq \frac{\zeta}{\zeta-1} g(\zeta)^n \zeta^{-N}.$$

These are exponential bounds on the probability that n copies of the variable X have a sum deviating substantially from the expected value. ◁

VIII. 8.2. Large powers: saddle-point analysis. The saddle-point bounds for large powers are technically shallow but useful, whenever only rough order of magnitude estimates are sought. In fact, the full saddle-point method is applicable under the very conditions of the preceding proposition.

Theorem VIII.8 (Saddle-point estimates of large powers). *Under the conditions of Proposition VIII.7, with $\lambda = N/n$, one has*

$$(84) \qquad [z^N]A(z) \cdot B(z)^n = A(\zeta) \frac{B(\zeta)^n}{\zeta^{N+1}\sqrt{2\pi n\xi}}(1+o(1)),$$

where ζ is the unique root of $\zeta B'(\zeta)/B(\zeta) = \lambda$ and

$$\xi = \frac{d^2}{d\zeta^2}\left(\log B(\zeta) - \lambda \log \zeta\right).$$

In addition, a full expansion in descending powers of n exists.

These estimates hold uniformly for λ in any compact interval of $(0, T)$, i.e., any interval $[\lambda', \lambda'']$ with $0 < \lambda' < \lambda'' < T$, where T is the spread.

Proof. We discuss the analysis corresponding to a fixed λ. For any fixed r such that $0 < r < R$, the function $|B(re^{i\theta})|$ is, by positivity of coefficients *and* aperiodicity, uniquely maximal at $\theta = 0$ (see The Daffodil Lemma on p. 266). It is also infinitely differentiable at 0. Consequently there exists a (small) angle $\theta_1 \in (0, \pi)$ such that

$$|B(re^{i\theta})| \le |B(re^{i\theta_1})| \qquad \text{for all } \theta \in [\theta_1, \pi],$$

and at the same time, $|B(re^{i\theta})|$ is strictly decreasing for $\theta \in [0, \theta_1]$ (it is given by a Taylor expansion without a linear term).

We carry out the integration along the saddle-point circle, $z = \zeta e^{i\theta}$, where the previous inequalities on $|B(z)|$ hold. The contribution for $|\theta| > \theta_1$ is exponentially negligible. Thus, up to exponentially small terms, the desired coefficient is given asymptotically by $J(\theta_1)$, where

$$J(\theta_1) = \frac{1}{2\pi}\int_{-\theta_1}^{\theta_1} A(\zeta e^{i\theta})B(\zeta e^{i\theta})^n e^{ni\theta}\, d\theta.$$

It is then possible to impose a *second* restriction on θ, by introducing θ_0 according to the general heuristic, namely, $n\theta_0^2 \to \infty$, $n\theta_0^3 \to 0$. We fix here

$$\theta_0 \equiv \theta_0(n) = n^{-2/5}.$$

By the decrease of $|B(\zeta e^{i\theta})|$ on $[\theta_0, \theta_1]$ and by local expansions, the quantity $J(\theta_1) - J(\theta_0)$ is of the form $\exp(-cn^{1/5})$ for some $c > 0$, that is, exponentially small.

Finally, local expansions are valid in the central range since θ_0 tends to 0 as $n \to \infty$. One finds for $z = \zeta e^{i\theta}$ and $|\theta| \le \theta_0$,

$$A(z)B(z)^n z^{-N} \sim A(\zeta)B(\zeta)^n \zeta^{-N} \exp(-n\xi\theta^2/2).$$

Then the usual process applies upon completing the tails, resulting in the stated estimate. A complete expansion in powers of $n^{-1/2}$ is obtained by extending the expansion of $\log B(z)$ to an arbitrary order (as in the case of Stirling's formula, p. 557). Furthermore, by parity, all the involved integrals of odd order vanish so that the expansion turns out to be in powers of $1/n$ (rather than $1/\sqrt{n}$). ∎

Example VIII.11. *Central binomials and trinomials, Motzkin numbers.* An automatic application of Theorem VIII.8 is to the central binomial coefficient $\binom{2n}{n} = [z^n](1 + z)^{2n}$. In the same way, one gets an estimate of the central trinomial number,

$$T_n := [z^n](1 + z + z^2)^n \qquad \text{satisfies} \qquad T_n \sim \frac{3^{n+1/2}}{2\sqrt{\pi n}}.$$

The Motzkin numbers count unary–binary trees,

$$M_n = [z^n]M(z) \qquad \text{where} \qquad M = z(1 + M + M^2).$$

The standard approach is the one seen earlier based on singularity analysis as the implicitly defined function $M(z)$ has an algebraic singularity of the $\sqrt{\ }$-type, but the Lagrange inversion formula provides an equally workable route. It gives

$$M_{n+1} = \frac{1}{n+1}[z^n](1+z+z^2)^{n+1},$$

which is amenable to saddle-point analysis via Theorem VIII.8, leading to

$$M_n \sim \frac{3^{n+1/2}}{2\sqrt{\pi n^3}}.$$

See below for more on this theme. .... ∎

We have opted for a basic formulation of the theorem with conditions on A and B that are not minimal. It is easily recognized that the estimates of Theorem VIII.8 continue to hold, provided that *the function $|B(re^{i\theta})|$ attains a unique maximum on the positive real axis, when $r \in (0, T)$ is fixed and θ varies on $[-\pi, \pi]$*. Also, in order for the statement to hold true, it is only required that *the function $A(z)$ does not vanish on $(0, T)$*, and $A(z)$ or $B(z)$ could then well be allowed to have negative coefficients: see Note VIII.36. Finally, if $A(\zeta) = 0$, then a simple modification of the argument still provides precise estimates in this vanishing case; see Note VIII.37 below.

▷ **VIII.34.** *Middle Stirling numbers.* The "middle" Stirling numbers of both kinds satisfy

$$\frac{n!}{(2n)!}\begin{bmatrix}2n\\n\end{bmatrix} \sim c_1 A_1^n n^{-1/2}\left(1 + O(n^{-1})\right), \qquad \frac{n!}{(2n)!}\begin{Bmatrix}2n\\n\end{Bmatrix} \sim c_2 A_2^n n^{-1/2}\left(1 + O(n^{-1})\right),$$

where $A_1 \doteq 2.45540$, $A_2 \doteq 1.54413$, and A_1, A_2 are expressible in terms of special values of the Cayley tree function. Similar estimates hold for $\begin{bmatrix}\alpha n\\\beta n\end{bmatrix}$ and $\begin{Bmatrix}\alpha n\\\beta n\end{Bmatrix}$. ◁

▷ **VIII.35.** *Integral points on high-dimensional spheres.* Let $L(n, \alpha)$ be the number of lattice points (i.e., points with integer coordinates) in n-dimensional space that lie *on* the sphere of radius $\sqrt{N}$, where $N = \alpha n$ is assumed to be an integer. Then,

$$L(n, \alpha) = [z^N]\Theta(z)^n, \qquad \text{where} \quad \Theta(z) := \sum_{m\in\mathbb{Z}} z^{m^2} = 1 + 2\sum_{m=1}^{\infty} z^{m^2}.$$

Mazo and Odlyzko [431] show that there exist computable constants C, D depending on α, such that $L(n, \alpha) \sim Cn^{-1/2}D^n$. The number of lattice points *inside* the sphere can be similarly estimated. (Such bounds are useful in coding theory, combinatorial optimization, especially the knapsack problem, and cryptography [393, 431].) ◁

▷ **VIII.36.** *A function with negative coefficients that is minimal along the positive axis.* Take $B(z) = 1 + z - z^{10}$. By design, $B(z)$ has both negative and positive Taylor coefficients. On the other hand, $|B(re^{i\theta})|$ for fixed $r \leq 1/10$ (say) attains its unique maximum at $\theta = 0$. For certain values of N, an estimate of $[z^N]B(z)^n$ is provided by (84): discuss its validity. ◁

▷ **VIII.37.** *Coalescence of a saddle-point with roots of the multiplier.* Fix ζ and take a multiplier $A(z)$ in Theorem VIII.8 such that $A(\zeta) = 0$, but $A'(\zeta) \neq 0$. The formula (84) is then to be modified as follows:

$$[z^N]A(z) \cdot B(z)^n = \left[A'(\zeta) + \zeta A''(\zeta)\right] \frac{B(\zeta)^n}{\zeta^{N+1}\sqrt{2\pi n^3 \xi^3}}(1 + o(1)).$$

Higher order cancellations can also be taken into account. ◁

Large powers: saddle-points versus singularity analysis. In general, the Lagrange inversion formula establishes an exact correspondence between two *a priori* different problems; namely,

> the estimation of coefficients of large order in large powers, and
> the estimation of coefficients of implicitly defined functions.

In one direction, the Lagrange Inversion Theorem has the capacity of bringing the evaluation of coefficients of implicit functions into the orbit of the saddle-point method. Indeed, let Y be defined implicitly by $Y = z\phi(Y)$, where ϕ is analytic at 0 and aperiodic. One has, by Lagrange,

$$[z^{n+1}]Y(z) = \frac{1}{n+1}[w^n]\phi(w)^{n+1},$$

which is of the type (84). Then, under the assumption that the equation $\phi(\tau) - \tau\phi'(\tau)$ has a positive root within the disc of convergence of ϕ, a direct application of Theorem VIII.8 yields

$$[z^n]Y(z) \sim \gamma\frac{\rho^{-n}}{2\sqrt{\pi n^3}}, \qquad \rho := \frac{\tau}{\phi(\tau)}, \qquad \gamma := \sqrt{\frac{2\phi(\tau)}{\phi''(\tau)}}.$$

This last estimate is equivalent to the statement of Theorem VII.2 (p. 453) obtained there by singularity analysis. (As we know from Chapter VII, this provides the number of trees in a simple variety, with ϕ being the degree generating function of the variety.) This approach is in a few cases more convenient to work with than singularity analysis, especially when explicit or uniform upper bounds are required, since constructive bounds tend to be more easily obtained on circles than on variable Hankel contours (Note VIII.38).

Conversely, the Lagrange Inversion Theorem makes it possible to approach problems relative to large powers by means of singularity analysis of an implicitly defined function[9]. This mode of operation can prove quite useful when there occurs a coalescence between saddle-points and singularities of the integrand (Note VIII.39).

▷ **VIII.38.** *An assertion of Ramanujan.* In his first letter to Hardy, Ramanujan (1913) announced that

$$\frac{1}{2}e^n = 1 + \frac{n}{1!} + \frac{n^2}{2!} + \cdots + \frac{n^{n-1}}{(n-1)!} + \frac{n^n}{n!}\theta,$$

$$\text{where} \qquad \theta = \frac{1}{3} + \frac{4}{135(n+k)},$$

and k lies between 8/45 and 2/21. Ramanujan's assertion indeed holds for all $n \geq 1$; see [237] for a proof based on saddle-points and effective bounds.   ◁

▷ **VIII.39.** *Coalescence between a saddle-point and a singularity.* The integral in

$$I_n := [y^n](1+y)^{2n}(1-y)^{-\alpha} = \frac{1}{2i\pi}\int_{0^+}\frac{(1+y)^{2n}}{(1-y)^\alpha}\frac{dy}{y^{n+1}},$$

---

[9]This is in essence an approach suggested by several sections of the original memoir of Darboux [137, §§3–5], in which "Darboux's method" discussed in Chapter VI was first proposed. It is also of interest to note that a Lagrangean change of variables transforms a saddle-point circle into a contour whose geometry is of the type used in singularity analysis.

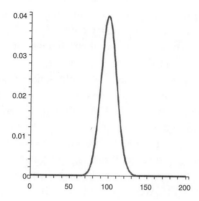

Figure VIII.10. The coefficients $[z^N]e^{nz}$, normalized by e^{-n}, when $n = 100$ is fixed and $N = 0..200$ varies, have a bell-shaped aspect.

can be treated directly, but this requires a suitable adaptation of the saddle-point method, given the coalescence between a saddle-point at 1 [the part without the $(1-y)^\alpha$ factor] and a singularity at that same point. Alternatively, it can be subjected to the change of variables $z = y/(1+y)^2$. Then y is defined implicitly by $y = z(1+y)^2$, so that

$$I_n = \frac{1}{2i\pi} \int_{0+} \frac{1+y}{(1-y)^{1+\alpha}} \frac{dz}{z^{n+1}} = [z^n] \frac{1+y}{(1-y)^{1+\alpha}}.$$

Since $y(z)$ has a square-root singularity at $z = 1/4$, the integrand is of type $Z^{-(1+\alpha)/2}$, and

$$I_n \sim \frac{2^{2n-\alpha}}{\Gamma(\frac{\alpha+1}{2})} n^{(\alpha-1)/2}.$$

In general, for $\phi(y)$ satisfying the assumptions (relative to B) of Theorem VIII.8, one finds, with $\tau : \phi(\tau) - \tau\phi'(\tau) = 0$),

$$\frac{1}{2i\pi} \int_{0+} \frac{\phi(y)^n}{(\phi(\tau) - \phi(y))^\alpha} \frac{dy}{y^n} \sim c \left(\frac{\phi(\tau)}{\tau}\right)^n \frac{n^{(\alpha-1)/2}}{\Gamma(\frac{\alpha+1}{2})}.$$

Van der Waerden discusses this problem systematically in [589]. See also Section VIII. 10 below for other coalescence situations. ◁

VIII. 8.3. Large powers: Gaussian forms.
Saddle-point analysis has consequences for multivariate asymptotics and it constitutes a direct way of establishing that many discrete distributions tend to the Gaussian law in the asymptotic limit. For large powers, this property derives painlessly from our earlier developments, especially Theorem VIII.8, by means of a "perturbation" analysis.

First, let us examine a particularly easy problem: *How do the coefficients of $[z^N]e^{nz}$ vary as a function of N when n is some large but fixed number?* These coefficients are

$$C_N^{(n)} = [z^N]e^{nz} = \frac{n^N}{N!}.$$

By the ratio test, they have a maximum when $N \approx n$ and are small when N differs significantly from n; see Figure VIII.10. The bell-shaped profile is also apparent on the figure and is easily verified by elementary real analysis. The situation is then parallel to what is already known of the binomial coefficients on the nth line of Pascal's triangle, corresponding to $[z^N](1+z)^n$ with N varying.

The asymptotically Gaussian character of coefficients of large powers is actually universal among a wide class of analytic functions. We prove this within the framework of large powers already investigated in Subsection VIII. 8.1 and consider the general problem of estimating the coefficients $[z^N](A(z) \cdot B(z)^n)$ as N varies. In accordance with the conditions on p. 586, we postulate the following: $(\mathbf{L_1})$: $A(z)$, $B(z)$ are analytic at 0, have non-negative coefficients, and are such that $B(0) \neq 0$; $(\mathbf{L_2})$: $B(z)$ is aperiodic; $(\mathbf{L_3})$ The radius of convergence R of $B(z)$ is a minorant of the radius of convergence of $A(z)$. We also recall that the *spread* has been defined as $T := \lim_{x \to R^-} x B'(x)/B(x)$.

Theorem VIII.9 (Large powers and Gaussian forms). *Consider the "large powers" coefficients:*

$$(85) \qquad C_N^{(n)} := [z^N]\left(A(z) \cdot B(z)^n\right).$$

Assume that the two analytic functions $A(z)$, $B(z)$ satisfy the conditions $(\mathbf{L_1})$, $(\mathbf{L_2})$, and $(\mathbf{L_3})$. Assume also that the radius of convergence of B satisfies $R > 1$. Define the two constants:

$$(86) \qquad \mu = \frac{B'(1)}{B(1)}, \qquad \sigma^2 = \frac{B''(1)}{B(1)} + \frac{B'(1)}{B(1)} - \left(\frac{B'(1)}{B(1)}\right)^2 \qquad (\sigma > 0).$$

Then the coefficients $C_N^{(n)}$ for fixed n as N varies admit a Gaussian approximation: for $N = \mu n + x\sqrt{n}$, there holds (as $n \to \infty$)

$$(87) \qquad \frac{1}{A(1)B(1)^n}C_N^{(n)} = \frac{1}{\sigma\sqrt{2\pi n}}e^{-x^2/(2\sigma^2)}\left(1 + O(n^{-1/2})\right),$$

uniformly with respect to x, when x belongs to a finite interval of the real line.

Proof. We start with a few easy observations that shed light on the global behaviour of the coefficients. First, since $R > 1$, we have the exact summation,

$$\sum_{N=0}^{\infty} C_N^{(n)} = A(1)B(1)^n,$$

which explains the normalization factor in the estimate (87). Next, by definition of the spread and since $R > 1$, one has

$$\mu = \frac{B'(1)}{B(1)} < T = \lim_{x \to R^-} \frac{x B'(x)}{B(x)},$$

given the general property that $x B'(x)/B(x)$ is increasing. Thus, the estimation of the coefficients in the range $N = \mu n \pm O(\sqrt{n})$ falls into the orbit of Theorem VIII.8 which expresses the results of the saddle-point analysis in the case of large powers.

Referring to the statement of Theorem VIII.8, the saddle-point equation is

$$\zeta \frac{B'(\zeta)}{B(\zeta)} = \frac{B'(1)}{B(1)} + \frac{x}{\sqrt{n}},$$

with ζ a function of x and n. For x in a bounded set, we thus have $\zeta \sim 1$ as $n \to \infty$. It then suffices to effect an asymptotic expansion of the quantities ζ, $A(\zeta)$, $B(\zeta)$, ξ in the saddle-point formula of Equation (84). In other words, the fact that N is close to μn induces for ζ a small perturbation with respect to the value 1. With $b_j := B^{(j)}(1)$, one finds mechanically

$$\zeta \quad = \quad 1 + \frac{b_0^2}{b_0 b_2 + b_0 b_1 - b_1^2} \frac{x}{\sqrt{n}} + O(n^{-1})$$

$$\frac{B(\zeta)}{\zeta^\mu} \quad = \quad b_0 + \frac{x^2}{2n} \frac{b_0^3}{b_0 b_2 + b_0 b_1 - b_1^2} + O(n^{-3/2}),$$

and so on. The statement follows. ∎

Take first $A(z) \equiv 1$. In the particular case when $B(z)$ is the probability generating function of a discrete random variable Y, one has $B(1) = 1$, and the coefficient $\mu = B'(1)$ is the mean of the distribution. The function $B(z)^n$ is then the probability generating function (PGF) of a sum of n independent copies of Y. Theorem VIII.9 describes a Gaussian approximation of the distribution of the sum near the mean. Such an approximation is called a *local limit law*, where the epithet "local" refers to the fact that the estimate applies to the coefficients themselves. (In contrast, an approximation of the partial sums of the coefficients by the Gaussian error function is known as a *central limit law* or, sometimes, as an *integral limit law*.) In the more general case in which $A(z)$ is also the PGF of a non-degenerate random variable (i.e., $A(z) \neq 1$), similar properties hold and one has:

Corollary VIII.3 (Local limit law for sums). *Let X be a random variable with probability generating function (PGF) $A(z)$ and $Y_1, \ldots, Y_n$ be independent variables with PGF $B(z)$, where it is assumed that X and the Y_j are supported on $\mathbb{Z}_{\geq 0}$. Assume that $A(z)$ and $B(z)$ are analytic in some disc that contains the unit disc in its interior and that $B(z)$ is aperiodic. Let the coefficients μ, σ be as in (86). Then the sum,*

$$S_n := X + Y_1 + Y_2 + \cdots + Y_n,$$

satisfies a local limit law of the Gaussian type: for t in any finite interval, one has

$$\mathbb{P}\left(S_n = \lfloor \mu n + t\sigma\sqrt{n} \rfloor\right) = \frac{e^{-t^2/2}}{\sqrt{2\pi n}}\left(1 + O(n^{-1/2})\right).$$

Proof. This is just a restatement of Theorem VIII.9, setting $x = t\sigma$ and taking into account $A(1) = B(1) = 1$. ∎

Gaussian forms for large powers admit many variants. As already pointed out in Section VIII. 4, the positivity conditions can be greatly relaxed. Furthermore, estimates for partial sums of the coefficients are possible by similar techniques. The asymptotic expansions can be extended to any order. Finally, suitable adaptations of Theorems VIII.8 and VIII.9 make it possible to allow x to tend slowly to infinity and manage what is known as a "moderate deviation" regime. We do not pursue these aspects here since we shall develop a more general framework, that of "Quasi-powers" in the next chapter.

▷ **VIII.40.** *An alternative proof of Corollary VIII.3.* The saddle-point ζ is near 1 when N is near the centre $N \approx \mu n$. It is alternatively possible to recover the $C_n^{(N)}$ by Cauchy's formula upon integrating along the circle $|z| = 1$, which is then only an *approximate* saddle-point contour. This convenient variant is often used in the literature, but one needs to take care of linear terms in expansions. Its origins go back to Laplace himself in his first proof of the local limit theorem (which was expressed however in the language of Fourier series as Cauchy's theory was yet to be born). See Laplace's treatise *Théorie Analytique des Probabilités* [402] first published in 1812 for much fascinating mathematics related to this problem.   ◁

VIII. 9. Saddle-points and probability distributions

Saddle-point methods are useful not only for estimating combinatorial counts, but also for extracting probabilistic characteristics of large combinatorial structures. In the previous section, we have already encountered the large powers framework, giving rise to Gaussian laws. In this section, we further examine the way a saddle-point analysis can serve to quantify properties of random structures.

VIII. 9.1. Moment analyses. Univariate applications of admissibility include the analysis of generating functions relative to moments of distributions, which are obtained by differentiation and specialization of corresponding multivariate generating functions. In the context of saddle-point analyses, the dominant asymptotic form of the mean value as well as bounds on the variance usually result, often leading to concentration of distribution (convergence in probability) properties. In what follows, we focus on the analysis of *first moments* (see also Subsection VII. 10.1, p. 532, for the "moment pumping" method developed in the context of singularity analysis).

The situation of interest here is that of a counting generating function $G(z)$, corresponding to a class $\mathcal{G}$, which is amenable to the saddle-point method. A parameter χ on $\mathcal{G}$ gives rise to a bivariate GF $G(z, u)$, which is a deformation of $G(z)$ when u is close to 1. Then the GFs

$$\partial_u G(z, u)|_{u=1}, \quad \partial_u^2 G(z, u)\Big|_{u=1}, \quad \dots$$

relative to successive (factorial) moments, are in many cases amenable to an analysis that closely resembles that of $G(z)$ itself. In this way, moments can be estimated asymptotically.

We illustrate the analysis of moments by two examples: (i) Example VIII.12 provides an analysis of the mean number of blocks in a random set partition by bivariate generating functions; (ii) Example VIII.13 estimates the mean number of increasing subsequences in a random permutation by a direct generating function construction. The first example foreshadows the full treatment of the corresponding limit distribution in the next chapter (Subsection IX. 8, p. 690).

***Example* VIII.12.** *Blocks in random set partitions.* The function

$$G(z, u) = e^{u(e^z - 1)}$$

is the bivariate generating function of set partitions, with u marking the number of blocks (or parts). We set $G(z) = G(z, 1)$ and define

$$M(z) = \frac{\partial}{\partial u} G(z, u) \Big|_{u=1} = e^{e^z - 1}(e^z - 1).$$

Thus, the quantity

$$\frac{m_n}{g_n} = \frac{[z^n] M(z)}{[z^n] G(z)}$$

represents the mean number of parts in a random partition of $[1..n]$. We already know that $G(z)$ is admissible and so is $M(z)$ by closure properties. The saddle-point for the coefficient integral of $G(z)$ occurs at ζ such that $\zeta e^\zeta = n$, and it is already known that $\zeta = \log n - \log \log n + o(1)$.

It would be possible to analyze $M(z)$ by means of Theorem VIII.4 directly: the analysis then involves a saddle-point $\widehat{\zeta} \neq \zeta$ that is relative to $M(z)$; an estimation of the mean then follows, albeit at the expense of some computational effort. It is however more transparent to appeal to Proposition VIII.5, p. 567, and analyse the coefficients of $M(z)$ at the saddle-point of $G(z)$.

Let $a(r), b(r)$ and $\widehat{a}(r), \widehat{b}(r)$ be the functions $\alpha_1(r), \alpha_2(r)$ of Equation (47), relative to $G(z)$ and $M(z)$, respectively:

$$
\begin{array}{llll}
\log G(z) &= e^z - 1 & \log M(z) &= e^z + z - 1 \\
a(r) &= re^r & \widehat{a}(r) &= re^r + r = a(r) + r \\
b(r) &= (r^2 + r)e^r & \widehat{b}(r) &= (r^2 + r)e^r + r = b(r) + r.
\end{array}
$$

Thus, estimating m_n by Proposition VIII.5 with the formula taken at $r = \zeta$, one finds

$$m_n = \frac{e^\zeta G(\zeta)}{\zeta^n \sqrt{2\pi \widehat{b}(\zeta)}} \left[\exp\left(-\frac{\zeta^2}{2\widehat{b}(\zeta)} \right) + o(1) \right],$$

while the corresponding estimate for g_n is

$$g_n = \frac{G(\zeta)}{\zeta^n \sqrt{2\pi b(\zeta)}} (1 + o(1)).$$

Given that $\widehat{b}(\zeta) \sim b(\zeta)$ and that ζ^2 is of smaller order than $\widehat{b}(\zeta)$, one has

$$\frac{m_n}{g_n} = e^\zeta (1 + o(1)) = \frac{n}{\log n} (1 + o(1)).$$

A similar computation applies to the second moment of the number of parts which is found to be asymptotic to $e^{2\zeta}$ (the computation involves taking a second derivative). Thus, the standard deviation of the number of parts is of an order $o(e^\zeta)$ that is smaller than the mean. This implies a concentration property for the distribution of the number of parts.

Proposition VIII.8. *The variable X_n equal to the number of parts in a random partition of the set $[1..n]$ has expectation*

$$\mathbb{E}\{X_n\} = \frac{n}{\log n}(1 + o(1)).$$

The distribution satisfies a "concentration" property: for any $\epsilon > 0$, one has

$$\mathbb{P}\left\{ \left| \frac{X_n}{\mathbb{E}\{X_n\}} - 1 \right| > \epsilon \right\} \to 0 \qquad as \ n \to +\infty.$$

The calculations are not especially difficult (see Note VIII.41 for the end result) but they require care in the manipulation of asymptotic expansions: for instance, Salvy and Shackell [530] who "do it right" report that two discrepant estimates (differing by a factor of e^{-1}) had been previously published regarding the value of the mean. .... ∎

▷ **VIII.41.** *Moments of the number of blocks in set partitions.* Let X_n be the number of blocks in a random partition of n elements. Then, one has

$$\mathbb{E}(X_n) = \frac{n}{\log n} + \frac{n \log \log n \, (1 + o(1))}{\log^2 n}, \quad \mathbb{V}(X_n) = \frac{n}{\log^2 n} + \frac{n(2 \log \log n - 1 + o(1))}{\log^3 n},$$

which proves concentration. The calculation is best performed in terms of the saddle-point ζ, then converted in terms of n. (See Salvy's étude [529] and the paper [530].)   ◁

▷ **VIII.42.** *The shape of random involutions.* Consider a random involution of size n, the EGF of involutions being $e^{z + z^2/2}$. Then the mean number of 1–cycles and 2-cycles satisfy

$$\mathbb{E}(\# \, 1\text{–cycles}) = \sqrt{n} + O(1), \qquad \mathbb{E}(\# \, 2\text{–cycles}) = \frac{1}{2}n - \frac{1}{2}\sqrt{n} + O(1).$$

In addition, the corresponding distributions are concentrated.   ◁

***Example* VIII.13.** *Increasing subsequences in permutations.* Given a permutation written in linear notation as $\sigma = \sigma_1 \cdots \sigma_n$, an increasing subsequence is a subsequence $\sigma_{i_1} \cdots \sigma_{i_k}$ which is in increasing order, i.e., $i_1 < \cdots < i_k$ and $\sigma_{i_1} < \cdots \sigma_{i_k}$. The question is: *What is the mean number of increasing subsequences in a random permutation?*

The problem has a flavour analogous to that of "hidden" patterns in random words, which was tackled in Chapter V, p. 315, and indeed similar methods are applicable here. Define a *tagged permutation* as a permutation together with one of its increasing subsequence distinguished. (We also consider the null subsequence as an increasing subsequence.) For instance,

$$7 \, | 3 \, 5 \, 2 \, | 6 \, 4 \, 1 \, | 8 \, 9$$

is a tagged permutation with the increasing subsequence **3 6 8** that is distinguished. The vertical bars are used to identify the tagged elements, but they may also be interpreted as decomposing the permutation into sub-permutation fragments. We let $\mathcal{T}$ be the class of tagged permutations, with $T(z)$ the corresponding EGF, and set $T_n = n![z^n]T(z)$. The *mean* number of increasing subsequences in a random permutation of size n is clearly $t_n = T_n/n!$.

In order to enumerate $\mathcal{T}$, we let $\mathcal{P}$ be the class of all permutations and $\mathcal{P}^+$ the subclass of non-empty permutations. Then, one has, up to isomorphism,

$$\mathcal{T} = \mathcal{P} \star \text{SET}(\mathcal{P}^+),$$

since a tagged permutation can be reconstructed from its initial fragment and the *set* of its fragments (by ordering the set according to increasing values of initial elements). This combinatorial argument gives the EGF $T(z)$ as

$$T(z) = \frac{1}{1 - z} \exp\left(\frac{z}{1 - z}\right).$$

The generating function $T(z)$ can be expanded, so that the quantity T_n admits a closed form,

$$T_n = \sum_{k=0}^{n} \binom{n}{k} \frac{n!}{k!}.$$

From this, it is possible to analyse T_n asymptotically by means of the Laplace method for sums, as was done by Lifschitz and Pittel in [407]. However, analytically, the function $T(z)$ is a mere variant of the EGF of fragmented permutations. Saddle-point conditions are again easily checked, either directly or via admissibility, to the effect that

$$(88) \qquad\qquad\qquad t_n \equiv \frac{T_n}{n!} \sim \frac{e^{-1/2}e^{2\sqrt{n}}}{2\sqrt{\pi}n^{1/4}}.$$

(Compare with the closely related estimate (45) on p. 562.)

The estimate (88) has the great advantage of providing information about an important and much less accessible parameter. Indeed, let $\lambda(\sigma)$ represent the *length of the longest increasing subsequence* in σ. With $I(\sigma)$ the number of increasing subsequences, one has the general inequality,

$$2^{\lambda(\sigma)} \leq I(\sigma),$$

since the number of increasing subsequences of σ is *at least* as large as the number of subsequences contained in the *longest* increasing subsequence. Let now ℓ_n be the expectation of λ over permutations of size n. Then, convexity of the function 2^x implies

$$(89) \qquad 2^{\ell_n} \leq t_n, \qquad \text{so that} \qquad \ell_n \leq \frac{2}{\log 2} \sqrt{n}(1 + o(1)).$$

In summary:

Proposition VIII.9. *The mean number of increasing subsequences in a random permutation of n elements is asymptotically*

$$\frac{e^{-1/2} e^{2\sqrt{n}}}{2\sqrt{\pi} n^{1/4}} (1 + o(1)).$$

Accordingly, the expected length of the longest increasing subsequence in a random permutation of size n satisfies the inequality

$$\ell_n \leq \frac{2}{\log 2} \sqrt{n}(1 + o(1)) \approx 2.89\sqrt{n}.$$

Note VIII.45 describes an elementary lower bound of the form $\ell_n \geq \frac{1}{2}\sqrt{n}$. In fact, around 1977, Logan and Shepp [411] and, independently, Vershik and Kerov [596] succeeded in establishing the much more difficult result

$$\ell_n \sim 2\sqrt{n}.$$

Their proof is based on a detailed analysis of the profile of a random Young tableau. (The bound obtained here by a simple mixture of saddle-point estimates and combinatorial approximations at least provides the right order of magnitude.) This has led in turn to attempts at characterizing the asymptotic *distribution* of the length of the longest increasing subsequence. The problem remained unsolved for two decades, despite many tangible steps forward. J. Baik, P. A. Deift, and K. Johansson [24] eventually obtained a solution, in 1999, by relating longest increasing subsequences to eigenvalues of random matrix ensembles (see Note VIII.45 for the end result). We regretfully redirect the reader to relevant presentations of the beautiful theory surrounding this sensational result, for instance [10, 148]. . . . . ∎

▷ **VIII.43.** *A useful recurrence.* A decomposition according to the location of n yields for t_n the recurrence

$$t_n = t_{n-1} + \frac{1}{n} \sum_{k=0}^{n-1} t_k, \qquad t_0 = 1.$$

Hence $T(z)$ satisfies the ordinary differential equation,

$$(1 - z)^2 \frac{d}{dz} T(z) = (2 - z)T(z), \qquad T(0) = 1,$$

which gives rise to the simpler recurrence

$$t_{n+1} = 2t_n - \frac{n}{n+1} t_{n-1}, \qquad t_0 = 0, \qquad t_1 = 2,$$

by which t_n can be computed efficiently in a linear number of operations. ◁

▷ **VIII.44.** *Related combinatorics.* The sequence $T_n = n! t_n$ starts as $1, 2, 7, 34, 209, 1546$, and is *EIS* **A002720**. The number T_n counts the following equivalent objects: (i) the $n \times n$ binary matrices with at most one entry 1 in each column; (ii) the partial matchings of the complete bipartite graph $K_{n,n}$; (iii) the injective partial mappings of $[1 . . n]$ to itself.   ◁

▷ **VIII.45.** *A simple probabilistic lower bound.* Elementary probability theory provides a simple lower bound on ℓ_n. Let $X_1, \ldots, X_n$ be independent random variables uniformly distributed over $[0, 1]$. Assume $n = m^2$. Partition $[0, 1[$ into m subintervals each of the form $[j - 1/m, j/m[$ and $X_1, \ldots, X_n$ into m blocks, each of the form $X_{(k-1)m+1}, \ldots, X_{km}$. There is a probability $1 - (1 - m^{-1})^m \sim 1 - e^{-1}$ that block numbered 1 contains an element of subinterval numbered 1, block numbered 2 contains an element of subinterval numbered 2, and so on. Then, with high probability, at least $m/2$ of the blocks contain an element in their matching subinterval. Consequently, $\ell_n \geq \frac{1}{2}\sqrt{n}$, for n large enough. (The factor $1/2$ can even be improved a little.) The crisp booklet by Steele [556] describes many similar as well as more advanced applications to combinatorial optimization. See also the book of Motwani and Raghavan [451] for applications to randomized algorithms in computer science.   ◁

▷ **VIII.46.** *The Baik–Deift–Johansson Theorem.* Consider the Painlevé II equation

$$u''(x) = 2u(x)^3 + xu(x)$$

and the particular solution $u_0(x)$ that is asymptotic to $-\operatorname{Ai}(x)$ as $x \to +\infty$, with $\operatorname{Ai}(x)$ the Airy function, which solves $y'' - xy = 0$. Define the Tracy–Widom distribution (arising in random matrix theory)

$$F(t) = \exp\left(\int_t^\infty (x - t) u_0(x)^2 \, dx \right).$$

The distribution of the length of the longest increasing subsequence, λ satisfies

$$\lim_{n \to \infty} \mathbb{P}\left(\lambda_n \leq 2\sqrt{n} + tn^{1/6} \right) = F(t),$$

for any fixed t. Thus the discrete random variable λ_n converges to a well-characterized distribution [24]. (An exact formula for associated GFs is due to Gessel; see p. 753.)   ◁

VIII. 9.2. Families of generating functions.

There is an extreme diversity of possible situations, which partly defy classification, when analysing a family of generating functions associated with an extremal parameter. Accordingly, we must content ourselves with the discussion of a single representative example relative to random allocations. (A good rule of thumb is once more that the saddle-point method is likely to succeed in cases involving some sort of exponential growth of GFs.) Problems of a true multivariate nature will be examined in the next chapter specifically dedicated to multivariate asymptotics and limit distributions.

Random allocations. The example that follows is relative to random allocations, occupancy statistics, and balls-in-bin models, as introduced in Subsection II. 3.2, p. 111.

Example **VIII.14.** *Capacity in occupancy problems.* Assume that n balls are thrown into m bins, uniformly at random. How many balls does the most filled bin contain? We shall examine the regime $n = \alpha m$ for some fixed α in $(0, +\infty)$; see Example III.10 (p. 177) for a first analysis and relations to the Poisson law. The size of the most filled bin is called the *capacity* and we let $C_{n,m}$ denote the random variable, when all m^n allocations are taken equally likely. Under our conditions a random bin contains on average a constant number, α, of balls. The proposition below proves that the most filled bin has somewhat more, as illustrated by Figure VIII.11. (We limit ourselves here to saddle-point bounds. The various regimes of the distribution are well covered in [388, pp. 94–115].)

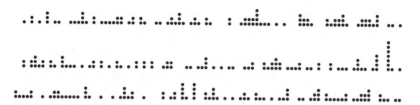

Figure VIII.11. Three random allocations of $n = 100$ balls in $m = 100$ bins.

Proposition VIII.10. *Let n and m tend simultaneously to infinity, with the constraint that $n/m = \alpha$ for some constant $\alpha > 0$. Then, the expected capacity satisfies*

$$\frac{1}{2}\frac{\log n}{\log\log n}(1 + o(1)) \le \mathbb{E}\{C_{n,m}\} \le 2\frac{\log n}{\log\log n}(1 + o(1)).$$

In addition, the probability of capacity to lie outside the interval determined by the lower and upper bounds tends to 0 as $m, n \to \infty$.

Proof. We detail the proof when $\alpha = 1$ and abbreviate $C_n = C_{n,m}$, the generalization to $\alpha \ne 1$ requiring only simple adjustments. From Chapter II, we know that

(90)
$$\begin{cases} \mathbb{P}\{C_n \le b\} &= \dfrac{n!}{n^n}[z^b](e_b(z))^n \\[2mm] \mathbb{P}\{C_n > b\} &= \dfrac{n!}{n^n}\left(e^{nz} - (e_b(z))^n\right), \end{cases}$$

where $e_b(z)$ is the truncated exponential:

$$e_b(z) = \sum_{j=0}^{b}\frac{z^j}{j!}.$$

The two equalities of (90) permit us to bound the left and right tails of the distribution. As suggested by the Poisson approximation of balls-in-bins model, we decide to adopt saddle-point bounds based on $z = 1$. This gives (cf Theorem VIII.2, p. 547):

(91)
$$\begin{cases} \mathbb{P}\{C_n \le b\} &\le \dfrac{n!e^n}{n^n}\left(\dfrac{e_b(1)}{e}\right)^n \\[2mm] \mathbb{P}\{C_n > b\} &\le \dfrac{n!e^n}{n^n}\left(1 - \left(\dfrac{e_b(1)}{e}\right)^n\right). \end{cases}$$

We set

(92)
$$\rho_b(n) = \left(\frac{e_b(1)}{e}\right)^n.$$

This quantity represents the probability that n Poisson variables of rate 1 all have value b or less. (We know from elementary probability theory that this should be a reasonable approximation of the problem at hand.) A weak form of Stirling's formula, namely, $n!e^n/n^n < 2\sqrt{\pi n}$, for $n \ge 1$, then yields an alternative version of (91),

(93)
$$\begin{cases} \mathbb{P}\{C_n \le b\} &\le 2\sqrt{\pi n}\rho_b(n) \\ \mathbb{P}\{C_n > b\} &\le 2\sqrt{\pi n}\,(1 - \rho_b(n)). \end{cases}$$

For fixed n, the function $\rho_b(n)$ increases steadily from e^{-n} to 1 as b varies from 0 to ∞. In particular, the "transition region" where $\rho_b(n)$ stays away from both 0 and 1 is expected to play a rôle. This suggests defining $b_0 \equiv b_0(n)$ such that

$$b_0! \leq n < (b_0 + 1)!,$$

so that

$$b_0(n) = \frac{\log n}{\log \log n}(1 + o(1)).$$

We also observe that, as $n, b \to \infty$, there holds

$$
\begin{aligned}
(94) \quad \rho_b(n) &= (e^{-1}e_b(1))^n = \left(1 - \frac{e^{-1}}{(b+1)!} + O(\frac{1}{(b+2)!})\right)^n \\
&= \exp\left(-\frac{ne^{-1}}{(b+1)!} + O(\frac{n}{(b+2)!})\right).
\end{aligned}
$$

Left tail. We take $b = \lfloor \frac{1}{2}b_0 \rfloor$ and a simple computation from (94) shows that for n large enough, $\rho_b(n) \leq \exp(-\sqrt[3]{n})$. Thus, by the first inequality of (93), the probability that the capacity be less than $\frac{1}{2}b_0$ is exponentially small:

$$(95) \qquad\qquad \mathbb{P}\{C_n \leq \frac{1}{2}b_0(n)\} \leq 2\sqrt{\pi n}\exp(-\sqrt[3]{n}).$$

Right tail. Take $b = 2b_0$. Then, again from (94), for n large enough, one has $1 - \rho_b(n) \leq 1 - \exp(-\frac{1}{n}) = \frac{1}{n}(1 + o(1))$. Thus, the probability of observing a capacity that exceeds $2b_0$ is vanishingly small, and is $O(n^{-1/2})$. Taking next $b = 2b_0 + r$ with $r > 0$, similarly gives the bound

$$(96) \qquad\qquad \mathbb{P}\{C_n > 2b_0(n) + r\} \leq 2\sqrt{\frac{\pi}{n}}\left(\frac{1}{b_0(n)}\right)^r.$$

The analysis of the left and right tails in Equations (95) and (96) now implies

$$
(97) \quad
\begin{cases}
\displaystyle \mathrm{E}\{C_n\} \leq 2b_0(n) + \sum_{r=0}^{\infty} 2\sqrt{\frac{\pi}{n}}(b_0(n))^{-r} = 2b_0(n)(1 + o(1)) \\
\displaystyle \mathrm{E}\{C_n\} \geq \sum_{r=0}^{\lfloor \frac{1}{2}b_0(n) \rfloor} \left[1 - 2\sqrt{\pi n}\exp(-\sqrt[3]{n})\right] = \frac{1}{2}b_0(n)(1 + o(1)).
\end{cases}
$$

This justifies the claim of the proposition when $\alpha = 1$. The general case ($\alpha \neq 1$) follows similarly from saddle-point bounds taken at $z = \alpha$. ∎

The saddle-point bounds described above are obviously not tight: with some care in derivations, one can show by the same means that the distribution is tightly concentrated around its mean, itself asymptotic to $\log n/\log \log n$. In addition, the saddle-point method may be used instead of crude bounds. These results, in the context of longest probe sequences in hashing, were obtained by Gonnet [301] under the Poisson model. Many key estimates regarding random allocations (including capacity) are to be found in the book by Kolchin *et al.* [388]. Analyses of this type are also useful in evaluating various dynamic hashing algorithms by means of saddle-point methods [217, 504]. . . . . ∎

VIII. 10. Multiple saddle-points

We conclude this chapter with a discussion of higher order saddle-points, accompanied by brief indications on what are known as phase transitions or critical phenomena in the applied sciences.

Multiple saddle-point formula. All the analyses carried out so far have been in terms of simple saddle-points, which represent by far the most common situation. In order to get a feel of what happens in the case of multiple saddle-points, consider first the problem of estimating the two *real* integrals,

$$ I_n := \int_0^1 (1 - x^2)^n \, dx, \qquad J_n := \int_0^1 (1 - x^3)^n \, dx. $$

(These examples are illustrative: as a check of the results, note that the integrals can be evaluated in closed form by way of the Beta function, Note B.10, p. 747.) The contribution of any interval $[x_0, 1]$ is exponentially small, and the ranges to be considered on the right of 0 are about $n^{-1/2}$ and $n^{-1/3}$, respectively. One thus sets

$$ x = \frac{t}{\sqrt{n}} \quad \text{for } I_n, \qquad x = \frac{t}{\sqrt[3]{n}} \quad \text{for } J_n. $$

Following the guidelines of the method of Laplace (Appendix B.6, p. 755), we proceed as follows: local expansions are applied, then tails are completed in the usual way, to the effect that

$$ I_n \sim \frac{1}{\sqrt{n}} \int_0^\infty e^{-t^2} \, dt, \qquad J_n \sim \frac{1}{\sqrt[3]{n}} \int_0^\infty e^{-t^3} \, dt. $$

The last integrals reduce to the Gamma function integral, which provides

$$ I_n \sim \frac{1}{2} \frac{\Gamma(\frac{1}{2})}{n^{1/2}}, \qquad J_n \sim \frac{1}{3} \frac{\Gamma(\frac{1}{3})}{n^{1/3}}. $$

The repeated occurrences of $\frac{1}{2}$ in the quadratic case and of $\frac{1}{3}$ in the cubic case stand out. The situation in the cubic case corresponds to the Laplace method for integrals, when a multiple critical point is present (Note B.23, p. 759).

What has been just encountered in the case of real integrals is typical of what to expect for *complex* integrals and saddle-points of higher orders, as we now explain. First, we briefly revisit the discussion of landscapes of analytic functions at the beginning of Section VIII. 1, p. 543. Consider, for simplicity, the case of a double saddle-point of an analytic function $F(z)$. At such a point ζ, we have $F(\zeta) \neq 0$, $F'(\zeta) = F''(\zeta) = 0$, and $F'''(\zeta) \neq 0$. Then, there are three steepest descent lines emanating from the saddle-point and three steepest ascent lines. Accordingly, one should think of the landscape of $|F(z)|$ as formed of three "valleys" separated by three mountains and meeting at the common point ζ. The characteristic aspect is that of a "monkey saddle" (comparable to a saddle with places for two legs and a tail) and is displayed in Figure VIII.12.

In order to avoid an unpleasant discussion of the combinatorics of valleys, we now discuss the case of a multiple saddle-point estimation of an integral $\int_A^B$ in the case where the starting point A coincides with the saddle-point ζ. By a painless surgery of paths, this entails no loss of generality. We can then enunciate a modified form of the saddle-point formula of Theorem VIII.3.

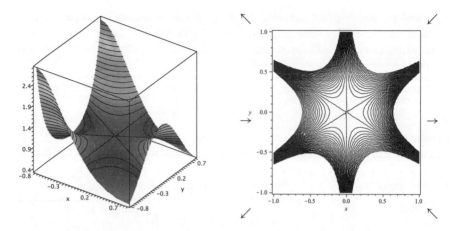

Figure VIII.12. A double saddle-point or "monkey saddle". *Left*: the surface $|\exp(z^3)|$ around the double saddle point $z = 0$; *right*: level curves with arrows pointing towards directions of increase. (Inward pointing arrows indicate valleys.)

Theorem VIII.10 (Double Saddle-point Algorithm). *Consider an integral $\int_\zeta^B F(z)\,dz$, where the integrand $F = e^f$ is an analytic function depending on a large parameter and ζ is a double saddle-point, which is a root of the saddle-point equations*

$$f'(\zeta) = 0, \qquad f''(\zeta) = 0$$

(or, equivalently, $F'(\zeta) = F''(\zeta) = 0$). The point B is supposed to lie inside one of the three valleys of the double saddle-point.

Assume that the contour C connecting ζ to B can be split into $C = C^{(0)} \cup C^{(1)}$ in such a way that the following conditions are satisfied: (i) the tail integral $\int_{C^{(1)}}$ is negligible; (ii) in the central domain $C^{(0)}$, a cubic approximation holds,

$$f(z) = f(\zeta) + \frac{1}{3!}f'''(\zeta)(z-\zeta)^3 + O(\eta_n),;$$

with $\eta_n \to 0$ as $n \to \infty$ uniformly; (iii) tails can be completed back. Then one has

(98)
$$\int_\zeta^B e^{f(z)}\,dz \sim \frac{\omega}{3}\Gamma\left(\frac{1}{3}\right)\frac{e^{f(\zeta)}}{\sqrt[3]{-f'''(\zeta)/3!}},$$

where ω is a cube root of unity ($\omega^3 = 1$), dependent upon the position of the valley of B.

Proof. The proof is a simple adaptation of that of Theorem VIII.3. The heart of the matter is now the integration of

$$\int_C \exp\left(\frac{1}{3!}f'''(\zeta)(z-\zeta)^3\right)\,dz,$$

with C composed of the half-line connecting ζ to a point at infinity in the valley of $f''(\zeta)(z-\zeta)^3$ that contains B. A linear change of variable finally reduces the integral to the canonical form $\int e^{-w^3}$. ∎

▷ **VIII.47.** *Higher-order saddle-points.* For a saddle-point of order $p + 1$, the saddle-point formula reads

$$\int_\zeta^B e^{f(z)} \, dz \sim \frac{\omega}{p} \Gamma\left(\frac{1}{p}\right) \frac{e^{f(\zeta)}}{\sqrt[p]{-f^{(p)}(\zeta)/p!}},$$

where $\omega^p = 1$. ◁

▷ **VIII.48.** *Vanishing multipliers and multiple saddle-points.* This note supplements Note VIII.47. For a saddle-point of order $p + 1$ and an integrand of the form $(z - \zeta)^b \cdot e^{f(z)}$, the saddle-point formula must be modified according to

$$\int_0^\infty x^b e^{-ax^p/p!} \, dx = \frac{1}{p} \Gamma\left(\frac{b+1}{p}\right) \left(\frac{p!}{a}\right)^{(b+1)/p}.$$

Thus, the argument of the Γ factor is changed from $1/p$ to $(b + 1)/p$, as is the exponent of $f^{(p)}(\zeta)$ and of n in the case of large power estimates. ◁

Forests and coalescence of saddle-points. We give below an application to the counting of forests of unrooted trees made of a large number of trees. The analysis precisely involves a double saddle-point in a certain critical region. The problem is in particular relevant to the analysis of random graphs during the phase where a giant component has not yet emerged.

Example VIII.15. *Forests of unrooted trees.* The problem here consists in determining the number $F_{m,n}$ of ordered forests, i.e., sequences, made of m (labelled, non-plane) unrooted trees and comprised of n nodes in total. The number of unrooted trees of size n is, by virtue of Cayley's formula, n^{n-2} and its EGF is expressed as $U = T - T^2/2$, where T is the Cayley tree function satisfying $T = ze^T$. Consequently, we have

$$\frac{1}{n!} F_{m,n} = [z^n] \left(T(z) - \frac{T(z)^2}{2}\right)^m = \frac{1}{2i\pi} \int_{0+} \left(T - \frac{T^2}{2}\right)^m \frac{dz}{z^{n+1}}.$$

The case of interest here is when m and n are linearly related. We thus set $m = \alpha n$, where *a priori* $\alpha \in (0, 1)$. Then, the integral representation of $F_{m,n}$ becomes

(99) $\quad \dfrac{1}{n!} F_{m,n} = \dfrac{1}{2i\pi} \int_C e^{nh_\alpha(t)} (1 - t) \dfrac{dt}{t}, \qquad h_\alpha(t) := \alpha \log\left(1 - \dfrac{t}{2}\right) + t + (\alpha - 1) \log t,$

where C encircles 0. This has the form of a "large power" integral. Saddle-points are found as usual as zeros of the derivative h'_α; there are two of them given by

$$\zeta_0 = 2 - 2\alpha, \qquad \zeta_1 = 1.$$

For $\alpha < 1/2$, one has $\zeta_0 > \zeta_1$ while for $\alpha > 1/2$ the inequality is reversed and $\zeta_0 < \zeta_1$. In both cases, a simple saddle-point analysis succeeds, based on the saddle-point nearer to the origin; see Note VIII.49 below. In contrast, when $\alpha = 1/2$, the points ζ_0 and ζ_1 coalesce to the common value 1. In this last case, we have $h'_{1/2}(1) = h''_{1/2}(1) = 0$ while $h'''_{1/2}(1) = -2$ is non-zero: there is a *double saddle-point* at 1.

The number of forests thus presents two different regimes depending on whether $\alpha < 1/2$ or $\alpha > 1/2$, and there is a discontinuity of the analytic form of the estimates at $\alpha = 1/2$ (see Figure VIII.13). The situation is reminiscent of "critical phenomena" and phase transitions (e.g., from solid to liquid to gas) in physics, where such discontinuities are encountered. This provides a good motivation to study what happens right at the "critical" value $\alpha = 1/2$.

As in the analytic proof of the Lagrange Inversion Theorem it proves convenient to adopt $t = T$ as an independent variable, so that $z = te^{-t}$ becomes a dependent variable. Since

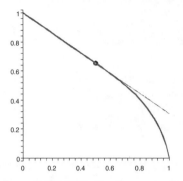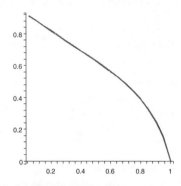

Figure VIII.13. The function $H(\alpha)$ governing the exponential rate of the number of forests exhibits a "phase transition" at $\alpha = 1/2$ (left); this is reflected by a plot of the quantity $\frac{1}{n}\log(F_{m,n}/n!)$, as a function of $\alpha = m/n$ for $n = 200$ (right).

$dz = (1 - t)e^{-t}$, this provides the integral representation, a special instance of (99):

$$\frac{1}{n!}F_{m,n} = \frac{1}{2i\pi}\int_{0+}\left(t - \frac{1}{2}t^2\right)^m e^{nt}(1 - t)\frac{dt}{t^{n+1}}.$$

We thus consider the special value $\alpha = 1/2$ and set $h \equiv h_{1/2}$. What is to be determined is therefore the number of forests of total size n that are made of $n/2$ trees, assuming naturally n even. Bearing in mind that the double saddle-point is at $\zeta = \zeta_0 = \zeta_1 = 1$, one has

$$h(z) = 1 - \frac{1}{3}(z - 1)^3 + O((z - 1)^4) \qquad (z \to 1).$$

Thus, upon neglecting the tails and localizing the integral to a disc centred at 1 with radius $\delta \equiv \delta(n)$ such that

$$n\delta^3 \to \infty, \qquad n\delta^4 \to 0$$

($\delta = n^{-3/10}$ is suitable), we have the asymptotic equivalence (with y representing $z - 1$)

$$(100) \qquad \frac{1}{n!}F_{m,n} = -\frac{e^{n(1 - \frac{1}{2}\log 2)}}{2i\pi}\int_D e^{-ny^3/3}y\,dy + \text{exponentially small},$$

where D is a certain (small) contour containing 0 obtained by transformation from C.

The discussion so far has left aside the choice of the contour C in (99), hence of the geometric aspect of D near 0, which is needed in order to fully specify (100). Because of the minus sign in the third derivative, $h'''(1) = -2$, the three steepest descent half-lines stemming from 1 have angles $0, e^{2i\pi/3}, e^{-2i\pi/3}$. This suggests the adoption, as original contour C in (99), of two symmetric segments stemming from 1 connected by a loop left of 0; see Figure VIII.14. Elementary calculations justify that the contour can be suitably dimensioned so as to remain always below level $h(1)$. See also the right-hand drawing of Figure VIII.14, in which the level curves of the valleys *below* the saddle-point are drawn, together with a legal contour of integration that winds about 0.

Once the original contour of integration has been fixed, the orientation of D in (100) is fully determined. After effecting the further change of variables $y = wn^{-1/3}$ and completing

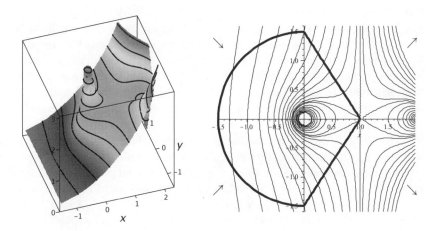

Figure VIII.14. Left: a plot of e^h with the double saddle-point at 1. Right: The level curves of e^h together with a legal integration contour through valleys.

the tails, we find

(101)
$$\frac{1}{n!}F_{m,n} \sim \frac{\lambda}{n^{2/3}}e^{n(1-\frac{1}{2}\log 2)}, \qquad \lambda = -\frac{1}{2i\pi}\int_E e^{-y^3/3}y\,dy,$$

where E connects $\infty e^{-2i\pi/3}$ to 0 then to $\infty e^{2i\pi/3}$. The evaluation of the integral giving λ is now straightforward (in terms of the Gamma function), which yields the following corollary.

Proposition VIII.11. *The number of forests of total size n comprised of $n/2$ unrooted Cayley trees satisfies*

$$\frac{1}{n!}F_{n/2,n} \sim 2 \cdot 3^{-1/3}\Gamma(2/3)e^{n(1-\frac{1}{2}\log 2)}n^{-2/3}.$$

The number *three* is characteristically ubiquitous in the formula. (Furthermore, the formula displays the exponent $2/3$ instead of $1/3$ in the general case (98) because of the additional factor $(1 - z)$ present in the integral representation (99), which vanishes at the saddle-point 1; see Note VIII.48.) .... ∎

The problem of analysing random forests composed of a large number of trees has been first addressed by the Russian School, most notably Kolchin and Britikov. We refer the reader to Kolchin's book [387, Ch. I] where nearly thirty pages are devoted to a deeper study of the number of forests and of associated parameters. Kolchin's approach is however based on an alternative presentation in terms of sums of independent random variables and stable laws of index $3/2$, so that it is limited to first order asymptotics. As it turns out there is a striking parallel with the analysis of the growth of the random graph in the critical region, when the random graph stops resembling a large collection of disconnected tree components.

An almost sure sign of (hidden or explicit) monkey saddles is the presence of $\Gamma(1/3)$ factors in the final formulae and cube roots in expressions involving n. It is in fact possible to go much further than we have done here with the analysis of forests (where we have stayed right at the critical point) and provide asymptotic expressions that describe the transition between regimes, here from $A^n n^{-1/2}$, to $B^n n^{-2/3}$, then to $C^n n^{-1/2}$. The analysis then appeals to the theory of coalescent saddle-points well developed by applied mathematicians (see, e.g., the presentation in [75, 465, 614]) and the already evoked rôle of the Airy function. We do not pursue this thread further since it properly belongs to multivariate asymptotics. It is developed in a detailed manner in an article of Banderier, Flajolet, Schaeffer, and Soria [28] relative to the size of the core in a random map, on which our presentation of forests has been modelled (see also Example IX.42, p. 713).

The results of several studies conducted towards the end of the previous millennium do suggest that, among threshold phenomena and phase changes, there is a fair amount of universality in descriptions of combinatorial and probabilistic problems by means of multiple and coalescing saddle-points. In particular $\Gamma(1/3)$ factors and the Airy function surface recurrently in the works of Flajolet, Janson, Knuth, Łuczak and Pittel [241, 354], which are relative to the Erdős–Renyi random graph model in its critical phase; see also [254] for a partial explanation. The occurrence of the Airy area distribution (in the context of certain polygon models related to random walks) can be related to this orbit of techniques, as first shown by Prellberg [496], and strong numerical evidence evoked in Chapter V (p. 365) suggests that this might extend to the difficult problem of self-avoiding walks [509]. Airy-related distributions also appear in problems relative to the satisfiability of random boolean expressions [77], the path length of trees (Proposition VII.15, p. 534 and [567, 565, 566]), as well as cost functionals of random allocations (Note VII.54, p. 534 and [249]). The reasons are sometimes well understood in separate contexts by probabilists, statistical physicists, combinatorialist, and analysts, but a global framework is still lacking.

▷ **VIII.49.** *Forests and simple saddle-points.*   When $0 < \alpha < 1/2$, the number of forests satisfies, for some computable $C_-(\alpha)$:

$$\frac{1}{n!} F_{n,m} \sim C_-(\alpha) \frac{e^{H_-(\alpha)}}{n^{1/2}}, \qquad H_-(\alpha) = 1 - \alpha \log 2.$$

When $1/2 < \alpha < 1$, the number of forests satisfies, for some computable $C_+(\alpha)$:

$$\frac{1}{n!} F_{n,m} \sim C_+(\alpha) \frac{e^{H_+(\alpha)}}{n^{1/2}}, \qquad H_+(\alpha) = \alpha \log \alpha + 2 - 2\alpha + (\alpha - 1) \log(2 - 2\alpha).$$

This results from a routine *simple saddle-point* analysis at ζ_1 and ζ_0, respectively.   ◁

VIII. 11. Perspective

One of the pillars of classical analysis, the saddle-point method plays a major rôle in analytic combinatorics. It provides an approach to coefficient asymptotics and can handle combinatorial classes that are not amenable to singularity analysis. The simplest case is that of urns, whose generating function e^z has no singularities at a finite distance. Similar functions commonly arise as composed SET constructions. Broadly speaking, for the class of generating functions that arise from the combinatorial constructions of Part A of this book, singularity analysis is effective for functions that have moderate growth at their singularities; the saddle-point method is effective otherwise.

The essential idea behind the saddle-point method is simple, and it is very easy to get good bounds on coefficient growth. In effect, for combinatorial generating functions, the Cauchy coefficient integral defines a surface with a well-defined saddle-point somewhere along the positive real axis, and choosing a circle centred at the origin and passing through the saddle-point already provides useful bounds by elementary arguments. The essence of the full saddle-point method is the development of more precise bounds, which are obtained by splitting the contour into two parts and balancing the associated errors.

Combinatorial classes that are amenable to saddle-point analysis have so far only been incorporated into relatively few schemas, compared to what we saw for singularity analysis. The consistency of the approach certainly argues for the existence of many more such schemas. A positive signal in that direction is the fact that several researchers have developed concepts of admissibility that serve to delineate classes of function for which the saddle-point method boils down to verifying simple conditions.

The saddle-point method also provides insights in more general contexts. Most notably, the general results on analysis of large powers lay the groundwork for distributional analyses and limit laws, which are the subject of the next chapter.

Bibliographic notes. Saddle-point methods take their sources in applied mathematics, one of them being the asymptotic analysis by Debye (1909) of Bessel functions of large order. (In fact, there are early signals of its use by Riemann in relation to hypergeometric functions [511] and to the zeta function, as noted by Edwards [186, p. 139], as well as traces of it in works of Cauchy published in 1827: see the scholarly study by Petrova and Solov'ev [483].) Saddle-point analysis is sometimes called steepest descent analysis, especially when integration contours strictly coincide with steepest descent paths. Saddle-points themselves are also called critical points (*i.e.*, points where a first derivative vanishes). Because of its roots in applied mathematics, the method is well covered by the literature in this area, and we refer to the books by Olver [465], Henrici [329], or Wong [614] for extensive discussions. A vivid introduction to the subject is to be found in De Bruijn's book [143]. We also recommend Odlyzko's impressive survey [460].

To a large extent, saddle-point methods were introduced into the world of combinatorial enumerations in the 1950s. Early combinatorial papers were concerned with permutations (involutions) or set partitions: this includes works by Moser and Wyman [448, 449, 450] that are mostly directed towards entire functions.

Hayman's approach [325] which we have expounded here (see also [614]) is notable in its generality as it envisions saddle-point analysis in an abstract perspective, which makes it possible to develop general closure theorems. A similar thread was followed by Harris and Schoenfeld who gave stronger conditions allowing for full asymptotic expansions [323]; Odlyzko and Richmond [462] were successful in connecting these conditions with Hayman admissibility. Another valuable work is Wyman's extension to non-positive functions [624].

Interestingly enough, developments that parallel the ones in analytic combinatorics have taken place in other regions of mathematics. Erwin Schrödinger introduced saddle-point methods in his lectures [535] at Dublin in 1944 in order to provide a rigorous foundation to some models of statistical physics that closely resemble balls-in-bins models. Daniels' publication [136] of 1954 is a historical source for saddle-point techniques in probability and statistics, in which refined versions of the central limit theorem can be obtained. (See for instance the description in Greene and Knuth's book [310].) Since then, the saddle-point method has proved a useful tool for deriving Gaussian limiting distributions. We have given here some idea of this approach which is to be developed further in Chapter IX, where we shall discuss some of Canfield's results [101]. Analytic number theory also makes a heavy use of saddle-point analysis. In additive number theory, the works by Hardy, Littlewood, and Ramanujan relative to integer partitions have been especially influential, see for instance Andrews' book [14] and Hardy's *Lectures* on Ramanujan [321] for a fascinating perspective. (In multiplicative number theory, generating functions take the form of Dirichlet series while Perron's formula replaces Cauchy's formula. For saddle-point methods in this context, we refer to Tenenbaum's book [576] and his seminar survey [575].)

A more global perspective on limit probability distributions and saddle-point techniques will be given in the next chapter, since there are strong relations to the quasi-powers framework developed there, to local limit laws, and to large deviation estimates. General references for some of these aspects of the saddle-point method are the articles of Bender–Richmond [45], Canfield [101], Gardy [280, 281, 282], and Gittenberger–Mandlburger [292]. With regard to multiple saddle-points and phase transitions, we refer the reader to references provided at the end of Section VIII. 10, on p. 605.

Part C

RANDOM STRUCTURES

Un problème relatif aux jeux du hasard,
proposé à un austère janseniste par un homme du monde,
a été à l'origine du Calcul des Probabilités[1].

— SIMÉON-DENIS POISSON

Analytic combinatorics concerns itself with the elucidation of properties of combinatorial structures in relation to algebraic and analytic properties of generating functions. The most basic cases are the enumeration of combinatorial classes and the analysis of moments of combinatorial parameters. These involve generating functions in one (formal or complex) variable as discussed extensively in previous chapters and represent essentially *univariate* problems.

Many applications, in various sciences as well as in combinatorics itself, require quantifying the behaviour of *parameters* of combinatorial structures. The corresponding problems are now of a *multivariate* nature, as one typically wants a way to estimate the number of objects in a combinatorial class having a fixed size *and* a given parameter value. Average-case analyses usually do not suffice, since it is often important to predict what is likely to be observed in simulations or on actual data that obey a given

---

[1] *"A problem relative to games of chance proposed to an austere Jansenist by a man of the world has been at the origin of the calculus of probabilities."* Poisson refers here to the fact that questions of betting and gambling posed by the Chevalier de Méré (who was both a gambler and a philosopher) led Pascal (an austere religious man) to develop some of the first foundations of probability theory.

randomness model, in terms of possible deviations from the mean—this signifies that information on *probability distributions* is wanted. Useful but crude estimates are derived from the moment inequalities developed in Section III. 2.2, p. 161. However, much more is usually true. Indeed, it is frequently observed that the histograms of the distribution of a combinatorial parameter (for varying size values) exhibit a common characteristic "shape", as the size of the random combinatorial structure tends to infinity. In this case, we say that there exists a *limit law*. Our goal in this chapter is precisely to introduce a methodology for distilling limit laws from combinatorial specifications.

In simpler cases, limit laws are *discrete* and, when this happens, they often turn out to be of the geometric or Poisson type. In many other situations, limit laws are *continuous*, a case of prime importance being the Gaussian law associated with the famous bell-shaped curve, which is found so often to occur in elementary combinatorial structures. This chapter develops a coherent set of analytic techniques dedicated to extracting such discrete and continuous laws by exploiting properties of bivariate generating functions. The starting point is provided by symbolic methods of Part A (especially Chapter III), which enable us to derive systematically *bivariate generating functions* for many natural parameters of combinatorial structures. The methods presented here then combine complex asymptotic techniques of Part B with a small selection of fundamental theorems from the analytic side of classical probability theory recalled in Appendix C (*Complements of Probability Theory*).

Under the theory to be expounded, bivariate generating functions are processed analytically as follows. The auxiliary variable marking the combinatorial parameter of interest is regarded as inducing a *deformation* of the (univariate) counting generating function. The way in which such deformations affect the type of singularity of the counting generating functions can then be studied: a *perturbation* of univariate singularity analysis is often sufficient to derive an asymptotic estimate of the probability generating function of a given parameter, when taken over objects of some large size. Continuity theorems from probability theory finally allow us to conclude on the existence of a limit law and characterize it.

An especially important component of this paradigm is the framework of "*quasi-powers*". Large powers tend to occur in the asymptotic form of coefficients of counting generating functions (think of radius of convergence bounds and ρ^{-n} factors). The collection of deformations of a counting generating function is then likely to induce for the corresponding coefficients a collection of approximations that also *asymptotically* involve large powers—technically, these are referred to as quasi-powers. From this, a Gaussian law is derived along lines that are somewhat reminiscent of the classical Central Limit Theorem of probability theory, which expresses the asymptotically Gaussian character of sums of independent random variables.

This chapter starts with an informal introduction to limit laws, either discrete or continuous (Section IX. 1). Sections IX. 2 and IX. 3 then present methods and examples relative to discrete laws in combinatorics. Continuous limit laws form the subject of Section IX. 4, dedicated to general methodology, and Section IX. 5 where the quasi-powers framework is introduced. Three sections, IX. 6, IX. 7, and IX. 8, then develop the extension of meromorphic asymptotics, singularity analysis, and saddle-point methods to the characterization of Gaussian limit laws in combinatorics. Additional properties, such as local limits and large deviations, form the subject of Sections IX. 9 and IX. 10, respectively. The chapter concludes with a discussion of non-Gaussian limits (in particular stable laws, Section IX. 11) and multivariate problems (Section IX. 12).

In the business of limit laws in combinatorics, as true elsewhere, the spirit is more important than the letter. That is, methods are often more important than theorems, whose statements may involve somewhat intricate technical conditions. We have made every effort to expound the former in a "conceptual" manner, but shall try our best to avoid the latter.

Within the perspective of analytic combinatorics, the direct relation that can be established between combinatorial specifications and asymptotic properties, in the form of limit laws, is striking and is a characteristic feature of the theory. In particular, all the *schemas* previously introduced in this book lead to well-characterized limit laws. As we shall see throughout this chapter, almost any basic law of probability theory and statistics is likely to occur somewhere in combinatorics; conversely, almost any simple combinatorial parameter is likely to be governed by a limit law.

IX. 1. Limit laws and combinatorial structures

What is given is a combinatorial class $\mathcal{F}$, labelled or unlabelled, and an integer valued combinatorial parameter χ. There results both a family of probabilistic models, namely for each n the uniform distribution over $\mathcal{F}_n$ that assigns to any $\gamma \in \mathcal{F}_n$ the probability

$$\mathbb{P}(\gamma) = \frac{1}{F_n}, \qquad \text{with} \quad F_n = \text{card}(\mathcal{F}_n),$$

and a corresponding family of random variables obtained by restricting χ to $\mathcal{F}_n$. Under the uniform distribution over $\mathcal{F}_n$, we then have

$$\mathbb{P}_{\mathcal{F}_n}(\chi = k) = \frac{1}{F_n} \text{card} \left\{ \gamma \in \mathcal{F}_n \mid \chi(\gamma) = k \right\}.$$

We write $\mathbb{P}_{\mathcal{F}_n}$ to indicate the probabilistic model relative to $\mathcal{F}_n$, but also freely abbreviate it to $\mathbb{P}_n$ or write the probability distribution as $\mathbb{P}(\chi_n = k)$, whenever $\mathcal{F}$ is clear from context.

As n increases, the histograms of the distribution of χ_n often share a common profile; see Example IX.1 and Figure IX.1 for two elementary parameters, one leading to a discrete law, the other to a continuous limit. It is from such observations that the notion of a *limit law* is abstracted.

***Example* IX.1.** *Binary words: elementary approach.* Consider the class $\mathcal{W}$ of binary words over $\{a, b\}$. We examine two parameters purposely chosen simple enough, so that explicit expressions are available for the probability distributions at stake. Define the parameters

$$\chi(w) := \text{number of initial } a \text{ in } w, \qquad \xi(w) := \text{total number of } a \text{ in } w,$$

and the corresponding counts,

$$W^\chi_{n,k} := \text{card}\{w \in \mathcal{W}_n \mid \chi(w) = k\}, \qquad W^\xi_{n,k} := \text{card}\{w \in \mathcal{W}_n \mid \xi(w) = k\}.$$

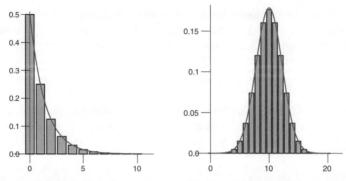

Figure IX.1. Histograms of probability distributions for the number of initial a in a random binary string for $n = 10$ (χ: left) and the total number of a for $n = 20$ (ξ: right). The histogram corresponding to χ is not normalized and direct convergence to a discrete geometric law is apparent; for ξ, the horizontal axis is scaled to n, and the histogram closely matches the bell-shaped curve that is characteristic of a continuous Gaussian limit.

Explicit expressions result from elementary combinatorics: for $0 \le k \le n$, we have

$$W^{\chi}_{n,0} = 2^{n-1}, \quad W^{\chi}_{n,1} = 2^{n-2}, \cdots, W^{\chi}_{n,n-1} = 1, \quad W^{\chi}_{n,n} = 1; \quad W^{\xi}_{n,k} = \binom{n}{k}.$$

The probability distributions are accordingly ($[\![\cdot]\!]$ is Iverson's notation for the indicator function):

$$\begin{cases} \mathbb{P}_{\mathcal{W}_n}(\chi = k) & = & \dfrac{1}{2^{k+1}}[\![0 \le k < n]\!] + \dfrac{1}{2^n}[\![k = n]\!], \\[3mm] \mathbb{P}_{\mathcal{W}_n}(\xi = k) & = & \dfrac{1}{2^n}\dbinom{n}{k}. \end{cases}$$

The probabilities relative to χ then resemble, in the asymptotic limit of large n, the geometric distribution. Indeed, one has, for each k,

$$\lim_{n\to\infty} \mathbb{P}_{\mathcal{W}_n}(\chi = k) = \frac{1}{2^{k+1}} \quad \text{and} \quad \lim_{n\to\infty} \mathbb{P}_{\mathcal{W}_n}(\chi \le k) = 1 - \frac{1}{2^{k+1}}.$$

We say that there is a *discrete limit law* of the geometric type for χ.

In contrast, the parameter ξ taken over $\mathcal{W}_n$ has mean $\mu_n := n/2$ and standard deviation $\sigma_n := \frac{1}{2}\sqrt{n}$. One should then centre and scale the parameter ξ, introducing the "standardized" (or "normalized") random variable

$$(1) \qquad\qquad X^{\star}_n := \frac{\xi_n - \mathbb{E}(\xi_n)}{\sqrt{\mathbb{V}(\xi_n)}} = \frac{\xi_n - n/2}{\frac{1}{2}\sqrt{n}}.$$

It then becomes possible to examine the (cumulative) distribution function $\mathbb{P}(X^{\star}_n \le y)$ for fixed values of y. In terms of ξ itself, we are considering $\mathbb{P}(\xi_n \le \mu_n + y\sigma_n)$ for real values of y. Then, the classical approximation of the binomial coefficients yields the approximation (Note IX.1):

$$(2) \qquad\qquad \lim_{n\to\infty} \mathbb{P}(\xi_n \le \mu_n + y\sigma_n) = \frac{1}{\sqrt{2\pi}}\int_{-\infty}^{y} e^{-t^2/2}\, dt.$$

We now say that there is a *continuous limit law* of the Gaussian type for ξ. .... ∎

▷ **IX.1.** *Local and central approximations of the binomial law.* Equation (2) is classically derived by summation from the "local" approximation,

$$(3) \qquad \frac{1}{2^n} \binom{n}{\frac{1}{2}n + \frac{1}{2}y\sqrt{n}} = \frac{e^{-y^2/2}}{\sqrt{\pi n/2}} \left(1 + O\left(\frac{y^3}{\sqrt{n}} \right) \right),$$

valid for $y = o(n^{1/6})$. A proof of (3) can be obtained by the method of De Moivre (1721), see Note III.3, p. 160, or by Stirling's formula.   ◁

Combinatorial distributions and limit laws. In accordance with the general notion of convergence in distribution (or weak convergence, see Appendix C.5: *Convergence in law*, p. 776), we shall say that a *limit law* exists for a parameter if there is *convergence* of the corresponding family of *cumulative distribution functions*. In virtually all cases[2] encountered in this book, there are, like in Example IX.1, two major types of convergence that the *a priori* discrete distribution of a combinatorial parameter may satisfy:

$$\boxed{\text{Discrete} \longrightarrow \text{Discrete}} \qquad \text{and} \qquad \boxed{\text{Discrete} \longrightarrow \text{Continuous}}.$$

Regarding the discrete-to-discrete case, convergence is established without standardizing the random variables involved. In the discrete-to-continuous case, the parameter is to be centred at its mean and scaled by its standard deviation, as in (1).

There is also interest in obtaining a *local limit law*, which, when available, quantifies individual probabilities (rather than the cumulative distribution functions). In the discrete-to-discrete case, the distinction between local and "global" limits is immaterial, since the existence of one type of law implies the other. In the discrete-to-continuous case, the local limit is expressed in terms of a fixed probability density, as in (3), and is technically more demanding to derive, since stronger analytic properties are required.

The *speed of convergence* in a limit law describes the way the finite combinatorial distributions approach their asymptotic limit. It provides useful information on the quality of asymptotic approximations for finite n models.

Finally, quantifying the "risk" of extreme configurations, far away from the mean, necessitates estimates on the *tails* of the distributions. Such estimates belong to the theory of *large deviation* and they constitute a useful complement to the study of central and local limits. These various notions are summarized in Figure IX.2.

Classical probability theory has elaborated highly useful tools for analysing limit distributions. For each of the major two types, a *continuity theorem* provides conditions under which convergence in law can be established from convergence of transforms. The transforms in question are *probability generating functions (PGFs)* for the discrete case, *characteristic functions* or *Laplace transforms* otherwise. Refinements, known as the Berry–Esseen inequalities relate speed of convergence of the combinatorial distributions to their limit on the one hand, and a distance between transforms on the other. Put otherwise, distributions are close if their transforms are close. Large deviation estimates are finally obtained by a technique of "shifting the mean", which is otherwise familiar in probability and statistics.

---

[2] See, however, the case of longest runs in words in Example V.4, p. 308, for a family of discrete distributions that need centring.

Limit law: An asymptotic approximation of the *cumulative distribution function* of a combinatorial parameter in terms of the cumulative distribution function of a fixed random variable, called the "limit". Thus one estimates $\mathbb{P}_n(\chi \leq k)$. Centring and scaling, a process called *standardization*, is needed in the case of a continuous limit.

Local limit law: A direct asymptotic estimate of "local values" of the combinatorial probabilities, $\mathbb{P}_n(\chi = k)$. In the discrete case, existence of basic and local limits are logically equivalent properties. In the continuous case, standardization is needed and the resulting estimate is expressed in terms of the *density* of a fixed continuous random variable.

Tail estimates and large deviations: For a given distribution, tail estimates are asymptotic estimates of the probability of deviating from the mean by a large quantity. Large deviation estimates quantify the tail probabilities of a family of distributions, when these decay at an exponential rate (in a suitable scale).

Speed of convergence: An upper bound on the error in asymptotic estimates.

Figure IX.2. An informal summary of the main notions of relevance to the analysis of combinatorial distributions.

Limit laws and bivariate generating functions. In this chapter, the starting point of a distributional analysis is invariably a bivariate generating function

$$F(z, u) = \sum_{n,k} f_{n,k} u^k z^n,$$

where $f_{n,k}$ represents (up to a possible normalization factor) the number of structures of size n in some class $\mathcal{F}$. What is sought is asymptotic information relative to the array of coefficients

$$f_{n,k} = [z^n u^k] F(z, u).$$

Thus, a *double coefficient extraction* is to be effected. This task could in principle be approached by an iterated use of Cauchy's coefficient formula,

$$[z^n u^k] F(z, u) = \left(\frac{1}{2i\pi}\right)^2 \int_\gamma \int_{\gamma'} F(z, u) \frac{dz}{z^{n+1}} \frac{du}{u^{k+1}},$$

but this approach is hard to carry out[3] and, under our current stage of knowledge, it appears to be less general than the path taken in this chapter.

Here is a broad outline of the principles behind the theory to be developed in the next few sections of this chapter. First, as we know all too well, the specialization at $u = 1$ of $F(z, u)$ gives the counting generating function of $\mathcal{F}$, that is, $F(z) = F(z, 1)$. Next, as seen repeatedly starting from Chapter III, the moments of the combinatorial distribution $\{f_{n,k}\}$ for fixed n and varying k are attainable through the partial derivatives at $u = 1$, namely

$$\text{first moment} \quad \leftrightarrow \quad \left.\frac{\partial}{\partial u} F(z, u)\right|_{u=1}, \qquad \text{second moment} \quad \leftrightarrow \quad \left.\frac{\partial^2}{\partial u^2} F(z, u)\right|_{u=1},$$

---

[3] A collection of recent works by Pemantle and coauthors [474, 475, 476] shows, however, that a well-defined class of bivariate asymptotic problems can be attacked by the theory of functions of several complex variables and a detailed study of the geometry of a singular variety.

Problem	GF	u-region	Reference
counting	$F(z, 1)$	$u = 1$	Ch. I and II
moments	$\left.\dfrac{\partial^r}{\partial u^r} F(z, u)\right\|_{u=1}$	$u = 1 \pm o(1)$	Ch. III
Discrete laws			
limit law	$F(z, u)$	$u \in \Omega \subseteq \{\|u\| \le 1\}$	Th. IX.1, p. 624
tails	$F(z, u)$	$\|u\| = r, \quad r > 1$	Th. IX.3, p. 627
Continuous laws			
limit law, Gaussian	$F(z, u)$	$u \in \Omega; \quad \Omega \subset \mathbb{C}, 1 \in \Omega$	Th. IX.8, p. 645
local Limit Law	$F(z, u)$	$u \in \Omega \cup \{\|u\| = 1\}$	Th. IX.14, p. 696
large deviations	$F(z, u)$	$u \in [1 - \delta, 1 + \delta']$	Th. IX.15, p. 700

Figure IX.3. A summary of the correspondence between analytic properties of bi-variate generating functions (BGFs) and probabilistic properties of combinatorial distributions.

and so on. In summary: *Counting is provided by the bivariate generating function $F(z, u)$ taken at $u = 1$; moments result from the bivariate generating function taken in an* infinitesimal *neighbourhood of $u = 1$.*

Our approach to limit laws will then be as follows. The goal is to estimate the "horizontal" generating function

$$f_n(u) := \sum_k f_{n,k} u^k \equiv [z^n] F(z, u),$$

which is proportional to the *probability generating function* of χ taken over $\mathcal{F}_n$, since $\mathbb{E}_{\mathcal{F}_n}(u^\chi) = f_n(u)/f_n(1)$. The problem is viewed as a single coefficient extraction (extracting the coefficient of z^n) but *parameterized* by u—see our paragraph on *"singularity perturbation"* below for a brief discussion. Thanks to the availability of continuity theorems, the following can then be proved for a great many cases of combinatorial interest: *The existence and the shape of the limit law are derived from an asymptotic estimate of $f_n(u)$, when u is taken in a* fixed *neighbourhood of 1, which estimate depends on the behaviour of the generating function $z \mapsto F(z, u)$, for $u \approx 1$.* This is the basic paradigm of analysis explored throughout most of the chapter.

In addition, thanks to Berry–Esseen inequalities, *the quality of a* uniform *asymptotic estimate for $f_n(u)$ translates into a speed of convergence estimate for the corresponding limit law.* Also, for the discrete-to-continuous case, as we shall see in Section IX. 9 based on the saddle-point method, *local limit laws* are derived from consideration of the generating function $z \mapsto F(z, u)$, when u is assigned *values on the unit circle*, $\|u\| = 1$. In that case, the secondary inversion, with respect to u, is effected by the saddle-point method, rather than by continuity theorems—the principles extend the analysis of large powers presented in Section VIII. 8, p. 585. Finally, *large deviation* estimates are found to arise from estimates of $f_n(u)$ when u is real and either $u < 1$ (left tail) or $u > 1$ (right tail), this property being simply a reflection of saddle-point bounds; see Section IX. 10.

The correspondence between analytic properties of bivariate generating functions and probabilistic properties of distributions is summarized in Figure IX.3; see also the diagram of Figure IX.9 (p. 649) specialized to continuous limit laws.

Singularity perturbation. As seen throughout Chapters IV–VIII, analytic combinatorics approaches the univariate problem of counting objects of size n starting from the Cauchy coefficient integral,

$$[z^n]F(z) = \frac{1}{2i\pi} \int_\gamma F(z) \frac{dz}{z^{n+1}}.$$

The singularities of $F(z)$ can be exploited, whether they are of a polar type (Chapters IV and V), algebraic–logarithmic of singularity analysis class (Chapters VI and VII) or essential and amenable to the saddle-point method (Chapter VIII).

From the discussion above, crucial information on combinatorial distributions is accessible from the bivariate generating function $F(z, u)$ when u varies in some domain containing 1. This suggests to consider $F(z, u)$ not so much as an analytic function of two complex variables, where z and u would play a symmetric rôle, but rather as a collection of functions of z indexed by a secondary parameter u. In other words, $F(z, u)$ is considered as a *deformation* of $F(z) \equiv F(z, 1)$ when u varies in a domain containing $u = 1$. Cauchy's coefficient integral gives

$$f_n(u) \equiv [z^n]F(z, u) = \frac{1}{2i\pi} \int_\gamma F(z, u) \frac{dz}{z^{n+1}}.$$

For $u = 1$, an asymptotic form of $f_n(1) = [z^n]F(z, 1)$ is obtained by suitable contour integration techniques of Part B. We can then examine the way the parameter u affects the asymptotic coefficient extraction process[4], with the goal of deriving an asymptotic estimate of $f_n(u)$, when u is close to 1. Such an approach is called a *singularity perturbation analysis*. For instance, a singularity of $F(z, 1)$ at $z = \rho$ typically implies for the coefficients of $F(z, 1)$ an estimate of the form $f_n(1) \approx \rho^{-n} n^\alpha$, and, in lucky cases (of which there are many, see Sections IX.6 and IX.7), this univariate analysis can be extended, resulting in an estimate of the form $f_n(u) \approx \rho(u)^{-n} n^\alpha$. Under such circumstances, the probability generating function of the parameter χ associated to $F(z, u)$ satisfies the estimate

$$(4) \qquad\qquad \mathbb{E}_{\mathcal{F}_n}(u^\chi) \equiv \frac{f_n(u)}{f_n(1)} \approx \left(\frac{\rho(u)}{\rho(1)}\right)^{-n}.$$

This analytical form is reminiscent of the central limit theorem of probability theory, according to which large powers of a fixed PGF (corresponding to sums of a large number of independent random variables) entail convergence to a Gaussian law[5]— such a law is indeed obtained here. In this chapter, we are going to see numerous applications of this strategy, which we now briefly illustrate by revisiting the case of binary words from Example IX.1.

---

[4]The essential feature of the analysis of coefficients of GFs by means of complex analytic techniques, as developed in Chapters IV–VIII, is to be robust: being based on contour integrals, it is usually amenable to smooth perturbations and provides *uniform* error terms.

[5]See also Section VIII.8, p. 585.

***Example* IX.2.** *Binary words: the BGF approach.* Regarding binary words and the two parameters χ (initial run of a's) and ξ (total number of a's), the general strategy of singularity perturbation starts from the BGFs,

$$\begin{cases} \mathcal{W}^{\chi} = \text{SEQ}(ua)\,\text{SEQ}(b\,\text{SEQ}(a)) & \implies \quad W^{\chi}(z, u) = \dfrac{1}{1 - uz}\dfrac{1}{1 - \frac{z}{1-z}} \\[3mm] \mathcal{W}^{\xi} = \text{SEQ}(ua + b) & \implies \quad W^{\xi}(z, u) = \dfrac{1}{1 - (zu + z)}, \end{cases}$$

and it instantiates as follows.

Consider the secondary variable u fixed at some value u_0. In the case of W^{χ}, there are two components in the BGF

$$W^{\chi}(z, u_0) = \frac{1}{1 - u_0 z} \cdot \boxed{\frac{1 - z}{1 - 2z}},$$

and the dominant singular part, with a simple pole at $z = 1/2$, arises from the second factor as long as $|u_0| < 2$. Accordingly, one has

$$W^{\chi}(z, u_0) \underset{z \to 1/2}{\sim} \frac{1/2}{1 - u_0/2} W(z) \qquad \text{implying} \qquad [z^n] W^{\chi}(z, u_0) \sim \frac{1/2}{1 - u_0/2} 2^n.$$

The probability generating function of χ over $\mathcal{W}_n$ is then obtained upon dividing by 2^{-n},

$$\mathbb{E}_{\mathcal{W}_n}\left(u_0^{\chi}\right) = \frac{1}{2^n}[z^n]W^{\chi}(z, u_0) \sim \frac{1/2}{1 - u_0/2} = \sum_{k=0}^{\infty} \frac{1}{2^{k+1}} u_0^k,$$

where the last expression is none other than the probability generating function of a discrete law, namely, the geometric distribution of parameter $1/2$. As we shall see in section IX. 2 where we enunciate a continuity theorem for probability generating functions, this is enough to conclude that the distribution of χ converges to a geometric law.

In the second case, that of W^{ξ}, the auxiliary parameter modifies the location of the singularity,

$$W^{\xi}(z, u_0) = \frac{1}{1 - z\boxed{(1 + u_0)}}.$$

Then, the (unique) singularity smoothly moves,

$$\rho(u_0) = \frac{1}{(1 + u_0)}$$

as u_0 varies, while the type of singularity (here a simple pole) remains the same—we thus encounter an extremely simplified form of (4). Accordingly, the coefficients $[z^n]W^{\xi}(z, u_0)$ are described by a "large power" formula (here of an exact type, as in Section VIII. 8, p. 585). As regards the probability generating function of ξ over $\mathcal{W}_n$, one has

$$\mathbb{E}_{\mathcal{W}_n}\left(u^{\xi}\right) = \frac{1}{2^n}[z^n]W^{\xi}(z, u_0) = \left(\frac{1}{2\rho(u_0)}\right)^n.$$

In the perspective of the present chapter, this last form (here especially simple) is amenable to continuity theorems for integral transforms (Section IX. 4). There results a continuous limit law of the Gaussian type in this case. .... ■

It is typical of the approach taken in this chapter that, once equipped with suitably general theorems, it is hardly more difficult to discuss the number of leaves in a non-plane unlabelled tree or the number of summands in a composition into primes.

$F(z, u)$ for $u \approx 1$	type of law	method and schemas	
Sing. + exp. fixed	Discrete limit	Subcritical composition	§IX. 3
	(Neg. bin., Poisson, ...)	Subcritical Seq., Set, ...	§IX. 3
Sing. moves, exp. fixed	Gaussian(n, n)	Supercritical composition	
—	—	Meromorphic perturb.	§IX. 6
—	—	(Rational functions)	§IX. 6
—	—	Sing. analysis perturb.	§IX. 7
—	—	(Alg., implicit functions)	§IX. 7.3
Sing. fixed, exp. moves	Gaussian($\log n, \log n$)	(Exp-log structures)	§IX. 7.1
—	—	(Differential eq.)	§IX. 7.4
Sing. + exp. move	Gaussian	[Gao–Richmond [277]]	
Essential singularity	often Gaussian	Saddle-point perturb.	§IX. 8
Discontinuous type	non-Gaussian	Various cases	§IX. 11
—	stable	Critical composition	§IX. 11.2

Figure IX.4. A rough typology of bivariate generating functions $F(z, u)$ and limit laws studied in this chapter, based on the way singularities and exponents evolve for $u \approx 1$.

The foregoing discussion rightly suggests that a "minor" perturbation of bivariate generating function that affects neither the location nor the nature of the singularity points to a discrete limit law. A "major" change, in location or in exponent, is conducive to a continuous limit law, of which the prime example is the normal distribution. Figure IX.4 outlines a typology of limit laws summarizing the spirit of this chapter: a bivariate generating function $F(z, u)$ is to be analysed; the deformation induced by u affects the type of singularity of $F(z, u)$ in various ways, and an adapted complex coefficient extraction provides corresponding limit laws.

IX. 2. Discrete limit laws

This section provides the basic analytic–probabilistic technology needed for the discrete-to-discrete situation, where the distribution of a (discrete) combinatorial parameter tends (without normalization) to a discrete limit. The corresponding notion of convergence is examined in Subsection IX. 2.1. Probability generating functions (PGFs) are important since, by virtue of a continuity theorem stated in Subsection IX. 2.2, convergence in distribution is implied by convergence of PGFs. At the same time, the fact that PGFs of two distributions are close implies that the original distribution functions are close. Finally, tail estimates for a distribution can be easily related to analytic continuation of the PGFs, a basic property discussed in Subsection IX. 2.3. This section organizes some general tools and accordingly we limit ourselves to a single combinatorial application, that of the number of cycles of some fixed size in a random permutation. The next section will provide a number of applications to random combinatorial structures.

This and the next section feature three classical discrete laws described in Appendix C.4: *Special distributions*, p. 774. For our reader's convenience, their definitions are recalled in Figure IX.5,

Distribution	probabilities	PGF
geometric (q)	$(1-q)q^k$	$\dfrac{1-q}{1-qu}$
negative binomial$[m]$ (q)	$\dbinom{m+k-1}{k}q^k(1-q)^m$	$\left(\dfrac{1-q}{1-qu}\right)^m$
Poisson (λ)	$e^{-\lambda}\dfrac{\lambda^k}{k!}$	$e^{\lambda(1-u)}$

Figure IX.5. The three major discrete laws of analytic combinatorics: the geometric, negative binomial, and Poisson laws.

IX. 2.1. Convergence to a discrete law. In order to specify precisely what a limit law is, we base ourselves on the general context described in Appendix C.5: *Convergence in law*, p. 776. The principles presented there provide for what must be the "right" notion convergence of a family of discrete distributions to a limit discrete distribution. Here is a self-standing definition, particularized to the cases of interest here.

Definition IX.1 (Discrete-to-discrete convergence). *The discrete random variables X_n supported by $\mathbb{Z}_{\geq 0}$ are said to* converge in law, *or* converge in distribution, *to a discrete variable Y supported by $\mathbb{Z}_{\geq 0}$, a property written as $X_n \Rightarrow Y$, if, for each $k \geq 0$, one has*

$$(5) \qquad \lim_{n \to \infty} \mathbb{P}(X_n \leq k) = \mathbb{P}(Y \leq k).$$

Convergence is said to take place at speed ϵ_n *if*

$$(6) \qquad \sup_k |\mathbb{P}(X_n \leq k) - \mathbb{P}(Y \leq k)| \leq \epsilon_n,$$

The condition in (5) can be expressed in terms of the distribution functions $F_n(k) = \mathbb{P}(X_n \leq k)$ and $G(k) = \mathbb{P}(Y \leq k)$ as

$$\lim_{n \to \infty} F_n(k) = G(k),$$

pointwise for each k, in which case it is written as $F_n \Rightarrow G$ and is known as *weak convergence*. One also says that the X_n (or the F_n) admit a *limit law* of type Y (or G).

In addition to limit laws in the sense of (5), there is also interest in examining the convergence of individual probability values. One says that there exists a *local limit law* if

$$(7) \qquad \lim_{n \to \infty} \mathbb{P}(X_n = k) = \mathbb{P}(Y = k),$$

for each $k \geq 0$, and δ_n is called a *local speed of convergence* if

$$\sup_k |\mathbb{P}(X_n = k) - \mathbb{P}(Y = k)| \leq \delta_n.$$

By differencing or summing, it is easily seen that the conditions (5) and (7) imply one another. In other words: *For the convergence of discrete random variables (RVs) to a discrete RV, there is complete equivalence between the existence of a limit law in the sense of* (5) *and of a local limit law* (7). Note IX.2 below shows elementarily that there always exists a speed of convergence that *tends to* 0 as n tends to infinity. In other words, plain convergence of distribution functions or of individual probabilities implies *uniform* convergence.

In the following, the random variables X_n are meant to represent a combinatorial parameter χ taken over some class $\mathcal{F}$ and restricted to $\mathcal{F}_n$, that is,

$$\mathbb{P}(X_n = k) := \mathbb{P}_{\mathcal{F}_n}(\chi = k).$$

The limit variable Y, i.e., its probability distribution G, is to be determined in each particular case. A highly plausible indication of the occurrence of a discrete law is the fact that the mean μ_n and variance σ_n^2 of X_n remain bounded, i.e., they satisfy $\mu_n = O(1)$ and $\sigma_n^2 = O(1)$. Examination of initial entries in the table of values of the probabilities will then normally permit one to detect whether a limit law holds.

Example IX.3. *Singleton cycles in permutations.* The case of the number of singleton cycles (cycles of length 1) in a random permutation of size n illustrates the basic notions, while it can be studied with minimal analytic apparatus. The exponential BGF is

$$(8) \qquad P = \mathrm{SET}(u\mathcal{Z} + \mathrm{CYC}_{\geq 2}(\mathcal{Z})) \qquad \Longrightarrow \qquad P(z, u) = \frac{\exp(z(u-1))}{1-z},$$

which determines the mean $\mu_n = 1$ (for $n \geq 1$) and the standard deviation $\sigma_n = 1$ (for $n \geq 2$). The table of numerical values of the probabilities $p_{n,k} := [z^n u^k] P(z, u)$ immediately tells what goes on:

	$k = 0$	$k = 1$	$k = 2$	$k = 3$	$k = 4$	$k = 5$
$n = 4$	0.375	0.333	0.250	0.000	0.041	
$n = 5$	0.366	0.375	0.166	0.083	0.000	0.008
$n = 10$	0.367	0.367	0.183	0.061	0.015	0.003
$n = 20$	0.367	0.367	0.183	0.061	0.015	0.003

The exact distribution is easily extracted from the bivariate GF,

$$(9) \qquad p_{n,k} \equiv [z^n u^k] P(z, u) = [z^n] \frac{z^k}{k!} \frac{e^{-z}}{1-z} = \frac{d_{n-k}}{k!},$$

where $n! d_n$ is the number of derangements of size n, that is,

$$d_n = [z^n] \frac{e^{-z}}{1-z} = \sum_{j=0}^{n} \frac{(-1)^j}{j!}.$$

Asymptotically, one has $d_n \sim e^{-1}$. Thus, for fixed k, we have a local form of a limit law:

$$\lim_{n \to \infty} p_{n,k} = p_k, \qquad \text{where} \quad p_k = \frac{e^{-1}}{k!}.$$

As a consequence: *the distribution of the number of singleton cycles in a random permutation of large size tends to a Poisson law of rate* $\lambda = 1$.

Convergence is quite fast. Here is a table of differences, $\delta_{n,k} = p_{n,k} - e^{-1}/k!$:

	$k = 0$	$k = 1$	$k = 2$	$k = 3$	$k = 4$	$k = 5$
$n = 10$	$2.3\,10^{-8}$	$-2.5\,10^{-7}$	$1.2\,10^{-6}$	$-3.7\,10^{-6}$	$7.3\,10^{-6}$	$1.0\,10^{-5}$
$n = 20$	$1.8\,10^{-20}$	$-3.9\,10^{-19}$	$3.9\,10^{-18}$	$-2.4\,10^{-17}$	$1.1\,10^{-16}$	$-3.7\,10^{-16}$

The speed of convergence is easily bounded. Indeed, one has $d_n = e^{-1} + O(1/n!)$ by the alternating series property, so that, uniformly,

$$
p_{n,k} = \frac{e^{-1}}{k!} + O\left(\frac{1}{k!\,(n-k)!}\right) = \frac{e^{-1}}{k!} + O\left(\frac{1}{n!}\binom{n}{k}\right) = \frac{e^{-1}}{k!} + O\left(\frac{2^n}{n!}\right).
$$

As a consequence, one obtains local (δ_n) and central (ϵ_n) speed estimates

$$
\delta_n = O\left(\frac{2^n}{n!}\right), \qquad \epsilon_n = O\left(\frac{n2^n}{n!}\right).
$$

These bounds are quite tight. For instance one computes that the best speed is $\delta_{50} \doteq 1.5\,10^{-52}$, while the quantity $2^n/n!$ evaluates to $3.7\,10^{-50}$. .... ∎

▷ **IX.2.** *Uniform convergence.* Local and global convergences to a discrete limit law are always *uniform*. In other words, there always exist speeds ϵ_n, δ_n tending to 0 as $n \to \infty$.
Proof. Set $p_{n,k} := \mathbb{P}(X_n = k)$ and $q_k := \mathbb{P}(Y = k)$. Assume simply the condition (5) and its equivalent form (7). Fix a small $\epsilon > 0$. First dispose of the tails: there exists a k_0 such that $\sum_{k \geq k_0} q_k \leq \epsilon$, so that $\sum_{k < k_0} q_k > 1 - \epsilon$. Now, by simple convergence, for all large enough $n \geq n_0$, there holds $|p_{n,k} - q_k| < \epsilon/k_0$, for each $k < k_0$. Thus, we have $\sum_{k < k_0} p_{n,k} > 1 - 2\epsilon$, hence $\sum_{k \geq k_0} p_{n,k} \leq 2\epsilon$. At this stage, we have proved that $\sum_{k \geq k_0} q_k$ and $\sum_{k \geq k_0} p_{n,k}$ are both in $[0, 2\epsilon]$. This shows that convergence of distribution functions is uniform, with speed $\epsilon_n \leq 3\epsilon$. Furthermore, a local speed exists, which satisfies $\delta_n \leq 2\epsilon$. ◁

▷ **IX.3.** *Speed in local and global estimates.* Let M_n be the spread of χ on $\mathcal{F}_n$ defined as $M_n := \max_{\gamma \in \mathcal{F}_n} \chi(\gamma)$. Then, a speed of convergence in (6) is given by

$$
\epsilon_n := M_n \delta_n + \sum_{k > M_n} q_k.
$$

(Refinements of these inequalities can be obtained from tail estimates detailed on p. 627.) ◁

▷ **IX.4.** *Total variation distance.* The *total variation distance* between X and Y is classically

$$
d_{TV}(X, Y) := \sup_{E \subseteq \mathbb{Z}_{\geq 0}} |\mathbb{P}_Y(E) - \mathbb{P}_X(E)| = \frac{1}{2} \sum_{k \geq 0} |\mathbb{P}(Y = k) - \mathbb{P}(X = k)|.
$$

(Equivalence between the two forms is established elementarily by considering the particular E for which the supremum is attained.) The argument of Note IX.2 shows that convergence in distribution also implies that the total variation distance between X_n and X tends to 0. In addition, by Note IX.3, one has $d_{TV}(X_n, X) \leq M_n \delta_n + \sum_{k > M_n} p_k$. ◁

▷ **IX.5.** *Escape to infinity.* The sequence X_n, where

$$
\mathbb{P}\{X_n = 0\} = 1/3, \quad \mathbb{P}\{X_n = 1\} = 1/3, \quad \mathbb{P}\{X_n = n\} = 1/3,
$$

does not satisfy a discrete limit law in the sense above, although $\lim_{n \to \infty} \mathbb{P}\{X_n = k\}$ exists for each k. Some of the probability mass escapes to infinity—in a way, convergence takes place in $\mathbb{Z} \cup \{+\infty\}$. ◁

IX. 2.2. Continuity theorem for PGFs.

A high level approach to discrete limit laws in analytic combinatorics is based on asymptotic estimates of the PGF $p_n(u)$ of a random variable X_n arising from a parameter χ over a class $\mathcal{C}_n$. If, for sufficiently many values of u, one has

$$
p_n(u) \to q(u) \qquad (n \to +\infty),
$$

one can infer that the coefficients $p_{n,k} = [u^k]p_n(u)$ (for any fixed k) tend to the limit $q_k = [u^k]q(u)$. A general *continuity theorem* for PGFs describes precisely the conditions under which convergence of PGFs to a limit entails convergence of coefficients to a limit, that is to say, the occurrence of a discrete limit law.

Theorem IX.1 (Continuity Theorem, discrete laws). *Let Ω be an arbitrary set contained in the unit disc and having at least one accumulation point in the interior of the disc. Assume that the probability generating functions $p_n(u) = \sum_{k\geq 0} p_{n,k}u^k$ and $q(u) = \sum_{k\geq 0} q_k u^k$ are such that there is convergence,*

$$\lim_{n\to+\infty} p_n(u) = q(u),$$

pointwise for each u in Ω. Then a discrete limit law holds in the sense that, for each k,

$$\lim_{n\to+\infty} p_{n,k} = q_k \qquad and \qquad \lim_{n\to+\infty} \sum_{j\leq k} p_{n,j} = \sum_{j\leq k} q_j.$$

Proof. The $p_n(u)$ are *a priori* analytic in $|u| < 1$ and uniformly bounded by 1 in modulus throughout $|u| \leq 1$. Vitali's Theorem, a classical result of analysis (see [577, p. 168] or [329, p. 566]), is as follows:

> **Vitali's theorem.** *Let $\mathcal{F}$ be a family of analytic functions defined in a region S (an open connected set) and uniformly bounded on every compact subset of S. Let $\{f_n\}$ be a sequence of functions of $\mathcal{F}$ that converges on a set $\Omega \subset S$ having a point of accumulation $q \in S$. Then $\{f_n\}$ converges in all of S, uniformly on every compact subset $T \subset S$.*

Here, we take S to be the open unit disc on which all the $p_n(u)$ are bounded (since $p_n(1) = 1$). The sequence in question is $\{p_n(u)\}$. By assumption, there is convergence of $p_n(u)$ to $q(u)$ on Ω. Vitali's theorem implies that this convergence is uniform in any compact subdisc of the unit disc, for instance, $|u| \leq 1/2$. Then, Cauchy's coefficient formula provides

$$
\begin{aligned}
(10) \qquad q_k &= \frac{1}{2i\pi} \int_{|u|=1/2} q(u)\, \frac{du}{u^{k+1}} \\
&= \lim_{n\to\infty} \frac{1}{2i\pi} \int_{|u|=1/2} p_n(u)\, \frac{du}{u^{k+1}} \\
&= \lim_{n\to\infty} p_{n,k},
\end{aligned}
$$

where uniformity granted by Vitali's theorem is combined with continuity of the contour integral (with respect to the integrand). ∎

Feller gives the sufficient set of conditions $p_n(u) \to q(u)$ pointwise for all real $u \in (0,1)$, which in our terminology corresponds to the special case $\Omega = (0,1)$; see [205, p. 280] for a proof that only involves elementary real analysis. It is perhaps surprising that very different sets Ω can be taken, for instance,

$$\Omega = \left[-\tfrac{1}{3}, -\tfrac{1}{2}\right], \qquad \Omega = \{\tfrac{1}{n}\}, \qquad \Omega = \left\{\tfrac{\sqrt{-1}}{2} + \tfrac{1}{2^n}\right\}.$$

The next statement relates a measure of distance between two PGFS, $p(u)$ and $q(u)$ to the distance between distributions. It is naturally of interest when quantifying speed of convergence to the limit in the discrete-to-discrete case.

Theorem IX.2 (Speed of convergence, discrete laws). *Consider two random variables supported by $\mathbb{Z}_{\geq 0}$, with distribution functions $F(x)$, $G(x)$ and probability generating functions $p(u)$, $q(u)$.*

(i) Assume the existence of first moments. Then, for any $T \in (0, \pi)$, one has,

(11)
$$\sup_k |F(k) - G(k)| \leq \frac{1}{4} \int_{-T}^{+T} \left| \frac{p(e^{it}) - q(e^{it})}{t} \right| dt + \frac{1}{2\pi T} \sup_{T \leq |t| \leq \pi} \left| p(e^{it}) - q(e^{it}) \right|.$$

(ii) Assume that $p(u)$ and $q(u)$ are analytic in $|u| < \rho$, for some $\rho > 1$. Then, for any r satisfying $1 < r < \rho$, one has

(12)
$$\sup_k |F(k) - G(k)| \leq \frac{1}{r - 1} \sup_{|u| = r} |p(u) - q(u)|.$$

Proof. *(i)* Observe first that $p(1) = q(1) = 1$, so that the integrand is of the form $\frac{0}{0}$ at $t = 0$, corresponding to $u \equiv e^t = 1$. By Appendix C.3: *Transforms of distributions*, p. 772, the existence of first moments, say μ for F and ν for G, implies that, for small t, one has $p(e^{it}) - q(e^{it}) = (\mu - \nu)t + o(t)$, so that the integral is indeed well defined.

For any given k, Cauchy's coefficient formula provides

(13)
$$F(k) - G(k) = \frac{1}{2i\pi} \int_\gamma \frac{p(u) - q(u)}{1 - u} \frac{du}{u^{k+1}} = \frac{1}{2\pi} \int_{-\pi}^{+\pi} \frac{p(e^{it}) - q(e^{it})}{1 - e^{it}} e^{-kit} \, dt,$$

where γ is taken to be the circle $|u| = 1$, and the trigonometric form results from setting $u = e^{it}$. (The factor $(1-u)^{-1}$ sums coefficients.) In the trigonometric integral, split the interval of integration according as $|t| \leq T$ and $|t| \geq T$. For $t \in [-\pi, \pi]$, one has elementarily

$$\left| \frac{t}{e^{it} - 1} \right| \leq \frac{\pi}{2}.$$

For $|t| \leq T$, this inequality makes it possible to replace $|1 - u|^{-1}$ by $1/|t|$, up to a constant multiplier and get as a majorant the first term on the right of (11). For $|t| \geq T$, trivial upper bounds provide the second term on the right of (11).

(ii) Start from the contour integral in (13), but now integrate along $|u| = r$. Trivial bounds provide (12). ∎

The first form holds with strictly minimal assumptions (existence of expectations); the second form is *a priori* only usable for distributions that have exponential tails, as discussed in Subsection IX. 2.3 below. The first form relates the distance on the unit circle between the PGF $p_n(u)$ of a combinatorial parameter and the limit PGF $q(u)$ to the speed of convergence to the limit law—it prefigures the Berry–Esseen inequalities discussed in the continuous context on p. 641.

Example IX.4. *Cycles of length m in permutations.* Let us first revisit the number χ of singleton cycles ($m = 1$) in this new light. The BGF $P(z, u) = e^{z(u-1)}/(1 - z)$, given by Equation (8) in Example IX.3, has for each u a simple pole at $z = 1$ and is otherwise analytic in $\mathbb{C} \setminus \{1\}$. Thus, a meromorphic analysis provides instantly, pointwise for any fixed u,

$$[z^n] P(z, u) = e^{(u-1)} + O(R^{-n}),$$

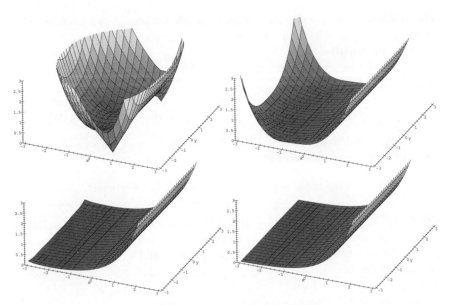

Figure IX.6. The PGFs of singleton cycles in random permutations of size $n = 4, 8, 12$ (left to right and top to bottom) illustrate convergence to the limit PGF of the Poisson(1) distribution (bottom right). The modulus of each PGF is displayed, for $|\Re(u)|, |\Im(u)| \leq 3$.

with any $R > 1$. This, by the continuity theorem, Theorem IX.1, implies convergence to a limit law, which is Poisson.

Next, in order to obtain a speed of convergence, one should estimate a distance between PGFs over the unit circle. One has, for $p_n(u)$ and $q(u)$, respectively, the PGF of χ over $\mathcal{P}_n$ and the PGF of a Poisson variable of parameter 1:

$$p_n(u) - q(u) = [z^n] \frac{e^{z(u-1)} - e^{(u-1)}}{1 - z}.$$

There is a removable singularity at $z = 1$. Thus, integration over the circle $|z| = 2$ in the z-plane is permissible, and

$$p_n(u) - q(u) = \frac{1}{2i\pi} \int_{|z|=2} \frac{e^{z(u-1)} - e^{u-1}}{1 - z} \frac{dz}{z^{n+1}}.$$

Trivial bounds applied to the last integral then yield

$$|p_n(u) - q(u)| \leq 2^{-n} \sup_{|z|=2} \left| e^{z(u-1)} - e^{(u-1)} \right| = O\left(2^{-n}|1 - u|\right),$$

uniformly for u in any compact set of $\mathbb{C}$. One can then apply Theorem IX.2, Part (i). The value $T = \frac{\pi}{2}$ is suitable, to the effect that a speed of convergence to the limit is found to be $O(2^{-n})$. (Any $O(R^{-n})$ is furthermore possible by a similar argument.) Numerical aspects of the convergence are illustrated in Figure IX.6.

This approach generalizes straightforwardly to the number of m–cycles in a random permutation (m kept *fixed*). The exponential BGF is

$$F(z, u) = \frac{e^{(u-1)z^m/m}}{1 - z}.$$

Then, singularity analysis of the meromorphic function of z (for u fixed) gives immediately

$$\lim_{n \to \infty} [z^n] F(z, u) = e^{(u-1)/m}.$$

The right-hand side of this equality is none other than the PGF of a Poisson law of rate $\lambda = 1/m$. The continuity theorem and the first form of the speed of convergence theorem then imply: *The number of m–cycles in a random permutation of large size converges in law to a Poisson distribution of rate $1/m$ with speed of convergence $O(R^{-n})$ for any $R > 1$.* This last result appreciably generalizes our previous observations on singleton cycles. .... ∎

▷ **IX.6.** *A quiz.* Figure IX.6 tacitly assumes that the property $|p_n(u)| \to |p(u)|$ suffices to conclude that $p_n(u) \to p(u)$. Can you justify it? [Hint: for an analytic function, if we know $|\phi(u)|$, we know $\log |\phi(u)| = \Re(\log \phi(u))$. But then we can reconstruct $\Im(\log \phi(u))$ by the Cauchy-Riemann equations (p. 742). Hence, we know $\log \phi(u)$, hence $\phi(u)$ itself.] ◁

▷ **IX.7.** *Poisson law for rare events.* Consider the binomial distribution with PGF $(q + pu)^n$. If p depends on n in such a way that $p = \lambda/n$ for some fixed λ, then the limit law of the binomial random variable is Poisson of rate λ. (This "law of small numbers" explains the Poisson character of activity in radioactive decay as well as the occurrence of accidental deaths of soldiers in the Prussian army resulting from the kick of a horse [Bortkiewicz, 1898].) ◁

IX. 2.3. Tail estimates. Tail estimates quantify the rate of decrease of probabilities away from the central part of the distribution. In the case of a discrete limit law having a finite mean, what one needs is information regarding $\mathbb{P}(X > k)$ as k gets large. A simple, but often effective, approach consists in appealing to saddle-point bounds. We give here a general statement which is nothing but a rephrasing of such bounds adapted to discrete probability distributions.

Theorem IX.3 (Tail bounds, discrete laws). *Let $p(u) = E(u^X)$ be a probability generating function that is analytic for $|u| \le r$ where r is some number satisfying $r > 1$. Then, the following "local" and "global" tail bounds hold:*

$$\mathbb{P}(X = k) \le \frac{p(r)}{r^k}, \qquad \mathbb{P}(X > k) \le \frac{p(r)}{r^k(r - 1)}.$$

Proof. The local estimate is a direct consequence of trivial bounds applied to Cauchy's integrals, namely

$$\mathbb{P}(X = k) = \frac{1}{2i\pi} \int_{|u|=r} p(u) \frac{du}{u^{k+1}} \le \frac{p(r)}{r^k}.$$

The cumulative bound is derived from the useful integral representation

$$\begin{aligned}
\mathbb{P}(X > k) &= \frac{1}{2i\pi} \int_{|u|=r} p(u) \left(1 + \frac{1}{u} + \frac{1}{u^2} + \cdots \right) \frac{du}{u^{k+2}} \\
&= \frac{1}{2i\pi} \int_{|u|=r} p(u) \frac{du}{u^{k+1}(u - 1)},
\end{aligned}$$

upon applying again trivial bounds. (Alternatively, summation from the local bounds can be used.) ∎

The bounds provided always exhibit a geometric decay in the value of k—this is both a strength and a limitation on the method. In accordance with the theorem and as is easily checked directly, the geometric and the negative binomial distributions have exponential tails; the Poisson law even has a "superexponential" tail, being $O(R^{-k})$ for any $R > 1$, since its PGF is entire. By their nature, the bounds can also be simultaneously applied to a whole family of probability generating functions, as shown by the characteristic example below. Hence their use in obtaining uniform estimates in the context of limit laws, in a way that prefigures the study of large deviations in Section IX. 10.

Example IX.5. *Permutations with a large number of singleton cycles.* The problem here is to quantify the probability that a permutation of size n has more than $k = \log n$ singleton cycles, a quantity that is far from the mean value 1. The elementary treatment of Example IX.3 is certainly applicable but it has the disadvantage of not easily generalizing to other situations. In the perspective of applying Theorem IX.3, we seek instead to bound $p_n(u)$ for $u > 0$, where $p_n(u) := [z^n]e^{z(u-1)}/(1 - z)$, by Equation (8). We have, for $u > 0$ and any $s \in (0, 1)$,

$$p_n(u) \equiv [z^n]e^{uz}\frac{e^{-z}}{1 - z} \leq e^{us}\frac{e^{-s}}{1 - s}s^{-n},$$

as found from saddle-point bounds (in the z–plane) applied to the BGF $P(z, u)$. Taking $s = 1 - 1/n$, which is suggested by the usual scaling of singularity analysis as well as by the saddle-point principles, gives the following bound on the PGF,

$$p_n(u) \leq 2ne^u,$$

valid for all $n \geq 2$. (Better estimates are available from the precise analysis of Example IX.4, but the improvement regarding tail bounds would be marginal.) Choosing now $r = \log n$ in the statement of Theorem IX.3 value provides an approximate saddle-point bound, and we get for $n \geq 10$ (say)

$$\sum_{j\geq\log n} p_{n,j} \leq \frac{2n^2}{n^{\log\log n}}.$$

Thus the probability of observing more than $\log n$ singleton cycles is asymptotically smaller than any inverse power of n. Note that, in this example, we have made use of Theorem IX.3, while opting to estimate the PGFs plainly by saddle-point bounds taken with respect to the principal variable z of the corresponding bivariate generating function. .... ∎

IX. 3. Combinatorial instances of discrete laws

In this section, we focus our attention on the general analytic schema based on compositions (p. 411), and more specifically on its subcritical case (Definition IX.2 below). It is such that the perturbations induced by the secondary variable (u) affects neither the location nor the nature of the basic singularity involved in the univariate counting problem. The limit laws are then of the discrete type. In particular, for the labelled universe and for subcritical sequences, sets, and cycles, these limit laws are invariably of the negative binomial, Poisson, and geometric type, respectively. Additionally, it is easy to describe the profiles of combinatorial objects resulting from such subcritical constructions.

Subcritical compositions. First, we consider the general *composition schema*,

$$\mathcal{F} = \mathcal{G} \circ (u\mathcal{H}) \qquad \Longrightarrow \qquad F(z, u) = g(uh(z)).$$

This schema expresses over generating functions the combinatorial operation $\mathcal{G} \circ \mathcal{H}$ of *substitution* of components $\mathcal{H}$ enumerated by $h(z)$ inside "templates" $\mathcal{G}$ enumerated by $g(z)$. (See Chapters I, p. 86 and II, p. 137, for the unlabelled and labelled versions, and Chapter III, p. 199, for the bivariate versions.) The variable z marks size as usual, and the variable u marks the size of the $\mathcal{G}$–template.

We assume globally that g and h have non-negative coefficients and that $h(0) = 0$ so that the composition $g(h(z))$ is well-defined. We let ρ_g and ρ_h denote the radii of convergence of g and h, and define

$$(14) \qquad \tau_g = \lim_{x \to \rho_g^-} g(x) \qquad \text{and} \qquad \tau_h = \lim_{x \to \rho_h^-} h(x).$$

The (possibly infinite) limits exist due to the non-negativity of coefficients. As already discussed in Section VI. 9, p. 411, three cases are to be distinguished.

Definition IX.2. *The composition schema* $F(z, u) = g(uh(z))$ *is said to be* subcritical *if* $\tau_h < \rho_g$, critical *if* $\tau_h = \rho_g$, supercritical *if* $\tau_h > \rho_g$.

In terms of singularities, the behaviour of $g(h(z))$ at its dominant singularity is dictated by the dominant singularity of h (subcritical case), or by the dominant singularity of g (supercritical case), or else it involves a mixture of the two (critical case). This section is concerned with the *subcritical* case[6].

Proposition IX.1 (Subcritical composition, number of components). *Consider the bivariate composition schema* $F(z, u) = g(uh(z))$. *Assume that* $g(z)$ *and* $h(z)$ *satisfy the* subcriticality *condition* $\tau_h < \rho_g$, *and that* $h(z)$ *has a unique singularity at* $\rho = \rho_h$ *on its disc of convergence, which, in a* Δ–*domain, is of the type*

$$h(z) = \tau - c \left(1 - \frac{z}{\rho} \right)^{\lambda} + o \left(\left(1 - \frac{z}{\rho} \right)^{\lambda} \right),$$

where $\tau = \tau_h$, $c \in \mathbb{R}^+$, $0 < \lambda < 1$. *Then, a discrete limit law holds for the number of* $\mathcal{H}$–*components: with* $f_{n,k} := [z^n u^k] F(z, u)$ *and* $f_n = [z^n] F(z, 1)$, *one has*

$$\lim_{n \to \infty} \frac{f_{n,k}}{f_n} = q_k, \qquad \text{where} \quad q_k = \frac{k g_k \tau^{k-1}}{g'(\tau)}.$$

The probability generating function of the limit distribution (q_k) *is*

$$q(u) = \frac{u g'(\tau u)}{g'(\tau)}.$$

Proof. First, we examine the univariate counting problem. Since $g(z)$ is analytic at τ, the function $g(h(z))$ is singular at ρ_h and is analytic in a Δ–domain. Its singular expansion is obtained by composing the regular expansion of $g(z)$ at τ with the singular

---

[6]By contrast with the discrete laws encountered here, the case of a supercritical composition leads to continuous limit laws of the Gaussian type (Section IX. 6). The critical case involves a confluence of singularities, which induces stable laws (Section IX. 11).

expansion of $h(z)$ at ρ_h:

$$F(z) \equiv g(h(z)) = g(\tau) - cg'(\tau)(1 - z/\rho)^\lambda (1 + o(1)).$$

Thus, $F(z)$ satisfies the conditions of singularity analysis, and

$$(15) \qquad f_n \equiv [z^n]F(z) = -\frac{cg'(\tau)}{\Gamma(-\lambda)}\rho^{-n}n^{-\lambda-1}(1 + o(1)).$$

By similar devices, the mean and variance of the distribution are found to be each $O(1)$.

Next, for the bivariate problem, fix any u with, say, $u \in (0,1)$. The BGF $F(z, u)$ is also seen to be singular at $z = \rho$, and its singular expansion obtained from $F(z, u) = g(uh(z))$ by composition, is

$$(16) \qquad \begin{aligned} F(z, u) = g(uh(z)) \quad &= \quad g(u\tau - cu(1 - z/\rho)^\lambda + o((1 - z/\rho)^\lambda)) \\ &= \quad g(u\tau) - cug'(u\tau)(1 - z/\rho)^\lambda + o((1 - z/\rho)^\lambda). \end{aligned}$$

Thus, singularity analysis implies immediately:

$$\lim_{n\to\infty} \frac{[z^n]F(z, u)}{[z^n]F(z, 1)} = \frac{ug'(u\tau)}{g'(\tau)}.$$

By the continuity theorem for PGFs, this is enough to imply convergence to the discrete limit law with PGF $ug'(\tau u)/g'(\tau)$, and the proposition is established. ∎

What stands out in the statement of Proposition IX.1 is the following general fact: *In a subcritical composition, the limit law is a direct reflection of the derivative of the outer function involved in the composition.*

▷ **IX.8.** *Tail bounds for subcritical compositions.* Under the subcritical composition schema, it is also true that the tails have a uniformly geometric decay. Let u_0 be any number of the interval $(1, \rho_g/\tau_h)$. Then the function $z \mapsto F(z, u_0)$ is analytic near the origin with a dominant singularity at ρ_h again obtained by composing the regular expansion of g with the singular expansion of h, and Equation (16) remains valid at $u = u_0$. There results the asymptotic estimate

$$p_n(u_0) = \frac{[z^n]F(z, u_0)}{[z^n]F(z, 1)} \sim g'(u_0\tau_h).$$

Thus, for some constant $K \equiv K(u_0)$, one has $p_n(u_0) < K$. It is also easy to verify that $p_n(u)$ is analytic at u_0, so that, by Theorem IX.3,

$$p_{n,k} \leq K(u_0) \cdot u_0^{-k}, \qquad \sum_{j>k} p_{n,j} \leq \frac{K(u_0)}{u_0 - 1} u_0^{-k}.$$

Therefore, the combinatorial distributions satisfy, uniformly with respect to n, a tail bound. In particular the probability that there are more than a logarithmic number of components satisfies

$$(17) \qquad \mathbb{P}_n(\chi > \log n) = O(n^{-\theta}) \qquad \text{and} \qquad \theta = \log u_0.$$

Such tail estimates may additionally serve to evaluate the speed of convergence to the limit law (as well as the total variation distance) in the subcritical composition schema. ◁

▷ **IX.9.** *Semi-small powers and singularity analysis.* Let $h(z)$ satisfy the stronger singular expansion

$$h(z) = \tau - c(1 - z/\rho)^\lambda + O(1 - z/\rho)^\nu,$$

for $0 < \lambda < \nu < 1$. Then, for $k \leq C \log n$ (some $C > 0$), the results of singularity analysis can be extended (under the form proved in Chapter VI, they are only valid for *fixed k*)

$$[z^n]h(z)^k = kc\rho^{-n}n^{-\lambda-1}\left(1 + O(n^{-\theta_1})\right),$$

for some $\theta_1 > 0$, uniformly with respect to k. [The proof recycles the Hankel contour Chapter VI, with some care needed in checking uniformity with respect to k; see also p. 709.] ◁

▷ **IX.10.** *Speed of convergence in subcritical compositions.* Combining the exponential tail estimate (17) and local estimates deriving from the singularity analysis of "semi-small" powers in the previous note, one obtains for the distribution functions associated with $p_{n,k}$ and p_k the speed estimate

$$\sup_k |F_n(k) - F(k)| \le \frac{L}{n^{\theta_2}}.$$

There, L and θ_2 are two positive constants. ◁

Subcritical constructions. The functional composition schema encompasses the sequence, set, and cycle constructions of the labelled universe. We state the following proposition.

Proposition IX.2 (Subcritical constructions, number of components). *Consider the labelled constructions of* sequence, set, *and* cycle. *Assume the subcriticality conditions of the previous proposition, namely* $\tau < 1$, $\tau < \infty$, $\tau < 1$, *respectively, where* τ *is the singular value of* $h(z)$. *Then, the distribution of the number* χ *of components determined by* $f_{n,k}/f_n$, *is such that* $\chi - 1$ *admits a discrete limit law that is of type, respectively, negative binomial* $NB[2]$, *Poisson, and geometric: the limit forms* $q_k = \lim_{n \to \infty} \mathbb{P}_n(\chi = k)$ *satisfy, respectively, for* $k \ge 0$,

$$q_{k+1}^{\text{SEQ}} = (1 - \tau)^2 (k + 1)\tau^k, \qquad q_{k+1}^{\text{SET}} = e^{-\tau} \frac{\tau^k}{k!}, \qquad q_{k+1}^{\text{CYC}} = (1 - \tau)\tau^k.$$

Proof. It suffices to take for the outer function g in the composition $g \circ h$ the quantities

$$(18) \qquad Q(w) = \frac{1}{1 - w}, \quad E(w) = e^w, \quad L(w) = \log \frac{1}{1 - w}.$$

According to Proposition IX.1 and Equation (18) above, the PGF of the discrete limit law involves the derivatives

$$Q'(w) = \frac{1}{(1 - w)^2}, \quad E'(w) = e^w, \quad L'(w) = \frac{1}{1 - w}.$$

By definition of the classical discrete laws in Figure IX.5, p. 621, it is seen that the last two cases precisely give rise to the classical Poisson and geometric law. The first case gives rise to the negative binomial law $NB[2]$, or equivalently the sum of two independent geometrically distributed random variables. ∎

The technical simplicity with which limit laws are extracted is worthy of note. Naturally, the statement also covers *unlabelled sequences*, since translation into GFs is the same in both universes. (Other unlabelled constructions usually lead to discrete laws, as long as they are subcritical; see Note IX.14 for a particular instance.) Also, subcriticality of a composition $g \circ h$ necessarily entails that τ_h is finite (since one has $\tau_h < \rho_g \le +\infty$, by definition). Primary cases of applications of Proposition IX.2 are thus in the realm of "treelike" structures, for which the GFs remain finite at their radius of convergence, as we have learnt in Chapter VII.

The example that follows illustrates the application of Proposition IX.1 to the analysis of root degrees in classical varieties of trees. It is especially interesting to observe the way limit laws directly reflect the combinatorial specifications. For instance, the root degree in a large random plane tree (a Catalan tree) is found to obey, in the asymptotic limit, a negative binomial ($NB[2]$) distribution, which, in a precise sense, echoes the sequence construction that expresses planarity. For labelled non-plane trees (Cayley tree), a Poisson law echoes the set construction attached to non-planarity.

Example IX.6. *Root degrees in trees.* Consider first the number of components in a sequence (ordered forest) of general Catalan trees. The bivariate OGF is

$$F(z, u) = \frac{1}{1 - uh(z)}, \quad h(z) = \frac{1}{2}\left(1 - \sqrt{1 - 4z}\right).$$

We have $\tau_h = 1/2 < \rho_g = 1$, so that the composition schema is subcritical. Thus, for a forest of total size n, the number X_n of tree components satisfies

$$\lim_{n\to\infty} \mathbb{P}\{X_n = k\} = \frac{k}{2^{k+1}} \quad (k \geq 1).$$

Since a tree is equivalent to a node appended to a forest, this asymptotic estimate also holds for the root degree of a general Catalan tree.

Consider next the number of components in a set (unordered forest) of Cayley trees. The bivariate EGF is

$$F(z, u) = e^{uh(z)}, \quad h(z) = ze^{h(z)}.$$

We have $\tau_h = 1 < \rho_g = +\infty$, again a subcritical composition schema. Thus the number X_n of tree components in a random unordered forest of size n admits the limit distribution

$$\lim_{n\to\infty} \mathbb{P}\{X_n = k\} = e^{-1}/(k-1)!, \quad (k \geq 1),$$

a shifted Poisson law of parameter 1; asymptotically, the same property also holds for the root degree of a random Cayley tree

The same method applies more generally to a simple variety of trees $\mathcal{V}$ (see Section VII.3, p. 452) with generator ϕ, under the condition of the existence of a root τ of the characteristic equation $\phi(\tau) - \tau\phi'(\tau) = 0$ at a point interior to the disc of convergence of ϕ. The BGF satisfies

$$V(z, u) = z\phi(uV(z)), \qquad V(z) = 1 - \gamma\sqrt{1 - z/\rho} + O(1 - z/\rho).$$

so that

$$V(z, u) \underset{z\to\rho}{\sim} \rho\phi(u\tau) - \gamma\frac{u\phi'(u\tau)}{\phi'(\tau)}\sqrt{1 - /z\rho}.$$

The PGF of the distribution of root degree is accordingly

$$\frac{u\phi'(\tau u)}{\phi'(\tau)} = \sum_{k\geq 1} \frac{k\phi_k \tau^k}{\phi'(\tau)} u^k.$$

This limit law was established under its local form in Chapter VII, p. 456, by means of univariate asymptotics; the present example shows the synthetic character of a derivation based on the continuity theorem for PGFs. .... ∎

A further direct application of the continuity of PGFs is the distribution of the number of $\mathcal{H}$–components of a fixed size m in a composition $\mathcal{G} \circ \mathcal{H}$ with GF $g(h(z))$, again under the *subcriticality* condition. In the terminology of Chapter III, we are thus characterizing the *profile* of combinatorial objects, as regards components of some fixed size. The bivariate GF is then

$$\mathcal{F} = \mathcal{G} \circ (\mathcal{H} \setminus \mathcal{H}_m + u\mathcal{H}_m) \qquad \Longrightarrow \qquad F(z, u) = g(h(z) + (u-1)h_m z^m),$$

with $h_m = [z^m]h(z)$. The singular expansion at $z = \rho$ is

$$F(z, u) = g(\tau + (u - 1)h_m\rho^m) - cg'(\tau + (u - 1)h_m\rho^m)(1 - z/\rho)^\lambda) + o((1 - z/\rho)^\lambda).$$

Thus, the PGF $p_n(u)$ for objects of size n satisfies

(19) $$\lim_{n\to\infty} p_n(u) = \frac{g'(\tau + (u - 1)h_m\rho^m)}{g'(\tau)}.$$

As before this calculation specializes to the case of sequences, sets, and cycles giving a result analogous to Proposition IX.1.

Proposition IX.3 (Subcritical constructions, number of fixed-size components). *Under the subcriticality conditions of Proposition IX.2, the number of components of a fixed size m in a labelled sequence, set, or cycle construction applied to a class with GF $h(z)$ admits a discrete limit law. Let $h_m := [z^m]h(z)$ and let ρ be the radius of convergence of $h(z)$, with $\tau := h(\rho)$. For sequences, sets, and cycles, the limit laws are, respectively, negative binomial $NB[2](a)$, Poisson(λ), and geometric(b), with parameters*

$$a = \frac{h_m\rho^m}{1 - \tau + h_m\rho^m}, \qquad \lambda = h_m\rho^m, \qquad b = \frac{h_m\rho^m}{1 - \tau + h_m\rho^m}.$$

Proof. Instantiate (19) with g, one of the three functions of (18). ∎

Example IX.7. *Root subtrees of size m.* In a Cayley tree, the number of root subtrees of some fixed size m has, in the limit, a Poisson distribution,

$$p_k = e^{-\lambda}\frac{\lambda^k}{k!}, \qquad \lambda := \frac{m^{m-1}e^{-m}}{m!}.$$

In a general Catalan tree, the distribution is a negative binomial $NB[2]$

$$p_k = (1 - a)^2(k + 1)a^k, \qquad a^{-1} := 1 + \frac{m2^{2m-1}}{\binom{2m-2}{m-1}}.$$

Generally, for a simple variety of trees under the usual conditions of existence of a solution to the characteristic equation, $V = z\phi(V)$, one finds *"en deux coups de cuillère à pot"*,

$$\begin{aligned}
V(z, u) &= z\phi(V(z) + V_m z^m(u - 1)) \\
V(z, u) &\sim \rho\phi(\tau + V_m\rho^m(u - 1)) - \rho\gamma\phi'(\tau + V_m\rho^m(u - 1))\sqrt{1 - z/\rho} \\
\text{limit PGF} &= \frac{\phi'(\tau + V_m\rho^m(u - 1))}{\phi'(\tau)}.
\end{aligned}$$

(Notations are the same as in Example IX.6.) . . . . ∎

We shall see later that similar discrete distributions (the Poisson and negative binomial law of Proposition IX.3) also arise in *critical* set constructions of the exp–log type (Example IX.23, p. 675), while *supercritical* sequences lead to Gaussian limits (Proposition IX.7, p. 652). Furthermore, given the generality of the methods and the analytic diversity of functional compositions, it should be clear that schemas leading to discrete limit laws can be listed *ad libitum*—in essence, conditions are that the auxiliary variable u does not affect the location nor the nature of the dominant singularity of $F(z, u)$. The notes below provide a small sample of the many extensions of the method that are possible.

▷ **IX.11.** *The product schema.* Define

$$F(z, u) = A(uz) \cdot B(z),$$

that corresponds to a product construction, $\mathcal{F} = \mathcal{A} \times \mathcal{B}$, with u marking the size of the $\mathcal{A}$–component in the product. Assume that the radii of convergence satisfy $\rho_A > \rho_B$ and that $B(z)$ has a unique dominant singularity of the algebraic–logarithmic type. Then, the size of the $\mathcal{A}$ component in a random $\mathcal{F}$–structure has a discrete limit law with PGF,

$$p(u) = \frac{A(\rho u)}{A(\rho)}.$$

The proof follows by singularity analysis. (Alternatively, an elementary derivation can be given under the weaker requirement that the $b_n = [z^n]B(z)$ satisfy $b_{n+1}/b_n \to \rho^{-1}$.)   ◁

▷ **IX.12.** *Bell number distributions.* Consider the "set-of-sets" schema

$$\mathcal{F} = \text{SET}(\text{SET}_{\geq 1}(\mathcal{H})) \qquad \Longrightarrow \qquad F(z, u) = \exp(e^{uh(z)} - 1),$$

assuming subcriticality. Then the number χ of components satisfies asymptotically a "derivative Bell" law:

$$\lim_{n \to +\infty} \mathbb{P}_n(\chi = k) = \frac{1}{K} \frac{k S_k \tau^k}{k!}, \qquad K = e^{-e^\tau - \tau - 1},$$

where $S_k = k![z^k]e^{e^z - 1}$ is a Bell number. There exist parallel results: for sequence-of-sets, involving the surjection numbers; for set-of-sequences involving the fragmented permutation numbers.   ◁

▷ **IX.13.** *High levels in Cayley trees.* The number of nodes at level 5 (i.e., at distance 5 from the root) in a Cayley tree has the nice PGF

$$u \frac{d}{du} \left(e^{-1 + e^{-1 + e^{-1 + e^{-1 + e^{-1 + u}}}}} \right),$$

so that the distribution involves "super-duper-hyper-Bell numbers".   ◁

▷ **IX.14.** *Root degree in non-plane unlabelled trees.* Discrete laws may also arise from an unlabelled set construction, but their form is complicated, reflecting the presence of Pólya operators. Consider the class of non-plane unlabelled trees (p. 71)

$$\mathcal{H} = \mathcal{Z} \times \text{MSET}(\mathcal{H}) \qquad \Longrightarrow \qquad H(z) = z \exp\left(\sum_{k \geq 1} \frac{1}{k} H(z^k) \right).$$

The OGF $H(z)$ is of singularity analysis class (Section VII. 5, p. 475), and $H(z) \sim 1 - \gamma(1 - z/\rho)^{1/2}$. Then the distribution with PGF

$$q(u) = u\rho \exp\left(\sum_{k \geq 1} \frac{u^k}{k} H(\rho^k) \right)$$

is the limit law of root degree in non-plane unlabelled trees.   ◁

Lattice paths. As a last example here, we discuss the length of the longest initial run of a's in random binary words satisfying various types of constraints. This discussion completes the informal presentation of Section IX. 1, Examples IX.1 and 2. The basic combinatorial objects are the set $\mathcal{W} = \{a, b\}^\star$ of binary words. A word $w \in \mathcal{W}$ can also be viewed as describing a walk in the plane, provided one interprets a and b as the vectors $(+1, +1)$ and $(+1, -1)$, respectively. Such walks in turn describe fluctuations in coin-tossing games, as described by Feller [205].

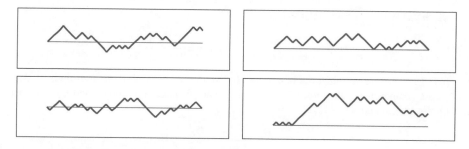

Figure IX.7. Walks, excursions, bridges, and meanders of Dyck type: from left to right and top to bottom, random samples of length 50.

The combinatorial decompositions of Section V. 4, p. 318, form the basis of our combinatorial treatment. What is especially interesting here is to observe the complete chain where a specific constraint leads in succession to a combinatorial decomposition, a specific analytic type of BGF, and a local singular structure that is eventually reflected by a particular limit law.

***Example* IX.8.** *Initial runs in random walks.* We consider here walks in the right half-plane that start from the origin and are made of steps $a = (1, 1)$, $b = (1, -1)$. According to the discussion of Chapter VII (p. 506), one can distinguish four major types of walks (Figure IX.7).

— *Unconstrained walks* ($\mathcal{W}$) corresponding to words and freely described by $\mathcal{W} = \mathrm{SEQ}(a, b)$;

— *Dyck paths* ($\mathcal{D}$), which always have a non-negative ordinate and end at level 0; the closely related class $\mathcal{G} = \mathcal{D}b$ represents the collection of gambler's ruin sequences. In probability theory, Dyck paths are also referred to as *excursions*.

— *Bridges* ($\mathcal{B}$), which are walks that may have negative ordinates but must finish at level 0.

— *Meanders* ($\mathcal{M}$), which always have a non-negative altitude and may end at an arbitrary non-negative altitude.

The parameter χ of interest is in all cases the length of the (longest) initial run of a's.

First, *unconstrained walks* obey the decomposition

$$\mathcal{W} = \mathrm{SEQ}(a)\,\mathrm{SEQ}(b\,\mathrm{SEQ}(a)),$$

already repeatedly employed. Thus, the BGF is

$$W(z, u) = \frac{1}{1 - zu}\frac{1}{1 - z(1 - z)^{-1}}.$$

By singularity analysis of the pole at $\rho = 1/2$, the PGF of χ on random words of $\mathcal{W}_n$ satisfies

$$p_n(u) \sim \frac{1/2}{1 - u/2},$$

for all u such that $|u| < 2$. This asymptotic value of the PGF corresponds to a limit law, which is a geometric with parameter $1/2$, in agreement with what was found in Examples IX.1 and IX.2.

Next, consider *Dyck paths*. Such a path decomposes into "arches" that are themselves Dyck paths encapsulated by a pair a, b, namely,

$$\mathcal{D} = \mathrm{SEQ}(a\mathcal{D}b),$$

which yields a GF of the Catalan domain,

$$D(z) = \frac{1}{1 - z^2 D(z)}, \qquad D(z) = \frac{1 - \sqrt{1 - 4z^2}}{2z^2}.$$

In order to extract the initial run of a's, we observe that a word whose initial a-run is a^k contains k components of the form bD. This corresponds to a decomposition in terms of the first traversals of altitudes $k - 1, \ldots, 1, 0$,

$$\mathcal{D} = \sum_{k \geq 0} a^k (b\mathcal{D})^k$$

(a special "first passage decomposition" in the sense of p. 321), illustrated by the following diagram:

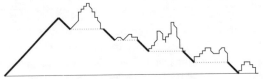

Thus, the BGF is

$$D(z, u) = \frac{1}{1 - z^2 u D(z)},$$

which is an even function of z. In terms of the singular element, $\delta = (1 - 4z)^{1/2}$, one finds

$$D(z^{1/2}, u) = \frac{2}{2 - u} - \frac{2u}{(2 - u)^2} \delta + O(\delta^2),$$

as $z \to 1/4$. Thus, the PGF of χ on random words of $\mathcal{D}_{2n}$ satisfies

$$p_{2n}(u) \sim \frac{u}{(2 - u)^2},$$

which is the PGF of a negative binomial $NB[2]$ of parameter $1/2$ shifted by 1. (Naturally, in this case, explicit expressions for the combinatorial distribution are available, as this counting is equivalent to the classical ballot problem.)

A *bridge* decomposes into a sequence of arches, either positive or negative,

$$\mathcal{B} = \text{SEQ}(a\mathcal{D}b + b\overline{\mathcal{D}}a),$$

where $\overline{\mathcal{D}}$ is like $\mathcal{D}$, but with the rôles of a and b interchanged. In terms of OGFs, this gives

$$B(z) = \frac{1}{1 - 2z^2 D(z)} = \frac{1}{\sqrt{1 - 4z^2}}.$$

The set $\mathcal{B}^+$ of non-empty walks that start with at least one a admits a decomposition similar to that of $\mathcal{D}$,

$$B^+(z) = \left(\sum_{k \geq 1} a^k b(Db)^{k-1} \right) \cdot \mathcal{B},$$

since the paths factor uniquely as a $\mathcal{D}$ component that hits 0 for the first time followed by a $\mathcal{B}$ oscillation. Thus,

$$B^+(z) = \frac{z^2}{1 - z^2 D(z)} B(z).$$

The remaining cases $\mathcal{B}^- = \mathcal{B} \setminus \mathcal{B}^+$ consist of either the empty word or of a sequence of positive or negative arches starting with a negative arch, so that

$$B^-(z) = 1 + \frac{z^2 D(z)}{1 - 2z^2 D(z)}.$$

The BGF results from these decompositions:

$$B(z, u) = \frac{z^2 u}{1 - z^2 u D(z)} B(z) + 1 + \frac{z^2 D(z)}{1 - 2z^2 D(z)}.$$

Again, the singular expansion is obtained mechanically,

$$B(z^{1/2}, u) = \frac{1}{(2 - u)} \frac{1}{\delta} + O(1), \qquad \text{where } \delta = (1 - 4z)^{1/2}.$$

Thus, the PGF of χ on random words of $\mathcal{B}_{2n}$ satisfies

$$p_{2n}(u) \sim \frac{1}{2 - u}.$$

The limit law is now geometric of parameter $1/2$.

A *meander* decomposes into an initial run a^k, a succession of descents with their companion (positive) arches in some number $\ell \le k$, and a succession of ascents with their corresponding (positive) arches. The computations are similar to the previous cases, more intricate but still "automatic". One finds that

$$M(z, u) = \left(\frac{XY}{(1 - X)(1 - Y)} - \frac{XY^2}{(1 - XY)(1 - Y)} \right) \frac{1}{1 - Y} + \frac{1}{1 - X},$$

with $X = zu$, $Y = zW_1(z)$, so that

$$M(z, u) = 2 \frac{1 - u - 2z + 2uz^2 + (u - 1)\sqrt{1 - 4z^2}}{(1 - zu)\left(1 - 2z - \sqrt{1 - 4z^2}\right)\left(2 - u + u\sqrt{1 - 4z^2}\right)}.$$

There are now two singularities at $z = \pm 1/2$, with singular expansions,

$$M(z, u) \underset{z \to 1/2}{=} \frac{u\sqrt{2}}{(2 - u)^2} \frac{1}{\sqrt{1 - 2z}} + O(1), \qquad M(z, u) \underset{z \to -1/2}{=} \frac{4 - u}{4 - u^2} + o(1),$$

so that only the singularity at $1/2$ matters asymptotically. Then, we have

$$p_n(u) \sim \frac{u}{(2 - u)^2},$$

and the limit law is a shifted negative binomial $NB[2]$ of parameter $1/2$. In summary:

Proposition IX.4. *The length of the initial run of a's in unconstrained walks and bridges is asymptotically distributed as a geometric; in Dyck excursions and meanders it is distributed as a negative binomial $NB[2]$.*

Similar analyses can be applied to walks with a finite set of step types [27]. .... ∎

▷ **IX.15.** *Left-most branch of a unary–binary (Motzkin) tree.* The class of unary–binary trees (or Motzkin trees) is defined as the class of unlabelled rooted plane trees where (out)degrees of nodes are restricted to the set $\{0, 1, 2\}$. The parameter equal to the length of the left-most branch has a limit law that is a negative binomial $NB[2]$. Find its parameter. ◁

IX. 4. Continuous limit laws

Throughout this chapter, our goal is to quantify sequences of random variables X_n that arise from an integer-valued combinatorial parameter χ defined on a combinatorial class $\mathcal{F}$. It is a fact that, when the mean μ_n and the standard deviation σ_n of X_n both tend to infinity as n gets large, then a limit law that is *continuous* usually holds. That limit law arises not directly from the X_n themselves (as was the case for discrete-to-discrete convergence in the previous section) but rather from their *standardized* versions:

$$X_n^\star = \frac{X_n - \mu_n}{\sigma_n}.$$

In this section, we provide definitions and major theorems needed to deal with such a discrete-to-continuous situation[7]. Our developments largely parallel those of Section IX. 2 relative to the discrete case, with integral transforms serving as the continuous analogue of probability generating functions.

IX. 4.1. Convergence to a continuous limit. A real random variable Y is in all generality specified by its *distribution function*,

$$\mathbb{P}\{Y \le x\} = F(x).$$

It is said to be *continuous* if $F(x)$ is continuous (see Appendix C.2: *Random variables*, p. 771). In that case, $F(x)$ has no jump, and there is no single value in the range of Y that bears a non-zero probability mass. If in addition $F(x)$ is differentiable, the random variable Y is said to have a *density*, $g(x) = F'(x)$, so that

$$\mathbb{P}(Y \le x) = \int_{-\infty}^{x} g(x)\, dx, \qquad \mathbb{P}\{x < Y \le x + dx\} = g(x)\, dx.$$

A particularly important case for us here is the standard *Gaussian* or *normal* $\mathcal{N}(0, 1)$ distribution function,

$$\Phi(x) = \frac{1}{\sqrt{2\pi}} \int_{-\infty}^{x} e^{-w^2/2}\, dw,$$

also called the *error function* (erf), the corresponding density being

$$\xi(x) \equiv \Phi'(x) = \frac{1}{\sqrt{2\pi}} e^{-x^2/2}.$$

This section and the next ones are relative to the existence of limit laws of the continuous type, with Gaussian limits playing a prominent rôle. The general definitions of convergence in law (or in distribution) and of weak convergence (see Appendix C.5: *Convergence in law*, p. 776) instantiate as follows.

Definition IX.3 (Discrete-to-continuous convergence). *Let Y be a continuous random variable with distribution function $F_Y(x)$. A sequence of random variables Y_n with*

---

[7]Probability theory has elaborated a unified way of dealing with discrete and continuous laws alike, as well as with mixed cases; see Appendix C.1: *Probability spaces and measure*, p. 769. For analytic combinatorics, it seems, however, preferable to develop the two branches of the theory in a parallel fashion.

distribution functions $F_{Y_n}(x)$ is said to converge in distribution to Y if, pointwise, for each x,

$$\lim_{n \to \infty} F_{Y_n}(x) = F_Y(x).$$

In that case, one writes $Y_n \Rightarrow Y$ and $F_{Y_n} \Rightarrow F_Y$. Convergence is said to take place with speed ϵ_n if

$$\sup_{x \in \mathbb{R}} \left| F_{Y_n}(x) - F_Y(x) \right| \le \epsilon_n.$$

The definition does not *a priori* impose uniform convergence. It is a known fact, however, that *convergence of distribution functions to a continuous limit is always uniform*. This uniformity property means that there always exists a speed ϵ_n that tends to 0 as $n \to \infty$.

IX. 4.2. Continuity theorems for transforms. Discrete limit laws can be established via convergence of PGFs to a common limit, as asserted by the continuity theorem for PGFs, Theorem IX.1, p. 624. In the case of continuous limit laws, one has to resort to integral transforms (see Appendix C.3: *Transforms of distributions*, p. 772), whose definitions we now recall.

— The *Laplace transform*, also called the *moment generating function*, $\lambda_Y(s)$ is defined by

$$\lambda_Y(s) := \mathbb{E}\{e^{sY}\} = \int_{-\infty}^{+\infty} e^{sx}\, dF(x).$$

— The *Fourier transform*, also called the *characteristic function*, $\phi_Y(t)$ is defined by

$$\phi_Y(t) := \mathbb{E}\{e^{itY}\} = \int_{-\infty}^{+\infty} e^{itx}\, dF(x).$$

(Integrals are taken in the sense of Lebesgue–Stieltjes or Riemann–Stieltjes; cf Appendix C.1: *Probability spaces and measure*, p. 769.)

There are two classical versions of the continuity theorem, one for characteristic functions, the other for Laplace transforms. Both may be viewed as extensions of the continuity theorem for PGFs. Characteristic functions always exist and the corresponding continuity theorem gives a *necessary and sufficient condition* for convergence of distributions. As they are a universal tool, characteristic functions are therefore often favoured in the probabilistic literature. In the context of this book, strong analyticity properties go along with combinatorial constructions so that both transforms usually exist and both can be put to good use (Figure IX.8).

Theorem IX.4 (Continuity of integral transforms). *Let Y, Y_n be random variables with Fourier transforms (characteristic functions) $\phi(t), \phi_n(t)$, and assume that Y has a continuous distribution function. A necessary and sufficient condition[8] for the convergence in distribution, $Y_n \Rightarrow Y$, is that, pointwise, for each real t,*

$$\lim_{n \to \infty} \phi_n(t) = \phi(t).$$

---

[8] The first part of this theorem is also known as *Lévy's continuity theorem* for characteristic functions.

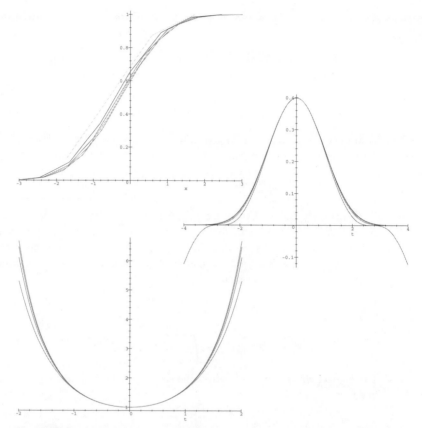

Figure IX.8. The standardized distribution functions of the binomial law (top), the corresponding Fourier transforms (right), and the Laplace transforms (bottom), for $n = 3, 6, 9, 12, 15$. The distribution functions centred around the mean $\mu_n = n/2$ and scaled according to the standard deviation $\sigma_n = \sqrt{n/4}$ converge to a limit which is the Gaussian error function, $\Phi(x) = \dfrac{1}{\sqrt{2\pi}} \displaystyle\int_{-\infty}^{x} e^{-w^2/2}\, dw$. Accordingly, the corresponding Fourier transforms (or characteristic functions) converge to $\phi(t) = e^{-t^2/2}$, while the Laplace transforms (or moment generating functions) converge to $\lambda(s) = e^{s^2/2}$.

Let Y, Y_n be random variables with Laplace transforms $\lambda(s), \lambda_n(s)$ that exist in a common interval $[-s_0, s_0]$, with $s_0 > 0$. If, pointwise for each real $s \in [-s_0, s_0]$,

$$\lim_{n \to \infty} \lambda_n(s) = \lambda(s),$$

then the Y_n converge in distribution to Y: $Y_n \Rightarrow Y$.

Proof. See Billingsley's book [68, Sec. 26] for Fourier transforms and [68, p. 408], for Laplace transforms. ∎

▷ **IX.16.** *Laplace transforms need not exists.* Let Y_n be a mixture of a Gaussian and a Cauchy distribution:

$$\mathbb{P}(Y_n \le x) = \left(1 - \frac{1}{n}\right) \int_{-\infty}^{x} \frac{e^{-w^2/2}}{\sqrt{2\pi}} \, dw + \frac{1}{\pi n} \int_{-\infty}^{x} \frac{dw}{1 + w^2}.$$

Then Y_n converges in distribution to a standard Gaussian limit Y, although $\lambda_n(s)$ only exists for $\Re(s) = 0$. ◁

In the discrete case, the continuity theorem for PGFs (Theorem IX.1, 624) eventually relies on continuity of the Cauchy coefficient formula that realizes the inversion needed in recovering coefficients from PGFs. In an analogous manner, the continuity theorem for integral transforms may be viewed as expressing the continuity of Laplace or Fourier inversion in the specific context of probability distribution functions.

The next theorem, called the *Berry–Esseen inequality*, is an effective version of the Fourier inversion theorem that proves especially useful for characterizing speeds of convergence. It bounds in a constructive manner the sup-norm distance between two distribution functions in terms of a special metric distance between their characteristic functions. Recall that $||f||_\infty := \sup_{x \in \mathbb{R}} |f(x)|$.

Theorem IX.5 (Berry–Esseen inequality). *Let F, G be distribution functions with characteristic functions $\phi(t), \gamma(t)$. Assume that G has a bounded derivative. There exist absolute constants c_1, c_2 such that for any $T > 0$,*

$$||F - G||_\infty \le c_1 \int_{-T}^{+T} \left| \frac{\phi(t) - \gamma(t)}{t} \right| \, dt + c_2 \frac{||G'||_\infty}{T}.$$

Proof. See Feller [206, p. 538] who gives

$$c_1 = \frac{1}{\pi}, \qquad c_2 = \frac{24}{\pi}$$

as possible values for the constants. ∎

This theorem is typically used with G being the limit distribution function (often a Gaussian for which $||G'||_\infty = (2\pi)^{-1/2}$) and $F = F_n$ a distribution that belongs to a sequence converging to G. The quantity T may be assigned an arbitrary value; the one giving the best bound in a specific application context is then naturally chosen.

▷ **IX.17.** *A general version of Berry–Esseen.* Let F, G be two distributions functions. Define Lévy's "concentration function", $Q_G(h) := \sup_x (G(x + h) - G(x))$, for $h > 0$. There exists an absolute constant C such that

$$||F - G||_\infty \le C Q_G(\frac{1}{T}) + C \int_{-T}^{+T} \left| \frac{\phi(t) - \gamma(t)}{t} \right| \, dt.$$

See Elliott's book [191, Lemma 1.47] and the article by Stef and Tenenbaum for a discussion [557]. The latter provides inequalities analogous to Berry–Esseen, but relative to Laplace transforms on the real line (distance bounds tend to be much weaker due to the smoothing nature of the Laplace transform). ◁

Large powers and the central limit theorem. Here is the simplest conceivable illustration of how to use the continuity theorem, Theorem IX.4. The unbiased binomial distribution $\mathrm{Bin}(n, 1/2)$ is defined as the distribution of a random variable X_n with PGF

$$p_n(u) \equiv \mathbb{E}(u^{X_n}) = \left(\frac{1}{2} + \frac{u}{2} \right)^n,$$

and characteristic function,

$$\phi_n(t) \equiv \mathbb{E}(e^{itX_n}) = p_n(e^{it}) = \frac{1}{2^n} \left(1 + e^{it}\right)^n.$$

The mean is $\mu_n = n/2$ and the variance is $\sigma_n^2 = n/4$. Therefore, the standardized variable $X_n^\star = (X_n - \mu_n)/\sigma_n$ has characteristic function

$$(20) \qquad \phi_n^\star(t) \equiv \mathbb{E}(e^{itX_n^\star}) = \left(\cosh \frac{it}{\sqrt{n}}\right)^n = \left(\cos \frac{t}{\sqrt{n}}\right)^n.$$

The asymptotic form is directly found by taking logarithms, and one gets

$$(21) \qquad \log \phi_n^\star(t) = n \log \left(1 - \frac{t^2}{2n} + \frac{t^4}{6n^2} + \cdots\right) = -\frac{t^2}{2} + O\left(\frac{1}{n}\right),$$

pointwise, for any fixed t, as $n \to \infty$. Thus, we have $\phi_n^\star(t) \to e^{-t^2/2}$, as $n \to \infty$. This establishes convergence to the Gaussian limit. In addition, upon choosing $T = n^{1/2}$, the Berry–Esseen inequalities (Theorem IX.5) show that the speed of convergence is $O(n^{-1/2})$.

▷ **IX.18.** *De Moivre's Central Limit Theorem.* Characteristic functions extend the normal limit law to biased binomial distributions with PGF $(p + qu)^n$, where $p + q = 1$. (Of course, the result is also accessible from elementary asymptotic calculus, which constitutes De Moivre's original derivation; see Note IX.1, p. 615.) ◁

The *Central Limit Theorem*, known as the *CLT* (the term was coined by Pólya in 1920, originally because of its "*zentralle Rolle*" [central rôle] in probability theory), expresses the asymptotically Gaussian character of sums of random variables. It was first discovered[9] in the particular case of binomial variables by De Moivre. The general version is due to Gauss (who, around 1809, had realized from his works on geodesy and astronomy the universality of the "Gaussian" law but had only unsatisfactory arguments) and to Laplace (in the period 1812–1820). Laplace in particular uses Fourier methods and his formulation of the CLT is highly general, although some of the precise validity conditions of his arguments only became apparent more than a century later.

Theorem IX.6 (Basic CLT). *Let T_j be independent random variables supported by $\mathbb{R}$ with a common distribution of (finite) mean μ and (finite) standard deviation σ. Let $S_n := T_1 + \cdots + T_n$. Then the standardized sum $S_n^\star$ converges to the standard normal distribution,*

$$S_n^\star \equiv \frac{S_n - \mu n}{\sigma \sqrt{n}} \Rightarrow \mathcal{N}(0, 1).$$

Proof. The proof is based on local expansions of characteristic functions, much like those in Equations (20) and (21). First, by a general theorem (see the summary in Figure B.2, p. 777 and [424, p. 22], for a proof), the existence of the first two moments implies that ϕ_{T_1} is twice differentiable at 0, so that

$$\phi_{T_1}(t) = 1 + i\mu t - \frac{1}{2}(\mu^2 + \sigma^2)t^2 + o(t^2), \qquad t \to 0.$$

---

[9]For a perspective on historical aspects of CLT, we refer to Hans Fischer's well-informed monograph [213].

By shifting, it suffices to consider the case of zero-mean variables ($\mu = 0$). We then have, pointwise for each t as $n \to \infty$,

$$(22) \qquad \phi_{T_1} \left(\frac{t}{\sigma \sqrt{n}} \right)^n = \left(1 - \frac{t^2}{2n} + o\left(\frac{t^2}{2n} \right) \right)^n \to e^{-t^2/2},$$

as in Equations (20) and (21). The conclusion follows from the continuity theorem. (This theorem is in virtually any basic book on probability theory, e.g., [206, p. 259] or [68, Sec. 27].) ∎

It is important to observe what happens if the T_j are discrete and given by their common PGF $p(u) \equiv p_{T_1}(u)$ (a case otherwise discussed in Subsection VIII. 8.3, p. 591, under a different angle). The proof above makes use of characteristic functions, that is, we set $u = e^{it}$, so that $u = 1$ corresponds to $t = 0$. Since there is a scaling of t by $1/\sqrt{n}$ in the crucial estimate (22), we only need information on $p(u)$ relatively to a small neighbourhood of $u = 1$. What this discussion brings is the following general fact: *in establishing continuous limit laws from discrete distributions, it is the behaviour near 1 of the discrete probability generating functions that matters.* We are going to make abundant use of this observation in the next section.

▷ **IX.19.** *Poisson distributions of large parameter.* Let X_λ be Poisson with rate λ. As λ tends to infinity, Stirling's formula provides easily convergence to a Gaussian limit. The error terms can then be compared to what the Berry–Esseen bounds provide. (In terms of speed of convergence, such Poisson variables of large parameters sometimes yield better approximations to combinatorial distributions than the standard Gaussian law; see Hwang's comprehensive study [341] for a general analytic approach.) ◁

▷ **IX.20.** *Extensions of the CLT.* The central limit theorem in the independent case is the subject of Petrov's comprehensive monographs [481, 482]. There are many extensions of the CLT, to variables that are independent but not necessarily identically distributed (the Lindeberg–Lyapunov conditions) or variables that are only dependent in some weak sense (mixing conditions); see the discussion by Billingsley [68, Sec. 27]. In the particular case where the Ts are discrete, a stronger "local" form of the Theorem results from the saddle-point method; see our earlier discussion in Section VIII. 8, p. 585, the classic treatment by Gnedenko and Kolmogorov [294], and extensions in Section IX. 9 below. ◁

IX. 4.3. Tail estimates.

Contrary to what happens with characteristic functions that are always defined, the mere existence of the Laplace transform of a distribution in a non-empty interval containing 0 implies interesting tail properties. We quote here:

Theorem IX.7 (Exponential tail bounds). *Let Y be a random variable such that its Laplace transform $\lambda(s) = \mathbb{E}(e^{sY})$ exists in an interval $[-a, b]$, where $-a < 0 < b$. Then the distribution of Y admits exponential tails, in the sense that, as $x \to +\infty$, there holds*

$$\mathbb{P}(Y < -x) = O(e^{-ax}), \qquad \mathbb{P}(Y > x) = O(e^{-bx}).$$

Proof. By symmetry (change Y to $-Y$), it suffices to establish the right-tail bounds. We have, for any s such that $0 \le s \le b$,

$$(23) \qquad \begin{aligned} \mathbb{P}(Y > x) &= \mathbb{P}(e^{sY} > e^{sx}) \\ &= \mathbb{P}\left[e^{sY} > \frac{e^{sx}}{\lambda(s)} \mathbb{E}(e^{sY}) \right] \\ &\le \lambda(s) e^{-sx}, \end{aligned}$$

where the last line results from Markov's inequality (Appendix A.3: *Combinatorial probability*, p. 727). It then suffices to choose $s = b$. ∎

Like its discrete counterpart, Theorem IX.3, this theorem is technically quite shallow but still useful, since it sets the stage for the ulterior development of large deviation estimates, in Section IX. 10.

IX. 5. Quasi-powers and Gaussian limit laws

The central limit theorem of probability theory admits a fruitful extension in the context of analytic combinatorics. As we show in this section, it suffices that the PGF of a combinatorial parameter *behaves* nearly *like a large power* of a fixed function to ensure convergence to a Gaussian limit—this is the *quasi-powers* framework. We first illustrate this point by considering the Stirling cycle distribution.

Example IX.9. *The Stirling cycle distribution.* The number χ of cycles in a permutation is described by the BGF

$$\mathcal{P} = \mathrm{SET}(u\,\mathrm{CYC}(\mathcal{Z})) \qquad \Longrightarrow \qquad P(z, u) = \exp\left(u \log \frac{1}{1-z}\right) = (1-z)^{-u}.$$

Let X_n be the random variable corresponding to χ taken over $\mathcal{P}_n$. The PGF of X_n is

$$p_n(u) = \binom{n+u-1}{n} = \frac{u(u+1)(u+2)\cdots(u+n-1)}{n!} = \frac{\Gamma(u+n)}{\Gamma(u)\Gamma(n+1)}.$$

We find for u near 1,

$$(24) \qquad p_n(u) \equiv \mathbb{E}(u^{X_n}) = \frac{n^{u-1}}{\Gamma(u)}\left(1 + O\left(\frac{1}{n}\right)\right) = \frac{1}{\Gamma(u)}\left(e^{(u-1)}\right)^{\log n}\left(1 + O\left(\frac{1}{n}\right)\right).$$

The last estimate results from Stirling's formula for the Gamma function (or from singularity analysis of $[z^n](1-z)^{-u}$, Chapter VI), with the error term being uniformly $O(n^{-1})$, provided u stays in a small enough neighbourhood of 1, for instance $|u - 1| \le 1/2$. Thus, as $n \to +\infty$, the PGF $p_n(u)$ *approximately* equals a large power of e^{u-1}, taken with exponent $\log n$ and multiplied by the fixed function, $\Gamma(u)^{-1}$. By analogy with the Central Limit Theorem, we may reasonably expect a Gaussian law to hold.

The mean satisfies $\mu_n = \log n + \gamma + o(1)$ and the standard deviation is $\sigma_n = \sqrt{\log n} + o(1)$. We then consider the standardized random variable,

$$X_n^\star = \frac{X_n - L - \gamma}{\sqrt{L}}, \qquad \text{where} \quad L := \log n.$$

The characteristic function of $X_n^\star$, namely $\phi_n^\star(t) = \mathbb{E}\left(e^{it X_n^\star}\right)$, then inherits the estimate (24) of $p_n(u)$:

$$\phi_n^\star(t) = \frac{e^{-it(L^{1/2}+\gamma L^{-1/2})}}{\Gamma(e^{it/\sqrt{L}})} \exp\left(L(e^{it/\sqrt{L}} - 1)\right)\left(1 + O\left(\frac{1}{n}\right)\right).$$

For fixed t, with $L \to \infty$, the logarithm is then found mechanically to satisfy

$$(25) \qquad \log \phi_n^\star(t) = -\frac{t^2}{2} + O\left((\log n)^{-1/2}\right),$$

so that $\phi_n^\star(t) \sim e^{-t^2/2}$. This is sufficient to establish a Gaussian limit law,

$$(26) \qquad \lim_{n \to \infty} \mathbb{P}\left\{ X_n \leq \log n + \gamma + x\sqrt{\log n} \right\} = \frac{1}{\sqrt{2\pi}} \int_{-\infty}^{x} e^{-w^2/2}\, dw.$$

Proposition IX.5 (Goncharov's Theorem). *The Stirling cycle distribution,* $\mathbb{P}(X_n = k) = \frac{1}{n!}\left[{n \atop k}\right]$, *describing the number of cycles (equivalently, the number of records) in a random permutation of size n is asymptotically normal.*

This result was obtained by Goncharov as early as 1944 (see [299]), albeit without an error term, as his investigations predate the Berry–Esseen inequalities. Our treatment quantifies the speed of convergence to the Gaussian limit as $O((\log n)^{-1/2})$, by virtue of Equation (25) and Theorem IX.5. .... ■

The cycle example is characteristic of the occurrence of Gaussian laws in analytic combinatorics. What happens is that the approximation (24) by a power with "large" exponent $\beta_n = \log n$ leads after normalization, to the characteristic function of a Gaussian variable, namely $e^{-t^2/2}$. From this, the limit distribution (26) results by the continuity theorem. This is in fact a very general phenomenon, as demonstrated by a theorem of Hsien-Kuei Hwang [337, 340] that we state next and that builds upon earlier statements of Bender and Richmond [44].

The following notations will prove especially convenient: given a function $f(u)$ analytic at $u = 1$ and assumed to satisfy $f(1) \neq 0$, we set

$$(27) \qquad \mathfrak{m}(f) = \frac{f'(1)}{f(1)}, \qquad \mathfrak{v}(f) = \frac{f''(1)}{f(1)} + \frac{f'(1)}{f(1)} - \left(\frac{f'(1)}{f(1)}\right)^2.$$

The notations $\mathfrak{m}, \mathfrak{v}$ suggest their probabilistic counterparts while neatly distinguishing between the analytic and probabilistic realms: If f is the PGF of a random variable X, then $f(1) = 1$ and $\mathfrak{m}(f)$, the mean, coincides with the expectation $\mathbb{E}(X)$; the quantity $\mathfrak{v}(f)$ then coincides with the variance $\mathbb{V}(X)$. Accordingly, we call $\mathfrak{m}(f)$ and $\mathfrak{v}(f)$, respectively, the *analytic mean* and *analytic variance* of function f.

Theorem IX.8 (Quasi-powers Theorem). *Let the X_n be non-negative discrete random variables (supported by $\mathbb{Z}_{\geq 0}$), with probability generating functions $p_n(u)$. Assume that,* uniformly *in a fixed complex neighbourhood of $u = 1$, for sequences $\beta_n, \kappa_n \to +\infty$, there holds*

$$(28) \qquad p_n(u) = A(u) \cdot B(u)^{\beta_n} \left(1 + O\left(\frac{1}{\kappa_n}\right)\right),$$

where $A(u), B(u)$ are analytic at $u = 1$ and $A(1) = B(1) = 1$. Assume finally that $B(u)$ satisfies the so-called "variability condition",

$$\mathfrak{v}(B(u)) \equiv B''(1) + B'(1) - B'(1)^2 \neq 0.$$

Under these conditions, the mean and variance of X_n satisfy

$$(29) \qquad \begin{aligned} \mu_n &\equiv \mathbb{E}(X_n) &=& \beta_n\, \mathfrak{m}(B(u)) + \mathfrak{m}(A(u)) + O\left(\kappa_n^{-1}\right) \\ \sigma_n^2 &\equiv \mathbb{V}(X_n) &=& \beta_n\, \mathfrak{v}(B(u)) + \mathfrak{v}(A(u)) + O\left(\kappa_n^{-1}\right). \end{aligned}$$

The distribution of X_n is, after standardization, asymptotically Gaussian, and the speed of convergence to the Gaussian limit is $O(\kappa_n^{-1} + \beta_n^{-1/2})$:

$$(30) \qquad \mathbb{P}\left\{ \frac{X_n - \mathbb{E}(X_n)}{\sqrt{\mathbb{V}(X_n)}} \le x \right\} = \Phi(x) + O\left(\frac{1}{\kappa_n} + \frac{1}{\sqrt{\beta_n}} \right),$$

where $\Phi(x)$ is the distribution function of a standard normal,

$$\Phi(x) = \frac{1}{\sqrt{2\pi}} \int_{-\infty}^{x} e^{-w^2/2}\, dw.$$

This theorem is a direct application of the following lemma, also due to Hwang [337, 340], that applies more generally to arbitrary discrete or continuous distributions (see also Note IX.22, p. 647), and is thus entirely phrased in terms of integral transforms.

Lemma IX.1 (Quasi-powers, general distributions). *Assume that the Laplace transforms $\lambda_n(s) = \mathbb{E}\{e^{sX_n}\}$ of a sequence of random variables X_n are analytic in a disc $|s| < \rho$, for some $\rho > 0$, and satisfy there an expansion of the form*

$$(31) \qquad \lambda_n(s) = e^{\beta_n U(s) + V(s)} \left(1 + O\left(\frac{1}{\kappa_n} \right) \right),$$

with $\beta_n, \kappa_n \to +\infty$ as $n \to +\infty$, and $U(s), V(s)$ analytic in $|s| \le \rho$. Assume also the variability condition, $U''(0) \ne 0$.

Under these assumptions, the mean and variance of X_n satisfy

$$(32) \qquad \begin{aligned} \mathbb{E}(X_n) &= \beta_n U'(0) + V'(0) + O(\kappa_n^{-1}), \\ \mathbb{V}(X_n) &= \beta_n U''(0) + V''(0) + O(\kappa_n^{-1}). \end{aligned}$$

The distribution of $X_n^\star := (X_n - \beta_n U'(0))/\sqrt{\beta_n U''(0)}$ is asymptotically Gaussian, the speed of convergence to the Gaussian limit being $O(\kappa_n^{-1} + \beta_n^{-1/2})$.

Proof. First, we estimate the mean and variance. The variable s is *a priori* restricted to a small neighbourhood of 0. By assumption, the function $\log \lambda_n(s)$ is analytic at 0 and it satisfies

$$\log \lambda_n(s) = \beta_n U(s) + V(s) + O\left(\frac{1}{\kappa_n} \right)$$

This asymptotic expansion carries over, with the same type of error term, to derivatives at 0 because of analyticity: this can be checked directly from Cauchy integral representations,

$$\frac{1}{k!} \frac{d^r}{ds^r} \log \lambda_n(s) \bigg|_{s=0} = \frac{1}{2i\pi} \int_{\gamma} \log \lambda_n(s) \frac{ds}{s^{r+1}},$$

upon using a small but fixed integration contour γ and taking advantage of the basic expansion of $\log \lambda_n(s)$. In particular, the mean and variance are seen to satisfy the estimates of (32).

Next, we consider the standardized variable,

$$X_n^\star = \frac{X_n - \beta_n U'(0)}{\sqrt{\beta_n U''(0)}}, \qquad \lambda_n^\star(s) = \mathbb{E}\{e^{sX_n^\star}\}.$$

We have

$$\log \lambda_n^\star(s) = -\frac{\beta_n U'(0)}{\sqrt{\beta_n U''(0)}}s + \log \lambda_n\left(\frac{s}{\sqrt{\beta_n U''(0)}}\right).$$

Local expansions to third order based on the assumption (31), with $\lambda_n(0) \equiv 1$, yield

$$(33) \qquad \log \lambda_n^\star(s) = \frac{s^2}{2} + O\left(\frac{|s| + |s|^3}{\beta_n^{1/2}}\right) + O\left(\frac{1}{\kappa_n}\right),$$

uniformly with respect to s in a disc of radius $O(\beta_n^{1/2})$, and in particular in any fixed neighbourhood of 0. This is enough to conclude as regards convergence in distribution to a Gaussian limit, by the continuity theorem of either Laplace transforms (restricting s to be real) or of Fourier transforms (taking $s = it$).

Finally, the speed of convergence results from the Berry–Esseen inequalities. Take $T \equiv T_n = c\beta_n^{1/2}$, where c is taken sufficiently small but non-zero, in such a way that the local expansion of $\lambda_n(s)$ at 0 applies. Then, the expansion (33) instantiated at $s = it$ entails that

$$\Delta_n := \int_{-T_n}^{T_n} \left|\frac{\lambda_n^\star(it) - e^{-t^2/2}}{t}\right| dt + \frac{1}{T_n}$$

satisfies $\Delta_n = O(\beta_n^{-1/2} + \kappa_n^{-1})$. The statement now follows from the Berry–Esseen inequality, Theorem IX.5. ∎

Theorem IX.8 under either form (28) or (31) can be read *formally* as expressing the distribution of a (pseudo)random variable

$$Z = Y_0 + W_1 + W_2 + \cdots + W_{\beta_n},$$

where Y_0 "corresponds" to $e^{V(s)}$ (or $A(u)$) and each W_j to $e^{U(s)}$ (or $B(u)$). However, there is no *a priori* requirement that β_n should be an integer, nor that $e^{U(s)}, e^{V(s)}$ be Laplace transforms of probability distribution functions (usually they *aren't*). In a way, the theorem recycles the intuition that underlies the classical proof of the central limit theorem and makes use of the analytic machinery behind it.

It is of particular importance to note that the conditions of Theorem IX.8 and Lemma IX.1 are purely local: *what is required is* local *analyticity of the quasi-power approximation at $u = 1$ for PGFs or, equivalently, $s = 0$ for Laplace–Fourier transforms.* This important feature ultimately owes to the standardization of random variables and the corresponding scaling of transforms that goes along with continuous limit laws

▷ **IX.21.** *Mean, variance and cumulants.* With the notations of (27), one has also

$$\mathfrak{m}(f) = \left.\frac{d}{dt}\log f(e^t)\right|_{t=0}, \qquad \mathfrak{v}(f) = \left.\frac{d^2}{dt^2}\log f(e^t)\right|_{t=0};$$

the higher order derivatives give rise to quantities known as *cumulants*. ◁

▷ **IX.22.** *Two equivalent forms of standardization.* By simple real analysis, one has also, under the assumptions of Lemma IX.1:

$$\mathbb{P}\left\{\frac{X_n - \mathbb{E}(X_n)}{\sqrt{\mathbb{V}(X_n)}} \le x\right\} = \Phi(x) + O\left(\frac{1}{\kappa_n} + \frac{1}{\sqrt{\beta_n}}\right).$$

Thus, main approximations in the convergence to the Gaussian limit are not affected by the way standardization is done, either with the *exact* values of the mean and variance of X_n or with their first-order asymptotic *approximations*. The same is true for Theorem IX.8.   ◁

▷ **IX.23.** *Higher moments under quasi-powers conditions.* Following Hwang [340], one has also, under the conditions of the Quasi-powers Theorem and for each fixed k,

$$\mathbb{E}(X_n^k) = \varpi_k(\beta_n) + O\left(\frac{1}{\kappa_n}\right), \qquad \varpi_k(x) := k![s^k]e^{xU(s)+V(s)}.$$

Thus, a polynomial ϖ_k, of exact degree k, describes the asymptotic form of higher moments. (Hint: make use of differentiability properties of asymptotic expansions of analytic functions, as in Subsection VI. 10.1, p. 418.)   ◁

Singularity perturbation and Gaussian laws. The main thread of this chapter is that of *bivariate generating functions*. In general, we are given a BGF $F(z, u)$ and aim at extracting a limit distribution from it. The quasi-power paradigm in the form (28) is what one should look for, when the mean and the standard deviation both tend to infinity with the size n of the combinatorial model.

We proceed heuristically in the following informal discussion, which expands on the brief indications of p. 618 relative to singularity perturbation—precise developments are given in the next sections. Start from a BGF $F(z, u)$ and consider u as a parameter. If a singularity analysis of sorts is applicable to the counting generating function $F(z, 1)$, it leads to an approximation,

$$f_n \approx C \cdot \rho^{-n} n^\alpha,$$

where ρ is the dominant singularity of $F(z, 1)$ and α is related to the critical exponent of $F(z, 1)$ at ρ. A similar type of analysis is often applicable to $F(z, u)$ for u near 1. Then, it is reasonable to hope for an approximation of the coefficients in the z-expansion of the bivariate GF,

$$f_n(u) \approx C(u)\rho(u)^{-n} n^{\alpha(u)}.$$

In this perspective, the corresponding PGF is of the form

$$p_n(u) \approx \frac{C(u)}{C(1)} \left(\frac{\rho(u)}{\rho(1)}\right)^{-n} n^{\alpha(u)-\alpha(1)}.$$

The strategy envisioned here is thus a perturbation analysis of singular expansions with the auxiliary parameter u being restricted to a small neighbourhood of 1.

In particular if only *the dominant singularity moves* with u, we have a rough form

$$p_n(u) \approx \frac{C(u)}{C(1)} \left(\frac{\rho(u)}{\rho(1)}\right)^{-n},$$

suggesting a Gaussian law with mean and variance that are both $O(n)$, by the Quasi-powers Theorem. If only *the exponent varies*, then

$$p_n(u) \approx \frac{C(u)}{C(1)} n^{\alpha(u)-\alpha(1)} = \frac{C(u)}{C(1)} \left(e^{\alpha(u)-\alpha(1)}\right)^{\log n},$$

suggests again a Gaussian law, but with mean and variance that are now both $O(\log n)$.

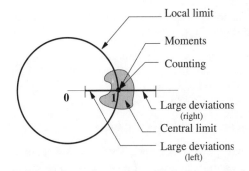

Region	Property
$u = 1$	Counting
$u = 1 \pm o(1)$	Moments
$u \in \mathcal{V}(1)$ (neighb.)	Central limit law
$|u| = 1$	Local limit law
$u \in [\alpha, \beta]$	Large deviations

Figure IX.9. The correspondence between regions of the u–plane when considering a combinatorial BGF $F(z, u)$ and asymptotic properties of combinatorial distributions.

These cases point to the fact that a rather simple perturbation of a univariate analysis is likely to yield a limiting Gaussian distribution. Each major coefficient extraction method of Chapters IV–VIII then plays a rôle, and the present chapter illustrates this important point in the following contexts.

— *Meromorphic analysis* for functions with polar singularities (Section IX. 6 below, based on a perturbation of methods of Chapters IV and V);

— *Singularity analysis* for functions with algebraic–logarithmic singularity (Section IX. 7 below, based on a perturbation of methods of Chapters VI and VII);

— *Saddle-point analysis* for functions with fast growth at their singularity (Section IX. 8 below, based on a perturbation of methods of Chapters VIII).

In essence, the decomposable character of many elementary combinatorial structures is reflected by strong analyticity properties of bivariate GFs that, after perturbation analysis, lead, via the Quasi-powers Theorem (Theorem IX.8), to Gaussian laws. The coefficient extraction methods being based on contour integration supply the necessary uniformity conditions.

We shall also see that several other properties often supplement the existence of Gaussian limit laws in combinatorics:

— *Local limit laws* [developed in Section IX. 9, p. 694 below] arise from quasi-power approximations, whenever these remain valid for all values of u *on the unit circle*. In that case, it is possible to express the combinatorial probability distribution directly in terms of the Gaussian density, by means of the saddle-point method (in a form similar to that of Section VIII. 8, p. 585, dedicated lo large powers) replacing the Continuity Theorem to effect the secondary coefficient extraction in $[u^k z^n] F(z, u)$.

— *Large deviation estimates* [developed in Section IX. 10, p. 699 below] quantify the probabilities of rare events, away from the mean value. As could be anticipated from Subsection IX. 4.3 relative to tail bounds, they are obtained by considering $[z^n] F(z, u_0)$ for some value of u_0 away from 1, via what are essentially saddle-point bounds applied to $[z^n] F(z, u_0)$.

The correspondence between u–domains and properties of combinatorial distributions is summarized in Figure IX.9. The next sections will copiously illustrate this paradigm for each of the main complex asymptotic methods of Part B.

IX. 6. Perturbation of meromorphic asymptotics

Once equipped with the general Quasi-powers Theorem, Theorem IX.8 (p. 645), it becomes possible to proceed and analyse broad classes of analytic schemas, along the lines of the principles of singularity perturbation informally presented in the previous section. We commence by investigating the effect of the secondary variable u on a bivariate generating function, whose univariate restriction $F(z, 1)$ can be subjected to a meromorphic analysis (Chapters IV and V), that is, its dominant singularities are poles. For basic parameters arising from the constructions examined there, Gaussian laws are the rule.

In what follows, we first examine supercritical compositions and sequences and establish the Gaussian character of the number of components. In this way, one gets precise information on the profile of supercritical sequences, which greatly refines the mean value estimates of Section V. 2, p. 293. We next enunciate a powerful statement widely applicable to meromorphic functions, with typical applications to runs in permutations, parallelogram polyominoes, and coin fountains. The section concludes with an investigation of the elementary perturbation theory of linear systems, whose applications are in the area of paths in graphs, finite automata, and transfer matrix models (Sections V. 5 and V. 6).

This section is largely based on works of Bender who, starting with his seminal article [35], was the first to propose abstract analytic schemas leading to Gaussian laws in analytic combinatorics. Our presentation also relies on subsequent works of Bender, Flajolet, Hwang, Richmond, and Soria [44, 258, 260, 337, 338, 339, 340, 547]. The essential philosophy here is that (almost) any univariate problem studied in Chapter V relative to rational and meromorphic asymptotics is susceptible to singularity perturbation, to the effect that limit Gaussian laws hold for basic parameters.

Supercritical compositions and sequences. Our first application of the quasi-powers framework is to supercritical compositions (p. 411), whenever the outer function has a dominant pole. This covers in particular supercritical sequences, for which asymptotic enumeration and moments have been worked out in Section V. 2, p. 293. In this way, we get access to distributions arising in surjections, alignments, and compositions of various sorts. Our reader is encouraged to study the proof that follows, since it constitutes the technically simplest, yet characteristic, instance of a *singularity perturbation process*.

Proposition IX.6 (Supercritical compositions). *Consider the bivariate composition schema $F(z, u) = g(uh(z))$. Assume that $g(z)$ and $h(z)$ satisfy the* supercriticality condition $\tau_h > \rho_g$, *that g is analytic in $|z| < R$ for some $R > \rho_g$, with a unique dominant singularity at ρ_g, which is a simple pole, and that h is aperiodic. Then the*

*number χ of $\mathcal{H}$–components in a random $\mathcal{F}_n$–structure, corresponding to the proba-
bility distribution $[u^k z^n] F(z, u)/[z^n] F(z, 1)$ has a mean and variance that are asymptotically proportional to n; after standardization, the parameter χ satisfies a limiting
Gaussian distribution, with speed of convergence $O(1/\sqrt{n})$.*

Proof. We start as usual with univariate analyses. Let ρ be such that $h(\rho) = \rho_g$
with $0 < \rho < \rho_h$. (Existence and unicity of ρ are guaranteed by the supercriticality
condition.) The expansions,

$$g(z) = \frac{C}{1 - z/\rho_g} + D + o(1), \quad h(z) = \rho_g + h'(\rho)(z - \rho) + \frac{1}{2}h''(\rho)(z - \rho)^2 + \cdots,$$

result from the hypotheses. Clearly, $F(z) \equiv F(z, 1)$ has a simple pole at $z = \rho$ and,
by composition of the expansions of g and h:

$$F(z) = \frac{C\rho_g}{\rho h'(\rho)} \frac{1}{1 - z/\rho} + O(1).$$

Aperiodicity of h also implies that ρ is the unique dominant singularity of $F(z, 1)$.
The usual process of meromorphic coefficient analysis then provides

$$[z^n] F(z) = \frac{C\rho_g}{\rho h'(\rho)} \rho^{-n}(1 + o(1)),$$

where $o(1)$ represents an exponentially small error term. Moments can be obtained
by differentiation, to the effect that the GF associated to the moment of order r has
a pole of order $(r + 1)$ and is amenable to singularity analysis. (This mimics the
univariate analysis of supercritical compositions in Section V. 2, p. 293.) However,
moment estimates also result from subsequent developments, so that this phase of the
analysis can be bypassed.

Now comes the singularity perturbation process. In what follows, we repeatedly
restrict u to a sufficiently small neighbourhood of 1. The equation in $\rho(u)$,

$$uh(\rho(u)) = \rho_g$$

admits a unique root near ρ, when u is sufficiently close to 1, and by the analytic
inversion lemma (Lemma IV.2, p. 275), the function $\rho(u)$ is analytic at $u = 1$. The
function $z \mapsto F(z, u)$ then has a simple pole at $z = \rho(u)$, and, by composition of
expansions, we obtain:

$$(34) \qquad F(z, u) \sim \frac{C\rho_g}{u\rho(u)h'(\rho(u))} \frac{1}{1 - z/\rho(u)} \qquad (z \to \rho(u)).$$

Next, for u again close enough to 1, we claim that the function $z \mapsto F(z, u)$
admits $\rho(u)$ as unique dominant singularity. The proof of this fact depends on the
aperiodicity of $h(z)$, which grants us the inequality $|h(z)| < h(\rho) = \rho_g$ for $|z| = \rho$,
$z \neq \rho$; also, for z near ρ, the equation $h(z) = \rho_g$ admits locally a unique solution, as
already seen above. Thus, there exists a quantity $r > \rho$ such that the equation $h(z) =
\rho$ admits in $|z| < r$ the unique solution $z = \rho$. But then, by keeping u close enough
to 1, one can find S with $\rho < S < r$, such that, in $|z| \leq S$, the unique solution to
the equation $uh(z) = \rho_g$ is $\rho(u)$ (see the continuity argument used in the proof of the
Analytic Inversion Theorem of Appendix B.5: *Implicit Function Theorem*, p. 753).

We can now conclude. Let us take S as in the previous paragraph and restrict u to a suitably small complex neighbourhood of 1, as the need arises. We then revisit the proof by contour integration of coefficient extraction in meromorphic functions, Theorem IV.10, p. 258. We have, by residues,

$$\frac{1}{2i\pi}\int_{|z|=S} F(z,u)\frac{dz}{z^{n+1}} = [z^n]F(z,u) + \text{Res}(g(uh(z))z^{-n-1}, z = \rho(u)),$$

and, since $F(z,u) = g(uh(z))$ is analytic, hence uniformly bounded, for $|z| = S$, we get via (34) the main *uniform* estimate

$$[z^n]F(z,u) = C(u) \cdot \rho(u)^{-n}\left(1 + O(K^{-n})\right), \qquad C(u) := \frac{C\rho_g}{u\rho(u)h'(\rho(u))},$$

for some $K > 1$. Thus, the PGF of χ over $\mathcal{F}_n$, which is $p_n(u) = [z^n]F(z,u)/[z^n]F(z,1)$ satisfies

$$p_n(u) = A(u) \cdot B(u)^n\left(1 + O(K^{-n})\right), \qquad A(u) = \frac{C(u)}{C(1)}, \qquad B(u) = \frac{\rho(1)}{\rho(u)}.$$

We are then precisely within the conditions of the Quasi-powers Theorem (Theorem IX.8, p. 645), and the statement follows. ■

A prime application of the last proposition is to supercritical sequences, where the properties elicited in Section V.2, p. 293, are seen to be supplemented by Gaussian laws.

Proposition IX.7 (Supercritical sequences). *Consider a sequence schema $\mathcal{F} = \text{SEQ}(u\mathcal{H}))$ that is supercritical, i.e., the value of h at its dominant positive singularity satisfies $\tau_h > 1$. Assuming h to be aperiodic and $h(0) = 0$, the number X_n of $\mathcal{H}$–components in a random $\mathcal{F}_n$–structure of large size n is, after standardization, asymptotically Gaussian with*

$$\mathbb{E}(X_n) \sim \frac{n}{\rho h'(\rho)}, \qquad \mathbb{V}(X_n) \sim n\frac{h''(\rho) + h'(\rho) - h'(\rho)^2}{\rho h'(\rho)^3},$$

where ρ is the positive root of $h(\rho) = 1$.

The number $X_n^{(m)}$ of components of some fixed size m is asymptotically Gaussian with mean $\sim \theta_m n$, where $\theta_m = h_m\rho^m/(\rho h'(\rho))$.

Proof. The first part is a direct consequence of Proposition IX.6 with $g(z) = (1-z)^{-1}$ and ρ_g replaced by 1. The second part results from the BGF

$$\mathcal{F} = \text{SEQ}(u\mathcal{H}_m + \mathcal{H} \setminus \mathcal{H}_m) \qquad \Longrightarrow \qquad F(z,u) = \frac{1}{1 - (u-1)h_m z^m - h(z)},$$

and from the fact that $u \approx 1$ induces a smooth perturbation of the pole of $F(z,1)$ at ρ, corresponding to $u = 1$. ■

The examples and notes that follow present two different types of applications of Propositions IX.6 and IX.7. The first batch deals with cases already encountered in Chapter V, namely, surjections (Example IX.10), alignments, and compositions— Figure V.1 (p. 297) and Figure IX.10 illustrate typical profiles of these structures. The second batch shows some purely probabilistic applications to closely related renewal problems (Example IX.11).

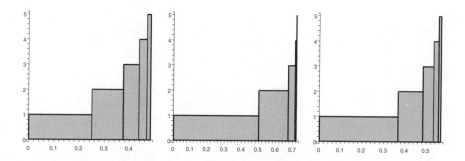

Figure IX.10. When components are sorted by size and represented by vertical segments of corresponding length, supercritical sequences present various profiles described by Proposition IX.7. The diagrams display the *limit* mean profiles of large compositions, surjections, and alignments, for component sizes ≤ 5.

Example IX.10. *The surjection distribution.* We revisit the distribution of image cardinality in surjections for which the concentration property has been established in Chapter V. This example serves to introduce bivariate asymptotics in the meromorphic case. Consider the distribution of image cardinality in surjections,

$$\mathcal{F} = \mathrm{SEQ}(u\,\mathrm{SET}_{\geq 1}(\mathcal{Z})) \qquad \Longrightarrow \qquad F(z, u) = \frac{1}{1 - u(e^z - 1)}.$$

Restrict u near 1, for instance $|u - 1| \leq 1/10$. The function $F(z, u)$, as a function of z, is meromorphic with singularities at

$$\rho(u) + 2ik\pi, \qquad \rho(u) = \log\left(1 + \frac{1}{u}\right).$$

The principal determination of the logarithm is used (with $\rho(u)$ near $\log 2$ when u is near 1). It is then seen that $\rho(u)$ stays within 0.06 from $\log 2$, for $|u - 1| \leq 1/10$. Thus $\rho(u)$ is the unique dominant singularity of F, the next nearest one being $\rho(u) \pm 2i\pi$ with modulus certainly larger than 5.

From the coefficient analysis of meromorphic functions (Chapter IV), the quantities $f_n(u) = [z^n]F(z, u)$ are estimated as follows,

$$
\begin{aligned}
(35) \qquad f_n(u) &= -\mathrm{Res}\left(F(z, u)z^{-n-1}\right)_{z=\rho(u)} + \frac{1}{2i\pi}\int_{|z|=5} F(z, u)\,\frac{dz}{z^{n+1}} \\
&= \frac{1}{u\rho(u)e^{\rho(u)}}\,\rho(u)^{-n} + O(5^{-n}).
\end{aligned}
$$

It is important to note that the error term is *uniform* with respect to u, once u has been constrained to (say) $|u - 1| \leq 0.1$. This fact is derived from the coefficient extraction method, since, in the remainder Cauchy integral of (35), the denominator of $F(z, u)$ stays bounded away from 0.

The second estimate in Equation (35), constitutes a prototypical case of application of the quasi-powers framework. Thus, the number X_n of image points in a random surjection of size n obeys in the limit a Gaussian law. The local expansion of $\rho(u)$,

$$\rho(u) \equiv \log(1 + u^{-1}) = \log 2 - \frac{1}{2}(u - 1) + \frac{3}{8}(u - 1)^2 + \cdots,$$

yields

$$\frac{\rho(1)}{\rho(u)} = 1 + \frac{1}{2\log 2}(u-1) - \frac{3\ln(2)-2}{8(\log 2)^2}(u-1)^2 + O\left((u-1)^3\right),$$

so that the mean and standard deviation satisfy

(36) $\qquad \mu_n \sim C_1 n, \quad \sigma_n \sim \sqrt{C_2 n}, \qquad C_1 := \frac{1}{2\log 2}, \quad C_2 := \frac{1-\log 2}{4(\log 2)^2}.$

In particular, the variability condition is satisfied. Finally, one obtains, with Φ the Gaussian error function,

$$\mathbb{P}\{X_n \leq C_1 n + x\sqrt{C_2 n}\} = \Phi(x) + O\left(\frac{1}{\sqrt{n}}\right).$$

This estimate can alternatively be viewed as a purely asymptotic statement regarding Stirling partition numbers.

Proposition IX.8. *The surjection distribution defined as $\frac{k!}{S_n}\begin{Bmatrix}n\\k\end{Bmatrix}$, with $S_n = \sum_k k!\begin{Bmatrix}n\\k\end{Bmatrix}$ a surjection number, satisfies* uniformly for all real x and C_1, C_2 given by (36):

$$\frac{1}{S_n} \sum_{k \leq C_1 n + x\sqrt{C_2 n}} k!\begin{Bmatrix}n\\k\end{Bmatrix} = \frac{1}{\sqrt{2\pi}} \int_{-\infty}^{x} e^{-w^2/2}\, dw + O\left(\frac{1}{\sqrt{n}}\right).$$

This result already appears in Bender's foundational study [35]. .... ∎

▷ **IX.24.** *Alignments and Stirling cycle numbers.* Alignments are sequences of cycles (Chapter II, p. 119), with exponential BGF given by

$$\mathcal{F} = \mathrm{SEQ}(u\,\mathrm{CYC}(\mathcal{Z})) \qquad \Longrightarrow \qquad F(z,u) = \frac{1}{1 - u\log(1-z)^{-1}}.$$

The function $\rho(u)$ is explicit, $\rho(u) = 1 - e^{-1/u}$, and the number of cycles in a random alignment is asymptotically Gaussian. This yields an asymptotic statement on Stirling cycle numbers: *Uniformly for all real x, with $O_n = \sum_k k!\begin{bmatrix}n\\k\end{bmatrix}$ the alignment number, there holds*

$$\frac{1}{O_n} \sum_{k \leq C_1 n + x\sqrt{C_2 n}} k!\begin{bmatrix}n\\k\end{bmatrix} = \frac{1}{\sqrt{2\pi}} \int_{-\infty}^{x} e^{-w^2/2}\, dw + O\left(\frac{1}{\sqrt{n}}\right),$$

where the two constants C_1, C_2 are $C_1 = \dfrac{1}{e-1}, C_2 = \dfrac{1}{(e-1)^2}.$ ◁

▷ **IX.25.** *Summands in constrained integer compositions.* Consider integer compositions where the summands are constrained to belong to a set $\Gamma \subseteq \mathbb{Z}_{\geq 1}$, and let X_n be the number of summands in a random composition of integer n. The ordinary BGF is

$$F(z,u) = \frac{1}{1 - uh(z)}, \qquad h(z) := \sum_{\gamma \in \Gamma} z^\gamma.$$

Assume that Γ contains at least two relatively prime elements, so that $h(z)$ is aperiodic. The radius of convergence of $h(z)$ can only be ∞ (when $h(z)$ is a polynomial) or 1 (when $h(z)$ comprises infinitely many terms but is dominated by $(1-z)^{-1}$). In all cases, the sequence construction is supercritical, so that the distribution of X_n is asymptotically normal. For instance, a Gaussian limit law holds for compositions into prime (or even twin-prime) summands enumerated in Chapter V (p. 297). ◁

***Example* IX.11.** *The Central Limit Theorem and discrete renewal theory.* Let $g(u)$ be any PGF ($g(1) = 1$) of a random variable supported by $\mathbb{Z}_{\geq 0}$ that is analytic at 1 and non-degenerate (i.e., $\mathfrak{v}(g) > 0$). Then

$$F(z, u) = \frac{1}{1 - zg(u)}$$

has a singularity at $\rho(u) := 1/g(u)$ that is a simple pole. Theorem IX.9 then applies to give the special form of the central limit theorem (p. 642) that is relative to discrete probability distributions with PGFs analytic at 1.

Under the same analytic assumptions on g, consider now the "dual" BGF,

$$G(z, u) = \frac{1}{1 - ug(z)},$$

where the rôles of z and u have been interchanged. In addition, we must impose for consistency that $g(0) = 0$. There is a simple probabilistic interpretation in terms of *renewal processes* of classical probability theory, when $g(1) = 1$. Assume a light bulb has a lifetime of m days with probability $g_m = [z^m]g(z)$ and is replaced as soon as it ceases to function. Let X_n be the number of light bulbs consumed in n days assuming independence, conditioned upon the fact that a replacement takes place on the nth day. Then the PGF of X_n is $[z^n]G(z, u)/[z^n]G(z, 1)$. (The normalizing quantity $[z^n]G(z, 1)$ is precisely the probability that a renewal takes place on day n.) Theorem IX.9 applies. The function G has a simple dominant pole at $z = \rho(u)$ such that $g(\rho(u)) = 1/u$, with $\rho(1) = 1$ since g is by assumption a PGF. One finds

$$\frac{1}{\rho(u)} = 1 + \frac{1}{g'(1)}(u - 1) + \frac{1}{2}\frac{g''(1) + 2g'(1) - 2g'(1)^2}{g'(1)^3}(u - 1)^2 + \cdots.$$

Thus the limit distribution of X_n is normal with mean and variance satisfying

$$\mathbb{E}(X_n) \sim \frac{n}{\mu}, \qquad \mathbb{V}(X_n) \sim n\frac{\sigma^2}{\mu^3},$$

where $\mu := \mathfrak{m}(g)$ and $\sigma^2 := \mathfrak{v}(g)$ are the mean and variance attached to g. (This calculation checks the variability condition *en passant*.) The mean value result certainly conforms to probabilistic intuition. .... ∎

▷ **IX.26.** *Renewals every day.* In the renewal scenario, no longer condition on the fact that a bulb breaks down on day n. Let Y_n be the number of bulbs consumed so far. Then the BGF of Y_n is found by expressing that there is a sequence of renewals followed by a last renewal that is to be credited to all intermediate epochs:

$$\sum_{n \geq 1} \mathbb{E}(u^{Y_n})z^n = \frac{1}{1 - ug(z)}\frac{g(u) - g(zu)}{1 - z}.$$

A Gaussian limit also holds for Y_n.   ◁

▷ **IX.27.** *A mixed CLT–renewal scenario.* Consider $G(z, u) = 1/(1 - g(z, u))$ where g has non-negative coefficients, satisfies $g(1, 1) = 1$, and is analytic at $(z, u) = (1, 1)$. This models the situation where bulbs are replaced but a random cost is incurred, depending on the duration of the bulb. Under general conditions, a limit law holds and it is Gaussian. This applies for instance to $H(z, u) = 1/(1 - a(z)b(u))$, where a and b are non-degenerate PGFs (a random repairman is called).   ◁

Singularity perturbation for meromorphic functions. The following analytic schema vastly generalizes the case of supercritical compositions.

Theorem IX.9 (Meromorphic schema). *Let $F(z, u)$ be a function that is bivariate analytic at $(z, u) = (0, 0)$ and has non-negative coefficients. Assume that $F(z, 1)$ is meromorphic in $z \leq r$ with only a simple pole at $z = \rho$ for some positive $\rho < r$. Assume also the following conditions.*

(*i*) Meromorphic perturbation: *there exists $\epsilon > 0$ and $r > \rho$ such that in the domain, $\mathcal{D} = \{|z| \leq r\} \times \{|u - 1| < \epsilon\}$, the function $F(z, u)$ admits the representation*

$$F(z, u) = \frac{B(z, u)}{C(z, u)},$$

where $B(z, u), C(z, u)$ are analytic for $(z, u) \in \mathcal{D}$ with $B(\rho, 1) \neq 0$. (Thus ρ is a simple zero of $C(z, 1)$.)

(*ii*) Non-degeneracy: *one has $\partial_z C(\rho, 1) \cdot \partial_u C(\rho, 1) \neq 0$, ensuring the existence of a non-constant $\rho(u)$ analytic at $u = 1$, such that $C(\rho(u), u) = 0$ and $\rho(1) = \rho$.*

(*iii*) Variability: *one has*

$$\mathfrak{v}\left(\frac{\rho(1)}{\rho(u)}\right) \neq 0.$$

Then, the random variable X_n with probability generating function

$$p_n(u) = \frac{[z^n] F(z, u)}{[z^n] F(z, 1)}$$

after standardization, converges in distribution to a Gaussian variable, with a speed of convergence that is $O(n^{-1/2})$. The mean and the standard deviation of X_n are asymptotically linear in n.

Proof. First we offer a few comments. Given the analytic solution $\rho(u)$ of the implicit equation $C(\rho(u), u) = 0$, the PGF $\mathbb{E}(u^{X_n})$ satisfies a quasi-power approximation of the form $A(u)(\rho(1)/\rho(u))^n$, as we prove below. The mean μ_n and variance σ_n^2 are then of the form

$$(37) \qquad \mu_n = \mathfrak{m}\left(\frac{\rho(1)}{\rho(u)}\right) n + O(1), \qquad \sigma_n^2 = \mathfrak{v}\left(\frac{\rho(1)}{\rho(u)}\right) n + O(1).$$

The variability condition of the Quasi-powers Theorem is precisely ensured by condition (*iii*). Set

$$c_{i,j} := \left. \frac{\partial^{i+j}}{\partial z^i \partial u^j} C(z, u) \right|_{(\rho, 1)}.$$

The numerical coefficients in (37) can themselves be solely expressed in terms of partial derivatives of $C(z, u)$ by series reversion,

(38)
$$\rho(u) = \rho - \frac{c_{0,1}}{c_{1,0}}(u - 1) - \frac{c_{1,0}^2 c_{0,2} - 2c_{1,0}c_{1,1}c_{0,1} + c_{2,0}c_{0,1}^2}{2c_{1,0}^3}(u - 1)^2 + O((u - 1)^3).$$

In particular the fact that $\rho(u)$ is non-constant, analytic, and is a simple root corresponds to $c_{0,1}c_{1,0} \neq 0$ (by the analytic Implicit Function Theorem). The variance condition is then computed to be equivalent to the cubic inequality in the $c_{i,j}$:

$$(39) \qquad \rho\, c_{1,0}{}^2 c_{0,2} - \rho\, c_{1,0}c_{1,1}c_{0,1} + \rho\, c_{2,0}c_{0,1}{}^2 + c_{0,1}{}^2 c_{1,0} + c_{0,1}c_{1,0}{}^2 \rho \neq 0.$$

We can now proceed with asymptotic estimates. Fix a u–domain $|u - 1| \leq \delta$ such that B, C are analytic. Then, one has

$$f_n(u) := [z^n]F(z, u) = \frac{1}{2i\pi} \oint F(z, u)\frac{dz}{z^{n+1}},$$

where the integral is taken along a small enough contour encircling the origin. We use the analysis of polar singularities described in Chapter IV, exactly as in (35). As $F(z, u)$ has at most one (simple) pole in $|z| \leq r$, we have

$$(40) \qquad f_n(u) = \mathrm{Res}\left(\frac{B(z, u)}{C(z, u)}z^{-n-1}\right)_{z=\rho(u)} + \frac{1}{2i\pi}\int_{|z|=r} F(z, u)\frac{dz}{z^{n+1}},$$

where we may assume u suitably restricted by $|u - 1| < \delta$ in such a way that $|r - \rho(u)| < \frac{1}{2}(r - \rho)$.

The modulus of the second term in (40) is bounded from above by

$$(41) \qquad \frac{K}{r^n} \qquad \text{where} \qquad K = \frac{\sup_{|z|=r,|u-1|\leq\delta}|B(z, u)|}{\inf_{|z|=r,|u-1|\leq\delta}|C(z, u)|}.$$

Since the domain $|z| = r, |u - 1| \leq \delta$ is closed, $C(z, u)$ attains its minimum that must be non-zero, given the unicity of the zero of C. At the same time, $B(z, u)$ being analytic, its modulus is bounded from above. Thus, the constant K in (41) is finite.

Trivial bounds applied to the integral of (40) then yield

$$f_n(u) = \frac{B(\rho(u), u)}{C'_z(\rho(u), u)}\rho(u)^{-n-1} + O(r^{-n}),$$

uniformly for u in a small enough fixed neighbourhood of 1. The mean and variance then satisfy (37), with the coefficient in the leading term of the variance term that is, by assumption, non-zero. Thus, the conditions of the Quasi-powers Theorem in the form (28), p. 645, are satisfied, and the law is Gaussian in the asymptotic limit. ∎

Some form of condition, such as those in (ii) and (iii), is a necessity. For instance, the functions

$$\frac{1}{1 - z}, \quad \frac{1}{1 - zu}, \quad \frac{1}{1 - zu^2}, \quad \frac{1}{1 - z^2u},$$

each fail to satisfy the non-degeneracy and the variability condition, the variance of the corresponding discrete distribution being identically 0. The variance is $O(1)$ for a related function such as

$$F(z, u) = \frac{1}{1 - z(u + 2) + 2z^2u} = \frac{1}{(1 - 2z)(1 - zu)},$$

which is excluded by the variability condition of the theorem—there, a discrete limit law (a geometric) is known to hold (p. 614). Yet another situation arises when considering

$$F(z, u) = \frac{1}{(1 - z)(1 - zu)}.$$

There is now a double pole at 1 when $u = 1$ that arises from "confluence" at $u = 1$ of two analytic branches $\rho_1(u) = 1$ and $\rho_2(u) = 1/u$. In this particular case, the limit law is continuous but non-Gaussian; in fact, this limit is the uniform distribution over the interval $[0, 1]$, since

$$F(z, u) = 1 + z(1 + u) + z^2(1 + u + u^2) + z^3(1 + u + u^2 + u^3) + \cdots.$$

In addition, for this case, the mean is $O(n)$ but the variance is $O(n^2)$. Such situations are examined in Section IX. 11, p. 703, at the end of this Chapter.

▷ **IX.28.** *Higher order poles.* Under the conditions of Theorem IX.9, a limit Gaussian law holds for the distributions generated by the BGF $F(z, u)^m$. More generally, the statement extends to functions with an mth order pole. See [35]. ◁

The next four applications of Theorem IX.9 are relative to runs in permutations, patterns in words, the perimeter of parallelogram polyominoes, and finally the analysis of Euclid's algorithm on polynomials. It is of interest to note that, for runs and patterns, the BGFs were each deduced in Chapter III by an inclusion–exclusion argument that involves sequences in an essential way.

Example IX.12. *Ascending runs in permutations and Eulerian numbers.* The exponential BGF of Eulerian numbers (that count runs in permutations) is, by Example III.25, p. 209,

$$F(z, u) = \frac{u(1 - u)}{e^{(u-1)z} - u},$$

where, for $u = 1$, we have $F(z, 1) = (1 - z)^{-1}$. The roots of the denominator are then

$$(42) \qquad \rho_j(u) = \rho(u) + \frac{2ij\pi}{u - 1}, \qquad \text{where} \quad \rho(u) := \frac{\log u}{u - 1},$$

and j is an arbitrary element of $\mathbb{Z}$. As u approaches 1, $\rho(u)$ is close to 1, whereas the other poles $\rho_j(u)$ with $j \neq 0$ escape to infinity. This fact is also consistent with the limit form $F(z, 1) = (1 - z)^{-1}$ which has only one (simple) pole at 1. If one restricts u to $|u| \leq 2$, there is clearly at most one root of the denominator in $|z| \leq 2$, given by $\rho(u)$. Thus, we have for u close enough to 1,

$$F(z, u) = \frac{1}{\rho(u) - z} + R(z, u),$$

with $z \mapsto R(z, u)$ analytic in $|z| \leq 2$, and

$$[z^n]F(z, u) = \rho(u)^{-n-1} + O(2^{-n}).$$

The variability conditions are satisfied since

$$\rho(u) = \frac{\log u}{(u - 1)} = 1 - \frac{1}{2}(u - 1) + \frac{1}{3}(u - 1)^2 + \cdots,$$

so that $\mathfrak{v}(1/\rho(u)) = \frac{1}{12}$ is non-zero.

Proposition IX.9. *The Eulerian distribution is, after standardization, asymptotically Gaussian, with mean and variance given by $\mu_n = (n + 1)/2$, $\sigma_n^2 = (n + 1)/12$. The speed of convergence is $O(n^{-1/2})$.*

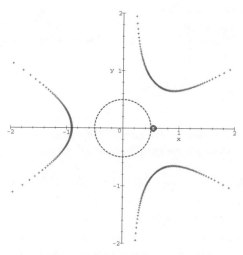

Figure IX.11. The diagram of poles of the BGF $z \mapsto F(z, u)$ associated to the pattern *abaa* with correlation polynomial $c(z) = 1 + z^3$, when u varies on the unit circle. The denominator is of degree 4 in z: one branch, $\rho(u)$ clusters near the dominant singularity $\rho = 1/2$ of $F(z, 1)$, whereas three other singularities stay away from the disc $|z| \le 1/2$ and escape to infinity as $u \to 1$.

This example is a famous one (see also our *Invitation*, p. 9) and our derivation follows Bender's paper [35]. The Gaussian character of the distribution has been known for a long time; it is for instance to be found in David and Barton's *Combinatorial Chance* [139] published in 1962. There are in this case interesting connections with elementary probability theory: if U_j are independent random variables that are uniformly distributed over the interval $[0, 1]$, then one has

$$[z^n u^k] F(z, u) = \mathbb{P}\{\lfloor U_1 + \cdots + U_n \rfloor < k\}.$$

Because of this fact, the normal limit is thus often derived as a consequence of the Central Limit Theorem, after one takes care of unimportant details relative to the integer part $\lfloor \cdot \rfloor$ function; see [139, 524]. .... ∎

Example IX.13. *Patterns in strings.* Consider the class $\mathcal{F}$ of binary strings (the "texts"), and fix a "pattern" w of length k. Let χ be the number of (possibly overlapping) occurrences of w. (The pattern w occurs if it is a factor, *i.e.*, if its letters occur contiguously in the text.) Let $F(z, u)$ be the BGF relative to the pair $(\mathcal{F}, \chi)$. The Guibas–Odlyzko correlation polynomial[10] relative to w is denoted by $c(z) \equiv c_w(z)$. We know, from Chapter I, that the OGF of words with pattern w excluded is

$$F(z, 0) = \frac{c(z)}{z^k + (1 - 2z)c(z)}.$$

By the inclusion–exclusion argument of Chapter III (p. 212), the BGF is

$$F(z, u) = \frac{1 - (c(z) - 1)(u - 1)}{1 - 2z - (u - 1)(z^k + (1 - 2z)(c(z) - 1))}.$$

---

[10]The correlation polynomial, as defined in Chapter I (p. 60), has coefficients in $\{0, 1\}$, with $[z^j]c(z) = 1$ iff w matches its image shifted to the right by j positions.

Let $D(z, u)$ be the denominator. Then $D(z, u)$ depends analytically on z, for u near 1 and z near $1/2$. In addition, the partial derivative $D'_z(1/2, 1)$ is non-zero. Thus, $\rho(u)$ is analytic at $u = 1$, with $\rho(1) = 1/2$ (see Figure IX.11). The local expansion of the root $\rho(u)$ of $D(\rho(u), u)$ follows from local series reversion,

$$2\rho(u) = 1 - 2^{-k}(u - 1) + (k2^{-2k} - 2^{-k}c(1/2))(u - 1)^2 + O\left((u - 1)^3\right).$$

Theorem IX.9 applies.

Proposition IX.10. *The number of occurrences of a fixed pattern in a large string is, after standardization, asymptotically normal. The mean μ_n and variance σ_n^2 satisfy*

$$\frac{n}{2^k} + O(1), \qquad \sigma_n^2 = \left(2^{-k}(1 + 2c(1/2)) + 2^{-2k}(1 - 2k)\right)n + O(1),$$

and the speed of convergence to the Gaussian limit is $O(n^{-1/2})$.

(The mean does not depend on the order of letters in the pattern, only the variance does.) Proposition IX.10 has been derived independently by many authors and it has been generalized in many ways, see for instance [43, 455, 506, 564, 603] and references therein. .... ∎

▷ **IX.29.** *Patterns in Bernoulli texts.* Asymptotic normality also holds when letters in strings are chosen independently but with an arbitrary probability distribution. It suffices to use the weighted correlation polynomial described in Note III.39, p. 213.   ◁

Example IX.14. *Parallelogram polyominoes.* Polyominoes are plane diagrams that are closely related to models of statistical physics, while having been the subject of a vast combinatorial literature. This example has the merit of illustrating a level of difficulty somewhat higher than in previous examples and typical of many "real-life" applications. Our presentation follows an early article of Bender [38] and a more recent paper of Louchard [419]. We consider here the variety of polyominoes called *parallelograms*. A parallelogram is a sequence of segments,

$$[a_1, b_1], [a_2, b_2], \ldots, [a_m, b_m], \qquad a_1 \le a_2 \cdots \le a_m, \; b_1 \le b_2 \le \cdots \le b_m,$$

where the a_j and b_j are integers with $b_j - a_j \ge 1$, and one takes $a_1 = 0$ for definiteness. A parallelogram can thus be viewed as a stack of segments (with $[a_{j+1}, b_{j+1}]$ placed on top of $[a_j, b_j]$) that leans smoothly to the right:

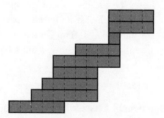

The quantity m is called the height, the quantity $b_m - a_1$ the width, their sum is called the (semi)perimeter, and the grand total $\sum_j (b_j - a_j)$ is called the area. (This instance has area 39, width 13, height 9, and perimeter $13 + 9 = 22$.) We examine here parallelograms of *fixed area* and investigate the *distribution of perimeter*.

The ordinary BGF of parallelograms, with z marking area and u marking perimeter is[11], as we shall prove momentarily

$$(43) \qquad F(z, u) = u \frac{J_1(z, u)}{J_0(z, u)},$$

where J_0, J_1 belong to the realm of "q–analogues" and generalize the classical Bessel functions,

$$J_0(q, u) := \sum_{n \geq 0} \frac{(-1)^n u^n q^{n(n+1)/2}}{(q; q)_n (uq; q)_n}, \qquad J_1(q, u) := \sum_{n \geq 1} \frac{(-1)^{n-1} u^n q^{n(n+1)/2}}{(q; q)_{n-1} (uq; q)_n},$$

with the "q–factorial" notation being used:

$$(a; q)_n = (1 - a)(1 - aq) \cdots (1 - aq^{n-1}).$$

Combinatorially, the BGF stated by (43), is obtained in a way that is reminiscent of Example III.22, p. 199. Its expression results from a simple construction: a parallelogram is either an interval, or it is derived from an existing parallelogram by stacking on top a new interval. Let $G(w) \equiv G(x, y, z, w)$ be the OGF with x, y, z, w marking width, height, area, and length of top segment, respectively. The GF of a parallelogram made of a single non-zero interval is

$$a(w) \equiv a(x, y, z, w) = \frac{xyzw}{1 - xzw}.$$

The operation of piling up a new segment on top of a segment of length m that is represented by a term w^m is described by

$$y \left(\frac{z^m w^m}{1 - xzw} + \cdots + \frac{zw}{1 - xzw} \right) = yzw \frac{1 - z^m w^m}{(1 - zw)(1 - xzw)}.$$

Thus, G satisfies the functional equation,

$$(44) \qquad G(w) = \frac{xyzw}{1 - xzw} + \frac{xyzw}{(1 - zw)(1 - xzw)} [G(1) - G(xzw)].$$

This is the method of "adding a slice" introduced in Chapter III, p. 199, which is reflected by the relation (44). Now, an equation of the form,

$$G(w) = a(w) + b(w)[G(1) - G(\lambda w)],$$

is solved by iteration:

$$
\begin{aligned}
G(w) &= a(w) + b(w)G(1) - b(w)G(\lambda w) \\
&= \left(a(w) - b(w)a(\lambda w) + b(w)b(\lambda w)a(\lambda^2 w) - \cdots \right) \\
&\quad + G(1) \left(b(w) - b(w)b(\lambda w) + b(w)b(\lambda w)b(\lambda^2 w) - \cdots \right).
\end{aligned}
$$

One then isolates $G(1)$ by setting $w = 1$. This expresses $G(1)$ as the quotient of two similar looking series (formed with sums of products of b values). Here, this gives $G(x, y, z, 1)$, from which the form (43) of $F(z, u)$ derives, since $F(z, u) = G(u, u, z, 1)$.

Analytically, one should first estimate $[z^n]F(z, 1)$, the number of parallelograms of size (i.e., area) equal to n. We have $F(z, 1) = J_1(z, 1)/J_0(z, 1)$, where the denominator is

$$J_0(z, 1) = 1 - \frac{z}{(1 - z)^2} + \frac{z^3}{(1 - z)^2(1 - z^2)^2} - \frac{z^6}{(1 - z)^2(1 - z^2)^2(1 - z^3)^2} + \cdots.$$

---

[11]Thus, $F(z, 1) = z + 2z^2 + 4z^3 + 9z^4 + 20z^5 + 46z^6 + \cdots$, corresponding to *EIS* **A006958** ("staircase polyominoes").

Clearly, $J_0(z, 1)$ and $J_1(z, 1)$ are analytic in $|z| < 1$, and it is not hard to see that $J_0(z, 1)$ decreases from 1 to about -0.24 when z varies between 0 and $1/2$, with a root at

$$\rho \doteq 0.43306\,19231\,29252,$$

where $J_0'(\rho, 1) \doteq -3.76 \neq 0$, so that the zero is simple[12]. Since $F(z, 1)$ is by construction meromorphic in the unit disc and $J_1(\rho, 1) \doteq 0.48 \neq 0$, the number of parallelograms satisfies

$$[z^n]F(z, 1) \sim \frac{J_1(\rho, 1)}{\rho J_0'(\rho, 1)} \left(\frac{1}{\rho}\right)^n = \alpha_1 \cdot \alpha_2^n,$$

where

$$\alpha_1 \doteq 0.29745\,35058\,07786, \quad \alpha_2 \doteq 2.30913\,85933\,31230.$$

As is common in meromorphic analyses, the approximation of coefficients is quite good; for instance, the relative error is only about 10^{-8} for $n = 35$.

We are now ready for bivariate asymptotics. Take $|z| \leq r = 7/10$ and $|u| \leq 11/10$. Because of the form of their general terms that involve $z^{n^2/2}u^n$ in the numerators while the denominators stay bounded away from 0, the functions $J_0(z, u)$ and $J_1(z, u)$ remain analytic there. Thus, $\rho(u)$ exists and is analytic for u in a sufficiently small neighbourhood of 1 (by Weierstrass preparation or implicit functions). The non-degeneracy conditions are easily verified by numerical computations. There results that Theorem IX.9 applies.

Proposition IX.11. *The perimeter of a random parallelogram polyomino of area n admits a limit law that is Gaussian with mean and variance that satisfy $\mu_n \sim \mu n$, $\sigma_n \sim \sigma \sqrt{n}$, with*

$$\mu \doteq 0.84176\,20156, \quad \sigma \doteq 0.42420\,65326.$$

This indicates that a random parallelogram is most likely to resemble a slanted stack of fairly short segments. .... ∎

▷ **IX.30.** *Width and height of parallelogram polyominoes are normal.* Similar perturbation methods show that the expected height and width are each $O(n)$ on average, again with Gaussian limit laws. ◁

▷ **IX.31.** *The base of a coin fountain.* A coin fountain (Example V.9, p. 330) is defined as a vector $v = (v_0, v_1, \ldots, v_\ell)$, such that $v_0 = 0$, $v_j \geq 0$ is an integer, $v_\ell = 0$ and $|v_{j+1} - v_j| = 1$. Take as size the *area*, $n = \sum v_j$. Then the distribution of the base length ℓ in a random coin fountain of size n is asymptotically normal. (This amounts to considering all ruin sequences of a fixed area as equally likely, and regarding the number of steps in the game as a random variable.) Similarly the number of "arches" is asymptotically Gaussian. ◁

Example IX.15. *Euclid's GCD Algorithm over polynomials.* We revisit the class $\mathcal{P} \subset \mathbb{F}_p[X]$ of monic polynomials in a variable X and coefficients in a prime field $\mathbb{F}_p$ (Example I.20, p. 90). Size of a polynomial is identified with degree. Euclidean division applies to any pair of polynomials (u, v), with $v \neq 0$: it provides a quotient (q) and a remainder (r), such that

$$u = vq + r, \quad \text{with} \quad r = 0 \quad \text{or} \quad \deg(r) < \deg v.$$

---

[12]As usual, such computations can be easily validated by carefully controlled numerical evaluations coupled with Rouché's theorem (see Chapter IV, p. 263).

Euclid's Greatest Common Divisor (GCD) Algorithm applies to any pair of polynomials (u_1, u_0) satisfying $\deg(u_1) < \deg(u_0)$, proceeding by successive divisions [379]:

$$(45) \qquad \begin{cases} u_0 & = & q_1 u_1 & + & u_2 \\ u_1 & = & q_2 u_2 & + & u_3 \\ \vdots & & \vdots & & \vdots \\ u_{h-2} & = & q_{h-1} u_{h-1} & + & u_h \\ u_{h-1} & = & q_h u_h & + & 0. \end{cases}$$

The number h is the *number of steps* of the algorithm. (It also corresponds to the *height* of the continued fraction representation of u_1/u_0: write $u_1/u_0 = 1/(q_1 + 1/ \cdots).)$ The quotient polynomials q_j, for $1 \le j \le h$ are each of degree at least 1 and one can always normalize things so that the u_j are monic. The last polynomial u_h is the gcd of the pair (u_1, u_0). (By convention, $\deg(0) = -\infty$, the gcd of $(0, u_0)$ is 1 and its height is 0.)

Together with the class $\mathcal{P}$, we introduce the class $\mathcal{G}$ of "general" (non-necessarily monic) polynomials and the subclass $\mathcal{G}^+$ of those of degree at least 1. The class $\mathcal{F}$ of *fractions* consists of all the pairs (u_1, u_0) such that: (i) the polynomial u_0 is monic; (ii) either $u_1 = 0$ or $\deg(u_1) < \deg(u_0)$. (View the pair as representing u_1/u_0.) The *size* of a fraction is by definition the degree of u_0. The corresponding OGF are instantly found to be:

$$(46) \qquad P(z) = \frac{1}{1 - pz}, \qquad G^+(z) = \frac{p(p-1)z}{1 - pz}, \qquad F(z) = \frac{1}{1 - p^2 z}.$$

The simple but startling fact that renders the analysis easy is the following: *Euclid's algorithm yields a combinatorial isomorphism between $\mathcal{F}$–fractions and pairs composed of a sequence of $\mathcal{G}^+$–polynomials (the quotients) and a $\mathcal{P}$–polynomial (the gcd).* In symbols:

$$(47) \qquad \mathcal{F} \cong \mathrm{SEQ}(\mathcal{G}^+) \times \mathcal{P}.$$

A direct consequence of (47) is the BGF of $\mathcal{F}$, with u marking the number of steps:

$$(48) \qquad F(z, u) = \frac{1}{1 - uG^+(z)} \cdot \frac{1}{1 - pz} = \frac{1}{1 - u \frac{p(p-1)z}{1-pz}} \cdot \frac{1}{1 - pz}.$$

Similarly, with u marking the number of quotients of some fixed degree k, one obtains the BGF

$$(49) \qquad \widehat{F}(z, u) = \frac{1}{1 - \frac{p(p-1)z}{1-pz} - z^k(u-1)p^k(p-1)} \cdot \frac{1}{1 - pz}.$$

Both cases give rise to direct applications of Theorem IX.9, p. 656, relative to the meromorphic schema. A simple computation then gives:

Proposition IX.12. *When applied to a random polynomial fraction of degree n, the number of steps of Euclid's algorithm is asymptotically normal with mean*

$$\mathbb{E}(\# \text{ steps}) = \frac{p-1}{p}n + O(1),$$

and variance $O(n)$. The number of quotients of a fixed degree k is also asymptotically Gaussian, with mean $\sim c_k n$ and variance $O(n)$, where $c_k = p^{-k-1}(p-1)^2$.

Similar considerations and the methods of Section IX. 2 show that the degree of the gcd itself is asymptotically geometric, with rate p^{-1}. Original analyses are due to Knopfmacher–Knopfmacher [371] and Friesen–Hensley [270]. In such a case, the transparent character of the analytic–combinatorial proofs is worthy of note. .... ∎

▷ **IX.32.** *Euclid's integer-gcd algorithm is Gaussian.* This spectacular and deep result is originally due to Hensley [331], with important improvements brought by Baladi–Vallée [25]. The reference set is now the pair of integers in $[1 .. n]$, to which Euclid's algorithm is applied. The number of steps has expectation

$$\frac{12 \log 2}{\pi^2} \log n + o(\log n),$$

as first established by Dixon [166] and Heilbronn [327]; see Knuth's book [379, pp. 356–373] for a good story. The proof of the Gaussian limit, following [25, 331], makes use of the *transfer operator* $\mathbf{G}_s$ associated with the transformation $x \mapsto \{1/x\} \equiv 1/x - \lfloor 1/x \rfloor$; namely,

$$\mathbf{G}_s[f](x) := \sum_{n=1}^{\infty} \frac{1}{(n+x)^{2s}} f\left(\frac{1}{n+x}\right).$$

It is then proved that a bivariate Dirichlet series describing the number of steps of Euclid's algorithm can be expressed in terms of the quasi-inverse $(\mathbf{I} - u\mathbf{G}_s)^{-1}$; compare with (48). Perturbation theory of the dominant eigenvalue $\lambda_1(s)$ of $\mathbf{G}_s$ in conjunction with the Mellin–Perron formula, an adapted form of singularity analysis, and the Quasi-powers Theorem (and hard work, as well) eventually yield the result. An operator analogue of (49) also holds, from which the frequency of quotient values can be quantified: the asymptotic frequency of k is $\log_2(1 + 1/(k(k+1)))$. See Vallée's surveys [583, 584], Hensley's book [332], and references therein for a review of these methods and many other applications.   ◁

Perturbation of linear systems. There is usually a fairly transparent approach to the analysis of BGFs defined implicitly as solutions of functional equations. One should start with the analysis at $u = 1$ and then examine the effect on singularities when u varies in a very small neighbourhood of 1. In accordance with what we have already seen many times, the process involves a perturbation analysis of the solution to a functional equation near a singularity, here one that *moves*.

We consider here functions defined implicitly by a *linear system* of positive equations, nonlinear systems being discussed in the next section. Positive linear systems arise in connection with problems specified by finite state devices, paths in graphs, and finite Markov chains, and transfer matrix models (Sections V.5, p. 336 and V.6, p. 356). The bivariate problem is then expressed by a linear equation

$$(50) \qquad Y(z, u) = V(z, u) + T(z, u) \cdot Y(z, u),$$

where $T(z, u)$ is an $m \times m$ matrix with entries that are polynomial in z, u with non-negative coefficients, $Y(z, u)$ is an $m \times 1$ column vector of unknowns, and $V(z, u)$ is a column vector of non-negative initial conditions.

Regarding the univariate problem,

$$(51) \qquad Y(z) = V(z) + T(z) \cdot Y(z),$$

where $Y(z) = Y(z, 1)$ and so on, we place ourselves under the assumptions of Corollary V.1, p. 358. This means that properness, positivity, irreducibility, and aperiodicity are assumed throughout. In this case (see the developments of Chapter V), Perron–Frobenius theory applies to the univariate matrix $T(z)$. In other words, the function

$$C(z) = \det(I - T(z))$$

has a unique dominant root $\rho > 0$ that is a simple zero. Accordingly, any component $F(z) = Y_i(z)$ of a solution to the system (50) has a unique dominant singularity at $z = \rho$ that is a simple pole,

$$F(z) = \frac{B(z)}{C(z)},$$

with $B(\rho) \neq 0$.

In the bivariate case, each component of the solution to the system (50) can be put under the form

$$F(z, u) = \frac{B(z, u)}{C(z, u)}, \qquad C(z, u) = \det(I - T(z, u)).$$

Since $B(z, u)$ is a polynomial, it does not vanish for (z, u) in a sufficiently small neighbourhood of $(\rho, 1)$. Similarly, by the analytic Implicit Function Theorem, there exists a function $\rho(u)$ locally analytic near $u = 1$, such that

$$C(\rho(u), u) = 0, \qquad \rho(1) = \rho.$$

Thus, it is sufficient that the variability conditions (38) be satisfied in order to infer a limit Gaussian distribution.

Theorem IX.10 (Positive rational systems). *Let $F(z, u)$ be a bivariate function that is analytic at $(0, 0)$ and has non-negative coefficients. Assume that $F(z, u)$ coincides with the component Y_1 of a system of linear equations in $Y = (Y_1, \ldots, Y_m)^T$,*

$$Y = V + T \cdot Y,$$

where $V = (V_1(z, u), \ldots, V_m(z, u))$, $T = \left(T_{i,j}(z, u)\right)_{i,j=1}^m$, and each of $V_j, T_{i,j}$ is a polynomial in z, u with non-negative coefficients. Assume also that $T(z, 1)$ is transitive, proper, and primitive, and let $\rho(u)$ be the unique solution of

$$\det(I - T(\rho(u), u)) = 0,$$

assumed to be analytic at 1, such that $\rho(1) = \rho$. Then, provided the variability condition,

$$\mathfrak{v}\left(\frac{\rho(1)}{\rho(u)}\right) > 0,$$

is satisfied, a Gaussian Limit Law holds for the coefficients of $F(z, u)$ with mean and variance that are $O(n)$ and speed of convergence that is $O(n^{-1/2})$.

Example IX.16. *Tilings.* (This prolongs the enumerative discussion of Example V.18, p. 360.) Take a $(2 \times n)$ chessboard of 2 rows and n columns, and consider coverings with "monomer tiles" that are (1×1)-pieces, and "dimer tiles" that are either of the horizontal (1×2) or vertical (2×1) type. The parameter of interest is the (random) number of tiles. Consider next the collection of all "partial coverings" in which each column is covered exactly, except possibly for the last one. The partial coverings are of one of four types and the legal transitions are described by a compatibility graph. For instance, if the previous column started with one horizontal dimer and contained one monomer, the current column has one occupied cell, and one free cell that may then be occupied either by a monomer or a dimer. This finite state description corresponds to a set of linear equations over BGFs (with z marking the area covered and u marking the total number of tiles), with the transition matrix found to be

$$T(z, u) = z \begin{pmatrix} u & u^2 & u^2 & u^2 \\ 1 & 0 & 0 & 0 \\ u & 0 & 0 & 0 \\ u & 0 & 0 & 0 \end{pmatrix}.$$

In particular, we have

$$\det(I - T(z, u)) = 1 - zu - z^2(u^2 + u^3).$$

Then, Theorem IX.10 applies: the number of tiles is asymptotically normal. The method clearly extends to $(k \times n)$ chessboards, for any fixed k (see Bender *et al.* [35, 46]). .... ∎

Example IX.17. *Limit theorem for Markov chains.* Assume that M is the transition matrix of an irreducible aperiodic Markov chain, and consider the parameter χ that records the number of passages through state 1 in a path of length n that starts in state 1. Then, Theorem IX.10 applies with

$$V = (1, 0, \ldots, 0)^T, \qquad T_{i,j}(z, u) = z M_{i,j} + z(u - 1) M_{i,1} \delta_{j,1}.$$

We therefore derive a classical limit theorem for Markov chains:

Proposition IX.13. *In an irreducible and aperiodic (finite) Markov chain, the number of times that a designated state is reached when n transitions are effected is asymptotically Gaussian.*

The conclusion also applies to paths in any strongly connected aperiodic digraph as well as to paths conditioned by their source and/or destination. .... ∎

▷ **IX.33.** *Sets of patterns in words.* This note extends Example IX.13 (p. 659) relative to the occurrence of a *single* pattern in a random text. Given the class $\mathcal{W} = \text{SEQ}(\mathcal{A})$ of words over a finite alphabet $\mathcal{A}$, fix a finite set of "patterns" $S \subset \mathcal{W}$ and define $\chi(w)$ as the total number of occurrences of members of S in the word $w \in \mathcal{W}$. It is possible to build finite automaton (essentially a digital tree built on S equipped with return edges) that records simultaneously the number of partial occurrences of each pattern. Then, the limit law of χ is Gaussian; see Bender and Kochman's paper [43], the papers [240, 263] for an approach based on the de Bruijn graph, [30, 457] for an inclusion–exclusion treatment, and [564] for a perspective.   ◁

▷ **IX.34.** *Constrained integer compositions.* Consider integer compositions where consecutive summands add up to at least 4. The number of summands in such a composition is asymptotically normal [46]. Similarly for a Carlitz composition (p. 201).   ◁

▷ **IX.35.** *Height in trees of bounded width.* Consider general Catalan trees of width less than a fixed bound w. (The width is the maximum number of nodes at any level in the tree.) In such trees, the distribution of height is asymptotically Gaussian.   ◁

IX.7. Perturbation of singularity analysis asymptotics

In this central section, we examine analytic–combinatorial schemas that arise when generating functions contain algebraic–logarithmic singularities. The underlying machinery is the method of singularity analysis detailed in Chapters VI and VII, on which suitable perturbative developments are grafted.

An especially important feature of the method of singularity analysis, stemming from properties of Hankel contours, is the fact that it preserves uniformity of expansions[13]. This feature is crucial in analysing bivariate generating functions, where we

---

[13]For instance, Darboux's method discussed in Section VI.11, p. 433, only provides *non-effective* error terms, since it is based on the Riemann–Lebesgue lemma, so that it cannot be conveniently employed for bivariate asymptotics. A similar comment applies to Tauberian theorems.

need to estimate *uniformly* a coefficient $f_n(u) = [z^n]F(z, u)$ that depends on the parameter u, given some (uniform) knowledge on the singular structure of $F(z, u)$, as a function of z. It is from such estimates that limit Gaussian laws can typically be derived via quasi-power approximations and the Quasi-powers Theorem (Theorem IX.8, p. 645).

In this section, we shall encounter two different types of situations, depending on the way the deformation induced by the secondary parameter affects the singularity of the function $z \mapsto F(z, u)$, when u is near 1. In accordance with the preliminary discussion of singularity perturbation and Gaussian laws, on p. 648, regarding the PGF $p_n(u) := f_n(u)/f_n(1)$, there is a fundamental dichotomy, depending on whether it is the singular exponent that varies or the dominant singularity that moves.

— *Variable exponent.* This corresponds to the case where the dominant singularity of $z \mapsto F(z, u)$ remains a constant ρ, but the singular exponent $\alpha(u)$ in the approximation $F(z, u) \approx (1 - z/\rho)^{-\alpha(u)}$ varies smoothly, to the effect that $p_n(u) \approx n^{\alpha(u) - \alpha(1)}$. We then have a Gaussian limit law in the scale of $\log n$ for the mean and the variance.

— *Movable singularity.* This is the case where the singular exponent retains a constant value α, but the dominant singularity $\rho(u)$ in the approximation $F(z, u) \approx (1 - z/\rho(u))^{-\alpha}$ moves smoothly with u, to the effect that $p_n(u) \approx (\rho(1)/\rho(u))^n$. There is again a Gaussian limit law, but a mean and variance that are now of the order of n.

The case of a variable exponent typically arises from the set construction, in the context of the exp–log schema introduced in Section VII. 2 (p. 445), which covers the cycle decomposition of permutations, connected components in random mappings, as well as the factorization of polynomials over finite fields. The mean value analyses of Chapter VII are then nicely supplemented by limit Gaussian laws, as we prove in Subsection IX. 7.1. Trees often lead to singularities that are of the square-root type and such a singular behaviour persists for a number of bivariate generating functions associated to additively inherited parameters (for instance the number of leaves). In that case, the singular exponent remains constant (equal to $1/2$), while the singularity moves. The basic technology adequate for such movable singularities is developed in Subsection IX. 7.2, where it is illustrated by means of simple examples relative to trees.

A notable feature of complex analytic methods is to be applicable to functions only known implicitly through a functional equation of sorts. We study implicit systems and algebraic functions in Subsection IX. 7.3: there, movable singularities are found, resulting in Gaussian limits in the scale of n. Differential systems display a broader range of singular behaviours, as discussed in Subsection IX. 7.4, to the effect that Gaussian laws can arise, both in the scale of $\log n$ and of n.

IX. 7.1. Variable exponents and the exp–log schema. The organization of this subsection is as follows. First, we state an easy but crucial lemma (Lemma IX.2) that takes care of the remainder terms in the expansions and hence enables the use of singularity analysis in a perturbed context. Then, we state a general theorem relative to the case of a fixed singularity and a variable exponent (Theorem IX.11). The major application is to the analysis of the exp–log schema as introduced in Section VII. 2, p. 445: Gaussian laws in the scale of $\log n$ are found to hold true for the number of components in several of the most classical structures of combinatorial theory.

Uniform expansions. The basis of the developments in this section is a uniformity lemma obtained from a simple re-examination of basic singularity analysis in the perspective of bivariate asymptotics.

Lemma IX.2 (Uniformity lemma, singularity analysis). *Let $f_u(z)$ be a family of functions analytic in a common Δ–domain Δ, with u a parameter taken in a set U. Suppose that there holds*

$$(52) \qquad |f_u(z)| \le K(u) \left| (1-z)^{-\alpha(u)} \right|, \qquad z \in \Delta, \quad u \in U,$$

where $K(u)$ and $\alpha(u)$ remain absolutely bounded: $K(u) \le K$ and $|\alpha(u)| \le A$, *for $u \in U$. Let B be such that $\Re(\alpha(u)) \le -B$. Then, there exists a constant λ (computable from A, B, Δ) such that*

$$(53) \qquad \left| [z^n] f_u(z) \right| < \lambda K n^{B-1}.$$

Proof. It suffices to revisit the proof of the Big-Oh Transfer Theorem (Theorem VI.3, p. 390), paying due attention to uniformity. The proof starts from Cauchy's formula,

$$f_{u,n} \equiv [z^n] f_u(z) = \frac{1}{2i\pi} \int_\gamma f_u(z) \frac{dz}{z^{n+1}},$$

where $\gamma = \bigcup_j \gamma_j$ is the Hankel contour displayed in Figure VI.6, p. 390. This contour is comprised of an inner circular arc (γ_1), an outer arc (γ_4), and two connecting linear parts (γ_2, γ_3); its half-angle is θ.

Decompose $\alpha(u)$ into its real and imaginary parts and set $\alpha(u) = \sigma(u) + i\tau(u)$. Also, set $z = 1 + t/n$, so that t lies on an image contour $\tilde{\gamma} = -1 + n\Delta$ and write $t = \rho e^{i\xi}$. We have

$$(54) \qquad \left| (1-z)^{-\alpha(u)} \right| = \left| (1-z)^{-\sigma(u)} \right| \cdot \left| \left(-\frac{t}{n} \right)^{-i\tau(u)} \right|,$$

with $|\tau(u)| \le A$. As t varies along $\tilde{\gamma}$, its argument ξ decreases continuously from $2\pi - \theta$ to θ. Thus, the second factor on the right of (54) remains bounded independently of n:

$$\left| \left(-\frac{t}{n} \right)^{-i\tau(u)} \right| \equiv \left| \left(-\frac{\rho e^{i\xi}}{n} \right)^{-i\tau(u)} \right| \le \lambda_1,$$

for some computable $\lambda_1 > 0$. In summary, we have found, for z on γ,

$$(55) \qquad \left| (1-z)^{-\alpha(u)} \right| \le \lambda_1 \left| (1-z)^{-\sigma(u)} \right|,$$

where $\sigma(u)$ is real and $-\sigma(u) \ge B$.

At this final stage, making use of (55), we can bound $[z^n] f_u(z)$ by a curvilinear integral:

$$\left| [z^n] f_u(z) \right| \le \frac{\lambda_1}{2\pi} \int_\gamma \left| (1-z)^{-\sigma(u)} \right| \frac{|dz|}{|z|^{n+1}}.$$

A direct application of the majorizations used in the proof of Theorem VI.3 then establishes the statement. ∎

▷ **IX.36.** *Uniformity in the presence of logarithmic multipliers.* Similar estimates hold when $f(z)$ is multiplied by a power of $L(z) = -\log(1-z)$: *if the condition* (52) *is replaced by*

$$|f_u(z)| \le K(u)\left|(1-z)^{-\alpha(u)}\right||L(z)|^{\beta},$$

for some $\beta \in \mathbb{R}$, then one has

$$\left|[z^n]f_u(z)\right| < \widetilde{\lambda}Kn^{B-1}(\log n)^{\beta},$$

for some $\widetilde{\lambda} = \widetilde{\lambda}(A, B, \Delta, \beta)$ (compare with (53)). ◁

The prototypical instance of a bivariate GF with a fixed singularity and a variable exponent is that of $F(z, u) := C(z)^{-\alpha(u)}$. We can in fact state a slightly more general result guaranteeing the presence of a Gaussian limit law in this and similar cases.

Theorem IX.11 (Variable exponent perturbation). *Let $F(z, u)$ be a bivariate function that is analytic at $(z, u) = (0, 0)$ and has non-negative coefficients. Assume the following conditions.*

(i) Analytic exponents. *There exist $\epsilon > 0$ and $r > \rho$ such that, with the domain $\mathcal{D}$ defined by*

$$\mathcal{D} = \left\{(z, u) \mid |z| \le r, |u - 1| \le \epsilon\right\},$$

the function $F(z, u)$ admits the representation

$$(56) \qquad F(z, u) = A(z, u) + B(z, u)C(z)^{-\alpha(u)}$$

where $A(z, u), B(z, u)$ are analytic for $(z, u) \in \mathcal{D}$. Suppose also that the function $\alpha(u)$ is analytic in $|u - 1| \le \epsilon$ with $\alpha(1) \notin \{0, -1, -2, \ldots\}$ and $C(z)$ is analytic for $|z| \le r$, with the equation $C(z) = 0$ having a unique root $\rho \in (0, r)$ in the disc $|z| \le r$ that is simple and such that $B(\rho, 1) \ne 0$.

(ii) Variability: *one has*

$$\alpha'(1) + \alpha''(1) \ne 0.$$

Then the variable with probability generating function

$$p_n(u) = \frac{[z^n]F(z, u)}{[z^n]F(z, 1)}$$

converges in distribution to a Gaussian variable with a speed of convergence $O((\log n)^{-1/2})$. The corresponding mean μ_n and variance σ_n^2 satisfy

$$\mu_n \sim \alpha'(1) \log n, \qquad \sigma_n^2 \sim (\alpha'(1) + \alpha''(1)) \log n.$$

Proof. Clearly, for the univariate problem, by singularity analysis, one has

$$(57) \qquad [z^n]F(z, 1) = B(\rho, 1)(-\rho C'(\rho))^{-\alpha(1)}\rho^{-n}\frac{n^{\alpha(1)-1}}{\Gamma(\alpha(1))}\left(1 + O\left(\frac{1}{n}\right)\right).$$

For the bivariate problem, the contribution to $[z^n]F(z, u)$ arising from $[z^n]A(z, u)$ is uniformly exponentially smaller than ρ^{-n}, since $A(z, u)$ is z–analytic in $|z| \le r$.

Write next

$$B(z, u) = (B(z, u) - B(\rho, u)) + B(\rho, u).$$

The first term satisfies

$$B(z, u) - B(\rho, u) = O((z - \rho)),$$

uniformly with respect to u, since

$$\frac{B(z, u) - B(\rho, u)}{z - \rho}$$

is analytic for $(z, u) \in \mathcal{D}$ (as seen by division of power series representations). Let A be an upper bound on $|\alpha(u)|$ for $|u - 1| \le \epsilon$. Then, by singularity analysis and its companion uniformity lemma,

$$(58) \qquad [z^n](B(z, u) - B(\rho, u))C(z)^{-\alpha(u)} = O(\rho^{-n} n^{A-2}).$$

By suitably restricting the domain of u, one may freely assume that $A < \alpha(1) + 1/2$ (say), ensuring that $A - 2 \le \alpha(1) - 3/2$. Thus, the contribution arising from (58) is uniformly polynomially small (by a factor $O(n^{-1/2})$).

It only remains to analyse

$$[z^n]B(\rho, u)C(z)^{-\alpha(u)}.$$

This is done exactly like in the univariate case: we have, uniformly for u in a small neighbourhood of 1,

$$(59) \qquad C(z)^{-\alpha(u)} = (-\rho C'(\rho))^{-\alpha(u)}(1 - z/\rho)^{-\alpha(u)} (1 + O(1 - z/\rho)),$$

and, taking once more advantage of the uniformity afforded by singularity analysis, we find by (58) and (59):

$$[z^n]F(z, u) = \frac{B(\rho, u)\rho^{-n}}{\Gamma(\alpha(u))}(-\rho C'(\rho))^{-\alpha(u)} n^{\alpha(u)-1} \left(1 + O(n^{-1/2})\right).$$

Thus, the Quasi-powers Theorem applies and the law is Gaussian in the limit. ∎

The exp–log schema. The next proposition covers the exponential–logarithmic ("exp–log") schema of Section VII. 2, p. 445, which is amenable to singularity perturbation techniques.

Proposition IX.14 (Sets of labelled logarithmic structures). *Consider the labelled set construction $\mathcal{F} = \mathrm{SET}(\mathcal{G})$. Assume that $G(z)$ has radius of convergence ρ and is Δ–continuable with a singular expansion of the form*

$$G(z) = \kappa \log \frac{1}{1 - z/\rho} + \lambda + O\left(\frac{1}{\log^2(1 - z/\rho)}\right).$$

Then, the limit law of the number of $\mathcal{G}$–components in a large $\mathcal{F}$–structure is asymptotically Gaussian with mean and variance each asymptotic to $\kappa \log n$ and with speed of convergence $O((\log n)^{-1/2})$.

Proof. Use the enhanced version of the uniformity lemma in Note IX.36. A quasipower approximation of the form $p_n(u) \approx n^{\alpha(u)-\alpha(1)}$, with $\alpha(u) \equiv \kappa u$, results from developments of the same type as in the proof of Theorem IX.11. ∎

Clearly, all the labelled structures of Section VII. 2 (p. 445) are covered by this proposition. A few examples, related to permutations, 2–regular graphs, and mappings, follow.

Example IX.18. *Cycles in derangements.* The bivariate EGF for *permutations* with u marking the number of cycles is given by the specification

$$\mathcal{F} = \text{SET}(u\,\text{CYC}(\mathcal{Z})) \quad\Longrightarrow\quad F(z,u) = \sum \begin{bmatrix} n \\ k \end{bmatrix} u^k \frac{z^n}{n!} = \exp\left(u\log\frac{1}{1-z} \right),$$

so that we are in the simplest case of an exp–log schema. Proposition IX.14 implies immediately that *the number of cycles in a random permutation of size* n *converges to a Gaussian limiting distribution.* (This classical result stating the asymptotically normality distribution of the Stirling cycle numbers could be derived directly in Proposition IX.5, p. 645, thanks to the explicit character of the horizontal generating functions—the Stirling polynomials—in this particular case.)

Similarly, the number of cycles is asymptotically normal in generalized derangements (Examples II.14, p. 122 and VII.1, p. 448) where a finite set S of cycle lengths are forbidden. This results immediately from Proposition IX.14, given the BGF

$$\mathcal{F} = \text{SET}(u\,\text{CYC}_{\mathbb{Z}_{\geq 1}\setminus S}(\mathcal{Z})) \quad\Longrightarrow\quad F(z,u) = \exp\left(u\left[\log\frac{1}{1-z} - \sum_{s\in S}\frac{z^s}{s} \right] \right).$$

The classical derangement problem corresponds to $S = \{1\}$. .... ∎

Example IX.19. *2–regular graphs.* A 2–regular graph is an undirected graph such that each vertex has degree exactly 2. Any 2–regular graph may be decomposed into a product of connected components that are *undirected* cycles of length at least 3 (Note II.22, p. 133 and Example VII.2, p. 449). Hence the bivariate EGF for 2–regular graphs, with u marking the number of connected components, is given by

$$\mathcal{F} = \text{SET}(u\,\text{UCYC}_{\geq 3}(\mathcal{Z})) \quad\Longrightarrow\quad F(z,u) = \exp\left(u\left[\frac{1}{2}\log\frac{1}{1-z} - \frac{z}{2} - \frac{z^2}{4} \right] \right).$$

By the logarithmic character of the function inside the exponential, the number of connected components in a 2–regular graph, has a Gaussian limit distribution. .... ∎

Example IX.20. *Connected components in mappings.* Mappings from a finite set to itself can be represented as labelled functional graphs. With u marking the number of connected components, the specification is (Subsection II. 5.2, p. 129 and Example VII.3, p. 449)

$$\mathcal{F} = \text{SET}(u\,\text{CYC}(\mathcal{T})) \quad\Longrightarrow\quad F(z,u) = \exp\left(u\log\frac{1}{1-T(z)} \right),$$

where $T(z)$ is the Cayley tree function defined implicitly by the relation $T(z) = z\exp(T(z))$. By the inversion theorem for implicit functions (Example VI.8, p. 403), we have a square-root singularity,

$$T(z) = 1 - \sqrt{2(1-ez)} + O(1-ez),$$

so that

$$F(z,u) = \exp\left(u\left[\frac{1}{2}\log\frac{1}{1-ez} + O((1-ez)^{1/2}) \right] \right).$$

From Proposition IX.14, we obtain a theorem originally due to Stepanov [559]: *The number of components in functional digraphs has a limiting Gaussian distribution.*

This approach extends to functional digraphs satisfying various degree constraints as considered in [18]. This analysis and similar ones are relevant to integer factorization, using Pollard's "rho" method [247, 379, 538]. .... ∎

Unlabelled constructions. In the unlabelled universe, the class of all finite multisets over a class $\mathcal{G}$ has ordinary *bivariate* generating function given by

$$\mathcal{F} = \mathrm{MSET}(u\mathcal{G}) \quad \Longrightarrow \quad F(z, u) = \exp\left(\frac{u}{1}G(z) + \frac{u^2}{2}G(z^2) + \frac{u^3}{3}G(z^3) + \cdots\right).$$

where u marks the number of $\mathcal{G}$–components (Chapter III).

The function $F(z, u)$ is consequently of the form $F(z, u) = e^{uG(z)}B(z, u)$, where $B(z, u)$ collects the contributions arising from $G(z^2)$, $G(z^3)$, …. If the radius of convergence ρ of $G(z)$ is assumed to be strictly less than 1, then, as it is easily checked, the function $B(z, u)$ is bivariate analytic in $|u| < 1 + \epsilon$, $|z| < R$ for some $\epsilon > 0$ and $R > \rho$. Here, we are interested in structures such that $G(z)$ has a logarithmic singularity, in which case the conclusions of Proposition IX.14 relative to the construction $\mathcal{F} = \mathrm{MSET}(u\mathcal{G})$ hold (this is verified by a simple combination of the proofs of Proposition IX.14 and Theorem IX.11). In summary:

> For the construction $\mathcal{F} = \mathrm{MSET}(\mathcal{G})$, under the assumption that $\rho < 1$ and $G(z)$ is logarithmic, the number of $\mathcal{G}$–components in a random $\mathcal{F}_n$ structure is asymptotically Gaussian in the scale of $\log n$, with speed $O((\log n)^{-1/2})$.

The same property also holds for the unlabelled powerset construction $\mathcal{F} = \mathrm{PSET}(\mathcal{G})$.

In what follows, we present two illustrations, one relative to the factorization of polynomials over finite fields, the other to unlabelled functional graphs.

Example IX.21. *Polynomial factorization.* Fix a finite field $\mathbb{F}_p$ and consider the class $\mathcal{P}$ of monic polynomials (having leading coefficient 1) in the polynomial ring $\mathbb{F}_p[z]$, with $\mathcal{I}$ the subclass of irreducible polynomials. The algebraic analysis has been performed in Example I.20, p. 90. One has $P_n = p^n$ and

$$P(z) = (1 - pz)^{-1}.$$

Because of the unique factorization property, a polynomial is a multiset of irreducible polynomials, whence the relation

$$P(z) = \exp\left(\frac{I(z)}{1} + \frac{I(z^2)}{2} + \frac{I(z^3)}{3} + \cdots\right).$$

The preceding relation can be inverted using Möbius inversion. With $L(z) = \log P(z)$, we have

$$I(z) = \sum_{k \geq 1} \mu(k)\frac{L(z^k)}{k} = \log\frac{1}{1 - pz} + \sum_{k \geq 2} \mu(k)\frac{L(z^k)}{k},$$

where μ is the Möbius function.

As it is apparent, $I(z)$ is logarithmic (it is indeed the sum of a logarithmic term and a function analytic for $|z| < p^{-1/2}$; see Example VII.4, p. 449). We have yet another instance of the exp–log schema (with $\kappa = 1$). Hence:

Proposition IX.15. *Let Ω_n be the random variable representing the number of irreducible factors of a random polynomial of degree n over $\mathbb{F}_p$, each factor being counted with its order of multiplicity. Then as n tends to infinity, we have, for any real x:*

$$\lim_{n \to +\infty} \mathbb{P}\{\Omega_n < \log n + x\sqrt{\log n}\} = \frac{1}{\sqrt{2\pi}}\int_{-\infty}^{x} e^{-t^2/2}\,dt.$$

This statement, which originally appears in [258], constitutes a counterpart of the famous Erdős–Kac Theorem (1940) for the number of prime divisors of natural numbers (with here $\log n$ that replaces $\log \log n$ when dealing with integers at most n; see [576]). The speed of convergence is once more $O((\log n)^{-1/2})$. Also, by the same devices, the same property holds for the parameter ω_n that represents the number of *distinct* irreducible factors in a random polynomial of degree n. .... ∎

It is perhaps instructive to re-examine this last example at an abstract level, in the light of general principles of analytic combinatorics.

> *A polynomial over a finite field is determined by the* sequence *of its coeffi-cients. Hence, the class of all polynomials, as a sequence class, has a polar singularity. On the other hand, unique factorization entails that a polyno-mial is also a* multiset *of irreducible factors ("primes"). Thus, the class of irreducible polynomials, that is implicitly determined, is logarithmic, since the multiset construction to be inverted is in essence an exponential oper-ator. As a consequence of the exp–log schema, the number of irreducible factors is asymptotically Gaussian.*

Example IX.22. *Unlabelled functional graphs (mapping patterns).* These are unlabelled di-rected graphs in which each vertex has outdegree equal to 1 (Chapter VII, p. 480). The specifi-cation of the class $\mathcal{F}$ of such digraphs is

$$\mathcal{F} = \mathrm{MSET}(\mathcal{L}), \qquad \mathcal{L} = \mathrm{CYC}(\mathcal{H}), \qquad \mathcal{H} = \mathcal{Z} \times \mathrm{MSET}(\mathcal{H}),$$

corresponding to multisets of cycles of rooted unlabelled trees $\mathcal{H}$.

Analytically, we know from Section VII. 5 (p. 475) relative to non-plane trees that $H(z)$ has a dominant square-root singularity:

$$H(z) = 1 - \gamma \sqrt{(1 - z/\eta)} + O(1 - z/\eta),$$

where $\eta \doteq 0.33832$ and γ is some positive constant. As a consequence, $L(z)$, which is obtained by translating an unlabelled cycle construction, is logarithmic with parameter $\kappa = 1/2$. Thus: *The number of components in a mapping pattern has a Gaussian limit distribution, with mean and variance each of the form $\frac{1}{2} \log n + O(1)$.* .... ∎

▷ **IX.37.** *Arithmetical semigroups.* Knopfmacher [370] defines an arithmetical semigroup as a semigroup with unique factorization, together with a size function (or degree) such that

$$|xy| = |x| + |y|,$$

and the number of elements of a fixed size is finite. If $\mathcal{P}$ is an arithmetical semigroup and $\mathcal{I}$ its set of 'primes' (irreducible elements), axiom $A^\#$ of Knopfmacher asserts the condition

$$\mathrm{card}\{x \in \mathcal{P} \ / \ |x| = n\} = cq^n + O(q^{\alpha n}) \quad (\alpha < 1),$$

with $q > 1$. It is shown in [370] that several algebraic structures forming arithmetical semi-groups satisfy axiom $A^\#$, and thus the conditions of Theorem IX.11 are automatically verified. Therefore, the results deriving from Theorem IX.11 fit into the framework of Knopfmacher's "abstract analytic number theory"—they provide general conditions under which theorems of the Erdős–Kac type must hold true. Examples of application mentioned in [370] are Galois polynomial rings (the case of polynomial factorization), finite modules or semi-simple finite algebras over a finite field $K = \mathbb{F}_q$, integral divisors in algebraic function fields, ideals in the principal order of a algebraic function field, finite modules, or semi-simple finite algebras over a ring of integral functions. ◁

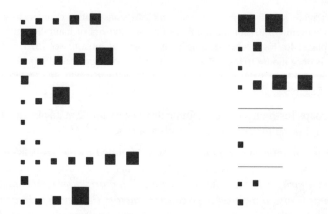

Figure IX.12. Small components of size ≤ 20 in random permutations (left) and random mappings (right) of size 1 000: each object corresponds to a line and each component is represented by a square of proportional area (for some of the mappings, such components may be lacking).

▷ **IX.38.** *A Central Limit Theorem on $GL_n(\mathbb{F}_q)$.* The title of this note is that of an article by Goh and Schmutz [297] who prove asymptotic normality for the number of irreducible factors that the characteristic polynomial of a random $n \times n$ matrix with entries in $\mathbb{F}_q$ has. [Some linear algebra relative to the canonical decomposition of matrices and due to Kung and Stong is needed.] The topic of random matrix theory over finite fields is blossoming: see Fulman's survey [272].   ◁

Number of fixed-size components in the exp–log schema. As we know all too well, the cycle structure of permutations is a typical instance of the exp–log schema, where everything is as explicit as can be. The Gaussian law for the total number of cycles actually summarizes information relative to the number of 1–cycles, 2–cycles, and so on. These can be analysed separately, and we learnt in Example IX.4 (p. 625) that, for m *fixed*, the number of m–cycles is asymptotically Poisson($1/m$)—in a way, the Gaussian law for cycles appears as the resultant of a large number of Poisson variables of slowly decreasing rates. As a matter of fact, similar properties hold true for any labelled class that belongs to the exp–log schema, namely, the number of m–components is in general asymptotically Poisson(λ_m), where the rate λ_m is computable and satisfies $\lambda_m = O(1/m)$; see Figure IX.12 for an illustration. (The alert reader may have noticed that we already obtained this property directly in Proposition VII.1 on p. 451, relative to profiles of exp–log structures, and that it is similar in spirit to what happens in subcritical constructions of Proposition IX.3, p. 633, although now the exp–log schema is *critical*!) Here we briefly indicate how such properties can be obtained by singularity perturbation: no quasi-power approximation is involved since a discrete-to-discrete convergence occurs, but the uniformity properties of the singularity analysis process, Lemma IX.2, p. 668, remains a central ingredient of the synthetic analysis to be developed below.

Example IX.23.   *Fixed-size components in sets of logarithmic structures*[14]. The number of components of some fixed size m in a set construction corresponds to the specification

$$\mathcal{F} = \text{SET}\left(u\mathcal{G}_m + (\mathcal{G} \setminus \mathcal{G}_m)\right) \quad \Longrightarrow \quad F(z, u) = \exp\left(G(z) + (u - 1)g_m z^m\right),$$

where $F(z, u)$ is an exponential BGF, $G(z)$ is an EGF, and $g_m := [z^m]G(z)$. As a consequence:

$$F(z, u) = \exp\left((u - 1)g_m z^m\right) F(z).$$

Under the assumption that $G(z)$ is logarithmic, one has, for u in a small neighbourhood of 1, as $z \to \rho$ in a Δ–domain,

$$F(z, u) = e^\lambda w(u)(1 - z/\rho)^{-\kappa}\left(1 + O(\log^{-2}(1 - z/\rho))\right), \qquad w(u) = \exp\left((u - 1)g_m\rho^m\right),$$

the uniformity of the expansion with respect to u being granted by the same argument as in Proposition IX.14. By singularity analysis, it is seen that

$$[z^n]F(z, u) = \frac{e^\lambda w(u)}{\Gamma(\kappa)}\rho^{-n}n^{\kappa-1}\left(1 + o(\log^{-1} n)\right).$$

Given the particular shape of $w(u)$, this last estimate tells us that *the number of m–components in a random $\mathcal{F}$–structure of large size tends to a Poisson distribution with parameter $\mu :=$ $g_m\rho^m$.*

This result applies for any m less than some arbitrary fixed bound B. In addition, truly multivariate methods evoked at the end of this chapter enable one to prove that the number of components of sizes $1, 2, \ldots, B$ are *asymptotically independent*. This gives a very precise model of the probabilistic profile of small components in random $\mathcal{F}$–objects as a product of independent Poisson laws of parameter $g_m\rho^m$ for $m = 1, \ldots, B$. Similar results hold for unlabelled multisets, but with the negative binomial law replacing the Poisson law. .... ∎

▷ **IX.39.** *Random mappings.* The number of components of some fixed size m in a large random mapping (functional graph) is asymptotically Poisson(λ) where $\lambda = K_m e^{-m}/m!$ and $K_m = m![z^m]\log(1 - T)^{-1}$ enumerates connected mappings. (There T is the Cayley tree function.) The fact that $K_m e^{-m}/m! \approx 1/(2m)$ explains the fact that small components are somewhat sparser for mappings than for permutations (Figure IX.12).   ◁

The last example concludes our detailed investigation of exp–log structures, and we may legitimately regard the most basic phenomena as well understood. Example IX.23 quantifies the distribution of the number of "small" components, whose presence is fairly sporadic (Figure IX.12) and for which an asymptotically independent Poisson structure prevails. Panario and Richmond [470] have further succeeded in proving that the size of the *smallest* component is asymptotically $O(\log n)$ *on average.* "Large" components also enjoy a rich set of properties. They cannot be independently distributed, since, for instance, a permutation can have only one cycle larger than $n/2$, two cycles larger than $n/3$, etc. As shown by Gourdon [305] under general exp–log conditions, the size of the largest component is $\Theta(n)$ on average and in probability, and the limit law involves the Dickman function otherwise known to describe the distribution of the largest prime divisor of a random integer over a large interval. A general probabilistic theory of the joint distribution of largest components in exp–log structures has been developed by Arratia, Barbour, and Tavaré [20], some of the initial developments of that theory drawing their inspiration from earlier

---

[14]This example revisits the analysis of Proposition VII.1, p. 451, under the perspective of continuity theorems for PGFs.

combinatorial–analytic studies. The joint distribution of large components appears to be characterized in terms of what is known as the Poisson–Dirichlet process.

IX. 7.2. Movable singularities. In accordance with the preliminary discussion offered at the beginning of the section (p. 666), we now examine BGFs $F(z, u)$ such that, for the function $z \mapsto F(z, u)$, the exponent at the singularity retains a constant value, while the location of the singularity $\rho(u)$ moves smoothly with u, for u kept in a sufficiently small neighbourhood of 1. A prototypical instance is a BGF involving a term $C(z, u)^{-\alpha}$, when $C(z, u)$ is bivariate analytic and $C(z, 1)$ has an isolated zero at the point $\rho \equiv \rho(1)$. The developments in the present subsection can then be seen as extending the perturbative analysis of meromorphic functions in Theorem IX.9 (p. 656), where the latter corresponds to exponents restricted to $\alpha = 1, 2, \dots$.

This subsection provides the general machinery for addressing such fixed-exponent movable-singularity situations, and it is once more based on the uniformity afforded by singularity analysis (Lemma IX.2, p. 668). We illustrate it by means of a few simple examples related to trees, where BGFs are explicitly known. (The next two subsections will explore further applications where BGFs are only accessible indirectly, via implicit analytic (especially, algebraic) equations and differential equations.) Our starting point is the following general statement, which parallels Theorem IX.9, p. 656.

Theorem IX.12 (Algebraic singularity schema). *Let $F(z, u)$ be a function that is bivariate analytic at $(z, u) = (0, 0)$ and has non-negative coefficients. Assume the following conditions:*

(*i*) Analytic perturbation: *there exist three functions A, B, C, analytic in a domain $\mathcal{D} = \{|z| \le r\} \times \{|u - 1| < \epsilon\}$, such that, for some r_0 with $0 < r_0 \le r$, and $\epsilon > 0$, the following representation*[15] *holds, with $\alpha \notin \mathbb{Z}_{\le 0}$,*

(60)
$$F(z, u) = A(z, u) + B(z, u)C(z, u)^{-\alpha};$$

   furthermore, assume that, in $|z| \le r$, there exists a unique root ρ of the equation $C(z, 1) = 0$, that this root is simple, and that $B(\rho, 1) \ne 0$.

(*ii*) Non-degeneracy: *one has $\partial_z C(\rho, 1) \cdot \partial_u C(\rho, 1) \ne 0$, ensuring the existence of a non-constant $\rho(u)$ analytic at $u = 1$, such that $C(\rho(u), u) = 0$ and $\rho(1) = \rho$.*

(*iii*) Variability: *one has*

$$\mathfrak{v}\left(\frac{\rho(1)}{\rho(u)}\right) \ne 0.$$

Then, the random variable with probability generating function

$$p_n(u) = \frac{[z^n] F(z, u)}{[z^n] F(z, 1)}$$

converges in distribution to a Gaussian variable with a speed of convergence that is $O(n^{-1/2})$. The mean μ_n and the standard deviation σ_n are asymptotically linear in n.

---

[15]By unicity of analytic continuation, the representation of $F(z, u)$ only needs to be established initially near $(z, u) = (0, 1)$, that is, for $|z| < r_0$, for some (arbitrarily small) positive r_0.

Proof. We start with the asymptotic analysis of the univariate counting problem. By the assumptions made, the function $F(z, 1)$ is analytic in $|z| < \rho$ and continuable to a Δ–domain. It admits a singular expansion of the form

(61)
$$
\begin{aligned}
F(z, 1) = {} & (a_0 + a_1(z - \rho) + \cdots) \\
& + (b_0 + b_1(z - \rho) + \cdots)\left(c_1(z - \rho) + c_2(z - \rho)^2 + \cdots\right)^{-\alpha}.
\end{aligned}
$$

There, the a_j, b_j, c_j represent the coefficients of the expansion in z of A, B, C for z near ρ when u is instantiated at 1. (We may consider $C(z, u)$ normalized by the condition that c_1 is positive real, and take, e.g., $c_1 = 1$.) Singularity analysis then implies the estimate

(62)
$$
[z^n]F(z, 1) = b_0(-c_1\rho)^{-\alpha}\rho^{-n}\frac{n^{\alpha-1}}{\Gamma(\alpha)}\left(1 + O\left(\frac{1}{n}\right)\right).
$$

All that is needed now is a *uniform* lifting of relations (61) and (62), for u in a small neighbourhood of 1.

First, we observe that, by the analyticity assumption on A, the coefficient $[z^n]A(z, u)$ is exponentially small compared to ρ^{-n}, for u close enough to 1. Thus, for our purposes, we may freely restrict attention to $[z^n]B(z, u)C(z, u)^{-\alpha}$. (The function A is only needed in some cases so as to ensure non-negativity of the first few coefficients of F.)

Next, we observe that there exists for u sufficiently near to 1, a unique simple root $\rho(u)$ near ρ of the equation

$$
C(\rho(u), u) = 0,
$$

which is an analytic function of u and satisfies $\rho(1) = \rho$. This results from the Analytic Implicit Function Theorem or, if one prefers, the Weierstrass Preparation Theorem: see Appendix B.5: *Implicit Function Theorem*, p. 753.

At this stage, due to the changing geometry of Δ–domains as u varies, it proves convenient to operate with a *fixed* rather than movable singularity. This is simply achieved by considering the normalized function

$$
\Psi(z, u) := B\left(z\rho(u), u\right)C\left(z\rho(u), u\right)^{-\alpha}.
$$

Provided u is restricted to a suitably small neighbourhood of 1 and z to $|z| < R$ for some $R > 1$, the functions $B(z\rho(u), u)$ and $C(z\rho(u), u)$ are analytic in both z and u (by composition of analytic functions), while $C(z\rho(u), u)$ now has a fixed (simple) zero at $z = 1$. There results that the function

$$
\frac{1}{1-z}C\left(z\rho(u), u\right)
$$

has a removable singularity at $z = 1$ (by division of series expansions) and hence is analytic in $|z| < R$ and $|u - 1| < \delta$, for some $\delta > 0$. In particular, near $z = 1$, Ψ satisfies an expansion of the form

(63)
$$
\Psi(z, u) = (1 - z)^{-\alpha}\sum_{n\geq 0}\psi_n(u)(1 - z)^n,
$$

that is convergent and such that each coefficient $\psi_j(u)$ is an analytic function of u for $|u - 1| < \delta$.

We can finally return to the analysis of $[z^n]F(z, u)$ and undo what has been done. We have

$$[z^n]F(z, u) = \rho(u)^{-n}[z^n]\Psi(z, u) + [z^n]A(z, u),$$

where the second term in the sum is (exponentially) negligible. Now, as we know from (63) and surrounding considerations, the function $z \mapsto \Psi(z, u)$ is analytic in a fixed Δ–domain, in which it admits a uniform singular approximation obtained by a simplification of (63),

$$\Psi(z, u) = \psi_0(u)(1 - z)^{-\alpha} + O\left((1 - z)^{\alpha-1}\right).$$

An application of the uniformity property of singularity analysis, Lemma IX.2, then provides the estimate

(64) $$[z^n]F(z, u) = \psi_0(u)\rho(u)^{-n}\frac{n^{\alpha-1}}{\Gamma(\alpha)}\left(1 + O\left(\frac{1}{n}\right)\right),$$

uniformly, for u restricted to a small neighbourhood of 1.

Equation (64) shows that $p_n(u) = f_n(u)/f_n(1)$, where $f_n(u) := [z^n]F(z, u)$, satisfies precisely the conditions of the Quasi-powers Theorem, Theorem IX.8. Therefore, the law with PGF $p_n(u)$ is asymptotically normal with a mean and a standard deviation that are both $O(n)$. Since the error term in (64) is $O(1/n)$, the speed of convergence to the Gaussian limit is $O(1/\sqrt{n})$.   ∎

The remarks following the statement of Theorem IX.9 apply. Accordingly, the mean μ_n and variance σ_n^2 are computable by the general formula (37), and the variability condition is expressible in terms of the values of C and its derivatives at $(\rho, 1)$ by means of Equation (39), p. 657.

▷ **IX.40.** *Logarithmic multipliers.* The conclusions of Theorem IX.12 extend to functions representable under the more general form ($k \in \mathbb{Z}_{\geq 0}$)

$$F(z, u) = A(z, u) + B(z, u)C(z, u)^{-\alpha}(\log C(z, u))^k.$$

(The proof follows the same pattern, based on Note IX.36, p. 669.)   ◁

In the remainder of this subsection, we illustrate the use of Theorem IX.12 by means of examples involving an explicit *fractional* power of a bivariate analytic function. Privileged cases of application of the theorem are the number of leaves in classical varieties of trees, such as Cayley trees, general or binary Catalan trees, and Motzkin trees, for which the GFs lead to an explicit square-root expression.

Example IX.24. *Leaves in general Catalan trees.* We revisit here under a complex asymptotic angle the analysis of the number of leaves in general Catalan trees $\mathcal{G}$, a problem already introduced in Example III.13, p. 182. The specification is

$$\mathcal{G} = \mathcal{Z}u + \mathcal{Z} \times \text{SEQ}_{\geq 1}(\mathcal{G}) \quad \Longrightarrow \quad G(z, u) = zu + \frac{zG(z, u)}{1 - G(z, u)},$$

with u marking the number of leaves. The solution of the implied quadratic equation then yields the explicit form

$$G(z, u) = \frac{1}{2}\left(1 + (u - 1)z - \sqrt{1 - 2(u + 1)z + (u - 1)^2z^2}\right),$$

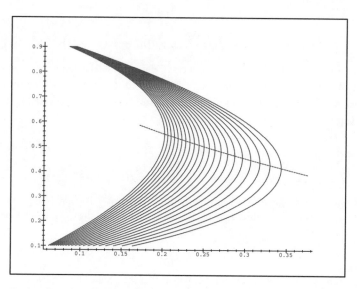

Figure IX.13. A display of the family of GFs $z \mapsto F(z, u)$ corresponding to leaves in general Catalan trees when $u \in [1/2, 3/2]$. It can be observed that the singularities are all of the square-root type, with a movable singularity at $\widetilde{\rho}(u) = (1 + u^{1/2})^{-2}$ (represented by the dashed line).

which is readily verified to be amenable to Theorem IX.12. Indeed, we have, in the notations of that theorem,

$$A(z, u) = \frac{1}{2}(1 + (u - 1)z), \quad B(z, u) \equiv -\frac{1}{2}, \quad C(z, u) = 1 - 2(u + 1)z + (u - 1)^2 z^2,$$

whose analyticity is obvious, together with the fixed exponent $\alpha = -1/2$. The factorization

$$C(z, u^2) = (1 - z(1 + u)^2) \cdot (1 - z(1 - u)^2),$$

implies that the zeros of $z \mapsto C(z, u)$ are at $(1 \pm \sqrt{u})^{-2}$. In particular, if $|u - 1| < 1/10$ (say), then the dominant singularity of $G(z, u)$ is at $\rho(u) = (1 + \sqrt{u})^{-2}$ and $\rho \equiv \rho(1) = 1/4$, as it should be.

The analytic perturbation assumption of Theorem IX.12 (Condition (i)) is then satisfied, with (say) $r = 1/3$. We next verify that $\partial_z C(\rho, 1) = -4$ and $\partial_u C(\rho, 1) = -1$, which ensures non-degeneracy (Condition (ii)). Finally, variability (Condition (iii)) is satisfied since $\mathfrak{v}(\rho(1)/\rho(u)) = 1/8$. Thus the theorem is applicable and the number of leaves is asymptotically normal.

The smooth displacement of singularities induced by the secondary variable u, which lies at the basis of such a Gaussian limit result, is illustrated in Figure IX.13. (Compare also with Figure 0.6 of our *Invitation*, p. 10.) .... ∎

Example IX.25. *Leaves in classical varieties of trees.* First, for leaves in binary Catalan trees, we have (Example III.14, p. 182)

$$\mathcal{B} = \mathcal{Z}u + 2(\mathcal{B} \times \mathcal{Z}) + (\mathcal{B} \times \mathcal{Z} \times \mathcal{B}) \quad \Longrightarrow \quad B(z, u) = z(u + 2zB(z, u) + B(z, u)^2),$$

so that

$$B(z, u^2) = \frac{1}{2z} \left(1 - 2z - \sqrt{(1 - 2z(1 + u))(1 - 2z(1 - u))} \right).$$

This is almost the same as the BGF of leaves in general Catalan trees. The dominant singularity of $z \mapsto B(z, u)$ is at $\rho(u) = \frac{1}{2(1+\sqrt{u})}$ and one finds $\upsilon(\rho(1)/\rho(u)) = 1/16$, so that the limit law is Gaussian. The asymptotic form of the mean and variance are also provided by $\rho(u)$: the number of leaves X_n in a binary Catalan tree of size n satisfies $E\{X_n\} = \frac{1}{4}n + O(1)$ and $\sigma\{X_n\} = \frac{1}{4}\sqrt{n} + O(n^{-1/2})$; the limit law is Gaussian.

Next, comes the case of Cayley trees (Note III.17, p. 183):

$$\mathcal{T} = \mathcal{Z}u + \text{SET}_{\geq 1}(\mathcal{T}) \qquad \Longrightarrow \qquad T(z, u) = z(u - 1 + e^{T(z,u)}).$$

(The distribution is closely related to the Stirling partition numbers.) By simple algebra, it is seen that the functional equation admits an explicit solution in terms of the Cayley tree function itself ($T = ze^T$): we find

$$T(z, u) = z(u - 1) + T(ze^{z(u-1)}).$$

As we know, the function $T(z)$ has a dominant singularity of the square-root type at e^{-1}, so

(65)
$$\rho(u) = \frac{1}{1 - u} T(e^{-1}(1 - u)),$$

and we get $\rho(1) = e^{-1}$, as we should. Accordingly, the function $z \mapsto T(z, u)$ has a singularity of the square-root type at $\rho(u)$, to which Theorem IX.12 can be applied. The expansion near $u = 1$ then comes automatically from (65):

$$\frac{\rho(u)}{\rho(1)} = 1 - e^{-1}(u - 1) + \frac{3}{2}e^{-2}(u - 1)^2 + O((u - 1)^3).$$

Hence the mean and the variance of the number X_n of leaves in a random tree of size n satisfy $E\{X_n\} \sim e^{-1}n \approx 0.36787\,n$ and $\sigma^2\{X_n\} \sim e^{-2}(e - 2)n \approx 0.09720\,n$, the limit law being Gaussian. .... ∎

Example IX.26. *Patterns in binary Catalan trees.* We present here a more sophisticated example that generalizes the problem of counting leaves in trees. It arises from the analysis of pattern matching and of compact representations of trees [257, 561]. The BGF of the number of (pruned) binary trees with z marking size and u marking the number of occurrences of a pattern of size m is

(66)
$$F(z, u) = \frac{1}{2z} \left(1 - \sqrt{1 - 4z - 4(u - 1)z^{m+1}} \right),$$

as seen in Note III.40 (p. 213) and Note III.41 (p. 214).

The quantity under the square-root in (66) has a unique root at $\rho = 1/4$ when $u = 1$, while it has $m + 1$ roots for $u \neq 1$. By general properties of implicit and, specifically, algebraic functions (Implicit Function Theorem, Weierstrass Preparation), as u tends to 1, one of these roots, call it $\rho(u)$ tends to 1/4, while all the others $\{\rho_j(u)\}_{j=1}^m$ escape to infinity. We have

$$H(z, u) := \frac{1 - 4z - 4z^{m+1}(u - 1)}{1 - z/\rho(u)} = \prod_{j=1}^{m} (1 - z/\rho_j(u)),$$

which is an analytic function in (z, u) for (z, u) in a complex neighbourhood of $(1/4, 1)$. (This results from the fact that the algebraic function $\rho(u)$ is analytic at $u = 1$.) The singular expansion of $G(z, u) = zF(z, u)$ is then given by

$$G(z, u) = \frac{1}{2} - \frac{1}{2}\sqrt{H(z, u)}\sqrt{1 - z/\rho(u)}.$$

Thus, we are under the conditions of Theorem IX.12. Accordingly, the number of occurrences taken over a random binary tree of size $n + 1$ has mean and variance given asymptotically by $\mathfrak{m}((4\rho(u))^{-1})n$ and $\mathfrak{v}((4\rho(u))^{-1})n$, respectively. The expansion of $\rho(u)$ at 1 is computed easily by iteration ("bootstrapping") from the defining equation,

$$z = \frac{1}{4} - z^{m+1}(u - 1) = \frac{1}{4} - \left(\frac{1}{4} - z^{m+1}(u - 1)\right)^{m+1}(u - 1) = \cdots,$$

to the effect that

$$\rho(u) = \frac{1}{4} - \frac{1}{4^{m+1}}(u - 1) + \frac{m + 1}{4^{2m+1}}(u - 1)^2 + \cdots.$$

Proposition IX.16. *The number of occurrences of a pattern of size m in a random Catalan tree of size $n + 1$ admits a Gaussian limit distribution, with mean μ_n and variance σ_n^2 that satisfy*

$$\mu_n \sim \frac{n}{4^m}, \qquad \sigma_n^2 \sim n\left(\frac{1}{4^m} - \frac{2m + 1}{4^{2m}}\right).$$

In particular, the probability of occurrence of a pattern at a random node of a random trees decreases fast (the factor of 4^{-m} in the estimate of averages) with the size of the pattern, a property that parallels the one already known for strings (p. 659). The paper of Steyaert and Flajolet [561] shows that similar properties hold for any simply generated family, at least in an expected value sense. Flajolet, Sipala, and Steyaert [257] build upon the foregoing analysis to show that the minimal "dag representation" of a random tree (where identical subtrees are "shared" and represented only once) is of average size $O(n(\log n)^{-1/2})$. .... ■

▷ **IX.41.** *Leaves in Motzkin trees.* The number of leaves in a unary–binary (Motzkin) tree is asymptotically Gaussian. ◁

▷ **IX.42.** *Patterns in classical varieties of trees.* Patterns in general Catalan trees and Cayley trees can be similarly analysed. ◁

IX. 7.3. Algebraic and implicit functions.

Under the univariate counting scenario, we have encountered in Chapter VII many analytic–combinatorial conditions leading to singular exponents that are non-integral. For instance, many implicitly defined functions, including important algebraic cases, have a dominant singularity that is of the square-root type (the exponent is $\alpha = -1/2$ in the notations of Theorem IX.12). If a corresponding specification is enriched by markers, there is a fair chance that the square-root singularity property will persist (as in Figure IX.13, p. 679) when the marking variable u remains close to 1, so that, by Theorem IX.12, a Gaussian law results in the scale of n. Similar comments apply to functions defined implicitly by systems of equations, including algebraic functions, provided suitable non-degeneracy conditions[16] are satisfied. Here, we only state a single proposition, which is meant to illustrate in a simple situation the type of treatment to which implicitly defined BGFs can be subjected.

---

[16] Subsection IX. 11.2 (p. 707) below examines cases where a confluence of singularities induces a stable law instead of the customary Gaussian distribution.

Proposition IX.17 (Perturbation of algebraic functions). *Let $F(z, u)$ be a bivariate function that is analytic at $(0, 0)$ and has non-negative coefficients. Assume that $F(z, u)$ is one of the solutions y of a polynomial equation*

$$y - \Phi(z, u, y) = 0,$$

where Φ is a polynomial of degree $d \geq 2$ in y, such that $\Phi(z, 1, y)$ satisfies the conditions of the smooth implicit function schema *of Section VII.4, p. 467, with $G(z, w) := \Phi(z, 1, w)$. Let ρ, τ be the solutions of the characteristic system (relative to $u = 1$), so that $y(z) := F(z, 1)$ is singular at $z = \rho$ and $y(\rho) = \tau$. Define the resultant polynomial (Appendix B.1: Algebraic elimination, p. 739),*

$$\Delta(z, u) = \mathbf{R}\left(y - \Phi(z, u, y), 1 - \frac{\partial}{\partial y}\Phi(z, u, y), y\right),$$

so that ρ is a simple root of $\Delta(z, 1)$. Let $\rho(u)$ be the unique root of the equation

$$\Delta(\rho(u), u),$$

analytic at 1, such that $\rho(1) = \rho$. Then, provided the variability condition

$$\mathfrak{v}\left(\frac{\rho(1)}{\rho(u)}\right) > 0,$$

is satisfied, a Gaussian Limit Law holds for the coefficients of $F(z, u)$.

Proof. By the developments of Theorem VII.3, p. 468, the function $y(z) = F(z, 1)$ has a square-root singularity at $z = \rho$. The polynomial $y - \Phi(\rho, 1, y)$ has a double (not triple) zero at $y = \tau$, so that

$$\left(\frac{\partial}{\partial y}\Phi(\rho, 1, y)\right)_{y=\tau} = 0, \qquad \left(\frac{\partial^2}{\partial y^2}\Phi(\rho, 1, y)\right)_{y=\tau} \neq 0.$$

Thus, the Weierstrass Preparation Theorem gives the local factorization

$$y - \Phi(z, u, y) = (y^2 + c_1(z, u)y + c_2(z, u))H(z, u, y),$$

where $H(z, u, y)$ is analytic and non-zero at $(\rho, 1, \tau)$ while $c_1(z, u), c_2(z, u)$ are analytic at $(z, u) = (\rho, \tau)$.

From the solution of the quadratic equation, we must have locally

$$y = \frac{1}{2}\left(-c_1(z, u) \pm \sqrt{c_1(z, u)^2 - 4c_2(z, u)}\right).$$

Consider first (z, u) restricted by $0 \leq z < \rho$ and $0 \leq u < 1$. Since $F(z, u)$ is real there, we must have $c_1(z, u)^2 - 4c_2(z, u)$ also real and non-negative. Since $F(z, u)$ is continuous and increasing with z for fixed u, and since the discriminant $c_1(z, u)^2 - 4c_2(z, u)$ vanishes at 0, the determination with the minus sign has to be constantly taken. In summary, we have

(67) $$F(z, u) = \frac{1}{2}\left(-c_1(z, u) - \sqrt{c_1(z, u)^2 - 4c_2(z, u)}\right).$$

Set $D(z, u) := c_1(z, u)^2 - 4c_2(z, u)$. The function $D(z, 1)$ has a simple real zero at $z = \rho$. Thus, by the Analytic Inverse Function Theorem (or Weierstrass preparation again), there is locally a unique analytic branch of the solution to $C(\rho(u), u) = 0$ such that $\rho(1) = \rho$, and $D(z, u)$ factorizes as

$$D(z, u) = (\rho(u) - z)K(z, u),$$

for some analytic K satisfying $K(\rho, 1) \neq 0$. The conditions of Theorem IX.12 therefore hold. The stated Gaussian law follows. ∎

The last proposition asserts that, under certain conditions, the only possible dominant singularity of the function $z \mapsto F(z, u)$ is a smooth lifting of the singularity of the univariate GF $F(z, 1)$, while the nature of the singularity does not change—it remains of the square-root type. Similar results, established by similar methods, hold true for more general equations and systems, under suitable non-degeneracy and variability conditions. Indeed, one can go all the way from algebraic functions defined by a single polynomial equation, as above, to functions implicitly defined by *systems of analytic equations*. This has been done by Drmota in an important paper [172]. For a system $\mathbf{y} = \Phi(z, u, \mathbf{y})$, the approach consists of looking at the Jacobian of the transformation, as in Subsection VII. 6.1 (p. 482) and imposing conditions that allow for a smooth singularity displacement. The Weierstrass Preparation Theorem normally provides the needed permanence of analytic relations that imply a persistent square-root singularity

The scope of Theorem IX.12, Proposition IX.17, and their derivative products is enormous—potentially, all the recursive combinatorial structures examined in Sections VII. 3–VII. 8 (pp. 452–518) are concerned. This includes trees of various sorts, mappings, lattice paths and their generalizations, planar maps, as well as languages and classes described by context-free specifications, to name a few.

Example IX.27. *A pot-pourri of Gaussian laws.* In the list that follows, all the mentioned parameters obey a Gaussian limit distribution in the scale of n. The proofs (omitted) involve in each case a precise investigation of the perturbation of univariate singular expansions induced by the secondary parameter, in a way similar to Theorem IX.12.

Simple varieties of trees, p. 452. The number of leaves is Gaussian (see Examples IX.24 and IX.25 above) and the property extends to the number of nodes of any fixed degree r as well as to the number of occurrences of any fixed pattern (see Example IX.26). This property also holds true for simple varieties of trees introduced in Section VII. 3, and it extends to unlabelled non-plane trees [121].

Mappings, p. 462. The number of points with r predecessors is Gaussian, as is the cardinality of the image set, the property being also true for mappings defined by degree restrictions [18, 247].

Irreducible context-free structures, p. 482. Examples given in the paper of Drmota [172] are the number of independent sets in a random tree and the number of patterns in a context-free language.

Non-crossing graphs, p. 485. The number of connected components and the number of edges in either forests or general non-crossing graphs is Gaussian [245]. (These properties are thus in sharp contrast with those of the usual random graph model of Erdős and Rényi [76].)

Walks in the discrete plane, p. 506. The number of steps of any fixed kind is Gaussian for walks, excursions, bridges and meanders. An extension of the known methods shows that the number of occurrences of any fixed pattern (made of contiguous letters) is also asymptotically normal. For instance, the number of occurrences of the pattern up-down-up-up-down in a random Dyck word (excursion) satisfies this property.

Planar maps, p. 513. The number of occurrences of any fixed submap is asymptotically Gaussian (see [278] for a proof based on moment methods). Thus, maps are like words and trees: any fixed collection of patterns occurs in a large enough random object with high probability (Borges' Theorem, p. 61). .... ■

IX. 7.4. Differential equations. We have encountered in this book sporadic combinatorial classes whose GFs are determined as solutions of *ordinary differential equations* (ODEs), and we have presented in Section VII. 9 (p. 518) several such structures that are amenable to singularity analysis. Basic parameters are then likely still to lead to ODEs, but ones that are now parameterized by the secondary variable u. (By contrast, *partial differential equations* have so far been only scarcely used in analytic combinatorics.) In such cases, a singularity perturbation analysis is often feasible. Both situations, that of a variable exponent and that of a movable singularity, can occur, as we now illustrate, largely by means of examples. The partial treatment given here should at least convey the spirit of the singularity perturbation process, in the context of differential equations.

Linear differential equations. ODEs in one variable, when *linear* and when having *analytic coefficients*, admit solutions whose singularities occur at well-defined places, namely those that entail a reduction of order (see Subsection VII. 9.1, p. 518, and Section VIII. 7, p. 581, for the so-called "regular and "irregular cases, respectively). The possible singular exponents of solutions are then obtained as roots of a polynomial equation, the indicial equation. Such ordinary differential equations are usually a reflection of a combinatorial decomposition of sorts, so that suitably parameterized versions open access to a number of combinatorial parameters. In the cases considered here, the ODE satisfied by a BGF $F(z, u)$ remains an ODE in the main variable z that records size, while the auxiliary variable u only affects coefficients. We start with a simple example, Example IX.28, relative to node levels in increasing binary trees, continue with a general statement, Proposition IX.18 relative to the case of a variable exponent in a linear ODE, and conclude with an application to node levels in quadtrees in Example IX.29.

Example IX.28. *Node levels in increasing binary trees.* Increasing binary trees are labelled (pruned) binary trees, such that any branch from the root has monotonically increasing labels. As explained in Example II.17 (p. 143), these trees are an important representation of permutations. Their specification, in terms of the boxed product of Chapter II, is

$$(68) \qquad \mathcal{F} = 1 + \left(\mathcal{Z}^{\square} \star \mathcal{F} \star \mathcal{F} \right) \qquad \Longrightarrow \qquad F(z) = 1 + \int_0^z F(t)^2 \, dt,$$

and, accordingly, their EGF is

$$F(z) = \frac{1}{1-z} = \sum_{n \geq 0} n! \frac{z^n}{n!},$$

Let $F(z, u)$ be the BGF of trees where u records the depth of external nodes. In other words, $f_{n,k} = [z^n u^k] F(z, u)$ is such that $\frac{1}{n+1} f_{n,k}$ represents the probability that a random external node in a random tree of size n is at depth k. (The probability space is then a product set of cardinality $(n + 1) \cdot n!$, as there are $n!$ trees each containing $(n + 1)$ external nodes. By a standard equivalence principle, the quantity $\frac{1}{n+1} f_{n,k}$ also give the probability that a random unsuccessful search in a random binary search tree of size n necessitates k comparisons.)

Since the depth of a node is inherited from subtrees, the function $F(z, u)$ satisfies the *linear* integral equation derived from (68) (see also Equation (VI.67), p. 429 in relation to the BST recurrence),

$$(69) \qquad F(z, u) = 1 + 2u \int_0^z F(t, u) \, \frac{dt}{1 - t},$$

or, after differentiation,

$$\frac{\partial}{\partial z} F(z, u) = \frac{2u}{1 - z} F(z, u), \qquad F(0, u) = 1.$$

This equation is nothing but a linear ODE, with u entering as a parameter in the *coefficients*,

$$\frac{d}{dz} y(z) - \frac{2u}{1 - z} y(z) = 0, \qquad y(0) = 0,$$

the solution of any such separable first-order ODE being obtained by quadratures:

$$F(z, u) = \frac{1}{(1 - z)^{2u}}.$$

From singularity analysis, provided u avoids $\{0, -1/2, -1, \ldots\}$, we have

$$f_n(u) := [z^n] F(z, u) = \frac{n^{2u-1}}{\Gamma(2u)} \left(1 + O \left(\frac{1}{n} \right) \right),$$

and a uniform approximation holds, provided (say) $|u-1| \le 1/4$. Thus, Theorem IX.11 applies, to the effect that *the distribution of the depth of a random external node in a random increasing binary tree*, with PGF $f_n(u)/f_n(1)$, *admits a Gaussian limit law*.

Naturally, explicit expressions are available in such a simple case,

$$\frac{f_n(u)}{f_n(1)} = \frac{2u \cdot (2u + 1) \cdots (2u + n - 1)}{(n + 1)!},$$

so a direct proof of the Gaussian limit in the line of Goncharov's theorem (p. 645) is clearly possible; see Mahmoud's book [429, Ch. 2], for this result originally due to Louchard. What is interesting here is the fact that $F(z, u)$ viewed as a function of z has a singularity at $z = 1$ that does not move and, in a way, originates in the combinatorics of the problem, through the EGF of permutations, $(1 - z)^{-1}$. The auxiliary parameter u appears here directly in the exponent, so that the application of singularity analysis or of the more sophisticated Theorem IX.11, (p. 669) is immediate.

A similar Gaussian law holds for levels of internal nodes, and is proved by similar devices. The Gaussian profile is even perceptible on single instance. In particular, Figure III.18 (p. 203) suggests a much stronger "functional limit theorem" for these objects (namely, almost all trees have an approximate Gaussian profile): this property, which seems currently beyond the scope of analytic combinatorics, has been proved by Chauvin and Jabbour [114] using martingale theory. .... ∎

Proposition IX.18 (Linear differential equations). *Let $F(z, u)$ be a bivariate generating function with non-negative coefficients that satisfies a linear differential equation*

$$a_0(z, u) \frac{\partial^r F}{\partial z^r} + \frac{a_1(z, u)}{(\rho - z)} \frac{\partial^{r-1} F}{\partial z^{r-1}} + \cdots + \frac{a_r(z, u)}{(\rho - z)^r} F = 0,$$

with $a_j(z, u)$ analytic at ρ, and $a_0(\rho, 1) \neq 0$. Let $f_n(u) = [z^n]F(z, u)$, and assume the following conditions:

- [Non-confluence] *The indicial polynomial*

(70) $$J(\alpha) = a_0(\rho, 1)(\alpha)_{(r)} + a_1(\rho, 1)(\alpha)_{(r-1)} + \cdots + a_r(\rho, 1)$$

  has a unique root $\sigma > 0$ which is simple and such that all other roots $\alpha \neq \sigma$ satisfy $\Re(\alpha) < \sigma$;
- [Dominant growth] $f_n(1) \sim C \cdot \rho^{-n} n^{\sigma - 1}$, *for some $C > 0$.*
- [Variability condition]

$$\sup \frac{\mathfrak{v}(f_n(u))}{\log n} > 0.$$

Then the coefficients of $F(z, u)$ admit a limit Gaussian law.

Proof. (See the paper by Flajolet and Lafforgue [243] for a detailed analysis and the books by Henrici [329] and Wasow [602] for a general treatment of singularities of linear ODEs.) We assume in this proof that no two roots of the indicial polynomial (70) differ by an integer. Consider first the univariate problem, for which we summarize the discussion started on p. 518. A differential equation,

(71) $$a_0(z)\frac{d^r F}{dz^r} + \frac{a_1(z)}{(\rho - z)}\frac{d^{r-1}F}{dz^{r-1}} + \cdots + \frac{a_r(z)}{(\rho - z)^r}F = 0,$$

with the $a_j(z)$ analytic at ρ and $a_1(\rho) \neq 0$ has a basis of local singular solutions obtained by substituting $(\rho - z)^{-\alpha}$ and cancelling the terms of maximum order of growth. The candidate exponents are thus roots of the *indicial equation*,

$$J(\alpha) \equiv a_0(\rho)(\alpha)_{(r)} + a_1(\rho)(\alpha)_{(r-1)} + \cdots + a_r(\rho) = 0.$$

If there is a unique (simple) root of maximum real part, α_1, then there exists a solution to (71) of the form

$$Y_1(z) = (\rho - z)^{-\alpha_1} h_1(\rho - z),$$

where $h_1(w)$ is analytic at 0 and $h_1(0) = 1$. (This results easily from a solution by indeterminate coefficients.) All other solutions are then of smaller growth and of the form

$$Y_j(z) = (\rho - z)^{-\alpha_j} h_j(\rho - z)(\log(z - \rho))^{k_j},$$

for some integers k_j and some functions $h_j(w)$ analytic at $w = 0$. Then, $F(z)$ has the form

$$F(z) = \sum_{j=1}^{r} c_j Y_j(z).$$

Then, provided $c_1 \neq 0$,

$$[z^n]F(z) = \frac{c_1}{\Gamma(\sigma)}\rho^{-n} n^{\alpha_1 - 1}(1 + o(1)).$$

Under the assumptions of the theorem, we must have $\sigma = \alpha_1$, and $c_1 \neq 0$. (The reality assumption on σ is natural for a series $F(z)$ that has real coefficients.)

When u varies in a neighbourhood of 1, we have a uniform expansion

(72) $$F(z, u) = c_1(u)(\rho - z)^{-\sigma(u)} H_1(\rho - z, u)(1 + o(1)),$$

for some bivariate analytic function $H_1(w, u)$ with $H_1(0, u) = 1$, where $\sigma(u)$ is the algebraic branch that is a root of

$$J(\alpha, u) \equiv a_0(\rho, u)(\alpha)_{(r)} + a_1(\rho, u)(\alpha)_{(r-1)} + \cdots + a_r(\rho, u) = 0,$$

and coincides with σ at $u = 1$. By singularity analysis, this entails

$$(73) \qquad [z^n]F(z, u) = \frac{c_1(u)}{\Gamma(\sigma)} \rho^{-n} n^{\sigma(u)-1}(1 + o(1)),$$

uniformly for u in a small neighbourhood of 1, with the error term being $O(n^{-a})$ for some $a > 0$. Thus Theorem IX.11 (p. 669) applies and the limit law is Gaussian.

The crucial point in (72) and (73) is the uniform character of expansions with respect to u. This results from two facts: (i) the solution to (71) may be specified by analytic conditions at a point z_0 such that $z_0 < \rho$ and there are no singularities of the equation between z_0 and ρ; and (ii) there is a suitable set of solutions with an analytic component in z and u and singular parts of the form $(\rho - z)^{-\alpha_j(u)}$, as results from the matrix theory of differential systems and majorant series. (This last point is easily verified if no two roots of the indicial equation differ by an integer; otherwise, see [243] for an alternative basis of solutions for u near 1, $u \neq 1$.) ∎

***Example* IX.29.** *Node levels in quadtrees.* Quadtrees defined in Example VII.23 (p. 522) are one of the most versatile data structures known for managing collections of points in multi-dimensional space. They are based on a recursive decomposition similar to that of binary search trees and increasing binary trees of the previous example.

This example is borrowed from [243]. We fix the dimension $d \geq 2$ of the ambient data space. Let $f_{n,k}$ be the number of external nodes at level k in a quadtree of size n grown by random insertions, and let $F(z, u)$ be the corresponding BGF. Two integral operators play an essential rôle,

$$\mathbf{I}\,g(z) = \int_0^z g(t)\, \frac{dt}{1-t} \qquad \mathbf{J}\,g(z) = \int_0^z g(t)\, \frac{dt}{t(1-t)}.$$

The basic equation that reflects the recursive splitting process of quadtrees is then (see [243] and Chapter VII, p. 522 for similar techniques)

$$(74) \qquad F(z, u) = 1 + 2^d u \mathbf{J}^{d-1} \mathbf{I}\, F(z, u).$$

The integral equation (74) satisfied by F then transforms into a differential equation of order d,

$$\mathbf{I}^{-1} \mathbf{J}^{1-d}\, F(z, u) = 2^d u F(z, u),$$

where

$$\mathbf{I}^{-1} g(z) = (1 - z)g'(z), \qquad \mathbf{J}^{-1} g(z) = z(1 - z)g'(z).$$

The linear ODE version of (74) has an indicial polynomial that is easily determined by examination of the reduced form of the ODE (74) at $z = 1$. There, one has

$$\mathbf{J}^{-1} g(z) = \mathbf{I}^{-1} g(z) - (z - 1)^2 g'(z) \approx (1 - z)g'(z).$$

Thus,

$$\mathbf{I}^{-1} \mathbf{J}^{1-d}(1 - z)^{-\theta} = \theta^d (1 - z)^{-\theta} + O((1 - z)^{-\theta+1}),$$

and the indicial polynomial is

$$J(\alpha, u) = \alpha^d - 2^d u.$$

In the univariate case, the root of largest real part is $\alpha_1 = 2$; in the bivariate case, we have

$$\alpha_1(u) = 2u^{1/d},$$

where the principal branch is chosen. Thus,

$$f_n(u) = \gamma(u) n^{\alpha_1(u)} (1 + o(1)).$$

By the combinatorial origin of the problem, $F(z, 1) = (1 - z)^{-2}$, so that the coefficient $\gamma(1)$ is non-zero. Thus, the conditions of Proposition IX.18 are satisfied: *The depth of a random external node in a randomly grown quadtree is Gaussian in the limit, with mean and variance*

$$\mu_n \sim \frac{2}{d} \log n, \qquad \sigma_n^2 \sim \frac{2}{d} \log n.$$

The same result applies to the cost of a (fully specified) random search, either successful or not, as shown in [243] by an easy combinatorial argument. . . . . ∎

From the global point of view of analytic combinatorics, it is of interest to place the last two examples in perspective. Simple varieties of trees, as considered in earlier subsections, are "square-root trees", where height and depth of a random node are each of order $\sqrt{n}$ (on average, in distribution), while the corresponding univariate GFs satisfy algebraic or implicit equations and have a square-root singularity. Trees that in some way arise from permutations (increasing trees, binary search trees, quadtrees) are "logarithmic trees": they are specified by order-constrained constructions that correspond to integro-differential operators, and their depth appears to be logarithmic with Gaussian fluctuations, as a reflection of a perturbative singularity analysis of ODEs.

Nonlinear differential equations. Although nonlinear differential equations defy classification in all generality, there are a number of examples in analytic combinatorics that can be treated by singularity perturbation methods. We detail here the typical analysis of "paging" in binary search trees (BSTs), or equivalently increasing binary trees, taken from [235]. The Riccati equation involved reduces, by classical techniques, to a linear second-order equation whose perturbation analysis is particularly transparent and akin to earlier analyses of ODEs. In this problem, the auxiliary parameter induces a movable singularity that leads to a Gaussian limit law in the scale of n.

Example IX.30. *Paging of binary search trees and increasing binary trees.* Fix a "page size" parameter $b \geq 2$. Given a tree t, its b–*index* is a tree constructed by retaining only those internal nodes of t which correspond to subtrees of size $> b$. As a computer data structure, such an index is well-suited to "paging", where one has a two-level hierarchical memory structure: the index resides in main memory and the rest of the tree is kept in pages of capacity b on peripheral storage, see for instance [429]. We let $\iota[t] = \iota_b[t]$ denote the size —number of nodes— of the b–index of t.

We consider here the analysis of paging in binary search trees, whose model is known to be equivalent to that of increasing binary trees. The bivariate generating function

$$F(z, u) := \sum_t \lambda(t) u^{\iota[t]} z^{|t|}$$

satisfies a Riccati equation that reflects the root decomposition of trees (see (68)),

$$(75) \qquad \frac{\partial}{\partial z} F(z, u) = u F(z, u)^2 + (1 - u) \frac{d}{dz} \left(\frac{1 - z^{b+1}}{1 - z} \right), \qquad F(0, u) = 1,$$

where the quadratic relation has to be adjusted in its low-order terms.

The GFs of moments are rational functions with a denominator that is a power of $(1 - z)$, as results from differentiation at $u = 1$. Mean and variance follow:

$$\mu_n = \frac{2(n+1)}{b+2} - 1, \qquad \sigma_n^2 = \frac{2}{3} \frac{(b-1)b(b+1)}{(b+2)^2} (n+1).$$

(The result for the mean is well-known, refer to quantity A_n in the analysis of quicksort on p. 122 of [378].)

Multiplying both sides of (75) by u now gives an equation satisfied by $H(z, u) := uF(z, u)$,

$$\frac{\partial}{\partial z} H(z, u) = H(z, u)^2 + u(1 - u) \frac{d}{dz} \left(\frac{1 - z^{b+1}}{1 - z} \right),$$

that may as well be taken as a starting point since $H(z, u)$ is the bivariate GF of parameter $1 + \iota_b$ (a quantity also equal to the number of external pages). The classical linearization transformation of Riccati equations,

$$H(z, u) = -\frac{X'_z(z, u)}{X(z, u)},$$

yields

(76) $\qquad \dfrac{\partial^2}{\partial z^2} X(z, u) + u(u - 1)A(z)X(z, u) = 0, \qquad A(z) = \dfrac{d}{dz} \left(\dfrac{1 - z^{b+1}}{1 - z} \right),$

with $X(0, u) = 1$, $X'_z(0, u) = -u$. By the classical existence theorem of Cauchy, the solution of (76) is an entire function of z for each fixed u, since the linear differential equation has no singularity at a finite distance. Furthermore, the dependency of X on u is also everywhere analytic; see the remarks of [602, §24], for which a proof derives by inspection of the classical existence property, based on indeterminate coefficients and majorant series. Thus, $X(z, u)$ is actually an entire function of *both* complex variables z and u. As a consequence, for any fixed u, the function $z \mapsto H(z, u)$ is a meromorphic function whose coefficients are amenable to singularity analysis.

In order to proceed further, we need to prove that, in a sufficiently small neighbourhood of $u = 1$, $X(z, u)$ has only one simple root, corresponding for $H(z, u)$ to a unique dominant and simple pole. This fact derives from the usual considerations surrounding the analytic Implicit Function Theorem and the Weierstrass Preparation Theorem (Appendix B.5: *Implicit Function Theorem*, p. 753). Here, we have $X(z, 1) \equiv 1 - z$. Thus, as u tends to 1, all solutions in z of $X(z, u) = 0$ must escape to infinity, except for one (analytic) branch $\rho(u)$ that satisfies $\rho(1) = 1$.

The argument is now complete: the BGF $F(z, u)$ and its companion $H(z, u) = uF(z, u)$ have a movable singularity at $\rho(u)$, which is a pole. Theorem IX.9 (p. 656) relative to the meromorphic case applies, and a Gaussian limit law results. .... ∎

As shown in [235], a similar analysis applies to patterns in binary search trees. The corresponding properties are (somewhat) related to the analysis of local order patterns in permutations, for which Gaussian limit laws have been obtained by Devroye [159] using extensions of the central limit theorem to weakly dependent random variables.

▷ **IX.43.** *Leaves in varieties of increasing trees.* Similar displacements of singularity arise for the number of nodes of a given type in varieties of increasing trees (Example VII.24, p. 526).

For instance, if $\phi(w)$ is the degree generator of a family of increasing trees, the nonlinear ODE satisfied by the BGF of the number of leaves is

$$\frac{\partial}{\partial z} F(z, u) = (u - 1)\phi(0) + \phi(F(z, u)).$$

Whenever ϕ is a polynomial, there is a spontaneous singularity at some $\rho(u)$ that depends analytically on u. Thus, the number of leaves is asymptotically Gaussian [49]. A similar result holds for nodes of any fixed degree r.   ◁

IX. 8. Perturbation of saddle-point asymptotics

The saddle-point method, which forms the subject of Chapter VIII, is also amenable to perturbation. For instance, we already know that the number of partitions of a domain of cardinality n into classes (set partitions enumerated by the nth Bell number) can be estimated by this method; a suitable perturbative analysis can then be developed, to the effect that the number of classes in a random set partition of large size is asymptotically Gaussian. Given the nature of univariate saddle-point expansions and their diversity (they do not reduce to the $\rho^{-n} n^{\alpha}$ paradigm), the Quasi-powers Theorem ceases to be applicable, and a more flexible framework is needed. In what follows, we base our brief discussion on a theorem taken from Sachkov's book [524].

Theorem IX.13 (Generalized quasi-powers). *Assume that, for u in a fixed neighbourhood Ω of 1, the generating function $p_n(u)$ of a non-negative discrete random variable (supported by $\mathbb{Z}_{\geq 0}$) X_n admits a representation of the form*

$$(77) \qquad\qquad p_n(u) = \exp(h_n(u))(1 + o(1)),$$

uniformly with respect to u, where each $h_n(u)$ is analytic in Ω. Assume also the conditions,

$$(78) \qquad h_n'(1) + h_n''(1) \to \infty \qquad and \qquad \frac{h_n'''(u)}{(h_n'(1) + h_n''(1))^{3/2}} \to 0,$$

uniformly for $u \in \Omega$. Then, the random variable

$$X_n^\star = \frac{X_n - h_n'(1)}{(h_n'(1) + h_n''(1))^{1/2}}$$

converges in distribution to a Gaussian with mean 0 and variance 1.

Proof. See [524, §1.4] for details. Set $\sigma_n^2 = h_n'(1) + h_n''(1)$, and expand the characteristic function of X_n at t/σ_n. Thanks to the form (77) and the conditions (78), inequalities implied by the Mean Value Theorem (Note IV.18, p. 249) give

$$h_n(e^{it/\sigma_n}) = h_n'(1)\frac{it}{\sigma_n} - \frac{t^2}{2} + o(1).$$

Thus, the characteristic function of $X_n^\star$ converges to the transform of a standard Gaussian. The statement follows from the continuity theorem of characteristic functions. ∎

▷ **IX.44.** *Real neighbourhoods.* The conditions of Theorem IX.13 can be relaxed by postulating only that Ω is a real interval containing $u = 1$. (Hint: use the continuity theorem for Laplace transforms of distributions.)   ◁

▷ **IX.45.** *Effective speed bounds.* When Ω is a complex neighbourhood of 1 (as stated in Theorem IX.13), a metric version of the theorem, with speed of convergence estimates, can be developed assuming effective error bounds in (77) and (78). (Hint: use the Berry–Esseen inequalities.)   ◁

The statement above extends the Quasi-powers Theorem, and, in order to stress the parallel, we have opted for a complex neighbourhood condition, which has the benefit of providing better error bounds in applications (Note IX.45). In effect, to see the analogy, note that if

$$h_n(u) = \beta_n \log B(u) + A(u),$$

then the second quantity in (78) is $O(\beta_n^{-1/2})$, uniformly. The application of this theorem to saddle-point integrals is in principle routine, although the manipulation of asymptotic scales associated with expressions involving the saddle-point value may become cumbersome. The fact that information for positive real values of u is sufficient (Note IX.44) may, however, help, since in applications, the GF $z \mapsto F(z, u)$ specialized for positive u stands a good chance of being an admissible function in the sense of Chapter VIII (p. 565), when $F(z, 1)$ is itself admissible. General conditions have been stated by Bender, Drmota, Gardy, and coauthors [174, 279, 280, 281]. Broadly speaking, such situations constitute the *saddle-point perturbation process.* Once more, uniformity of expansions is an issue, which can be technically demanding (one needs to revisit the dependency of univariate analyses on the secondary parameter $u \approx 1$), but is not conceptually difficult.

We first detail here the case of singletons in random involutions for which the saddle-point is an explicit algebraic function of n and u. Then, we prove the Gaussian character of the Stirling partition numbers, which is a classic result first obtained by Harper [322] in 1967. We continue with a pot-pourri of Gaussian laws, which can be obtained by the saddle-point method, and conclude with a note that provides brief indications on BGFs only indirectly accessible through functional equations,

Example IX.31. *Singletons in random involutions.* The exponential BGF of involutions, with u marking the number of singleton cycles, is given by

$$\mathcal{F} = \text{SET}\,(u\,\text{CYC}_1(\mathcal{Z}) + \text{CYC}_2(\mathcal{Z})) \qquad \Longrightarrow \qquad F(z, u) = \exp\left(zu + \frac{z^2}{2}\right).$$

The saddle-point equation (Theorem VIII.3, p. 553) is then

$$\frac{d}{dz}\left(uz + \frac{z^2}{2} - (n+1)\log z\right)_{z=\zeta} = 0.$$

This defines the saddle-point $\zeta \equiv \zeta(n, u)$,

$$\begin{aligned}
\zeta(n, u) &= -\frac{u}{2} + \frac{1}{2}\sqrt{4n + 4 + u^2} \\
&= \sqrt{n} - \frac{u}{2} + \frac{u^2 + 4}{8}\frac{1}{\sqrt{n}} + O(n^{-1}),
\end{aligned}$$

where the error term is uniform for u near 1. By the saddle-point formula, one has

$$[z^n]F(z, u) \sim \frac{1}{\sqrt{2\pi D(n, u)}} F(\zeta(n, u), u)\zeta(n, u)^{-n}.$$

The denominator is determined in terms of second derivatives, according to the classical saddle-point formula (p. 553),

$$D(n, u) = \frac{\partial^2}{\partial z^2} \left(uz + \frac{z^2}{2} - (n+1)\log z \right)\Bigg/_{z=\zeta(n,u)},$$

and its main asymptotic order does not change when u varies in a sufficiently small neighbourhood of 1,

$$D(n, u) = 2n - u\sqrt{n} + O(1),$$

again uniformly. Thus, the PGF of the number of singleton cycles satisfies

(79) $$p_n(u) = \frac{F(\zeta(n, u), u)}{F(\zeta(n, 1), 1)} \left(\frac{\zeta(n, u)}{\zeta(n, 1)} \right)^{-n} (1 + o(1)).$$

This is of the form

$$p_n(u) = \exp(h_n(u)) (1 + o(1)),$$

and local expansions then yield the centring and scaling constants

$$a_n := h_n'(1) = \sqrt{n} - \frac{1}{2} + O(n^{-1/2}), \qquad b_n^2 := h_n'(1) + h_n''(1) = \sqrt{n} - 1 + O(n^{-1/2}).$$

Uniformity in (79) can be checked by returning to the original Cauchy coefficient integral and to bounds relative to the saddle-point contour. Theorem IX.13 then applies to the effect that the variable $\frac{1}{b_n}(X_n - a_n)$ is asymptotic to a standard normal. (With a little additional care, one can verify that the mean μ_n and the standard deviation σ_n are asymptotic to a_n and b_n, respectively.) Therefore:

Proposition IX.19. *The number of singletons in a random involution of size n has mean $\mu_n \sim n^{1/2}$ and standard deviation $\sigma_n \sim n^{1/4}$; it admits a limit Gaussian law.*

A random involution thus has, with high probability, a small number of singletons. .... ∎

***Example* IX.32.** *The Stirling partition numbers.* The numbers $\left\{ {n \atop k} \right\}$ correspond to the BGF

$$\mathcal{F} = \text{SET}\left(u \, \text{SET}_{\geq 1}(\mathcal{Z}) \right) \qquad \Longrightarrow \qquad F(z, u) = \exp\left(u(e^z - 1) \right).$$

The saddle-point $\zeta \equiv \zeta(n, u)$ is determined as the positive root near $n/\log n$ of the equation $\zeta e^\zeta = (n+1)/u$. The derivatives occurring in the saddle-point approximation are computed as derivatives of inverse functions in a standard way. The conditions of Theorem IX.13, together with the required uniformity, can then be checked. Hence:

Proposition IX.20. *The Stirling partition distribution defined by $\frac{1}{S_n}\left\{ {n \atop k} \right\}$, with S_n a Bell number, is asymptotically normal, with mean and variance that satisfy*

$$\mu_n \sim \frac{n}{\log n}, \qquad \sigma_n^2 \sim \frac{n}{(\log n)^2}.$$

(See also p. 594 for first moments.) We refer once more to Sachkov's book [524, 526] for computational details. .... ∎

▷ **IX.46.** *Harper's analysis of Stirling behaviour.* Harper's original derivation [322] of Proposition IX.20 is of independent interest. Consider the Stirling polynomials defined by $\sigma_n(u) := n![z^n] \exp(u(e^z - 1))$. Each such polynomial has roots that are real, distinct, and non-positive. Then, for some positive $\beta_{n,k}$, one has

$$\sigma_n(u) = u \prod_{k=1}^{n-1} \left(1 + \frac{u}{\beta_{n,k}} \right).$$

Thus, $\sigma_n(u)/\sigma_n(1)$ can be viewed as the PGF of the sum of a large number of independent (but not identical) Bernoulli variables. One then can conclude by a suitable version of the Central Limit Theorem. $\lhd$

Example IX.33. *A pot-pourri of saddles and Gaussian laws.* Theorem IX.13 combined with a uniformly controlled use of the saddle-point method yields Gaussian laws for most of the structures examined in Chapter VIII. We leave the following cases as exercises to the reader.

Section VIII. 4 (p. 558) has examined three classes, (involutions, set partitions, and fragmented permutations), of which the first two have already been identified as leading to Gaussian laws. Fragmented permutations (p. 562) also have a number of components (fragments) that is Gaussian in the asymptotic limit. In this case, we have a singularity at a finite distance, which is of the exponential-of-a-pole type. (This last result can be rephrased as the fact that the coefficients of the classical Laguerre polynomials are asymptotically normal.)

Saddle-point perturbation applies to the field of exponentials-of-polynomials (p. 568), which vastly generalizes the case of involutions: this field has been pioneered by Canfield [101] in 1977. The number of components is Gaussian in permutations of order p, permutations with longest cycle $\leq p$, and set partitions with largest block $\leq p$, with p a fixed parameter. The number of connected components in idempotent mappings (p. 571) is also Gaussian.

Integer partitions have been asymptotically enumerated in VIII. 6 (p. 574). As regards unconstrained integer partitions, the Gaussian law for the number of summands is originally due to Erdős and Lehner [194]. By contrast, the number of summands in partitions with *distinct* summands is *not* Gaussian (it is a double-exponential distribution [194]). Subtle phenomena are at stake in these cases, which involve Pólya operators and functions having the unit circle as a natural boundary. . . . . ∎

▷ **IX.47.** *Saddle-points and functional equations.* The average-case analysis of the number of nodes in random digital trees or "tries" can be carried out using the Mellin transform technology. The corresponding distributional analysis is appreciably harder and due to Jacquet and Régnier [344]. A complete description is offered in Section 5.4 of Mahmoud's book which we follow. What is required is to analyse the BGF

$$F(z, u) = e^z T(z, u),$$

where the Poisson generating function $T(z, u)$ satisfies the nonlinear difference equation,

$$T(z, u) = uT\left(\frac{z}{2}, u\right)^2 + (1 - u)(1 + z)e^{-z}.$$

This equation is a direct reflection of the problem specification. At $u = 1$, one has $T(z, 1) = 1$, $F(z, 1) = e^z$. The idea is thus to analyse $[z^n]F(z, u)$ by the saddle-point method.

The saddle-point analysis of F requires asymptotic information on $T(z, u)$ for $u = e^{it}$ (the original treatment of [344] is based on characteristic functions). The main idea is to quasi-linearize the problem, setting

$$L(z, u) = \log T(z, u),$$

with u a parameter. This function satisfies the approximate relation $L(z, u) \approx 2L(z/2, u)$, and a bootstrapping argument shows that, in suitable regions of the complex plane, $L(z, u) = O(|z|)$, uniformly with respect to u. The function $L(z, u)$ is then expanded with respect to $u = e^{it}$ at $u = 1$, *i.e.*, $t = 0$, using a Taylor expansion, its companion integral representation, and the bootstrapping bounds. The moment-like quantities,

$$L_j(z) = \left.\frac{\partial^j}{\partial t^j} L(z, e^{it})\right|_{t=0},$$

can be subjected to Mellin analysis for $j = 1, 2$ and bounded for $j \geq 3$. In this way, it is shown that

$$L(z, e^{it}) = L_1(z)t + \frac{1}{2}L_2(z)t^2 + O(zt^3),$$

uniformly. The Gaussian law under a Poisson model immediately results from the continuity theorem of characteristic functions. Under the original Bernoulli model, the Gaussian limit follows from a saddle-point analysis of

$$F(z, e^{it}) = e^z e^{L(z,e^{it})}.$$

An even more delicate analysis has been carried out by Jacquet and Szpankowski [345] by means of analytic depoissonization (Subsection VIII. 5.3, 572). It is relative to path length in digital search trees and involves the formidable nonlinear bivariate difference-differential equation

$$\frac{\partial}{\partial z} F(z, u) = F\left(\frac{z}{2}, u\right)^2.$$

See Szpankowski's book [564] for this and similar results that play an important rôle in the analysis of data compression algorithms (the Lempel–Ziv schemes).   ◁

At this stage, by making use of the material expounded in Sections IX. 5–IX. 8, we can avail ourselves of a fairly large arsenal of techniques dedicated to extracting Gaussian limit laws from BGFs. For instance, we now have the property that *all four Stirling distributions*,

$$(80) \qquad \frac{1}{n!}\begin{bmatrix} n \\ k \end{bmatrix}, \quad \frac{k!}{O_n}\begin{bmatrix} n \\ k \end{bmatrix}, \quad \frac{1}{S_n}\begin{Bmatrix} n \\ k \end{Bmatrix}, \quad \frac{k!}{R_n}\begin{Bmatrix} n \\ k \end{Bmatrix},$$

associated with permutations, alignments, set partitions, and surjections are, after standardization, asymptotically Gaussian. The method is in each case a reflection of the underlying combinatorics. Typically, for the four cases of (80), we have used, respectively: (*i*) singularity analysis perturbation (the exp–log schema for the SET ∘ CYC construction of permutations); (*ii*) meromorphic perturbation (for alignments that are of type SEQ ∘ CYC); (*iii*) saddle-point perturbation (for set partitions that are of type SET ∘ SET and whose BGF is entire); (*iv*) meromorphic perturbation again (for surjections that are of type SEQ ∘ SET).

IX. 9. Local limit laws

The occurrence of continuous limit laws has been examined so far from the angle of convergence of (cumulative) distribution functions. Combinatorially, regarding the random variable X_n that represents some parameter χ taken over a class $\mathcal{F}_n$, we then quantify the *sums*

$$\sum_{j \leq k} F_{n,j}.$$

Specifically, we have focused our attention in previous sections on the case in which these sums (once normalized by $1/F_n$) are approximated by the Gaussian "error function", i.e., the (cumulative) distribution function of a standard normal variable. Combinatorialists would often rather have a direct estimate of the *individual* counting quantities, $F_{n,k}$, which is then a true bivariate asymptotic estimate.

Assume that we have already obtained the existence of a convergence in law, $X_n \Rightarrow Y$, and the standard deviation σ_n of X_n tends to infinity while the distribution

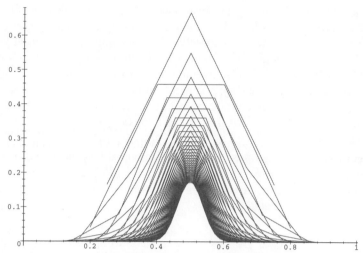

Figure IX.14. The histogram of the Eulerian distribution scaled to $(n + 1)$ on the horizontal axis, for $n = 3 \mathinner{\ldotp\ldotp} 60$. (The distribution is seen to quickly converge to a bell-shaped curve corresponding to the Gaussian density $e^{-x^2/2}/(2\pi)^{1/2}$.)

of Y admits a *density* $g(x)$. (Here, typically, $g(x)$ will be the Gaussian density.) If the $F_{n,k}$ vary smoothly enough, one may expect each of them to share about $1/\sigma_n$ of the total probability mass, and, in addition, somehow anticipate that their profile could resemble the curve $x \mapsto g(x)$. In that case, we expect an approximation of the form

$$F_{n,k} \approx \frac{1}{\sigma_n} g(x), \qquad \text{where} \quad x := \frac{k - \mu_n}{\sigma_n},$$

and μ_n is the expectation of X_n. Informally speaking, we say that a *Local Limit Law* (LLL) holds in this case.

We examine here the occurrence of local limit laws of the *Gaussian type*, which means convergence of a discrete probability distribution to the *Gaussian density function*. Figure IX.14 reveals that, at least for the Eulerian distribution (rises in permutations), such a local limit law holds, and we know, from De Moivre's original Central Limit Theorem (Note IX.1, p. 615) that a similar property holds for binomial coefficients as well. As a matter of fact, for reasons soon to be presented, virtually all the Gaussian limit laws obtained in Sections IX. 5–IX. 8 admit a local version.

Definition IX.4. *A sequence of discrete probability distributions, $p_{n,k} = \mathbb{P}\{X_n = k\}$, with mean μ_n and standard deviation σ_n is said to obey a* local limit law *of the Gaussian type if, for a sequence $\epsilon_n \to 0$,*

$$(81) \qquad \sup_{x \in \mathbb{R}} \left| \sigma_n p_{n, \lfloor \mu_n + x\sigma_n \rfloor} - \frac{1}{\sqrt{2\pi}} e^{-x^2/2} \right| \leq \epsilon_n.$$

The local limit law is said to hold with speed ϵ_n.

Note carefully, that a local limit law does not logically follows from a convergence in distribution in the usual sense, upon taking differences (the individual probabilities appear as differences at nearly identical points of values of a distribution function, hence they are "hidden" behind the error terms). Some additional regularity assumptions are needed. Here, we are naturally concerned with distilling local limit laws from BGFs $F(z, u)$. It turns out, rather nicely, that the Quasi-powers Theorem (Theorem IX.8, p. 645) can be amended by imposing constraints on the way the secondary variable affects the asymptotic approximation of $[z^n]F(z, u)$, *when u varies* globally *on the whole of the unit circle* (rather than just in a complex neighbourhood of 1). In that case, the *saddle-point method* is effective to effect the inversion with respect to the secondary variable u.

Theorem IX.14 (Quasi-powers, Local Limit Law). *Let X_n be a sequence of non-negative discrete random variables with probability generating function $p_n(u)$. Assume that the $p_n(u)$ satisfy the conditions of the Quasi-powers Theorem, in particular, the quasi-power approximation,*

$$p_n(u) = A(u) \cdot B(u)^{\beta_n} \left(1 + O\left(\frac{1}{\kappa_n} \right) \right),$$

holds uniformly in a fixed complex neighbourhood Ω of 1. Assume in addition the existence of a uniform bound,

$$(82) \qquad\qquad |p_n(u)| \leq K^{-\beta_n},$$

for some $K > 1$ and all u in the intersection of the unit circle and the complement $\mathbb{C} \setminus \Omega$. Under these conditions, the distribution of X_n satisfies a local limit law of the Gaussian type with speed of convergence $O(\beta_n^{-1/2} + \kappa_n^{-1})$.

Proof. Note first that the Quasi-powers Theorem (Theorem IX.8, p. 645) provides the mean and variance of the distribution of X_n as quantities asymptotically proportional to β_n. Furthermore, the standardized version of X_n converges to a standard Gaussian (in the sense of cumulative distribution functions).

The idea is to use Cauchy's formula and *integrate along the unit circle*. We have

$$(83) \qquad\qquad p_{n,k} \equiv [u^k]p_n(u) = \frac{1}{2i\pi} \int_{|u|=1} p_n(u) \frac{du}{u^{k+1}}.$$

We propose to appeal to the saddle-point method as a replacement for the continuity theorem of integral transforms used in the case of the central limit law (p. 645).

We first estimate $p_{n,k}$ when k is at a fixed number of standard deviations from the mean μ_n, namely, $k = \mu_n + x\sigma_n$, and accordingly restrict x to some arbitrary compact set of the real line. We can then import *verbatim* the treatment of large powers given in Section VIII. 8, p. 585. The integration circle in (83) is split into the "central range", near the real axis, where $|\arg(u)| \leq \theta_0$ with $\theta_0 = n^{-2/5}$, and the remainder of the contour. The remainder integral is exponentially small, as is verified by the arguments of the proof of Theorem VIII.8, p. 587 *and* the condition (82). The perturbative analysis conducted in Theorem IX.14 then shows the existence of a *uniform* local Gaussian approximation (in the sense of (81)), with β_n replacing n in the statement of Theorem IX.14.

We are almost done. It suffices to observe that, as x increases unboundedly, both the $p_{n,k}$ and the Gaussian density are fast decreasing functions of x, that is, the tails of the combinatorial distribution and of the limit Gaussian distribution, are both small. (For the $p_{n,k}$, this results from the Large Deviation Theorem, Theorem IX.15 below.) Thus Equation (81) actually holds when the supremum is taken over *all* real x (not just values of x restricted to compact sets). A careful revisitation of the arguments used in the proof then shows that the speed of convergence is, like in the central limit case, of the order of $\kappa_n^{-1} + \beta_n^{-1/2}$. ∎

This theorem applies in particular to the case of a movable singularity in a BGF $F(z, u)$, whenever the dominant singularity $\rho(u)$, of the function $z \mapsto F(z, u)$, as u ranges over the unit circle $|u| = 1$, uniquely attains its minimum modulus at $u = 1$. Given the positivity inherent in combinatorial GFs, we may expect this situation to occur frequently. Indeed, for a BGF $F(z, u)$ with non-negative coefficients, we already know that the property $|\rho(u)| \leq \rho(1)$ holds for $u \neq 1$ and u on the unit circle—only a strengthening to the strict inequality $|\rho(u)| < \rho(1)$ is needed. Similar comments apply to the case of variable exponents (where $\Re(\alpha(u))$ should be uniquely minimal) and, with adaptation, to the generalized quasi-powers framework of Theorem IX.13 (p. 690), which is suitable for the saddle-point method. These are the ultimate reasons why essentially all our previous central limit results can be supplemented by a local limit law.

Example IX.34. *Local limit laws for sums of discrete random variables.* The simplest application is to the binomial distribution, for which $B(u) = (1 + u)/2$. In a precise technical sense, the local limit arises from the BGF, $F(z, u) = 1/(1 - z(1 + u)/2)$, because the dominant singularity $\rho(u) = 2/(1 + u)$ exists on the whole of the unit circle, $|u| = 1$, and attains uniquely its minimum modulus at $u = 1$, so that $B(u) = \rho(1)/\rho(u)$ is uniquely maximal at $u = 1$.

More generally, Theorem IX.14 applies to any sum $S_n = T_1 + \cdots + T_n$ of independent and identically distributed *discrete* random variables whose maximal span is equal to 1 and whose PGF is analytic on the unit circle. In that case, the BGF is

$$F(z, u) = \frac{1}{1 - zB(u)},$$

the PGF of S_n is a pure power, $p_n(u) = B(u)^n$, and the fact that the minimal span of the X_j is 1 entails that $B(u)$ attains uniquely its maximum at 1 (by the Daffodil Lemma IV.1, p. 266). Such cases have been known for a long time in probability theory. See Chapter 9 of [294]. . ∎

Example IX.35. *Local limit law for the Eulerian distribution.* This example relative to Eulerian numbers shows the case of a movable singularity, subjected to a meromorphic analysis on p. 658, which we now revisit. The approximation obtained there is

$$p_n(u) = B(u)^{-n-1} + O(2^{-n}),$$

when u is close enough to 1, with

$$B(u) = \rho(u)^{-1} = \frac{u - 1}{\log u}.$$

A rendering of the function $|B(u)|$ when u ranges over the unit circle is given in Figure IX.15.

The analysis leading to (42), p. 658, also characterizes the complete set of poles $\rho_j(u)\}_{j \in \mathbb{Z}}$ of the associated BGF $F(z, u)$. From it, we can deduce, by simple complex geometry, that $\rho(u)$ is the unique dominant singularity, when $\Re(u) \geq 0$. the other ones remaining at distance at least $\pi/\sqrt{8} \doteq 1.110721$. Also, it is not hard to see that all the poles, including the dominant

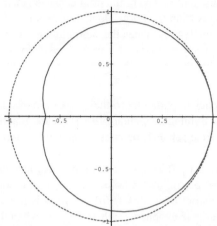

Figure IX.15. The values of the function $|B(u)|$ relative to the Eulerian distribution when $|u| = 1$, as represented by a polar plot of $|B(e^{i\theta})|$ on the ray of angle θ. (The dashed contour represents the unit circle, for comparison.) The maximum is uniquely attained at $u = 1$, where $B(1) = 1$, which entails a local limit law.

one, remain in the region $|z| > 11/10$, when $\Re(u) < 0$ and $|u| = 1$. Thus, $p_n(u)$ satisfies an estimate which is either of the quasi-powers type (when $\Re(u) \geq 0$) or of the form $O((10/11)^{-n})$ (when $\Re(u) \leq 0$). As a consequence: *a local limit law of the Gaussian type holds for the Eulerian distribution.* (This result appears in [35, p. 107].) . . . . ∎

▷ **IX.48.** *Congruence properties associated to runs.* Fix an integer $d \geq 2$. Let $P_n^{(j)}$ be the number of permutations whose number of runs is congruent to j modulo d. Then, there exists a constant $K > 1$ such that, for all j, one has: $|P_n^{(j)} - n!/d| \leq K^{-n}$. Thus, the number of runs is in a strong sense almost uniformly distributed over all residue classes modulo d. [Hint: use properties of the BGF for values of $u = \omega^d$, with ω a primitive dth root of unity.]   ◁

***Example* IX.36.** *A pot-pourri of local limit laws.* The following combinatorial distributions admit a local limit law (LLL).

The number of components in random surjections (p. 653) corresponds to the array of Stirling$_2$ numbers $k!\left\{{n \atop k}\right\}$. In that case, we have a movable singularity at $\rho(u) = \log(1 + u^{-1})$, all the other singularities remaining at distance at least 2π, and escaping to infinity as $u \to -1$. This ensures the validity of condition (82), hence an LLL (with $\beta_n = n$). Similarly for alignments (p. 654) associated to the array of Stirling$_1$ numbers $k!\left[{n \atop k}\right]$, various types of constrained compositions (p. 654), and more generally, the number of components in supercritical compositions, including compositions into prime summands.

Variable exponents also lead to an LLL under normal circumstances. Prototypically, the Stirling cycle distribution (p. 671) associated to the array $\left[{n \atop k}\right]$ satisfies

$$p_n(u) \sim \frac{e^{(u-1)\log n}}{\Gamma(u)},$$

and a suitably uniform version results from the Uniformity Lemma (p. 668), hence an LLL (this fact was already observed in [35, p. 105]). The property extends to the exp–log schema including the number of components in mappings (p. 671) and the number of irreducible factors in polynomials over finite fields (p. 672).

Cases of structures amenable to singularity perturbation with a movable singularity include leaves in Catalan and other classical varieties of trees (p. 678), patterns in binary trees (p. 680), as well as the mean level profile of increasing trees (p. 684), whose BGF is given by a differential equation.

Finally, central limit laws resulting from the saddle-point method and Theorem IX.13 (p. 690) can often be supplemented by an LLL. An important case is that of the number of blocks in set partitions, which is associated to the Stirling$_2$ array $\{{n \atop k}\}$. (The result appears in Bender's paper [35, p. 109], where it is derived from log-concavity considerations.) .... ∎

▷ **IX.49.** *Non-existence of a local limit.* Consider a binomial RV conditioned to assume only even values, so that $p_{n,2k} = 2^{1-n} \binom{n}{2k}$ and $p_{n,2k+1} = 0$. The BGF

$$F(z, u) = \frac{1}{2} \frac{1}{1 - z(1 + u)/2} + \frac{1}{2} \frac{1}{1 - z(1 - u)/2}$$

has two poles, namely $\rho_1(u) = 2/(1 + u)$ and $\rho_2(u) = 2/(1 - u)$, and it is simply not true that a single one dominates throughout the domain $|u| = 1$. Accordingly, the PGF satisfies

$$p_n(u) = 2^{-n} \left[(1 + u)^n + (1 - u)^n \right],$$

and smallness away from the positive real line cannot be guaranteed all along the unit circle (one has for instance $p_n(1) = p_n(-1)$). ◁

IX. 10. Large deviations

The term *large deviation principle*[17] is loosely defined as an exponentially small bound on the probability that a collection of random variables deviate substantially from their mean value. It thus quantifies rare events in an appropriate scale. Moment inequalities, although useful in establishing concentration of distribution (Subsection III. 2.2, p. 161), usually fall short of providing such exponentially small estimates, and the improvement over Chebyshev inequalities afforded by the methods presented here can be dramatic. For instance, for runs in permutations (the Eulerian distribution), the probability of deviating by 10% or more from the mean appears to be of the order of 10^{-6} for $n = 1\,000$ and 10^{-65} for $n = 10\,000$, with a spectacular 10^{-653} for $n = 100\,000$. (By contrast, the Chebyshev inequalities would only bound from above the last probability by a quantity about 10^{-3}.) Figure IX.16 provides a plot of the logarithms of the individual probabilities associated to the Eulerian distribution, which is characteristic of the phenomena at stake here.

Definition IX.5. *Let β_n be a sequence tending to infinity and ξ a non-zero real number. A sequence of random variables (X_n) having $\mathbb{E}(X_n) \sim \xi\beta_n$, satisfies a large deviation property relative to the interval $[x_0, x_1]$ containing ξ if a function $W(x)$ exists, such that $W(x) > 0$ for $x \neq \xi$ and, as n tends to infinity:*

$$(84) \quad \begin{cases} \dfrac{1}{\beta_n} \log \mathbb{P}(X_n \leq x\beta_n) &= -W(x) + o(1) \quad x_0 \leq x \leq \xi \quad (left\ tails) \\ \dfrac{1}{\beta_n} \log \mathbb{P}(X_n \geq x\beta_n) &= -W(x) + o(1) \quad \xi \leq x \leq x_1 \quad (right\ tails). \end{cases}$$

---

[17]Large deviation theory is introduced nicely in the book of den Hollander [153].

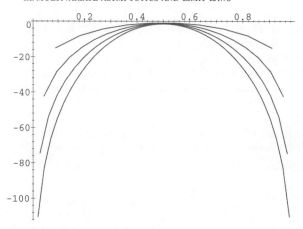

Figure IX.16. The quantities $\log p_{n,xn}$ relative to the Eulerian distribution illustrate an extremely fast decay away from the mean, which corresponds to $\xi \equiv \frac{1}{2}$. Here, the diagrams are plotted for $n = 10, 20, 30, 40$ (top to bottom). The common shape of the curves indicates a large deviation principle.

The function $W(x)$ is called the rate function *and β_n is the* scaling factor.

Figuratively, a large deviation property, in the case of left tails ($x < \xi$), expresses an exponential approximation of the rough form

$$\mathbb{P}(X_n \leq x\beta_n) \approx e^{-\beta_n W(x)},$$

for the probability of being away from the mean, and similarly for right tails. Under the conditions of the Quasi-powers Theorem, a large deviation principle invariably holds, a fact first observed by Hwang in [338].

Theorem IX.15 (Quasi-powers, large deviations). *Consider a sequence of discrete random variables (X_n) with PGF $p_n(u)$. Assume the conditions of the Quasi-powers Theorem (Theorem IX.8, p. 645); in particular, there exist functions $A(u)$, $B(u)$, which are analytic over some interval $[u_0, u_1]$ with $0 < u_0 < 1 < u_1$, such that, with $\kappa_n \to \infty$, one has*

(85)
$$p_n(u) = A(u)B(u)^{\beta_n}\left(1 + O(\kappa_n^{-1})\right),$$

uniformly. Then the X_n satisfy a large deviation property, relative to the interval $[x_0, x_1]$, where $x_0 = u_0 B'(u_0)/B(u_0)$, $x_1 = u_1 B'(u_1)/B(u_1)$; the scaling factor is β_n and the large deviation rate $W(x)$ is given by

(86)
$$W(x) = -\min_{u\in[u_0,u_1]} \log\left(\frac{B(u)}{u^x}\right).$$

Proof. We examine the case of the left tails, $\mathbb{P}(X_n \leq x\beta_n)$ with $x < \xi$ and $\xi = B'(1)$, the case of right tails being similar. It proves instructive to start with a simple inequality that suggests the physics of the problem, then refine it into an equality by a classical technique known as "shifting of the mean".

Inequalities. The basic observation is that, if $f(u) = \sum_k f_k u^k$ is a function analytic in the unit disc with non-negative coefficients at 0, then, for positive $u \le 1$, we have

$$(87) \qquad \sum_{j \le k} f_j \le \frac{f(u)}{u^k},$$

which belongs to the broad category of saddle-point bounds (see also our discussion of tail bounds on p. 643). The combination of (87), applied to $p_n(u) := \mathbb{E}(u^{X_n})$, and of assumption (85) yields

$$(88) \qquad \mathbb{P}(X_n \le x\beta_n) \le O(1) \left(\frac{B(u)}{u^x} \right)^{\beta_n},$$

which is usable *a priori* for any fixed $u \in [u_0, 1]$. In particular the value of u that minimizes $B(u)/u^x$ can be used, provided that this value of u exists, is less than 1, and also the minimum itself is less than 1.

The required conditions are granted by developments closely related to Boltzmann models and associated convexity properties, as developed in Note IV.46, p. 280, which we revisit here. Simple algebra with derivatives shows that

$$(89) \quad \frac{d}{du}\left(\frac{B(u)}{u^x}\right) = \left[\frac{uB'(u)}{B(u)} - x\right]\frac{B(u)}{u^{x+1}}, \qquad \frac{d}{du}\left(\frac{uB'(u)}{B(u)}\right) = \frac{1}{u}\mathfrak{v}_t(B(ut)),$$

where by $\mathfrak{v}_t(B(ut))$ is meant the analytic variance of the function $t \mapsto B(ut)$: u is treated as a parameter and $\mathfrak{v}(f)$ is taken in the sense of (27), p. 645. From the non-negativity of variances, we see by the second relation of (89) that the function $uB'(u)/B(u)$ is increasing. This grants us the existence of a root of the equation $uB'(u)/B(u) = x$, at which point, by the first relation of (89), the quantity $B(u)/u^x$ attains its minimum. Since $B(1) = 1$, that minimum is itself strictly less than 1, so that an inequality,

$$(90) \qquad \log \mathbb{P}(X_n \le x\beta_n) \le -\beta_n W(x) + O(1),$$

results, with $W(x)$ as stated in (86).

Equalities. The family $X_{n,\lambda}$ of random variables, with PGF

$$p_{n,\lambda}(u) := \frac{p_n(\lambda u)}{p_n(\lambda)},$$

when λ varies, is known as an *exponential family* (or as a family of exponentially shifted versions of X_n). Fix now λ to be the particular value of u at which the minimum of $B(u)/u^x$ is attained, so that $\lambda B'(\lambda)/B(\lambda) = x$. The PGFs $p_{n,\lambda}(u)$ satisfy a quasi-power approximation

$$(91) \qquad p_{n,\lambda}(u) = \frac{A(\lambda u)}{A(\lambda)}\left(\frac{B(\lambda u)}{B(\lambda)}\right)^{\beta_n}\left((1 + O(\kappa_n^{-1}))\right),$$

so that a central limit law (of Gaussian type) holds for these specific $X_{n,\lambda}$. By elementary calculus, we have $\mathbb{E}(X_{n,\lambda}) = x\beta_n + O(1)$. Thus, by the Quasi-powers Theorem applied to the centre of the Gaussian distribution, we find

(92)
$$\lim_{n\to\infty} \mathbb{P}(X_{n,\lambda} \leq x\beta_n) = \frac{1}{2}.$$

Fix now an arbitrary $\epsilon > 0$. We have a useful refinement of (92):

(93)
$$\mathbb{P}((x - \epsilon)\beta_n < X_{n,\lambda} \leq x\beta_n) = \frac{1}{2} + o(1).$$

We can then write

$$
\begin{aligned}
\mathbb{P}(X_n \leq x\beta_n) &\geq & \mathbb{P}((x - \epsilon)\beta_n < X_n \leq x\beta_n) \\
&\geq & \frac{p_n(\lambda)}{\lambda^{(x-\epsilon)\beta_n}}\mathbb{P}((x - \epsilon)\beta_n < X_{n,\lambda} \leq x\beta_n) \\
&\geq & \left(\frac{1}{2} + o(1)\right)\frac{B(\lambda)^{\beta_n}}{\lambda^{(x-\epsilon)\beta_n}}A(\lambda)\,(1 + o(1)),
\end{aligned}
$$
(94)

where the second line results from the definition of exponential families and the third from (93) and the quasi-powers assumption. Then, since the last line of (94) is valid for any $\epsilon > 0$, we get, in the limit $\epsilon \to 0$, the desired lower bound:

(95)
$$\log \mathbb{P}(X_n \leq x\beta_n) \geq -\beta_n W(x) + O(1),$$

Hence, Equation (95) combined with its converse (90), yields the statement relative to left tails.   ∎

The proof above yields an explicit algorithm to *compute* the rate function $W(x)$ from $B(u)$ and its derivatives. Indeed, the quantity $\lambda \equiv \lambda(x)$ is obtained by inversion of $uB'(u)/B(u)$,

(96)
$$\lambda(x)\frac{B'(\lambda(x))}{B(\lambda(x))} = x,$$

and the large deviation rate function is

(97)
$$W(x) = -\log B(\lambda(x)) - x\log \lambda(x).$$

▷ **IX.50.** *Extensions.* Speed of convergence estimates can be developed by making use of the Quasi-powers Theorem, with error terms. Also "local" forms of the large deviation principle (concerning $\log p_{n,k}$) can be derived under additional properties similar to those of Theorem IX.14 (p. 696) relative to local limit laws. (Hint: see [338, 339].)   ◁

Example IX.37. *Large deviations for the Eulerian distribution.* In this case, the BGF has a unique dominant singularity for u with $\epsilon < u < 1/\epsilon$, and any $\epsilon > 0$. Thus, there is a quasi-power expansion with

$$B(u) = \frac{u - 1}{\log u},$$

valid on any compact subinterval of the positive real line. Then, $\lambda(x)$ is computable as the inverse function of

$$h(u) = \frac{u}{u - 1} - \frac{1}{\log u}.$$

(The function $h(u)$ maps increasingly $\mathbb{R}_{>0}$ to the interval $(0, 1)$, so that its inverse function is always defined.) The function $W(x)$ is then computable by (96) and (97). Figure IX.17 presents a plot of $W(x)$ that explains the data of Figure IX.16, p. 700, as well as the estimates given in the introduction of this section. .... ∎

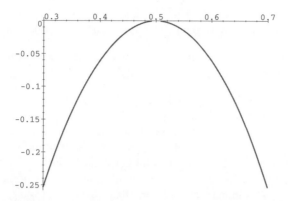

Figure IX.17. The large deviation rate function $-W(x)$ relative to the Eulerian distribution, for $x \in [0.3, 0.7]$, with scaling sequence $\beta_n = n$ and $\xi = 1/2$.

All the distributions mentioned in previous *pot-pourris* (Example IX.27, p. 683 and IX.36, p. 698) that result either from meromorphic perturbation or from singularity perturbation satisfy a large deviation principle, as a consequence of Theorem IX.15. For distributions amenable to the saddle-point method (Example IX.33, p. 693) tail probabilities also tend to be very small: their approximations are not expressed as simply as in Definition IX.5, but depend on the particulars of the asymptotic scale at play in each case. The interest of large deviation estimates in probability theory stems from their robustness with respect to changes in randomness models or under composition with non-mass-preserving transformations. In combinatorics, they have been most notably used to analyse depth and height in several types of increasing trees and search trees by Devroye and his coauthors [95, 160, 161].

IX. 11. Non-Gaussian continuous limits

Previous sections of this chapter have stressed two basic paradigms for bivariate asymptotics:

— a "minor" change in singularities, leading to discrete laws, which occurs when the nature and location of the dominant singularity remains unaffected by small changes in the values of the secondary parameter u;

— a "major" singularity perturbation mode leading to the Gaussian law, which arises from a variable exponent and/or a movable singularity.

However, it has been systematically the case, so far, that the collection of singular expansions parameterized by the auxiliary variable all belong to a sufficiently gentle analytic type (eventually leading to a quasi-power approximation) and, in particular, exhibit no sharp discontinuity when the secondary parameter traverses the special value $u = 1$. In this section we first illustrate, by means of examples, the way discontinuities in singular behaviour induce non-Gaussian laws (Subsection IX. 11.1), then examine a fairly general case of confluence of singularities, corresponding to the critical composition schema (Subsection IX. 11.2). The discontinuities observed in such situations are reminiscent of what is known as *phase-transition* phenomena in statistical physics, and we have found it suggestive to import this terminology here.

IX. 11.1. Phase-transition diagrams.

Perhaps the simplest case of discontinuity in singular behaviour is provided by the BGF,

$$F(z, u) = \frac{1}{(1-z)(1-zu)},$$

where u records the parameter equal to the number of initial occurrences of a in a random word of $\mathcal{F} = \mathrm{SEQ}(a)\,\mathrm{SEQ}(b)$. Clearly the distribution is uniform over the discrete set of values $\{0, 1, \ldots n\}$. The limit law is then continuous: it is the *uniform distribution* over the *real* interval $[0, 1]$. From the point of view of the singular structure of $z \mapsto F(z, u)$, summarized by a formula of the type $(1 - z/\rho(u))^{-\alpha(u)}$, three distinct cases arise, depending on the values of u:

— $u < 1$: simple pole at $\rho(u) = 1$, corresponding to $\alpha(u) = 1$;
— $u = 1$: double pole at $\rho(1) = 1$, corresponding to $\alpha(u) = 2$;
— $u > 1$: simple pole at $\rho(u) = 1/u$, corresponding to $\alpha(u) = 1$.

Here, both the location of the singularity $\rho(u)$ and the singular exponent $\alpha(u)$ experience a non-analytic transition at $u = 1$. This situation arises from a collapsing of two singular terms, when $u = 1$.

In order to visualize such cases, it is useful to introduce a simplified diagram representation, called a *phase-transition diagram* and defined as follows. Write $Z = \rho(u) - z$ and summarize the singular expansion by its dominant singular term $Z^{\alpha(u)}$. Then, the diagram corresponding to $F(z, u)$ is

$u = 1 - \epsilon$	$u = 1$	$u = 1 + \epsilon$
$\rho(u) = 1$	$\rho(1) = 1$	$\rho(u) = 1/u$
Z^{-1}	Z^{-2}	Z^{-1}

$\qquad Z := \rho(u) - z.$

A complete classification of such discontinuities is lacking (see, however, Marianne Durand's thesis [181] for several interesting schemas), and is probably beyond reach given the vast diversity of situations to be encountered in a combinatorialist's practice. We provide here two illustrations: the first example is relative to the classical theory of coin-tossing games (the arcsine distribution); the second one is relative to area under excursions and path length in trees (the Airy distribution of the area type). Both are revisited here under the perspective of phase transition diagrams, which provide a useful way to approach and categorize non-Gaussian limits.

Example **IX.38.** *Arcsine law for unbiased random walks.* This problem is studied in detail by Feller [205, p. 94] who notes, regarding gains in coin-tossing games: "Contrary to intuition, the maximum accumulated gain is much more likely to occur towards the very beginning or the very end of a coin-tossing game than somewhere in the middle." See Figure IX.18.

We let χ be the time of first occurrence of the maximum in a random game (that is, a walk with ± 1 steps) and write X_n for the RV representing χ restricted to the set $\mathcal{W}_n$ of walks of duration n. The BGF $W(z, u)$, where u marks χ, results from the standard decomposition of positive walks. Essentially, there is a sequence of steps ascending to the (non-negative) maximum accompanied by "arches" (the left factor) followed by a mirror excursion back to

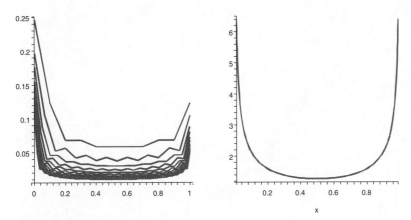

Figure IX.18. Histograms of the distribution of the location of the maximum of a random walk for $n = 10 .. 60$ (left) and the density of the arcsine law (right).

the maximum, followed by a sequence of descending steps with their companion arches. This construction translates directly into an equation satisfied by the BGF $W(z, u)$ of the location of the first maximum

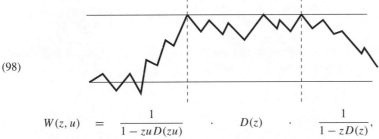

(98)

$$W(z, u) = \frac{1}{1 - zu D(zu)} \cdot D(z) \cdot \frac{1}{1 - z D(z)},$$

which involves the GF of a gambler's ruin sequences (equivalently Dyck excursions, Example IX.8, p. 635), namely,

$$(99) \qquad D(z) = \frac{1 - \sqrt{1 - 4z^2}}{2z^2}.$$

In such a simple case, explicit expressions are available from (98), when we expand first with respect to u, then to z. We obtain in this way the ultra-classical result that the probability that X_n equals either $k = 2r$ or $k = 2r + 1$ is $\frac{1}{2} u_{2r} u_{2v-2r}$, where $u_{2v} := 2^{-2v} \binom{2v}{v}$. The usual approximation of central binomial coefficients, $u_{2v} \sim (\pi v)^{-1/2}$, followed by a summation then leads to the following statement.

Proposition IX.21 (Arcsine law). *For any $x \in (0, 1)$, the position X_n of the first maximum in a random walk of even length n satisfies a limit* arcsine *law:*

$$\lim_{n \to \infty} \mathbb{P}_n(X_n < xn) = \frac{2}{\pi} \arcsin \sqrt{x}.$$

It is instructive to compare this to the way singularities evolve as u crosses the value 1. The dominant positive singularity is at $\rho(u) = 1/2$ if $u < 1$, while $\rho(u) = 1/(2u)$, if $u > 1$.

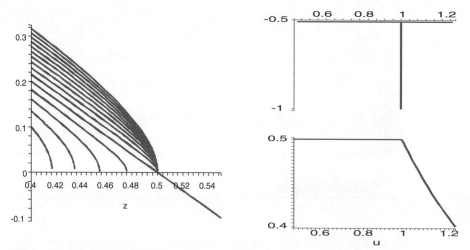

Figure IX.19. A plot of $1/W(z, u)$ for $z \in [0.4, 0.55]$ when u is assigned values between $1/2$ and $5/4$ (left); The exponent function $\alpha(u)$ (top right) and the singular value $\rho(u)$ (bottom right), for $u \in [0.5, 0.55]$.

Local expansions show that, with $c_<(u)$, $c_>(u)$ two computable functions, there holds:

$$W(z, u) \sim c_<(u)\frac{1}{\sqrt{1 - 2z}}, \qquad W(z, u) \sim c_>(u)\frac{1}{\sqrt{1 - 2zu}}.$$

Naturally, at $u = 1$, all sequences are counted and $W(z, 1) = 1/(1 - 2z)$. Thus, the corresponding phase-transition diagram is (see Figure IX.19):

$u = 1 - \epsilon$	$u = 1$	$u = 1 + \epsilon$
$\rho(u) = 1/2$	$\rho(1) = 1/2$	$\rho(u) = 1/(2u)$
$Z^{-1/2}$	Z^{-1}	$Z^{-1/2}$

The point to be made here is that the arcsine law could be expected when a similar phase-transition diagram occurs. There is indeed universality in this singular view of the arcsine law, which extends to walks with zero drift (Chapter VII). This analytic kind of universality is a parallel to the universality of Brownian motion, which is otherwise familiar to probabilists. ∎

▷ **IX.51.** *Number of maxima and other stories.* The construction underlying (98) also serves to analyse; (i) the number of times the maximum is attained. (ii) the difference between the maximum and the final altitude of the walk; (iii) the duration of the period following the last occurrence of the maximum.   ◁

***Example* IX.39.** *Path length in trees.* A final example is the distribution of path length in trees, whose non-Gaussian limit law has been originally characterized by Louchard and Takács [416, 417, 567, 569]. The distribution is recognized *not* to be asymptotically Gaussian, as it is verified from a computation of the first few moments. In the case of general Catalan trees, the analysis is equivalent to that of area under Dyck paths (Examples V.9, p. 330, and VII.26, p. 533) and is closely related to our discussion of coin fountains and parallelogram polyomino models, earlier in this chapter (p. 662). It reduces to that of the functional equation

$$F(z, u) = \frac{1}{1 - zF(zu, u)},$$

which determines $F(z, u)$ as a formal continued fraction, and setting $F(z, u) = A(z, u)/B(z, u)$, we found (p. 331)

$$B(z, u) = 1 + \sum_{n=1}^{\infty} (-1)^n \frac{u^{n(n-1)} z^n}{(1 - u)(1 - u^2) \cdots (1 - u^n)},$$

with a very similar expression for $A(z, u)$. Because of the quadratic exponent involved in the powers of u, the function $z \mapsto F(z, u)$ has radius of convergence 0 when $u > 1$, and is thus non-analytic. By contrast, when $u < 1$, the function $z \mapsto B(z, u)$ is an entire function, so that $z \mapsto F(z, u)$ is meromorphic. Hence the singularity diagram:

$u = 1 - \epsilon$	$u = 1$	$u = 1 + \epsilon$
$\rho(u) > \frac{1}{4}$	$\rho(1) = \frac{1}{4}$	$\rho(u) = 0$
Z^{-1}	$Z^{1/2}$	—

The limit law is the *Airy distribution of the area type* [244, 352, 416, 417, 567, 569], which we have encountered in Chapter VII, p. 533. By an analytical *tour de force*, Prellberg [496] has developed a method based on contour integral representations and coalescing saddle-points (Chapter VIII, p. 603) that permits us to make precise the phase transition diagram above and obtain uniform asymptotic expansions in terms of the Airy function. Since similar problems occur in relation to connectivity of random graphs under the Erdős–Rényi model [254], and conjecturally in self-avoiding walks (p. 363), future years might see more applications of Prellberg's methods. . . . . ■

IX. 11.2. Semi-large powers, critical compositions, and stable laws. We conclude this section by a discussion of critical compositions that typically involve confluences of singularities and lead to a general class of continuous distributions closely related to *stable laws* of probability theory. We start with an example where everything is explicit, that of zero contacts in random bridges, then state a general theorem on "semi-large" powers of functions of singularity analysis type, and finally return to combinatorial applications, specifically trees and maps.

Example IX.40. Zero-contacts in bridges. Consider once more fluctuations in coin tossings, and specifically bridges, corresponding to a conditioning of the game by the fact that the final gain is 0 (negative capitals are allowed). These are sequences of arbitrary positive or negative "arches", and the number of arches in a bridge is exactly equal to the number of intermediate steps at which the capital is 0. From the arch decomposition, it is found that the ordinary BGF of bridges with z marking length and u marking zero-contacts is

$$B(z, u) = \frac{1}{1 - 2uz^2 D(z)}$$

with $D(z)$ as in (99), p. 705. Analysing this function is conveniently done by introducing

$$F(z, u) \equiv B\left(\frac{1}{2}\sqrt{z}, u\right) = \frac{1}{1 - u(1 - \sqrt{1 - z})}.$$

The phase-transition diagram is then easily found to be:

$u = 1 - \epsilon$	$u = 1$	$u = 1 + \epsilon$
$\rho(u) = 1$	$\rho(1) = 1$	$\rho(u) = 1 - (1 - u^{-1})^2$
$Z^{1/2}$	$Z^{-1/2}$	Z^{-1}

Thus, there are discontinuities, both in the location of the singularity and the exponent, but of a different type from that which gives rise to the arcsine law of random walks.

The problem of the limit law is here easily solved since explicit expressions are provided by the Lagrange Inversion Theorem. One finds:

(100)
$$
\begin{aligned}
[u^k][z^n] F(z, u) &= [z^n]\left(1 - \sqrt{1 - z}\right)^k \\
&= \frac{k}{n}[w^{n-k}](2 - w)^{-n} = 2^{k-2n}\frac{k}{n}\binom{2n - k - 1}{n - 1}.
\end{aligned}
$$

A random variable with density and distribution function given by

(101)
$$
r(x) = \frac{x}{2}e^{-x^2/4}, \qquad R(x) = 1 - e^{-x^2/4},
$$

is called a *Rayleigh law*. Then Stirling's formula easily provides the following proposition.

Proposition IX.22. *The number X_n of zero-contacts of a random bridge of size $2n$ satisfies, as $n \to \infty$ a local limit law of the Rayleigh type:*

$$
\lim_{n\to\infty} \mathbb{P}(X_n = x\sqrt{n}) = \frac{x}{2\sqrt{n}}e^{-x^2/4}.
$$

The explicit character of (100) makes the analysis transparently simple. .... ∎

$\triangleright$ **IX.52.** *The number of cyclic points in mappings.* The number of cyclic points in mappings has exponential BGF $(1 - uT(z))^{-1}$, with T the Cayley tree function. The singularity diagram is of the same form as in Example IX.40. Explicit forms are derived from Lagrange inversion: the limit law is again Rayleigh. This property extends to the number of cyclic points in a simple variety of mappings (e.g., mappings defined by a finite constraint on degrees, as in Example VII.10, p. 464): see [18, 175, 176]. $\triangleleft$

Both Example IX.40 and Note IX.52 above exemplify the situation of an analytic composition scheme of the form $(1 - uh(z))^{-1}$ which is critical, since in each case h assumes value 1 at its singularity. Both can be treated elementarily since they involve powers that are amenable to Lagrange inversion, eventually resulting in a Rayleigh law. As we now explain, there is a family of functions that appear to play a universal rôle in problems sharing similar singular types. What follows is largely borrowed from an article by Banderier *et al.* [28].

We first introduce a function S that otherwise naturally surfaces in the study of *stable*[18] distributions in probability theory. For any parameter $\lambda \in (0, 2)$, define the entire function

---

[18]In probability theory, stable laws are defined as the possible limit laws of sums of independent identically distributed random variables. The function S is a trivial variant of the density of the stable law of index λ; see Feller's book [206, p. 581–583]. Valuable informations regarding stable laws may be found in the books by Breiman [93, Sec. 9.8], Durrett [182, Sec. 2.7], and Zolotarev [629].

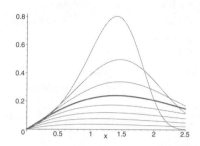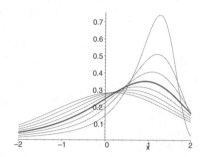

Figure IX.20. The S-functions for $\lambda = 0.1 \ldots 0.8$ (left; from bottom to top) and for $\lambda = 1.2 \ldots 1.9$ (right; from top to bottom); the thicker curves represent the Rayleigh law (left, $\lambda = 1/2$) and the Airy map law (right, $\lambda = 3/2$).

$$(102) \quad S(x, \lambda) := \begin{cases} \dfrac{1}{\pi} \displaystyle\sum_{k \geq 1} (-1)^{k-1} x^k \dfrac{\Gamma(1 + \lambda k)}{\Gamma(1 + k)} \sin(\pi k \lambda) & (0 < \lambda < 1) \\[2em] \dfrac{1}{\pi x} \displaystyle\sum_{k \geq 1} (-1)^{k-1} x^k \dfrac{\Gamma(1 + k/\lambda)}{\Gamma(1 + k)} \sin(\pi k/\lambda) & (1 < \lambda < 2) \end{cases}$$

The function $S(x; 1/2)$ is a variant of the Rayleigh density (101). The function $S(x; 3/2)$ constitutes the density of the "Airy map distribution" found in random maps as well as in other coalescence phenomena, as discussed below; see (109).

Theorem IX.16 (Semi-large powers). *The coefficient of z^n in a power $H(z)^k$ of a Δ-continuable function $H(z)$ with singular exponent λ admits the following asymptotic estimates.*

(i) For $0 < \lambda < 1$, that is, $H(z) = \sigma - h_\lambda (1 - z/\rho)^\lambda + O(1 - z/\rho)$, and when $k = xn^\lambda$, with x in any compact subinterval of $(0, +\infty)$, there holds

$$(103) \qquad [z^n] H^k(z) \sim \sigma^k \rho^{-n} \frac{1}{n} S\left(\frac{x h_\lambda}{\sigma}, \lambda \right).$$

(ii) For $1 < \lambda < 2$, that is, $H(z) = \sigma - h_1(1 - z/\rho) + h_\lambda(1 - z/\rho)^\lambda + O((1 - z/\rho)^2)$, when $k = \frac{\sigma}{h_1} n + xn^{1/\lambda}$, with x in any compact subinterval of $(-\infty, +\infty)$, there holds

$$(104) \qquad [z^n] H^k(z) \sim \sigma^k \rho^{-n} \frac{1}{n^{1/\lambda}} (h_1/h_\lambda)^{1/\lambda} S\left(\frac{x h_1^{1+1/\lambda}}{\sigma h_\lambda^{1/\lambda}}, \lambda \right).$$

(iii) For $\lambda > 2$, a Gaussian approximation holds. In particular, for $2 < \lambda < 3$, that is, $H(z) = \sigma - h_1(1 - z/\rho) + h_2(1 - z/\rho)^2 - h_\lambda(1 - z/\rho)^\lambda + O((1 - z/\rho)^3)$, when $k = \frac{\sigma}{h_1} n + x\sqrt{n}$, with x in any compact subinterval of $(-\infty, +\infty)$, there holds

$$(105) \quad [z^n] H^k(z) \sim \sigma^k \rho^{-n} \frac{1}{\sqrt{n}} \frac{\sigma/h_1}{a\sqrt{2\pi}} e^{-x^2/2a^2} \qquad \text{with } a = 2(\frac{h_2}{h_1} - \frac{h_1}{2\sigma})\sigma^2/h_1^2.$$

The term "semi-large" refers to the fact that the exponents k in case (i) are of the form $O(n^\theta)$ for some $\theta < 1$ chosen in accordance with the region where an "interesting" renormalization takes place and dependent on each particular singular exponent. When the interesting region reaches the $O(n)$ range in case (iii), the analysis of large powers, as detailed in Chapter VIII (p. 591), takes over and Gaussian forms result.

Proof. The proofs are somewhat similar to the basic ones in singularity analysis, but they require a suitable adjustment of the geometry of the Hankel contour and of the corresponding dimensioning.

Case (i). A classical Hankel contour, with the change of variable $z = \rho(1 - t/n)$, yields the approximation

$$[z^n]H^k(z) \sim -\frac{\sigma^k \rho^{-n}}{2i\pi n} \int e^{t - \frac{h_\lambda x}{\sigma} t^\lambda} dt$$

The integral is then simply estimated by expanding $\exp(-\frac{h_\lambda x}{\sigma} t^\lambda)$ and integrating termwise

$$(106) \qquad [z^n]H^k(z) \sim -\frac{\sigma^k \rho^{-n}}{n} \sum_{k \geq 1} \frac{(-x)^k}{k!} \left(\frac{h_\lambda}{\sigma}\right)^k \frac{1}{\Gamma(-\lambda k)},$$

which is equivalent to Equation (103), by virtue of the complement formula for the Gamma function.

Case (ii). When $1 < \lambda < 2$, the contour of integration in the z-plane is chosen to be a positively oriented loop, made of two rays of angle $\pi/(2\lambda)$ and $-\pi/(2\lambda)$ that intersect on the real axis at a distance $1/n^{1/\lambda}$ left of the singularity. The coefficient integral of H^k is rescaled by setting $z = \rho(1 - t/n^{1/\lambda})$, and one has

$$[z^n]H^k(z) \sim -\frac{\sigma^k \rho^{-n}}{2i\pi n^{1/\lambda}} \int e^{\frac{h_\lambda}{h_1} t^\lambda} e^{-\frac{xh_1}{\sigma} t} dt.$$

There, the contour of integration in the t-plane comprises two rays of angle π/λ and $-\pi/\lambda$, intersecting at -1. Setting $u = t^\lambda h_\lambda / h_1$, the contour transforms into a classical Hankel contour, starting from $-\infty$ over the real axis, winding about the origin, and returning to $-\infty$. So, with $\alpha = 1/\lambda$, one has

$$[z^n]H^k(z) \sim -\frac{\sigma^k \rho^{-n}}{2i\pi n^\alpha} \alpha \left(\frac{h_1}{h_\lambda}\right)^\alpha \int e^u e^{-\frac{xh_1^{\alpha+1}}{\sigma h_\lambda^\alpha} u^\alpha} u^{\alpha-1} du.$$

Expanding the exponential, integrating termwise, and appealing to the complement formula for the Gamma function finally reduces this last form to (104).

Case (iii). This case is only included here for comparison purposes, but, as recalled before the proof, it is essentially implied by the developments of Chapter VIII based on the saddle-point method. When $2 < \lambda < 3$, the angle ϕ of the contour of integration in the z–plane is chosen to be $\pi/2$, and the scaling is $\sqrt{n}$: under the change of variable $z = \rho(1 - t/\sqrt{n})$, the contour is transformed into two rays of angle $\pi/2$ and $-\pi/2$ (*i.e.*, a vertical line), intersecting at -1, and

$$[z^n]H^k(z) \sim -\frac{\sigma^k \rho^{-n}}{2i\pi \sqrt{n}} \int e^{pt^2 - \frac{h_1 x}{\sigma} t} dt,$$

with $p = \frac{h_2}{h_1} - \frac{h_1}{2\sigma}$. Complementing the square, and letting $u = t - \frac{h_1 x}{2p\sigma}$, we get

$$[z^n]H^k(z) \sim -\frac{\sigma^k \rho^{-n}}{2i\pi\sqrt{n}} e^{-\frac{h_1^2}{4p\sigma^2}x^2} \int e^{pu^2}\,du\,,$$

which gives Equation (105). By similar means, such a Gaussian approximation can be shown to hold for any non-integral singular exponent $\lambda > 2$. ∎

▷ **IX.53.** *Zipf distributions.* Zipf's law, named after the Harvard linguistic professor George Kingsley Zipf (1902–1950), is the observation that, in a language like English, the frequency with which a word occurs is roughly inversely proportional to its rank—the kth most frequent word has frequency proportional to $1/k$. The *generalized Zipf distribution* of parameter $\alpha > 1$ is the distribution of a random variable Z such that

$$\mathbb{P}(Z = k) = \frac{1}{\zeta(\alpha)} \frac{1}{k^\alpha}.$$

It has infinite mean for $\alpha \le 2$ and infinite variance for $\alpha \le 3$. It was proved in Chapter VI (p. 408) that polylogarithms are amenable to singularity analysis. Consequently, the sum of a large number of independent Zipf variables satisfies a local limit law of the stable type with index $\alpha - 1$ (for $\alpha \ne 2$). ◁

Example **IX.41.** *Mean level profiles in simple varieties of trees.* Consider the RV equal to the depth of a random node in a random tree taken from a simple variety $\mathcal{Y}$ that satisfies the smooth inverse-function schema (Definition VII.3, p. 453). The problem of quantifying the corresponding distribution is equivalent to that of determining the *mean* level profile, that is the sequence of numbers $M_{n,k}$ representing the mean number of nodes at distance k from the root. (Indeed, the probability that a random node lies at level k is $M_{n,k}/n$.) The first few levels have been characterized in Example VII.7 (p. 458) and the analysis of Chapter VII can now be completed thanks to Theorem IX.16. (The problem was solved by Meir and Moon [435] in an important article that launched the analytic study of simple varieties of trees. Meir and Moon base their analysis on a Lagrangean change of variable and on the saddle-point method, along the lines of our remarks in Chapter VIII, p. 590.) As usual, we let $\phi(w)$ be the generator of the simple variety $\mathcal{Y}$, with $Y(z)$ satisfying $Y = z\phi(Y)$, and we designate by τ the positive root of the characteristic equation:

$$\tau\phi'(\tau) - \phi(\tau) = 0.$$

It is known from Theorem VII.3 (p. 468) that the GF $Y(z)$ has a square root singularity at $\rho = \tau/\phi(\tau)$. For convenience, we also assume aperiodicity of ϕ. Meir and Moon's major result (Theorem 4.3 of [435]) is as follows

Proposition IX.23 (Mean level profiles). *The mean profile of a large tree in a simple variety obeys a Rayleigh law in the asymptotic limit: for $k/\sqrt{n}$ in any bounded interval of $\mathbb{R}_{\ge 0}$, the mean number of nodes at altitude k satisfies asymptotically*

$$M_{n,k} \sim Ake^{-Ak^2/(2n)},$$

where $A = \tau\phi''(\tau)$.

The proof goes as follows. For each k, define $Y_k(z, u)$ to be the BGF with u marking the number of nodes at depth k. Then, the root decomposition of trees translates into the recurrence:

$$Y_k(z, u) = z\phi(Y_{k-1}(z, u)), \qquad Y_0(z, u) = zu\phi(Y(z)) = uY(z).$$

By construction, we have

$$M_{n,k} = \frac{1}{Y_n}[z^n]\left(\frac{\partial}{\partial u}Y_k(z, u)\right)_{u=1}.$$

On the other hand, the fundamental recurrence yields

$$\left(\frac{\partial}{\partial u} Y_k(z, u)\right)_{u=1} = \left(z\phi'(Y(z))\right)^k Y(z).$$

Now, $\phi'(Y)$ has, like Y, a square-root singularity. The semi-large powers theorem applies with $\lambda = \frac{1}{2}$, and the result follows. .... ∎

▷ **IX.54.** *The width of trees.* The expectation of the width W of a tree in a simple variety satisfies

$$C_1 \sqrt{n} \le \mathbb{E}_{\mathcal{Y}_n}(W) \le C_2 \sqrt{n \log n},$$

for some $C_1, C_2 > 0$. (This is due to Odlyzko and Wilf [463], a possible approach consisting in suitably bounding the level profile of random trees. Better bounds are known, now that $W_n/\sqrt{n}$ has been recognized to be related to Brownian excursion. In particular, the expected width is $\sim c\sqrt{n}$; see Example V.17, p. 359 and the references there.)

◁

Critical compositions. Theorem IX.16 provides useful information on compositions of the form

$$F(z, u) = G(uH(z)),$$

provided $G(z)$ and $H(z)$ are of singularity analysis class. As we know, combinatorially, this represents a substitution between structures, $\mathcal{F} = \mathcal{G} \circ \mathcal{H}$, and the coefficient $[z^n u^k] F(z, u)$ counts the number of $\mathcal{M}$–structures of size n whose $\mathcal{G}$–component, also called *core* in what follows, has size k. Then the probability distribution of core-size X_n in $\mathcal{F}$–structures of size n is given by

$$\mathbb{P}(X_n = k) = \frac{[z^k] G(z)}{[z^n] G(H(z))} [z^n] H(z)^k.$$

The case where the schema is critical, in the sense that $H(r_H) = r_G$ with r_H, r_G the radii of convergence of H, G, follows as a direct consequence of Theorem IX.16. What comes out is the following informally stated general principle (details would closely mimic the statement of Theorem IX.16 and are omitted: see [28]).

Proposition IX.24 (Critical compositions). *In a composition schema $F(z, u) = G(uH(z))$ where H and G have singular exponents λ, λ' with $\lambda' \le \lambda$:*

(i) *for $0 < \lambda < 1$, the normalized core-size X_n/n^λ is spread over $(0, +\infty)$ and it satisfies a local limit law whose density involves a stable law of index λ; in particular, $\lambda = \frac{1}{2}$ corresponds to a Rayleigh law.*

(ii) *for $1 < \lambda < 2$, the distribution of X_n is bimodal and the "large region" $X_n = cn + xn^{1/\lambda}$ involves a stable law of index λ;*

(iii) *for $2 < \lambda$, the standardized version of X_n admits a local limit law that is of Gaussian type.*

Similar phenomena occur when $\lambda' > \lambda$, but with a greater preponderance of the "small" region. Many instances have already appeared scattered in the literature. especially in connection with rooted trees. For instance, this proposition explains well the occurrence of the Rayleigh law ($\lambda = 1/2$) as the distribution of cyclic points in random mappings and of zero-contacts in random bridges. The case $\lambda = 3/2$ appears in forests of unrooted trees (see the discussion in Chapter VIII, p. 603, for an alternative approach based on coalescing saddle-points) and it is ubiquitous in planar maps, as attested by the article of Banderier *et al.* on which this subsection is largely based [28]. We detail one of the cases in the following example, which explains the meaning of the term "large region" in Proposition IX.24.

Example IX.42. *Biconnected cores of planar maps.* The OGF of rooted planar maps, with size determined by the number of *edges*, is, by Subsection VII.8.2 (p. 513),

$$(107) \qquad M(z) = -\frac{1}{54z^2}\left(1 - 18z - (1 - 12z)^{3/2}\right),$$

with a characteristic $3/2$ exponent. Define a separating vertex or *articulation point* in a map to be a vertex whose removal disconnects the graph. Let $\mathcal{C}$ denote the class of non-separable maps, that is, maps without an articulation point (also known as biconnected maps). Starting from the root edge, any map decomposes into a non-separable map, called the "core" on which are grafted arbitrary maps, as illustrated by the following diagram:

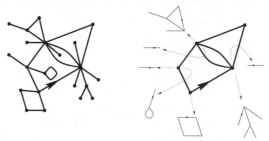

There results the equation:

$$(108) \qquad M(z) = C(H(z)), \qquad H(z) = z(1 + M(z))^2.$$

Since we know M, hence H, this last relation gives by inversion the OGF of non-separable maps as an algebraic function of degree 3 specified implicitly by the equation

$$C^3 + 2C^2 + (1 - 18z)C + 27z^2 - 2z = 0,$$

with expansion at the origin (*EIS* **A000139**):

$$C(z) = 2z + z^2 + 2z^3 + 6z^4 + 22z^5 + 91z^6 + \cdots, \qquad C_{k+1} = 2\frac{(3k)!}{(k+1)!(2k+1)!}.$$

(The closed form results from a Lagrangean parameterization.) Now the singularity of C is also of the $Z^{3/2}$ type as seen by inversion of (108) or from the Newton diagram attached to the cubic equation. We find in particular

$$C(z) = \frac{1}{3} - \frac{4}{9}(1 - 27z/4) + \frac{8\sqrt{3}}{81}(1 - 27z/4)^{3/2} + O((1 - 27z/4)^2),$$

which is reflected by the asymptotic estimate,

$$C_k \sim \frac{2}{27}\frac{\sqrt{3}}{\pi}\left(\frac{17}{4}\right)^k k^{-5/2}.$$

The parameter considered here is the distribution of the size X_n of the core (containing the root) in a random map of size n. The composition relation is $\mathcal{M} = \mathcal{C} \circ \mathcal{H}$, where $\mathcal{H} = \mathcal{Z}(1 + \mathcal{M})^2$. The BGF is thus $M(z, u) = C(uH(z))$ where the composition $C \circ H$ is of the singular type $Z^{3/2} \circ Z^{3/2}$. What is peculiar here is the "bimodal" character of the distribution of core-size (see Figure IX.21 borrowed from [28]), which we now detail.

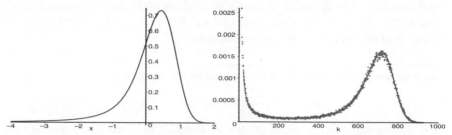

Figure IX.21. Left: The standard "Airy map distribution". Right: Observed frequencies of core-sizes $k \in [20; 1000]$ in 50 000 random maps of size 2 000, showing the bimodal character of the distribution.

First straight singularity analysis shows that, for *fixed k*,

$$\mathbb{P}(X_n = k) = C_k \frac{[z^n]H(z)^k}{M_n} \underset{n \to \infty}{\sim} k C_k h_0^{k-1},$$

where $h_0 = 4/27$ is the value of $H(z)$ at its singularity. In other words, there is local convergence of the probabilities to a fixed *discrete* law. The estimate above can be proved to remain uniform as long as k tends to infinity sufficiently slowly. We shall call this the "small range" of k values. Now, summing the probabilities associated to this small range gives the value $C(h_0) = 1/3$. Thus, *one-third of the probability mass of core-size arises from the small range, where a discrete limit law is observed.*

The other part of the distribution constitutes the "large range" to which Theorem IX.16 applies. It contains asymptotically $2/3$ of the probability mass of the distribution of X_n. In that case, the limit law is related to a stable distribution with density $S(x; 3/2)$ and is also known as the "Airy map" distribution: one finds for $k = \frac{1}{3}n + xn^{2/3}$, the local limit approximation:

$$(109) \quad \mathbb{P}(X_n = k) \sim \frac{1}{3n^{2/3}} \mathcal{A}\left(\frac{3}{4}2^{2/3}x\right), \qquad \mathcal{A}(x) := 2e^{-2x^2/3}\left(x \, \text{Ai}(x^2) - \text{Ai}'(x^2)\right).$$

There $\text{Ai}(x)$ is the Airy function (defined in the footnote on p. 534) and $\mathcal{A}(x)$ specifies the Airy map distribution displayed in Figure IX.21.

The bimodal character of the distribution of core-sizes can now be better understood [28]. A random map decomposes into biconnected components and the largest biconnected component has, with high probability, a size that is $O(n)$. There are also a large number ($O(n)$) of "dangling" biconnected components. In a rooted map, the root is in a sense placed "at random". Then, with a fixed probability, it either lies in the large component (in which case, the distribution of that large component is observed, this is the continuous part of the distribution given by the Airy map law), or else one of the small components is picked up by the root (this is the discrete part of the distribution). .... ∎

▷ **IX.55.** *Critical cycles.* The theory adapts to logarithmic factors. For instance the critical composition $F(z, u) = -\log(1 - ug(z))$ leads to developments similar to those of the critical sequence. In this way, it becomes possible for instance to analyse the number of cyclic points in a random connected mapping. ◁

▷ **IX.56.** *The base of supertrees.* Supertrees defined in Chapter VI (p. 412) are trees grafted on trees. Consider the bicoloured variant $\mathcal{K} = \mathcal{G}(2\mathcal{Z}\mathcal{G})$, with $\mathcal{G}$ the class of general Catalan trees. Then, the law of the external $\mathcal{G}$–component is related to a stable law. ◁

IX. 12. Multivariate limit laws

Combinatorics can take advantage of the enumeration of objects with respect to a whole collection of parameters. The symbolic methods of Part A are well suited and we have seen in Chapter III ways to solve problems like: how many permutations are there of size n with n_1 singleton cycles and n_2 cycles of length 2? In combinatorial terms we are seeking information about a multivariate (rather than plainly bivariate) sequence, say F_{n,k_1,k_2}. In probabilistic terms, we aim at characterizing the *joint distribution*, say $(X_n^{(1)}, X_n^{(2)})$, of a family of random variables. Methods developed in this chapter adapt well to multivariate situations. Typically, there exist natural extensions of continuity theorems, both for PGFs and for integral transforms and the most abstract aspects of the foregoing discussion regarding central and local limit laws as well as tail estimates and large deviations can be recycled.

Consider for instance the joint distribution of the numbers χ_1, χ_2 of singletons and doubletons in random permutations. Then, the parameter $\chi = (\chi_1, \chi_2)$ has a trivariate EGF

$$F(z, u_1, u_2) = \frac{\exp((u_1 - 1)z + (u_2 - 1)z^2/2)}{1 - z}.$$

Thus, the bivariate PGF satisfies, by meromorphic analysis,

$$p_n(u_1, u_2) = [z^n]F(z, u_1, u_2) \sim e^{(u_1 - 1)} e^{(u_2 - 1)/2},$$

uniformly when the pair (u_1, u_2) ranges over a compact set of $\mathbb{C} \times \mathbb{C}$. As a result, the joint distribution of (χ_1, χ_2) is a product of a Poisson(1) and a Poisson(1/2) distribution; in particular χ_1 and χ_2 are asymptotically independent.

Consider next the joint distribution of $\chi = (\chi_1, \chi_2)$, where χ_j is the number of summands equal to j in a random integer composition. Each parameter individually obeys a limit Gaussian law, since the sequence construction is supercritical. The trivariate GF is

$$F(z, u_1, u_2) = \frac{1}{1 - z(1 - z)^{-1} - (u_1 - 1)z - (u_2 - 1)z^2}.$$

By meromorphic analysis, a higher dimensional quasi-power approximation may be derived:

$$[z^n]F(z, u_1, u_2) \sim c(u_1, u_2)\rho(u_1, u_2)^{-n},$$

for some third-degree algebraic function $\rho(u_1, u_2)$. In such cases, multivariate versions of the continuity theorem for integral transforms can be applied. (See the book by Gnedenko and Kolmogorov [294] and especially the treatment of Bender and Richmond in [44].) As a result, the joint distribution is, in the asymptotic limit, a bivariate Gaussian distribution with a covariance matrix that is computable from $\rho(u_1, u_2)$. Such generalizations are typical and involve essentially no radically new concept, just natural technical adaptations.

A highly interesting approach to multivariate problems is that of *functional limit theorems*. The goal is now to characterize the joint distribution of an unbounded collection of parameters. The limit process is then a *stochastic process*, essentially an object that lives in some *infinite-dimensional space*. For instance, the joint distribution of all altitudes in random walks is accounted for by Brownian motion. The joint distribution of all cycle lengths in random permutations is described explicitly by Cauchy's formula (p. 188) and DeLaurentis and Pittel [149] have shown a convergence to the standard Brownian motion process, after a suitable renormalization. A rather spectacular application of this circle of ideas was provided in 1977 by Logan, Shepp, Vershik and Kerov [411, 596]. These authors established that the shape of the pair of Young tableaux associated to a random permutation conforms, in the asymptotic limit and with high probability, to a deterministic trajectory defined as the solution to a variational problem. In particular, the width of a Young tableau associated to a permutation gives the length of the longest increasing sequence of the permutation. By specializing their results, the authors were then able to show that the expected length in a random permutation of size n is asymptotic to $2\sqrt{n}$, a long-standing conjecture at the time (see also our remarks on p. 597 for subsequent developments). There is currently a flurry of activity on these questions, with methods ranging from purely probabilistic to purely analytic.

Among extensions of the standard approach presented in this book to analytic combinatorics, we single out a few, which seem especially exciting. Lalley [397] has extended the framework of the important Drmota–Lalley–Woods Theorem (p. 489) to certain infinite systems of equations, by appealing to Banach space theory—this has applications in the theory of random walks on groups. Vallée and coauthors (see Note IX.32, p. 664, and the survey [584]) have developed a broad theory based on transfer operators from dynamical systems theory, where generating operators replace generating functions and operate on certain infinite dimensional functional spaces—there are surprising applications both in information theory and in analytic number theory (e.g., the analysis of Euclidean algorithms). McKay [432] has shown how to extend the one-dimensional saddle-point theory presented in Chapter VIII in a highly non-trivial way in order to treat certain counting problems where a problem of size n is represented by a $d(n)$-dimensional integral, with $d(n)$ tending to infinity with n—this is especially important since a great many hard combinatorial problems can be represented in this manner, including for instance the celebrated random SAT-problem [77, 486].

We hope that the fairly complete treatment of standard aspects of the theory offered in this book will help our reader to master and enrich a field, which is extremely vast, blooming, and pregnant with fascinating problems at the crossroads of discrete and continuous mathematics.

IX. 13. Perspective

The study of parameters of combinatorial structures ideally culminates in an understanding of the distribution of the parameter's values, typically under the assumption that each instance of a given size in a combinatorial class appears with equal likelihood.

First, as we have already seen in Chapter III, we can extend the basic combinatorial constructions of Chapters I and II to include bivariate generating functions

(BGFs) whose second variable carries information about the parameter. Our combinatorial constructions then provide a systematic way to develop succinct BGFs for a broad range of combinatorial classes and parameters, which are of interest in combinatorics, computer science, and other applied sciences.

Next, the various methods considered in Chapters IV–VIII (Part B) of this book can be extended to develop asymptotic results for BGFs by studying slight perturbations of the singularities, controlled by the second variable. The uniform precision of the asymptotic results that we develop in Part B is a critical component in our ability to do this, by contrast with other classical methods for coefficient asymptotics (Darboux's method and Tauberian theorems) which are, to a large extent, non-constructive.

These asymptotic results take the form of *limit laws*: the distribution governing the behaviour of parameters converge to a fixed discrete distribution, or appropriately scaled, to a continuous distribution. Whereas BGFs are purely formal objects, to determine whether the distribution is discrete or continuous requires analysis of them as functions of complex variables. In a preponderance of cases, the limit laws say that parameter values approach a single distribution, the well-known Gaussian (normal) distribution. The well-known central limit theorem is but one example (not the explanation) of this phenomenon, whose breadth is truly remarkable. For example, we have encountered numerous examples where the occurrence of a given fixed pattern in a large random object is almost certain, with the number of occurrences governed by Gaussian fluctuations. This property holds true for strings, uniform tree models, and increasing trees. The associated BGFs are rational functions, algebraic functions, and solutions to nonlinear differential equations, respectively, but the approach of extending the methods of Part B to study local perturbations of singularities is effective in each case—the proofs eventually reduce to establishing an extremely simple property, a singularity that smoothly moves.

Such studies are an appropriate conclusion to this book, because they illustrate the power of analytic combinatorics. We are able to use formal methods to develop succinct formal objects that encapsulate the combinatorial structure (BGFs), then, treating those BGFs as objects of analysis (functions of one, then two complex variables) we are able to obtain wide encompassing asymptotic information about the original combinatorial structure. Such an approach has serendipitous consequences. Combinatorial problems can then be organized into broad *schemas*, covering infinitely many combinatorial types and governed by simple asymptotic laws—the discovery of such schemas and of the associated *universality* properties constitutes the very essence of analytic combinatorics.

Bibliographic notes. This chapter is primarily inspired by the studies of Bender and Richmond [35, 44, 46], Canfield [101], Flajolet, Soria, and Drmota [171, 172, 175, 176, 258, 260, 547] as well as Hwang [337, 338, 339, 340]. Bender's seminal study [35] initiated the study of bivariate analytic schemes that lead to Gaussian laws and the paper [35] may rightly be considered to be at the origin of the field. Canfield [101], building upon earlier studies showed the approach to extend to saddle-point schemas.

Tangible progress was next made possible by the development of the singularity analysis method [248]. Earlier research was mostly restricted to methods based on subtraction of singularities, as in [35], which is in particular effective for meromorphic cases. The extension to algebraic–logarithmic singularities was, however, difficult given that the classical method of Darboux does not provide for uniform error terms. In contrast, singularity analysis *does* apply to classes of analytic functions, since it allows for uniformity of estimates. The papers by Flajolet and Soria [258, 260] were the first to make clear the impact of singularity analysis on bivariate asymptotics. Gao and Richmond [277] were then able to extend the theory to cases where both a singularity and its singular exponent are allowed to vary.

From there, Soria developed the framework of schemas considerably in her doctorate [547]. Hwang extracted the very important concept of "quasi-powers" in his thesis [337] together with a wealth of properties such as full asymptotic expansions, speed of convergence, and large deviations. Drmota established general existence conditions leading to Gaussian laws in the case of implicit, especially algebraic, functions [171, 172]. The "singularity perturbation" framework for solutions of linear differential equations first appears under that name in [243]. Finally, the books by Sachkov, see [525] and especially [526] (based on the 1978 edition [524]) offer a modern perspective on bivariate asymptotics applied to classical combinatorial structures.

וְיֹתֵר מֵהֵמָּה בְּנִי הִזָּהֵר עֲשׂוֹת סְפָרִים
הַרְבֵּה אֵין קֵץ וְלַהַג הַרְבֵּה יְגִעַת בָּשָׂר:

*("But beyond this, my son, be warned: the writing of many books
is endless; and excessive devotion to books is wearying to the body."))*

— Tanakh (The Bible), Qohelet (Ecclesiastes) 12:12.

Part D

APPENDICES

APPENDICES

APPENDIX A

Auxiliary Elementary Notions

We combine in the three appendices definitions and theorems related to key mathematical concepts not covered directly in the text. Generally, the entries in the appendices are independent, intended for reference while addressing the main text. Our own *Introduction to the Analysis of Algorithms* [538] is a gentle introduction to many of the concepts underlying analytic combinatorics at a level accessible to any college student and is reasonable preparation for undergraduates or anyone undertaking to read this book for self-study.

This appendix contains entries that are arranged in alphabetical order, regarding the following topics:

> *Arithmetical functions; Asymptotic notations; Combinatorial probability; Cycle construction; Formal power series; Lagrange inversion; Regular languages; Stirling numbers; Tree concepts.*

The corresponding notions and results are used throughout the book, and especially in Part A relative to *Symbolic Methods*. Accessible introductions to the subject of this appendix are the books by Graham–Knuth–Patashnik [307], and Wilf [608], regarding combinatorial enumeration, and De Bruijn's vivid booklet [142], regarding asymptotic analysis. Reference works in combinatorial analysis are the books by Comtet [129], Goulden–Jackson [303], and Stanley [552, 554].

A.1. Arithmetical functions

A general reference for this section is Apostol's book [16]. First, the *Euler totient function* $\varphi(k)$ intervenes in the unlabelled cycle construction (pp. 27, 84, 165, as well as 729 below). It is defined as the number of integers in $[1 \mathinner{..} k]$ that are relatively prime to k. Thus, one has $\varphi(p) = p - 1$ if $p \in \{2, 3, 5, \ldots\}$ is a prime. More generally when the prime number decomposition of k is $k = p_1^{\alpha_1} \cdots p_r^{\alpha_r}$, then

$$\varphi(k) = p_1^{\alpha_1 - 1}(p_1 - 1) \cdots p_r^{\alpha_r - 1}(p_r - 1).$$

A number is squarefree if it is not divisible by the square of a prime. The *Möbius function* $\mu(n)$ is defined to be 0 if n is not squarefree and otherwise is $(-1)^r$ if $n = p_1 \cdots p_r$ is a product of r distinct primes.

Many elementary properties of arithmetical functions are easily established by means of a *Dirichlet generating functions* (DGF). Let $(a_n)_{n \geq 1}$ be a sequence; its DGF is formally defined by

$$\alpha(s) = \sum_{n=1}^{\infty} \frac{a_n}{n^s}.$$

In particular, the DGF of the sequence $a_n = 1$ is the Riemann zeta function, $\zeta(s) = \sum_{n \geq 1} n^{-s}$. The fact that every number uniquely decomposes into primes is reflected by Euler's formula,

$$(1) \qquad \zeta(s) = \prod_{p \in \mathcal{P}} \left(1 - \frac{1}{p^s}\right)^{-1},$$

where p ranges over the set $\mathcal{P}$ of all primes. (As observed by Euler, the fact that $\zeta(1) = \infty$ in conjunction with (1) provides a simple analytic proof that there are infinitely many primes! See Note IV.1, p. 228)

Equation (1) implies that the DGF of the Möbius function satisfies

$$(2) \qquad M(s) := \sum_{n \geq 1} \frac{\mu(n)}{n^s} = \prod_{p \in \mathcal{P}} \left(1 - \frac{1}{p^s}\right) = \frac{1}{\zeta(s)}.$$

(Verification: expand the infinite product and collect the coefficient of $1/n^s$.)

Finally, if $(a_n), (b_n), (c_n)$ have DGF $\alpha(s), \beta(s), \gamma(s)$, then one has the equivalence

$$\alpha(s) = \beta(s)\gamma(s) \quad \Longleftrightarrow \quad a_n = \sum_{d \mid n} b_d c_{n/d}.$$

In particular, taking $c_n = 1$ ($\gamma(s) = \zeta(s)$) and solving for $\beta(s)$ shows (using (2)) the implication

$$(3) \qquad a_n = \sum_{d \mid n} b_d \quad \Longleftrightarrow \quad b_n = \sum_{d \mid n} \mu(d) a_{n/d},$$

which is known as *Möbius inversion*. This relation is used in the enumeration of irreducible polynomials (Section I.6.3, p. 88).

A.2. Asymptotic notations

Let $\mathbb{S}$ be a set and $s_0 \in \mathbb{S}$ a particular element of $\mathbb{S}$. We assume a notion of neighbourhood to exist on $\mathbb{S}$. Examples are $\mathbb{S} = \mathbb{Z}_{>0} \cup \{+\infty\}$ with $s_0 = +\infty$, $\mathbb{S} = \mathbb{R}$ with s_0 any point in $\mathbb{R}$; $\mathbb{S} = \mathbb{C}$ or a subset of $\mathbb{C}$ with $s_0 = 0$, and so on. Two functions ϕ and g from $\mathbb{S} \setminus \{s_0\}$ to $\mathbb{R}$ or $\mathbb{C}$ are given.

— *O–notation*: write

$$\phi(s) \underset{s \to s_0}{=} O(g(s))$$

if the ratio $\phi(s)/g(s)$ stays bounded as $s \to s_0$ in $\mathbb{S}$. In other words, there exists a neighbourhood $\mathcal{V}$ of s_0 and a constant $C > 0$ such that

$$|\phi(s)| \leq C \, |g(s)|, \qquad s \in \mathcal{V}, \quad s \neq s_0.$$

One also says that "ϕ *is of order at most* g", or "ϕ *is big–Oh of* g" (as s tends to s_0).

— $\sim$–*notation*: write

$$\phi(s) \underset{s \to s_0}{\sim} g(s)$$

if the ratio $\phi(s)/g(s)$ tends to 1 as $s \to s_0$ in $\mathbb{S}$. One also says that "ϕ *and* g *are asymptotically equivalent*" (as s tends to s_0).

— *o–notation*: write

$$\phi(s) \underset{s \to s_0}{=} o(g(s))$$

if the ratio $\phi(s)/g(s)$ tends to 0 as $s \to s_0$ in $\mathbb{S}$. In other words, for any (arbitrarily small) $\varepsilon > 0$, there exists a neighbourhood $\mathcal{V}_\varepsilon$ of s_0 (depending on ε), such that

$$|\phi(s)| \leq \varepsilon \, |g(s)|, \qquad s \in \mathcal{V}_\varepsilon, \quad s \neq s_0.$$

One also says that "*ϕ is of order smaller than g, or ϕ is little–oh of g*" (as s tends to s_0).

These notations are due to Bachmann and Landau towards the end of the nineteenth century. See Knuth's note for a historical discussion [381, Ch. 4].

Related notations, of which, however, we only make a scant use, are

— *Ω-notation*: write

$$\phi(s) \underset{s \to s_0}{=} \Omega(g(s))$$

if the ratio $\phi(s)/g(s)$ stays bounded from below in modulus by a non-zero quantity, as $s \to s_0$ in $\mathbb{S}$. One then says that ϕ is *of order at least g*.

— *Θ-notation*: if $\phi(s) = O(g(s))$ and $\phi(s) = \Omega(g(s))$, write

$$\phi(s) \underset{s \to s_0}{=} \Theta(g(s)).$$

One then says that ϕ is *of order exactly g*.

For instance, one has as $n \to +\infty$ in $\mathbb{Z}_{>0}$:

$$\sin n = o(\log n); \quad \log n = O(\sqrt{n}); \quad \log n = o(\sqrt{n});$$
$$\binom{n}{2} = \Omega(n\sqrt{n}); \quad \pi n + \sqrt{n} = \Theta(n).$$

As $x \to 1$ in $\mathbb{R}_{\leq 1}$, one has

$$\sqrt{1 - x} = o(1); \quad e^x = O(\sin x); \quad \log x = \Theta(x - 1).$$

We take as granted in this book the elementary asymptotic calculus with such notations (see, e.g., [538, Ch. 4] for a smooth introduction close to the needs of analytic combinatorics and de Bruijn's classic [143] for a beautiful presentation.). We shall retain here in particular the fact that Taylor expansions (Note A.6, p. 726) imply asymptotic expansions; for instance, the convergent expansions, all valid for $|u| < 1$,

$$\log(1 + u) = \sum_{k=1}^{\infty} \frac{(-1)^{k-1}}{k} u^k, \quad \exp(u) = \sum_{k \geq 0} \frac{1}{k!} u^k, \quad (1 - u)^{-\alpha} = \sum_{k \geq 0} \binom{k + \alpha - 1}{k} u^k,$$

imply (as $u \to 0$)

$$\log(1 + u) = u + O(u^2), \quad \exp(u) = 1 + u + \frac{u^2}{2} + O(u^3), \quad (1 - u)^{1/2} = 1 - \frac{u}{2} + O(u^2),$$

and so forth. Consequently, as $n \to +\infty$, one has:

$$\log\left(1 + \frac{1}{n}\right) = \frac{1}{n} + O\left(\frac{1}{n^2}\right), \quad \left(1 - \frac{1}{\log n}\right)^{1/2} = 1 - \frac{1}{2 \log n} + o\left(\frac{1}{\log n}\right).$$

Two important asymptotic expansions are Stirling's formula for factorials and the harmonic number approximation, valid for $n \geq 1$,

(4)
$$n! = n^n e^{-n} \sqrt{2\pi n} \left(1 + \epsilon_n\right), \qquad 0 < \epsilon_n < \tfrac{1}{12n}$$
$$H_n = \log n + \gamma + \frac{1}{2n} - \frac{1}{12n^2} + \eta_n \quad \eta_n = O\left(n^{-4}\right), \quad \gamma \doteq 0.57721,$$

that are commonly established as consequences of the Euler–Maclaurin summation formula that relates sums to integrals (see Note A.7, p. 726, references [143, 538], as well as Appendix B.7: *Mellin transform*, p. 762).

▷ **A.1.** *Simplification rules for the asymptotic calculus.* Some of them are

$$
\begin{array}{lll}
O(\lambda f) & \longrightarrow & O(f) \qquad\qquad (\lambda \neq 0) \\
O(f) \pm O(g) & \longrightarrow & O(|f| + |g|) \\
& \longrightarrow & O(f) \qquad\qquad \text{if } g = O(f) \\
O(f \cdot g) & \longrightarrow & O(f)O(g).
\end{array}
$$

Similar rules apply for $o(\cdot)$.   ◁

Asymptotic scales. An important notion due to Poincaré is that of an *asymptotic scale*. A sequence of functions $\omega_0, \omega_1, \ldots$ is said to constitute an asymptotic scale if all functions ω_j exist in a common neighbourhood of $s_0 \in \mathbb{S}$ and if they satisfy there, for all $j \geq 0$:

$$\omega_{j+1}(s) = o(\omega_j(s)), \qquad \text{i.e.,} \qquad \lim_{s \to s_0} \frac{\omega_{j+1}(s)}{\omega_j(s)} = 0.$$

Examples at 0 are the scales: $u_j(x) = x^j$; $v_{2j}(x) = x^j \log x$ and $v_{2j+1}(x) = x^j$; $w_j(x) = x^{j/2}$. Examples at infinity are $t_j(n) = n^{-j}$, and so on. Given a scale $\Phi = (\omega_j(s))_{j \geq 0}$, a function f is said to admit an *asymptotic expansion* in the scale Φ if there exists a family of complex coefficients (λ_j) (the family is then necessarily unique) such that, for each integer m:

(5)
$$f(s) = \sum_{j=0}^{m} \lambda_j \omega_j(s) + O(\omega_{m+1}(s)) \qquad (s \to s_0).$$

In this case, one writes

(6)
$$f(s) \sim \sum_{j=0}^{\infty} \lambda_j \omega_j(s), \qquad (s \to s_0)$$

with an extension of the symbol '$\sim$'. (Some authors prefer the notation '$\approx$', but in this book, we reserve it to mean informally "approximately equal" or "of the rough form".)

The scale may be finite and in most cases, we do not need to specify it as it is clear from context. For instance, one can write

$$H_n \sim \log n + \gamma + \frac{1}{12n}, \qquad \tan x \sim x + \frac{1}{3}x^3 + \frac{2}{15}x^5.$$

In the first case, it is understood that $n \to \infty$ and the scale is $\log n, 1, n^{-1}, n^{-2}, \ldots$. In the second case, $x \to 0$ and the scale is $x, x^3, x^5, \ldots$. Note carefully that in the case of a complete expansion (6), convergence of the infinite sum is *not* in any way implied: the relation is to be interpreted literally, in the sense of (5); namely, as a collection of more and more precise descriptions of f when s becomes closer and closer to s_0. (As a matter of fact, almost all the asymptotic expansions of number sequences developed in this book, starting with Stirling's formula, are divergent.)

▷ **A.2.** *Harmonics of harmonics.* The harmonic numbers are readily extended to non-integral index by (cf also the ψ function p. 746)

$$H_x := \sum_{k=1}^{\infty} \left(\frac{1}{k} - \frac{1}{k+x} \right).$$

For instance, $H_{1/2} = 2 - 2\log 2$. This extension is related to the Gamma function [604], and it can be proved that the asymptotic estimate (4), with x replacing n, remains valid as $x \to +\infty$. A typical asymptotic calculation shows that

$$H_{H_n} = \log\log n + \gamma + \frac{\gamma + \frac{1}{2}}{\log n} + O\left(\frac{1}{\log^2 n} \right).$$

What is the shape of an asymptotic expansion of $H_{H_{H_n}}$? ◁

▷ **A.3.** *Stackings of dominos.* A stock of dominos of length 2cm is given. It is well known that one can stack up dominos in a harmonic mode:

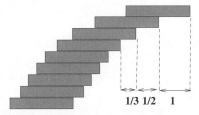

1/3 1/2 1

Estimate within 1% the minimal number of dominos needed to achieve a horizontal span of 1m (=100cm). (Hint: about $1.50926\, 10^{43}$ dominos!) Set up a scheme to evaluate this integer exactly, and do it! ◁

▷ **A.4.** *High precision fraud.* Why is it that, to forty decimal places, one finds

$$4 \sum_{k=1}^{500\,000} \frac{(-1)^{k-1}}{2k-1} \doteq 3.14159\underline{0}653589793\underline{2}404626433832\underline{6}9502884197$$

$$\pi \doteq 3.141592653589793238462643383279502884197,$$

with only four "wrong" digits in the first sum? (Hint: consider the simpler problem

$$\frac{1}{9801} \doteq 0.00\,01\,02\,03\,04\,05\,06\,07\,08\,09\,10\,11\,12\,13\,14\,15\,16\,17\,18\,19\,20\,21\,22\,23\,24\,25\cdots.)$$

Many fascinating facts of this kind are found in works by Jon and Peter Borwein [79, 80]. ◁

Uniform asymptotic expansions. The notions previously introduced allow for *uniform* versions in the case of families dependent on a secondary parameter [143, pp. 7–9]. Let $\{f_u(s)\}_{u \in U}$ be a family of functions indexed by U. An asymptotic equivalence like

$$f_u(s) = O(g(s)) \qquad (s \to s_0),$$

is said to be *uniform with respect to u* if there exists an absolute constant K (independent of $u \in U$) and a fixed neighbourhood $\mathcal{V}$ of s_0 such that

$$\forall u \in U, \ \forall s \in \mathcal{V}: \qquad |f_u(s)| \le K|g(s)|.$$

This definition in turn gives rise to the notion of a uniform asymptotic expansion: it suffices that, for each m, the O error term in (5) be uniform. Such notions are central for the determination of limit laws in Chapter IX, where a uniform expansion of a class of generating functions near a singularity is usually required.

▷ **A.5.** *Examples of uniform asymptotics.* One has *uniformly*, for $u \in \mathbb{R}$ and $u \in [0, 1]$ respectively:

$$\sin(ux) \underset{x \to \infty}{=} O(1), \qquad \left(1 + \frac{1}{n}\right)^u \underset{n \to \infty}{=} 1 + \frac{u}{n} + O\left(\frac{1}{n^2}\right).$$

However, the second expansion no longer holds uniformly with respect to u when $u \in \mathbb{R}$ (take $u = \pm n$), although it holds *pointwise* (non-uniformly) for any fixed $u \in \mathbb{R}$. What about the assertion $\left(1 + \frac{1}{n}\right)^u \underset{n \to \infty}{=} 1 + \frac{u}{n} + O\left(\frac{u^2}{n^2}\right)$ for $u \in \mathbb{R}$?   ◁

▷ **A.6.** *Taylor expansions.* Let (ϕ_k) be a sequence of polynomials such that $\phi_0 = 1$ and $\phi'_{k+1} = \phi_k$, for all $k \geq 0$. A repeated use of integration by parts shows that, for a function f assumed to be sufficiently smooth, one has ($[h]_A^B$ denotes the variation $h(B) - h(A)$)

$$
(7) \qquad \int_0^1 f(t)\phi_0(t)\, dt = [f\phi_1]_0^1 - [f'\phi_2]_0^1 + \cdots + (-1)^{m-1}[f^{(m-1)}\phi_m]_0^1 \\
+ (-1)^m \int_0^1 f^{(m)}(t)\, \phi_m(t)\, dt.
$$

Choosing $\phi_k(t) = (t - 1)^k/k!$ yields the *basic Taylor expansion with remainder*:

$$(8) \qquad \int_0^1 f(t)\, dt = \sum_{k=0}^{m-1} \frac{f^{(k)}(0)}{(k+1)!} + \frac{1}{m!} \int_0^1 f^{(m)}(t)\, (1-t)^m\, dt.$$

If $|f^{(m)}(t)|$ is less than $m!A^{-m}$ for some $A > 1$, then a convergent representation follows. Setting $f(t) = xg'(xt)$ then yields the classical *Taylor expansion with remainder*

$$(9) \qquad g(x) = \sum_{k=0}^{m} g^{(k)}(0)\frac{x^k}{k!} + \frac{1}{m!} \int_0^x g^{(m+1)}(t)\, (x - t)^m\, dt,$$

and a convergent infinite series can be deduced under suitable growth assumptions on the derivatives of g. (Complex analytic methods of Chapter IV and Appendix B develop a powerful theory by which one can avoid explicitly determining and bounding derivatives.)   ◁

▷ **A.7.** *Euler–Maclaurin summation.* Choose now $\phi_k(t) = [z^n]ze^{tz}/(e^z - 1)$. The ϕ_k are, up to normalization, *Bernoulli polynomials* and their coefficients involve the Bernoulli numbers (p. 268): $\phi_0(t) = 1$, $\phi_1(t) = t - \frac{1}{2}$, $\phi_2(t) = t^2/2 - t/2 + 1/12$, and so on. Equation (7) then yields the *basic Euler–Maclaurin expansion with remainder*:

$$\int_0^1 f(t)\, dt = \frac{f(0) + f(1)}{2} - \sum_{k=1}^{M} \frac{B_{2k}}{(2k)!}[f^{(2k-1)}]_0^1 + \int_0^1 f^{(2M)}(t)\phi_{2M}(t)\, dt.$$

From here, a formula results by summation (with $\{x\} := x - \lfloor x \rfloor$), which serves to compare sums and integrals:

$$\int_0^n f(t)\, dt = \frac{f(0) + f(n)}{2} + \sum_{j=1}^{n-1} f(j) - \sum_{k=1}^{M} \frac{B_{2k}}{(2k)!}[f^{(2k-1)}]_0^n + \int_0^n f^{(2M)}(t)\phi_{2M}(\{t\})\, dt.$$

The asymptotic expansions of (4), p. 724, can finally be developed: use $f(t) = \log(t + 1)$ and $f(t) = 1/(t + 1)$. (Hint: see [142, §3.6], [465, pp. 281–289], or [538, §4.5].) The fine characterisation of the "Euler–Maclaurin constants" (Euler's constant γ for H_n, Stirling's constant $\sqrt{2\pi}$ for Stirling's approximation) is in general non-obvious: see pp. 238, pp. 410, and pp. 766 for complex-analytic alternatives. ◁

A.3. Combinatorial probability

This entry gathers elementary concepts from probability theory specialized to the discrete case and used in Chapter III. A more elaborate discussion of probability theory forms the subject of Appendix C.

Given a finite set $\mathcal{S}$, the *uniform probability measure* assigns to any $\sigma \in \mathcal{S}$ the probability mass

$$\mathbb{P}(\sigma) = \frac{1}{\text{card}(\mathcal{S})}.$$

The probability of any set, also known as *event*, $\mathcal{E} \subseteq \mathcal{S}$, is then measured by

$$\mathbb{P}\{\mathcal{E}\} := \frac{\text{card}(\mathcal{E})}{\text{card}(\mathcal{S})} = \sum_{\sigma \in \mathcal{E}} \mathbb{P}(\sigma)$$

("the number of favorable cases over the total number of cases").

Given a combinatorial class $\mathcal{A}$, we make extensive use of this notion with the choice of $\mathcal{S} = \mathcal{A}_n$. This defines a probability model (indexed by n), in which elements of size n in $\mathcal{A}$ are taken with equal likelihood. For this uniform probabilistic model, we write

$$\mathbb{P}_n \qquad \text{and} \qquad \mathbb{P}_{\mathcal{A}_n},$$

whenever the size and the type of combinatorial structure considered need to be emphasized.

Next consider a parameter χ, which is a function from $\mathcal{S}$ to $\mathbb{Z}_{\geq 0}$. We regard such a parameter as a *random variable*, determined by its probability distribution,

$$\mathbb{P}(\chi = k) = \frac{\text{card}(\{\sigma \mid \chi(\sigma) = k\})}{\text{card}(\mathcal{S})}.$$

The notions above extend gracefully to non-uniform probability models that are determined by a family of non-negative numbers $(p_\sigma)_{\sigma \in \mathcal{S}}$ which add up to 1:

$$\mathbb{P}(\sigma) = p_\sigma, \quad \mathbb{P}(\mathcal{E}) := \sum_{\sigma \in \mathcal{E}} p_\sigma, \quad \mathbb{P}(\chi = k) = \sum_{\chi(\sigma) = k} p_\sigma.$$

Moments. Important information on a distribution is provided by its *moments*. We state here the definitions for an arbitrary discrete random variable supported by $\mathbb{Z}$ and determined by its probability distribution, $\mathbb{P}(X = k) = p_k$ where the $(p_k)_{k \in \mathbb{Z}}$ are non-negative numbers that add up to 1. The *expectation* of $f(X)$ is defined as the linear functional

$$\mathbb{E}(f(X)) = \sum_k \mathbb{P}\{X = k\} \cdot f(k).$$

In particular, the (power) *moment* of order r is defined as the expectation:

$$\mathbb{E}(X^r) = \sum_k \mathbb{P}\{X = k\} \cdot k^r.$$

Of special importance are the first two moments of the random variable X. The expectation (also mean or average) $\mathbb{E}(X)$ is

$$\mathbb{E}(X) = \sum_k \mathbb{P}\{X = k\} \cdot k.$$

The second moment $\mathbb{E}(X^2)$ gives rise to the *variance*,

$$\mathbb{V}(X) = \mathbb{E}\left((X - \mathbb{E}(X))^2\right) = \mathbb{E}(X^2) - \mathbb{E}(X)^2,$$

and, in turn, to the *standard deviation*

$$\sigma(X) = \sqrt{\mathbb{V}(X)}.$$

The mean deserves its name as first observed by Galileo Galilei (1564–1642): if a large number of draws are effected and values of X are observed, then the arithmetical mean of the observed values will normally be close to the expectation $\mathbb{E}(X)$. The standard deviation measures in a mean quadratic sense the dispersion of values around the expectation $\mathbb{E}(X)$.

▷ **A.8.** *The weak law of large numbers.*   Let (X_k) be a sequence of mutually independent random variables with a common distribution. If the expectation $\mu = \mathbb{E}(X_k)$ exists, then for every $\epsilon > 0$:

$$\lim_{n \to \infty} \mathbb{P}\left(\left|\frac{1}{n}(X_1 + \cdots + X_n) - \mu\right| > \epsilon\right) = 0.$$

(See [205, Ch X] for a proof.) Note that the property does not require finite variance.   ◁

Probability generating function. The *probability generating function* (PGF) of a discrete random variable X, with values in $\mathbb{Z}_{\geq 0}$, is by definition:

$$p(u) := \sum_k \mathbb{P}(X = k)u^k,$$

and an alternative expression is $p(u) = \mathbb{E}(u^X)$. Moments can be recovered from the PGF by differentiation at 1, for instance:

$$\mathbb{E}(X) = \left.\frac{d}{du}p(u)\right|_{u=1}, \qquad \mathbb{E}(X(X-1)) = \left.\frac{d^2}{du^2}p(u)\right|_{u=1}.$$

More generally, the quantity,

$$\mathbb{E}(X(X-1)\cdots(X-k+1)) = \left.\frac{d^k}{du^k}p(u)\right|_{u=1},$$

is known as the kth factorial moment.

▷ **A.9.** *Relations between factorial and power moments.* Let X be a discrete random variable with PGF $p(u)$; denote by $\mu_r = \mathbb{E}(X^r)$ its rth moment and by ϕ_r its factorial moment. One has

$$\mu_r = \left.\partial_t^r p(e^t)\right|_{t=0}, \qquad \phi_r = \left.\partial_u^r p(u)\right|_{u=1}.$$

Consequently, with $\left\{{n \atop k}\right\}$ and $\left[{n \atop k}\right]$ the Stirling numbers of both kinds (Appendix A.8: *Stirling numbers*, p. 735),

$$\phi_r = \sum_j (-1)^{r-j}\left[{r \atop j}\right]\mu_j; \qquad \mu_r = \sum_j \left\{{r \atop j}\right\}\phi_j.$$

(Hint: for $\phi_r \to \mu_r$, expand the Stirling polynomial defined in (17), p. 736; in the converse direction, write $p(e^t) = p(1 + (e^t - 1))$.) ◁

Markov–Chebyshev inequalities. These are fundamental inequalities that apply equally well to discrete and to continuous random variables (see Appendix C for the latter).

Theorem A.1 (Markov–Chebyshev inequalities). *Let X be a* non-negative *random variable and Y an* arbitrary *real random variable. One has for an arbitrary $t > 0$:*

$$\mathbb{P}\{X \geq t\mathbb{E}(X)\} \qquad \leq \quad \frac{1}{t} \qquad \textit{(Markov inequality)}$$

$$\mathbb{P}\{|Y - \mathbb{E}(Y)| \geq t\sigma(Y)\} \quad \leq \quad \frac{1}{t^2} \qquad \textit{(Chebyshev inequality)}.$$

Proof. Without loss of generality, one may assume that X has been scaled in such a way that $\mathbb{E}(X) = 1$. Define the function $f(x)$ whose value is 1 if $x \geq t$, and 0 otherwise. Then

$$\mathbb{P}\{X \geq t\} = \mathbb{E}(f(X)).$$

Since $f(x) \leq x/t$, the expectation on the right is less than $1/t$. Markov's inequality follows. Chebyshev's inequality then results from Markov's inequality applied to $X = |Y - \mathbb{E}(Y)|^2$. ∎

Theorem A.1 informs us that the probability of being much larger than the mean must decay (Markov) and that an upper bound on the decay is measured in units given by the standard deviation (Chebyshev).

Moment inequalities are discussed for instance in Billingsley's reference treatise [68, p. 74]. They are of great importance in discrete mathematics where they have been put to use in order to show the *existence* of surprising configurations. This field was pioneered by Erdős and is often known as the "probabilistic method" [in combinatorics]; see the book by Alon and Spencer [13] for many examples. Moment inequalities can also be used to estimate the probabilities of complex events by reducing the problems to moment estimates for occurrences of simpler configurations—this is one of the bases of the "first and second moment methods", again pioneered by Erdős, which are central in the theory of random graphs [76, 355]. Finally, moment inequalities serve to design, analyse, and optimize randomized algorithms, a theme excellently covered in the book by Motwani and Raghavan [451].

A.4. Cycle construction

The unlabelled cycle construction is introduced in Chapter I and is classically obtained within the framework of Pólya theory (Note I.58, p. 85 and [129, 488, 491]). The derivation given here is based on an elementary use of symbolic methods that follows [259]. It relies on bivariate GFs developed in Chapter III, with z marking size and u marking the number of components. Consider a class $\mathcal{A}$ and the sequence class $\mathcal{S} = \text{SEQ}_{\geq 1}(\mathcal{A})$. A sequence $\sigma \in \mathcal{S}$ is primitive (or aperiodic) if it is not the repetition of another sequence (e.g., $\alpha\beta\beta\alpha\alpha$ is primitive, but $\alpha\beta\alpha\beta = (\alpha\beta)^2$ is not). The class

$\mathcal{PS}$ of primitive sequences is determined implicitly,

$$S(z, u) \equiv \frac{u A(z)}{1 - u A(z)} = \sum_{k \geq 1} P S(z^k, u^k),$$

which expresses that every sequence possesses a "root" that is primitive. Möbius inversion (Equation (3), p. 722) then gives

$$P S(z, u) = \sum_{k \geq 1} \mu(k) S(z^k, u^k) = \sum_{k \geq 1} \mu(k) \frac{u^k A(z^k)}{1 - u^k A(z^k)}.$$

A cycle is primitive if all of its linear representations are primitive. There is an exact one-to-ℓ correspondence between primitive ℓ–cycles and primitive ℓ–sequences. Thus, the BGF $PC(z, u)$ of primitive cycles is obtained by effecting the transformation $u^\ell \mapsto \frac{1}{\ell} u^\ell$ on $PS(z, u)$, which means

$$PC(z, u) = \int_0^u P S(z, v) \frac{dv}{v},$$

giving after term-wise integration,

$$PC(z, u) = \sum_{k \geq 1} \frac{\mu(k)}{k} \log \frac{1}{1 - u^k A(z^k)}.$$

Finally, cycles can be composed from arbitrary repetitions of primitive cycles (each cycle has a primitive "root"), which yields for $\mathcal{C} = \text{CYC}(A)$:

$$C(z, u) = \sum_{k \geq 1} P C(z^k, u^k).$$

The arithmetical identity $\sum_{d \mid k} \mu(d)/d = \varphi(k)/k$ gives eventually

(10) $$C(z, u) = \sum_{k \geq 1} \frac{\varphi(k)}{k} \log \frac{1}{1 - u^k A(z^k)}.$$

Formula (10) is reduced to the formula that appears in the translation of the cycle construction in the unlabelled case (Theorem I.1, p. 27), upon setting $u = 1$; this formula also coincides with the statement of Proposition III.5, p. 171, regarding the number of components in cycles, and it yields the general multivariate version (Theorem III.1, p. 165) by a simple adaptation of the argument.

A.5. Formal power series

Formal power series [330, Ch. 1] extend the usual algebraic operations on polynomials to infinite series of the form

(11) $$f = \sum_{n \geq 0} f_n z^n,$$

where z is a formal indeterminate. The notation $f(z)$ is also employed. Let $\mathbb{K}$ be a ring of coefficients (usually we shall take one of the fields $\mathbb{Q}, \mathbb{R}, \mathbb{C}$); the ring of formal power series is denoted by $\mathbb{K}[[z]]$ and it is the set $\mathbb{K}^{\mathbb{N}}$ of infinite sequences of elements of $\mathbb{K}$, written as infinite sums (11), endowed with the operations of sum and product:

$$\left(\sum_n f_n z^n\right) + \left(\sum_n g_n z^n\right) \quad := \quad \sum_n (f_n + g_n)\, z^n$$

$$\left(\sum_n f_n z^n\right) \times \left(\sum_n g_n z^n\right) \quad := \quad \sum_n \left(\sum_{k=0}^n f_k g_{n-k}\right) z^n.$$

A topology, known as the *formal topology*, is put on $\mathbb{K}[[z]]$ by which two series f, g are "close" if they coincide to a large number of terms. First, the valuation of a formal power series $f = \sum_n f_n z^n$ is the smallest r such that $f_r \neq 0$ and is denoted by $\mathrm{val}(f)$. (One sets $\mathrm{val}(0) = +\infty$.) Given two power series f and g, their distance $d(f, g)$ is then defined as $2^{-\mathrm{val}(f-g)}$. With this distance (in fact an ultrametric distance), the space of all formal power series becomes a *complete metric space*. The limit of a sequence of series $\{f^{(j)}\}$ exists if, for each n, the coefficient of order n in $f^{(j)}$ eventually stabilizes to a fixed value as $j \to \infty$. In this way *formal convergence* can be defined for infinite sums: it suffices that the general term of the sum should tend to 0 in the formal topology, i.e., the valuation of the general term should tend to ∞. Similarly for infinite products, where $\prod(1 + u^{(j)})$ converges as soon as $u^{(j)}$ tends to 0 in the topology of formal power series.

It is then a simple exercise to prove that the sum $Q(f) := \sum_{k\geq 0} f^k$ exists (the sum converges in the formal topology) whenever $f_0 = 0$; the quantity then defines the *quasi-inverse* written $(1 - f)^{-1}$, with the implied properties with respect to multiplication (namely, $Q(f)(1 - f) = 1$). In the same way one defines formally logarithms and exponentials, primitives and derivatives, etc. Also, the composition $f \circ g$ is defined whenever $g_0 = 0$ by substitution of formal power series. More generally, any process on series that involves only finitely many operations at each coefficient is well-defined and is accordingly a continuous functional in the formal topology.

It can then be verified that the usual functional properties of analysis extend to formal power series provided they make sense formally; for instance, the logarithm and the exponential of formal power series, as defined by their usual expansions, are inverses of one another (e.g., $\log(\exp(zf)) = zf$; $\exp(\log(1 + zf)) = 1 + zf$). The extension to multivariate formal power series follows along entirely similar lines.

▷ **A.10.** *The OGF of permutations.* The ordinary generating function of permutations,

$$P(z) := \sum_{n=0}^{\infty} n!z^n = 1 + z + 2z^2 + 6z^3 + 24z^4 + 120z^5 + 720z^6 + 5040z^7 + \cdots$$

exists as an element of $\mathbb{C}[[z]]$, although the series has radius of convergence 0. The quantity $1/P(z)$ is well-defined (via the quasi-inverse) and one can effectively compute $1 - 1/P(z)$ whose coefficients enumerate indecomposable permutations (p. 90). The formal series $P(z)$ can even be made sense of, analytically, but as an *asymptotic series* (Euler [198]), since

$$\int_0^{\infty} \frac{e^{-t}}{1 + tz}\, dt \sim 1 - z + 2!z^2 - 3!z^3 + 4!z^4 - \cdots \quad (z \to 0+).$$

Thus, the OGF of permutations is also representable as the (formal, divergent) asymptotic series associated to an integral.   ◁

A.6. Lagrange inversion

Lagrange inversion (Lagrange, 1770) relates the coefficients of the compositional inverse of a function to coefficients of the powers of the function itself (see [129, §3.8] and [330, §1.9]). It thus establishes a fundamental correspondence between functional composition and standard multiplication of series. Although the proof is technically simple, the result is altogether non-elementary.

The inversion problem $z = h(y)$ consists in expressing y as a function of z; it is solved by the Lagrange series given below. It is assumed that $[y^0]h(y) = 0$, so that inversion is formally well defined, and $[y^1]h(y) \neq 0$. The problem is then conveniently standardized by defining $\phi(y) = y/h(y)$.

Theorem A.2 (Lagrange Inversion Theorem). *Let $\phi(u) = \sum_{k\geq 0} \phi_k u^k$ be a power series of $\mathbb{C}[[u]]$ with $\phi_0 \neq 0$. Then, the equation $y = z\phi(y)$ admits a unique solution in $\mathbb{C}[[z]]$ whose coefficients are given by (Lagrange form)*

$$(12) \qquad y(z) = \sum_{n=1}^{\infty} y_n z^n, \qquad \text{where} \quad y_n = \frac{1}{n}[u^{n-1}]\phi(u)^n.$$

Furthermore, one has for $k > 0$ (Bürmann form)

$$(13) \qquad y(z)^k = \sum_{n=1}^{\infty} y_n^{(k)} z^n, \qquad \text{where} \quad y_n^{(k)} = \frac{k}{n}[u^{n-k}]\phi(u)^n.$$

By linearity, a form equivalent to Bürmann's (13), with H an arbitrary function, is

$$(14) \qquad [z^n]H(y(z)) = \frac{1}{n}[u^{n-1}]\left(H'(u)\phi(u)^n\right).$$

Proof. The method of indeterminates coefficients provides a system of polynomial equations for $\{y_n\}$ that is seen to admit a unique solution:

$$y_1 = \phi_0, \qquad y_2 = \phi_0\phi_1, \qquad y_3 = \phi_0\phi_1^2 + \phi_0^2\phi_2, \ldots .$$

Since y_n only depends polynomially on the coefficients of $\phi(u)$ till order n, one may assume without loss of generality, in order to establish (12) and (13), that ϕ is a polynomial. Then, by general properties of analytic functions, $y(z)$ is analytic at 0 (see Chapter IV and Appendix B.2: *Equivalent definitions of analyticity*, p. 741 for definitions) and it maps conformally a neighbourhood of 0 into another neighbourhood of 0. Accordingly, the quantity $ny_n = [z^{n-1}]y'(z)$ can be estimated by Cauchy's coefficient formula:

$$(15) \qquad \begin{aligned} ny_n &= \frac{1}{2i\pi}\int_{0+} y'(z)\frac{dz}{z^n} \qquad &\text{(Direct coefficient formula for } y'(z)) \\ &= \frac{1}{2i\pi}\int_{0+} \frac{dy}{(y/\phi(y))^n} \qquad &\text{(Change of variable } z \mapsto y) \\ &= [y^{n-1}]\phi(y)^n \qquad &\text{(Reverse coefficient formula for } \phi(y)^n). \end{aligned}$$

In the context of complex analysis, this useful result appears as nothing but an avatar of the change-of-variable formula. The proof of Bürmann's form is similar. ∎

There exist instructive (but longer) combinatorial proofs based on what is known as the "cyclic lemma" or "conjugacy principle" [503] for Łukasiewicz words. (See Note I.47, p. 75 and the remarks surrounding Proposition III.7, p. 194.) Another classical proof due to Henrici relies on properties of iteration matrices [330, §1.9]; see also Comtet's book for related formulations [129].

Lagrange inversion serves most notably to develop explicit formulae for simple varieties of trees (Chapters I, p. 66, and II, p. 128), mappings (Subsection II.5.2, p. 129), planar maps (Chapter VII, p. 516) and more generally for problems involving coefficients of powers of functions.

▷ **A.11.** *Lagrange–Bürmann inversion for fractional powers.* The formula

$$[z^n]\left(\frac{y(z)}{z}\right)^\alpha = \frac{\alpha}{n+\alpha}[u^n]\phi(u)^{n+\alpha}$$

holds for any real or complex exponent α, and hence generalizes Bürmann's form. One can similarly expand $\log(y(z)/z)$.   ◁

▷ **A.12.** *Abel's identity.* By computing in two different ways the coefficient

$$[z^n]e^{(\alpha+\beta)y} = [z^n]e^{\alpha y} \cdot e^{\beta y},$$

where $y = ze^y$ is the Cayley tree function, one derives a collection of identities

$$(\alpha+\beta)(n+\alpha+\beta)^{n-1} = \alpha\beta\sum_{k=0}^{n}\binom{n}{k}(k+\alpha)^{k-1}(n-k+\beta)^{n-k-1},$$

known as *Abel's identities*.   ◁

▷ **A.13.** *A variant of Lagrange inversion.* If $y(z)$ satisfies $y = z\phi(y)$, then one has $zy' = y/(1 - z\phi'(y))$. Hence, for a function $a(y)$, the chain

$$[z^n]\frac{ya(y)}{1-z\phi'(y)} = [z^{n-1}]y'a(y) = n[z^n]A(y),$$

where A is such that $A' = a$. This, by (14), yields the general evaluation:

$$[z^n]\frac{ya(y)}{1-z\phi'(y)} = [u^{n-1}]a(u)\phi(u)^n.$$

In particular, for $\phi(u) = e^u$, we have $y \equiv T$ (the Tree function), and $[z^n]T/(1-T) = n^n$, which gives back the number of mappings of size n.   ◁

A.7. Regular languages

A *language* is a set of words over some fixed alphabet $\mathcal{A}$. The structurally simplest (yet non-trivial) languages are the *regular languages* that, as asserted on p. 57, can be defined in several equivalent ways (see [6, Ch. 3] or [189]): by regular expressions, either ambiguous or not, and by finite automata, either deterministic or non-deterministic. Our definitions of S–regularity (S as in specification) and A–regularity (A as in automaton) from Section I.4, p. 49, correspond to definability by *unambiguous* regular expression and *deterministic* automaton, respectively.

Regular expressions and ambiguity. Here is first the classical definition of a
regular expression in formal language theory.

Definition A.1. *The category* RegExp *of regular expressions is defined inductively
by the property that it contains all the letters of the alphabet ($a \in \mathcal{A}$) as well as the
empty symbol ϵ, and is such that, if $R_1, R_2 \in$ RegExp, then the formal expressions
$R_1 \cup R_2$, $R_1 \cdot R_2$ and $R_1^\star$ are regular expressions.*

Regular expressions are meant to denote *languages*. The language $\mathbf{L}(R)$ associated to R is obtained by interpreting '$\cup$' as set-theoretic union, '$\cdot$' as catenation
product extended to sets and '$\star$' as the star operation: $\mathbf{L}(R^\star) := \{\epsilon\} \cup \mathbf{L}(R) \cup$
$(\mathbf{L}(R) \cdot \mathbf{L}(R)) \cup \cdots$. These operations, since they rely on set-theoretic operations,
place no condition on multiplicities (a word may be obtained in several different
ways). Accordingly, the notions of a regular expression and a regular language are
useful when studying structural properties of languages, but they must be adapted for
enumeration purposes, where unambiguous specifications are needed.

A word $w \in \mathbf{L}(R)$ may be parsable in several ways according to R: the *ambiguity
coefficient* (or *multiplicity*) of w with respect to the regular expression R is defined[1]
as the number of parsings and written $\kappa(w) = \kappa_R(w)$.

A regular expression R is said to be *unambiguous* if for all w, we have $\kappa_R(w) \in
\{0, 1\}$, ambiguous otherwise. In the unambiguous case, if $\mathcal{L} = \mathbf{L}(R)$, then $\mathcal{L}$ is S–
regular in the sense of Chapter I, and a specification is obtained by the translation
rules

$$(16) \qquad\qquad \cup \mapsto +, \qquad \cdot \mapsto \times, \qquad ( )^\star \mapsto \text{SEQ},$$

so that the translation mechanism afforded by Proposition I.2 p. 52 applies. (Use of
the general mechanism (16) in the ambiguous case would imply that we enumerate
words with multiplicities [ambiguity coefficients] taken into account.)

A–regularity implies S–regularity. This construction is due to Kleene [367]
whose interest had its origin in the formal expressive power of nerve nets. Within
the classical framework of the theory of regular languages, it produces from an automaton (possibly non-deterministic) a regular expression (possibly ambiguous).

For our purposes, let a deterministic automaton $\mathfrak{a}$ (as defined in Subsection I.4.2,
p. 56) be given, with alphabet $\mathcal{A}$, set of states Q, with q_0 and $\overline{Q}$ the initial state
and the set of final states respectively (Definition I.11, p. 56). The idea consists in
constructing inductively the family of languages $\mathcal{L}_{i,j}^{(r)}$ of words that connect state q_i to
state q_j passing only through states $q_0, \ldots, q_r$ in between q_i and q_j. We initialize the
data with $\mathcal{L}_{i,j}^{(-1)}$ to be the singleton set $\{a\}$ if the transition $(q_i \circ a) = q_j$ exists, and
the emptyset ($\emptyset$) otherwise. The fundamental recursion

$$\mathcal{L}_{i,j}^{(r)} = \mathcal{L}_{i,j}^{(r-1)} + \mathcal{L}_{i,r}^{(r-1)} \, \text{SEQ}(S)\{\mathcal{L}_{r,r}^{(r-1)}\}\mathcal{L}_{r,j}^{(r-1)},$$

incrementally takes into account the possibility of traversing the "new" state q_r.
(The unions are clearly disjoint and the segmentation of words according to passages

---

[1] For instance if $R = (a \cup aa)^\star$ and $w = aaaa$, then $\kappa(w) = 5$ corresponding to the five parsings:
$a \cdot a \cdot a \cdot a$, $\quad a \cdot a \cdot aa$, $\quad a \cdot aa \cdot a$, $\quad aa \cdot a \cdot a$, $\quad aa \cdot aa$.

$$S\text{–regularity} \quad \equiv \quad \begin{array}{c} \text{Unambiguous} \\ \text{RegExp} \end{array} \quad \longrightarrow \quad \begin{array}{c} \text{General} \\ \text{RegExp} \end{array}$$

$$\uparrow \mathbf{K} \qquad\qquad\qquad \downarrow \mathbf{I}$$

$$A\text{–regularity} \quad \equiv \quad \begin{array}{c} \text{Deterministic} \\ \text{FA} \end{array} \quad \overset{\mathbf{RS}}{\longleftarrow} \quad \begin{array}{c} \text{Non-deterministic} \\ \text{FA} \end{array}$$

Figure A.1. Equivalence between various notions of regularity: **K** is Kleene's construction; **RS** is Rabin–Scott's reduction; **I** is the inductive construction of the text.

through state q_r is unambiguously defined, hence the validity of the sequence construction.) The language $\mathcal{L}$ accepted by $\mathfrak{a}$ is then given by the regular specification

$$\mathcal{L} = \sum_{q_j \in \overline{Q}} \mathcal{L}_{0,j}^{\|Q\|},$$

that describes the set of all words leading from the initial state q_0 to any of the final states while passing freely through any intermediate state of the automaton.

S–regularity implies A–regularity. An object described by a regular specification $\mathfrak{r}$ can be first encoded as a word, with separators indicating the way the word should be parsed unambiguously. These encodings are then describable by a regular expression using the correspondence of (16). Next any language described by a regular expression is recognizable by an automaton (possibly non-deterministic) as shown by an inductive construction. (We only state the principles informally here.) Let $\rightarrow\boxed{\mathfrak{r}}\mapsto$ represent symbolically the automaton recognizing the regular expression $\mathfrak{r}$, with the initial state represented by an incoming arrow on the left and the final state(s) by an outgoing arrow on the right. Then, the rules are schematically

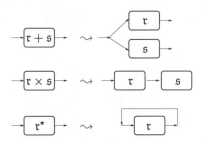

Finally, a standard result of the theory, the Rabin–Scott theorem, asserts that any non-deterministic finite automaton can be emulated by a deterministic one. (Note: this general reduction produces a deterministic automaton whose set of states is the powerset of the set of states of the original automaton; it may consequently involve an exponential blow-up in the size of descriptions.)

A.8. Stirling numbers.

These numbers count among the most famous ones of combinatorial analysis. They appear in two kinds:

- the *Stirling cycle number* (also called 'of the first kind') $\begin{bmatrix} n \\ k \end{bmatrix}$ enumerates permutations of size n having k cycles;
- the *Stirling partition number* (also called 'of the second kind') $\begin{Bmatrix} n \\ k \end{Bmatrix}$ enumerates partitions of an n-set into k non-empty equivalence classes.

The notations $\begin{bmatrix} n \\ k \end{bmatrix}$ and $\begin{Bmatrix} n \\ k \end{Bmatrix}$ proposed by Knuth (himself anticipated by Karamata) are nowadays most widespread; see [307].

The most natural way to define Stirling numbers is in terms of the "vertical" EGFs when the value of k is kept fixed:

$$\sum_{n\geq 0} \begin{bmatrix} n \\ k \end{bmatrix} \frac{z^n}{n!} = \frac{1}{k!}\left(\log\frac{1}{1-z}\right)^k$$

$$\sum_{n\geq 0} \begin{Bmatrix} n \\ k \end{Bmatrix} \frac{z^n}{n!} = \frac{1}{k!}\left(e^z - 1\right)^k.$$

From here, the bivariate EGFs follow straightforwardly:

$$\sum_{n,k\geq 0} \begin{bmatrix} n \\ k \end{bmatrix} u^k \frac{z^n}{n!} = \exp\left(u\log\frac{1}{1-z}\right) = (1-z)^{-u}$$

$$\sum_{n,k\geq 0} \begin{Bmatrix} n \\ k \end{Bmatrix} u^k \frac{z^n}{n!} = \exp\left(u(e^z - 1)\right).$$

Stirling numbers and their cognates satisfy a host of algebraic relations. For instance, the differential relations of the EGFs imply recurrences reminiscent of the binomial recurrence

$$\begin{bmatrix} n \\ k \end{bmatrix} = \begin{bmatrix} n-1 \\ k-1 \end{bmatrix} + (n-1)\begin{bmatrix} n-1 \\ k \end{bmatrix}, \qquad \begin{Bmatrix} n \\ k \end{Bmatrix} = \begin{Bmatrix} n-1 \\ k-1 \end{Bmatrix} + k\begin{Bmatrix} n-1 \\ k \end{Bmatrix}.$$

By techniques akin to Lagrange inversion or by expanding the powers in the vertical EGF of the Stirling partition numbers, one finds explicit forms

$$\begin{bmatrix} n \\ k \end{bmatrix} = \sum_{0\leq j\leq h\leq n-k} (-1)^{j+h}\binom{h}{j}\binom{n-1+h}{n-k+h}\binom{2n-k}{n-k-h}\frac{(h-j)^{n-k+h}}{h!}$$

$$\begin{Bmatrix} n \\ k \end{Bmatrix} = \frac{1}{k!}\sum_{j=0}^{k}\binom{k}{j}(-1)^j(k-j)^n.$$

Although comforting, these forms are not too useful in general, due to their sign alternation. (The one relative to Stirling cycle numbers was obtained by Schlömilch in 1852; see [129, p. 216].)

An important relation is that of the generating polynomials of the $\begin{bmatrix} n \\ r \end{bmatrix}$ for fixed n,

$$(17) \qquad P_n(u) \equiv \sum_{r=0}^{n}\begin{bmatrix} n \\ r \end{bmatrix}u^r = u\cdot(u+1)\cdot(u+2)\cdots(u+n-1),$$

which nicely parallels the OGF for the $\left\{{n \atop r}\right\}$, for fixed r:

$$\sum_{n=0}^{\infty} \left\{{n \atop r}\right\} z^n = \frac{z^r}{(1-z)(1-2z)\cdots(1-rz)}.$$

▷ **A.14.** *Schlömilch's formula.* It is established starting from

$$\frac{k!}{n!}\left[{n \atop k}\right] = \frac{1}{2i\pi}\oint \log^k \frac{1}{1-z}\frac{dz}{z^{n+1}},$$

via the change of variable *a la* Lagrange: $z = 1 - e^{-t}$. See [129, p.216] and [251].   ◁

A.9. Tree concepts

In the abstract graph-theoretic sense, a *forest* is an acyclic (undirected) graph and a *tree* is a forest that consists of just one connected component. A *rooted tree* has a specific node is distinguished, the *root*. Rooted trees are drawn with the root either below (the mathematician's and botanist's convention) or on top (the genealogist's and computer scientist's convention), and in this book, we employ both conventions interchangeably. Here then are two planar representations of the same rooted tree

(18)

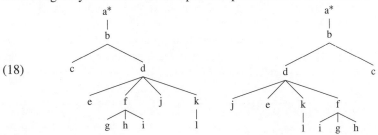

where the star distinguishes the root. (Tags on nodes, a, b, c, etc, are not part of the tree structure but only meant to discriminate nodes here.) A tree whose nodes are labelled by distinct integers then becomes a *labelled tree*, this in the precise technical sense of Chapter II. Size is defined by the number of nodes (vertices). Here is for instance a labelled tree of size 9:

(19)

In a rooted tree, the *outdegree* of a node is the number of its descendants; with the sole exception of the root, outdeegree is thus equal to degree (in the graph-theoretic sense, i.e., the number of neighbours) minus 1. Once this convention is clear, one usually abbreviates "outdegree" by "degree" when speaking of rooted trees. A *leaf* is a node without descendant, that is, a node of (out)degree equal to 0. For instance the tree in (19) has five leaves. Non-leaf nodes are also called internal nodes.

Many applications from genealogy to computer science require superimposing an additional structure on a graph-theoretic tree. A *plane tree* (sometimes also called

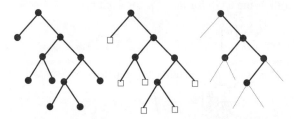

Figure A.2. Three representations of a binary tree.

a *planar tree*) is defined as a tree in which subtrees dangling from a common node are ordered between themselves and represented from left to right in order. Thus, the two representations in (18) are equivalent as graph-theoretic trees, but they become distinct objects when regarded as plane trees.

   Binary trees play a very special role in combinatorics. These are rooted trees in which every non-leaf node has degree 2 exactly as, for instance, in the first two drawings of Figure A.2. In the second case, the leaves have been distinguished by '□'. The *pruned binary tree* (third representation) is obtained from a regular binary tree by removing the leaves—such a tree then has unary branching nodes of either one of two possible types (left- or right-branching). A binary tree can be fully reconstructed from its pruned version, and a tree of size $2n + 1$ always expands a pruned tree of size n.

   A few major classes are encountered throughout this book. Here is a summary[2].

general plane trees (Catalan trees)	$\mathcal{G} = \mathcal{Z} \times \mathrm{SEQ}(\mathcal{G})$	(unlabelled)
binary trees	$\mathcal{A} = \mathcal{Z} + (\mathcal{Z} \times \mathcal{A} \times \mathcal{A})$	(unlabelled)
non-empty pruned binary trees	$\mathcal{B} = \mathcal{Z} + 2(\mathcal{Z} \times \mathcal{B}) + (\mathcal{Z} \times \mathcal{B} \times \mathcal{B})$	(unlabelled)
pruned binary trees	$\mathcal{C} = \mathbf{1} + (\mathcal{Z} \times \mathcal{B} \times \mathcal{B})$	(unlabelled)
general non-plane trees (Cayley trees)	$\mathcal{T} = \mathcal{Z} \times \mathrm{SET}(\mathcal{T})$	(labelled)

The corresponding GFs are, respectively,

$$G(z) = \frac{1 - \sqrt{1 - 4z}}{2}, \quad A(z) = \frac{1 - \sqrt{1 - 4z^2}}{2z}, \quad B(z) = \frac{1 - 2z - \sqrt{1 - 4z}}{2z},$$
$$C(z) = \frac{1 - \sqrt{1 - 4z}}{2z}, \quad T(z) = z e^{T(z)},$$

being of type OGF for the first four and EGF for the last one. The corresponding counts are

$$G_n = \frac{1}{n}\binom{2n - 2}{n - 1}, \quad A_{2n+1} = \frac{1}{n + 1}\binom{2n}{n}, \quad B_n = \frac{1}{n + 1}\binom{2n}{n} \quad (n \geq 1),$$
$$C_n = \frac{1}{n + 1}\binom{2n}{n}, \quad T_n = n^{n-1}.$$

The common occurrence of the Catalan numbers, $(C_n = B_n = A_{2n+1} = G_{n+1})$ is explained by pruning and by the rotation correspondence described on p. 73.

---

[2] The term "general" refers to the fact that no degree constraint is imposed.

APPENDIX B

Basic Complex Analysis

This appendix contains entries arranged in alphabetical order regarding the following topics:

> *Algebraic elimination; Equivalent definitions of analyticity; Gamma function; Holo-*
> *nomic functions; Implicit Function Theorem; Laplace's method; Mellin transform;*
> *Several complex variables.*

The corresponding notions and results are used starting with Part B, which is relative to *Complex Asymptotics*. The present entries, together with the first sections of Chapter IV, should enable a reader, previously unacquainted with complex analysis but with a fair background in basic calculus, to follow the main developments of analytic combinatorics. There are a number of excellent classic presentations of complex analysis: the books by Dieudonné [165], Henrici [329], Hille [334], Knopp [373], Titchmarsh [577], and Whittaker–Watson [604] are of special interest, given their concrete approach to the subject (see also our comments on p. 286).

B.1. Algebraic elimination

Auxiliary quantities can be eliminated from systems of polynomial equations. In essence, elimination is achieved by suitable combinations of the equations themselves. One of the best strategies is based on Gröbner bases and is presented in the excellent book of Cox, Little, and O'Shea [135]. This entry develops a more elementary approach based on *resultants*. It is necessitated by the analysis of algebraic curves, function, and systems (Sections VII. 6, p. 482, and VII. 7, p. 493), with a general applicability to context-free structures introduced on p. 79.

Resultants. Consider a field of coefficients $\mathbb{K}$, which may be specialized as $\mathbb{Q}, \mathbb{C}, \mathbb{C}(z), \ldots$, as the need arises. A polynomial of degree d in $\mathbb{K}[x]$ has at most d roots in $\mathbb{K}$ and exactly d in the algebraic closure $\overline{\mathbb{K}}$ of $\mathbb{K}$. Given two polynomials,

$$P(x) = \sum_{j=0}^{\ell} a_j x^{\ell-j}, \qquad Q(x) = \sum_{k=0}^{m} b_k x^{m-k},$$

their *resultant* (with respect to the variable x) is the determinant of order $(\ell + m)$,

$$(1) \qquad \mathbf{R}(P, Q, x) = \det \begin{vmatrix} a_0 & a_1 & a_2 & \cdots & 0 & 0 \\ 0 & a_0 & a_1 & \cdots & 0 & 0 \\ \vdots & \vdots & \vdots & \ddots & \vdots & \vdots \\ 0 & 0 & 0 & \cdots & a_{\ell-1} & a_\ell \\ b_0 & b_1 & b_2 & \cdots & 0 & 0 \\ 0 & b_0 & b_1 & \cdots & 0 & 0 \\ \vdots & \vdots & \vdots & \ddots & \vdots & \vdots \\ 0 & 0 & 0 & \cdots & b_{m-1} & b_m \end{vmatrix},$$

also called the Sylvester determinant. By its definition, the resultant is a polynomial form in the coefficients of P and Q. The main properties of resultants are the following: (*i*) *ff $P(x), Q(x) \in \mathbb{K}[x]$ have a common root in the algebraic closure $\overline{\mathbb{K}}$ of*

$\mathbb{K}$, then $\mathbf{R}(P(x), Q(x), x) = 0$; (ii) conversely, if $\mathbf{R}(P(x), Q(x), x) = 0$ holds, then either $a_0 = b_0 = 0$ or else $P(x)$, $Q(x)$ have a common root in $\overline{\mathbb{K}}$. (The idea of the proof of (i) is as follows. Let S be the matrix in (1). Then the homogeneous linear system $Sw = 0$ admits a solution $w = (\xi^{\ell+m-1}, \ldots, \xi^2, \xi, 1)$ in which ξ is a common root of P and Q; this is only possible if $\det(S) \equiv \mathbf{R}$ vanishes.) See especially van der Waerden's crisp treatment in [590] and Lang's treatise [401, V.10] for a detailed presentation of resultants

Equating the resultant to 0 thus provides a *necessary* condition for the existence of common roots, but not always a sufficient one. This has implications in situations where the coefficients a_j, b_k depend on one or several parameters. In that case, the condition $\mathbf{R}(P, Q, x) = 0$ will certainly capture all the situations in which P and Q have a common root, but it may also include some situations where there is a reduction in degree, although the polynomials have no common root. For instance, take $P(x) = tx - 2$ and $Q(x) = tx^2 - 4$ (with t a parameter); the resultant with respect to x is

$$\mathbf{R} = 4t(1 - t).$$

Indeed, the condition $\mathbf{R} = 0$ corresponds to either a common root ($t = 1$ for which $P(2) = Q(2) = 0$) or to some degeneracy in degree ($t = 0$ for which $P(x) = -2$ and $Q(x) = -4$ have no common zero).

Systems of equations. Given a system

(2) $$\{P_j(z, y_1, y_2, \ldots, y_m) = 0\}, \qquad j = 1 \mathinner{.\,.} m,$$

defining an algebraic curve, we can then proceed as follows in order to extract a single equation satisfied by one of the indeterminates. By taking resultants with P_m, eliminate all occurrences of the variable y_m from the first $m-1$ equations, thereby obtaining a new system of $m-1$ equations in $m-1$ variables (with z kept as a parameter, so that the base field is $\mathbb{C}(z)$). Repeat the process and successively eliminate $y_{m-1}, \ldots, y_2$. The strategy (in the simpler case where variables are eliminated in succession exactly one at a time) is summarized in the skeleton procedure **Eliminate**:

> **procedure** Eliminate $(P_1, \ldots, P_m, y_1, y_2, \ldots y_m)$;
> {*Elimination of $y_2, \ldots, y_m$ by resultants*}
> $(A_1, \ldots, A_m) := (P_1, \ldots, P_m)$;
> **for** j **from** m **by** -1 **to** 2 **do**
> **for** k **from** $j - 1$ **by** -1 **to** 1 **do**
> $\quad A_k := \mathbf{R}(A_k, A_j, y_j)$;
> **return**(A_1).

The polynomials obtained need not be minimal, in which case, one should appeal to multivariate polynomial factorization in order to select the relevant factors at each stage. (Gröbner bases provide a neater alternative to these questions, see [135].)

Computer algebra systems usually provide implementations of both resultants and Gröbner bases. The complexity of elimination is, however, exponential in the worst-case: degrees essentially multiply, which is somewhat intrinsic. For example, y_0 in the quadratic system of k equations

$$y_0 - z - y_k = 0, \quad y_k - y_{k-1}^2 = 0, \quad \ldots, \quad y_1 - y_0^2 = 0$$

(determining the OGF of regular trees of degree 2^k) represents an algebraic function of degree 2^k and no less.

▷ **B.1.** *Resultant and roots.* Let $P, Q \in \mathbb{C}[x]$ have roots $\{\alpha_j\}$ and $\{\beta_k\}$, respectively. Then

$$\mathbf{R}(P, Q, x) = a_0^\ell b_0^m \prod_{i=1}^{\ell} \prod_{j=1}^{m} (\alpha_i - \beta_j) = a_0^\ell \prod_{i=1}^{m} Q(\alpha_i).$$

The *discriminant* of P classically defined by $D(P) := a_0^{-1} \mathbf{R}(P(x), P'(x), x)$ satisfies

$$D(P) \equiv a_0^{-1} \mathbf{R}(P(x), P'(x), x) = a_0^{2\ell-2} \prod_{i \neq j} (\alpha_i - \alpha_j).$$

Given the coefficients of P and the value of $D(P)$, an effectively computable bound on the minimal separation distance δ between any two roots of P can be found. (Hint. Let $A = 1 + \max_j(|a_j/a_0|)$. Then each α_j satisfies $|\alpha_j| < mA$. Set $L = \binom{\ell}{2}$. Then $\delta \geq |a_0|^{2-2\ell} |D(P)| (2A)^{L-1}$.) ◁

B.2. Equivalent definitions of analyticity

Two parallel notions are introduced at the beginning of Chapter IV: analyticity (defined by power series expansions) and holomorphy (defined as complex differentiability). As is known from any textbook on complex analysis, these notions are equivalent. Given their importance for analytic combinatorics, this appendix entry sketches a proof of the equivalence, which is summarized by the following diagram:

$$\begin{array}{ccc} & [A] & \\ \text{Analyticity} & \overset{\longrightarrow}{\underset{[C]}{\longleftarrow}} & \mathbb{C}\text{-differentiability} \\ & & \downarrow [B] \\ & & \text{Null integral Property} \end{array}$$

A. Analyticity implies complex-differentiability. Let $f(z)$ be analytic in the disc $D(z_0; R)$. We may assume without loss of generality that $z_0 = 0$ and $R = 1$ (else effect a linear transformation on the argument z). According to the definition of analyticity, the series representation

$$f(z) = \sum_{n=0}^{\infty} f_n z^n, \tag{3}$$

converges for all z with $|z| < 1$. Elementary series rearrangements first entail that $f(z)$ given by this representation is analytic at any z_1 interior to $D(0; 1)$; similar techniques then show the existence of the derivative as well as the fact that the derivative can be obtained by term-wise differentiation of (3). See Note B.2 for details.

▷ **B.2.** *Proof of [A]: Analyticity implies differentiability.* Formally, the binomial theorem provides

$$\begin{aligned} f(z) &= \sum_{n \geq 0} f_n z^n &= \sum_{n \geq 0} f_n (z_1 + z - z_1)^n \\ &= \sum_{n \geq 0} \sum_{k=0}^{n} \binom{n}{k} f_n z_1^k (z - z_1)^{n-k} \\ &= \sum_{m \geq 0} c_m (z - z_1)^m, \qquad c_m := \sum_{k \geq 0} \binom{m+k}{k} f_{m+k} z_1^k. \end{aligned} \tag{4}$$

Let r_1 be any number smaller than $1 - |z_1|$. We observe that (4) makes analytic sense. Indeed, one has the bound $|f_n| \leq CA^n$, valid for any $A > 1$ and some $C > 0$. Thus, the terms in (4) are dominated in absolute value by those of the double series

$$(5) \qquad \sum_{n \geq 0} \sum_{k=0}^{n} \binom{n}{k} CA^n |z_1|^k r_1^{n-k} = C \sum_{n \geq 0} A^n (|z_1| + r_1)^n = \frac{C}{1 - A(|z_1| + r_1)},$$

which is absolutely convergent as soon as A is chosen such that $A < (|z_1| + r_1)^{-1}$.

Complex differentiability of at any $z_1 \in D(0; 1)$ is derived from the analogous calculation, valid for small enough δ,

$$
\begin{aligned}
(6) \qquad \frac{1}{\delta}(f(z_1 + \delta) - f(z_1))) &= \sum_{n \geq 0} n f_n z_1^{n-1} + \delta \sum_{n \geq 0} \sum_{k=2}^{n} \binom{n}{k} f_n z_1^k \delta^{n-k-2} \\
&= \sum_{n \geq 0} n f_n z_1^{n-1} + O(\delta),
\end{aligned}
$$

where boundedness of the coefficient of δ results from an argument analogous to (5).   ◁

The argument of Note B.2 has shown that the derivative of f at z_1 is obtained by differentiating termwise the series representing f. More generally derivatives of all orders exist and can be obtained in a similar fashion. In view of this fact, the equalities of (4) can also be interpreted as the *Taylor expansion* (by grouping terms according to values of k first)

$$(7) \qquad f(z_1 + \delta) = f(z_1) + \delta f'(z_1) + \frac{\delta^2}{2!} f''(z_1) + \cdots,$$

which is thus generally valid for analytic functions.

B. Complex differentiability implies the "Null Integral" Property. The Null Integral Property relative to a domain Ω is the property:

$$\int_{\lambda} f(z)\, dz = 0 \qquad \text{for any loop } \lambda \subset \Omega.$$

(A loop is a closed path that can be contracted to a single point in the domain Ω.) Its proof results from the Cauchy–Riemann equations and Green's formula.

▷ **B.3.** *Proof of* [*B*]: *the Null Integral Property.* This starts from the *Cauchy–Riemann equations*. Let $P(x, y) = \Re f(x + iy)$ and $Q(x, y) = \Im f(x + iy)$. By adopting successively in the definition of complex differentiability $\delta = h$ and $\delta = ih$, one finds $P'_x + i Q'_x = Q'_y - i P'_y$, implying the Cauchy–Riemann equations:

$$(8) \qquad \frac{\partial P}{\partial x} = \frac{\partial Q}{\partial y} \qquad \text{and} \qquad \frac{\partial P}{\partial y} = -\frac{\partial Q}{\partial x},$$

(The functions P and Q satisfy the partial differential equations $\Delta f = 0$, where Δ is the two-dimensional *Laplacian* $\Delta := \frac{\partial^2}{\partial x^2} + \frac{\partial^2}{\partial y^2}$; such functions are known as *harmonic functions*.) The Null Integral Property, given differentiability, results from the Cauchy–Riemann equations, upon taking into account Green's theorem of multivariate calculus,

$$\int_{\partial K} A\, dx + B\, dy = \int \int_{K} \left(\frac{\partial B}{\partial x} - \frac{\partial A}{\partial y} \right) dx\, dy,$$

which is valid for any (compact) domain K enclosed by a simple curve ∂K.   ◁

C. Complex differentiability implies analyticity. The starting point is the formula

(9)
$$f(a) = \frac{1}{2i\pi} \int_\gamma \frac{f(z)}{z-a}\, dz,$$

knowing only differentiability of f and its consequence, the Null Integral Property, but precisely *not* postulating the existence of an analytic expansion (here γ is a simple positive loop encircling a inside a region in which f is analytic).

▷ **B.4.** *Proof of* [*C*]: *the integral representation.* The proof of (9) is obtained by decomposing $f(z)$ in the original integral as $f(z) = f(z) - f(a) + f(a)$. Define accordingly $g(z) = (f(z) - f(a))/(z-a)$, for $z \neq a$, and $g(a) = f'(a)$. By the differentiability assumption, g is continuous and holomorphic (differentiable) at any point other than a. Its integral is thus 0 along γ. On the other hand, we have

$$\int_\gamma \frac{1}{z-a}\, dz = 2i\pi,$$

by a simple computation: deform γ to a small circle around a and evaluate the integral directly by setting $z - a = re^{i\theta}$. ◁

Once (9) is granted, it suffices to write, e.g., for an expansion at 0,

$$
\begin{aligned}
f(z) &= \frac{1}{2i\pi} \int_\gamma f(t)\, \frac{dt}{t-z} \\
&= \frac{1}{2i\pi} \int_\gamma f(t) \left(1 + \frac{z}{t} + \frac{z^2}{t^2} + \cdots\right) \frac{dt}{t} \\
&= \sum_{n \geq 0} f_n z^n, \qquad f_n := \frac{1}{2i\pi} \int_\gamma f(t)\, \frac{dt}{t^{n+1}}.
\end{aligned}
$$

(Exchanges of integration and summation are justified by normal convergence.) Analyticity is thus proved from the Null Integral Property.

▷ **B.5.** *Cauchy's formula for derivatives.* One has

$$f^{(n)}(a) = \frac{n!}{2i\pi} \int_\gamma \frac{f(z)}{(z-a)^{n+1}}\, dz.$$

This follows from (9) by differentiation under the integral sign. ◁

▷ **B.6.** *Morera's Theorem.* Suppose that f is continuous [but not *a priori* known to be differentiable] in an open set Ω and that its integral along any triangle in Ω is 0. Then, f is analytic (hence holomorphic) in Ω. (For details, see, e.g, [497, p. 68].) This theorem is useful for disposing of *apparent* (or "removable") singularities, as in $(\cos(z) - 1)/\sin(z)$. ◁

B.3. Gamma function

The formulae of singularity analysis in Chapter IV involve the *Gamma function* in an essential manner. The Gamma function extends to non-integral arguments the factorial function. We collect in this appendix a few classical facts regarding it. Proofs may be found in classic treatises like Henrici's [329] or Whittaker and Watson's [604].

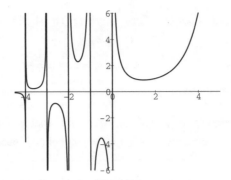

Figure B.1. A plot of $\Gamma(s)$ for real s.

Basic properties. Euler introduced the Gamma function as

$$(10) \qquad \Gamma(s) = \int_0^\infty e^{-t} t^{s-1} \, dt,$$

where the integral converges provided $\Re(s) > 0$. Through integration by parts, one immediately derives the basic functional equation of the Gamma function,

$$(11) \qquad \Gamma(s+1) = s\Gamma(s).$$

Since $\Gamma(1) = 1$, one has $\Gamma(n+1) = n!$, so that the Gamma function serves to extend the factorial function for non-integral arguments. The special value,

$$(12) \qquad \Gamma\left(\frac{1}{2}\right) := \int_0^\infty e^{-t} \frac{dt}{\sqrt{t}} = 2 \int_0^\infty e^{-x^2} \, dx = \sqrt{\pi},$$

proves to be quite important. It implies in turn $\Gamma(-\tfrac{1}{2}) = -2\sqrt{\pi}$.

From (11), the Gamma function can be analytically continued to the whole of $\mathbb{C}$ with the exception of poles at $0, -1, -2, \ldots$ indeed, the functional equation used backwards yields

$$\Gamma(s) \sim \frac{(-1)^m}{m!} \frac{1}{s+m} \qquad (s \to -m),$$

so that the residue of $\Gamma(s)$ at $s = -m$ is $(-1)^m/m!$. Figure B.1 depicts the graph of $\Gamma(s)$ for real values of s.

▷ **B.7.** *Evaluation of the Gaussian integral.* Define $J := \int_0^\infty e^{-x^2} \, dx$. The idea is to evaluate J^2:

$$J^2 = \int_0^\infty \int_0^\infty e^{-(x^2+y^2)} \, dx \, dy.$$

Going to polar coordinates, $(x^2 + y^2)^{1/2} = \rho$, $x = \rho \cos\theta$, $y = \rho \sin\theta$ yields, via the standard change of variables formula:

$$J^2 = \int_0^\infty \int_0^{\frac{\pi}{2}} e^{-\rho^2} \rho \, d\rho \, d\theta.$$

The equality $J^2 = \pi/4$ results.   ◁

Hankel contour representation. Euler's integral representation of $\Gamma(s)$ used in conjunction with the functional equation permits us to continue $\Gamma(s)$ to the whole of the complex plane. A direct approach due to Hankel provides an alternative integral representation valid for all values of s.

Theorem B.1 (Hankel's contour integral). *Let $\int_{+\infty}^{(0)}$ denote an integral taken along a contour starting at $+\infty$ in the upper plane, winding counterclockwise around the origin, and proceeding towards $+\infty$ in the lower half-plane. Then, for all $s \in \mathbb{C}$,*

$$(13) \qquad \frac{1}{\pi} \sin(\pi s) \Gamma(1-s) = \frac{1}{\Gamma(s)} = -\frac{1}{2i\pi} \int_{+\infty}^{(0)} (-t)^{-s} e^{-t} \, dt.$$

In (13), $(-t)^{-s}$ is assumed to have its principal determination when t is negative real, and this determination is then extended uniquely by continuity throughout the contour. The integral then closely resembles the definition of $\Gamma(1-s)$. The first form of (13) can also be rewritten as $\frac{1}{\Gamma(s)}$, by virtue of the complement formula given below.

▷ **B.8.** *Proof of Hankel's representation.* We refer to volume 2 of Henrici's book [329, p. 35] or Whittaker and Watson's treatise [604, p. 245] for a detailed proof.

A contour of integration that fulfills the conditions of the theorem is typically the contour $\mathcal{H}$ that is at distance 1 of the positive real axis comprising three parts: a line parallel to the positive real axis in the upper half-plane; a connecting semi-circle centered at the origin; a line parallel to the positive real axis in the lower half-plane. More precisely, $\mathcal{H} = \mathcal{H}^- \cup \mathcal{H}^+ \cup \mathcal{H}^\circ$, where

$$(14) \qquad \begin{cases} \mathcal{H}^- &= \{z = w - i, \ w \geq 0\} \\ \mathcal{H}^+ &= \{z = w + i, \ w \geq 0\} \\ \mathcal{H}^\circ &= \{z = -e^{i\phi}, \ \phi \in [-\frac{\pi}{2}, \frac{\pi}{2}]\}. \end{cases}$$

Let ϵ be a small positive real number, and denote by $\epsilon \cdot \mathcal{H}$ the image of $\mathcal{H}$ by the transformation $z \mapsto \epsilon z$. By analyticity, for the integral representation, we can equally well adopt as integration path the contour $\epsilon \cdot \mathcal{H}$, for any $\epsilon > 0$. The main idea is then to let ϵ tend to 0.

Assume momentarily that $s < 0$. (The extension to arbitrary s then follows by analytic continuation.) The integral along $\epsilon \cdot \mathcal{H}$ decomposes into two parts:

1. The integral along the semi-circle is 0 if we take the circle of a vanishing small radius, since $-s > 0$.
2. The combined contributions from the upper and lower lines give, as $\epsilon \to 0$

$$\int_{+\infty}^{(0)} (-t)^{-s} e^{-t} \, dt = (-U + L) \int_0^\infty t^{-s} e^{-t} \, dt$$

where U and L denote the determinations of $(-1)^{-s}$ on the half-lines lying in the upper and lower half-planes respectively.

By continuity of determinations, $U = (e^{-i\pi})^{-s}$ and $L = (e^{+i\pi})^{-s}$. Therefore, the right-hand side of (13) is equal to

$$-\frac{(-e^{i\pi s} + e^{-i\pi s})}{2i\pi} \Gamma(1-s) = \frac{\sin(\pi s)}{\pi} \Gamma(1-s),$$

which completes the proof of the theorem. ◁

Expansions. The Gamma function has poles at the non-positive integers but has no zeros. Accordingly, $1/\Gamma(s)$ is an entire function with zeros at $0, -1, \ldots$, and the position of the zeros is reflected by the product decomposition,

$$(15) \qquad \frac{1}{\Gamma(s)} = se^{\gamma s} \prod_{n=1}^{\infty}\left[(1+\frac{s}{n})e^{-s/n}\right]$$

(of the so-called Weierstrass type). There $\gamma = 0.57721$ denotes Euler's constant

$$\gamma = \lim_{n\to\infty}(H_n - \log n) \equiv \sum_{n=1}^{\infty}\left[\frac{1}{n} - \log(1+\frac{1}{n})\right].$$

The logarithmic derivative of the Gamma function is classically known as the psi function and is denoted by $\psi(s)$:

$$\psi(s) := \frac{d}{ds}\log\Gamma(s) = \frac{\Gamma'(s)}{\Gamma(s)}.$$

In accordance with (15), $\psi(s)$ admits a partial fraction decomposition

$$(16) \qquad \psi(s+1) = -\gamma - \sum_{n=1}^{\infty}\left[\frac{1}{n+s} - \frac{1}{n}\right].$$

From (16), it can be seen that the Taylor expansion of $\psi(s+1)$, and hence of $\Gamma(s+1)$, involves values of the Riemann zeta function, $\zeta(s) = \sum_{n=1}^{\infty}\frac{1}{n^s}$, at the positive integers: for $|s| < 1$,

$$\psi(s+1) = -\gamma + \sum_{n=2}^{\infty}(-1)^n\zeta(n)s^{n-1}.$$

so that the coefficients in the expansion of $\Gamma(s)$ around any integer are polynomially expressible in terms of Euler's constant γ and values of the zeta function at the integers. For instance, as $s \to 0$,

$$\Gamma(s+1) = 1 - \gamma s + \left(\frac{\pi^2}{12} + \frac{\gamma^2}{2}\right)s^2 + \left(-\frac{\zeta(3)}{3} - \frac{\pi^2\gamma}{12} - \frac{\gamma^3}{6}\right)s^3 + O(s^4).$$

Another direct consequence of the infinite product formulae for $\Gamma(s)$ and $\sin\pi s$ is the complement formula for the Gamma function,

$$(17) \qquad \Gamma(s)\Gamma(-s) = -\frac{\pi}{s\sin\pi s},$$

which directly results from the factorization of the sine function (due to Euler),

$$\sin s = s\prod_{n=1}^{\infty}\left(1 - \frac{s^2}{n^2\pi^2}\right).$$

In particular, Equation (17) gives back the special value (cf (12)): $\Gamma(1/2) = \sqrt{\pi}$.

▷ **B.9.** *The duplication formula.* This is

$$2^{2s-1}\Gamma(s)\Gamma(s+1/2) = \pi^{1/2}\Gamma(2s),$$

which provides the expansion of Γ near $1/2$:

$$\Gamma(s+1/2) = \pi^{1/2} - (\gamma + 2\log 2)\,\pi^{1/2}s + \left(\frac{\pi^{5/2}}{4} + \frac{(\gamma + 2\log 2)^2\,\pi^{1/2}}{2}\right)s^2 + O(s^3).$$

The coefficients now involve $\log 2$ as well as zeta values. ◁

Finally, a famous and absolutely fundamental asymptotic formula is Stirling's approximation, familiarly known as "Stirling's formula":

$$\Gamma(s+1) = s\Gamma(s) \sim s^s e^{-s}\sqrt{2\pi s}\left[1 + \frac{1}{12s} + \frac{1}{288s^2} - \frac{139}{51840s^3} + \cdots\right].$$

It is valid for (large) real $s \in \mathbb{R}_{>0}$, and more generally for all $s \to \infty$ in $|\arg(s)| < \pi - \delta$ (any $\delta > 0$). For the purpose of obtaining effective bounds, the following quantitative relation [604, p. 253] often proves useful,

$$\Gamma(s+1) = s^s e^{-s}(2\pi s)^{1/2}e^{\theta/(12s)}, \qquad \text{where } 0 < \theta \equiv \theta(s) < 1,$$

an equality that holds now for all $s \geq 1$. Stirling's formula is usually established by appealing to the method of Laplace applied to the integral representation for $\Gamma(s+1)$, see Appendix B.6: *Laplace's method*, p. 755, or by Euler–Maclaurin summation (Note A.7, p. 726). It is derived by Mellin transforms in Appendix B.7, p. 762.

▷ **B.10.** *The Eulerian Beta function.* It is defined for $\Re(p), \Re(q) > 0$ by any of the following integrals,

$$B(p,q) := \int_0^1 x^{p-1}(1-x)^{q-1}\,dx = \int_0^\infty \frac{y^{p-1}}{(1+y)^{p+q}}\,dy = 2\int_0^{\frac{\pi}{2}} \cos^{2p-1}\theta\,\sin^{2q-1}\theta\,d\theta,$$

where the last form is known as a Wallis integral. It satisfies:

$$B(p,q) = \frac{\Gamma(p)\Gamma(q)}{\Gamma(p+q)}.$$

[See [604, p. 254] for a proof generalizing that of Note B.7.] ◁

▷ **B.11.** *Facts about the Riemann zeta function* (ζ). Here are a few properties of this function, whose elementary theory centrally involves the Gamma function. It is initially defined by

$$\zeta(s) := \sum_{n\geq 1}\frac{1}{n^s}, \qquad \Re(s) > 1,$$

and it admits a meromorphic expansion to the whole of $\mathbb{C}$, with only a pole at $s = 1$, where $\zeta(s) = 1/(s-1) + \gamma + \cdots$ and γ is Euler's constant. Special values for $k \in \mathbb{Z}_{\geq 1}$ are

$$\zeta(2k) = \frac{2^{2k-1}|B_{2k}|}{(2k)!}\pi^{2k}, \qquad \zeta(-2k+1) = -\frac{B_{2k}}{2k}, \qquad \zeta(-2k) = 0,$$

with B_{2k} a Bernoulli number. Other interesting values are $\zeta(0) = -\frac{1}{2}$, $\zeta'(0) = -\log\sqrt{2\pi}$. The functional equation admits many forms, among which the reflection formula:

$$\Gamma\left(\frac{s}{2}\right)\pi^{-s/2}\zeta(s) = \Gamma\left(\frac{1-s}{2}\right)\pi^{-(1-s)/2}\zeta(1-s).$$

The proofs make an essential use of Mellin transforms (Appendix B.7, p. 762, and especially Equation (46), p. 764) as well as Hankel contours. Accessible introductions are to be found in [186, 578, 604]. ◁

B.4. Holonomic functions

Doron Zeilberger [626] has introduced discrete mathematicians to a powerful framework, the *holonomic framework*, which takes its roots in classical differential algebra [72, 133] and has found innumerable applications in the theory of special functions and symbolic computation [480], combinatorial identities, and combinatorial enumeration. In these pages, we can only offer a (too) brief orientation tour of this wonderful theory. Major contributions in the perspective of *Analytic Combinatorics* are due to Stanley [551], Zeilberger [626], Gessel [289], and Lipshitz [409, 410]. As we shall see there is a chain of growing generality and power,

$$\text{rational} \quad \to \quad \text{algebraic} \quad \to \quad \text{holonomic}.$$

The associated asymptotic problems are examined in Subsection VII. 9.1, p. 518 ("regular" singularities) and Section VIII. 7, p. 581 ("irregular" singularities).

Univariate holonomic functions. Holonomic functions[1] are solutions of linear differential equations or systems whose coefficients are rational functions. The univariate theory is elementary.

Definition B.1. *A formal power series (or function)* $f(z)$ *is said to be* holonomic *if it satisfies a linear differential equation,*

$$(18) \qquad c_0(z)\frac{d^r}{dz^r}f(z) + c_1(z)\frac{d^{r-1}}{dz^{r-1}}f(z) + \cdots + c_r(z)f(z) = 0,$$

where the coefficients $c_j(z)$ *lie in the field* $\mathbb{C}(z)$ *of rational functions. Equivalently,* $f(z)$ *is holonomic if the vector space over* $\mathbb{C}(z)$ *spanned by the set of all its derivatives* $\{\partial^j f(z)\}_{j=0}^{\infty}$ *is finite dimensional.*

By clearing denominators, we can assume, if needed, the quantities $c_j(z)$ in (18) to be polynomials. It then follows that the coefficient sequence (f_n) of a holonomic $f(z)$ satisfies a recurrence,

$$(19) \qquad \widehat{c}_s(n)f_{n+s} + \widehat{c}_{s-1}(n)f_{n+s-1} + \cdots + \widehat{c}_0(n)f_n = 0,$$

for some polynomials $\widehat{c}_j(n)$, provided $n \geq n_0$ (some n_0). Such a recurrence (19) is known as a *P–recurrence*. (The two properties of sequences, to be the coefficients of a holonomic function and to be *P*–recursive, are equivalent.)

Functions such as $e^z, \log z, \cos(z), \arcsin(z), \sqrt{1+z},$ and $\mathrm{Li}_2(z) := \sum_{n\geq 1} z^n/n^2$ are holonomic. Formal power series like $\sum z^n/(n!)^2$ and $\sum n!z^n$ are holonomic. Sequences like $\frac{1}{n+1}\binom{2n}{n}, 2^n/(n^2+1)$ are coefficients of holonomic functions and are *P*–recursive. However, sequences like $\sqrt{n}, \log n$ are not *P*–recursive, a fact that can be proved by an examination of singularities of associated generating functions [232]. For similar reasons, $\tan z, \sec z,$ and $\Gamma(z)$ that have infinitely many singularities are *not* holonomic.

Holonomic functions enjoy a rich set of closure properties. Define the Hadamard product of two functions $h = f \odot g$ to be the termwise product of series: $[z^n]h(z) = ([z^n]f(z)) \cdot ([z^n]g(z))$. We have the following theorem.

---

[1]A synonymous name is ∂-finite or *D*-finite.

Theorem B.2 (Univariate holonomic closure). *The class of univariate holonomic functions is closed under the following operations: sum ($+$), product ($\times$), Hadamard product ($\odot$), differentiation (∂_z), indefinite integration ($\int^z$), and algebraic substitution ($z \mapsto y(z)$ for some algebraic function $y(z)$).*

Proof. An exercise in vector space manipulations. For instance, let $VS(\partial^\star f)$ be the vector space over $\mathbb{C}(z)$ spanned by the derivative $\{\partial_z^j f\}_{j \geq 0}$. If $h = f + g$ (or $h = f \cdot g$), then $VS(\partial^\star h)$ is finite dimensional since it is included in the direct sum $VS(\partial^\star f) \oplus VS(\partial^\star g)$ (respectively, the tensor product $VS(\partial^\star f) \otimes VS(\partial^\star g)$). For Hadamard products, if $h_n = f_n g_n$, then a system of P–recurrences can be obtained for the quantities $h_n^{(i,j)} = f_{n+i} g_{n+j}$ from the recurrences satisfied by f_n, g_n, and then a single P–recurrence can be obtained. Closure under algebraic substitution results from the methods of Note B.12. See Stanley's historic paper [551] and his book chapter [554, Ch. 6] for details. ∎

▷ **B.12.** *Algebraic functions are holonomic.* Let $y(z)$ satisfy $P(z, y(z)) = 0$, with P a polynomial. Any non-degenerate rational fraction $Q(z, y(z))$ can be expressed as a polynomial in $y(z)$ with coefficients in $\mathbb{C}(z)$. [Proof: let D be the denominator of Q; the Bezout relation $AP - BD = 1$ (in $\mathbb{C}(x)[y]$), obtained by a gcd calculation between polynomials (in y), expresses $1/D$ as a polynomial in y.] Then, all derivatives of y live in the space spanned over $\mathbb{C}(z)$ by $1, y, \ldots, y^{d-1}$, with $d = \deg_y P(z, y)$. (The fact that algebraic functions are holonomic was known to Abel [1, p. 287], and an algorithm has been described in recent times by Comtet [128].) The closure under algebraic substitutions ($y \mapsto y(z)$) asserted in Theorem B.2 can be established along similar lines. ◁

Zeilberger observed that holonomic functions with coefficients in $\mathbb{Q}$ can be specified by a *finite amount of information*. Equality in this subclass is then a decidable property, as the following skeleton algorithm suggests (detailed validity conditions are omitted).

> **Algorithm Z**: *Decide whether two holonomic functions $A(z)$, $B(z)$ are equal*
> Let Σ, T be holonomic descriptions of A, B (by equations or systems);
> Compute a holonomic differential equation Υ for $h := A - B$;
> Let e be the order of Υ.
> Output 'equal' iff $h(0) = h'(0) = \cdots = h^{(e-1)}(0) = 0$, with e the order of Υ.

The book titled "$A = B$" by Petkovšek, Wilf, and Zeilberger [480] abundantly illustrates the application of this method to combinatorial and special function identities. Interest in the approach is reinforced by the existence of powerful symbolic manipulation systems and algorithms: Salvy and Zimmermann [531] have implemented univariate algebraic closure operations; Chyzak and Salvy [120, 123] have developed algorithms for multivariate holonomicity discussed below.

***Example* B.1.** *The Euler–Landen identities for dilogarithms.* Let as usual $\mathrm{Li}_\alpha(z) := \sum_{n \geq 1} z^n / n^\alpha$ represent the polylogarithm function (p. 408). Around 1760, Landen and Euler discovered the dilogarithmic identity [52, p. 247],

$$(20) \qquad \mathrm{Li}_2 \left(-\frac{z}{1-z} \right) = -\frac{1}{2} \log^2 (1 - z) - \mathrm{Li}_2(z),$$

which corresponds to the (easy) identity on coefficients (extract $[z^n]$)

$$\text{(21)} \qquad \sum_{k=1}^{n} \binom{n-1}{k-1} \frac{(-1)^k}{k^2} = -\frac{1}{n^2} - \sum_{k=1}^{n-1} \frac{1}{k(n-k)},$$

and specializes (at $z = 1/2$) to the infinite series evaluation

$$\text{Li}_2 \left(\frac{1}{2} \right) \equiv \sum_{n \geq 1} \frac{1}{n^2 2^n} = \frac{\pi^2}{12} - \frac{1}{2} \log^2 2.$$

Write A and B for the left and right sides of (20), respectively. The differential equations for A, B are built in stages, according to closure properties:

$$\text{(22)} \qquad \begin{array}{ll} \text{Li}_1(z): & (1-z)\partial^2 y - \partial y = 0 \\ \text{Li}_1(z)^2: & (1-z)^2\partial^3 y + 3(1-z)\partial^2 y + \partial y = 0 \\ \text{Li}_2(z): & z(1-z)\partial^3 y + (2-3z)\partial^2 y - \partial y = 0 \\ B(z): & z^3(36z^5 + \cdots)(1-z)^6 \partial^9 y + \cdots - 48(225z^5 + \cdots)\partial y = 0 \\ A(z): & z(1-z)^2\partial^3 y + (1-z)(2-5z)\partial^2 y - (3-4z)\partial y = 0 \end{array}$$

Thus, $A - B$ lives *a priori* in a vector space of dimension $12 = 3 + 9$. It thus suffices to *check* the coincidence of the expansions of both members of (20) up to order 12 in order to *prove* the identity $A = B$. (An upper bound on the dimension of the vector space is actually enough.) Equivalently, given the automatic computations of (22), it suffices to *verify* sufficiently many cases of the identity (21) in order to have a complete *proof* of it. .... ∎

▷ **B.13.** *Holonomic functions as solutions of systems.* (This is a simple outcome of Note VII.48, p. 522.) A holonomic function $y(z)$ which satisfies a linear differential equation of order m with coefficients in $\mathbb{C}(z)$ is also the first component of a first-order differential system of dimension m with rational coefficients: $y(z) = Y_1(z)$, where

$$\text{(23)} \qquad \begin{cases} \dfrac{d}{dz} Y_1(z) &= a_{11}(z)Y_1 + \cdots + a_{1m}(z)Y_m(z) \\ \quad \vdots & \quad \vdots \qquad \qquad \vdots \\ \dfrac{d}{dz} Y_m(z) &= a_{m1}(z)Y_1 + \cdots + a_{mm}(z)Y_m(z), \end{cases}$$

where each $a_{i,j}(z)$ is a rational function. Conversely, any solution of a system (23) with the $a_{i,j} \in \mathbb{C}(z)$ is holonomic in the sense of Definition B.1.   ◁

▷ **B.14.** *The Laplace transform.* Let $f(z) = \sum_{n \geq 0} f_n z^n$ be a formal power series. Its (formal) *Laplace transform* $g = \mathcal{L}[f]$ is defined as the formal power series:

$$\mathcal{L}[f](x) = \sum_{n=0}^{\infty} n! f_n x^n.$$

(Thus Laplace transforms convert EGFs into OGFs.) Under suitable convergence conditions, the Laplace transform is analytically representable by

$$\mathcal{L}[f](x) = \int_0^{\infty} f(xz)e^{-z}\, dz.$$

The following property holds: *A series is holonomic if and only if its Laplace transform is holonomic.* [Hint: use P–recurrences (19).]   ◁

▷ **B.15.** *Hypergeometric functions.* It is customary to employ the notation $(a)_n$ for representing the falling factorial $a(a-1)\cdots(a-n+1)$. The function of one variable, z, and three parameters, a, b, c, defined by

$$(24) \qquad F[a, b; c; z] = 1 + \sum_{n=1}^{\infty} \frac{(a)_n (b)_n}{(c)_n} \frac{z^n}{n!},$$

is known as a *hypergeometric function*. It satisfies the differential equation

$$(25) \qquad z(1-z)\frac{d^2 y}{dz^2} + (c - (a+b+1)z)\frac{dy}{dz} - aby = 0,$$

and is consequently a holonomic function. An accessible introduction appears in [604, Ch XIV].

The generalized hypergeometric function (or series) depends on $p + q$ parameters $a_1, \ldots, a_p$ and $c_1, \ldots, c_q$, and is defined by

$$(26) \qquad {}_pF_q[a_1, \ldots, a_p; c_1, \ldots, c_q; z] = 1 + \sum_{n=1}^{\infty} \frac{(a_1)_n \cdots (a_p)_n}{(c_1)_n \cdots (c_q)_n} \frac{z^n}{n!},$$

so that F in (24) is a ${}_2F_1$. Hypergeometric functions satisfy a rich set of identities [193, 542], many of which can be verified (though not *discovered*) by Algorithm Z. ◁

Multivariate holonomic functions. Let $\mathbf{z} = (z_1, \ldots, z_m)$ be a collection of variables and $C(\mathbf{z})$ the field of all rational fractions in the variables $\mathbf{z}$. For $\mathbf{n} = (n_1, \ldots, n_m)$, we define $\mathbf{z}^{\mathbf{n}}$ to be $z_1^{n_1} \cdots z_m^{n_m}$ and let $\partial^{\mathbf{n}}$ represent $\partial_{z_1^{n_1}} \cdots \partial_{z_m^{n_m}}$.

Definition B.2. *A multivariate formal power series (or function) $f(\mathbf{z})$ is said to be* holonomic *if the vector space over $\mathbb{C}(\mathbf{z})$ spanned by the set of all derivatives $\{\partial^{\mathbf{n}} f(\mathbf{z})\}$ is finite dimensional.*

Since the partial derivatives $\partial_{z_1}^j f$ are bound, a multivariate holonomic function satisfies a differential equation of the form

$$c_{1,0}(\mathbf{z})\frac{\partial^{r_1}}{\partial z_1^{r_1}} f(\mathbf{z}) + \cdots + c_{1,r_1}(\mathbf{z}) f(\mathbf{z}) = 0,$$

and similarly for $z_2, \ldots, z_m$. (Any system of equations with possibly mixed partial derivatives that allows one to determine all partial derivatives in terms of a finite number of them serves to define a multivariate holonomic function.) Denominators can be cleared, upon multiplication by the l.c.m of all the denominators that figure in the system of defining equations. There results that coefficients of multivariate holonomic functions satisfy particular systems of recurrence equations with polynomial coefficients, which are characterized in [410].

Given $f(\mathbf{z})$ viewed as a function of z_1, z_2 (the remaining variables being parameters) and abbreviated as $f(z_1, z_2)$, the *diagonal* with respect to variables z_1, z_2 is

$$\mathrm{Diag}_{z_1, z_2}[f(z_1, z_2)] = \sum_{\nu} f_{\nu, \nu} z_1^{\nu}, \qquad \text{where} \quad f(z_1, z_2) = \sum_{n_1, n_2} f_{n_1, n_2} z_1^{n_1} z_2^{n_2}.$$

The Hadamard product is defined, as in the univariate case, with respect to a specific variable (e.g., z_1).

Theorem B.3 (Multivariate holonomic closure). *The class of multivariate holonomic functions is closed under the following operations: sum (+), product (×), Hadamard* *product ($\odot$), differentiation (∂), indefinite integration ($\int$), algebraic substitution, specialization (setting some variable to a constant), and diagonal.*

An elementary proof of this remarkable theorem (in the sense that it does not appeal to higher concepts of differential algebra) is given by Lipshitz in [409, 410]. The closure theorem and its companion algorithms [120, 570] make it possible to prove, or verify, automatically identities, many of which are non-trivial. For instance, in his proof of the irrationality of the number $\zeta(3) = \sum_{n \geq 1} 1/n^3$, Apéry introduced the combinatorial sequence,

$$(27) \qquad A_n = \sum_{k=0}^{n} \binom{n}{k}^2 \binom{n+k}{k}^2,$$

for which a proof was needed [588] of the fact that it satisfies the recurrence

$$(28) \qquad (n+1)^3 B_n + (n+2)^3 B_{n+2} - (2n+3)(17n^2 + 51n + 39) B_{n+1} = 0,$$

with $B_1 = 5$, $B_2 = 73$. Obviously, the generating function $B(z)$ of the sequence (B_n) as defined by the P–recurrence (28) is univariate holonomic. Repeated use of the multivariate closure theorem shows that the ordinary generating function $A(z)$ of the sequence A_n of (28) is holonomic. (Indeed, start from the explicit

$$\sum_{n_1, n_2} \binom{n_1}{n_2} z_1^{n_1} z_2^{n_2} = \frac{1}{1 - z_1(1 + z_2)}, \qquad \sum_{n_1, n_2} \binom{n_1 + n_2}{n_2} z_1^{n_1} z_2^{n_2} = \frac{1}{1 - z_1 - z_2},$$

and apply suitable Hadamard products and diagonal operations.) This gives an ordinary differential equation satisfied by $A(z)$. The proof is then completed by checking that A_n and B_n coincide for enough initial values of n.

Holonomic functions in infinitely many variables. Let f be a power series in infinitely many variables $x_1, x_2, \ldots$. Let $S \subset \mathbb{Z}_{\geq 1}$ be a subset of indices. We write f_S for the specialization of f in which all the variables whose indices do not belong to S are set to 0. Following Gessel [289], we say that the series f is holonomic if, for each finite S, the specialization f_S is holonomic (in the variables x_s for $s \in S$). Gessel has developed a powerful calculus in the case of series f that are *symmetric functions*, with stunning consequences for combinatorial enumeration.

An undirected graph is called k–regular if every vertex has exact degree k. A *standard Young tableau* is the Ferrers diagram of an integer partition, filled with consecutive integers in a way that is increasing along rows and columns. The classical Robinson–Schensted–Knuth correspondence establishes a bijection between permutations of size n and pairs of Young tableaux of size n having the same shape. The common height of the tableaux in the pair associated to a permutation σ coincides with the length of the longest increasing subsequence of σ. A $k \times n$ *Latin rectangle* is a $k \times n$ matrix with elements in the set $\{1, 2, \ldots, n\}$ such that entries in each row and column are distinct. (It is thus a k–tuple of "discordant" permutations.)

Gessel's calculus [288, 289] provides a unified approach for establishing the holonomic character of many generating functions of combinatorial structures, such as: Young tableaux, permutations of uniform multisets, increasing subsequences in permutations, Latin rectangles, regular graphs, matrices with fixed row and column sums, and so on. For instance: *the generating functions of Latin rectangles and Young tableaux of height at most k, of k–regular graphs, and of permutations with longest increasing subsequence of length k are* holonomic *functions.* In particular, the number $Y_{n,k}$ of permutations of size n with longest increasing subsequence $\leq k$ satisfies

$$(29) \qquad \sum_{n \geq 0} Y_{n,k} \frac{z^{2n}}{(n!)^2} = \det\left[I_{|i-j|}(2z)\right]_{1 \leq i,j \leq k}, \quad \text{where } I_\nu(2z) = \sum_{n=0}^{\infty} \frac{x^{2n+\nu}}{n!(n+\nu)!},$$

that is, a corresponding GF is expressible as a determinant of Bessel functions. Other applications are described in [122, 444].

The asymptotic problems relative to the holonomic framework are examined in Subsection VII. 9.1, p. 518 and Section VIII. 7, p. 581.

B.5. Implicit Function Theorem

In its real-variable version, the Implicit Function Theorem asserts that, for a sufficiently smooth function $F(z, w)$ of two variables, a solution to the equation $F(z, w) = 0$ exists in the vicinity of a solution point (z_0, w_0) (therefore satisfying $F(z_0, w_0) = 0$) provided the partial derivative satisfies $F'_w(z_0, w_0) \neq 0$. This theorem admits a complex extension, which is essential for the analysis of recursive structures.

Without loss of generality, one restricts attention to $(z_0, w_0) = (0, 0)$. We consider here a function $F(z, w)$ that is analytic in two complex variables in the sense that it admits a convergent representation valid in a polydisc,

$$(30) \qquad F(z, w) = \sum_{m,n \geq 0} f_{m,n} z^m w^n, \qquad |z| < R, \quad |w| < S.$$

for some $R, S > 0$ (cf Appendix B.8: *Several complex variables*, p. 767).

Theorem B.4 (Analytic Implicit Functions). *Let F be bivariate analytic near $(0, 0)$. Assume that $F(0, 0) \equiv f_{0,0} = 0$ and $F'_w(0, 0) \equiv f_{0,1} \neq 0$. Then, there exists a unique function $f(z)$ analytic in a neighbourhood $|z| < \rho$ of 0 such that $f(0) = 0$ and*

$$F(z, f(z)) = 0, \qquad |z| < \rho.$$

▷ **B.16.** *Proofs of the Implicit Function Theorem.* See Hille's book [334] for details.

(i) Proof by residues. Make use of the principle of the argument and Rouché's Theorem to see that the equation $F(z, w)$ has a unique solution near 0 for $|z|$ small enough. Appeal then to the result, based on the residue theorem, that expresses the sum of the solutions to an equation as a contour integral: with C a small enough contour around 0 in the w–plane, one has

$$(31) \qquad f(z) = \frac{1}{2i\pi} \int_C w \frac{F'_w(z, w)}{F(z, w)} \, dw$$

(Note IV.39, p. 270), which is checked to represent an analytic function of z.

(ii) Proof by majorant series. Set $G(z, w) := w - f_{0,1}^{-1} F(z, w)$. The equation $F(z, w) = 0$ becomes the fixed-point equation $w = G(z, w)$. The bivariate series G has its coefficients dominated termwise by those of

$$\widehat{G}(z, w) = \frac{A}{(1 - z/R)(1 - w/S)} - A - A\frac{w}{S}.$$

The equation $w = \widehat{G}(z, w)$ is quadratic. It admits a solution $\widehat{f}(z)$ analytic at 0,

$$\widehat{f}(z) = A\frac{z}{R} + \frac{A(A^2 + AS + S^2)}{S^2} \frac{z^2}{R^2} + \cdots,$$

whose coefficients dominate termwise those of f.

(iii) Proof by Picard's method of successive approximants. With G as before, define the sequence of functions

$$\phi_0(z) := 0; \quad \phi_{j+1}(z) = G(z, \phi_j(z)),$$

each analytic in a small neighbourhood of 0. Then $f(z)$ can be obtained as

$$f(z) = \lim_{j \to \infty} \phi_j(z) \equiv \phi_0(z) - \sum_{j=0}^{\infty} \left(\phi_j(z) - \phi_{j+1}(z) \right),$$

which is itself checked to be analytic near 0 by the geometric convergence of the series. ◁

Weierstrass Preparation. The Weierstrass Preparation Theorem (WPT) also familiarly known as *Vorbereitungssatz* is a useful complement to the Implicit Function Theorem.

Given a collection $\mathbf{z} = (z_1, \ldots, z_m)$ of variables, we designate as usual by $\mathbb{C}[[\mathbf{z}]]$ the ring of formal power series in indeterminates $\mathbf{z}$. We let $C\{\mathbf{z}\}$ denote the subset of these that are convergent in a neighbourhood of $(0, \ldots, 0)$, i.e., analytic (cf Appendix B.8: *Several complex variables.*, p. 767).

Theorem B.5 (Weierstrass Preparation). *Let $F = F(z_1, \ldots, z_m)$ in $\mathbb{C}[[\mathbf{z}]]$ (respectively, $\mathbb{C}\{\mathbf{z}\}$) be such that $F(0, \ldots, 0) = 0$ and F depends on at least one of the z_j with $j \geq 2$ (i.e., $F(0, z_2, \ldots, z_m)$ is not identically 0). Define a Weierstrass polynomial to be a polynomial of the form*

$$W(z) = z^d + g_1 z^{d-1} + \cdots + g_d,$$

where $g_j \in \mathbb{C}[[z_2, \ldots, z_m]]$ (respectively, $g_j \in \mathbb{C}\{z_2, \ldots, z_m\}$), with $g_j(0, \ldots, 0) = 0$. Then, F admits a unique factorization

$$F(z_1, z_2, \ldots, z_m) = W(z_1) \cdot X(z_1, \ldots, z_m),$$

where $W(z)$ is a Weierstrass polynomial and X is an element of $\mathbb{C}[[z_1, \ldots, z_m]]$ (respectively, $\mathbb{C}\{z_1, \ldots, z_m\}$) satisfying $X(0, 0 \ldots, 0) \neq 0$.

▷ **B.17.** *Weierstrass Preparation: sketch of a proof.* An accessible proof and a discussion of the formal algebraic result are found in Abhyankar's lecture notes [2, Ch. 16].

The analytic version of the theorem is the one of use to us in this book. We prove it in the representative case where $m = 2$ and write $F(z, w)$ for $F(z_1, z_2)$. First, the number of roots of the equation $F(z, w) = 0$ is given by the integral formula

$$(32) \qquad \frac{1}{2i\pi} \int_\gamma \frac{F'_w(z, w)}{F(z, w)} \, dw,$$

where γ is a small contour encircling 0 in the w-plane. There exists a sufficiently small open set Ω containing 0 such that the quantity (32), which is an analytic function of z while being an integer, is constant, and thus necessarily equal to its value at $z = 0$, which we call d. The quantity d is the multiplicity of 0 as a root of the equation $F(0, w) = 0$. In other words, we have shown that if $F(0, w) = 0$ has d roots equal to 0, then there are d values of w near 0 (within γ) such that $F(z, w) = 0$, provided z remains small enough (within Ω).

Let $y_1, \ldots, y_d$ be these d roots. Then, we have for the power sum symmetric functions,

$$y_1^r + \cdots + y_r^d = \frac{1}{2i\pi} \int_\gamma \frac{F'_w(z, w)}{F(z, w)} w^r \, dw,$$

which are analytic functions of z when z is sufficiently near to 0. There results from relations between symmetric functions (Note III.64, p. 88) that $y_1, \ldots, y_r$ are the solutions of a polynomial equation with analytic coefficients, W, which is a uniquely defined Weierstrass polynomial. The factorization finally results from the fact that F/W has removable singularities. ◁

In essence, by Theorem B.5, functions implicitly defined by a transcendental equation (an equation $F = 0$) are locally of the same nature as algebraic functions (corresponding to the equation $W = 0$). In particular, for $m = 2$, when the solutions have singularities, these singularities can only be branch points and companion Puiseux expansions hold (Section VII. 7, p. 493). The theorem acquires even greater importance when perturbative singular expansions (corresponding to $m \geq 3$) become required for the purpose of extracting limit laws in Chapter IX.

▷ **B.18.** *Multivariate implicit functions.* The following extension of Theorem B.4 is important, with regard to the solution of *systems* of equations (Section VII. 6, p. 482). Its statement [104, §IV.5] makes use of the notion of analytic functions of several variables (Appendix B.8, p. 767). **Theorem B.6** (Multivariate implicit functions). *Let* $f_i(x_1, \ldots, x_m; z_1, \ldots, z_p)$, *with* $i = 1, \ldots, m$, *be analytic functions in the neighbourhood of a point* $x_j = a_j$, $z_k = c_k$. *Assume that the* Jacobian determinant *defined as*

$$J := \det \left(\frac{\partial f_i}{\partial x_j} \right)$$

is non-zero at the point considered. Then the equations (in the x_j)

$$y_i = f_i(x_1, \ldots, x_m; z_1, \ldots, z_p), \qquad i = 1, \ldots, m,$$

admit a solution with the x_j *near to the* a_j, *when the* z_k *are sufficiently near to the* c_k *and the* y_i *near to the* $b_i := f_i(a_1, \ldots, a_m; c_1, \ldots, c_p)$: *one has*

$$x_j = g_j(y_1, \ldots, y_m; z_1, \ldots, z_p),$$

where each g_j *is analytic in a neighbourhood of the point* $(b_1, \ldots, b_m; c_1, \ldots, c_p)$.

The basic idea is that the linear approximations expressed by the *Jacobian matrix* $\left(\frac{\partial f_i}{\partial x_j} \right)$ can be inverted. Hence the x_j depend locally linearly on the y_i, z_k; hence they are analytic. ◁

B.6. Laplace's method

The method of Laplace serves to estimate asymptotically *real* integrals depending on a large parameter n (which may be an integer or a real number). Although it is primarily a real analysis technique, we present it in detail, given its relevance to the saddle-point method, which deals instead with *complex* contour integrals.

Case study: a Wallis integral. In order to demonstrate the essence of the method, consider first the problem of estimating asymptotically the Wallis integral

$$(33) \qquad\qquad I_n := \int_{-\pi/2}^{\pi/2} (\cos x)^n \, dx,$$

as $n \to +\infty$. The cosine attains its maximum at $x = 0$ (where its value is 1), and since the integrand of I_n is a large power, the contribution to the integral outside any fixed segment containing 0 is exponentially small and can consequently be discarded for all asymptotic purposes. A glance at the plot of $\cos^n x$ as n varies (Figure B.2) also suggests that the integrand tends to conform to a bell-shaped profile near the centre as n increases. This is not hard to verify: set $x = w/\sqrt{n}$, then a local expansion yields

$$(34) \qquad \cos^n x \equiv \exp(n \log \cos(x)) = \exp \left(-\frac{w^2}{2} + O(n^{-1}w^4) \right),$$

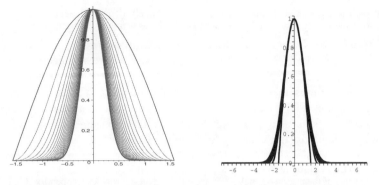

Figure B.2. Plots of $\cos^n x$ [left] and $\cos^n (w/\sqrt{n})$ [right], for $n = 1 \mathinner{.\,.} 20$.

the approximation being valid as long as $w = O(n^{1/4})$. Accordingly, we choose (somewhat arbitrarily)

$$\kappa_n := n^{1/10},$$

and define the central range by $|w| \leq \kappa_n$. These considerations suggest to rewrite the integral I_n as

$$I_n = \frac{1}{\sqrt{n}} \int_{-\pi\sqrt{n}/2}^{+\pi\sqrt{n}/2} \left(\cos \frac{w}{\sqrt{n}} \right)^n dw,$$

and expect under this new form an approximation by a Gaussian integral arising from the central range.

Laplace's method proceeds in three steps.
  (*i*)   *Neglect the tails of the original integral.*
 (*ii*)   *Centrally approximate the integrand by a Gaussian.*
(*iii*)   *Complete the tails of the Gaussian integral.*

In the case of the cosine integral (33), the chain is summarized in Figure B.3. Details of the analysis follow.

  (*i*) *Neglect the tails of the original integral*: By (34), we have

$$\cos^n \left(\frac{\kappa_n}{\sqrt{n}} \right) \sim \exp \left(-\frac{1}{2} n^{1/5} \right),$$

and, since the integrand is unimodal, this exponentially small quantity bounds the integrand throughout $|w| > \kappa_n$, that is, on a large part of the integration interval. This gives

$$(35) \qquad I_n = \int_{-\kappa_n/\sqrt{n}}^{+\kappa_n/\sqrt{n}} \cos^n x \, dx + O \left(\exp \left(-\frac{1}{2} \kappa_n^2 \right) \right),$$

and the error term is of the order of $\exp(-\frac{1}{2} n^{1/5})$.

$$
\begin{aligned}
\int_{-\pi/2}^{\pi/2} \cos^n x \, dx \quad &= \quad \frac{1}{\sqrt{n}} \int_{-\frac{\pi}{2}\sqrt{n}}^{\frac{\pi}{2}\sqrt{n}} \left(\cos \frac{w}{\sqrt{n}} \right)^n dw \quad && \text{Set } x = w/\sqrt{n}; \text{ choose } \kappa_n = n^{1/10} \\[2mm]
&\sim \quad \frac{1}{\sqrt{n}} \int_{-\kappa_n}^{\kappa_n} \left(\cos \frac{w}{\sqrt{n}} \right)^n dw \quad && [\textit{Neglect the tails}] \\[2mm]
&\sim \quad \frac{1}{\sqrt{n}} \int_{-\kappa_n}^{\kappa_n} e^{-w^2/2} dw \quad && [\textit{Central approxim.}] \\[2mm]
&\sim \quad \frac{1}{\sqrt{n}} \int_{-\infty}^{\infty} e^{-w^2/2} dw \quad && [\textit{Complete the tails}] \\[2mm]
&\sim \quad \sqrt{\frac{2\pi}{n}}.
\end{aligned}
$$

Figure B.3. A typical application of the Laplace method.

(*ii*) *Centrally approximate the integrand by a Gaussian*: In the central region, we have

$$
\begin{aligned}
(36) \qquad I_n^{(1)} \quad &:= \quad \int_{-\kappa_n/\sqrt{n}}^{+\kappa_n/\sqrt{n}} \cos^n x \, dx \\[2mm]
&= \quad \frac{1}{\sqrt{n}} \int_{-\kappa_n}^{+\kappa_n} e^{-w^2/2} \exp\left(O(n^{-1}w^4) \right) dw \\[2mm]
&= \quad \frac{1}{\sqrt{n}} \int_{-\kappa_n}^{+\kappa_n} e^{-w^2/2} \left(1 + O(n^{-1}w^4) \right) dw \\[2mm]
&= \quad \frac{1}{\sqrt{n}} \int_{-\kappa_n}^{+\kappa_n} e^{-w^2/2} \, dw + O(n^{-3/5}),
\end{aligned}
$$

given the uniformity of approximation (34) for w in the integration interval.

(*iii*) *Complete the tails of the Gaussian integral*: The incomplete Gaussian integral in the last line of (36) can be easily estimated once it is observed that its tails are small. Precisely, one has, for $W \geq 0$,

$$
\int_W^\infty e^{-w^2/2} \, dw \leq e^{-W^2/2} \int_0^\infty e^{-h^2/2} \, dh \equiv \sqrt{\frac{\pi}{2}} e^{-W^2/2}
$$

(by the change of variable $w = W + h$). Thus,

$$
(37) \qquad \int_{-\kappa_n}^{+\kappa_n} e^{-w^2/2} \, dw = \int_{-\infty}^{+\infty} e^{-w^2/2} \, dw + O\left(\exp\left(-\frac{1}{2}\kappa_n^2 \right) \right).
$$

It now suffices to collect the three approximations, (35), (36), and (37): we have obtained in this way.

$$
(38) \qquad I_n = \frac{1}{\sqrt{n}} \int_{-\infty}^{+\infty} e^{-w^2/2} \, dw + O(n^{-3/5}) \equiv \sqrt{\frac{2\pi}{n}} + O(n^{-3/5}).
$$

These three steps comprise Laplace's method.

▷ **B.19.** *A complete asymptotic expansion.* In the asymptotic scale of the problem, the exponentially small errors in the tails can be completely neglected; the main error in (38) then arises from the central approximation (34), and its companion $O(w^4 n^{-1})$ term. This can easily be improved and it suffices to appeal to further terms in the expansion of $\log \cos x$ near 0. For instance, one has (with $x = w/\sqrt{n}$):

$$\cos^n x = e^{-w^2/2}\left(1 - w^4/12n + O(n^{-2}w^8)\right).$$

Proceeding as before, we find that a further term in the expansion of I_n is obtained by considering the additive correction

$$\epsilon_n := -\frac{1}{\sqrt{n}} \int_{-\infty}^{+\infty} e^{-w^2/2}\left(\frac{w^4}{12n}\right)dw \equiv -\sqrt{\frac{\pi}{8n^3}},$$

so that

$$I_n = \sqrt{\frac{2\pi}{n}} - \sqrt{\frac{\pi}{8n^3}} + O(n^{-17/10}).$$

A complete asymptotic expansion in the scale $n^{-1/2}, n^{-3/2}, n^{-5/2}, \ldots$ can easily be obtained in this way. ◁

▷ **B.20.** *Wallis integrals, central binomials, and the squaring of the circle.* The integral I_n is an integral considered by John Wallis (1616–1703). It can be evaluated through partial integration or by its relation to the Beta integral (Note B.10, p. 747) as $I_n = \Gamma(\frac{1}{2})\Gamma(\frac{n}{2} + \frac{1}{2})/\Gamma(\frac{n}{2} + 1)$. There results ($n \mapsto 2n$):

$$\binom{2n}{n} \sim \frac{2^{2n}}{\sqrt{\pi n}}\left(1 - \frac{1}{8n} + \frac{1}{128n^2} + \frac{5}{1024n^3} - \cdots\right),$$

which is yet another avatar of Stirling's formula. Wallis' evaluation, when combined with its asymptotic estimate, is, in Euler's terms, a formula for "squaring the circle"

$$\frac{\pi}{4} = \frac{2 \cdot 4 \cdot 4 \cdot 6 \cdot 6 \cdot 8 \cdot 8 \cdot 10 \cdot 10}{3 \cdot 3 \cdot 5 \cdot 5 \cdot 7 \cdot 7 \cdot 9 \cdot 9 \cdot 11}\&c,$$

albeit one that cannot be finitely implemented with ruler and compass. ◁

General case of large powers. Laplace's method applies under general conditions to integrals involving large powers of a fixed function.

Theorem B.7 (Laplace's method). *Let f and g be indefinitely differentiable real-valued functions defined over some compact interval I of the real line. Assume that $|g(x)|$ attains its maximum at a unique point x_0 interior to I and that $f(x_0), g(x_0), g''(x_0) \neq 0$. Then, the integral*

$$I_n := \int_I f(x)g(x)^n\, dx$$

admits a complete asymptotic expansion:

$$(39) \qquad I_n \sim \sqrt{\frac{2\pi}{\lambda n}} f(x_0)g(x_0)^n\left(1 + \sum_{j \geq 1}\frac{\delta_j}{n^j}\right), \qquad \lambda := -\frac{g''(x_0)}{g(x_0)}.$$

▷ **B.21.** *Proof of Laplace's method.* Assume first that $f(x) \equiv 1$. Then, one chooses κ_n as a function tending slowly to infinity like before ($\kappa_n = n^{1/10}$ is suitable). It suffices to expand

$$I_n^{(1)} := \int_{x_0-\kappa_n/\sqrt{n}}^{x_0+\kappa_n/\sqrt{n}} e^{n\log g(x)}\, dx,$$

as the difference $I_n - I_n^{(1)}$ is exponentially small. Set first $x = x_0 + X$ and

$$L(X) := \log g(x_0 + X) - \log g(x_0) + \lambda\frac{X^2}{2},$$

so that, with $w = X\sqrt{n}$, the central contribution becomes:

$$I_n^{(1)} = \frac{g(x_0)^n}{\sqrt{n}} \int_{-\kappa_n}^{\kappa_n} e^{-\lambda w^2/2} e^{nL(w/\sqrt{n})} \, dw.$$

Then, expanding $L(X)$ to any order M,

$$L(X) = \sum_{j=3}^{M-1} \ell_j X^j + O(X^M),$$

shows that $e^{nL(w/\sqrt{n})}$ admits a full expansion in descending powers of $\sqrt{n}$:

$$e^{nL(w/\sqrt{n})} \sim 1 + \frac{\ell_3 w^3}{\sqrt{n}} + \frac{2\ell_4 w^4 + \ell_3^2 w^6}{2n} + \cdots.$$

There, by construction, the coefficient of $n^{-k/2}$ is a polynomial $E_k(w)$ of degree $3k$. This expression can be truncated to any order, resulting in

$$I_n^{(1)} = \frac{g(x_0)^n}{\sqrt{n}} \int_{-\kappa_n}^{\kappa_n} e^{-\lambda w^2/2} \left(1 + \sum_{k=1}^{M-1} \frac{E_k(w)}{n^{k/2}} + O\left(\frac{1+w^{3M}}{n^{M/2}} \right) \right) dw.$$

One can then complete the tails at the expense of exponentially small terms since the Gaussian tails are exponentially small.

The full asymptotic expansion is revealed by the following device: for any power series $h(w)$, introduce the Gaussian transform,

$$\mathfrak{G}[f] := \int_{-\infty}^{\infty} e^{-w^2/2} f(w) \, dw,$$

which is understood to operate by linearity on integral powers of w,

$$\mathfrak{G}[w^{2r}] = 1 \cdot 3 \cdots (2r-1)\sqrt{2\pi}, \qquad \mathfrak{G}[w^{2r+1}] = 0.$$

Then, the complete asymptotic expansion of I_n is obtained by the formal expansion

(40) $$\frac{g(x_0)^n}{\sqrt{n\lambda}} \cdot \mathfrak{G}\left[\exp\left(\lambda^{-3/2} w^3 y \tilde{L}(\lambda^{-1/2} wy) \right) \right], \qquad \tilde{L}(X) := \frac{1}{X^3} L(X), \quad y \mapsto \frac{1}{\sqrt{n}}.$$

The addition of the prefactor $f(x)$ (omitted so far) induces a factor $f(x_0)$ in the main term of the final result and it affects the coefficients in the smaller order terms in a computable manner. Details are left as an exercise to the reader. $\triangleleft$

▷ **B.22. *The next term?*** One has (with $f_j := f^{(j)}(x_0)$, etc):

$$\frac{I_n \sqrt{\lambda n}}{\sqrt{2\pi} g(x_0)^n} = f_0 + \frac{-9\lambda^3 f_0 + 12\lambda^2 f_2 + 12\lambda f_1 g_3 + 3\lambda f_0 g_4 + 5 g_3^2 f_0}{24\lambda^3 n} + O(n^{-2}),$$

which is best determined using a symbolic manipulation system. $\triangleleft$

The method is amenable to a large number of extensions. Roughly it requires a point where the integrand is maximized, which induces some sort of exponential behaviour, local expansions then allowing for a replacement by standard integrals.

▷ **B.23. *Special cases of Laplace's method.*** When $f(x_0) = 0$, the integral normalizes to an integral of the form $\int w^2 e^{-w^2/2}$. If $g'(x_0) = g''(x_0) = g^{(iii)}(x_0) = 0$ but $g^{(iv)}(x_0) \neq 0$ then a factor $\Gamma(1/4)$ replaces the characteristic $\sqrt{\pi} \equiv \Gamma(1/2)$. [Hint: $\int_0^\infty \exp(-w^\beta) w^\alpha \, dw = \beta^{-1}\Gamma((\alpha+1)\beta^{-1})$.] If the maximum is attained at one end of the interval $I = [a, b]$ while $g'(x_0) = 0$, $g''(x_0) \neq 0$, then the estimate (39) must be multiplied by a factor of $1/2$. If the maximum is attained at one end of the interval I while $g'(x_0) \neq 0$, then the right normalization is $w = x/n$ and the integrand is reducible to an exponential e^{-w}. Here are some dominant asymptotic terms:

$x_0 \neq a, b \quad g''(x_0) \neq 0, f(x_0) = 0$	$\sqrt{\frac{\pi}{2\lambda^5 n^3}} g(x_0)^n (\lambda f''(x_0) + f'(x_0) g'''(x_0))$
$x_0 \neq a, b \quad g''(x_0) = 0, g^{(iv)}(x_0) \neq 0$	$\Gamma(\frac{1}{4}) \sqrt[4]{\frac{3}{2\lambda^* n}} f(x_0) g(x_0)^n \quad \left(\lambda^* = -\frac{g^{(iv)}(x_0)}{g(x_0)} \right)$
$x_0 = a \quad f(x_0) \neq 0, g'(x_0) \neq 0$	$-\frac{1}{ng'(x_0)} f(x_0) g(x_0)^{n+1}$.

A similar analysis is employed in Section VIII. 10, p. 600, when we discuss coalescence cases of the saddle-point method. ◁

Example B.2. *Stirling's formula via Laplace's method.* Start from an integral representation involving $n!$, namely,

$$I_n := \int_0^\infty e^{-nx} x^n \, dx = \frac{n!}{n^{n+1}} \, .$$

This is a direct case of application of the theorem, except for the fact that the integration interval is not compact. The integrand attains its maximum at $x_0 = 1$ and the remainder integral $\int_2^\infty$ is accordingly exponentially small as proved by the chain

$$
\begin{aligned}
\int_2^\infty e^{-nx} x^n \, dx &= (2e^{-2})^n \int_0^\infty \left(1 + \frac{x}{2}\right)^n e^{-nx} \, dx && [x \mapsto x + 2] \\
&< (2e^{-2})^n \int_0^\infty e^{nx/2} e^{-nx} \, dx = \frac{2}{n}(2e^{-2})^n && [\log(1 + x/2) < x/2].
\end{aligned}
$$

Then the integral from 0 to 2 is amenable to the standard version of Laplace's method as stated in Theorem B.7 to the effect that

$$n! = n^n e^{-n} \sqrt{2\pi n} \left(1 + O\left(\frac{1}{n}\right)\right).$$

The asymptotic expansion of I_n is derived from (40) and involves the combinatorial GF

$$(41) \qquad H(z, u) := \exp\left(u \left(\log(1 - z)^{-1} - z - \frac{z^2}{2}\right)\right).$$

The noticeable fact is that $H(z, u)$ is the exponential BGF of generalized derangements involving no cycles of length 1 or 2, with z marking size and u marking the number of cycles:

$$H(z, u) = \sum_{n,k \geq 0} h_{n,k} u^k \frac{z^n}{n!} = 1 + \frac{1}{3} u z^3 + \frac{1}{4} u z^4 + \frac{1}{5} u z^5 + (\frac{1}{6} u + \frac{1}{18} u^2) z^6 + (\frac{1}{7} u + \frac{1}{12} u^2) z^7 + \cdots.$$

Then, a complete asymptotic expansion of I_n is obtained by applying the Gaussian transform $\mathfrak{G}$ to $H(wy, -y^{-2})$ (with $y = n^{-1/2}$), resulting in

$$n! \sim n^n e^{-n} \sqrt{2\pi n} \left(1 + \frac{1}{12n} + \frac{1}{288n^2} - \frac{139}{51840n^3} - \cdots\right).$$

Proposition B.1 (Stirling's formula). *The factorial function admits the asymptotic expansion:*

$$x! \equiv \Gamma(x + 1) \sim x^x e^{-x} \sqrt{2\pi x} \left(1 + \sum_{q \geq 1} \frac{c_q}{x^q}\right) \qquad (x \to +\infty).$$

The coefficients satisfy $c_q = \sum_{k=1}^{2q} \frac{(-1)^k}{2^{q+k}(q + k)!} h_{2q+2k,k}$, *where* $h_{n,k}$ *counts the number of permutations of size n having k cycles, all of length ≥ 3.*

The derivation above is due to Wrench (see [129, p. 267]). .... ∎

The scope of the method goes much beyond the case of integrals of large powers. Roughly, what is needed is a localization of the main contribution of an integral to a smaller range ("Neglect the tails") where local approximations can be applied ("Centrally approximate"). The approximate integral is then finally estimated by completing back the tails ("Complete the tails").

The Laplace method is excellently described in books by de Bruijn [143] and Henrici [329]. A thorough discussion of special cases and multidimensional integrals is found in the book by Bleistein and Handelsman [75]. Its principles are fundamental to the development of the saddle-point method in Chapter VIII.

▷ **B.24.** *The classical proof of Stirling's formula.* This proceeds from the integral

$$J_n := \int_0^\infty e^{-x} x^n \, dx \quad (= n!)$$

The maximum is at $x_0 = n$ and the central range is now $n \pm \kappa_n \sqrt{n}$. Reduction to a Gaussian integral follows, but the estimate is no longer a direct application of Theorem B.7. ◁

Laplace's method for sums. The basic principles of the method of Laplace (for integrals) can often be recycled for the asymptotic evaluation of discrete sums. Take a finite or infinite sum S_n defined by

$$S_n := \sum_k t(n, k).$$

A preliminary task consists in working out the general aspect of the family of numbers $\{t(n, k)\}$ for fixed (but large) n as k varies. In particular, one should locate the value $k_0 \equiv k_0(n)$ of k for which $t(n, k)$ is maximal. In a vast number of cases, tails can be neglected; a central approximation $\widehat{t}(n, k)$ of $t(n, k)$ for k in the "central" region near k_0 can be determined, frequently under the form [remember that we use in this book '$\approx$' in the loose sense of "approximately equal"]

$$\widehat{t}(n, k) \approx s(n)\phi\left(\frac{k - k_0}{\sigma_n}\right),$$

where ϕ is some smooth function while $s(n)$ and σ_n are scaling constants. The quantity σ_n indicates the range of the asymptotically significant terms. One may then expect

$$S_n \approx s(n) \sum_k \phi\left(\frac{k - k_0}{\sigma_n}\right).$$

Then provided $\sigma_n \to \infty$, one may further expect to approximate the sum by an integral, which after completing the tails, gives

$$S_n \approx s(n)\sigma_n \int_{-\infty}^\infty \phi(t) \, dt.$$

Example B.3. *Sums of powers of binomial coefficients.* Here is, in telegraphic style, an application to sums of powers of binomial coefficients:

$$S_n^{(r)} = \sum_{k=-n}^{+n} \binom{2n}{n + k}^r.$$

The largest term arises at $k_0 = 0$. Furthermore, one has elementarily

$$\frac{\binom{2n}{n+k}}{\binom{2n}{n}} = \frac{\left(1 - \frac{1}{n}\right) \cdots \left(1 - \frac{k-1}{n}\right)}{\left(1 + \frac{1}{n}\right) \cdots \left(1 + \frac{k}{n}\right)}.$$

By the exp–log transformation and the expansion of $\log(1 \pm x)$, one has

(42)
$$\frac{\binom{2n}{n+k}}{\binom{2n}{n}} = \exp\left(-\frac{k^2}{n} + O(k^3 n^{-2})\right).$$

This approximation holds for $k = o(n^{2/3})$, where it provides a Gaussian approximation ($\phi(x) = e^{-rx^2}$) with a span of $\sigma_n = \sqrt{n}$. Tails can be neglected, so that

$$\frac{1}{\binom{2n}{n}^r} S_n^{(r)} \sim \sum_k \exp\left(-r\frac{k^2}{n}\right),$$

say with $|k| < n^{1/2}\kappa_n$ where $\kappa_n = n^{1/10}$. Then approximating the Riemann sum by an integral and completing the tails, one gets

$$S_n^r \sim \binom{2n}{n}^r \sqrt{n} \int_{-\infty}^{\infty} e^{-rw^2}\, dw, \quad \text{that is,} \quad S_n^r \sim \frac{2^{2rn}}{\sqrt{r}} (\pi n)^{-(r-1)/2},$$

which is our final estimate. .... ∎

▷ **B.25.** *Elementary approximation of Bell numbers.* The Bell numbers counting set partitions (p. 109) are

$$B_n = n![z^n]e^{e^z - 1} = e^{-1} \sum_{k=0}^{\infty} \frac{k^n}{k!}.$$

The largest term occurs for k near e^u where u is the positive root of the equation $ue^u = n+1$; the central terms are approximately Gaussian. There results the estimate,

(43) $B_n = n!e^{-1}(2\pi)^{-1/2}(1 + u^{-1})^{-1/2} \exp\left(e^u(1 - u\log u) - \frac{1}{2}u\right)(1 + O(e^{-u})).$

This alternative to saddle-point asymptotics (p. 560) is detailed in [143, p. 108].   ◁

B.7. Mellin transforms

The Mellin transform[2] of a function f defined over $\mathbb{R}_{>0}$ is the complex-variable function $f^{\star}(s)$ defined by the integral

(44)
$$f^{\star}(s) := \int_0^{\infty} f(x)x^{s-1}\, dx.$$

This transform is also occasionally denoted by $\mathcal{M}[f]$ or $\mathcal{M}[f(x); s]$. Its importance devolves from two properties: (*i*) it *maps* asymptotic expansions of a function at 0 and $+\infty$ to singularities of the transform; (*ii*) it *factorizes* harmonic sums (defined below). The conjunction of the mapping property and the harmonic sum property makes it possible to analyse asymptotically rather complicated sums arising from a

---

[2]In the context of this book, Mellin transforms are useful in analyses relative the longest run problem (p. 311), the height of trees (p. 329) polylogarithms (p. 408), and integer partitions (p. 576). They also serve to establish fundamental asymptotic expansions, as in the case of harmonic and factorial numbers (below).

linear superposition of models taken at different scales. Major properties are summarized in Figure B.4. In this brief review, detailed analytic conditions must be omitted: see the survey [234] as well as comments and references at the end of this entry.

It is assumed that f is locally integrable. Then, the two conditions,

$$f(x) \underset{x \to 0^+}{=} O(x^u), \qquad f(x) \underset{x \to +\infty}{=} O(x^v),$$

guarantee that f^* exists for s in a *strip*,

$$s \in \langle -u, -v \rangle, \qquad \text{i.e.,} \qquad -u < \Re(s) < -v.$$

Thus existence of the transform is granted provided $v < u$. The prototypical Mellin transform is the Gamma function discussed earlier in this appendix:

$$\Gamma(s) := \int_0^\infty e^{-x} x^{s-1} \, dx = \mathcal{M}[e^{-x}; s], \qquad 0 < \Re(s) < \infty.$$

Similarly $f(x) = (1+x)^{-1}$ is $O(x^0)$ at 0 and $O(x^{-1})$ at infinity, and hence its transform exists in the strip $\langle 0, 1 \rangle$; it is in fact $\pi / \sin \pi s$, as a consequence of the Eulerian Beta integral. The Heaviside function defined by $H(x) := [\![0 \leq x < 1]\!]$ exists in $\langle 0, +\infty \rangle$ and has transform $1/s$.

Harmonic sum property. The Mellin transform is a linear transform. In addition, it satisfies the simple but important rescaling rule:

$$f(x) \overset{\mathcal{M}}{\mapsto} f^*(s) \qquad \text{implies} \qquad f(\mu x) \overset{\mathcal{M}}{\mapsto} \mu^{-s} f^*(s),$$

for any $\mu > 0$. Linearity then entails the derived rule

$$(45) \qquad \sum_k \lambda_k f(\mu_k x) \overset{\mathcal{M}}{\mapsto} \left(\sum_k \lambda_k \mu_k^{-s} \right) \cdot f^*(s),$$

valid *a priori* for any finite set of pairs (λ_k, μ_k) and extending to infinite sums whenever the interchange of $\int$ and $\sum$ is permissible. A sum of the form (45) is called a *harmonic sum*, the function f is the "base function", the λ values are the "amplitudes" and the μ values the "frequencies". Equation (45) then yields the "harmonic sum rule": *The Mellin transform of a harmonic sum factorizes as the product of the transform of the base function and a generalized Dirichlet series associated to amplitudes and frequencies.* Harmonic sums surface recurrently in the context of analytic combinatorics and Mellin transforms are a method of choice for coping with them.

Here are a few examples of application of the harmonic sum rule (45):

$$\sum_{k \geq 1} e^{-k^2 x^2} \underset{\Re(s) > 1}{\mapsto} \tfrac{1}{2} \Gamma(s/2) \zeta(s) \qquad \sum_{k \geq 0} e^{-x 2^k} \underset{\Re(s) > 0}{\mapsto} \frac{\Gamma(s)}{1 - 2^{-s}}$$

$$\sum_{k \geq 0} (\log k) e^{-\sqrt{k} x} \underset{\Re(s) > 2}{\mapsto} -\zeta'(s/2) \Gamma(s) \qquad \sum_{k \geq 1} \frac{1}{k(k+x)} \underset{0 < \Re(s) < 1}{\mapsto} \zeta(2 - s) \frac{\pi}{\sin \pi s}.$$

▷ **B.26.** *Connection between power series and Dirichlet series.* Let (f_n) be a sequence of numbers with at most polynomial growth, $f_n = O(n^r)$, and with OGF $f(z)$. Then, one has

$$\sum_{n \geq 1} \frac{f_n}{n^s} = \frac{1}{\Gamma(s)} \int_0^\infty f(e^{-x}) x^{s-1} \, dx, \qquad \Re(s) > r + 1.$$

Function $(f(x))$	Mellin transform $(f^\star(s))$	
$f(x)$	$\displaystyle\int_0^\infty f(x)x^{s-1}\,dx$	**definition**, $s \in \langle -u, -v \rangle$
$\displaystyle\frac{1}{2i\pi}\int_{c-i\infty}^{c+i\infty} f^\star(s)x^{-s}\,ds$ $f^\star(s)$		**inversion th.**, $-u < c < -v$
$\displaystyle\sum_i \lambda_i f_i(x)$	$\displaystyle\sum_i \lambda_i f_i^\star(s)$	linearity
$f(\mu x)$	$\mu^{-s} f^\star(s)$	scaling rule $(\mu > 0)$
$x^\rho f(x^\theta)$	$\displaystyle\frac{1}{\theta}f^\star\left(\frac{s+\rho}{\theta}\right)$	power rule
$\displaystyle\sum_i \lambda_i f(\mu_i x)$	$\displaystyle\left(\sum_i \lambda_i \mu_i^{-s}\right)\cdot f^\star(s)$	**harmonic sum rule** $(\mu_i > 0)$
$\displaystyle\int_0^\infty \lambda(t)f(tx)\,dt$	$\displaystyle\int_0^\infty \lambda(t)t^{-s}\,dt \cdot f^\star(s)$	harmonic integral rule
$f(x)\log^k x$	$\partial_s^k f^\star(s)$	diff. I, $k \in \mathbb{Z}_{\geq 0}$, $\partial_s := \frac{d}{ds}$
$\partial_x^k f(x)$	$\displaystyle\frac{(-1)^k\Gamma(s)}{\Gamma(s-k)}f^\star(s-k)$	diff. II, $k \in \mathbb{Z}_{\geq 0}$, $\partial_x := \frac{d}{dx}$
$\displaystyle\mathop{\sim}_{x\to 0} x^\alpha(\log x)^k$	$\displaystyle\mathop{\sim}_{s\to -\alpha}\frac{(-1)^k k!}{(s+\alpha)^{k+1}}$	**mapping**: $x \to 0$, left poles
$\displaystyle\mathop{\sim}_{x\to +\infty} x^\beta(\log x)^k$	$\displaystyle\mathop{\sim}_{s\to -\beta}\frac{(-1)^{k-1}k!}{(s+\beta)^{k+1}}$	**mapping**: $x \to \infty$, right poles

Figure B.4. A summary of major properties of Mellin transforms.

For instance, one obtains the Mellin pairs

$$(46) \qquad \frac{e^{-x}}{1-e^{-x}} \stackrel{\mathcal{M}}{\mapsto} \zeta(s)\Gamma(s) \quad (\Re(s) > 1), \qquad \log\frac{1}{1-e^{-x}} \stackrel{\mathcal{M}}{\mapsto} \zeta(s+1)\Gamma(s) \quad (\Re(s) > 0).$$

These serve to analyse sums or, conversely, deduce analytic properties of Dirichlet series. $\lhd$

Mapping properties. Mellin transforms map asymptotic terms in the expansions of a function f at 0 and $+\infty$ onto singular terms of the transform $f^\star$. This property stems from the basic Heaviside function identities

$$H(x)x^\alpha \stackrel{\mathcal{M}}{\mapsto} \frac{1}{s+\alpha} \ (s \in \langle -\alpha, +\infty\rangle), \qquad (1-H(x))x^\beta \stackrel{\mathcal{M}}{\mapsto} -\frac{1}{s+\beta} \ (s \in \langle -\infty, -\beta\rangle),$$

as well as what one obtains by differentiation with respect to α, β.

The converse mapping property also holds. Like for other integral transforms, there is an *inversion formula*: if f is continuous in an interval containing x, then

$$(47) \qquad\qquad f(x) = \frac{1}{2i\pi}\int_{c-i\infty}^{c+i\infty} f^\star(s)x^{-s}\,ds,$$

where the abscissa c should be chosen in the "fundamental strip" of f; for instance any c satisfying $-u < c < -v$ with u, v as above is suitable.

In many cases of practical interest, $f^\star$ is continuable as a meromorphic function to the whole of $\mathbb{C}$. If the continuation of $f^\star$ does not grow too fast along vertical lines, then one can estimate the inverse Mellin integral of (47) by residues. This corresponds to shifting the line of integration to some $d \neq c$ and taking poles into account by the residue theorem. Since the residue at a pole s_0 of $f^\star$ involves a factor of x^{-s_0}, the contribution of s_0 will give useful information on $f(x)$ as $x \to \infty$ if s_0 lies to the right of c, and on $f(x)$ as $x \to 0$ if s_0 lies to the left. Higher order poles introduce additional logarithmic factors. The "dictionary" is simply

$$(48) \qquad \frac{1}{(s - s_0)^{k+1}} \xrightarrow{\mathcal{M}^{-1}} \pm \frac{(-1)^k}{k!} x^{-s_0} (\log x)^k,$$

where the sign is '+' for a pole on the left of the fundamental strip and '−' for a pole on the right.

Mellin asymptotic summation. The combination of mapping properties and the harmonic sum property constitutes a powerful tool of asymptotic analysis, as shown by the examples and the notes below.

***Example* B.4.** *Asymptotics of a simple harmonic sum.* Let us first investigate the pair

$$F(x) := \sum_{k \geq 1} \frac{1}{1 + k^2 x^2}, \qquad F^\star(s) = \frac{1}{2} \frac{\pi}{\sin \frac{1}{2} \pi s} \zeta(s),$$

where $F^\star$ results from the harmonic sum rule and has fundamental strip $\langle 1, 2 \rangle$. The function $F^\star$ is continuable to the whole of $\mathbb{C}$ with poles at the points $0, 1, 2$ and $4, 6, 8, \ldots$. The transform $F^\star$ is small towards infinity, so that application of the dictionary (48) is justified. One finds

$$F(x) \underset{x \to 0^+}{\sim} \frac{\pi}{2x} - \frac{1}{2} + O(x^M), \qquad F(x) \underset{x \to +\infty}{\sim} \frac{\pi^2}{6x^2} - \frac{\pi^4}{90x^4} + \cdots,$$

where the expansion at 0 is valid for any $M > 0$. .... ∎

***Example* B.5.** *Asymptotics of a dyadic sum.* A particularly important quantity in analytic combinatorics is the following harmonic sum, stated here together with its Mellin transform:

$$\Phi(x) := \sum_{k=0}^{\infty} \left(1 - e^{-x/2^k}\right); \qquad \Phi^\star(s) = -\frac{\Gamma(s)}{1 - 2^s}, \qquad s \in \langle -1, 0 \rangle.$$

It occurs for instance in the analysis of longest runs in words (p. 311). The transform of $e^{-x} - 1$ is also $\Gamma(s)$, but in the shifted strip $\langle -1, 0 \rangle$. The singularities of $\Phi^\star$ are at $s = 0$, where there is a double pole, at $s = -1, -2, \ldots$ which are simple poles, but also at the complex points

$$\chi_k = \frac{2ik\pi}{\log 2}.$$

The Mellin dictionary (48) can still be applied provided one integrates along a long rectangular contour that passes in-between poles. The salient feature is here the presence of fluctuations induced by the imaginary poles, since $x^{-\chi_k} = \exp\left(-2ik\pi \log_2 x\right)$, and each pole induces a Fourier element. All in all, one finds (any $M > 0$):

$$(49) \qquad \begin{cases} \Phi(x) \underset{x \to +\infty}{\sim} \log_2 x + \dfrac{\gamma}{\log 2} + \dfrac{1}{2} + P(x) + O(x^M) \\[2mm] \qquad P(x) := \dfrac{1}{\log 2} \displaystyle\sum_{k \in \mathbb{Z} \setminus \{0\}} \Gamma\left(\dfrac{2ik\pi}{\log 2}\right) e^{-2ik\pi \log_2 x}. \end{cases}$$

The analysis for $x \to 0$ yields, in this particular case, $\Phi(x) \underset{x \to 0}{\sim} \sum_{n \geq 1} \frac{(-1)^{n-1}}{1 - 2^{-n}} \frac{x^n}{n!}$, which would also result from expanding $\exp(-x/2^k)$ in $\Phi(x)$ and reorganizing the terms. .... ∎

Example B.6.   *Euler–Maclaurin summation via Mellin analysis.* Let f be continuous on $(0, +\infty)$ and satisfy $f(x) =_{x \to +\infty} O(x^{-1-\delta})$, for some $\delta > 0$, and

$$f(x) \underset{x \to 0^+}{\sim} \sum_{k=0}^{\infty} f_k x^k.$$

The summatory function $F(x)$ satisfies

$$F(x) := \sum_{n \geq 1} f(nx), \qquad F^\star(s) = \zeta(s) f^\star(s),$$

by the harmonic sum rule. The collection of (trimmed) singular expansions of $f^\star$ at $s = 0, -1, -2, \ldots$ is summarized by the formal expansion, conventionally represented by $\asymp$:

$$f^\star(s) \asymp \left(\frac{f_0}{s} \right)_{s=0} + \left(\frac{f_1}{s+1} \right)_{s=1} + \left(\frac{f_2}{s+2} \right)_{s=1} + \cdots .$$

Thus, by the mapping properties, provided $F^\star(s)$ is small towards $\pm i\infty$ in finite strips, one has

$$F(x) \underset{x \to 0}{\sim} \frac{1}{x} \int_0^\infty f(t)\, dt + \sum_{j=0}^{\infty} f_j \zeta(-j) x^j,$$

where the main term is associated to the singularity of $F^\star$ at 1 and arises from the pole of $\zeta(s)$, with $f^\star(1)$ giving the integral of f. The interest of this approach is that it is very versatile and allows for various forms of asymptotic expansions of f at 0 as well as multipliers like $(-1)^k$, $\log k$, and so on; see [234] for details and Gonnet's note [300] for alternative approaches. .. ∎

▷ **B.27.** *Mellin-type derivation of Stirling's formula.* One has the Mellin pair

$$L(x) = \sum_{k \geq 1} \log\left(1 + \frac{x}{k}\right) - \frac{x}{k}, \qquad L^\star(s) = \frac{\pi}{s \sin \pi s} \zeta(-s), \quad s \in \langle -2, -1 \rangle.$$

Note that $L(x) = \log(e^{-\gamma x} / \Gamma(1+x))$. Mellin asymptotics provides

$$L(x) \underset{x \to +\infty}{\sim} -x \log x - (\gamma - 1)x - \frac{1}{2} \log x - \log \sqrt{2\pi} - \frac{1}{12x} + \frac{1}{360x^3} - \frac{1}{1260x^5} + \cdots,$$

where one recognizes Stirling's expansion of $x!$:

$$\log x! \underset{x \to +\infty}{\sim} \log\left(x^x e^{-x} \sqrt{2\pi x}\right) + \sum_{n \geq 1} \frac{B_{2n}}{2n(2n-1)} x^{1-2n}$$

(the B_n are the Bernoulli numbers).   ◁

▷ **B.28.** *Mellin-type analysis of the harmonic numbers.* For $\alpha > 0$, one has the Mellin pair:

$$K_\alpha(x) = \sum_{k \geq 1} \left(\frac{1}{k^\alpha} - \frac{1}{(k+x)^\alpha} \right), \qquad K_\alpha^\star(s) = -\zeta(\alpha - s) \frac{\Gamma(s)\Gamma(\alpha - s)}{\Gamma(\alpha)}.$$

This serves to estimate harmonic numbers and their generalizations, for instance,

$$H_n \underset{n \to \infty}{\sim} \log n + \gamma - \frac{1}{2n} - \sum_{k \geq 2} \frac{B_k}{k} n^{-k} \sim \log n + \gamma + \frac{1}{2n} - \frac{1}{12n^2} + \frac{1}{120n^4} - \cdots,$$

since $K_1(n) = H_n$.   ◁

General references on Mellin transforms are the books by Doetsch [168] and Widder [605]. The term "harmonic sum" and some of the corresponding technology originates with the abstract [253]. This brief presentation is based on the survey article by Flajolet, Gourdon, and Dumas [234] to which we refer for a detailed treatment; see also the self-contained treatment by Butzer and Jansche [100]. Mellin analysis of "harmonic integrals" is a classical topic of applied mathematics for which we refer to the books by Wong [614] and Paris–Kaminski [472]. Valuable accounts of properties of use in discrete mathematics and analysis of algorithms appear in the books by Hofri [335], Mahmoud [429], and Szpankowski [564].

B.8. Several complex variables

The theory of analytic (or holomorphic) functions of one complex variables extends non-trivially to several complex variables. This profound theory has been largely developed in the course of the twentieth century. Here we shall only need the most basic *concepts*, not the deeper results, of the theory.

Consider the space $\mathbb{C}^m$ endowed with the metric

$$|\mathbf{z}| = |(z_1, \ldots, z_m)| = \sum_{j=1}^m |z_j|^2,$$

under which it is isomorphic to the Euclidean space $\mathbb{R}^{2m}$. A function f from $\mathbb{C}^m$ to $\mathbb{C}$ is said to be analytic at some point $\mathbf{a}$ if in a neighbourhood of a it can be represented by a convergent power series,

$$(50) \qquad f(\mathbf{z}) = \sum_{\mathbf{n}} f_{\mathbf{n}}(\mathbf{z} - \mathbf{a})^{\mathbf{n}} \equiv \sum_{n_1, \ldots, n_m} f_{n_1, \ldots, n_m}(z_1 - a_1)^{n_1} \cdots (z_m - a_m)^{n_m}.$$

There and throughout the theory, extensive use is made of the multi-index convention, as encountered in Chapter III, p. 165.

An expansion (50) converges in a polydisc $\prod_j \{|z_j - a_j| < r_j\}$, for some $r_j > 0$. A convergent expansion at $(0, \ldots, 0)$ has its coefficients majorized in absolute value by those of a series of the form

$$\prod_{j=1}^m \frac{1}{1 - z_j/R_j} = \sum_{\mathbf{n}} \mathbf{R}^{-\mathbf{n}} \mathbf{z}^{\mathbf{n}} \equiv \sum_{n_1, \ldots, n_m} R_1^{-n_1} \cdots R_m^{-n_m} z_1^{n_1} \cdots z_m^{n_m}.$$

Closure of analytic functions under sums, products, and compositions results from standard manipulations of majorant series (see p. 250 for the univariate case). Finally, a function is analytic in an open set $\Omega \subseteq \mathbb{C}^m$ iff it is analytic at each $\mathbf{a} \in \Omega$.

A remarkable theorem of Hartogs asserts that $f(\mathbf{z})$ with $\mathbf{z} \in \mathbb{C}^m$ is analytic *jointly* in all the z_j (in the sense of (50)) if it is analytic *separately* in each variable z_j. (The version of the theorem that postulates *a priori* continuity is elementary.)

As in the one-dimensional case, analytic functions can be equivalently defined by means of differentiability conditions. A function is $\mathbb{C}$-differentiable or holomorphic at $\mathbf{a}$ if, as $\Delta \mathbf{z} \to 0$ in $\mathbb{C}^m$, one has

$$f(\mathbf{a} + \Delta \mathbf{z}) - f(\mathbf{a}) = \sum_{j=1}^{m} c_j \Delta z_j + o\left(|\Delta \mathbf{z}|\right).$$

The coefficients c_j are the partial derivatives, $c_j = \partial_{z_j} f(\mathbf{a})$. The fact that this relation does not depend on the way $\Delta \mathbf{z}$ tends to 0 implies the Cauchy–Riemann equations. In a way that parallels the single variable case, it is proved that two conditions are equivalent: f is analytic; f is complex-differentiable.

Iterated integrals are defined in the natural way and one finds, by a repeated use of calculus in a single variable,

$$(51) \qquad f(\mathbf{z}) = \frac{1}{(2i\pi)^m} \int_{C_1} \cdots \int_{C_m} \frac{f(\zeta)}{(\zeta_1 - z_1) \cdots (\zeta_m - z_m)} \, d\zeta_1 \cdots d\zeta_m,$$

where C_j is a small circle surrounding z_j in the z_j–plane. By differentiation under the integral sign, Equation (51) also provides an integral formula for the partial derivatives of f, which is the analogue of Cauchy's coefficient formula. Iterated integrals are independent of details of the "polypath" on which they are taken, and uniqueness of analytic continuation holds.

The theory of functions of several complex variables develops in the direction of an integral calculus that is much more powerful than the iterated integrals mentioned above; see, for instance, the book by Aïzenberg and Yuzhakov [8] for a multidimensional residue approach. Egorychev's monograph [187] develops systematic applications of the theory of functions of one or several complex variables to the evaluation of combinatorial sums. Pemantle together with several coauthors [474, 475, 476] has launched an ambitious research programme meant to extract the coefficients of meromorphic multivariate generating functions by means of this theory, with the ultimate goal of obtaining systematically asymptotics from multivariate generating functions. By contrast, see especially Chapter IX, we can limit ourselves to developing a perturbative theory of one-variable complex function theory.

In the context of this book, the basic notion of analyticity in several complex variables serves to confer a *bona fide* analytic meaning to multivariate generating functions. Basic definitions are also needed in the context of functions f defined implicitly by functional relations of the form $H(z, f) = 0$ or $H(z, u, f) = 0$, where analytic functions of two or more complex variables make an appearance. (See in particular the discussion of the analytic Implicit Function Theorem and the Weierstrass Preparation Theorem in this appendix, p. 753.)

APPENDIX C

Concepts of Probability Theory

This appendix contains entries arranged in logical order regarding the following topics:

> *Probability spaces and measure; Random variables; Transforms of distributions; Special distributions; Convergence in law.*

In this book we start from probability spaces that are finite, since they arise from objects of a fixed size in some combinatorial class (see Chapter III and Appendix A.3: *Combinatorial probability*, p. 727 for elementary aspects), then need basic properties of continuous distributions in order to discuss asymptotic limit laws. The entries in this appendix are related principally to Chapter IX of Part C (*Random Structures*). They present a unified framework that encompasses discrete and continuous probability distributions alike. For further study, we recommend the superb classics of Feller [205, 206], given the author's concrete approach, and of Billingsley [68], whose coverage of limit distributions is of great value for analytic combinatorics.

C.1. Probability spaces and measure

An axiomatization of probability theory[1] was discovered in the 1930s by Kolmogorov. A *measurable space* consists of a set Ω, called the set of elementary events or the sample set and a σ-algebra $\mathcal{A}$ of subsets of Ω called events (that is, a collection of sets containing $\emptyset$ and closed under complement and denumerable unions). A *measure space* is a measurable space endowed with a measure $\mu : \mathcal{A} \mapsto \mathbb{R}_{\geq 0}$ that is additive over finite or denumerable unions of disjoint sets; in that case, elements of $\mathcal{A}$ are called measurable sets. A *probability space* is a measure space for which the measure satisfies the further normalization $\mu(\Omega) = 1$; in that case, we also write $\mathbb{P}$ for μ. Any set $S \subseteq \Omega$ such that $\mu(S) = 1$ is called a *support* of the probability measure. These definitions given above cover several important cases.

(*i*) *Finite sets with the uniform measure* (also known as "counting" measure). In this case, Ω is finite, all sets are in $\mathcal{A}$ (i.e., are measurable), and ($\| \cdot \|$ denotes cardinality)

$$\mu(E) := \frac{\|E\|}{\|S\|}.$$

Non-uniform measures over a finite set Ω are determined by assigning a non-negative weight $p(\omega)$ to each element of Ω (with $\sum_{\omega \in \Omega} p(\omega) = 1$) and setting

$$\mu(E) := \sum_{e \in E} p(e).$$

(We also write $\mathbb{P}(e)$ for $\mathbb{P}(\{e\}) \equiv \mu(\{e\}) = p(e)$.) In this book, Ω is usually the subclass $\mathcal{C}_n$ formed by the objects of size n in some combinatorial class $\mathcal{C}$. The uniform measure is usually assumed, although suitably weighted models often prove to be of

---

[1] For this entry we refer to the vivid and well-motivated presentation in Williams' book [609] or to many classical treatises such as those by Billingsley [68] and Feller [205].

interest: see for instance in Chapter III the discussion of weighted word models and Bernoulli trials as well as the case of weighted tree models and branching processes.

(*ii*) *Discrete probability measures over the integers* (supported by $\mathbb{Z}$ or $\mathbb{Z}_{\geq 0}$). In this case the measure is determined by a function $p : \mathbb{Z} \mapsto \mathbb{R}_{\geq 0}$ and

$$\mu(E) := \sum_{e \in E} p(e),$$

with $\mu(\mathbb{Z}) = 1$. (All sets are measurable.) More general discrete measures supported by denumerable sets of $\mathbb{R}$ can be similarly defined.

(*iii*) *The real line* $\mathbb{R}$ equipped with the σ-algebra generated by the open intervals constitutes a standard example of a measurable space; in that case, any member of the σ-algebra is known as a Borel set. The measure, denoted by λ, that assigns to an interval (a, b) the value $\lambda(a, b) = b - a$ (and is extended non-trivially to all Borel sets by additivity) is known as the Lebesgue measure. The interval $[0, 1]$ endowed with λ is a probability space. The line $\mathbb{R}$ itself is not a probability space since $\lambda(\mathbb{R}) = +\infty$.

In the measure-theoretic framework, a *random variable* is a mapping X from a probability space Ω (equipped with its σ-algebra $\mathcal{A}$ and its measure $\mathbb{P}_\Omega$) to $\mathbb{R}$ (equipped with its Borel sets $\mathcal{B}$) such that the preimage $X^{-1}(B)$ of any $B \in \mathcal{B}$ lies in $\mathcal{A}$. For $B \in \mathcal{B}$, the probability that X lies in B is then defined as

$$\mathbb{P}(X \in B) := \mathbb{P}_\Omega(X^{-1}(B)).$$

Since the Borel sets can be generated by the semi-infinite intervals $(-\infty, x]$, this probability is equivalently determined by the function

$$F(x) := \mathbb{P}(X \leq x),$$

which is called the *distribution function* or *cumulative distribution function* of X. It is then possible to introduce random variables directly by means of distribution functions, see the entry below, *Random variables*.

Integration. The next step is to go from measures of sets to integrals of (real-valued) functions. Lebesgue integrals are constructed, first for indicator functions of intervals, then for simple (staircase) functions, then for non-negative functions, finally for integrable functions. One defines in this way, for an arbitrary measure μ, the Lebesgue integral

$$(1) \qquad \int f d\mu, \qquad \text{also written} \qquad \int f(x) d\mu(x) \quad \text{or} \quad \int f(x)\mu(dx),$$

where the last notation is often preferred by probabilists. The basic idea is to decompose the domain of *values* of f into finitely many measurable sets (A_i) and, for a positive function f, consider the supremum over all finite decompositions (A_i)

$$(2) \qquad \int f d\mu := \sup_{(A_i)} \sum_i \left[\inf_{\omega \in A_i} f(\omega) \right] \mu(A_i).$$

(Thus Riemann integration proceeds by decomposing the domain of the function's *arguments* while Lebesgue integrals decomposes the domain of *values* and appeals to a richer notion of measure for point sets.)

In (1) and (2), the possibility exists that μ assigns a non-zero measure to certain individual points. In such a context, the integral is sometimes referred to as the *Lebesgue-Stieltjes* integral. It suitably generalizes the *Riemann-Stieltjes* integral which, given a real valued function M, defines the following extension of the standard Riemann integral:

$$(3) \qquad \int f(x)\, dM(x) = \lim_{(B_k)} \sum_k f(x_k) \Delta_{B_k}(M).$$

There the B_k form a finite partition of the domain in which the argument of f ranges, the limit is taken as the largest B_k tends to 0, each x_k lies in B_k, and $\Delta_{B_k}(M)$ is the variation of M on B_k.

The great advantage of Stieltjes (hence automatically of Lebesgue) integrals is to unify many of the formulae relative to discrete and continuous probability distributions while providing a simple framework adapted to mixed cases.

C.2. Random variables

A real random variable X is fully characterized by its (cumulative) distribution function

$$F_X(x) := \mathbb{P}(X \le x),$$

which is a non-decreasing right-continuous function satisfying $F(-\infty) = 0$, $F(+\infty) = 1$.

A variable is *discrete* if it is supported by a finite or denumerable set. Almost all discrete distributions in this book are supported by $\mathbb{Z}$ or $\mathbb{Z}_{\ge 0}$. (An interesting exception is the collection of distributions occurring in longest runs of words, Chapter IV, p. 308.)

A variable X is *continuous* if it assigns zero probability mass to any finite or denumerable set. In particular, it has no jump. An easy theorem states that any distribution function can be decomposed into a discrete and a continuous part,

$$F(x) = c_1 F^{\mathrm{d}}(x) + c_2 F^{\mathrm{c}}(x), \qquad c_1 + c_2 = 1.$$

(The jumps must sum to at most 1, hence their set is at most denumerable.) A variable is *absolutely continuous* if it assigns zero probability mass to any Borel set of measure 0. In that case, the Radon–Nikodym Theorem asserts that there exists a function w such that

$$F_X(x) = \int_{-\infty}^{x} w(y)\, dy.$$

(There, in all generality, the Lebesgue integral is required but the Riemann integral is sufficient for all practical purposes in this book.) The function $w(x)$ is called a *density* of the random variable X (or of its distribution function). When F_X is differentiable everywhere it admits the density

$$w(x) = \frac{d}{dx} F_X(x),$$

by the Fundamental Theorem of Calculus.

▷ **C.1.** *The Lebesgue decomposition theorem.* It states that any distribution function $F(x)$ decomposes as

$$F(x) = c_1 F^d(x) + c_2 F^{ac} + c_3 F^s(x), \qquad c_1 + c_2 + c_3 = 1,$$

where F^d is discrete, F^{ac} is absolutely continuous, and F^s is continuous but *singular*, i.e., it is supported by a Borel set of Lebesgue measure 0. Singular random variables are constructed, e.g., from the Cantor set. ◁

In this book, all combinatorial distributions are by nature discrete (and then supported by $\mathbb{Z}_{\geq 0}$). All continuous distributions obtained as limits of discrete ones are, in our context, absolutely continuous and the qualifier "absolutely" is globally understood when discussing continuous distributions.

If X is a random variable, the *expectation* of a function $g(X)$ is defined as

$$\mathbb{E}\,(g(X)) = \int_{\mathbb{R}} g(x)dF(x),$$

which involves the distribution function F of X. In particular the *expectation* or *mean* of X is $\mathbb{E}(X)$, and generally its *moment* of order r is

$$\mu^{(r)} = \mathbb{E}(X^r).$$

(These quantities may not exist for $r \neq 0$.)

▷ **C.2.** *Alternative formulae for expectations.* If X is supported by $\mathbb{R}_{\geq 0}$:

$$\mathbb{E}(X) = \int_0^{\infty} (1 - F(x))\,dx.$$

If X is supported by $\mathbb{Z}_{\geq 0}$:

$$\mathbb{E}(X) = \sum_{k \geq 0} \mathbb{P}(X > k).$$

Proofs are by partial integration and summation: for instance with $p_k = \mathbb{P}(X = k)$,

$$\mathbb{E}(X) = \sum_{k \geq 1} k p_k = (p_1 + p_2 + p_3 + \cdots) + (p_2 + p_3 + \cdots) + (p_3 + \cdots) + \cdots .$$

Similar formulae hold for higher moments. ◁

C.3. Transforms of distributions

The *Laplace transform* of X (or of its distribution function F) is defined by

$$\lambda_X(s) := \mathbb{E}\left(e^{sX}\right) = \int_{-\infty}^{+\infty} e^{sx}\,dF(x).$$

(If F has a discrete component, then integration is to be taken in the sense of Lebesgue–Stieltjes or Riemann–Stieltjes.) The Laplace transform is also known as the *moment generating function* (see below for an existential discussion). The *characteristic function* is defined by

$$\phi_X(t) = \mathbb{E}\left(e^{itX}\right) = \int_{-\infty}^{+\infty} e^{itx}\,dF(x),$$

and it is a Fourier transform. Both transforms are formal variants of one another and $\phi_X(t) = \lambda_X(it)$.

If X is discrete and supported by $\mathbb{Z}$, then its *probability generating function* (PGF) is, as defined as in Appendix A.3: *Combinatorial probability*, p. 727:

$$P_X(u) := \mathbb{E}(u^X) = \sum_{k \in \mathbb{Z}} \mathbb{P}(X = k) u^k.$$

As an analytic object this always exists when X is non-negative (supported by $\mathbb{Z}_{\geq 0}$), in which case the PGF is analytic at least in the open disc $|u| < 1$. If $X \in \mathbb{Z}$ assumes arbitrarily large negative values, then the PGF certainly exists on the unit circle, but sometimes not on a larger domain. The precise domain of existence of the PGF as an analytic function depends on the geometric rate of decay of the left and right tails of the distribution, that is, of $\mathbb{P}(X = k)$ as $k \to \pm\infty$. The characteristic function of the variable X (and of its distribution function F_X) is

$$\phi_X(t) := \mathbb{E}(e^{itX}) = P_X(e^{it}) = \sum_{k \in \mathbb{Z}} \mathbb{P}(X = k) e^{ikt}.$$

It exists for *all real values* of t. The Laplace transform of the discrete variable X is

$$\lambda_X(s) := \mathbb{E}(e^{sX}) = P_X(e^s) = \sum_{k \in \mathbb{Z}} \mathbb{P}(X = k) e^{ks}.$$

If X is a continuous random variable with distribution function $F(x)$ and density $w(x)$, then the characteristic function is expressed as

$$\phi_X(t) := \mathbb{E}(e^{itX}) = \int_{\mathbb{R}} e^{itx} w(x) \, dx.$$

and the Laplace transform is

$$\lambda_X(s) := \mathbb{E}(e^{sX}) = \int_{\mathbb{R}} e^{sx} w(x) \, dx.$$

The Fourier transform always exists for real arguments (by integrability of the Fourier kernel e^{it} whose modulus is 1). The Laplace transform, when it exists in a strip, extends analytically the characteristic function via the equality $\phi_X(t) = \lambda_X(it)$. The Laplace transform is also called the *moment generating function* since an alternative formulation of its definition, valid for discrete and continuous cases alike, is

$$\lambda_X(s) := \sum_{k \geq 0} \mathbb{E}(X^k) \frac{s^k}{k!},$$

which indeed represents the exponential generating function of moments. (We avoid this terminology in the text, because of a possible confusion with the many other types of generating functions employed in this book.)

The importance of the transforms is due to the existence of *continuity theorem* by which convergence of distributions can be established via convergence of transforms.

▷ **C.3.** *Centring, scaling, and standardization.* Let X be a random variable. Define $Y = \frac{X - \mu}{\sigma}$. The representations as expectations of the Laplace transform and of the characteristic function make it obvious that

$$\phi_Y(t) = e^{-\mu it} \phi_X\left(\frac{t}{\sigma}\right), \qquad \lambda_Y(s) = e^{-\mu s} \lambda_X\left(\frac{s}{\sigma}\right).$$

One says that Y is obtained from X by centring (by a shift of μ) and scaling (by a factor of σ). If μ and σ are the mean and standard deviation of X, then one says that Y is a *standardized version* of X.   ◁

▷ **C.4.** *Moments and transforms.* The moments are accessible from either transform,

$$\mu^{(r)} := E\{Y^r\} = \left.\frac{d^r}{ds^r}\lambda(s)\right|_{s=0} = (-i)^r \left.\frac{d^r}{dt^r}\phi(t)\right|_{t=0}.$$

In particular, we have

$$
\begin{aligned}
\mu &= \left.\frac{d}{ds}\lambda(s)\right|_{s=0} = -i\left.\frac{d}{dt}\phi(t)\right|_{t=0} \\[2mm]
(4)\qquad \mu^{(2)} &= \left.\frac{d^2}{ds^2}\lambda(s)\right|_{s=0} = -\left.\frac{d}{dt}\phi(t)\right|_{t=0} \\[2mm]
\sigma^2 &= \left.\frac{d^2}{ds^2}\log\lambda(s)\right|_{s=0} = -\left.\frac{d^2}{dt^2}\log\phi(t)\right|_{t=0}.
\end{aligned}
$$

The direct expression of the standard deviation in terms of $\log\lambda(s)$, called the *cumulant generating function*, often proves computationally handy.   ◁

▷ **C.5.** *Mellin transforms of distributions.* The quantity $M(s) := \mathbb{E}(X^{s-1})$ is the Mellin transform of X or of its distribution function F, when X is supported by $\mathbb{R}_{\geq 0}$ (see Appendix B.7: *Mellin transform*, p. 762). In particular, if X admits a density, then this notion coincides with the usual definition of a Mellin transform. When it exists, the value of the Mellin transform at an integer $s = k$ provides the moment of order $k - 1$; at other points, it provides moments of fractional order.   ◁

▷ **C.6.** *A "symbolic" fragment of probability theory.* Consider discrete random variables supported by $\mathbb{Z}_{\geq 0}$. Let $X, X_1, \ldots$ be independent random variables with PGF $p(u)$ and let Y have PGF $q(u)$. Then, certain natural operations admit a translation into PGFs.

Operation		*PGF*
switch	$(\mathrm{Bern}(\lambda) \Rightarrow X \mid Y)$	$\lambda p(u) + (1 - \lambda)q(u)$
sum	$X + Y$	$p(u) \cdot q(u)$
	$X_1 + \cdots + X_n$	$p(u)^n$
random sum	$X_1 + \cdots + X_Y$	$q(p(u))$
size bias	∂X	$\dfrac{u p'(u)}{p'(1)}$

("Bern" means a Bernoulli $\{0, 1\}$ variable B, with $\mathbb{P}(1) = \lambda$; the switch is interpreted as $BX + (1 - B)Y$. Size-biased distributions occur in Chapter VII.)   ◁

C.4. Special distributions

A compendium of special probability distributions of frequent occurrence in analytic combinatorics is provided by Figure C.1.

A *Bernoulli trial* of parameter q is an event such that it has probability p of having value 1 (interpreted as "success") and probability q of having value 0 (interpreted as "failure"), with $p + q = 1$. Formally, this is the set $\Omega = \{0, 1\}$ endowed with the probability measure $\mathbb{P}(0) = q$, $\mathbb{P}(1) = p$. (By extension, we also refer to independent experiments with finitely many possible outcomes as Bernoulli trials. In

	Distribution	Prob. (D), density (C)	PGF(D), Char. f. (C)
D	Binomial (n, p)	$\binom{n}{k} p^k (1-p)^{n-k}$	$(q + pu)^n$
D	Geometric (q)	$(1-q)q^k$	$\dfrac{1-q}{1-qu}$
D	Neg. binomial$[m]$ (q)	$\binom{m+k-1}{k} q^k (1-q)^m$	$\left(\dfrac{1-q}{1-qu}\right)^m$
D	Log. series (λ)	$\dfrac{1}{-\log(1-\lambda)} \dfrac{\lambda^k}{k!}$	$\dfrac{\log(1-\lambda u)}{\log(1-\lambda)}$
D	Poisson (λ)	$e^{-\lambda} \dfrac{\lambda^k}{k!}$	$e^{\lambda(1-u)}$
C	Gaussian or Normal, $\mathcal{N}(0, 1)$	$\dfrac{e^{-x^2/2}}{\sqrt{2\pi}}$	$e^{-t^2/2}$
C	Exponential	e^{-x}	$\dfrac{1}{1-it}$
C	Uniform $[-1/2, +1/2]$	$[\![-1/2 \le x \le +1/2]\!]$	$\dfrac{\sin(t/2)}{(t/2)}$

Figure C.1. A list of commonly encountered discrete (D) and continuous (C) probability distributions: type, name, probabilities or density, probability generating function or characteristic function.

that sense, the model of words of some fixed length over a finite alphabet and non-uniform letter weights (or probabilities) belongs to the category of Bernoulli models; see Chapter III.) The *binomial distribution* of parameters n, q is the random variable that represents the number of successes in n independent Bernoulli trials. This is the probability distribution associated with the game of heads-and-tails. The *geometric distribution* is the distribution of a random variable X that records the number of failures till the first success is encountered in a potentially arbitrarily long sequence of Bernoulli trials. The *negative binomial* distribution of index m (written $NB[m]$) and parameter q corresponds to the number of failures before m successes are encountered. We have found in Chapter VII that it is systematically associated with the number of r–components in an unlabelled multiset schema $\mathcal{F} = \mathfrak{M}(\mathcal{G})$ whose composition of singularities is of the exp–log type. The geometric distribution appears in several schemas related to sequences while the logarithmic series distribution is closely tied to cycles (Chapter V). indexlogarithmic-series distribution

The *Poisson distribution* counts among the most important distributions of probability theory. Its essential properties are recalled in Figure C.1. It occurs for instance in the distribution of singleton cycles and of r–cycles in a random permutation and more generally in labelled composition schemes (Chapter IX).

In this book all probability distributions arising directly from combinatorics are *a priori* discrete as they are defined on finite sets—typically a certain subclass $\mathcal{C}_n$ of a combinatorial class $\mathcal{C}$. However, as the size n of the objects considered grows, these finite distributions usually approach a continuous limit. In this context, by far the most important law is the *Gaussian law* also known as *normal law*, which is defined by its density and its distribution function:

$$(5) \qquad g(x) = \frac{e^{-x^2/2}}{\sqrt{2\pi}}, \qquad \Phi(x) = \frac{1}{\sqrt{2\pi}} \int_{-\infty}^{x} e^{-y^2/2} \, dy.$$

The corresponding Laplace transform is then evaluated by completing the square,

$$\lambda(s) = \frac{1}{\sqrt{2\pi}} \int_{-\infty}^{+\infty} e^{-y^2/2 + sy} \, dy. = e^{s^2/2},$$

and, similarly, the characteristic function is $\phi(t) = e^{-t^2/2}$. The distribution of (5) is referred to as the *standard* normal distribution, $\mathcal{N}(0, 1)$; if X is $\mathcal{N}(0, 1)$, the variable $Y = \mu + \sigma X$ defines the normal distribution with mean μ and standard deviation σ, denoted $\mathcal{N}(\mu, \sigma)$.

Among other continuous distributions appearing in this book, we mention the theta distributions associated with the height of trees and Dyck paths (Chapter V) and the stable laws, which surface in Chapter IX.

C.5. Convergence in law

The central notion, which is of the greatest interest for analytic combinatorics, is the notion of *convergence in law*, also known as *weak convergence*.

Definition C.1. *Let F_n be a family of distribution functions. The F_n are said to converge weakly to a distribution function F if pointwise there holds*

$$(6) \qquad \lim_n F_n(x) = F(x),$$

at every continuity point x of F. This is expressed by writing $F_n \Rightarrow F$ as well as $X_n \Rightarrow X$, if X_n, X are random variables corresponding to F_n, F. We say that X_n converges in distribution *or* converges in law *to X.*

This definition has the merit of covering discrete and continuous distributions alike. For discrete distributions supported by $\mathbb{Z}$, an equivalent form of (6) is $\lim_n F_n(k) = F(k)$ for each $k \in \mathbb{Z}$; for continuous distributions, Equation (6) just means that $\lim_n F_n(x) = F(x)$ for all $x \in \mathbb{R}$. Although in all generality anything can tend to anything else, due to the finite nature of combinatorics, we only need in this book the convergences

Discrete $\Rightarrow$ Discrete,   Discrete $\Rightarrow$ Continuous (after standardization).

Three major tools can be used to establish convergence in law: characteristic functions, Laplace transforms, and moment convergence theorems.

Characteristic functions and limit laws. Properties of random variables are reflected by probabilities of characteristic functions, in accordance with general principles of Fourier analysis—Figure C.2 offers an aperçu. Most important for us is the *Continuity Theorem* for characteristic functions due to Lévy and used extensively in Chapter IX, starting on p. 639, through the Quasi-powers Theorem of p. 645.

Characteristic function $(\phi(t))$	*distribution function* $(F(x))$
$\phi(0) = 1$	$F(-\infty) = 0$, $F(+\infty) = 1$
$\|\phi(t_0)\| = 1$ for some $t_0 \neq 0$	Lattice distribution, span $\frac{2\pi}{t_0}$
$\phi(t) \underset{t \to 0}{=} 1 + i\mu t + o(t)$	$\mathbb{E}(X) = \mu < \infty$
$\phi(t) \underset{t \to 0}{=} 1 + i\mu t - \nu\frac{t^2}{2} + o(t^2)$	$\mathbb{E}(X^2) = \nu < \infty$
$\log \phi(t) = -\frac{t^2}{2}$	$X \overset{d}{=} \mathcal{N}(0, 1)$
$\phi(t) \to 0$ as $t \to \infty$	X is continuous
$\phi(t)$ integrable (is in $\mathcal{L}_1$)	X is absolutely continuous
	density is $w(x) = \dfrac{1}{2\pi} \displaystyle\int_{-\infty}^{+\infty} e^{-itx} \phi(t)\, dt$
$\lambda(s) := \phi(-is)$ exists in $\alpha < \Re(s) < \beta$	Exponential tails
$\lim_{T \to \infty} \frac{1}{2T} \int_{-T}^{+T} \|\phi(t)\|^2\, dt$	equals $\sum_i (p_i)^2$; the p_i are the jumps
$\phi_n(t) \to \phi(t)$ (point conv.)	$F_n \Rightarrow F$ (weak conv.)
	$X_n \Rightarrow X$ (conv. in distribution)
ϕ_n "close" to ϕ	F_n "close" to F (Berry–Esseen)

Figure C.2. The correspondence between properties of the distribution function (F) of a random variable (X) and properties of its characteristic function (ϕ).

Theorem C.1 (Continuity theorem for characteristic functions). *Let Y, Y_n be random variables with characteristic functions ϕ, ϕ_n. A necessary and sufficient condition for the weak convergence $Y_n \Rightarrow Y$ is that $\phi_n(t) \to \phi(t)$ for each t.*

For a proof, see [68, §26]. What is notable is that the theorem provides a *necessary and sufficient condition*. In addition, the Berry–Esseen inequalities stated in Chapter IX, p. 641, lie at the origin of precise speed of convergence estimates to asymptotic limits.

Laplace transforms and limit laws. The continuity theorem for Laplace transforms is stated in Chapter IX, p. 639. In principle, it is of a more restricted scope than Theorem C.1 since Laplace transforms need not exist. Also, error bounds derived from Laplace transform can be exponentially worse than those resulting from Berry–Esseen inequalities [557]. For these reasons, the rôle of Laplace transforms in this book is mostly confined to large deviation estimates (Section IX. 10, p. 699).

The method of moments. For the purpose of establishing limit laws in combinatorics, it is may be convenient (sometimes even necessary) to access distributions by moments. One then attempts to deduce convergence of distributions from convergence of moments. This approach requires conditions under which a distribution is uniquely characterized by its moments—finding these is known as the *moment problem* in analysis. A lucid discussion is offered by Billingsley in [68, §30], which we follow.

A distribution function $F(x)$, with $x \in \mathbb{R}$, is characterized by its moments if the sequence of real numbers

$$\mu_k = \int_{\mathbb{R}} x^k \, dF(x), \qquad k = 0, 1, 2, \ldots,$$

uniquely determines F (that is: $\int x^k dF = \int x^k dG$ for all k implies $F = G$). The following basic conditions are known to be *sufficient* for such a property to hold: (i) F has finite support; (ii) the exponential generating function of (μ_k) is analytic at 0, that is, for some $R > 0$, one has

$$(7) \qquad \qquad \mu_k \frac{R^k}{k!} \to 0, \qquad k \to \infty.$$

(The first case is proved by appealing to Weierstrass' theorem to the effect that polynomials are dense among continuous functions over a finite interval with respect to the uniform norm; the second case results from the continuity theorem of Laplace transforms, which are none other than exponential generating functions of moments.) Clearly, the uniform distribution over $[0, 1]$, the exponential distribution, and the Gaussian distribution are characterized by their moments.

Equation (7) expresses the fact that a distribution is characterized by its moments provided they do not grow too fast, which indicates that its tails decay sufficiently rapidly. Other useful sufficient conditions for $F(x)$ to be characterized by moments are [157, XIV.2]:

$$(8) \quad \begin{cases} Carleman: & \displaystyle\sum_{k=0}^{\infty} \mu_{2k}^{-1/(2k)} = +\infty & (\text{support}(F) \subset \mathbb{R}) \\[2ex] \text{---}: & \displaystyle\sum_{k=0}^{\infty} \mu_k^{-1/(2k)} = +\infty & (\text{support}(F) \subset \mathbb{R}_{\geq 0}) \\[2ex] Krein: & \displaystyle\int_{-\infty}^{\infty} \log(f(x)) \frac{dx}{1+x^2} = -\infty & (F'(x) = f(x)). \end{cases}$$

One has the following theorem.

Theorem C.2 (Moment Convergence Theorem). *Let F be determined by its moments and assume that a sequence of distribution functions $F_n(x)$, $x \in \mathbb{R}$ satisfies for each $k = 0, 1, 2 \ldots,$*

$$\lim_{n \to \infty} \int_R x^k \, dF_n(x) = \int_{\mathbb{R}} x^k \, dF(x).$$

Then weak convergence holds: $F_n \Rightarrow F$.

For a proof, see [68, §30]. In this book, moment methods are used to validate the moment pumping method expounded in Chapter VII, p. 532.

▷ **C.7.** *The log–normal distribution.* As its name indicates, this is the distribution of the exponential of a standard normal, with density $f(x) = e^{-(\log x)^2/2}/(x\sqrt{2\pi})$, for $x > 0$. The distribution with density $f(x)(1 + \sin(2\pi \log x))$ has the same moments (Stieltjes, 1895). ◁

Bibliography

[1] ABEL, N. H. *Oeuvres complètes. Tome II.* Éditions Jacques Gabay, Sceaux, 1992. Suivi de "Niels Henrik Abel: sa vie et son action scientifique" par C.-A. Bjerknes. [Followed by "Niels Henrik Abel: his life and his scientific activity" by C.-A. Bjerknes] (1884), Edited and with notes by L. Sylow and S. Lie, Reprint of the second (1881) edition.

[2] ABHYANKAR, S.-S. *Algebraic geometry for scientists and engineers.* American Mathematical Society, 1990.

[3] ABRAMOWITZ, M., AND STEGUN, I. A. *Handbook of Mathematical Functions.* Dover, 1973. A reprint of the tenth National Bureau of Standards edition, 1964.

[4] ABRAMSON, M., AND MOSER, W. Combinations, successions and the n-kings problem. *Mathematics Magazine 39*, 5 (November 1966), 269–273.

[5] AHO, A. V., AND CORASICK, M. J. Efficient string matching: an aid to bibliographic search. *Communications of the ACM 18* (1975), 333–340.

[6] AHO, A. V., AND ULLMAN, J. D. *Principles of Compiler Design.* Addison-Wesley, 1977.

[7] AIGNER, M., AND ZIEGLER, G. *Proofs from THE BOOK.* Springer-Verlag, 2004.

[8] AĬZENBERG, I. A., AND YUZHAKOV, A. P. *Integral representations and residues in multidimensional complex analysis*, vol. 58 of *Translations of Mathematical Monographs.* American Mathematical Society, 1983.

[9] ALDOUS, D. J. Deterministic and stochastic models for coalescence (aggregation and coagulation): a review of the mean-field theory for probabilists. *Bernoulli 5*, 1 (1999), 3–48.

[10] ALDOUS, D. J., AND DIACONIS, P. Longest increasing subsequences: from patience sorting to the Baik-Deift-Johansson theorem. *American Mathematical Society. Bulletin. New Series 36*, 4 (1999), 413–432.

[11] ALDOUS, D. J., AND FILL, J. A. *Reversible Markov Chains and Random Walks on Graphs.* 2003. Book in preparation; manuscript available electronically.

[12] ALDOUS, D. J., AND PITMAN, J. The asymptotic distribution of the diameter of a random mapping. *C. R. Math. Acad. Sci. Paris 334*, 11 (2002), 1021–1024.

[13] ALON, N., AND SPENCER, J. H. *The probabilistic method.* John Wiley & Sons Inc., 1992.

[14] ANDREWS, G. E. *The Theory of Partitions*, vol. 2 of *Encyclopedia of Mathematics and its Applications.* Addison–Wesley, 1976.

[15] ANDREWS, G. E., ASKEY, R., AND ROY, R. *Special Functions*, vol. 71 of *Encyclopedia of Mathematics and its Applications.* Cambridge University Press, 1999.

[16] APOSTOL, T. M. *Introduction to Analytic Number Theory.* Springer-Verlag, 1976.

[17] APOSTOL, T. M. *Modular functions and Dirichlet series in number theory.* Springer-Verlag, 1976. Graduate Texts in Mathematics, No. 41.

[18] ARNEY, J., AND BENDER, E. A. Random mappings with constraints on coalescence and number of origins. *Pacific Journal of Mathematics 103* (1982), 269–294.

[19] ARRATIA, R., BARBOUR, A. D., AND TAVARÉ, S. Random combinatorial structures and prime factorizations. *Notices of the American Mathematical Society 44*, 8 (1997), 903–910.

[20] ARRATIA, R., BARBOUR, A. D., AND TAVARÉ, S. *Logarithmic Combinatorial Structures: a Probabilistic Approach.* EMS Monographs in Mathematics. European Mathematical Society (EMS), 2003.

[21] ATHREYA, K. B., AND NEY, P. E. *Branching processes.* Springer-Verlag, New York, 1972. Die Grundlehren der mathematischen Wissenschaften, Band 196.

[22] AYOUB, R. *An introduction to the analytic theory of numbers.* Mathematical Surveys, No. 10. American Mathematical Society, 1963.

[23] BACH, E. Toward a theory of Pollard's rho method. *Information and Computation 90*, 2 (1991), 139–155.

[24] BAIK, J., DEIFT, P., AND JOHANSSON, K. On the distribution of the length of the longest increasing subsequence of random permutations. *Journal of the American Mathematical Society 12*, 4 (1999), 1119–1178.

[25] BALADI, V., AND VALLÉE, B. Euclidean algorithms are Gaussian. *Journal of Number Theory 110* (2005), 331–386.

[26] BANDERIER, C., BOUSQUET-MÉLOU, M., DENISE, A., FLAJOLET, P., GARDY, D., AND GOUYOU-BEAUCHAMPS, D. Generating functions of generating trees. *Discrete Mathematics 246*, 1-3 (Mar. 2002), 29–55.

[27] BANDERIER, C., AND FLAJOLET, P. Basic analytic combinatorics of directed lattice paths. *Theoretical Computer Science 281*, 1-2 (2002), 37–80.

[28] BANDERIER, C., FLAJOLET, P., SCHAEFFER, G., AND SORIA, M. Random maps, coalescing saddles, singularity analysis, and Airy phenomena. *Random Structures & Algorithms 19*, 3/4 (2001), 194–246.

[29] BARBOUR, A. D., HOLST, L., AND JANSON, S. *Poisson approximation*. The Clarendon Press Oxford University Press, 1992. Oxford Science Publications.

[30] BASSINO, F., CLÉMENT, J., FAYOLLE, J., AND NICODÈME, P. Counting occurrences for a finite set of words: an inclusion-exclusion approach. *Discrete Mathematics & Theoretical Computer Science Proceedings* (2007). 14 pages. Proceedings of the AofA07 (Analysis of Algorithms) Conference. In press.

[31] BEARDON, A. F. *Iteration of Rational Functions*. Graduate Texts in Mathematics. Springer Verlag, 1991.

[32] BÉKESSY, A., BÉKESSY, P., AND KOMLÓS, J. Asymptotic enumeration of regular matrices. *Studia Scientiarum Mathematicarum Hungarica 7* (1972), 343–355.

[33] BELL, J. P., BURRIS, S. N., AND YEATS, K. A. Counting rooted trees: The universal law $t(n) \sim C\rho^{-n}n^{-3/2}$. *Electronic Journal of Combinatorics 13*, R63 (2006), 1–64.

[34] BELLMAN, R. *Matrix Analysis*. S.I.A.M. Press, 1997. A reprint of the second edition, first published by McGraw-Hill, New York, 1970.

[35] BENDER, E. A. Central and local limit theorems applied to asymptotic enumeration. *Journal of Combinatorial Theory 15* (1973), 91–111.

[36] BENDER, E. A. Asymptotic methods in enumeration. *SIAM Review 16*, 4 (Oct. 1974), 485–515.

[37] BENDER, E. A. The asymptotic number of non-negative integer matrices with given row and column sums. *Discrete Mathematics 10* (1974), 217–223.

[38] BENDER, E. A. Convex n–ominoes. *Discrete Mathematics 8* (1974), 219–226.

[39] BENDER, E. A., AND CANFIELD, E. R. The asymptotic number of labeled graphs with given degree sequences. *Journal of Combinatorial Theory*, Series A *24* (1978), 296–307.

[40] BENDER, E. A., CANFIELD, E. R., AND MCKAY, B. D. Asymptotic properties of labeled connected graphs. *Random Structures & Algorithms 3*, 2 (1992), 183–202.

[41] BENDER, E. A., GAO, Z., AND WORMALD, N. C. The number of labeled 2-connected planar graphs. *Electronic Journal of Combinatorics 9*, 1 (2002), Research Paper 43, 13 pp.

[42] BENDER, E. A., AND GOLDMAN, J. R. Enumerative uses of generating functions. *Indiana University Mathematical Journal* (1971), 753–765.

[43] BENDER, E. A., AND KOCHMAN, F. The distribution of subword counts is usually normal. *European Journal of Combinatorics 14* (1993), 265–275.

[44] BENDER, E. A., AND RICHMOND, L. B. Central and local limit theorems applied to asymptotic enumeration II: Multivariate generating functions. *Journal of Combinatorial Theory*, Series A *34* (1983), 255–265.

[45] BENDER, E. A., AND RICHMOND, L. B. Multivariate asymptotics for products of large powers with application to Lagrange inversion. *Electronic Journal of Combinatorics 6* (1999), R8. 21pp.

[46] BENDER, E. A., RICHMOND, L. B., AND WILLIAMSON, S. G. Central and local limit theorems applied to asymptotic enumeration. III. Matrix recursions. *Journal of Combinatorial Theory*, Series A *35*, 3 (1983), 264–278.

[47] BENTLEY, J., AND SEDGEWICK, R. Fast algorithms for sorting and searching strings. In *Eighth Annual ACM-SIAM Symposium on Discrete Algorithms* (1997), SIAM Press.

[48] BERGE, C. *Principes de combinatoire*. Dunod, 1968.

[49] BERGERON, F., FLAJOLET, P., AND SALVY, B. Varieties of increasing trees. In *CAAP'92* (1992), J.-C. Raoult, Ed., vol. 581 of *Lecture Notes in Computer Science*, pp. 24–48. Proceedings of the 17th Colloquium on Trees in Algebra and Programming, Rennes, France, February 1992.

[50] BERGERON, F., LABELLE, G., AND LEROUX, P. *Combinatorial species and tree-like structures*. Cambridge University Press, 1998.

[51] BERLEKAMP, E. R. *Algebraic Coding Theory*. Mc Graw-Hill, 1968. Revised edition, 1984.

[52] BERNDT, B. C. *Ramanujan's Notebooks, Part I*. Springer Verlag, 1985.

[53] BERSTEL, J. Sur les pôles et le quotient de Hadamard de séries *n*-rationnelles. *Comptes–Rendus de l'Académie des Sciences 272*, Série A (1971), 1079–1081.

[54] BERSTEL, J., Ed. *Séries Formelles*. LITP, University of Paris, 1978. (Proceedings of a School, Vieux–Boucau, France, 1977).

[55] BERSTEL, J., AND PERRIN, D. *Theory of codes*. Academic Press Inc., 1985.

[56] BERSTEL, J., AND REUTENAUER, C. Recognizable formal power series on trees. *Theoretical Computer Science 18* (1982), 115–148.

[57] BERSTEL, J., AND REUTENAUER, C. *Les séries rationnelles et leurs langages*. Masson, 1984.

[58] BERTOIN, J., BIANE, P., AND YOR, M. Poissonian exponential functionals, *q*-series, *q*-integrals, and the moment problem for log-normal distribution. In *Proceedings Stochastic Analysis, Ascona* (2004), vol. 58 of *Progress in Probability*, Birkhäuser Verlag, pp. 45–56.

[59] BERTONI, A., CHOFFRUT, C., GOLDWURM, M. G., AND LONATI, V. On the number of occurrences of a symbol in words of regular languages. *Theoretical Computer Science 302*, 1–3 (2003), 431–456.

[60] BESSIS, D., ITZYKSON, C., AND ZUBER, J.-B. Quantum field theory techniques in graphical enumeration. *Advances in Applied Mathematics 1* (1980), 109–157.

[61] BETREMA, J., AND PENAUD, J.-G. Modèles avec particules dures, animaux dirigés et séries en variables partiellement commutatives. ArXiv Preprint, 2001. arXiv:math/0106210.

[62] BHARUCHA-REID, A. T. *Elements of the Theory of Markov Processes and Their Applications*. Dover, 1997. A reprint of the original McGraw-Hill edition, 1960.

[63] BIANE, P. Permutations suivant le type d'excédance et le nombre d'inversions et interprétation combinatoire d'une fraction continue de Heine. *European Journal of Combinatorics 14* (1993), 277–284.

[64] BIANE, P., PITMAN, J., AND YOR, M. Probability laws related to the Jacobi theta and Riemann zeta functions, and Brownian excursions. *Bulletin of the American Mathematical Society (N.S.) 38*, 4 (2001), 435–465.

[65] BIEBERBACH, L. *Lehrbuch der Funktionentheorie*. Teubner, 1931. In two volumes. Reprinted by Johnson, 1968.

[66] BIGGS, N. L. *Algebraic Graph Theory*. Cambridge University Press, 1974.

[67] BIGGS, N. L., LLOYD, E. K., AND WILSON, R. *Graph Theory, 1736–1936*. Oxford University Press, 1974.

[68] BILLINGSLEY, P. *Probability and Measure*, 2nd ed. John Wiley & Sons, 1986.

[69] BINGHAM, N. H., GOLDIE, C. M., AND TEUGELS, J. L. *Regular variation*, vol. 27 of *Encyclopedia of Mathematics and its Applications*. Cambridge University Press, 1989.

[70] BIRKHOFF, G. D. Formal theory of irregular linear difference equations. *Acta Mathematica 54* (1930), 205–246.

[71] BIRKHOFF, G. D., AND TRJITZINSKY, W. J. Analytic theory of singular difference equations. *Acta Mathematica 60* (1932), 1–89.

[72] BJÖRK, J. E. *Rings of Differential Operators*. North Holland P. C., 1979.

[73] BŁASIAK, P., A., H., PENSON, K. A., DUCHAMP, G. H. E., AND SOLOMON, A. I. Boson normal ordering via substitutions and Sheffer-type polynomials. *Physical Letters A 338* (2005), 108.

[74] BŁASIAK, P., A., H., PENSON, K. A., SOLOMON, A. I., AND DUCHAMP, G. H. E. Combinatorics and Boson normal ordering: A gentle introduction. arXiv:0704.3116v1 [quant-ph], 8 pages.

[75] BLEISTEIN, N., AND HANDELSMAN, R. A. *Asymptotic Expansions of Integrals*. Dover, New York, 1986. A reprint of the second Holt, Rinehart and Winston edition, 1975.

[76] BOLLOBÁS, B. *Random Graphs*. Academic Press, 1985.

[77] BOLLOBÁS, B., BORGS, C., CHAYES, J. T., KIM, J. H., AND WILSON, D. B. The scaling window of the 2-SAT transition. *Random Structures & Algorithms 18*, 3 (2001), 201–256.

[78] BORWEIN, D., RANKIN, S., AND RENNER, L. Enumeration of injective partial transformations. *Discrete Mathematics 73* (1989), 291–296.

[79] BORWEIN, J. M., AND BORWEIN, P. B. Strange series and high precision fraud. *American Mathematical Monthly 99*, 7 (Aug. 1992), 622–640.

[80] BORWEIN, J. M., BORWEIN, P. B., AND DILCHER, K. Pi, Euler numbers and asymptotic expansions. *American Mathematical Monthly 96*, 8 (1989), 681–687.

[81] BOURDON, J., AND VALLÉE, B. Generalized pattern matching statistics. In *Mathematics and computer science, II (Versailles, 2002)*, B. Chauvin *et al.*, Ed. Birkhäuser, 2002, pp. 249–265.

[82] BOUSQUET-MÉLOU, M. A method for the enumeration of various classes of column-convex poly-gons. *Discrete Math. 154*, 1-3 (1996), 1–25.

[83] BOUSQUET-MÉLOU, M. Limit laws for embedded trees: Applications to the integrated SuperBrow-nian excursion. *Random Structures and Algorithms 29* (2006), 475–523.

[84] BOUSQUET-MÉLOU, M. Rational and algebraic series in combinatorial enumeration. In *Proceedings of the International Congress of Mathematicians* (2006), pp. 789–826.

[85] BOUSQUET-MÉLOU, M., AND GUTTMANN, A. J. Enumeration of three-dimensional convex poly-gons. *Annals of Combinatorics 1* (1997), 27–53.

[86] BOUSQUET-MÉLOU, M., AND PETKOVŠEK, M. Linear recurrences with constant coefficients: the multivariate case. *Discrete Mathematics 225*, 1-3 (2000), 51–75.

[87] BOUSQUET-MÉLOU, M., AND RECHNITZER, A. Lattice animals and heaps of dimers. *Discrete Mathematics 258* (2002), 235–274.

[88] BOUTTIER, J. *Physique statistique des surfaces aléatoires et combinatoire bijective des cartes planaires*. Ph.D. Thesis, Universté Paris 6, 2005.

[89] BOUTTIER, J., DI FRANCESCO, P., AND GUITTER, E. Census of planar maps: from the one-matrix model solution to a combinatorial proof. *Nuclear Physics B 645*, 3 (2002), 477–499.

[90] BOUTTIER, J., DI FRANCESCO, P., AND GUITTER, E. Geodesic distance in planar graphs. *Nuclear Physics B 663*, 3 (2003), 535–567.

[91] BOUTTIER, J., DI FRANCESCO, P., AND GUITTER, E. Statistics of planar graphs viewed from a vertex: A study via labeled trees. *Nuclear Physics B 675*, 3 (2003), 631–660.

[92] BRASSARD, G., AND BRATLEY, P. *Algorithmique: conception et analyse*. Masson, Paris, 1987.

[93] BREIMAN, L. *Probability*. Society for Industrial and Applied Mathematics (SIAM), 1992. Corrected reprint of the 1968 original.

[94] BRÉZIN, É., ITZYKSON, C., PARISI, G., AND ZUBER, J.-B. Planar diagrams. *Communications in Mathematical Physics 59* (1978), 35–51.

[95] BROUTIN, N., AND DEVROYE, L. Large deviations for the weighted height of an extended class of trees. *Algorithmica 46* (2006), 271–297.

[96] BROWN, W. G., AND TUTTE, W. T. On the enumeration of rooted non-separable planar maps. *Canadian Journal of Mathematics 16* (1964), 572–577.

[97] BURGE, W. H. An analysis of binary search trees formed from sequences of nondistinct keys. *Jour-nal of the ACM 23*, 3 (July 1976), 451–454.

[98] BURRIS, S. N. *Number theoretic density and logical limit laws*, vol. 86 of *Mathematical Surveys and Monographs*. American Mathematical Society, 2001.

[99] BURRIS, S. N. Two corrections to results in the literature on recursive systems. Unpublished memo. 8 pages., January 2008.

[100] BUTZER, P. L., AND JANSCHE, S. A direct approach to the Mellin transform. *The Journal of Fourier Analysis and Applications 3*, 4 (1997), 325–376.

[101] CANFIELD, E. R. Central and local limit theorems for the coefficients of polynomials of binomial type. *Journal of Combinatorial Theory, Series A 23* (1977), 275–290.

[102] CANFIELD, E. R. Remarks on an asymptotic method in combinatorics. *Journal of Combinatorial Theory, Series A 37* (1984), 348–352.

[103] CARLITZ, L. Permutations, sequences and special functions. *S.I.A.M. Review 17* (1975), 298–322.

[104] CARTAN, H. *Théorie élémentaire des fonctions analytiques d'une ou plusieurs variables complexes*. Hermann, 1961.

[105] CARTIER, P., AND FOATA, D. *Problèmes combinatoires de commutation et réarrangements*, vol. 85 of *Lecture Notes in Mathematics*. Springer Verlag, 1969. (New free web edition, 2006).

[106] CATALAN, E. Note sur une équation aux différences finies. *Journal de Mathématiques Pures et Appliquées 3* (1838), 508–516. Freely accessible under the Gallica-MathDoc site.

[107] CATALAN, E. Addition à la note sur une équation aux différences finies, insérée dans le volume précédent, page 508. *Journal de Mathématiques Pures et Appliquées 4* (1839), 95–99. Freely acces-sible under the Gallica-MathDoc site.

[108] CATALAN, E. Solution nouvelle de cette question: Un polygone étant donné, de combien de manières peut-on le partager en triangles au moyen de diagonales? *Journal de Mathématiques Pures et Appliquées 4* (1839), 91–94. Freely accessible under the Gallica-MathDoc site.

[109] CAZALS, F. Monomer-dimer tilings. *Studies in Automatic Combinatorics 2* (1997). Electronic pub-lication http://algo.inria.fr/libraries/autocomb/autocomb.html.

[110] CHABAUD, C. *Séries génératrices algébriques:asymptotique et applications combinatoires*. PhD thesis, Université Paris VI, 2002.

[111] CHASSAING, P., AND MARCKERT, J.-F. Parking functions, empirical processes, and the width of rooted labeled trees. *Electronic Journal of Combinatorics 8*, 1 (2001), Research Paper 14, 19 pp. (electronic).

[112] CHASSAING, P., MARCKERT, J.-F., AND YOR, M. The height and width of simple trees. In *Mathematics and computer science (Versailles, 2000)*, Trends Math. Birkhäuser Verlag, 2000, pp. 17–30.

[113] CHASSAING, P., AND SCHAEFFER, G. Random planar lattices and integrated superBrownian excursion. *Probability Theory and Related Fields 128* (2004), 161–212.

[114] CHAUVIN, B., DRMOTA, M., AND JABBOUR-HATTAB, J. The profile of binary search trees. *The Annals of Applied Probability 11*, 4 (2001), 1042–1062.

[115] CHAUVIN, B., FLAJOLET, P., GARDY, D., AND GITTENBERGER, B. And/Or Trees Revisited. *Combinatorics, Probability and Computing 13*, 4–5 (2004), 501–513. Special issue on Analysis of Algorithms.

[116] CHERN, H.-H., AND HWANG, H.-K. Partial match queries in random quadtrees. *SIAM Journal on Computing 32*, 4 (2003), 904–915.

[117] CHERN, H.-H., HWANG, H.-K., AND TSAI, T.-H. An asymptotic theory for Cauchy-Euler differential equations with applications to the analysis of algorithms. *Journal of Algorithms 44*, 1 (2002), 177–225.

[118] CHIHARA, T. S. *An Introduction to Orthogonal Polynomials*. Gordon and Breach, 1978.

[119] CHOMSKY, N., AND SCHÜTZENBERGER, M. P. The algebraic theory of context–free languages. In *Computer Programing and Formal Languages* (1963), P. Braffort and D. Hirschberg, Eds., North Holland, pp. 118–161.

[120] CHYZAK, F. Gröbner bases, symbolic summation and symbolic integration. In *Gröbner Bases and Applications*, B. Buchberger and F. Winkler, Eds., vol. 251 of *London Mathematical Society Lecture Notes Series*. Cambridge University Press, 1998, pp. 32–60. In *Proceedings of the Conference 33 Years of Gröbner Bases*.

[121] CHYZAK, F., DRMOTA, M., KLAUSNER, T., AND KOK, G. The distribution of patterns in random trees. *Combinatorics, Probability and Computing 17* (2008), 21–59.

[122] CHYZAK, F., MISHNA, M., AND SALVY, B. Effective scalar products of D-finite symmetric functions. *Journal of Combinatorial Theory, Series A 112*, 1 (2005), 1–43.

[123] CHYZAK, F., AND SALVY, B. Non-commutative elimination in Ore algebras proves multivariate identities. *Journal of Symbolic Computation 26*, 2 (1998), 187–227.

[124] CLÉMENT, J., FLAJOLET, P., AND VALLÉE, B. Dynamical sources in information theory: A general analysis of trie structures. *Algorithmica 29*, 1/2 (2001), 307–369.

[125] COMPTON, K. J. A logical approach to asymptotic combinatorics. I. First order properties. *Advances in Mathematics 65* (1987), 65–96.

[126] COMPTON, K. J. A logical approach to asymptotic combinatorics. II. Second–order properties. *Journal of Combinatorial Theory*, Series A *50* (1987), 110–131.

[127] COMPTON, K. J. 0–1 laws in logic and combinatorics. In *Proceedings NATO Advanced Study Institute on Algorithms and Order* (Dordrecht, 1988), I. Rival, Ed., Reidel, pp. 353–383.

[128] COMTET, L. Calcul pratique des coefficients de Taylor d'une fonction algébrique. *Enseignement Mathématique. 10* (1964), 267–270.

[129] COMTET, L. *Advanced Combinatorics*. Reidel, 1974.

[130] CONRAD, E. V. F., AND FLAJOLET, P. The Fermat cubic, elliptic functions, continued fractions, and a combinatorial excursion. *Séminaire Lotharingien de Combinatoire 54*, B54g (2006), 1–44.

[131] CORLESS, R. M., GONNET, G. H., HARE, D. E. G., JEFFREY, D. J., AND KNUTH, D. E. On the Lambert W function. *Advances in Computational Mathematics 5* (1996), 329–359.

[132] CORMEN, T. H., LEISERSON, C. E., AND RIVEST, R. L. *Introduction to Algorithms*. MIT Press, 1990.

[133] COUTINHO, S. C. *A primer of algebraic D-modules*, vol. 33 of *London Mathematical Society Student Texts*. Cambridge University Press, 1995.

[134] COVER, T. M., AND THOMAS, J. A. *Elements of information theory*. John Wiley & Sons Inc., 1991. A Wiley-Interscience Publication.

[135] COX, D., LITTLE, J., AND O'SHEA, D. *Ideals, Varieties, and Algorithms: an Introduction to Computational Algebraic Geometry and Commutative Algebra*, 2nd ed. Springer, 1997.

[136] DANIELS, H. E. Saddlepoint approximations in statistics. *Annals of Mathematical Statistics 25* (1954), 631–650.

[137] DARBOUX, G. Mémoire sur l'approximation des fonctions de très grands nombres, et sur une classe étendue de développements en série. *Journal de Mathématiques Pures et Appliquées* (Feb. 1878), 5–56,377–416.

[138] DAVENPORT, H. *Multiplicative Number Theory*, revised by H. L. Montgomery, second ed. Springer-Verlag, 1980.

[139] DAVID, F. N., AND BARTON, D. E. *Combinatorial Chance*. Charles Griffin, 1962.

[140] DE BRUIJN, N. G. A combinatorial problem. *Nederl. Akad. Wetensch., Proc. 49* (1946), 758–764. Also in *Indagationes Math.* **8**, 461–467 (1946).

[141] DE BRUIJN, N. G. On Mahler's partition problem. *Indagationes Mathematicae X* (1948), 210–220. Reprinted from Koninklijke Nederlansche Akademie Wetenschappen.

[142] DE BRUIJN, N. G. A survey of generalizations of Pólya's enumeration theory. *Nieuw Archief voor Wiskunde XIX* (1971), 89–112.

[143] DE BRUIJN, N. G. *Asymptotic Methods in Analysis*. Dover, 1981. A reprint of the third North Holland edition, 1970 (first edition, 1958).

[144] DE BRUIJN, N. G., AND KLARNER, D. A. Multisets of aperiodic cycles. *SIAM Journal on Algebraic and Discrete Methods 3* (1982), 359–368.

[145] DE BRUIJN, N. G., KNUTH, D. E., AND RICE, S. O. The average height of planted plane trees. In *Graph Theory and Computing* (1972), R. C. Read, Ed., Academic Press, pp. 15–22.

[146] DE SEGNER, A. Enumeration modorum quibus figurae planae rectlineae per diagonales dividuntur in triangula. *Novi Commentarii Academiae Scientiarum Petropolitanae 7* (1758/59), 203–209.

[147] DÉCOSTE, H., LABELLE, G., AND LEROUX, P. Une approche combinatoire pour l'itération de Newton-Raphson. *Advances in Applied Mathematics 3* (1982), 407–416.

[148] DEIFT, P. Integrable systems and combinatorial theory. *Notices of the American Mathematical Society 47*, 6 (2000), 631–640.

[149] DELAURENTIS, J. M., AND PITTEL, B. G. Random permutations and brownian motion. *Pacific Journal of Mathematics 119*, 2 (1985), 287–301.

[150] DELEST, M.-P., AND VIENNOT, G. Algebraic languages and polyominoes enumeration. *Theoretical Computer Science 34* (1984), 169–206.

[151] DELLNITZ, M., SCHÜTZE, O., AND ZHENG, Q. Locating all the zeros of an analytic function in one complex variable. *J. Comput. Appl. Math. 138*, 2 (2002), 325–333.

[152] DEMBO, A., VERSHIK, A., AND ZEITOUNI, O. Large deviations for integer partitions. *Markov Processes and Related Fields 6*, 2 (2000), 147–179.

[153] DEN HOLLANDER, F. *Large deviations*. American Mathematical Society, 2000.

[154] DENEF, J., AND LIPSHITZ, L. Algebraic power series and diagonals. *Journal of Number Theory 26* (1987), 46–67.

[155] DERSHOWITZ, N., AND ZAKS, S. The cycle lemma and some applications. *European Journal of Combinatorics 11* (1990), 35–40.

[156] DEVANEY, R. L. *A first course in chaotic dynamical systems*. Addison-Wesley Studies in Nonlinearity. Addison-Wesley Publishing Company Advanced Book Program, 1992. Theory and experiment, With a separately available computer disk.

[157] DEVROYE, L. A note on the height of binary search trees. *Journal of the ACM 33* (1986), 489–498.

[158] DEVROYE, L. Branching processes in the analysis of the heights of trees. *Acta Informatica 24* (1987), 277–298.

[159] DEVROYE, L. Limit laws for local counters in random binary search trees. *Random Structures & Algorithms 2*, 3 (1991), 302–315.

[160] DEVROYE, L. Universal limit laws for depths in random trees. *SIAM Journal on Computing 28*, 2 (1999), 409–432.

[161] DEVROYE, L. Laws of large numbers and tail inequalities for random tries and patricia trees. *Journal of Computational and Applied Mathematics 142* (2002), 27–37.

[162] DHAR, D., PHANI, M. K., AND BARMA, M. Enumeration of directed site animals on two-dimensional lattices. *Journal of Physics A: Mathematical and General 15* (1982), L279–L284.

[163] DI FRANCESCO, P. Folding and coloring problems in mathematics and physics. *Bulletin of the American Mathematical Society (N.S.) 37*, 3 (2000), 251–307.

[164] DIENES, P. *The Taylor Series*. Dover, New York, 1958. A reprint of the first Oxford University Press edition, 1931.

[165] DIEUDONNÉ, J. *Calcul Infinitésimal*. Hermann, Paris, 1968.

[166] DIXON, J. D. The number of steps in the Euclidean algorithm. *Journal of Number Theory 2* (1970), 414–422.

[167] DIXON, J. D. Asymptotics of generating the symmetric and alternating groups. *Electronic Journal of Combinatorics 12*, R56 (2005), 1–5.

[168] DOETSCH, G. *Handbuch der Laplace-Transformation,* Vol. 1–3. Birkhäuser Verlag, Basel, 1955.

[169] DOMB, C., AND BARRETT, A. Enumeration of ladder graphs. *Discrete Mathematics 9* (1974), 341–358.

[170] DOYLE, P. G., AND SNELL, J. L. *Random walks and electric networks*. Mathematical Association of America, 1984.

[171] DRMOTA, M. Asymptotic distributions and a multivariate Darboux method in enumeration problems. Manuscript, Nov. 1990.

[172] DRMOTA, M. Systems of functional equations. *Random Structures & Algorithms 10*, 1–2 (1997), 103–124.

[173] DRMOTA, M., AND GITTENBERGER, B. On the profile of random trees. *Random Structures & Algorithms 10*, 4 (1997), 421–451.

[174] DRMOTA, M., GITTENBERGER, B., AND KLAUSNER, T. Extended admissible functions and Gaussian limiting distributions. *Mathematics of Computation 74*, 252 (2005), 1953–1966.

[175] DRMOTA, M., AND SORIA, M. Marking in combinatorial constructions: Generating functions and limiting distributions. *Theoretical Computer Science 144*, 1–2 (June 1995), 67–99.

[176] DRMOTA, M., AND SORIA, M. Images and preimages in random mappings. *SIAM Journal on Discrete Mathematics 10*, 2 (1997), 246–269.

[177] DUCHON, P., FLAJOLET, P., LOUCHARD, G., AND SCHAEFFER, G. Boltzmann samplers for the random generation of combinatorial structures. *Combinatorics, Probability and Computing 13*, 4–5 (2004), 577–625. Special issue on Analysis of Algorithms.

[178] DUFFIN, R. J. Ruble's universal differential equation. *Proceedings of the National Academy of Sciences USA 78*, 8 (1981), 4661–4662.

[179] DUMAS, P., AND FLAJOLET, P. Asymptotique des récurrences mahleriennes: le cas cyclotomique. *Journal de Théorie des Nombres de Bordeaux 8*, 1 (June 1996), 1–30.

[180] DUQUESNE, T., AND LE GALL, J.-F. Random Trees, Levy Processes and Spatial Branching Processes. arXiv:math.PR/0509558, 2005.

[181] DURAND, M. *Combinatoire analytique et algorithmique des ensembles de données*. Ph.D. Thesis, École Polytechnique, France, 2004.

[182] DURRETT, R. *Probability: theory and examples*, second ed. Duxbury Press, 1996.

[183] DUTOUR, I., AND FÉDOU, J.-M. Object grammars and random generation. *Discrete Mathematics and Theoretical Computer Science 2* (1998), 47–61.

[184] DVORETZKY, A., AND MOTZKIN, T. A problem of arrangements. *Duke Mathematical Journal 14* (1947), 305–313.

[185] EDELMAN, A., AND KOSTLAN, E. How many zeros of a random polynomial are real? *Bulletin of the American Mathematical Society (N.S.) 32*, 1 (1995), 1–37.

[186] EDWARDS, H. M. *Riemann's Zeta Function*. Academic Press, 1974.

[187] EGORYCHEV, G. P. *Integral representation and the computation of combinatorial sums*, vol. 59 of *Translations of Mathematical Monographs*. American Mathematical Society, 1984. Translated from the Russian by H. H. McFadden, Translation edited by Lev J. Leifman.

[188] EHRENFEST, P., AND EHRENFEST, T. Über zwei bekannte Einwände gegen das Boltzmannsche *H*-Theorem. *Physikalische Zeitschrift 8*, 9 (1907), 311–314.

[189] EILENBERG, S. *Automata, Languages, and Machines*, vol. A. Academic Press, 1974.

[190] ELIZALDE, S., AND NOY, M. Consecutive patterns in permutations. *Advances in Applied Mathematics 30*, 1-2 (2003), 110–125.

[191] ELLIOTT, P. D. T. A. *Probabilistic number theory. I*, vol. 239 of *Grundlehren der Mathematischen Wissenschaften [Fundamental Principles of Mathematical Science]*. Springer-Verlag, 1979.

[192] ENTING, I. G. Generating functions for enumerating self-avoiding rings on the square lattice. *Journal of Physics A: Mathematical and General 18* (1980), 3713–3722.

[193] ERDÉLYI, A. *Higher Transcendental Functions*, second ed., vol. 1-2-3. R. E. Krieger publishing Company, Inc., 1981.

[194] ERDŐS, P., AND LEHNER, J. The distribution of the number of summands in the partitions of a positive integer. *Duke Mathematical Journal 8* (1941), 335–345.

[195] ERDŐS, P., AND RÉNYI, A. On a classical problem of probability theory. *Magyar Tud. Akad. Mat. Kutató Int. Közl. 6* (1961), 215–220.

[196] EULER, L. Letter to Goldbach, dated September 4, 1751. Published as "Lettre CXL, Euler à Goldbach". In *Leonhard Euler Briefwechsel*, Vol. I, p. 159, letter 868.

[197] EULER, L. Enumeration modorum, quibus figurae planae rectlineae per diagonales dividuntur in triangula, auct. i. a. de segner. *Novi Commentarii Academiae Scientiarum Petropolitanae 7* (1758/59), 13–14. Report by Euler on de Segner's note [146].

[198] EULER, L. De seriebus divergentibus. *Novi Commentarii Academiae Scientiarum Petropolitanae 5* (1760), 205–237. In *Opera Omnia*: Series 1, Volume 14, pp. 585–617. Available on the Euler Archive as E247.

[199] EULER, L. Observationes analyticae. *Novi Commentarii Acad. Sci. Imper. Petropolitanae 11* (1765), 124–143.

[200] EVEREST, G., VAN DER POORTEN, A., SHPARLINSKI, I., AND WARD, T. *Recurrence sequences*, vol. 104 of *Mathematical Surveys and Monographs*. American Mathematical Society, 2003.

[201] FARKAS, H. M., AND KRA, I. *Riemann surfaces*, second ed., vol. 71 of *Graduate Texts in Mathematics*. Springer-Verlag, 1992.

[202] FAYOLLE, G., AND IASNOGORODSKI, R. Two coupled processors: the reduction to a Riemann-Hilbert problem. *Zeitschrift für Wahrscheinlichkeitstheorie und Verwandte Gebiete 47*, 3 (1979), 325–351.

[203] FAYOLLE, G., IASNOGORODSKI, R., AND MALYSHEV, V. *Random walks in the quarter-plane*. Springer-Verlag, 1999.

[204] FAYOLLE, J. An average-case analysis of basic parameters of the suffix tree. In *Mathematics and Computer Science III: Algorithms, Trees, Combinatorics and Probabilities* (2004), M. Drmota *et al.*, Ed., Trends in Mathematics, Birkhäuser Verlag, pp. 217–227.

[205] FELLER, W. *An Introduction to Probability Theory and its Applications*, third ed., vol. 1. John Wiley, 1968.

[206] FELLER, W. *An Introduction to Probability Theory and Its Applications*, vol. 2. John Wiley, 1971.

[207] FILL, J. A. On the distribution of binary search trees under the random permutation model. *Random Structures & Algorithms 8*, 1 (1996), 1–25.

[208] FILL, J. A., FLAJOLET, P., AND KAPUR, N. Singularity analysis, Hadamard products, and tree recurrences. *Journal of Computational and Applied Mathematics 174* (Feb. 2005), 271–313.

[209] FILL, J. A., AND JANSON, S. Approximating the limiting quicksort distribution. *Random Structures & Algorithms 19* (2001), 376–406.

[210] FILL, J. A., AND KAPUR, N. Limiting distributions for additive functionals on Catalan trees. *Theoretical Computer Science 326*, 1–3 (2004), 69–102.

[211] FINCH, S. *Mathematical Constants*. Cambridge University Press, 2003.

[212] FINKEL, R. A., AND BENTLEY, J. L. Quad trees, a data structure for retrieval on composite keys. *Acta Informatica 4* (1974), 1–9.

[213] FISCHER, H. *Die verschiedenen Formen und Funktionen des zentralen Grenzwertsatzes in der Entwicklung von der klassischen zur modernen Wahrscheinlichkeitsrechnung*. Shaker Verlag, 2000. 318 p. (ISBN: 3-8265-7767-1).

[214] FLAJOLET, P. Combinatorial aspects of continued fractions. *Discrete Mathematics 32* (1980), 125–161. Reprinted in the 35th Special Anniversary Issue of *Discrete Mathematics*, Volume 306, Issue 10–11, Pages 992-1021 (2006).

[215] FLAJOLET, P. *Analyse d'algorithmes de manipulation d'arbres et de fichiers*, vol. 34–35 of *Cahiers du Bureau Universitaire de Recherche Opérationnelle*. Université Pierre et Marie Curie, Paris, 1981. 209 pages.

[216] FLAJOLET, P. On congruences and continued fractions for some classical combinatorial quantities. *Discrete Mathematics 41* (1982), 145–153.

[217] FLAJOLET, P. On the performance evaluation of extendible hashing and trie searching. *Acta Informatica 20* (1983), 345–369.

[218] FLAJOLET, P. Approximate counting: A detailed analysis. *BIT 25* (1985), 113–134.

[219] FLAJOLET, P. Elements of a general theory of combinatorial structures. In *Fundamentals of Computation Theory* (1985), L. Budach, Ed., vol. 199 of *Lecture Notes in Computer Science*, Springer Verlag, pp. 112–127. Proceedings of FCT'85, Cottbus, GDR, September 1985 (Invited Lecture).

[220] FLAJOLET, P. Analytic models and ambiguity of context–free languages. *Theoretical Computer Science 49* (1987), 283–309.

[221] FLAJOLET, P. Mathematical methods in the analysis of algorithms and data structures. In *Trends in Theoretical Computer Science*, E. Börger, Ed. Computer Science Press, 1988, ch. 6, pp. 225–304. (Lecture Notes for *A Graduate Course in Computation Theory*, Udine, 1984).

[222] FLAJOLET, P. Constrained permutations and the principle of inclusion-exclusion. *Studies in Automatic Combinatorics II* (1997). Available electronically at http://algo.inria.fr/libraries/autocomb.

[223] FLAJOLET, P. Singularity analysis and asymptotics of Bernoulli sums. *Theoretical Computer Science 215*, 1-2 (1999), 371–381.

[224] FLAJOLET, P. Counting by coin tossings. In *Proceedings of ASIAN'04 (Ninth Asian Computing Science Conference)* (2004), M. Maher, Ed., vol. 3321 of *Lecture Notes in Computer Science*, pp. 1–12. (Text of Opening Keynote Address.).

[225] FLAJOLET, P., DUMAS, P., AND PUYHAUBERT, V. Some exactly solvable models of urn process theory. *Discrete Mathematics & Theoretical Computer Science (Proceedings) AG* (2006), 59–118.

[226] FLAJOLET, P., FRANÇON, J., AND VUILLEMIN, J. Sequence of operations analysis for dynamic data structures. *Journal of Algorithms 1* (1980), 111–141.

[227] FLAJOLET, P., FUSY, E., GOURDON, X., PANARIO, D., AND POUYANNE, N. A hybrid of Darboux's method and singularity analysis in combinatorial asymptotics. *Electronic Journal of Combinatorics 13*, 1:R103 (2006), 1–35.

[228] FLAJOLET, P., FUSY, É., AND PIVOTEAU, C. Boltzmann sampling of unlabelled structures. In *Proceedings of the Ninth Workshop on Algorithm Engineering and Experiments and the Fourth Workshop on Analytic Algorithmics and Combinatorics* (2007), D. A. *et al.*, Ed., SIAM Press, pp. 201–211. Proceedings of the New Orleans Conference.

[229] FLAJOLET, P., GABARRÓ, J., AND PEKARI, H. Analytic urns. *Annals of Probability 33*, 3 (2005), 1200–1233. Available from ArXiv:math.PR/0407098.

[230] FLAJOLET, P., GAO, Z., ODLYZKO, A., AND RICHMOND, B. The distribution of heights of binary trees and other simple trees. *Combinatorics, Probability and Computing 2* (1993), 145–156.

[231] FLAJOLET, P., GARDY, D., AND THIMONIER, L. Birthday paradox, coupon collectors, caching algorithms, and self–organizing search. *Discrete Applied Mathematics 39* (1992), 207–229.

[232] FLAJOLET, P., GERHOLD, S., AND SALVY, B. On the non-holonomic character of logarithms, powers, and the nth prime function. *Electronic Journal of Combinatorics 11(2)*, A1 (2005), 1–16.

[233] FLAJOLET, P., GONNET, G., PUECH, C., AND ROBSON, J. M. Analytic variations on quadtrees. *Algorithmica 10*, 7 (Dec. 1993), 473–500.

[234] FLAJOLET, P., GOURDON, X., AND DUMAS, P. Mellin transforms and asymptotics: Harmonic sums. *Theoretical Computer Science 144*, 1–2 (June 1995), 3–58.

[235] FLAJOLET, P., GOURDON, X., AND MARTÍNEZ, C. Patterns in random binary search trees. *Random Structures & Algorithms 11*, 3 (Oct. 1997), 223–244.

[236] FLAJOLET, P., GOURDON, X., AND PANARIO, D. The complete analysis of a polynomial factorization algorithm over finite fields. *Journal of Algorithms 40*, 1 (2001), 37–81.

[237] FLAJOLET, P., GRABNER, P., KIRSCHENHOFER, P., AND PRODINGER, H. On Ramanujan's Q–function. *Journal of Computational and Applied Mathematics 58*, 1 (Mar. 1995), 103–116.

[238] FLAJOLET, P., AND GUILLEMIN, F. The formal theory of birth-and-death processes, lattice path combinatorics, and continued fractions. *Advances in Applied Probability 32* (2000), 750–778.

[239] FLAJOLET, P., HATZIS, K., NIKOLETSEAS, S., AND SPIRAKIS, P. On the robustness of interconnections in random graphs: A symbolic approach. *Theoretical Computer Science 287*, 2 (2002), 513–534.

[240] FLAJOLET, P., KIRSCHENHOFER, P., AND TICHY, R. F. Deviations from uniformity in random strings. *Probability Theory and Related Fields 80* (1988), 139–150.

[241] FLAJOLET, P., KNUTH, D. E., AND PITTEL, B. The first cycles in an evolving graph. *Discrete Mathematics 75* (1989), 167–215.

[242] FLAJOLET, P., LABELLE, G., LAFOREST, L., AND SALVY, B. Hypergeometrics and the cost structure of quadtrees. *Random Structures & Algorithms 7*, 2 (1995), 117–144.

[243] FLAJOLET, P., AND LAFFORGUE, T. Search costs in quadtrees and singularity perturbation asymptotics. *Discrete and Computational Geometry 12*, 4 (1994), 151–175.

[244] FLAJOLET, P., AND LOUCHARD, G. Analytic variations on the Airy distribution. *Algorithmica 31*, 3 (2001), 361–377.

[245] FLAJOLET, P., AND NOY, M. Analytic combinatorics of non-crossing configurations. *Discrete Mathematics 204*, 1-3 (1999), 203–229. (Selected papers in honor of Henry W. Gould).

[246] FLAJOLET, P., AND ODLYZKO, A. M. The average height of binary trees and other simple trees. *Journal of Computer and System Sciences 25* (1982), 171–213.

[247] FLAJOLET, P., AND ODLYZKO, A. M. Random mapping statistics. In *Advances in Cryptology* (1990), J.-J. Quisquater and J. Vandewalle, Eds., vol. 434 of *Lecture Notes in Computer Science*, Springer Verlag, pp. 329–354. Proceedings of EUROCRYPT'89, Houtalen, Belgium, April 1989.

[248] FLAJOLET, P., AND ODLYZKO, A. M. Singularity analysis of generating functions. *SIAM Journal on Algebraic and Discrete Methods 3*, 2 (1990), 216–240.

[249] FLAJOLET, P., POBLETE, P., AND VIOLA, A. On the analysis of linear probing hashing. *Algorithmica 22*, 4 (Dec. 1998), 490–515.

[250] FLAJOLET, P., AND PRODINGER, H. Level number sequences for trees. *Discrete Mathematics 65* (1987), 149–156.

[251] FLAJOLET, P., AND PRODINGER, H. On Stirling numbers for complex argument and Hankel contours. *SIAM Journal on Discrete Mathematics 12*, 2 (1999), 155–159.

[252] FLAJOLET, P., AND PUECH, C. Partial match retrieval of multidimensional data. *Journal of the ACM 33*, 2 (1986), 371–407.

[253] FLAJOLET, P., RÉGNIER, M., AND SEDGEWICK, R. Some uses of the Mellin integral transform in the analysis of algorithms. In *Combinatorial Algorithms on Words* (1985), A. Apostolico and Z. Galil, Eds., vol. 12 of *NATO Advance Science Institute Series*. Series F: Computer and Systems Sciences, Springer Verlag, pp. 241–254. (Invited Lecture).

[254] FLAJOLET, P., SALVY, B., AND SCHAEFFER, G. Airy phenomena and analytic combinatorics of connected graphs. *Electronic Journal of Combinatorics 11*, 2:#R34 (2004), 1–30.

[255] FLAJOLET, P., SALVY, B., AND ZIMMERMANN, P. Automatic average–case analysis of algorithms. *Theoretical Computer Science 79*, 1 (Feb. 1991), 37–109.

[256] FLAJOLET, P., AND SEDGEWICK, R. Mellin transforms and asymptotics: finite differences and Rice's integrals. *Theoretical Computer Science 144*, 1–2 (June 1995), 101–124.

[257] FLAJOLET, P., SIPALA, P., AND STEYAERT, J.-M. Analytic variations on the common subexpression problem. In *Automata, Languages, and Programming* (1990), M. S. Paterson, Ed., vol. 443 of *Lecture Notes in Computer Science*, pp. 220–234. Proceedings of the 17th ICALP Conference, Warwick, July 1990.

[258] FLAJOLET, P., AND SORIA, M. Gaussian limiting distributions for the number of components in combinatorial structures. *Journal of Combinatorial Theory*, Series A *53* (1990), 165–182.

[259] FLAJOLET, P., AND SORIA, M. The cycle construction. *SIAM Journal on Discrete Mathematics 4*, 1 (Feb. 1991), 58–60.

[260] FLAJOLET, P., AND SORIA, M. General combinatorial schemas: Gaussian limit distributions and exponential tails. *Discrete Mathematics 114* (1993), 159–180.

[261] FLAJOLET, P., AND STEYAERT, J.-M. A complexity calculus for classes of recursive search programs over tree structures. In *Proceedings of the 22nd Annual Symposium on Foundations of Computer Science* (1981). IEEE Computer Society Press, pp. 386–393.

[262] FLAJOLET, P., AND STEYAERT, J.-M. A complexity calculus for recursive tree algorithms. *Mathematical Systems Theory 19* (1987), 301–331.

[263] FLAJOLET, P., SZPANKOWSKI, W., AND VALLÉE, B. Hidden word statistics. *Journal of the ACM 53*, 1 (Jan. 2006), 147–183.

[264] FLAJOLET, P., ZIMMERMAN, P., AND VAN CUTSEM, B. A calculus for the random generation of labelled combinatorial structures. *Theoretical Computer Science 132*, 1-2 (1994), 1–35.

[265] FOATA, D. *La série génératrice exponentielle dans les problèmes d'énumération*. S.M.S. Montreal University Press, 1974.

[266] FOATA, D., LASS, B., AND HAN, G.-N. Les nombres hyperharmoniques et la fratrie du collectionneur de vignettes. *Seminaire Lotharingien de Combinatoire 47*, B47a (2001), 1–20.

[267] FOATA, D., AND SCHÜTZENBERGER, M.-P. *Théorie Géométrique des Polynômes Euleriens*, vol. 138 of *Lecture Notes in Mathematics*. Springer Verlag, 1970. Revised edition of 2005 freely available from D. Foata's web site.

[268] FORD, W. B. *Studies on divergent series and summability and the asymptotic developments of functions defined by Maclaurin series*, 3rd ed. Chelsea Publishing Company, 1960. (From two books originally published in 1916 and 1936.).

[269] FRANÇON, J., AND VIENNOT, G. Permutations selon leurs pics, creux, doubles montées et doubles descentes, nombres d'Euler et de Genocchi. *Discrete Mathematics 28* (1979), 21–35.

[270] FRIESEN, C., AND HENSLEY, D. The statistics of continued fractions for polynomials over a finite field. *Proceedings of the American Mathematical Society 124*, 9 (1996), 2661–2673.

[271] FROBENIUS, G. Über Matrizen aus nicht negativen Elementen. *Sitz.-Ber. Akad. Wiss., Phys-Math Klasse, Berlin* (1912), 456–477.

[272] FULMAN, J. Random matrix theory over finite fields. *Bulletin of the American Mathematical Society 39*, 1 (2001), 51–85.

[273] FULTON, W. *Algebraic Curves*. W.A. Benjamin, Inc., 1969.

[274] FUSY, ÉRIC. Transversal structures on triangulations: combinatorial study and straight-line drawing. In *Graph Drawing* (2006), P. Healy and N. S. Nikolov, Eds., vol. 3843 of *Lecture Notes in Computer Science*, pp. 177–188. Proceedings of 13th International Symposium, GD 2005, Limerick, Ireland, September 12-14, 2005.

[275] GÁL, A., AND MILTERSEN, P. B. The cell probe complexity of succinct data structures. In *Automata, Languages and Programming* (2003), vol. 2719 of *Lecture Notes in Computer Science*, Springer Verlag, pp. 332–344. Proceedings of ICALP 2003.

[276] GANTMACHER, F. R. *Matrizentheorie*. Deutscher Verlag der Wissenschaften, 1986. A translation of the Russian original Teoria Matriz, Nauka, 1966.

[277] GAO, Z., AND RICHMOND, L. B. Central and local limit theorems applied to asymptotic enumerations IV: Multivariate generating functions. *Journal of Computational and Applied Mathematics 41* (1992), 177–186.

[278] GAO, Z., AND WORMALD, N. C. Asymptotic normality determined by high moments and submap counts of random maps. *Probability Theory and Related Fields 130*, 3 (2004), 368–376.

[279] GARDY, D. On coefficients of powers of functions. Tech. Rep. CS-91-53, Brown University, Aug. 1991.

[280] GARDY, D. Méthode de col et lois limites en analyse combinatoire. *Theoretical Computer Science 92*, 2 (1992), 261–280.

[281] GARDY, D. Normal limiting distributions for projection and semijoin sizes. *SIAM Journal on Discrete Mathematics 5*, 2 (1992), 219–248.

[282] GARDY, D. Some results on the asymptotic behaviour of coefficients of large powers of functions. *Discrete Mathematics 139*, 1-3 (1995), 189–217.

[283] GARDY, D. Random Boolean expressions. In *Computational Logic and Applications (CLA'05)* (2005), vol. AF of *Discrete Mathematics and Theoretical Computer Science Proceedings*, pp. 1–36.

[284] GASPER, G., AND RAHMAN, M. *Basic Hypergeometric Series*, vol. 35 of *Encyclopedia of Mathematics and its Applications*. Cambridge University Press, 1990.

[285] GESSEL, I., AND WANG, D. L. Depth-first search as a combinatorial correspondence. *Journal of Combinatorial Theory*, Series A *26*, 3 (1979), 308–313.

[286] GESSEL, I. M. A factorization for formal Laurent series and lattice path enumeration. *Journal of Combinatorial Theory* Series A *28*, 3 (1980), 321–337.

[287] GESSEL, I. M. A noncommutative generalization and q–analog of the Lagrange inversion formula. *Transactions of the American Mathematical Society 257*, 2 (1980), 455–482.

[288] GESSEL, I. M. Enumerative applications of symmetric functions. In *Actes du 17ième Séminaire Lotharingien de Combinatoire* (Strasbourg, 1988), P. IRMA, Ed., pp. 5–21.

[289] GESSEL, I. M. Symmetric functions and P–recursiveness. *Journal of Combinatorial Theory*, Series A *53* (1990), 257–285.

[290] GIMÉNEZ, O., AND NOY, M. The number of planar graphs and properties of random planar graphs. *Discrete Mathematics and Theoretical Computer Science Proceedings AD* (2005), 147–156.

[291] GIMÉNEZ, O., AND NOY, M. Counting planar graphs and related families of graphs. Preprint, August 2008.

[292] GITTENBERGER, B., AND MANDLBURGER, J. Hayman admissible functions in several variables. *Electronic Journal of Combinatorics 13*, R106 (2006), 1–29.

[293] GLASSER, M. L. A Watson sum for a cubic lattice. *Journal of Mathematical Physics 13* (1972), 1145–1146.

[294] GNEDENKO, B. V., AND KOLMOGOROV, A. N. *Limit Distributions for Sums of Independent Random Variables*. Addison-Wesley, 1968. Translated from the Russian original (1949).

[295] GODSIL, C. D. *Algebraic Combinatorics*. Chapman and Hall, 1993.

[296] GODSIL, C. D., AND MCKAY, B. D. Asymptotic enumeration of Latin rectangles. *Journal of Combinatorial Theory*, Series B *48* (1990), 19–44.

[297] GOH, W. M. Y., AND SCHMUTZ, E. A central limit theorem on $GL_n(F_q)$. *Random Structures & Algorithms 2*, 1 (1991).

[298] GOLDWURM, M., AND SANTINI, M. Clique polynomials have a unique root of smallest modulus. *Information Processing Letters 75*, 3 (2000), 127–132.

[299] GONCHAROV, V. On the field of combinatory analysis. *Soviet Math. Izv., Ser. Math. 8* (1944), 3–48. In Russian.

[300] GONNET, G. H. Notes on the derivation of asymptotic expressions from summations. *Information Processing Letters 7*, 4 (1978), 165–169.

[301] GONNET, G. H. Expected length of the longest probe sequence in hash code searching. *Journal of the ACM 28*, 2 (1981), 289–304.

[302] GOOD, I. J. Random motion and analytic continued fractions. *Proceedings of the Cambridge Philosophical Society 54* (1958), 43–47.

[303] GOULDEN, I. P., AND JACKSON, D. M. *Combinatorial Enumeration*. John Wiley, New York, 1983.

[304] GOULDEN, I. P., AND JACKSON, D. M. Distributions, continued fractions, and the Ehrenfest urn model. *Journal of Combinatorial Theory. Series A 41*, 1 (1986), 21–31.

[305] GOURDON, X. Largest component in random combinatorial structures. *Discrete Mathematics 180*, 1-3 (1998), 185–209.

[306] GOURDON, X., AND SALVY, B. Asymptotics of linear recurrences with rational coefficients. Tech. Rep. 1887, INRIA, Mar. 1993. To appear in Proceedings FPACS'93.

[307] GRAHAM, R. L., KNUTH, D. E., AND PATASHNIK, O. *Concrete Mathematics*. Addison Wesley, 1989.

[308] GREENE, D. H. *Labelled formal languages and their uses*. Ph.D. Thesis, Stanford University, June 1983. Available as Report STAN-CS-83-982.

[309] GREENE, D. H., AND KNUTH, D. E. *Mathematics for the analysis of algorithms*. Birkhäuser, 1981.

[310] GREENE, D. H., AND KNUTH, D. E. *Mathematics for the analysis of algorithms*, second ed. Birkhauser, Boston, 1982.

[311] GUIBAS, L. J., AND ODLYZKO, A. M. Long repetitive patterns in random sequences. *Zeitschrift für Wahrscheinlichkeitstheorie und Verwandte Gebiete 53*, 3 (1980), 241–262.

[312] GUIBAS, L. J., AND ODLYZKO, A. M. Periods in strings. *Journal of Combinatorial Theory,* Series A 30 (1981), 19–42.

[313] GUIBAS, L. J., AND ODLYZKO, A. M. String overlaps, pattern matching, and nontransitive games. *Journal of Combinatorial Theory. Series A 30*, 2 (1981), 183–208.

[314] GUILLEMIN, F., ROBERT, P., AND ZWART, B. AIMD algorithms and exponential functionals. *Annals of Applied Probability 14*, 1 (2004), 90–117.

[315] HABSIEGER, L., KAZARIAN, M., AND LANDO, S. On the second number of Plutarch. *American Mathematical Monthly 105* (1998), 446–447.

[316] HADAMARD, J. *The Psychology of Invention in the Mathematical Field*. Princeton University Press, 1945. (Enlarged edition, Princeton University Press, 1949; reprinted by Dover.).

[317] HALMOS, P. Applied mathematics is bad mathematics. In *Mathematics Tomorrow*, L. Steen, Ed. Springer-Verlag, 1981, pp. 9–20. Excerpted in *Notices of the AMS* , 54:9, (2007), pp. 1136–1144.

[318] HANSEN, J. C. A functional central limit theorem for random mappings. *Annals of Probability 17*, 1 (1989), 317–332.

[319] HARARY, F., AND PALMER, E. M. *Graphical Enumeration*. Academic Press, 1973.

[320] HARARY, F., ROBINSON, R. W., AND SCHWENK, A. J. Twenty-step algorithm for determining the asymptotic number of trees of various species. *Journal of the Australian Mathematical Society* (Series A) *20* (1975), 483–503.

[321] HARDY, G. H. *Ramanujan: Twelve Lectures on Subjects Suggested by his Life and Work*, third ed. Chelsea Publishing Company, 1978. Reprinted and Corrected from the First Edition, Cambridge, 1940.

[322] HARPER, L. H. Stirling behaviour is asymptotically normal. *Annals of Mathematical Statistics 38* (1967), 410–414.

[323] HARRIS, B., AND SCHOENFELD, L. Asymptotic expansions for the coefficients of analytic functions. *Illinois Journal of Mathematics 12* (1968), 264–277.

[324] HARRIS, T. E. *The Theory of Branching Processes*. Dover Publications, 1989. A reprint of the 1963 edition.

[325] HAYMAN, W. K. A generalization of Stirling's formula. *Journal für die reine und angewandte Mathematik 196* (1956), 67–95.

[326] HECKE, E. *Vorlesungen über die Theorie der algebraischen Zahlen*. Akademische Verlagsgesellschaft, Leipzig, 1923.

[327] HEILBRONN, H. On the average length of a class of continued fractions. In *Number Theory and Analysis* (New York, 1969), P. Turan, Ed., Plenum Press, pp. 87–96.

[328] HENNEQUIN, P. Combinatorial analysis of quicksort algorithm. *Theoretical Informatics and Applications 23*, 3 (1989), 317–333.

[329] HENRICI, P. *Applied and Computational Complex Analysis*, vol. 2. John Wiley, 1974.

[330] HENRICI, P. *Applied and Computational Complex Analysis*, vol. 1. John Wiley, 1974.

[331] HENSLEY, D. The number of steps in the Euclidean algorithm. *Journal of Number Theory 49*, 2 (1994), 142–182.

[332] HENSLEY, D. *Continued fractions*. World Scientific Publishing, 2006.

[333] HICKERSON, D. Counting horizontally convex polyominoes. *Journal of Integer Sequences 2* (1999). Electronic.

[334] HILLE, E. *Analytic Function Theory*. Blaisdell Publishing Company, Waltham, 1962. 2 Volumes.

[335] HOFRI, M. *Analysis of Algorithms: Computational Methods and Mathematical Tools*. Oxford University Press, 1995.

[336] HOWELL, J. A., SMITH, T. F., AND WATERMAN, M. S. Computation of generating functions for biological molecules. *SIAM Journal on Applied Mathematics 39*, 1 (1980), 119–133.

[337] HWANG, H.-K. *Théorèmes limites pour les structures combinatoires et les fonctions arithmetiques*. Ph.D. Thesis, École Polytechnique, Dec. 1994.

[338] HWANG, H.-K. Large deviations for combinatorial distributions. I. Central limit theorems. *The Annals of Applied Probability 6*, 1 (1996), 297–319.

[339] HWANG, H.-K. Large deviations of combinatorial distributions. II. Local limit theorems. *The Annals of Applied Probability 8*, 1 (1998), 163–181.

[340] HWANG, H.-K. On convergence rates in the central limit theorems for combinatorial structures. *European Journal of Combinatorics 19*, 3 (1998), 329–343.

[341] HWANG, H.-K. Asymptotics of Poisson approximation to random discrete distributions: an analytic approach. *Advances in Applied Probability 31*, 2 (1999), 448–491.

[342] IMMINK, G. K. *Asymptotics of Analytic Difference Equations*. No. 1085 in Lecture Notes in Mathematics. 1980.

[343] ISMAIL, M. E. H. *Classical and Quantum Orthogonal Polynomials in One Variable*. No. 98 in Encyclopedia of Mathematics and its Applications. Cambridge University Press, 2005.

[344] JACQUET, P., AND RÉGNIER, M. Trie partitioning process: Limiting distributions. In *CAAP'86* (1986), P. Franchi-Zanetacchi, Ed., vol. 214 of *Lecture Notes in Computer Science*, pp. 196–210. Proceedings of the 11th Colloquium on Trees in Algebra and Programming, Nice France, March 1986.

[345] JACQUET, P., AND SZPANKOWSKI, W. Asymptotic behavior of the Lempel-Ziv parsing scheme and digital search trees. *Theoretical Computer Science 144*, 1–2 (1995), 161–197.

[346] JACQUET, P., AND SZPANKOWSKI, W. Analytical de-Poissonization and its applications. *Theoretical Computer Science 201*, 1-2 (1998), 1–62.

[347] JACQUET, P., AND SZPANKOWSKI, W. Analytic approach to pattern matching. In *Applied Combinatorics on Words* (2004), M. Lothaire, Ed., vol. 105 of *Encycl. of Mathematics and Its Applications*, Cambridge University Press, pp. 353–429. Chapter 7.

[348] JANSON, S. The Wiener index of simply generated random trees. *Random Structures & Algorithms 22*, 4 (2003), 337–358.

[349] JANSON, S. Functional limit theorems for multitype branching processes and generalized Pólya urns. *Stochastic Processes and Applications 110*, 2 (2004), 177–245.

[350] JANSON, S. Random cutting and records in deterministic and random trees. Technical Report, 2004. *Random Structures & Algorithms*, 42 pages, to appear.

[351] JANSON, S. Limit theorems for triangular urn schemes. *Probability Theory and Related Fields 134* (2006), 417–452.

[352] JANSON, S. Brownian excursion area, Wright's constants in graph enumeration, and other Brownian areas. *Probability Surveys 3* (2007), 80–145.

[353] JANSON, S. Sorting using complete subintervals and the maximum number of runs in a randomly evolving sequence. arXiv:math/0701288, January 2007. 31 pages. Extended abstract, *DMTCS Proceedings* (AofA 2007 Conference).

[354] JANSON, S., KNUTH, D. E., ŁUCZAK, T., AND PITTEL, B. The birth of the giant component. *Random Structures & Algorithms 4*, 3 (1993), 233–358.

[355] JANSON, S., ŁUCZAK, T., AND RUCINSKI, A. *Random graphs*. Wiley-Interscience, 2000.

[356] JENSEN, I. A parallel algorithm for the enumeration of self-avoiding polygons on the square lattice. *Journal of Physics A: Mathematical and General 36* (2003), 5731–5745.

[357] JOHNSON, N. L., AND KOTZ, S. *Urn Models and Their Application*. John Wiley, 1977.

[358] JONES, W. B., AND MAGNUS, A. Application of Stieltjes fractions to birth-death processes. In *Padé and rational approximation* (New York, 1977), E. B. Saff and R. S. Varga, Eds., Academic Press Inc., pp. 173–179. Proceedings of an International Symposium held at the University of South Florida, Tampa, Fla., December 15-17, 1976.

[359] JOYAL, A. Une théorie combinatoire des séries formelles. *Advances in Mathematics 42*, 1 (1981), 1–82.

[360] JUNGEN, R. Sur les séries de Taylor n'ayant que des singularités algébrico-logarithmiques sur leur cercle de convergence. *Commentarii Mathematici Helvetici 3* (1931), 266–306.

[361] KAC, M. Random walk and the theory of Brownian motion. *American Mathematical Monthly 54* (1947), 369–391.

[362] KARLIN, S., AND MCGREGOR, J. The classification of birth and death processes. *Transactions of the American Mathematical Society 86* (1957), 366–400.

[363] KARLIN, S., AND TAYLOR, H. *A First Course in Stochastic Processes*, second ed. Academic Press, 1975.

[364] KEMP, R. Random multidimensional binary trees. *Journal of Information Processing and Cybernetics (EIK) 29* (1993), 9–36.

[365] KIRWAN, F. *Complex Algebraic Curves*. No. 23 in London Mathematical Society Student Texts. Cambridge University Press, 1992.

[366] KLAMKIN, M. S., AND NEWMAN, D. J. Extensions of the birthday surprise. *Journal of Combinatorial Theory 3* (1967), 279–282.

[367] KLEENE, S. C. Representation of events in nerve nets and finite automata. In *Automata studies*. Princeton University Press, 1956, pp. 3–41.

[368] KNOPFMACHER, A., ODLYZKO, A. M., PITTEL, B., RICHMOND, L. B., STARK, D., SZEKERES, G., AND WORMALD, N. C. The asymptotic number of set partitions with unequal block sizes. *Electronic Journal of Combinatorics 6*, 1 (1999), R2:1–37.

[369] KNOPFMACHER, A., AND PRODINGER, H. On Carlitz compositions. *European Journal of Combinatorics 19*, 5 (1998), 579–589.

[370] KNOPFMACHER, J. *Abstract Analytic Number Theory*. Dover, 1990.

[371] KNOPFMACHER, J., AND KNOPFMACHER, A. The exact length of the Euclidean algorithm in $F_q[X]$. *Mathematika 35* (1988), 297–304.

[372] KNOPFMACHER, J., AND KNOPFMACHER, A. Counting irreducible factors of polynomials over a finite field. *Discrete Mathematics 112* (1993), 103–118.

[373] KNOPP, K. *Theory of Functions*. Dover Publications, 1945.

[374] KNUTH, D. E. Mathematical analysis of algorithms. In *Information Processing 71* (1972), North Holland Publishing Company, pp. 19–27. Proceedings of IFIP Congress, Ljubljana, 1971.

[375] KNUTH, D. E. The average time for carry propagation. *Indagationes Mathematicae 40* (1978), 238–242.

[376] KNUTH, D. E. Bracket notation for the 'coefficient of' operator. E-print `arXiv:math/9402216`, Feb. 1994.

[377] KNUTH, D. E. *The Art of Computer Programming*, 3rd ed., vol. 1: Fundamental Algorithms. Addison-Wesley, 1997.

[378] KNUTH, D. E. *The Art of Computer Programming*, 2nd ed., vol. 3: Sorting and Searching. Addison-Wesley, 1998.

[379] KNUTH, D. E. *The Art of Computer Programming*, 3rd ed., vol. 2: Seminumerical Algorithms. Addison-Wesley, 1998.

[380] KNUTH, D. E. Linear probing and graphs. *Algorithmica 22*, 4 (Dec. 1998), 561–568.

[381] KNUTH, D. E. *Selected papers on analysis of algorithms*. CSLI Publications, Stanford, CA, 2000.

[382] KNUTH, D. E., MORRIS, JR., J. H., AND PRATT, V. R. Fast pattern matching in strings. *SIAM Journal on Computing 6*, 2 (1977), 323–350.

[383] KNUTH, D. E., AND PITTEL, B. A recurrence related to trees. *Proceedings of the American Mathematical Society 105*, 2 (Feb. 1989), 335–349.

[384] KNUTH, D. E., AND SCHÖNHAGE, A. The expected linearity of a simple equivalence algorithm. *Theoretical Computer Science 6* (1978), 281–315.

[385] KNUTH, D. E., AND VARDI, I. Problem 6581 (the asymptotic expansion of $2n$ choose n). *American Mathematical Monthly 95* (1988), 774.

[386] KOLCHIN, V. F. *Random Mappings*. Optimization Software Inc., 1986. Translated from *Slučajnye Otobraženija*, Nauka, Moscow, 1984.

[387] KOLCHIN, V. F. *Random Graphs*, vol. 53 of *Encyclopedia of Mathematics and its Applications*. Cambridge University Press, 1999.

[388] KOLCHIN, V. F., SEVASTYANOV, B. A., AND CHISTYAKOV, V. P. *Random Allocations*. John Wiley and Sons, 1978. Translated from the Russian original *Slučajnye Razmeščeniya*.

[389] KOREVAAR, J. *Tauberian theory*, vol. 329 of *Grundlehren der Mathematischen Wissenschaften [Fundamental Principles of Mathematical Sciences]*. Springer-Verlag, 2004.

[390] KRASNOSELSKII, M. *Positive solutions of operator equations*. P. Noordhoff, 1964.

[391] KRATTENTHALER, C. Advanced determinant calculus. *Seminaire Lotharingien de Combinatoire 42* (1999). Paper B42q, 66 pp.

[392] LABELLE, J., AND YEH, Y. N. Generalized Dyck paths. *Discrete Mathematics 82* (1990), 1–6.

[393] LAGARIAS, J. C., AND ODLYZKO, A. M. Solving low-density subset sum problems. *JACM 32*, 1 (1985), 229–246.

[394] LAGARIAS, J. C., ODLYZKO, A. M., AND ZAGIER, D. B. On the capacity of disjointly shared networks. *Computer Networks and ISDN Systems 10*, 5 (1985), 275–285.

[395] LALLEY, S. P. Finite range random walk on free groups and homogeneous trees. *The Annals of Probability 21*, 4 (1993), 2087–2130.

[396] LALLEY, S. P. Return probabilities for random walk on a half-line. *Journal of Theoretical Probability 8*, 3 (1995), 571–599.

[397] LALLEY, S. P. Random walks on regular languages and algebraic systems of generating functions. In *Algebraic methods in statistics and probability (Notre Dame, IN, 2000)*, vol. 287 of *Contemporary Mathematics*. American Mathematical Society, 2001, pp. 201–230.

[398] LALLEY, S. P. Algebraic systems of generating functions and return probabilities for random walks. In *Dynamics and randomness II*, vol. 10 of *Nonlinear Phenom. Complex Systems*. Kluwer Academic Publishers, 2004, pp. 81–122.

[399] LAMÉ, G. Extrait d'une lettre de M. Lamé à M. Liouville sur cette question: Un polygone convexe étant donné, de combien de manières peut-on le partager en triangles au moyen de diagonales? *Journal de Mathématiques Pures et Appliquées 3* (1838), 505–507. Accessible under the Gallica-MathDoc site.

[400] LANDO, S. K. *Lectures on generating functions*, vol. 23 of *Student Mathematical Library*. American Mathematical Society, 2003. Translated from the 2002 Russian original by the author.

[401] LANG, S. *Linear Algebra*. Addison-Wesley, 1966.

[402] LAPLACE, P.-S. *Théorie analytique des probabilités. Vol. I, II*. Éditions Jacques Gabay, 1995. Reprint of the 1819 and 1820 editions.

[403] LAWLER, G. F. *Intersections of random walks*. Birkhäuser Boston Inc., 1991.

[404] LE GALL, J. F. The topological structure of scaling limits of large planar maps. ArXiv, 2006. arXiv:math/0607567v2, 45 pages.

[405] LEFMANN, H., AND SAVICKÝ, P. Some typical properties of large AND/OR Boolean formulas. *Random Structures & Algorithms 10* (1997), 337–351.

[406] LEWIN, L., Ed. *Structural Properties of Polylogarithms*. American Mathematical Society, 1991.

[407] LIFSCHITZ, V., AND PITTEL, B. The number of increasing subsequences of the random permutation. *Journal of Combinatorial Theory*, Series A *31* (1981), 1–20.

[408] LINDELÖF, E. *Le calcul des résidus et ses applications à la théorie des fonctions*. Collection de monographies sur la théorie des fonctions, publiée sous la direction de M. Émile Borel. Gauthier-Villars, Paris, 1905. Reprinted by Gabay, Paris, 1989.

[409] LIPSHITZ, L. The diagonal of a *D*-finite power series is *D*-finite. *Journal of Algebra 113* (1988), 373–378.

[410] LIPSHITZ, L. *D*-finite power series. *Journal of Algebra 122* (1989), 353–373.

[411] LOGAN, B. F., AND SHEPP, L. A. A variational problem for random Young tableaux. *Advances in Mathematics 26* (1977), 206–222.

[412] LORENTZEN, L., AND WAADELAND, H. *Continued Fractions With Applications*. North-Holland, 1992.

[413] LOTHAIRE, M. *Combinatorics on Words*, vol. 17 of *Encyclopedia of Mathematics and its Applications*. Addison–Wesley, 1983.

[414] LOTHAIRE, M. *Applied combinatorics on words*. Encyclopedia of Mathematics and its Applications. Cambridge University Press, Cambridge, 2005. (A collective work edited by Jean Berstel and Dominique Perrin).

[415] LOUCHARD, G. The Brownian motion: a neglected tool for the complexity analysis of sorted table manipulation. *RAIRO Theoretical Informatics 17* (1983), 365–385.

[416] LOUCHARD, G. The Brownian excursion: a numerical analysis. *Computers and Mathematics with Applications 10*, 6 (1984), 413–417.

[417] LOUCHARD, G. Kac's formula, Lévy's local time and Brownian excursion. *Journal of Applied Probability 21* (1984), 479–499.

[418] LOUCHARD, G. Random walks, Gaussian processes and list structures. *Theoretical Computer Science 53*, 1 (1987), 99–124.

[419] LOUCHARD, G. Probabilistic analysis of some (un)directed animals. *Theoretical Computer Science 159*, 1 (1996), 65–79.

[420] LOUCHARD, G. Probabilistic analysis of column-convex and directed diagonally-convex animals. *Random Structures & Algorithms 11*, 2 (1997), 151–178.

[421] LOUCHARD, G., AND PRODINGER, H. Probabilistic analysis of Carlitz compositions. *Discrete Mathematics & Theoretical Computer Science 5*, 1 (2002), 71–96.

[422] LOUCHARD, G., SCHOTT, R., TOLLEY, M., AND ZIMMERMANN, P. Random walks, heat equations and distributed algorithms. *Journal of Computational and Applied Mathematics 53* (1994), 243–274.

[423] LUCAS, E. *Théorie des Nombres*. Gauthier-Villard, Paris, 1891. Reprinted by A. Blanchard, Paris 1961.

[424] LUKACS, E. *Characteristic Functions*. Griffin, 1970.

[425] LUM, V. Y., YUEN, P. S. T., AND DODD, M. Key to address transformations: A fundamental study based on large existing format files. *Communications of the ACM 14* (1971), 228–239.

[426] LYNCH, J. F. Probabilities of first-order sentences about unary functions. *Transactions of the American Mathematical Society 287*, 2 (Feb. 1985), 543–568.

[427] MACKAY, D. J. C. *Information theory, Inference and Learning Algorithms*. Cambridge University Press, 2003.

[428] MACMAHON, P. A. *Introduction to combinatory analysis*. Chelsea Publishing Co., New York, 1955. A reprint of the first edition, Cambridge, 1920.

[429] MAHMOUD, H. M. *Evolution of Random Search Trees*. John Wiley, 1992.

[430] MARTÍNEZ, C., AND MOLINERO, X. A generic approach for the unranking of labeled combinatorial classes. *Random Structures & Algorithms 19*, 3-4 (2001), 472–497. Analysis of algorithms (Krynica Morska, 2000).

[431] MAZO, J. E., AND ODLYZKO, A. M. Lattice points in high-dimensional spheres. *Monatshefte für Mathematik 110*, 1 (1990), 47–61.

[432] MCKAY, B. D. The asymptotic numbers of regular tournaments, Eulerian digraphs and Eulerian oriented graphs. *Combinatorica 10*, 4 (1990), 367–377.

[433] MCKAY, B. D., BAR-NATAN, D., BAR-HILLEL, M., AND KALAI, G. Solving the bible code puzzle. *Statistical Science 14* (1999), 150–173.

[434] MEINARDUS, G. Asymptotische Aussagen über Partitionen. *Mathematische Zeitschrift 59* (1954), 388–398.

[435] MEIR, A., AND MOON, J. W. On the altitude of nodes in random trees. *Canadian Journal of Mathematics 30* (1978), 997–1015.

[436] MEIR, A., AND MOON, J. W. On random mapping patterns. *Combinatorica 4*, 1 (1984), 61–70.

[437] MEIR, A., AND MOON, J. W. Recursive trees with no nodes of out-degree one. *Congressus Numerantium 66* (1988), 49–62.

[438] MEIR, A., AND MOON, J. W. Erratum: "On an asymptotic method in enumeration". *Journal of Combinatorial Theory, Series A 52*, 1 (1989), 163.

[439] MEIR, A., AND MOON, J. W. On an asymptotic method in enumeration. *Journal of Combinatorial Theory, Series A 51*, 1 (1989), 77–89.

[440] MEIR, A., AND MOON, J. W. The asymptotic behaviour of coefficients of powers of certain generating functions. *European Journal of Combinatorics 11* (1990), 581–587.

[441] MERLINI, D., SPRUGNOLI, R., AND VERRI, M. C. The tennis ball problem. *Journal of Combinatorial Theory, Series A 99*, 2 (2002), 307–344. lattice paths.

[442] MILLER, S. D., AND VENKATESAN, R. Spectral analysis of Pollard rho collisions. Arxiv:math.NT/0603727, 2006.

[443] MILNOR, J. *Dynamics in one complex variable*. Friedr. Vieweg & Sohn, 1999.

[444] MISHNA, M. Automatic enumeration of regular objects. Preprint available on ArXiv, 2005. ArXiv:CO/0507249.

[445] MOON, J. W. Counting labelled trees. In *Canadian Mathematical Monographs N.1* (1970), William Clowes and Sons.

[446] MORRIS, M., SCHACHTEL, G., AND KARLIN, S. Exact formulas for multitype run statistics in a random ordering. *SIAM Journal on Discrete Mathematics 6*, 1 (1993), 70–86.

[447] MOSER, L., AND WYMAN, M. An asymptotic formula for the Bell numbers. *Transactions of the Royal Society of Canada XLIX* (June 1955).

[448] MOSER, L., AND WYMAN, M. On the solution of $x^d = 1$ in symmetric groups. *Canadian Journal of Mathematics 7* (1955), 159–168.

[449] MOSER, L., AND WYMAN, M. Asymptotic expansions. *Canadian Journal of Mathematics 8* (1956), 225–233.

[450] MOSER, L., AND WYMAN, M. Asymptotic expansions II. *Canadian Journal of Mathematics* (1957), 194–209.

[451] MOTWANI, R., AND RAGHAVAN, P. *Randomized Algorithms*. Cambridge University Press, 1995.

[452] MYERSON, G., AND VAN DER POORTEN, A. J. Some problems concerning recurrence sequences. *The American Mathematical Monthly 102*, 8 (1995), 698–705.

[453] NEBEL, M. Combinatorial properties of RNA secondary structures. *Journal of Computational Biology 9*, 3 (2002), 541–574.

[454] NEWMAN, D. J., AND SHEPP, L. The double dixie cup problem. *American Mathematical Monthly 67* (1960), 58–61.

[455] NICODÈME, P., SALVY, B., AND FLAJOLET, P. Motif statistics. *Theoretical Computer Science 287*, 2 (2002), 593–617.

[456] NIJENHUIS, A., AND WILF, H. S. *Combinatorial Algorithms*, 2nd ed. Academic Press, 1978.

[457] NOONAN, J., AND ZEILBERGER, D. The Goulden-Jackson cluster method: Extensions, applications, and implementations. *Journal of Difference Equations and Applications 5*, 4 & 5 (1999), 355–377.

[458] NÖRLUND, N. E. *Vorlesungen über Differenzenrechnung*. Chelsea Publishing Company, New York, 1954.

[459] ODLYZKO, A. M. Periodic oscillations of coefficients of power series that satisfy functional equations. *Advances in Mathematics 44* (1982), 180–205.

[460] ODLYZKO, A. M. Explicit Tauberian estimates for functions with positive coefficients. *Journal of Computational and Applied Mathematics 41* (1992), 187–197.

[461] ODLYZKO, A. M. Asymptotic enumeration methods. In *Handbook of Combinatorics*, R. Graham, M. Grötschel, and L. Lovász, Eds., vol. II. Elsevier, 1995, pp. 1063–1229.

[462] ODLYZKO, A. M., AND RICHMOND, L. B. Asymptotic expansions for the coefficients of analytic generating functions. *Aequationes Mathematicae 28* (1985), 50–63.

[463] ODLYZKO, A. M., AND WILF, H. S. Bandwidths and profiles of trees. *Journal of Combinatorial Theory*, Series B *42* (1987), 348–370.

[464] ODLYZKO, A. M., AND WILF, H. S. The editor's corner: *n* coins in a fountain. *American Matematical Monthly 95* (1988), 840–843.

[465] OLVER, F. W. J. *Asymptotics and Special Functions*. Academic Press, 1974.

[466] OTTER, R. The number of trees. *Annals of Mathematics 49*, 3 (1948), 583–599.

[467] PAINLEVÉ, P. *Analyse des travaux scientifiques jusqu'en 1900*. Albert Blanchard, 1967. (A summary by Painlevé of his scientific works until 1900, published posthumously.).

[468] PANARIO, D., PITTEL, B., RICHMOND, B., AND VIOLA, A. Analysis of Rabin's irreducibility test for polynomials over finite fields. *Random Structures Algorithms 19*, 3-4 (2001), 525–551. Analysis of algorithms (Krynica Morska, 2000).

[469] PANARIO, D., AND RICHMOND, B. Analysis of Ben-Or's polynomial irreducibility test. In *Proceedings of the Eighth International Conference "Random Structures and Algorithms" (Poznan, 1997)* (1998), vol. 13, pp. 439–456.

[470] PANARIO, D., AND RICHMOND, B. Smallest components in decomposable structures: Exp-log class. *Algorithmica 29* (2001), 205–226.

[471] PANHOLZER, A., AND PRODINGER, H. An analytic approach for the analysis of rotations in fringe-balanced binary search trees. *Annals of Combinatorics 2* (1998), 173–184.

[472] PARIS, R. B., AND KAMINSKI, D. *Asymptotics and Mellin-Barnes integrals*, vol. 85 of *Encyclopedia of Mathematics and its Applications*. Cambridge University Press, 2001.

[473] PEITGEN, H.-O., AND RICHTER, P. H. *The beauty of fractals*. Springer-Verlag, 1986. Images of complex dynamical systems.

[474] PEMANTLE, R. Generating functions with high-order poles are nearly polynomial. In *Mathematics and computer science (Versailles, 2000)*. Birkhäuser, 2000, pp. 305–321.

[475] PEMANTLE, R., AND WILSON, M. C. Asymptotics of multivariate sequences. I. Smooth points of the singular variety. *Journal of Combinatorial Theory. Series A 97*, 1 (2002), 129–161.

[476] PEMANTLE, R., AND WILSON, M. C. Asymptotics of multivariate sequences, Part II: Multiple points of the singular variety. *Combinatorics, Probability and Computing 13*, 4–5 (2004), 735–761.

[477] PERCUS, J. K. *Combinatorial Methods*, vol. 4 of *Applied Mathematical Sciences*. Springer-Verlag, 1971.

[478] PERRON, O. Über Matrizen. *Mathematische Annalen 64* (1907), 248–263.

[479] PERRON, O. *Die Lehre von der Kettenbrüchen*, vol. 1. Teubner, 1954.

[480] PETKOVŠEK, M., WILF, H. S., AND ZEILBERGER, D. $A = B$. A. K. Peters Ltd., 1996.

[481] PETROV, V. V. *Sums of Independent Random Variables*. Springer-Verlag, 1975.

[482] PETROV, V. V. *Limit theorems of probability theory*, vol. 4 of *Oxford Studies in Probability*. The Clarendon Press Oxford University Press, New York, 1995. Sequences of independent random variables, Oxford Science Publications.

[483] PETROVA, S. S., AND SOLOV'EV, A. D. The origin of the method of steepest descent. *Historia Mathematica 24* (1997), 361–375.

[484] PITTEL, B. Normal convergence problem? Two moments and a recurrence may be the clues. *The Annals of Applied Probability 9*, 4 (1999), 1260–1302.

[485] PIVOTEAU, C., SALVY, B., AND SORIA, M. Boltzmann oracle for combinatorial systems. *Discrete Mathematics & Theoretical Computer Science Proceedings* (2008). *Mathematics and Computer Science Conference*. In press, 14 pages.

[486] PLAGNE, A. On threshold properties of k-SAT: An additive viewpoint. *European Journal of Combinatorics 27*, 3 (2006), 1186–1198 1186–1198 1186–1198.

[487] POLLARD, J. M. A Monte Carlo method for factorization. *BIT 15*, 3 (1975), 331–334.

[488] PÓLYA, G. Kombinatorische Anzahlbestimmungen für Gruppen, Graphen und chemische Verbindungen. *Acta Mathematica 68* (1937), 145–254.

[489] PÓLYA, G. On picture-writing. *American Mathematical Monthly 63*, 10 (1956), 689–697.

[490] PÓLYA, G. On the number of certain lattice polygons. *Journal of Combinatorial Theory, Series A 6* (1969), 102–105.

[491] PÓLYA, G., AND READ, R. C. *Combinatorial Enumeration of Groups, Graphs and Chemical Compounds*. Springer Verlag, 1987.

[492] PÓLYA, G., AND SZEGŐ, G. *Aufgaben und Lehrsätze aus der Analysis*, 4th ed. Springer Verlag, 1970.

[493] PÓLYA, G., TARJAN, R. E., AND WOODS, D. R. *Notes on Introductory Combinatorics*. Progress in Computer Science. Birkhäuser, 1983.

[494] POSTNIKOV, A. G. *Tauberian theory and its applications*, vol. 144 of *Proceedings of the Steklov Institute of Mathematics*. American Mathematical Society, 1980.

[495] POUYANNE, N. On the number of permutations admitting an mth root. *Electronic Journal of Combinatorics 9*, 1:R3 (2002), 1–12.

[496] PRELLBERG, T. Uniform q-series asymptotics for staircase polygons. *Journal of Physics A: Math. Gen. 28* (1995), 1289–1304.

[497] PRIESTLEY, H. A. *Introduction to Complex Analysis*. Oxford University Press, 1985.

[498] PRODINGER, H. Approximate counting via Euler transform. *Mathematica Slovaka 44* (1994), 569–574.

[499] PRODINGER, H. A note on the distribution of the three types of nodes in uniform binary trees. *Séminaire Lotharingien de Combinatoire 38* (1996). Paper B38b, 5 pages.

[500] PROSKUROWSKI, A., RUSKEY, F., AND SMITH, M. Analysis of algorithms for listing equivalence classes of k-ary strings. *SIAM Journal on Discrete Mathematics 11*, 1 (1998), 94–109 (electronic).

[501] QUISQUATER, J.-J., AND DELESCAILLE, J.-P. How easy is collision search? Application to DES. In *Proceedings of EUROCRYPT'89* (1990), vol. 434 of *Lecture Notes in Computer Science*, Springer-Verlag, pp. 429–434.

[502] RAINS, E. M., AND SLOANE, N. J. A. On Cayley's enumeration of alkanes (or 4-valent trees). *Journal of Integer Sequences 2* (1999). Article 99.1.1; available electronically.

[503] RANEY, G. N. Functional composition patterns and power series reversion. *Transactions of the American Mathematical Society 94* (1960), 441–451.

[504] RÉGNIER, M. Analysis of grid file algorithms. *BIT 25* (1985), 335–357.

[505] RÉGNIER, M. A limiting distribution for quicksort. *Theoretical Informatics and Applications 23*, 3 (1989), 335–343.

[506] RÉGNIER, M., AND SZPANKOWSKI, W. On pattern frequency occurrences in a Markovian sequence. *Algorithmica 22*, 4 (1998), 631–649.

[507] RÉNYI, A., AND SZEKERES, G. On the height of trees. *Australian Journal of Mathematics 7* (1967), 497–507.

[508] RÉVÉSZ, P. Strong theorems on coin tossing. In *Proceedings of the International Congress of Mathematicians (Helsinki, 1978)* (Helsinki, 1980), Acad. Sci. Fennica, pp. 749–754.

[509] RICHARD, C. Scaling behaviour of two-dimensional polygon models. *Journal of Statistical Physics 108*, 3/4 (2002), 459–493.

[510] RICHARD, C. On q-functional equations and excursion moments. ArXiv:math/0503198, 2005.

[511] RIEMANN, B. Sullo svolgimento delquoziente di due serie ipergeometriche in frazione continua infinita. In *Bernhard Riemann's Gesammelte mathematische Werke und wissenschaftlicher Nachlass*, H. Weber and R. Dedekind, Eds. Teubner, 1863. Fragment of a manuscript (# XXIII), posthumously edited by H. A. Schwarz.

[512] RIORDAN, J. *Combinatorial Identities*. Wiley, 1968.

[513] RIORDAN, J. *Combinatorial Identities*. Dover Publications, 2002. A reprint of the Wiley edition 1958.

[514] RIVIN, I. Growth in free groups (and other stories). ArXiv, 1999. arXv:math.CO/9911076v2, 31 pages.

[515] ROBERT, P. *Réseaux et files d'attente: méthodes probabilistes*, vol. 35 of *Mathématiques & Applications*. Springer, Paris, 2000.

[516] ROGERS, L. J. On the representation of certain asymptotic series as convergent continued fractions. *Proceedings of the London Mathematical Society (Series 2) 4* (1907), 72–89.

[517] RÖSLER, U. A limit theorem for quicksort. *RAIRO Theoretical Informatics and Applications 25*, 1 (1991), 85–100.

[518] ROTA, G.-C. *Finite Operator Calculus*. Academic Press, 1975.

[519] ROTH, K. F., AND SZEKERES, G. Some asymptotic formulae in the theory of partitions. *Quarterly Journal of Mathematics, Oxford Series 5* (1954), 241–259.

[520] ROURA, S., AND MARTÍNEZ, C. Randomization of search trees by subtree size. In *Algorithms—ESA'96* (1996), J. Diaz and M. Serna, Eds., no. 1136 in Lecture Notes in Computer Science, pp. 91–106. Proceedings of the Fourth European Symposium on Algorithms, Barcelona, September 1996.

[521] RUBEL, L. A. Some research problems about algebraic differential equations. *Transactions of the American Mathematical Society 280*, 1 (1983), 43–52.

[522] RUBEL, L. A. Some research problems about algebraic differential equations II. *Illinois Journal of Mathematics 36*, 4 (1992), 659–680.

[523] RUDIN, W. *Real and complex analysis*, 3rd ed. McGraw-Hill Book Co., 1987.

[524] SACHKOV, V. N. *Verojatnostnye Metody v Kombinatornom Analize*. Nauka, 1978.

[525] SACHKOV, V. N. *Combinatorial Methods in Discrete Mathematics*, vol. 55 of *Encyclopedia of Mathematics and its Applications*. Cambridge University Press, 1996.

[526] SACHKOV, V. N. *Probabilistic methods in combinatorial analysis*. Cambridge University Press, Cambridge, 1997. Translated and adapted from the Russian original edition, Nauka,1978.

[527] SALOMAA, A., AND SOITTOLA, M. *Automata-Theoretic Aspects of Formal Power Series*. Springer, 1978.

[528] SALVY, B. *Asymptotique automatique et fonctions génératrices*. Ph.D. Thesis, École Polytechnique, 1991.

[529] SALVY, B. Asymptotics of the Stirling numbers of the second kind. *Studies in Automatic Combinatorics II* (1997). Published electronically.

[530] SALVY, B., AND SHACKELL, J. Symbolic asymptotics: Multiseries of inverse functions. *Journal of Symbolic Computation 27*, 6 (June 1999), 543–563.

[531] SALVY, B., AND ZIMMERMANN, P. GFUN: a Maple package for the manipulation of generating and holonomic functions in one variable. *ACM Transactions on Mathematical Software 20*, 2 (1994), 163–167.

[532] SAMET, H. *The Design and Analysis of Spatial Data Structures*. Addison–Wesley, 1990.

[533] SCHAEFFER, G. *Conjugaison d'arbres et cartes combinatoires aléatoires*. Ph.D. Thesis, Université de Bordeaux I, Dec. 1998.

[534] SCHMITT, W. R., AND WATERMAN, M. S. Linear trees and RNA secondary structure. *Discrete Applied Mathematics. Combinatorial Algorithms, Optimization and Computer Science 51*, 3 (1994), 317–323.

[535] SCHRÖDINGER, E. *Statistical thermodynamics*. A course of seminar lectures delivered in January-March 1944, at the School of Theoretical Physics, Dublin Institute for Advanced Studies. Second edition, reprinted. Cambridge University Press, 1962.

[536] SEDGEWICK, R. Quicksort with equal keys. *SIAM Journal on Computing 6*, 2 (June 1977), 240–267.

[537] SEDGEWICK, R. *Algorithms*, second ed. Addison–Wesley, Reading, Mass., 1988.

[538] SEDGEWICK, R., AND FLAJOLET, P. *An Introduction to the Analysis of Algorithms*. Addison-Wesley Publishing Company, 1996.

[539] SHARP, R. Local limit theorems for free groups. *Mathematische Annalen 321* (2001), 889–904.

[540] SHEPP, L. A., AND LLOYD, S. P. Ordered cycle lengths in a random permutation. *Transactions of the American Mathematical Society 121* (1966), 340–357.

[541] SHPARLINSKI, I. E. *Finite fields: theory and computation*, vol. 477 of *Mathematics and its Applications*. Kluwer Academic Publishers, Dordrecht, 1999. The meeting point of number theory, computer science, coding theory and cryptography.

[542] SLATER, L. J. *Generalized Hypergeometric Functions*. Cambridge University Press, 1966.

[543] SLOANE, N. J. A. *The On-Line Encyclopedia of Integer Sequences*. 2006. Published electronically at www.research.att.com/~njas/sequences/.

[544] SLOANE, N. J. A., AND PLOUFFE, S. *The Encyclopedia of Integer Sequences*. Academic Press, 1995.

[545] SLOANE, N. J. A., AND WIEDER, T. The number of hierarchical orderings. *Order 21* (2004), 83–89.

[546] SOITTOLA, M. Positive rational sequences. *Theoretical Computer Science 2* (1976), 317–322.

[547] SORIA-COUSINEAU, M. *Méthodes d'analyse pour les constructions combinatoires et les algorithmes*. Doctorat ès sciences, Université de Paris–Sud, Orsay, July 1990.

[548] SPENCER, J. Enumerating graphs and Brownian motion. *Communications on Pure and Applied Mathematics 50* (1997), 293–296.

[549] SPRINGER, G. *Introduction to Riemann surfaces*. Addison-Wesley Publishing Company, 1957. Reprinted by Chelsea.

[550] STANLEY, R. P. Generating functions. In *Studies in Combinatorics,* M.A.A. Studies in Mathematics, Vol. 17. (1978), G.-C. Rota, Ed., The Mathematical Association of America, pp. 100–141.

[551] STANLEY, R. P. Differentiably finite power series. *European Journal of Combinatorics 1* (1980), 175–188.

[552] STANLEY, R. P. *Enumerative Combinatorics*, vol. I. Wadsworth & Brooks/Cole, 1986.

[553] STANLEY, R. P. Hipparchus, Plutarch, Schröder and Hough. *American Mathematical Monthly 104* (1997), 344–350.

[554] STANLEY, R. P. *Enumerative Combinatorics*, vol. II. Cambridge University Press, 1999.

[555] STARK, H. M., AND TERRAS, A. A. Zeta functions of finite graphs and coverings. *Advances in Mathematics 121*, 1 (1996), 124–165.

[556] STEELE, J. M. *Probability theory and combinatorial optimization*. Society for Industrial and Applied Mathematics (SIAM), 1997.

[557] STEF, A., AND TENENBAUM, G. Inversion de Laplace effective. *Ann. Probab. 29*, 1 (2001), 558–575.

[558] STEIN, P. R., AND WATERMAN, M. S. On some new sequences generalizing the Catalan and Motzkin numbers. *Discrete Mathematics 26*, 3 (1979), 261–272.

[559] STEPANOV, V. E. On the distribution of the number of vertices in strata of a random tree. *Theory of Probability and Applications 14*, 1 (1969), 65–78.

[560] STEYAERT, J.-M. *Structure et complexité des algorithmes*. Doctorat d'état, Université Paris VII, Apr. 1984.

[561] STEYAERT, J.-M., AND FLAJOLET, P. Patterns and pattern-matching in trees: an analysis. *Information and Control 58*, 1–3 (July 1983), 19–58.

[562] STIELTJES, T. Sur la réduction en fraction continue d'une série procédant selon les puissances descendantes d'une variable. *Annales de la Faculté des Sciences de Toulouse 4* (1889), 1–17.

[563] SZEGŐ, G. *Orthogonal Polynomials*, vol. XXIII of *American Mathematical Society Colloquium Publications*. A.M.S, Providence, 1989.

[564] SZPANKOWSKI, W. *Average-Case Analysis of Algorithms on Sequences*. John Wiley, 2001.

[565] TAKÁCS, L. A Bernoulli excursion and its various applications. *Advances in Applied Probability 23* (1991), 557–585.

[566] TAKÁCS, L. Conditional limit theorems for branching processes. *Journal of Applied Mathematics and Stochastic Analysis 4*, 4 (1991), 263–292.

[567] TAKÁCS, L. On a probability problem connected with railway traffic. *Journal of Applied Mathematics and Stochastic Analysis 4*, 1 (1991), 1–27.

[568] TAKÁCS, L. Random walk processes and their application to order statistics. *The Annals of Applied Probability 2*, 2 (1992), 435–459.

[569] TAKÁCS, L. The asymptotic distribution of the total heights of random rooted trees. *Acta Scientifica Mathematica (Szeged) 57* (1993), 613–625.

[570] TAKAYAMA, N. An approach to the zero recognition problem by Buchberger algorithm. *Journal of Symbolic Computation 14*, 2-3 (1992), 265–282.

[571] TANGORA, M. C. Level number sequences of trees and the lambda algebra. *European Journal of Combinatorics 12* (1991), 433–443.

[572] TAURASO, R. The dinner table problem: The rectangular case. *INTEGERS 6*, A11 (2006), 1–13.

[573] TEMPERLEY, H. N. V. On the enumeration of the Mayer cluster integrals. *Proc. Phys. Soc. Sect. B. 72* (1959), 1141–1144.

[574] TEMPERLEY, H. N. V. *Graph theory and applications*. Ellis Horwood Ltd., Chichester, 1981.

[575] TENENBAUM, G. La méthode du col en théorie analytique des nombres. In *Séminaire de Théorie des Nombres, Paris 1985–1986* (1988), C. Goldstein, Ed., Birkhauser, pp. 411–441.

[576] TENENBAUM, G. *Introduction to analytic and probabilistic number theory*. Cambridge University Press, 1995. Translated from the second French edition (1995) by C. B. Thomas.

[577] TITCHMARSH, E. C. *The Theory of Functions*, second ed. Oxford University Press, 1939.

[578] TITCHMARSH, E. C., AND HEATH-BROWN, D. R. *The Theory of the Riemann Zeta-function*, second ed. Oxford Science Publications, 1986.

[579] TUTTE, W. T. A census of planar maps. *Canadian Journal of Mathematics 15* (1963), 249–271.

[580] TUTTE, W. T. On the enumeration of planar maps. *Bull. Amer. Math. Soc. 74* (1968), 64–74.

[581] TUTTE, W. T. On the enumeration of four-colored maps. *SIAM Journal on Applied Mathematics 17* (1969), 454–460.

[582] TUTTE, W. T. Planar enumeration. In *Graph theory and combinatorics (Cambridge, 1983)*. Academic Press, 1984, pp. 315–319.

[583] VALLÉE, B. Dynamical sources in information theory: Fundamental intervals and word prefixes. *Algorithmica 29*, 1/2 (2001), 262–306.

[584] VALLÉE, B. Euclidean dynamics. *Discrete and Continuous Dynamical Systems 15*, 1 (2006), 281–352.

[585] VAN CUTSEM, B. Combinatorial structures and structures for classification. *Computational Statistics & Data Analysis 23*, 1 (1996), 169–188.

[586] VAN CUTSEM, B., AND YCART, B. Indexed dendrograms on random dissimilarities. *Journal of Classification 15*, 1 (1998), 93–127.

[587] VAN DER HOEVEN, J. Majorants for formal power series. Preprint, 2003. 29 pages. Available from author's webpage.

[588] VAN DER POORTEN, A. A proof that Euler missed ... Apéry's proof of the irrationality of $\zeta(3)$. *Mathematical Intelligencer 1* (1979), 195–203.

[589] VAN DER WAERDEN, B. L. On the method of saddle points. *Applied Scientific Research 2* (1951), 33–45.

[590] VAN DER WAERDEN, B. L. *Algebra. Vol. I.* Springer-Verlag, New York, 1991. Based in part on lectures by E. Artin and E. Noether, Translated from the seventh German edition.

[591] VAN LEEUWEN, J., Ed. *Handbook of Theoretical Computer Science*, vol. A: Algorithms and Complexity. North Holland, 1990.

[592] VAN RENSBURG, E. J. J. *The statistical mechanics of interacting walks, polygons, animals and vesicles.* Oxford University Press, 2000.

[593] VAUGHAN, R. C. On the number of partitions into primes. *Ramanujan Journal 15*, 1 (2008), 109–121.

[594] VEIN, R., AND DALE, P. *Determinants and Their Applications in Mathematical Physics*, vol. 134 of *Applied Mathematical Sciences*. Springer-Verlag, 1998.

[595] VERSHIK, A. M. Statistical mechanics of combinatorial partitions, and their limit configurations. *Funktsional'nyĭ Analiz i ego Prilozheniya 30*, 2 (1996), 19–39.

[596] VERSHIK, A. M., AND KEROV, S. V. Asymptotics of the Plancherel measure of the symmetric group and the limiting form of Young tables. *Soviet Mathematical Doklady 18* (1977), 527–531.

[597] VIENNOT, G. X. Heaps of pieces, I: basic definitions and combinatorial lemmas. In *Combinatoire énumérative* (1986), G. Labelle and P. Leroux, Eds., vol. 1234 of *Lecture Notes in Mathematics*, Springer-Verlag.

[598] VITTER, J. S., AND FLAJOLET, P. Analysis of algorithms and data structures. In *Handbook of Theoretical Computer Science*, J. van Leeuwen, Ed., vol. A: Algorithms and Complexity. North Holland, 1990, ch. 9, pp. 431–524.

[599] VON ZUR GATHEN, J., AND GERHARD, J. *Modern computer algebra*. Cambridge University Press, 1999.

[600] VUILLEMIN, J. A unifying look at data structures. *Communications of the ACM 23*, 4 (Apr. 1980), 229–239.

[601] WALL, H. S. *Analytic Theory of Continued Fractions*. Chelsea Publishing Company, 1948.

[602] WASOW, W. *Asymptotic Expansions for Ordinary Differential Equations*. Dover, 1987. A reprint of the John Wiley edition, 1965.

[603] WATERMAN, M. S. *Introduction to Computational Biology*. Chapman & Hall, 1995.

[604] WHITTAKER, E. T., AND WATSON, G. N. *A Course of Modern Analysis*, fourth ed. Cambridge University Press, 1927. Reprinted 1973.

[605] WIDDER, D. V. *The Laplace Transform*. Princeton University Press, 1941.

[606] WILF, H. S. Some examples of combinatorial averaging. *American Mathematical Monthly 92* (Apr. 1985), 250–261.

[607] WILF, H. S. *Combinatorial Algorithms: An Update*. No. 55 in CBMS–NSF Regional Conference Series. Society for Industrial and Applied Mathematics, Philadelphia, 1989.

[608] WILF, H. S. *Generatingfunctionology*. Academic Press, 1990.

[609] WILLIAMS, D. *Probability with martingales*. Cambridge Mathematical Textbooks. Cambridge University Press, Cambridge, 1991.

[610] WIMP, J. Current trends in asymptotics: Some problems and some solutions. *Journal of Computational and Applied Mathematics 35* (1991), 53–79.

[611] WIMP, J., AND ZEILBERGER, D. Resurrecting the asymptotics of linear recurrences. *Journal of Mathematical Analysis and Applications 111* (1985), 162–176.

[612] WINKLER, P. Seven puzzles you think you must not have heard correctly. Preprint, 2006. Paper presented at the Seventh Gathering for Gardner (in honour of Martin Gardner).

[613] WOESS, W. *Random walks on infinite graphs and groups*, vol. 138 of *Cambridge Tracts in Mathematics*. Cambridge University Press, Cambridge, 2000.

[614] WONG, R. *Asymptotic Approximations of Integrals*. Academic Press, 1989.

[615] WONG, R., AND WYMAN, M. The method of Darboux. *J. Approximation Theory 10* (1974), 159–171.

[616] WOODS, A. R. Coloring rules for finite trees, and probabilities of monadic second order sentences. *Random Structures Algorithms 10*, 4 (1997), 453–485.

[617] WRIGHT, E. M. Asymptotic partition formulae: I plane partitions. *Quarterly Journal of Mathematics, Oxford Series II* (1931), 177–189.

[618] WRIGHT, E. M. The coefficients of a certain power series. *Journal of the London Mathematical Society 7* (1932), 256–262.

[619] WRIGHT, E. M. On the coefficients of power series having exponential singularities. *Journal of the London Mathematical Society 24* (1949), 304–309.

[620] WRIGHT, E. M. The number of connected sparsely edged graphs. *Journal of Graph Theory 1* (1977), 317–330.

[621] WRIGHT, E. M. The number of connected sparsely edged graphs. II. Smooth graphs. *Journal of Graph Theory 2* (1978), 299–305.

[622] WRIGHT, E. M. The number of connected sparsely edged graphs. III. Asymptotic results. *Journal of Graph Theory 4* (1980), 393–407.

[623] WRIGHT, R. A., RICHMOND, B., ODLYZKO, A., AND MCKAY, B. Constant time generation of free trees. *SIAM Journal on Computing 15*, 2 (1985), 540–548.

[624] WYMAN, M. The asymptotic behavior of the Laurent coefficients. *Canadian Journal of Mathematics 11* (1959), 534–555.

[625] YANG, Y. Partitions into primes. *Transactions of the American Mathematical Society 352*, 6 (2000), 2581–2600.

[626] ZEILBERGER, D. A holonomic approach to special functions identities. *Journal of Computational and Applied Mathematics 32* (1990), 321–368.

[627] ZEILBERGER, D. Symbol-crunching with the transfer-matrix method in order to count skinny physical creatures. *Integers 0* (2000), Paper A9. Published electronically at http://www.integers-ejcnt.org/vol0.html.

[628] ZIMMERMANN, P. *Séries génératrices et analyse automatique d'algorithmes*. Ph. d. thesis, École Polytechnique, 1991.

[629] ZOLOTAREV, V. M. *One-dimensional stable distributions*. American Mathematical Society, 1986. Translated from the Russian by H. H. McFaden, Translation edited by Ben Silver.

[630] ZVONKIN, A. K. Matrix integrals and map enumeration: An accessible introduction. *Mathematical and Computer Modelling 26*, 8–10 (1997), 281–304.

Index